DIESEL EXHAUST AFTERTREATMENT 2000–2007

Other SAE books of interest:

Diesel Emissions and Their Control
By Magdi K. Khair and W. Addy Majewski
(Order No. R-303)

Diesel Particulate Filter Technology
Edited by Timothy V. Johnson
(Order No. PT-124)

DIESEL EXHAUST AFTERTREATMENT 2000–2007

PT-126

Edited by
Magdi Khair
Federico Millo

Published by
SAE International
400 Commonwealth Drive
Warrendale, PA 15096-0001
U.S.A.
Phone: (724) 776-4841
Fax: (724) 776-5760
www.sae.org
April 2008

For permission and licensing requests contact:

SAE Permissions
400 Commonwealth Drive
Warrendale, PA 15096-0001-USA
Email: permissions@sae.org
Fax: 724-776-3036
Tel: 724-772-4028

Global Mobility Database®

All SAE papers, standards, and selected books are abstracted and indexed in the Global Mobility Database.

green press INITIATIVE

SAE International is committed to preserving ancient forests and natural resources. We elected to print *PT-126 Diesel Exhaust Aftertreatment 2000-2007* on 30% post consumer recycled paper, processed chlorine free. As a result, for this printing, we have saved:

6 Trees (40' tall and 6-8" diameter)
2,375 Gallons of Wastewater
955 Kilowatt Hours of Electricity
262 Pounds of Solid Waste
514 Pounds of Greenhouse Gases

SAE International made this paper choice because our printer, Thomson-Shore, Inc., is a member of Green Press Initiative, a nonprofit program dedicated to supporting authors, publishers, and suppliers in their efforts to reduce their use of fiber obtained from endangered forests.

For more information, visit www.greenpressinitiative.org

For multiple print copies contact:

SAE Customer Service
Tel: 877-606-7323 (inside USA and Canada)
Tel: 724-776-4970 (outside USA)
Fax: 724-776-0790
Email: CustomerService@sae.org

ISBN-978-0-7680-1709-0

Library of Congress Catalog Card Number: 2005933884
SAE/PT-126

SAE Order No. PT-126

Printed in USA

Preface

The diesel engine is well established as a prime mover for heavy-duty commercial applications worldwide. It has earned this position because of its superior fuel economy and durability. Recent market studies indicate that the diesel engine is rapidly increasing its dominance in light-duty applications such as passenger cars, especially in Europe. Its share of the European passenger car market has risen to more than 50 percent, and currently there is growing interest in light-duty diesel applications in the United States.

The exceptional success of the diesel engine has been made possible through improved performance and lower emissions from a new generation of engines. Many, if not most, of the improvements obtained in these new engines can be attributed to the advances in the fuel injection and air charging systems. Thanks to these two technological advances, the diesel engine has attained high levels of specific power outputs that approach those of its rival gasoline-fueled engine. In the meantime, the diesel engine has maintained its superior fuel economy and high torque at low engine speeds for improved drivability and responsiveness and has achieved significant improvements in its noise, vibration, and harshness (NVH) quality.

The increased popularity of the diesel engine invariably has raised public awareness and has led to increased concern regarding the emissions profile of the diesel engine. This concern was translated into a new set of extremely restrictive limits on emissions from diesel engines. Of special note are nitric oxides (NOx) and particulate matter (PM) emissions, which combine to form the Achilles' heel of the diesel engine. In contrast with the relaxed limits imposed on diesel engines in the early 2000s, the U.S. Environmental Protection Agency (EPA) and the European Commission both are imposing much stricter limits for light-duty applications through their Tier 2 and Euro V and VI regulations, respectively. Note that the new limits in the United States are fuel neutral and apply equally to gasoline, diesel, and a host of other fuels considered for future power plants. The same level of tight regulations also is planned for Japan. Figure 1 illustrates the standards for the regulations in these various locations.

New and extremely demanding regulations also are expected by 2010 for heavy-duty applications, with EPA 2010 standards and Euro V and Euro VI, as shown in Figure 2.

Through the early 2000s, diesel emission controls were well managed by in-cylinder technologies. However, further reductions in diesel engine emissions will require highly efficient and reliable exhaust aftertreatment technologies. Diesel oxidation catalysts (DOC) have been the state of the art for diesel engines for a number of years. In Europe, diesel particulate filters (DPF) have been used for almost a decade and continue to increase in popularity and application. Although NOx aftertreatment systems have spread throughout Japan and Europe, application of these systems in diesel engines in the United States is being considered for introduction in the 2010 model year.

Diesel Exhaust Aftertreatment 2000–2007 has been compiled to help guide engineers in selecting the most appropriate and effective tools to achieve the above-mentioned demanding low-emission targets. This book features a collection of more than 50 papers, selected from the extensive SAE International literature of the past seven years. The aim of this compilation is to summarize the most representative studies in the field of diesel exhaust aftertreatment, highlighting the recent findings, developments, and technology trends for both light-duty and heavy-duty applications.

The first section of this book addresses the *regulation framework* and highlights the implications in terms of aftertreatment device requirements for different markets and applications. Moreover, some of the selected papers provide a meaningful and comprehensive review of in-cylinder measures for emissions control, thus defining in a more complete way the "boundary conditions" for exhaust aftertreatment systems.

The next section includes fundamental studies and recent achievements in *PM control.* This section pays particular attention to the effectiveness of PM abatement techniques, not only from a gravimetric point of view but also as far as particle number is concerned. Particle size, number, surface area, and other related technologies are receiving more emphasis in Europe at this time and will be included in future regulations. Different techniques to enhance the ash storage capacity of particulate wall-flow filters and to improve their regeneration behavior (both through the use of fuel-borne catalysts and through catalyzed filters) are presented. Finally, some examples of flow-through filters also are provided.

In the third section, *NOx control technologies* are reviewed, starting with selective catalytic reduction (SCR) and including fundamental studies as well as application examples for both heavy-duty and light-duty applications. Control issues related to the urea dosage also are addressed. Subsequently, lean NOx traps (LNT) and lean NOx catalysts (LNC) are examined. This section concludes with some examples of combined systems.

In the fourth section, *integrated and combined systems for simultaneous PM and NOx control* are presented, starting with combinations of SCR and DPFs for heavy-duty applications and then progressing to more integrated solutions. One example is the diesel particulate—NOx reduction (DPNR) system for simultaneous PM and NOx reduction in light-duty applications.

The fifth section covers the *effects of fuel and lubricants* on engine emissions and their impact on diesel aftertreatment systems. Special focus is placed on alternative fuels and on lubricants, whose influence will become increasingly important as the sulfur content in fuels approaches minimum levels.

Finally, the last section reviews the *modeling of diesel aftertreatment systems*, including filtration and regeneration processes in diesel particulate filters, as well as numerical simulation and control issues concerning LNT and SCR systems for NOx control.

We hope this compilation will serve as a valuable reference to all engineers involved in reducing emissions from diesel engines and minimizing the potentially adverse impact of those emissions on the environment. The goal that we share with technical staffs that are pursuing a cleaner environment is ultimately to remove the diesel engine from the environmental debate by having it produce zero or near zero emissions for a better tomorrow.

Magdi Khair
Federico Millo

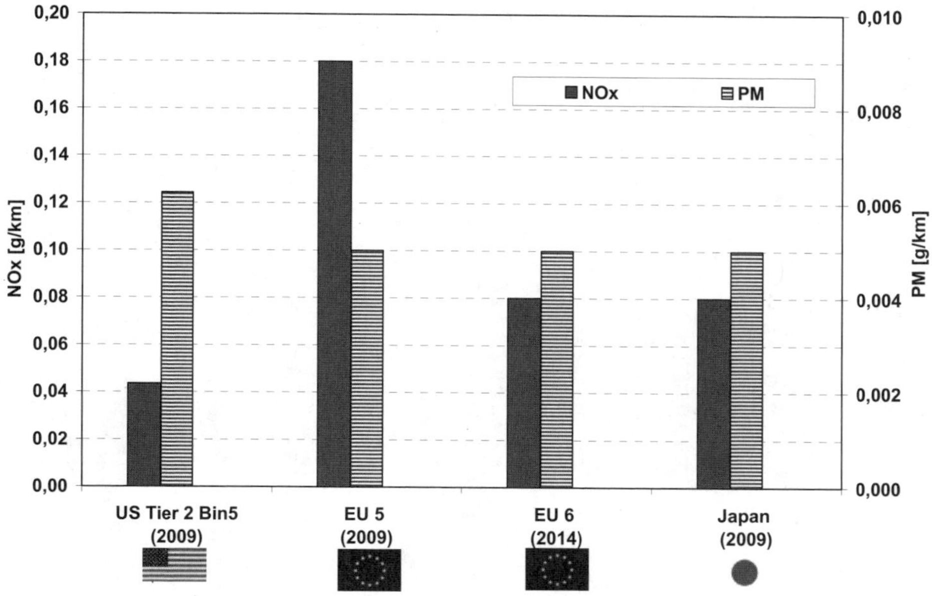

Figure 1. A comparison of future standards for light-duty diesels in the United States, European Union, and Japan (all limits in g/km, with no correction for different driving cycles).

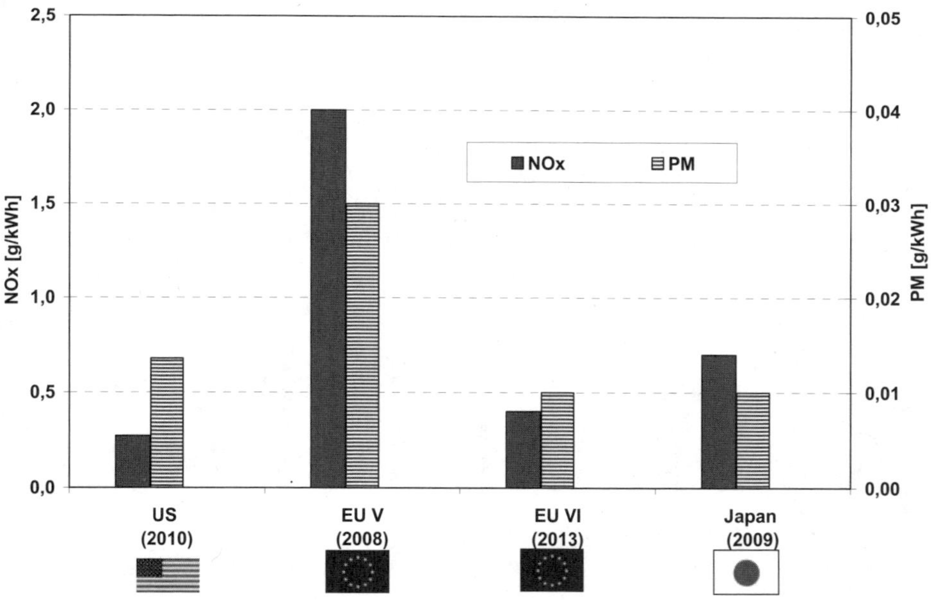

Figure 2. A comparison of future standards for heavy-duty diesels in the United States, European Union, and Japan (transient test cycles, all limits in g/kWh, no correction for different test cycles).

Table of Contents

II.2 Substrate geometry effects

II.3 Technologies

III. NO_x Control

III.1 SCR systems

III.2 LNT, LNC systems

III.3 Combined systems

IV. Integrated and combined systems for simultaneous PM and NO_x control

IV.1 Heavy Duty

IV.2 Light Duty

V. Fuel & Lubricants effects

VI. Modelling

I. Introduction: Regulations, Emission Control Requirements, Technology Trends

2003-01-0040

Heavy Duty Diesel Emission Regulations – Past, Present, and Future

David F. Merrion
Detroit Diesel Corporation

ABSTRACT

Exhaust emissions from heavy-duty truck and bus diesel engines have been legislated from 1970 to 2010. Smoke emissions continue to be controlled but exhaust odor regulations were never promulgated. Gaseous emissions (oxides of nitrogen, carbon monoxide and hydrocarbons) were regulated in 1973 and particulate matter first regulated in 1988. Throughout this period there has been cooperation, lawsuits, statement of principles, Consent Decrees and now The Clean Diesel Independent Review Panel. If the 2010 rulemaking remains in place, diesel emissions of NOx and PM will have been reduced 99% from uncontrolled levels.

INTRODUCTION

The exhaust emissions from diesel engines have been a concern since Rudolf Diesel invented the diesel cycle in 1882. The diesel has been known for its low carbon monoxide emissions and in the 1940s was declared an engine with low toxic emissions. The emissions of concern were visible smoke and exhaust odor, and these were both considered nuisance emissions.

The diesel engine's exhaust emissions have been regulated since the early 1960s and now have regulations up to year 2010. In this early period the diesel was threatened by standards that would regulate it out of the market (particularly in California) and by other engines such as: steam engines (William P. Lear's antipollution steam engine circa 1970), gas turbines (Detroit Diesel and Ford's gas turbine for trucks and buses 1967-1975), rotary diesel engines (John Deere Program), Stirling engines (Stirling Thermal Motors), adiabatic diesels (ceramic components), alternate fuel engines (ethanol, methanol, natural gas), and now fuel cells.

The emission regulations by both U.S. EPA and California are discussed for the period from the 1960s to 2010, including the impact of the 1998 Consent Decree signed by seven diesel engine manufacturers.

DIESEL EXHAUST SMOKE

The smoke from a diesel engine can be black (carbon), blue (lube oil) or white (unburned fuel). The greatest concern has been with the black smoke since it is the most displeasing to the eye and contains small particles. The diesel cycle operates on the excess air principle but when too much fuel is injected into the air containing combustion chamber, there is not enough time for all the fuel to find air and some of the fuel is only partially oxidized.

One interesting aspect of diesel exhaust smoke is the technique for measuring it. Over the years the most popular way to measure smoke was with visual techniques using arbitrary scales of light, medium, heavy, or dark, black, etc. Back in the late 1800's Ringelmann [1] developed a scale and a printed card system for comparison rating. (Fig. 1)

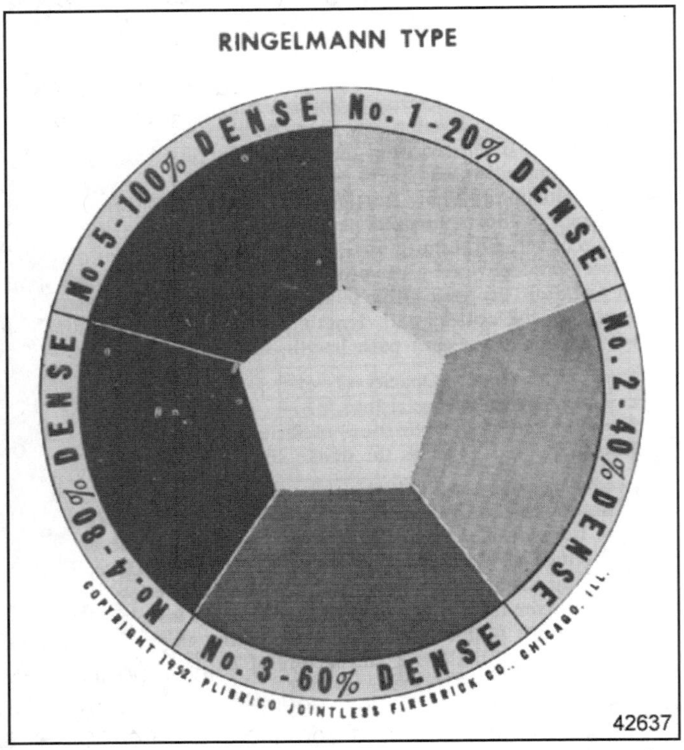

Figure 1 Ringelmann Type Smoke Guide

Derivatives of this system have been developed, such as the EMA smoke rating guide (Fig. 2), but none were widely accepted and smoke meters became the preferred method of measurement.

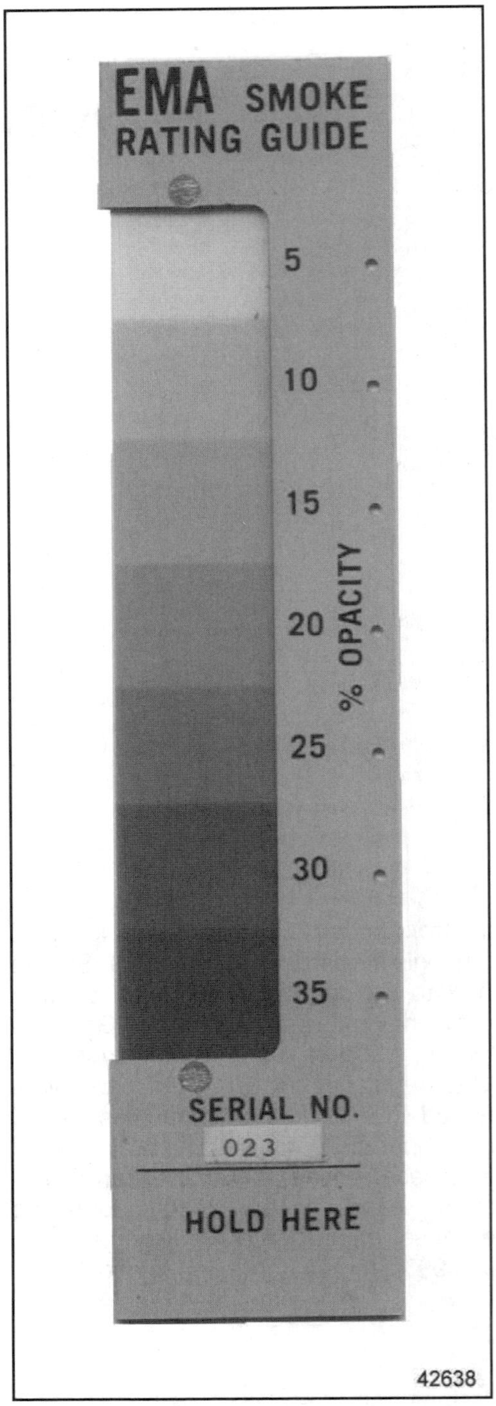

Figure 2 EMA Smoke Guide

Early smoke meters were the Von Brand, Robert Bosch and Leslie Hartridge meters. Papers have been written about the correlation of smoke measurement devices [2, 24] and the subject is presented in the SAE handbook. It wasn't until the Public Health Services (PHS) Smoke-meter (Fig. 3) was developed in the late 1960's and exhaust smoke legislation proposed in 1968, did smoke measurement move from the subjective, or questionable correlation, to the objective or quantifiable.

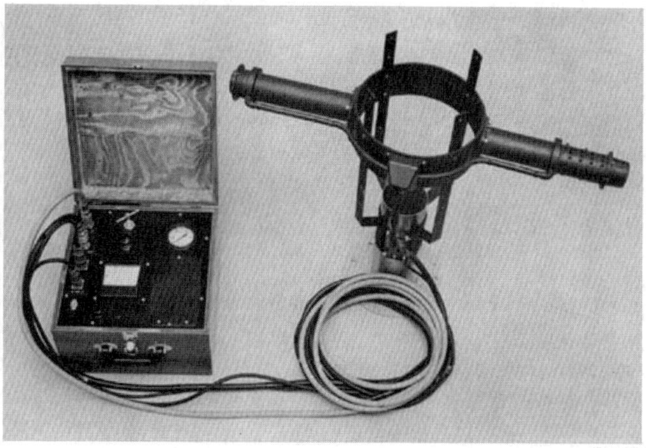

42639

Figure 3 Public Health Service Smoke Meter

DIESEL EXHAUST ODOR

In the early years, diesel odor was more scientifically investigated than exhaust smoke since it was more difficult to measure, to quantify, and to reduce. References to research on diesel odor go back to 1942 [3] but the most significant early paper was by GM Research titled "Diesel Exhaust Odor: Its Evaluation and Relation to Exhaust Gas Composition" presented at the SAE National Diesel Engine Meeting November 2, 1956. [4] This paper outlines measurement techniques, (using the human nose), (Fig. 4) the effects of design variables such as speed, load, engine type (diesel, gasoline, propane, two-stroke vs. four stroke), fuels, lubricating oil, chemical composition (aldehydes, hydrocarbons, oxides of nitrogen, carbon monoxide), and outlines methods to reduce odor.

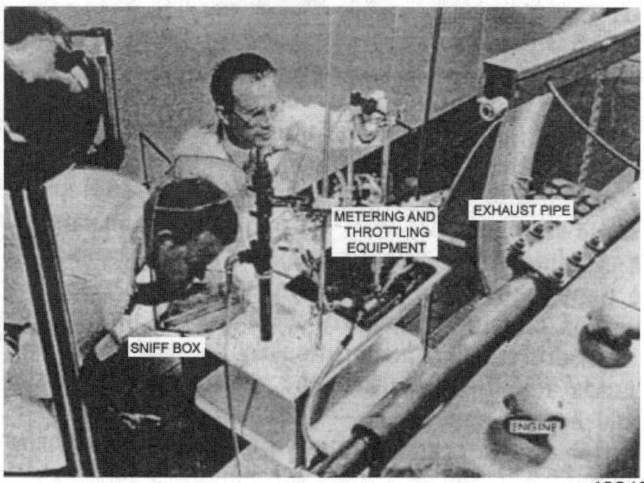

42640

Figure 4 GM Research Odor Room

On June 10, 1966, the California Board of Public Health adopted standards for exhaust odor and irritation as follows:

1. "The average intensity of odor as determined by subjective appraisal shall be less than the intensity from diesel vehicles with horizontal exhaust pipes representative of the diesels in use in 1966 and whose odorant concentrations have been reduced by at least 80%.
2. There shall be no detectable eye, nasal or throat irritation to at least 75% of the panel.
3. Exhaust odors that are different in quality from characteristic diesel odor shall be less objectionable to the panel than the odor from diesel vehicles with horizontal exhaust pipes representative of the diesels in use in 1966 and whose odorant concentrations have been reduced by at least 80%.
4. The conditions for the appraisal are:
   a. The odor and irritation panel shall consist of not less than 10 persons.
   b. Appraisal of odor and irritation shall be made on a vertical plane ten feet distant from the exhaust outlet to either side of the motor vehicle parallel to the longitudinal axis. For vehicles with more than one exhaust outlet, the appraisal shall be made on a vertical plane parallel to the longitudinal axis at a horizontal distance ten feet from midpoint of the exhaust outlets.
   c. The exhaust gas shall be evaluated during the modes of idle and full throttle acceleration.
   d. Other methods of odor appraisal or measurement may be used if approved by the Department of Public Health."

These odor regulations stimulated more odor research that continued at General Motors, since they were the manufacturers of locomotives, trucks, and urban buses, all contributors of odor. The 1965 amendment to the Clean Air Act of 1963 charged the Department of Health Education and Welfare with the responsibility for research on the problem of emissions from diesel engines. Accordingly, the Motor Vehicle Research and Development Program of the National Centers for Air Pollution Control, Public Health Service, undertook a broad diesel emission program. This program included several odor contracts including: Scott Research Laboratories to study diesel exhaust chemistry, odor analysis and odor chemistry; Dr. Amos Turk, odor consultant, Dr. Dean Foster, research psychologist and Dr. Ralph Bradley, statistical consultant to develop odor evaluation procedures involving human panels; U.S. Bureau of Mines Petroleum Center to map emissions from diesel engines including odor; and Southwest Research Institute to study odor from diesel powered trucks and buses. This government funded work started in 1965 and continued until 1977. The studies of diesel odor are interesting, in part because it was break through science for measurement and control, but, also because diesel odor was never regulated by either

California or U.S. EPA. A summary of this work is as follows:

1. General Motors Research used the dilution technique for studies (Fig. 5) where the technique of rating the odor intensity at various dilution ratios was used as well as the technique diluting exhaust to the just detectable level. This basic research resulted in techniques that could be used on full-scale buses. (Fig. 6) The GM airplane hanger at Detroit City Airport was used to drive buses into the hanger, close the doors and have a panel measure the odor or state if they detected an odor. With this technique they were able to evaluate changes to the GM Detroit Diesel engine as well as evaluating exhaust pipe position on a GMC transit bus including a vertical exhaust.[6, 7]

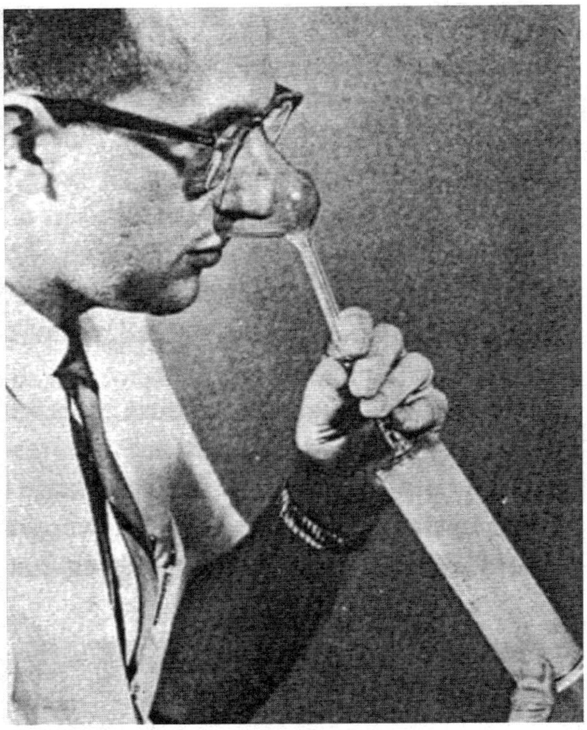

42641

Figure 5 GM Odor Measurement Using Dilution Technique

Figure 6 GM Odor Measurement of Full-Scale Buses

2. Detroit Diesel Engine Division of GM also conducted studies of diesel odor (Fig. 7) and evaluated the difference between engine designs. The bus engine designated 6V-71E, was compared to a 6V-71N and significant odor reduction was achieved with the 6V-71N engine. The "N" engine had higher compression ratio and needle valve injectors, with much lower uncontrolled volume of fuel below the control valve in the injector tip. It was later discovered that the uncontrolled volume of fuel below the injector control valve was a major contributor to exhaust hydrocarbons and odor.[8, 9]

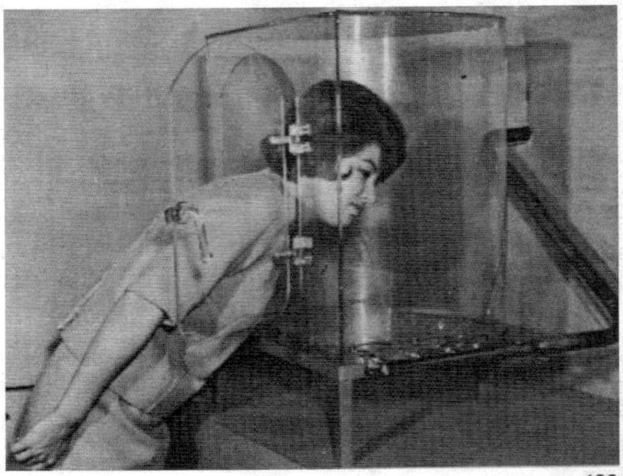

42643

Figure 7 Detroit Diesel Odor Measurement

3. The Public Health Service contracts resulted in a Turk Odor Kit (Fig. 8) for standards for odor quality and intensity Southwest Research Institute (SwRI) used this kit to measure odor from trucks and buses running on a chassis dynamometer using an odor panel of ten women (Fig. 9). The SwRI studies of diesel exhaust odor were the most extensive of any odor studies covering a period over 12 years from 1965 to 1977. Those studies resulted in methods to

measure odor, methods to qualify and quantify the odor and methods to reduce odor.[10, 11]

Figure 8 Turk Odor Kit

Figure 9 Southwest Research Institute Odor Measurement

In retrospect, the government spent millions of dollars to research diesel odor but never legislated its control. The reason was two-fold:

1. Detroit Diesel discovered the cause of hydrocarbons in diesel exhaust as being related to the volume of fuel below the fuel injector control valve. Reducing this volume to near zero significantly reduced hydrocarbons (including the aldehydes) and significantly reduced the odor.
2. The GM studies showed the vertical exhaust pipes on urban buses diluted the exhaust an additional seven times compared to horizontal exhaust pipes. GMC Truck and Coach Division changed from horizontal exhaust to vertical exhaust.

The combination of these two control methods made significant odor reductions and made regulation unnecessary.

DIESEL ENGINE INDUSTRY

In this period of the mid 1960's it became obvious that the U.S. Department of Health Education and Welfare (HEW), the State of California and various States for vehicles in-service would regulate the diesel industry. The diesel industry gathered together under the umbrella of the old Internal Combustion Engine Institute and in 1968 formed the Engine Manufacturers Association, EMA. The visionaries at this time were Tom Head from Cummins, Ken Hulsing from Detroit Diesel, Jules Loefler, Roger Ringham, Bill Wallace from International Harvester, and Chuck Leber from Caterpillar. EMA became the interface organization for both HEW and California. The concept within EMA was to sponsor research through the Coordinating Research Council (CRC), write procedures and recommended practices through the Society of Automotive Engineers (SAE) and have the governments set the standards. Future discussion will illustrate how well this concept worked.

1968 DIESEL EXHAUST SMOKE REGULATION FOR 1970

On January 4, 1968, smoke standards were proposed by HEW [12] using the Public Health Service (PHS) smoke meter that that been developed by Andy Rose at the National Center for Air Pollution Control in Cincinnati (Fig. 3). It is interesting that even though the smoke emissions were considered a nuisance, the first regulations stated "The proposed standards represent the application of current technology to the control of such motor vehicle emissions which, in the judgment of the Secretary, cause or contribute to, air pollution which endangers the health or welfare of any person". Input was sought from EMA regarding the smoke cycle and SwRI did the development work for both a chassis test and an engine dyno test. The standards were proposed as "shall not reduce the transmission of a beam of light by more than 20 percent except that reduction of light transmission of not more than 40 percent for a total of not more than 5 seconds..." The test was a 5-minute idle, acceleration for 25 seconds and a lug down for 35 seconds (Fig. 10). The engines tested were the highest rating in the family and the fuel was a "worst case" type fuel. The final rule was issued on June 4, 1968, [13] and was basically the same as the proposed rule, which took effect with engines produced after January 1, 1970. In 1971 it was proposed to revise the smoke standards effective with 1973 engine production. The standard change was 20% during the acceleration mode 15% during the engine lugging mode and 50% during the peaks in either mode. This smoke standard continues on the books today using the same test procedure and smoke measurement device.

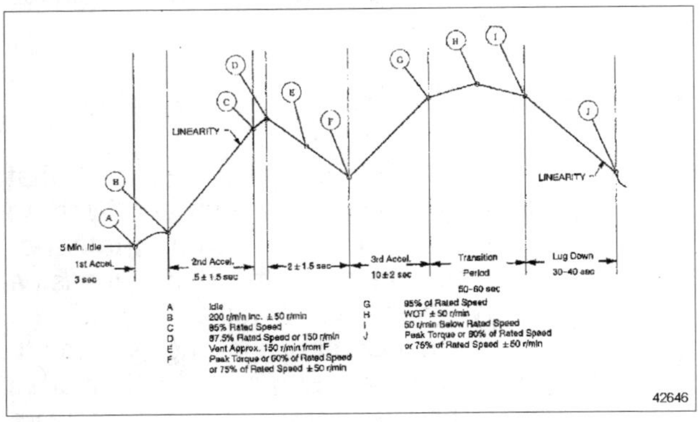

Figure 10 EPA Diesel Smoke Cycle

1970 CALIFORNIA GASEOUS EMISSION REGULATION FOR 1973

In 1967, merging the California Motor Vehicle Pollution Control Board and the Bureau of Air Sanitation created the California Air Resources Board (CARB). Dr. Arie Haagen-Smit (the discoverer of the cause of photochemical smog) was named its first Chairman.

Diesel gaseous emissions (hydrocarbons, carbon monoxide, oxides of nitrogen) had been discussed in the literature for several years, for instance:

- 1943, Bureau of Mines Report "Diesel Engines Underground IV, Effect on Composition of Exhaust Gas Variables Influencing Fuel Injection" [14]
- 1949, Bureau of Mines "Sampling and Determination of Aldehydes in Diesel Engine Exhaust Gas and in Mine Air" [15]
- 1950, General Motors "Composition of Diesel Exhaust Gas" [16]
- 1955, General Motors "Composition of Exhaust Gases from Diesel, Gasoline and Propane Powered Motor Coaches" [17]
- 1956, General Motors "Diesel Exhaust Odor –
- Its Evaluation and Relation to Exhaust Gas Composition" [18]
- 1963, Caterpillar "Nitrogen Oxides and Variables in Precombustion Chamber Type Diesel Engines" [19]
- 1965, Bureau of Mines "Air Pollution Inventory Enter the Diesel" [20]
- 1966, Cummins "Exhaust Characteristics of the Automotive Diesel" [21]
- 1966, A Report to the California Legislative California Department of Public Health "The Diesel Vehicle and Its Role in Air Pollution" [22]
- 1967, American Oil "Atmospheric Pollution from Diesel Engines" [23]
- 1967, General Motors "Measuring The Total Hydrocarbons in Diesel Exhaust" [24]

But most significant was the series of papers on diesel gaseous emissions presented in 1968 by Bureau of Mines,[25, 26] International Harvester,[27] Cummins Engine Co.,[28] Caterpillar Tractor Co.,[29] and Detroit Diesel Engine Division of General Motors.[30] These six papers and three papers by General Motors Research, Scott Research laboratory and Southwest Research Institute on diesel odor[31, 32, 5] served as a status report on diesel emissions in the 1960s.

In 1968 the California legislature revised the state's laws on air pollution by enacting Assembly Bill 357. This bill required the Air Resources Board (ARB) to establish emissions standards for carbon monoxide, oxides of nitrogen and hydrocarbons from diesel-powered vehicles. These standards had to be established by January 1, 1971 and were effective January 1, 1973. These "first" gaseous emission standards for diesels resulted in the industry becoming very cooperative with the government agencies (ARB and HEW). In fact Roger Bascom of Cummins (representing EMA) and Gay Hass of CARB wrote an SAE paper in August 1970, which outlines this cooperation.[33] The organizations involved were EMA, SAE, CRC, CARB, HEW and their contractor SwRI. HEW was involved since they were required to grant a Federal preemption before the California standard could become effective. This cooperation resulted in the 13-mode steady state test procedure (Fig. 11). The concept was, that the emission map could then be weighted differently for different applications and cycles for various applications could be derived using different weighting factors. The final regulation from California resulted in weighting factors of 8% for each of the 10 intermediate and rated speed modes and 20% for the three idle modes. The standards were set as follows:

HC + NOx	16 g/hp-hr
CO	40 g/hp-hr

(The schedule will go to 5 g/hp-hr NOx in 1975.)

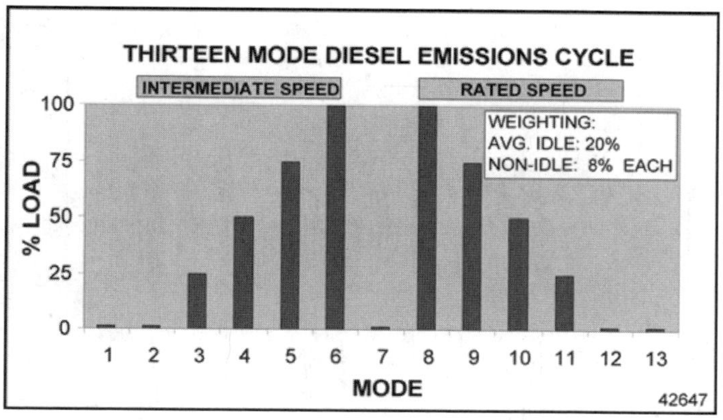

Figure 11 EPA Gaseous Emission 13-mode Test Procedure

1971 FEDERAL GASEOUS EMISSION REGULATION FOR 1974

In 1963 the Federal Clean Air Act (CAA) was passed and in 1965 the Motor Vehicle Air Pollution Control Act amended the CAA and HEW was directed to establish auto emission standards. In 1967 the Federal Air Quality Act was enacted and in 1970 the U.S. Environmental Protection Agency (U.S. EPA) was created and the Federal Clean Air Act Amendments of 1970 were enacted. All this activity resulted in a October 5, 1971 Federal Register [34] proposing the following diesel standard:

HC	3.0 g/hp-hr *
CO	7.5 g/hp-hr *
NOx	12.5 g/hp-hr *

* Using the CARB/SAE 13-mode steady state emission test

The final U.S. EPA rules were published on November 15, 1972 [35] and set the standards the same as California, 16 g/hp-hr NOx + HC and 40 g/hp-hr CO, on the 13-mode steady state cycle but effective January 1, 1974 versus 1973 for California. Part of the Federal Test Procedure made reference to the SAE recommended practice but no explanation was given why the HC and NOx standards were combined and why the CO standard was relaxed from the proposed rule. Thus, in 1974, the California and Federal standards were the same and the NOx + HC standard of 16 g/hp-hr was set at a level that some engines could meet, but some needed NOx control with retarded injector timing. (Note: uncontrolled diesel engines varied in NOx from 3 to 19 g/hp-hr and HC from 0.2 to 3.5 g/hp-hr so the highest uncontrolled engine could have been as high as 22.5 g/hp-hr NOx + HC). [36] Using the Detroit Diesel product line as an example of the technology impact on one diesel engine manufacturer, in 1974 Detroit Diesel continued to offer naturally aspirated engines, and out of 15 engine models offered for sale only four were turbocharged.

CALIFORNIA GASEOUS EMISSIONS – 1973 TO 1979

The standard proposed in 1970 of 5 g/hp-hr NOx + HC for 1975, was revised to 10 g/hp-hr for 1975 and 1976 and the 5 g/hp-hr NOx + HC stayed on the books for 1977 to 1979 with an option for 7.5 g/hp-hr NOx and 1.0 g/hp-hr HC. The impact on Detroit Diesel was in 1975/1976; naturally aspirated engines were still offered but the turbocharged engines used jacket water aftercooling to achieve the NOx level. In 1977 to 1979, emission certifying to the 7.5 + 1.0 g/hp-hr standards, only one naturally aspirated model remained with turbocharging and aftercooling being used for NOx control.

1976 EPA PROPOSED REVISION FOR 1979

The 1976 Federal Register [37] contained several revisions to the rules for 1979.

1. Oxides of nitrogen will be measured by chemiluminescence (cl) which will measure NO + NO_2.
2. All diesel engine test procedures are in the regulation with no reference to SAE practices.
3. Revised smoke opacity standards
   - 20% acceleration mode
   - 15% lugging mode
   - 35% peaks in either mode
4. Revised gaseous emission standards
   - HC = 1.5 g/hp-hr
   - CO = 25 g/hp-hr
   - HC + NOx = 10 g/hp-hr

On September 8, 1977 [38] EPA published the final rules for 1979 that, based on comments from industry, the only change was to drop the 35% peak smoke change since it would affect vehicle acceleration and the air quality benefit was "not readily identifiable." It is interesting to note at this point in time, 1977, that the cooperation between industry and EPA was decreasing and EPA was adopting more of a "control" approach to regulating.

The most important issues raised by the engine and vehicle manufacturers are:

> "The proposed standards will force discontinuance of naturally aspirated diesel engines and require the use of intercoolers and turbochargers."

EPA's response was:

"EPA believes that the existing trend toward diesel engines that use intercoolers and turbochargers is only marginally related to the emission standards. The trend toward constant speed (55 mph) heavy-duty vehicles equipped with an engine with significant torque backup is in response to the desire for improved fuel economy, not emission control. The manufacturers did not submit data that would demonstrate that turbochargers and intercoolers are required to meet the emission standards. Most importantly, a maximum increase of only 4 percent in the use of the turbochargers may occur. Therefore, EPA views the costs of intercoolers and turbochargers as not being properly chargeable to the emission control standards."

Also note that just over one year lead-time was given from final regulation to effectiveness. The 1979 10 g/hp-hr HC + NOx requirement for EPA and the 7.5 g/hp-hr NOx and 1.0 g/hp-hr HC requirement for CARB resulted in the following Detroit Diesel models being offered:

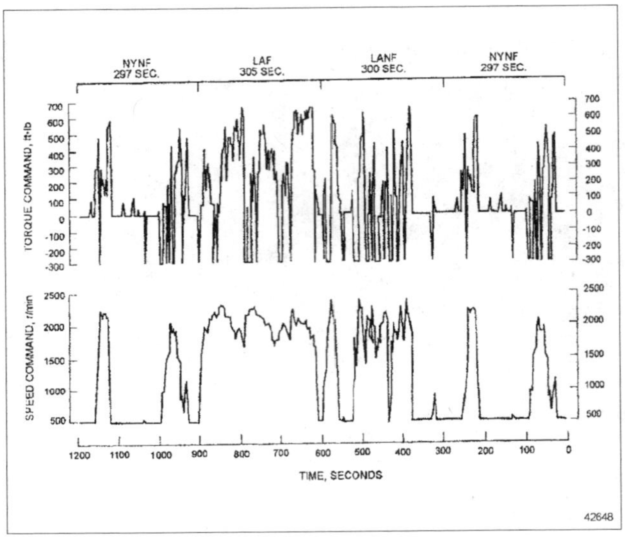

Figure 12 EPA Diesel Gaseous and Particulate 1200 Second Transient Test

Federal	15 models	4 naturally aspirated
		4 turbocharged
		7 turbocharged and aftercooled
California	12 models	1 naturally aspirated
		2 turbocharged
		9 turbocharged and aftercooled

1980 EPA PROPOSED REVISIONS FOR 1984

The 1980 proposed regulations for 1984 [39] were significant to the diesel industry since EPA changed from the 13-mode steady state test procedure to a 1200 second transient test for gaseous emissions. (Fig. 12) The transient test was 5 years in development with little participation by industry. EPA reasons for the new test procedure were:

1. EPA plans to soon propose and promulgate regulations governing particulate emissions from heavy-duty diesel engines.
2. Currently, modal engine tests are used which subject the engines to a sequence of steady-state (i.e. unchanging) speed and torque combinations which have little relationship to actual patterns of use. Since the modal tests sample only selected operating modes, they are easier to defeat than the transient test.

This transient test also required electric dynamometers, constant volume samplers, dilution systems, new measurement techniques, hot and cold cycles etc. The emission standards were set after running correlation tests with the steady-state procedure, observing twelve heavy-duty gasoline engines and several diesels on the new test procedures, as follows:

HC	1.3 g/hp-hr
CO	15.5 g/hp-hr
NOx	10.7 g/hp-hr

EPA did allow one year, 1984, where the diesel engine manufacturers had the option to certify to the steady state standards as follows:

HC	0.5 g/hp-hr
CO	15.5 g/hp-hr
NOx	9.0 g/hp-hr

In January 1983, this 1984 requirement was delayed to 1985 due to insufficient lead-time.

1981 EPA PROPOSED PARTICULATE STANDARD FOR 1986

In January 1981 [40], the EPA proposed a 0.25 g/hp-hr particulate (PM) standard to be effective in January 1986. The CAA required the Administrator to prescribe regulations for emissions of particulate matter for vehicles manufactured during and after model year 1981. EPA did not meet this statutory requirement. This proposal was based on an average uncontrolled PM level of 0.41 g/hp-hr and a 60% reduction in engine out PM from the application of particulate trap-oxidizers. Also mentioned, was methanol as a fuel and the 75%

reduction in NOx from uncontrolled gasoline engine levels (1.7 g/hp-hr NOx).

The Natural Resources Defense Council (NRDC), an environmental group, filed suit against EPA in 1984, challenging the agency's failure to promulgate the NOx and particulate regulations described above. NRDC was granted summary judgment on September 14, 1984, by the U.S. District Court for the District of Columbia, which found that EPA had failed to fulfill non-discretionary statutory duties under the Clean Air Act Amendments of 1977. The court ordered EPA to publish proposed heavy-duty engine NOx and particulate standards no later than October 15, 1984, and the corresponding final regulations no later than March 15, 1985.

1983 EPA REVISIONS FOR 1985

On November 16, 1983 EPA published a final rule [41] for useful life as follows

Light-duty diesel	110,000 mi
Medium-heavy diesel	185,000 mi
Heavy-heavy diesel	290,000 mi

1984 EPA REVISIONS FOR 1987 AND 1990

As mentioned earlier, EPA had two statutory requirements, (CAA), that included a 75% reduction in NOx, (1.7 g/hp-hr NOx), from uncontrolled heavy duty gasoline engines and a particulate matter standard for 1981 model year. In the October 15, 1984 Federal Register [42] there was considerable discussion of the situation, including the following:

> "Section 202(a)(3)(A)(iii) of the Act directs the Administrator of EPA to establish standards for the emissions of particulate matter from heavy-duty diesel engines. According to that section, regulations are to require "the greatest degree of emission reduction achievable through the application of technology which the Administrator determines will be available giving appropriate consideration to the cost and to noise, energy, and safety factors associated with the application of such technology."
>
> "This statutory description corresponds to what are generally referred to as "technology-forcing" standards. That is, they are to be based upon that technology which the Administrator determines *will* be available, not necessarily that technology which is *already* available."
>
> "A final note on the statute concerns the model year provision in this section. This clearly presents a dilemma to EPA, since the original compliance date of 1981 has already passed while standards have yet to be promulgated. EPA sees its present task as implementing particulate standards as soon as possible, while allowing manufacturers the necessary amount of time to reach compliance."
>
> "Provisions for the control of NOx emissions from heavy-duty engines are also found in Section 202(a)(3)(A) of the Act."
>
> "In summary, then, the Act calls for a 75 percent NOx reduction standards unless temporarily revised due to cost or fuel economy reasons, or changed due to studies on the effect of emission on the public health and welfare, or other such information. The Act required such standards to be promulgated, with four years of lead-time, for the 1985 model year. Clearly, while four years of lead-time could still be provided, it is no longer possible to have revised standards in place for vehicles or engines manufactured in 1985."

As such then EPA <u>proposed</u> the following standards:

1987	NOx	6.0 g/hp-hr
	PM	0.60 g/hp-hr
1990	NOx	4.0 g/hp-hr
	PM	0.25 g/hp-hr
1990	PM	Averaging

It is interesting to note that EPA commented that the January 7, 1981 [43] proposal for NOx… "Strictly speaking, neither of these standards corresponded to the 75 percent reduction requirement of section (a)(3)(A)(ii)…" "Publication of these two notices produced considerable degree of public response, and while <u>no</u> further regulatory action has been taken until today's proposal, EPA has remained active…"

MARCH 1985 EPA ANNOUNCEMENT FOR 1988 AND 1991

EPA was under court order to publish <u>final</u> standards by March 15, 1985 [44] and they met this date as follows:

NOx	6.0 g/hp-hr for 1988-1990
NOx	5.0 g/hp-hr for 1991 and later
PM	0.60 g/hp-hr for 1988-1990
PM	0.25 g/hp-hr for 1991-1993
PM	0.10 g/hp-hr for 1991-1993 urban buses
PM	0.10 g/hp-hr for 1994 and later

Note the changes from the previous proposed regulations: i.e.: 6.0 g/hp-hr NOx in 1988 vs. 1987 and 5.0 g/hp-hr NOx in 1991 vs. 4.0 g/hp-hr NOx in 1990.

On April 23, 1985 the Natural Resources Defense Council, Inc. (NRDC) sent letters to interested parties asking them to join NRDC in a lawsuit against EPA. The basis for the lawsuit was "EPA ignored the Clean Air Act's explicit command to base the standards on the capabilities of the best performing engines. Rather, EPA deliberately set standards at levels well below this level, in order to accommodate engines that lag behind the best performers." On November 7, 1986 the United States Court of Appeals for the District of Columbia

Circuit handed down a decision, which rejected the NRDC, argument that the NOx and PM standards should have been set based on the capability of the "technological leader." The court also upheld the industry position regarding the requirement for four-year lead-time. Following this, the schedule for emission compliance was as follows:

1988	10.7 g/hp-hr NOx	0.60 g/hp-hr PM
1990	6.0 g/hp-hr NOx	0.60 g/hp-hr PM
1991	5.0 g/hp-hr NOx	0.25 g/hp-hr PM
1994	5.0 g/hp-hr NOx	0.10 g/hp-hr PM

After this flurry of regulatory activity and lawsuits in the 1980s, the regulatory activity settled into business as usual. Regulations were set or proposed as follows:

1988	Non-conformance penalties
1988	Certification procedures
1989	Program for banking and trading emissions
1989	Warranty regulations and voluntary aftermarket part certification
1989	Standards for emissions from methanol-fueled motor vehicles
1989	Regulation of fuels and fuel additives
1989	Selective enforcement audit
1990	Certification procedures
1991	Heavy duty engine rebuild study
1992	Revision to durability procedures
1993	California diesel fuel aromatics reduced to 10%
1993	Federal and California diesel fuel sulfur reduced to 500 ppm

In 1991 [45] the American Public Transit Association (APTA) petitioned EPA to delay the 0.10 g/hp-hr urban bus PM standard until the 1994 model year to coincide with the truck PM standard. Before EPA could take action on the APTA petition, Congress took up the issue, as part of the CAA amendments, setting the PM standard for urban buses at 0.10 g/hp-hr for 1993.

In this period of time, 1987 to 1991, Detroit Diesel had developed a new four-stroke cycle truck diesel engine but continued to emission certify and sell their two-stroke cycle diesel engines. A significant event was the Detroit Diesel use of electronic controlled unit injectors on its two-cycle engines starting in 1985 and on its four-cycle engine in 1987. The four-cycle engine, the Series 60, was only available as a turbocharged and air-air charge cooled engine with electronic unit injectors. The two-stroke engines continued with various models (turbocharged, turbocharged and jacket water aftercooled, mechanical and electronic unit injector) until the 1991 0.25 g/hp-hr PM standards.

In 1991 only three two-cycle diesel engine models were offered (all turbocharged, aftercooled and electronic unit injectors), plus a methanol fueled transit coach model. In this period of the early 1990's, many transit bus properties were looking for cleaner exhaust from their buses with Detroit Diesel two-cycle engines.

The California initiative for methanol fuel started in 1983 and resulted in an emission certified methanol engine for Los Angeles in 1991. Houston, Texas wanted a natural gas engine and Detroit Diesel provided several varieties i.e.: dual fuel (diesel pilot), direct injection of natural gas and spark ignited natural gas. New York City did not want alternate fuel and insisted on a dual particulate trap system developed by Donaldson Co. All these technologies (methanol, ethanol, CNG, LNG and particulate traps) failed to be effective emission reducers over the useful life of a bus. When the PM standard was further reduced for buses in 1994, the Detroit Diesel two-cycle bus engine models were removed from the market.

1993 FINAL EPA RULE FOR 1998

In March 1993 [46] EPA finalized the rules and regulations for 1998 as follows:

HC	1.3 g/hp-hr
CO	15.5 g/hp-hr
NOx	4.0 g/hp-hr
PM	0.10 g/hp-hr
PM (buses)	0.05 g/hp-hr*

*(0.07 g/hp-hr for in-use enforcement)

1995 STATEMENT OF PRINCIPLES

The next significant event in the history of diesel emission rules and regulations coincided with an administration effort on behalf of the federal government to work with regulated industries toward negotiated regulations (reg-neg). This regulatory approach was initiated during the early Clinton administration but was specifically brought about in diesel emission regulations by a proposed California State Implementation Plan (SIP) in 1995. California intended to adopt a 2.5 g/hp-hr NOx standard for on-highway engines in 2002 and EPA was committed to proposing a 2.0 g/hp-hr NOx standard for 2004. EPA and CARB approached several engine manufacturers (Caterpillar, Cummins, Detroit Diesel and Navistar) about a "cooperative approach" to implementing such a standard. This started the discussions that would result in a "Statement of Principles" (SOP). After much negotiation between April and July 1995, when the industry tried to get improved diesel fuel quality into the SOP, a final SOP was signed in July 1995 with the following announcement taken from an EPA "Environmental Fact Sheet".

"EPA, California Air Resources Board, and Manufacturers of Heavy-Duty Engines Sign "Statement Of Principles"--

The Environmental Protection Agency (EPA), the California Air Resources Board (CARB), and the leading manufacturers of heavy-duty engines signed a historic agreement to reduce engine emissions, bringing the nation closer to the goal of clean, healthy air. In this "Statement of Principles," the signatories agree to pursue new standards that will cut oxides of nitrogen (NOx) emissions from new trucks and buses in half. This agreement is an example of the type of private/public and federal/state partnership approach to environmental regulation that EPA is pursuing.

The SOP also covered the following major points:

- Reduce NOx to 2.0 g/hp-hr in 2004 (NMHC +
- NOx = 2.5 g/hp-hr and NMHC cap of 0.5 g/hp-hr).
- EPA will review technological feasibility in 1999.
- California will harmonize with the EPA.
- Agree to a research agreement to reduce NOx and PM to 1.0 g/hp-hr and 0.05 g/hp-hr respectively.

It is interesting to note that during these "negotiations" that the subject of "off-cycle emissions" was brought up by EPA stating that they were testing engines and were concerned about one-manufacturer emissions being high off-cycle.

EPA RULEMAKING FOR 2004

Even with the SOP being signed in July 1995, EPA published an advanced notice of proposed rulemaking (ANPRM) in the Federal Register on August 31, 1995, that incorporated the standards from the SOP i.e.: 2.4 g/hp-hr NMHC + NOx or 2.5 g/hp-hr NMHC + NOx with a limit of 0.5 g/hp-hr NMHC. After comments were received they published a notice of proposed rulemaking (NPRM) [47] on June 27, 1996. This proposed rule restated the above emission standards, proposed a 1999 technology review, extended the useful life for heavy-heavy engines to 435,000 miles (from 290,000 miles) and stated some relaxed emission levels if modified diesel fuel was required technically, but EPA never acts to require the fuel for this standard. On October 21, 1997, [48] EPA issued the final rule, with no mention of off-cycle emissions or supplemental tests. At an EPA / CARB Symposium in early October, 1997, the director of EPA's engine programs and compliance division in EPA's Office of Mobile Sources made the following statements: [59]

- Off cycle NOx control puts upward pressure on fuel consumption.
- EPA transient test cycle designed as an urban test, with PM focus, is not representative of regional scale driving patterns.
- Control strategies are based on the transient test cycle "that is the regulatory requirement."
- Looking toward the future, EPA together with ARB, is focusing effort on development of low NOx technology without big fuel economy impacts.

It was in November 1997 that the previously cooperative relationship with EPA started to deteriorate. The issue of increased NOx emission under "off-cycle" conditions was examined by EPA Office of Mobile Sources (OMS) in conjunction with the EPA Office of Enforcement and Compliance Assurance (OECA) and the U.S. Department of Justice (DOJ). Meetings were held with individual companies and with industry groups that ultimately resulted in seven companies signing consent decrees with EPA, DOJ and CARB. Historically, the understanding of the regulatory view of off-cycle emissions control was based on a number of significant events that included:

1. 1985, EPA abandons the steady-state emission test and adopts a transient emission test, mainly to represent urban driving and to measure and control particulate emissions.
2. 1985, Detroit Diesel (DDC) introduces electronic control of fuel injection and other companies follow in subsequent years.
3. 1987, Detroit Diesel uses advanced injection timing when a truck is in cruise control.
4. 1987 and forward, Detroit Diesel continues to develop algorithms that sense when a truck is running steady-state and advances the injection timing to achieve better fuel economy (and related NOx increases). These algorithms are active both on the Federal Test Procedures (FTP) and under typical driving conditions.
5. 1993, EPA publishes a non-road NPRM proposing an option to use the FTP or the 8-mode non-road steady state test since results are "comparable".
6. 1993, John Deere submits data to EPA on Cummins and Detroit Diesel engines showing that the injection timing is advanced under steady state conditions and NOx is increased thus the FTP and steady state test are not comparable.
7. 1994, Europe's ACEA releases a report on the development of new test cycles and discusses what is described as-- the problem of "cycle beating" the U.S. transient test.
8. 1994, EPA publishes the California FIP NPRM in which EPA cites the Deere Testimony: "[They] had received information that certain on-highway heavy-duty diesel engines are using 'transient sensoring algorithms' that have the effect of retarding the timing during transient engine operating conditions and advancing the timing during certain steady state operating conditions."
9. In 1994, 1995, 1996 and 1997: EPA issues certificates of conformity to diesel engine manufacturers without any mention of "defeat devices". Also such issues were not raised during the 1995 SOP negotiations.
10. June 1997, EPA tests a 1994 Detroit Diesel Series 60 engine and makes subsequent requests for the source code used in the electronic controls.

From these events stated above, it would appear that some people at EPA were aware of the algorithms used by heavy duty manufacturers for at least four years, but EPA did not start to take any action until late 1997.

CONSENT DECREE

The Consent Decree negotiations took place over a year and a half during which time the government and the affected companies prepared their legal arguments regarding the regulation of "off-cycle" emissions. As is often the case in complex regulatory disputes, the parties felt that the risks of litigation and the need to continue receiving certification and operating their business weighed in favor of a negotiated outcome. Seven engine manufacturers signed consent decrees in July 1999. But for one company, the terms of the consent decrees were largely identical and contained the following main points:

1. The engine manufacturers agreed to pull ahead the 2004 requirements by 15 months, October 1, 2002.
2. The engine manufacturers agreed to supplemental emissions tests (Fig. 13) under steady-state conditions with not-to-exceed (NTE) requirements (Fig. 14).

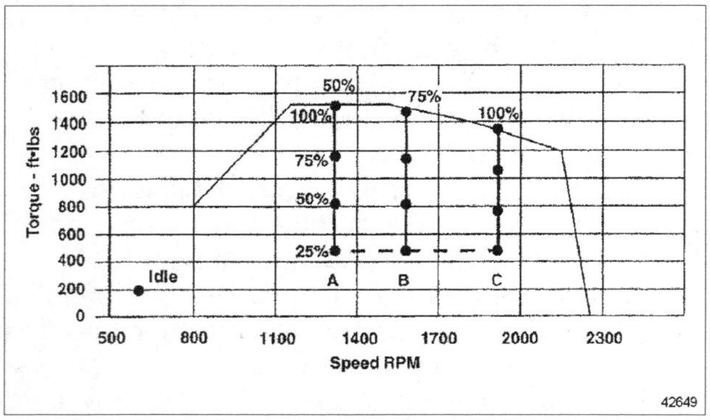

Figure 13 EPA Steady State Euro III Type Test

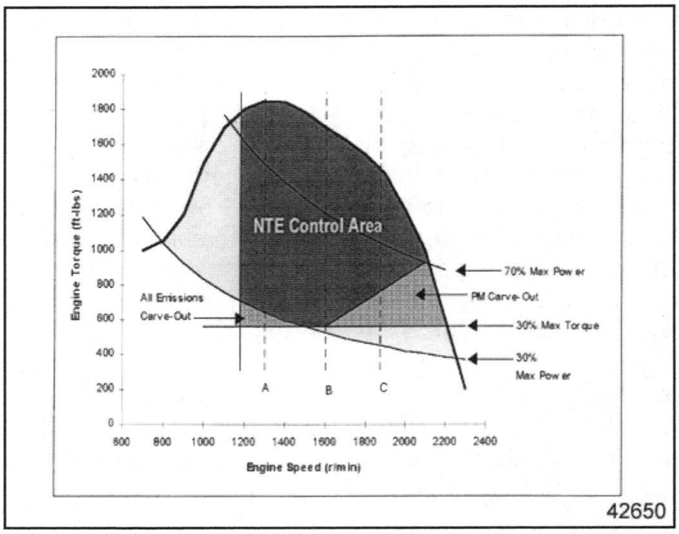

Figure 14 EPA Steady State NTE Test Area

3. The engine manufacturers agreed to meet a 4.0 g/hp-hr NOx requirement under transient conditions and 6.0 g/hp-hr NOx under steady state conditions until October 1, 2002.
4. EPA agreed to allow the use of certain contested auxiliary emission control devices (AECDs) until October 2002.
5. The engine manufacturers agreed to implement a program to reduce NOx emissions from older engines/vehicles through software reprogramming at the time of rebuild.
6. The engine manufacturers agreed to pay fines, to spend defined amounts for emission development toward compliance in 2002, and conduct specific emission related special projects.

The standard of NOx + NMHC of 2.5 g/hp-hr was technology forcing with most manufacturers going to exhaust gas recirculation (EGR) to meet the NOx requirement. (This included variable nozzle turbochargers, EGR coolers, EGR valves, sensors, software, etc.) The truck fleets wanted pilot models to test at least twelve months before October 2002, but manufacturers did not have pilots for field test until early 2002. Tests that fleets did participate in showed that the new systems were not very reliable and they were taking a 3 to 9 % fuel economy loss. This resulted in fleets buying their 2002/2003 requirements in the first nine months of 2002. An indication of EPA's potential dissatisfaction with the certification testing and supporting data was that only conditional certificates were issued to all manufacturers for the period from October 1 to December 31, 2002. Only Caterpillar did not use EGR for October 2002. They will introduce new technology in early 2003, called Advanced Combustion Emission Reduction Technology (ACERT). In the interim they will produce "bridge engines" using internal EGR and will pay non-conformance penalties (NCP).

Several disputes have arisen since the signing of the Consent Decree:

1. In early 2001, the Truck Manufacturers Association petitioned the court administrating the Consent Decrees, to intervene in the Consent Decree process. May 7, 2001, the Court denied their motion.
2. In 2001, many letters were written between manufacturers and EPA regarding AECDs (letters from EPA dated January 19, 2001 and August 17, 2001). Finally Caterpillar, on October 22, 2001, submitted a Memorandum in Support of Motion for Joint Status Conference.[49]
3. The NCP rules also came under fire because the NCP formula was for 1998 and later model year diesel engines with a relatively low cost-of-compliance value. On January 16, 2002,[50] EPA proposed NCPs as high as $15,000 per engine with a cost-of-compliance about ten times higher than the original rules that were approved by the Office of Management and Budget (OMB). On August 8, 2002,[51] EPA issued the final NCP rule and lowered the maximum penalty to $12,000 per year. The issue of EPAs cost of compliance numbers being off by an order of magnitude was never resolved.
4. Others started getting involved in the protest of the Consent Decree engines. Transport Topics reported in their December 17, 2001 issue that Schneider National, Yellow Freight, and Swift Transportation management said "most fleets won't buy 2003 trucks until they have time to evaluate them especially since they can buy low mileage used trucks."
    - On July 1, 2002, Transport Topics reported, "American Trucking Association said it had sent President Bush a letter signed by 345 carriers that are ATA members. In it they warned that unless something is done quickly, truck-related industries likely would be forced to lay off significant number of employees."
    - On July 22, 2002, Transport Topics reported, "American Trucking Association has asked the U.S. District Court for the District of Columbia to modify Consent Decrees…"
    - On August 5, 2002, Transport Topics reported
    - "The White House last week said it would not delay tougher federal engine emission standards scheduled to take effect October 1…"
5. In February 2002, both Caterpillar and DDC issued letters to EPA and DOJ attempting to trigger the "dispute resolution provision" of the Consent Decree.
    - The dispute was with the October 1, 2002 compliance date due to new information on cost of compliance (NCP rule) and fleet buy ahead concern.
    - May 10, 2002, DDC submits Statement of Position of DDC on dispute regarding October 1, 2002 requirements.
    - July 3, 2002, DDC petitions Honorable Henry H. Kennedy, Jr., United States District Judge, with a "Memorandum in Support of Motion to Review Dispute and Amend consent Decree."
    - September 5, 2002, Henry H. Kennedy, Jr. denies DDCs and Caterpillar's motion for review and modification of the consent Decree.

EPA 1999 RULE FOR 2004

On October 29, 1999, EPA published a NPRM [52] that proposed to reaffirm the technical and economic feasibility of the 2004 model year diesel NOx + NMHC standard. This proposal also included the requirement for all engine manufacturers to test under the supplemental tests of the Consent Decrees. On October 6, 2000, EPA issued a final rule that did affirm the feasibility of the NOx + NMHC standard and identified costs of complying with the rule.[53]

EPA 1999 / 2000 RULES FOR 2007

In May 1999, EPA issued an ANPRM [54] and a NPRM in June 2000 [55] proposing the following for 2007 through 2010:

1. NOx = 0.20 g/hp-hr by 2010 with a 50% phase-in for 2007, 2008 and 2009 (50% phase-in means NOx = 1.1 g/hp-hr)
2. HC = 0.14 g/hp-hr by 2010 with a 50% phase-in for 2007, 2008 and 2009
3. PM = 0.01 g/hp-hr in 2007 (no phase-in)
4. Diesel fuel sulfur limit reduced to 15 ppm maximum by mid-2006

In January 2001 a final rule was published [56] with a discussion of the technology required to make the 90% reduction in NOx (2.0 g/hp-hr to 0.2 g/hp-hr) and the 90% reduction in PM (0.1 g/hp-hr to 0.01 g/hp-hr). The preferred technology, by EPA, is NOx absorbers and catalyzed particulate traps. A Clean Diesel Independent Review Panel has been established by EPA under the Clean Air Act Advisory Committee to assess the progress of:

1. Manufacturers of diesel engines and emission control systems in developing technology to reduce engine exhaust pollutants
2. The fuels industry in developing and demonstrating technologies to effectively lower the sulfur level of highway diesel fuel

The EPA limited the scope of the Panel to addressing only these two areas with four specific questions to be answered.

The Panel issued their required report on October 30, 2002, [57] stating that..."The Panel found that significant progress is being made to develop NOx absorbers and catalyzed particulate filter systems for use in diesel engines in 2007"..."Technical challenges remain to be resolved"..."while we do not know the solutions now, none is considered to be insurmountable at this time."..."The next twelve months of development will be extremely important."

The National Petrochemical and Refiners Association brought suit against EPA challenging the diesel fuel rule but the U.S. Court of Appeals for the District of Columbia Court upheld the EPA regulations.

CONCLUSIONS / SUMMARY

In summary:

1. The diesel engine odor emissions were not regulated because the industry stepped up to the issues and significantly reduced odorant concentrations.
2. The gaseous emissions of HC, CO, NOx, continue to be reduced, especially NOx, which has been reduced from uncontrolled levels of approximately 20 g/hp-hr to 2 g/hp-hr (90% reduction) and will be reduced to 0.2 g/hp-hr by 2010 (99% reduction).
3. The particulate matter (PM) emissions have been reduced from approximately 1.0 g/hp-hr uncontrolled to 0.1 g/hp-hr (90% reduction) and will be reduced to 0.01 g/hp-hr in 2007 (99% reduction).
4. The periods of cooperation, the lawsuits, the "Statement of Principles", the Consent Decree, the clean Diesel Independent Review Panel, are all indications of the roller coaster relationship between industry, the regulators, and the environmentalists.

AUTHOR'S NOTES:

It is interesting to review the Federal Registers over the period of this paper. Back in the 60's and 70's very little was put in the preamble describing EPA's thoughts, logic or process. Now in the 2007 rulemaking there is voluminous material on process, health effects, technology, ambient air quality, cost, etc. This paper has been limited to heavy-heavy duty diesel engines but could have included much more when considering diesel engines, or gasoline engines, in trucks over 8500 lbs. (light-heavy and medium-heavy). Plus, I have not included any discussion of the toxicity of diesel exhaust, which is fully discussed in the Federal Register and other sources. It is a fact that California has declared diesel particulate as a human carcinogen and EPA published in the September 3, 2002 Federal Register,[58] that diesel particulate matter is a mobile source air toxic and a likely human carcinogen. It is well known that the U.S. emission standards are a major factor in the lack of dieselization in passenger cars in the U.S. As dieselization approaches 50% in European passenger cars, because of good fuel economy and low CO_2 emissions, we will see how the future emission standards will affect the market for diesel engines in trucks over 8500 lbs. GVW. A summary of the diesel gaseous emission regulations from 1973 to 2010 is shown in Fig. 15.

Diesel Emission Standards U.S. Federal and California vs. Years

	U.S. FEDERAL						CALIFORNIA					
Year	CO	HC	NOx	HC+NOx	PM Truck	PM Bus	CO	HC	NOx	HC+NOx	PM Truck	PM Bus
1973	-		-	-	-	-	40	-	-	16	-	-
1974	40		-	16	-	-	40	-	-	16	-	-
1975	40	-	-	16	-	-	30	-	-	10	-	-
1976	40	-	-	16	-	-	30	1.0	7.5	5	-	-
1979	25	1.5	-	10	-	-	30	1.0	7.5	5	-	-
1980	25	1.5	-	10	-	-	30	-	-	6	-	-
1984[1]	25	1.5	-	10	-	-	.5	1.3	5.1	-	-	-
1985[2]	15.5	1.3	10.7	-			15.5	1.3	5.1	-	-	-
1988	15.5	1.3	10.7	-	0.60	0.60	15.5	1.3	6.0	-	0.60	0.60
1990	15.5	1.3	6.0	-	0.60	0.60	15.5	1.3	6.0	-	0.60	0.60
1991	15.5	1.3	5.0	-	0.25	0.25	15.5	1.3	5.0	-	0.25	0.10
1993	15.5	1.3	5.0	-	0.25	0.10	15.5	1.3	5.0	-	0.25	0.10
1994	15.5	1.3	5.0	-	0.25	0.07	15.5	1.3	5.0	-	0.10	0.07
1996	15.5	1.3	5.0	-	0.10	0.05	15.5	1.3	5.0	-	0.10	0.05
1998	15.5	1.3	4.0	-	0.10	0.05	15.5	1.3	4.0	-	0.10	0.05
2002[3]	15.5	-	-	2.5	0.10	0.05	15.5	-	-	2.5	0.10	0.05
2004	15.5	-	-	2.5	0.10	0.05	15.5	-	-	2.5	0.10	0.05
2007[4]	15.5	-	-	1.1	0.01	0.01	15.5	-	-	1.1	0.01	0.01
2010	15.5	-	-	0.2	0.01	0.01	15.5	-	-	0.2	0.01	0.01

(1) Transient for California, (2) Transient for Federal and California,
(3) Consent Decree Companies, (4) 15 ppm S fuel

42651

Figure 15 Diesel Emission Standards: U.S. Federal and California vs. Years

ACKNOWLEDGMENTS

I would like to acknowledge people from industry and government who made significant contributions to the state-of-the-art of diesel emission in the early days of emission measurement and emission control.

Stan Ford, Dick Hames, Detroit Diesel

Tom Head, Al Carey, and Roger Bascom: Cummins

Ernie Landen and Joe Perez: Caterpillar

John Johnson: International Harvester

Joe Colucci, Gerry Barnes, and Howard Pearsall: General Motors

Tom Young: Engine Manufacturers Association

Karl Springer: Southwest Research Institute

Dick Hurn: Bureau of Mines

Ralph Stahman and George Kittredge: U.S. Public Health Service

Gay Hass, Miles Brubacher: State of California

I would also like to acknowledge the following people from Detroit Diesel who helped me with this paper: Fred Rushman, Barbara Lauter, Joan Mannino, Michele Moss, John Farmer, Josh Yaker, Don Keski-Hynnilla, and Joanna Vardas.

REFERENCES

1) SAE 670224 "Smoke Production in Diesel Engines" by A. W. Carey, Jr.
2) SAE 710484 "Design Factors That Affect Diesel Emissions" by R. C. Bascom, L. C. Broeing and D. E. Wulfhorst
3) SAE Transactions, Vo. 50, December 1942, pp. 509-520 "Effect of Diesel Fuel on Exhaust Smoke Odor" by R. S. Wetmiller and L. E. Endsley, Jr.
4) SAE 570057 "Diesel Exhaust Odor – Its Evaluation and Relation to Exhaust Gas Composition" by F. G. Rounds and H. W. Pearsall
5) SAE 660548 "Reduction of Diesel Smoke in California" by M. L. Brubacher
6) SAE 680445 "Relation of Lean Combustion Limits in Diesel Engines to Exhaust Odor Intensity" by G. J. Barnes
7) SAE 700105 "Evaluation of Vehicle Exhaust Gas Odor Intensity Using Natural Dilution" by J. M. Colucci and G. J. Barnes
8) SAE 700734 "Reducing Hydrocarbons and Odor in Diesel Exhaust by Fuel Injector Design" by H. S. Ford, D. F. Merrion and R. J. Hames.
9) SAE 710671 "Some Effects of Fuel Injection System Parameters on Diesel Exhaust Emissions" by R. J. Hames, D. F. Merrion and H. S. Ford
10) Natural Petroleum Refiners Association September 1966 "An Investigation of Diesel Powered Vehicle Odor and Smoke" by K. J. Springer and R. C. Stahman
11) ASME 69 – WA / APC-3 "Four Years of Diesel Odor and Smoke Control Technology Evaluations – A Summary" by K. J. Springer and C. T. Hare
12) Federal Register Vol. 33, No. 2, Thursday, January 4, 1968
13) Federal Register Vol. 33, No. 108, Tuesday, June 4, 1968
14) Bureau of Mines Report on "Diesel Engines Underground IV, Effect on Composition of Exhaust Variables Influencing Fuel Injection" by H, C. Holtz, L. B. Berger, M.A. Elliott and H. H. Schrank
15) Bureau of Mines "Sampling and Determination of Aldehydes in Diesel Engine Exhaust Gas and in Mine Air" by H. W. Busch and L. B. Berger
16) SAE Quarterly Transactions, Vol. 4, "Composition of Diesel Exhaust Gas" by M. A. Elliott and R. F. Davis
17) APCA Journal, Vol. 5, "Composition of Exhaust Gases from Diesel, Gasoline and Propane Powered Motor Coaches" by M. A. Elliott, G. J. Nebel and F. G. Rounds
18) SAE 570050 "Diesel Exhaust Odor – its Evaluation and Relation to Exhaust Gas Composition" " by F. G. Rounds and H. W. Pearsall
19) SAE 630167 "Nitrogen Oxides and Variables in Precombustion Chamber Type Diesel Engines" by E. W. Landen
20) American Petroleum Institute 30th Mid Year Meeting Pre-print 24-65 "Air Pollution Inventory Enter the Diesel" by R. W. Hurn and D. E. Selzinger
21) SAE 660550 "Exhaust Characteristics of Automotive Diesel" by R. C. Schmidt, A. W. Carey and Roy Kamo
22) A Report to the California Legislative California Department of Public Health "The Diesel Vehicle and Its Role in Air Pollution" by J. R. Kinosian, J. A. Maga and J. R. Goldsmith, M.D
23) Seventh World Petroleum Congress "Atmospheric Pollution from Diesel Engines" by H. R. Talifarro, J. O. Becker and T. O Wagner
24) SAE 670089 "Measuring The Total Hydrocarbons in Diesel Exhaust"
25) SAE 680418 "Techniques for Diesel Emission Measurement" by R. W. Hurn and W. F. Marshall
26) SAE 680528 "Factors Influencing Diesel Emissions" by W. F. Marshall and R. W. Hurn
27) SAE 680419 "A Flame Ionization Technique for Measuring Total Hydrocarbons in Diesel Exhaust" by J.H. Johnson, E. J. Sienicki and O. F. Zeck
28) SAE 680420 "Exhaust Emission Characteristics of Four-Stroke, Direct Injection, Compression Ignition Engines" by V. S. Yumlu and A. W. Carey, Jr.
29) SAE 680421 "Exhaust Emission Characteristics of Precombustion Chamber Engines" by J. W. Perez and E. W. Landen
30) SAE 680422 "Effect of Design Revisions on Two-Stroke Cycle Diesel Engine Exhaust" by D.F. Merrion
31) SAE 680433 "Smoke and Odor Control for Diesel-Powered Trucks and Buses" by R. C. Stahman, G. D. Kittredge and K. J. Springer
32) SAE 680444 "Diesel Exhaust Odor Measurement Using Human Panels" by L. R. Reckner and R. E. Squires
33) SAE 700671 "A Status Report on the Development of the 1973 California Diesel Emissions Standards" by R. C. Bascom and G. C. Hass
34) Federal Register Volume 36, Number 193, Tuesday, October 5, 1971
35) Federal Register Volume 37, Number 221, Wednesday, November 15, 1972

36) University of Missouri, Air Pollution Conference, November 19, 1969 “Diesel and Turbine Driven Vehicles and Air Pollution” by D. F. Merrion
37) Federal Register Volume 41, Number 101, Monday, May 20, 1976
38) Federal Register Volume 42, Number 174, Thursday, September 8, 1977
39) Federal Register Volume 45, Number 14, Monday, January 21, 1980
40) Federal Register Volume 46, Number 4, Wednesday, January 7, 1981
41) Federal Register Volume 48, Number 222, Wednesday, November 16, 1983
42) Federal Register Volume 49, Number 200, Monday, October 15, 1984
43) Federal Register [46FR6838] January 19, 1981
44) Federal Register Volume 50, Number 51, Friday, March 15, 1985
45) Federal Register Volume 56, Number 185, Tuesday, September 24, 1991
46) Federal Register Volume 58, Number 55, Wednesday, March 24, 1993
47) Federal Register Volume 61, Number 125, Thursday, June 27, 1996
48) Federal Register Volume 62, Number 203, Tuesday, October 21, 1997
49) In the United States District Court for the District of Columbia Civil action No. 98-2544 HHK.
50) Federal Register Volume 67, Number 11, Wednesday, January 16, 2002
51) Federal Register Volume 67, Number 153, August 8, 2002
52) Federal Register 65FR58472, October 29, 1999
53) Federal Register 65FR59896, October 6, 2000
54) Federal Register 64FR26142, May 13, 1999
55) Federal Register 65FR35430, June 2, 2000
56) Federal Register 66FR5001 January 18, 2001
57) Final Report: Meeting Technology Challenges for the 2007 Heavy-Duty Highway Diesel Rule. Clean Diesel Independent Review Subcommittee Clean Air Act Advisory Committee, October 30, 2002.
58) Federal Register Volume 67, Number 170, September 3, 2002.
59) EPA/CARB Symposium, Sacramento, California, October 1997 (from DDC files)

2003-01-3753

Developments In Diesel Emission Aftertreatment Technology

Philip G. Blakeman
Andrew F. Chiffey
Paul R. Phillips
Martyn V. Twigg
Andrew P. Walker
Johnson Matthey

ABSTRACT

The modern Diesel engine is one of the most versatile power sources available for mobile applications. The high fuel economy and torque of the Diesel engine has long resulted in global application for heavy-duty applications. Moreover, the high power and excellent driveability of today's turbo-charged small high-speed Diesel engines, coupled with their low CO_2 emissions, has resulted in an increasing demand for Diesel powered light-duty vehicles.

However, the demand for Diesel vehicles can only be realised if their exhaust emissions meet the increasingly stringent emissions legislation being introduced around the world. In the USA, light-duty Diesel (LDD) vehicles will have to meet the same emissions legislation as gasoline vehicles from 2004 onwards, while in Europe a similar target is expected when European Stage 5 legislation is introduced. In practice, such targets mean very high reductions (up to 90%) of nitrogen oxides (NOx) and particulate matter (PM) emissions may be required from today's levels. Drastically reduced NOx and PM emissions from heavy-duty Diesel (HDD) engines are also required in Europe and the USA in a similar time frame.

This paper reviews the developments in Diesel exhaust emissions control devices. The application of Diesel oxidation catalysts, particulate filters, and NOx control catalysts (NOx adsorber catalysts and Selective Catalytic Reduction systems) to help meet both light- and heavy-duty legislation is discussed. An overview of likely catalyst system designs to achieve high levels of PM and NOx conversion is given.

INTRODUCTION

Future legislation around the world for Diesel vehicles is becoming increasingly stringent, with the emphasis placed on lower PM and NOx emissions. Advances in engine design and control has shown, and continues to show, significant reductions in emissions. At the same time further advances in the high torque and excellent driveability of the Diesel engine has fuelled expansion of the Diesel sector.

To help achieve the tighter emissions requirements, advanced aftertreatment technology from oxidation catalysts to systems capable of high conversion of all four pollutants is being developed.

For passenger cars in Europe, Stage 4 legislation will be introduced in 2005, with significantly lower emissions limits than the currently applied Stage 3 legislation (Table 1). For Stage 4, NOx and PM emissions have been halved from the Stage 3 limits, while CO and HC emissions require more moderate reductions. In addition, the durability requirement over which the emission standards have to be met has increased from 80,000 km for Stage 3 to 100,000 km for Stage 4.

While LDD Stage 5 has yet to be set, it is widely expected that a further reduction of NOx and PM limits will occur, to give emissions levels comparable to those required from gasoline cars (0.08 g/km NOx in Stage 4), and similar to the standards already set in the USA.

Future USA legislation for LDD vehicles also focuses on NOx and PM emissions. For passenger vehicles, Tier 2 limits phase in between 2004 and 2007 with the limits for Diesel vehicles being the same as gasoline-powered vehicles. By 2007, the fleet average NOx emissions must meet 0.07 g/mile for all passenger vehicles up to 8500 lb. Within Tier 2, there are a number of emissions levels (bins) which vehicles can be certified to – the 0.07 g/mile fleet average NOx value occurs in bin 5 (Table 2). The emissions must be durable over 120,000 miles.

Emissions legislation for HDD engines in Europe is also forcing lower NOx and PM emissions. Table 3 shows the emissions limits for the European steady-state cycle (ESC) test procedure. Stage 4 (commencing in 2005) has a significant drop in PM and a more moderate NOx decrease. Stage 5 (commencing in 2008) introduces a further drop in NOx.

The steady decrease of NOx emissions for HDD engine legislation in Europe is in contrast to that in the USA. In the 2007 – 2010 timeframe, the NOx and PM emissions for HDD engines in the US are reduced ten-fold from the 2004 limits

(Table 4). The PM legislation is effective immediately in 2007, while the NOx and NMHC limits will be gradually phased in between 2007 and 2010. In addition, not-to-exceed limits of 150% of the legislation have been set, and these apply to a large proportion of the engine map in an effort to ensure high levels of emission control occur under a whole range of driving conditions. The emissions have to be durable for up to 435,000 miles.

Table 1: European LDD emissions legislation (g/km).

Stage (Year)	*CO*	*HC+NOx*	*NOx*	*PM*
3 (2000)	0.64	0.56	0.5	0.05
4 (2005)	0.5	0.3	0.25	0.025

Table 2: USA Tier 2 LDD emissions legislation (g/mile) to be phased in from 2004.

Bin	*CO*	*NMOG*	*NOx*	*PM*
8	4.2	0.125	0.2	0.02
7	4.2	0.09	0.15	0.02
6	4.2	0.09	0.1	0.01
5	4.2	0.09	0.07	0.01
4	2.1	0.07	0.04	0.01
3	2.1	0.055	0.03	0.01
2	2.1	0.01	0.02	0.01
1	0.0	0.0	0.0	0.0

Table 3: European HDD emissions legislation (g/kW-hr) for the ESC test.

Stage (Year)	*CO*	*HC*	*NOx*	*PM*
3 (2000)	2.1	0.66	5.0	0.1
4 (2005)	1.5	0.46	3.5	0.02
5 (2008)	1.5	0.46	2.0	0.02

In Brazil, Diesel engines are used in trucks and busses, however, and legislation for these has been based on the European standards. Currently, legislation equivalent to European Stage 2 is in force (ECE R49 test, 4.0 g/kW-hr CO, 1.1 g/kW-hr HC, 7.0 g/kW-hr NOx and 0.15 g/kW-hr PM). In 2006 the legislation will become equivalent to European Stage 3, and in 2009 it will be equivalent to European Stage 4 (Table 3).

Table 4: USA HDD emissions legislation (g/bhp-hr).

Year	*CO*	*HC / NMHC[a]*	*NOx*	*PM*
1998	15.5	1.3	4.0	0.1
2004[b]	15.5	0.5	2.0	0.1
2007 – 2010	15.5	0.14	0.2	0.01

a) HC applies to 1998 legislation b)
An alternative limit of NOx + NMHC = 2.5 g/bhp-hr also exisits

Legislation has been set for light vehicles (< 1700 kg) to be effective in 2007 and 2009. The test cycle is to be the FTP-75 cycle, but the legislated values are not equivalent to those in the USA legislation (Table 5). As in the rest of the world, there is a significant tightening of the NOx legislation by 2009.

Table 5: Brazilian LDD emissions legislation (g/km).

Level (Year)	*CO*	*NMHC*	*NOx*	*PM*
4 (2007)	2.0	0.16	0.6	0.05
5 (2009)	2.0	0.05	0.25	0.05

Such legislation is forcing the use of exhaust aftertreatment systems of greater complexity. To help ensure prolonged activity and durability of these systems, Diesel fuel sulfur levels are being reduced. For Stage 4 in Europe, 50 ppm sulfur fuel is required (down from 350 ppm sulfur for Stage 3), while in the US 15 ppm sulfur fuel is required in 2006 (down from 500 ppm today). In Brazil, 500 ppm sulfur fuel is the target for 2009 (although 50 ppm is proposed for metropolitan regions), and this could limit the types of aftertreatment technology applicable in this time frame.

The main aftertreatment devices for use in Diesel exhaust are discussed below.

DIESEL OXIDATION CATALYSTS (DOC)

Today, the main market for LDD vehicles is in Europe, and the majority of these vehicles currently meet Stage 3 legislation with a DOC as the only aftertreatment device. The DOC will remain a principal aftertreatment device for future vehicles, either used alone or, as discussed later, as part of more advanced emissions control systems.

The DOC principally controls HC, CO and the volatile organic fraction (VOF) of PM emissions. The legislation for these pollutants is tighter for Stage 4 than Stage 3, and there is a general trend towards lower exhaust gas temperatures and higher CO and HC emissions from LDD engines, mainly as the result of advanced engine control methods that are used to minimise NOx emissions. These factors place greater emphasis on the need for high catalyst activity at low temperatures [1, 2].

High temperature durability is important in order to withstand maximum exhaust temperatures and exotherm events that may occur on the catalyst surface (the latter usually caused by HC that can be adsorbed on the catalyst surface at low temperatures). High durability of the catalyst with respect to extended low temperature ageing is also required. Sulfur, heavy HC, and carbonaceous deposits can build-up on the catalyst surface at low temperatures and poison the catalytic activity. This ageing mechanism is particularly important because of the general trend towards lower exhaust gas temperatures on modern LDD aplications.

Figure 1 illustrates the progress that has been made in improving low temperature performance. The engine bench CO light-off activity of various catalysts is shown after extended low temperature ageing (< 320°C). A typical Stage 3 DOC demonstrates 50% CO conversion at 180°C (the T50 value) on this test. By contrast, a typical DOC suitable for Stage 4 applications has a T50 of 164°C, clearly a significant improvement. The latest developments in DOC technology demonstrate an even lower T50 value of 154°C – a 26°C improvement in light-off activity compared to a current Stage 3 DOC technology.

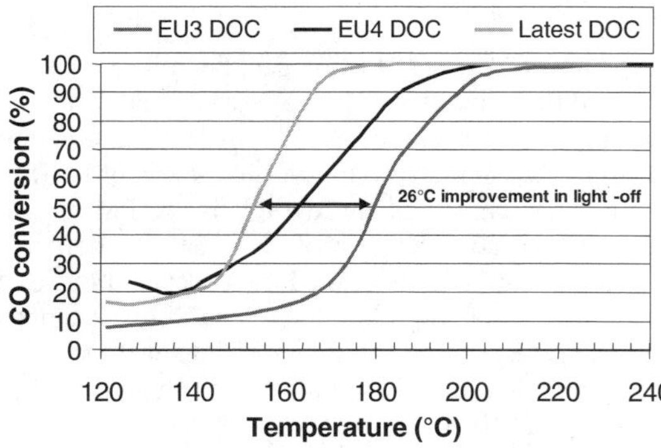

Figure 1: Engine bench CO light-off of various DOC technologies after extended low temperature ageing.

The improvement in vehicle emissions between a Stage 3 and Stage 4 DOC after extended low temperature ageing is shown in Figure 2. The improved low temperature activity of the Stage 4 DOC results in a faster light-off on the European test cycle. The improved activity enables lower emissions to be achieved with almost half of the platinum content than that used in the Stage 3 catalyst (50 g/ft^3 on the Stage 4 DOC and 90 g/ft^3 on the Stage 3 DOC). Thus the improved activity and durability of the Stage 4 DOC demonstrated on the engine bench translates to an emissions improvement at significantly lower platinum cost on a vehicle.

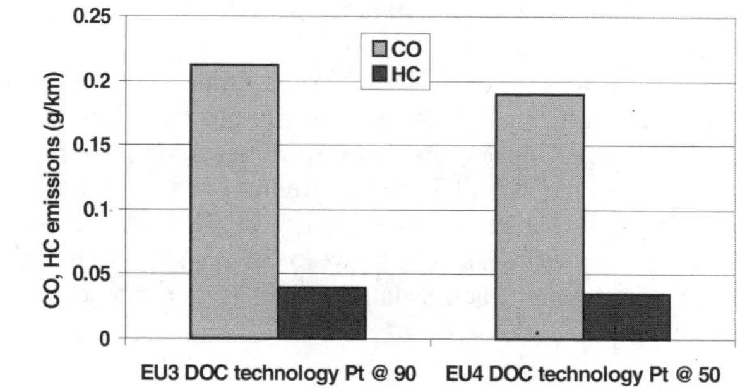

Figure 2: CO and HC emissions over the European Stage 3 test cycle (on a Stage 3 vehicle) for a Stage 3 DOC with a Pt loading of 90 g/ft^3 and a Stage 4 DOC with a Pt loading of 50 g/ft^3 after extended low temperature ageing.

DEVICES FOR CONTROL OF PM EMISSIONS

A number of methods of achieving high levels of removal of PM emissions in the exhaust of Diesel vehicles have been developed that are mainly based on a filtration technique for removal of soot from the exhaust gas.

The most common and well-known filtration devices are wall-flow filters. These filters typically consist of a honeycomb substrate and may be made from cordierite or silicon carbide (although other materials have also been employed) [3, 4, 5]. Typically, half of the cells of the honeycomb substrate are sealed at the inlet face in a checkerboard pattern, and the remaining cells are sealed at the outlet face of the substrate. When placed in the exhaust, the gas cannot pass straight through the device, but instead is forced by the seals to pass through the walls of the cells. This process results in the removal of soot (and other particles) suspended in the gas. The filtration efficiency of the wall-flow filters can be extremely high, with greater than 95% removal of particulate mass possible. This type of wall-flow filter is already fitted to a number of passenger cars in Europe [6].

Other filtration devices have also been developed. Sintered metal, ceramic foams and compacted fibres have also been demonstrated to remove PM [7, 8]. In addition, filter designs have been developed with partial flow through the filter material to give partial filtration [9].

Whatever the type of PM control device employed, the common system control issue is that soot removal by the device cannot continue indefinitely without the soot itself being removed from the system, typically by combustion. The implications of soot regeneration for LDD and HDD applications are now discussed.

CONTROL OF LDD PARTICULATE EMISSIONS

In Europe, advanced engine design and combustion control has been successfully employed in combination with low temperature light-off oxidation catalysts to reduce soot and heavy HC emissions from LDD vehicles. The contribution of sulfate emissions to total PM emissions has also been lowered by the reduction of the sulfur level in the fuel.

The further lowering of PM and NOx limits however has increased the need for PM filtration devices to meet European Stage 4 limits particularly for heavier vehicles, and for LDD vehicles to meet USA '07 limits. Improvements in engine technology can further reduce PM and NOx emissions, however, the general trade-off between NOx and PM engine-derived emissions necessitates the requirement for aftertreatment devices.

The effect on PM emissions of fitting a wall-flow filter device to a Stage 3 passenger car is shown in Figure 3. As expected, highly efficient removal of particulate is demonstrated.

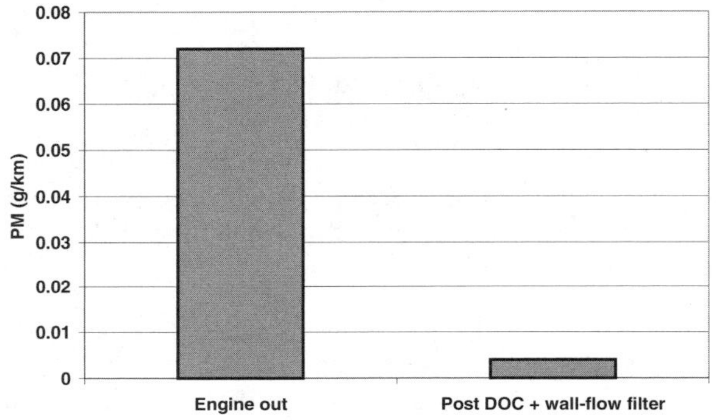

Figure 3: Effect on PM emissions over the European Stage 3 test cycle of fitting a DOC + wall-flow filter device to a Stage 3 passenger car.

In order to maintain functionality of the filter system, the filter device has to be regenerated. Combustion of soot within the filter can be achieved either by reaction with oxygen (Reaction 1), or by reaction with nitrogen dioxide (NO_2, Reaction 2). In the case of Reaction 2, a catalyst can be used to oxidise engine-out NO to NO_2 (Reaction 3), as shown in Figure 4, and the NO_2 can then react with the collected soot.

$$PM + O_2 \rightarrow CO_2 \quad (1)$$

$$PM + NO_2 \rightarrow CO_2 + NO \quad (2)$$

$$NO + O_2 \rightarrow NO_2 \quad (3)$$

Figure 5 shows the temperatures at which soot combustion is achieved with either O_2 or NO_2 gas. The combustion of soot with NO_2 clearly occurs at significantly lower temperatures than with O_2.

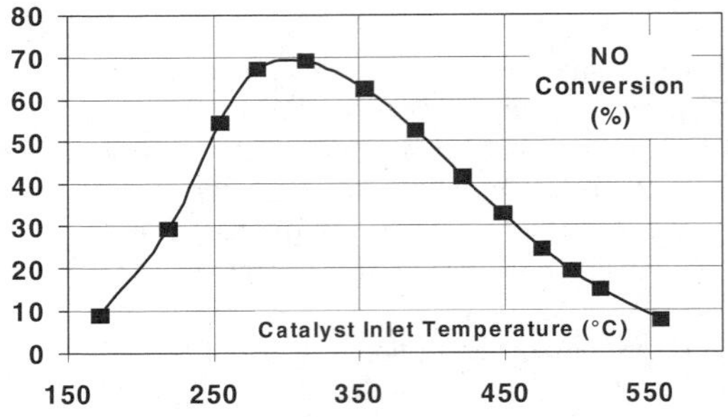

Figure 4: Temperature window for oxidation of NO to NO_2 over an oxidation catalyst.

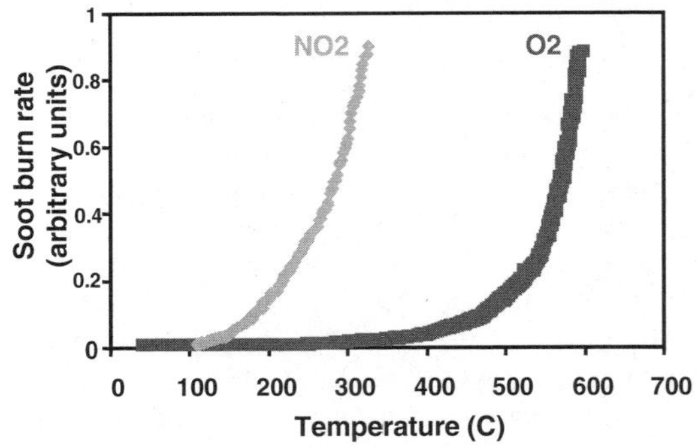

Figure 5: Combustion of Diesel soot in oxygen and nitrogen dioxide gas feeds.

In practice, the minimum temperature required to ensure a significant soot combustion rate with NO_2 is around 270°C. The soot combustion rate will also depend on the ratio of PM and NO_2 available for a given engine condition. In general, a weight ratio of around 16:1 NOx:PM is required to ensure continuous removal of soot. Under operating conditions that have sufficient temperature, combustion of the filtered soot with NO_2 can either provide continuous, or at least partial removal of soot (depending on the NOx:PM ratio), and the optimised design of the catalyst and overall system layout can maximise the efficiency of this low temperature soot combustion.

Whilst NO_2 combustion of soot can be utilised in certain driving conditions, modern LDD vehicles are characterised by low exhaust gas temperatures and low NOx concentrations emitted from the engine. The use of high levels of EGR mean that engine NOx emissions are low relative to PM emissions, and therefore the concentration of NO_2 that can be made by the oxidation catalyst is also low. Low concentrations of NO_2 result in low rates of soot burn, even if sufficient temperature is achieved in the exhaust gas. Figure 6 illustrates the effect of NOx concentration on soot combustion (monitored by the backpressure of the filter, and initial and final soot mass within the filter) on a LDD engine with and without EGR enabled at 350°C. Without EGR, there is 4.2 times higher mass flow of NOx than with EGR enabled, and a high rate of soot combustion occurred. With a high EGR rate enabled (*i.e.* as found on advanced engine applications), there is only half the amount of soot removed in the same time. Whilst complete regeneration of the filter would take a relatively long time under the high EGR rate condition, it can be seen that the balance point is exceeded such that the filter loses soot mass. This NO_2-based passive regeneration can be used to minimise the build-up of soot, even when the engine-out NOx/PM ratio is too low to allow totally passive regeneration. This soot combustion via passive regeneration can significantly reduce the frequency of active regeneration.

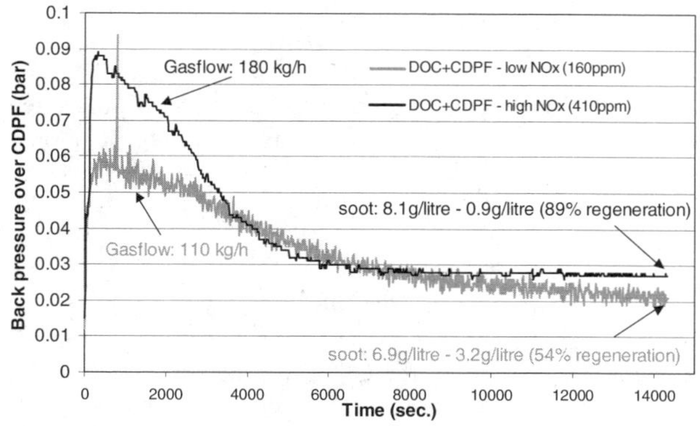

Figure 6: Effect of NOx mass flow (EGR on or off) at 350°C on soot combustion with NO_2.

In addition to the low temperature NO_2 soot combustion mechanism, in order to ensure the filter can be regenerated in all operating conditions, an active strategy is required to periodically combust soot from the filter. For example, combustion of the soot can be achieved by raising the temperature of the exhaust gas to that required for combustion with oxygen. The excess oxygen conditions present in the Diesel exhaust enables a fast rate of soot combustion to take place. Thus, the higher temperature conditions for oxygen regeneration only need to be maintained for a relatively short time.

The two key steps involved in an oxygen-based combustion of soot are raising the exhaust gas temperature in the filter to initiate soot combustion, and then controlling the combustion under transient operation once soot burning has been initiated.

Raising the exhaust gas temperature within the filter can be done through modifications to the engine calibration to create higher temperatures. Fuel can be injected into the cylinder after the main fuel combustion event, and this late injected fuel can be combusted in-cylinder to increase the gas temperatures. Alternatively, excess fuel can be injected very late after the main injection into the cylinder, or directly into the exhaust gas. This fuel can then be combusted over an oxidation catalyst to create a certain exotherm. A combination of both techniques can also be employed.

Once soot combustion is initiated, it is important that the regeneration is maintained and controlled under transient operating conditions. Following the initiation of soot combustion, the gas flow rate may substantially change to a lower flow, for example, as expected under engine idling conditions. The resulting lower gas flow will be less effective at removing the heat generated by soot combustion, and therefore a significant rise in filter bed temperature can occur. The effects of this type of transient gas flow conditions during soot regeneration of a SiC filter has been modelled to show the influence of different oxygen concentrations under the low flow condition, as shown in Figure 7. The simulation is based on a 650°C inlet gas condition to initiate the soot combustion of a filter at an exhaust flow equivalent to idle conditions. At high oxygen concentrations, a maximum bed temperature of 1800°C is obtained. Lowering the oxygen concentration in the exhaust (for example, by increasing the EGR rate) significantly slows the rate of soot combustion and the peak temperatures within the filter are much lower.

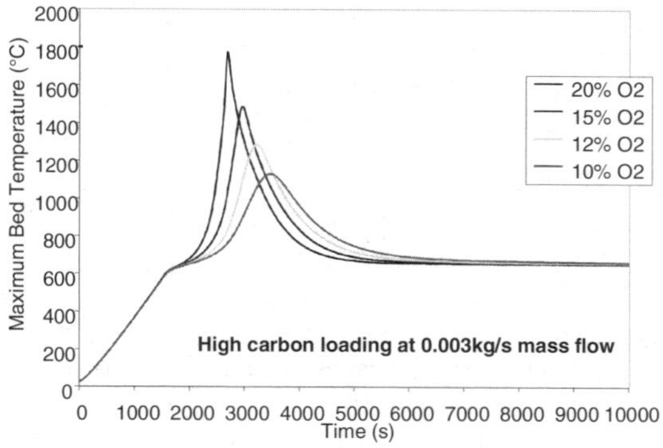

Figure 7: Modelling results showing maximum predicted filter bed temperatures for a SiC wall-flow filter under idle conditions.

There are three different system architectures possible for a filter system, shown schematically in Figure 8.

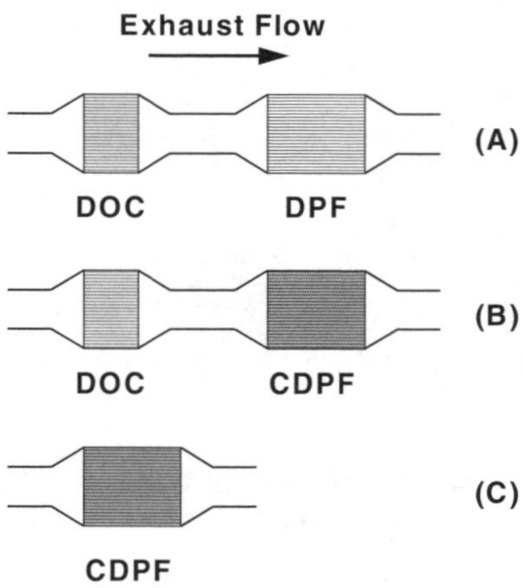

Figure 8: Three possible DPF systems (DOC = Diesel Oxidation Catalyst, DPF = Diesel Particulate Filter, CDPF = Catalysed Diesel Particulate Filter).

System A

This system utilises the Diesel oxidation catalyst (DOC) placed upstream of the DPF to create the exotherm needed for soot combustion. The temperature required at the rear of the DOC during an active regeneration may be significantly higher than the temperature at the inlet to the DPF because of heat losses between the DOC and DPF. Therefore the oxidation catalyst must have high durability to high temperature conditions. The upstream oxidation catalyst can also be used to maximise the passive soot combustion via the oxidation of NO to NO_2, and subsequent reaction of NO_2 with soot. A variation on system A is already in production in Europe, where a fuel additive is used to lower the temperature required to combust soot with oxygen to around 450°C [6].

System B

System B also utilises the upstream DOC to create exotherms to aid the active regeneration strategy, and to generate NO_2. In addition, the filter is catalysed to improve soot combustion. This system provides improved NO_2-based regeneration, since NO formed via reaction 2 can be reoxidised on the CDPF and re-used to react further with soot in the filter. In this way, the system makes better use of the available NOx.

Figure 9 compares the low temperature performance of System A and System B over a low temperature cycle, in which the inlet temperature is always below 270°C. The backpressure was used to provide an indication of the extent of regeneration. The backpressure of a bare filter is also shown, to demonstrate the rate of backpressure increase from soot accumulation in the absence of any regeneration. The backpressure of System B is significantly below that of System A, revealing that System B has better low temperature passive regeneration [10].

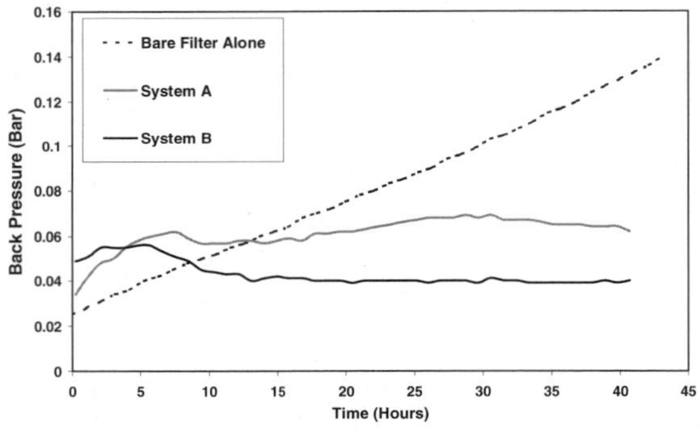

Figure 9: Passive regeneration of Systems A and B over a low temperature cycle.

In addition, CO can be formed by incomplete oxidation of soot or the HC used for the exotherm generation, and the CDPF is very efficient at removing this (Figure 10). In addition, the exotherm does not need to be created 100% by the DOC. The catalyst on the filter can also be utilised, thus creating extra heat in the same location as the soot.

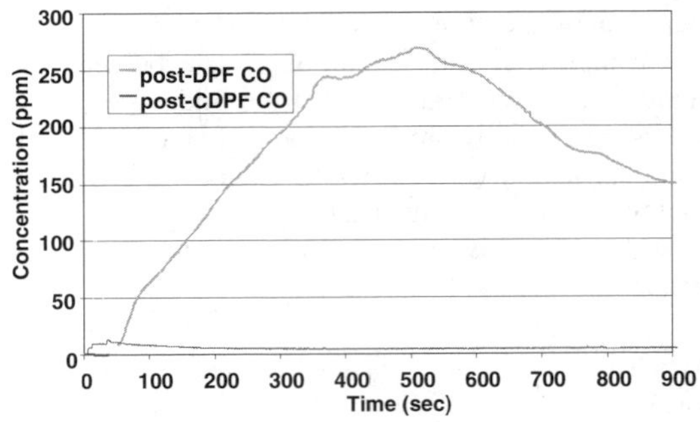

Figure 10: Comparison of tailpipe CO emissions from System A and System B during an oxygen-based soot regeneration at 620°C.

System C

System C is the most thermally efficient of the systems because there is only one substrate to heat-up, and the exotherm created during the CDPF regeneration is created at the same location as the soot. As there is no separate flow-through DOC, all CO and HC emissions must be controlled by the CDPF, and therefore this arrangement places very high demands on the catalyst coating. The catalyst must have very high thermal durability to keep high activity over repeated soot regenerations where severe temperature conditions may be encountered. In addition, the catalyst has a high exposure to inorganic ash deposits, such as calcium, zinc and phosphorus, to which it must also be very tolerant.

The common element to all three systems is the extreme conditions to which the DOC and/or CDPF are exposed. New specialised catalyst formulations have been developed for these applications.

CONTROL OF HDD PARTICULATE EMISSIONS

In Europe, the five-fold reduction in PM emissions required by the incoming Stage 4 legislation is not accompanied by a large reduction in the NOx emission limit with respect to the Stage 3 level. In principle there are two ways to meet the Stage 4 limits. EGR may be applied to reduce the NOx level, but this will lead to an increase in PM emissions, so it is expected that a filter will need to be added to control PM emissions when using this strategy. The alternative approach is to advance the injection timing, which leads to low PM emissions (within the Stage 4 limits) but high NOx emissions (outside the Stage 4 limits). Selective Catalytic Reduction (SCR) will then be applied to control the NOx (see below). The latter strategy is expected to be the principal one used since advancing the timing leads to a significant fuel economy benefit.

A similar argument applies to European Stage 5 limits, as the PM limit is currently expected to be unchanged from the Stage 4 level. The NOx limit for European Stage 5 is lower than Stage 4. However, the possibility still remains that European Stage 5 could be achieved by a combination of engine-out PM control and SCR.

In the USA, many HDD engines are already being fitted with EGR to meet the USA 2004 legislation. This strategy, combined with the large reduction in both NOx and PM emissions required to meet the US 2007 legislation, means that one possible strategy to meet these limits is to combine EGR with a particulate filter [11].

The same operational considerations discussed above for LDD also apply to HDD particulate control devices. However, due to the higher temperatures generally prevalent in HDD applications, the extent of NO_2-based regeneration is likely to be higher in HDD systems. Indeed, there is substantial experience with retrofitting such devices to existing vehicles and relying entirely upon NO_2 to regenerate the filter [12, 13].

Nevertheless an oxygen-based regeneration must still be used to ensure the system works in all situations. Figure 11 shows an oxygen regeneration of System A (with a cordierite wall-flow filter) conducted on a HDD engine. Diesel fuel was injected directly into the exhaust to create a temperature of 600°C after the DOC, with an engine-out temperature of 250°C. The filter regeneration was monitored by measuring the backpressure of the system. Once regeneration starts, it is rapid and controlled, with the backpressure dropping from 0.1 bar to 0.03 bar in about 4 minutes.

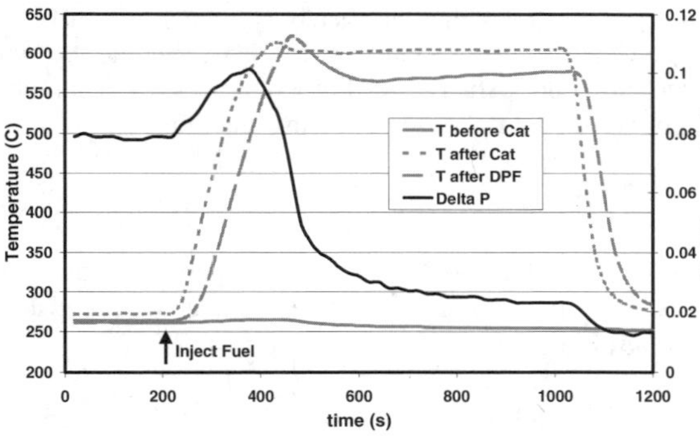

Figure 11: System backpressure and temperatures during an oxygen-based soot regeneration with System A on a HDD engine.

DIESEL NO_X CONTROL AFTERTREATMENT

Both LD and HD Diesels in the USA will have to meet extremely stringent NOx emissions legislation in the 2007 – 2010 time frame, almost certainly necessitating the use of aftertreatment for NOx control.

In Europe, catalytic treatment of NOx emissions is also assumed to be required on LDD vehicles for Stage 5 limits (when they are set). In addition, HDD engines are likely to use NOx control for Stage 4 onwards.

There are two main aftertreatment technologies for high levels of NOx conversion – NOx Adsorber Catalysts (NACs) and Selective Catalytic Reduction (SCR).

NO_X ADSORBER CATALYSTS (NAC)

The chemical principles behind NOx adsorber catalysts have been well documented [14, 15]. Under lean conditions (the normal operating mode of a Diesel engine), the catalyst promotes NOx adsorption as illustrated by reactions 4 and 5, where M is the NOx adsorbing element, MO is the stable oxidized form of the element, and MNO_3 is the stable NOx containing compound formed by element M.

$$2NO + O_2 \rightarrow 2NO_2 \quad (4)$$

$$NO_2 + MO \rightarrow MNO_3 \quad (5)$$

Under fuel-rich conditions, the catalyst promotes the reverse reaction *i.e.* decomposition of the nitrate phase to release the stored NOx (reaction 6). Finally, the catalyst reduces the NOx to form N_2 (reaction 7).

$$MNO_3 \rightarrow MO + \tfrac{1}{2}O_2 + NO \quad (6)$$

$$NO + CO / HC \rightarrow \tfrac{1}{2}N_2 + CO_2 \quad (7)$$

Fuel rich conditions on Diesel engines can be achieved in-cylinder via engine calibration techniques [16, 17], or with fuel injection directly into the exhaust gas [18 - 21].

For most vehicles requiring the use of a NAC, the use of PM control devices is also expected to be required. Additionally, if the operation of NAC systems is through the use of in-cylinder techniques, engine-out PM emissions during rich operation may increase such that a PM control device is necessary [22].

A possible exhaust arrangement for a NAC + CDPF system is illustrated in Figure 12. The NAC is shown upstream of the CDPF because of thermal considerations – the high thermal mass of filter devices (especially silicon carbide wall-flow filters) would significantly delay the light-off of a catalyst placed behind it. This thermal lag is especially important in cold-start situations.

Also illustrated in Figure 12 is a DOC upstream of the NAC. This catalyst is optional, but can bring some benefits to the system, such as better oxidation activity for CO and HC, and enhancing the NAC activity. Figure 13 shows the NOx conversion from engine bench evaluation of a NAC versus a DOC + NAC system. An improvement of 10 – 30 % in NOx conversion is seen across a wide temperature range.

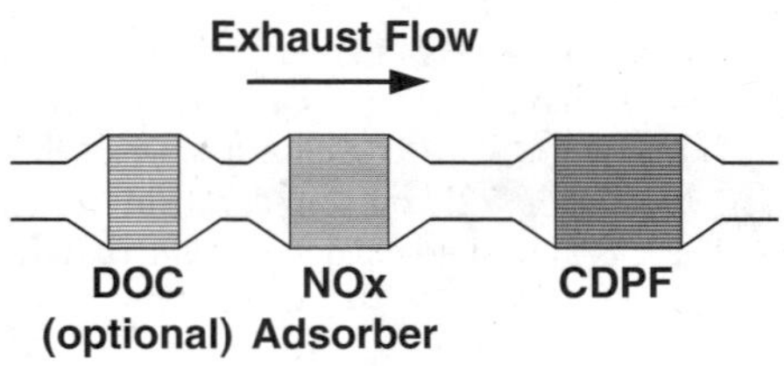

Figure 12: Schematic of a possible exhaust architecture for a NAC system.

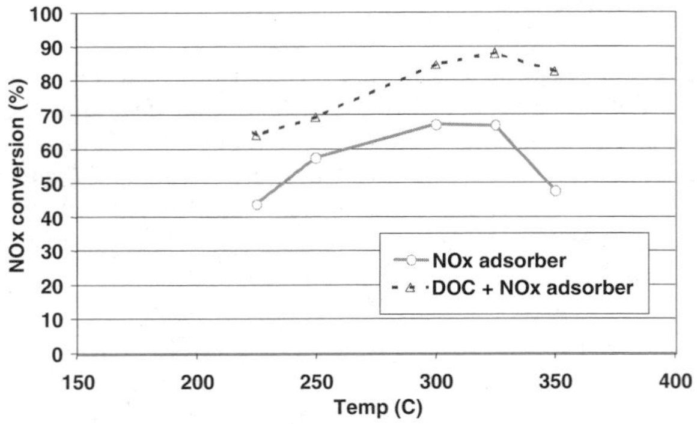

Figure 13: Engine bench NOx conversions of a NAC with and without an upstream DOC. The engine was cycled 60 seconds lean, 4 seconds rich at each temperature.

Because of the higher loads generally experienced in HDD applications, it can be difficult to control a rich pulse in the cylinder. In this case, fuel injection directly into the exhaust may be used to achieve rich gas conditions. However, the high exhaust gas mass flow from large engines means that very high quantities of fuel would be needed in order to create rich exhaust over the catalyst. To overcome this difficulty, it has been proposed [18] that a dual stream exhaust is employed for HDD applications. Valves can be employed to divert the majority of exhaust gas through one catalyst system, while a significantly smaller gas flow is directed through the other catalyst system and fuel is injected to regenerate the catalyst under low flow conditions.

High thermal durability of the NAC is required to ensure effective performance throughout the vehicle lifetime. As discussed above, the filter device may require periodic cleaning of soot by combustion in oxygen, which requires the generation of temperatures of around 600°C at the filter. Higher temperatures would be expected over the NAC whether it is situated upstream or downstream of the filter device.

Another important issue is the response of the NAC to sulfur exposure. NOx adsorbing materials are also very efficient at adsorbing sulfur oxides. The sulfur then blocks the NOx adsorption sites, and the catalyst efficiency decreases. Under rich conditions, the adsorbed sulfur can be desorbed (in the same way as NOx) [14, 23], but this requires a higher temperature than for the NOx release because of the higher stability of the sulfur compounds with respect to the corresponding nitrate compounds. NACs optimised for application on Diesel engines that meet the thermal and sulfur durability criteria have been extensively developed [14]. Figure 14 shows the effect of ageing a state-of-the-art NOx adsorber at 750°C for 72 hours under lean / rich cycling conditions. There is almost no change in the activity window.

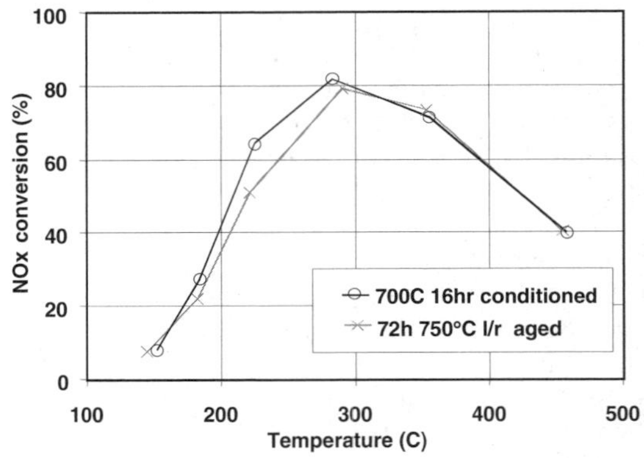

Figure 14: Activity of a conditioned and aged NOx adsorber (laboratory test cycling 60s lean, 5 s rich).

SELECTIVE CATALYTIC REDUCTION (SCR)

As mentioned previously, vehicles targeting HDD European Stage 4 legislation (and possibly Stage 5) may utilise a low PM / high NOx engine calibration, coupled with an SCR aftertreatment system to reduce NOx to below the required limits.

An SCR system incorporates a strategy for the injection of ammonia into the exhaust gas. Sources of ammonia can be the compressed gas, or compounds such as ammonium carbamate or urea, which readily decompose to give ammonia gas when injected into a hot exhaust gas. For example, urea decomposes to ammonia according to the following reaction:

$$(NH_2)_2CO + H_2O \rightarrow 2\,NH_3 + CO_2 \qquad (8)$$

Reaction 8 represents the thermal decomposition and hydrolysis of urea to generate ammonia, and occurs above temperatures of about 160°C.

Over an SCR catalyst, ammonia reacts with NOx according to the following reactions:

$$4NH_3 + 4NO + O_2 \rightarrow 4N_2 + 6H_2O \qquad (9)$$

$$2NH_3 + NO + NO_2 \rightarrow 2N_2 + 3H_2O \qquad (10)$$

$$4NH_3 + 2NO_2 + O_2 \rightarrow 3N_2 + 6H_2O \qquad (11)$$

Of these three reactions, reaction 10 is more facile and therefore occurs at lower reaction temperatures than either reaction 9 or 11. Thus, if the NOx is present in a 1:1 ratio of NO:NO_2, the SCR system will perform with the highest efficiency at low temperatures.

A possible SCR system architecture is shown schematically in Figure 15.

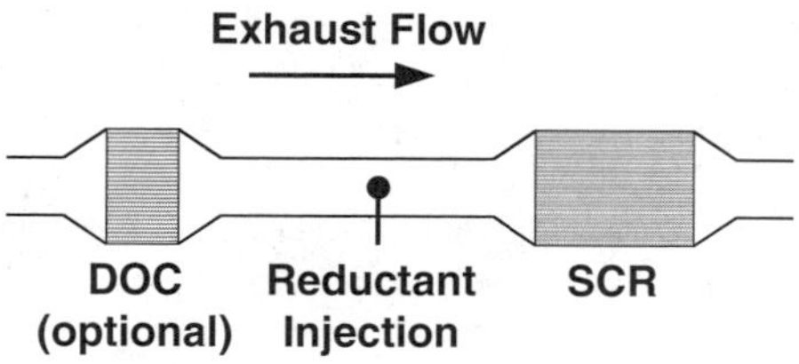

Figure 15: Schematic of a possible exhaust architecture for an SCR system for application on a Diesel engine.

Figure 16 shows the laboratory light-off activity of a typical SCR catalyst in a gas feed with NO as the only NOx species, and the effect of using $NO:NO_2$ in a 1:1 ratio with NO (i.e. simulating the use of an upstream oxidation catalyst).

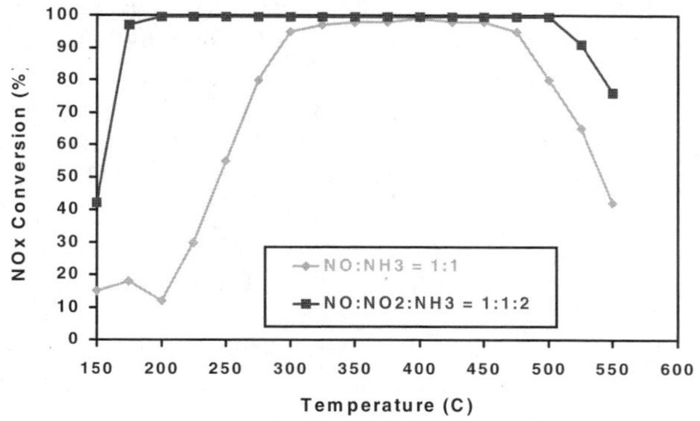

Figure 16: SCR light off activity with 100% NO and with $NO:NO_2$ 1:1 as the NOx species.

A very wide temperature window for NOx conversion is achieved by the catalyst system. Table 6 shows the NOx conversion obtained over the legislated HDD ESC test. High conversion is achieved for the SCR system alone, however, for cold applications or for very high levels of NOx conversion, a pre-oxidation catalyst can further improve system performance.

Although SCR systems may be employed without an additional PM aftertreatment, NOx and PM technologies can be combined. Because the SCR system is most likely to be employed on HDD applications with relatively high engine-out NOx levels, it can be advantageous to fit the PM device upstream of the SCR system (despite the thermal delay this can cause in the SCR catalysts). By placement of the PM device upstream of the SCR catalyst, the PM device can regenerate soot with NO_2, and the SCR can subsequently remove NOx. By correct design of the PM system, the ratio of $NO:NO_2$ exiting the PM device can enhance the SCR performance at low temperatures as discussed earlier. Such systems have been shown to achieve very high NOx conversions [24, 25].

SCR catalysts have been shown to be extremely durable both thermally and with respect to poisons [26].

Table 6: ESC NOx conversions of various SCR systems on a 10 litre HDD engine.

System	*ESC NOx conversion (%)*
SCR (volume 8.5 litres)	82
SCR (volume 17 litres)	88
DOC + SCR **(SCR volume 8.5 litres)**	92
DOC + SCR **(SCR volume 17 litres)**	96

CONCLUSIONS

The future emissions limits for Diesel vehicles are likely to require the use of aftertreatment for control of both PM and NOx. These systems are likely to require advanced Diesel oxidation catalysts to work in conjuction with the PM and NOx aftertreatment systems.

Filter systems to remove soot are expected to be required on engines which use EGR to control NOx. They need to be coupled with an active regeneration strategy to ensure collected soot is removed. This requires a high level of control to ensure that the integrity of the filter is maintained during transient driving conditions. Such systems are already in production on a number of passenger cars in Europe, and their use is expected to expand to meet Stage 4 legislation in 2005. In addition, in the US, they are expected to phased in on passenger vehicles from 2004, and may be used on heavy duty vehicles in 2007.

For NOx control, either NAC or SCR systems can be used. Both have difficulties associated with their use – the complexity of control needed for a NAC system, and the secondary reductant needed for an SCR system. Nevertheless, SCR is expected to be employed on a number of HDD vehicles in Europe for Stage 4, and the durability and high activity of SCR systems has been demonstrated.

NACs are favoured in Europe and the US for LDD, and also for HDD in the US. The properties required by the NAC are severe, particularly with respect to durability in a system. However, catalysts have been developed to meet the requirements of these applications.

ACKNOWLEDGMENTS

The authors wish to thank colleagues at JM sites in Sweden, the United Kingdom and North America for their help in conducting this work. We are grateful to Johnson Matthey plc for permission to publish this work.

REFERENCES

1. A. J. J. Wilkins, M. V. Twigg, G. R. Chandler, P. R. Phillips, "Advances in Catalyst Technology for Light Duty Diesels for South American and European Emission Levels", SAE paper 982950 (1998).
2. P. R. Phillips, G. R. Chandler, D. M. Jollie, A. J. J. Wilkins, M. V. Twigg, "Development of Advanced Diesel Oxidation Catalysts", SAE paper 1999-01-3075 (1999).
3. S. Ichikawa, Y. Uchida, A. Otsuka, T. Harada, T. Hamanaka, K. Nagashima, M. Nagata, "Material development of High Porous SiC for Catalysed Diesel Particulate Filters", SAE paper 2003-01-0380 (2003).
4. K. Ogyu, A. Kudo, Y. Oshimi, H. Sato, K. Ohno, "Characterisation of Thin Wall SiC-DPF", SAE paper 2003-01-0377 (2003).
5. N. Miyakawa, H. Maeno, H. Takahashi, "Characteristics and Evaluation of Porous Silicon Nitride DPF", SAE paper 2003-01-0386 (2003).
6. O. Salvat, P. Marez, G. Belot, "Passenger Car Serial Application of a Particulate Filter System on a Common Rail Dierct Injection Diesel Engine", SAE paper 2000-01-0473 (2000).
7. A. Setiabudi, M. Makkee, J. A. Moulijn, "An Optimal Usage of NOx in a Combined Pt/Ceramic Foam and a Wall-Flow Monolith Filter for an Effective NOx-Assisted Diesel Soot Oxidation", SAE paper 2003-01-0379 (2003).
8. R. K. Miller, W. C. Haberkamp, K. M. Badeau, Z. G. Liu, R. C. Shirk, T. Wood, "Design, Development and Performance of a Composite Diesel Particulate Filter", SAE paper 2002-01-0323 (2002).
9. F. Diefke, M. Lundgren, P. Nilsson, R. Brück, C. Kruse, S. Schaper, "New Diesel Catalyst Systems to Achieve European 2005 Legislation – Tested on a Volvo S60 Passenger Car", Vienna Motor Symposium, 2003.
10. R. Allansson, P. G. Blakeman, B. J. Cooper, H. Hess, P. J. Silcock, A. P. Walker, "Optimising the Low Temperature Performance and Regneration of the Continuously Regenerating Diesel Particulate Filter (CR-DPF) System", SAE paper 2002-01-0428 (2002).
11. S. Chatterjee, R. Conway, S. Viswanathan, M. Blomquist, B. Klüsener, S. Andersson, "NOx and PM Control From Heavy-Duty Diesel Engines Using a Combination of Low Pressure EGR and Continuously Regenerating Diesel Particulate Filter", SAE paper 2003-01-0048 (2003).
12. S. Chatterjee, C. McDonald, R. Conway, H. Windawi, K. Vertin, C. A. LeTavec, N. N. Clark, M. Gautam, "Performance and Durability Evaluation of Continuously Regenerating Particulate Filters on Diesel Powered Urban Buses at NY City Transit", SAE paper 2001-01-0512 (2001).
13. R. Allansson, B. J. Cooper, J. E. Thoss, A. J. Uusimaki, A. P. Walker, J. P. Warren, "European Experience of High Mileage Durability of Continuously Regenerating Diesel Particulate Filter Technology", SAE paper 2000-01-0480 (2000).
14. P. G. Blakeman, P. J. Andersen, H.-Y. Chen, J. D. Jonsson, P. R. Philips, M. V. Twigg, "Performance of NOx Adsorber Emissions Control Systems for Diesel Engines", SAE paper 2003-01-0045 (2003).
15. N. Miyoshi, S. Matsumoto, K. Katoh, T. Tanaka, J. Harada, N. Takahashi, K. Yokota, M. Sigiura, K. Kasahara, "Development of New-Concept Three-Way Catalyst for Automotive Lean-Burn Engines", SAE paper 950809 (1995).
16. H. Klein, H. Böhnke, M. Hoffmann, E. Lox, T. Cartus, K. Neunteufl, L. Burgler, P. Herzog, "The problems of NOx Treatment for Diesel Passenger Cars Solved? Desulphurisation of NOx Storage Catalysts", Vienna Motor Symposium, 2001.
17. S. Michon, C. Gérentet, L. Pierron, T. Colliou, J.-B. Dementhon, B. Martin, "Application of a NOx trap for emission Reduction of a Heavy-Duty Engine", Vienna Motor Symposium, 2003.
18. C. Schenk, J. McDonald, B. Olson, "High-Efficiency NOx and PM Exhaust Emission Control for Heavy-Duty On-Highway Diesel Engines", SAE paper 2001-01-1351 (2001).
19. C. Schenk, J. McDonald, B. Olson, "High-Efficiency NOx and PM Exhaust Emission Control for Heavy-Duty On-Highway Diesel Engines – Part Two", SAE paper 2001-01-3619 (2001).
20. C. Schenk, C. Laroo, "NOx adsorber Aging on a Heavy-Duty On-Highway Diesel Engine – Part One", SAE paper 2003-01-0042 (2003).
21. J. Tahara, Y. Ohtsubo, T. Aoyama, T. Ito, H. Matsuoka, T. Fujimura, "Development of an Exhaust Aftertreatment System for Diesel Passenger Cars", JSAE Annual Congress Proceedings No. 23-02, Pages 9 –12 (2002).
22. T. Colliou, J. Lavy, B. Martin, J. B. Dementhon, G. Pichon, K. Chandes, L. Pierron, " Coupling of a NOx-trap and a DPF for Emission Reduction of a 6-Cylinder HD Engine", Vienna Motor Symposium, 2003.
23. S. Geckler, D. Tomazic, V. Scholz, M. V. Whalen, D. McKinnon, J. Orban, R. A. Gorse, O. Bailey, J. C. Hoelzer, "Development of a Desulfurization Strategy for a NOx Adsorber Catalyst System", SAE paper 2001-01-0510 (2001).
24. A. P. Walker, R. Allansson, P. G. Blakeman, M. Lavenius, S. Erkfeldt, H. Landalv, B. Ball, P. Harrod, D. Manning, L. Bernegger, "The Development and Performance of the Compact SCR-Trap System: A 4-Way Diesel Emission Control System", SAE paper 2003-01-0778 (2003).
25. G. R. Chandler, B. J. Cooper, J. P. Harris, J. E. Thoss, A. J. Uusimaki, A. P. Walker, J. P. Warren, "An Integrated SCR and Continuously Regenerating Trap System to Meet Future NOx and PM Legislation", SAE paper 2000-01-0188 (2000).
26. E. Bardasz, D. Mackney, N. Britton, G. Kleinschek, K. Olofsson, I. Murray, A. P. Walker, "Investigations of the Interactions Between Lubricant-Derived Species and Aftertreatment Systems on a State-of-the-Art Heavy Duty Diesel Engine", SAE paper 2003-01-1963 (2003).

2004-01-1452

Passenger Vehicle Diesel Engines for the U.S.

Chris Cowland
AVL Powertrain Engineering, Inc.

Peter Gutmann and Peter L. Herzog
AVL List, GmbH

ABSTRACT

HSDI diesel engines have become increasingly popular in Europe over the last 20 years to the extent that market penetration now exceeds 40%. The situation is very different in the US where diesel accounts for only 4.5% of the light vehicle total at present. This paper provides a perspective on the technological, economic and societal issues associated with the large scale introduction of passenger vehicle diesels to the US.

INTRODUCTION

With the amount of miles driven by the average US consumer rising coupled with a decrease in vehicle fuel economy the usage of fossil fuels is at an all time high for the US. A number of alternative propulsion systems are being evaluated by vehicle OEM's, Government bodies and research establishments. Fuel cells, Hybrid Electric Vehicles and Hydrogen IC engines are all being presented as the solution to help reduce the dependence on oil. In Europe, where reductions in fuel consumption have been taken on board voluntarily by the car companies, the diesel engine is currently the main solution to the fuel equation. Why then, has this technology not been adopted for use in US passenger vehicles?

US VS EUROPE

The modern High Speed Direct Injection (HSDI) diesel engine is becoming increasingly popular as the power plant of choice in Europe. Diesel penetration for all of Europe is currently exceeding 40 % and in some countries such as Austria the figure is about 70 %. Whereas diesels were traditionally selected solely on the fuel saving potential the new generation of engines offer fuel economy and refinement as well as increased driving pleasure due to the high torque available. FIG 1 shows the tremendous advances in specific power and torque which these engines have achieved since their introduction in the mid-eighties. (Gasoline engine advances for the same period are provided for comparison). The diesel achievement is even more exceptional when one considers that legislation over the same period has called for a ~90% reduction in NOx and Particulate emissions.

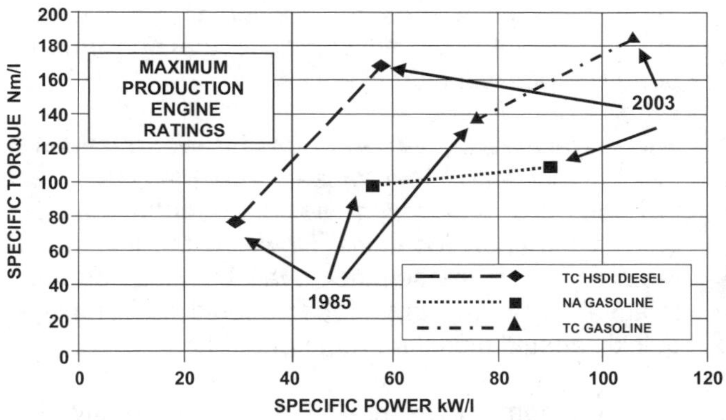

FIG 1 MAXIMUM PRODUCTION ENGINE RATINGS 1985-2003

The situation in the US is very different. In 2003 only one vehicle manufacturer offered diesel passenger cars for the US. This situation is set to change with diesel passenger cars being offered by Daimler Chrysler - in addition to VW - for 2004. The majority of the diesel engines in the US are currently fitted into heavy duty pick up trucks where the market penetration is expected nearly 6% in 2003. In these applications consumers clearly appreciate the added utility of diesel and are willing to pay premiums of up to $5000 for the diesel powered vehicles.

DIESEL BARRIERS

So, why when diesel penetration is rapidly increasing in Europe, is there a very limited diesel growth in the US? There are perhaps 5 main reasons to consider:

1. Emissions Legislation. Future Tier II and US 2007/10 emissions legislation presents a technical challenge to diesels which will require both the use of advanced combustion and aftertreatment systems. Whilst the required technology will be available in the required timeframe cost issues remain.

2. Product Cost. The level of technology required to meet the future emissions legislation may result in engines which are significantly more expensive to produce than gasoline engines of today. The major cost contributors on the advanced diesel engine are the Fuel-, Turbomachinery-, EGR- and Aftertreatment systems.

3. Diesel Perception. Many consumers still regard diesel as "dirty, noisy, smelly and slow" even though advanced diesel technologies available today are very different from engines in the early eighties. The main challenge is to get the consumer to test drive a modern diesel powered vehicle – the performance benefits soon win over skeptics.

4. Fuel Infrastructure. Around 30% of filling stations in the US currently offer diesel fuel. Many of these locations are truck stops with the pumps located in non user friendly locations. However the investment required to increase the availability is probably relatively slow and this process could be completed quickly should diesel market penetration increase.

5. Fuel Economy Pressure. There is currently little pressure to improve the fuel consumption of passenger vehicles. CAFE legislation has been relatively ineffective at promoting a reduction in fuel consumption, in fact, average fuel consumption has actually increased in the US which is in stark contrast to the trend seen in European countries such as Germany. See FIG 2. The cost of fuel is a large driver for the reduction of fuel consumption in Europe where typical fuel costs can be 3 times that of the US. See FIG 3. All car companies are also committed to the voluntary reduction of CO2 emissions under the ACEA agreement. The situation in the US may change in the near future due to the governments stated desire to reduce the dependence on oil. This may either promote more interest in oil consumption reduction or a change to alternative fuels such as hydrogen. In the near term the former is probably the most likely scenario due to the infrastructure required for the Hydrogen option.

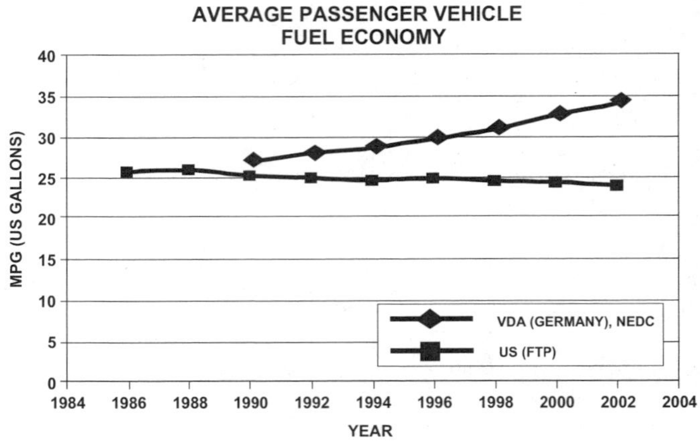

FIG 2 FUEL CONSUMPTION TRENDS

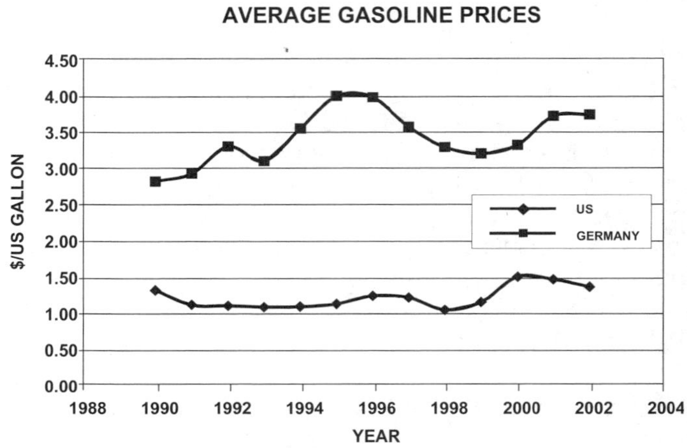

FIG 3 GASOLINE PRICE COMPARISON

Many recent marketing studies have been conducted to investigate the potential market for diesels in the US market [e.g. 1][1]. These studies have indicated that up to two thirds of consumers would consider "clean diesel" for their next vehicle and penetration predictions as high as 20% have been presented.

LEGISLATION

Before investigating the future emissions legislation it is first important to highlight the issue of the two differing sets of future legislation. Depending upon the vehicle weight class diesel equipped passenger car vehicles can be certified either as in-vehicle on a chassis dyno (light duty - FTP 75), or as a loose engine (heavy duty - US HDDTC). Light duty emissions limits are in g/mile whereas heavy duty are g/bhphr. The break point for vehicle class is > 8500lb GVWR. FIGS 4 and 5 show both the light duty and heavy duty emissions legislation.

[1] Number in parentheses designate the references at the end of the paper

Year of Implementation [1]	Emission Standards in g/BHP-hr NMHC		CO	NMHC + NO_x	NO_x	PM	Test Cycles & Comments	Fuel Sulphur Content
2004 (2002 for Consent Decree-Companies) [2) 3)]	Option 1	0.5	15.5	2.5	2.0	0.10 (0.05)	HDDTC; ESC, NTE-Areas; LRT for reference only	500 mg/kg
	Option 2	--	15.5	2.4	--	0.10 (0.05)		
2007 – 2009 Option 1 [4) 5)]	$\left(\frac{0.14+0.5}{2}\right)=$ 0.32		15.5	--	$\left(\frac{0.2+2.0}{2}\right)=$ 1.1	0.01	↑	15 mg/kg from June 2006
2007 – 2009 Option 2 [4) 5)]				$\left(\frac{0.2+2.4}{2}\right)=1.3$				
2010 [5)]	0.14		15.5	(0.34)	0.20	0.01	↑	15 mg/kg

Status 06/2002

Notes:
HDDTC = Heavy-Duty Diesel Transient Cycle (also known as FTP = Federal Test Procedure)
ESC = European Steady State Cycle (also known as SET = Supplemental Emission Test Cycle)
NTE = Not-to-Exceed area = defined operating area where NMHC, NOx, and PM may exceed HDDTC Standard by 1.5 times, CO-standard 1.25 times
LRT = Load Response Test, for determining smoke puffs at 5 engine speeds, for Reference only, no Standard set sofar
1) = Exh. Emission Standards in this table apply primarily to the HDDTC-Test Procedure
2) = 2004 Standards have to be met in 2002 by those companies having signed the Consent Decree. These companies need also to certify along the ESC, NTE and LRT; all other companies have to meet 2004 Standards in 2004 by certifying along the HDDTC only
3) = for 2004 (2002) there are 2 options: 1st option: (NMHC + NOx) ≤ 2.5 g/BHP-hr, with a NMHC cap of 0.5 g/BHP-hr; 2nd option: (NMHC + NOx) ≤ 2.4 g/BHP-hr, no NMHC standard. PM for Buses is 0.05 g/BHP-hr
4) = Actually, 2010 standards shown in the table apply already for 2007. However, for Model Years 2007 - 2009 due to phasing in, 50:50 averaging of NMHC, (NMHC + NOx) and NOx standards of 2004 and 2010 is allowed resulting in the effective standards shown for 2007 - 2009. If averaging is not made use of, the 2010 Standards would become effective already in 2007.
5) = From 2007 no crankcase emissions shall be discharged directly into the ambient atmosphere

FIG 4 HEAVY DUTY EMISSIONS LEGISLATION

The heavy duty emission scenario for vehicles above 8500lb (except MDPV's) is slightly confused by the phase in period between 2007 and 2010 for the 0.2 g/bhphr NOx limit. The 50% phase in of 0.2 g in 2007 offers the manufactures the option of either having 50% of engines at 2004 and 50% at 2010 levels for 2007 or having an average fleet emissions of not higher than the total NOx that this scenario would produce. Hence it is most likely that the majority of manufacturers will certify 100% of engines to a figure of around 1.1 g.

Bin No.	Tier 2 Light Duty Full Useful Life Emission Standards , FTP 75, [g / mi NOx	NMOG	CO	HCHO	PM	Comments
11	0.9	0.280	7.3	0.032	0.12	a,c
10	0.6	0.156 / 0.230	4.2 / 6.4	0.018 / 0.027	0.08	a, b, d
9	0.3	0.090 / 0.180	4.2	0.018	0.06	a, b, e
The above temporary bins expire in 2006 (for LDVs and LLDTs) and 2008 (for HLDTs and MDPVs)						
8	0.20	0.125 / 0.156	4.2	0.018	0.02	b, f
7	0.15	0.090	4.2	0.018	0.02	
6	0.10	0.090	4.2	0.018	0.01	
5	0.07	0.090	4.2	0.018	0.01	
4	0.04	0.070	2.1	0.011	0.01	
3	0.03	0.055	2.1	0.011	0.01	
2	0.02	0.010	2.1	0.004	0.01	
1	0.00	0.000	0.0	0.000	0.00	

Comments:
a. Bin deleted at end of 2006 model year (2008 for HLDTs and MDPVs).
b. The higher NMOG, CO and HCHO values apply for HLDTs and MDPVs only.
c. This bin is only for MDPVs.
d. Optional NMOG standard of 0.280 g/mi applies for qualifying LDT4s and qualifying MDPVs only.
e. Optional NMOG standard of 0.130 g/mi applies for qualifying LDT2s only.
f. Higher NMOG standard deleted at end of 2008 model year.

Interim vehicles and Tier 2 vehicles (except for those Tier 2 vehicles in the lowest bins) would also have to meet intermediate useful life standards that apply for 5 years or 50,000 miles.

FIG 5 LIGHT DUTY EMISSIONS LEGISLATION

The heavy duty standards are applicable to 435,000 miles or the life of the engine, whichever is sooner. This very long emissions compliance period also presents significant challenges due to degradation of aftertreatment and combustion systems over time.

From a light duty perspective the majority of OEM's are targeting BIN 5 emissions due to the fact that achievement of this bin matches the corporate average NOx requirement and hence no penalties would be associated with the sale of diesels. If engines are certified at higher bin levels then credits will be required from gasoline engines running at lower bin levels hence effectively limiting the number of potential diesel sales or increasing the cost to produce lower bin gasoline engines.

The light duty emissions legislation is further complicated by differing classes of vehicles. The main effect of the vehicle classes is the change in effective emissions life. This leads to a longer life requirement for the heavier vehicles which will result in the requirement for larger deterioration factors.

The implications of the two differing sets of emissions are considerable with respect to engine technology and rating strategy. FIG 6 shows a comparison of the heavy and light duty emissions cycles. The cycles have been reduced to representative modes to illustrate the contribution of each mode and the relative engine speed and load. The US heavy duty cycle has a large proportion of the emissions cycle at very high engine speeds and loads. The US HDDTC also includes the 13 mode ESC test - not shown in Fig. 6 - and also not to exceed areas (NTE) where the emissions at any selected point must not exceed 1.25 (or 1.5 depending on NOx level) the legislated values. By comparison the light duty FTP cycle involves running at much lower speed and load points in the engine map. The major differences in these cycles drives the need for completely different engine hardware (EGR system), calibration and potentially aftertreatment systems. The durability requirements of the engines are also significantly different.

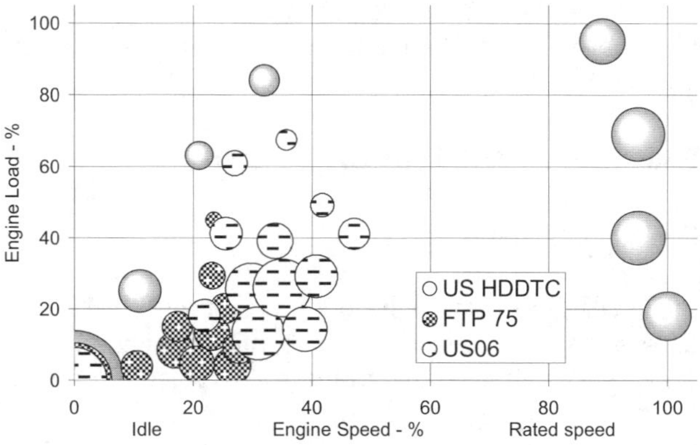

FIG 6 US CYCLE MODAL COMPARISON

FIG 7 shows a comparison of the parameters for typical light and heavy duty certified engines and vehicles used in the US. The widely varying parameters clearly show the challenges associated with the different classes and also demonstrate that developing base engines or families which span differing engine classes can create considerable technical challenges.

	HD	HD	LD
PRIMARY USE	ON-HIGHWAY TRUCK CLASS 4 - 8	PICK-UP/ VAN > 8500 LB GVWR	PASSENGER CAR/LIGHT DUTY TRUCK < 8500LB*
US ENGINE MANUFACTURERS	CAT, CUMMINS, DDC, MACK, ISUZU,	INTERNATIONAL, ISUZU, CUMMINS	???
TYPICAL CONFIGURATION	I6	V8, I6	V6, I6, I5, I4
DISPLACEMENT LITRES	7-18	5.5 - 7	1.9 – 4.5
DISP/CYL LITRES	1 - 3	0.7 – 0.85	0.5 – 0.8
RATED SPEED RPM	1800 - 2200	2800 - 3600	3800 - 4200
MAX SPECIFIC RATING kW/L	< 30	< 43	< 58
PEAK BMEP BAR	23	18.5**	21
CERTIFICATION	HD(ENGINE) US HDDTC,ESC, NTE	HD(ENGINE) US HDDTC, ESC, NTE	LD(IN-CHASSIS) FTP75, US06, SC03
FUTURE LEGISLATION	2007/2010	2007/2010	2004 - 2009
DURABILITY MILES	500K (1M class 8)	300K	175K
ENGINE VOLUMES PER ENGINE MODEL/YEAR	<100K (< 20K = class 8)	>150K	???

* EXCEPT MDPV = < 10000 lb
** GENERALLY TRANSMISSION LIMITED

VALUES IN TABLE ARE FOR GUIDELINES ONLY. SOME APPLICATIONS MAY LIE OUTSIDE STATED VALUES.

FIG 7 US DIESEL ENGINE CLASSES

The typical exhaust temperatures seen in the 2 classes of applications are also somewhat different which leads to differing aftertreatment regeneration challenges.

Hence from the above discussion we can see that the challenges associated with the light duty engines are very similar to those experienced in Europe. The heavy duty engines for the medium duty truck market in the US are in a segment of their own, running higher engine speeds and at higher ratings than traditional class 8 truck engines, but being certified to the same levels and conditions.

BREAKING DOWN THE BARRIERS

ENGINE TECHNOLOGY

Taking a vehicle mass of 4000 lbs over the FTP 75 cycle it is interesting to examine the potential displacement of an engine to suit the US market. In typical European applications the engine selected for this vehicle mass would be approximately 2.5-2.7 liters. However by examining the plots in FIG 8 it is possible to concur that the best compromise (when considering the US 06 acceleration-/emissions- and FTP emissions- and fuel consumption- tradeoff) is around 3.5L displacement for US applications. This displacement naturally leads us to a V6 engine configuration in the interests of cylinder size, packaging and cost.

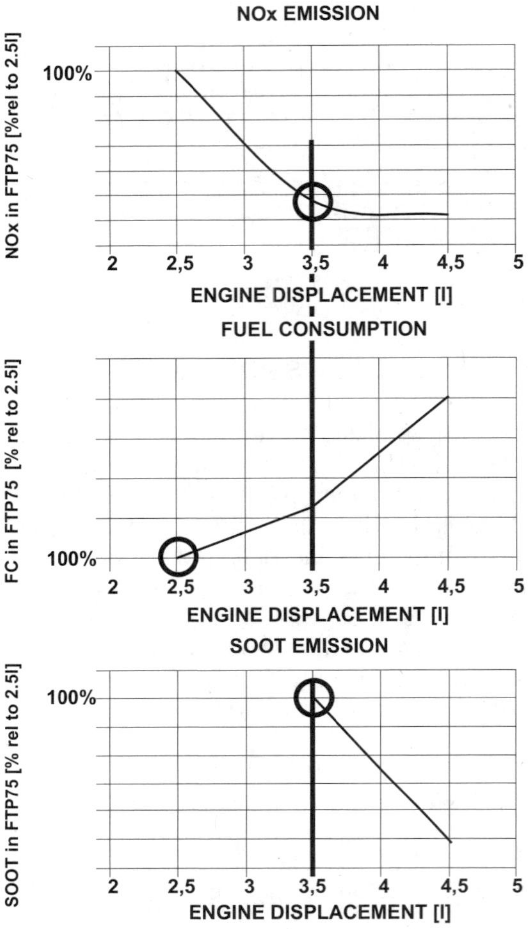

FIG 8 DISPLACEMENT COMPROMISE

For heavier, light duty certified SUV applications engine displacements up to 4.5 liters can be envisaged but again a V6 engine configuration is the natural choice. One of the challenges associated with this approach for SUV's is the relatively high mean piston speed which will be needed assuming a rated speed of around 4000 rpm and slightly undersquare bore/stroke ratio. Mean piston speeds as high as 13.5 m/s may be required which leads to emissions, fuel consumption and heat rejection challenges. The alternatives of oversquare bore/stoke ratio or increased number of cylinders are not attractive from efficiency and cost reasons respectively.

For the heavy duty certified engines used in the over 8500 lb vehicle class 5.5 - 6.5 liter engines appear to make the most sense. However the cylinder size cannot realistically be increased based on the smaller SUV engines due to the mean piston speed constraint and hence these engines will probably be of V8 configuration with rated speeds in the 3100-3600 rpm range. The lowered piston and engine speeds help both from a durability and emissions point of view.

From a performance perspective the diesel engine clearly has much to offer the end user. FIG 9 shows a direct comparison between two identical European vehicles equipped with 3.0L gasoline and diesel engines producing a similar power level. As can be seen from the comparison the superior torque of the diesel engine gives a significant improvement with respect to in-gear acceleration times, but what the data does not show is the improvement in driveability and flexibility this offers the driver. In this example the diesel powered vehicle is also over 100 kg heavier than the 3.0L gasoline due to the additional weight from the engine, transmission and noise package. These benefits are achieved with a combined 39% improvement in fuel economy for the combined European city and highway cycle. For additional comparison an identical 4.4 liter gasoline powered vehicle is also included which still does not achieve the in-gear acceleration of the 3.0L diesel powered car.

	Power kW	Torque Nm	V Max mph	0-62 mph s	50-75 mph s 4th	50-75 mph s 5th	NEDC MPG (US gal)
3.0L Diesel	160	500	153	7.1	5.4	6.7	34.7
3.0L Gas	170	300	155	6.9	7.3	10.0	25.0
4.4L Gas	245	447	155	5.8	5.8	7.4	21.8

Note: 4,4L Gasoline engine is cut off at max speed of 155mph

FIG 9 DIESEL/GASOLINE PERFORMANCE

Using the 0 – 62.5 mph acceleration performance it can easily be shown that for the same vehicle acceleration, diesel is around 40% more fuel efficient, or at the same fuel economy the diesel car accelerates around 2 seconds faster. See FIG 10 .These results are notwithstanding the fact that the standstill acceleration test does not particularly suit the turbocharged diesel vehicles.

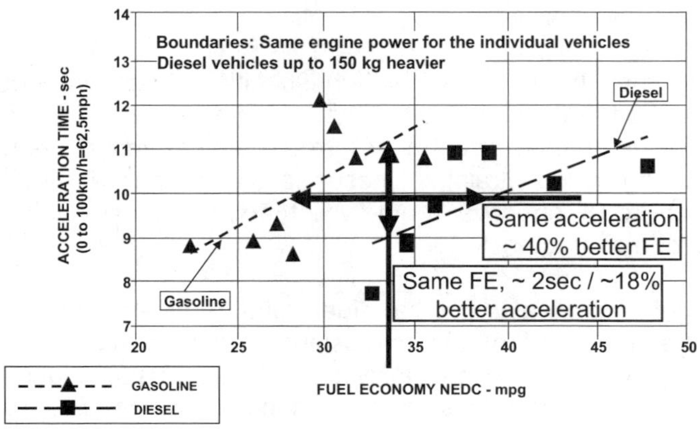

FIG 10 ACCELERATION/FC TRADEOFF

Using averaged data from a number of tests conducted on a wide range of gasoline and diesel vehicles, the noise comparison at full load is shown in FIG 11. The objective measure of noise does not necessarily show the whole picture and the subjective rating of diesel vs gasoline is often viewed differently. This is mainly due to the different perception of mechanical or combustion related noise. With modern FIE and noise reduction technology it is possible however to make the diesel competitive with the gasoline from a NVH perspective.

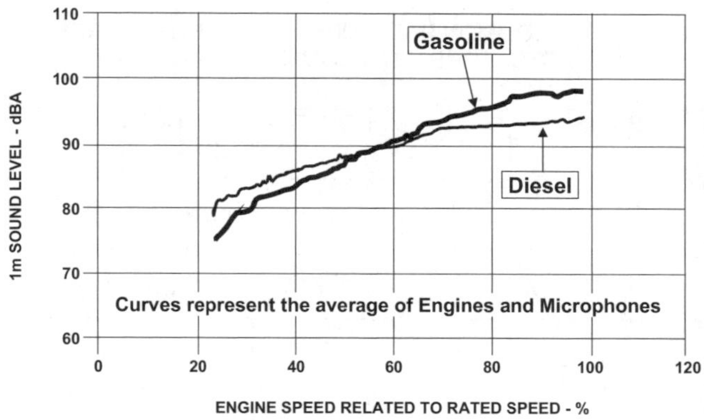

FIG 11 FULL LOAD ACCELERATION VEHICLE NOISE COMPARISON

EMISSIONS TECHNOLOGY

As discussed previously the emissions challenges for LD and HD certified engines are considerably different. As a result of these differences the technology packages for future engines may utilize alternative approaches.

A number of technologies available today are likely to form the base emission package used on future US passenger vehicle diesels. Many of the advancements made recently focus on the increased flexibility of fuel and air systems hence allowing further optimization of combustion systems and also the control of advanced aftertreatment technologies.

ADVANCED FIE – Generation 2 Common Rail FIE allows the possibility of both high injection pressure capability and also flexibility in injection. Injection pressures of 1800 bar will be common place with 2000 bar a possibility for the future. Up to 5 injections per cycle are offered with very close separation being feasible. This added flexibility offers the opportunity to utilize advanced combustion systems to deliver low NOx and soot in the emission relevant map areas. Many suppliers of FIE are now investigating alternative architectures to improve injection control. One of the most critical issues is the rapid control of needle opening and closing. To this end a number of injectors with the control element close to the needle are available which results in reduced hydraulic volumes and mechanical masses for the needle control hence giving faster response. Piezo actuation may also offer benefits over solenoid although the architecture of the injector is equally important as the actuation method. NOx and soot emissions up to 25 % lower have been demonstrated using advanced injectors in back to back tests with the same nozzle geometry. See FIG 12.

2 DIFFERENT COMMON RAIL INJECTOR ARCHITECTURES AND ACTUATORS WITH COMMON NOZZLES

BASIS	IMPROVED NEEDLE CONTROL + PIEZO
100%	+ 4% peak power
100%	-25% NOx @ constant soot
100%	-25% soot @ constant NOx
100%	-15% NOx/soot

FIG 12 ADVANCED INJECTOR COMPARISON

VARIABLE SWIRL - Variable swirl offers further combustion optimization possibilities. On fixed swirl combustion systems generally the combustion is compromised with a too low swirl number at low engine speeds and too high swirl at high speeds. FIG 13 shows the benefit of variable swirl on an HSDI diesel engine.

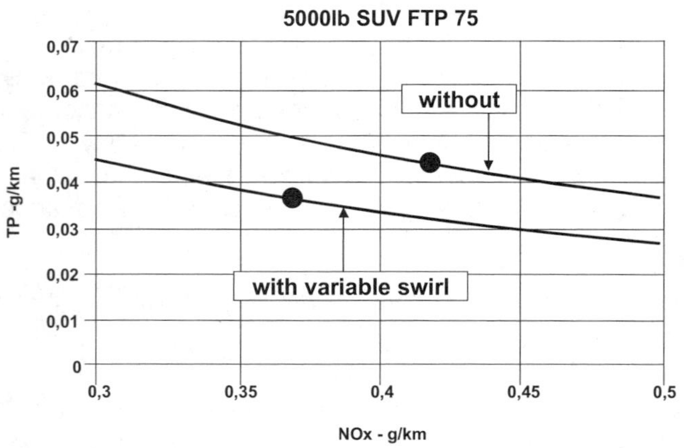

FIG 13 VARIABLE SWIRL BENEFITS

In addition to the emissions benefits variable swirl also offers lower peak heat rejection and improved cold startability.

Traditionally, variable swirl systems suffered from limited air flow capacity due to the nature of achieving variable swirl via closing off one of the inlet ports on a 4 valve engine. To overcome this problem AVL has developed a "split port" design which achieves the desired flexibility without impacting volumetric efficiency considerably [2]. FIG 14 shows a comparison of traditional and "split port" performance.

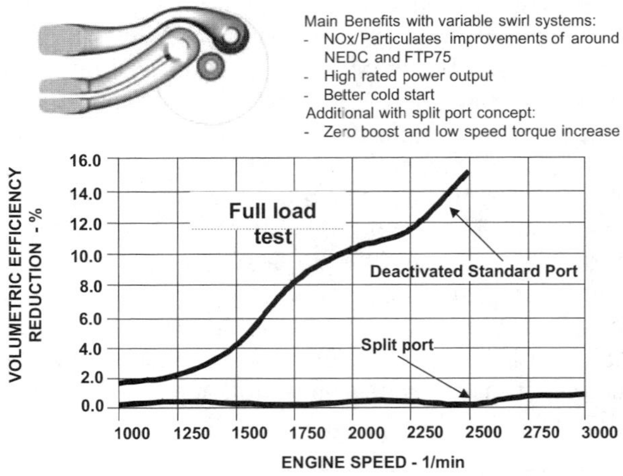

FIG 14 AVL SPLIT PORT PERFORMANCE

ADVANCED AIR SYSTEMS – In achieving the 60 kW/liter ratings the HSDI diesel engines are beginning to come to the limit of current turbocharger technology for single stage systems. Compressor pressure ratios of around 3.5 are required and with current technology it is not possible to get acceptable compressor map widths at these high pressure ratios to allow attainment of required engine performance both at low and high speed. For these reasons engines running at specific powers > 60 kW/liter will require the use of variable geometry compressors or 2 stage turbocharging systems. Electrically assisted turbos or E-booster systems are another means of achieving acceptable boosting system match across the operating range of the engine. At present work is on-going looking into the feasibility of variable compressors similar to the variable turbines used on many engines today. These variable compressor turbos should extend the width of the compressor maps considerably and allow single stage turbocharging to be used for higher ratings than are currently feasible.

For the US where such high specific ratings are not required or desirable, either single or twin VGT arrangements will be the preferred route depending upon packaging constraints. These solutions offer a much more cost effective route whilst achieving acceptable torque characteristics.

Light and Heavy duty air fuel ratio requirements are different due to the emissions requirements. This coupled with the EGR system design issue places different requirements on the air handling system. FIG 15 provides a comparison of LD and HD lambda values in the load/speed map for a couple of typical engines.

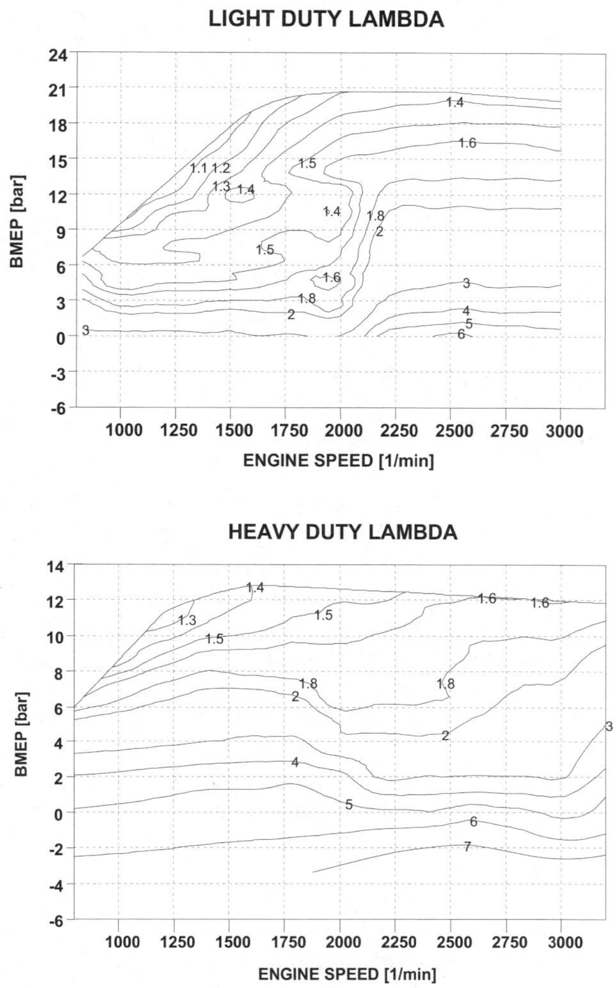

FIG 15 LAMBDA COMPARISON LD vs HD

When considering the vehicle cooling challenge of the diesel engine compared to a gasoline engine the additional cooling load from the intercooler and EGR must be considered. The total cooling load for light duty applications is however, very similar between diesel and gasoline for the same power rating due to the improved thermal efficiency of the diesel engine. Hence the additional space required for the intercooler can be traded against the smaller radiator package required for the diesel. In heavy duty applications where EGR must be applied across the torque curve the considerable increases in heat rejection due to EGR and intercooling must be contained. This represents a significant challenge with respect to vehicle packaging.

EGR SYSTEMS – Precise metering of large quantities of EGR (up to 65%) is a prerequisite for achieving low emissions in LD applications. Future emissions legislation will demand high efficiency cooled EGR with a cooler bypass to avoid overcooling the EGR under some operating conditions or for critical engine / vehicle combinations

In heavy duty certified applications, EGR requirements are lower at around 20-25% but they are applied in totally different regions of the engine map. At extremely high ratings one of the problems is that the pressure differential between exhaust and inlet is not high enough to drive the required quantity of EGR and supple-mentary means may be necessary.

FIG 16 shows a comparison of LD and HD EGR maps for a typical engine calibration.

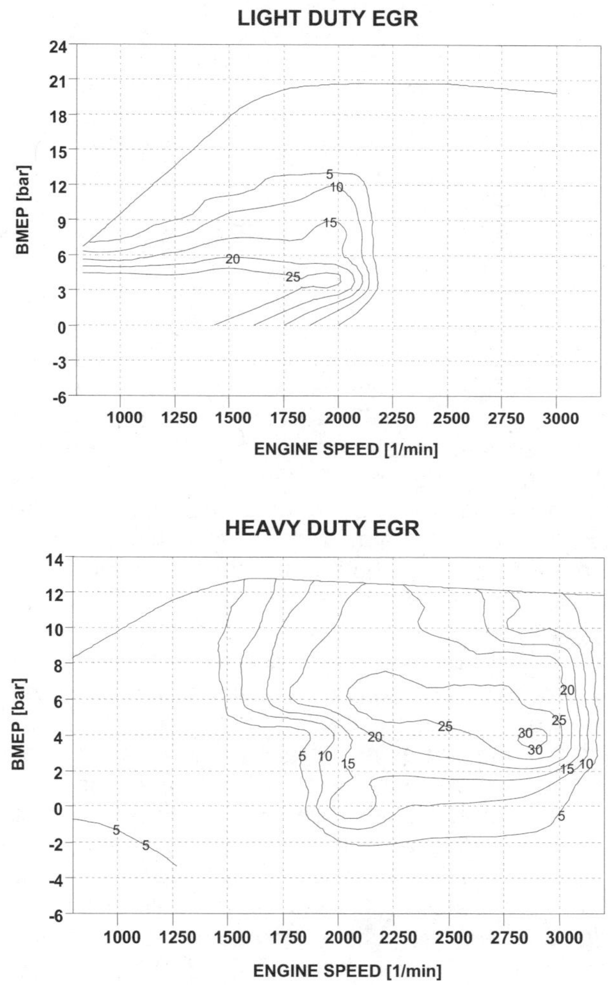

FIG 16 EGR MAPS LD vs HD

Increased EGR cooling is a very effective method of improving the NOx/soot tradeoff. Using modal test results and applying them to a drive cycle, a 40deg C reduction in EGR temp has been proven to offer a 30% reduction in emissions. See FIG 17.

Reductions in EGR temperature can be realized not only by increases in cooler size but by careful optimization of the cooler design and water flow. In the future twin circuit systems may be beneficial but the cost/performance tradeoff must be investigated and the packaging issues addressed. At the lowest emissions levels individual cylinder introduction of EGR may also offer benefits but once again the cost/value tradeoff needs to be closely analyzed.

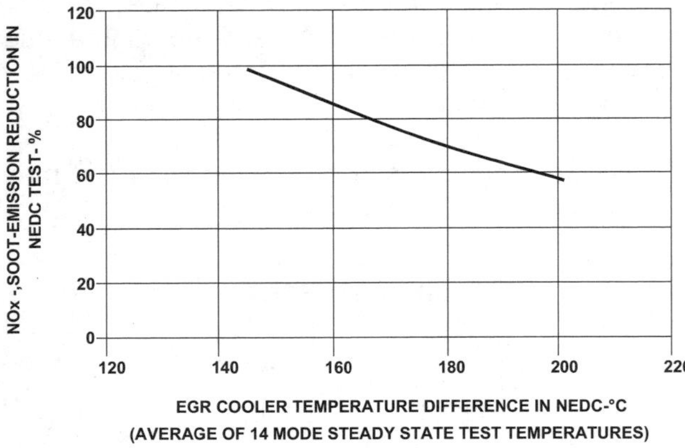

FIG 17 EFFECT OF EGR TEMPERATURE ON EMISSIONS

REDUCED COMPRESSION RATIO – Traditionally, HSDI diesel engines employed compression ratios in the range of 18 to 19 : 1. These high compression ratios generate very high peak firing pressures at the highest engine ratings presenting considerable challenges with respect to structural design, engine mass and durability. Cold startability was the main driver for maintaining high compression ratio's, but with the introduction of advanced Common Rail FIE and significant improved glow plug systems, compression ratio's can be reduced considerably. In Europe, compression ratio's as low as 15 : 1 are being developed. Typically a 1 ratio drop in compression reduces the peak firing pressure by around 8 -10 bar for the same engine rating. For North American use it may not be feasible to drop compression ratio quite as far as in Europe due to the stricter cold start requirements and lower cetane fuel. Compression ratio's of 16.0 – 16.5 : 1 are, however, still considered feasible. Careful optimization of cold start calibration and spray/glow plug interaction is required to effectively employ low compression ratio's. FIG 18 shows the cold start behavior of an HSDI engine with a CR of 16.

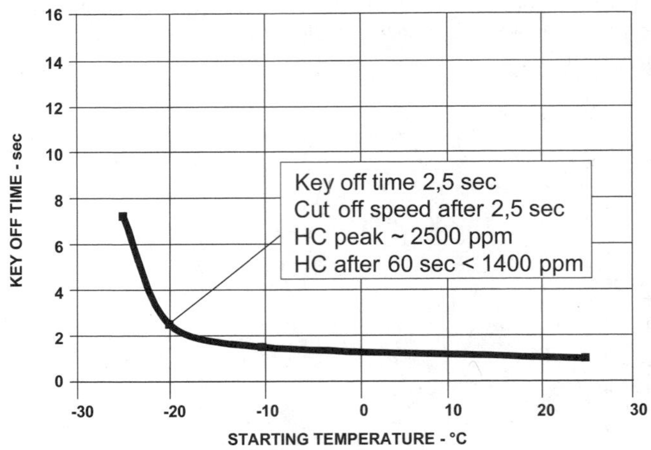

FIG 18 COLD START BEHAVIOR AT 16:1 CR

AFTERTREATMENT

Looking at the exhaust gas legislation in the US, the legal emission limits cannot be achieved by combustion measures alone, even if alternative combustion processes are used. This particularly refers to heavy vehicles and high engine displacements. Therefore, efficient exhaust gas aftertreatment is indispensable. Diesel particulate filters are already used in series production and various announcements exist that other OEM's will follow shortly. However, besides particulates, NOx emissions have to be reduced simultaneously. The challenge is to combine DPF and NOx catalysts to a 4-way catalyst (NOx, particulates, HC, CO) as already described in [3].

Various combinations/systems of NOx-cat and DPF have been tested in-house including e.g. real world driving, different vehicles, DPF regeneration aids, DeNOx-purge events etc. [4]. FIG 19 presents the particulates-NOx-conversion rates for different emission test cycles for a NOx-adsorber/catalysed DPF combination. A simultaneous NOx/TP-reduction of about 60/90% can be achieved for both the NEDC and the FTP 75. The US 06 shows a significantly lower NOx conversion rate of about 35%. Nevertheless, this rate can be increased up to 60% by measures not tested in the case in question.

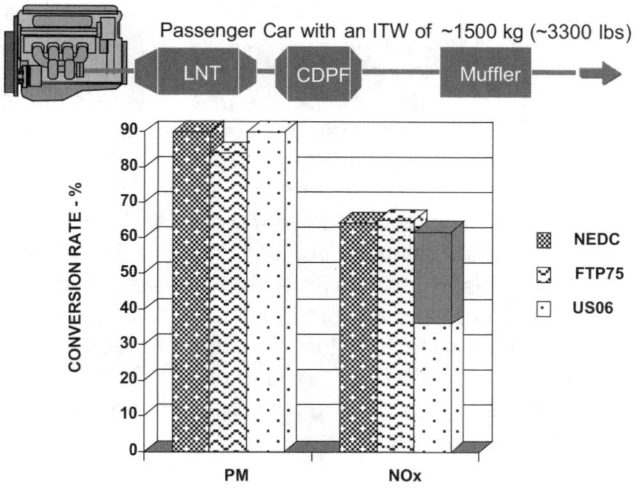

FIG 19 EFFICIENCY OF A 4 WAY DIESEL CATALYST

ALTERNATIVE COMBUSTION

Alternative combustion strategies are ideally suited to light duty certified engines where the majority of the emissions cycle is spent at low speed and load. A variety of strategies can be applied to achieve low NOx and soot levels. FIG 20 demonstrates the part load results with AVL´s alternative combustion [5] relative to current HSDI engine data. These results were obtained without the need for fundamental hardware changes to the base engine. Some combustion related specification- and calibration- changes are used to achieve the desired characteristics.

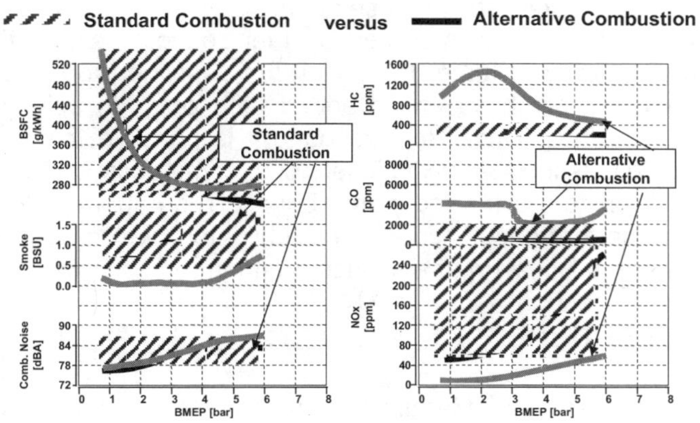

FIG 20 ALTERNATIVE vs STD COMBUSTION

The bmep limitation of alternative combustion strategies does not lend itself to application on heavy duty certified engines. As can be seen from the results above at bmep levels exceeding 5 bar a significant degradation in fuel consumption occurs and since this is the area which heavy duty engines operate the application of this technology is not desirable.

COST ISSUES

If the cost comparison between diesel and gasoline is based on current US gasoline technology vs 2007 diesel technology the diesel engine production cost could potentially be up to 200% higher including the necessary aftertreatment. However, if we project that fuel consumption reduction for the gasoline engines is a prerequisite based on potential future CAFE or CO2 standard's the equation becomes a little more palatable. The addition of VVT, GDI and turbocharging technology to gasoline engines to improve fuel consumption will increase the base engine cost considerably. Perhaps in it's ultimate form, with fully flexible valvetrain system, the Gasoline engine may be able to compete with a light duty diesel engine but the cost and complexity are significant and the diesel then becomes an attractive option.

The cost tradeoffs discussed above are based on European design HSDI diesels and a further optimization of this cost may be feasible for the US market if US specific designs are required. Many of the European designs have evolved due to the very high specific power ratings required for HSDI diesel engines. If, for the US market alone specific ratings in the range of 40-50 kW/l are deemed to be suitable substantial savings on base engine cost can be considered. These can be achieved via the use of 2 valve per cylinder head designs rather than 4 valve. With the advent of high pressure FIE and the mandatory use of DPF's the prerequisite for 4 valves per cylinder can now be challenged, particularly in the US. Since pushrod technology is also acceptable from a marketing perspective in the US, the adoption of this approach for a vee engine offers considerable base engine cost reduction potential. Further benefits such as mass, packaging and friction reductions are also possible via this approach.

Hence in order to fully exploit the potential for HSDI's in the US, particularly for the very significant SUV and light truck market, the diesel engines are likely to be custom designs for the US rather than transplanted European technology. These custom designs will allow the cost gap to be narrowed between gasoline and diesel whilst still offering all the benefits of driveability, reduced fuel consumption, low noise etc. For smaller passenger car applications where 4 cylinder engines are likely to dominate, applying European engines into US vehicles is considered more feasible. The additional volume for this class of engine will also reduce the cost of the European engines.

CONCLUSIONS

The technology is available today to achieve light duty Tier 2 Bin 5 emissions levels in vehicles up to at least 6000lb inertia test weight. The approach required to meet these levels includes increased engine flexibility via the use of advanced air and fuel systems. These systems allow the optimization of low engine out NOx and soot as well as the flexibility to control aftertreatment devices.

Variable swirl offers substantial benefits from an emissions, torque curve and heat rejection perspective and will become common place on future engines. Utilization of concepts such as the AVL Split Port design allows the implementation of variable swirl without the usually associated reduction in volumetric efficiency when applying the higher swirl stage.

From an aftertreatment perspective DPF will be mandatory for all engine classes. DPF application is now commonplace in Europe and the technology is relatively mature. NOx aftertreatment is much less mature and presents many technical challenges. For passenger cars under ~4000 lbs the achievement of Tier 2 Bin 5 appears feasible without the need for NOx aftertreatment. This approach is possible due to the reduced emissions of the lighter vehicles and the fact that alternative combustion can be applied for a wider range of the emission relevant speeds and loads. For heavier vehicles NOx aftertreatment is required and the development of suitable systems is still underway. However, it is likely that in the Tier 2 timeframe suitable NOx aftertreatment devices will be available. In the future DPF and NOx aftertreatment devices are likely to be integrated into "4 way catalysts" such as the Toyota DPNR system (REF) offering the possibility for cost reduction.

Ratings for the US market will probably be lower than in Europe due to the emissions tradeoffs. Hence engines will be of larger displacement to achieve the desired power output. Specific ratings of around 40-50 kW/liter are expected with cylinder displacements in the 0.6 – 0.7 liter range.

V6 engines are likely to dominate the market based on the above mentioned rating and displacement issues. The V6 also offers competitive packaging for the majority of US applications and can be used in both longitudinal and transverse applications which offers the potential for higher base engine volumes.

From a heavy duty passenger vehicle perspective the 2007 emissions levels appear achievable without the need for NOx aftertreatment but with the application of a DPF. However the 2010 regulations are a different matter and at present these will require significant development in both NOx aftertreatment and combustion technology to ensure emissions compliance.

These engines will have a similar cylinder displacement to the light duty engines although displacements in the 5.5 – 6.5 liter range are expected. Hence V8 configurations will dominate. It is feasible that the LD V6 and HD V8 engines will be derivatives will be of the same base engine design although challenges exist with respect to bank angle selection and the differing durability requirements of the two engine classes. The requirements of the two EGR systems will also represent a design commonality issue.

The biggest factors determining the fate of diesels in the US passenger vehicle market are not technical but more a blend of economic and social.

At present there is no significant driver to reduce fuel consumption and until this situation changes there is little consumer pressure to improve fuel consumption. The desire by the US government to reduce oil dependence may stimulate a change in this direction but legislation is probably required. Whether taxation on fuel is a politically acceptable means to drive down fuel consumption is a still unanswered question.

The perception of diesel is also an issue. However, the main challenge here largely perception rather than truth. Modern HSDI diesels are fuel efficient, clean, quiet and offer the end user improved driveability over the gasoline engine. Once end users experience these engines the negative perception will quickly be removed. The major challenge may be getting the consumers to test drive the vehicles in the first place. This then represents a marketing, rather than technical challenge, which must be addressed by the auto manufacturers.
Last, and very significantly, the increased cost of the diesel engine must be considered. Unless the OEM's can convince the end user of the added benefits of diesel it will be difficult to charge a premium for the vehicle and hence retain profit. As aftertreatment system development and production volumes increase the on-cost for these systems should reduce. Also fuel system costs, which account for around 40% of total diesel cost, should reduce with increased competition and production volumes. If value is placed on fuel economy either via fuel taxation or future legislation the gasoline production costs will climb based on added technology to achieve fuel savings. Under this scenario it is possible that the on-cost associated with diesel will become acceptable. In addition to the fuel economy benefit it is important that the performance aspect is highlighted which allows further justification of any price premium.

REFERENCES

1. Walter S. McManus: Consumer Demand for Clean Diesel-Powered Vehicles. SAE Executive Management Briefing: Diesels and Their Future in the U.S. May 28, 2003. Troy, Michigan
2. AVL: Österreichisches Gebrauchsmuster GM 600/98
3. Tanaka, T.: Simultaneous Reduction of PM and NOx - A New After-Treatment System. 22nd Int. Wiener Motorensymposium. Vienna, Austria April 26-27, 2001
4. Pfahl, U.; Hirtler, W.; Cartus, Th.: Passenger Car Investigations of NOx-Adsorber & DPF Combination to Fulfill Future Diesel Emission Limits. SAE 2003-01-0043

5. Weissbäck, M.; Csató, J.; Glensvig, M.; Sams, Th.; Herzog, P.: Alternative Combustion - An Approach for Future HSDI Diesel Engines. MTZ (9) 2003
6. Herzog, P.: U.S. Light Duty Diesel Technology. Global Powertrain Congress 2003, September 23-25, 2003, Ann Arbor, Michigan

CONTACT

Chris Cowland / chris.cowland@avlna.com
Peter Gutmann / peter.gutmann@avl.com
Peter L. Herzog / peter.herzog@avl.com

www.avl.com

DEFINITIONS, ACRONYMS, ABBREVIATIONS

BMEP : Brake Mean Effective Pressure

CAFE : Corporate Average Fuel Economy

CR : Compression Ratio

DPF: Diesel Particulate Filter

EGR : Exhaust Gas Recirculation

FIE : Fuel Injection Equipment

FTP : Federal Test Procedure

GDI : Gasoline Direct Injection

GVWR : Gross Vehicle Weight Rating

HDDTC : Heavy Duty Diesel Transient Cycle

HSDI : High Speed Direct Injection

MDPV : Medium Duty Passenger Vehicle

NEDC : New European Driving Cycle

NTE : Not To Exceed

PM : Particulate Matter

SUV : Sport Utility Vehicle

VVA : Variable Valve Actuation

2005-01-3628

Developing Diesel Engines to Meet Ultra-low Emission Standards

Stephen J. Charlton
Cummins Inc.

ABSTRACT

The modern diesel engine is used around the world to power applications as diverse as passenger cars, heavy-duty trucks, electrical power generators, ships, locomotives, agricultural and industrial equipment. The success of the diesel engine results from its unique combination of fuel economy, durability, reliability and affordability – which drive the lowest total cost of ownership.

The diesel engine has been developed to meet the most demanding on-highway emission standards, through the introduction of advanced technologies such as: electronic controls, high pressure fuel injection, and cooled exhaust gas recirculation. The standards to be introduced in the U.S. in 2007 will see the introduction of the *Clean Diesel* which will achieve near-zero NO_x and particulate emissions, while retaining the customer values outlined above. The progress toward near-zero emissions has involved the development of:

- Advanced engines using new technologies
- Advanced aftertreatment systems
- Availability of ultra-low sulfur diesel fuels

This paper describes the technology, tools and processes used to develop ultra-low emission diesel engines for on-highway heavy-duty applications. While meeting emissions is an important goal, it is vital to develop engines that retain traditional customer values. The primary purpose of this paper is to provide a set of ideas and practical tools that allow engine design and development to focus on the end-user, while satisfying regulatory requirements. The on-highway heavy-duty (HD) diesel engine is used to illustrate the processes; however the general principles may be applied to other diesel engine applications, such as off-highway, marine or power generation.

The paper underlines the importance of selecting the right technology and system architecture for the application, executing the design of hardware and controls software, and optimizing the performance, fuel economy and emissions through controls calibration using the best available tools. Understanding customer requirements is the fundamental foundation for effective engine design, as is the integration of the engine system with the machine it will drive – be it a truck, a ship or a power generating set.

The paper will describe the modern HD diesel engine, as used in on-highway applications in the US, and will discuss the technologies that make these engines possible. The evolution of emissions requirements in the U.S. and worldwide will be discussed. The important processes of system integration and product definition will be discussed briefly and put in context with the demands of developing new engine systems to meet proposed ultra-low emissions standards. The paper will close with a look forward to advanced technologies that will be candidates for meeting future stringent emission standards.

LIST OF ACRONYMS

AFR	Air:Fuel Ratio
BMEP	Brake Mean Effective Pressure
BSFC	Brake Specific Fuel Consumption
DegCA	Degrees Crank Angle
CAD	Computer-Aided Design
CFD	Computational Fluid Dynamics
CO	Carbon Monoxide
CO_2	Carbon Dioxide
DOC	Diesel Oxidation Catalyst
DPF	Diesel Particulate Filter
ECM	Electronic Control Module
EGR	Exhaust Gas Recirculation
FEA	Finite Element Analysis
FGT	Fixed Geometry Turbocharger
HC	Unburned Hydrocarbons
HD	Heavy-Duty
LNT	Lean NO_x Trap
LNC	Lean NO_x Catalyst
NAC	NO_x Adsorber Catalyst
NO	Nitric Oxide
NO_2	Nitrogen Dioxide

NO_x	Oxides of Nitrogen
PM	Particulate Matter
SCR	Selective Catalytic Reduction
SOx	Oxides of Sulfur
TWC	Three-Way Catalyst
WGT	Wastegated Turbocharger

1. INTRODUCTION

This paper describes the technology, tools and processes used to develop ultra-low emission diesel engines for on-highway heavy-duty applications. While meeting emissions is an important goal, it is vital to develop engines that retain traditional customer values, such as fuel economy, performance, reliability and durability. Therefore, the primary purpose of this paper is to provide a set of ideas and practical tools that allow engine design and development to focus on the end-user, while satisfying regulatory requirements. The on-highway heavy-duty (HD) diesel engine is used to illustrate the processes; however the general principles may be applied to other diesel engine applications, such as off-highway, marine or power generation.

The modern HD diesel engine operates as a part, albeit an important part, of a highly integrated system, Figure 1-1. Taking the typical class 8 truck as an example, with gross combined weights (GCW) of up to 150,000 lbs - the truck design at a system level is a thoughtful balance between engine and aftertreatment performance and packaging, cooling system design, aerodynamic efficiency, rolling resistance, transmission characteristics and electronics integration. The integrated product should be easy and safe to operate at peak efficiency, under a variety of duty-cycles, have low cost of ownership, be easy to maintain and have exceptional uptime. Today, the demands placed on engine and truck design are such that a highly integrated engineering program is required to produce optimum results.

- Aerodynamic Design
- Engine Performance
- Engine Fuel Efficiency
- Cooling Capacity
- Rolling Resistance
- Transmission System
- Braking System
- Service Tools
- Electronics Systems Integration
- Engine & Aftertreatment Packaging

Figure 1-1, Today's HD truck is a highly integrated system.

DESIGN AND OPTIMIZATION OF THE MODERN HD DIESEL ENGINE

The on-highway HD diesel engine is used extensively throughout the developed world to haul commercial freight and deliver consumer goods. The diesel engine is successful in this application primarily because of its outstanding fuel economy, durability, reliability and performance, and because it is environmentally friendly. In recent years the term *Clean Diesel* [1] has been used to describe the new generation of engines to be launched in 2007, which will have very low NO_x and particulate emissions. *Clean Diesel* combines advances in diesel engine design, cleaner burning fuels and emission control systems to produce engine systems that have near-zero particulate and NO_x emissions.

This generation of diesel engine will feature particulate filter technology (DPF) which effectively eliminates smoke and odor from the exhaust, and will have emission reductions of 98% compared with unregulated engines. The diesel engine, by virtue of its thermal efficiency, also emits the least carbon dioxide (CO_2) of any available internal combustion engine per unit of work.

Figure 1-2 shows the general layout of a HD on-highway diesel engine. Such engines are turbocharged and use air-to-air charge air cooling to increase charge density and reduce NO_x. They feature high pressure injection systems (1600-2400bar) and full-authority electronic control. At the prevailing emission standards in the U.S. (EPA 04), these engines typically have cooled Exhaust Gas Recirculation (EGR) technology to control NO_x emissions. The electronic control system typically features several sensors to enable control of engine performance and emissions under a wide range of operational and ambient conditions – see Figure 2-5.

Optimization of the modern HD diesel engine starts at the design stage. The ability of an engine system to meet customer requirements begins with the selection of the system architecture, followed closely by the execution of the design itself – including controls software, and finally depends on the controls calibration. A successful engine system will require attention to all aspects:

- System Architecture
- System Design
- System Controls Calibration

CUSTOMER REQUIREMENTS

The fuel economy of the modern HD diesel engine derives from a number of design factors. First is the thermodynamic cycle which allows for high indicated thermal efficiency by virtue of a high compression and expansion ratio. Second is turbocharging, which allows high power density, and a consequent improvement in mechanical efficiency since parasitic losses become a smaller fraction of the indicated power of the engine. Third is the application of cooled Exhaust Gas Recirculation (EGR) as a key component of the emission control system. Cooled EGR will be discussed in greater detail in a later section. Cooled EGR controls flame temperature through oxygen dilution, which allows combustion to be optimized for fuel economy and performance.

The modern HD diesel engine operates at specific power levels between 30 and 40 HP/liter, and maximum BMEP levels of 20-24 bar. The rate of increase in specific power levels has slowed in recent years with the added requirements associated with design for ultra-low emissions. However, with the introduction of turbocharging and charge air cooling in the 1970s and 1980s respectively, the diesel engine today achieves power levels 3-4 times higher than were available 40 years ago.

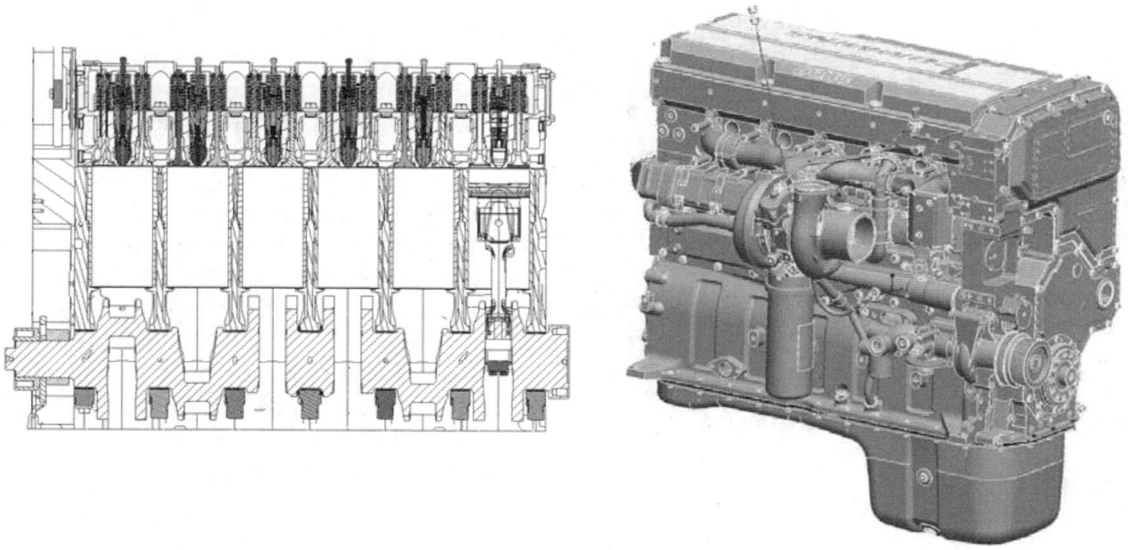

Figure 1-2, On-highway heavy-duty diesel engine. (Courtesy of Cummins Inc.)

The durability of the HD diesel engine is well known, and results from the robust design of the base engine – crankshaft, cylinder block, and cylinder head, which are designed to withstand the high cylinder pressures associated with the diesel engine. HD diesel engines are designed for remanufacture, to allow extended life. Design for remanufacturing requires provision of excess material on critical surfaces to allow subsequent re-machining, availability of oversized parts, and modular design of critical components such as removable cylinder liners. Today's HD diesel engines are designed for 1,000,000 mile life, with many engines operating beyond this mileage with the help of remanufacturing.

A key requirement for the on-highway HD diesel is extremely high reliability and uptime. Reliability defines an engine's ability to avoid failures between service events – in other words the ability to avoid unexpected failures. *Uptime* is both a function of the engine's reliability and the *downtime* associated with effecting repairs. Design and development for reliability is one of the more demanding tasks in new engine product development. One of the major challenges with reliability development is being able to measure or assess the reliability of an engine design during the early design and subsequent development phases.

On-Road Mobile Sources	Non-Road Mobile Sources	Stationary Sources
Cars	Construction Equipment	Electrical Power Generators
Light Trucks	Farm Equipment	Oil and Gas Operations
Heavy Trucks	Locomotives	Industrial Plant & Equipment
Buses	Marine Vessels	Waste Incineration Plant

Table 1-1, Stationary and Mobile Sources regulated by the Clean Air Act of 1970.

EMISSION REGULATION AND CONTROL IN THE U.S.

In the late 1960s concern for air quality began to emerge in the US, and especially in California due to air quality concerns in the Los Angeles basin and San Joaquin Valley. These concerns were caused by a combination of the unique geographic and climatic conditions in those regions and the rapidly increasing number of vehicles on the roads. This led to the creation of the California Air Resources Board (CARB) in 1967. In 1970 the US Congress passed amendments to the Clean Air Act, which led to the creation of National Ambient Air Quality Standards (NAAQs) and statutory deadlines for compliance [2]. The 1970 amendments also led to the creation of the Environmental Protection Agency (EPA) and began the regulation of various emission sources – including stationary and mobile sources – see Table 1-1.

The National Ambient Air Quality Standards (NAAQs) govern so-called criteria pollutants, which include:

- 1-Hour Ozone
- 8-Hour Ozone
- Carbon Monoxide
- Nitrogen Dioxide
- Sulfur Dioxide
- Particulate Matter PM-10
- Particulate Matter PM-2.5

Local Authorities and States are required by law to measure and record criteria pollutants, and are subject to standards based on public health criteria. Areas that fail to comply may be designated as 'Non-Attainment' areas, and become subject to more stringent emission standards until attainment is achieved. In general, the standards for mobile and stationary sources are designed to allow NAAQs to be met in light of increasing vehicle miles traveled (VMT), increasing energy consumption and industrial production, and better understanding of public health effects.

Regulation of HD diesel engines effectively began in 1978 with the introduction of NO_x, Particulate, HC and CO standards. Compliance with these standards was established through certification using a steady-state 13 mode emissions test.

	1978	2007	Test Cycle
NO_x	10.0g/HP-hr	0.2g/HP-hr	Steady-State
Particulate	1.00g/HP-hr	0.01g/HP-hr	Transient & Steady-State

Table 1-2, Comparison between the 1978 and 2007 US EPA on-highway standards

In 1985 a transient test cycle was introduced as shown in Figure 1-3. This cycle was designed to test engines under real-world driving conditions, and contained additional features such as periods of motoring, hard accelerations, urban and highway driving segments, not previously captured in the 13 mode steady-state test. For a comparison of standards in 1978 with 2007, see Table 1-2.

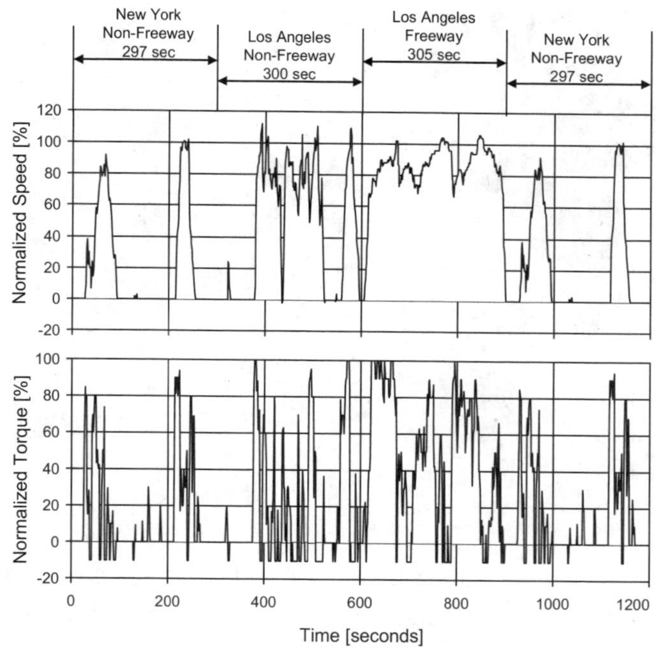

Figure 1-3, EPA transient emissions engine dynamometer test cycle introduced in 1985 vehicles greater than 8500lbs GVW.

Figure 1-4 shows the evolution of emission standards for US EPA HD on-highway vehicles of greater than 8500 lbs Gross Vehicle Weight (GVW). It will be readily apparent that the industry has made considerable progress, and that by 2010 emission of NO_x and Particulate will be at near-zero levels. The next section of this paper will present and discuss the evolution of technology that has made this dramatic progress possible.

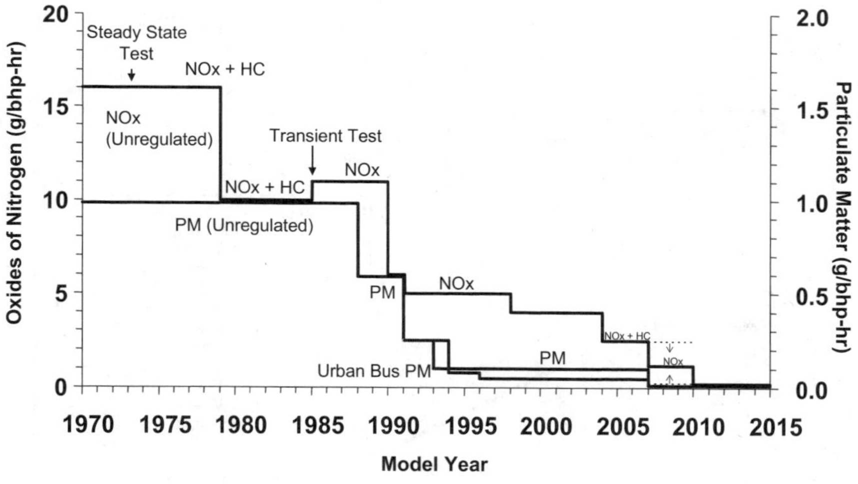

Figure 1-4, Evolution of HD engine emission standards – U.S. on-highway.

EVOLUTION OF WORLD-WIDE EMISSIONS REQUIREMENTS

The concern for clean air, and the regulation of engine emissions that began in the US in the 1960s, were closely followed by regulatory programs in Europe and Japan. Figure 1-5 shows the evolution of actual and expected worldwide emission standards through 2010.

It will readily be seen that while the US, Europe and Japan lead with the most stringent standards; other areas are following closely behind. In many developing countries, the trend is toward a combination of national or regional standards, and local or city standards to deal with local air quality issues. Examples of local regulation can be found in New Delhi, India and Beijing, China.

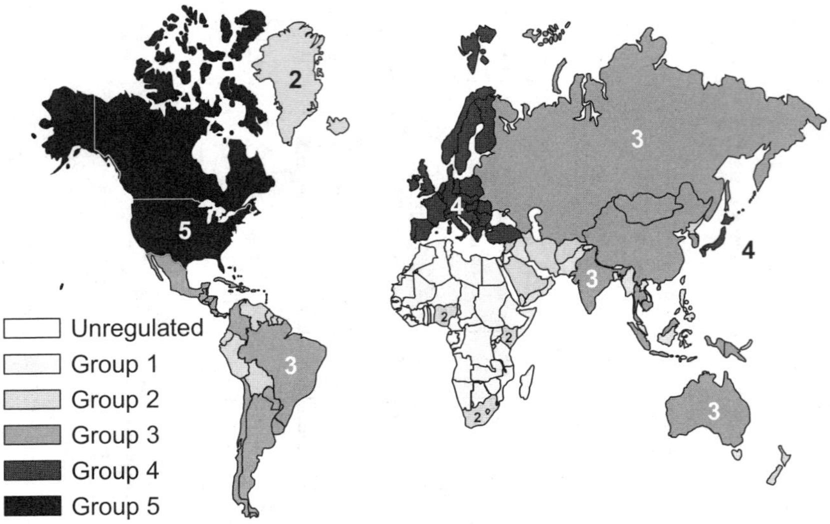

Figure 1-5, Evolution of worldwide emission standards – projection to 2010. Groups 1 through 5 represent increasing emission control stringency.

2. EVOLUTION OF EMISSIONS TECHNOLOGY TO MEET CURRENT STANDARDS

The challenge posed by the stringent emission standards introduced since 1978 has led to the development and introduction of new technologies - all of which form a key part of today's low emission engine designs:

- Combustion System Development
- Air-to-Air Charge Cooling
- Electronic Controls
- VG Turbocharger
- Cooled EGR
- Electronic Fuel Systems
- Aftertreatment – Diesel Oxidation Catalyst

Figure 2-1 illustrates the introduction of technology to support improved customer value while meeting increasingly stringent emission standards.

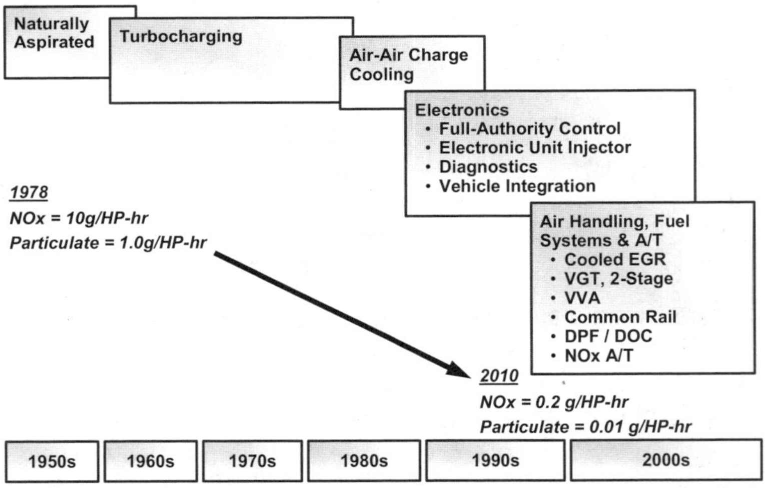

Figure 2-1, Evolution of key technologies for HD on-highway diesel engines.

COMBUSTION SYSTEM DEVELOPMENT

In the modern diesel engine the charge of air, or EGR and air, is compressed through a relatively high compression ratio (typically between 16.5:1 and 19:1) to raise the charge temperature prior to fuel injection. Although many schemes exist for early pilot, pilot, post and late-post injection events, the main fuel injection event occurs close to piston top-dead-center (TDC). The high pressure fuel systems in use today operate at fuel pressures between 1600bar and 2400bar, to create a highly atomized fuel spray and short injection duration (less than 30 degCA). The fuel rapidly vaporizes in the hot charge in the cylinder, and begins to mix with the surrounding charge. The introduction of swirl (ordered charge rotation inside the cylinder) in the early 1990s allowed a step improvement in engine-out particulate emission and removed the need for the early introduction of PM filter technology.

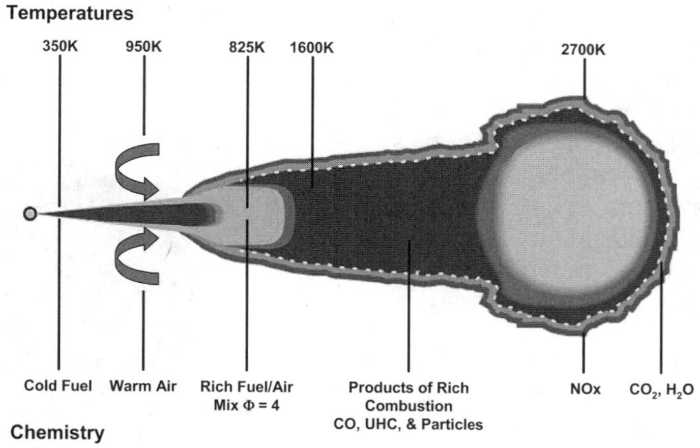

Figure 2-2, Sandia / Cummins model of diffusion flame combustion [3].

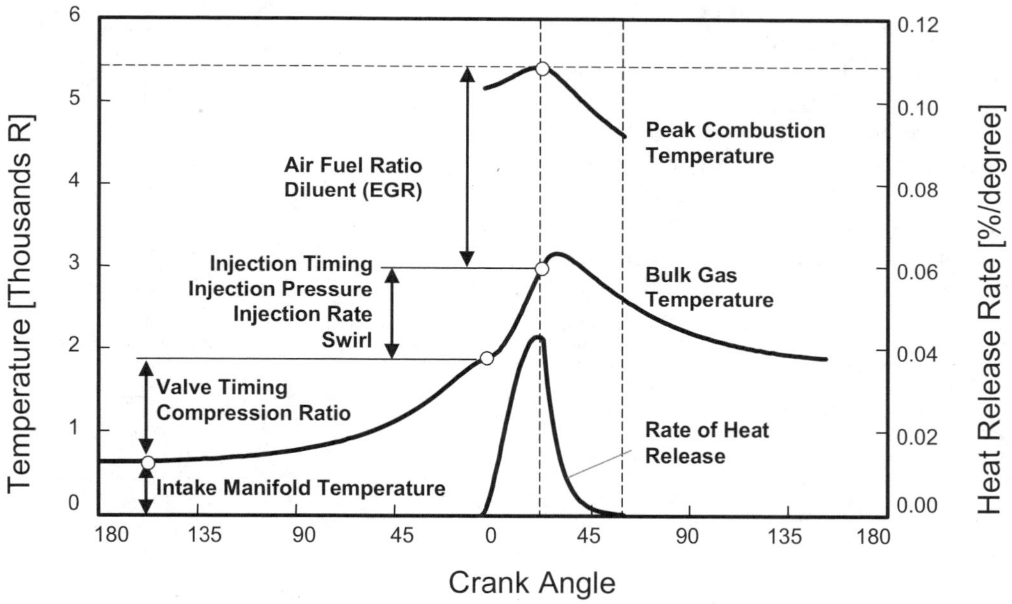

Figure 2-3, Factors affecting flame temperature and NO_x production.

Diesel combustion has been characterized by Dec et al [3] as shown in Figure 2-2, which shows the various phases of combustion development from the raw fuel at the injector nozzle, to the edge of the combustion plume.

NO_x is formed in the high temperature diffusion flame region at the interface between the fuel rich plume and the surrounding fresh charge. Since NO_x production is an exponential function of flame temperature, in-cylinder control of NO_x has been accomplished through control of factors which directly influence flame temperature, illustrated in Figure 2-3. Several of these factors will be described in the following sections.

AIR-TO-AIR CHARGE COOLING

In the 1980s air-to-air charge cooling was introduced in order to reduce charge temperatures (hence flame temperatures), and to increase charge density, thereby providing more engine torque and power. Air-to-air charge cooling can reduce intake manifold temperatures almost to ambient temperature – following compression heating in the turbocharger, producing very large reductions in NO_x, of the order of 50% from unregulated levels. The charge air cooler is a cooling matrix installed between the turbocharger compressor and intake manifold. The charge cooler is usually installed at the front of the truck – in front of the engine radiator so that it is cooled by the ambient air either by natural ram air from the forward motion of the vehicle, or with cooling fan assistance.

ELECTRONIC CONTROLS

In the late 1980s full-authority electronic control systems were introduced. Electronic controls were introduced initially to allow control of a new generation of electronic fuel systems. Subsequently, electronics has been used to control many aspects of the engine – turbochargers and EGR systems being notable examples. Electronics are also used to diagnose problems with the engine, provide protection under extreme operating conditions, and to support on-board diagnostics (OBD).

Electronic controls allow accurate control of all critical functions of the engine, under the widest range of conditions. The evolution of electronic controllers is summarized in Table 2-1. First generation controllers used relatively few sensors and actuators, and featured relatively simple algorithms that made extensive use of a limited number of look-up tables to produce outputs. Examples include controlling outputs such as injected fuel quantity and timing in direct response to inputs such as engine speed and accelerator pedal position. With growing requirements second generation controllers were developed that used several sensors and controlled a small number of actuators, such as fuel injectors and EGR valves. Such controllers made use of multiple look-up tables, featured limited closed-loop control and often made corrections to control outputs based on ambient temperature or altitude. Examples include EGR control based on EGR control valve position with feedback, open-loop control of fuel system quantity and timing. Although the fuel system itself is effectively controlled open-loop, all engine controllers control engine speed or vehicle speed by means of a governor or cruise control.

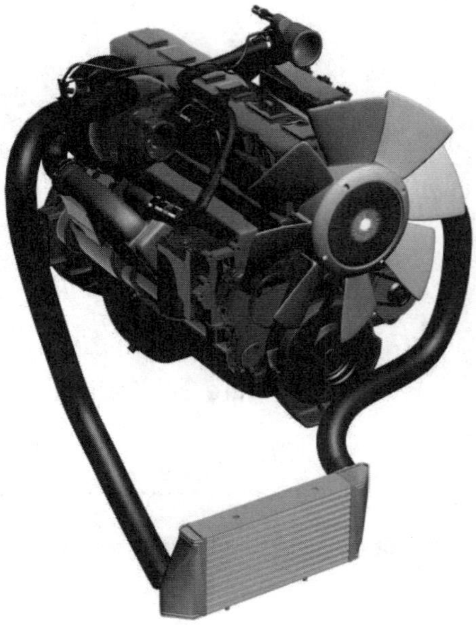

Figure 2-4, Air-to-air after-cooler.

First Generation Diesel Controller	Second Generation Diesel Controller	Third Generation Diesel Controller
• Fuel System Control • Few Sensors • Few Actuators • Simple Look-Up Tables • Limited Ambient Correction • Limited Service Diagnostics • Limited Vehicle Integration	• Fuel, EGR and Turbocharger • Many Sensors • Few Actuators • Multiple Look-Up Tables • Some Closed-Loop Compensation • Some Correction for Ambient Conditions • Service Diagnostics • Powertrain Integration	• Engine & A/T Control • Many Sensors • Virtual Sensors • Physical Models • Many Actuators • Multiple Look-Up Tables • Closed-Loop Control Critical Variables • Extensive Correction for Ambient Temperature and Altitude • Service Diagnostics • On-Board Diagnostics • Vehicle Integration

Table 2-1, Evolution of electronic diesel control

Third generation controllers use a rich set of inputs – determined directly from engine and vehicle mounted sensors, and indirectly from physical models often known as Virtual Sensors, see Figure 2-5. They feature closed-loop control of most critical variables – such as EGR flow rate or air flow rate – which is necessary to limit the amount of variation in performance and emissions from engine-to-engine, and with variation in climate and duty-cycle. Third generation controllers have smart actuators where the electronic control module (ECM) communicates with the actuator over a serial data-link. Smart actuators feature local micro-controllers which can decode control inputs from the data-link and execute control and diagnostics algorithms local to the actuator. Examples include smart EGR valves and smart turbochargers.

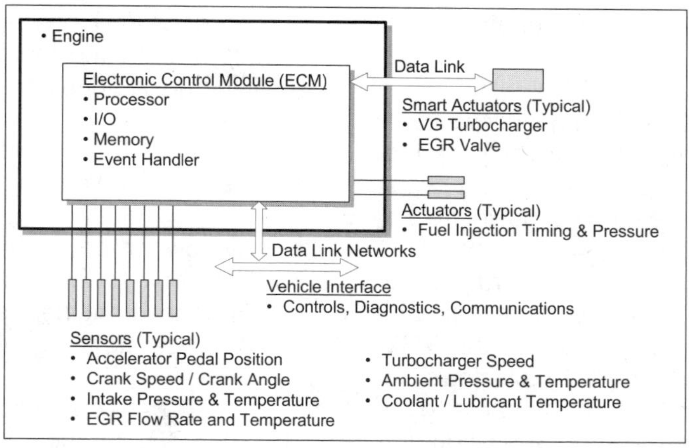

Figure 2-5, Typical third generation engine control system.

VARIABLE GEOMETRY (VG) TURBOCHARGER

The variable geometry turbocharger was first conceived in the mid-1980s, and after a period of development and optimization was introduced into series production in the late 1990s [4]. A conventional turbocharger consists of a turbine stage in the exhaust stream of the engine which drives a compressor stage installed in the intake system.

The primary goal of the turbocharger is to increase charge density (boost) and hence allow increased engine power. The combination of charge compression by a turbocharger and air-to-air charge cooling has allowed 3 to 4 fold increases in diesel engine power density over the past 40 years. The amount of intake air compression achieved depends on the characteristics of the turbine stage. The turbine develops power by expanding exhaust gases through a stator or nozzle ring and rotor. The two key parameters of the turbine are its swallowing capacity and isentropic efficiency. Swallowing capacity is a term used to describe the overall flow restriction caused by the turbine, and hence the upstream pressure head created. The higher the pressure head, the higher the expansion ratio through the stage, and the higher the turbine power developed. The isentropic efficiency determines how much of the available turbine power is actually achieved in the stage.

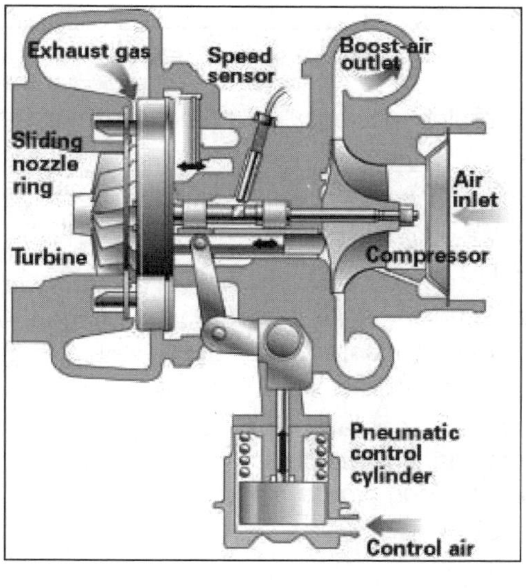

Figure 2-6, Variable geometry (VGT) turbocharger with sliding nozzle ring (Courtesy of Holset).

Given that the swallowing capacity of the turbine is a critical parameter in determining turbocharger power, interest in variable geometry turbocharging was a natural evolution from the initial 'fixed' geometry turbocharger (FGT) designs. The first evolution of variable geometry was the development of the turbine wastegate. The wastegate is essentially a controlled by-pass path around the turbine stage. By allowing part of the exhaust gas to by-pass the turbine rotor, the power of the turbine stage could be modulated from its maximum power by opening the wastegate. The wastegate allowed turbine stages with small swallowing capacities, which could produce high boost pressures at low engine speeds with the wastegate closed. At higher engine speeds, the wastegate would be opened to prevent over-boosting and protect the engine from damage.

A truly variable geometry turbocharger allows the turbine swallowing capacity to be adjusted during engine operation, without the need to by-pass any of the exhaust gas. This allows the turbine power to be modulated in response to real-time operating conditions, for example, during a hard acceleration the VGT can be closed to speed up the compressor and produce excess boost. At low engine speeds the VGT can be closed down to generate high air flows to provide high engine torque levels. At higher engine speeds the VGT can be opened to reduce air flow for optimum engine operation without risk of damage.

A number of VGT concepts have been proposed, which include swing vane designs and sliding nozzle designs. A sliding nozzle design is shown in Figure 2-6.

Today, the VGT serves several purposes in engine performance and emission control:

- Air flow for torque and performance
- EGR delivery and control
- Enhanced engine braking
- Aftertreatment regeneration

The VGT can be used to modulate the delivery of cooled EGR, through real-time control of swallowing capacity. A typical VGT swallowing capacity versus turbine efficiency relationship is shown in Figure 2-7.

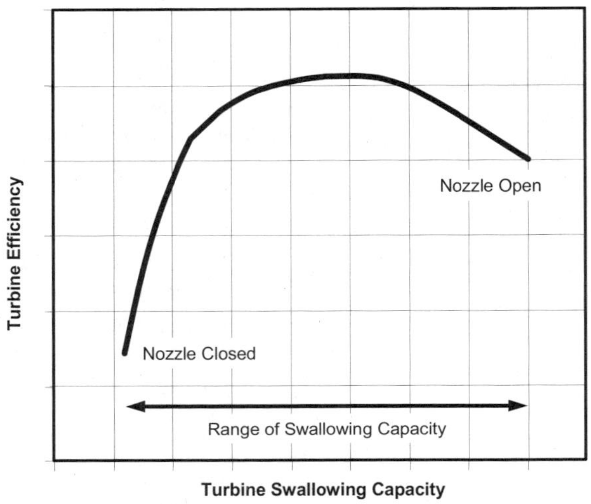

Figure 2-7, Typical VG turbine efficiency as a function of turbine swallowing capacity.

COOLED EGR

Cooled EGR was first introduced in US HD on-highway applications in 2002, to meet the 2.5g/HP-hr NO_x +HC standard. Cooled EGR is a highly effective NO_x control, capable of almost total elimination of NO_x when used at high levels. Figure 2-8 shows an engine schematic, where EGR is drawn from the exhaust manifold and returned to the intake manifold – so-called *Short-Route EGR*.

Exhaust Gas Recirculation involves returning a proportion of the exhaust gas to the engine cylinder. This may be achieved by taking gas from the exhaust manifold and routing it through a metering valve and heat exchanger – known as an EGR cooler, Figure 2-9. The cooler is designed to reduce the exhaust gas temperature to limit thermal loading in the engine, further reduce NO_x (since NO_x is a function of temperature) and to allow high power density.

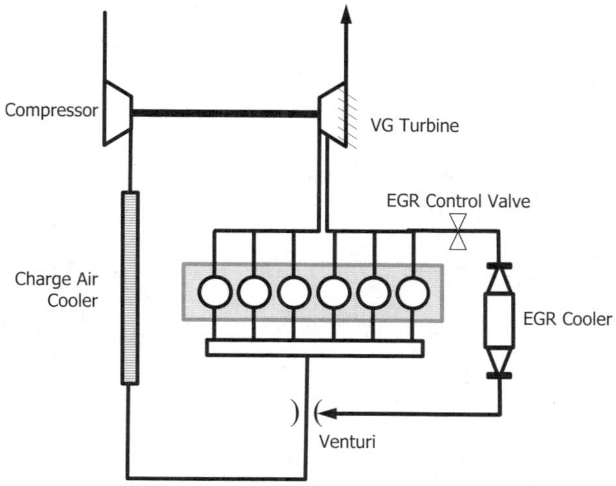

Figure 2-8, Engine schematic showing *short-loop* cooled EGR circuit.

Cooled EGR reduces NO_x by diluting the oxygen content of the combustion charge with inert gases – mostly nitrogen, such that when combustion occurs the resulting flame temperature is greatly reduced. The inert gases absorb part of the heat of combustion and limit the overall temperature rise. Again, since NO_x kinetics are an exponential function of temperature, cooled EGR has the effect of reducing NO_x. A characteristic NO_x reduction curve is shown in Figure 2-10. This figure shows that as EGR is increased, the NO_x reduction is very significant, approaching 90% reduction as the intake charge oxygen mass fraction approaches 18%.

Electrically Actuated EGR Valve

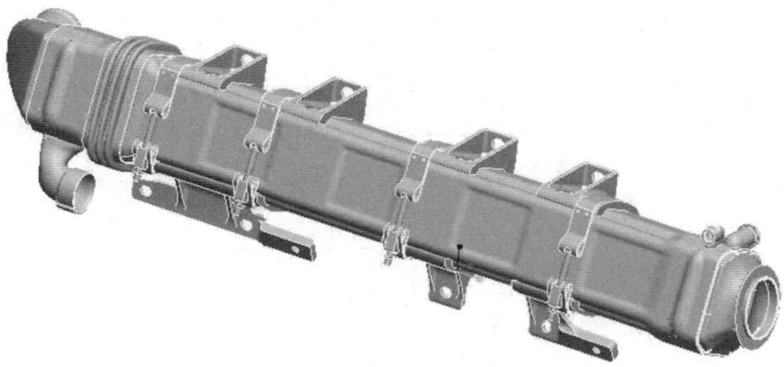

Heavy-Duty EGR Cooler

Figure 2-9, EGR components. (Courtesy of Eaton Corporation and Behr)

Figure 2-10 also shows the effect of fresh air:fuel ratio on the effectiveness of cooled EGR. Fresh air:fuel ratio is the ratio of the mass of incoming fresh air and the mass of injected fuel. Diesel engines typically operate with excess air, such that fresh air:fuel ratio is in the range of 20:1 to 30:1 at high engine loads. It can be seen that as the fresh air:fuel ratio is reduced, the effectiveness of the recirculated exhaust gas is increased. This occurs because the oxygen content of the recirculated gas is reduced with lower fresh air:fuel ratios.

Through the 1990s engine fuel economy was improved through efforts to improve turbocharger turbine efficiency by attention to the detail design of the turbine stage. Higher turbine efficiency meant that turbine expansion ratio could be reduced, thereby reducing back-pressure on the engine and engine pumping work. Prior to the introduction of cooled EGR, the conventional (FGT) turbocharger was a significant contributor to engine fuel economy, with intake boost pressures exceeding exhaust pressures by as much as 0.5 to 1.0 bars at high loads.

With the introduction of cooled EGR it was essential to raise the average exhaust pressure to allow exhaust gas to flow from exhaust manifold to intake manifold. Although this led to an increase in engine pumping work, the overall benefits of cooled EGR on the combustion process – i.e. the placement of heat release - ensured little impact on fuel consumption.

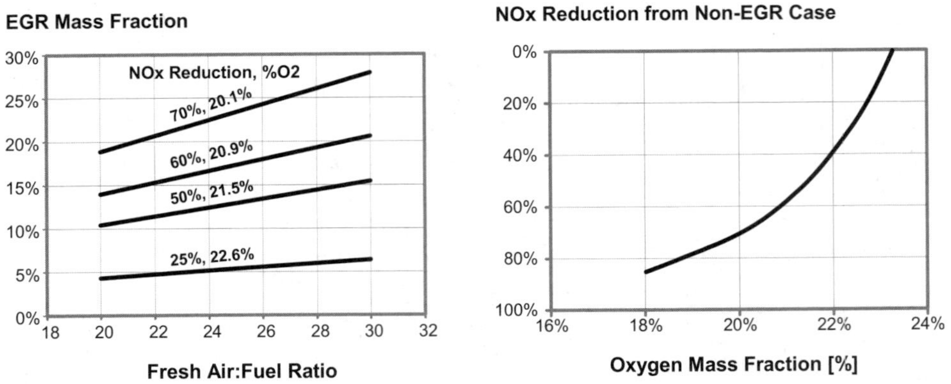

Figure 2-10, NO_x reduction efficiency for cooled EGR with variation in fresh air:fuel ratio.

ELECTRONIC FUEL SYSTEMS

From the earliest days of the diesel engine, the fuel system has been one of the most critical and challenging engine sub-systems. In recent years the evolution of fuel systems has accelerated with the development and introduction of a generation of advanced unit injector and common rail fuel systems. Figure 2-11 classifies the range of fuel systems in widespread use today.

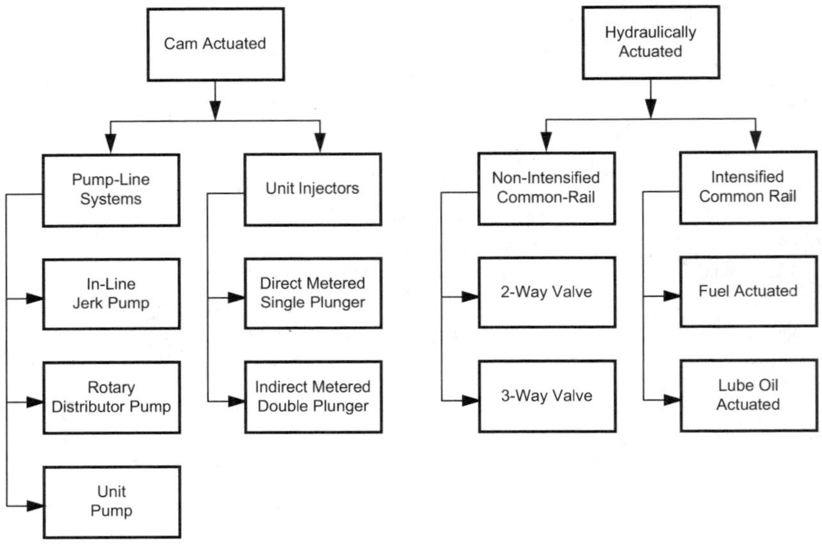

Figure 2-11, Fuel system classification based on the method of fuel pressurization.

Broadly, fuel systems are either mechanically actuated or hydraulically actuated. Mechanically (cam) actuated systems include in-line jerk pump designs, rotary designs and unit injectors. Mechanically controlled unit injectors were used in the U.S. on HD engines for many years [5]. A unit injector effectively combines the pumping element (plunger) and the nozzle into a single injector body, or unit, Figure 2-12. The force required to pressurize the trapped fuel for injection is applied by a cam mechanism acting on the top of the plunger. The unit injector design is sufficiently compact that it can be installed in the cylinder head between the engine valves. The design of the unit injector was later developed for electronic control, and became known as the electronic unit injector (EUI). The advantages of the EUI are very high injection pressure capability – typically up to 2400 bar, high mean-effective injection pressure leading to high injection rates and compact design compared with pump-line-nozzle systems. While EUIs have excellent pressure capability, they do not have the flexibility of common rail systems. Recent developments of the EUI have added a measure of flexibility through independent control of trapped fuel volume and hence injection pressure, combined with needle control for provision of limited pilot- and post-injection capability. The Delphi E3 unit injector is an example of two-solenoid unit injector technology that allows increased injection pressures at lower engine speeds and limited multiple injection capability [6].

Hydraulically actuated fuel systems use a high pressure fluid (engine oil or fuel) to effect the injection process. In the Caterpillar HEUI system, high pressure engine oil acts on the injection plunger to pressurize the fuel and produce an injection event. In the high pressure common rail (HPCR) injection system, fuel is pressurized in a rail that feeds all injectors. A high-speed actuator is used to release the pressure from the top of the needle valve causing it to lift and allow injection – so-called needle control. The design of the HPCR allows multiple injections to occur, of varying length – at any time during the engine cycle. Thus, although injection rate-shaping of the injection event is more difficult with common rail, this disadvantage can be partially off-set by the flexibility of multiple injection events in any given cycle. The common rail has allowed quiet operation of the diesel engine through injection of a small pilot charge ahead of the main injection event – see Figure 2-13.

Figure 2-12 and Figure 2-14 show typical high pressure EUI and common rail (HPCR) fuel systems in use currently.

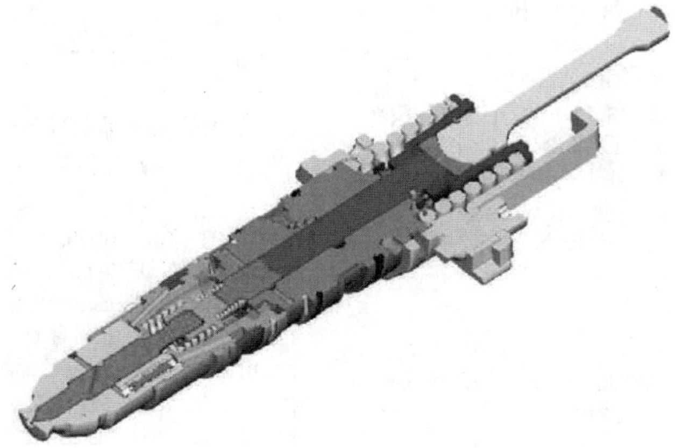

Figure 2-12, Cross-section of an Electronic Unit Injector (EUI). (Courtesy of Cummins Inc.)

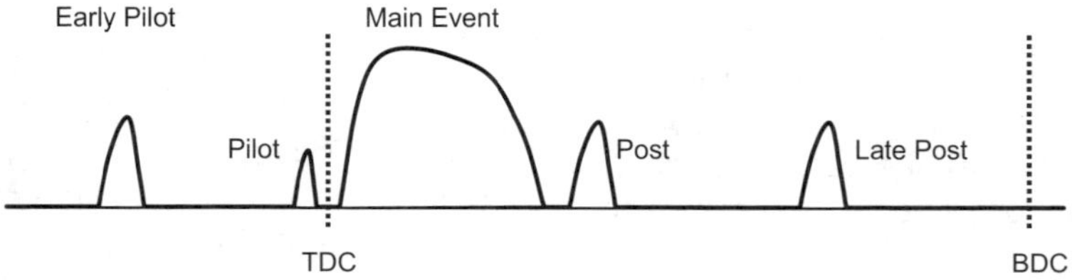

Figure 2-13, Multiple injection events possible with the common rail fuel system.

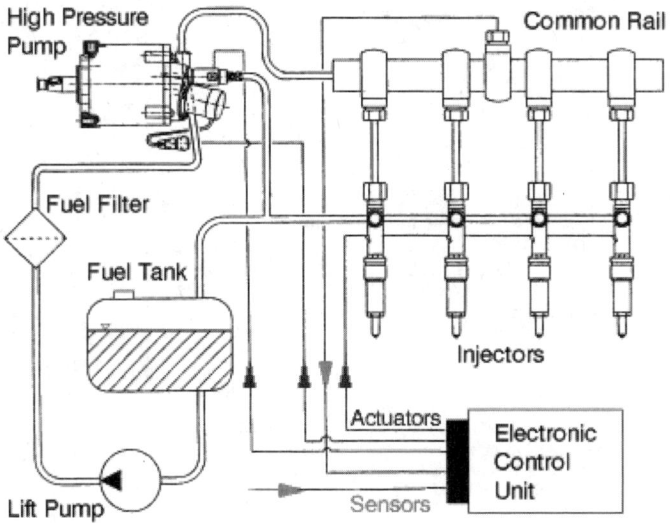

Figure 2-14, High Pressure Common Rail (HPCR) fuel system (Courtesy of Robert Bosch)

DIESEL OXIDATION CATALYST (DOC)

2007 will see the introduction of Active Particulate Filters for controlling exhaust particulate emissions to very low levels. Prior to 2007 a number of engine designs have used oxidation catalysts to control total particulate as part of an integrated engine-aftertreatment system. As the name suggests an oxidation catalyst, will oxidize unburned hydrocarbons emitted from the engine and reduce total particulate.

As a side-effect DOCs will also oxidize sulfur dioxide (SO_2) to sulfur trioxide (SO_3) which is emitted from the exhaust as particulate matter. Thus DOCs designed to operate with Low Sulfur Diesel Fuel (500PPM) are a careful balance of precious metals and base metal oxides to promote hydrocarbon oxidation and inhibit formation of sulfates.

In the diesel engine, unburned hydrocarbons normally make up 10-30% of the total particulate, and may be reduced by 80-90% by a well-designed DOC. The sources of unburned hydrocarbon include incomplete combustion – due to over-mixing of the vaporized fuel spray prior to combustion, fuel being trapped in crevices such as the top land of the piston, and fuel being trapped in the injection nozzle tip. The lubricating oil may also act as a source of unburned hydrocarbons, as a result of imperfect oil control in the cylinder, due to sealing issues at the exhaust or intake valve guides or turbocharger, or through crankcase ventilation in closed systems.

The DOC is a passive device installed in the engine exhaust system through which all of the engine exhaust flows. It consists of a metal outer casing which contains a core or substrate, usually a ceramic honeycomb. The substrate forms a matrix of flow passages parallel to the direction of flow, having as many as 400 cells/sq inch. The object of the cellular matrix is to provide a large surface area exposed to the exhaust gases, thereby increasing the probability of contact and reaction between exhaust components. The substrate is coated with a washcoat consisting of base metal oxides such as ceria or alumina, and in some cases small quantities of precious metals – typically a combination of platinum, palladium and rhodium. The combination of the base metals and precious metals act as the catalyst.

3. ENGINE AND VEHICLE SYSTEMS INTEGRATION

System Integration is defined here as the synthesis and evaluation of a relatively complex system from its sub-systems, subassemblies and components [7]. System Integration is first a creative process which involves the synthesis of new ideas to form new or improved products that satisfy a set of functional requirements. However, success in system integration will ultimately result from the application of system engineering tools which provide the discipline and structure that allow large teams to synchronize their work and their schedules.

System engineering tools allow top-level requirements to be driven down into the sub-systems and components. They provide measures that can be used to track compliance, and they allow complex interfaces to be defined and managed. In system engineering it is critical to define system boundaries very early in the product development cycle – recognizing important inputs, outputs and interactions across internal boundaries, see Figure 3-1.

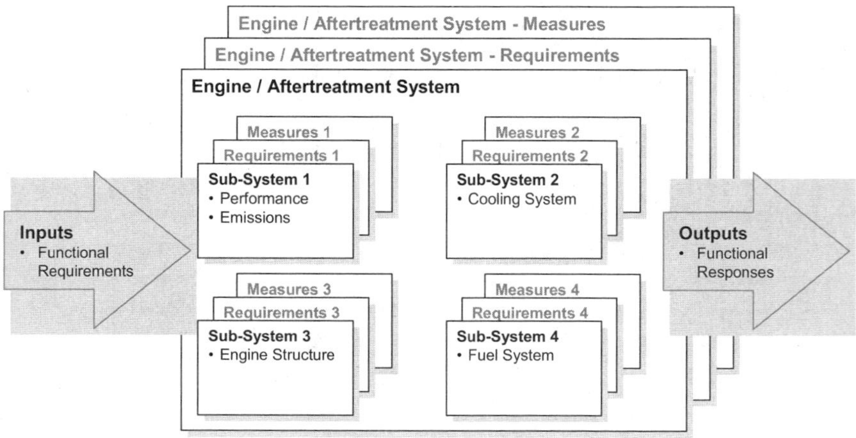

Figure 3-1, System engineering drives the definition of the system and its sub-systems, and provides tools to manage the interactions at the boundaries between sub-systems.

System integration involves optimization of the total system to better meet customer expectations, expressed as functional requirements – which will involve cost of ownership, performance, maintenance, reliability and durability. Engine systems are assemblies of sub-systems: cooling and lubricating system, aftertreatment system, fuel system, structure, air handling system, combustion system etc. Therefore, engine design optimization is a systems integration and system engineering challenge. The engine, however, is itself part of a larger system – in this case, the vehicle system. As such the engine is a sub-system that has to be optimized as part of the vehicle along with other vehicle sub-systems, see Figure 3-2. Vehicle sub-systems include: chassis system, powertrain system, braking system, cab system, electronic systems etc.

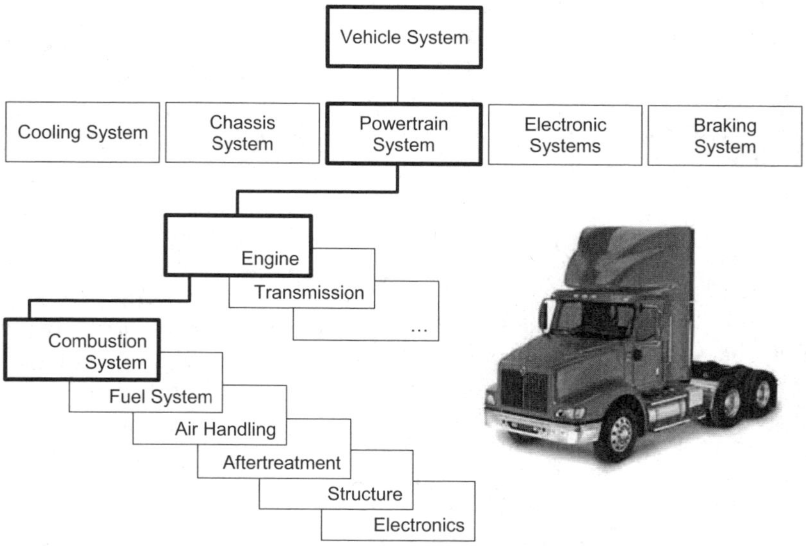

Figure 3-2, System Integration – The engine is a sub-system of the vehicle system.

System integration is both critical to the success of a project, and challenging, for a number of reasons: overall complexity of vehicle and engine systems, cross-functional requirements, multiple teams located at multiple sites and communication barriers. A number of tools exist to improve the effectiveness of system integration, which will be explored here:

- System Requirements Definition
- Sub-System Technical and Economic Requirements Flow-Down
- Technology Selection
- Platform Architecture
- Virtual Prototyping

SYSTEM REQUIREMENTS DEFINITION

System requirements definition is a step that forms the essential foundation for product definition. Requirements should be obtained from all stakeholders, including end-user customers, service, manufacturing, original equipment makers (OEMS), suppliers, marketing and sales, see Figure 3-3. Many tools exist for gathering and organizing requirements into usable information [7].

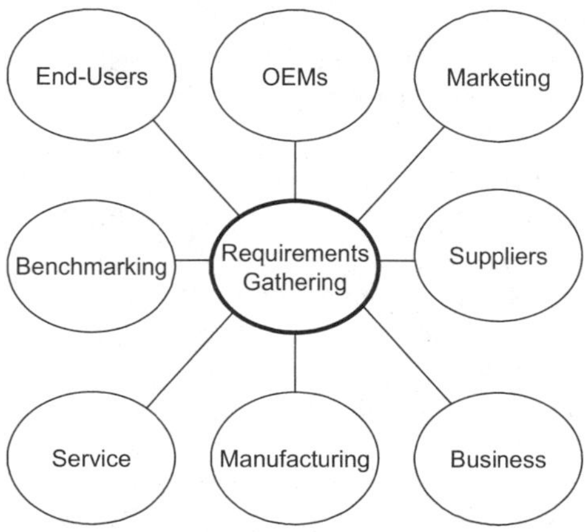

Figure 3-3, Diagram showing requirements sources.

Requirements gathering is a team activity, and should include representatives of the many functions and skills that will be used in executing the project: engineering, service, marketing, manufacturing, electronics for example. Time spent in requirements gathering is in many ways the most valuable investment, since it will create a technical and economic direction for the project, help set priorities for project goals, and remind the team who they are working for – the end-user customer.

SUB-SYSTEM TECHNICAL AND ECONOMIC REQUIREMENTS FLOW-DOWN

The key task in system integration is the flow-down of requirements from the system level to the sub-system level. This requires skilled system integrators who understand the system from both technical and economic perspectives. System models and empirical data are invaluable at this stage of the development process. Models allow trade-offs and optimization, and 'what-if' scenarios to be explored, prior to setting technical and economic targets for each sub-system. When system and sub-system requirements are defined it is essential to develop a set of measures against which progress can be tracked toward the goal – which is to fully satisfy the functional requirements.

TECHNOLOGY SELECTION

New product development usually involves design challenges that cannot be met by current product technologies. The investment in new products is being made specifically to meet new requirements that cannot be met by current products. Examples of new requirements would include increased power levels from an existing engine design, or low emission levels not previously achieved, or an entirely new engine design for a new application. Whatever the new requirements driving the need for a new product, the need to consider new technologies will inevitably arise and should be considered explicitly within the overall product development plan.

In considering new technologies it is important to consider a number of factors, which include:

- Readiness for product development
- Time and resources needed to launch the new technology
- Life-cycle costs

Readiness is an important concept in product development, and one that is not easily measured or assessed quantitatively. A common definition of technology readiness is that the technology is well understood, can be modeled mathematically, has been demonstrated in real world conditions, and has production processes and costs that are stable and understood. The time and resources needed to bring a technology to production is a critical consideration, and one that needs careful consideration in the technology selection phase.

Life-cycle costs should always be assessed and used to drive technology selections that are both technically and economically the right solution for the customer. Factors that affect life-cycle costs include:

- Product material costs
- Maintenance costs
- Operating costs
- Downtime / reliability costs
- Residual value

PLATFORM ARCHITECTURE

Platform architecture is the top-level design of the entire product line. Not only does it encompass the product line options for a given platform, but it should also describe the relationships between platforms, and in particular how they leverage commonality and standardization, Figure 3-4.

For example, within an engine platform it is important to define at the outset all of the anticipated applications for the platform, such as, automotive, industrial equipment, rail, marine and power generation. This leads to a superset of requirements such as maximum torque and power, maximum fuel quantities, package and weight constraints, range of fuel and lubricant quality etc.

Optimizing an individual engine platform design across an engine product line requires a higher level appreciation of the opportunities to reduce cost by re-use, common technology, common design and standardization. Taking time at the formative stage of the project to create a cross-platform architecture strategy will provide major opportunities for control of life-cycle costs later.

Platform architecture involves optimizing the product line for:

- Re-Use of stable components and sub-systems
- Common technology
- Modular design & standardization
- Common standards
- Reduced proliferation
- Managed obsolescence

VIRTUAL PROTOTYPING

Virtual prototyping is a cost-effective means to advance system integration using computer models and analysis – such as CAD solid models, finite element analysis, control system simulation, performance and emissions simulation. Virtual prototypes are also a valuable project management tool that allow assessment of overall project readiness. In addition, virtual prototypes are a relatively low cost means to synthesize the product, evaluate product performance, estimate product costs, check for interference between components etc – all without having to resort to expensive prototype hardware.

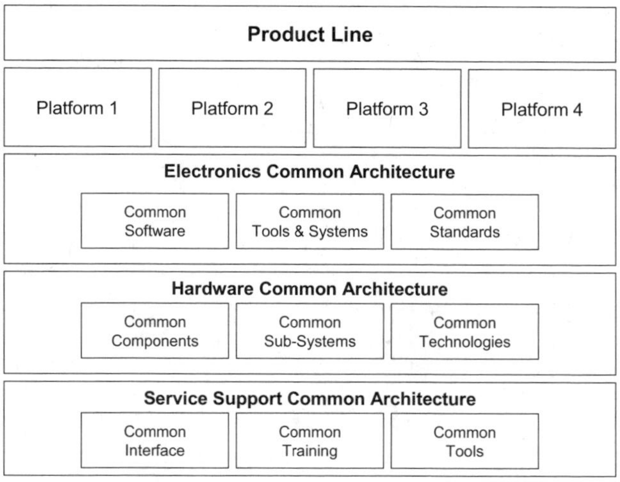

Figure 3-4, Example of platform architecture structure.

4. TOOLS FOR DIESEL ENGINE DESIGN AND DEVELOPMENT

The complexity of modern diesel engines is such that optimization of the design is only possible with the aid of computational tools, such as:

- Total System Simulation
- Engine Performance Simulation
- Computational Fluid Dynamics (CFD)
- Finite Element Analysis (FEA)
- Virtual Prototype - Fully Integrated Mechanical System Modeling

Other important computational tools include:

- Computer-Aided Design (CAD)
- Automated Test Control, Data Acquisition and Multi-Variable Optimization
- Controls Simulation and Software Code Generation

Such tools not only serve to define and deliver improved designs, but contribute significantly to the productivity of technical resources.

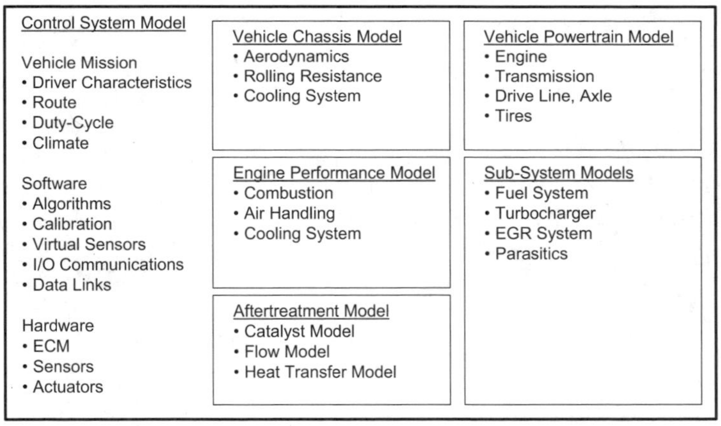

Figure 4-1, Total system simulation models contain all dynamic components of the system including driver characteristics, vehicle and engine dynamics.

TOTAL SYSTEM SIMULATION

Total system simulation is used to model all important aspects of a system for the purpose of robust controls development, see Figure 4-1. Typically such models will include a plant model (for example, a vehicle model including engine, aftertreatment and transmission models), and a control system model complete with controls software, sensor and actuator models. Such models are used to:

- Develop control strategies
- Design control algorithms
- Develop control system calibrations
- Establish system stability and response
- Establish system accuracy relative to control commands
- Generate embedded controls software

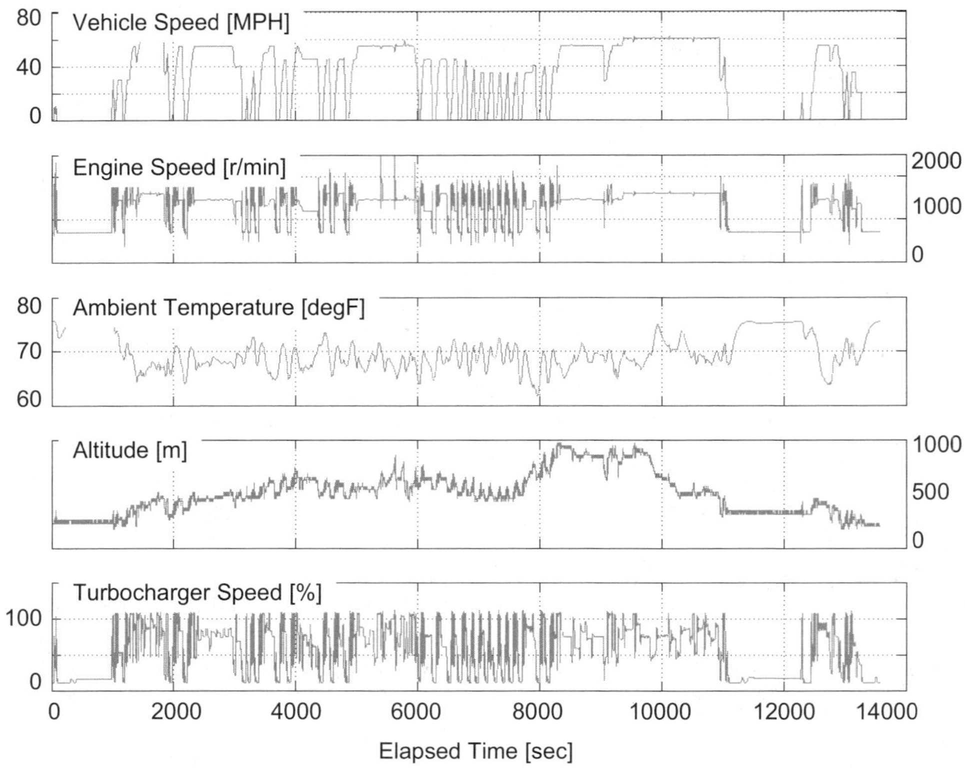

Figure 4-2, Example of total system simulation – Class 8 truck mission with 80k LB GVW.

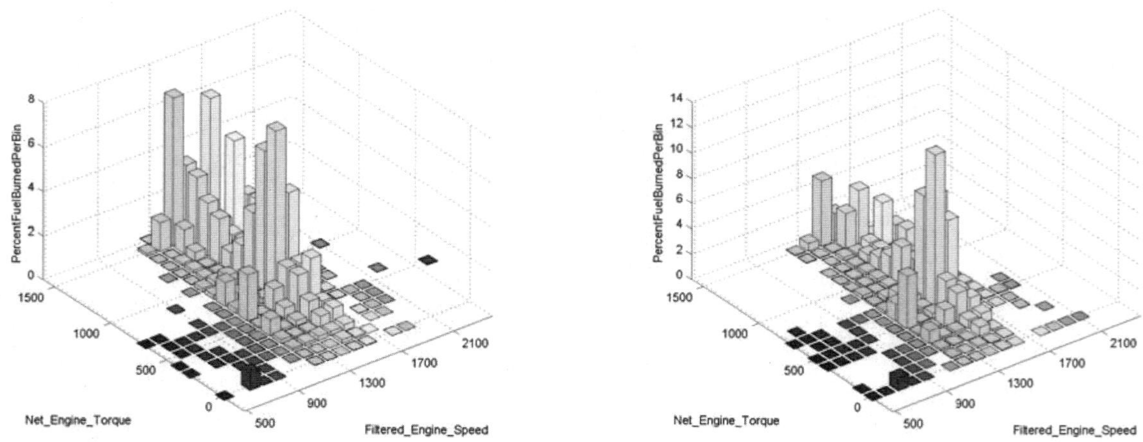

Figure 4-3, Example of total system simulation – Class 8 truck duty-cycle.
Left - 120k LB GVW, Right - 80k LB GVW.

Total system simulations are typically used to model the interactions between sub-systems, and to study the design and performance of control systems.

Total system simulations may contain physical hardware – so-called hardware-in-the-loop (or hybrid) simulations, for example an engine control module complete with embedded software and calibration could be combined with an engine simulation and a vehicle simulation. Such hybrid models have a number of advantages and as a result are widely used to solve difficult engine and control system performance problems.

Figure 4-2 and Figure 4-3 show the results from a Total System Simulation of a Class 8 HD Truck. In this example the truck was simulated over a mission that involved variation of vehicle speed, ambient temperature and altitude. The simulation was used assess the impact of GVW on the operation of the engine and aftertreatment. Figure 4-3 shows the distribution of fuel burned as a function of engine load and speed for 80k and 120k LB GVW cases. This information is used in the design of the engine system, to ensure reliable and durable operation under real-world conditions.

ENGINE SYSTEM PERFORMANCE MODELING

Engine performance models are used for a variety of design studies:

- Torque and power, fuel consumption
- Manifold design and tuning
- Transient performance and torque response
- Turbocharger matching
- Combustion and emissions
- EGR system design

Such models have progressed rapidly with advances in the cost and processing power of computer work stations, and with progress in understanding the fundamentals of key engine processes such as combustion and catalysis. Originally these models, known as cycle simulations, concentrated on the thermodynamics of the engine, however today such models of necessity include models of aftertreatment systems, such as particulate filter and oxidation catalysts.

Today it is common to use comprehensive engine cycle simulation models that feature 1-dimensional gas dynamics in the intake and exhaust systems, quasi-dimensional combustion models, 1 dimensional heat transfer models and 2- or 3-dimensional models of catalyst systems. Such models have the advantages of rapid run times, ease of use and good predictive capability within the range of the validation. They allow system interactions to be studied – for example turbocharger component matching, catalyst matching and cooling system design and optimization. A typical engine model representation is shown in Figure 4-4, showing both the engine-level model and exhaust system sub-model.

The extension of cycle simulation to include more detailed quasi-dimensional combustion models has allowed study of the effects of design variables such as injection timing, pressure and rate, EGR fraction, and air-fuel ratio on performance and emissions. Since NO_x production is largely a function of local flame temperature within the combustion process, such cycle simulations have proved both accurate and valuable in engine design optimization studies. They allow trade-offs to be explored analytically, saving development time and money, and improving the final result.

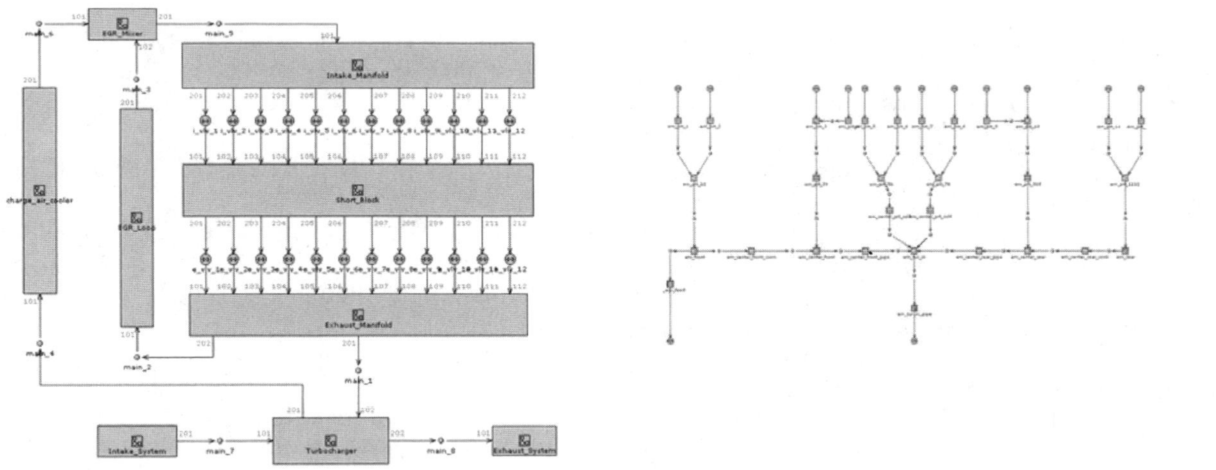

Engine Model

Exhaust Manifold Sub-Model

Figure 4-4, Engine performance simulation model (GT-Power).

Exhaust aftertreatment integration with the engine system is an area that can be modeled quite well with modern cycle simulation tools. The performance of almost all aftertreatment systems depends on the gas flow rate through the catalyst, the composition of the exhaust gas, the temperature of the catalyst bed as well as the exhaust gas temperature, and the make-up of the catalyst washcoat surface itself.

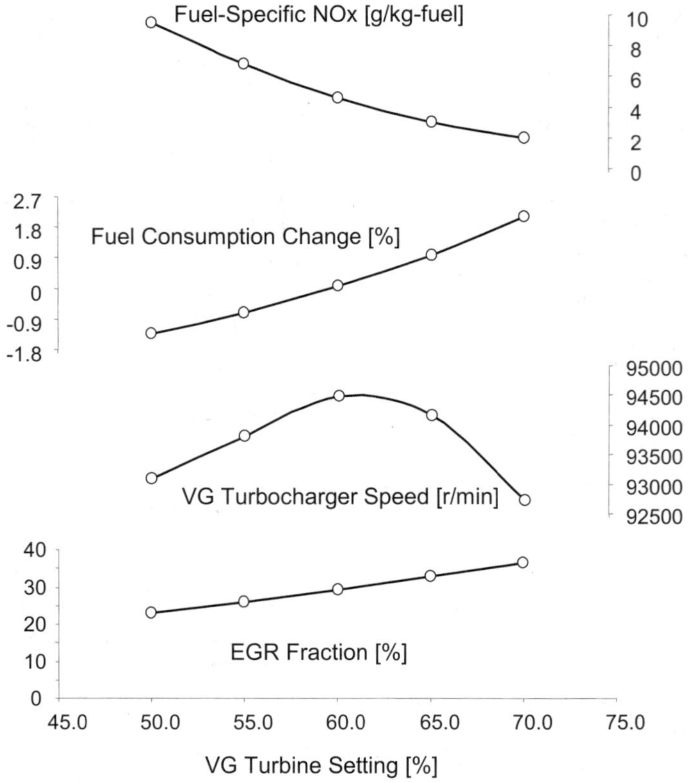

Figure 4-5, Example of engine performance simulation using GT Power.

The Figure 4-5 shows an example of the application of engine performance simulation. The figure shows typical relationships between the critical variables – in this case the response of engine-out NO_x and BSFC to EGR fraction in response to variation in VG turbine setting.

Catalyst bed temperature may be effectively modeled as a function of the thermal properties of the catalyst brick, the heat transfer to the surroundings, the exhaust gas temperature and any exothermic reactions proceeding within the catalyst.

Today, cycle simulations run sufficiently fast that transient investigations of relatively complex systems are feasible. Examples include modeling an engine and aftertreatment system over the transient test cycles used in engine certification, or modeling catalyst warm-up from a cold start. Without such models the task of engine development for low emissions would be extremely costly and inefficient.

An increasing trend is to integrate engine and aftertreatment simulation models with other analysis tools, such as combustion analysis (CFD), manifold and port flow analysis (CFD), and detailed fuel injection analysis.

COMPUTATIONAL FLUID DYNAMICS (CFD)

Computational fluid dynamics (CFD) programs are used widely throughout the engine industry to investigate a number of problems:

- Combustion and emissions modeling
- Cooling system performance
- Turbocharger stage design
- EGR system design

In recent years the development of multi-dimensional CFD models has advanced rapidly. CFD models are now used routinely to study reacting flows such as diesel engine combustion and non-reacting flows such as coolant flow or EGR flow around the engine. Figure 4-6 shows results from a CFD analysis of coolant flow and heat transfer in the jacket of a HD diesel engine cylinder block. The model was used to optimize cooling between cylinders prior to hardware testing.

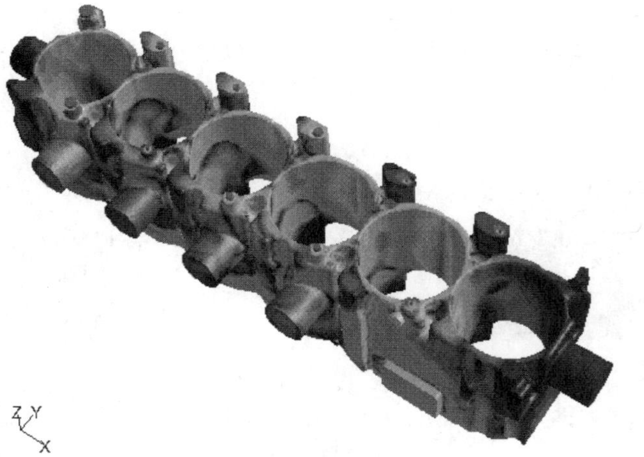

Figure 4-6, Example of CFD applied to a diesel engine cylinder block cooling jacket.

CFD models use fine grids – see Figure 4-7 - to numerically solve the Navier-Stokes equations that represent viscous, turbulent, compressible fluid flow. Additional sub-models are required to model important phenomena such as turbulence, heat transfer and fuel injection, atomization and mixing. The CFD models used to investigate diesel engine combustion comprise both analytical and empirical models. The effective application of such models requires accurate boundary and initial conditions, accurate characterization of the fuel injection process, and detailed chemical kinetics for the reactions included in the model. Such detail puts heavy demand on computer resources and ultimately forces a trade-off between model complexity and model productivity. Of paramount importance in combustion modeling is the calibration and validation of the model for its intended purpose. In general, detailed experimental data is required to calibrate and validate the various sub-models in the area of interest, before the model is used to make predictions beyond the experimental data base. Figure 4-8 lists some of the experimental tools used to validate CFD combustion models.

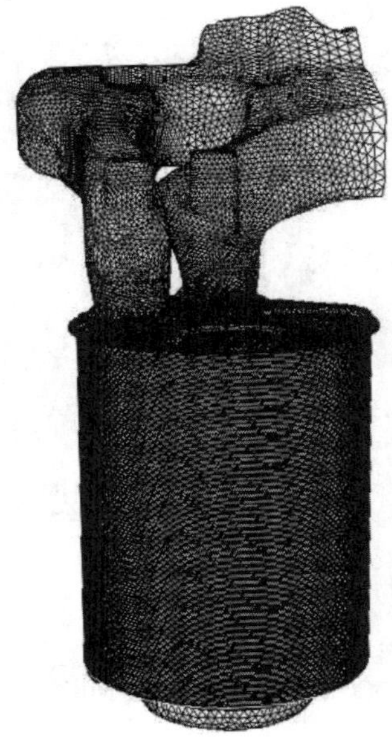

Figure 4-7, CFD computational grid, for engine cylinder, intake and exhaust ports.

Numerous physical phenomena must be modeled to accurately represent diesel combustion, as illustrated in Figure 4-9. Over the past decade, considerable advancements have been achieved in the area of diesel spray modeling with primary atomization, collision and coalescence, and aerodynamic breakup for a wide variety of injector nozzle geometries [8]. Another significant accomplishment in the area of combustion modeling is the incorporation of detailed chemical kinetics [9]. Figure 4-10 shows a comparison between measured and computed OH distributions at 21% and 8% oxygen concentration, which shows a good qualitative correlation. Historically, unacceptably long simulation times have prohibited the use of detailed kinetics. Improved numerical techniques and high performance computers allow large chemical mechanisms to be used to represent diesel fuel. A typical mechanism consists of approximately 2000 reactions and 200 species. Detailed soot and NO_x chemistry is employed in order to track the large number of chemical species throughout the in-cylinder volume.

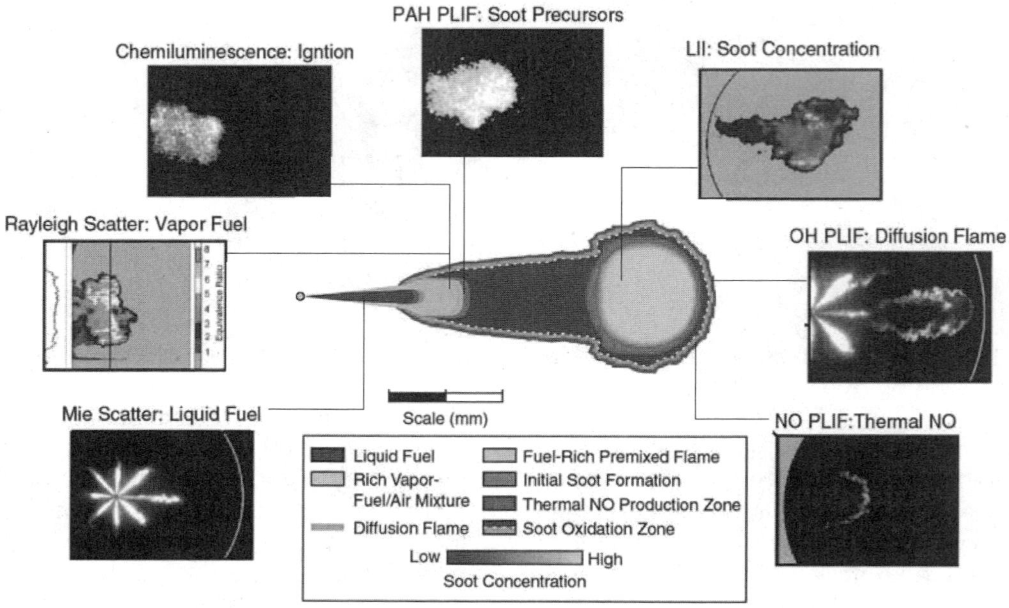

Figure 4-8, In-cylinder laser diagnostic techniques used to validate CFD models. (Courtesy of Sandia National Laboratory)

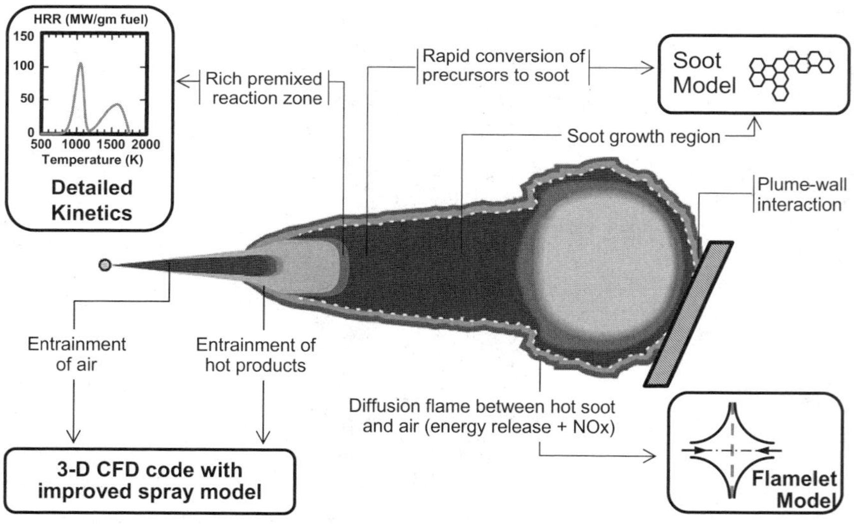

Figure 4-9 Physical processes modeled with advanced CFD codes.

With the improved modeling capability, the use of CFD as a predictive design tool has been greatly advanced. Many of the analysis tools are now coupled to provide a detailed assessment of the engine system including interactions between the combustion process and the fuel system, air handling system, and aftertreatment as illustrated in Figure 4-11. The combination of tools is used to improve engine design by exploring a larger design space compared with older cut-and-try methods. An example of the predictive capability of the CFD code is shown in Figure 4-12 for three different engine designs. This example demonstrates the degree of fidelity that can be achieved, leading to reduced development cycle times and expenses, and improved design solutions.

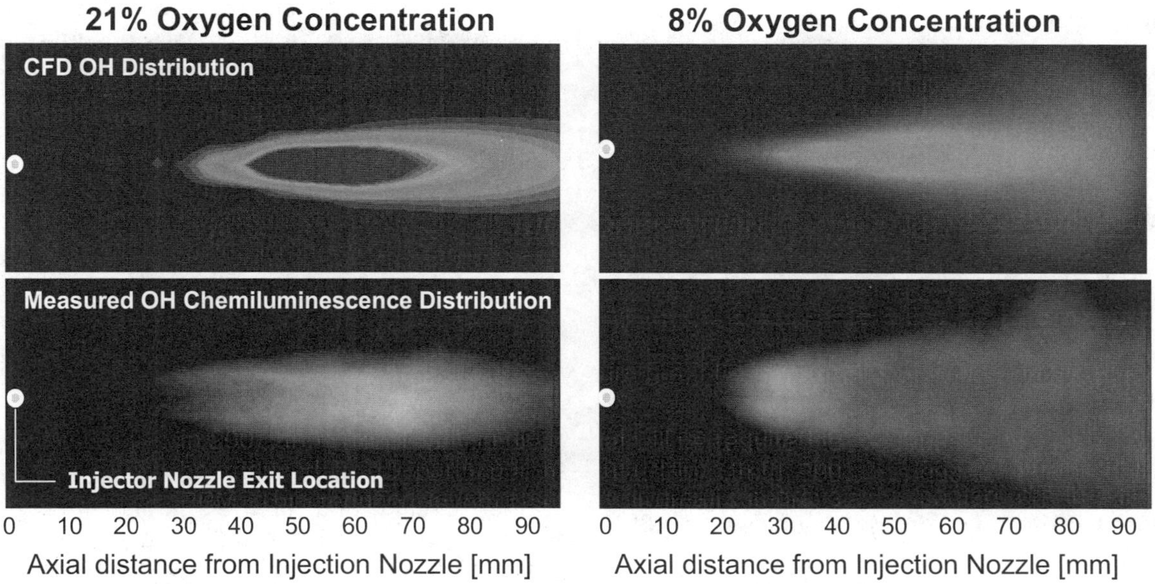

Figure 4-10, Comparison of computed and measured spatial distribution of OH.

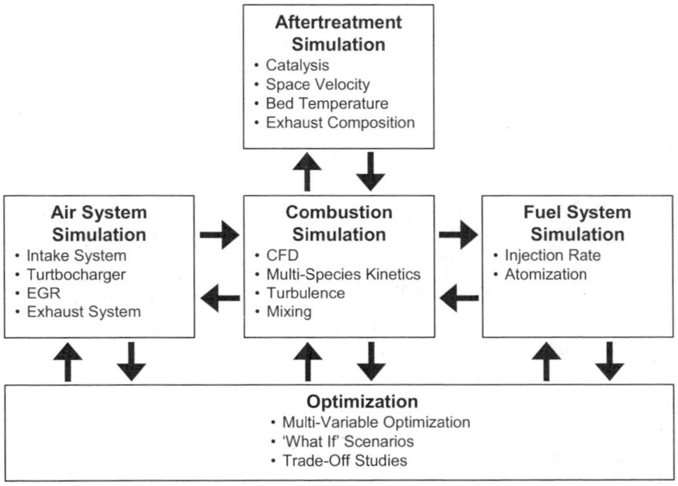

Figure 4-11, Coupling of the analysis tools for engine design.

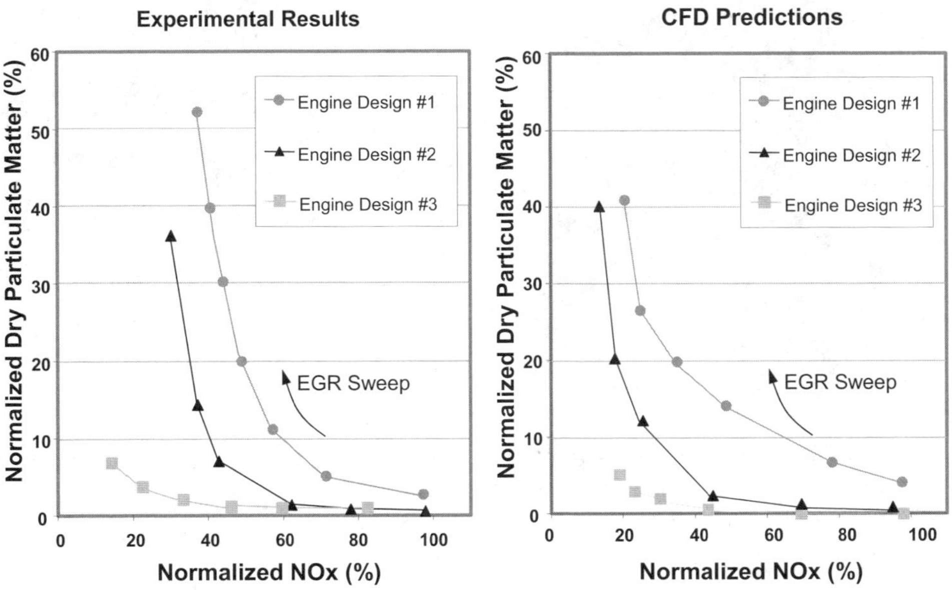

Figure 4-12, Comparison of predicted and measured emissions results for different engine designs.

STRUCTURAL AND DYNAMIC ANALYSIS

The mechanical design of the engine – the cylinder block, head, head gasket, crankshaft and bearings – is the critical foundation for a reliable and durable product. Many tools have been developed to allow computer analysis and optimization of these components and sub-systems:

- Finite Element Analysis (FEA)
- Multi-body dynamic analysis
- Hydrodynamic bearing analysis

Of these, Finite element analysis has become the most broadly applied computational method for predicting engine reliability and durability. Finite element analysis is used extensively in engine design, examples include:

- Engine structural analysis – stress, fatigue, bore distortion
- Thermal analysis of piston, cylinder liner, cylinder head
- Vibration modes and natural frequencies for the primary structure and for components attached to the engine

Finite element analysis is used to model linear and non-linear behaviors in complex systems within the engine. Non-linear analysis is required to accurately model a variety of complex physical phenomena. Non-linearities that can be modeled include material properties with strain or temperature dependence and complex contact at component interfaces. In engine design, such non-linearities are common and have to be accounted for if the analysis is to be sufficiently accurate for decision making. Finite element analysis techniques have progressed significantly, and are now a stable technology used routinely for structural analysis.

FEA was first used in engine design in the late 1970s as a stand-alone analysis tool used by specialists for solving a limited number of critical problems in engine design. More recently the use of FEA has evolved such that it is increasingly integrated within the overall design work flow. Today there is a tight integration between CAD and FEA tools – usually including bi-directional associativity and parametric optimization. A further improvement in the integration of FEA into the design process is the development of seamless interfaces with other design and analysis tools – such as CFD, acoustic and electromagnetic analysis. Such integration allows solutions for problems that are strongly 'coupled' across two or more physical phenomena – for example, fluid flow, heat transfer and surface temperatures in the coolant passage of an engine cylinder head. The latest step in the integration of FEA with other tools is the Virtual Prototype, discussed in a later section.

Figure 4-13 shows a finite element model of a diesel engine cylinder block. The block is a complete bottom-end assembly with bearing shells and bearing caps. Also shown in Figure 4-13 are the stress results predicted by the analysis. The results were used in this case to improve the design prior to investing in expensive prototype parts and committing to a prototype testing program.

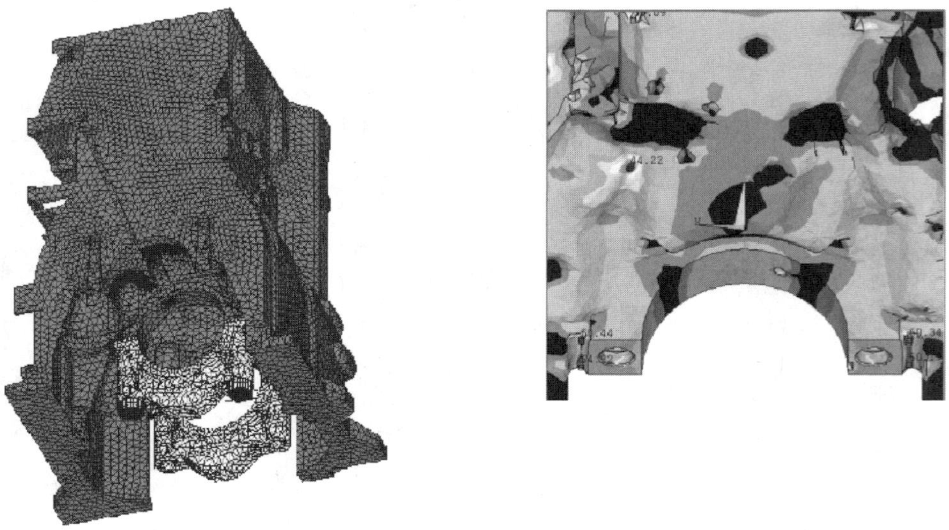

Figure 4-13, Finite element model of a diesel engine cylinder block assembly shown left, predicted stress distribution shown right.

VIRTUAL PROTOTYPE - FULLY INTEGRATED MECHANICAL SYSTEM MODELING

Virtual prototyping is an emerging technology that allows the integration of design, analysis and test into a seamless environment, allowing efficient use of resources and improved product design, see Figure 4-14.

The Virtual Prototype has emerged in recent years as a means to build and test system concepts and designs in the computer long before a commitment is made to expensive hardware prototypes. Such Virtual Prototypes involve the seamless integration of an array of powerful design and analysis tools such as Computer Aided Design (CAD), Finite Element Analysis (FEA), Multi-Body Dynamic Modeling, Computational Fluid Dynamics (CFD) and Optimization tools.

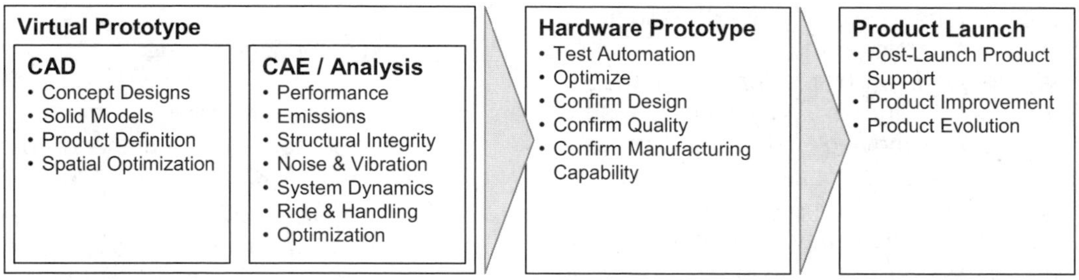

Figure 4-14, The role of a virtual prototype is to optimize and improve the quality of the final design through integration of design, analysis and test.

Increasingly the many tools used to design, analyze and optimize a design are being integrated into seamless computer environments where data can flow rapidly and efficiently between applications and through the product development process. This has many benefits, not least the reduction of non-value added engineering effort in transferring or translating data between applications. A further benefit is the data integrity and longevity that results from integrating design and analysis (and test) data in a single unified data base.

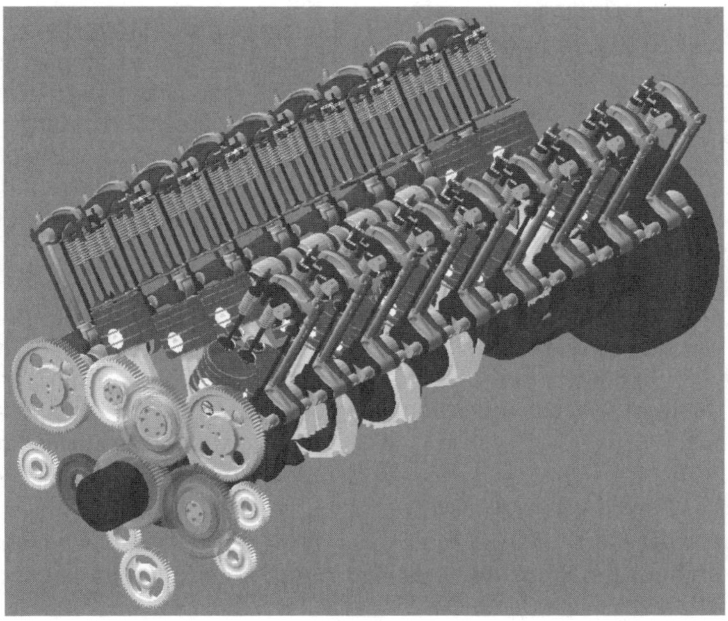

Figure 4-15, Engine system dynamic model includes crankshaft system, power cylinder components, gear train, valve train and flywheel.

Virtual Prototyping allows analysis tools to be brought into the design process, rather than being an external, after-the-fact component-level analysis. Having analysis as an integral part of the design synthesis reduces design re-iteration by forcing intelligent design iteration during the design process.

The field of modal analysis and structural dynamics has expanded rapidly in recent years due to the development of improved techniques for system modeling. These techniques have become an important part of the product development cycle for designing cost effective structures. Such analyses allow engineers to simulate the performance of mechanical systems in terms of structural integrity, noise and vibration, durability, system dynamics, ride and handling and other attributes.

For a mechanical system, the process typically begins with the development of a CAD solid model of the system components. The system is next assembled with a variety of kinematic joints. Forces between components may be modeled using springs, dampers, tires, friction, or contact forces. A system dynamic model for all of the major moving parts of an engine is shown in Figure 4-15. Solid modeling combined with material density provides the component masses and inertia prior to analysis, which may include kinematic, dynamic, inverse dynamic, quasi-static and pre-load analyses. Output from such analysis will include forces, displacements, velocities, and accelerations for the system components, thereby enabling design optimization for performance, structural integrity and durability. Virtual prototypes also may provide opportunity for animation, illustrating key effects in ways that improve understanding.

5. CONTROL TECHNOLOGIES FOR FUTURE ULTRA-LOW EMISSION STANDARDS

In the US, Europe and Japan, HD diesel emission standards for NO_x and particulate will be near-zero by the early years of the next decade. In the U.S. the 2010 on-highway NO_x standard is 0.2g/HP-hr, and the particulate standard is 0.01g/HP-hr. These standards represent greater than 98% reductions from unregulated levels.

The trend in developing countries is to adopt standards from Europe or the U.S. and to phase these in several years behind the developed countries. Nonetheless by the end of the next decade it is expected that almost every nation will have relatively strict emission standards for HD diesel engines.

In the U.S. it is possible that the focus will move away from NO_x and Particulate to in-use compliance tools such as in-use testing and on-board diagnostics. It is also likely that emphasis will be placed on CO_2 emissions with growing concern over global warming and the greenhouse effect.

The diesel engine has been developed to meet the most demanding on-highway emission standards, through the introduction of advanced technologies such as electronic controls, high pressure fuel injection, and cooled exhaust gas recirculation. Progress toward near-zero emissions has involved the development of:

- Advanced engines using new technologies
- Advanced aftertreatment systems
- Availability of ultra-low sulfur diesel fuels

Meeting tough future emission standards will require further advances in technology, such as, advanced combustion technologies, aftertreatment technologies, air handling technologies and electronic controls. This section will describe the these advanced technologies.

ADVANCED COMBUSTION

<u>Premixed combustion systems</u>

In the past decade there has been a strong interest in the development of advanced diesel combustion systems that control both NO_x and particulate in-cylinder to very low levels through control of local equivalence ratio AND local flame temperature. Equivalence ratio is the actual fuel/air ratio normalized by the stoichiometric fuel/air ratio

$$\phi = \frac{(Fuel/Air)_{actual}}{(Fuel/Air)_{stoichiometric}}$$

[ϕ = 1.0 is the stoichiometric condition. Conditions where there is an excess of air present in the mixture are "lean", ϕ < 1.0. Mixtures with an excess of fuel are "rich", ϕ > 1.0]

In conventional diffusion combustion systems, shown in Figure 2-2 for example, the greater part of the heat release occurs in the diffusion flame at the interface between the fresh charge and the injection plume. In the diffusion mode, the equivalence ratio in the flame is close to the stoichiometric ratio – which gives rise to high local temperatures and high NO_x

In the 1990s a number of research programs were focused on homogeneous-charge compression ignition systems (HCCI) [10,11]. The early work generally used a premixed charge, prepared in the intake manifold or very early in the compression stroke, and relied on compression in the cylinder to ignite the charge. While such a system could produce very low NO_x and particulate, it suffered the drawbacks of wide variation in the combustion parameters from event to event, high noise levels, and transient control challenges.

Conceptually, the diesel combustion process can be represented by an equivalence ratio and gas temperature plot shown in Figure 5-1. A number of regions have been identified on this plot. First, the soot and NO_x production regions are shown. These regions have been created using detailed chemical kinetic analysis of diesel fuel along with limited shock tube experiments. For typical diesel combustion, the process is dominated by mixing of the fuel and air as illustrated by the mixing line from A to B. Once the fuel is injected, the equivalence ratio decreases due to mixing, and the mixture temperature increases due to the compression process until ignition starts. The gas temperature of the combustion plume increases from B to C due to the energy release, and in the rich core of the plume soot formation begins. The soot is oxidized in the region of the diffusion flame where the soot and fresh charge mix. NO_x is formed in this high temperature zone as indicated by point D. The gas temperature eventually decreases due to the expansion process as indicated by point E.

More recent research has focused on creating a diesel injection and combustion event that avoids both the NO_x and particulate producing regions. Such combustion systems require that the fuel be injected rapidly prior to the start of combustion – effectively during the ignition delay period. This allows the fuel to vaporize and form a premixed charge in the cylinder prior to the onset of combustion [12]. In such systems, the NO_x is controlled by high levels of cooled EGR (>40%), and the particulate is controlled through the pre-mixing of fuel and air – thus avoiding the rich combustion states that form soot.

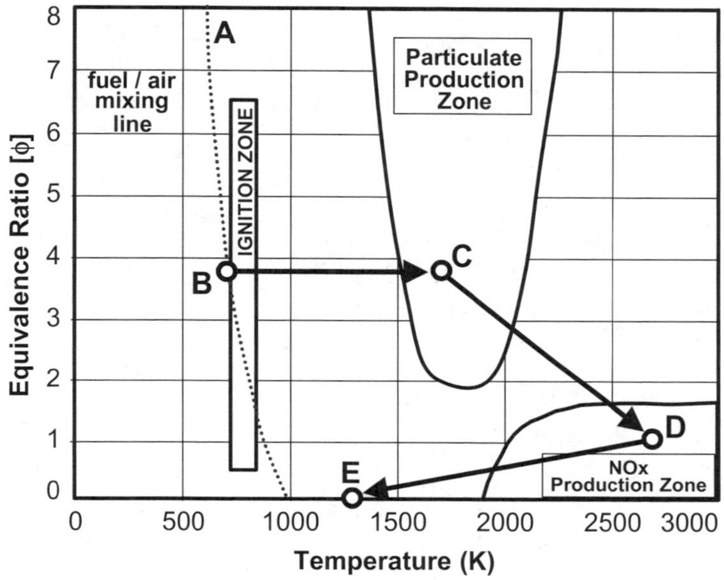

Figure 5-1, The diesel combustion process can be represented by an equivalence ratio and gas temperature plot.

ADVANCED AFTERTREATMENT TECHNOLOGIES

Effects of Sulfur

Sulfur is present in both diesel fuel and lubricating oils used by engines. It originates in the crude oils used to produce refined products. It has been recognized for many years that sulfur is potentially harmful to engines and aftertreatment, and is emitted as sulfur dioxide (SO_2) which can lead to the formation of acid rain. Sulfur dioxide is a colorless acidic gas produced in the combustion process within the engine. Wall and Hoekmann [13] studied the effects of fuel properties on exhaust emissions from diesel engines, and developed an understanding of the effects of sulfur, and the linkage to sulfate formation in the exhaust. Sulfur levels in diesel fuel have been reduced over the past decade from 2000-3000 PPM to 500PPM today, with further reduction to 15PPM planned for 2007.

The removal of sulfur from diesel fuel is a key enabler for ultra-low emission engine systems that use aftertreatment. The performance of oxidation catalysts, used in a variety of applications, is impaired by the presence of sulfur. Sulfur can form a barrier on the surface of the catalyst and reduce its performance significantly – especially at low catalyst bed temperatures. As temperatures are increased the sulfur will normally be released from the surface, restoring performance. The presence of sulfur in the fuel will lead to the emission of sulfates which contribute to total particulate emissions, and can produce acid droplets which form in the atmosphere.

The NO_x adsorber catalyst (NAC) is poisoned by sulfur, as discussed later in the paper. The NO_x adsorber contains base metal oxides designed to store nitrogen dioxide, an acid gas. Unfortunately the storage mechanism also applies to the sulfur dioxide in the exhaust stream such that over time, even at very low sulfur levels, the base metal oxide sites will be occupied by sulfates, causing a loss of NO_x conversion efficiency.

Urea-SCR

Selective Catalytic Reduction (SCR) was first introduced over 20 years ago for stationary application in power plants and other industrial process plants. More recently Siemens [14], and others have demonstrated SCR for mobile on-highway applications – specifically aimed at European heavy-duty diesel truck applications. In the US Ford has pursued SCR for light-duty diesel applications to meet the tough Tier 2, Bin 5 standards [15]. In Japan, Nissan recently introduced SCR applied to a heavy-duty diesel for their Quon truck [16].

The SCR process exploits the reactions between ammonia (NH_3) and NO_X over a catalyst to form nitrogen and water vapor. A number of catalyst materials can be used to achieve the required reactions, including precious metal catalysts such as platinum, and other materials such as vanadia and zeolites. Each of these materials has trade-offs in terms of material cost, processing costs, performance versus bed temperature, ammonia storage characteristics and durability. Under tightly controlled steady-state conditions it is possible to approach 100% NO_X conversion efficiency using SCR. Steady-state performance is a function of catalyst size – which determines space-velocity (or residence time), catalyst temperature and NO_X /NH_3 ratio. If ammonia is made available at the stoichiometric ratio then under these ideal conditions, NO_X can be eliminated and the supply of ammonia is entirely reacted with the NO_x. If conditions depart from ideal, the efficiency will fall, and either NO_X will be slipped through the catalyst, or ammonia will be slipped. A typical SCR system schematic is shown in Figure 5-2.

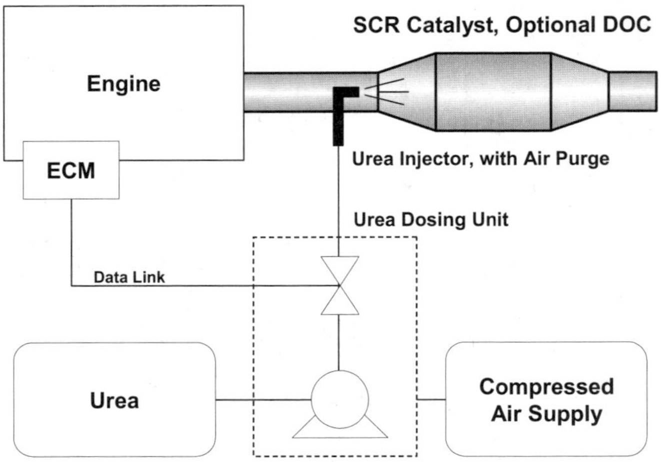

Figure 5-2, Typical mobile SCR system schematic.

In real-world heavy-duty on-highway operation, the engine operating state is constantly changing – causing sharp changes in exhaust flow, exhaust temperature and exhaust NO_X flow rate. These disturbances make control of the SCR reactions extremely demanding [17, 18, 19, 20, 21]. A further complication is the dynamic storage of ammonia within the catalyst. The storage potential is generally highest at low bed temperatures, such that ammonia is released as the bed temperature rises during a load increase. Each of these disturbances leads to a negative impact on the ability of the SCR system to achieve high conversion efficiencies. As a result the maximum in-use NO_X performance of practical SCR systems is limited to between 80% and 90% depending on the sophistication of the control system. To achieve the highest performance, it is necessary to use closed-loop control with NO_X sensing and embedded models for ammonia storage. A further consideration is the aging of the SCR catalyst due to thermal effects such as sintering or migration and poisoning effects caused by lubricant additives plating on the catalyst surface. Operation of the SCR catalyst is most effective above 300 degC bed temperature and is greatly diminished at lower temperatures, however, this can be improved with a high percentage of NO_X being converted to NO_2. Under light operating conditions or low ambient temperature conditions, the engine exhaust temperature is generally lower than the minimum required for effective NO_X conversion.

The chemistry of SCR is generally well understood, and consists of several reaction steps:

- Partial oxidation:
  $2\ NO + O_2 > 2\ NO_2$
- Urea hydrolysis:
  $(NH_2)2CO + H2O > 2NH_3 + CO_2$
- Selective reduction:
  $NO + NO_2 + 2NH_3 > 2N_2 + 3H_2O$

Urea is a crystalline substance, which is produced synthetically in large amounts from ammonia and carbon dioxide. It is used mainly in agriculture as a fertilizer and feed additive, but also in the plastic, textile, pharmaceutical and particle board industries. Urea as well as its aqueous solutions (e. g. AdBlue) are of low acute toxicity and do not act as a skin or eye irritant; sensitizing, carcinogenic or mutagenic effects are unknown. Therefore, urea is not listed as a dangerous substance according to the Hazardous Substances Ordinance and does not require any labeling according to the Chemicals Act [22, 23].

SCR has been deployed by a number engine and truck manufacturers as part of an integrated emission control strategy, combined with other technologies aimed at meeting customer requirements. For EURO IV, SCR or cooled EGR may be used and in fact both solutions will be available in the market place in 2006. SCR has the advantage that it requires a minimal amount of engine development, since it can be applied to a EURO III engine almost without modification. SCR has the disadvantages of requiring the operator to replenish the urea tank on a regular basis and it requires a distribution system for urea. In the U.S. SCR was considered for the 2004 and 2007 standards, and rejected primarily because the end-user economics were unfavorable compared with cooled EGR. A further hurdle for the introduction of SCR in the U.S. are the regulatory expectations: primarily that the system is operational and effective at all times and that urea is widely available. SCR will again be considered for introduction to meet the 2010 standards – however, the earlier questions related to end-user economics, regulatory compliance and urea availability will need to be addressed.

The economics of SCR depend on a number of factors:

- Annual mileage
- Urea pricing and availability
- Diesel fuel pricing
- Service intervals
- Tail pipe NO_x requirement
- Engine design capability

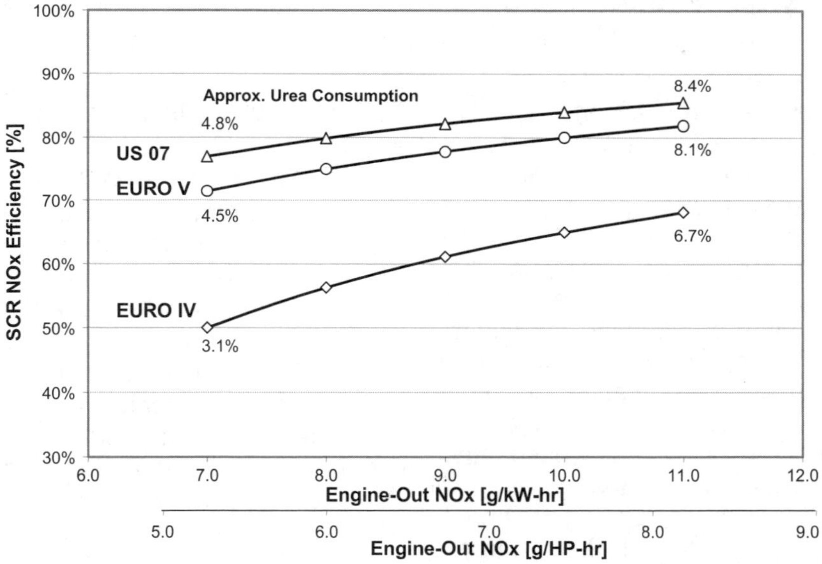

Figure 5-3, Urea-SCR system performance for European and US emissions compliance, showing engine-out NO_x and approximate urea consumption as a percentage of fuel consumption.

In regions where the fuel price is high, for example Europe, an argument may be made for SCR versus other approaches, since the engine itself can be operated at higher engine-out NO_x and at better fuel economy. However, the counter-balancing factor is the consumption of urea – which will be between 3% and 7% at EURO IV NO_x levels depending on the required SCR performance, see Figure 5-3.

NOx Adsorbers

The NO_x Adsorber Catalyst was first demonstrated by Toyota in the late 1990s [24,25,26] as a means to remove NO_x from gasoline direct injection (GDI) and light-duty diesel engines – both of which operate with an overall excess of oxygen in the exhaust stream. The excess exhaust oxygen associated with these combustion systems prevents the effective use of three-way catalyst systems (TWC) so prevalent in stoichiometric gasoline engine applications. TWCs are operated at, or close to, the stoichiometric ratio, such that the very low oxygen concentration in the exhaust allows a set of three-way reactions to take place, over a precious metal catalyst, between the hydrocarbons (HC), the carbon monoxide (CO) and the oxides of nitrogen (NO_x) – producing nitrogen, carbon dioxide and water vapor as by-products.

Since the diesel engine operates with an excess of oxygen the TWC is ineffective, and as a result special measures have to be taken to achieve NO_X control in the exhaust. The NO_X Adsorber Catalyst (NAC) operates in a cycle, wherein NO_X is stored for a period of time on the catalyst surface as a nitrate, and is later released from the catalyst surface under locally rich conditions, and reduced to nitrogen, see Figures 5.4 and 5.5.

Storage Cycle / Lean Operation

$NO + 1/2O_2 => NO_2$

$BaO + NO_2 + 1/2O_2 => Ba(NO_3)_2$

$SO_2 + 1/2O_2 => SO_3$ }
$BaO + SO_3 => BaSO_4$ } Unwanted Reactions

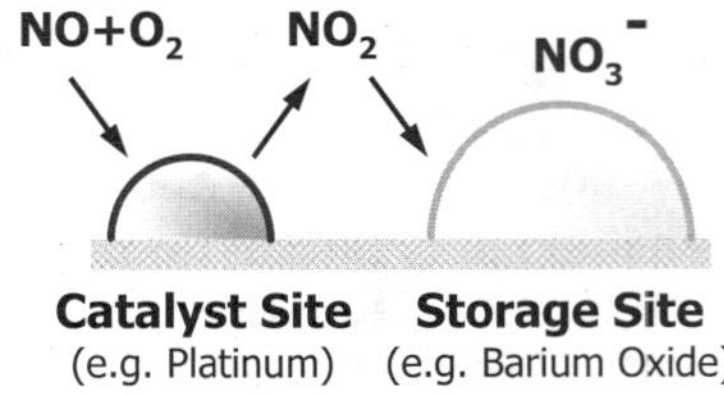

Reduction Cycle / Rich Operation

$Ba(NO_3)_2 => BaO + 2NO + 3/2O_2$

$Ba(NO_3)_2 => BaO + 2NO_2 + 1/2O_2$

3-Way Catalyst NOx Reduction ... example

$NO + CO => 1/2N_2 + CO_2$

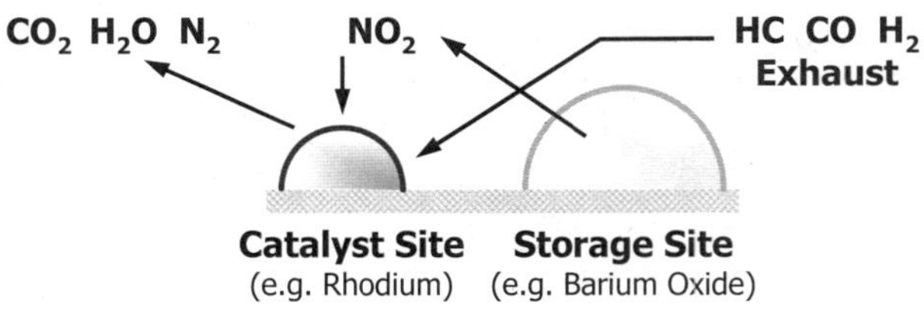

Figure 5-4, NO_X Adsorber Catalyst (NAC) chemical reactions for NO_X control.

Figure 5-5 shows the operation of the NAC as a repeating cycle of processes. The first process converts nitric oxide (NO) to nitrogen dioxide (NO_2) over a precious metal catalyst, for example platinum. The conversion to NO_2 is required for the subsequent process which is the storage of NO_x onto the catalyst surface as a nitrate. In this step, the NO_2 reacts with the base metal oxide components of the catalyst to form a stable nitrate compound on the surface – effectively storing the NO_x. Eventually all available base metal storage sites will be occupied by nitrates and NO_x will start to 'slip' through the catalyst, leading to excess emissions. When this point is reached, the exhaust is enriched - either by control of engine parameters – such as air flow, EGR rate and fuel injection parameters, or by enrichment in the exhaust through fuel dosing, either in the full exhaust flow or a by-pass flow. This period of enrichment causes the nitrates to become unstable and to release into the exhaust as NO_2. Under these rich conditions a three-way reaction is established over the precious metal catalyst, similar to that described earlier for the gasoline TWC, which allows the NO_X to be reduced, provided an adequate supply of hydrocarbons and carbon monoxide is available – so-called reductants. These reductants will generally be produced naturally as dissociation products of rich combustion.

Since the NAC is capable of storing NO_X as nitrates during normal lean operations, it is unfortunately also capable of storing sulfates. The presence of sulfur in the fuel or in the lubricating oil provides a ready source of sulfur, which will occupy base metal oxide sites and reduce the overall efficiency of the catalyst – effectively 'poisoning' the catalyst. The countermeasures for this poisoning are to:

- Reduce sulfur levels in the fuel and lubricant to very low levels
- Develop a de-sulfation cycle to remove sulfur from the catalyst surface

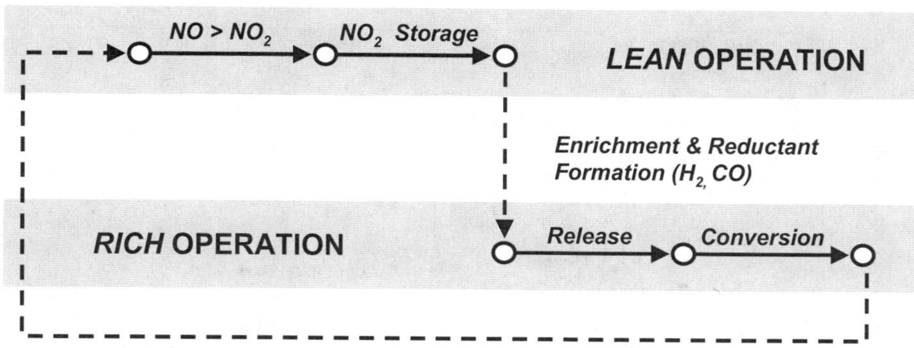

Figure 5-5, NO_x Adsorber Catalyst cycle of operation.

In 2007, the sulfur levels in diesel fuel will be required to be reduced to 15 parts per million (PPM), from 500 PPM. This reduced level of sulfur will greatly improve the feasibility of NAC technology for diesel engine applications, but would not remove the need for periodic on-board de-sulfation.

On-board de-sulfation has been demonstrated by many researchers [27]. It requires operation of the catalyst at relatively high bed temperatures, typically greater than 600 degC, under rich conditions for an extended time period.

Such conditions could conceivably be achieved through manipulation of engine parameters during high load operations – however, the challenge would be to make these operations effective and transparent to the operator.

Active Particulate Filter Systems

Over the past 5 years there have been many programs in the U.S., Europe, and Japan to promote the application of particulate filter technologies for control of particulates from older vehicles already in service. Such systems, known as Retro-Fit Particulate Filters, generally consist of a wall-flow ceramic filter in the exhaust stream which removes in excess of 90% of exhaust carbon particles, see Figure 5-6. Generally these Retro-Fit systems rely on the heat of the exhaust gases to oxidize the filtered carbon to carbon dioxide gas – with the help of a precious metal catalyst either coated on the filter itself, or on to a DOC placed upstream of the filter. Carbon may be oxidized directly by oxygen in the exhaust stream, or indirectly by nitrogen dioxide (NO_2), see Table 5-1. The oxidation of carbon by NO_2 is especially effective since it is active at relatively low temperatures, 280-320 degC being typical. These retro-fit systems are generally applied to engines with higher NO_x levels (4-6 g/HP-hr being typical), and total particulates of 0.1g/HP-hr. The predominant form of NO_x emitted from the engine is nitric oxide (NO), rather than NO_2, which is required for the oxidation of carbon. However, with the use of a catalyst placed upstream of the particulate filter NO can be readily oxidized to NO_2 [28]. Such particulate filter systems are known as *Passive Systems* because they do not involve any form of active control of either the engine or the aftertreatment itself.

The 2007 U.S. EPA standards for particulate described earlier in this paper are sufficiently stringent to require the application of particulate filter technologies for all HD on-highway engines. A typical ceramic wall flow filter is shown in Figure 5-7. In this design, a ceramic honeycomb of typically 200 cells/sq inches is used with alternate channels blocked at opposite ends of the filter, such that the exhaust gases have to pass through the porous ceramic walls.

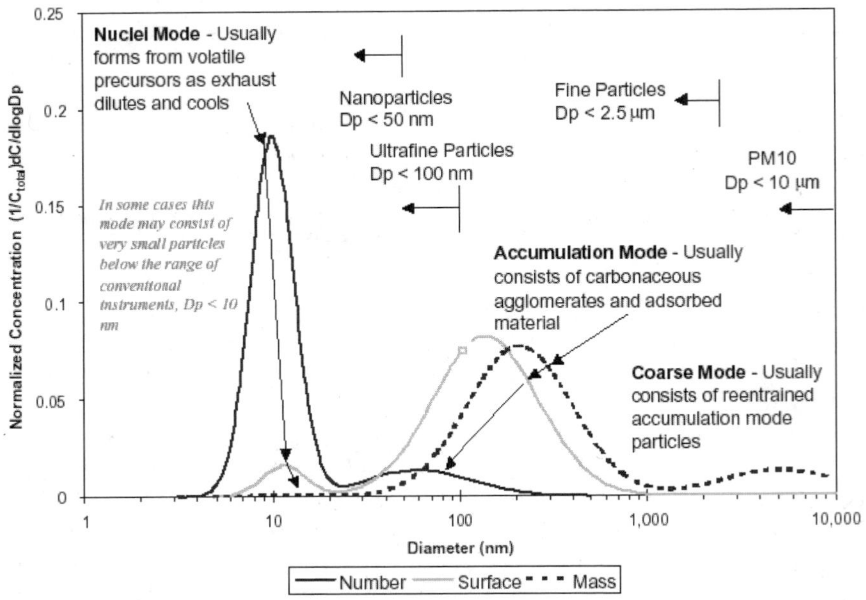

Figure 5-6, Typical particle size distribution in raw diesel exhaust. [Kittelson, 29]

The Active Particulate Filter system, shown in Figure 5-8, is comprised of elements necessary to perform the particulate removal as well as ensuring trouble free operation.

From Figure 5-6, it is clear that if 90% of the mass is to be eliminated, in order to meet future standards, then all particles greater than approximately 60 nm must be removed.

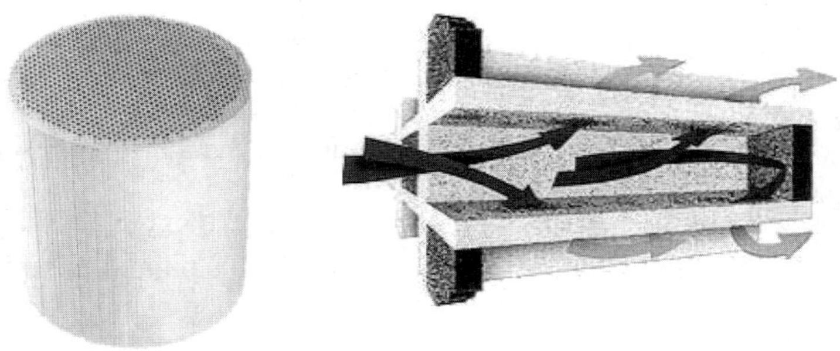

Figure 5-7, Typical wall-flow filter substrate.

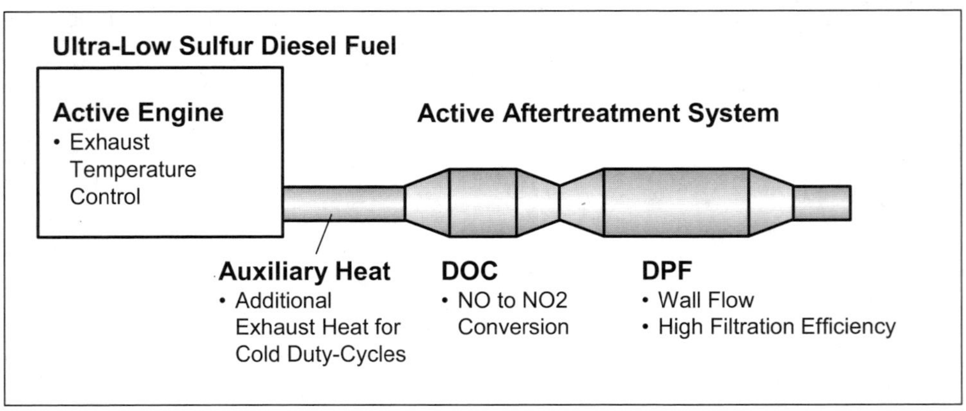

Figure 5-8, Active diesel particulate filter system.

Candidate filter materials should have the following characteristics:

- High filtration efficiency for soot particles
- Low variation in manufacture – pore size etc
- High soot and ash capacity
- High allowable material temperature

- Low pressure drop with increasing soot load
- Long term chemical stability
- Reliability and durability
- Low thermal capacity to aid rapid catalyst light-off

Currently, there are limited structures that can meet these requirements. The monolithic ceramic wall flow filter shows the greatest promise, Figure 5-7. Candidate materials include:

- Ceramic monoliths (Cordierite, SiC)
- Ceramic foam
- Ceramic fiber
- Sintered metal
- Metal fiber
- Pleated paper

Given the high trapping efficiency required, the filter will rapidly plug if soot is not continuously oxidized. Being primarily carbon, the soot is consumed by three mechanisms as illustrated in Table 5-1.

Temperature Range	**280 – 400°C**	**300 – 500°C**	**500°C and above**
Primary Oxidation Mechanism	NO_2	Catalytic O2 radicals	O_2
Regeneration Regime	Passive	Passive or Active	Active

Table 5-1, Soot oxidation mechanisms with effective temperature ranges in lean exhaust conditions.

Since the filter media is designed to trap soot particles of the order of 60nm in diameter, the media is also capable of trapping ash which derives primarily from the lubricant additive package used in the engine. In 2007 new lubricants will be introduced with reduced ash-forming components – which is expected to reduce ash levels from approximately 1.4% to 0.9%. Kittelson et al [30] have studied the effects of metallic ash derived from lubricating oil on the oxidation of soot in the exhaust. They suggest that the presence of metallic ash in the filter may modify the oxidation kinetics.

Ash deposits – typically zinc, magnesium, calcium and phosphorous – cannot be removed from the filter by passive or active regeneration. With prolonged engine service back pressure will increase and the filter will require cleaning to remove the ash and restore the filter to its original condition. This can be achieved in purpose-designed cleaning machines that reverse flush the filter with compressed air for a period of 10-20 minutes.

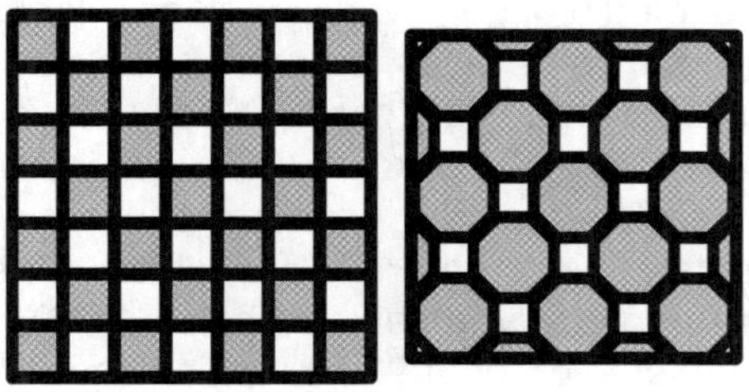

Figure 5-9, Particulate Filter Substrate designs – Left regular grid, Right asymmetric grid for increased ash capacity [31,32].

Recent research has shown that ash capacity may be increased through asymmetric channel design in the filter substrate [31,32]. In such designs, the inlet channels are arranged to be larger in cross-section than the outlet channels, thereby exposing a greater surface area and pore volume to the incoming (ash bearing) flow.

DPF Regeneration Strategies

Experience from passive soot filter applications has shown that many applications spend significant periods of time below the lowest temperature range shown in Table 5-1. For this reason an active regeneration system that can raise the temperature of the filter on demand, to the levels needed to oxidize carbon in exhaust gas is required.

Active systems can increase oxidation rates through the following approaches:

- Heating the soot directly
- Heating the catalyst bed directly
- Increasing exhaust gas temperatures

A number of concepts have been proposed and developed, at least to a demonstration or concept level:

- Active engine control
- Electrical resistance heating
- Microwave heating
- Fuel burner systems
- Low temperature fuel oxidation systems

Active Engine Control

In Active Engine Control schemes the controls flexibility of the engine is used to increase exhaust gas temperature during regeneration. An example of such control is late cycle post injection using the flexibility of high pressure common rail fuel systems, Figure 2-13. Other examples include increased idle speeds and the use of intake or exhaust throttles to restrict air flow and apply load to the engine.

Electrical Resistance Heating Systems
A number of active regeneration schemes have been proposed that exploit electrical resistive heating. Many propose a separate heating element placed upstream of the PM filter to effect and increase in exhaust gas temperature. Other systems have been proposed that involve heating the substrate directly by using materials with electrical resistance properties, thereby increasing the filter bed temperature – causing regeneration of the filtered soot [33]. Advantages of this approach are that the heat is applied directly to the filter bed where it is required. Disadvantages can include the additional electric alternator capacity required to support regeneration, and the weight of the filter compared with light weight ceramic substrates such as cordierite.

Microwave Heating Systems
Microwave heating can be used to heat particulate matter directly because of the microwave absorption properties of diesel soot. The substrate is required to be transparent to the microwave radiation. A number of research programs have shown the theoretical possibilities [34], however at this time no microwave systems are in series production. The technical challenges include: non-uniform heating of the soot leading to local exotherms, and electro-magnetic interference due to microwave leakage.

Fuel Burner Systems
Fuel burners systems have been proposed which take diesel fuel from the vehicle fuel tanks and burn it in a carefully designed combustor or burner installed in the vehicle exhaust system upstream of the particulate filter. Such burners are typically capable of generating temperatures of 650 degC to 900 degC. Burners provide the temperature levels required to support direct soot oxidation by the oxygen in the exhaust stream. Burner designs can be divided into two broad types: full function burners that can operate without operator intervention at almost any engine condition, and single point burners that operate only at idle conditions under the control of the operator.

Zeuna Starker and later ArvinMeritor [35] have proposed fuel burner systems that use diesel fuel directly to generate heat in the exhaust. Such systems typically consist of a flame holder and ignition system, often with auxiliary air supplies to promote a stable flame under all operating conditions.

Low Temperature Oxidation Systems
Diesel fuel may be oxidized over a precious metal catalyst at temperatures above approximately 300 degC. The resulting mixture temperatures produced can be readily controlled by the amount of fuel injected relative to the exhaust mass flow rate and oxygen concentration. Given typical diesel exhaust conditions, it is possible to heat the exhaust to temperatures in the range 400 to 600 degC, which is ideal for rapid PM filter regeneration

ADVANCED ELECTRONIC CONTROLS

The evolution of software engineering tools in recent years has made it possible to create computer models of integrated vehicle, engine and aftertreatment systems. Such models provide a rich development environment in which control strategies can be developed and tested. More recently these tools have allowed a degree of automation, such that the validated system model can be used to produce software code for use in the control module directly, leading to significant productivity and quality gains.

In Table 2-1, three generations in the evolution of electronic control systems were described. The table described progress from the initial systems introduced in the mid-1980s through to the present day. A fourth generation of electronic controls may be envisioned which features mathematical models rather than look-up tables, closed-loop control using emissions sensors for direct control of emissions, with extensive use of smart sensors and actuators. Deploying physical models in the controller itself changes the nature of engine controls calibration from entering tens of thousands of parameters into tables, to one of calibrating and validating the embedded models over the range of operation of the controller. Such controllers will be more efficient to develop, and provide improved performance. As costs come down, it is also reasonable to expect wireless communication within the engine, system, eliminating or simplifying the engine wiring harness.

ON-BOARD DIAGNOSTICS (OBD)

On-board diagnostics (OBD) refers to regulations in the U.S. and Europe to define the in-service monitoring of emission control systems for gasoline, diesel and alternate fueled engines for passenger cars, light duty trucks and medium duty vehicles. The Medium Duty Vehicle (MDV) is a class of vehicle defined by the California Air Resources Board (CARB) between 2720 kg (6000 lbs) and 6350 kg (14000 lbs) in which the emissions can be tested on either a chassis or engine dynamometer. For other U.S. states the OBD regulations apply only to vehicles below 3850 kg (8500 lbs). At present, on-board diagnostics regulations do not apply to heavy duty vehicles above these classifications.

The primary goal of the OBD regulations is to enable malfunctions in the emission control system to be detected rapidly, in order to reduce the time over which increased emissions are produced prior to repair. The California regulations, known as OBD-2, are defined in *The California Code of Regulations* [36]. The EPA (49 State) US regulations, known as OBD, are defined in

The Code of Federal Regulations [37]. The European Union (EU) is developing a set of regulations for on-board diagnostics for emissions controls that will be similar to those defined for the U.S. It is expected that common standards will be used to define the EU OBD requirements such that engines developed for the U.S. and Europe will comply for OBD in both domains.

On-board diagnostics monitoring requirements are listed in simplified form in Table 5-2, with comments regarding the applicability to diesel engines. Table 5-2 shows 10 'monitors' which are intended to verify that each listed sub-system is performing its prescribed function for the useful life of the vehicle. If a malfunction is detected that would cause emissions to exceed the regulated standard by more than 50%, a Malfunction Indicator Light (MIL) must be lit on the dash-panel to warn the operator that repair is required. The 'comprehensive component' monitor in Table 5-2 refers to any component that directly or indirectly affects emissions, for example, coolant or intake manifold temperature sensors. If such components affect emissions they must be monitored. In the case of sensors, monitoring for open or short circuit is readily achieved by checking for an out-of-range signal.

Emission System Monitors	Applicability to Diesel Engines
1. Catalyst	Applicable but no practical means of detecting malfunction is known
2. Heated Catalyst	Applicable but no practical means of detecting malfunction is known
3. Misfire	Malfunctions can be detected
4. Evaporative System	Not applicable
5. Secondary Air System	Not applicable
6. Air Conditioning System	Not applicable if non-CFC refrigerant is used
7. Fuel System	Malfunctions can be detected
8. Oxygen Sensor	Not applicable
9. EGR System	Malfunctions can be detected
10. 'Comprehensive Component'	Malfunctions can be detected

Table 5-2, OBD-2 Emission System Monitoring Requirements

Figure 6-1 shows a simplified flow chart representation of an OBD-2 system for a diesel engine. The flow chart shows the general operating principles. Not all monitors are active on a continuous basis, for example the EGR system may be monitored only once per trip. Assuming the conditions are met for a sub-system to be monitored, the controller will initiate monitoring. If the conditions are met for a malfunction to be declared, the fault is stored in the ECU for later retrieval during servicing. The MIL light will be illuminated after two consecutive occurances of the malfunction. If a malfunction does not recur in the next 3 trips, the MIL may be extinguished. If a malfunction does not recur for 40 trips the fault code may be erased from the ECU.

6. SUMMARY AND CONCLUSIONS

The *Clean Diesel* engine is an integral and important part of the world economy – moving freight across continents with outstanding fuel efficiency, reliability and with low emissions today, and near-zero emissions by the end of the decade with the introduction of the 2007 EPA standards.

Meeting projected ultra-low emission standards will require the development of:

- Advanced engine technologies
- Advanced aftertreatment systems, and
- Ultra-low sulfur diesel fuels

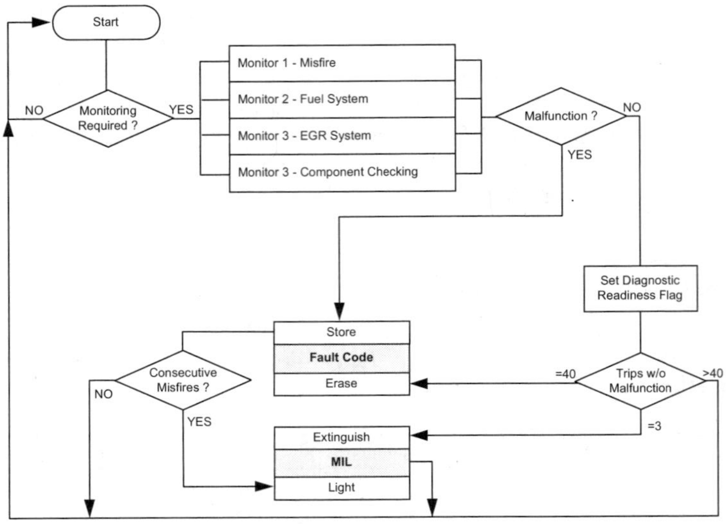

Figure 6-1, OBD II logic tree.

The paper has discussed the techniques used to develop modern ultra-low emission diesel engines as used in on-highway applications in the U.S. The principles presented may however be applied to the development of low emission diesel engines for non-road applications and for regions other than the U.S.

Emphasis is placed on the need to *choose the right technology* for any given market and application. It is essential in any technology or product development program to invest heavily in understanding the customer expectations from the outset. Time spent understanding the customer needs will be re-payed handsomely as the product is developed and taken to market.

ACKNOWLEDGEMENTS

This paper and the accompanying lecture would not have been possible without the support and encouragement of many colleagues from Cummins Inc. Technical Center, Columbus Indiana. They have provided guidance and counsel throughout. I would also like to recognize the generous support of Cummins Inc. in allowing me to publish the results of work undertaken as part of technology and product development programs at Cummins.

REFERENCES

1 Walsh, M.P., Clean Diesel Implementation Workshop – EPA Sponsored Discussion Panel, Chicago August 7th 2003.

2 Rogers P.G., "The Clean Air Act of 1970", EPA Journal - January/February 1990.

3 Flynn, P.F., R.P. Durrett, G.L. Hunter, A. Zurloye, O.C. Akinyemi, J.E. Dec, C.K. Westbrook. "Diesel Combustion: An Integrated View Combining Laser Diagnostics, Chemical Kinetics, and Empirical Validation," SAE 1999-01-0509. 1999.

4 Caddy SJ, 'Variable Geometry Turbocharger Control', Patent Application, 1996

5 Cummins, C.L., US Patent 1,561,913 "Oil Engine" (Mechanical Unit Injector), November 17, 1925

6 "Delphi E3 Diesel Electronic Unit Injector", http://www.delphi.com/pdf/commercial/e3_eui.pdf

7\. Creveling C.M., Slutsky J.L. and Antis D., 'Design for Six Sigma in Technology and Product Development', Prentice Hall ISBN 0-13-009223-1, Pearson Education 2003.

8 Ricart, L., J Xin, G.R. Bower, R.D. Reitz "In-Cylinder Measurement and Modeling of Liquid Fuel Spray Penetration in a Heavy-Duty Diesel Engine," SAE 971591. 1997.

9 Barths, H., C. Antoni, N. Peters. "Three-Dimensional Simulation of Pollutant Formation in a Di Diesel Engine Using Multiple Interactive Flamelets," SAE 982459. 1998.

10 Hikosaka, N., 1997. "A View of The Future of Automotive Diesel Engines", SAE 972682

11 Aceves, S.M., D.L. Flowers, J. Martinez-Frias, J.R. Smith, R. Dibble, M. Au and J. Girard, 2001. "HCCI Combustion: Analysis and Experiments", SAE 2001-01-2077,

12 Neely, G.D., S. Sasaki, and J.A. Leet. Experimental Investigation of PCCI-DI Combustion on Emissions in a Light-Duty Diesel Engine. Presented at the SAE World Congress and Exhibition, Detroit, Michigan, March 2004. Paper No. SAE 2004-01-0121, 2004

13 Wall, J.C. and S.K. Hoekman (1984). "Fuel Composition Effects on Heavy-Duty Diesel Particulate Emissions". Soc. Auto. Eng. Paper No. 841364.

14 Cobb, D., et al., 1991. "Application of Selective Catalytic Reduction (SCR) Technology for NOx Reduction From Refinery Combustion Sources", Environmental Progress, 10, pg. 49

15 Hammerle, R., 2003. "Urea SCR and DPF System for Diesel Sport Utility Vehicle Meeting Tier II Bin 5", US DOE, 9th Diesel Engine Emissions Reduction Conference (DEER), Newport, RI, August 2003

16 Hirata, K., N. Masaki, H. Ueno, and H. Akagawa, 2005. "Development of Urea-SCR System for a Heavy-Duty Commercial Vehicles", SAE 2005-01-1860

17 Walker, A.P., P.G. Blakeman, T. Ilkenhans, B. Magnusson, A.C. McDonald, P. Kleijwegt, F. Stunnenberg, M. Sanchez, 2004a. "The Development and In-Field Demonstration of Highly Durable SCR Catalyst Systems", SAE 2004-01-1289

18 Walker, A.P., et al., 2003. "The Development and Performance of the Compact SCR-Trap System: A 4-Way Diesel Emission Control System", SAE 2003-01-0778

19 Havenith, C., R.P. Verbeek, 1997. "Transient Performance of a Urea DeNOx Catalyst for Low Emissions Heavy-Duty Diesel Engines", SAE 970185

20 Miller W.R., et al., 2000. "The Development of Urea-SCR Technology for US Heavy Duty Trucks", SAE 2000-01-0190

21 Block, M.C., N. Clark, W.S. Wayne, R. Nine, W. Miller, 2005. "An Investigation into the Emissions Reduction Performance of an SCR System Over Two Years' In-Use Heavy-Duty Vehicle Operation", SAE 2005-01-1861

22 DMGK, 2003. "AdBlue as a Reducing Agent for the Decrease of NOx Emissions from Diesel Engines of Commercial Vehicles", Research Report 616-1, DGMK, Hamburg, Germany, September 2003

23 DMGK, 2005. "AdBlue as a Reducing Agent for the Decrease of NOx Emissions from Diesel Engines of Commercial Vehicles. Part 2: Laboratory and Field Testing of AdBlue Logistics", Research Report 616-2, DGMK, Hamburg, Germany

24 Tanaka, T., 2001. "Simultaneous Reduction of PM and NOx - A New After-Treatment System", 22. Internationales Wiener Motorsymposium, Fortschritt-Berichte VDI (VDI Verlag, Duesseldorf), Reihe 12, Nr. 455, Band 2, pg. 216-228

25 Yamaguchi, J., 2001. "Toyota diesel catalytic converter", Automotive Engineering, February 2001, pg. 81-84

26 Hachsuka I., et al., "Deactivation Mechanism of NOx Storage-Reduction Catalyst and Improvement of Its Performance", SAE 2000-01-1196

27 DECSE, 2000. "Phase II Summary Report: NOx Adsorber Catalysts", U.S. DOE, October 2000

28 Cooper, B.J., J.E. Thoss, 1989. "Role of NO in Diesel Particulate Emission Control", SAE 890404

29 Kittelson, D.B., W.F. Watts, J. Johnson, 2002. "Diesel Aerosol Sampling Methodology - CRC E-43: Final Report", University of Minnesota, Report for the Coordinating Research Council, 19 August 2002

30 Heejung Jung, David B. Kittelson and Michael R. Zachariah, “The Influence of Engine Lubricating Oil on Diesel Nanoparticle Emissions and Kinetics of Oxidation”, SAE Paper 2003-01-3179

31 Young, D.M., D.L. Hickman, G. Bhatia and N. Gunasekaran, 2004. "Ash Storage Concept for Diesel Particulate Filters", SAE 2004-01-0948

32 Ogyu, K., K. Ohno, S.e Hong and T. Komori, 2004. "Ash Storage Capacity Enhancement of Diesel Particulate Filter", SAE 2004-01-0949

33 DePetrillo, F., et al., 2002. "Rypos Trap Active Diesel Particulate Filter System: Field Demonstration", US DOE, 8th Diesel Emissions Reduction Conference (DEER), San Diego, CA, August 2002

34 Gautam M., S. Popuri, B. Rankin and M. Seehra, 1999. "Development of A Microwave Assisted Regeneration System for A Ceramic Diesel Particulate System", SAE 1999-01-3565

35 Zelenka, P., C. Telford, D. Pye, N. Birkby, 2002. "Development of a Full-Flow Burner DPF System for Heavy Duty Diesel Engines", SAE 2002-01-2787

36 California Code of Regulations (CCR [CARB]), Title 13, Section 1968.1 - Malfunction and Diagnostic Requirements - 1994 and Subsequent Model Year Passenger Cars, Light Duty Trucks, and Medium Duty Vehicles and Engines (OBD-2).

37 Code of Federal Regulations (CFR [EPA]), Title 40, Part 86, Section 86.094-17 - Control of Air Pollution from New Motor Vehicles and New Motor Vehicle Engines; Regulations Requiring On-Board Diagnostic Systems on 1994 and Later Model Year Light Duty Vehicles and Light Duty Trucks (OBD)

2006-01-0030

Diesel Emission Control in Review

Timothy V. Johnson
Corning Incorporated

ABSTRACT

The paper summarizes the key developments in diesel emission control, generally for 2005. Regulatory targets for the next 10 years and projected advancements in engine technology are used to estimate future emission control needs. Recent NOx control developments on selective catalytic reduction (SCR), lean NOx traps (LNT) and lean NOx catalysts (LNC) are then summarized. Likewise, the paper covers important recent developments on diesel particulate filters (DPFs), summarizing regeneration strategies, new filter and catalyst materials, ash management, and PM measurement. Recent developments in diesel oxidation catalysts are also briefly summarized. Finally, the paper discusses examples of how it is all pulled together to meet the tightest future regulations.

INTRODUCTION

Controlling diesel engine emissions is one of the most important aspects of modern air quality management. The field is very active with upwards of perhaps 1000 papers and presentations annually on the health effects of diesel exhaust, new fuels, engine technologies, and emission control technologies. This paper will offer a review of a narrow aspect of this field, diesel exhaust emission control.

As in the past (1), the review is not intended to be all-encompassing. Rather, the objective is to summarize representative studies that show the key trends in the industry. An emphasis is placed on reports from 2005. First, the regulatory issues are addressed, followed by a quick overview of engine technologies as a means of estimating the exhaust emission control requirements. Then the author will review NOx, PM (particulate matter), and hydrocarbon control developments, and close with some examples of integrated systems.

REGULATIONS, ENGINES, AND GENERAL EMISSION CONTROL REQUIREMENTS

It is important to put emission control technologies in the perspective of regulations, which are the primary driver for advancements, and what engine technologies can reasonably deliver. The difference between the two represents the challenge to the aftertreatment technologies. Following is an attempt summarize these needs.

REGULATIONS

Before summarizing the tailpipe regulations, it is worthy to note some recent regulatory and related developments that will have future impacts.

In the US, the EPA finalized their proposal in December 2005 to tighten the National Ambient Air Quality Standards for PM (1). The proposal, which has to be met across the US by 2015, or regions could lose highway funding, calls for tightening of the PM2.5 (PM smaller than 2.5 μm) 24-hour ambient air levels to 35 μg/m^3 from 65 μg/m^3. The PM2.5 annual average standard would remain unchanged at 15 μg/m^3. Further, a new standard for particles sized between 2.5 and 10 μm would be established at 70 μg/m^3, the intent of which is to better capture anthropogenic urban emissions.

Similarly, the European Commission finalized their Clean Air for Europe (CAFE) thematic strategy in September 2005 (2). The strategy will return about €42-130 billion (0.3-1% GDP) to society in health benefits in 2020 at a cost of €7-8 billion. On PM2.5, the daily limit is 25 μg/m^3, to be phased-in from 2010-15.

The tightening of these ambient air quality standards in the US and Europe are important because significant regions are now in nonattainment to current standards. These new levels could result in further tightening of tailpipe standards and increases of diesel retrofits

Moving into current tailpipe regulations, for the purpose here, diesel tailpipe emission regulations are categorized by light-duty, on-road heavy duty, and non-road heavy duty applications.

Light-Duty Diesel

Figure 1 shows general relationships between the future light-duty diesel (LDD) regulations for the US, Europe, and Japan (no adjustment for test cycle differences). The US regulations were finalized in 1999, and the Japanese 2009 regulations were finalized in March 2005. Although the European regulations have been in discussion for a few years, the European Commission just submitted their proposal to the Parliament in December 2005. (For reference, today's Euro IV standards are at 0.025 g/km PM and 0.25 g/km NOx. Japan 2005 standards are 0.013 g/km PM and 0.14 g/km NOx)

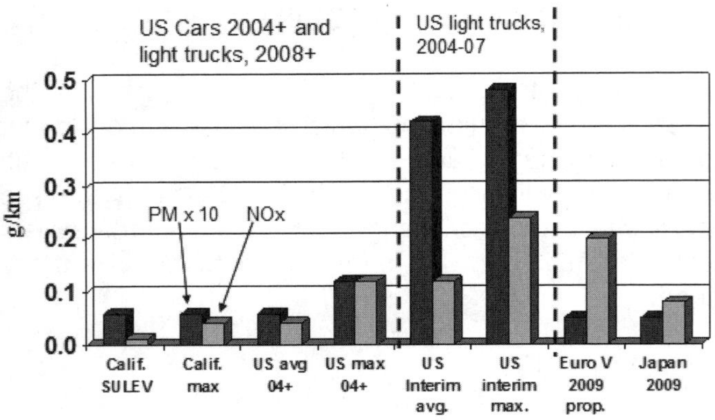

Figure 1. General comparison of light-duty diesel standards in the US, Europe, and Japan. Left bar is PM (X 10).

The US standards are based on fleet average NOx levels, while California uses a fleet average non-methane hydrocarbon (NMHC) basis. The others are maximums. Note that the maximum allowed NOx level in California is equal to the US fleet average requirement ("Bin 5"), meaning that to sell cars in all 50 states, the California maximum level (0.07 g/mi or 0.042 g/km NOx) can't be exceeded. Also noteworthy, this level of NOx is nominally 80% lower than the Euro V proposed standard. The US will require unique engine and aftertreatment technology from Europe, or even Japan.

In addition to these regulations, California finalized their CO_2 regulation, dropping greenhouse-gas CO_2 equivalent emissions 30% by 2016. This results in a similar fleet CO_2 emission as from the European commitment of 140 g/km CO_2, but four years later. Canada has adopted the California standards, and several states are considering following. Most significant, California and followers are combining the tightest tailpipe and greenhouse gas standards in the world. This presents a huge challenge to the industry, and is already sparking significant developments in powertrain technology, especially for gasoline vehicles.

Also of significance, the European Commission signaled in their proposal to Parliament the desire to add a particulate number based standard (numbers of particles per km), pending successful verification of the UN-ECE PMP (Particulate Measurement Program). The main intent is to better tie emissions standards to health impacts by regulating solid ultrafine particles.

Heavy-duty diesel

The on-road heavy-duty diesel (HDD) standards are shown in Figure 2, as are estimates of engine emissions performance.

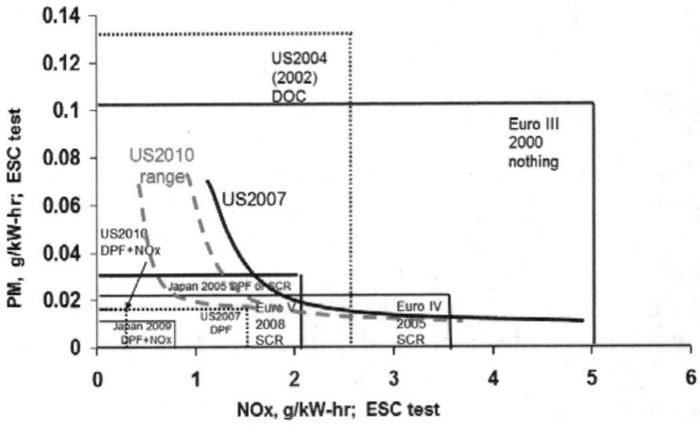

Figure 2. General comparison of on-road HDD standards in the US, Japan, and Europe. Estimated engine-out emissions for 2007 and 2010 (range) are shown. Steady-state cycle.

The European Commission is anticipating a Euro VI proposal to the Parliament in 2006 (3). It is widely anticipated that Euro VI will bring the emission levels to the US2010 and Japan 2009 ranges (<1.0 g/kW-hr NOx and <0.01 g/kW-hr PM).

Although meeting the steady-state and transient dynamometer test requirements are at the heart of regulatory compliance in Japan and Europe, and thus determines engine and aftertreatment technologies, in the US the Not-to-Exceed (NTE) part of the regulation represents the greatest challenge. In Europe and Japan, peak emissions (for example at full load) are unlimited as long as the test cycle average emissions are less than the limit values. With NTE, peak emissions are limited to 1.5X the limit values for any 30 second steady-state in-use operating point as measured with portable emission monitors, with some exceptions for operating point and ambient conditions. Thus, NTE demands high efficiency at high load, the most challenging condition.

Notable future trends relate to NTE. Europe and Japan are considering adapting the requirements, and if the US program is successful, the author speculates that only NTE compliance will be required in the future. Also in discussion (4) are worldwide harmonization of tests and regulations, an increased emphasis on fuel economy, and a look at number-based PM regulations, at least in Europe.

Non-Road

Figure 3 shows the US and European non-road diesel regulations (5), which are essentially harmonized. Japan is also moving in that direction. The regulations emerging in 2011-15 are about double the on-road standards regarding absolute levels, as well as four years behind. This is because the test cycles are more demanding, the applications so varied, and there are different fuel quality issues. As such, it is anticipated that similar technologies will be applied in this sector as in the on-road sector.

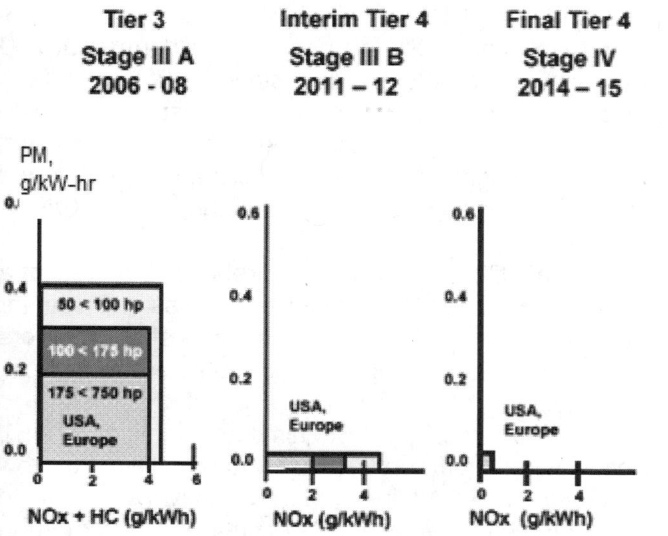

Figure 3. Future US and European non-road regulations for 50 to 750 hp engines (6).

ENGINE TECHNOLOGIES AND RESULTANT EMISSION CONTROL REQUIREMENTS

Traditional diffusion flame diesel combustion is benefiting from the continuous evolution of fuel injection equipment, air handling, exhaust gas recirculation (EGR), combustion chamber design, and sensors and control. The advancements on research engines are very impressive, even pointing towards hitting the US2010 NOx requirements without NOx aftertreatment (6, 7, 8, 9). Developing and transferring these technologies into production will impact emissions. Also, transient operation is much more difficult (10) than the steady state results shown here.

Another key trend is the advancement of cool-flame, pre-mixed "advanced combustion". There are several versions, with homogeneous charge compression ignition (HCCI) perhaps being the most common. All are typified by managing combustion via EGR, mixing, and fuel injection to keep flames out of PM and NOx formation regimes. Figure 4 illustrates the principle (11) for early and late injections at different EGR levels. Although NOx and PM are rather low, hydrocarbon and CO levels are high, often an order of magnitude higher than with traditional diesel combustion. Fuel penalties are less than 3%, and often less than 1%.

Advanced combustion is difficult to control at high load. As such, "mixed-mode" combustion is emerging, wherein advanced combustion is used at low load and traditional diesel combustion is used at higher loads. Depending on the transition point between the two modes, mixed-mode combustion can significantly reduce the need for low-temperature NOx control and for low-load regeneration of diesel particulate filters (DPF).

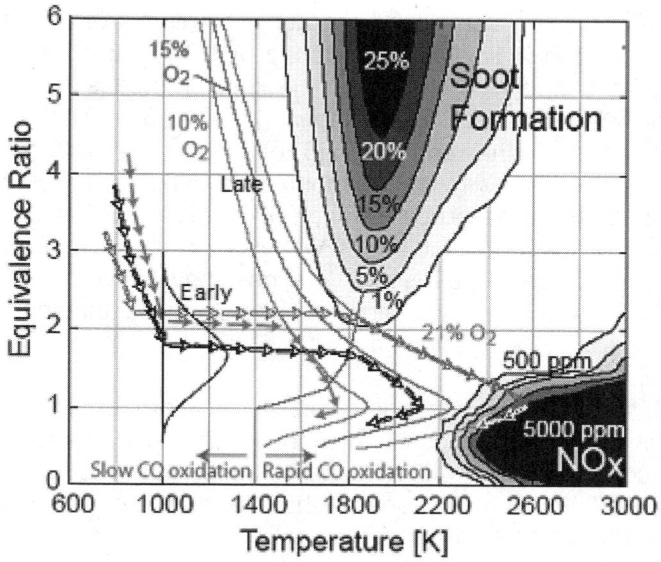

Figure 4. Principle of advanced combustion, wherein the flame temperature and equivalency ratio, , are kept away from NOx and PM formation regimes using injection timing and EGR (12).

Given that the filter technology has been commercial for five years, and NOx regulations are the most challenging to hit, engine status and aftertreatment requirements are focused on NOx.

Light-duty diesel engine status and emission control requirements

Reports on traditional diesel combustion show the potential to get down to about 0.09 to 0.14 g/km NOx on the European test cycle (12, 13). For example, a 1600 kg prototype vehicle is hitting Bin 8 (0.2 g/mi or 0.12 g/km) on the US test cycle (14). Using these engines, no NOx aftertreatment will be needed to hit the proposed Euro V requirements (even if the NOx limit is cut in half), and given reasonable advancements in technology, very little if any NOx aftertreatment will be needed to hit the Japan 2009 levels for all but the HD vehicles (>3500 kg). US Bin 5 requires nominally 70% NOx efficiency.

Given the very tight NOx requirements for LDD in the US, mixed-mode combustion developments are generally aimed at this market. The US test cycles mostly operate at less than 50% load, so most of the cycle can be run in advanced combustion mode with low NOx (9). As such, some are even hitting Bin 5 without NOx aftertreatment on research engines (14, 15). The first mixed-mode engines in the US for heavier applications will need modest NOx treatment, about 30-50% efficiency at the high load test points, but it would appear that 1600 kg vehicles can do without such by about 2009 (14).

Heavy-duty engine status

NOx emissions using traditional diesel combustion at steady-state high load operating points have been reported (16, 17, 18) in the range of 1.0 to 1.3 g/kW-hr (0.6-0.8 g/bhp-hr). Advanced combustion modes at <40 to 60% load are delivering (16, 17, 18, 19) NOx emissions in the range of 0.26 to 0.50 g/kW-hr (0.2-0.4 g/bhp-hr). A minimum threshold average emission value has been proposed by Parche (20) of around 0.6 g/kW-hr NO and 0.015-0.020 g/kW-hr PM. If the emissions performance of the research engines shown here can be transferred into production, it appears these threshold emissions will be attained.

Given that there is another year or more of engine development and engineering before freezing technologies for the Japan 2009 and US2010 regulations, it is reasonable that the above engine-out levels might be commercially realized. If so, to meet US NTE, about 50-70% NOx treatment will be needed at 500-520C. To hit Japan 2009 at 0.7 g/kW-hr NOx it appears only about 20-30% NOx control, if any, is required. However, as NOx systems can deliver fuel economy gains and are being implemented today (Euro IV and Japan 2005), HD engines in Japan 2009 will likely have such.

For non-road applications, it is difficult to predict emissions from these engines for the 2011-14 timeframe. However, it is reasonable that engine companies that make similar engines for both on-road and non-road applications will try to maximize synergies. As such, we could see Japan 2009 and US2010 engines show up in 2014 non-road applications and probably sooner to take advantages of early introduction incentives. In 2011, when generally only DPFs are expected too be needed, we will see NOx targets being met with advanced fuel injection, air handling systems, and some EGR, albeit lack of ram air limits the range of cooled EGR.

NOx CONTROL

Given the tight NOx emission regulations in the US and Japan, and the fuel economy impacts NOx aftertreatment can have, NOx control technologies will play a key role going into the future. As the state of NOx aftertreatment is more dynamic than that of PM or HC and CO control, the author elected to include it first.

Following is an assessment of the state-of-the-art on selective catalytic reduction (SCR), lean NOx traps (LNT), and lean NOx catalysts (LNC).

SELECTIVE CATALYTIC REDUCTION

Although NOx control was not required to meet the Euro IV or Japan 2005 HDD regulations (beginning October 2005), SCR was selected by several truck manufacturers because the high NOx efficiencies in the tailpipe allow the engine to be run at higher NOx levels for better efficiency and lower PM, thus delivering competitive fuel economy and eliminating the need for a DPF. In the US, the EPA generally laid out its requirements for SCR: no operation without urea, and conveniently located urea filling stations (21). The EPA is looking for a proposal from the industry on how to accomplish this. One reasonable approach to the first requirement might be to use interlocks that are engaged if the urea:fuel ratio is out of balance right after fueling.

Hirata, et al, (22) provides a comprehensive review of a modern SCR system as developed by Nissan Diesel for Japan. Because the Japanese test cycle strongly favors low-load operation, exhaust temperatures are rather low. They chose to use a DOC in front of the system to generate NO_2, which, with equimolar quantities of NO, reacts most efficiently with ammonia at low temperatures. In adding the DOC, they were able to increase the NOx removal efficiency at 200C from 40 to up to 70%. Further, converse to Europe, a zeolite SCR catalyst was chosen over the traditional vanadia-based catalyst because of better performance and smaller size. Figure 5 shows the efficiency curves for several tested catalysts. Simulated fuel economy gains from SCR over EGR/DPF systems were estimated to be net 4% at 90 km/hr cruise, after the 5% urea consumption (assumed 50% the cost of fuel) is taken into account. The authors also elaborated on the urea and delivery system, OBD (on-board diagnostics), and on the urea infrastructure.

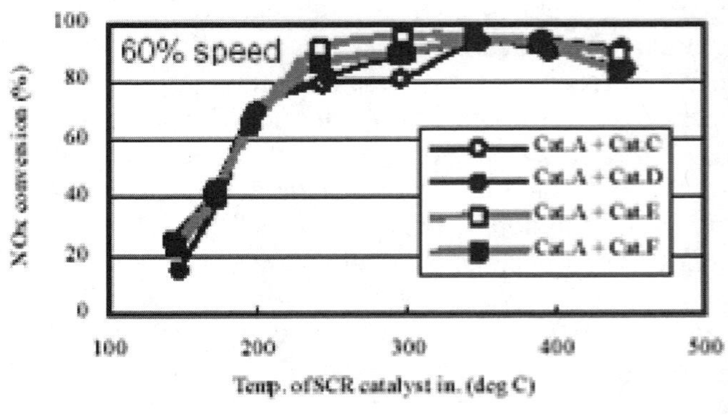

Figure 5. NOx removal efficiency curves for various SCR catalyst configurations (22). "Catalyst A" is an oxidation catalyst to promote the fast $NO_2/NO/NH_3$ reaction. "Catalyst C" is vanadia-based, and the remaining catalysts are zeolite (25% smaller than the vanadia system).

Regarding catalyst options, several reports show the advantages of zeolite for SCR applications (23, 24, 25, 26, 27, 28). For low temperature efficiency, Walker (27) shows better low temperature performance with zeolite than with vanadia, and more tolerance to non-ideal NO_2 levels. Excessive NO_2, that is NO_2>NO, is difficult to reduce using ammonia. Holcomb (28) reports more sensitivity to this effect at about 240C than at 310C. He also shows DOC length and catalyzed DPF design can be used to optimize NO_2 formation for SCR. More importantly, zeolite exhibits much better high-temperature durability. This is important when SCR systems are added directly after DPF systems, wherein DPF exit temperatures can approach 650C during regeneration. Although vanadia systems show signs of aging at temperatures less than 600C, zeolite shows impressive durability at 650C (26) or even 700C (25, 27).

SCR operation in very cold climates can cause problems, as the standard urea solution freezes at about -10C. Ammonium formate has the same ammonia content as urea but freezes at -20 to -30C, depending on formulation, and behaves similarly in SCR systems (29). An alternative is to heat the tank using engine cooling water and to use heating elements for the lines (22).

Some work is beginning to expand on earlier reports (30, 31) of possible secondary and trace emissions from SCR systems. Sluder, et al. (32), show that at 205C and with ideal NO_2/NO levels, steady state reduction of NO over zeolite catalysts doesn't begin until after 300 seconds, and for NO_2 not until after 500 seconds, independent of whether urea or ammonia is used. Reductant storage in the catalyst was hypothesized as the cause. N_2O, which is a very strong greenhouse gas, is the preferred reduction product if there is excess NO_2. Given this, it is not surprising that longer catalysts result in higher N_2O in these conditions. On the other hand, HCN emission, which occurs at similar levels (low ppm) regardless if ammonia or urea is used, was slightly higher at higher space velocities, Figure 6. Finally, small amounts of nitromethane, nitroethane, and nitropropane (<250 ppb) were detected upstream and downstream of the SCR catalyst. In the case of HCN and the nitro-HCs, the authors expect that a clean-up oxidation catalyst will virtually eliminate them.

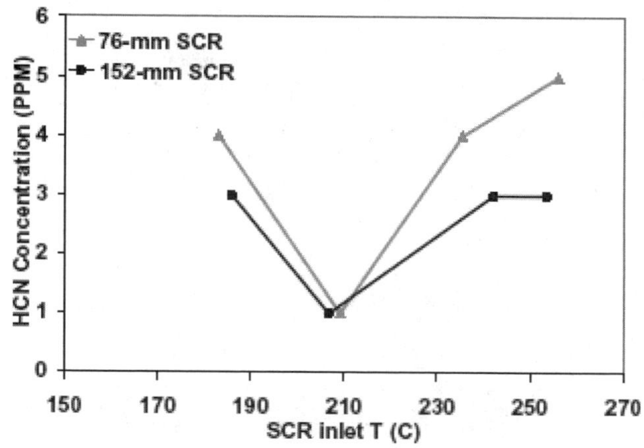

Figure 6. HCN formation from a reaction with NH_3, measured after an SCR system. HCN/NH_3 ratios are constant up to 6 ppm HCN (32).

Finally, on SCR and OBD, the European Commission finalized their requirements in September 2005 (3). It primarily calls for urea level detection and urea quality detection, combined with urea consumption correlation with engine NOx production. Alternatively, NOx sensors combined with level detection is preferred. If things are not right, torque limiting begins after the first stop.

LEAN NOx TRAPS

Lean NOx traps offer an attractive NOx solution for light-duty applications and those HD applications in which an extended urea infrastructure might be problematic. In US HD applications the challenge for LNT is acceptable high temperature efficiency to meet the NTE requirements (50-70% control at 500-520C), and do it over the useful life of the vehicle (185,000 miles for medium HDD and 435,000 miles for heavy HDD). On the light-duty side, efficiencies need to be high enough to hit Tier 2 Bin 5 (70% using traditional diesel combustion; 30-50% in the 50 to 70% load range using mixed mode combustion) over the useful life of the vehicle (120,000 miles).

HT LNT systems for lean burn gasoline systems have been developing for a number of years. Figure 7 shows the efficiency-temperature relationship for light-duty and heavy-duty applications reported by Roth (33). In light-duty applications (chassis certified) using traditional diesel combustion, low temperature NOx control is important.

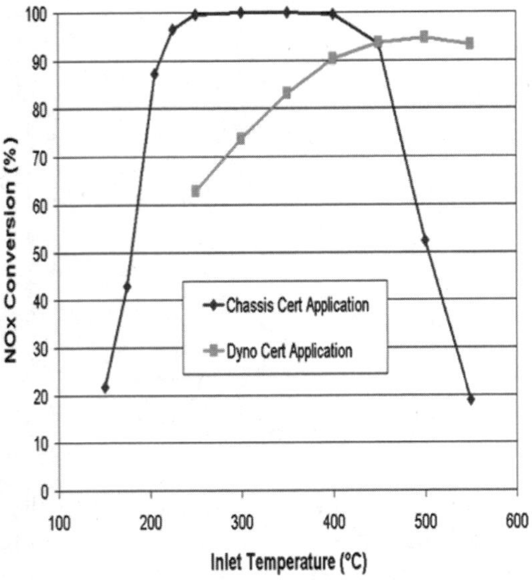

Figure 7. LNT efficiency for LDD (chassis certified) and HDD (dyno certified) applications (33). LT performance is critical for most LDD applications, whereas HT efficiency under high NOx fluxes is needed to meet HD NTE requirements. Hydrothermal aged 5 hours at 850C, 30,000/hr and 120,000/hr space velocities, respectively.

Hinz, et al. (34) also report on a HT LNT system for HD applications in which a bypass system is used during NOx regeneration. The system achieves at least 75% efficiency at all full load points, at a fuel penalty of only 2 to 3% at all but the low-load points. The swept volume ratio (SVR; ratio of LNT volume to the engine displacement) is 1.4.

Small amounts of hydrogen in the exhaust stream enhance the low temperature performance of LNTs. Fuel reformers are emerging to accomplish this task (35, 36). If a few percents hydrogen are present in the exhaust during rich regeneration, the LT efficiency markedly improves, even down to 150C. One can envision such a system for HD applications in which NTE is addressed by placing the LNT further back from the engine (35).

LNT durability problems surface when desulfating the LNT under rich and HT conditions. Reports of several years ago showed upwards of 50% loss of NOx storage capacity. Later results (37) are showing 10 to 20% capacity loss, with minimal decline after about 20 desulfations (generally 100,000 km). HT LNT formulations require HT desulfation, generally up to 700C, which would be expected to result in more deterioration. Wu (36) shows that temperature might be better controlled using exotherms in the LNT from combustion of reformer gas, thus minimizing deterioration.

As more is learned about sulfur loading and desulfation, deterioration issues are being addressed. In this regard, Takahashi, et al. (38) show that both CO and hydrogen in the regeneration gas are important, and most of the sulfur is present in the inlet regions. Sulfur at the exit is very difficult to remove.

Concerning cost, platinum group metal (PGM) loadings are coming down, saving on cost. In a very thorough and well-designed experimental plan, Theis, et al. (39) explain the positive and negative impacts and trade-offs of high PGM loadings in a gasoline application. Figure 8 is one example. At low temperature, higher PGM loading aids NO_2 formation, thus improving efficiency. However, at high temperatures, the PGM causes faster nitrate decomposition, thus dropping efficiency. This works in the direction of NTE and mixed mode requirements. In the case of mixed mode combustion, wherein little NOx is formed at LT, much reduced PGM loadings would appear possible.

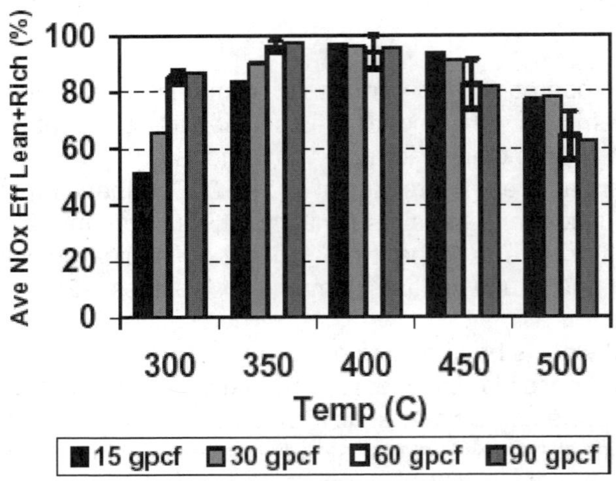

Figure 8. At lower temperatures, high PGM loadings are preferred, but at higher temperatures, lower loading are better (39).

Finally, Lambert (26) put an SCR catalyst behind the LNT to make use of ammonia that comes out of the LNT during rich periods to take up NOx regeneration spikes. Figure 9 shows the results. As the rate controlling step in LNTs is the reduction of NOx coming off the LNT during rich spikes under many conditions, it follows that the SCR combination can result in significant PGM savings.

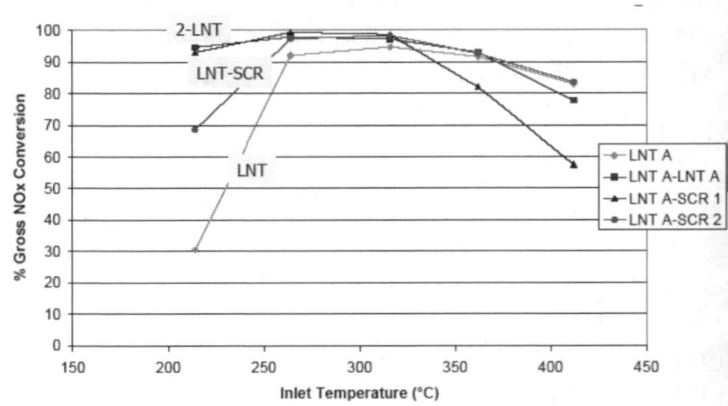

Figure 9. When an SCR is added to an LNT, system efficiencies go up (26).

LEAN NOx CATALYSTS

Lean NOx catalysts are attractive because they use fuel as a reductant, and potentially cheap catalyst (silver, zeolite, others). The problem is that the reductant has to be used continually, unlike with an LNT, and performance has been only in the 10-30% range with relatively high fuel penalties (6%).

Advancements are forthcoming. The KNOWNOX consortium in Europe reported (24) that small amounts of hydrogen (<2%) can markedly increase silver-based LNC reduction efficiency over much of the temperature range (200-400C). Further, a US Department of Energy consortium has developed a rapid LNC formulation screening process involving computational chemistry, synthesis, and rapid testing (40). More than 4500 compositions have been evaluated. Figure 10 shows an example of a new formulation that is getting close to SCR HT efficiencies.

One can envision combining the two discoveries: Hydrogen additions to improve LT performance, and the new LNC formulations for HT performance.

<u>On-Board Diagnostics (OBD)</u>

NOx OBD is centered on NOx sensors. A recent report (41) summarizes the performance and durability after 6000 hours of testing as part of the US DOE APBF (US Dept. of Energy Advanced Petroleum-Based Fuels) program on HD SCR. Figure 11 summarizes the results. At low NOx levels, as would be experienced after a NOx control system, accuracy and durability are impressive, but if placed ahead of the NOx control system, durability issues arose, albeit they appear predictable.

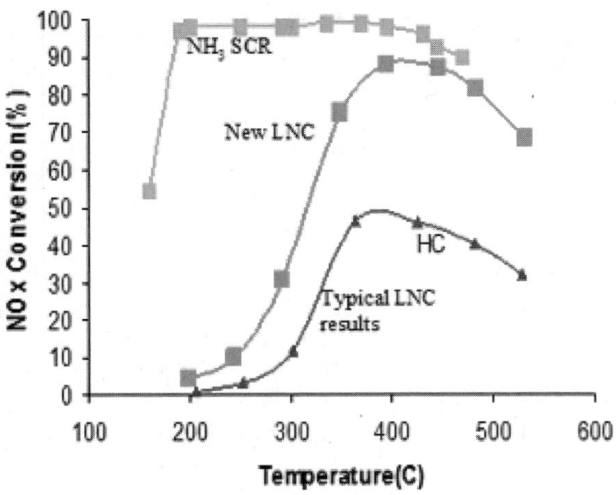

Figure 10. New LNC formulations discovered using a rapid screening process show improved HT performance compared to typical formulations (40).

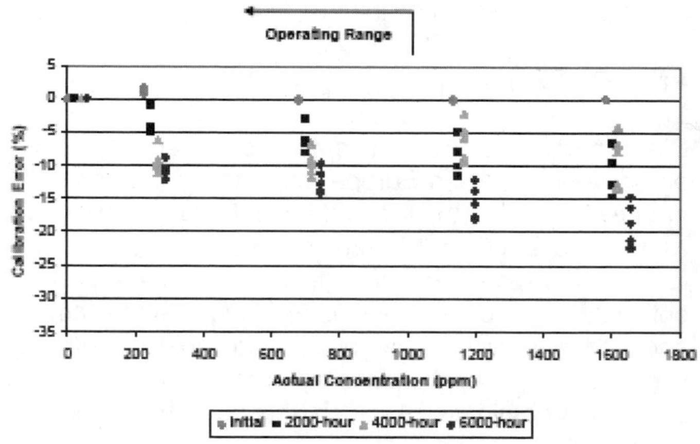

Figure 11. Sensor calibration error as a function of NOx concentration and exposure time to exhaust for one type of NOx sensor. Performance is consistent and durable at low NOx levels, as would be observed for an OBD sensor after the NOx control system (41).

NOx SUMMARY

For LDD, traditional diesel combustion will meet foreseeable European and Japanese needs without NOx aftertreatment. In the US, upwards of 70% NOx control is needed. Both SCR and LNTs apparently can meet this requirement. SCR has significant infrastructure issues (45,000+ diesel filling stations), and LNTs will require significant control. LNT PGM costs are coming down. Perhaps the combination of LNT+SCR can do it more economically and easily. However, as we move into mixed mode combustion (2009+), LNT costs will come down significantly simply because the NOx burden is lower, but more importantly, with reduced low-load NOx this challenge is lessened PGM savings can be realized.

For HD, the issues are different. Fuel economy and current approaches (and urea infrastructure) in Japan and Europe will keep SCR in play for the coming years. However in the US, although urea infrastructure is more challenging, NTE HT requirements place a huge demand on LNTs. SCR shows better HT efficiencies and is developing a track record. LNTs can hit the required HT efficiency, but desulfation issues with the HT formulations also raise durability questions not yet addressed in the literature. For the medium and light heavy duty vehicles, despite needing to hit NTE, fuel economy considerations are not as significant as in the heavy sector, and SCR infrastructure for these urban applications is more significant. LNT is a more plausible option here than for long haul applications.

If SCR is chosen for US2010 applications, we need to expect that in following years, hydrocarbon based NOx solutions like LNT and LNC will advance faster than SCR optimization. The rapid developments on LNC are especially interesting for less fuel-consumption sensitive applications. It is conceivable that HC-based deNOx could be more attractive than SCR in later years.

PM CONTROL

Diesel particulate filters have been commercially applied on passenger cars in Europe for more than 6 years, but are just beginning to go commercial for HD applications, with Japan 2005 and US2007 regulations coming into force. Despite that the field is generally in an optimization and cost reduction stage, developments and understanding is still advancing rapidly. Although the LD and HD applications have their differences, the similarities dominate. As such, this section consolidates LD and HD developments.

REGENERATION

Zink (42) provided a comprehensive review of active DPF regeneration approaches and systems. The numerous examples for LDD applications (43, 44, 45, 46, 47) show common characteristics, summarized in Figure 12:

- Estimation of DPF soot loading using engine and back pressure models, and fuel consumption.
- Preheat the catalyst to ensure that injected hydrocarbons can ignite and heat up the filter.
- Increase of exhaust hydrocarbon levels via in-cylinder or supplemental fuel injection, for burning on a catalyst.
- Control and monitoring of the regeneration as a function of operating point and conditions.
- Recalculation of pertinent models to account for ash build-up.

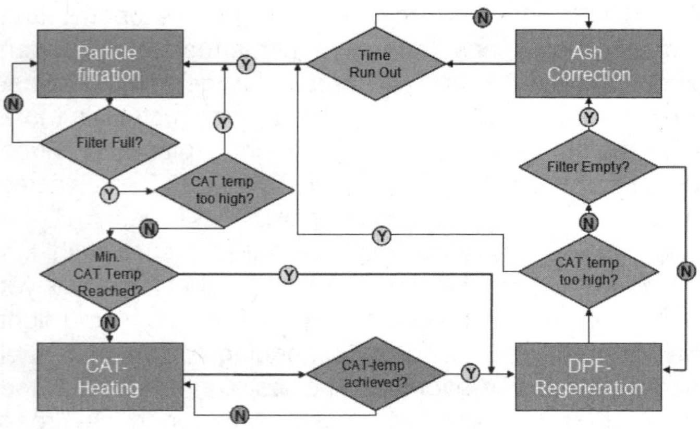

Figure 12. Example of LDD DPF regeneration plan. (45)

Each aspect of the plan is specific to the system and manufacturer, and can be rather detailed and complex. Key control parameters are fuel injection timing and quantity, air flow using throttling and EGR, and filter type and catalyst. The approaches are also pertinent to heavy duty and non-road applications.

Manns, et al. (48) do a very thorough review of LDD system layout and implications to calibration. They conclude that close coupled catalyzed filters, with or without a DOC are preferred due to better thermal management, less oil dilution by fuel, and better hydrocarbon conversion. They also discuss heat losses, thermal gradients in the filter, and post- and late-injection strategies.

For HDD applications, Kodama, et al. (49) described the system Mitsubishi FUSO will be using. One aspect of HDD filter regeneration that can be especially difficult is achieving temperatures high enough to burn fuel in the exhaust system under low load and/or cold ambient conditions. Key aspects of their approach is to use both intake and exhaust throttling to achieve DOC temperatures of 250C to enable fuel burning. They can accomplish this at speeds as low at 14 km/hr and in temperatures as low at -10C. Also noteworthy of the Mitsubishi approach is that instead of limiting regeneration duration by time, they chose to use total oxygen delivered to the filter, Figure 13. They save 20% on regeneration fuel consumption at 44 km/hr, but this savings diminishes as the speed goes down.

Other solutions that aid the low load problem involve exhaust system insulation, close placement to the engine, and improvements in catalyst formulation. In the latter case, Holcomb (28) was able to drop the DOC ignition temperature from 240C down to 215C by converting from a platinum to a platinum/palladium catalyst formulation.

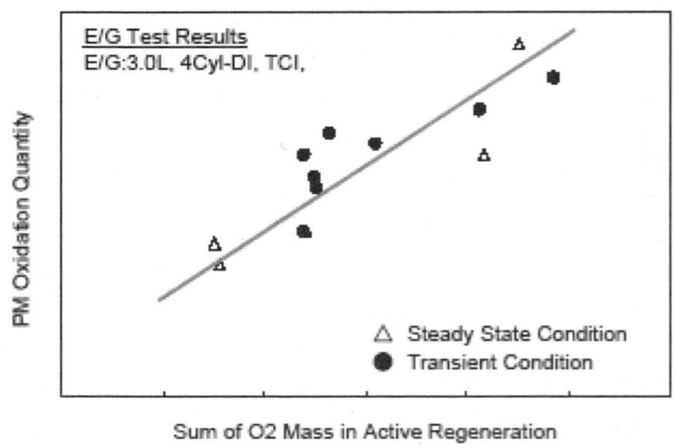

Figure 13. Total oxygen mass delivered to the filter during regeneration is used instead of time to determine when to stop DPF regeneration (49).

Mitsubishi is using SiC filters in their HDD application. Craig, et al. (50) describe the various regeneration properties of cordierite. As shown in Figure 14, peak filter temperatures and gradients, and completeness of regeneration are a function of DPF inlet temperature. Flow rate, soot loading, catalysts, and filter thermal mass also have significant impacts. The authors suggest that to increase soot loadings and improve regeneration efficiency, initial inlet temperatures of 550C might be used, which are then increased as soot loading decreases.

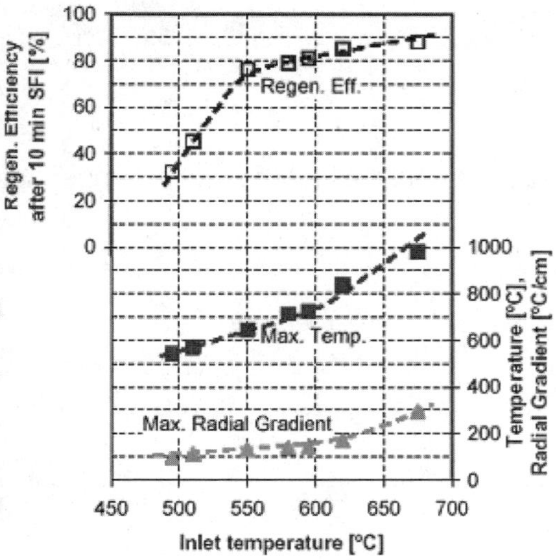

Figure 14. DPF inlet temperatures of >550C are needed to efficiently regenerate catalyzed cordierite filters with a 4 g/l soot loading. To further enhance regeneration, the inlet temperature can be increased later in the process (50).

Other items of note on recent DPF regeneration reports include:

- By moving the filter system from an underfloor to a close coupled position, the amount of regeneration fuel goes down because the system is inherently hotter. As such, the optimum soot loading, which balances fuel consumption due to increased pressure drop with regeneration frequency, drops from 10-11 g/liter down to 6-8 g/liter for SiC filters (51).
- Regeneration fundamentals using oxygen and NO_2 are investigated for catalyzed filters. NO_2 formation and utilization for passive regeneration is a complex relationship between NOx flux and temperature (for NO_2 formation and for oxidizing soot). Maximum passive regeneration occurs at medium to high speed and medium load conditions (52).
- It takes 65 to 70% of the total regeneration fuel to burn the last 25% of soot (53).
- Description of a supplementary fuel vaporizer that improves fuel distribution in the exhaust (54).

FILTER MATERIALS AND CATALYSTS

A new filter material besides SiC is being used in a high volume series production application (55). In an earlier report, Ogunwumi, et al. (56) describe the fundamental aspects of the aluminum titanate based composition, and later Heibel, et al. (57) describe engine and vehicle testing. In the latter paper, back pressure with soot and ash, regeneration properties as a function of soot loading, long term durability testing, and ash loading properties are described. Back pressure is lower due to tight pore size distribution and cell geometry, specifically, using an asymmetric cell structure (inlet cells larger than exit) and minor changes in cell density and wall thickness. Figure 15 shows some back pressure comparisons for fresh and loaded filters. Even though the filter has low thermal conductivity, the high heat capacity and physical properties enable an unsegmented design with maximum soot loadings of about 8 g/liter. This compares with cordierite in the 4 g/liter range and SiC in the 10-15 g/liter range.

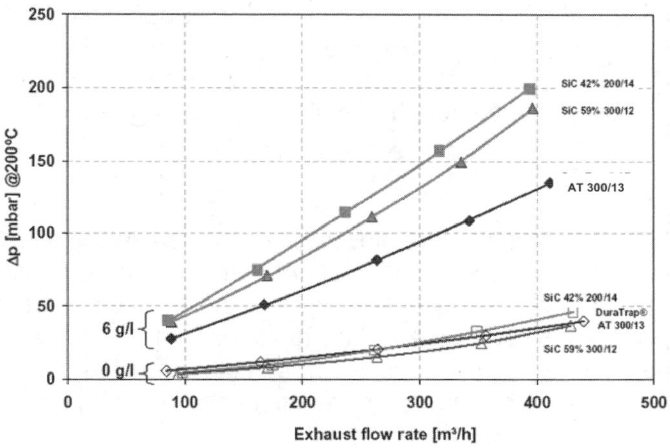

Figure 15. A new aluminum titanate filter that is now in series production has comparable unloaded back pressure as an SiC standard, but is 25% lower at 6 g/liter soot loading (57).

On other alternative filter materials, Mao, et al. (58) gave an update on a mullite filter with a novel whisker microstructure. The high porosity (60%) 25 μm pore structure allows low back pressure at high catalyst loadings (15% lower than SiC despite 3X washcoat loadings – 150 g/l). There was no deterioration in the 95% filtration efficiency after 100 regenerations at 8 g/liter soot loadings. On other designs, although metal "open" filters, wherein gas can either pass through filter media or go through unfiltered, are in both a LD and HD series applications they may have a niche in some retrofit applications (59). Presti, et al. (60) show 40% filtration efficiency for an unloaded open filter in a LDD application, but this drops to 25% efficiency as the filter loads up. Konstandopolous, (53) did a comprehensive paper on DPF design and performance. They show that depth filters, with an average pore size of about 40 µm, result in lower pressure drop and better soot penetration into the wall. This results in lower regeneration temperatures due to more intimate contact between soot and catalyst.

Improving catalyst behavior on DPFs will be important as engines move into mixed mode combustion. It is expected that oxidizing the high concentration of HCs from these engines at low load on the DPF will also continually burn soot (6). Ido, et al. (61) show that much catalyst within the wall might be under-utilized. Despite using DPF substrates with widely different pore size distributions, after washcoating the distributions all broadened. This will result in gas preferentially flowing through the large pores. Catalyst in the small pores thus is under-utilized. This phenomenon might contribute to the observation that system CO removal is higher if the PGM is distributed more on the DOC than a catalyzed DPF (62). On the other hand, catalyzed filters can give better regeneration behavior than a comparable system with a DOC+ catalyzed DPF (63). The filters, with an improved coating, burned more soot at 450 and 600C than if the HC oxidation process occurs primarily on a DOC. Instead of mainly residing on the wall, the improved coated penetrated into the wall.

ASH MANAGEMENT

MECA (Manufacturers of Emission Controls Association) published a paper on the state of ash management (64). The reader is directed there for a much more comprehensive review of the topic than can be covered here. The report covers sources and composition of ash, filter designs, filter maintenance intervals, and cleaning procedures.

Soeger, et al., (65) show that, although high filter back pressure is one reason to clean filters of ash, performance can be adversely affected. Figure 16 shows that the balance point temperature (temperature at which soot accumulation and oxidation are balanced) decreases from 369C for a filter aged 75,000 km in a long haul application, to 349C after ash cleaning. Similar results are available for NO_2 generation and HC oxidation.

Plumley (66) shows at 25 and 50% load, lube oil consumption rates are similar, as they are for higher loads at lower engine speeds. However, as shown in Figure 17, at 50 and 75% load consumption markedly increases. Given that low load applications, like refuse trucks and transit buses, require more frequent ash cleaning (around 48,000 to 90,000 km) than high load applications, like fuel transporters (600,000 km), these results indicate that perhaps hours of operation rather than mileage might be a better index of ash cleaning interval. Plumley also shows that soot is about 10% ash.

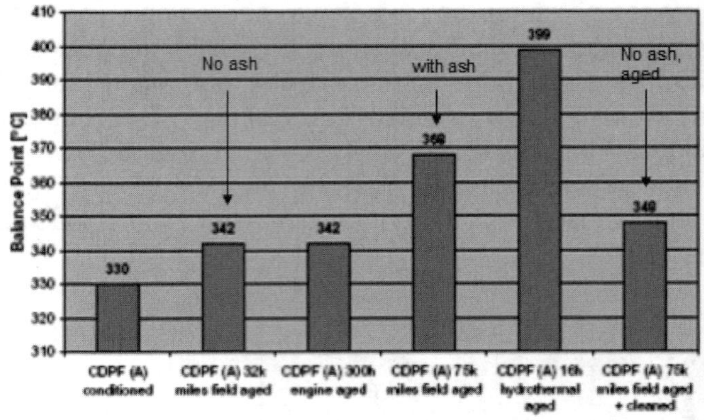

Figure 16. Ash can affect balance point temperature (65).

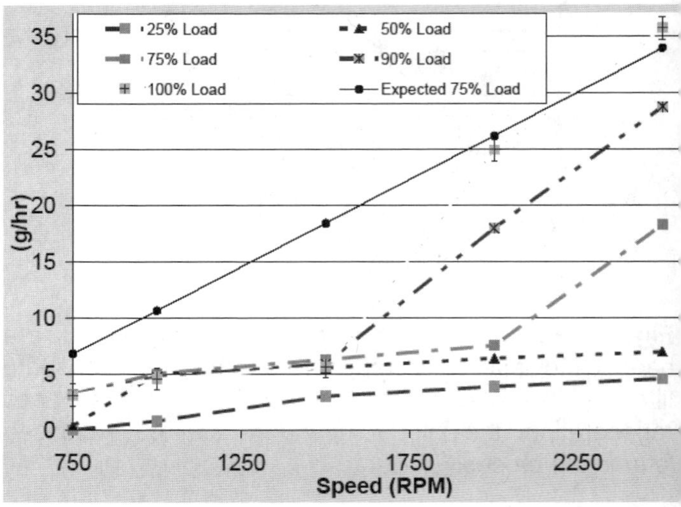

Figure 17. Lube oil consumption rates at 50, 75, and 90% load are similar up to about 1500 RPM, after which the consumption at 90% load increases (66). 2002 Cummins ISB, 5.9 liter engine.

Finally, researchers investigating ash effects have proposed a puzzle. Bardasz, et al. (67) did a mass balance on the lube oil consumed and the ash collected on a high-efficiency DPF, and found that only 46% of the phosphorous, 37% of the calcium, 31% of the magnesium, and 5% of the boron from the lube oil are captured on the filter. Others have found similar results to varying degrees, and no component is collected at >50%. One might jump to the conclusion that this ash goes through the filter, but no one has shown such, and these filters generally have high efficiency (80+%) in the size range of the fundamental ash particle (10-30 nm), and 99%+ efficiency for agglomerates 2-3X this size.

PM MEASUREMENT AND FILTER PERFORMANCE

Measuring the PM emissions when DPFs are used is quite challenging. Emissions are relatively undetectable compared to those if filters are not used. As shown in Figure 18, Khalek, et al., (68) shows that qualified analytical filters can result in a 3.5X range of PM emissions, only depending on filter type. The error is not on the low weight filters letting PM through, but rather those filters that collect more PM may have flaws that allow non-PM aerosols to accumulate. Similar results on dilution ratio variability and other parameters also are shown to impact the measurement (69, 70).

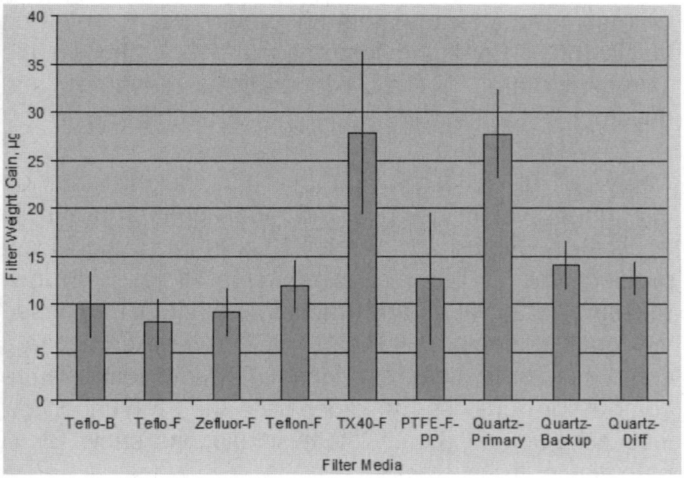

Figure 18. The choice of analytical filter can impact PM emissions results for systems using DPFs. Here two approved filters have 3.5X more emissions on them than other approved types (68).

To alleviate some of these problems, the European Union is investigating a number-based protocol instead of a mass-based one. Figure 19 shows preliminary results (71). Although there is more variability in the number-based measurement for filtered diesel, the method is able to differentiate multi-port gasoline (MPI) from direct injection gasoline (GDI) much better than the mass-based method.

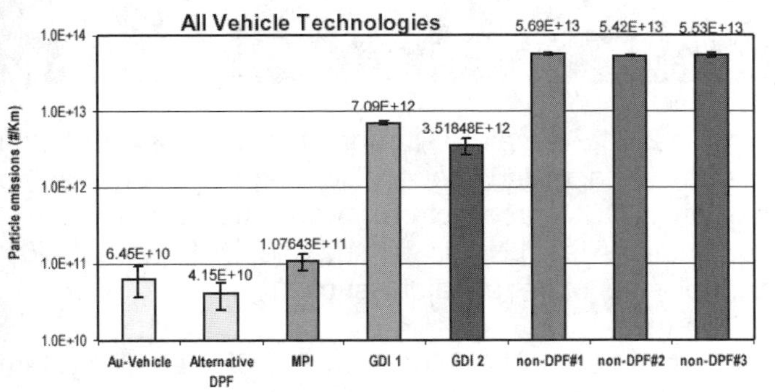

Figure 19. Preliminary results from the European PMP test program to develop a number-based PM measurement method (71).

Regarding performance, an earlier report on Paris taxis (72) shows ultrafine numbers increasing 10X after 80,000 km (but still removing 99% of them). A more recent AECC (Association of Emission Control by Catalysts) shows (70) little additional deterioration after 160,000 km of aging, and these emissions are still only 10% that of similar Euro IV gasoline vehicles. In fact, this and other studies (57) show PM mass efficiency improvement with age due to ash build-up.

Finally, DPF systems are not without issue. A key problem that is gaining attention is the generation of NO_2 from filter systems that use the compound to oxidize soot. Not included are systems that primarily use a fuel borne catalyst and/or heat from excess fuel to oxidize all the collected soot (no NO_2 regeneration). Mayer (73) shows that tailpipe NO_2 concentrations increase 2.5 to 6X at 25 to 50% load across the filter systems, depending on vehicle and system. However, one has to consider that NOx flux is low under these conditions. Even so, at full load NO_2 from modern filter systems can go up 20 to 60%. California is considering an NO_2 limit, even though modeling shows no significant increase in ground level ozone in conservative scenarios, and personal exposure is well below occupational health limits. Better system engineering, and some new system designs (74, 75) are beginning to address the problem.

SUMMARY – PM CONTROL

Although we may see LD and HD engines meet the tight US and Japanese standards without NOx treatment, it is doubtful we will see the regulations being met without filters. Given the growing evidence on the adverse health effects of fine particles, it is reasonable that diesel particulate filters will propagate through all countries, just like we are seeing with the catalytic converter. As we learn more about regeneration control, and as the materials (substrate and catalyst) improve, costs will continue to decline and performance will get better.

In that regard, filter regeneration strategies are reviewed, with new methods described for controlling regeneration duration, efficiency, and safety. The performance of new filter materials are summarized, as well as some new catalysts and coating methods. A summary of the ash management is provided, while presenting a question on ash deposition and the mass balance. Challenges in measuring PM at the very low emission levels are summarized, as well as that of tackling the NO2 emission issue. More results on performance and durability were summarized.

HYDROCARBON AND CO CONTROL

Although diesel oxidation catalysts (DOC) are a relatively mature technology, having been placed on 10's of millions LD diesels in Europe, and on HD applications in series production in the late 1990's, two recent developments are very noteworthy. The first is the developments of an oxidation catalyst that can operate with high-sulfur fuel (2000 ppm) without significant SO_2 formation (76). Figure 20 shows that with the new catalyst sulfur-based particulate is the same as that coming out of the engine, compared to a

3X increase with other typical formulations. Such a development has interesting applications in retrofit applications in developing countries with high sulfur fuel and older diesel engines.

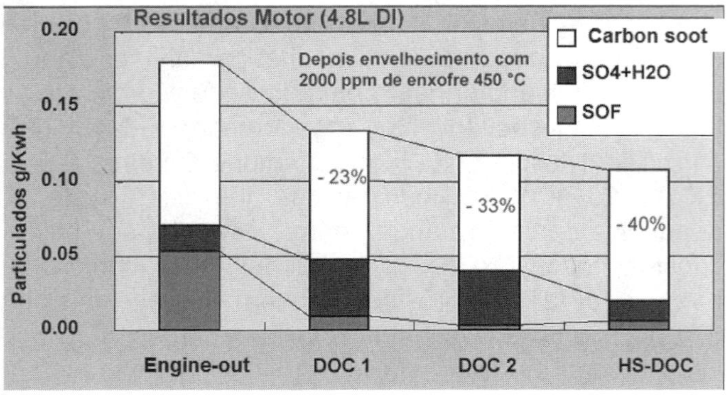

Figure 20. A new high sulfur DOC (HS-DOC) does ot form sulfates in 2000 ppm sulfur fuel, despite high temperatures (76).

The second development (77) is a low temperature DOC that has application in both retrofitting of older vehicles with cold exhaust, Figure 21, and possibly for the emerging mixed mode engines, wherein highly efficient oxidation catalysts will be needed to treat cool exhaust with high HC and CO levels. For the case of older engines at idle, 80-90% of the soluble organic fraction (SOF) is removed even though the temperatures are only 105-163C.

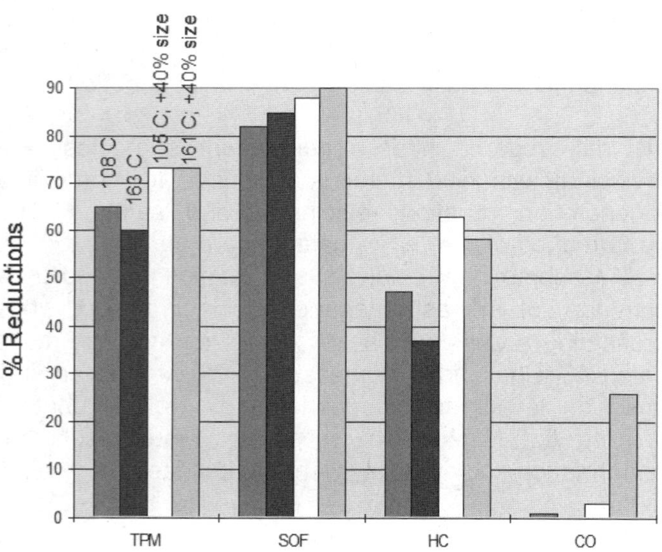

Figure 21. Results on a new LT DOC on an older engine. 80-90% SOF reductions given 60-70% total PM reductions under idle conditions (105-161C). Reference 77.

INTEGRATED NOX AND PM CONTROL

The first applications of integrated NOx and PM control will likely be in US Tier 2 Bin 5 LD applications. At the 2006 Detroit Auto Show, DaimlerChrysler announced an E-class that will use a DOC followed by an LNT, DPF, and an SCR system (78) to meet the Bin 5 standards. VW officials have made a less formal announcement of a Bin 5 Jetta for 2006 introduction (79), although the system was not described. On a related matter, the US EPA showed a Ford minivan prototype that uses high EGR, boost, and advanced fuel injection and a DPF without NOx aftertreatment to obtain Bin 5 emission levels (80). Ford and International Truck and Engine Company are in a cooperative research and development agreement with the EPA on this engine.

Neely, et al. (81), provide the first insights into an integrated system using mixed mode control in a LDD application, Figure 22. As LT NOx is not a significant problem, the LNT is placed after the DPF. Both low-pressure (after DPF) and traditional HP EGR are used for engine control. Rich PCCI (premixed charge compression ignition) and rich LTC (low temperature combustion) are used to regenerate the LNT under low load conditions. Steady state simulations show Bin 3 potential for the system.

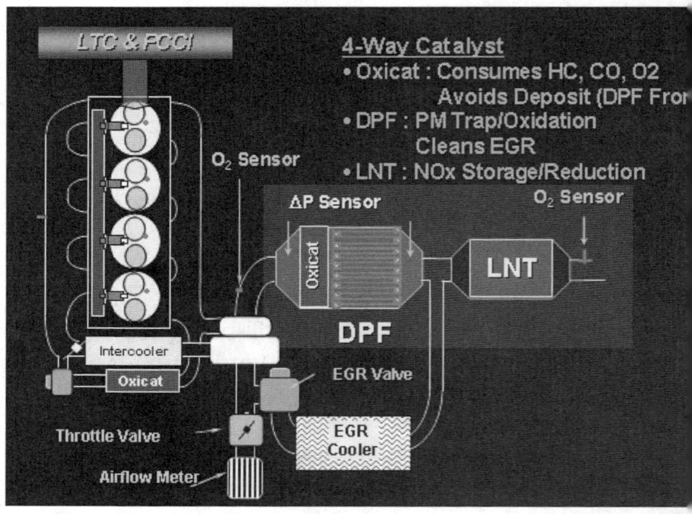

Figure 22. Example of an integrated emission control system for a mixed mode LDD engine. Converse to typical LDD architecture with traditional diesel combustion, the LNT is behind the DPF as LT NOx control is no longer a major issue (81).

Looking towards US2010, a DPF+SCR retrofit system on a 12 liter Caterpillar engine retrofit with an EGR system (82) hit the required NOx levels at 0.18 g/bhp-hr and PM levels at 5 mg/bhp-hr, with no deterioration after 6000 hours of testing. The more challenging NTE performance was missed at one steady-state load point.

Finally, although the tight US and European non-road regulations are five years off (2011), Baumgard, et al., (83) showed the first insights into how a US Tier 4 2011 system might look, Figure 23. They used a high pressure common rail system and cooled EGR to take a Tier 2 John Deere 6068H 6.8 liter engine to Tier 4 interim NOx levels. A DPF got them to the PM levels.

The new configuration results in a 3 to 5% fuel penalty versus Tier 2.

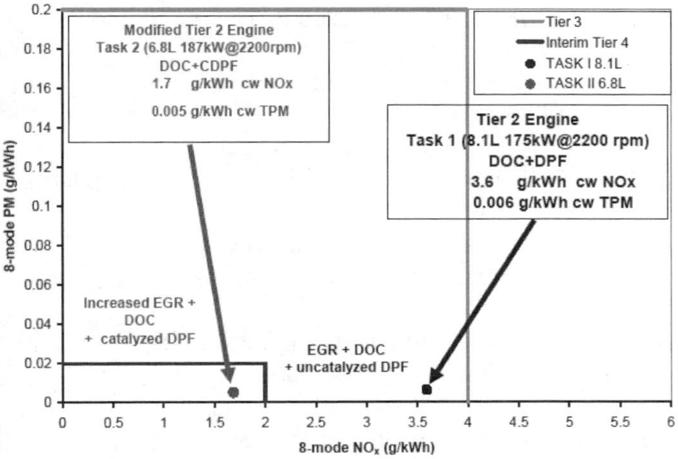

Figure 23. A interim Tier 4 (2011) non-road system, using high-pressure common rail fuel injection and cooled EGR to hit the NOx levels and a DPF to hit the PM levels. Fuel penalty is 3-5%. John Deere 6068H Tier 2 engine (83).

CONCLUSIONS

The regulatory framework for LDD, HDD, and non-road applications through 2014 are described, as well as emerging engine technologies. This helps define the future aftertreatment requirements. In general, for LDD using incremental advances in traditional diesel combustion, although DPF will be needed, nominally US Tier 2 Bin 8 (0.12 g/km) can be achieved today with no NOx aftertreatment, as well as Euro V, and Japan 2009 will likely be hit in the future. US Bin 5 will need 70% NOx control and DPFs. As mixed mode combustion develops, Bin 5 passenger cars will likely require no NOx aftertreatment in 2009, followed by the same status for heavier LD vehicles in 2011. For HDD The Japan 2009 regulations show good chances of being hit with advanced combustion, at least as indicated by today's research engines, but NOx control will likely be extended from Japan 2005 to offer better fuel economy. For US 2007, 50% NOx control at 500-520C is indicated by today's research engines to hit the tough NTE requirements.

For NOx control, LNTs will meet the LDD requirements, and SCR leads the HD field. The middle sized applications could go either way. SCR is moving towards zeolite catalysts, and reports are surfacing on secondary emissions. For LNTs, HT formulations to address HD NTE are being reported, and PGM loadings are dropping.

Filters are continuing the movement towards cost reduction and performance improvement. DPF regeneration strategies are summarized and quite sophisticated. New filter materials are emerging, and for the first time in six years, a wall-flow filter of a new material (aluminum titanate) is in series production. Substrate materials are improving, as are catalysts, and reports are offering insights for improvements. Ash management is emerging as a key concerns, and PM measurement in clean exhaust has its challenges.

Regarding HC and CO control, two recent developments are significant, especially for retrofit applications in developing countries: a high sulfur DOC is reported that does not form sulfates, and a LT DOC that works well at 105C is available.

On integrated systems, the US Tier 2 Bin 5 application will be the first high production application for integrating both PM and NOx control. We may see two vehicles meeting this requirement by the end of 2006. The first mixed mode system for LD application is also described. US2010 appears within reach, and the first paper on hitting the interim US Tier 4 non-road regulations for 2011 is summarized.

CONTACT

JohnsonTV@Corning.com

REFERENCES

1. http://www.epa.gov/ttn/naaqs/standards/pm/s_pm_cr_sp.html
2. http://europa.eu.int/comm/environment/air/cafe/general/pdf/cafe_lot1.pdf
3. R. Schulte-Brauchs, SAE HHD Emission Control Symposium, Gothenburg, September 2005.
4. Regulatory topics panel discussion, SAE HHD Emission Control Symposium, Gothenburg, September 2005.
5. Courtesy of John Deere.
6. Hogan, M., Gray, C., US Department of Energy Diesel Engine Emissions Reductions (DEER) Conference, Chicago, September 2005. http://www.eere.energy.gov/vehiclesandfuels/resources/proceedings/index.shtml
7. Ryan, T., "Combustion Targets for Low Emissions and High Efficiency", US Department of Energy Diesel Engine Emissions Reductions (DEER) Conference, Chicago, September 2005, http://www.eere.energy.gov/vehiclesandfuels/resources/proceedings/index.shtml.
8. Idicheria, C. A. , Pickett, L., "Soot Formation in Diesel Combustion Under High-EGR Conditions", SAE 2005-01-3834.

9. Stanton, D., "Analysis-Lead Design for Engine Systems Development to Hit US2010 Emission Standards", University of Wisconsin Engine Research Center Symposium, June 2005.
10. Moser, F. X., "The Challenge of Transient Operation in Various HDDE Emission Test Cycles Base Layout and Transient Control", SAE HHD Emission Control Symposium, Gothenburg, September 2005.
11. Miles, P., "The role of flow structures and mixing on low-temperature diesel combustion", 9th International Conference on Present and Future Engines for Automobiles", June 2005; San Antonio
12. Hammer, J., "Evolution of the Common Rail Technology", Ispa, Italy, ATA conference, October 2004.
13. Imarisio, R., et al, "Diesel Strategies Towards Fuel Neutral European Emissions Standards", ATA Conference, Ispa, Italy, October 2004.
14. Greaney, A., "Ricardo's ACTION Strategy — an Enabling Light Duty Diesel Technology for the U.S. Market", US Department of Energy Diesel Engine Emissions Reductions (DEER) Conference, Chicago, September 2005, http://www.eere.energy.gov/vehiclesandfuels/resources/proceedings/index.shtml.
15. Pfahl, U., "Combustion, Aftertreatment and Control Key Elements for Emission Reduction of US HSDI Diesel of US HSDI Diesel Engines University of Wisconsin Engine Research Center Symposium, June 2005.
16. Aneja, R., "Integrated Engine and Aftertreatment Technology Roadmap for 2010 Heavy-Duty Emissions Regulations", US Department of Energy Diesel Engine Emissions Reductions (DEER) Conference, Chicago, September 2005, http://www.eere.energy.gov/vehiclesandfuels/resources/proceedings/index.shtml.
17. Shimoda, M., "Japanese Perspective on Clean Heavy-Duty Diesel Engines", SAE HHD Emission Control Symposium, Gothenburg, September 2005.
18. Edwards, S., "Commercial Vehicle Diesel Technology to Meet the Global Market Requirements of 2010+", SAE HHD Emission Control Symposium, Gothenburg, September 2005.
19. Khair, M., "US Perspective on HDDE Controls to Meet 2010 Emissions Standards", SAE HHD Emission Control Symposium, Gothenburg, September 2005.
20. Parche, M., "Expected Trends in Fuel Injection Systems", SAE HHD Emission Control Symposium, Gothenburg, September 2005.
21. Charmley, W., US Department of Energy Diesel Engine Emissions Reductions (DEER) Conference, Chicago, September 2005, http://www.eere.energy.gov/vehiclesandfuels/resources/proceedings/index.shtml.
22. Hirata, K., et al., "Development of Urea SCR System for HDD Commercial Vehicles", SAE 2005-01-1860.
23. Gieshoff, J., et al., "Advanced Urea SCR Catalysts for Automotive Applications", SAE 2001-01-0514
24. Cider, L., et al., "RESULTS FROM KNOWNOX - A EUROPEAN COOPERATION PROJECT", paper number F2004V184, FISITA 2004, Barcelona.
25. Engelhard, October 2004, private communications.
26. Lambert, C., et al., "Urea SCR and DPF System for Diesel LDT/SUV Meeting Tier 2 Bin 5", US Department of Energy Diesel Engine Emissions Reductions (DEER) Conference, Chicago, September 2005, http://www.eere.energy.gov/vehiclesandfuels/resources/proceedings/index.shtml.
27. Walker, A., "Heavy-Duty Emissions Control Systems – 2010 and Beyond", SAE HHD Emission Control Symposium, Gothenburg, September 2005.
28. Holcomb, T., Engelhard Symposium, Plant Opening, Nieburg, Germany, September 2005.
29. Solla, A., et al., "Effect of Ammonium Formate and Mixtures of Urea and Ammonium Formate on Low-Temperature Activity of SCR Systems", SAE 2005-01-1856.
30. Ball, J., "A Toxicological Evaluation of Potential Thermal Degradation Products of Urea", SAE 2001-01-3621
31. Koebel, M. et al. "Urea-SCR: A Promising Technique to Reduce NOx Emissions from Automotive Diesel Engines", Catalysis Today, 59, 2000, pp. 335-345.
32. Sluder, C. S., et al., "Low-Temperature Urea Decomposition and Scr Performance", SAE 2005-01-1858.
33. Roth, S., "Challenges with Catalyst System – Contrasting US and EU Issues", SAE HHD Emission Control Symposium, Gothenburg, September 2005.
34. Hinz, A., et al., "The Application of a NOx Absorber Catalyst System on a Heavy-Duty Diesel Engine", SAE 2005-01-1084.
35. Dalla Betta, R. et al., "Transient Dynamometer Testing of a Single Leg NOx Adsorber Combined with a Fuel Processor for Enhanced NOx Control", US Department of Energy Diesel Engine Emissions Reductions (DEER) Conference, Chicago, September 2005, http://www.eere.energy.gov/vehiclesandfuels/resources/proceedings/index.shtml.
36. Wu, M. C., "Experimental Evaluation of Reformate-Assisted Diesel NOx Trap Desulfation", SAE 2005-01-3878.
37. Hallstrom, K., "Catalyst Based Diesel Emission Control Technology", Asian Vehicle Emission Control Conference, Beijing, June 2004, WWW.MECA.org.
38. Takahashi, Y., et al., "Development of NOx Trap System for Commercial Vehicle - Basic Characteristics and Effects of Sulfur Poisoning", SAE 2004-01-0580.
39. Theis, J. R., et al. " The Effects of Aging Temperature and PGM Loading on the NOx Storage Capacity of a Lean NOx Trap", SAE 2005-01-1117.

40. Blint, R. J., "Discovery of New NOx Reduction Catalysts for CIDI Engines Using Combinatorial Techniques", Combinatorial & High Throughput Materials Science, Gordon Conference, August 15-19, 2005
41. Orban, J. E., et al., "Long-Term Aging of NOx Sensors in Heavy-Duty Engine Exhaust", SAE 2005-01-3793.
42. Zink, U., Johnson, T. V., "DPF Regeneration Management: A Review of European LD Concepts – Implications for HD Systems",US Department of Energy Diesel Engine Emissions Reductions (DEER) Conference, Chicago, September 2005, http://www.eere.energy.gov/vehiclesandfuels/resources/proceedings/index.shtml.
43. Boretto, G., et al.; "Serial Application of a Catalyzed Particulate Filter on Common Rail DI Diesel Engines for Passenger Cars", 2004V068, FISITA Conference, Barcelona May 2004
44. Mattes, W. et al.; MTZ 7-8/2004 Vol 65
45. Sommers, J., et al., "Das neue Mercedes Benz Dieselpartikelfilterkonzept fuer PKW in Verbindung mit der Abgasstufe EU4"; 25. Wiener Motorensymposium, April 2004
46. Zink, U., et al. "Regeneration Control – Key to successful application of new DPF systems", Proceedings of 13. Aachen Kolloquium Fahrzeug- und Motorentechnik, October 2004.
47. Wonkun, K., et al., "Application of Catalyzed Diesel Particulate Filter System (CPF) To Small SUVs for Euro IV", 4th HMC Diesel Symposium, October 21, 2004.
48. Manns, J., "Diesel Particulate filters for Passenger Cars – Interaction Between System Layout and Requirements for Calibration", Global Powertrain Conference, Ann Arbor, Michigan, September 2005.
49. Kodama, K., et al., "Development of DPF System for Commercial Vehicles: (Second Report) - Active Regenerating Function in Various Driving Condition", SAE 2005-01-3694.
50. Craig, A., et al. "Performance aspects of Cordierite Diesel Particulate filters in HD Applications", SAE Commercial Vehicle Conference, Chicago, October 2005.
51. Oya, T., et al., "Performance Evaluation of SiC-DPF Sintered with Sintering Additive", SAE 2005-01-0579.
52. Soeger, N., et al., "Impact of Aging and NOx/Soot Ratio on the Performance of a Catalyzed Particulate Filter for Heavy-Duty Diesel Applications", SAE 2005-01-0663.
53. Konstandopolous, A., et al., "Catalytic Filter Systems With Direct and Indirect Soot Oxidation Activity", SAE 2005-01-0670.
54. Chiew, L., et al., "Diesel Vaporiser: An Innovative Technology for Reducing Complexity and Costs Associated with DPF Regeneration", SAE 2005-01-0671.
55. Volkswagen press release, November 2005.
56. Ogunwumi, S., et al., "Aluminum Titanate Compositions for Diesel Particulate Filters", SAE 2005-01-0583
57. Heibel, A., et al., "Performance and Durability of the New Corning DuraTrap® AT Diesel Particulate Filter – Performance and Durability Evaluation Results from Engine Bench and Vehicle Tests", Aachen Kolloquium, October 2005.
58. Mao, F., et al., "Performance Validation of an Advanced Diesel Particulate Filter With High Catalyst Loading Capacity", SAE 2005-01-3696
59. Kassel, R., Clean Air Initiative - Asia BAQ 2004, Agra, India, December 2004.
60. Presti, M., et al., "Metal Supported Particulate Matter-Cat, A Low Impact and Cost Effective Solution for a 1.3 Euro IV Diesel Engine", SAE 2005-01-0471
61. Ido, T., et al., "Study on the Filter Structure of SiC-DPF with Gas Permeability for Emission Control", SAE 2005-01-0578.
62. Kwon, C., et al., "Control of Diesel Catalyzed Particulate Filter System II (The Optimization of Catalyst in the CPF System)", 2005-01-0672
63. Pfeifer, M., et al., "The Second Generation of Catalyzed Diesel Particulate Filter Systems for Passenger Cars - Particulate Filters With Integrated Oxidation Catalyst Function", SAE 2005-01-1756.
64. "Diesel Particulate Filter Maintenance: Current Practices and Experience", www.MECA.org, June 2005.
65. Soeger, N., et al., "Impact of Aging and NOx/Soot Ratio on the Performance of a Catalyzed Particulate Filter for Heavy-Duty Diesel Applications", SAE 2005-01-0663.
66. Plumley, M., "Lubricant Formulation and Consumption Effects on Diesel Exhaust Ash Emissions: Measurements and Sample Analyses from a HD Diesel Engine", US Department of Energy Diesel Engine Emissions Reductions (DEER) Conference, Chicago, September 2005, http://www.eere.energy.gov/vehiclesandfuels/resources/proceedings/index.shtml.
67. Bardasz, E., et al., "Future Engine Fluids Technologies: Durable, Fuel-Efficient, and Emissions-Friendly", 9th International Conference on Present and Future Engines for Automobiles", June 2005; San Antonio, and SAE 2005-01-2168
68. Khalek, I., "Overview of 2007 Diesel Particulate Measurement Research (Project E-66)", 9th ETH Conference on Combustion Generated Nanoparticles, Zurich, Switzerland, August 15-18, 2005.
69. Shimpi, S., "Update on 2007 Diesel Particle Measurement Research (Project E-66)", US Department of Energy Diesel Engine Emissions Reductions (DEER) Conference, Chicago, September 2005, http://www.eere.energy.gov/vehiclesandfuels/resources/proceedings/index.shtml.

70. Automobile Emissions Control by Catalysts (AECC) Association, April 2005, LDD Vehicle Report.
71. PMP, Aaqius and Aaqius Conference, Frankfurt Auto Show, September 2005.
72. Jeuland, N., et al., "Performances and Durability of DPF (Diesel Particulate Filter) Tested on a Fleet of Peugeot 607 Taxis: Final Results", SAE 2004-01-0073
73. Mayer, A., Aaqius and Aaqius Conference, Frankfurt Auto Show, September 2005.
74. Allansson, R., et al., "The Development and In-Field Performance of Highly Durable Particulate Control Systems", SAE 2004-01-0072.
75. Private communications, A. Gorel, SST Emtec, Janaury 2005.
76. Blumrich, S., SAE Spring Fuels and Lubes Conference, Rio de Janeiro, May 2005.
77. Hallstrom, K., Engelhard Symposium, Plant Opening, Nieburg, Germany, September 2005.
78. DaimlerChrysler Press Release, Detroit Auto Show, January 2006.
79. Diesel Fuel News, September 12, 2005
80. US EPA demonstration vehicle at SAE Government and Industry Meeting, Washington, D.C., May 2005.
81. Neely, G. D., et al., "New Diesel Emission Control Strategy to Meet Us Tier 2 Emissions Regulations", SAE 2005-01-1091.
82. Khair M., Sharp, C., "Update on Progress of APBF-DEC EGR/DPF/SCR Demonstration Program An Update on Progress of APBF-DEC EGR/DPF/SCR Demonstration Program at SwRI", US DOE Diesel Engine Emissions Reduction Conference, August 2004.
83. Baumgard, K., "Exhaust Aftertreatment and Low Pressure Loop EGR Applied to an Off-Highway Engine", US Department of Energy Diesel Engine Emissions Reductions (DEER) Conference, Chicago, September 2005,

2006-01-1145

Advanced Diesel Technology to Achieve Tier 2 Bin 5 Emissions Compliance in US Light-Duty Diesel Applications

Brian Cooper, Ian Penny, Matt Beasley, Adrian Greaney and Jackie Crump
Ricardo Consulting Engineers

ABSTRACT

Future growth of the US light-duty diesel market is highly dependent on achieving cost effective and robust emissions solutions. For this reason, ACTION (Advanced Combustion Technology to Improve engine-Out NOx) is being developed by Ricardo. This research is delivering reduced emissions through the application of Highly Premixed Cool Combustion (HPCC). NOx reduction is primarily achieved through the reduction of charge oxygen concentration, accomplished by reducing air/fuel ratio and increasing exhaust gas recirculation. This approach coupled with enhanced combustion system efficiency delivers a practical approach to meet future emissions compliance while minimizing aftertreatment requirement.

The paper will examine key engine technology developments to deliver US Tier 2 Bin 5 emissions in Light Duty Diesel applications. In particular, Bin 5 emissions compliance will be possible in passenger car applications. ACTION level 2 technology will be sufficiently mature for 2007 and will require NOx aftertreatment solutions with 60-80% conversion capability. The paper describes progress to further apply the HPCC approach and lower engine-out NOx emissions while controlling fuel consumption and soot. The results show the potential to minimise the requirement for NOx aftertreatment. The paper describes key areas of development being undertaken by Ricardo research projects. In particular the importance of advanced air/EGR technology and combustion control is described.

INTRODUCTION

The Light Duty Diesel engine is the preferred powertrain choice in European passenger cars. The growth in diesel car sales has been driven by the application of new technology. Turbocharging, fuel injection and control technology in particular have significantly improved the performance, drivability and refinement of passenger cars. These characteristics coupled with a 20-25% typical fuel consumption advantage compared to gasoline cars has convinced European consumers.

Although European Light Duty Diesel engine sales are exceptional, globally there is a different position. Major opportunity exists for diesel sales growth in Japan, China and the US. In the US, where the total market for light duty vehicles was 16.9 million in 2004, the current percentage diesel sales is less than 0.3%. For global growth, future engines must concurrently meet cost, emissions, fuel consumption/CO_2, refinement and drivability demands. Reducing the compromise between these factors is at the heart of Ricardo Light Duty Diesel research.

The most severe emissions challenge has been defined by US legislators. In order to sell vehicles in all states, US Tier 2 passenger car regulations require 0.07 g/mi (0.044 g/km) NOx at 120k miles full life on the FTP-75 cycle. Refer to Figure 1.

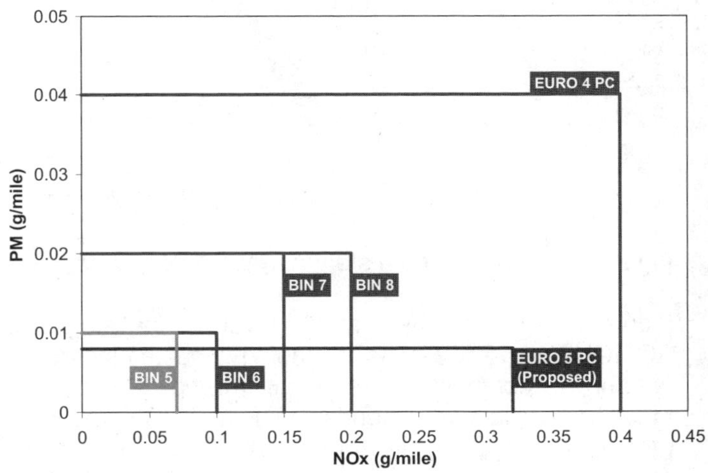

Figure 1: Overview of US Passenger Car Emissions Challenge compared to Europe

Furthermore, the US06 cycle will demand NOx reduction at much higher loads than current European legislation. This legislation will require the NOx emissions of typical

Euro 4 engines to be reduced by over 90%. To meet this requirement, NOx aftertreatment is under development. However, major issues must be solved before such technology can be applied in the US. In regard to SCR technology, the introduction of Urea infrastructure by 2010 is unlikely. This will inhibit the introduction of in-service refill of Urea. Lean NOx Trap (LNT) technology has robustness, efficiency and durability issues to overcome. Whichever NOx aftertreatment technology succeeds globally, the reduction of engine-out emissions is forecast to be the most efficient approach.

ACTION STATUS

To target NOx and soot emissions reduction at source, extensive research has been dedicated to the investigation of so-called "alternative combustion strategies". In this respect, many researchers have suggested that a radical change in technology is necessary to lower emissions. However Ricardo experience suggests there is a progressive roadmap of technologies to enhance pre-mixed and lower temperature combustion. These trends have already started and will continue to advance. Ricardo describes this roadmap as "Advanced Combustion Technology for Improved engine-Out NOx" ACTION. This is an integrated approach which combines air, EGR, fuel system, combustion system design and control developments. This roadmap has been defined through internal research and a number of advanced engineering prototype vehicles have been successfully completed.

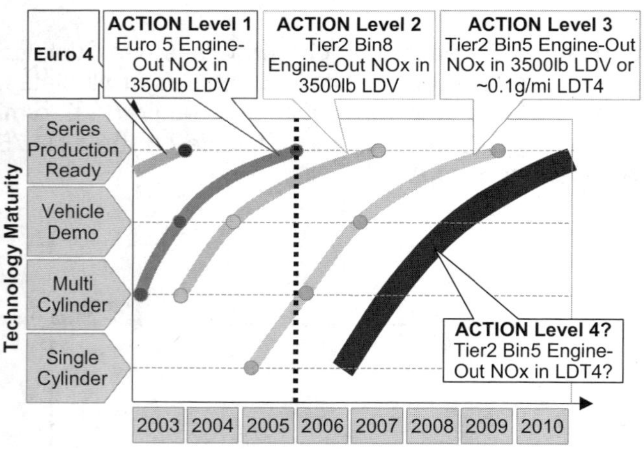

Figure 2: Overview of the status of ACTION technology

Figure 2 illustrates the maturity level for the stages of ACTION technology. The technology enables increased control of combustion while minimizing the cost of meeting legislative demands. The ACTION roadmap produces a synergistic effect, resulting in reduced emissions and high performance engine output at moderate peak cylinder pressure levels.

Figures 3 and 4 illustrate the progressive improvement in emissions achievable with technology application in relation to critical US Tier 2 Light Duty Diesel emissions challenges.

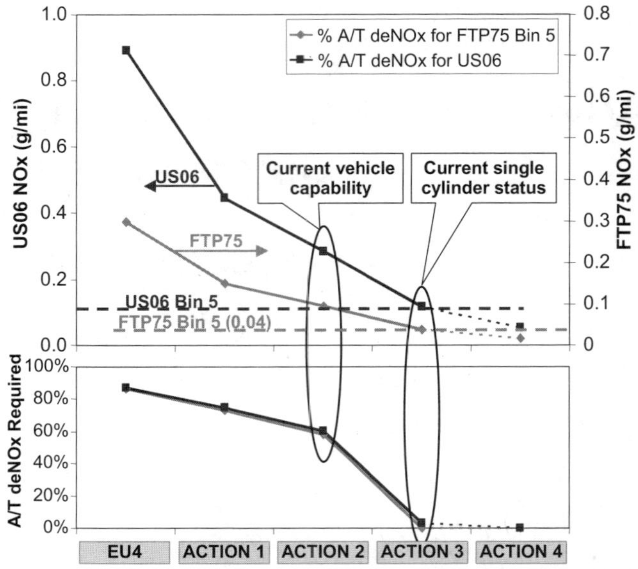

Figure 3: US ACTION status for LDV

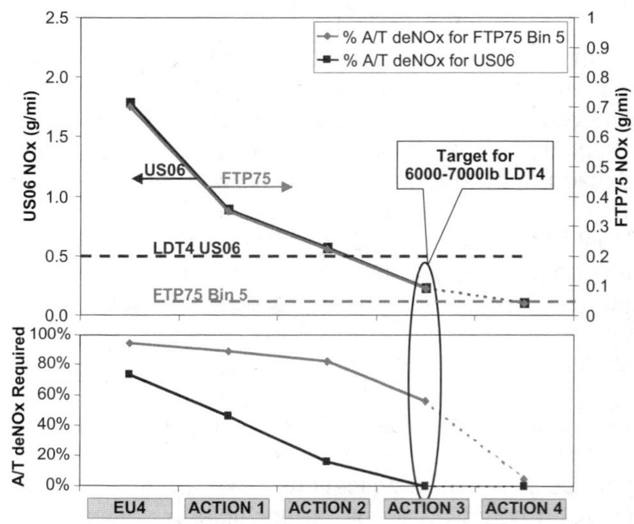

Figure 4: US ACTION Status for LDT4

The plots show that production ready solutions are becoming available for Bin 5 requirements to be met in LDV applications. The reduction of engine out NOx will enable practical NOx aftertreatment solutions with approximately 60-65% conversion performance. LDV4 applications however have a greater technical challenge and within the 2007 timeframe, higher levels of NOx aftertreatment control is expected to be necessary. In this regard, engine technology is advancing rapidly and research results are showing the potential to further reduce NOx at source. This will provide reduced risk and cost in regard to the aftertreatment system.

The ACTION level 3 solutions suggested by the authors aims to minimise engine cost. Figure 5 gives an overview of the technical elements, benefits and cost estimated to meet future requirements. This example shows the cost impact for a 2 liter diesel engine. At all levels, the cost and complexity of solutions are minimized. Level 2 technology requires a new combustion system design with reduced compression ratio. Also fuel system, EGR, air and control system developments will be production ready for 2007. Very expensive solutions such as Variable Valve Actuation or radical fuel injection solutions are covered by ACTION Level 4 technologies.

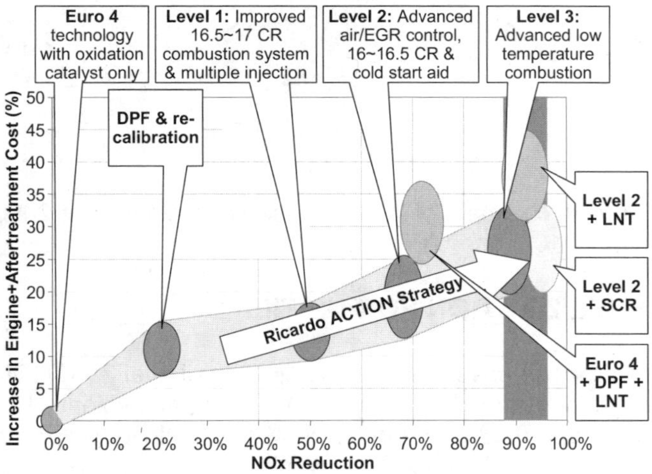

Figure 5: ACTION cost and technology overview

The chart also shows a Ricardo prediction for the costs of alternative approaches to NOx control. Ricardo suggest that engine-out technology alone will become a viable alternative to NOx aftertreatment by 2010 for some applications. Such technology availability will enable the optimum solution to be selected for an application and total costs to be minimised. The following sections describe the results achieved in greater detail.

THE HPCC TECHNICAL APPROACH

Major international research has been dedicated to the early injection strategy known as Homogeneous Charge Compression Ignition (HCCI). [1] [2]. This approach faces significant challenges and operation has been limited to very light load conditions. A more practical approach has been suggested by Kimura et al. [3] The Nissan MK system highlighted the strategy of promoting air and fuel mixing during an extended ignition delay period. In this respect "HCCI" research and technical papers have grown to encompass a wide variety of injection strategies and research approaches.

The practical approach to lower NOx emissions is by limiting combustion temperature through the reduction of charge oxygen concentration. This can be achieved by reducing air/fuel ratio (AFR) and increasing exhaust gas recirculation (EGR) levels. Figure 6 shows a comparison of this trend with ACTION technology and illustrates the primary route to ultra low NOx. The plot shows a compilation of 2000 rev/min steady state key point results achieved on a 0.5 liter single cylinder research engine Results from the three levels of the Ricardo approach are shown at the key point conditions.

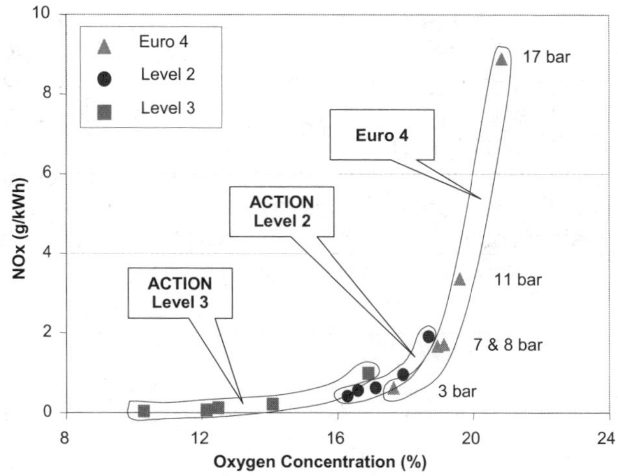

Figure 6: Progressive reduction in charge Oxygen concentration and NOx with ACTION technology

The dramatic reduction of NOx is primarily a function of reduced Oxygen concentration. Limiting the impact to efficiency and soot emissions is therefore a key issue. This requires enhanced combustion system operation. In particular improved air/fuel mixing and increased control of pre-mixture preparation is essential. Additional secondary benefits arise from the operation of lean pre-mixed combustion enabled by reduced pre-combustion temperatures. This is achieved by reducing inlet temperatures and compression ratio in combination with optimized combustion phasing. Reference [7] These operating characteristics are the reason the authors describe the combustion approach as Highly-Premixed Cool Combustion (HPCC).

An important factor of the ACTION technology is increased flexibility and control of the factors that control combustion. This flexibility enables the most appropriate air and fuel mixing to be provided for combustion. Such flexibility combined with advanced development techniques offers the potential to optimize the hardware and calibration to a heightened capability. Figure 7 illustrates the terminology applied in Figure 8. In particular, "Ignition Delay" is defined as the period between the actual start of needle lift and the 5% burn angle.

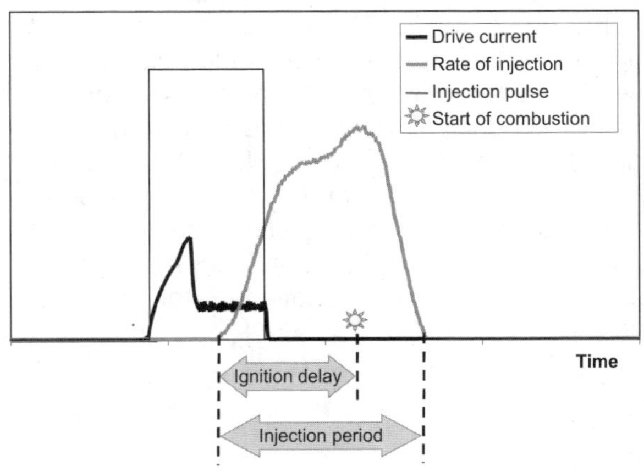

Figure 7: Illustration of Terminology

Figure 8 shows the change in combustion timing that is observed with the HPCC approach. With technology application, more pre-mixture preparation is achieved with increased ignition delay and reduced injection period (needle opening time). At light load conditions it is more practical to achieve completely pre-mixed combustion but at higher loads, diffusion combustion must be increasingly optimized. The result is NOx reduction while minimizing impact to fuel consumption.

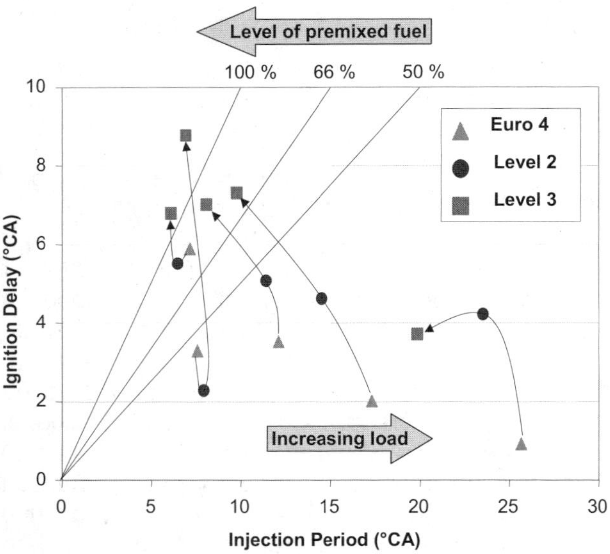

Figure 8: Pre-mix combustion trend with ACTION

The ACTION strategy increases ignition delay by the combination of increased EGR rate, reduced AFR, reduced compression ratio, intake temperature control and injection timing. Also injection period is reduced by increases in rail pressure and nozzle flow selection. Such a strategy results in initially slow combustion controlled by the low Oxygen concentration. The timing of combustion must be optimized to control combustion efficiency.

As load increases, achieving the combination of low NOx, soot, noise and fuel consumption becomes more challenging. Figure 9 shows an example of ACTION Level 3 single cylinder results achieved at 2000 rpm, 10.8 bar gross IMEP. This is a critical condition for FTP and US06 drive cycles. In this example, the optimized Level 3 engine achieved 93% NOx reduction relative to the Euro 4 baseline. The figure shows initial slow and retarded combustion followed by rapid heat release. Deterioration in combustion efficiency is limited to 4% by optimization of the combustion period. Such an approach however results in increased combustion noise. This is a key challenge for ultra low NOx emissions.

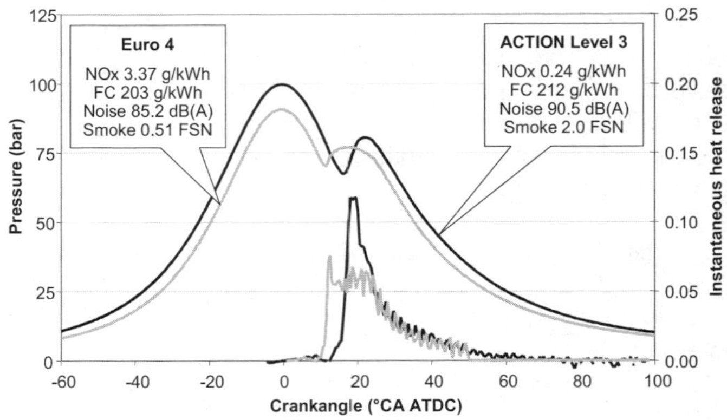

Figure 9: Example of heat release characteristic at 2000 rev/min, 10.8 bar IMEP

LEVEL 2 ACTION FOR US TIER 2

In European passenger car applications Level 2 ACTION technology has demonstrated gasoline equivalent NOx emissions (0.08 g/km) over the MVEG-B drive cycle with no fuel consumption or combustion noise penalty. Furthermore, the engine specific power of 50 kW/L and vehicle driveability is not compromised relative to Euro 4 vehicles.

The following figures show the results achieved for a 3500lb Passenger Car with 6 speed manual transmission using ACTION level 2 technology. The following technologies have been applied in this example:

- A 16.5 compression ratio combustion system with conventional cone angle nozzle (~150°)
- Optimized intake port geometry with fixed swirl
- Solenoid actuated Common Rail fuel system with maximum rail pressure of 1600 bar
- EGR cooler bypass control
- Ricardo air/EGR system and transient control algorithms

Figures 10 and 11 show the potential for ACTION to meet Bin 8 emissions targets over the FTP drive cycle. In this case an additional 60% NOx reduction would be necessary to meet Bin 5 requirements.

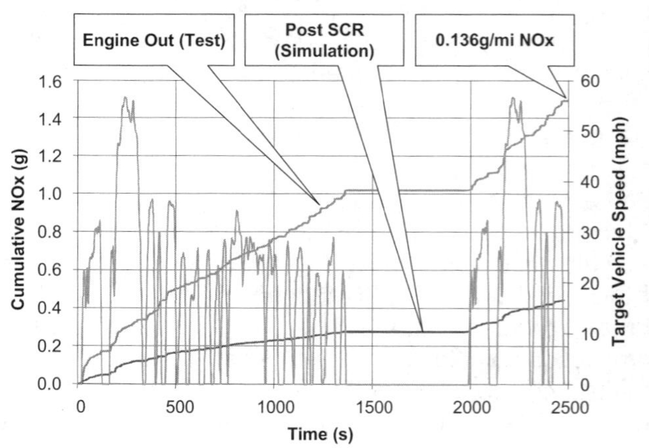

Figure 10: FTP drive cycle results for 3500lb vehicle

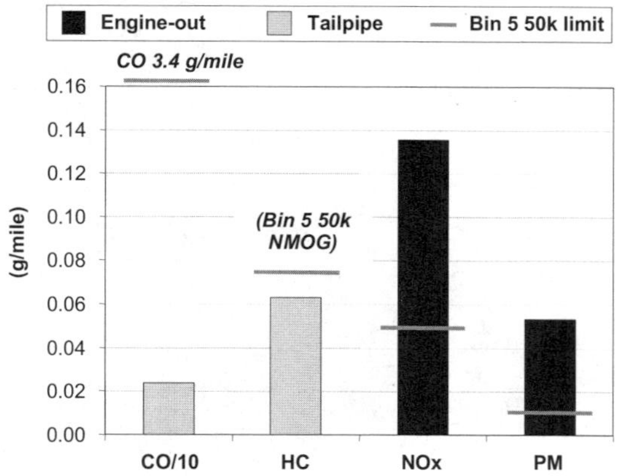

Figure 11 – Overview of FTP emissions results

The testing has also identified the fuel consumption impact. Fuel consumption of 40.2mpg(US) was achieved over FTP-75 cycle, representing a penalty of 3% relative to the baseline Euro 4 vehicle subject to the FTP drive cycle. Furthermore, Figures 12 and 13 show the results achieved over the stringent US06 drive cycle.

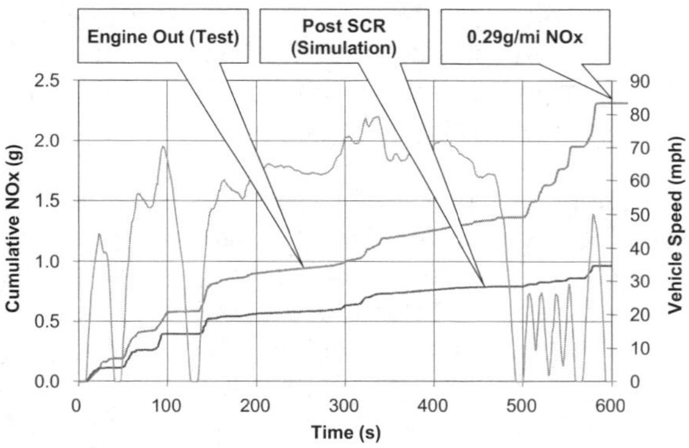

Figure 12 – US06 drive cycle results for 3500lb vehicle

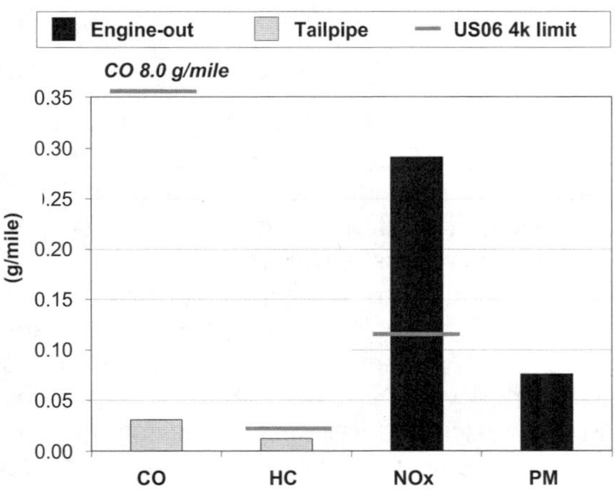

Figure 13 – Overview of US06 emissions results

This example shows the potential for available technology to control engine out emissions in combination with Diesel Particulate Filter technology. Additional NOx aftertreatment reduction of 60-70% is anticipated for LDV applications to meet Bin 5 requirements with engineering margin. The control of engine out emissions to such levels offers the potential for SCR technology to be selected. Service interval refill of Urea becomes a practical solution where no infrastructure is available. In this example, it is estimated that a 5-6L Urea tank would provide twice the volume of Urea necessary for typical passenger car driving over a service interval period of 12,000 miles. Where Lean NOx Trap technology is selected, engine-out NOx reduction offers to limit the critical issues of high system cost, durability sensitivity and fuel/oil dilution.

LEVEL 3 ACTION COMBUSTION TECHNOLOGY

Research is progressing to minimize and potentially eliminate NOx aftertreatment. To achieve this HPCC operation must be enhanced, particularly at the higher load conditions of the US06 drive cycle. The following results illustrate the operation of ultra low NOx on a single cylinder research engine of 0.5 liter capacity. The following engine technology has been applied in this research example:

- Ricardo 16:1 compression ratio combustion system
- Solenoid actuated Common Rail fuel system with maximum rail pressure of 1600 bar
- Optimized intake port geometry with variable swirl control
- Air and EGR system conditions predicted from 1-D engine simulation of new air/EGR system design

The test bed setup used a Ricardo HYDRA single cylinder engine. The installation was capable of controlling intake pressure, temperature and EGR rate. Initial validation of the engine operation was undertaken to match multicylinder test data. In particular, attention was paid to optimization of exhaust orifice and back pressure valve settings. Instrumentation included KISLER 6125 pressure measurement in combination with "Adaptive-CAS" high speed data processing software. Emissions measurement was by HORIBA MEXA 9100DEGR and AVL 415 Variable sampling Smoke Meter. Combustion noise was examined with an AVL 450 Combustion Noise Meter. The paper shows EGR rate calculated from CO_2 measured in the intake and exhaust manifold and Air/Fuel ratio is calculated using dry Spindt [6].

$$EGR = \frac{[CO_2]_{inlet} - [CO_2]_{ambient}}{[CO_2]_{exhaust} - [CO_2]_{ambient}}$$

Note: Where $[CO_2]_x$ is the dry volumetric concentration of CO_2 in the location *x*.

Research has shown that highly pre-mixed combustion can be achieved by two generally distinct approaches. Firstly by an early injection approach, generally described as "HCCI" or a retarded combustion strategy with extended ignition delay. The following section compares these strategies with the hardware described.

Figure 14 shows results at 1500 rpm, 3.0 bar Gross Indicated Mean Effective Pressure (IMEP). Three calibration strategies are shown. In each case an EGR swing is performed from 45 to 55% EGR rate at fixed air/fuel ratio. It should be noted that the hardware and calibration are not fully optimized in this example. The base calibration in this condition is representative of a Euro 4 calibration and operates with AFR of 30:1 and shows the lowest sensitivity to changes in EGR rate.

Two alternative injection strategies have been assessed at this light load condition; an early injection "HCCI" approach and a "Pilot + main" injection strategy. Both tests are performed at a fixed AFR of 20.5:1.

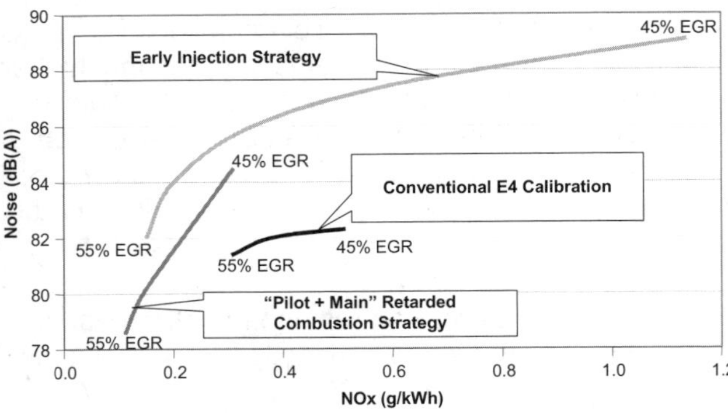

Figure 14: EGR sensitivity with alternative combustion strategies

With the highly pre-mixed "Pilot +main" injection strategy, reduced AFR and retarded combustion enable low NOx to be achieved. However, there is increased noise and NOx sensitivity relative to a typical Euro 4 calibration approach. In this single injection case the early injection strategy results in increased combustion noise and is very sensitive to EGR variation.

The early injection strategy is therefore more sensitive to variations in EGR rate. This is important on a multi-cylinder engine where individual cylinders experience differences in trapped air and EGR levels. Also of concern with such a strategy is the risk of fuel/oil dilution due to fuel impingement on the cylinder walls.

As load increases, it becomes increasingly difficult to apply an early injection strategy. Figure 15 shows an example at 1500 rev/min, 6.6 bar IMEP. At such a condition, a typical Euro 4 engine operates with approximately 26:1 AFR and 16% EGR. The figure shows timing response for two alternative air/fuel ratio levels of 17:1 and 19:1 with fixed EGR rate of 43%. In each case a single injection is applied. The figure shows results at these fixed AFR and EGR conditions. This is an extreme condition, which demands enhanced air and EGR technology to be applied. The two cases illustrate that reduced AFR and retarded timing are necessary to suppress NOx. The result was NOx levels of 0.1 g/kWh, which are approximately 95% lower than production Euro 4 levels. As seen at light load conditions, soot emissions rise with timing advance and subsequently fall. Even with a high EGR rate of 43% and low AFR, testing beyond −22° ATDC resulted in unacceptable cylinder pressures and noise levels (>97dB).

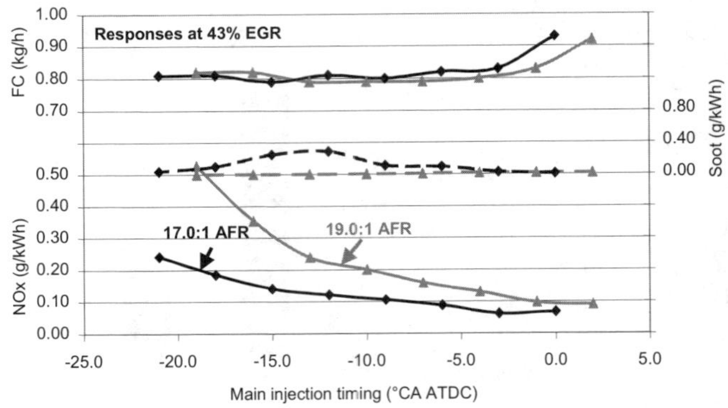

Figure 15: Effect of injection timing at 1500 rpm, 6.6 bar IMEP

The results show that with the conventional hardware applied, the early injection "HCCI" approach is limited to loads below 5 bar BMEP. Additionally the early injection strategy forces the application of a more complex control system to manage variation and perform calibration switching. This is due to the necessity of applying a conventional injection strategy at increased loads. The "pilot + main" injection strategy has been shown to offer improved combustion noise characteristics and requires a less complex control approach.

The operating range for HPCC combustion and the early injection strategy can be improved with the application of multiple injection, reduced cone angle nozzles, reduced compression temperature and pressure. However the following results show the potential without such complex developments. Further development of the combustion system hardware and calibration was performed on a single cylinder engine with a compression ratio of 16:1. Calibration optimization was undertaken using advanced Design of Experiment techniques. Figure 16 shows a summary of the optimum results achieved with the selected hardware.

The results are shown relative to steady state NOx targets for Tier 2 Bin 5 and US06 emissions to be achieved with a 2.0 liter engine powering a 3500 lb (1590 kg) LDV application. The steady state emissions targets at the key point conditions were established using a quasi-steady state emissions prediction. In this method vehicle emissions were predicted using emissions maps applied at expected instantaneous speed and loads during the drive cycle. This characteristic was established from vehicle test results. Transient and cold compensation factors were applied based on Ricardo guidelines for the appropriate drive cycles. The figure shows the target NOx levels predicted to be required to achieve 0.05g/mi and 0.126g/mi engine-out cumulative NOx for the FTP-75 and US06 drive cycles respectively. The results at four key points are shown in terms of gross IMEP (360° crank revolution) and therefore indicate the impact to combustion efficiency. The conversion of Brake targets to Gross has been achieved by applying Friction and Pumping losses predicted by air system simulation using Ricardo WAVE simulation. This incorporates a new air system design which enables the required EGR rates to be achieved.

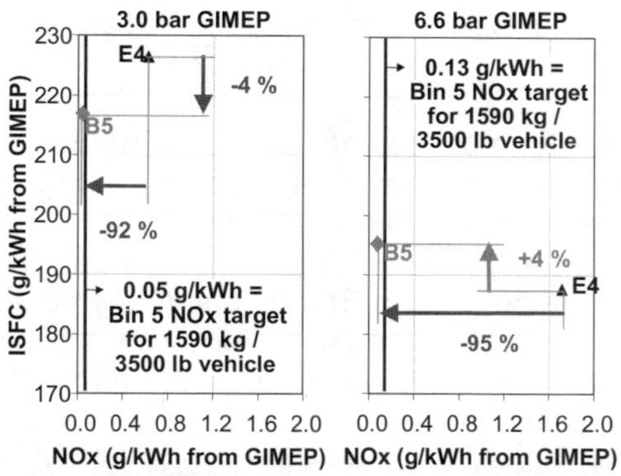

	Euro 4 (3.0 bar)	Bin 5 (3.0 bar)	Euro 4 (6.6 bar)	Bin 5 (6.6 bar)
Soot (g/kWh)	0.10	0.00	0.04	0.03
HC (g/kWh)	0.6	4.3	0.2	0.6
CO (g/kWh)	7.7	21.3	1.9	6.1
Noise (dBa)	83.6	80.0	83.5	91.0

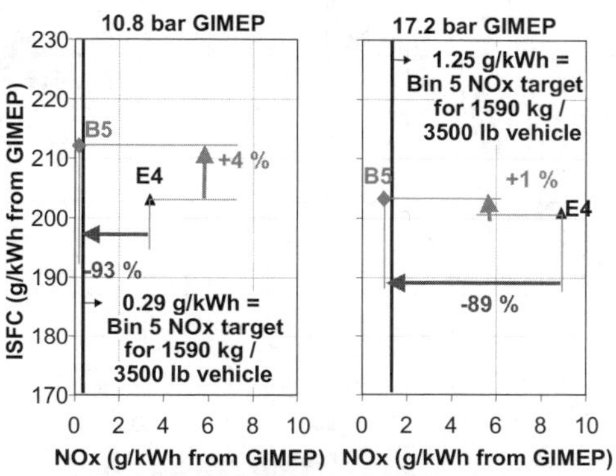

	Euro 4 (10.8 bar)	Bin 5 (10.8 bar)	Euro 4 (17.2 bar)	Bin 5 (17.2 bar)
Soot (g/kWh)	0.03	0.15	0.01	0.03
HC (g/kWh)	0.2	0.2	0.1	0.1
CO (g/kWh)	1.0	1.7	0.5	0.7
Noise (dBa)	85.2	90.5	85.1	88.0

Figure 16: Optimized Bin 5 single cylinder results

The charts show that target NOx levels were achieved. At light load, NOx and soot enhancement was possible. At very light load improvement in combustion efficiency was also possible. However as load increased, efficiency and noise became more challenging. In combination with over 90% NOx reduction, HC and CO levels were significantly increased with the suppression of combustion temperatures.

Figure 17 shows the full load results possible with the improved hardware and combustion system design within a maximum cylinder pressure of 150 bar and a maximum rail pressure of 1600 bar. As well as reducing the soot emissions at part load, the combustion system demonstrated the capability of achieving 60 kW/liter rated power. The smoke emissions were also below 1.5 FSN. Of importance is the relative improvement in the trade-off characteristic between smoke and AFR. This is a function of detailed optimization of the reduced compression ratio combustion system design. This enhancement reduces the boost pressure demand for the engine and enables air system optimization for part load conditions. The combination of improved combustion system performance and air system efficiency are fundamental to delivering the test results observed. The ACTION strategy therefore enables both emissions and performance to be achieved while limiting maximum cylinder pressure and engine weight.

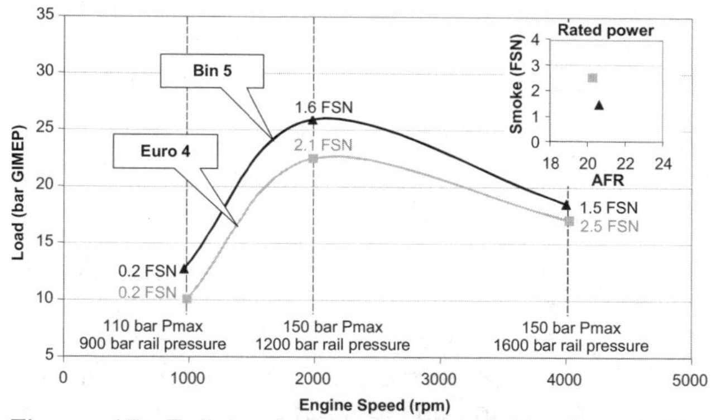

Figure 17: Full load comparison of Bin 5 and Euro 4 engines

KEY TECHNOLOGIES UNDER DEVELOPMENT

Fuel System Technology

This paper shows results achieved with fuel systems of limited capability. New fuel system technology offers improved pressure capability and control of injection events. It is predicted that re-optimization will result in improved part load results without compromising full load performance.

The following developments offer exciting potential and are under assessment for addition to ACTION level 3 and 4 solutions:

- 1800-2000 bar Piezo FIE (level 3)
- Hydraulically amplified fuel systems (level 3)
- Variable nozzle technology (level 4)
- Variable cone angle (level 4)

Air and EGR System Technology

Of fundamental importance to achieve ultra low NOx is the delivery of increased EGR while maintaining lean operation. Figure 18 shows an example of the trend with low NOx technology. Air and EGR system developments are delivering increased EGR while combustion system enhancements enable efficient, clean combustion at low levels of oxygen concentration. The combined effects is low temperature and highly pre-mixed combustion.

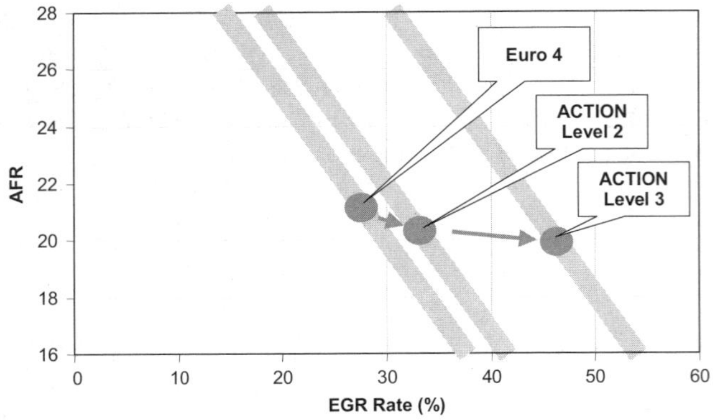

Figure 18: Overview of approach to lowering Oxygen concentration

The air/EGR requirements are driving innovative air and EGR solutions. Increased EGR delivery must be achieved while limiting parasitic friction and pumping losses. Air and EGR systems are being developed to meet the demands of the single cylinder results shown. Multi-cylinder test programmes are now progressing and vehicle development will follow. Figure 19 summarizes simulation results and illustrates how advanced air systems can be tuned to deliver the best compromise between rated power and EGR delivery. In summary, Ricardo expect the trend to be from single stage to two stage air systems. Also the demand for increased EGR cooling will encourage the application of long route EGR technology.

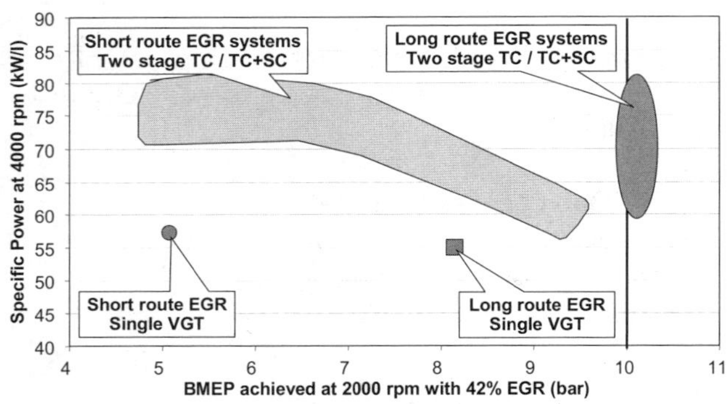

Figure 19: Overview of Air/EGR system capability to enhance EGR delivery and performance

Control System Technology

While combustion system characteristics are continuing to advance, to fully capitalize on this performance requires control system enhancements. As previously described, advanced combustion concepts will be more sensitive to calibration and hardware variations. Fundamental to delivering production solutions will be improved control strategies. The following key development areas are being progressed:

- Improved transient control of air and EGR systems
- Combustion control
- Model based control of engine parameters
- Adaptive engine control with changes to hardware

The most exciting area of current development is closed loop combustion control through the application of new cylinder pressure sensors and increased ECU processing power. The application of such innovations opens new possibilities for the optimization of diesel engines.

In this regard, Ricardo has developed CPEMS (Cylinder Pressure Engine Management System). Figure 20 shows one element of what is possible with closed loop combustion control. In this multi-cylinder engine, balancing of the 50% burn (B50) event in each individual cylinder has been implemented. Following activation of the closed loop combustion controller, the main injection timing is adjusted on each cylinder to target the B50 set point. In particular the figure shows how the cylinder to cylinder variation in B50 is reduced using closed loop control.

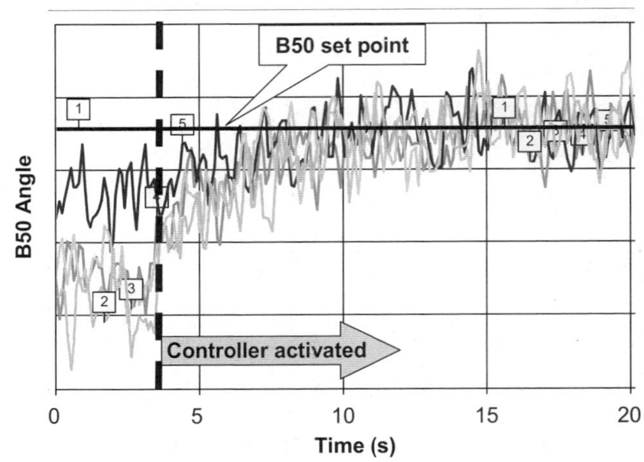

Figure 20: B50 control using Ricardo CPEMS

Figure 21 shows one potential benefit of such capability. Two injection timing swings are plotted, with and without CPEMS active. In the closed loop case the B50 set-point is varied. The results show how in this simple case, the soot is reduced at the same NOx level. Through such an approach improved matching of the combustion in each cylinder can enable improved emissions, fuel economy and noise optimization.

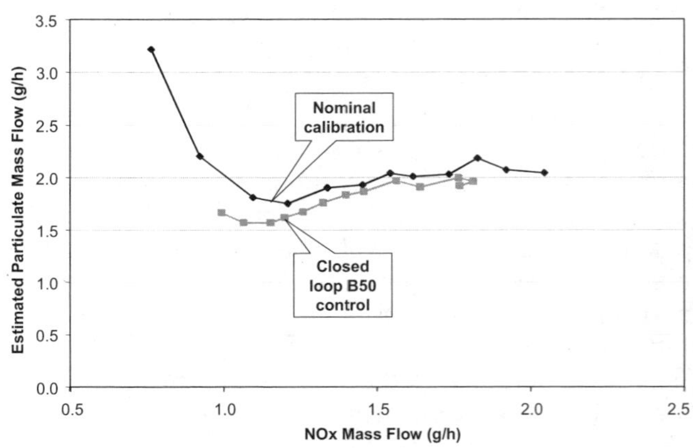

Figure 21: B50 control at 2500 rpm, 8 bar BMEP

CONCLUSIONS

The results reported in the paper has led to the following conclusions:

- A practical roadmap of technologies is available that will progressively lower diesel engine emissions at source
- Highly Premixed Cool Combustion offers to significantly lower NOx emissions while controlling combustion efficiency and soot emissions. Combustion noise and increased products of incomplete combustion become increasingly challenging at ultra low NOx levels.
- By 2007, Bin 5 emissions will be achieved through the application of Level 2 ACTION technology in combination with 60-70% NOx reducing aftertreatment. Service interval refill of Urea becomes practical with SCR systems.
- Technology will advance to deliver Tier 2 Bin 5 emissions without NOx aftertreatment in LDV applications. Reduced compression ratio combustion system, advanced air/EGR system and combustion control will be key enabling technologies.

ACKNOWLEDGMENTS

The Authors would like to thank the Directors of Ricardo for permission to publish this paper and their colleagues for the work that generated the results. The authors would also like to thank the following for their technical contribution to the paper: Olivier Laguitton, Richard Cornwell, Anthony Truscott and Tomasz Salamon.

ABBREVIATIONS

ACTION	Advanced Combustion Technology to Improve engine-Out NOx
AFR	Air Fuel Ratio
ATDC	After Top Dead Center
A/T	Aftertreatment
B50	50% Burn
BMEP	Brake Mean Effective Pressure
CA	Crank Angle
CO	Carbon Monoxide
CPEMS	Cylinder Pressure Engine Management System
CR	Compression Ratio
DPF	Diesel Particulate Filter
EGR	Exhaust Gas Recirculation
FC	Fuel Consumption
FIE	Fuel Injection Equipment
GIMEP	Gross Indicated Mean Effective Pressure
HC	Hydrocarbons
HCCI	Homogeneous Charge Compression Ignition
HPCC	Highly Premixed Cool Combustion
IMEP	Indicated Mean Effective Pressure
ISFC	Indicated Specific Fuel Consumption
LDT	Light Duty Truck
LDV	Light Duty Vehicle
LNT	Lean NOx Trap
NMHC	Non Methyl Hydrocarbons
NOx	Nitrogen Oxides
PC	Passenger Car
Rs	Swirl Ratio
SC	Supercharger
SCR	Selective Catalyst Reduction
SOI	Start Of Injection
TC	Turbocharger
VGT	Variable Geometry Turbocharger

REFERENCES

1. NAJT, P.M., FOSTER, D.E. "Compression-Ignited Homogeneous Charge Combustion" SAE 830264
2. THRING, R. H. "Homogeneous-Charge Compression Ignition (HCCI) Engines", SAE 1989-2068
3. KIMURA, S.; AOKI, O.; KITAHARA, Y.; AIJOSHIZAWA, E. "Ultra-clean Combustion Technology Combining a Low-Temperature and Premixed Combustion Concept for Meeting Future Emission Standards", SAE paper 2001-01-0200
4. SEABROOK, J.; COOPER, B.G.; PATTERSON, J.; EDWARDS, S. "Optimisation Developments in Diesel Engine Calibration". Design of Experiments in der Motorentwicklung, Berlin, Dec 2001.
5. DOWNES, T.A., RAWLINS, D.J., "Development of HSDI Combustion Systems Using CFD Analysis Techniques", Fuel Injection Systems – I Mech E Conference Proceedings, November 2002

6 SPINDT, R.S. "Air-Fuel Ratios from Exhaust Gas Analysis" SAE 650507

7 WHELAN, S., COOPER, B.G. A New Integrated Approach for Diesel Engines to Deliver High Performance and Low Emissions for 2010, , 9th EAEC International Congress, Paris, C109, June 2003.

8 COOPER, B.G. PENNY, I.J., WHELAN, S. "A New Diesel System Methodology to Meet the Demands of Low Engine-out NOx and High Performance for Future Global Markets" 20th International VDI/VW Conference, Wolfsburg, VDI-Berichte Nr. 1808, 4th December 2003.

9 COOPER, B., JACKSON, N., PENNY, I.J., RAWLINS, D., TRUSCOTT, A, SEABROOK, J. "Advanced Combustion System Design, Control and Optimisation to Meet the Requirements of Future European Passenger cars" Thiesel 2004, Valencia, September 2004

10 RINDONE, G., KEENAN, M., JACKSON, N., GORDON R.L., BEASLEY, M. "Achieving Gasoline Euro 4 Emissions with a Diesel Passenger Car ", JSAE 2005-5431.

11 GREANEY, A. "Ricardo ACTION Strategy. An Enabling Technology for the US Market", 11th DEER conference, Chicago, Aug 2005

2007-01-1128

Combined Particulate Matter and NOx Aftertreatment Systems for Stringent Emission Standards

Stefan Pischinger, Thomas Körfer, Andreas Wiartalla and Jürgen Schnitzler
FEV Motorentechnik GmbH

Dean Tomazic and Marek Tatur
FEV Engine Technology, Inc.

ABSTRACT

The HSDI Diesel engine contributes substantially to the decrease of fleet fuel consumption thus to the reduction of CO_2 emissions. This results in the rising market acceptance which is supported by desirable driving performance as well as greatly improved NVH behavior. In addition to the above mentioned requirements on driving performance, fuel economy and NVH behavior, continuously increasing demands on emissions performance have to be met.

From today's view the Diesel particulate trap presents a safe technology to achieve the required reduction of the particle emission of more than 95%. However, according to today's knowledge a further, substantial NOx engine-out emission reduction for the Diesel engine is counteracts with the other goal of reduced fuel consumption.

To comply with current and future emission standards, Diesel engines will require DeNOx technologies. The NH_3-SCR and Lean NOx-Trap (LNT) technologies are currently the most promising solutions to meet stringent NOx standards. While the NH_3-SCR shows benefits considering fuel consumption penalty, the usage of an additional reduction component is considered as a backdraw.

Introduction of DeNOx technologies in combination with DOC/DPF application requires a detailed and thorough analysis of exhaust system layout at the very beginning of the engine development. To define the appropriate technology for different applications, modeling and simulation of emission and fuel consumption are mandatory.

This paper will discuss the simulation results for aftertreatment components such as DOC, DPF, SCR and LNT and the impact of different exhaust system layouts on emission reduction.

INTRODUCTION

Due to its superior fuel economy in combination with excellent driving performance, good acoustics as well as exhaust emissions, the Diesel engine has achieved market shares of about 50% in Europe in the recent past and is also raising increasing interest in the U.S. These significant improvements have been achieved by applying technologies such as common-rail, turbo-charging, cooled EGR as well as sophisticated control algorithms. In order to further establish the image of a clean engine, the particulate filter is becoming a state-of-the-art technology for Diesel engines.

Nevertheless significant further improvements on the engine-out emissions as well as exhaust aftertreatment technologies will be required in order to meet future stringent emission standards such as EU6 or U.S. Tier2. In addition to the state-of-the-art Diesel oxidation catalyst (DOC), the Diesel particulate filter (DPF) as well as high-efficient NOx aftertreatment technologies will become mandatory in the future.

Thereby the reduction of the nitrogen oxide emissions (NOx) of Diesel engines will become one of the greatest development challenges. Prime goal is to maintain the Diesel engine as a propulsion source with highest fuel economy for the future. From todays point of view only lean NOx trap (LNT) and SCR technology present promising capabilities to achieve the required NOx reduction targets. SCR technology has been recently introduced into the market for HD applications in Europe as well as Japan /1/, /2/ and is also announced for market introduction for passenger cars and light-duty trucks in Europe as well as U.S. LNT technology has been introduced for lean-burn gasoline engines in Europe and Japan as well as for Diesel engines as the integrated DPNR technology approach by Toyota /3/. Besides pure LNT and SCR system concepts currently also combinations of LNT and SCR technology are published, such as by DaimlerChrysler for the E320 Tier

2 Bin 8 BLUETEC concept /4/ or Honda with regard to their Tier 2 Bin 5 aftertreatment concept /5/, /6/.

In the course of its rising market acceptance the high-speed DI (HSDI) Diesel engine contributes substantially to the decrease of fleet fuel consumption thus in the reduction of CO_2. Any typical internal measures in order to reduce exhaust emissions – especially NOx emissions – inherit increasing fuel consumption without compensation measures and thus counteract the realization of CO_2 emission goals. Figure 1 points out, that NOx reduction only by internal measure (such as increased EGR or retarded injection timing) is directly combined with significant drawbacks in fuel consumption. Therefore an optimized integration of improved engine-out emissions as well as a highly efficient aftertreatment technology is indispensable for a tailored integrated approach in order to achieve future stringent emission standards while maintaining excellent fuel economy.

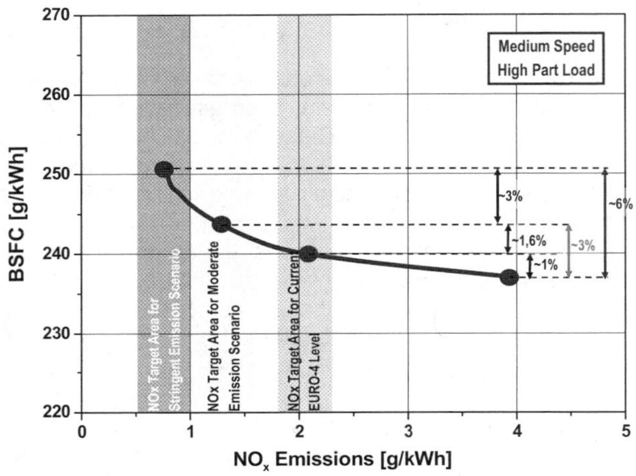

Fig. 1: Impact of conventional internal NOx reduction measures on BSFC

NOX AFTERTREATMENT TECHNOLOGIES

From today´s point of view only lean NOx trap (LNT) and SCR technology represent promising solutions to achieve future NOx reduction targets. The operating principle as well as the specific pros and cons of the two different technologies will be described in more detail in the following chapters.

LNT technology

The lean NOx trap (LNT) or NOx adsorber catalyst (NAC) is a discontinuously operating aftertreatment technology and is characterized by the following operating modes (Figure 2):

- NOx storage during lean engine operation
- NOx reduction during rich operation phases
- LNT desulfurization under rich conditions and high temperatures

The most challenging operating modes of a LNT under real transient conditions are the rich operation of the Diesel engine, the transitions between lean and rich operation, the desulfurization process as well as control of the LNT system including sensors.

Due to its operating principle the mixture-controlled (quality-controlled) Diesel engine operates with a high amount of excess oxygen, particularly under part-load, which is most relevant especially for light-duty applications. A thorough optimization of the Diesel engine air and fuel management is required in order to reduce the high amount of excess oxygen in the rich operating condition. The removal of oxygen is essential for a successful regeneration of the NOx adsorber catalyst /7/, /8/. Therefore efficient calibration methodologies based on DoE (Design of Experiments) approaches, which consider aspects like oil dilution, component limits, CO/HC-ratio, black smoke emissions, combustion noise, sensor requirements as well as the requirement for smooth maps and map transitions during the calibration process are necessary.

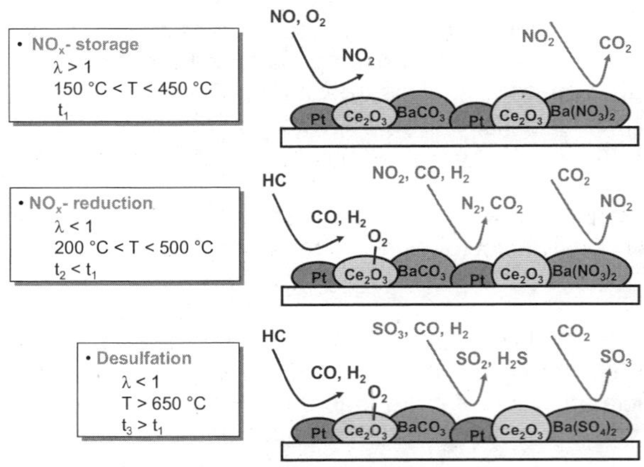

Fig. 2: LNT operating modes

Furthermore sophisticated and robust control functionalities for the transition between lean and rich engine operation have to be developed and implemented. Besides the definition of a suitable control strategy, an appropriate sensor setup represents an important task. This includes detailed assessment of the sensor signal quality with regard to the demands for LNT system control in order to enable a control strategy targeting for high NOx-conversion efficiencies with lowest fuel consumption penalty and also enabling the detection of thermal aging as well as sulfur poisoning of the LNT /19/.

Efficient desulfurization of current state-of-the-art LNT technologies require high temperatures of at least 650°C in combination with locally rich exhaust gas conditions within the catalyst. To avoid severe thermal aging of the LNT, sophisticated temperature control strategies have to be developed. The major portion of sulphur will be released as H_2S from the LNT during an efficient desulfurization event. Because significant amounts of

H_2S cannot be emitted into the atmosphere, a post-oxidation of the H_2S has to be ensured. This can be achieved by a suitable management of oxygen downstream of the LNT; e.g. using a control strategy based on the oxygen storage capacity of any catalyst placed downstream LNT in the exhaust line.

The main challenges of LNT technology can be summarized as follows:

- DeNOx regeneration by engine internal measures in terms of drivability and driver transparency
- Limited DeNOx regeneration operation area
- Sulphur poisoning / desulfurization
- Reliable desulfurization strategy
- Long-term stability / thermal aging
- DeNOx and DeSOx management / complexity of aftertreatment control

SCR Technology

SCR technology is based on reduction of NOx by ammonia, which has to be generated on-board from a suitable reductant (Figure 3), since a transportation of pure ammonia is not desired or advisable. Today the usage of urea/water solution ("Adblue") is most common for SCR systems. Aside of the liquid urea also alternative system concepts based on ammonia carbamate /9/ or solid urea /10/ are under development.

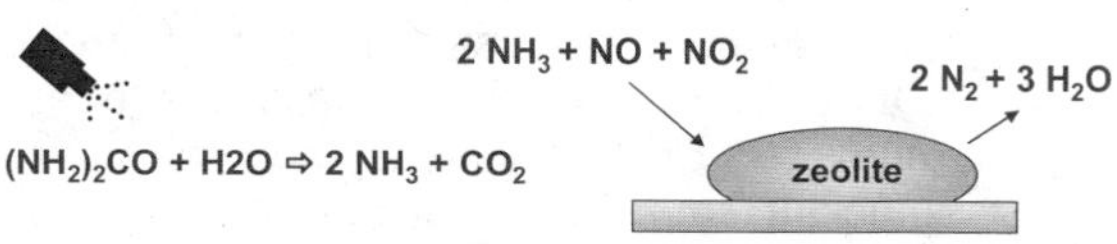

Fig. 3: SCR operating modes

For passenger car and light-duty applications current concepts are utilizing non-air assist urea injection systems and zeolite type catalysts. For the urea dosing, a reliable operation of the injector and a uniform distribution of urea in the exhaust gas stream has to be achieved under all relevant operating conditions. Furthermore the dependency of low temperature activity on NO_2 content in the exhaust as well as the formation of exhaust components such as N_2O, NH_3, formic acid or iso-cyanuric acid has to be examined carefully. SCR catalysts are characterized by complex processes of ammonia storage, NOx conversion as well as ammonia slip and therefore detailed characterisation of the SCR catalyst as well as sophisticated control algorithms are required in order to achieve high NOx conversion rates while maintaining low ammonia slip under transient operating conditions.

The main challenges of SCR technology can be summarized as follows:

- Reliable urea injection
- Uniform ammonia distribution in the exhaust
- NOx neutral SCR-catalyst heating-up strategy
- Dosing strategy
- Ammonia slip
- Vehicle package
- System costs

In the following chapter the specific attributes of SCR and LNT technology considering the arrangement in the exhaust line will be pointed out.

Comparison of LNT and SCR Technology

In order to achieve lowest NOx tailpipe emissions over the desired emission certification test cycles, the NOx conversion efficiency as a function of exhaust temperature is one of the most determining factors. Nevertheless, the specific boundary conditions and limitations of either LNT as well as SCR technology have to be considered in order to allow a fair comparison of the two aftertreatment technologies.

Systems configurations with a DeNOx system placed upstream of a CDPF, which are expected to be more beneficial with regard to NOx reduction performance under lower engine-out temperature conditions, will be more common in light-duty vehicle applications. A more detailed assessment of combined NOx aftertreatment and DPF systems will be performed in the following section.

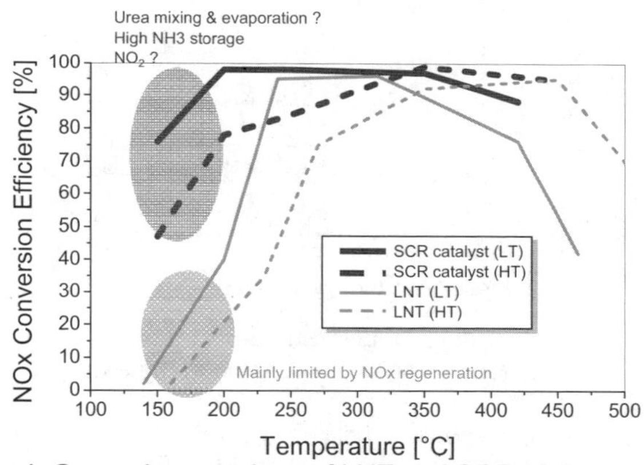

Fig. 4: General comparison of LNT and SCR technology

Figure 4 shows typical idealized NOx conversion efficiency curves for LNT and SCR systems. For both technologies a more ´low temperature´ oriented coating technology as well as a more ´high temperature´ oriented coating technology are depicted. Considering the temperature dependency of the NOx conversion efficiencies, there is a clear advantage of the SCR technology, as SCR can achieve highest efficiencies at low temperature as well as high temperature areas. However with regard to low temperature activity some important differences between SCR and LNT technology have to be pointed out: High NOx conversion efficiencies can only be achieved using the SCR technology with NO_2 ratios of about 50% in the exhaust as well as significant ammonia storage on the SCR catalyst. This requires a sophisticated control strategy to avoid ammonia slip. Besides that, the evaporation and mixing of the liquid urea/water solution has to be guaranteed under such low exhaust temperature conditions. In contrast to that, the

LNT technology is mainly limited with regard to NOx regeneration in the low temperature region, whereas LNT catalysts can store significant amounts of NOx even at low temperatures. This first consideration already points out, that a more detailed assessment of the different technologies is required in order to select a suitable technology for a specific application.

Furthermore it has to be taken into account, that an SCR catalyst cannot be positioned really close-coupled to the turbocharger outlet, as there is always a certain length in the exhaust line required for the urea injection and appropriate mixing. In contrast to that the LNT catalyst can be positioned comparably close-coupled to the turbine outlet position if the vehicle package conditions provide the necessary volume /20,21,22/. Due to this fact the exhaust temperature at SCR catalyst position will always be less compared to the close-coupled LNT position, especially during the warm-up phase after cold start. In Figure 5 this effect is considered by a temperature decrease of about 60°C between SCR and LNT catalyst position. As a result it becomes obvious, that the close-coupled LNT and the under-floor SCR show nearly the same low temperature performance.

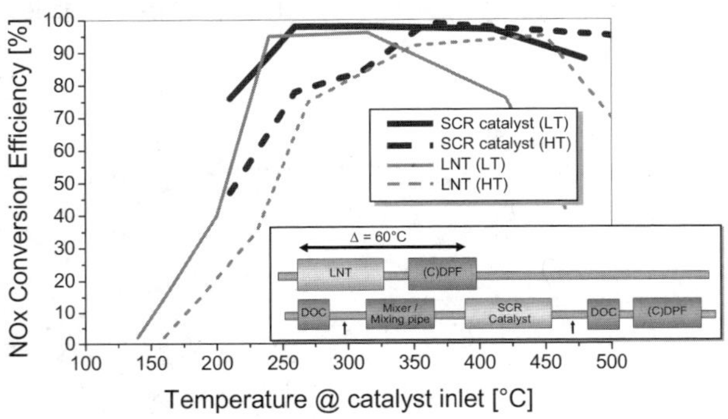

Fig. 5: Comparison of LNT and SCR technology considering possible position in the exhaust

The following scenario utilizes a split LNT design which is not be possible for the SCR system. As pointed out in Figure 6, such an approach provides the potential to extend the operating window of the LNT concept. In this configuration the close-coupled LNT brick is used to convert the NOx in the low temperature area (e.g. over the FTP75), whereas the under-floor LNT provides the required efficiencies under high temperature operating conditions (e.g. the US06). Further improvements might be achieved with an optimized coating for both bricks.

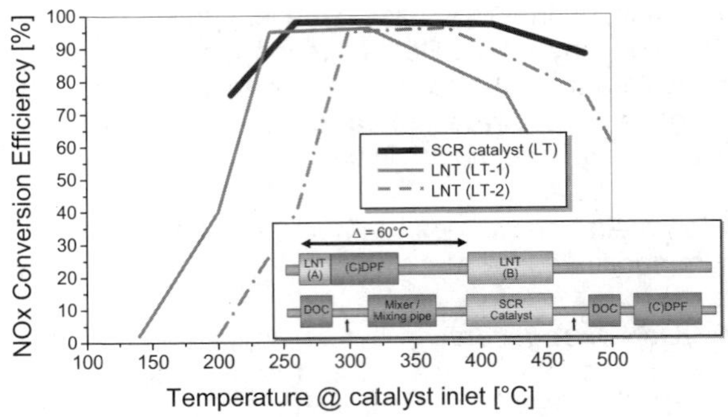

Fig. 6: Comparison of LNT and SCR technology considering split of LNT catalyst into two bricks

Heating strategies

Both NOx aftertreatment technologies require a heating strategy in order to improve the low temperature performance and achieve high NOx conversion rates a short time after engine start as well as for low-load operating. These operating conditions are dominant during the European (NEDC) as well as US (FTP75) test cycles. As discussed above, the current state-of-the-art SCR catalysts show comparably low NOx conversion for exhaust temperatures below 200°C and the evaporation of urea and mixture formation in the exhaust remains difficult under these conditions as well. Therefore heating directly after engine start-up is required. In contrast to that, LNT technology mainly requires a heating strategy for DeNOx regeneration in order to provide a sufficient amount of reductants and to avoid excessive HC slip. In this case heating will be mainly required in case of a DeNOx demand. Thus the demands for heating and also the impact on fuel consumption will be completely different for SCR compared to LNT technology. With regard to the overall NOx reduction performance, such heating strategies have to be tailpipe emission neutral in any case and should not contribute to any significant oil dilution.

Based on these fundamental considerations it can be stated, that LNT/SCR catalyst conversion efficiency as a function of temperature cannot be taken as the decisive factor for selection of a suitable NOx aftertreatment technology. Further aspects like the combination of DeNOx and DPF technology, the specific vehicle boundary conditions as well as the desired test cycles have to be taken more into account as well.

An assessment of the different NOx aftertreatment approaches for combined DPF and NOx aftertreatment systems will be performed in the following section. A more detailed comparison of SCR and LNT technologies based on real vehicle measurement data and simulation of temperatures in the exhaust line as well as tailpipe NOx emissions will be presented and discussed later within this paper.

COMBINED PARTICULATE AND NOX AFTERTREATMENT SYSTEMS

For future applications any NOx aftertreatment system will have to be considered in combination with a DPF which can be placed up- or downstream the NOx aftertreatment unit. In contrast to SCR systems, the LNT provides the potential to be placed the in close-coupled location to the engine and/or to split the catalyst into two separate bricks. The different system concepts are layed out in Figure 7.

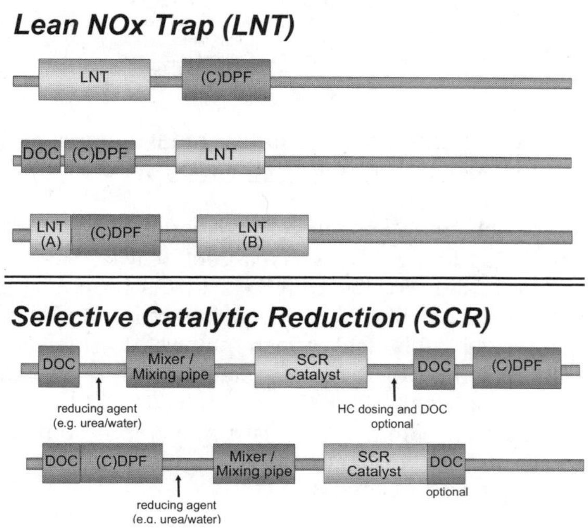

Fig. 7: Concepts for combined DPF and NOx aftertreatment systems

Figure 8 shows an assessment of the different concepts with regard to important factors such as:

- Low / high temperature activity
- Effort for active DPF regeneration
- Possibility for soot oxidation by NO_2
- Risk and degree of thermal aging
- Risk of sulfur poisoning / effort for desulfurization
- Potential for NOx reduction during DPF regeneration (with regard to emission adjustment factors)
- Packaging demands / system complexity
- Potential with regard to fuel consumption

It has to be mentioned, that cost as well as infrastructure related issues are not considered in this assessment.

A comparison and assessment of the different system concepts shows, that there are specific pros and cons for each system. Different approaches result in the optimum choice for each individual application. The LNT technology is favoured with regard to packaging conditions and therefore for small vehicle applications. LNT systems have to be considered as bearing significantly higher risk with respect to thermal aging and sulfur poisoning. In general the LNT provides advantages during low temperature operation, whereas a system layout with SCR downstream CDPF is the only concept enabling significant NOx conversion during DPF regenerations. With regard to DPF related aspects, the LNT has to be considered as more beneficial compared to SCR technology. Thereby SCR is the only technology which even can offer potential for fuel consumption benefits in case of limited NOx reduction demands.

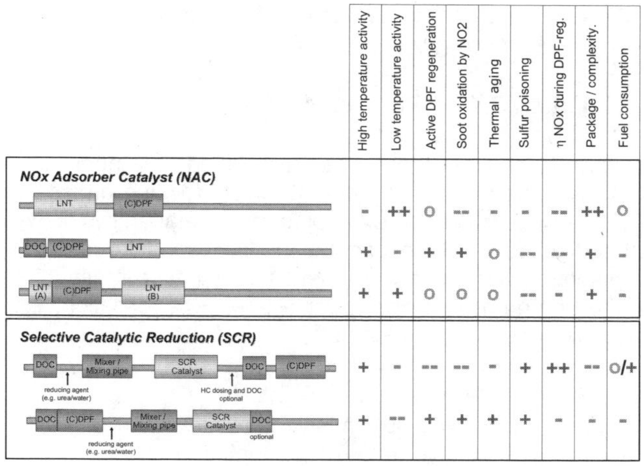

Fig. 8: Assessment of concepts for combined DPF and NOx aftertreatment systems

As a conclusion, any aftertreatment system layout has to be addressed with specific weighting factors for each individual vehicle application.

Once the required NOx conversion rates, the specific boundaries of the legislative test cycles as well as the available space in the vehicle are quantified as decisive factors for the selection of the appropriate NOx aftertreatment technology, the following general trend can be observed: The SCR system is proposed as first choice for large vehicles which require high NOx conversion efficiencies over high vehicle mileage (such as SUVs for US Tier 2 Bin 5 emission standards). The LNT technology is considered as an attractive alternative for smaller vehicles with lower NOx reduction efficiency demand (e.g. for EU5 and post EU5 legislation), see Figure 9. Therefore the selection of the aftertreatment technology as well as exhaust system layout is required for each specific application, which has to be supported by simulation of the exhaust system including temperatures in the exhaust system as well as emissions in different driving cycles.

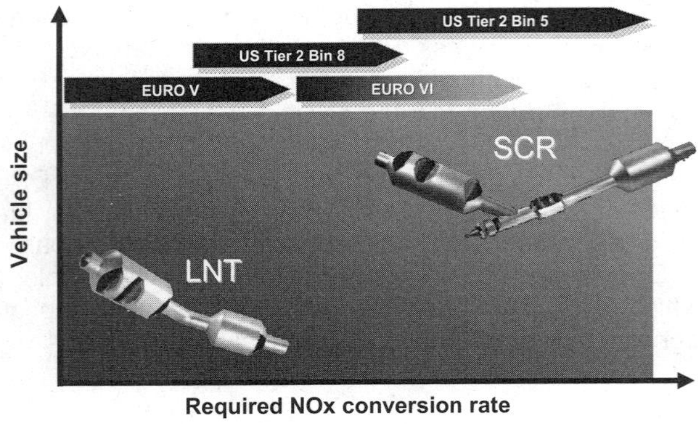

Fig. 9: Development trends for LNT and SCR technology

SIMULATON APPROACH

In the following a tool for fast simulation of temperatures in the exhaust line as well as vehicle tailpipe emissions will be presented and examples for simulation results will be discussed. Such simulations are considered to be essential with regard to selection of a suitable NOx aftertreatment technology for a specific application, e.g. considering factors such as:

- Vehicle inertia weight
- Engine-out emission profile
- Vehicle packaging situation
- Driving cycle

Simulation of the temperature profile along an exhaust system is critical to calculate the reduction performance of DeNOx systems. The tool used for the calculations within this paper is based on a Simulink model, which provides the possibility for fast and easy abstraction of an exhaust aftertreatment design (Figure 10).

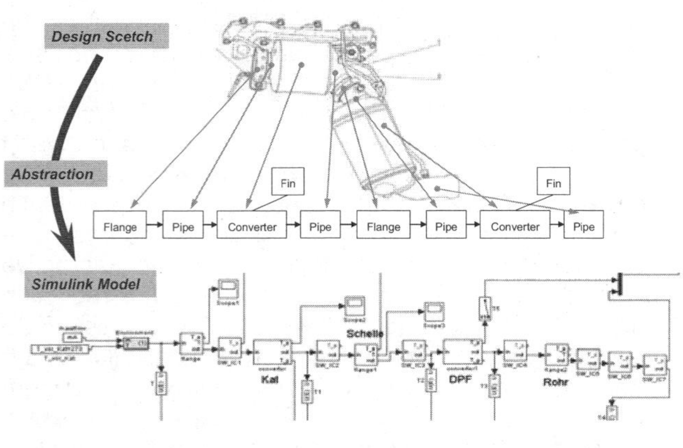

Fig. 10: Aftertreatment simulation - Simulink temperature model

Figure 11 depicts the high level of agreement between measured and simulated temperatures along the exhaust systems especially also at the beginning of the FTP75 driving cycle. Heat releases by chemical reactions as well adsorption and desorption of gaseous compounds on the catalyst surface are implemented by parameterized functions dependent on brick temperature and gaseous matrix.

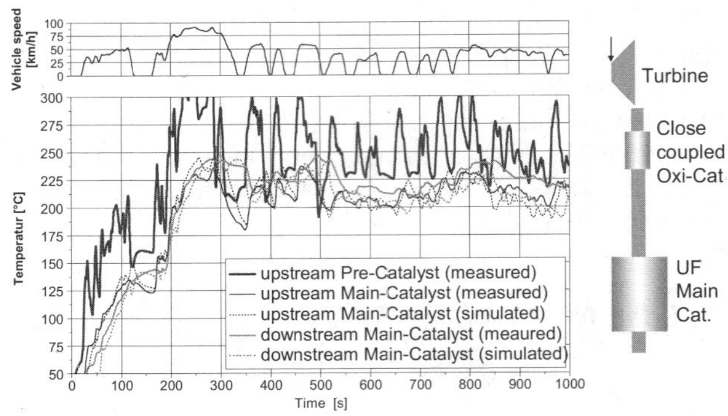

Fig. 11: Comparison of temperature measurement and simulation during FTP75-driving cycle

Overall each catalytic brick in the exhaust system can be divided in up to 3 zones with different activity. The basis of the simulation for the DeNOx systems are laboratory experiments on FEV's own engineered catalyst test bench. Within this test bench catalyst bricks with a maximum length of 8" and a diameter of 2" at space velocities between 10,000 and 125,000 1/h can be evaluated with different exhaust gas compositions. It is possible to run validation tests within a temperature range between 25°C and 800°C.

The evaluation results of a typical fresh (de-greened) low temperature LNT are used as an input for the driving cycle simulations, such as NOx storage capacity as a function of temperature, stored NOx and NO_2/NOx ratio /14/, /15/, /17/. The regeneration efficiency for the LNT is simulated as a function of temperature. The management of DeNOx (rich exhaust gas) regeneration is simulated by limitation of maximum fuel consumption penalty (in this case 3.5% are used) and of the maximum amount of stored NOx (in this case 0.3 g are used). The temperature dependent NOx storage capacity of the LNT used for the calculations within this paper is shown in Figure 12.

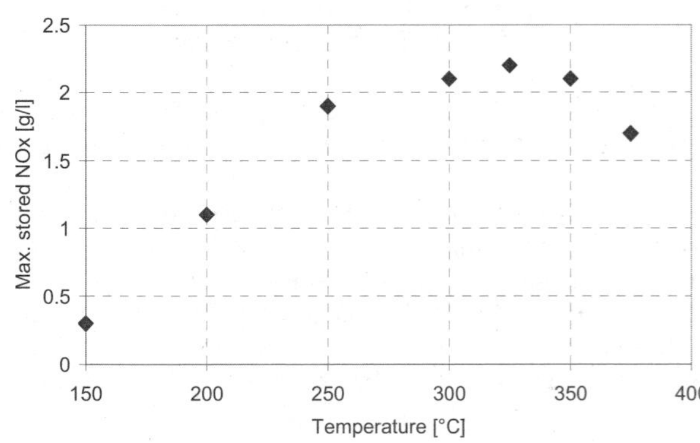

Fig. 12: NOx storage capacity of the fresh LNT as a function of temperature, SV 50,000 1/h

As an input for the SCR zeolite based catalyst the NH_3 storage behavior as a function of temperature and stored NH_3 as well the NOx conversion efficiency as a function of temperature and stored NH_3 are used in the model simulation /16/, /18/. Figure 13 illustrates the maximum NOx conversion efficiencies at SV 50,000 1/h.

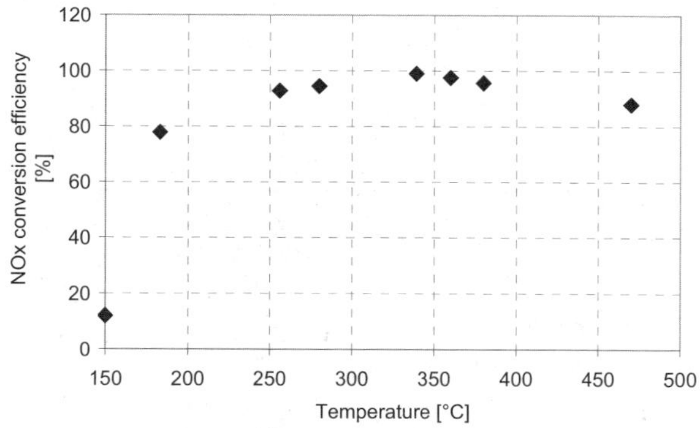

Fig. 13: Maximum NOx conversion efficiency of the fresh SCR catalyst as a function of temperature, SV 50,000 1/h, Feed ratio = 1.0, at max. NH_3 storage

SIMULATON RESULTS

The simulation is performed for the FTP75 driving cycle for different vehicle applications:

Curb weight [kg]	Engine displacement [l]	Engine-out NOx over FTP 75 [g]
1470	2.0	2.51 (0.23 g/mi)
1700	2.0	3.31 (0.30 g/mi)
2270	3.0	3.62 (0.33 g/mi)

The engine-out NOx emissions are based on combustion system developments for European applications and reflect what can be achieved without complex internal NOx reduction measures, such as low pressure EGR. These emission levels can be seen as an intermediate step towards U.S. Tier 2 Bin 5 combustion system developments and should be considered only as the basis for this simulation work. U.S. Tier 2 Bin 5 combustion system developments will most likely further reduce the NOx engine-out emission levels.

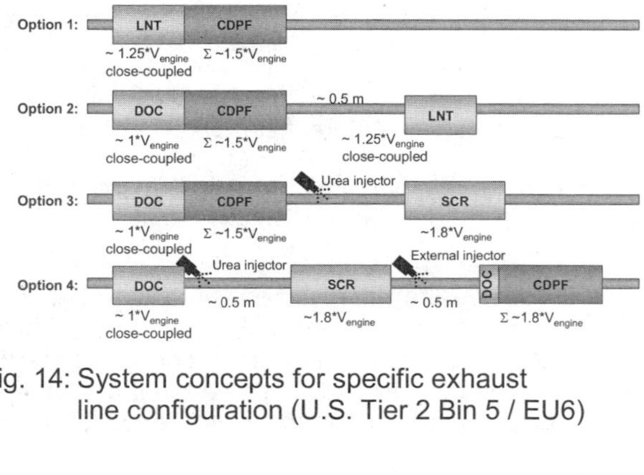

Fig. 14: System concepts for specific exhaust line configuration (U.S. Tier 2 Bin 5 / EU6)

DeNOx conversion performance is simulated for the different vehicle applications with the exhaust system layouts shown in Figure 14.

Figure 15 and 16 show two examples of the simulated temperature and accumulated NOx emission profiles during the FTP75 cycle for a vehicle with a curb weight of 1470 kg. The soaking time of 10 minutes after 1379 s is not shown in the graphs, but considered within the simulation.

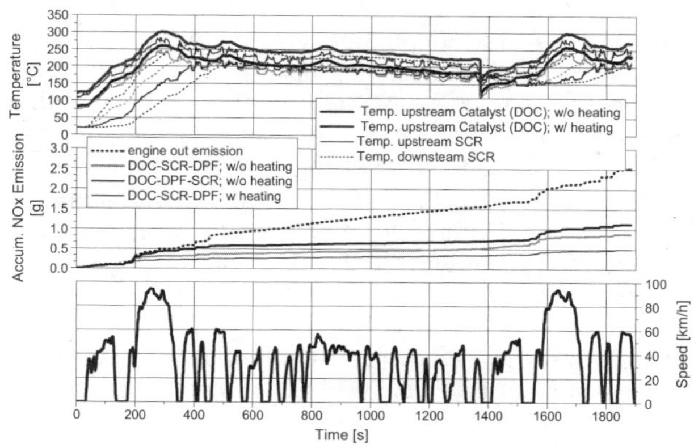

Fig. 15: Emission and temperature profile during FTP75 cycle with different SCR-catalyst configuration, curb weight 1470 kg

The impact of exhaust system configuration on the NOx tailpipe emission profile can be seen clearly. For both DeNOx systems the NOx tailpipe emission can be heavily reduced, when the DeNOx catalyst is positioned upstream of the DPF.

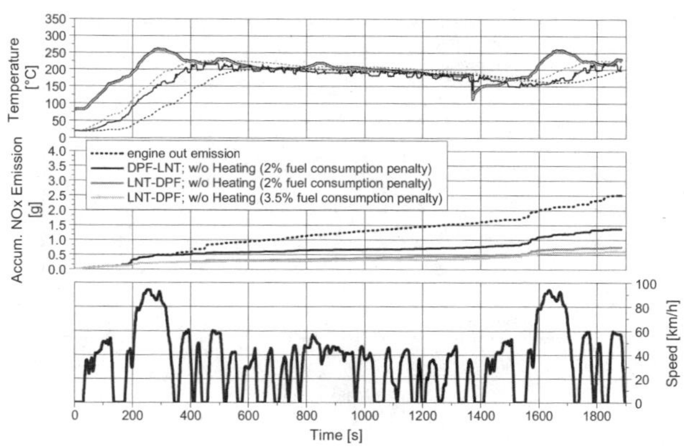

Fig. 16: Emission and temperature profile during FTP75 cycle with different LNT-catalyst configuration, curb weight 1470 kg

An increase of fuel economy penalty for the LNT system (a result of more frequent LNT regenerations) leads to a further decrease in NOx tailpipe emissions. The NOx tailpipe emissions during the first 300 seconds of the cycle account for more than 50% of the overall NOx tailpipe emissions. They can be reduced for all DeNOx systems through sophisticated fast heat-up strategies, which will be discussed later in more detail.

IMPACT OF INERTIA WEIGHT / POWER OUTPUT RATIO AND EXHAUST SYSTEM CONFIGURATION

Looking at the NOx conversion efficiency as a function of the vehicle inertia for different aftertreatment technologies and configurations over the FTP75 cycle, the following statements can be made:

For all SCR system concepts the NOx conversion efficiency significantly increases with increasing vehicle weight (Figure 17). This effect can be explained by the strong temperature dependency of the NOx conversion rate especially in the low temperature operating range. With increasing vehicle inertia weight the engine load in the test cycle and thus the exhaust temperature increases faster and catalyst light-off temperature for the is achieved earlier. Due to the higher engine load higher NOx engine-out emission level is anticipated (which is considered within the simulations) and thus also the required NOx conversion efficiency is higher.

Furthermore the achievable NOx conversion efficiency for a system configuration with SCR located downstream DPF is significantly lower compared to a configuration with SCR located upstream DPF. This again can be explained by a reduced temperature level caused by temperature losses in the exhaust line as well as the heat capacity of the DPF, which delay the heating-up phase of the SCR catalyst. But it has to be taken into account, that DPF regenerations downstream SCR might not be possible without external injector in order to ensure DPF regeneration under all conditions and to avoid excessive oil dilution.

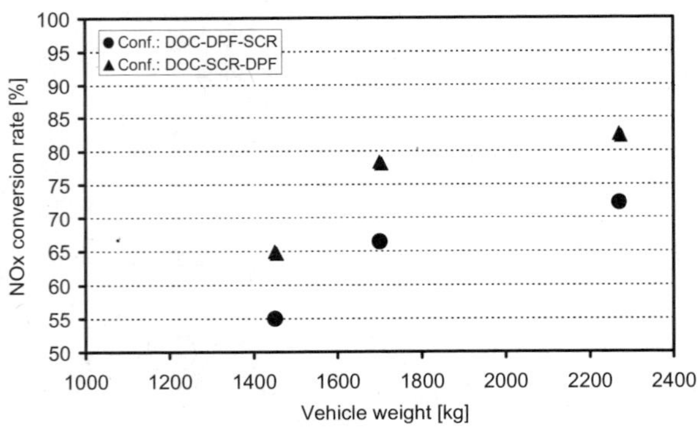

Fig. 17: NOx conversion rate as a function of vehicle weight and SCR exhaust system configurations

The simulation results for LNT technology show less pronounced conversion efficiency increases with vehicle weight (Figure 18). The reason for this is the fact that LNT catalysts are able to store NOx even under low temperature conditions and thus the impact of vehicle weight in the light-off phase is less pronounced. If the LNT is located downstream DPF, NOx conversion efficiency is significantly reduced, because the temperature level in the beginning of the test cycle becomes too low for NOx storage in the LNT.

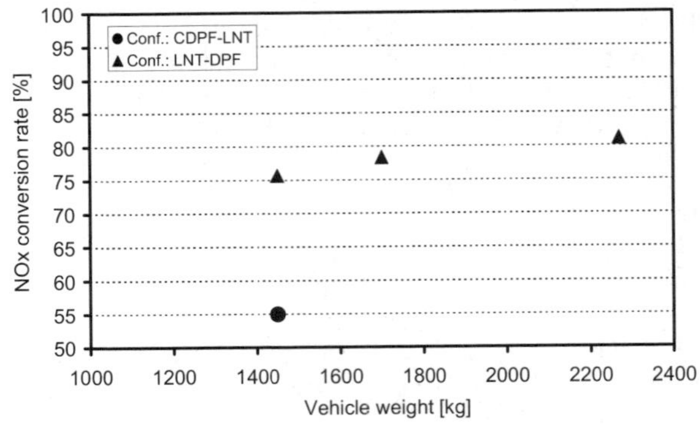

Fig. 18: NOx conversion rate as a function of vehicle weight and LNT exhaust system configurations

Comparing the calculated NOx conversion efficiencies for the most beneficial configuration between SCR and LNT technology focusing on NOx conversion rate (Figure 19), both technologies show very similar NOx conversion efficiencies for higher vehicle weights. For low vehicle weights there is a clear advantage for the LNT technology.

It has to be taken into account that all simulations are performed with de-greened - but not aged - catalysts. Furthermore a 3.5% fuel consumption penalty is to be considered for LNT technology. No heating strategy for the SCR technology – which will lead to improved NOx conversion efficiency on the one hand but fuel consumption penalty on the other hand – is considered so far. The potential of heating strategies for both LNT and SCR will be addressed in the following section.

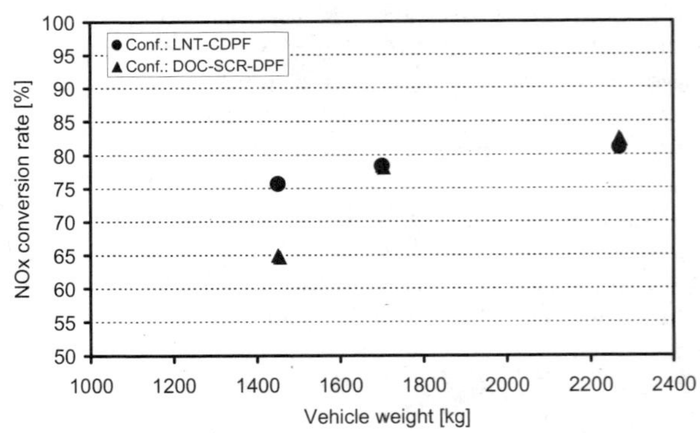

Fig. 19: Comparison of different DeNOx technologies

IMPACT OF HEATING ON NOX CONVERSION RATE

Figure 20 shows the improvement potential of heating strategies on NOx conversion efficiency for the lowest vehicle weight. For both SCR system configurations the efficiency can be significantly improved, but overall NOx conversion efficiency is still limited to about 80%. The explanation for this effect can be found in the reduced NOx conversion efficiency of SCR technology for low temperatures: Figure 21 depicts that even with a heating mode and corresponding temperature increase, the NOx tailpipe emissions in the first phase of the test cycle still contribute with a considerable amount to the overall tailpipe emissions. Thus a very fast heat-up of the SCR catalyst in order to achieve fast light-off in combination with lowest engine-out emissions in the early phase of the FTP75 cycle are mandatory to meet the stringent Tier 2 Bin 5 emission standards.

Hereby it is emphasized, that the simulations are based on de-greened - but not aged – catalysts. Considering desulfurization and thermal aging as well as the requirement for 120,000 miles emission compliance in the U.S., the SCR technology in combination with sophisticated heating strategies and adapted engine calibration is considered as the superior solution for heavy vehicle applications, but might also be applied to smaller vehicles from a technological point of view.

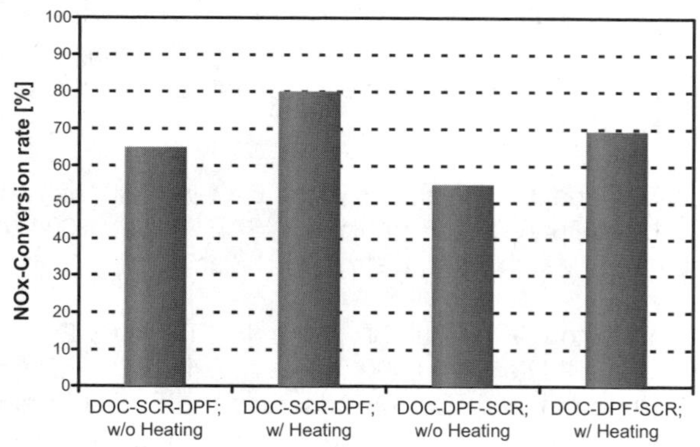

Fig. 20: Influence of heating on DeNOx performance for SCR applications, 1450 kg vehicle weight

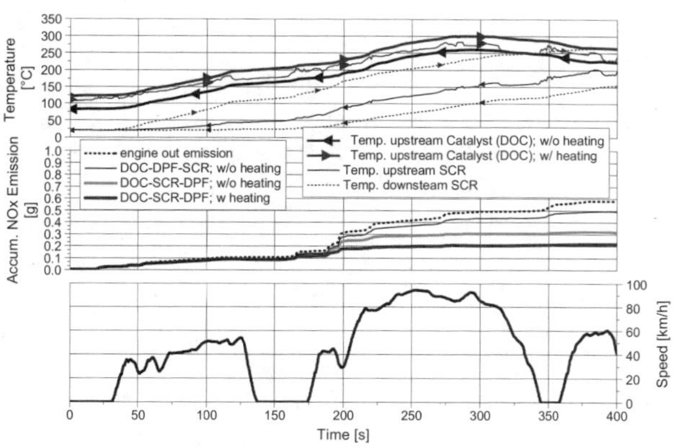

Fig. 21: Tailpipe emission during first phase of FTP-cycle for different SCR applications, 1450 kg vehicle weight

CONCLUSION

For current and future stringent exhaust emission standards SCR as well as LNT applications will be developed for market introduction. Both technologies provide specific pros and cons and based on the individual boundaries such as:

- Package conditions
- Vehicle size
- Ratio between engine displacement and vehicle mass
- Test cycle
- Required NOx conversion efficiency

One of both technologies will be the optimum choice for each selected vehicle application. General considerations as well as simulations for specific application examples show that SCR is the preferred technical solution for heavy vehicle applications targeting the U.S. Tier 2 Bin 5 emission limits, whereas LNT technology is an alternative option for lighter vehicles.

With regard to future developments the desulfurization and thermal aging and thus the long-term stability remains a major challenge for LNT technology. On the other hand the system packaging in the vehicle including required SCR catalyst and tank volume as well as the low temperature activity will be important issues to be solved for the SCR technology.

The infrastructure for the urea distribution especially in the U.S. remains a considerable challenge. EPA's concerns regarding this technology remain and have to be addressed by each manufacturer that attempts to launch a Diesel vehicle in the U.S. using SCR exhaust aftertreatment.

REFERENCES

/1/ T. Braun, K.-J. Marquardt, W. Müller : „Erfüllung zukünftiger Abgasgrenzwerte mittels NH3-geregelter SCR-Technologie“, 2. FAD-Konferenz, 10.-11.11.2004, Dresden

/2/ "Nissan launching Quon Truck with Urea-SCR system", www.dieselnet.com, November 2004

/3/ T. Tanaka : "Simultaneous Reduction of PM and NOx – A New After-Treatment System"; 22nd International Vienna Motor Symposium, 2001

/4/ J. Schommers, C. Enderle, H. Breitbach, et.al.: „World Wide Emission Strategy with BLUETEC for Diesel Passenger Cars“, 15. Aachener Kolloquium Fahrzeug- und Motorentechnik 2006

/5/ N. Satoh, H. Ohno, T. Nakatsuji : A NOx Reduction System Using Ammonia Storage-Selective Catalytic Reduction in Rich and Lean Operation“,15. Aachener Kolloquium Fahrzeug- und Motorentechnik 2006

/6/ „Honda Previews Next-Generation Engine and Power Technologies", www.greencarcongress.com, 25. September 2006

/7/ S. Pischinger, H.O. Herrmann, V. Scholz, J. Schnitzler : „Untersuchung zum Einsatz eines NOx-Speicherkatalysators im Pkw-Dieselmotor", 10. Aachener Kolloquium Fahrzeug- und Motorentechnik, 2001

/8/ S. Pischinger, J. Schnitzler, V. Scholz, A. Wiartalla : "Investigation of a NOx-Adsorber Catalyst used with a Passenger Car Diesel Engine", MTZ Worldwide 03/2003

/9/ M. Krüger, P. Nisius, V. Scholz, A. Wiartalla : „A compact Solid-SCR System for NOx-Reduction in Passenger Cars and Light Duty Trucks", MTZ Worldwide 06/2003

/10/ W. Müller, A. Herr, S. Käfer, A. Lacroix, E. Jacob: „SCR using solid urea", 3. Internationales Forum ´Abgas und Partikelemissionen´, Sinsheim, 14./15.9. 2004

/11/ Jacob, E.: Perspektiven der mobilen SCR-Technik, 15. Aachener Kolloquium Fahrzeug- und Motorentechnik 2006

/12/ I. Grißtede, S. Bremm, W. Müller, et.al.: „NOX-Nachbehandlungssysteme bei Dieselmotoren zur Serieneinführung in Nordamerika", 15. Aachener Kolloquium Fahrzeug- und Motorentechnik 2006

/13/ Breitbach, H.; Schön, C.; Leyrer, J.: „Potenzial und Grenzen der Abgasnachbehandlung durch NOx-Speicherkatalysatoren", 14. Aachener Kolloquium Fahrzeug- und Motorentechnik 2005

/14/ Hepburn, J. S.; Thanasiu, E.; Dobson, D. A.; Watkins, W. L. Experimental and Modelling Investigations of NOx Trap Performance, SAE 962051, SAE 962051, 1996.

/15/ Kobayashi, Yamada, Kayano Study of NOx Trap Reaction by Thermodynamic Calculation, SAE-970745SP1248, SAE, 1997.

/16/ Waltz, Ch. NOx-Minderung nach dem SCR-Verfahren:Untersuchungen zum Einfluß des NO2-Anteils, Diss., Karlsruhe, 2000.

/17/ Ketfi-Cherif, A.; Wissel, D.; Beurthey, St. Modeling and Control of a NOx Trap Catalyst, SAE 2000-01-1199

/18/ Schär, C.M.; Onder, C.H.; Geering, H.P. Control-Oriented Model of an SCR Catalytic, 2004-01-0153, SEA, 2004.

/19/ Tomazic D., Tatur M., Thornton M. "Development of a Diesel Passenger Car Meeting Tier 2 Emissions Levels," SAE Paper 2004-01-0581, March 2004, Detroit, Michigan.

/20/ Tatur M., Tomazic D. Thornton M. "Tier 2 Useful Life (120,000 miles) Exhaust Emission Results for a NOx Adsorber and Diesel Particle Filter Equipped Light-Duty Diesel Vehicle" SAE Paper 2006-01-0424

/21/ Tatur M., Tomazic D. Thornton M. "Desulfurization Effects on a Light-Duty Diesel Vehicle NOx Adsorber Exhaust Emission Control System" SAE Paper 2006-01-0423

/22/ Tatur M., Tyrer H., Tomazic D., Thornton M., McDonald, J. "Tier 2 Intermediate Useful Life (50,000 miles) and 4000 Mile Supplemental Federal Test Procedure (SFTP) Exhaust Emission Results for a NO_x Adsorber and Diesel Particle Filter Equipped Light-Duty Diesel Vehicle," SAE Paper 2005-01-1755

/23/ Tomazic D., Tatur M., Thornton M. "APBF-DEC NO_x Adsorber/DPF Project: Light-Duty Passenger Car Platform," DEER Paper, August 2003, Newport, Rhode Island.

/24/ McDonald, J. "Progress in the Development of Tier 2 Light-Duty Diesel Vehicles," SAE Paper 2004-01-1791, March 2004, Detroit, Michigan.

/25/ Thornton M., Tatur M., Tyrer H., Tomazic D. "Full Useful Life (120,000 miles) Exhaust Emission Performance of a NO_x Adsorber and Diesel Particle Filter Equipped Passenger Car and Medium-Duty Engine in Conjunction with Ultra Low Sulfur Fuel," DEER 2005..

CONTACT

Jürgen Schnitzler, FEV Motorentechnik GmbH, Senior Engineer Diesel Aftertreatment, E-Mail: schnitzler@fev.com

II. PM Control

II.1 FUNDAMENTAL STUDIES

2000-01-1016

Fundamental Studies of Diesel Particulate Filters: Transient Loading, Regeneration and Aging

Athanasios G. Konstandopoulos
Foundation for Research and Technology – Hellas/Chemical Process Engineering Research Institute (FORTH/CPERI); ME, EM Department, Michigan Technological University

Margaritis Kostoglou, Evangelos Skaperdas, Eleni Papaioannou, Dimitrios Zarvalis and Evdoxia Kladopoulou
Foundation for Research and Technology – Hellas/Chemical Process Engineering Research Institute (FORTH/CPERI)

ABSTRACT

Compliance with future emission standards for diesel powered vehicles is likely to require the deployment of emission control devices, such as particulate filters and $DeNO_x$ converters. Diesel emission control is merging with powertrain management and requires deep knowledge of emission control component behavior to perform effective system level integration and optimization. The present paper focuses on challenges associated with a critical component of diesel emission control systems, namely the diesel particulate filter (DPF), and provides a fundamental description of the transient filtration/loading, catalytic/NO_2-assisted regeneration and ash-induced aging behavior of DPF's. The derived models (some in analytic form) are based on rigorous descriptions of the underlying physicochemical processes and they are shown to be in excellent agreement with experimental data (collected in-house and from the literature) as well as by comparison to, 'presumably exact", 3-dimensional Computational Fluid Dynamics calculations. The validated models thus provide a robust core to build upon/interface system level simulation and optimization tools.

INTRODUCTION

The global automotive industry is currently facing complex challenges to provide solutions that reduce transportation related environmental impact, conserve energy and advance social welfare without adverse effects on economic growth, employment and competitiveness. Significant reductions in CO_2 emissions and the associated greenhouse-related environmental impacts of internal combustion engines, can be effected through increased deployment of the modern, highly efficient, direct injection (DI) diesel engine. This requires that engine manufacturers meet increasingly tightened particulate matter (PM) and NO_x emissions legislation [1-3]. The use of diesel particulate filter (DPF) systems (traps) appears inescapable for complying with heavy-duty engine emission standards in the next five years. The evolution of light-duty diesel emission standards is likely to make necessary the introduction of particulate emission control systems before the end of the decade, on both sides of the Atlantic [1]. Efficient DPF's can quickly accumulate significant amounts of particulates. These collected particles have to be removed from the filter by oxidation in order to prevent excessively high pressure drop in the exhaust system, which would affect the engine operation. The in-situ particulate oxidation process is termed regeneration of the filter. The availability of reliable, and efficient particulate reduction technology in combination with appropriate engine management/tuning and $DeNO_x$ technology, would then enable the diesel engine to become the automotive power plant of choice for the future.

It is becoming widely appreciated in the last years, that exhaust emission reduction devices should be an integral part of the powertrain. A shift in paradigm thus is becoming evident, from the "black-box converter" fitted to the exhaust of an already developed engine/vehicle, towards a framework of "concurrent engineering and integrated design" at the system level. This requires deep knowledge of emission control component behavior to perform effective system-level integration and optimization. A long-term research program carried out at the Aerosol & Particle Technology (APT) Laboratory of FORTH/CPERI is concerned with contributions to this framework, that would allow fast optimization of diesel emission control systems, as well as assist in the development of control strategies and assessment of novel concepts. In the present paper we discuss critical aspects of the diesel particulate emission control, with emphasis on the fundamentals of the filtration/loading, regeneration and aging behavior of DPF's.

The structure of the paper is as follows. At first a brief review of DPF technology is given to set the stage for the subsequent developments. Then analytical and numerical models for the transient loading and filtration process of a wall-flow DPF are derived and validated with experimental data and CFD simulations for different filter materials. The regeneration behavior of DPF's is further studied extending a mathematical model, developed previously by the authors, to cover the combined influence of NO_2 and catalytic coatings on diesel particulate oxidation. The new model is validated against experimental data collected in a dedicated pilot-scale filter testing unit, connected to a diesel engine bench. Finally the aging process of the filter pressure drop due to ash accumulation is simulated and validated against experimental data from the literature.

REVIEW OF DPF TECHNOLOGIES

Filters separate the typically sub-μm diesel particulate from the exhaust with a variety of transport mechanisms. Main deposition mechanisms are those of Brownian diffusion and direct interception while thermophoresis can be important in the presence of temperature gradients [4]. In the last 15 years many DPF system concepts have appeared, incorporating different filter media and geometric configurations, regeneration technologies and control/monitoring options. The interested reader should consult the recent literature [1-3] as well as on-line databases [5]. Filter media and geometric configuration is a key element in the DPF system and the selection process deserves careful balancing of different criteria including pressure drop performance, particulate collection efficiency, ease of regeneration, durability and cost. A partial list of available commercially or as prototypes filter media is shown in Table 1.

Table 1. Available Media for Diesel Particulate Filters

Ceramic

- Cordierite (extrudates)
- Mullite (extrudates, foams, fibers, infiltrates)
- Silicon Carbide (extrudates, foams, felts, infiltrates)
- Tialite (extrudates)
- Mullite/zirconia (extrudates, foams)
- Mullite/tialite (extrudates, foams)

Metallic

- Sintered metal
- Wire mesh
- Metal foam

Geometric configurations of filters are listed in Table 2 and some are schematically illustrated in Figure 1.

The wall-flow monolith honeycomb design originally introduced in 1981 [6] still remains the most popular configuration, since it exhibits a very compact arrangement, exhibiting thus a low pressure drop without sacrifice in filtration rate or available space.

Table 2. Geometric Configurations for DPF's

- Extruded honeycomb wall-flow monoliths
- Assembled parallel plate wall-flow elements
- Cylindrical Cartridges (based on fibrous structures)
- Foam monolithic blocks and plates
- Concentric tubular wall-flow elements

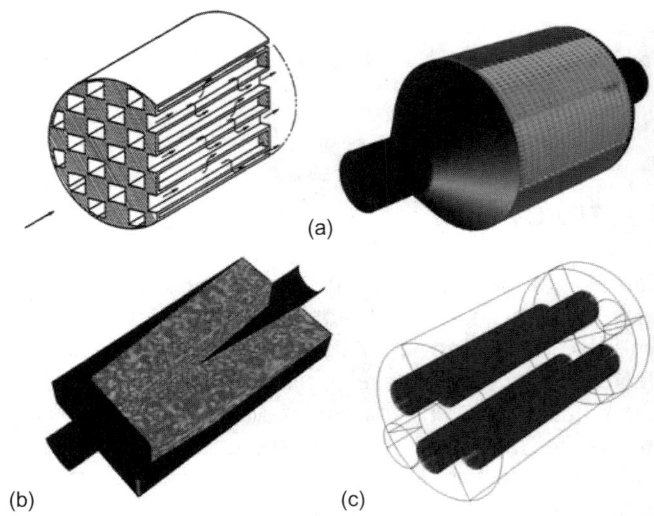

Figure 1. Geometric configurations of some commercially available DPF's : (a) honeycomb wall-flow monolith, (b) V-shaped foam filter, and (c) cylindrical fibrous cartridge.

Since the majority of filter configurations uses some sort of wall-flow arrangement we will focus our investigation for the remaining of the paper on the wall-flow geometry and honeycomb structures. It should be noted though that the developments/results can be extended straightforwardly to the other geometries.

Regeneration of diesel traps requires temperatures in excess of 550 °C. Such exhaust temperatures do not occur often in practical applications of the diesel engine. Especially in light-duty diesel applications the exhaust gas temperature can be as low as 100-200°C, for most of the driving route of the vehicle, especially in cities. Successful regeneration of diesel traps should achieve fast, reliable and cost effective conversion of the accumulated particulate mass, before the trap backpressure rises to unacceptably high levels.

It is convenient to classify regeneration measures [1-3,5] as active (employing external or engine means, see Table 3) or passive (usually employing catalytic means although catalytic means can also be applied in active systems, see Table 4). Combinations of measures are also common.

Table 3. Active Regeneration Measures

External means
• Fuel burners (full and partial flow)
• Electric heating (upstream or embedded in the filter)
• Micro-wave heating
• Injection of combustibles (e.g. fuel) in the exhaust
• Injection of catalytic and/or reactive species in the exhaust (e.g. H_2O_2 [7])
• Generation of reactive species (e.g. non-thermal plasma [8])
• Electrochemical filter reactor [9]
Engine means
• Exhaust gas recirculation
• Post-injection of fuel
• Decrease of boost pressure
• Intercooler by-pass
• Injection timing retard

Table 4. Passive Regeneration Measures

Catalytic means
• Fuel-borne catalysts
• Catalytic filter coatings
• Reactive species generation (e.g. NO_2 [10])

Effective catalytic oxidation of diesel particulate relies on the establishment of good contact between the soot particles and active catalyst sites [11], where oxidation takes place. Catalyzed diesel particulate filters involve the deposition of catalytic coatings on the filter media, while fuel-borne catalysts are continuously supplied with the fuel. Applications combining fuel-borne catalysts and catalytically coated filters have also been suggested [12]. The particulate oxidation mechanism with fuel catalytic additives is similar to that in a catalyzed trap. Commercially available regeneration promoting fuel-borne catalysts are based on Cerium (Ce), Copper (Cu), Iron (Fe), Platinum (Pt), Strontium (Sr) and Sodium (Na) fuel-soluble compounds and their combinations [1-3, 5]. In fleet operations, the catalytic compounds can be dopped in a central fuel tank, while for application to individual vehicles a reliable on-board, controlled dosing system is required. The combined use of a fuel-borne catalyst and temperature-raising engine measures, are employed in a recent particulate emission control system marketed by a European car manufacturer [13].

In other recent developments, diesel particulates are removed through oxidation by NO_2 [10]. Initially the technology required low-fuel sulphur content (<10 ppm), but it is believed that it can operate with up to 50 ppm sulphur content in the fuel, provided there is a favorable NO_x/C ratio in the engine out emissions.

Finally, non-thermal plasma based technologies [5] can be exploited to promote diesel particulate oxidation at low temperatures [8], through the creation of highly reactive species (including NO_2) and radicals. However, the actual potential of such technologies integrated with modern diesel engines remains to be systematically investigated and assessed against performance, reliability and cost (Table 5).

Table 5. Assessment Aspects for DPF Systems

Performance
• Filtration efficiency (mass, number, size-specific)
• Effects on other emissions
• Pressure drop at rated speed/full load (fresh, regeneration limit, safety limit)
• Regeneration efficiency
• Noise reduction
Reliability
• Monitoring, control and safety warning algorithms
• Life expectancy, aging performance
• Inspection and maintenance verification protocols
Cost
• Capital cost (including system integration, optimization and tampering resistance)
• Operating cost (including fuel penalty)
• Maintenance cost

In summary, diesel particulate filter regeneration technologies aim at increasing the reactivity of the accumulated carbonaceous particulate either by raising (directly or indirectly) the exhaust gas temperature, or by making the particulate and/or the environment in the filter intrinsically more reactive, or both. Lately, heat recovery approaches are under investigation for extending the operating range of regeneration strategies. These involve heat recovery by reciprocating flow operation [14] as well as through the use of heat recuperators [15].

It is convenient to distinguish between two regimes of DPF operation namely filtration/loading and regeneration. However for some technologies the boundary between these regimes can be blur or even non-existent (e.g. in continuously regenerating systems). In the following we study these two regimes.

FILTRATION AND LOADING BEHAVIOR

OVERVIEW OF THEORETICAL DEVELOPMENTS – The first applications and validations of filtration theory to diesel particulate filters were done in [16] for fibrous structures and in [4] for wall-flow extruded filters. In these studies it was shown that classic filtration theory employing the concept of the "unit collector" (spheres for extruded filters and cylinders for fiber and foamy structures) can give a good estimate of the size-specific collection efficiency of "clean" DPF's with respect to solid particles. The first study that provided validated design equations for sizing wall-flow filters based on fundamental principles of fluid mechanics and flow through porous media (Darcy's law) was by Konstandopoulos and Johnson [4] using a one-dimensional (perimeter averaged) model of the trap channels. In that study analytical solutions were obtained for the flow fields and pressure drop of wall-flow monoliths in terms of the filter media microstructure and geometric configuration, that were validated experimentally for a particular extruded monolith design. The analytical model (extended for non-Darcian flow effects) was recently shown to be in excellent agreement with 3-D Computational Fluid Dynamics (CFD) simulations and was further validated against a larger variety of filter media [17]. Furthermore, explicit design and optimization criteria for a systematic selection of the filter material microstructural characteristics and filter configuration (wall thickness, cell density, length) were obtained, for routine sizing of wall-flow monoliths with special emphasis on filters regenerating continuously [17]. The Konstandopoulos and Johnson [4] model was very recently extensively tested against a variety of filter samples, and it is reported to give excellent *a priori* predictions of the pressure drop, opening new development possibilities [18,19].

Appendix I presents the extension of the Konstandopoulos et al. [4, 17] trap pressure drop model to the case of arbitrary filter loadings. An analytical expression derived for the trap pressure drop as a function of particulate loading (expressed in terms of a uniform, effective deposit thickness, w on the filter wall) is given below:

$$\Delta P = \frac{\mu Q}{2V_{trap}}(\alpha + w_s)^2\left[\frac{w_s}{k_o\alpha} + \frac{1}{2k_{soot}} ln\left(\frac{\alpha}{\alpha - 2w}\right) + \right. \\ \left. + \frac{4FL^2}{3}\left(\frac{1}{(\alpha - 2w)^4} + \frac{1}{\alpha^4}\right)\right] \tag{1}$$

The symbols appearing in Eq. (1) are defined in a separate section at the end of the paper.

For coated filters, temperature dependent slip-flow effects cause the pressure drop to deviate from the continuum form of Darcy's law, however these can be accounted as shown in Appendix II through a simple procedure, see also [4]. Figure 2 illustrates how the pressure drop dependence of a membrane-coated cordierite filter [20] is accurately described in terms of the slip-flow corrected permeability (Appendix II) that depends on the local value of the Knudsen number of the flow through the coating.

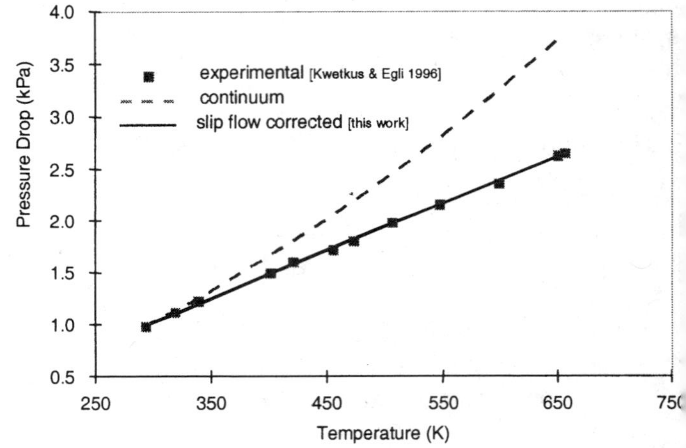

Figure 2. Effect of temperature on pressure drop of a membrane coated filter, experiment [20] and theory (see Appendix II for slip-flow corrected theory).

Depending on their microstructure, filters can exhibit a transition from the so-called deep bed filtration regime, to the cake (or surface) filtration regime [21], characterized by a linear dependence of the pressure drop on the accumulated particulate mass in the filter. DPF's are no exception and for some filters such behavior has been documented experimentally in the literature [22, 23]. In Appendix III we describe the extension of the clean filter model of Konstandopoulos and Johnson [4], based on the concept of the "unit collector", into the domain of transient filtration/loading. The derived model is based on first principles that describe the spatial deposition of particles inside the porous filter wall, and simulates in a natural way the deep-bed-to-cake transition through the progressive blocking of the filter wall. The model equations cannot be solved analytically, but a very efficient numerical code (*FilSim* v.1.0) has been developed for their solution. The transient filtration algorithm is shown in the flow chart of Figure 3 and is briefly described below.

A prescribed particle size distribution (particle size loop in Figure 3) is tracked through the filter medium, which is discretized in slabs where the local collection efficiency is computed using the concept of the "unit collector" (spatial efficiency loop in Figure 3). The locally deposited mass along the filter thickness modifies the microstructure (spatial microstructure loop in Figure 3) -porosity, collector diameter- of each slab which in turn determines the new local collection efficiency and pressure drop. The procedure is repeated for every time step (temporal loop in Figure 3) to compute the evolution of the filter behavior in terms of collection efficiency and pressure drop. Eventually a cake emerges from the wall, and deposit growth continuous on the surface of the filter cake.

The *FilSim* code can accommodate any type of "unit cell" collector model (here it is shown for spherical collectors) and arbitrary spatial variations of filter microstructure including gradients in porosity/pore size as for example in the case of coated filters or multilayer structured filters (e.g. fibrous cartridges).

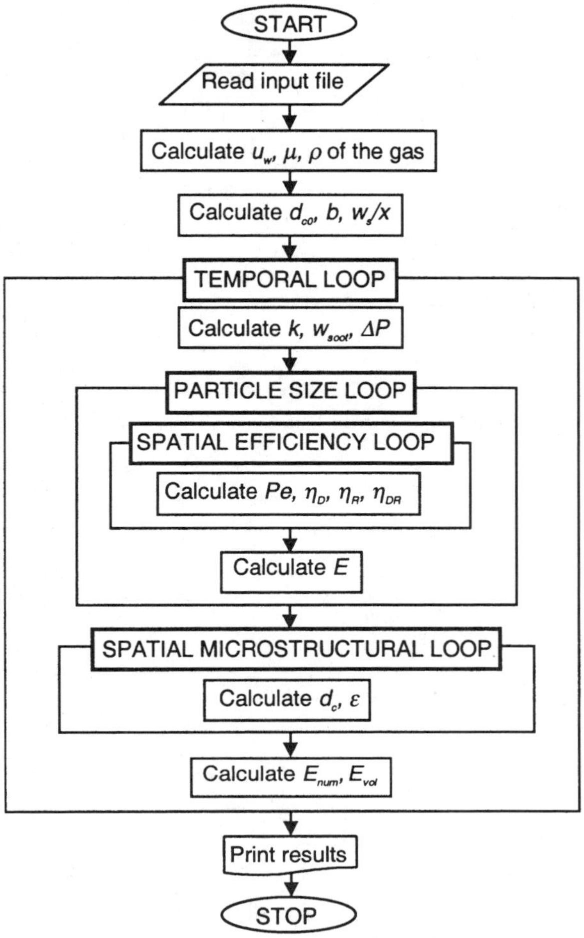

Figure 3. Flow chart of transient filter loading model. See Appendix III for details.

FILTRATION AND LOADING RESULTS – Eq. (1) is compared against three-dimensional (3-D) Computational Fluid Dynamics calculations in Figure 4. In addition, Eq. (1) is validated with experimental data [24] obtained fitting two types of filters in the exhaust of a single cylinder diesel engine, providing a 70 kg/hr flow at 280 °C and a 3.2 gr/hr particulate emission rate. Filter A had a filtration area of 2.1 m^2 and a cell opening of 1.44 mm (Figure 4), while filter B had a 1.53 m^2 filtration area and a cell opening of 2.11 mm (Figure 5). It is evident that there is excellent agreement between all three sets of data (analytical, numerical and experimental), establishing thus confidence in the use of Eq. (1) for filter pressure drop calculations at arbitrary loading.

A characteristic of the present filter pressure drop model is that it can accurately describe the non-linear behavior of the pressure drop trace for heavily loaded filters (particularly evident for filters with smaller cell openings, as in Figure 4).

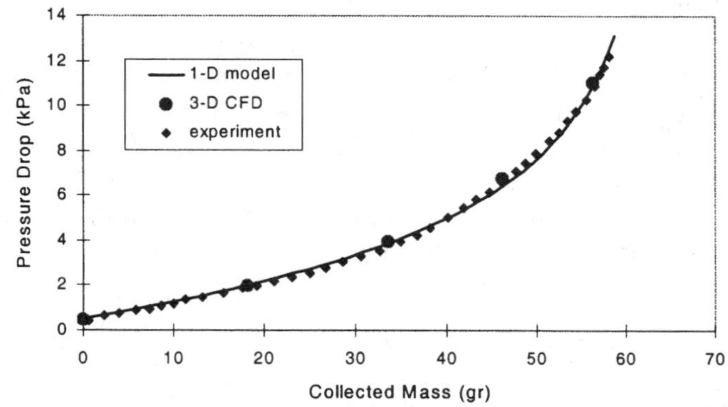

Figure 4. Experimental pressure drop curve for filter A [24], compared with analytical model, Eq. (1) and CFD results.

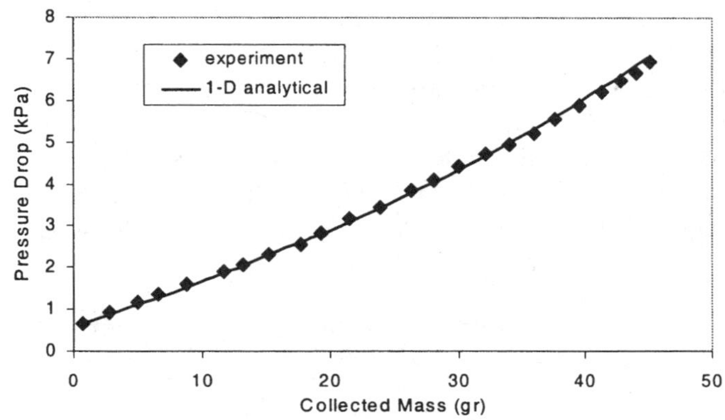

Figure 5. Experimental pressure drop curve for filter B [24], compared with analytical model, Eq. (1).

The particulate layer packing density $\rho_{soot,c}$ governs the non-linear behavior of the pressure drop, as the trap channel starts to be blocked, while the initial slope of the pressure drop rise curve depends mainly on the product of the particulate layer packing density and its Darcy permeability, k_{soot}. Pressure drop measurements of deliberately heavily loaded filters can be used in conjunction with the developed model to extract microstructural characteristics of the particulate layer grown in the filter. The present model provides thus for the first time a tool for the independent estimation of the particulate deposit packing density and permeability. For the particular examples shown in Figures 4 and 5 these characteristics are given in Table 6 and are very similar for the particulate layers grown on both filters.

Table 6. Particulate Microstructural Characteristics

Filter	Permeability (m^2)	packing density (kg/m^3)
A	3.2 x 10^{-14}	97
B	3.3 x 10^{-14}	110

It should be stated here that for accurate estimation of $\rho_{soot,c}$ it is necessary to have sufficient experimental data in the non-linear regime of pressure drop versus collected mass. Based on the analysis of such experiments a database of particulate layer characteristics (packing density and permeability) grown on many filter media under various operating conditions is developed at our laboratory to assist in rational filter selection and design.

Figure 6 depicts the simulation of the transient evolution of the size specific collection efficiency of a 10.5"x12" 100 cpi cordierite filter, exposed to the exhaust of a heavy duty diesel engine (0.236 Nm^3/s) at 260 °C and a 18 gr/hr particulate mass emission rate. A typical diesel particle lognormal distribution was used as input to *FilSim*. It is evident that very quickly the "clean" filter collection efficiency increases to high values. The spatial distribution of collected particulate mass normalized with the total mass in the filter wall is shown in Figure 7. We observe an exponential distribution of the collected particulate mass along the filter wall thickness, that is expected to have significant implications in the case of catalyzed filters, where the local contact and particulate to catalyst ratio are important.

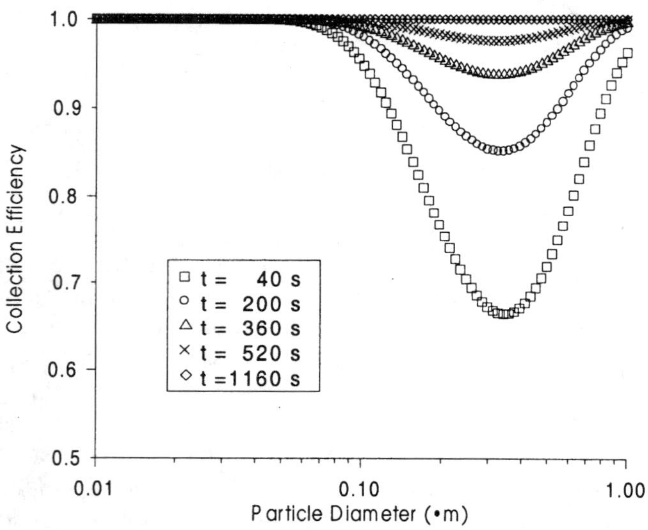

Figure 6. Simulation of the evolution of size specific collection efficiency of a cordierite filter.

Experimental validation of the filtration aspect of the model is provided in Figures 8 and 9 where the size distributions downstream of a corderite [25] and a sintered metal filter [26] are compared with model predictions. In the case of the sintered metal filter the comparison indicates that some post filter nuclei-mode particle formation [4,27] is taking place. Overall the filtration model is in good agreement with the data for the accumulation mode particles as was already found in [4] for another type of filter.

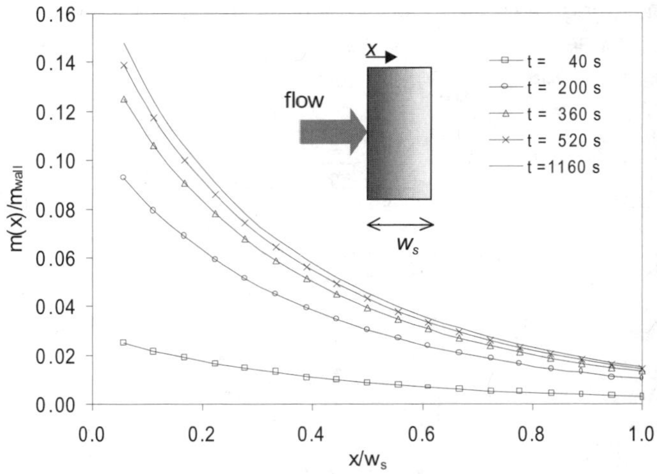

Figure 7. Simulation of the evolution of collected mass along the wall thickness of a cordierite filter.

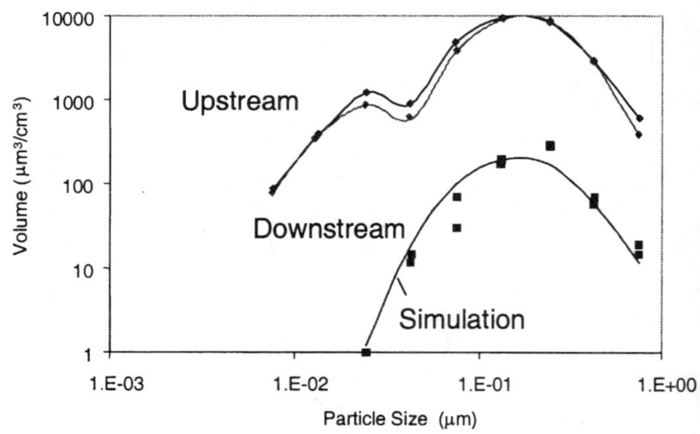

Figure 8. Particle size distributions measured upstream and downstream of a cordierite wall-flow filter [25], together with theoretical prediction.

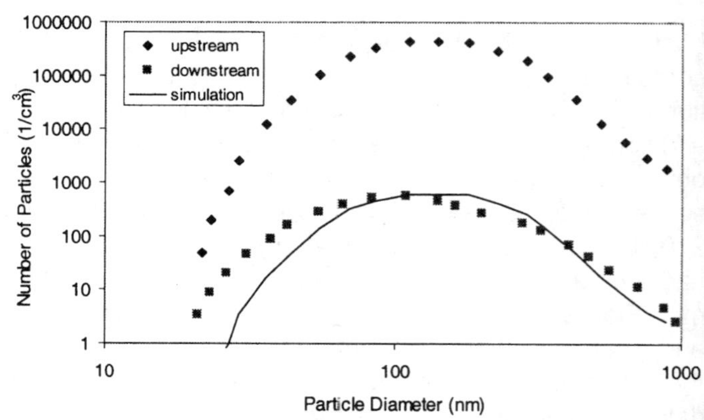

Figure 9. Particle size distributions measured upstream and downstream of a sintered metal filter [25], together with theoretical prediction.

The transient filtration model is used to simulate the experimental data in [22] on the pressure drop of various cordierite filter types (Figure 15 in [22]). The filters were exposed to the exhaust of a heavy duty diesel engine at 0.236 Nm^3/s and 260 °C. A particulate mass flow of 18 gr/hr as given in the paper was used for the simulation. Since it was referred to as "soot" in the paper [22] we assumed that it corresponded to the emission rate of the solid particulate that would be actually entering the filter. All DPF's had dimensions of 10.5"X12.0", cell density of 100 cpi and a wall thickness of 0.017 in. The reader is referred to the original paper [22] for additional experimental details. The simulation of these data was performed with the parameter values shown in Table 7. All parameters except the last three in Table 7 were fixed to the indicated values. The first two parameters in Table 7 are nominal values from the filter manufacturer, the clean filter permeability was estimated from the first experimental data point of Figures 10-13, and the cake packing density value was estimated based on our experience with heavily loaded filters (see also Table 5). A non-linear regression code using the Levenberg-Marquardt method [28] was written to estimate the values of the last three parameters in Table 7 from the experimental data.

Table 7. Particulate Microstructural Characteristics

FILTER:	EX-47	EX-54	EX-66	EX-80
d_{pore} (μm)	13.4	24.4	34.1	12.5
ε (%)	48	50	50	48
k_0 (m^2)	2.4x10^{-13}	3.3x10^{-13}	3.5x10^{-13}	1.8 x10^{-13}
$\rho_{soot,\ c}$	91	91	91	91
k_{soot} (m^2)	1.8 x10^{-14}	1.9 x10^{-14}	2.1 x10^{-14}	2.2x10^{-14}
$\rho_{soot,\ w}$	14.10	12.24	12.22	8.26
ψ	0.9203	0.9246	0.9201	0.8651

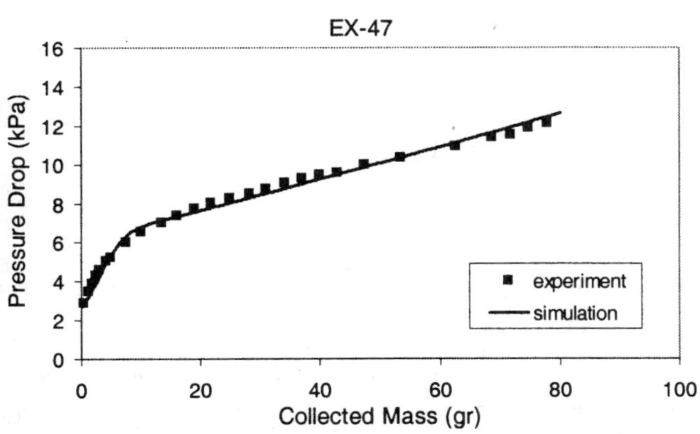

Figure 10. Comparison of transient filtration model pressure drop prediction against experiments [22] for a Corning EX-47 filter.

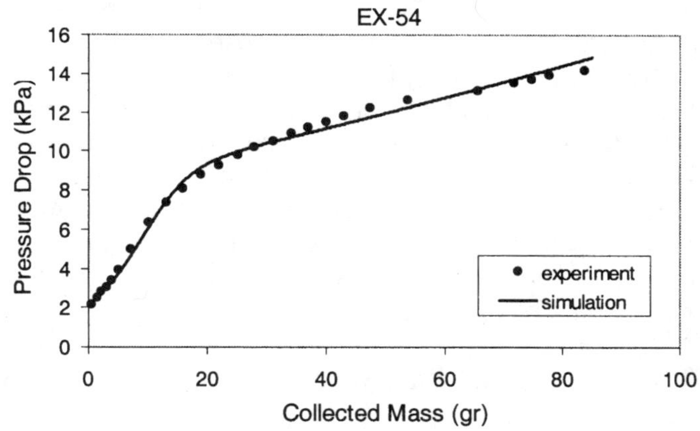

Figure 11. Comparison of transient filtration model pressure drop prediction against experiments [22] for a Corning EX-54 filter.

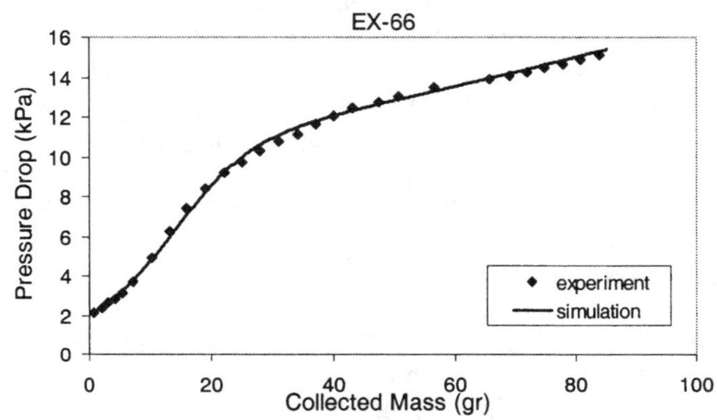

Figure 12. Comparison of transient filtration model pressure drop prediction against experiments [22] for a Corning EX-66 filter.

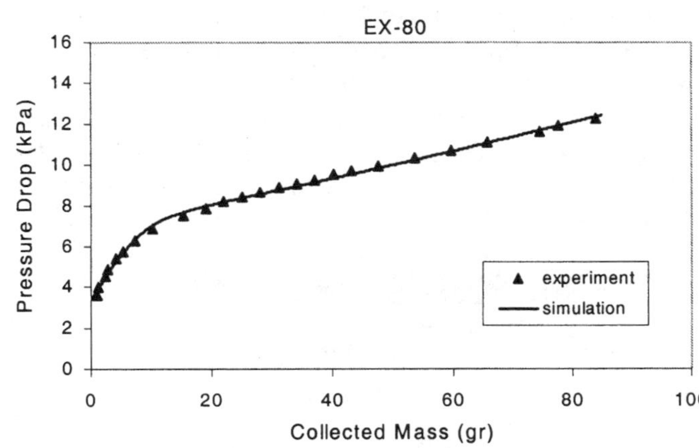

Figure 13. Comparison of transient filtration model pressure drop prediction against experiments [22] for a Corning EX-80 filter.

Examining Figures 10-13, we observe that the developed transient filtration model describes very well the entire transition from deep bed-to-cake behavior for all tested filters. In addition the estimated parameters of all particulate layers (Table 7) are in close agreement with each other, establishing thus the internal consistency of the model. It is interesting to notice the low packing density determined for the particulate collected inside the filter wall vs. that of the particulate cake. This is consistent with the notion of *percolation*, where small "islands" of particulate dispersed into the (deep-bed like) filter wall, can cause substantial increase of pressure drop by pore bridging/blocking.

REGENERATION BEHAVIOR

THEORETICAL FRAMEWORK – There are a lot of aspects that must be included in a comprehensive model for the regeneration behavior of a particulate filter. In general, there are three coexisting scales ("levels") which the modeller of particulate traps must take into account. The first level refers to the phenomena occurring across the particulate layer and the porous filter walls, the second level to the phenomena occurring along the filter channels and the third level refers to the phenomena at the macroscopic scale of the trap. Submodels at each level must be developed and properly combined in order to derive a model of the entire process. In the following, these three levels are discussed in more detail.

Level I. Filter wall scale – The exhaust stream can be assumed to be always in a quasi-steady state, since the residence time of the exhaust gas as it transverses the filter wall, is very small compared to the characteristic scale of variations of the filter inlet conditions (velocity, temperature, oxygen content). The fluid mechanics at this level is trivial (just a constant filtration velocity). A separate differential mass balance must be written for each active species (all species which take part in a reaction in the particulate layer, e.g. O_2, NO_2, gaseous and adsorbed hydrocarbons). In addition, a transient energy balance equation must be derived for the temperature distribution across the particulate layer and porous wall.

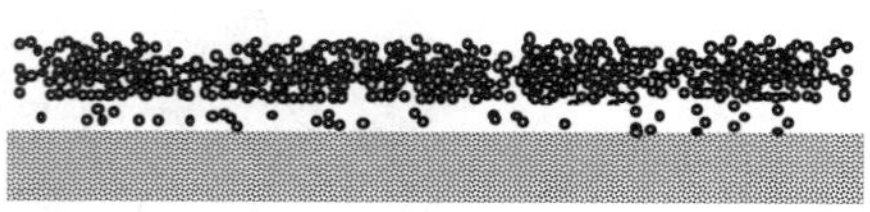

Figure 14. Schematic of filter wall.

A complication arises from the oxidation of the particles trapped in the pores and on the wall surface, that necessitates the solution of a moving boundary problem. Bissett addressed the above problem [29, 30] and showed using a singular perturbation expansion valid for small thermal Peclet numbers (a very good assumption) that the temperature can be assumed uniform across the particle layer and porous wall. This leads to a great simplification, since only the species mass balances must be integrated across the particle layer and filter thickness. This integration can be done analytically for some simple cases (e.g. for the case of O_2 or O_2/NO_2 particulate oxidation), while in more complex cases the integration has to be performed numerically. Application of the methodology to a case simulating our experiments with catalytic coatings and coupled reactions among the active species (described later) is given in Appendix IV.

Structural considerations for the particulate deposit and the porous filter wall (possibly serving as a substrate for a catalytic coating) must also be taken into account in Level I (Figure 14). Recently, Konstandopoulos and Kostoglou developed [31] and applied [14, 32] a mesoscopic microstructural filter model (the so-called two-layer model) to describe the effect of the catalytic coating of the porous walls, supplemented with a general "mixture-model" for the carbonaceous particulate-catalyst promoted oxidation kinetics, taking explicitly into account the particulate-catalyst contact. Appendix IV describes the extension of that model to account for the presence of NO_2 in the filter inlet, to investigate NO_2-regenerative technologies (currently practiced with uncatalyzed filters) in combination with catalyzed filters. Such combinations could lower the dependence of NO_2-regenerative technologies to high engine-out NO_x concentrations, as NO_x emission standards become tighter.

The relevant reactions that are assumed to describe the process are:

Thermal particulate oxidation

$$C+(1-f_{CO}^{the}/2)O_2 \rightarrow f_{CO}^{the}CO+(1-f_{CO}^{the})CO_2 \qquad (2)$$

Catalytic particulate oxidation

$$C+(1-f_{CO}^{cat}/2)O_2 \rightarrow f_{CO}^{cat}CO+(1-f_{CO}^{cat})CO_2 \qquad (3)$$

NO_2-assisted particulate oxidation

$$C+(2-g_{CO})NO_2 \rightarrow g_{CO}CO+ (1-g_{CO})CO_2+(2-g_{CO})NO \qquad (4)$$

NO to NO_2 catalytic oxidation

$$2NO+O_2 \rightarrow 2NO_2 \qquad (5)$$

Where the selectivities f_{CO}^{i}, g_{CO} (*i* = *the* or *cat*) of the reactions are given explicitly in Appendix IV.

In the general case the model of Level I serves to provide a "constitutive relation" for the local reaction rate to the Level II model.

Level II. Filter channel scale – If the transient behavior of the filter is slow with respect to the residence time of the exhaust gas as it travels across the filter, then the quasi-steady state approximation for the exhaust flow in the channels can be used.

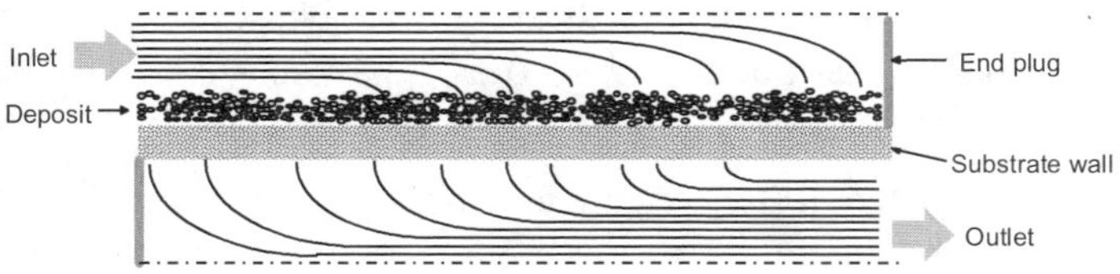

Figure 15. Schematic of filter channel.

A rigorous approach for this scale was first provided by Bissett [33]. His one-dimensional model includes the mass and momentum balances for a representative inlet and outlet channel (Figure 15) plus a Forcheimer modified Darcy's law for the porous filter wall. As discussed in the previous section on the loading behavior, this approach for the fluid mechanics has been validated against 3-D CFD and experiments, see also [4,17,19]. The reduction in dimensionality of the problem possible with this approach, in connection with its proven accuracy, support its choice as the best available, for a fluid mechanical submodel for the trap channels.

To model the particulate oxidation process, additional equations must be solved coupled with the fluid mechanical submodel. These include: quasi-steady state energy balances for the exhaust gas stream in the inlet and outlet channel, transient energy balance for the system porous wall-particulate layer and an evolution equation for the particulate layer mass (or thickness). If the inlet conditions to the trap can be assumed to be spatially (radially) uniform, solving the model for two representative channels (inlet and outlet) is equivalent to solving it for the entire trap (assumed radially homogeneous). This approach was successfully validated against experimental results in [14, 32]. In the general case the Level II model serves to provide the laws of phase interactions for the model of Level III.

Level III. Trap scale – This is the macroscopic scale of the DPF. The need for a model at this scale arises from the possible existence of inlet radial non-uniformities. Radial variations of the exhaust gas stream velocity and temperature at the inlet of the filter have been assessed by simulating the exhaust system with CFD (see [17]). Some previous attempts for modeling at this level are based on the approximation of the radial non-uniformities through a number of "zones" (collections of several channels) along the radius of the filter with thermal coupling between adjacent "zones" [34]. This approach requires closure for the "zone" interactions.

A self-consistent continuum model for the entire porous filter, is currently under development at our laboratory following the principles of volume averaging in multiphase flows [35]. The three phases in the trap are the exhaust gas in the inlet channels, the exhaust gas in the outlet channels and the solid phase, whereas diesel particulate is modeled as a species component distributed in all three phases. The advantage of the multiphase continuum formalism is that all relevant phenomena can be described by a compact set of Partial Differential Equations, easily integrated into the framework of CFD codes (in-house or commercial). Figures 16-18 depict 3-D flow simulation of several types of diesel particulate filters (wall-flow honeycomb, foam and fibrous cartdridge filters) at the macroscopic scale

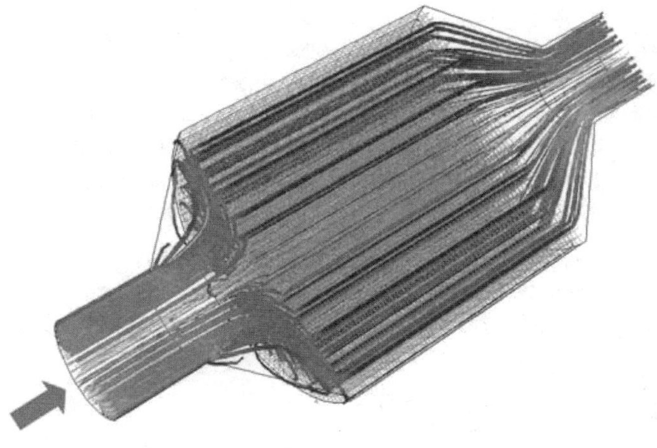

Figure 16. Flow simulation of honeycomb DPF.

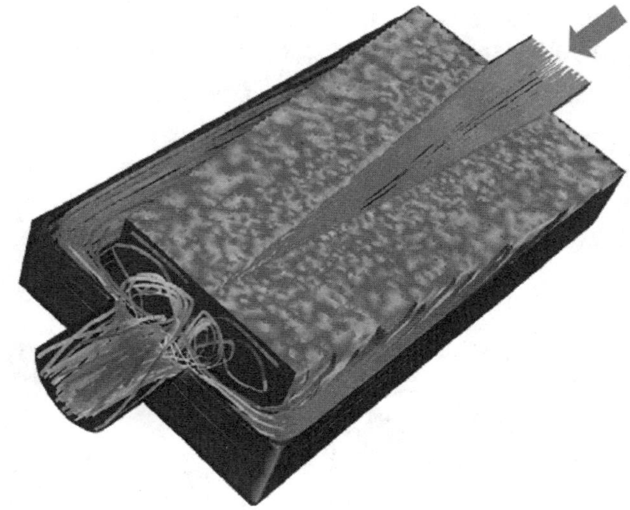

Figure 17. Flow simulation of V-shaped foam DPF.

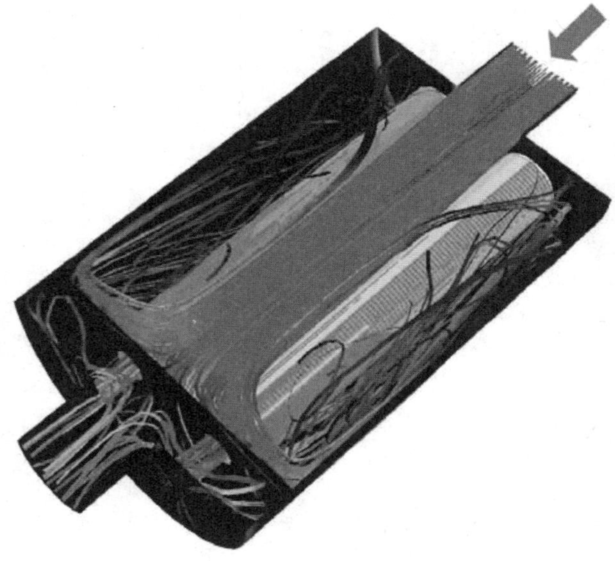

Figure 18. Flow simulation of fibrous cartridge DPF.

NO$_2$-ASSISTED PARTICULATE OXIDATION

Experimental – NO$_2$ oxidation of diesel particulate was studied on prototype catalyzed ceramic filter disks (of diameter 60 mm and thickness about 1 mm) loaded in a special pilot scale filter testing unit (Figure 19), at the Aerosol & Particle Technology Laboratory of FORTH/ CPERI [36]. The filters were exposed to a constant exhaust flow from a passenger car turbo-charged direct injection diesel engine with a 1.9 Lit displacement (rated at 66 kW/4000 rpm) operated at 2400 rpm and 50% load, coupled to a hydraulic bolt-on dynamometer.

The filter disk is held in place sandwiched between two gaskets cut from heat expanding fibrous mat in a stainless-steel holder, housed in a heated stainless-steel reactor. The exposed filter diameter to the flow is 40 mm. As shown in the diagram of Figure 19, the unit is instrumented with thermocouples and incorporates a by-pass valve that permits isolation of the filter from the exhaust flow under an inert (N$_2$) atmosphere. A heated tube before the reactor heats the exhaust gases to the desired temperature. The unit sits on a bench that incorporates sampling pumps and ancillary controllers and transducers to maintain constant volume flow through the filter. A dedicated data acquisition computer collects temperature, pressure and exhaust gas analysis data from the unit.

To study the NO$_2$-particulate reaction on the catalyzed filters, the filters were first loaded with particulate from the engine exhaust. Subsequently they were in-situ exposed under a temperature ramp (3 °C/min), in the range of 300 °C to 500 °C, to a synthetic exhaust flow of 3 NLit/min. In these conditions the presence of the catalyst can promote the re-oxidation of NO (resulting from NO$_2$ reduction by the particulate) back into NO$_2$ and accelerate thus the rate of particulate oxidation.

Results – Particulate oxidation was followed by monitoring the concentration of CO$_2$, NO and NO$_2$ behind the filter as well as the pressure drop evolution across the filter. To clarify the influence of NO$_2$ three types of experiments were performed.

- Oxidation by a 10% O$_2$ in N$_2$ stream (Figure 20)
- Oxidation by a 400 ppm NO$_2$ in N$_2$ stream (Figure 21)
- Oxidation by a 400 ppm NO$_2$/10% O$_2$ in N$_2$ stream (Figure 22)

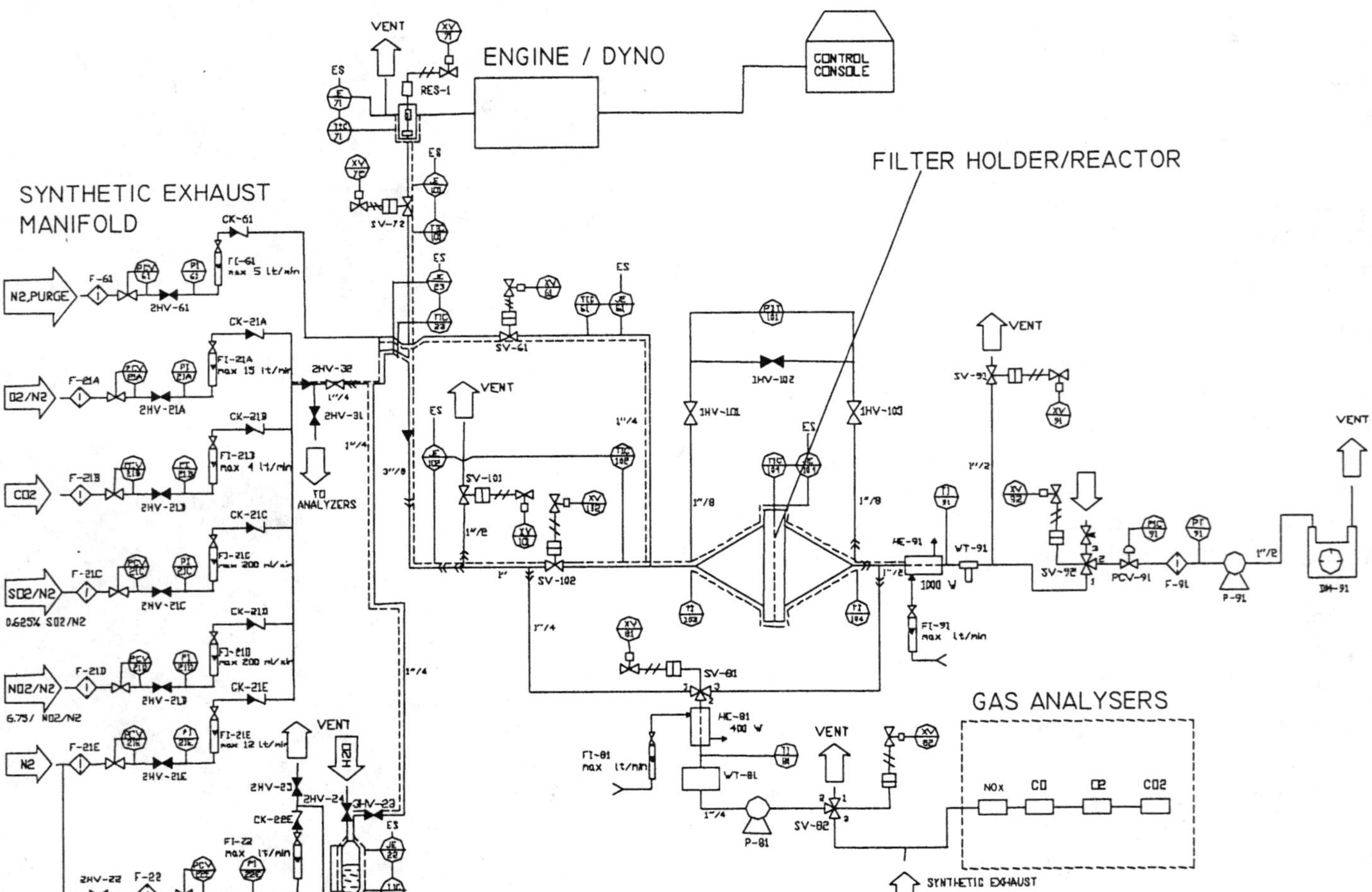

Figure 19. Flow chart of the experimental set-up used for loading prototype particulate filters from diesel exhaust and studying their regeneration under controlled conditions of feed composition and temperature.

The dotted lines in Figures 20-22 are based on the application of the NO_2 assisted particulate oxidation model described in Appendix IV. The NO_2-particulate oxidation kinetics were adjusted to fit the NO_2 evolution profile and then the CO_2 profile was computed without any other adjustment to assess the validity of the assumed reaction mechanism. A high degree of consistency is observed between model and experiments building confidence in the use of the model to extract kinetic data from these and similar experiments. The activation energies of reactions (Eqs. 2, 4, 5) extracted from these experiments are shown in Table 8. The data imply that the catalytic coating has minimal effect on the direct catalytic oxidation of the particulate, however its presence makes possible the acceleration of the particulate oxidation rate through the oxidation of NO to NO_2 which in turn oxidizes the particulate. This is evident comparing Figure 21 to Figure 22. It is seen in Figure 22 that in the presence of O_2, there is significantly higher amount of NO_2 remaining due to the action of the catalytic coating towards NO_2 production. On the contrary, in Figure 21 NO_2 is almost consumed entirely when the particulate oxidation reaches its highest rate.

Table 8. Activation Energies of Eqs. (2), (4) and (5) on Catalyzed Filter

Reaction	E/R in K
$C + O_2$	18040.
$C + NO_2$	14650.
$NO + ½ O_2$	10900.

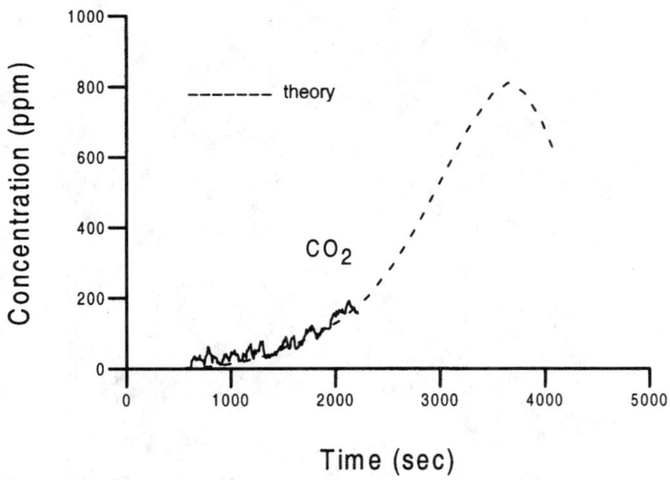

Figure 20. O_2-particulate reaction on catalyzed filter.

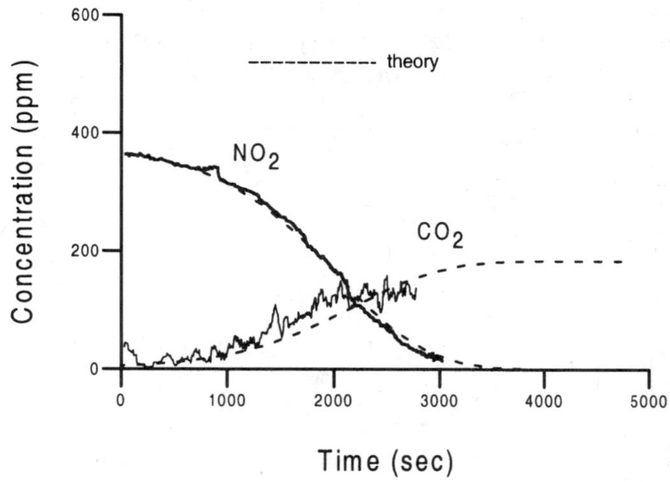

Figure 21. NO_2-particulate reaction on catalyzed filter in the absence of O_2.

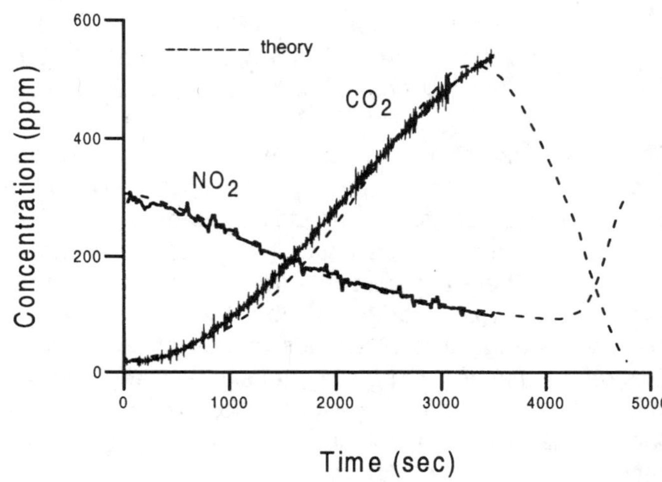

Figure 22. NO_2-particulate reaction on catalyzed filter in the presence of O_2.

AGING OF FILTER PRESSURE DROP DUE TO ASH

A critical factor that affects the expected lifetime of DPF's is ash accumulation after repeated regenerations. The origin of the ash is the lubricating oil additives, engine wear as well as the fuel, especially if it contains additives for regeneration promotion. The problem of filter plugging by ash accumulation is stated as one of the most important problems by many diesel engine manufacturers [37]. It is therefore interesting to develop and validate appropriate models for the aging behavior of the filter pressure drop, to assist the life cycle analysis and optimization of DPF's for a particular application. The transient filter loading model developed in this paper is indispensable in this respect, since it can describe not only the collection of diesel particulate but also the accumulation of ash particles inside the filter, provided that appropriate ash deposit microstructural properties are employed.

Figure 23 depicts the application of the developed pressure drop model to the experimental data of [38] in order to estimate ash layer microstructural properties.

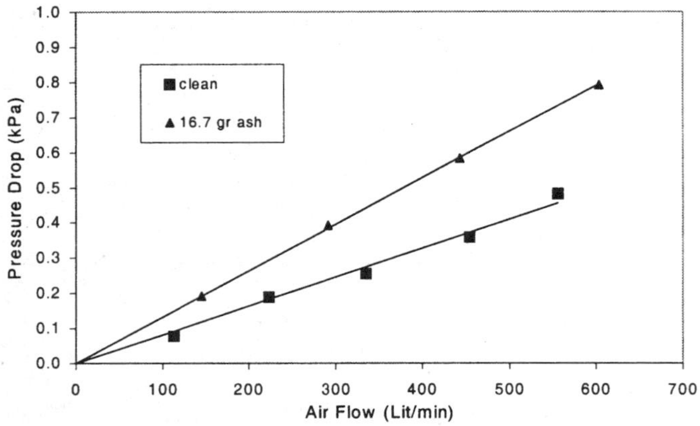

Figure 23. Experimental points of pressure drop for a 100 cpi cordierite 2.66"x12" filter [38] compared with theory (continuous lines) for the clean filter [4] and after accumulation of 16.7 gr of ash [present model based on Eq. (1)].

Based on first principles, one would expect the ash packing density to be of O(10^3) kg/m^3, while its permeability for μm-sized particles to be of O(10^{-14}) m^2. Ash packing density was estimated by an independent experiment where ash collected from a regenerated trap, was poured into a graduated tube and its volume and mass were measured at various stages of compaction. This resulted in an ash packing density between 400 and 1000 kg/m^3. The loading model for these values of packing density then estimates ash layer permeability between 2.77 and 7.4 x 10^{-14} m^2. Applying the Kuwabara model for the permeability [4, 21, 39] one can estimate that these permeabilities are consistent with an "effective" ash particle size between 0.5 and 2 μm. Measurements of ash particle size distribution were conducted with two techniques available at the APT Laboratory: an aerodynamic time-of-flight individual particle count, technique with a minimum detection limit of 0.2 μm (TSI Aerosizer) and a laser ensemble diffraction technique (Malvern Mastersizer). Figure 24 shows the ensemble diffraction based ash particle size distribution, while Figure 25 shows the time-of-flight measured size distribution (by number, area and volume). Considering that the optical technique measures some large agglomerates (peak around 100 μm in Figure 24) which are broken down by the dispersion in the hypersonic flow field of the Aerosizer, we conclude that both techniques indicate a substantial amount of ash particle volume to lie between 1 and 10 μm.

Ash particles were also observed in the Scanning Electron Microscope (Figure 26). It can be seen that ash particles are aggregates (mainly in the 1-10 μm range) of smaller units. The smaller units have characteristic dimensions that broadly agree with the permeability-based estimates of "ash effective size".

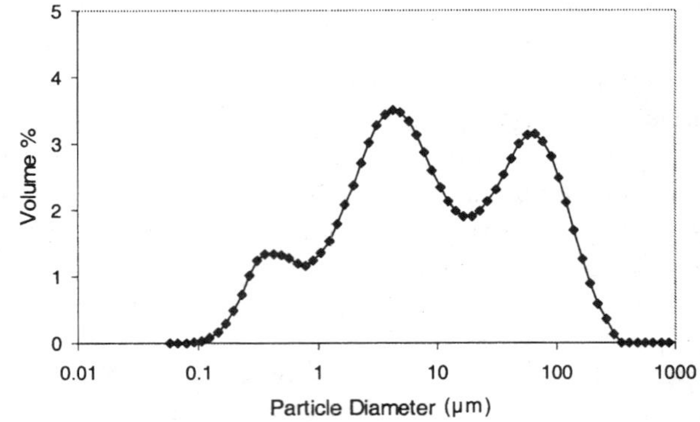

Figure 24. Size distribution of ash measured with the Mastersizer.

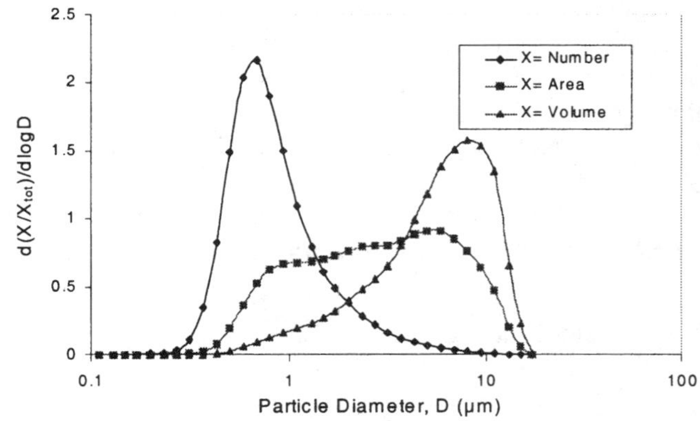

Figure 25. Size distribution of ash measured with the Aerosizer.

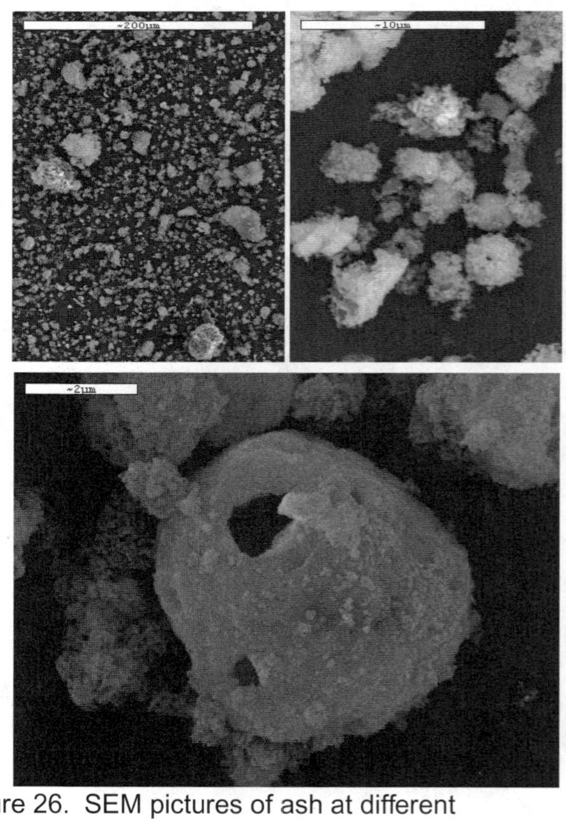

Figure 26. SEM pictures of ash at different magnifications.

In the case of fuel-borne catalysts, the size distribution of the resulting ash most likely depends on the catalyst dose. Burtcher et al. [40] presented interesting experimental data on this phenomenon (Figure 27), for a Cerium fuel-borne catalyst. In that study, a locus of catalyst dosage is seen to exist below which the Cerium is scavenged by the soot particles, and above which Cerium nucleates homogeneously into (oxide) nano-particles giving rise to a distinct mode in the size distribution [40, 41].

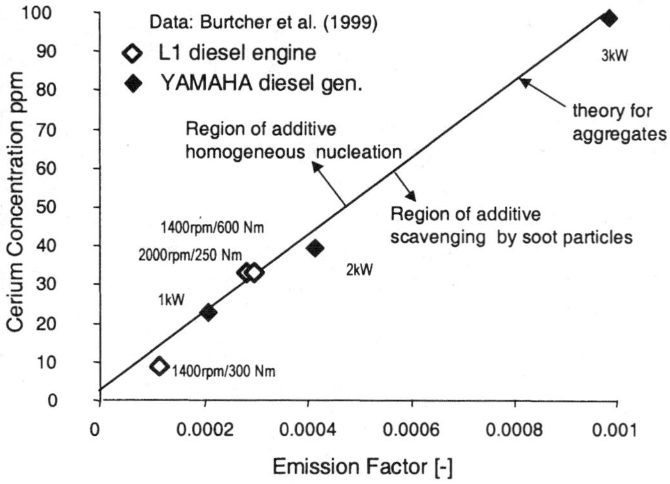

Figure 27. Cerium fuel-borne catalyst homogeneous nucleation locus. Experiments [40, 41] and theory [42].

Assuming that the scavenging rate is proportional to the available particulate area it can be shown [42] that for soot aggregates the points on Figure 27 should fall on a straight line. As seen in Figure 27 this is well supported by the experiments. We expect the activity of the fuel-borne catalyst to become less above the line in Figure 27, as it will exist in the form of individual particles that would exhibit poorer contact with the diesel particulate than the catalyst scavenged directly and incorporated inside the particulate. We anticipate thus that any dose above the locus in Figure 27 would contribute more into the ash aging of the filter, rather than to the improvement of its regeneration and therefore it should be avoided, to prevent fast aging of the filter pressure drop. Konstandopoulos and Kostoglou [32] have found analyzing the data of Tan et al. [43] for a Copper fuel-borne catalyst, that a dosage of 33 ppm was already in contact with 95% of the available soot surface area and therefore increasing the dose to 66 ppm would induce no further enhancement, a fact supported by the experiments of [43].

Combining the available information on ash properties a simulation of ash accumulation into 9"x12" cordierite filters of 100 and 200 cpi exposed to the exhaust of a diesel engine at 0.33 actual m^3/sec and 310 °C was performed for two values of ash emission rates, namely 7 and 14 gr/1000km.

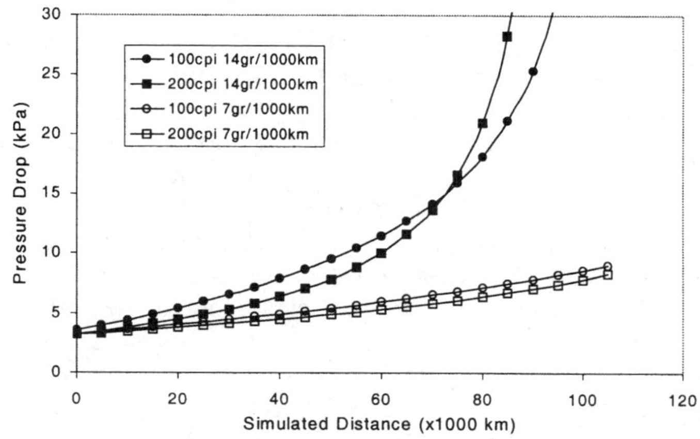

Figure 28. Pressure drop aging simulation of a 9"x12" cordierite wall-flow filter.

It is interesting to note in Figure 28 for the higher ash emission simulations, that after about 75000 km the 200 cpi filter loses its better aging behavior (lower pressure drop rise due to ash accumulation) over the 100 cpi filter, as both filters start to become severely plugged with ash. The results shown in Figure 28 illustrate the utility of the aging model for performing Life Cycle Analysis as well as in supporting the selection process of diesel particulate filters.

CONCLUSIONS

Optimized diesel particulate emission control system design encompasses selection of porous media, filter geometric configuration and regeneration strategies. The process can be significantly aided through the use of validated mathematical models based on rigorous theoretical descriptions of the fundamental underlying physicochemical processes. In the present paper we have outlined the development of state-of-the-art, experimentally validated analytical and numerical models for three critical aspects of the operation of diesel particulate emission control systems, namely transient filter loading, catalyzed/NO_2-assisted filter regeneration and filter pressure drop aging due to ash accumulation. The major findings in each of these areas are stated below:

TRANSIENT FILTER LOADING

1. By extending the clean filter pressure drop model of [17], an analytical expression has been derived Eq. (1), for the pressure drop of arbitrarily loaded filters that operate in the cake filtration regime. The expression was shown to be in excellent agreement with "presumably exact" 3-D CFD simulations and with experimental data collected with honeycomb filters fitted in the exhaust of a diesel engine. The simplicity of the model makes it a valuable tool for optimized preliminary diesel particulate filter selection and design.

2. To describe the initial deep-bed filtration regime of some filters a transient filtration simulator was developed (*FilSim* v. 1.0) applying the notion of "unit cell" filtration theory, locally across the filter wall. The simulator predicts the transient collection efficiency and pressure drop of diesel particulate honeycomb filters with arbitrary wall structure and was shown to describe with excellent accuracy transient loading experiments, available in the literature for a variety of filter grades. The validated simulator can be used for designing tailored filter structures for particular applications, e.g. for catalyzed filter substrates with optimized diesel particle-catalyst contact (currently in progress).

FILTER REGENERATION

1. The regeneration process of prototype small scale catalyzed filter disks, loaded in-situ from the exhaust of a diesel engine was experimentally studied in the presence of NO_2 in a specially constructed testing unit. The results indicate the potential benefits that catalyzed filters can bring to NO_2-regenerative technologies, through the in-situ re-oxidation of NO (produced by the carbon-NO_2 reaction) back to NO_2 alleviating the dependence on high engine-out NO_X concentrations.
2. A previously published regeneration model by the authors [14, 31, 32] was extended to account for the presence of NO_2 on particulate oxidation and was shown to describe very well the collected experimental data. The model is very useful for the design of next-generation NO_2-regenerative technologies employing catalyzed filters.

FILTER PRESSURE DROP AGING DUE TO ASH ACCUMULATION – The transient filtration model was adapted to describe the process of ash accumulation after repeated regenerations of the filter. Using filter pressure drop aging data from the literature and simple characterizations of ash particles collected from regenerated filters, a validated filter pressure drop aging model was developed. The aging model can be used for performing filter Life Cycle Analysis as well as in supporting the sizing process of diesel particulate filters for applications where extended life time is required (e.g. in heavy-duty trucks).

The integration of the derived models into the practice of engine/vehicle manufacturers and their suppliers (currently in progress) will lead to a systematic and hence more efficient, concurrent engineering procedure for optimized design of emission reduction components and their interface into complete systems.

ACKNOWLEDGMENTS

Our work in this area has been supported in part by the General Secretariat of Research and Technology (GSRT) of the Hellenic Republic through OPRT-II, Project Grant No. 127 and the European Commission through BRITE-EURAM III Projects DIDTREAT and CERFIL. We thank our respective consortium partners for their permission to publish these results.

We are also grateful to our colleagues in the APT Laboratory: Mr. C. Altiparmakis and Mr. K. Moustakeas (for assistance with engine runs), Dr. M. Nikitidis (for the data acquisition interface), Dr. I. Mavroidis (for assistance with CFD and graphics), Mrs. G. Macheridou and Mrs K. Karadimitra (for ash characterizations).

We also thank Dr. J. A. Martens (K. Univ. Leuven) for providing a catalyst powder sample, Dr. J. Warren (Johnson Matthey) for providing an ash sample and discussions regarding ash properties, Dr. G. Elfinger and Mr. H. Bussmann (Zeuna Staerker, GmbH) for the data in Figures 4 and 5, Mr. P. Stobbe (formerly of Notox International) for discussions regarding porous structures and filter samples, Dr. J. H. Johnson (Michigan Tech) and Dr. M. Masoudi (Corning, Inc.) for sharing pre-prints of their work. The provision of the engine by Volkswagen A. G. and the support of Dr. A. Koening and Dr. K. Weidmann is deeply appreciated.

REFERENCES

1. **Diesel Exhaust Aftertreatment**, SAE Special Publication, SP-1414, 1999
2. **Diesel Exhaust Aftertreatment**, SAE Special Publication, SP-1313, 1998.
3. **Diesel Exhaust Aftertreatment**, SAE Special Publication, SP-1227, 1997.
4. Konstandopoulos, A. G., Johnson, J. H., "Wall-flow diesel particulate filters-their pressure drop and collection efficiency" *SAE Trans.* **98** sec. 3 (*J. Engines*) *Paper No. 890405*, p. 625, 1989.
5. http://www.dieselnet.com
6. Howitt, J., Montierth, M., "Cellular ceramic diesel particulate filter", *SAE Paper No. 81104*, 1981.
7. Roth, P., Eckhardt, T., Franz, B., Patschull, J. "H_2O_2-assisted regeneration of diesel particulate traps at typical exhaust gas temperatures" *Combust. Flame* **115**, p. 28, 1998.
8. Lueders, H., Stommel, P., Geckler, S. "Diesel exhaust treatment-new approaches to ultra low emission diesel vehicles", *SAE Paper No. 1999-01-0108*, 1999.
9. Christensen, H., Dinesen, J., Engell, H., Hansen, K., "Electrochemical reactor for exhaust gas purification" *SAE Paper No 1999-01-0472*, 1999.
10. Hawker, P., Myers, N., Huthwohl, G., Vogel, H., Bates, B., Magnusson, L., Bronneberg, P., "Experience with a new particulate trap technology in Europe" *SAE Paper No 970182*, 1997.
11. Neeft, J., **Catalytic oxidation of soot; Potential for the reduction of diesel particulate emissions**, Ph.D. Thesis, Tech. Univ. Delft, 1995.
12. Jelles, S., Makkee, M. Moulijn, J., Acres, G., Peter-Hoblyn, J., "Diesel particulate control. Application of

an activated particulate trap in combination with fuel additives at an ultra low dose rate", *SAE Paper No 1999-01-0113*, 1999.

13. http://www.dieselnet.com/news/9904peugeot.html
14. Konstandopoulos, A., Kostoglou, M., "Periodically reversed flow regeneration of diesel particulate traps" *SAE Paper No 1999-01-0469*, 1999.
15. Page D. L., MacDonald J. R., Edgar B. L., "The QuadCAT^{TM} Four-Way Catalytic Converter: An Integrated Aftertreatment System for Diesel Engines", *SAE Paper No 1999-01-2924*, 1999.
16. Oh, S., McDonald, J., Vaneman, G., Hegedus, L., "Mathematical modeling of fibrous filters for diesel particulate-Theory and experiments", *SAE Paper No 810113*, 1981.
17. Konstandopoulos, A., Skaperdas, E., Warren, J., Allansson, R., "Optimized filter design and selection criteria for continuously regenerating diesel particulate traps", *SAE Paper No 1999-01-0468*, 1999.
18. *Auto Emissions Magazine*, Corning Inc., No. 14, 1999.
19. Masoudi, M., Zinc, U., Then, P. M., Cash, T., Thompson, D., Heibel, A., "Predicting pressure drop of diesel particulate filters: modeling and experiment", *SAE Paper No 2000-01-0184*, 2000.
20. Kwetkus, B. A., Egli, W. "The ceramic monolithic filter module: filtration properties and DeNox potential", in **High Temperature Gas Cleaning** (Eds. Schmidt et al.), p.278, Institut fuer Mechanische Verfahrenstechnik und Mechanik, Karlsruhe, 1996.
21. Tien C., **Granular Filtration of Aerosols and Hydrosols**, Butterworths, New York, (1989)
22. Murtagh, M. J., Sherwood, D. L., Socha, L. S. Jr. "Development of diesel particulate filter composition and its effect on thermal durability and filtration performance", *SAE Paper No 940235*, 1994.
23. Suresh, A., Khan, A., Johnson, J. H. "Pressure Drop and Permeability of Clean and Particulate Loaded Filters", *SAE Paper No 2000-01-0476*, 2000.
24. Final Report to the European Commission on BRITE–EURAM III Project No. 1531-BE95-1984, March 1999.
25. Harvey, G. D., Baumgard, K. J. Johnson, J. H. Gratz, L. D., Bagley, S. T., Leddy, D. G. "Effects of a Ceramic Particle trap and Copper Fuel Additive on Heavy-Duty Diesel Emissions", *SAE Paper No 942068*,1994.
26. Rogers, T., "Solving Ultrafine Particulate Emissions", *Engine Technology International* '98 Annual Review p. 17, 1998.
27. Abdul-Khalek, I. S., Kittelson, D. B., Graskow, B. R., Wei Q., Brear F. "Diesel Exhaust Particle Size: Measurement Issues and Trends", *SAE Paper No 980525*, 1998.
28. Press, W.H., Teukolsky, S.A., Vetterling, W.T. and Flannery, B.P., **Numerical Recipes. The art of scientific computing**, Second Edition, Cambridge University Press, Cambridge, 1992.
29. Bissett, E., "Thermal regeneration of particle filters with large conduction" *Math. Model.* **8**, p.1, 1985.
30. Bissett, E., Shadman, F., "Thermal regeneration of diesel particulate monolithic filters" *AIChE J.* **31**, p.753, 1985.
31. Konstandopoulos, A., Kostoglou, M., "A mathematical model of soot oxidation on catalytically coated ceramic filters" *Advances in Vehicle Control and Safety*, Amiens, France, p.137, 1998.
32. Konstandopoulos, A.G., Kostoglou, M. "Reciprocating flow regeneration of soot filters", *Comb. and Flame*, in press, 2000.
33. Bissett, E., "Mathematical model of the thermal regeneration of a wall-flow monolith diesel particulate filter" *Chem. Eng. Sci.* **39**, p.1232, 1984.
34. Aoki, H., Asano, A., Kurazono, K., Kobashi, K., Sami, H., "Numerical simulation model for the regeneration process of a wall-flow monolith particulate filter", *SAE Paper No 930364*, 1993.
35. Kuo, K. **Principles of combustion**, John Wiley & Sons, New York, 1986.
36. Papaioannou, E., Zarvalis, D., Konstandopoulos A. G, "Influence of NO_2 on soot particles oxidation" *Proc. Second Panhellenic Symposium of Chemical Engineers*, p. 325, 1999.
37. Minutes of European Commission *PREMTECH Network Workshop on Aftertreatment*, Oct. 14-15, 1999, FORTH/CPERI, Thessaloniki.
38. Sachdev R., Wong V. and Shahed S., "Effect of ash accumulation of diesel exhaust particulate traps" (SAE 830182)
39. Lee K. W., Gieseke J. A., "Collection of aerosol particles by packed beds", *Env. Sci. Tech.* **13** (4), p.1761, 1978.
40. Burtscher H, Matter U., Skillas G.,"The effect of fuel additives on diesel engine particulate emissions" *J. Aerosol Sci.*, **30**, p. S851, 1999
41. Burtscher H, Skillas G., Baltensperger, U., Matter U.,"Particle Formation due to Fuel Additives" *Proc. 3rd Int. ETH Workshop on Nanoparticle Measurement*, August 9-10, Zurich, 1999.
42. Konstandopoulos, A.G., Kostoglou, M. "A Population Balance Study of the Simultaneous Evolution of Soot and Fuel Additive-Originating Particles", in preparation, 2000.
43. Tan, J., Opris, C., Baumgard, K., Johnson, J. H. "A study of the regeneration process in diesel particulate traps using a Copper fuel additive", *SAE Paper No 960136*, 1996.
44. *VDI-Waermeatlas*, 6. Aufl., VDI-Verlag Duesseldorf, 1992
45. Fluent Inc., **Fluent 5 User's Guide**, July 1998.
46. Friedlander S. K., **Smoke, dust and haze: Fundamentals of aerosol behavior**, John Wiley & Sons, New York, 1977.
47. Konstandopoulos, A. G., "Deposit growth dynamics: particle sticking and scattering phenomena", *J. Powder Tech.*, in press, 1999.

48. Tognotti, L., Longwell J. P. and Sarofim, A. F., *23rd Int. Symp. Comb.*, p.1207, 1990.

49. Cooper B.J., Thoss J.E. "Role of NO in Diesel Particulate Emission Control" *SAE Paper No. 890404*, 1989.

DEFINITIONS, ACRONYMS, ABBREVIATIONS

A: total filtration area
A_o: filtration area of a single channel
b: "unit cell" diameter
$C_{p,1}$: particle layer heat capacity
$C_{p,g}$: gas phase heat capacity
$C_{p,s}$: filter wall heat capacity
d_c: loaded collector diameter
d_{c0}: clean collector diameter
d_p: particle diameter
d_{pore}: pore diameter
D: monolithic trap diameter
D_p: particle diffusion coefficient : $\frac{k_B \cdot T}{3ð \cdot ì \cdot d_p} \cdot SCF$
E: grade collection efficiency
E_{cat}: activation energy for catalytic soot oxidation
E_{f1},E_{f2},E_{f3}: parameters in CO selectivity Eqs.(A.42)-(A.44)
E_{NO}: activation energy for NO oxidation
E_{NO_2}: activation energy for carbon oxidation by NO_2
E_{num}: total collection efficiency by number
E_{th}: activation energy for thermal carbon oxidation
E_{vol}: total collection efficiency by volume
$f(\varepsilon)$: geometric function by Kuwabara, see Eq. (A.39)
f_{CO}^{the}: thermal CO selectivity for carbon oxidation
f_{CO}^{cat}: catalytic CO selectivity for carbon oxidation
F: factor equal to 28.454
$g(\varepsilon)$: geometric function by Kuwabara, see Eq. (A.27)
g_{CO}: CO selectivity for carbon oxidation by NO_2
k: loaded filter wall permeability
k_0: clean filter wall permeability
k_B: Boltzman's constant 1.38×10^{-23} J/moleÞK
k_{f1},k_{f2},k_{f3}: parameters in CO selectivity Eqs. (A.42)-(A.44)
$k_{o,cat}$: frequency factor for catalytic carbon oxidation
$k_{o,th}$: frequency factor for thermal carbon oxidation
$k_{o,NO}$: frequency factor for NO oxidation
k_{o,NO_2}: frequency factor for carbon oxidation with NO_2
k_{soot}: particulate layer permeability
Kn_p: particle Knudsen number, $2\lambda/d_p$
L: effective channel length
$\dot{m}$: exhaust gas mass flow rate
$\dot{m}_p$: particulate mass flow rate per unit collector
m_c: deposited particulate mass per unit collector
m_1: particulate mass in Layer I
$m_{1,0}$: initial particulate mass in Layer I
m_2: particulate mass in Layer II
$m_{2,0}$: initial particulate mass in Layer II
m_b: initial total particulate mass
m_{soot}: particulate mass collected in DPF
m_w: particulate mass in the filter wall
M_c: molecular weight of carbon deposit
M_{NO_2}: molecular weight of NO_2
M_{O_2}: molecular weight of oxygen
MW: gas molecular weight
n: exponent in Eq. (A.47)
N_{cells}: number of inlet cells of wall-flow filter
N_R: interception parameter, Eq. (A.30)
Pe: Peclet number, Eq. (A.24)
Q: exhaust volumetric flow rate
Q_o: exhaust volumetric flow rate of a single channel
R: Universal gas constant
Re: trap entrance Reynolds number, $U\alpha / \nu$
R_j^i: depletion rate of species *j* in Layer *i*
s: interception exponent, Eq. (A.29)
S_p: specific area of deposit layer
T: temperature of the filter wall (used in Appendix IV)
T: gas temperature
T_I: temperature of exhaust gases (used in Appendix IV)
U: inlet channel entrance velocity
U_I: interstitial or pore velocity, Eq. (A.25)
u_w: filter wall/approach velocity
V_{trap}: filter volume
w: particulate layer thickness
w_s: filter wall thickness
x: filter wall thickness coordinate
y_I: mole fraction of species *i* in the exhaust gas
$y_{w,I}$: local mole fraction of species *i* in the wall
Y_I: mass fraction of species *i* in the exhaust gas
$Y_{w,I}$: local mass fraction of species *i* in the wall

Greek letters

α: honeycomb filter cell size
β: particulate-catalyst contact fraction
γ: exponent in Eq. (A.47)
$\Delta H_{NO_2}^{CO}$: specific heat of CO formation (oxidation by NO_2)
$\Delta H_{O_2}^{CO}$: specific heat of CO formation (oxidation by O_2)
$\Delta H_{O_2}^{CO_2}$: specific heat of CO_2 formation (oxidation by O_2)
$\Delta H_{NO_2}^{CO_2}$: specific heat of CO_2 formation (oxidation by NO_2)
ΔP: pressure drop across the filter
ε: loaded filter porosity
ε_0: clean filter porosity
λ: exhaust gas-mean-free-path
μ: exhaust dynamic viscosity
μ_1,μ_2,μ_3: parameters in CO selectivity Eqs. (A.42)-(A.44)
ν: exhaust kinematic viscosity
ξ: interaction parameter between Layers I and II [14, 31]
ρ: exhaust gas density
$\rho_{soot,c}$: particulate packing density in cake deposit
$\rho_{soot,w}$: particulate packing density inside filter wall
ρ_{sub}: bulk density of filter substrate
σ: honeycomb filter cell density
φ: initial soot mass fraction in Layer I
ψ: "unit cell" bridging onset parameter in Eq.(A.38)

APPENDIX I

LOADED FILTER MODEL

Analytical – The clean filter flow model of Konstandopoulos and Johnson [4] is here extended for high particulate loading effects. The loading is assumed to be uniform across the filter channels since the flow would self-adjust to follow the path of least resistance as the filter gets loaded. This is also supported by separate deposit growth simulations into the filter channels (see also [14]). The basic equation for the pressure drop of a clean wall flow filter is [4,17]:

$$\Delta P_{clean} = \frac{\mu}{k_o}\frac{U\alpha}{4L} w_s + \frac{2\mu F}{3\alpha^2} UL \qquad \text{(A1)}$$

where the first term represents losses through the porous walls and the second term frictional losses along the filter channel.

Expressing the channel inlet velocity in terms of the total exhaust gas volume flow Q :

$$U = \frac{8Q}{\pi D^2 \sigma \alpha^2} \qquad \text{(A2)}$$

$$\text{where} \quad \sigma = \frac{1}{(\alpha + w_s)^2} \qquad \text{(A3)}$$

Eq. (A1) takes the form :

$$\Delta P_{clean} = \frac{\mu Q}{2V_{trap}} (\alpha + w_s)^2 \left[\frac{w_s}{k_o \alpha} + \frac{8FL^2}{3\alpha^4} \right] \qquad \text{(A4)}$$

The clean filter model is extended to the arbitrary loaded filter case employing the *ansatz* that the second term of Eq. (A1) can be split as:

$$\Delta P = \underbrace{\frac{\mu}{k_o}\frac{U\alpha}{4L} w}_{wall} + \underbrace{\frac{1\mu F}{3\alpha^2} UL}_{inlet\ channel} + \underbrace{\frac{1\mu F}{3\alpha^2} UL}_{outlet\ channel} =$$

$$= \frac{\mu Q}{2V_{trap}} (\alpha + w_s)^2 \left[\underbrace{\frac{w_s}{k_o \alpha}}_{wall} + \underbrace{\frac{4FL^2}{3\alpha^4}}_{inlet\ channel} + \underbrace{\frac{4FL^2}{3\alpha^4}}_{outlet\ channel} \right] \qquad \text{(A5)}$$

and that the splitting remains valid for a loaded filter as well. The pressure drop of a loaded filter then consists of four contributions:

$$\Delta P = \Delta P_{filter\ wall} + \Delta P_{soot\ layer} + \Delta P_{inlet\ channel} + \Delta P_{outlet\ channel} \qquad \text{(A6)}$$

where the first two components obey Darcy's law and the other two are obtained taking into account the real flow area and its effects on axial velocities (see below Eqs. (A11) and (A12)), separately for each channel.

The first term is given as in Eq. (A4) by:

$$\Delta P_{filter\ wall} = \frac{\mu Q}{2V_{trap}} (\alpha + w_s)^2 \left(\frac{w_s}{k_o \alpha} \right) \qquad \text{(A7)}$$

The second term of Eq. (A6) can be evaluated by integrating Darcy's law across the deposit thickness, to account for the variation of the cross sectional flow area, A_o which is significant for high deposit loadings (see Figure A1):

$$\Delta P_{soot\ layer} = \frac{\mu}{k_{soot}} \int_0^w u(x)dx = \frac{\mu}{k_{soot}} \int_o^w \frac{Q_o}{A_o} dx = \frac{\mu Q_o}{4Lk_{soot}} \int_0^w \frac{1}{(\alpha - 2w + 2x)} dx = \frac{\mu Q (\alpha + w_s)^2}{L\pi D^2 k_{soot}} ln\left(\frac{\alpha}{\alpha - 2w} \right) \qquad \text{(A8)}$$

For a loaded filter the volume of particulate mass deposited in one inlet channel (see Figure A1) is given by:

$$V_{soot} = \frac{m_{soot}}{N_{cells}\rho_{soot,c}} = \left[\alpha^2 L - (\alpha - 2w)^2 L \right] = 4\alpha wL\left(1 - \frac{w}{\alpha} \right) \qquad \text{(A9)}$$

Re-arranging gives the particulate layer thickness in terms of the deposited mass:

$$w = \frac{\alpha - \sqrt{\alpha^2 - \dfrac{m_{soot}}{N_{cells} L \rho_{soot,c}}}}{2} \qquad \text{(A10)}$$

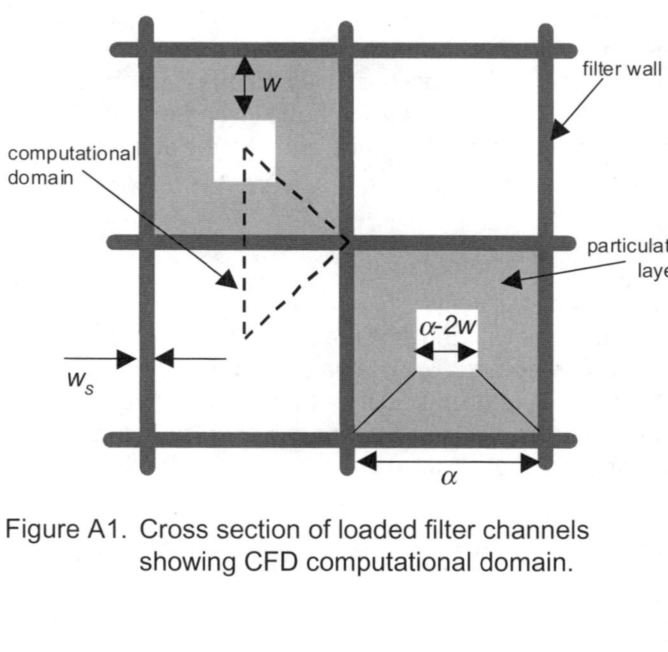

Figure A1. Cross section of loaded filter channels showing CFD computational domain.

For the inlet (loaded) channel the effective channel width is given by:

$$\alpha^* = \alpha - 2w \tag{A11}$$

resulting in the following value for inlet channel axial velocity:

$$U_1^* = \frac{8Q}{\pi D^2 \sigma (\alpha - 2w)^2} \tag{A12}$$

Applying the previously mentioned *ansatz*, the loaded inlet channel losses are then written from Eq. (A 5), as:

$$\Delta P_{inlet\ channel} = \frac{\mu Q}{2V_{trap}}(\alpha + w_s)^2 \left(\frac{4FL^2}{3(\alpha - 2w)^2} \right) \tag{A13}$$

while the outlet channel losses remain the same as for a clean filter:

$$\Delta P_{outlet\ channel} = \frac{\mu Q}{2V_{trap}}(\alpha + w_s)^2 \left(\frac{4FL^2}{3\alpha^2} \right) \tag{A14}$$

The final equation for the pressure drop is obtained by summing up Eqs. (A7, A8, A13, A14):

$$\Delta P = \frac{\mu Q}{2V_{trap}}(\alpha + w_s)^2 \left[\frac{w_s}{k_o \alpha} + \frac{1}{2k_{soot}} ln\left(\frac{\alpha}{\alpha - 2w} \right) + \frac{4FL^2}{3}\left(\frac{1}{(\alpha - 2w)^4} + \frac{1}{\alpha^4} \right) \right] \tag{A15}$$

The effect of non-Darcian flow in the porous wall can be straightforwardly included in the model as shown in [17]. It should be noted though that it is not desirable, from a practical point of view, to operate DPF's in the non Darcian regime, as there is a penalty of increased pressure drop for the same useful filtration area.

Note that in the above derivations the entrance and exit pressure losses due to the contraction and expansion of the flow into and out of the monolith channels have not been taken into account, as their contribution is negligible. This is shown by the following analysis. The pressure drop at the entrance and at the exit are given by [44]:

$$\Delta P_{entrance} = \zeta_E \frac{\rho U^2}{2} \tag{A16}$$

ζ_E is a factor varying from 0.2 to 0.8 according to area blockage ratio and *Re*. The values for these components of the pressure drop are of $O(10^{-2}\text{-}10^{-3})$ compared to the total pressure drop of the loaded filter, hence they are considered negligible.

<u>3-D CFD Validation</u> – To assess the validity of the 1-D analytical model, 3-D CFD simulations were performed on a single channel of a DPF (Table A1), for various particulate loadings. The study was performed using a commercial CFD package [45]. The flow is steady and laminar as the Reynolds number based on the channel width is around 300. Taking advantage of the symmetry of the geometry of the problem, and of the unstructured mesh capabilities of the code, only the triangular section shown in Figure A1 is simulated.

Table A1. Input Data for CFD Simulations

Inlet Velocity U (m/sec)	5.85 – 29.62
Gas density, ρ (kg/m^3)	0.64
Viscosity, μ (kg/m/sec)	2.87x10^{-5}
Wall permeability, k_o (m^2)	5.3x10^{-13}
Deposit permeability (m^2)	3.2x10^{-14}
Channel width, α (mm)	1.44
Channel length, L (mm)	144
Wall thickness, w_s (mm)	0.356
Deposit thickness w (mm)	0.0 – 0.4

The 3-D unstructured grid shown in Figure A2, consists of 105,000 hexahedral cells, with 300 of them distributed along the channel length. Nodes are clustered towards the porous walls to represent better the boundary layer, and towards the inlet and outlet sides to capture the velocity variations. Each CFD simulation point in Figure A4 takes about 100 minutes of CPU time on a Digital 500 series Alphastation with 320 Mb of RAM.

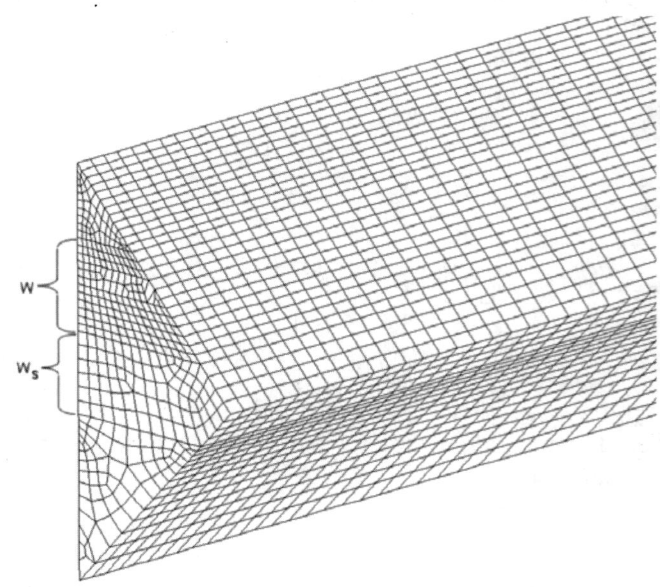

Figure A2. Front part of the 105,000 cells unstructured grid.

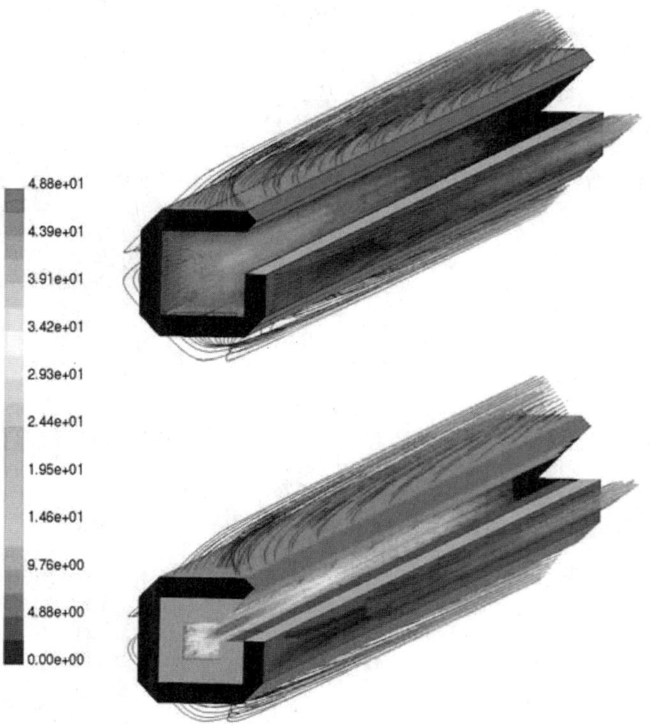

Figure A3. Streamlines colored by velocity magnitude (m/sec) for a clean (top) and a loaded (bottom) channel of a 5.66"x6" 200 cpi filter.

Figure A3 displays the flow streamlines in the inlet channel of the loaded filter of Figure 4. The streamlines of the clean filter are also shown for comparison.

The contributions of each term of Eq. (A6) are plotted in Figure A4, and are shown to be in very good agreement with the 1-D analytical model.

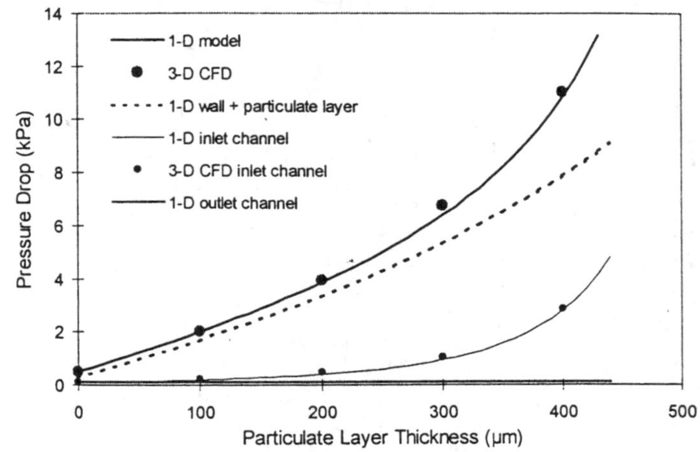

Figure A4. Diagram showing the contributions of the pressure drop as predicted by the analytical model and the CFD simulations.

APPENDIX II

SLIP-FLOW CORRECTION OF PERMEABILITY

When continuum fluid mechanics is applicable, the permeability coefficient entering Darcy's law is an intrinsic property of the porous medium. In this case the influence of temperature on filter pressure drop is due only to the variation of the exhaust gas viscosity and density with temperature. This translates into a power-law dependence of the filter pressure drop on temperature.

When the mean-free-path of the gas molecules approaches the pore size of the porous medium, it is necessary to account for slip and transitional flow effects, which depend on the magnitude of the local Knudsen, *Kn* number:

$$Kn = \frac{2\lambda}{d_{pore}} \tag{A17}$$

with the mean-free-path of the gas given by:

$$\lambda = \nu \sqrt{\frac{\pi MW}{2RT}} \tag{A18}$$

Exploiting the similar problem for the non-continuum drag of a spherical particle [46] we employ the Stokes-Cunningham Factor (*SCF*) to correct for slip-flow effects on the permeability:

$$SCF = 1 + Kn\left(1.257 + 0.4e^{-1.1/Kn}\right) \tag{A19}$$

$$k_{slip\ flow} = SCF \cdot k_{continuum} \tag{A20}$$

As seen in Figure 2 this approach captures very well the experimental data [20] for a membrane coated filter. In that case the membrane Knudsen number varies from 0.3 to 0.7 as the temperature varies from 20 °C to 400 °C.

APPENDIX III

TRANSIENT FILTRATION MODEL

The transient filtration model is based on applying locally across the filter wall, "unit collector" filtration theory in a self-consistent manner. The filter wall is divided into slabs (Figure A5) and the calculation of the variation of the local properties ("unit collector" diameter, porosity, permeability etc.) is performed by tracking the amount of particulate mass accumulated in each slab. The "unit collector" model used here, is that of spherical collectors [4,39] since it is closer to the granular structure of extruded ceramic and sintered metal filters. However any other "unit collector" model can be employed for other filter structures (e.g. fibers).

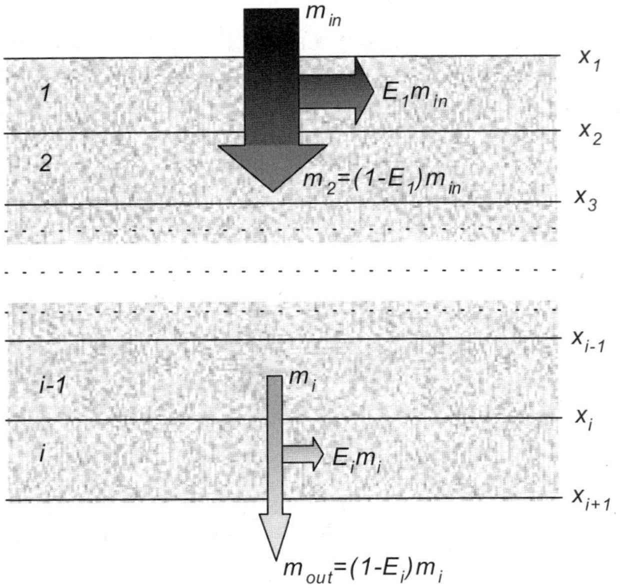

Figure A5. Schematic representation of filter wall discretization into slabs of "unit cell/ collectors".

Representing the microstructure of the filter by spherical "unit cells" of size b, each hosting a collector of size d_c, the "unit collector" is assumed to collect particles through the mechanisms of Brownian diffusion and interception [4,39] with combined efficiency denoted by η_{DR}. The collection efficiency E for the clean filter can then be related to η_{DR} by integrating across the filter wall thickness (see [4,39] and references therein):

$$E = 1 - exp\left[-\frac{3\eta_{DR} \cdot (1-\varepsilon_0) \cdot w_s}{2\varepsilon_0 d_{c0}} \right] \tag{A21}$$

The "unit collector" size is related to the "unit cell" size through the porosity according to:

$$\frac{b^3}{d_{c_0}^3} = (1-\varepsilon_0) \tag{A22}$$

The single spherical collector efficiency for Brownian diffusion is given by (see [4,39] and references therein):

$$\eta_D = 3.5 \cdot g(\varepsilon) \cdot Pe^{-2/3} \tag{A23}$$

where the Peclet number is given by:

$$Pe = \frac{U_i \cdot d_c}{D_p} \tag{A24}$$

and U_i represents the "interstitial" or "pore" velocity related to the approach velocity u_w by :

$$U_i = \frac{u_w}{\varepsilon} \tag{A25}$$

The unit collector diameter d_{c0} for a clean filter is given by [4,39]:

$$d_{c0} = \frac{3}{2} \cdot \frac{(1-\varepsilon_0)}{\varepsilon_0} \cdot d_{pore} \tag{A26}$$

and the geometric function $g(\varepsilon)$ for the Kuwabara "unit cell" is [4,39]:

$$g(\varepsilon) = \left[\frac{\varepsilon}{2 - \varepsilon - \frac{9}{5} \cdot (1-\varepsilon)^{1/3} - \frac{1}{5} \cdot (1-\varepsilon)^2} \right]^{1/3} \tag{A27}$$

while the efficiency due to interception is given by [4,39]:

$$\eta_R = 1.5 \cdot N_R^2 \cdot \frac{[g(\varepsilon)]^3}{(1+N_R)^s} \tag{A28}$$

with the exponent s defined as :

$$s = \frac{3-2\varepsilon}{3\varepsilon} \tag{A29}$$

and the interception parameter N_R computed as [4,39]:

$$N_R = \frac{d_p}{d_c} \tag{A30}$$

Assuming the two collection mechanisms to be independent of each other, the combined collection efficiency η_{DR} can be expressed as [4,39]:

$$\eta_{DR} = \eta_D + \eta_R - \eta_D \cdot \eta_R \tag{A31}$$

The particulate mass deposited in the filter m_c satisfies:

$$\frac{\partial \dot{m}_p}{\partial x} + \frac{\partial m_c}{\partial t} = 0 \tag{A32}$$

$$\frac{\partial \dot{m}_p}{\partial x} = -e_f(m_c)\dot{m}_p \tag{A33}$$

The above hyperbolic system is solved numerically by discretizing the filter depth into slabs as shown in Figure A5. As mass enters the filter, it is distributed in it according to the local collection efficiency in each slab E_i.

The mass deposited at each slab is modifying the local "unit collector" size according to:

$$d_c(i,t) = 2 \cdot \left[\frac{3}{4\pi} \cdot \frac{m_c(i,t)}{\rho_{soot,w}} + \left(\frac{d_{c_0}}{2} \right)^3 \right]^{1/3} \tag{A34}$$

Respectively the local porosity ε of the filter as a function of time is given by:

$$\varepsilon(i,t) = 1 - \left(\frac{d_c(i,t)}{d_{c_0}} \right)^3 \cdot (1-\varepsilon_0) \tag{A35}$$

and the slab collection efficiency evolution $E(i,t)$ is computed from:

$$E(i,t) = \int_{x_i}^{x_{i+1}} e_f(m_c(x,t))dx =$$

$$1 - exp\left[-\frac{3\eta_{DR}(i,t)\cdot(1-\varepsilon(i,t))\cdot(x_{i+1}-x_i)}{2\varepsilon(i,t)d_c(i,t)} \right] \quad \text{(A36)}$$

The particles are collected initially inside the filter wall but eventually, they form a cake on its frontal area. It is therefore useful to define a partition coefficient Φ ($0 \leq \Phi \leq 1$) that determines the fraction of mass collected in the first slab that contributes to cake formation. Based on geometrical grounds this partition coefficient is expected to depend on the dimensionless blocked-area fraction at the scale of the unit "collector". Φ is assumed to be given by :

$$\Phi(t) = \left(\frac{(d_c(1,t))^2 - d_{c_0}^2}{(\psi \cdot b)^2 - d_{c_0}^2} \right) \quad \text{(A37)}$$

where ψ is a dimensionless "percolation" control constant ($0<\psi<1$), that determines the onset of pore bridging, which has to be estimated from experimental data or detailed discrete particle dynamics simulations using the methods of [47].

The local permeability of the filter wall changes as the particles deposit on each "unit collector". The local permeability k_i of the loaded filter is related to the permeability of the clean filter k_0 by:

$$\frac{k(i,t)}{k_0} = \left(\frac{d_c(i,t)}{d_{c_0}} \right)^2 \cdot \frac{f(\varepsilon(i,t))}{f(\varepsilon_0)} \quad \text{(A38)}$$

where $f(\varepsilon)$ is a geometric function defined as follows for the Kuwabara "unit cell" [4,39]:

$$f(\varepsilon) = \frac{2}{9} \cdot \frac{\left[2 - \frac{9}{5}\cdot(1-\varepsilon)^{1/3} - \varepsilon - \frac{1}{5}\cdot(1-\varepsilon)^2 \right]}{(1-\varepsilon)} \quad \text{(A39)}$$

APPENDIX IV

NO_2 SOOT OXIDATION

In this Appendix the incorporation of NO_2-assisted particulate oxidation into the regeneration model framework of Konstandopoulos and Kostoglou [14, 31, 32] is performed.

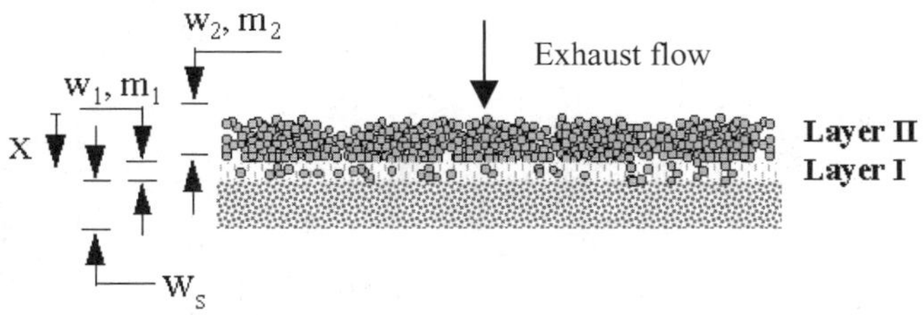

Figure A6. Schematic illustrating the filter wall, catalytic coating influenced Layer I and overlaid particulate deposit Layer II.

With reference to Figure A6, exhaust gases pass perpendicular to the wall of the filter with a specified mass flux imposed by the exhaust flow rate (which in general is time dependent), $\dot{m}$(t) and the total filtration area of the filter, A, satisfying the following balance equation :

$$\frac{d}{dx}(\rho u_w) = 0 \quad \text{(A40)}$$

which upon integration along the particle/wall layer (Figure A6) becomes:

$$\rho u_w = \frac{\dot{m}}{A} \quad \text{(A41)}$$

For analytic simplicity, particulate oxidation [see Eqs. (2), (3)] is assumed to follow a first-order heterogeneous oxidation both in the presence and absence of catalyst but with, in principle, different reaction CO selectivity for the non-catalyst and catalyst influenced reactions, denoted as f_{CO}^{the} and f_{CO}^{cat} respectively, and assumed to have a temperature dependence of the following form [48]:

$$f_{CO}^{the} = \frac{1}{1 + k_{f1}\, y_{O_2}^{\mu_1}\, e^{E_{f1}/RT}} \quad \text{(A42)}$$

$$f_{CO}^{cat} = \frac{1}{1 + k_{f2}\, y_{O_2}^{\mu_2}\, e^{E_{f2}/RT}} \quad \text{(A43)}$$

The CO selectivity g_{CO} for the first-order [49] oxidation reaction of soot by NO_2, is assumed to have also a similar to Eqs. (A42)-(A43) temperature dependence:

$$g_{CO} = \frac{1}{1 + k_{f3}\, y_{O_2}^{\mu_3}\, e^{E_{f3}/RT}} \quad \text{(A44)}$$

The oxygen and NO_2 balance equations in terms of the local oxygen mass fraction Y_{w,O_2} and local NO_2 mass fraction Y_{w,NO_2} for each layer can then be written as follows:

Layer II – In this layer soot oxidation with oxygen occurs only through the non-catalytic path.

O_2 mass balance

$$\frac{d}{dx}(\rho u_w Y_{w,O_2}) = -S_p \rho Y_{w,O_2} k_{th}(T)(1 - f_{CO}^{the}/2) \tag{A45}$$

NO_2 mass balance

$$\frac{d}{dx}(\rho u_w Y_{w,NO_2}) = -S_p \rho Y_{w,NO_2} k_{NO_2}(T)(2 - g_{CO}) \tag{A46}$$

Layer I – In this layer occurs additional production of NO_2 from NO and O_2. A general kinetic expression is assumed for this reaction (see [49] for the reaction orders):

$$\frac{dy_{NO_2}}{dt} = k_{NO} y_{O_2}^{\gamma} y_{NO}^{n} \tag{A47}$$

with an Arrhenius form for the rate constant:

$$k_{NO}(T) = k_{o,NO} e^{-(\frac{E_{NO}}{RT})} \tag{A48}$$

O_2 mass balance

$$\frac{d}{dx}(\rho u_w Y_{w,NO_2}) = -S_p \rho Y_{w,O_2} \times$$
$$[(1-\beta)k_{th}(T)(1 - f_{CO}^{the}/2) + \beta k_{th}(T)(1 - f_{CO}^{the}/2)]$$
$$-\frac{1}{2}\frac{k_{NO}(T)}{M_{O_2}} y_{w,NO}^{n} y_{w,O_2}^{\gamma} \tag{A49}$$

NO_2 mass balance

$$\frac{d}{dx}(\rho u_w Y_{w,NO_2}) = -S_p \rho Y_{w,NO_2} \times$$
$$k_{NO_2}(T)(2 - g_{CO}) + \frac{k_{NO}(T)}{M_{NO_2}} y_{w,NO}^{n} y_{w,NO_2}^{\gamma} \tag{A50}$$

The local NO mole fraction $y_{w,NO}$ is obtained from the total nitrogen atoms conservation:

$$y_{w,NO} = y_{NO} + y_{NO_2} - y_{w,NO_2} \tag{A51}$$

In Eq. (A49) it has been assumed that inside the catalytically affected layer, carbon oxidation proceeds according to two paths, a catalytic path over a certain fraction of the carbon specific area, β and a non-catalytic (thermal) path over the rest. This is the simplest way to account for the effect of incomplete contact between particles and catalytic sites. Each path has in principle its own activation energy and frequency factor assumed to be given by a modified-Arrhenius form [14, 30]:

$$k_{th}(T) = k_{o,th} T e^{-(\frac{E_{th}}{RT})} \tag{A52}$$

$$k_{cat}(T) = k_{o,cat} T e^{-(\frac{E_{cat}}{RT})} \tag{A53}$$

The rate constant for carbon oxidation by NO_2 is assumed to have an Arrhenius form:

$$k_{NO_2}(T) = k_{o,NO_2} e^{-(\frac{E_{NO2}}{RT})} \tag{A54}$$

The exhaust gas density is related to temperature via the ideal gas law. Upon integration across the respective layer thicknesses the total rate of oxygen and NO_2 depletion per layer can be derived as shown below. In Layer I we distinguish between oxygen depleted in the non-catalytic and catalytic paths:

O_2 and NO_2 depletion rates

$$R_{O_2}^{II} = \rho u_w Y_{O_2}(1 - e^{\frac{-S_p k_{th}(T)(1-f_{CO}^{the}/2)w_2}{u_w}}) \tag{A55}$$

$$R_{NO_2}^{II} = \rho u_w Y_{NO_2}(1 - e^{\frac{-S_p k_{NO_2}(T)(1-f_{CO}^{the}/2)w_2}{u_w}}) \tag{A56}$$

$$R_{O_2}^{I,th} = \rho S_p(1-\beta)k_{th}(T)(1 - f_{CO}^{the}/2)\int_{w_2}^{w_1+w_2} Y_{w,O_2}dx \tag{A57}$$

$$R_{O_2}^{I,cat} = \rho S_p \beta k_{th}(T)(1 - f_{CO}^{cat}/2)\int_{w_2}^{w_1+w_2} Y_{w,O_2}dx \tag{A58}$$

$$R_{NO_2}^{I} = \rho S_p k_{NO_2}(T)(2 - g_{CO})\int_{w_2}^{w_1+w_2} Y_{w,NO_2}dx \tag{A59}$$

Introducing the following decompositions of the total particulate mass m_{soot} and layer thickness w (where the subscript 0 denotes initial conditions):

$$m_{soot} = m_1 + m_2$$
$$m_{1,0} = \phi m_b$$
$$m_{2,0} = (1-\phi)m_b \tag{A60}$$

$$w = w_1 + w_2 \tag{A61}$$

and invoking a quasi-steady approximation for the rate of particulate deposit depletion, we can write the following evolution equations for the particulate mass in each layer:

$$\frac{1}{A}\frac{dm_1}{dt} = -(1-\xi)(\frac{M_c}{M_{O_2}}\frac{1}{1-f_{CO}^{the}/2}R_{O_2}^{I,th} +$$
$$\frac{M_c}{M_{O_2}}\frac{1}{1-f_{CO}^{cat}/2}R_{O_2}^{I,cat} + \frac{M_c}{M_{NO_2}}\frac{1}{2-g_{CO}}R_{NO_2}^{I}) \tag{A62}$$

$$\frac{1}{A}\frac{dm_2}{dt} = -(\frac{M_c}{M_{O_2}}\frac{1}{1-f_{CO}^{the}/2}R_{O_2}^{II} +$$
$$\frac{M_c}{M_{NO_2}}\frac{1}{2-g_{CO}}R_{NO_2}^{II})$$
$$-\xi(\frac{M_c}{M_{O_2}}\frac{1}{1-f_{CO}^{the}/2}R_{O_2}^{I,th} +$$
$$\frac{M_c}{M_{O_2}}\frac{1}{1-f_{CO}^{cat}/2}R_{O_2}^{I,cat} + \frac{M_c}{M_{NO_2}}\frac{1}{2-g_{CO}}R_{NO_2}^{I}) \tag{A63}$$

In writing these equations we have introduced a "mixing parameter", x to account for the interaction between the two layers [14]. In the above developments we have assumed that regeneration occurs so fast, that it is permissible to assume that no further deposit growth happens as the soot laden, exhaust gases pass through the filter. It is straightforward though to include a particulate deposit growth term on the right hand side of Eq. (A63), see [14].

Applying a small-Peclet number perturbation expansion similar to that in [29, 30] we find that the filter temperature evolution is given by:

$$\frac{dT}{dt} = \frac{\dot{m}c_{pg}}{(\rho_{soot}c_{p1}w + \rho_{sub}c_{ps}w_s)}\left(\frac{A}{\dot{m}c_{pg}}Q_r + T_i - T\right) \tag{A64}$$

where $T_i(t)$ is the filter inlet temperature (in general a function of engine operating point, hence time).

The oxidation reactions Eqs.(2)-(4) generate heat that is given by:

$$\begin{aligned} Q_r = {} & [\Delta H_{O_2}^{CO} f_{CO}^{the} + \Delta H_{O_2}^{CO_2}(1 - f_{CO}^{the})] \times \\ & \frac{R_{O_2}^{I,th} + R_{O_2}^{II}}{M_{O_2}(1 - f_{CO}^{the}/2)} + [\Delta H_{O_2}^{CO} f_{CO}^{cat} + \\ & \Delta H_{O_2}^{CO_2}(1 - f_{CO}^{cat})]\frac{R_{O_2}^{I,cat}}{M_{O_2}(1 - f_{CO}^{cat}/2)} + \\ & [\Delta H_{NO_2}^{CO} g_{CO} + \Delta H_{NO_2}^{CO_2}(1 - g_{CO})] \times \\ & \frac{R_{NO_2}^{I} + R_{NO_2}^{II}}{M_{NO_2}(2 - g_{CO})} + \end{aligned} \tag{A65}$$

where the respective heats of reaction have been computed based on the stoichiometry of the reactions, accounting for finite CO selectivity.

The mass balances in Layer II are integrated analytically to give the depletion rates in Eqs. (A55) and (A56) and the initial conditions for the integration of mass balances in Layer I

At $x = w_2$ we find:

$$Y_{w,O_2} = Y_{O_2}\left(1 - e^{\frac{-S_p k_{th}(T)(1 - f_{CO}^{the}/2)w_2}{u_w}}\right) \tag{A66}$$

$$Y_{w,NO_2} = Y_{NO_2}\left(1 - e^{\frac{-S_p k_{NO_2}(T)(1 - f_{CO}^{the}/2)w_2}{u_w}}\right) \tag{A67}$$

The mass balances in Layer I cannot be integrated analytically. Eqs. (A57)-(A59) are differentiated with respect to *x* and are combined with Eqs. (A49)-(A50) to form a system of 5 ODE's in the spatial domain. This system is solved with an explicit Runge-Kutta method with constant step (spatial integration). For the time integration of Eqs. (A62) and (A63) an explicit integrator with prespecified accuracy and automatically adapted step is used [28]. Results from application of the model are shown in Figures 20-22.

2007-01-0313

Multi-Instrumental Assessment of Diesel Particulate Filters

Athanasios G. Konstandopoulos, Dimitrios Zarvalis and Ioannis Dolios
Aerosol and Particle Technology Laboratory, CERTH/CPERI

ABSTRACT

As different Diesel Particulate Filter (DPF) designs and media are becoming widely adopted, research efforts in the characterization of their influence on particle emissions intensify. In the present work the influence of a Diesel Oxidation Catalyst (DOC) and five different Diesel Particulate Filters (DPFs) under steady state and transient engine operating conditions on the particulate and gaseous emissions of a common-rail diesel engine are studied. An array of particle measuring instrumentation is employed, in which all instruments simultaneously measure from the engine exhaust. Each instrument measures a different characteristic/metric of the diesel particles (mobility size distribution, aerodynamic size distribution, total number, total surface, active surface, etc.) and their combination assists in building a complete characterization of the particle emissions at various measurement locations: engine-out, DOC-out and DPF-out. The results provide useful guidelines for the selection of various filter media and measuring methodologies. In this paper, among other themes, the inter-comparison of SMPS and EEPS measurements is discussed since these instruments exhibit small but systematic differences. In addition, the evolution of the collection efficiency of each filter medium (evaluated with respect to each of the characteristic metrics measured by each instrument) under steady state and transient engine operation conditions, as a function of soot load of the filter and the particle emissions during filter regeneration is investigated.

INTRODUCTION

Due to the more stringent diesel engine exhaust emissions regulations the use of Diesel Particulates Filters (DPFs) has increased in the automotive industry in the recent years. Every year innovative DPFs appear in the market with the potential to offer high diesel exhaust particulate filtration efficiency without imposing significant penalty to the vehicle fuel consumption [1-5]. On the other hand, even if today's particulate emission regulations are concerned only with the gravimetrically measured particulate matter (PM) emitted, numerous diesel particulate emission analyzers appear in the market offering the ability to measure particle concentration and size distribution based on different measurement principles. Each one of the available instruments shows several advantages as well as disadvantages regarding the ability to measure particle concentration and size distribution as far as reliability, accuracy, time resolution and amount of information are concerned.

In the present study, five diesel exhaust particle analyzers have been used. The engine exhaust layout incorporated a Diesel Oxidation Catalyst (DOC) upstream of the DPF; a typical layout for most automotive applications. Five different DPFs have been evaluated. These DPFs were of different material and filter structures. The range of DPF types examined are representative of what is presently been considered in the modern automotive industry. The scope of the study was two-fold: First, to evaluate the particle analyzers ability to measure at different engine exhaust points; engine-out, DOC-out and DPF-out. Second, to assess the DPFs filtration efficiency under steady state and under transient engine operation conditions.

The paper structure is the following: Initially, the test set-up and testing procedures are presented. Then the steady state particle analyzers results are presented for the inter-comparison of the instrument measurements and the determination of the effect of the DOC and the DPF on the engine exhaust particle emissions. Then the transient results are presented. These results offer the opportunity for the filtration efficiency assessment of the different DPFs employed, as well as the comparison between the different measurement methodologies under transient conditions. Finally the ability of the employed instrumentation to measure soot particle concentration during filter regeneration conditions is addressed.

EXPERIMENTAL METHODOLOGY

ENGINE AND EXHAUST SETUP

A common-rail diesel engine (displacement 1.9L, rated power 60 KW), coupled to a servo-controlled dynamometer was used. The engine exhaust system was able to host several DPFs of different dimensions as well as a DOC upstream of the DPF. Special built exhaust modules for particle and gas sampling were placed upstream of the DOC and the DPF and downstream of the DPF. The system had the capability to by-pass the exhaust flow at the beginning of the test to allow the engine to reach steady state operation conditions (Figure 1). It also incorporated the necessary pressure and temperature sensors as well as a hydrocarbon port injection system (in-house development [6]) for DPF regeneration purposes.

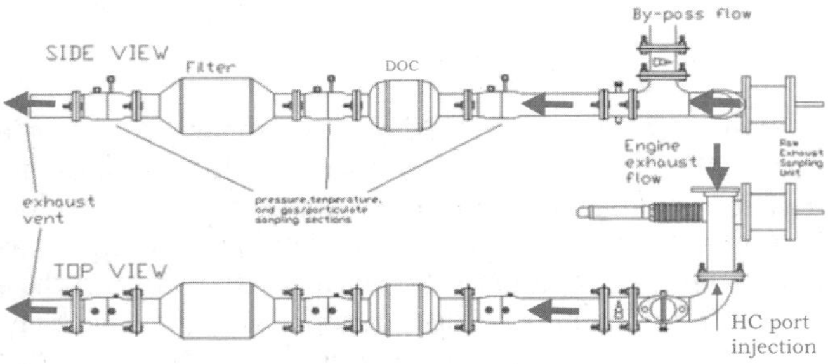

Figure 1. Exhaust setup for particle measurements and DPF assessment.

PARTICLE MEASUREMENT INSTRUMENTATION

The particle measurement instruments were the following:

1. Electric Low Pressure Impactor (ELPI). This is an aerodynamic method based instrument. It measures in real-time particle size and number concentration of particles. It is connected to a 2-stage mini diluter and provides the number concentration of particles with size in the range of 30 nm to 8 μm [7].

2. Condensation Particle Counter (Standalone CPC). This is a laser based particle counter. It is connected to a 3-stage mini diluter system and it is used for real-time total particle number concentration measurements [8].

3. Scanning Mobility Particle Sizer (SMPS). This is an electrical mobility method based instrument. It is also connected to a 3-stage mini diluter system and consists of a Differential Mobility Analyser (DMA) combined with an Ultrafine Condensation Particle Counter (U-CPC). It provides the number concentration of each particle size in the range of 10 to 430 nm [9]. The integration of the number concentration over the entire particle size range provides the total particle concentration. SMPS can also be set to continuously measure (in real time) particle number concentration at a specific particle size (e.g. 80 nm).

4. Nanoparticle Metrology system (NANOMET). This is a composite instrument sampling through a rotary diluter [10]. It consists of two sensors:

    - The DC (Diffusion Charger) sensor provides the total accessible surface of the diesel aerosol ($\mu m^2/cm^3$).

    - The PAS (Photoelectric Aerosol Sensor) provides a signal (in fA) that correlates with the particles' elemental carbon content.

5. Engine Exhaust Particle Sizer (EEPS). The EEPS spectrometer measures the size distribution of engine-exhaust particle emissions in the range from 5.6 to 560 nm with a response time in the range of 1 sec (real time). It displays measurements in 32 channels total (16 channels per decade [11]). EEPS was connected to the 3-stage mini diluter system (same as CPC and SMPS units).

The 3-stage mini dilution system (for the CPC, the SMPS and the EEPS) was heated to 160 C and operated at a Dilution Ratio (DR) range of 100 – 1600 depending on the sampling point. Its sampling flow rate was 3 SLPM. The ELPI two stage mini dilution system used dried, filtered and heated at 160 C air to accomplish DRs also in the range of 100 - 1600. Its sampling flow rate was 8 SLPM. The NANOMET rotary diluter operated at 160 C, 8 SLPM sampling flow rate and DRs in the range of 130 to 260.

DIESEL PARTICULATE FILTERS STUDIED

A set of DPFs was studied. This set was chosen to be representative of current and future DPF structures. The DPFs differ from each other regarding the filter type (wall-flow monolith or other), cell density and filtration area (100-200 cpsi, filtration area: 0.08 to 2.00 m^2), filter material (non-oxide or oxide ceramics), porosity (42-85 %), dimensions and permeability (2.0E-11 – 4.1E-13 m^2). Table 1 shows the characteristics of the DPFs examined.

Table 1. Characteristics of DPFs under evaluation.

DPF	Type/ Material	Dimensions Diameter x Length (in)	Filtration area (m^2)	Porosity (%)	Permeability (m^2)
1	Non-oxide ceramic wall-flow monolith	5.66x6	1.90	42	5.4 E-13
2	Oxide ceramic wall-flow monolith	5.66x6	2.00	60	4.1E-13
3	Oxide ceramic wall-flow monolith	5.66x6	1.13	48	8.4E-13
4	Non-oxide ceramic wall-flow monolith	5.66x8.4	1.70	42	3.7E-12
5	Fibrous structure	5 x 3	0.08	85	2.0 E-11

ENGINE OPERATION CONDITIONS

The engine was operated under steady state or transient conditions. The steady state engine points used are shown in Table 2. The New European Driving Cycle NEDC) was employed as representative of engine transient operation. Figure 2 provides the evolution of engine speed and torque during the NEDC.

Table 2. Steady state engine operation points used.

Engine point	Speed (rpm)	BMEP (bar)	Soot mass conc. (mg/m^3)	Exhaust mass flow (kg/h)	Exhaust temp. (C)
1	1500	3	50.2	75	259
2	1500	4	70.0	77	292
3	1500	5	59.3	86	327
4	2350	3.8	25.4	126	307
5	2250	6.7	25.9	157	381
6	2450	8.8	35.0	207	448
7	2500	5	28.4	153	348
8	1700	5.5	62.8	104	351

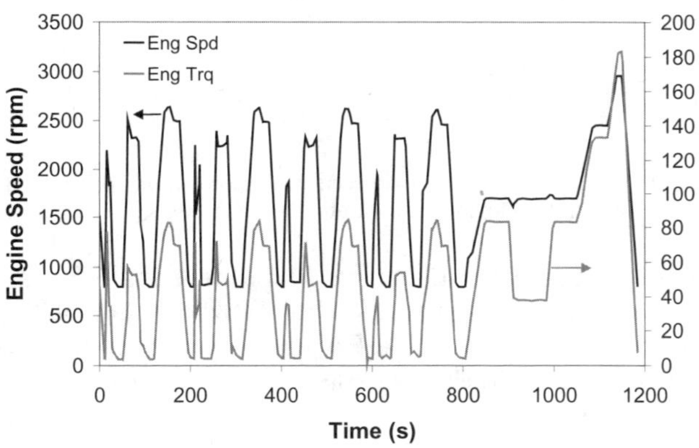

Figure 2. Engine speed and torque during the NEDC

STEADY STATE ENGINE OPERATION MEASUREMENTS

The target during the steady state engine operation tests was to accomplish an inter-instrument comparison and evaluate the effect of the DOC and the DPF on particulate emissions.

INSTRUMENT COMPARISON

The comparison between the results of each measuring methodology was performed both at the engine-out and DOC-out emissions. The trends established were equivalent in the two cases. The SMPS and the CPC data correlate reasonably well with the EEPS data. In Figure 3 and in Figure 4 the EEPS measurements (DOC-out emissions) are plotted over the SMPS and CPC measurements. The line of perfect agreement (y=x) is also plotted in both cases.

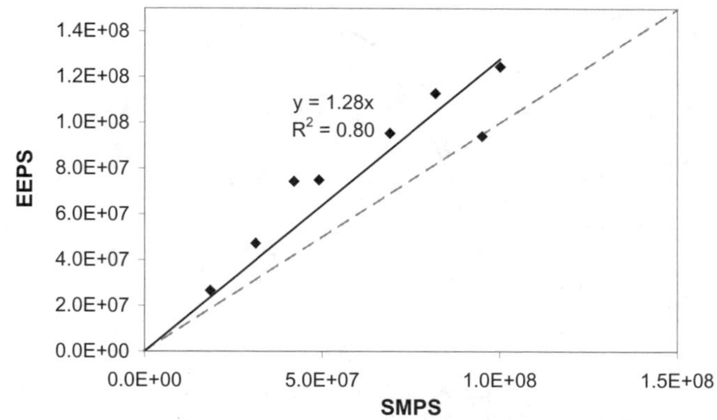

Figure 3 The EEPS total particle number concentration versus the SMPS total particle number concentration (#/cm^3).

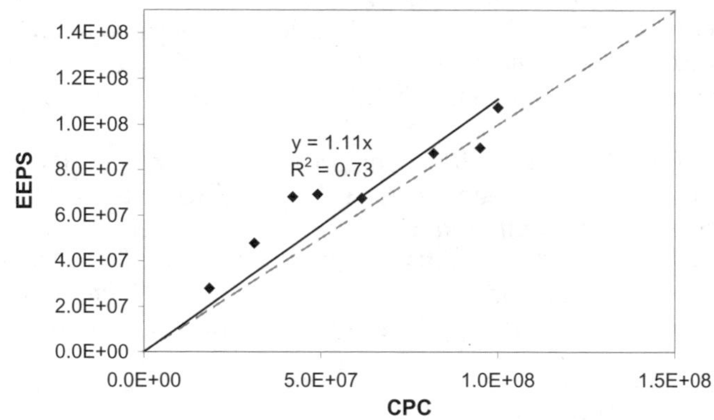

Figure 4. The EEPS total particle number concentration versus the CPC total particle number concentration (#/cm^3)

The EEPS total particle number concentration is consistently slightly higher than the total number concentration as measured by the CPC and the SMPS. The same outcome is true if one examines the particle size distribution curves measured by the EEPS and the SMPS at the same engine operation points. As an example the particle size distributions given by the EEPS and the SMPS measured at the engine-out during the engine operation points 5 and 8 are shown in Figure 5 and Figure 6, respectively.

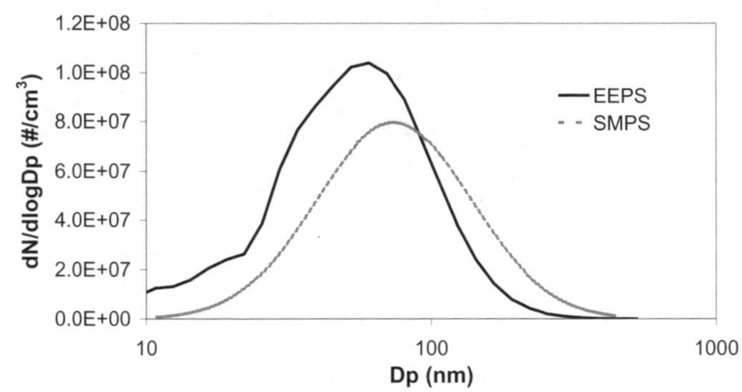

Figure 5. Engine-out EEPS and SMPS particle size distributions for engine operation at 2250 RPM, 6.7 bar.

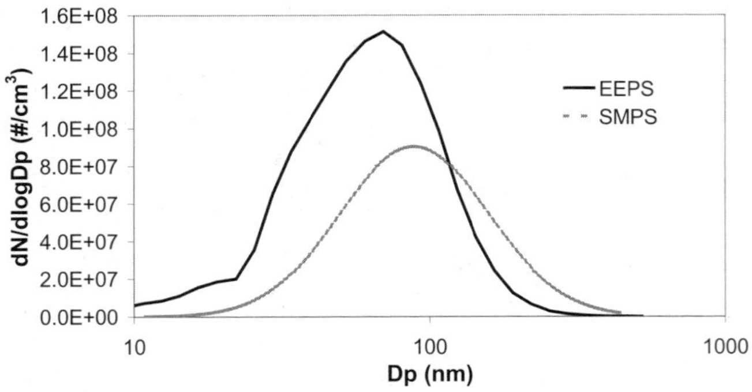

Figure 6. Engine-out EEPS and SMPS particle size distributions for engine operation at 1700 RPM, 5.5 bar.

From Figure 5 and Figure 6 it is apparent that the EEPS not only measures higher total number soot particle concentrations than the SMPS but also results in smaller mean particle diameters. A shift to the left of the particle size axis of approximately 20 nm is evident in all cases.

Figure 7 presents the comparison between the SMPS and the CPC measurements. The SMPS total number particle concentration is about 13% smaller than the CPC, a result that is justified by the fact that the SMPS measures particles in the range of 10 to 430 nm while the CPC measures particles over the entire size range.

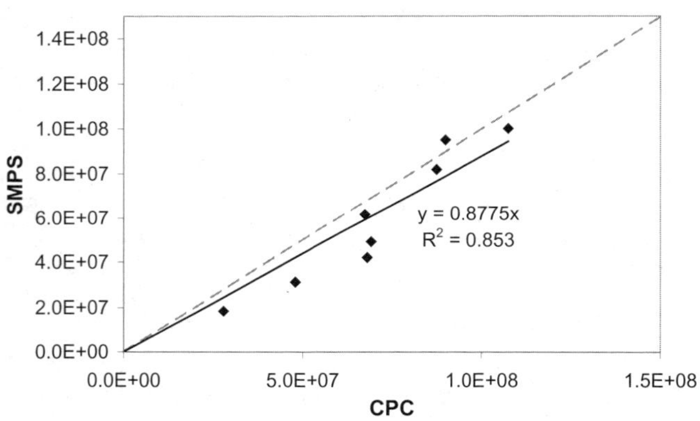

Figure 7. DOC-out SMPS total particle number concentration versus the CPC total particle number concentration (#/cm^3)

Figure 8 and Figure 9 compare the ELPI measurements with the SMPS and the EEPS measurements respectively. The EPLI measures a significantly lower total particle concentration compared to the other two instruments.

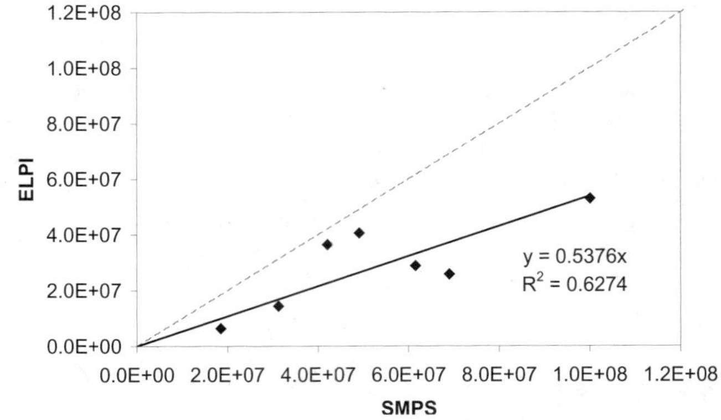

Figure 8. DOC-out ELPI total particle number concentration versus the SMPS total particle number concentration (#/cm^3).

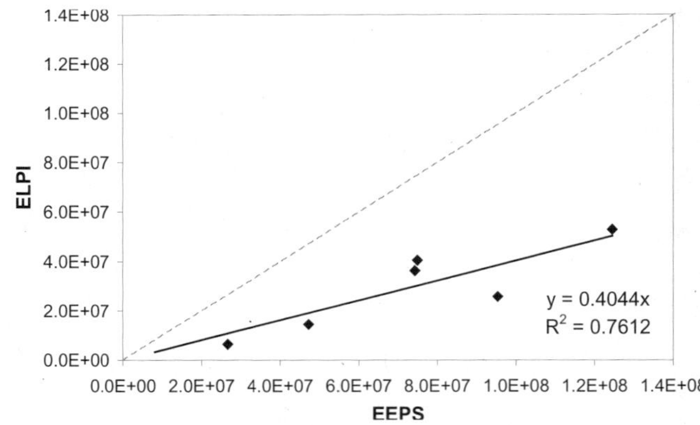

Figure 9. DOC-out ELPI total particle number concentration versus the EEPS total particle number concentration (#/cm^3).

Finally, Figure 10 and Figure 11 present the PAS and DC measurements versus the SMPS total particle number concentration. The PAS and DC signal correlate linearly with the SMPS measurements.

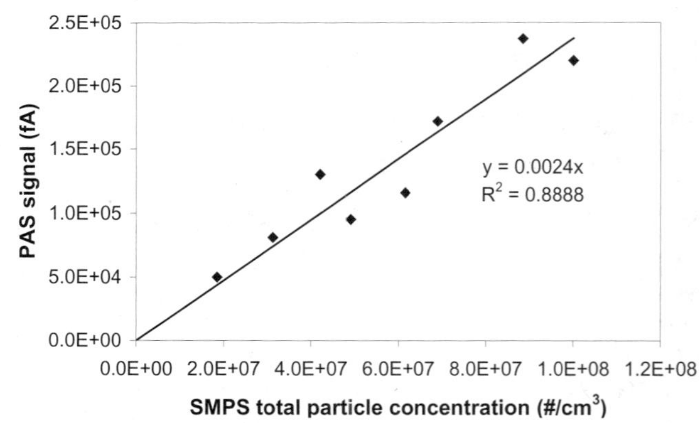

Figure 10. DOC-out PAS signal versus the SMPS total number particle concentration (#/cm^3).

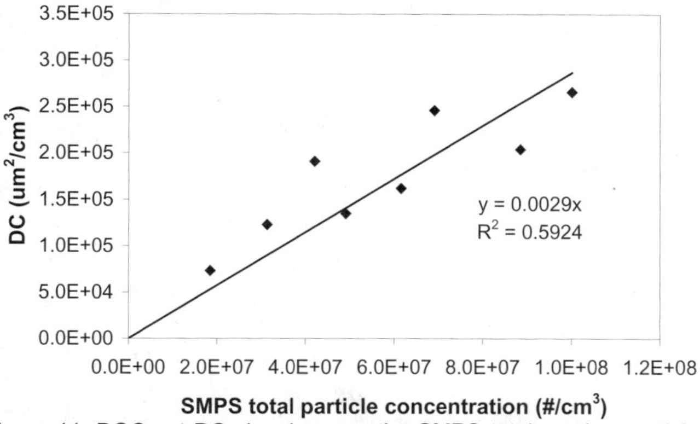

Figure 11. DOC-out DC signal versus the SMPS total number particle concentration (#/cm^3).

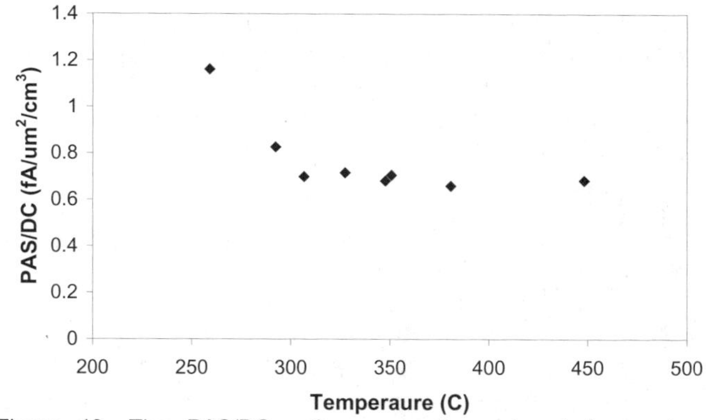

Figure 13. The PAS/DC ratio versus the exhaust temperature downstream of the DOC.

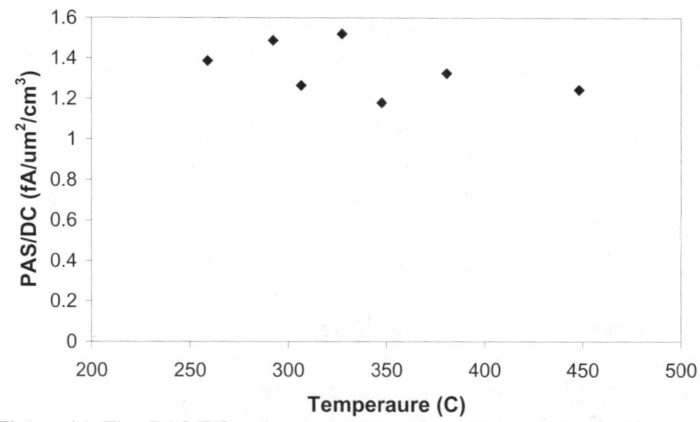

Figure 14. The PAS/DC ratio versus the exhaust temperature upstream of the DOC.

EFFECT OF DOC ON ENGINE-OUT EMISSIONS

Figure 12 shows the DC signal versus the PAS signal for the case of Nanomet sampling downstream of the DOC. While there is a quite strong linear relationship between the PAS and the DC for 6 engine points there are two outliners in this graph. Interestingly enough, these outliners correspond to engine points 1 and 2 with exhaust temperatures less than 300 C. Figure 13 presents the evolution of the PAS/DC ratio with the exhaust temperature. For exhaust temperatures higher than 300 C the PAS/DC ratio is always less than 0.8 while for temperatures less than 300 C the same ratio is greater than 0.8. The same data are plotted in Figure 14 for the case of Nanomet sampling upstream of the DOC. In this case the PAS/DC ratio is always higher than 1. The above indicates that the DOC has a substantial effect on the surface photoionization potential of the soot particles for exhaust temperatures higher than 300 C.

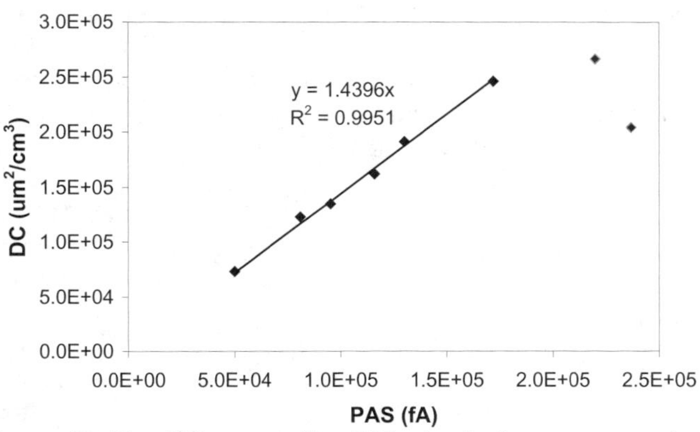

Figure 12. The DC versus the PAS signal. Sensors measuring downstream of the DOC.

EFFECT OF DPF ON ENGINE-OUT EMISSIONS

The effect of DPFs on engine out particle emissions as well as on the particle analyzers ability to monitor the filtration efficiency of the DPFs during steady state engine operation was examined during this part of the experimental work. The filters were placed in the engine exhaust in the absence of a DOC to avoid any NO_2-assisted regeneration. The 1st engine point of Table 2 was used. The engine exhaust soot emission rate was measured gravimetrically to allow the estimation of the amount of soot that had entered the filter from the test start. This quantity hereafter will be referred as the challenge mass load (g/m^2) and is equivalent to the actual soot mass load that would be collected on an absolute filter (100 % efficiency) divided by the actual surface area of the DPF under examination.

Before the actual start of each test, measurements with all available instruments were performed upstream of the DPF to establish the actual concentration and particle size distribution of the soot entering the DPF. During the test the particle instruments measured downstream of the DPFs in order to establish the soot particle concentration and size distribution exiting the filters. From the above measurements the evolution of the filtration efficiency with the challenge mass load was calculated. All instruments demonstrated the ability to measure the filtration efficiency in real time for the entire range of DPFs examined (from the more efficient to the less efficient one). Figure 15 presents the evolution of filtration efficiency and the pressure drop for DPF 1.

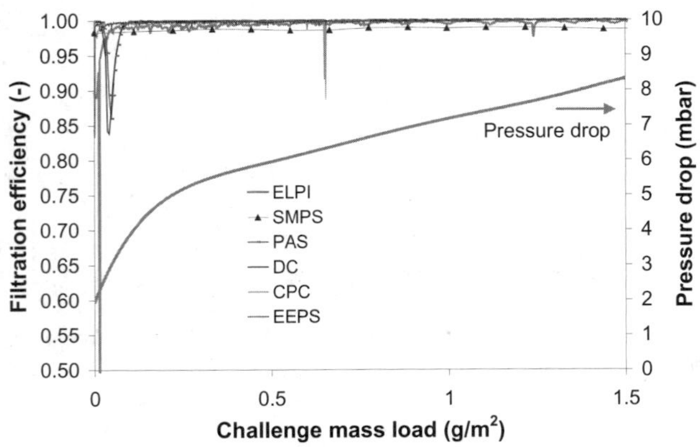

Figure 15. Pressure drop and filtration efficiency for DPF 1.

From Figure 15, it is evident that the filtration efficiency attains a very high value very rapidly as the DPF accumulates the soot particles. All instruments determine nearly the same excellent filtration efficiency.

Figure 16 shows the evolution of the soot particle size distribution measured by the SMPS during the filter soot loading test. On the left y-axis the probability density function of the particle size distributions downstream of the DPF is plotted while on the right y-axis the one upstream of the DPF is plotted.

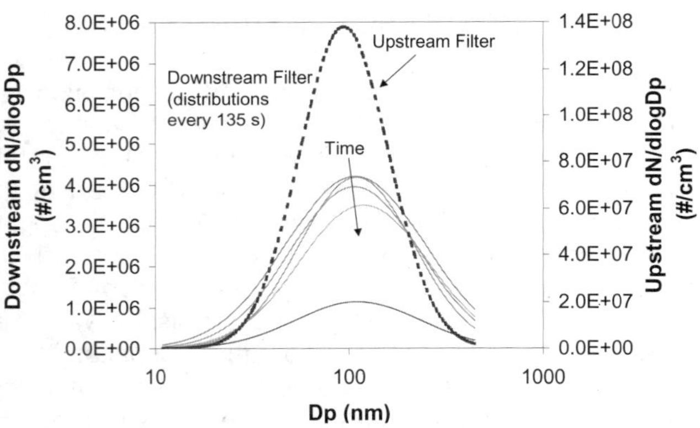

Figure 16. SMPS particle size distributions upstream and downstream DPF 1.

Interestingly enough, there is a shift of the mean particle diameter to higher values for the soot particles downstream of the DPF. This fact reveals that filtration is diffusion controlled with the smaller particles filtered with higher efficiency [12, 13]. The same is also verified by the ELPI measurements as shown in Figure 19.

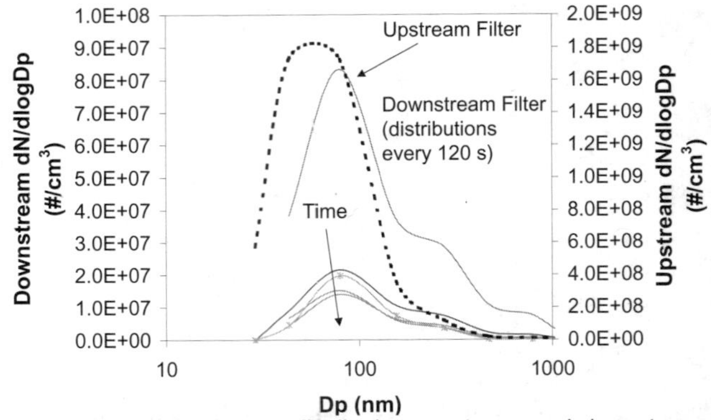

Figure 17. ELPI particle size distributions upstream and downstream DPF 1.

The same behavior was also established for the high porosity wall-flow DPF 2. Figure 20 reveals the evolution of filtration efficiency and pressure drop for DPF 2.

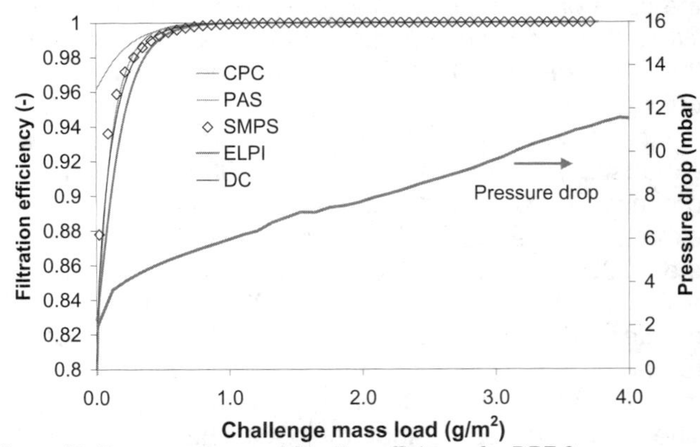

Figure 18. Pressure drop and filtration efficiency for DPF 2.

DPF 2 is not as fast as DPF 1 in reaching high values of filtration efficiency. Nevertheless, the filtration efficiency reaches values larger than 0.98 for values of challenge mass load greater than 0.5 g/m^2. One can interpret the same to DPF 1 diffusion-controlled filtration mechanism from the SMPS data of Figure 21.

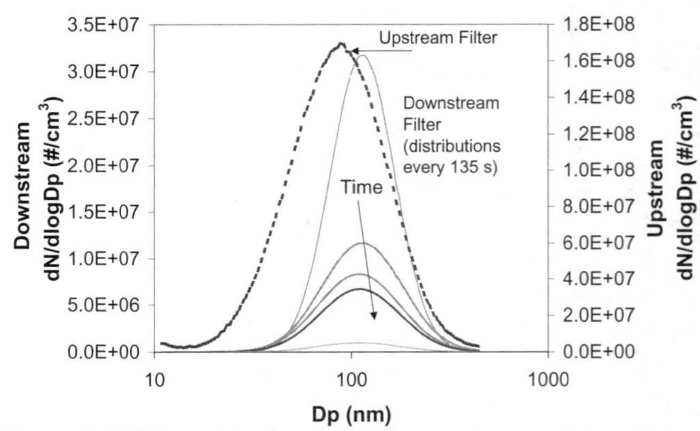

Figure 19. SMPS particle size distributions upstream and downstream DPF 2.

A different filtration behavior than in the wall-flow monolith was found for the fibrous filter (DPF 5, Figure 22). It is important to note that the test was performed with a filtration velocity of 35 cm/s (engine point 1) due to the very small filtration area of the fibrous filter. As the filter starts collecting soot the pressure drop increases at a very small rate. Filtration efficiency was low at the beginning of the test, starting from 0.2. As the filter's pores start blocking the pressure drop begins to increase linearly reaching very high values. At the same time, the filtration efficiency attains values greater than 0.9.

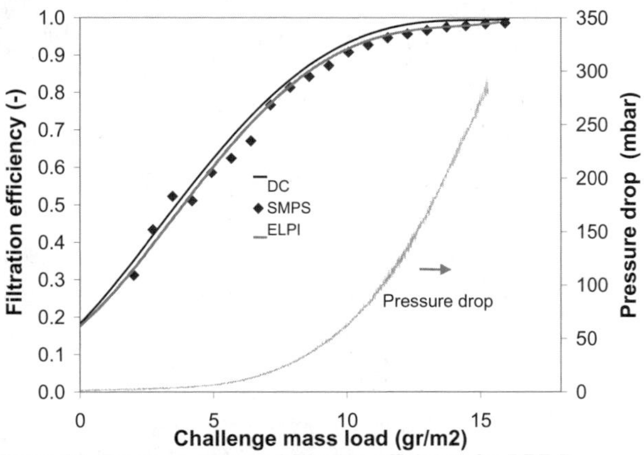

Figure 20. Pressure drop and filtration efficiency for DPF 5.

Figure 21 shows the evolution of the SMPS size distribution of soot particles exiting the filter. In this graph the particle size distributions are not shifted to the right as in the case of the monolithic filters.

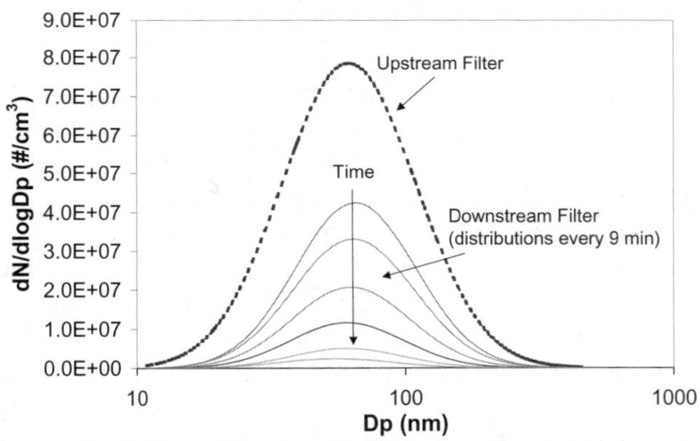

Figure 21. SMPS particle size distributions upstream and downstream DPF 5.

Figure 22 shows the ELPI size distributions of soot particles exiting the DPF 5 during the same soot loading test.

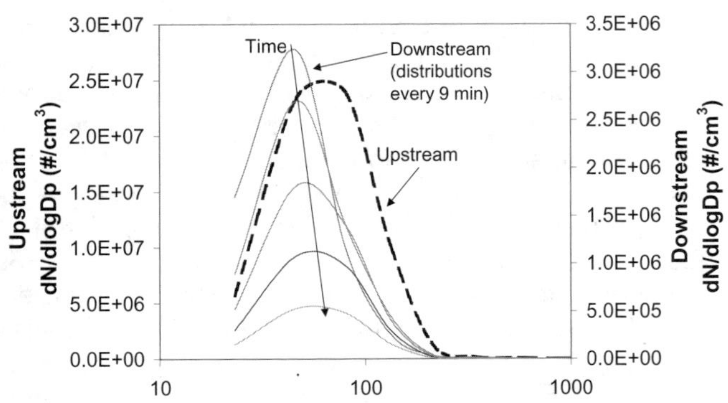

Figure 22. ELPI particle size distributions upstream and downstream DPF 5.

In Figure 22 the shift of the mean Dp to smaller sizes is obvious, indicating that interception and inertial phenomena play a major role in the filtration mechanism of the fibrous filter.

TRANSIENT ENGINE OPERATION MEASUREMENTS

With the exception of the SMPS all other particle analyzers are able to monitor the exhaust soot particles in real time (time resolution of 1 s) during transient engine operation. The SMPS can also be set to continuously measure particle number concentration but only for a given particle size (e.g. 80nm) denoted hereafter as SMPS-80. During the transient tests the ability of the analyzers to measure at the various points on the engine exhaust was assessed together with the filtration efficiency of the DPFs.

INSTRUMENT COMPARISON

Figure 23 and Figure 23 show the EEPS, SMPS-80 and ELPI real time measurements during the NEDC. The CPC, PAS and DC measurements are not presented due to space limitations. Through the tests it became obvious that every analyzer examined was capable to monitor the exhaust soot particle concentration and to account for the relative changes with time.

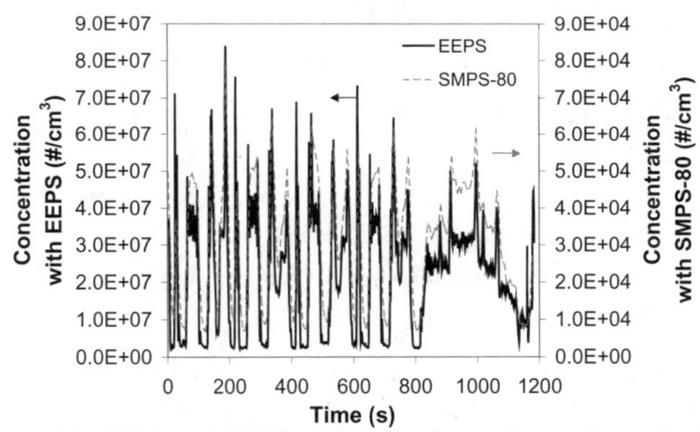

Figure 23. EEPS and SMPS measurements on the engine exhaust during the NEDC.

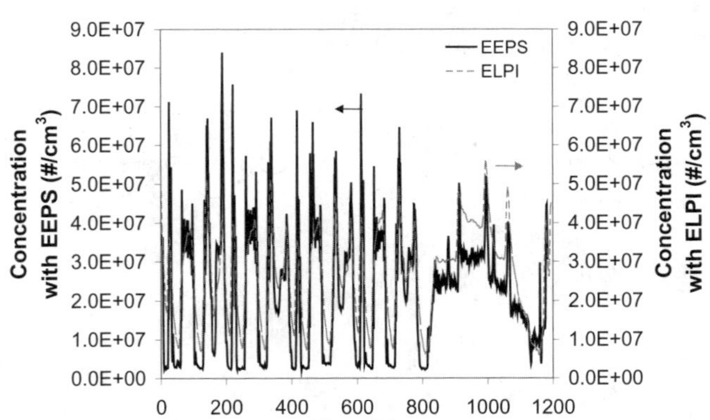

Figure 24. EEPS and ELPI measurements on the engine exhaust during the NEDC.

EFFECT OF DOC

In Figure 25 and Figure 26 the EEPS and ELPI measurements are presented for the engine-out and DOC-out exhaust points. No significant effect of the DOC on the total particle number concentrations is observed.

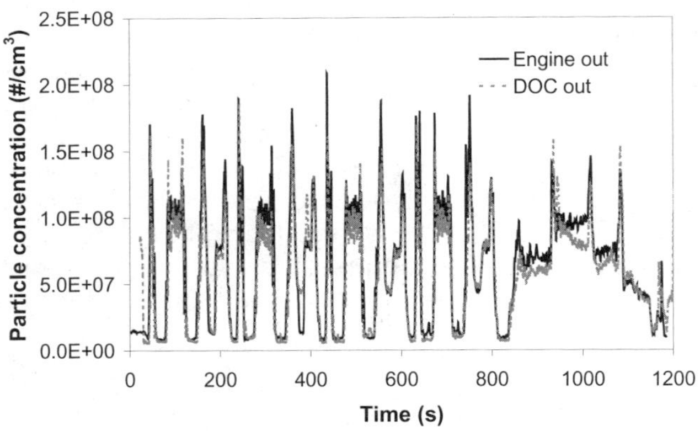

Figure 25. EEPS measurements at the engine-out and DOC-out emissions.

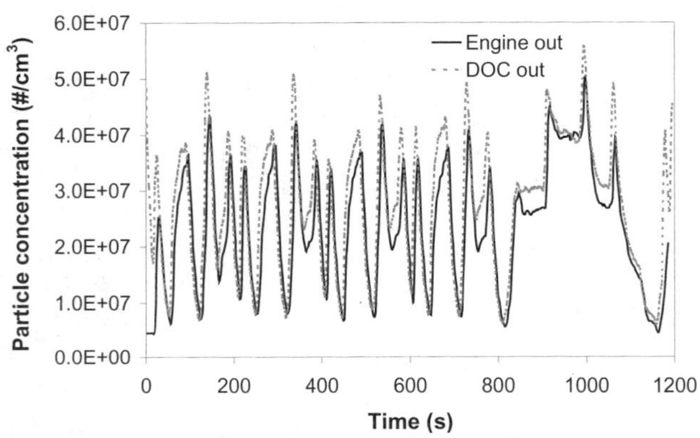

Figure 26. ELPI measurements at the engine-out and DOC-out emissions.

Figure 27 reveals a significant decrease of the PAS/DC ratio for the DOC-out emissions compared to the engine out emissions. The DOC effect on the surface photoionization potential of the soot particles is more evident as the cycle evolves due to the higher exhaust temperatures attained.

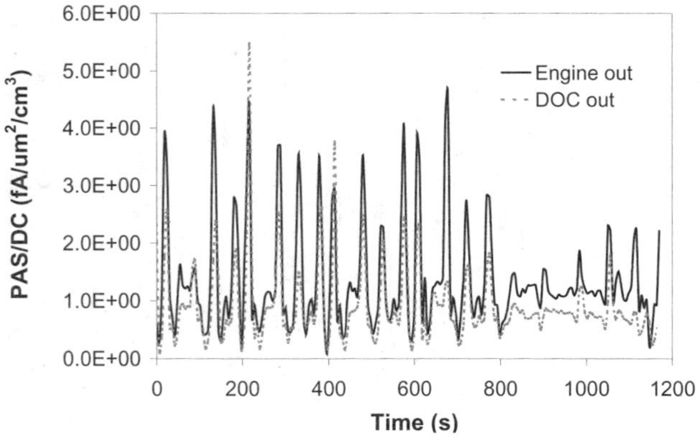

Figure 27. PAS/DC ratio for engine-out and DOC-out exhaust points.

The above is inline with what was found during the tests under steady state engine operation. Figure 28 depicts the PAS/DC plotted against the temperatures occurring during the NEDC. For comparison, data from the steady state tests are also included.

Apparently the variation in the residence time of the soot particles in the DOC also plays a role at lower cycle temperatures which become less pronounced at higher temperatures.

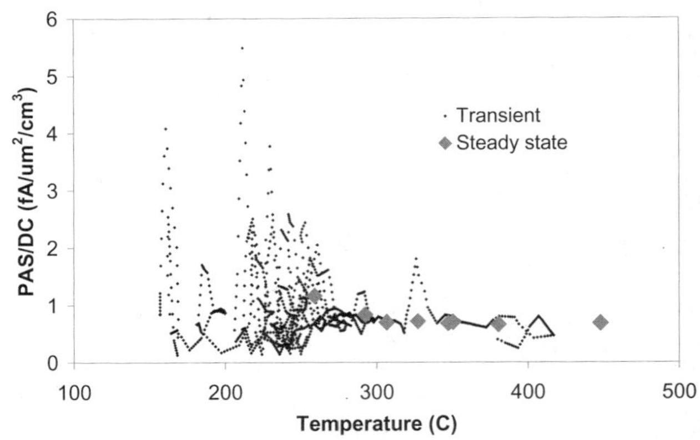

Figure 28. PAS/DC ratio plotted against exhaust temperature for transient and steady state tests.

EFFECT OF DPF ON ENGINE-OUT EMISSIONS

The effect of DPFs on engine out particle emissions under transient engine operation was examined during this part of the experimental work. The filters were placed in the engine exhaust in the absence of a DOC to avoid any NO_2-assisted regeneration. Before the test start, two subsequent NEDCs were run with exhaust by-passed from the DPF to take baseline (engine-out) measurements with all available instruments. Then, the exhaust flow was directed to the DPF and the engine was operated for two subsequent NEDCs with an interval at idle of approximately 450 s. The filter pressure drop and the filter inlet and filter outlet temperatures (Tin, Tout) were monitored in each case. The challenge mass load of each DPF during the test was calculated with an in-house developed virtual soot sensor [6,14-18], which takes into consideration the exhaust conditions, the filter pressure drop, the filter geometry and the soot microstructural properties. During the transient test the particle instruments measured downstream of the DPFs in order to establish the soot particle concentration and size distribution exiting the filters. From the above measurements the evolution of the filtration efficiency with the challenge mass load was calculated. As in the case of the steady state engine operation all instruments demonstrated the ability to measure the filtration efficiency in real time for the entire range of DPFs examined (from the more efficient to the less efficient one). For reasons of space and clarity the filtration efficiency evolution for each DPF is presented only for the EEPS, CPC and ELPI.

Figure 29 shows the pressure drop and temperature evolution for DPF 1, while Figure 30 shows again the pressure drop and challenge mass load evolution.

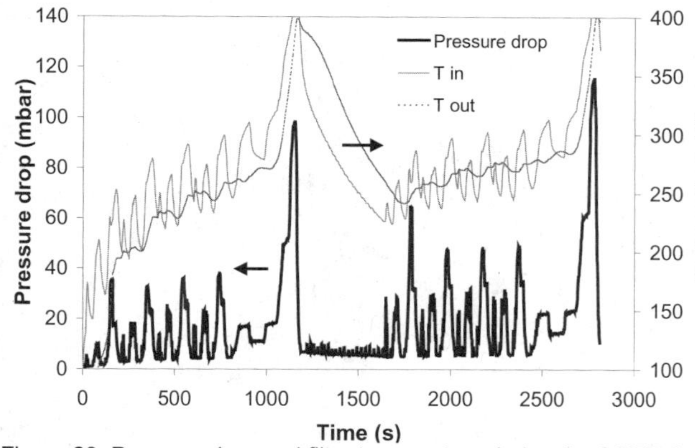

Figure 29. Pressure drop and filter temperature during the 2 NEDC tests for DPF 1.

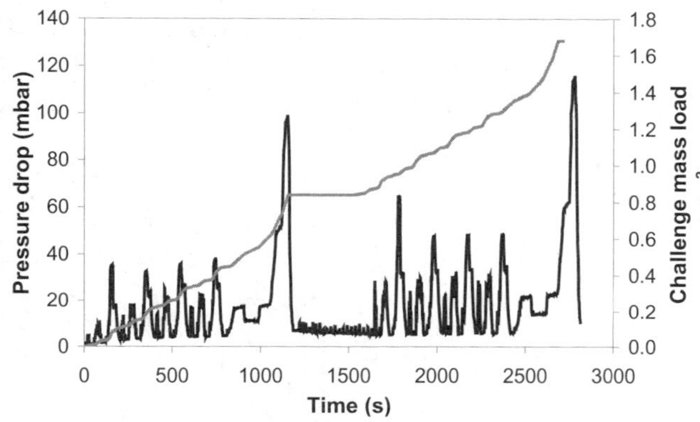

Figure 30. Pressure drop and filter challenge mass load during the 2 NEDC tests for DPF 1.

Figure 31 presents the evolution of filtration efficiency with the challenge mass load for the same filter. All instruments exhibited excellent filtration efficiencies during the 2 NEDC tests and only during the last part of each NEDC a drop in the filtration efficiency occurred bellow 0.9, as measured by the ELPI and the EEPS. The CPC did not display such a decrease. The last part of the NEDC is where the speed and torque is increased and consequently high values of filtration velocity occur as can be seen in Figure 32.

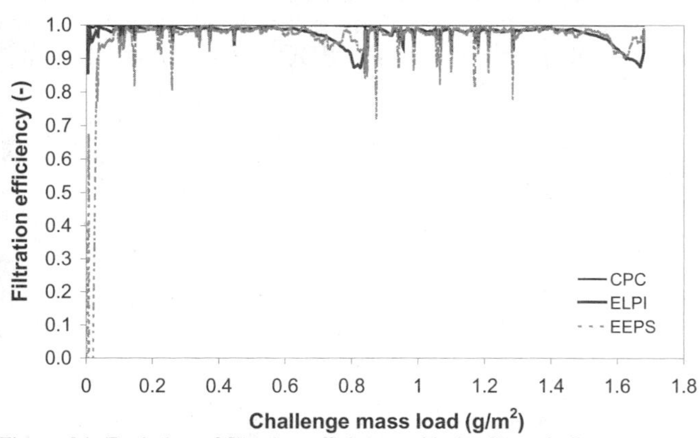

Figure 31. Evolution of filtration efficiency with the filter challenge mass load during the 2 NEDC tests for DPF 1.

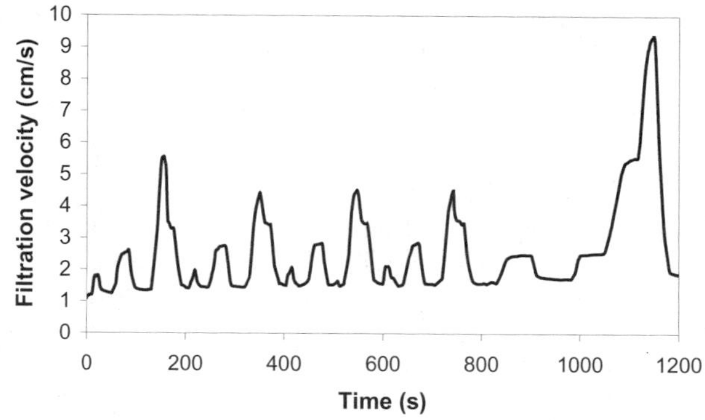

Figure 32. Evolution of filtration velocity during the NEDC for DPF 1.

The same conclusions concerning the filtration efficiency can be drawn if one compares the size distributions of soot particles entering and exiting the filter. Figure 33 shows the soot particle size distribution upstream of the DPF and Figure 34 shows the same downstream of the DPF sampling location. It is apparent that after a few seconds of soot accumulation the number of particles exiting the filter is very small for all particle sizes.

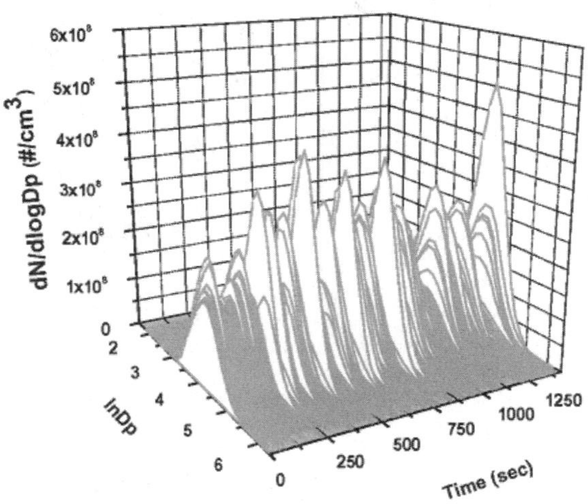

Figure 33. EEPS particle size distributions upstream of DPF 1 during the NEDC test.

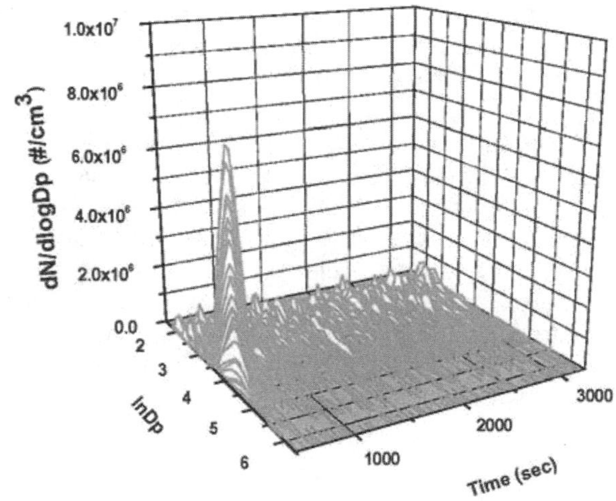

Figure 34. EEPS measurements downstream of DPF 1 during the NEDC test.

Figure 35 presents the pressure drop and challenge mass load evolution during the 2 NEDC tests for DPF 2.

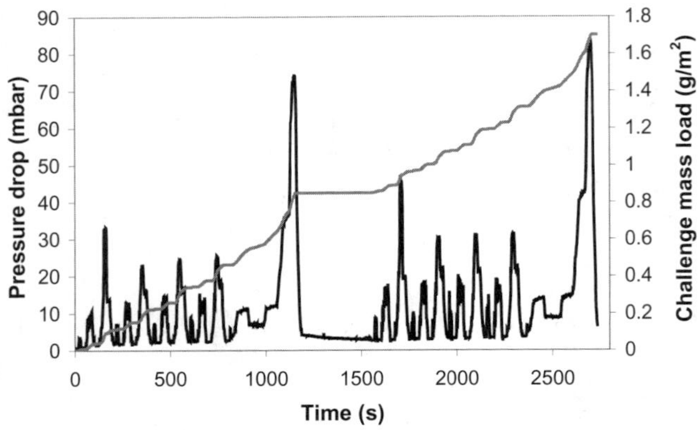

Figure 35. Pressure drop and filter challenge mass load during the 2 NEDC tests for DPF 2.

Figure 36 presents the filtration efficiency versus the challenge mass load for DPF 2.

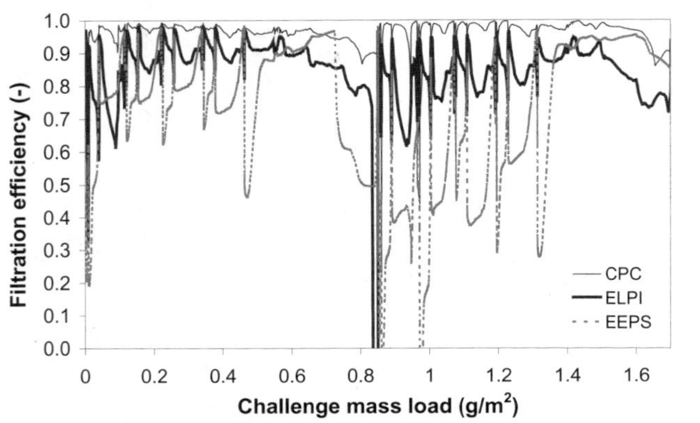

Figure 36. Evolution of filtration efficiency with the filter challenge mass load during the 2 NEDC tests for DPF 2.

In the case of DPF 2, the filtration efficiency plot is somewhat different than in the case of DPF 1. CPC filtration efficiency is higher than 0.9 with the exception of high filtration velocities (end of NEDC cycles). EEPS and ELPI filtration efficiencies start from quite low values to evolve to higher ones as soot is accumulated in the filter. After the first NEDC, and at 0.85 g/m^2 challenge mass load, the filtration efficiency fails to remain at high values despite the low filtration velocities during the first part of the second NEDC. Filtration velocity starts rising again with soot accumulation at 1.2 g/m^2 challenge mass load. The above phenomenon could be attributed to partial blow-off of soot particles i.e. a number of soot particles deposited on the filter are carried away from the filter structure. This is also verified by the EEPS graph of particle size distribution exiting the DPF as Figure 37 shows.

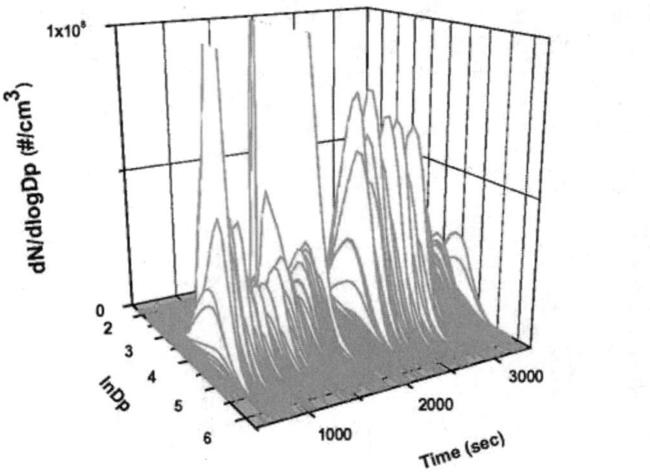

Figure 37. EEPS measurements downstream of DPF 2 during the NEDC test.

Figure 43 shows the pressure drop and challenge mass load evolution during the 2 NEDC tests for DPF 3. Figure 39 shows the filtration efficiency evolution with the challenge mass load for the same filter.

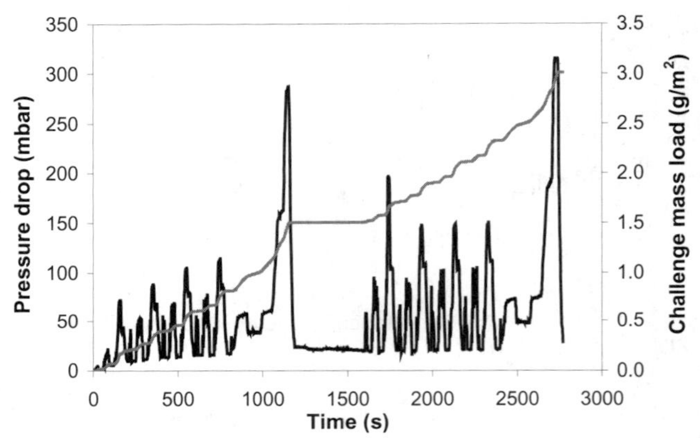

Figure 38. Pressure drop and filter challenge mass load during the 2 NEDC tests for DPF 3.

In Figure 39 the CPC filtration efficiency is excellent for the entire test with the exception of high filtration velocities where some small dips occur. The ELPI measures smaller values of filtration efficiency and reveals a larger dip at the end of NEDC 1. The EEPS filtration efficiency remains at substantially lower values compared to the two other instruments reaching zero values at points in the NEDC with higher filtration velocities.

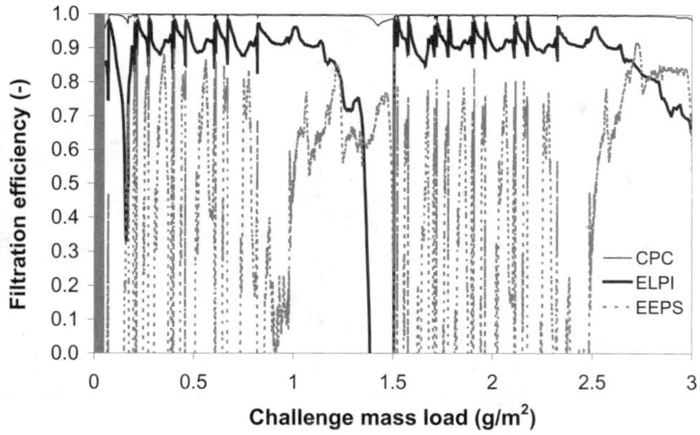

Figure 39. Evolution of filtration efficiency with the filter challenge mass load during the 2 NEDC tests for DPF 3.

Figure 40 presents the evolution of the size distribution of particles downstream of DPF 3 as obtained by the EEPS. Indications of blow-off phenomena exist here also but to a limited extent.

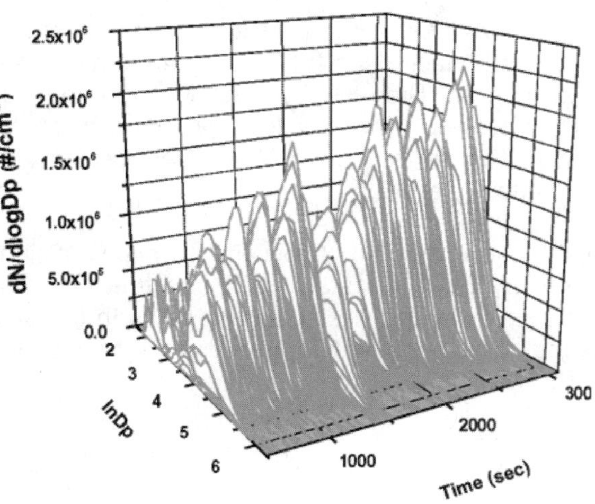

Figure 40. EEPS measurements downstream of DPF 3 during the NEDC test.

Figure 41 and Figure 42 show the test results for DPF 4

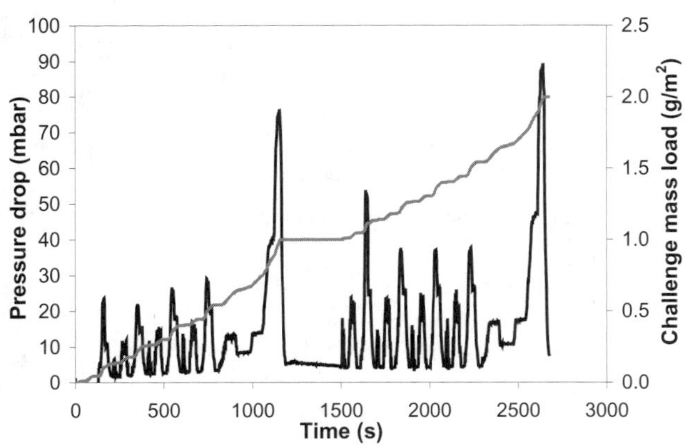

Figure 41. Pressure drop and filter challenge mass load during the 2 NEDC tests for DPF 4.

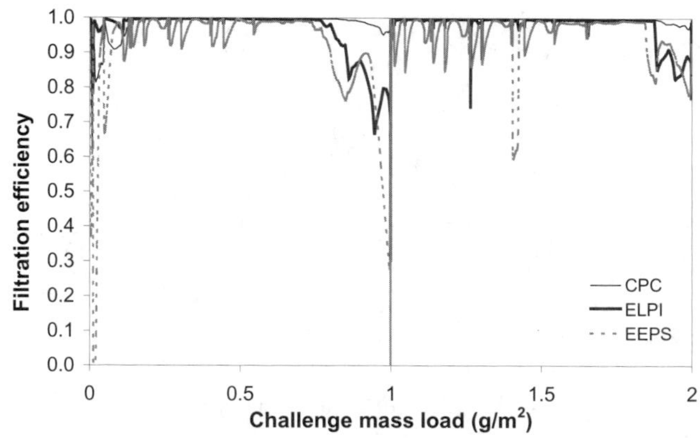

Figure 42. Evolution of filtration efficiency with the filter challenge mass load during the 2 NEDC tests for DPF 4.

For DPF 4 filtration efficiency is very good and becomes excellent (> 0.9) with soot accumulation. As in the case of the previous filters filtration efficiency is decreased for higher filtration velocities. The CPC measures the highest values of filtration efficiency and the EEPS the smallest. Figure 43 presents the EEPS size distribution of soot particles downstream of DPF 4. Some indications of blow-off phenomena during the second NEDC exist.

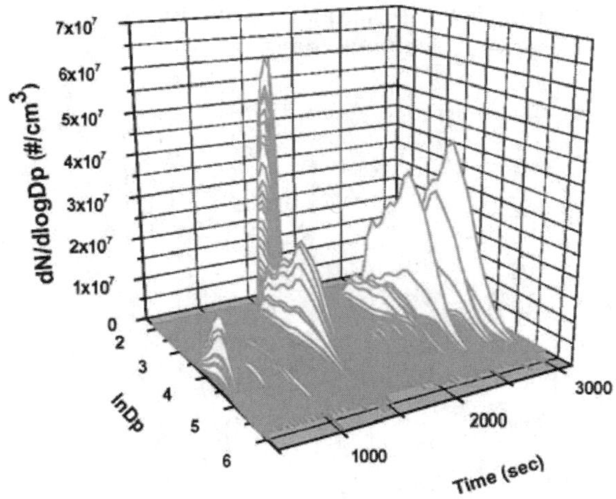

Figure 43. EEPS measurements downstream of DPF 4 during the NEDC test.

Figure 44 and Figure 45 show the test results for DPF 5.

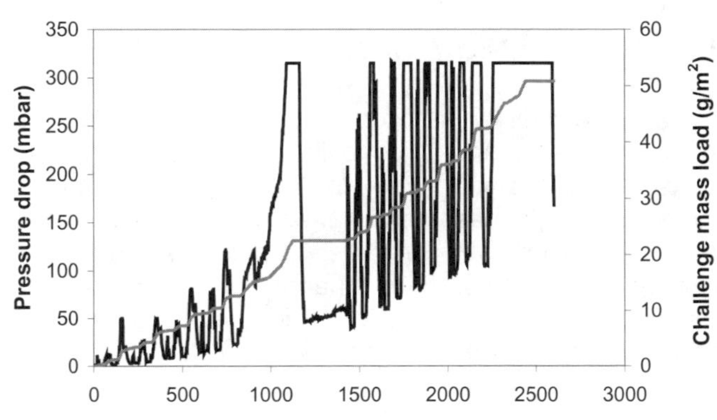

Figure 44. Pressure drop and filter challenge mass load during the 2 NEDC tests for DPF 5.

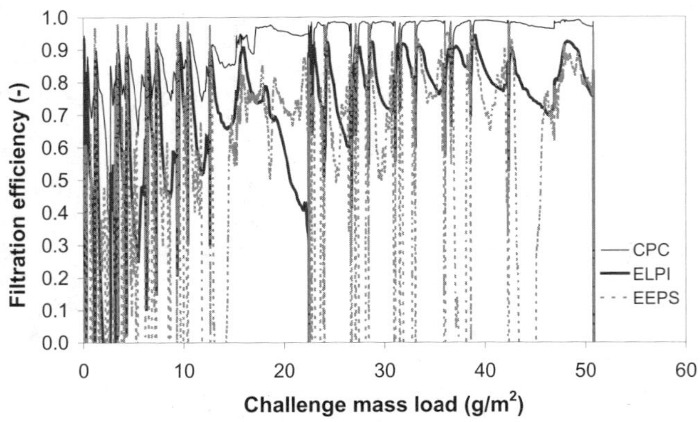

Figure 45. Evolution of filtration efficiency with the filter challenge mass load during the 2 NEDC tests for DPF 5

In Figure 45 only the values for pressure drop, which are less than 315 mbar, are plotted due to the limits of the pressure transducer used. It is obvious from the graph that even higher pressure drop values are realized. Filtration efficiency is quite low for the clean filter but as the filter is clogged with soot particles it gets higher values. Again, the CPC measures the highest values of filtration efficiency and the EEPS the smallest. The EEPS size distributions (Figure 46) verify the above findings with particles exiting the DPF decreasing with time and soot mass accumulated in the filter.

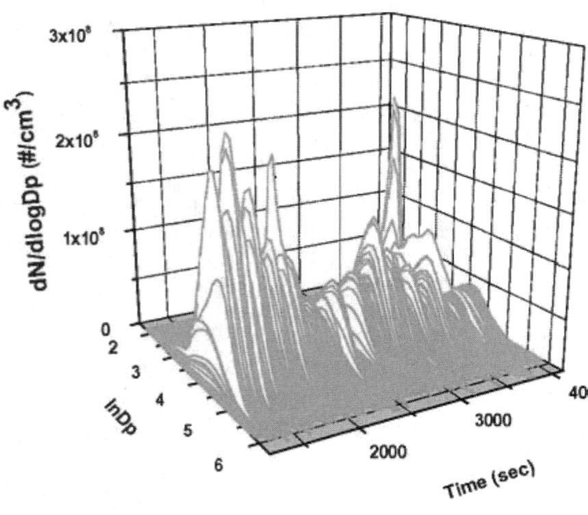

Figure 46. EEPS measurements downstream of DPF 5.

FILTER REGENERATION

Filtration efficiency monitoring can also be performed under filter regeneration conditions. For the regeneration tests the DPFs were first loaded with soot up to a moderate soot mass load. Then a DOC was placed upstream of the DPF. In order to increase the exhaust temperature, a predefined quantity of engine fuel was injected upstream of the DOC. An example regeneration test (DPF 2) is shown in Figure 47.

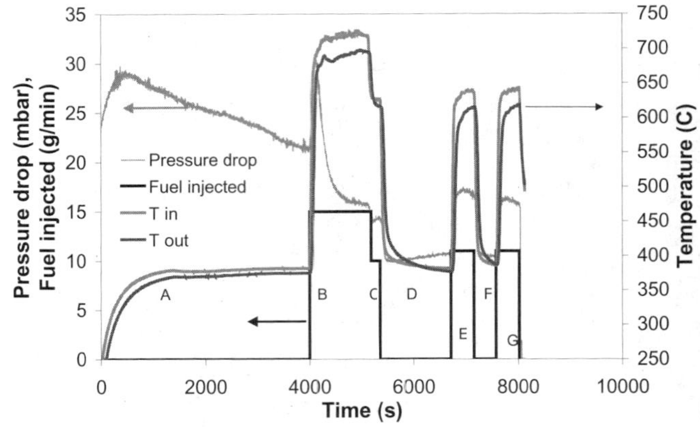

Figure 47. Pressure drop and temperature evolution during the DPF 2 regeneration test.

The procedure followed included seven steps. During step A (0 to 4000 s) no fuel injection was performed. Due to the DOC existence upstream of the DPF, an NO_2-assisted partial filter regeneration occurred. At 4000 s, fuel was injected upstream of the DOC at a constant rate of 15 g/min. This raised the filter inlet temperature to 720 C. This step lasted for 1100 s and led to a complete filter regeneration (Step B). The fuel injection continued for 100 s more, but at a smaller fuel injection rate (10 g/min; step C). Further fuel injections were performed in following steps (steps E and G).

Figure 48 and Figure 49 show the filtration efficiency (measured by ELPI, PAS and DC) and the filter mass load (calculated by the virtual soot sensor). In the same graphs the corresponding filter temperature to each step is also shown.

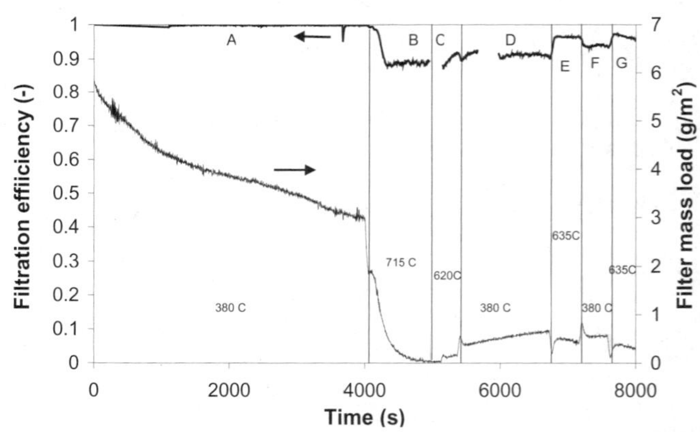

Figure 48. ELPI filtration efficiency and filter mass load during the regeneration test of DPF 2.

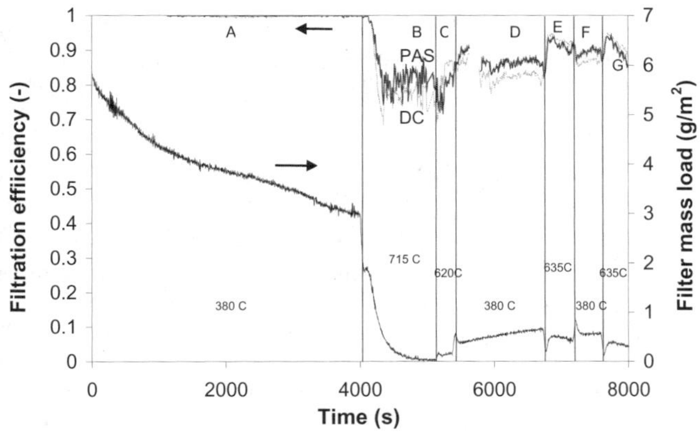

Figure 49. PAS and DC filtration efficiency and filter mass load during the regeneration test of DPF 2.

During step A the filtration efficiency is high because the filter is loaded with soot. From the start of the regeneration (step B), the filtration efficiency (ELPI-based) decreases down to 0.90 corresponding to the clean filter filtration efficiency. As step C starts, the filtration efficiency is decreased initially due to the temperature decrease [19,20] and then starts increasing again (from 0.87 to 0.92) with soot accumulation. The same changes in the filtration efficiency occur during the rest of the test according to the filter temperature and whether soot is accumulated or oxidized in the filter. The same pattern of changes in the filtration efficiency is recorded when PAS and the DC are employed.

CONCLUSIONS

An extended testing campaign was carried out with the employment of many state of the art particle analyzers and five different DPF technologies, in an effort to evaluate the possible correlations between the different measurement methodologies and assess their ability to measure the filtration efficiency under different exhaust conditions.

Small but systematic differences were found between the EEPS and the SMPS under steady state engine operating conditions. Generally, the EEPS total particle number concentration was higher than that of the SMPS and CPC. The ELPI also correlated linearly to the above instruments but measured significantly lower particle concentrations. The particle number concentration measuring instruments correlated linearly with the PAS and the DC. The DOC caused a decrease in the PAS/DC ratio in a temperature dependent fashion.

All instruments were able to monitor the exhaust particle concentration during transient engine operation. The response of the instruments to the variations of the exhaust conditions was approximately the same.

All instruments were able to provide filtration efficiency measurements during transient engine operation, though they resulted in systematic differences. The CPC provided the less noisy filtration efficiency values and the highest ones. The EEPS values for filtration efficiency were the noisiest and the lowest.

All instruments were able to pick reduced filtration efficiency events during the transient engine operation that were due to higher filtration velocities or particle blow-off phenomena.

The different filter structure of the examined DPFs resulted in differences in the attained pressure drop and filtration efficiency. The fibrous filter was the least effective in backpressure and filtration efficiency terms. For the monolithic DPF with the higher porosity (DPF 2) particle blow-off phenomena occurred over the transient cycle. Only the EEPS showed that this might be the case also for DPF 3 and DPF 4, but to a smaller extent.

According to the filter structure of the DPF, the particles downstream of it may be shifted to larger or smaller sizes in relation to the dominating filtration mechanism (diffusion or inertial interception controlled).

The instruments are able to measure the DPF filtration efficiency during filter regeneration conditions. A testing methodology to this effect was demonstrated.

ACKNOWLEDGMENTS

This work has been supported in part by the European Commission through the projects MAAPHRI, IMITEC and STYFF-DEXA. We thank Mr. Oliver Bischoff (TSI) for making available the EEPS. We thank our APT Lab colleague Mrs Eleni Papaioannou for her contribution in the data analysis.

REFERENCES

1. Ogyu K., Ohno K., Hong S., Komori T., "Ash Storage Capacity Enhancement of Diesel Particulate Filter", *SAE Tech. Paper* No. 2004-01-0949, 2004.
2. Young D. M., Hickman D. L., Bhatia G., Gunasekaran N., "Ash Storage Concept for Diesel Particulate Filters", *SAE Tech. Paper* No. 2004-01-0948, 2004.
3. Bardon S., Bouteiller B., Bonnail N., Girot P., Gleize V., Oxarango L., Higelin P., Michelin J., Schuerholz S., Terres F., "Asymmetrical Channels to Increase DPF Lifetime", *SAE Tech. Paper* No. 2004-01-0950, 2004.
4. Boretto G., Imarisio R., Rellecati P., Barucchi E., Sanguedolce A., "Serial Application of a Catalyzed Particulate Filter on Common Rail DI Diesel Engines For Passenger Cars", F2004V068, FISITA World Automotive Congress, Barcelona, Spain, May 23-27, 2004, 2004.
5. Konstandopoulos A. G., Zarvalis D., Papaioannou E., Vlachos N. D., Boretto G., Pidria M. F., Faraldi

P., Piacenza O., Prenninger P., Cartus T., Schreier H., Brandstatter W., Wassermayr C., Lepperhof G., Scholz V., Luers B., Schnitzler J., Claussen M., Wollmann A., Maly M., Tsotridis G., Vaglieco B. M., Merola S. S., Webster D., Bergeal D., Gorsmann C., Obernosterer H., Fino D., Russo N., Saracco G., Specchia V., Moral N., D'Anna A., D'Alessio A., Zahoransky R., Laile E., Schmidt S., Ranalli M., "The Diesel Exhaust Aftertreatment (DEXA) Cluster: A Systematic Approach to Diesel Particulate Emission Control in Europe", *SAE Tech. Paper No.* 2004-01-0694 (SP-1861), 2004.

6. Konstandopoulos A. G., Papaioannou E., Zarvalis D., Skopa S., Baltzopoulou P., Kladopoulou E., Kostoglou M, Lorentzou S., (2005) "Catalytic Filter Systems with Direct and Indirect Soot Oxidation Activity", SAE Tech. Paper No. 2005-01-0670 (SP-1942).
7. DEKATI. (2001). Electrical Low Pressure Impactor, Operation Manual, Dekati Ltd, Tampere, Finland.
8. TSI. (1998). Condensation Particle Counter Model 2022 Operation Manual, TSI inc., Aachen, Germany.
9. TSI. (1998). Scanning Mobility Particle Sizer Model 3936 Operation Manual, TSI inc., Aachen, Germany.
10. Matter Engineering. (2000). Nanoparticle Meterology System Operation Manual, Matter Engineering AG, Wohlen, Switzerland.
11. TSI. (2004). Engine Exhaust Particle Sizer Model 3090 Operation Manual, TSI inc., Aachen, Germany.
12. Konstandopoulos A. G., Kostoglou M., "Theory and Practice of Soot Filter Operation", *Proceedings of Global Powertrain Congress 99* (invited paper), Powertrain Int., Stuttgart, October 5-7, 1999.
13. Konstandopoulos A. G., Kostoglou M., Housiada P., Vlachos N., Zarvalis D., "Multichannel Simulation of Soot Oxidation in Diesel Particulate Filters", *SAE Tech. Paper* No. 2003-01-0839 (SP-1755), 2003.
14. Konstandopoulos A. G., Skaperdas E., Masoudi M., "Microstructural Properties of Soot Deposits in Diesel Particulate Traps", *SAE Tech. Paper* No. 2002-01-1015 (SP-1673), 2002.
15. Konstandopoulos A. G., Kostoglou M., Housiada P., Vlachos N., Zarvalis D., "Multichannel Simulation of Soot Oxidation in Diesel Particulate Filters", *SAE Tech. Paper* No. 2003-01-0839 (SP-1755), 2003.
16. Konstandopoulos A. G., Kostoglou M., "Microstructural Aspects of Soot Oxidation in Diesel Particulate Filters", *SAE Tech. Paper* No. 2004-01-0693 (SP-1861), 2004.
17. Skaperdas E., Konstandopoulos A. G., "Prediction of Diesel Particulate Filter Loading Behavior for System Control Applications", CDAuto01A3018, CD AUTO 01, *3rd International Conference on Control and Diagnostics in Automotive Applications* Sestri-Levante, Italy, 4-6 July, 2001.
18. Konstandopoulos A. G., Kladopoulou E., "A Virtual Sensor for On-Board Diagnostics and Control of Diesel Particulate Filters", CDAuto03A2028, CD AUTO 03, *4th International Conference on Control and Diagnostics in Automotive Applications*, Sestri-Levante, Italy, 18-20 June, 2003.
19. Konstandopoulos A. G., and Johnson J. H., (1989) "Wall-Flow Diesel Particulate Filters-Their Pressure Drop and Collection Efficiency", SAE Tech. Paper No. 890405, SAE Transactions 1998, Sec. 3 (J. Engines).
20. Konstandopoulos A. G., Kostoglou M., Skaperdas E., Papaioannou E., Zarvalis D., and Kladopoulou E., "Fundamental Studies of Diesel Particulate Filters: Transient Loading, Regeneration and Aging", *SAE Tech. Paper No.* 2000-01-1016 (SP-1497), 2000.

CONTACT

Dr. Athanasios G. Konstandopoulos, Head of Aerosol & Particle Technology Laboratory, CERTH/CPERI, P.O. Box 361, Thermi, Thessaloniki 57001, Greece. e-mail: agk@cperi.certh.gr, web: http://apt.cperi.certh.gr

DEFINITIONS, ACRONYMS, ABBREVIATIONS

CPC: Condensation Particle Counter

DPF: Diesel Particulate Filter

DOC: Diesel Oxidation Catalyst

EEPS: Engine Exhaust Particle Sizer

ELPI: Electric Low Pressure Impactor

NANOMET: Nanoparticle Metrology System

NEDC: New European Driving Cycle

PAS: Photoelectric Aerosol Sensor

2007-01-0921

Filtration Behavior of Diesel Particulate Filters (1)

E. Ohara, Y. Mizuno, Y. Miyairi, T. Mizutani, K. Yuuki, Y. Noguchi, T. Hiramatsu, M. Makino, A. Takahashi and H. Sakai
NGK Insulators, Ltd.

M. Tanaka, A. Martin, S. Fujii and P. Busch
NGK Automotive Ceramics USA, Inc.

T. Toyoshima, T. Ito, I. Lappas and C. D. Vogt
NGK Europe GmbH

ABSTRACT

This paper is Part-1 of two papers discussing the filtration behavior of diesel particulate filters. Results of the fundamental study are presented in Part-1, and test results for real size DPFs are reported in the supplement, Part-2. In this paper, a fundamental experimental study was performed on the effect of pore size and pore size distribution on the PM filtration efficiency of the ceramic, wall-flow Diesel Particulate Filter (DPF). Small round plates of various average mean pore sizes (4.6, 9.4, 11.7, 17.7 micro-meters) with a narrow pore size distribution were manufactured for the tests. During the DPF filtration efficiency tests, $ZnCl_2$ particles in the range of 10 nm to 500 nm were used instead of PM from actual diesel engine exhaust. $ZnCl_2$ particles were made using an infrared furnace and separated into monodisperse particles by DMA (Differential Mobility Analyzer). Various $ZnCl_2$ particle sizes were introduced upstream of the test plates and particle numbers were counted at the upstream and the downstream positions to determine filtration efficiency. In addition, theoretical calculations were made for the collection mechanisms of i) diffusion, ii) interception and iii) inertia. The theoretical calculations were compared with the experimental test results. On-engine testing of a DPF, with optimum pore size distribution, demonstrated overall filtration efficiency of >90% during preconditioning operation of the NEDC test cycle. The potential of the deep-bed filtration model for predicting filtration efficiency of wall-flow type ceramic DPFs was confirmed. Additional vehicle tests were performed on DPFs with various pore structures. The results are shown and discussed in the Part-2.

INTRODUCTION

The high thermal efficiency characteristic of diesel engines yields excellent fuel economy and less CO_2, which contributes to global warming, when compared to gasoline engines. Conversely, diesel engines produce a large amount of PM (Particulate Matter) due to its diffusion combustion characteristic, and it is difficult to reduce NOx with emission control devices due to the high air/fuel ratio. PM and NOx emission regulations for diesel engines have lagged gasoline engines. However, in response to community health concerns, Japan, Europe and US have begun to tighten PM and NOx restrictions, moving toward the same level as gasoline engines. The current weight limit basis for PM emission regulations is in question and there is growing pressure to restrict the fine PM particle count, which has the greatest impact on human health.

The leading technology in reducing diesel PM is the DPF. This wall-flow type filter, based on a monolithic substrate with alternately plugged inlets and outlets channels, exhibits minimum backpressure, excellent PM filtration efficiency and a large filtration area in a relatively small volume. Immediate oxidation of the trapped PM is most desirable to minimize the maximum DPF temperature and system backpressure. Some emission system designs feature a catalyzed substrate in front of the filter, or have the filter itself support a catalyst in order to continuously oxidize the PM.

Porous ceramic materials, such as cordierite, SiC and the like, are typically used for the DPF material. For these materials, the mean pore size and pore size distribution need to be optimized in order to obtain adequate particulate filtration efficiency. Experiments and modeling were conducted to determine the DPF pore size necessary for achieving adequate PM filtration efficiency, taking into account the worldwide trend to regulate PM particle count.

FUNDAMENTAL EXPERIMENTS ON DPF WALL MATERIAL FILTRATION PROPERTIES

PARTICULATE COLLECTION MECHANISMS

Extensive research has been conducted on the various collection mechanisms (i.e.: diffusion, interception, inertia, gravity, electrostatics and thermophoretic [References 2-11]. Utilizing a granular bed model, Otani et al. performed basic experiments to determine the collection efficiency of diffusion, interception, inertia, and gravity mechanisms, for a wide range of aerosol particle diameters and filtration speeds. From this model, empirical equations were developed relating each collection mechanism and its efficiency [Reference 1].

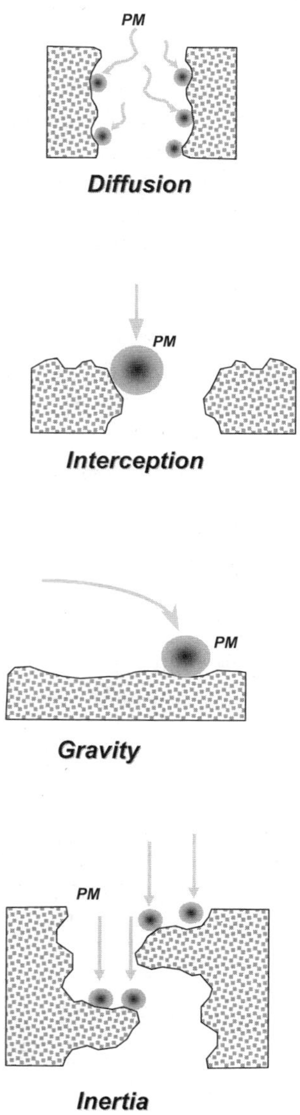

Figure 1: Filtration Mechanism

As shown in Figure 1, the diffusion collection mechanism is defined as aerosol particles that are made to deviate from their line of flow due to Brown's diffusion movement, and are collected by coming into contact with the filter material. In terms of Otani et al.'s empirical equation, the diffusion collection efficiency around a single sphere is shown as Equation 1, and the smaller the aerosol particles, the easier they are to be collected.

Diffusion

$$\left.\begin{aligned} \eta_D &= 8.0 Pe^{-2/3} \quad \text{for } Re \to 0 \\ \eta_D &= 2.1 Pe^{-1/2} \quad \text{for } Re \to \infty \end{aligned}\right\} \text{Equation (1)}$$

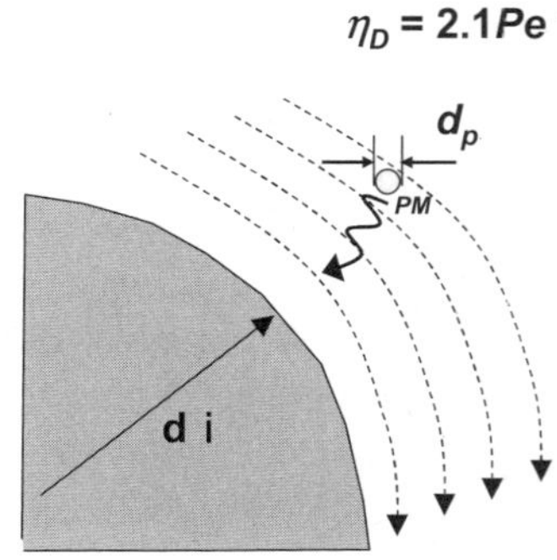

Pe : Peclet number $= ud_i / D$
u : flow velocity
d_i : Granule diameter
D : Brown diffusivity $= C_c kT/(3\pi\mu d_p)$
C_c : Cunningham slip correction factor
k : Boltzmann factor
d_p : Particle diameter
μ : Viscosity of gas

The interception collection mechanism is defined as particles that follow along the flow line and are collected by coming into contact with the filter material. The larger the aerosol particles, the easier they are to be collected. The empirical equation for a single sphere is shown as Equation 2.

Interception

$$\left.\begin{aligned} \eta_R &= 16 R^2 \quad \text{for } Re \to 0 \\ \eta_R &= 16 R \quad \text{for } Re \to \infty \end{aligned}\right\} \text{Equation (2)}$$

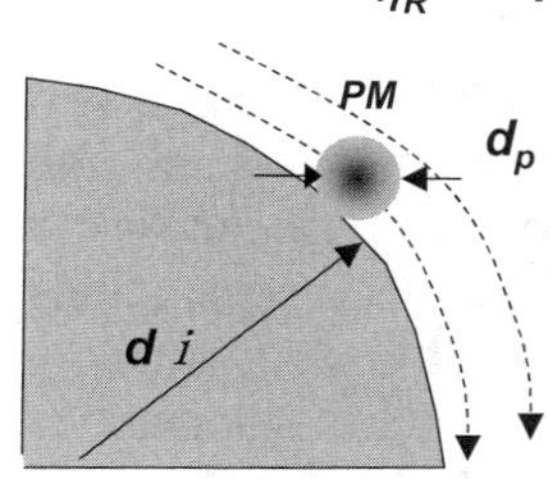

$R = dp/di$
d_p : PM(particle matter) diameter
d_i : Granule diameter

The inertia collection mechanism is the condition whereby due to a rapid change in flow angle, the aerosol particles deviate from their flow line as a result of their inertia and are collected by colliding with the filter material. Along with this method, the gravity and electrostatic collection mechanisms also assume aerosol particles deviate from their flow line due to the respective forces, and particles are collected after coming into contact with the filter material. The filtration efficiency of the granular bed (Figure 2), which is a collective unit of single spheres (granules), can be determined from single sphere collection efficiencies using Equation 3.

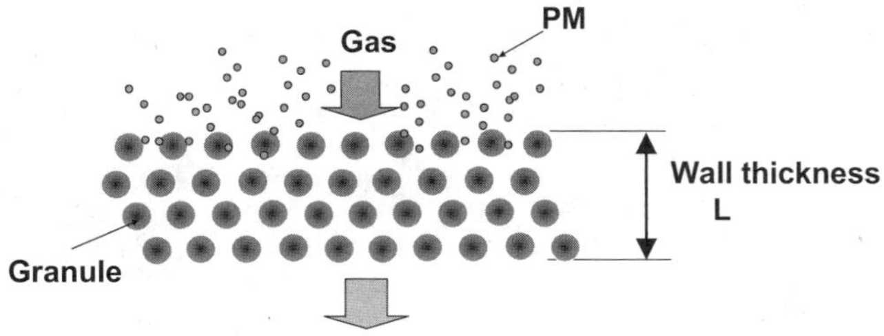

$$E = 1 - \exp[(-3/2)\eta L(1-\varepsilon)/d_i] \quad \text{(Equation 3)}$$

E: Total bed efficiency
η: Single sphere efficiency
ε: Porosity
L: wall thickness
d_i: Granule diameter

Figure 2: Granular Bed Model

This model can be considered useful for making predictions about the PM filtration efficiency of DPF walls. However, There is a problem in translating the measured pore size to the granule diameter in the model. In Otani et al.'s equtions (Equation 1, 2), the principal scale factor of the filter structure is the granule diameter of the granular bed. On the other hand, the pore diameter distribution of ceramic porous material is generally measured by the mercury penetration method. The difference between the effective pore diameter for diffusion collection and the measured pore diameter by mercury penetration is illustrated in Figure 3. The measured diameter by the mercury penetration method (d_m) corresponds to a minimum diameter in the pore channel. However, the effective diameter for diffusion is larger. In contrast, for interception, minimum diameter is directly effective.

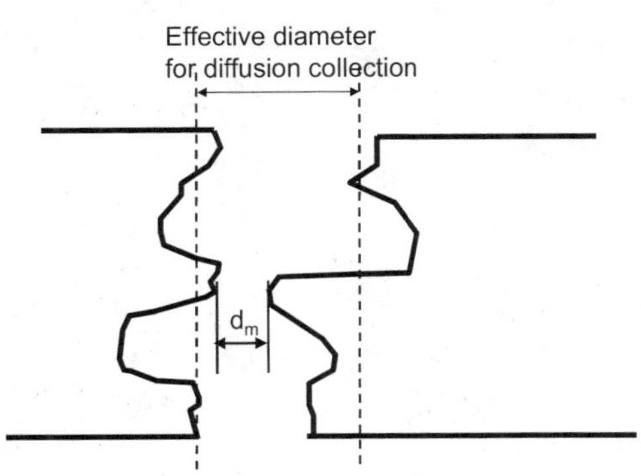

Figure 3: Relationship between the effective pore diameter for diffusion and the pore diameter by the mercury penetration method

It is necessary to identify a coefficient to convert the pore diameter from the mercury penetration method to the granule diameter of a granular bed with equivalent filtration efficiency. We assumed linear coefficients between the pore diameter using the mercury penetration method (d_m) and the granule diameter of the granular bed (d_s) that has equivalent filtration efficiency.

$$d_s = C_1 d_m \quad \text{(for Interception)} \qquad (4)$$

$$d_s = C_2\, d_m \quad \text{(for Diffusion)} \qquad (5)$$

C_1 and C_2 were tuned so that the measurement and prediction trends agree, providing the following relationships:

Interception trapping: $C_1=1 \quad d_s=d_m$ (6)

Diffusion trapping: $C_2=4 \quad d_s=4d_m$ (7)

Using Otani et al.'s equations, the relationship between PM diameter and single sphere collection efficiency in a granular bed with a granule diameter of 17.7 μm, at 300 °C and a filtration speed of 0.05 m/s, is shown in Figure 4. When the PM particle size is smaller than 200 nm, diffusion collection is predominant, and for larger PM sizes, interception collection becomes the predominant mechanism. It is evident that gravity and inertia mechanisms have a negligible influence.

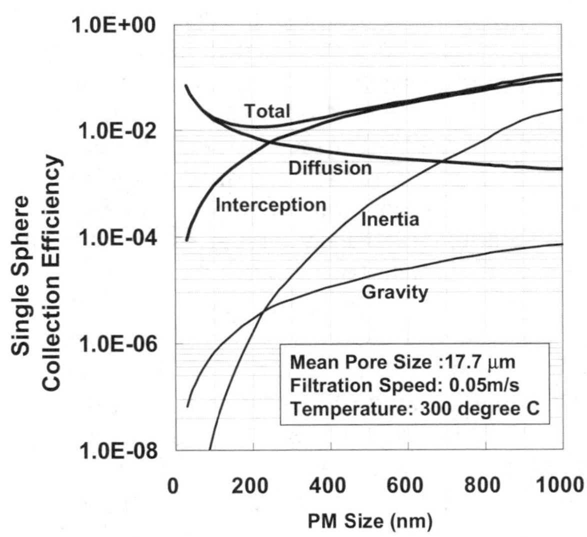

Figure 4: Single Sphere Collection Efficiency for Each Mechanism

Thus far, the collection mechanisms of the filter material itself have been described. However, during actual DPF usage, when PM loading reaches the 0.3 g/L to 0.5 g/L range, a layer is formed on the wall surface. This PM layer acts as a filter and filtration efficiency improves. Figure 5 shows the typical results of an experiment evaluating the relationship between PM loading and filtration efficiency using an actual DPF. The PM filtration efficiency during the period in which filtration relies solely on the wall material, was 55 %. When the PM loading density increased to 0.3 g/L or more, a filtration efficiency of 90 % was measured, due to the filtering effect of the PM layer. [Reference: 12]

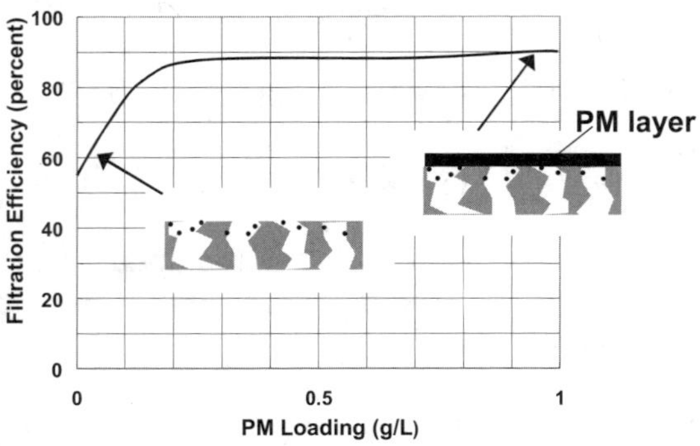

Figure 5: Filtration Efficiency at Various PM Loadings

FILTER MATERIAL FILTRATION EFFICIENCY EXPERIMENTS

Test Conditions

As described above, current diesel particulate filters are able to achieve a >90% filtration efficiency due to the filtering effect of the PM layer. However, during the initial PM loading of a new filter and immediately after conducting a complete filter regeneration, it is necessary to rely on the filtration efficiency of the material alone. Further, in order to meet future regulations in terms of PM count, it is desirable that the material itself have adequate filtration efficiency. Filtration efficiency is influenced by each of the mechanisms described above and also changes according to PM diameter. Basic experiments were performed to determine the influence of various filter wall pore diameters on PM filtration efficiency. The equipment used for the experiments is shown in Figure 6. We used discs with a 25 mm diameter and 0.305 mm (12 mil) thickness for the filter wall samples. This thickness is identical to diesel particulate filter wall thickness used in current and future diesel engines. Cordierite-bonded SiC samples were prepared with narrow pore size distributions. The mean pore diameters were 4.6μm, 9.4μm, 11.7μm, and 17.7μm. The pore size distribution of each test sample, as well as the production DPF material, is shown in Figure 7. SEM photographs of the test samples are shown in Figure 8. For the aerosol particles, $ZnCl_2$ was heated in an infrared ray furnace enabling evaporation. The particles that formed by condensation were electrically neutralized by a radioactive source of ^{241}Am. A DMA (Differential Mobility Analyzer) was used to isolate $ZnCl_2$ particles with a specific particle diameter. After removing the charge with ^{241}Am, the isolated particles were introduced upstream of the filter sample. The count concentration of $ZnCl_2$ aerosol particles was measured upstream and downstream of the sample using a CNC (Condensation Nuclei Counter), from which the aerosol particle filtration efficiency of the sample material was determined. Five (5) different particle diameters between 20 nm and 300 nm, and 3 filtration speeds from 0.025 to 0.1 m/s were tested. This range of filtration speeds represents the range of PM filter filtration speeds used in DPF-equipped systems of current diesel engines.

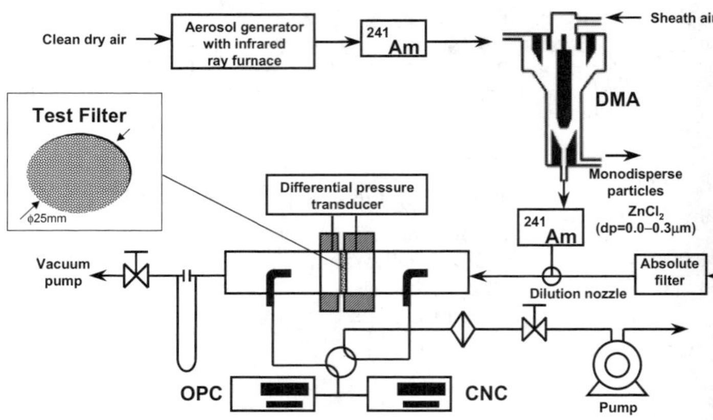

Figure 6: Test Equipment

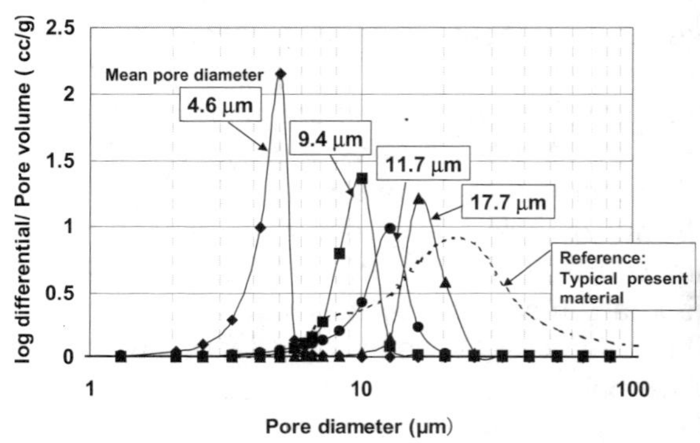

Figure 7: Pore Size Distribution of Each Test Sample

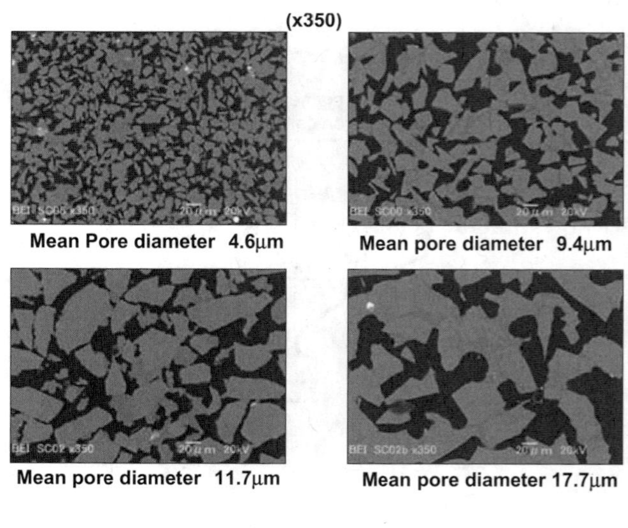

Figure 8: SEM Photograph of Test Samples

The Difference between $ZnCl_2$ Particles and Diesel Exhaust Particles

There are two main differences between $ZnCl_2$ particles and actual diesel exhaust particles. The first is specific density. Zn and Cl are heavier than carbon. The second difference is the shape of the particles. The shape of the $ZnCl_2$ particle is a sphere. The nuclei mode particle, of diesel exhaust particles, also has a spherical shape. However, larger particles in diesel exhaust are agglomerates of small particles. The shape difference of the agglomerated particles may have an influence on the interception collection efficiency. The difference in the specific density, however, does not affect the results of this study because particle diameter is defined by mobility(the sized particles are obtained by DMA). Brownian diffusivity, *D,* is a function of mobility as shown in Equation 8.

$$D = kTB \qquad \text{Equation (8)}$$

B : mobility

$$B = C_c/(3\pi\mu d_p) \qquad \text{Equation (9)}$$

T: Temperature

D : Brownian diffusivity

k : Boltzmann factor

μ : Viscosity of gas

C_c : Cunningham slip factor

$$C_c = 1 + 2.514(l/d_p) + 0.80(l/d_p)\exp[-0.55(d_p/l)] \qquad \text{Equation (10)}$$

d_p : Particle diameter

l : Average free distance of gas

These equations show that the specific density of the particles does not affect mobility and Brownian diffusivity. The basis for this study is that the filtration efficiency of diesel particulates can be evaluated using $ZnCl_2$ particles of equivalent mobility particle size.

Test Results

The filtration efficiency experiment results are shown in the graphs measured at flow speeds of 0.025, 0.05, and 0.1 m/s (see Figures 9, 10, and 11). The particle diameters are shown on the horizontal axis. The solid lines are the experimental results and the dotted lines are predictions.

As shown in Figures 9-11, at each of the filtration speeds, the graphs are concave and the filtration efficiency is lowest between particle diameters of 100 nm - 200 nm. This is attributed to the diffusion collection mechanism that is predominant in the region of < 100 nm diameter particles. Whereas once 200 nm is exceeded, interception collection dominates. In the interim range (100 nm - 200 nm), no single collection mechanism is dominant. The greater the filtration speed, the smaller the particle diameters are, at which filtration efficiency reaches a minimum. Considering the differences in mean pore size of the filter materials in each graph (Figures 9-11), it is evident that there is a large difference in the minimum filtration efficiency values. At the highest filtration speed of 0.1 m/s (Figure 11), the filter with a 4.6 μm mean pore diameter had a 74% minimum filtration efficiency, compared to 7% measured on the filter with a 17.7 μm mean pore diameter. This shows that the smaller the mean pore diameter, the greater the filtration efficiency. Comparing the effect of filtration speed in Figures 9-11, the higher the filtration speed, the lower the filtration efficiency. The same tendency is evident regarding smaller mean pore diameters having greater filtration efficiency.

In Figures 9-11, the dotted line shows the calculated filtration efficiency values based on the granular bed model, corrected with Equation's (9) and (10) noted above. The calculated results confirmed the trends: 1) filtration efficiency will have a minimum value with regard to aerosol particle diameter, 2) there will be a difference in filtration efficiency relative to the pore diameter of the material and 3) the filtration speed.

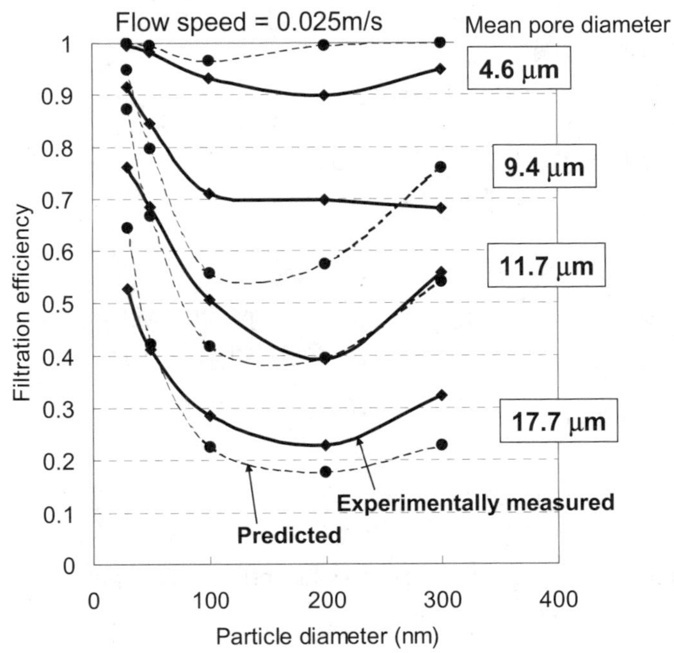

Figure 9: Filtration Efficiency of Each Test Sample at a Flow Speed of 0.025 m/s

Figure 12 shows pressure drop through the disc samples at three filtration speeds. Smaller mean pore diameters and higher filtration speeds result in greater pressure drop. When the mean pore diameter of a DPF wall is less than 10 µm, it produces a pressure drop that cannot be ignored with respect to the total DPF pressure drop.

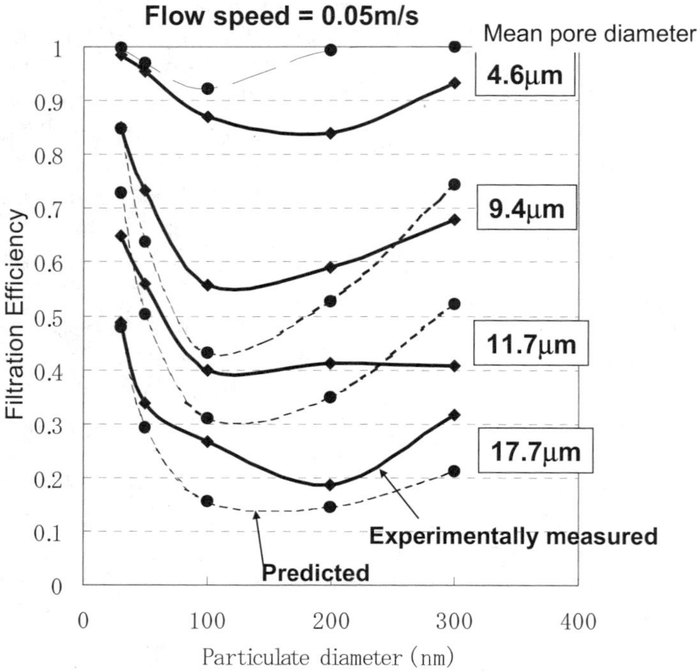

Figure 10: Filtration Efficiency of Each Test Sample at a Flow Speed of 0.05 m/s

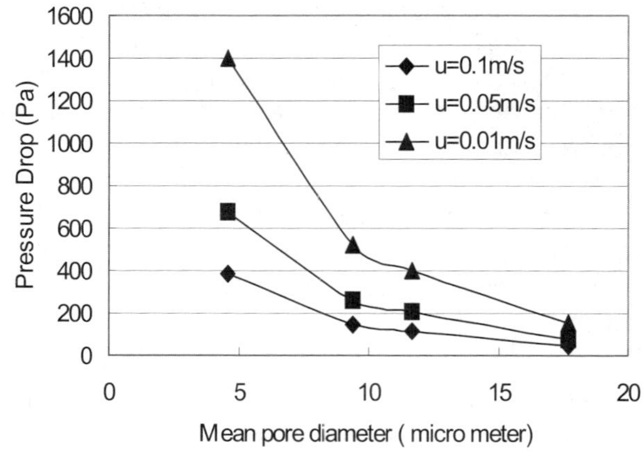

Figure 12: Pressure Drop Through the Wall

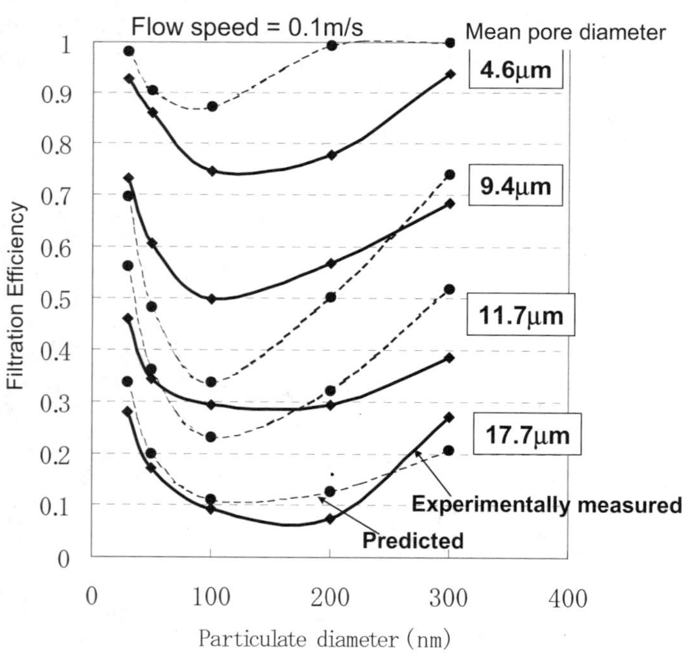

Figure 11: Filtration Efficiency of Each Test Sample at a Flow Speed of 0.1 m/s

Comparison Between $ZnCl_2$ Particles and Combustion Particles

As mentioned earlier, the filtration efficiency of diesel particulates can be evaluated by using $ZnCl_2$ particles of equivalent mobility particle size. To validate this concept, filtration efficiency test results were compared using $ZnCl_2$ particles and diesel combustion particles. Cast-2F was used as a combustion soot generator. A test specimen with a mean pore size of 9.4 µm was used for this validation test. Figure 13 shows the test equipment.

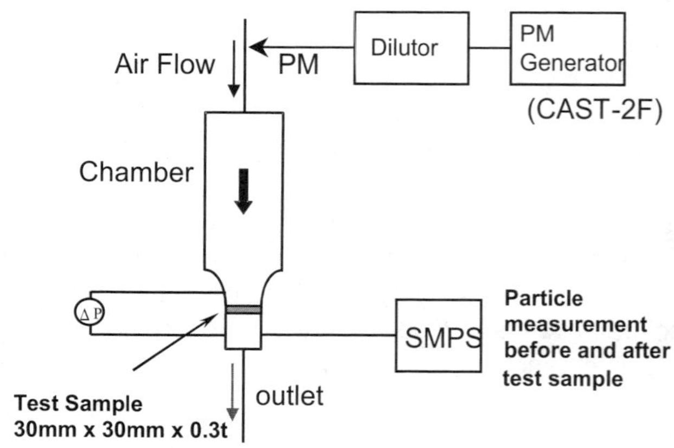

Figure 13: Test Equipment with Combustion Particles

Using three different PM size distribution: (1) distribution with a mean particulate size of 34 nm, (2) distribution with a mean particulate size of 156nm, and (3) distribution with peaks of 30 nm and 150 nm, particle size/count measurements were performed before and after the specimen using SMPS. In Figure 14, SMPS data is shown. For each particle size, penetration and filtration efficiency were calculated. The results are shown in Figure 15. Figure 15 shows good agreement of the results between $ZnCl_2$ particles and combustion particles.

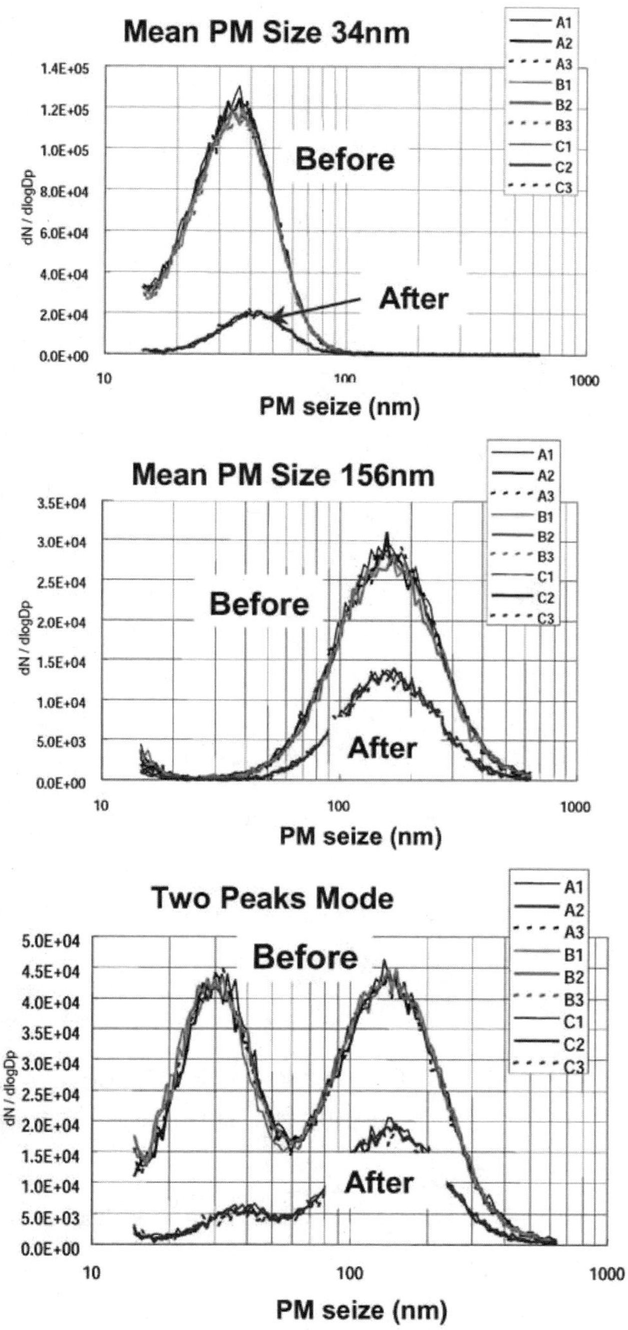

Figure 14: SMPS Measurement Before the Test Specimen and After the Test Specimen. - Test specimen: Mean Pore Size = 9.4 μm, Filtration speed: 0.05m/s, A1-A3 and B1-B3 are repeated measurement results for inlet. C1-C3 are repeated measurement results for outlet.

The PM concentration generated in Cast-2F was set to approximately 12 x 10^7 1/cm^3. The dilution ratio downstream of Cast-2F was set to 300. The combustion gas was diluted again in the mixing chamber to a ratio of 10. Therefore, the total dilution ratio was 3000 and the PM concentration of the primary exhaust stream was maintained at approximately 4 x 10^4 1/cm^3. The primary exhaust stream was directly introduced into SMPS. The SMPS used an impactor size of 0.457mm, a sheath flow rate of 3 liter/min, and a sampling flow rate of 0.3 liter/min. The diameter of the sampling tube was 6 mm and the Reynolds numbers of the sampling flow and the primary flow were almost the same: 70 – 80. The measurement interval was 3 min for the PM size range of 14.3 nm to 649 nm with 106 steps. Measurements were taken 3 times.

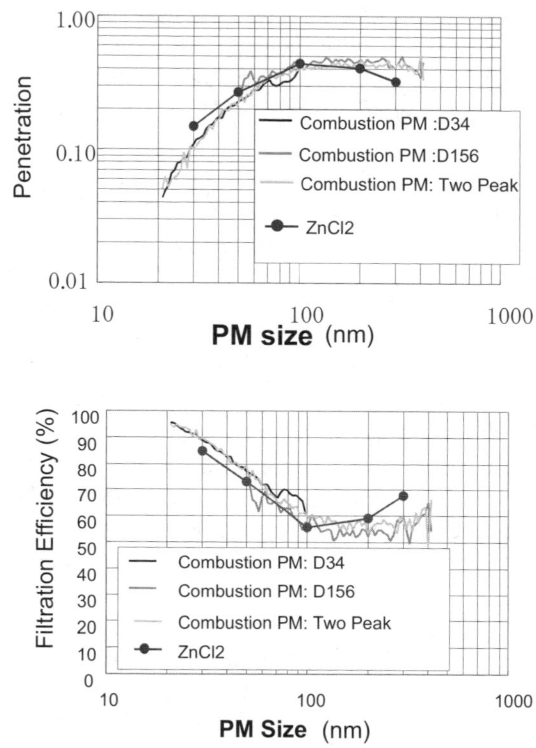

Figure 15: Penetration and Filtration Efficiency of Combustion Particles and $ZnCl_2$ Particles. - Test Condition: Mean Pore Diameter = 17.7μm, Thickness =0.304mm, Filtration Speed = 0.05m/s, Temperature = 20 degree C, Combustion Soot: D34 has a mean PM size of 34 nm, D156 has a mean PM size of 156 nm, the Two Peak Mode has PM distribution peaks of 30 nm and 150 nm, $ZnCl_2$: Monodisperse particles of 30 nm, 50 nm, 100 nm, 200 nm and 300 nm.

Filtration Efficiency Predictions

Using Equation 3, the filtration efficiency of an actual DPF without a PM layer was calculated. The pore distribution, mean pore diameters, and distribution ranges were varied as shown in Figure 16. σ is the logarithmic standard deviation. It was assumed that pore diameter does not change with depth into the wall surface. The wall area was divided and assumed that each area has different pore diameter in proportion to the pore diameter distribution. The ratio of the amount of gas flowing in each region was determined from the pressure drop balance using Ergun's pressure drop equation. We calculated the filtration efficiency using different filtration speeds and pore diameters in each flow region. Based on SMPS results of actual diesel exhaust, we assumed the pore size distribution shown in Figure 17 and obtained the overall filtration efficiency from those measured at each PM particle diameter. To determine the gas flow rate, we performed predictions for the weighted load mean gas flow of PM emissions in a 2-liter engine on the NEDC drive cycle. A 3-liter DPF volume and a 12 mil / 300 cpsi cell structure were assumed.

The modeling presented here is not capable of predicting filtration efficiency under soot-loaded condition. Only under the initial clean condition, the model can predict the filtration efficiency.

The calculation results are plotted in Figure 18. The results suggest the following relationships: (1) the smaller the mean pore diameter, the greater the filtration efficiency, and (2) the narrower the pore size distribution, the greater the filtration efficiency. A filter with a 20 μm pore diameter, with a logarithmic standard deviation of $\sigma = 0.5$ μm, has a filtration efficiency of approximetely 40%. With a smaller, uniform pore diameter (15μm, σ=0.0μ), the filtration efficiency reaches 70%. The filtration efficiency was measured at the fresh, initial state, assuming no PM layer on the wall surface. Since, in real world usage, a PM layer exists most of the time, higher filtration efficiency is can be expected versus the predicted.

The last graph in Figure 18 shows the change in DPF pressure drop due to differences in pore diameter. The lower curve corresponds to the calculated pressure drop of a bare DPF and the upper curve is the measured pressure drop of an actual catalyzed DPF. For the pressure drop calculation, uniform pore size was assumed. Mean pore diameters of 12μm or smaller, resulted in insufficient wall permeability due to plugging of the small pores with catalyst. Consequently, there was a radical increase in pressure drop. With mean pore diameters larger than 12μm, the contribution of wall flow pressure drop to overall filter pressure drop is minimal. Since initial pressure drop is primarily composed of contributions from the cell channels and contraction/expansion of the channels at the inlets and outlets, changes in DPF pressure drop due to differences in pore diameter are minor. [Ref: 12]

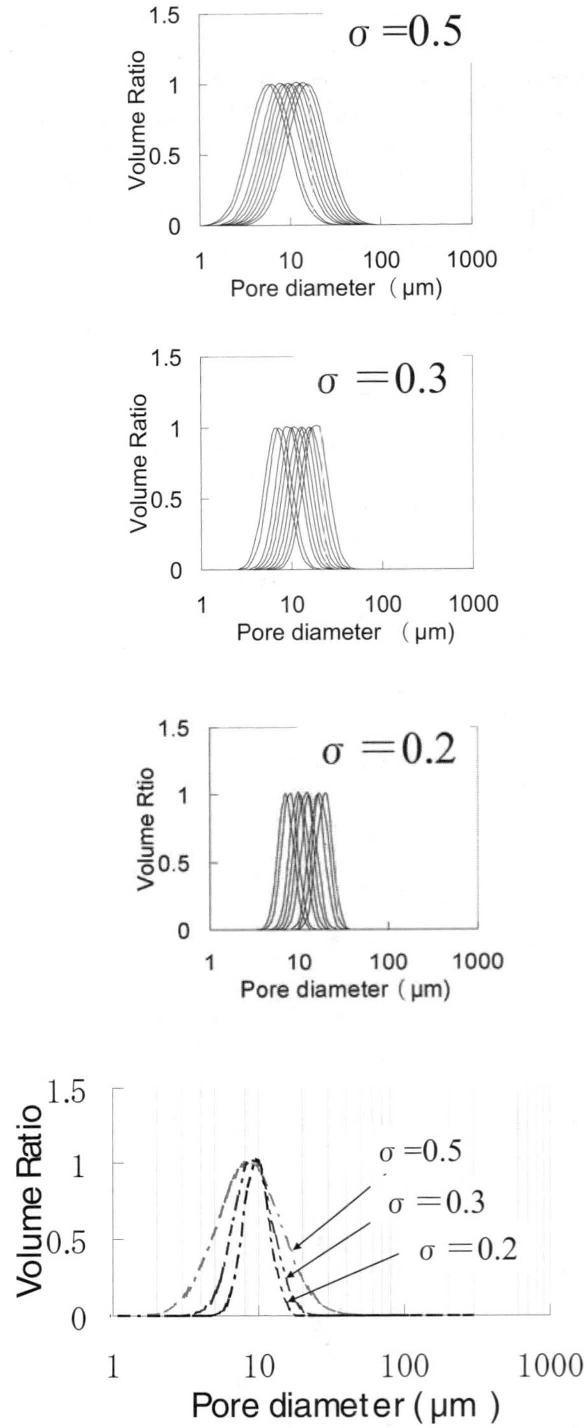

Figure 16: Pore Size Distributions for Model Calculation

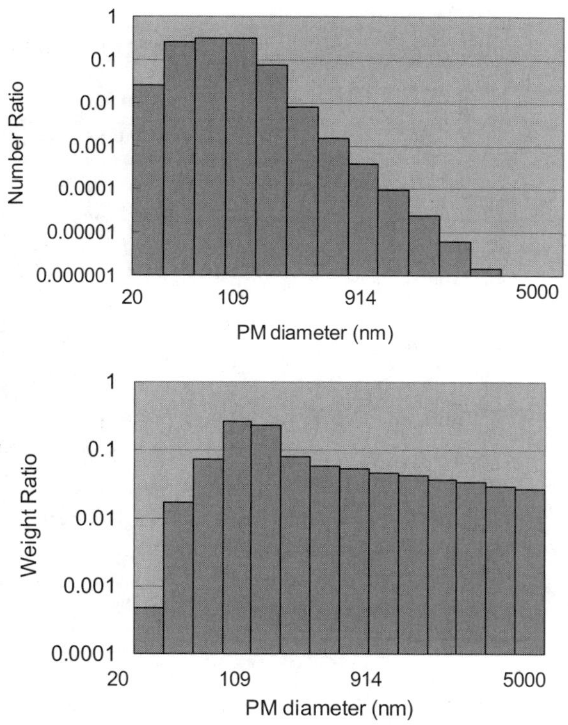

Figure 17: PM Size Distribution for Model Calculation

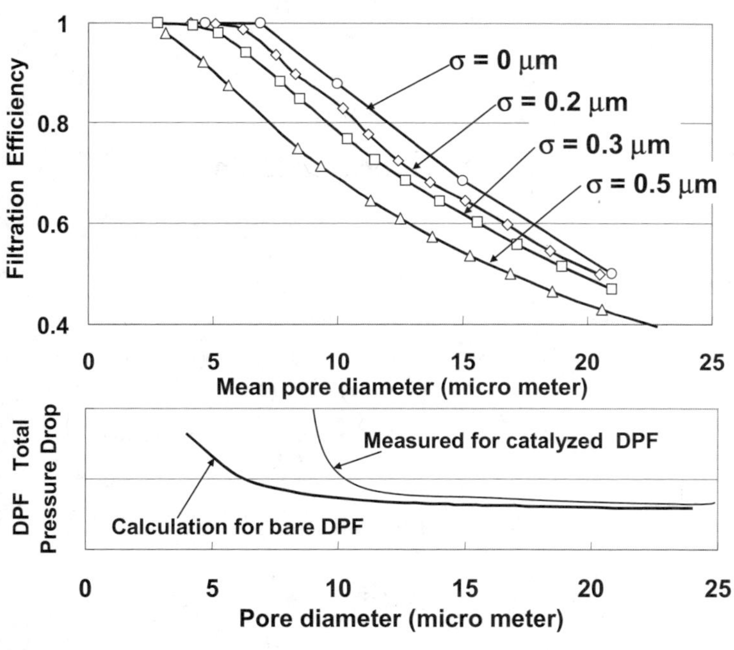

Figure 18: Influence of Pore Size Distribution on DPF Filtration Efficiency Without a PM Layer

VEHICLE EXPERIMENTS WITH A DPF

DESCRIPTION

Vehicle PM emissions with two (2) cordierite DPF materials, having different mean pore diameters, were measured on the NEDC test cycle. The experiment equipment used in the experiment is outlined in Figure 19. The DPFs had a volume of 2.5 liters and a cell structure of 12 mil/ 300 cpsi. In consideration of the performance trade-off's of filtration efficiency versus pressure drop, the mean pore diameters of 25 μm and 15 μm were selected for DPF A and DPF B respectively. Both DPF A and DPF B have pore distribution standard deviations of $\sigma = 0.4$ μm. The vehicle was driven on the NEDC test cycle by a robot while the exhaust gas was directed into a full dilution tunnel, PM was sampled from the gas line using filter paper and weighed. At the same time, the vehicle out PM count was continuously measured using ELPI. The test procedure included a preconditioning phase of running 3 repetitions of the last portion of the NEDC mode, after which, the particulate matter on the filter paper was weighed. ELPI measurements over the 3 cycles were continuously recorded. The vehicle was calibrated to meet Euro III regulations. For these tests, an oxidation catalyst was installed in front of the un-catalyzed DPF.

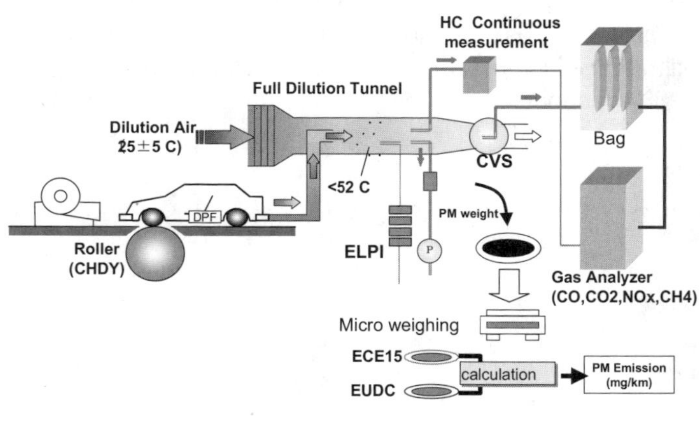

Figure 19 Chassis Dynamometer and Diesel Emission Test Equipment

TEST RESULTS

Figure 20 shows the real-time measurement results of the PM emissions count using ELPI for DPF A and DPF B, for both the preconditioning and NEDC drive cycles. The particulate matter collected on the filter paper during the emission test cycle weighed 3 mg/km for DPF A and 1 mg/km for DPF B.

Since ELPI measurement involves sampling exhaust gas that has been diluted and cooled to below 52°C in the dilution tunnel, the sample also includes nuclei mode particulates formed by the concentration of hydrocarbons in the exhaust gas. In Figure 20, a portion of the particles with diameters less than 70 nm, counted during the first preconditioning cycle, can be inferred to be nuclei mode particulates generated because the oxidation catalyst in front of the DPF did not reach its activation temperature. The same oxidation catalyst was used for both the DPF A and DPF B experiments. Therefore, it can be assumed that the nuclei mode particulate emissions were the same for A and B. Consequently, the difference between DPF A and DPF B results is deemed to be solely due to the difference in carbon particle count. After the preconditioning cycle of DPF A, the PM emissions count decreases for successive EUDC cycles, as a result of the PM layer formed on the wall surface. However, at the end of the third cycle, a leak, although small, can be identified. By contrast, even before the PM layer is formed within DPF B, the PM emissions count is much smaller than that of DPF A. This is evidence that the DPF B material has higher PM filtration efficiency. Engine out PM emission was 50 mg/km. The overall filtration efficiency of DPF A during preconditioning was only 80%. The overall filtration efficiency of DPF B during preconditioning was improved to 92%. During the Emission Test Mode, after preconditioning, overall filtration efficiency of DPF A and DPF B were 94% and 98%, respectively.

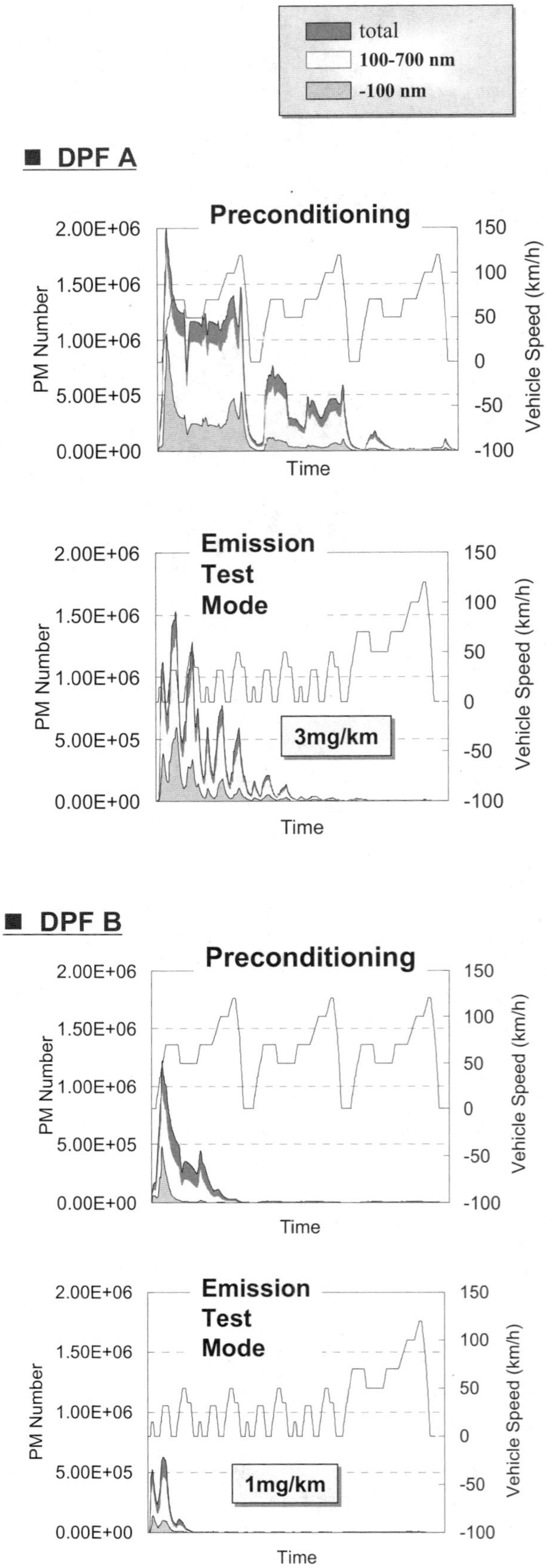

Figure 20: Chassis Dynamometer Test Results

For reference, filtration efficiency test results using a diesel fuel burner are shown in Figure 21. The weight of the particulate matter at a fixed mass flow, upstream and downstream of DPF A and DPF B, were measured by filter paper sampling. Figure 21 shows that the filtration efficiency of DPF A without a soot layer is only 55%, while changes to the wall material characteristics of DPF B resulted in an improved initial efficiency of 70 %.

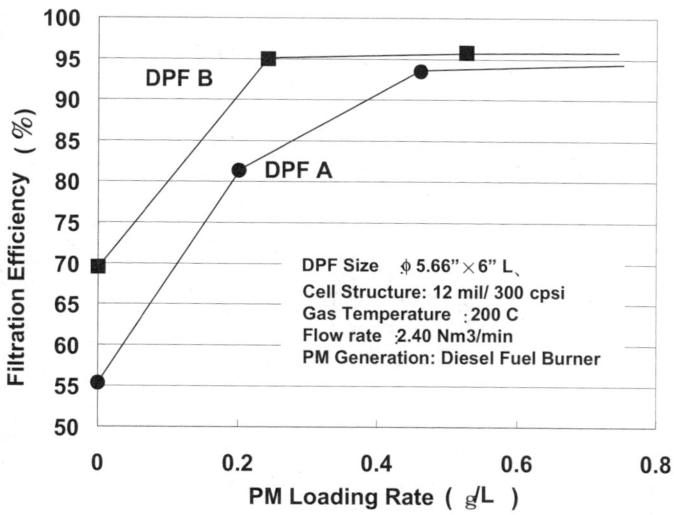

Figure 21: Filtration Efficiency Test Results Using a Diesel Fuel Burner

CONCLUSIONS

1) The effects of the DPF mean pore diameter on PM filtration efficiency were examined. Fundamental experiments were conducted to measure $ZnCl_2$ aerosol particle filtration efficiency of plate samples with the same thickness as diesel particulate filters. The following results were obtained:

a) In the mean pore diameter range of 4.6 - 17.7 μm and filtration speed of 0.025 - 0.1 m/s, the PM filtration efficiency of a filter with a 0.305 mm wall thickness reached a minimum value for particle diameters in the 100 nm – 200 nm range. High filtration efficiency was attributed to the diffusion collection mechanism for small diameter particles (<100 nm) and to the interception collection mechanism for large particles (>200nm).

b) A reduction in pore diameter and flow speed always raised the filtration efficiency.

c) Filtration efficiency of $ZnCl_2$ particles is well matched with that of combustion particles with the same mobility diameter.

2) The following findings were obtained by comparing filtration efficiency calculations using the granular bed model against the experimental results:

a) Applying the diffusion collection mechanism analysis, filtration efficiency could be predicted by assuming a granule diameter (d_s) that was 4 times the pore diameter (d_m) obtained by the mercury penetration method.

b) For the interception collection mechanism, filtration efficiency could be predicted by setting $d_m = d_s$.

3) Filtration efficiency predictions during the mean flow rate conditions of the EUDC test cycle were conducted using the granular bed model for various mean pore diameter distributions, assuming no soot layer. The following results were obtained:

a) The smaller the mean pore diameter, the greater the filtration efficiency. Narrower pore size distributions exhibited higher filtration efficiency.

b) A filter with a 20μm mean pore diameter, with a logarithmic standard deviation of σ = 0.5μm has a filtration efficiency of approximately 40%. With a smaller, uniform pore diameter (15μm, σ=0.0μm), the filtration efficiency reaches 70%.

c) The smaller the mean pore diameter, the greater the filtration efficiency. Small mean pore diameters, less than 12μm, result in a dramatic increase in pressure drop, due to pores being plugged with catalyst. A mean pore size of 12μm to 18μm seems to be optimal.

4) Vehicle PM emissions measurements on a chassis dynamometer, equipped with cordierite diesel particulate filters, with mean pore diameters of 25 μm and 15 μm, obtained the following results:

There was a significant difference in PM emissions between the 25μm mean pore diameter filter and the 15 μm filter during preconditioning. Overall filtration efficiency during preconditioning was improved from 80% to 92% by changing mean pore diameter from 25 μm to 15 μm.

5) When PM loading rate becomes more than 0.3 g/l, filtration efficiency with optimized pore size will be >95% (see Figure 21).

REFERENCE

1. Otani et.al.,"Experimental Study of Aerosol Filtration by the Granular Bed Over a Wide Range of Reynolds Numbers", Aerosol Sience and Technology 10.463-474(1989)

2. Tardos,G.T.,Abuaf,N.,and Gutfinger,C.,"Dust Deposition in Granular Bed Filters:Theories and Experiments", J.Air Pollut. Control Assoc.28:354.(1978)

3. Parctsky,L.,Theodore,L.,Pfelfer,R. and Squires,A.M.,"Panel Bed Filters for Simultaneous Removal of Flt Ash and Sulfer Diaxide: II. Filtration of Dilute Aerosols by Sand Beds", ,J.Air pollut.Control assoc.21:204.(1971)

4. Pendse,J.and Tien,C.,"General Correlation of the Initial Collection Efficiency of Granular Filter Beds", AIChE J.28.677.(1982)

5. Mann,L..A.,and goren,S.L.,"Aerosol Capture in Granular Beds in the Sedimentation and Diffusion Dominated Regimes", Aerosol Sci. Technol.3:195, (1984)

6. Knettig,P.,and Beeckmans,A.M.,"Inertial Capture of Aerosol Particles by Swarms of Accelerating Spheres", J.Aerosol. Sci.5:225, (1974)

7. Kennard,M.L..,and Miesen.A.,"Aerosol Collection in Granular Beds", 2nd World Filtration Congr., pp.229, (1979)

8. Gebhart,J.,Roth.C.,and Stahlhofen,W.,"Filtration Properties of Glass Bed media for Aerosol Particles in the 0.1-2μm Size Range", J. Aeosol Aci. 4:355, (1973)

9. Gal,E.,Tardos,G.,and Pfeffer,R.,"A Study of Inertial Effects in Granular Bed Filtration", AIChE J.31:1093. (1985)

10. Ranz,W.E.,and Wong, "Impaction of Dust and Smoke Particles", J.B.Ind.Eng.Chem.44:1371.(1952)

11. Kanaoka,C.,Yoshioka,N.,Iinoya,K.,and Emi,H.,"Inertial Capture of Aerosol Particles by Single Sphere", Kagaku Kougaku 36:104. (1972)

12. Hashimoto,S., Miyairi,Y., Hamanaka,T., et.al.,"SiC and Cordierite Diesel Particulate Filters Designed for Low Pressure Drop and Catalyzed, Uncatalyzed Systems", SAE Paper :2002-01-0322

APPENDIX

1) Single Sphere Gravity Collection Efficiency [1]

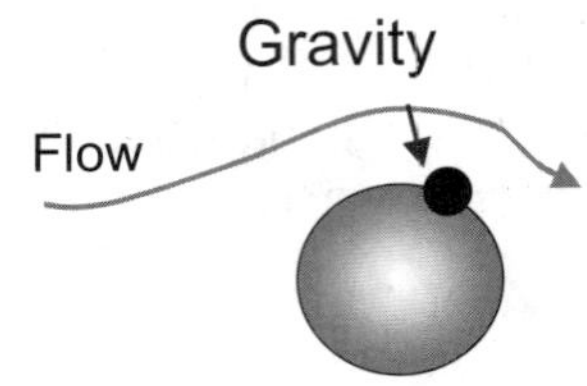

$$\eta_g = G/(1+G)$$

$$G = \rho_p\, d_p^{\,2}\, g\, /\, (18\mu u)$$

ρ_p: Specific Density of PM
μ: Gas Viscosity
g : Gravity Acceleration
u : Flow Speed

2) Single Sphere Inertia Collection Efficiency [1]

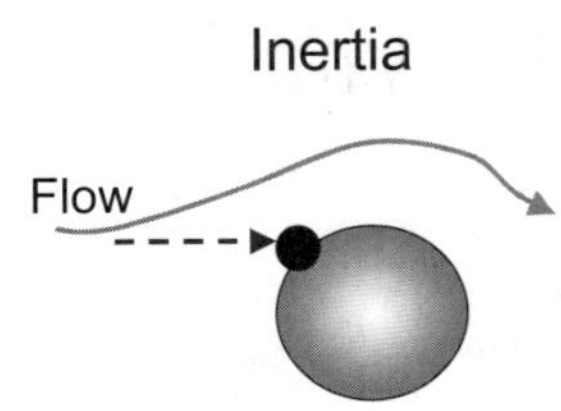

$$\eta_I = St_{eff}^{\;3} / (1.4E\text{-}02 + St_{eff}^{\;3})$$

$$St_{eff} = St\ (\ 1 + 1.75\varepsilon\ Re\ /\ 150\ (\ 1- \varepsilon\ )$$

$$St = \ \rho_p\ /(9\rho_f)\ Re\ (\ d_p/d_i)^2$$

Re: Reynolds Number
ρ_p: Specific Density of PM
ρ_f: Gas Density
ε : Porosity
d_p: PM Diameter
d_i : Granule Diameter

2007-01-0923

Filtration Behavior of Diesel Particulate Filters (2)

T. Mizutani, A. Kaneda, S. Ichikawa, Y. Miyairi, E. Ohara, A. Takahashi, K. Yuuki, H. Matsuda and H. Kurachi
NGK Insulators, Ltd.

T. Toyoshima, T. Ito, I. Lappas and C. D. Vogt
NGK Europe GmbH

M. Tanaka, A. Martin, S. Fujii and P. Busch
NGK Automotive Ceramics USA, Inc.

ABSTRACT

Due to its better fuel efficiency and low CO_2 emissions, the number of diesel engine vehicles is increasing worldwide. Since they have high Particulate Matter (PM) emissions, tighter emission regulations will be enforced in Europe, the US, and Japan over the coming years. The Diesel Particulate Filter (DPF) has made it possible to meet the tighter regulations and Silicon Carbide and Cordierite DPF's have been applied to various vehicles from passenger cars to heavy-duty trucks. However, it has been reported that nano-size PM has a harmful effect on human health. Therefore, it is desirable that PM regulations should be tightened.

This paper will describe the influence of the DPF material characteristics on PM filtration efficiency and emissions levels, in addition to pressure drop.

INTRODUCTION

Diesel engines have low CO_2 emissions and good fuel efficiency, however, they have had high PM and NOx emissions, which were big disadvantages. Common-rail fuel injection systems and DPF's made it possible to reduce PM and NOx emissions, achieving almost the same level as gasoline engines.

As an example, it was confirmed in this paper that current Si-SiC (Silicon-Bonded Silicon Carbide) DPF's [1]-[4] and Cordierite DPF's (shown in Figure 1) can meet the proposed new European emission standard, Euro-5 [5] (PM: 5mg/km, NOx: 200mg/km), which is scheduled to be enforced from 2009 for new models and from 2011 for all models.

However, it is expected that PM regulation levels, including those for gasoline engines, will be tightened again in the next phase, since it has been reported that nano-particles are harmful to human health.

Some papers have analyses of the basic filtration performance and emission levels of DPF's, but only a few examinations of the materials and pore structures of DPF's have been offered [6]-[8]. Therefore, the influence of DPF pore structures on filtration performance was investigated using a disc-type filter with different pore size distributions in Part 1 [9] of this two part, simultaneously published, series. Part 1 contains an examination of the emission, filtration performance and pressure drop of DPF's with different materials and pore structures, measured on a chassis dynamometer and engine bench.

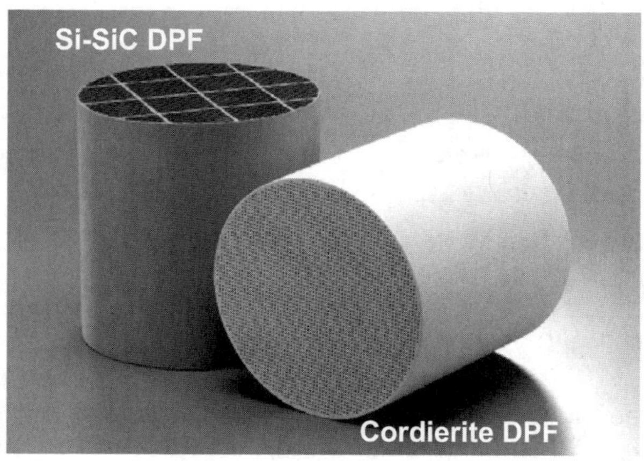

Figure 1. Examples of Silicon-Bonded Silicon Carbide DPF (Si-SiC DPF) and Cordierite DPF

EMISSION AND FILTRATION PERFORMANCE

The filtration performance of several DPF materials was evaluated. The emission level was determined by vehicle testing and the filtration efficiency was measured by engine bench testing.

EMISSION TEST

Experimental Set-up

A Volkswagen Passat (2.0L TDI, 103kW, Euro-4 application, single Close-Coupled CSF with integrated DOC function) was used for the emission test. The emission measurement system is shown in Figure 2.

The European regulation mode (New European Driving Cycle [NEDC]) was used. The emission test procedure is shown in Figure 3. As reported in several papers[9], filtration efficiency rapidly increases to more than 90%. As soot accumulates inside of the DPF, the pores begin to fill and behave like secondary filters. Therefore, accumulated soot inside the DPF was burned out using an electric furnace before each emission test. This pre-test step allowed the filtration efficiency to be evaluated under the most severe conditions.

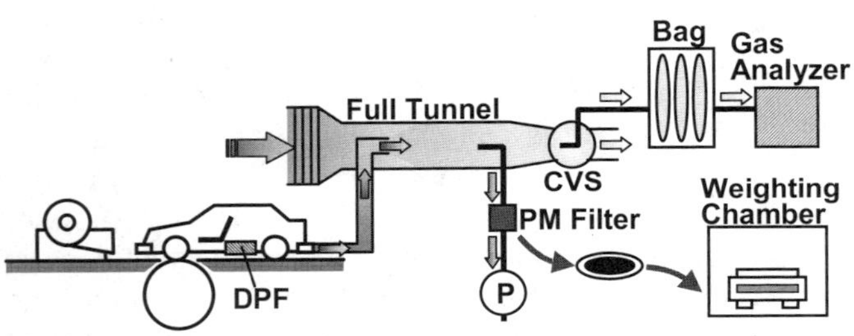

Figure 2. Emission Measurement System

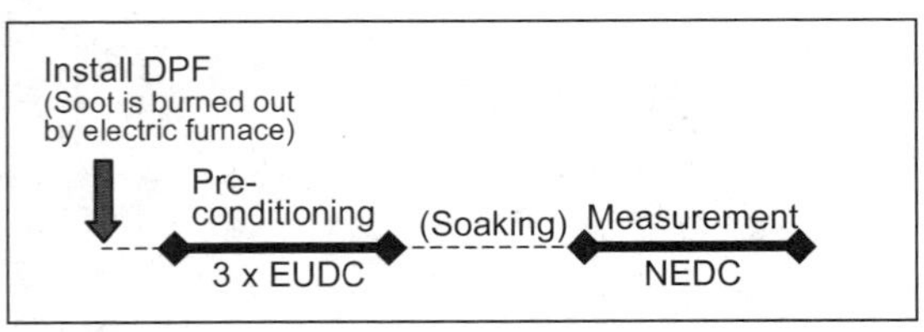

Figure 3. Test Mode and Test Procedure

DPF Test Sample Description

Three SiC DPF's with different porosities and mean pore sizes (MPS) and one Cordierite (Cd) DPF were evaluated (SiC.A, SiC.B, SiC.C, Cd.A, listed in Table 1). The porosity and MPS data described in Table 1 are based on uncoated conditions. All of the DPF's had identical outer dimensions (oval) and catalyst coating.

Table 1. Emission Test DPF Description

Material	SiC.A	SiC.B	SiC.C	SiC.D	Cd.A
Porosity	58%	52%	48%	43%	50%
MPS	23um	21um	14um	18um	14um
Cell Structure	12mil /300cpsi (Wall Thickness: 0.31mm, Cell Pitch: 1.47mm)				
Catalyst	Coated				

Results and Discussion

The PM emission test results, shown in Figure 4, prove that the emissions of all of the DPF's are less than 1mg/km, which is far below the proposed 5mg/km Euro-5 regulation value. It is presumed that the PM layer, accumulated during preconditioning, filters the soot generated during the measurement phase. Consequently, all of the DPF's showed virtually equivalent emissions levels, even though they all have different porosities and MPS.

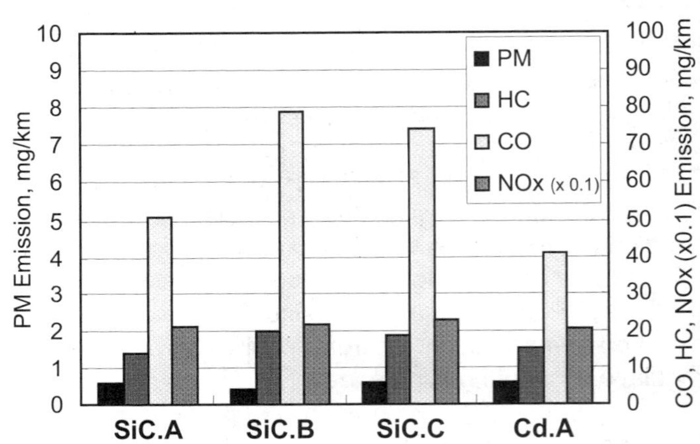

Figure 4. NEDC Emission Test Results

ENGINE BENCH TEST

Introduction

DPF filtration performance was also evaluated on an engine bench. The performance was evaluated at high load conditions and steady state lower load conditions. The engine used was different from the one used in the emission test (2.0L TDI with common-rail, Euro-3 application, Underfloor DOC and CSF).

DPF Test Sample Description

The four DPF's listed in Table 1 were evaluated. The DPF's were all ϕ5.66 inches x 6 inches length (ϕ144 x 152mm length). All of the DPF's were coated with the same catalyst.

Smoke Test During Full Load

[Experimental Set-up]

To evaluate filtration performance during high load conditions, the smoke level, during the full load condition with different engine speeds, 1200 → 2000 → 2500 → 3000 → 3500 → 4000 RPM (3 minutes for each step), was measured using an AVL Smoke Meter (AVL 415S). As was done in the emission test procedure, the soot accumulated within the DPF's was burned out before each test. In addition, the interval between engine ignition and the 1st step (1200rpm x full load) was controlled to within 1 minute to minimize the influence of accumulated soot on filtration efficiency.

[Results and Discussion]

Results of the smoke test during full load conditions, with and without a DPF, are shown in Figure 5. The smoke levels of all of the DPF's versus engine speed are shown in Figure 6. Because the PM concentration upstream of the DPF is high and there is no soot layer, the smoke level at 1200 RPM is higher. At engine speeds above 1200 RPM, the smoke level becomes very low as soot begins to clog pores, increasing filtration efficiency, while engine out PM emissions also decrease. At engine speeds from 3000 to 4000 RPM, the smoke level downstream of the DPF increases as accumulated soot starts to burn, allowing small particles to pass through the cell walls, while the engine out PM emissions also increase.

The smoke level at 1200 RPM is primarily influenced by the MPS. The filtration efficiency was calculated using the engine out smoke levels shown in Figure 7 and Figure 8, respectively. The pore sizes were measured using mercury porosimetry. The heat capacity and thermal conductivity of the DPF affects the rate of soot burning, which, in turn, affects the smoke level at high speeds. At low speeds, however, the DPF thermal characteristics can be ignored, allowing the filtration performance evaluated independently. Therefore, the smoke level at low speeds was utilized in the calculations. Figure 7 and Figure 8 show that MPS smaller than 15um results in zero smoke emission and 100% filtration efficiency, respectively in this test.

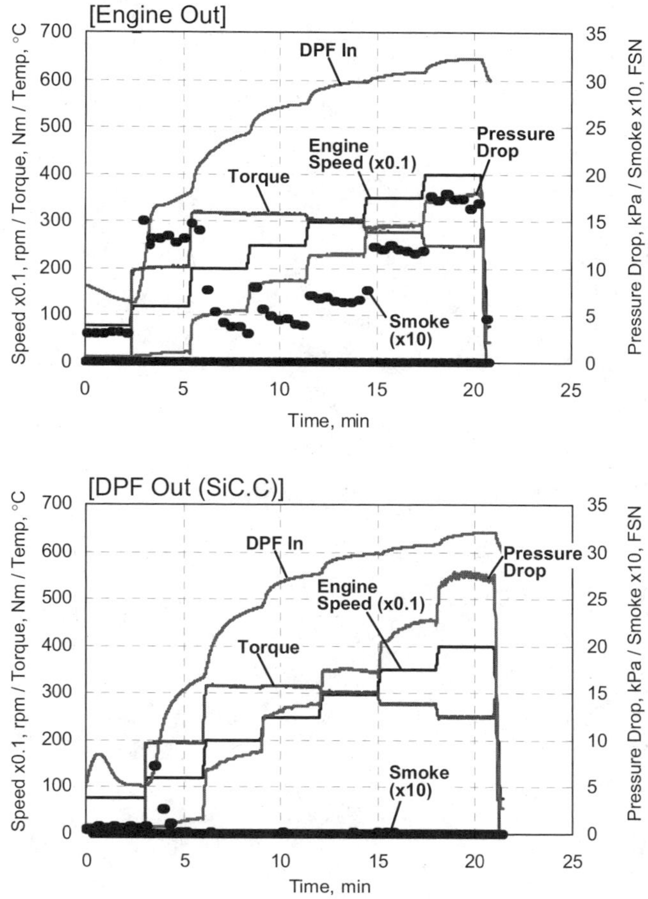

Figure 5. Full Load Smoke Test [Engine Out / DPF Out]

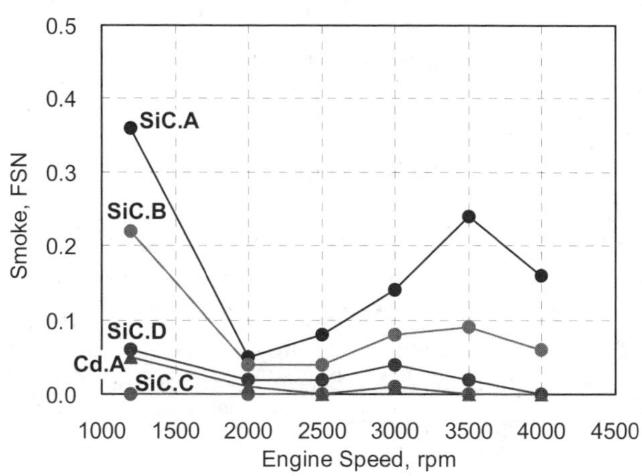

Figure 6. Smoke Level versus Engine Speed

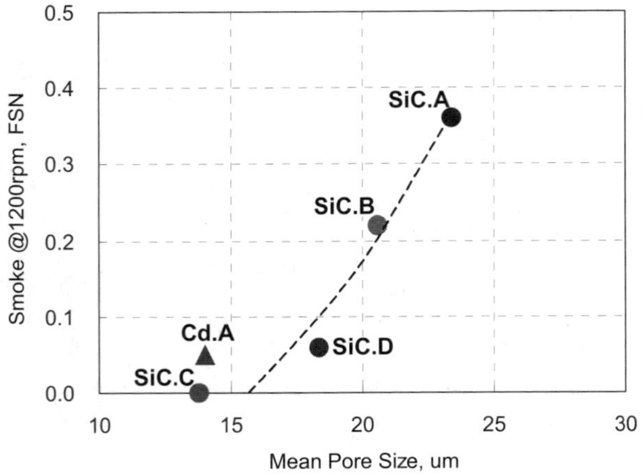

Figure 7. Smoke Level versus MPS

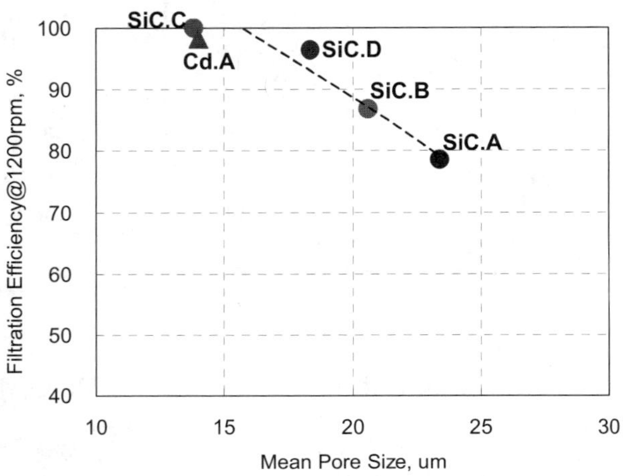

Figure 8. Filtration Efficiency versus MPS

Steady-State SMPS Test

[Experimental Set-up]

To investigate the filtration performance during steady-state, normal driving conditions (2000rpm x 50Nm), filtration efficiency was measured using a Scanning Mobility Particle Sizer (SMPS). A Micro Tunnel, MDLT-1300T, was used as a diluter.

[Experimental Set-up]

Like the full load smoke test, accumulated soot was burned out with an electric furnace before each test. The engine was warmed up with a dummy DPF. After warm-up, the dummy DPF was replaced with a test sample and engine conditions were changed from idle to 2000 RPM x 50Nm in 10 seconds. The first measurement was made after 30 seconds and repeated 4 times. The measurement point located downstream of the DPF was evaluated first, followed by the upstream location. The test procedure is illustrated in Figure 9.

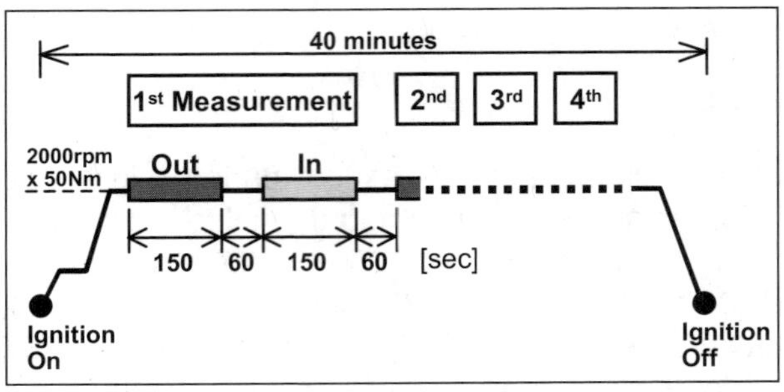

Figure 9. Steady-State SMPS Test Procedure

[Results and Discussion]

The PM particle number and filtration efficiency distribution as a function of the PM particle size obtained from SiC.C are shown in Figure 10. The uppermost figures are results of the first measurement and the lower two figures are results of the second measurement. Although the duration between the first and second measurements is only eight minutes, the filtration efficiency increases drastically. As described in the first paper [9], this increase is caused by pore clogging, i.e., PM accumulates at the necks of the pores and filters accumulating PM. A schematic of this behavior is illustrated in Figure 11.

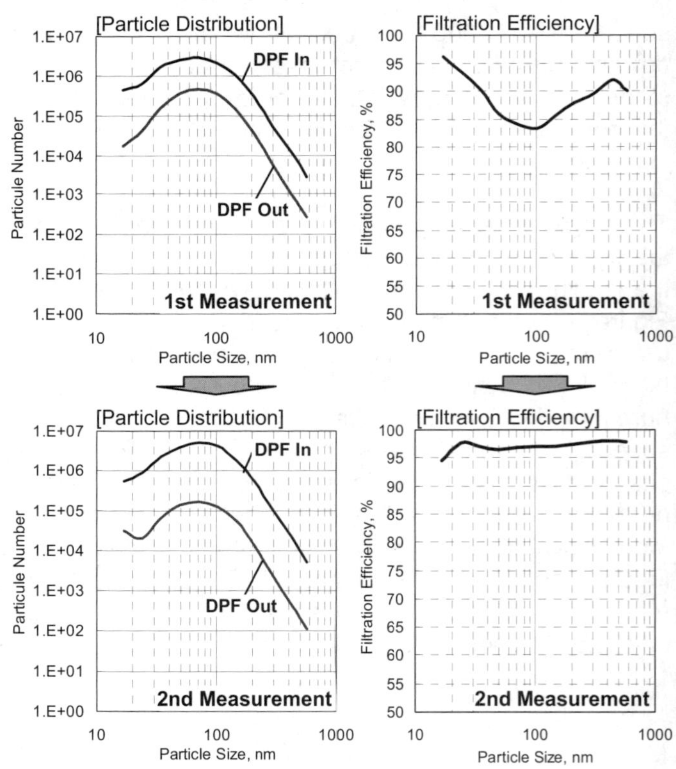

Figure 10. PM Particle Number and Filtration Efficiency Distributions

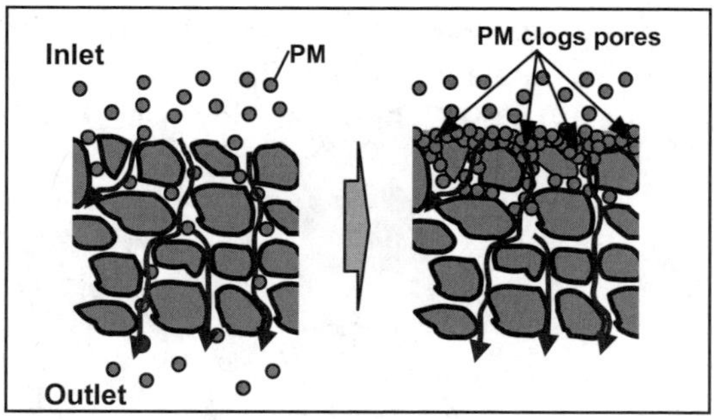

Figure 11. Schematic of Filtration Behavior

The first measurement of filtration efficiency was calculated by integrating from 15nm to 500nm and was influenced by the MPS as shown in Figure 12. The calculations revealed the same relationship as the full load smoke test results; smaller MPS have higher filtration efficiency.

The influence of pore volume (<10um, 10-20um, 20-30um, 30-40um, >40um) of each DPF is illustrated in Figure 13. "Pore volume (10-20um)" is defined as the integrated pore volume between 10um and 20um pore sizes. The investigation shows that a pore volume above 20um results in a reduction of filtration efficiency. Also, pores smaller than 20um have almost no correlation with filtration efficiency. Therefore, it seems that pores smaller than 20um have no impact on filtration efficiency, i.e., filtration efficiency would not decrease no matter how much the volume of pores smaller than 20um is increased.

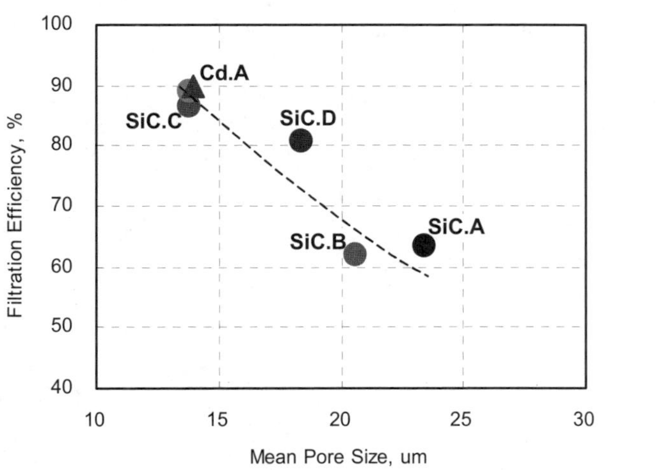

Figure 12. Filtration Efficiency versus MPS (First measurement)

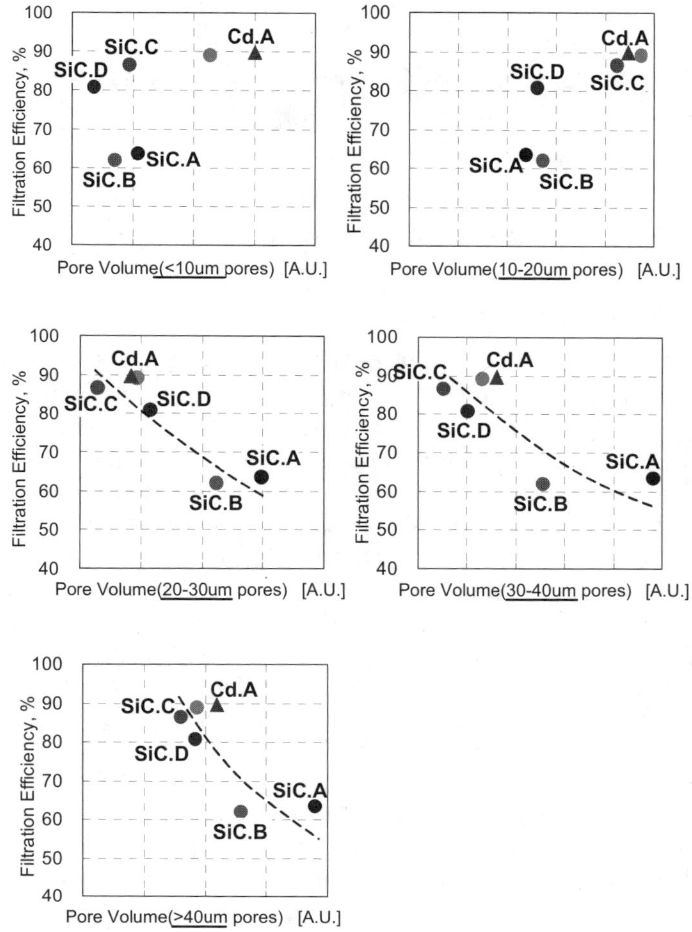

Figure 13. Filtration Efficiency versus Pore Volume

PRESSURE DROP PERFORMANCE

Introduction

The pressure drop performance with and without soot loading was evaluated. Since the purpose of this study was to investigate the influence of pore size on pressure drop, only the pressure drop of SiC DPF's were evaluated to avoid inherent differences between segmented and monolithic designs. Pressure drop performance is highly affected by catalyst loading conditions, such as washcoat distributions inside pores, catalyst particle size, and so forth. Therefore, four SiC DPF's listed in Table 1 were evaluated without catalyst coatings. The DPF size was ϕ5.66inch x 6inch length (ϕ144 x 152mm length).

Pressure Drop Without Soot Loading

[Experimental Set-up]

Pressure drop without soot loading was measured using a cold flow bench (Figure 14) at room temperature.

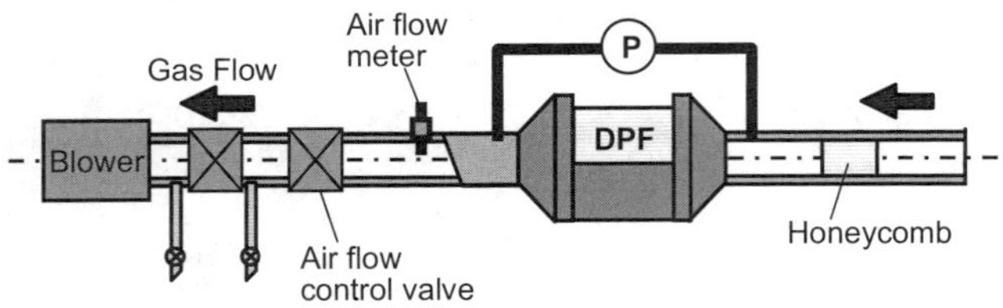

Figure 14. Sketch of Cold Flow Bench

[Results and Discussion]

The pressure drop at 9Nm^3/min (normalized at 25°C) was influenced by the MPS as shown in Figure 15. This figure was rearranged by MPS based on Figure 3 in the previous paper[2], and results of three SiC DPF's were added. As described in the previous paper[6], the predominant factors in the pressure drop without soot loading are the plug loss (contraction loss at the inlet side and expansion loss at the outlet side) and channel flow loss (friction loss through a narrow duct). The influence of the wall flow loss (determined by the wall material properties) is very minor. However, if the permeability of the DPF material is less than 1 x 10^{-12}m^2, the pressure drop without soot loading drastically increases[2].

It had been reported in other publications that permeability correlates with porosity and the square of the MPS [6, 11], i.e., when the total area of the flow passage is very small, the pressure drop drastically increases. Because MPS is squared, the effect is large. The influence of MPS on the pressure drop without soot loading is shown in Figure 15. There is a considerable increase in pressure drop when the MPS is below 10um. The rate of pressure drop decrease is almost negligible when the MPS is more than 10um.

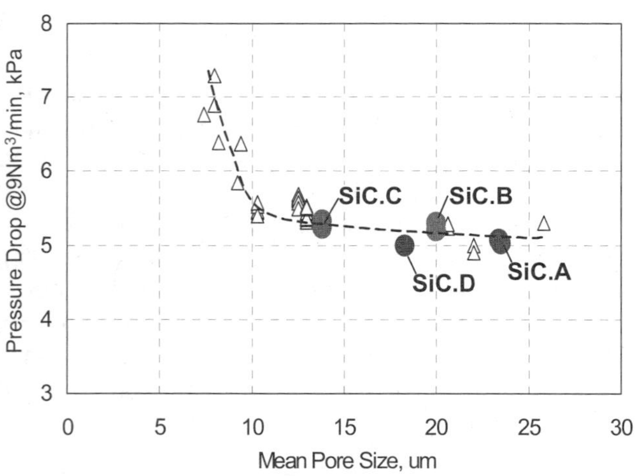

Figure 15. Pressure Drop Without Soot versus MPS

Pressure Drop With Soot Loading

[Experimental Set-up]

A gas burner based Soot Generator, shown in Figure 16, was used for the pressure drop test with soot loading. The generator produces soot as a result of incomplete combustion of diesel fuel. The temperature in front of the DPF was set to 200°C and gas flow rate was set to 2.27Nm^3/min.

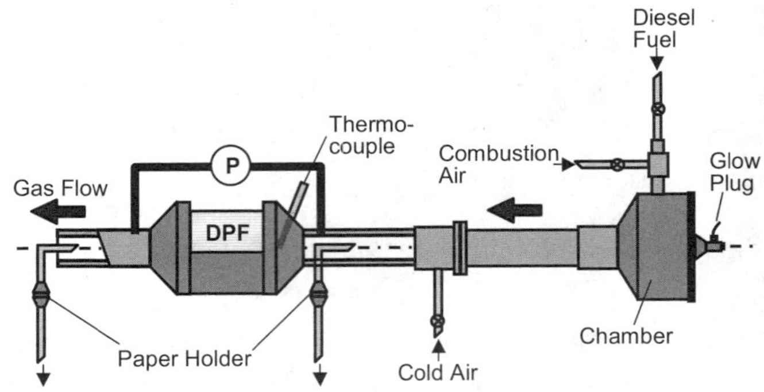

Figure 16. Sketch of Soot Generator

[Results and Discussion]

The pressure drop of all of the DPF's versus soot density is shown in Figure 17. The influences of porosity and pore volume on pressure drop are shown in Figure 18 and Figure 19, respectively.

In the initial phase, the pressure drop slope has a large positive gradient. In the second phase, it changes to a smaller gradient after an inflection point as shown in Figure 17. Soot clogs the pores in the initial phase, resulting in larger pressure drop increases. Then, soot accumulates on the cell wall and generates a soot cake layer resulting in a smaller pressure drop increase. The gradient of the slope in the second phase is mostly determined by the cell structure. Because all of the DPF's have the same cell structure, the gradients in the second phase are almost the same.

Figure 18 shows that pressure drop with soot loading depends on porosity, i.e., the higher the porosity, the lower the pressure drop. Since pore volume has no strong correlation with pressure drop with soot as shown in Figure 19, the pore size has very little affect on pressure drop with soot.

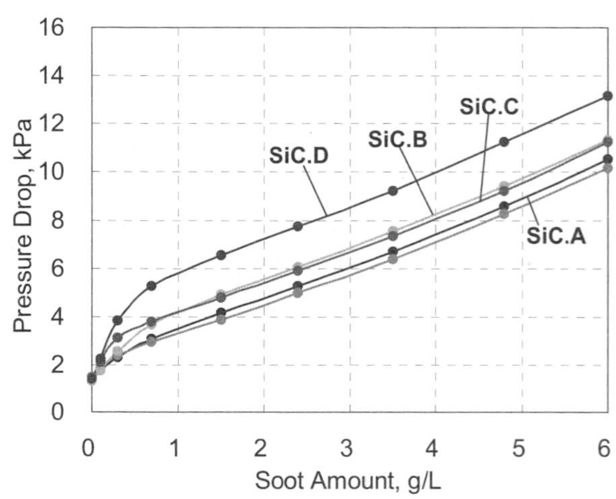

Figure 17. Pressure Drop versus Soot Density

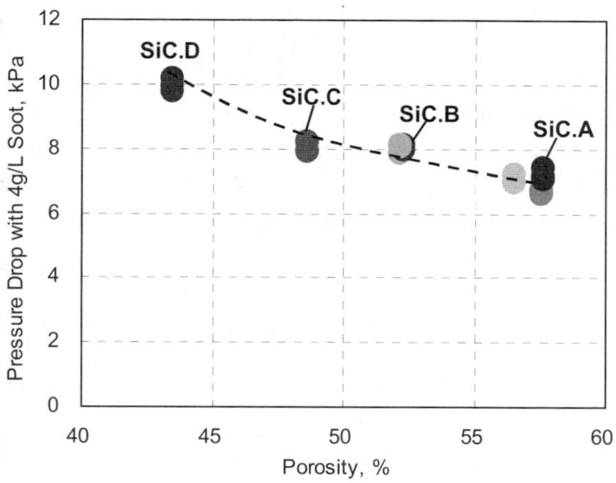

Figure 18. Pressure Drop With 4g/L Soot versus Porosity

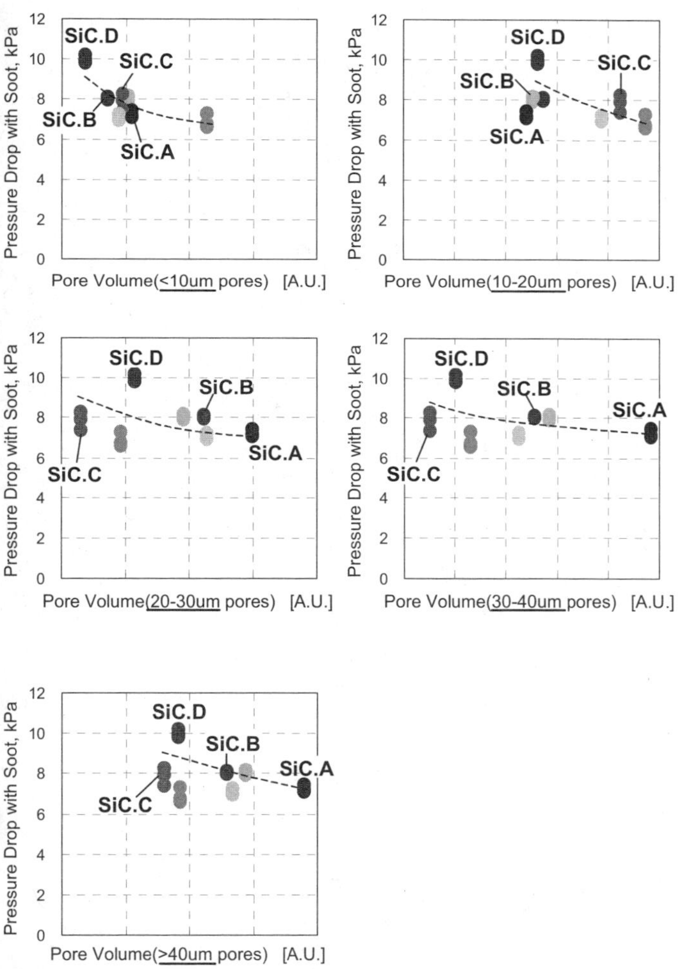

Figure 19. Pressure Drop versus Pore Volume With 4g/L Soot

CONCLUSION

The emission levels, filtration performances, and pressure drops of Si-bonded-SiC and Cordierite DPF's were evaluated and the following results were obtained:

1. The emission levels of three SiC DPF's and one Cordierite DPF, with 48 to 58% porosity and 14 to 23um MPS, were evaluated under the NEDC test mode. It was confirmed that PM emissions from all of the DPF's were less than 1mg/km and can thus achieve the Euro-5 PM emissions level (5mg/km).
2. Filtration performance was evaluated under full load and steady-state lower load conditions on the engine bench. It was found that there is a correlation between MPS and filtration efficiency: the larger the volume of pores greater than 20um, the lower the filtration efficiency.
3. Pressure drop without soot loading was evaluated on bare DPF's. The study revealed that pressure drop drastically increases when MPS is less than 10um.
4. When pressure drop with soot loading was tested, it was confirmed that the higher the porosity, the lower the pressure drop.
5. SiC.C (48% porosity and 14um MPS) has a good balance of material characteristics because it has better filtration efficiency than the other candidates and acceptable pressure drop performance.

ACKNOWLEDGMENTS

The authors wish to express our appreciation to BASF and Johnson Matthey Catalysts for providing the DPF catalyst coating. We also wish to thank our NGK Insulators, Ltd. associates who supported this study.

REFERENCES

1. Y. Uchida, et. al., “Durability Study on Si-SiC Material for DPF”, SAE 2003-01-0384
2. K. Yuuki, et. al., “The Effect of SiC Properties on the Performance of Catalyzed Diesel Particulate Filter (DPF)”, SAE 2003-01-0383
3. S. Ichikawa, et. al., “Durability Study on Si-SiC Material for DPF (2)”, SAE 2004-01-0951
4. M. Harada, et. al., “Durability Study on Si-SiC Material for DPF (3)”, SAE 2005-01-0582
5. http://ec.europa.eu/, “Preliminary draft proposal for a Regulation of the European Parliament and of the Council relating to emissions of atmospheric pollutants from motor vehicles (Euro 5)”

6. S. Hashimoto, et. al., "SiC and Cordierite Diesel Particulate Filters Designed for Low Pressure Drop and Catalyzed, Uncatalyzed Systems", SAE 2002-01-0322
7. T. Mizuno, et. al., "Development of a New DPNR Catalyst", SAE 2004-01-0578
8. Y. Kasai, et. al., "New Cordierite Diesel Particulate Filter Material for the Diesel Particulate -NOx Reduction System", SAE 2004-01-0953
9. E. Ohara, et. al., "Filtration Behavior of Diesel Particulate Filters (1)", SAE 2007-01-0923
10. T. V. Johnson, "Diesel Emission Control in Review", SAE 2006-01-0030
11. A. G. Konstandopoulos, "Flow Resistance Descriptors for Diesel Particulate Filters: Definitions, Measurements and Testing", SAE 2003-01-0846

2007-01-0320

Detailed Diesel Exhaust Particulate Characterization and Real-Time DPF Filtration Efficiency Measurements During PM Filling Process

Ekathai Wirojsakunchai, Eric Schroeder, Christopher Kolodziej and David E. Foster
University of Wisconsin – Engine Research Center

Niklas Schmidt and Thatcher Root
University of Wisconsin–Madison

Terunao Kawai
National Traffic Safety and Environmental Laboratory

Toshiyuki Suga
Honda R&D, Inc.

Tim Nevius and Takeshi Kusaka
Horiba Instrument, Inc.

ABSTRACT

An experimental study was performed to investigate diesel particulate filter (DPF) performance during filtration with the use of real-time measurement equipment. Three operating conditions of a single-cylinder 2.3-liter D.I. heavy-duty diesel engine were selected to generate distinct types of diesel particulate matter (PM) in terms of chemical composition, concentration, and size distribution. Four substrates, with a range of geometric and physical parameters, were studied to observe the effect on filtration characteristics. Real-time filtration performance indicators such as pressure drop and filtration efficiency were investigated using real-time PM size distribution and a mass analyzer. Types of filtration efficiency included: mass-based, number-based, and fractional (based on particle diameter). In addition, time integrated measurements were taken with a Rupprecht & Patashnick Tapered Element Oscillating Microbalance (TEOM), Teflon and quartz filters.

Initial breakthrough phenomenon was observed for all clean DPFs. As the soot cake layer forms on top the substrate, the filtration efficiency is improved. In reality, the DPF substrate itself has poor filtration efficiency. It acts more as a facilitator to form a soot cake for the filtration of diesel particulates. A lower pressure drop filter (with approximately 70% larger filtration area than the others tested) showed considerably longer initial breakthrough. Results show different engine operating conditions resolve into distinct PM concentration, chemical composition, size distribution, and filtration velocity. Differences in the particulate composition and filtration characteristics appear to influence DPF pressure drop and filtration performance. Although the coating of the substrate contributes to higher pressure drop, it does not influence the onset of the different stages of filling. Comparing substrates with different geometrical and physical properties, a shift of the most penetrating particle range (Greenfield Gap) was observed.

INTRODUCTION

US EPA 2007 heavy-duty diesel emissions standards will see a 90% reduction in PM from 2002/04 levels. Diesel particulate filters (DPFs) are designed to remove PM from the exhaust stream. Latest researches have shown that DPF filtration efficiency can be as high as 99% in the case of monolithic wall flow filters [1, 2, 3] when a soot cake is formed. In these filters, alternate channels are plugged and exhaust gases are forced through the porous walls, filtering the particulate from the exhaust stream (Figure 1). This filtration process takes place in two major steps: PM is first deposited in the porous walls (wall loading) and later builds up a cake layer (soot cake loading) on the inlet channel walls. Periodically the DPF needs to be cleaned to remove the captured PM from the filter in a process referred to as regeneration. The substrate temperature is raised to the PM ignition temperature to oxidize PM to CO_2.

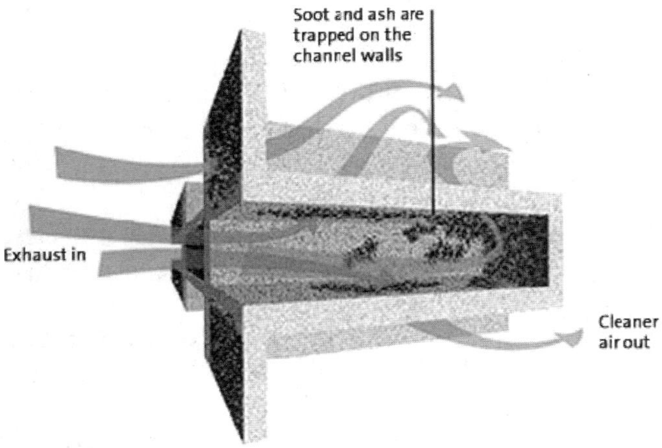

Figure 1. Illustration of a single channel of a DPF (Illustration Copyright 2001, Corning Incorporated. All right reserved. Reprinted by permission.)

Variation in filter properties (such as porosity, pore size distribution, pore connectivity, tortuosity, and washcoat and catalyst formulation) and engine operating conditions (including volumetric flow rate, PM size distribution, and PM chemical composition) affect pressure drop and filtration performance during DPF filling [2, 3, 4]. Regeneration performance depends on filter properties (such as heat capacity, thermal diffusivity, and catalyst formulation), exhaust characteristics (including temperature, volumetric flow rate, and oxygen concentration), and PM characteristics (including mass, chemical composition, and PM deposition location) [5, 6, 7, 8]. By gaining a better understanding of the effects of these parameters on the mechanisms by which PM is deposited and oxidized, DPF performance can be optimized with the engine to maximize the system efficiency.

The objective of this work is to better understand the effect of different engine operating conditions and filter physical and chemical properties on the filtration and regeneration mechanisms which occur inside the DPF. This paper is one of two which is being released by the University of Wisconsin – Madison DPF research group. The second publication on regeneration can be found in [6].

EXPERIMENTAL METHOD

LABORATORY SETUP

A single-cylinder heavy-duty diesel research engine adapted from an inline six-cylinder Cummins N14-series engine was used in this study. The specifications of the engine are shown in Table 1.

Table 1. Cummins N14 engine specification

Base Engine	Single Cylinder Cummins N14
Piston Geometry	Shallow Dish Mexican Hat
Number of Intake Valves	2
Number of Exhaust Valves	2
Compression Ratio	16:1
Swirl Ratio	1.4
Displacement	2.3 L
Bore x Stroke	140 mm x 152 mm
Piston Bowl Diameter	~94 mm
Injection System	Direct Injection (D.I.)
Injector	Electronic Unit Injector
Nozzle Holes	$8 \times \phi$ 0.2 mm
Hole Length/Dia, L/D	4.1
Spray Angle	152°

A schematic of the research engine and exhaust system is shown in Figure 2.

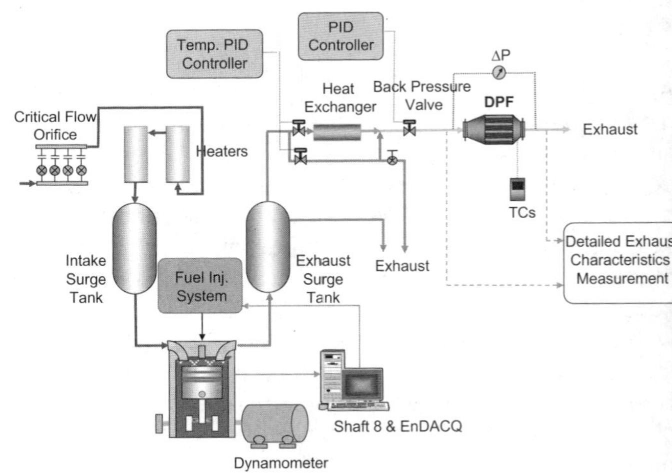

Figure 2. Schematic diagram of engine layout

In order to simulate turbocharged operating conditions, the intake boost pressure and exhaust back pressure are controlled by a series of critical flow orifices and a back pressure valve, respectively. Building-supplied compressed air is heated with twin air heaters mounted in series. The air flows through the intake surge tank and into the cylinder head. After combustion, the engine exhaust flows through an exhaust surge tank in order to minimize pressure pulsations. After the surge tank, the exhaust is routed into two parallel exhaust lines, one with and one without a heat exchanger. Gate valves installed upstream of these two exhaust routes are operated with a PID controller to maintain the desired DPF inlet temperature, thereby simulating the enthalpy extraction of a turbocharger turbine. A DPF is placed further downstream of the back pressure valve. A differential pressure transducer was installed across the DPF to monitor the pressure drop during the experiments. Thermocouples are placed at the upstream and downstream ends of the DPF inlet and outlet cones to measure the DPF inlet and outlet temperatures. Nine thermocouples were inserted into the DPF outlet channels to measure DPF internal temperatures.

EXHAUST GAS DILUTION SYSTEM

A Rupprecht & Patashnick Model 6100 two-stage exhaust gas partial dilutor was used (Figure 3). The exhaust is rapidly and isothermally diluted to minimize particle sampling artifacts. Dilution is carried out in two stages. The first dilution is conducted as close to the source as possible in a perforated sample probe located just outside of the exhaust system. The second stage is an ejector-type diluter, located downstream from the primary dilution stage. A series of magnetic valves control the flow and dilution ratio for the primary and secondary dilutions. Thermodynamic states of the particles entering the tunnel can be accurately controlled to minimize artifacts from the measurements. The software for the diluter also controls a real-time PM mass monitoring device (TEOM 1105: Tapered Element Oscillating Microbalance) from the same manufacturer. A more complete description of this sampling system can be found in [9].

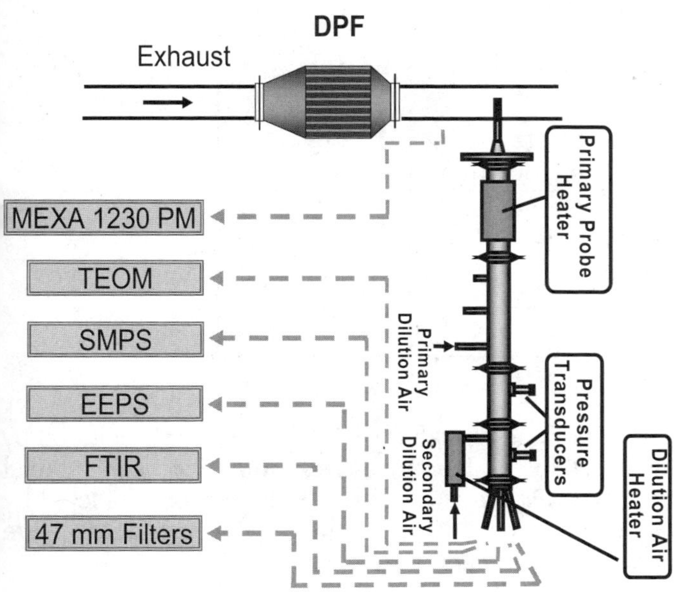

Figure 3. Microdiluter (Model 6100) and instruments

Dilution Ratios

In the course of this work, the following dilution ratios found in Table 2 were used. These dilution ratios were verified using gas-phase concentration measurements of CO_2 in raw and diluted exhaust. Past work at the University of Wisconsin-Madison, Engine Research Center, has shown this dilution system to have good correlation between the software readings and verified dilution ratios from gaseous measurement at dilution ratios between 20 and 35 for raw engine exhaust measurement. Lower dilution ratios allow changes in nuclei-mode particles (<50 nm diameter) to occur. Higher dilution ratios cause oscillations of sampled flow into the dilution system. A lower dilution ratio was used to detect low PM concentrations downstream of the DPF.

Table 2. Dilution ratios used in this study

Valve Comb.	PDAT [°C]	SDAT [°C]	DR (software)	Verified DR
2/3/7	230±7	100±4	22 [a]	21.2±1
2/3	230±7	100±4	7 [b]	7.1±0.2

DR: Dilution Ratio

PDAT: Primary Dilution Air Temperature

SDAT: Secondary Dilution Air Temperature

a: DR used at Upstream of DPF

b: DR used at Downstream of DPF

EEPS

An Engine Exhaust Particle Sizer (TSI EEPS Model 3090) Spectrometer was used to measure the size distribution and total concentration of PM in the range from 5.6 to 560 nm with 1 Hz time resolution. Its operating principle is detailed in [10] and has been previously applied to PM measurements downstream of particulate filters [11].

SMPS

For comparison, a Scanning Mobility Particle Sizer (SMPS, TSI Model 3936) was used when conditions were closer to steady-state. The system consists of an impactor nozzle, electrostatic classifier, aerosol neutralizer, differential mobility analyzer (DMA), and condensation particle counter (CPC). In this configuration, the system can measure particle mobility diameters from 7 to 290 nm.

FTIR

Gas phase emissions from the exhaust of the engine were measured by infrared spectroscopy using a Thermo Nicolet Nexus Fourier Transform Infrared Spectrometer (FTIR) with a 2 meter gas cell. For the current study, the FTIR was used to monitor the engine-out emissions for each mode of engine operation and sampling system dilution ratio.

47 mm Filters

Gravimetric analyses were performed using 47mm Teflon (Pall Teflo 2 μm pore size) and quartz (Pall Tissuquartz 2500QAO-UP) filters. Quartz filters were analyzed for elemental carbon (EC) and organic carbon (OC) according to the National Institute of Occupational Safety and Health (NIOSH 5040) method with a Sunset Laboratories Carbon Analyzer [12].

MEXA 1230 PM

The Horiba MEXA (Motor Exhaust Gas Analyzer) 1230 PM is an instrument that measures real-time soot and SOF (soluble organic fraction) concentrations in engine exhaust (see Figure 4). The equipment measures soot by using a diffusion charger and SOF by using a dual FID method. Its principal operation is detailed in [13].

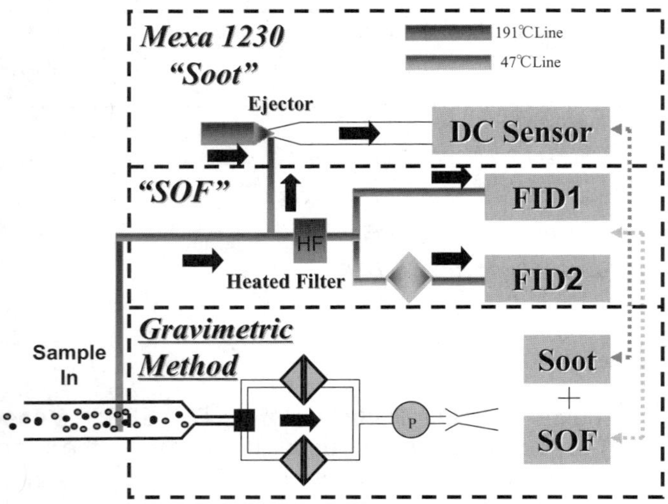

Figure 4. MEXA 1230 PM measurement method compared with traditional gravimetric method [13]

This method has the advantage of measuring PM at high temperatures which is more representative of the conditions in the exhaust and DPF. The fast response also allows it to measure soot and SOF in real time during transient events.

DPF SPECIFICATIONS

Two different substrates, typical Cordierite (Filters A, B, and C) and commercial SiC (Filter D) DPFs, were used in this study (Table 3). Cordierite DPFs including bare (A), washcoated† (B), and catalyzed (C) DPFs were tested to study the effects of substrate coatings during the filling process. The substrates were packaged and fitted with a pair of inlet/outlet cones designed to promote an even flow distribution as shown in Figure 5.

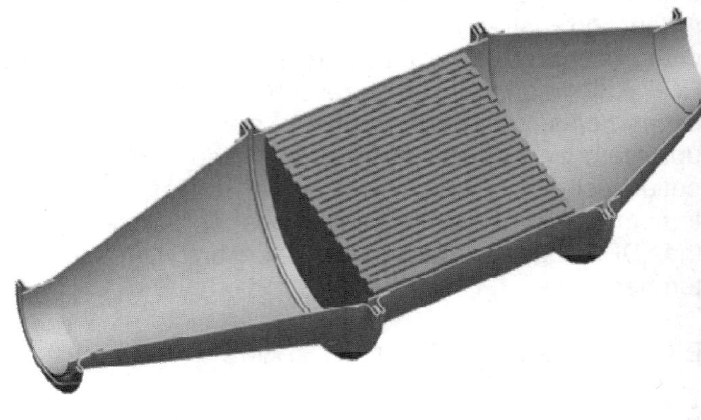

Figure 5. DPF packaging design

A summary of important parameters for the examined filters is given in Table 3.

Table 3. DPF specifications

Designation	A	B	C	D
Substrate		Cordierite		Silicon-Carbide
Cell Density, N (CPSI)		88		302
Wall Thickness (mil)		17		12
Geometric Filter Area (m^2/l)		0.62		1.08
Substrate Porosity, ε (%)	48.8	45.7	45.1	49.1
Median Pore Diameter, d50 (μm)	12.8	11.8	11.8	26.6
Substrate Volume, V_s (l)		2.44		3.46
Washcoat	no	yes	yes	yes
Pt Loading (g/l)	0	0	3	1

The internal temperatures of the DPF were monitored by inserting nine thermocouples into the DPF outlet channels (Figure 6). These locations were selected by assuming that DPF internal temperatures are concentrically uniform.

† *A washcoat is a high-surface-area inorganic material, i.e. Alumina (Al_2O_3) and Zeolites, in which catalysts like Pt and Pd can be dispersed.*

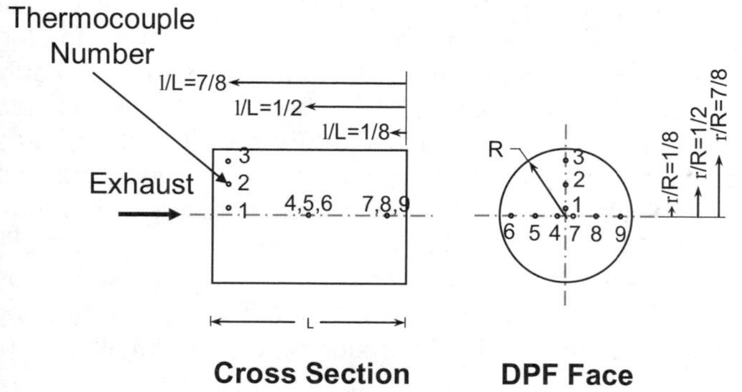

Figure 6. Layout of thermocouple locations

For the current study, ultra-low sulfur diesel fuel (ULSD) was used to avoid catalyst deactivation (Table 4).

Table 4. Properties of diesel fuel used in this study

Manufacturer		BP-Amoco
Unit		ECD-1
API Gravity @ 15.6°C		39.1 (°API)
Kinematic Viscosity @ 40°C		2.352 (cst)
Distillation D-86 (°C)	IBP	178
	10%	202
	50%	250
	90%	311
	FBP	344
Cetane Number		53.7
Sulfur Content		7.5 ppm
Cloud Point		-10°C
PM Flash		71°C

EXPERIMENTAL DESIGN AND RUN CONDITIONS

Engine Operating Conditions

Experiments were performed for three engine modes selected from the California Air Resources Board (CARB) 8 modes, namely Mode 3, 4, and 5. Previous work has shown significant differences in the chemical composition and size distribution of the PM created by these engine operating conditions [14]. A detailed characterization of the emissions for these engine operating conditions is described in the following.

Exhaust Gas Emissions

Gaseous emissions were measured for each engine operating condition using the FTIR. The emissions index for selected gas-phase emissions are shown in Table 5.

Table 5. Exhaust gas emissions index for three investigated engine modes

	Mode 3	Mode 4	Mode 5
CO_2 [g/kg.fuel]	9,200	8,952	10,021
CO [g/kg.fuel]	12	11	78
UHC [g/kg.fuel]	2.5	3.5	0.026
NO_x [g/kg.fuel]	34	37	28

PM Characterization

A detailed study was carried out to characterize the PM properties for each of the three investigated engine operating conditions. A summary is given in Table 6.

Table 6. PM properties for examined engine operating conditions

Engine Mode	Mode 3	Mode 4	Mode 5
Speed [rpm]	1800	1200	1200
Torque [N-m]	81	61	258
IMEP [bar]	7.3	5.7	14.2
Equivalence Ratio	0.32	0.2	0.76
Exhaust Flow Rate (kg/s)	0.051	0.039	0.036
Exhaust Temperature at Surge Tank (°C)	390	245	710
Soot Conc. [mg/m^3] (Teflon @ 47°C)	**14.5**	**5.1**	**225**
Filtration Velocity [cm/s]* Filter A,B,C/D	5.0/2.0	3.8/1.5	3.6/1.4
Geometric Mean Particle Size [nm]**	75	68	127
EC/OC ratio (quartz @ 47°C)	0.36	0.23	0.85

** These numbers are based on the exhaust temperature at 230°C. The exhaust from all engine modes was cooled to avoid passive regeneration in a catalyzed DPF.*

*** Data from SMPS*

Results for particle size distribution measurements from the SMPS for raw exhaust are shown in Figure 7.

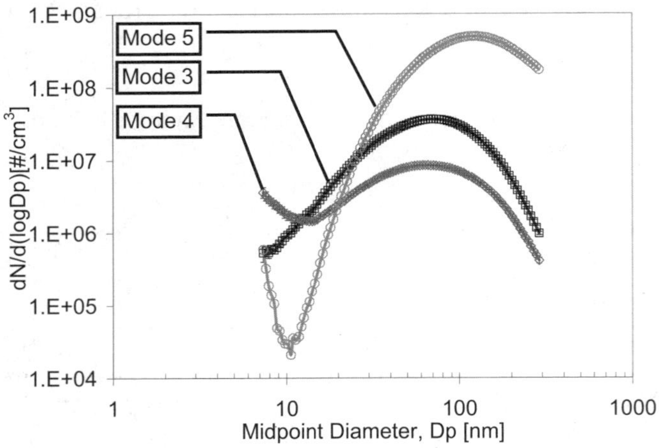

Figure 7. Particle size distribution for raw exhaust

To account for differences in total concentration of different engine modes, a concentration-normalized plot is shown in Figure 8 [15]. The vertical lines represent the geometric mean particle size for each mode of engine operation as shown in Table 6. Each condition gave very different particle size characteristics.

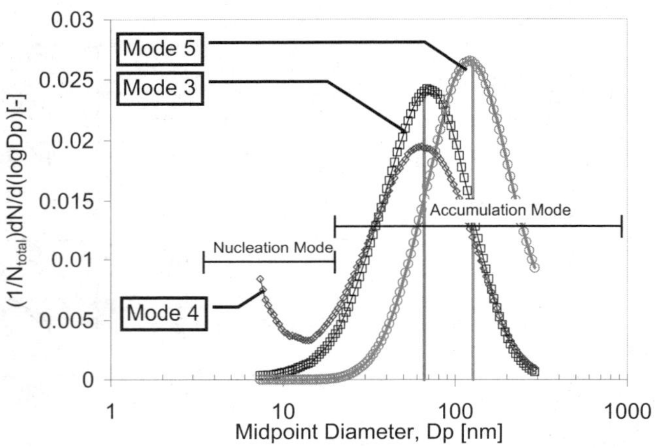

Figure 8. Normalized particle size distribution

Since Mode 5 is a peak torque condition, high concentrations of accumulation mode particles and CO emissions were observed. Mode 4 was run at low load. Higher concentration in nuclei-mode particles were seen along with increased hydrocarbon emissions. Particles larger than 100 nm contribute significantly to the PM total mass, so Mode 5 had the highest total mass and Mode 4 had the lowest (Table 6). Mode 3 has similar PM characteristics to Mode 4, but has higher particle concentrations and exhaust flow rates.

Comparison of Measurement Techniques

TEOM, MEXA total particulate concentrations, and the sum of EC and OC concentration from the NIOSH analysis are compared to values obtained by the Teflon filter gravimetric analysis (x axis) across a range of particle concentrations. Figure 9 shows the correlation between the different measurement methods of the three engine operating conditions. Correlation for the MEXA and EC+OC methods are very good and only slightly over estimated the total particulate concentration. This is expected for the EC+OC case as positive artifacts of OC tend to cause an over prediction in particulate mass [16]. However, measurements from the TEOM under predict the particulate mass but are still very linear. This under prediction has been observed by other researchers and is suspected to be caused by differences in volatile absorption on the TEOM filter [17].

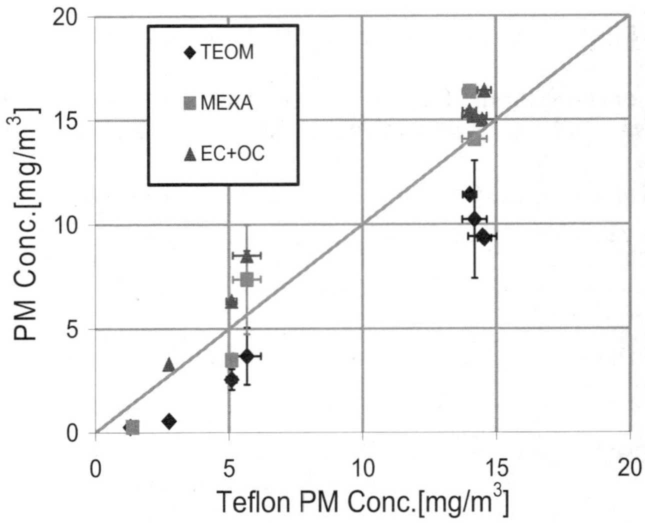

Figure 9. Comparison of total PM concentration from various measurement methods

RESULTS AND DISCUSSIONS

DPF FILLING – PRESSURE DROP

In order to investigate the effect of engine operating condition and DPF properties on the filtration mechanisms in the DPF, the filters were loaded with particulate from each of the three engine operating conditions. The engine exhaust was allowed to reach steady-state conditions in each mode before being routed to the DPF. Exhaust temperature for all engine modes was controlled to 230°C at the DPF inlet. During filling, the DPF thermocouple temperatures, pressure drop, and downstream particulate concentration were recorded. Filling times were selected to collect equivalent PM loads for the different engine operating conditions. After finishing the loading process, the DPF was regenerated in an electric furnace to return it to a clean state for the next experiment. Details on the furnace regeneration can be found in [6].

DPF FILTRATION MECHANISMS

The deposition mechanisms of particles in a porous DPF wall are described using the unit collector filtration theory via diffusion, impaction, and interception [18], the same forces which cause particle loss in the exhaust and sampling system. Thermophoresis, electrostatic forces, and sedimentation (settling due to gravity) are not often considered in current DPF filtration models because their contributions to the overall filtration efficiency are minimal.

Diffusion deposition occurs due to a gradient in particle concentrations between the bulk flow and the walls of the pores (Brownian motion). The Peclet number (Pe) is used to characterize this phenomenon. Pe number can be seen as the ratio of convective to diffusive transport, where high Pe number systems favor convective transport and low Pe number systems favor diffusion. Pe number can be used as a qualitative parameter only, as it does not take into account all DPF microstructure properties (for example, tortuosity and pore size distribution), or size range of particles. However, Pe number is effective in comparing DPF wall face velocity and average particle size. The characteristic length scale of the unit collector can be either pore diameter (wall loading stage) or primary soot particle (soot layer stage) depending on what particular filtration stage is of interest [2, 4]. In this analysis, Pe number is calculated based on clean DPF pore diameter to qualify the characteristics of the initial filtration mechanisms.

Inertial impaction occurs when aerosol particles are unable to follow the streamlines of the flow and are impacted on the collecting body. The characteristic number for this mechanism is the Stokes number (Stk). The Stk number can be viewed as the ratio of the stopping distance of a particle to a characteristic dimension of the collecting body.

The particle may become intercepted and deposited in the case where particles follow the streamlines but pass within one particle radius of the collecting body. The interception mechanism is commonly characterized by the interception parameter (N_r). It is the ratio between the particle diameter and the characteristic length of the collecting body.

CONCEPTUAL MODEL

Two major filtration steps as discussed previously can be broken down into four stages of DPF filtration as shown in Figure 10. Stage 1 shows a clean DPF. Stage 2 illustrates PM being collected on the inside walls of the substrate pores. In stage 3, PM accumulates to cause pore bridging. Finally, stage 4 is reached when a thin "cake layer" of PM begins to develop on the inlet channel wall.

A conceptual model of the pressure drop is shown in Figure 11. The pressure drop rises quickly as the pores become plugged in stages 1 to 3 [19]. As we can see in Figure 11, transitioning from stage 1 to 2 has less effect on the pressure drop than the transition from stage 2 to 3. It is also shown that pressure drop during stages 1 to 3 increases non-linearly. But once stage 4 is reached, there is a linear and slower rise in the pressure drop curve as the cake layer thickens.

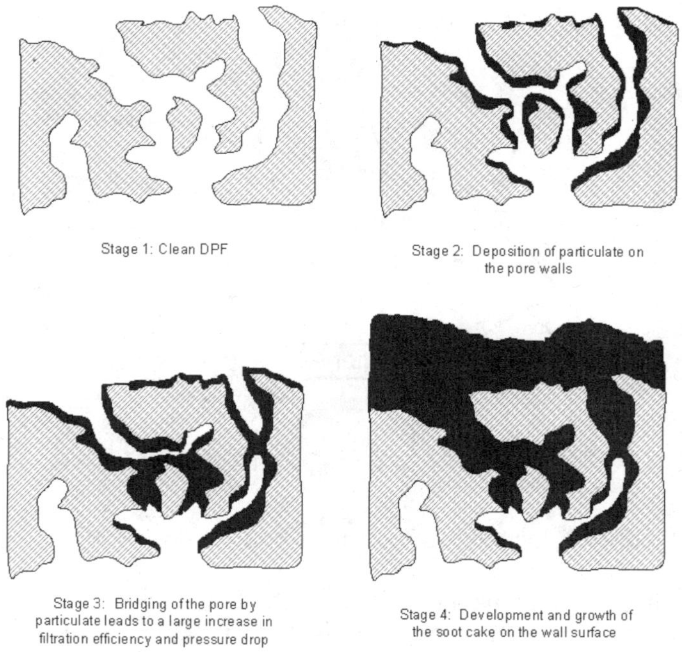

Figure 10. Conceptual model of DPF filling

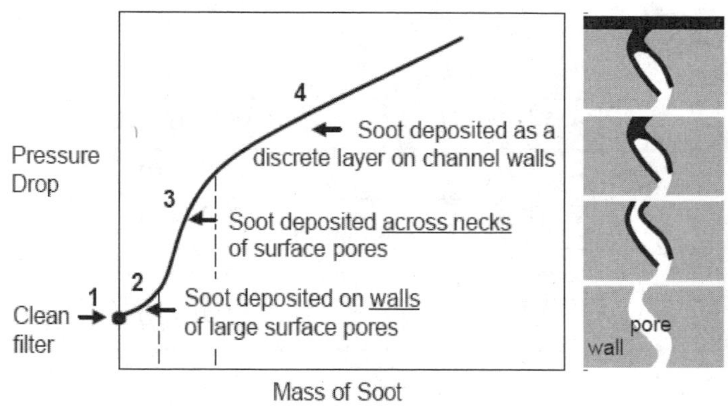

Figure 11. Conceptual model of the pressure drop during DPF filling [19]

INFLUENCE OF SUBSTRATE COATING

Figure 12 shows the pressure drop curves of Filters A, B, and C, which were made of the same substrate material but have different substrate coatings. The numbers on the pressure drop curve refer to the filling stages, as shown in Figure 11. Results show stages 1 through 3 happen very quick in these filters. Because of this, only stages 3 and 4 can be clearly named in Figure 12. The results show that the coating of the substrate does not change the shape of the pressure drop curve during filling. Although the coatings contribute to the amount of the pressure drop, they do not influence the onset of the different stages of filling for these DPFs. Once the soot cake (stage 4) has been reached, all pressure drop curves show the same slope. At this point (soot cake layer is formed), the pressure drop evolution becomes independent of the underlying substrate coating and is controlled by the deposition of the particulate in the cake layer [20].

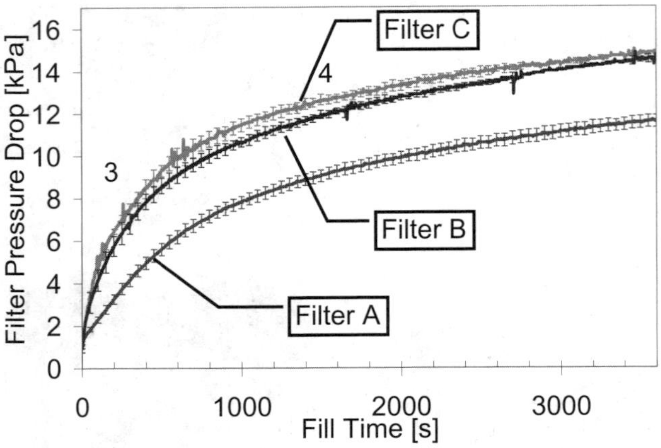

Figure 12. Influence of substrate coating (Mode 3)

The offset between the pressure drop curves can be traced back to the differences in DPF substrate permeability, as shown in Table 7. DPF permeability was measured with cold flow tests on a flow bench as proposed by Konstandopoulos [21].

Table 7. Permeability of substrate (from cold flow testing)

DPF Designation	Permeability, k_{global} [m^2]
A	3.96×10^{-13}
B	2.42×10^{-13}
C	2.37×10^{-13}
D	1.10×10^{-13}

INFLUENCE OF SUBSTRATE

Because of different filter suppliers, Filters C and D examined in Figure 13 had approximately 70 % difference in filtration area. These differences are shown in Table 3. Accordingly, the filtration velocities of the Filter C are ~1.7 times higher than that of Filter D at the same engine operating condition. In Figure 13, the difference in magnitude of the pressure drop for Filters C and D is not only caused by the differences in filtration velocity, but also the physical properties of each filter. The qualitative differences of the pressure drop curves are explained by the variation in substrate characteristics. With Filter D, the transition through stages 2, 3, and 4 is much more apparent and prolonged. As mentioned, Filter C fills so quickly that stages 1 and 2 are unobserved.

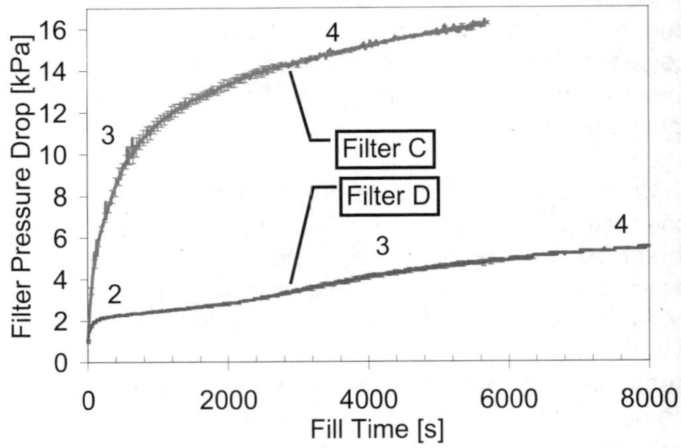

Figure 13. Influence of filter substrate (Mode 3)

The substrate of Filter C has been shown to have higher tortuosity due to its material properties [3]. In addition material properties, Filter C also had lower porosity (Table 3). With higher filtration velocities and pore tortuosity, and lower porosity, Filter C had an increase in convective filtration. These effects lead to quicker pore bridging and faster soot cake formation with higher initial filtration efficiency [22, 23, 24]. However, the substrate of Filter D had lower tortuosity [25] and larger mean pore size (Table 3) than Filter C, which caused lower initial pressure drop and lower initial filtration efficiency so diesel particulate could more easily flow through the highly connected walls of Filter D.

Figure 14 shows the pressure drop curve for Filter D from Figure 13 in more detail. The pressure drop curve closely follows the conceptual model as proposed in Figure 11.

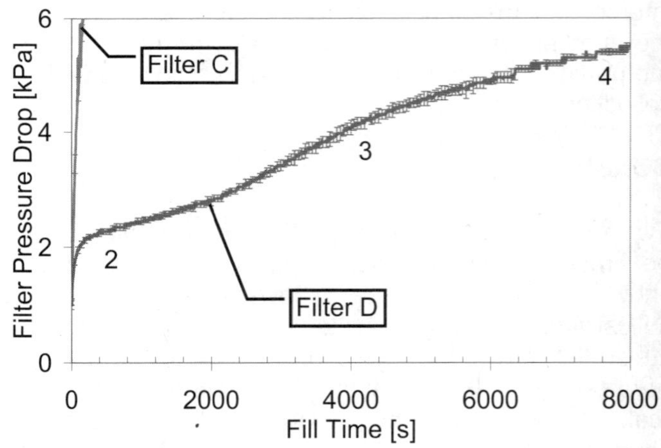

Figure 14. Influence of filter substrate (Mode 3)

INFLUENCE OF ENGINE OPERATING CONDITION

To account for the differences in exhaust mass flow rate and PM concentration, the measured pressure drop is normalized (delta P/exhaust mass flow rate) and the time axis is scaled to reflect the PM mass fed to the DPF. The results of this analysis can be seen in Figure

15. Each plot is an average of several runs and error bars represent one standard deviation. Mode 4 required several hours of filling and was only performed once.

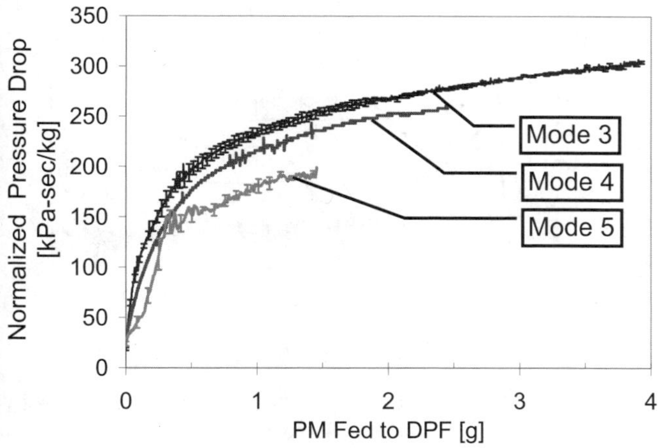

Figure 15. Influence of engine operating condition on filter pressure drop (Filter C)

Although the curves do not fall directly on top of each other, which would suggest identical PM-substrate interaction for each engine operating condition, the slope of the pressure drop curve and PM loading is similar during the cake layer regime. This means the component of pressure drop due to the cake layer is very similar over a wide range of engine operating conditions. Thereby, the cake layer is likely to have similar permeability over a wide range of engine operating conditions for this examined DPF and research engine.

However, large differences in pressure drop are noticed during the wall-loading stage for different engine operating conditions. Even though Mode 4 and 5 have approximately the same exhaust mass flow rate, Mode 5 reaches the soot cake stage of the pressure drop curve sooner (in term of PM mass loaded into a DPF) and its normalized pressure drop is less than Mode 4.

The Peclet numbers, Interception Parameters, and Stokes numbers for the analyzed engine operating conditions for clean Filter C (Cordierite DPF substrate) based on the unit collector model as proposed by Konstandopoulos [23] are shown in Table 8. Note that Peclet number can be viewed as a means of comparing convective forces per diffusive forces.

Table 8. Peclet number, Interception parameter, and Stokes number of investigated engine modes for clean Filter C

Engine Mode	Mode 3	Mode 4	Mode 5
Pe Number	**1074**	**673**	**2005**
N_r Number ($x10^{-3}$)	**3.5**	**3.2**	**5.9**
Stk Number ($x10^{-4}$)	**7.5**	**5.1**	**9.7**
Filtration Velocity [cm/s]	**5.0**	**3.8**	**3.6**
Geometric Mean Particle Size [nm]	**75**	**68**	**127**

Mode 5 contained bigger particles and the filtration mechanism was less diffusive while the impaction and interception mechanism was more pronounced as we see from the highest Pe, N_r and Stk numbers among all examined engine modes. The motion of Mode 5 particles was more deterministic than diffusive which likely lead particles to deposit in the pores (phase 2 and 3 from Figure 10). The pores become clogged due to big particles, which lead to fast formation of the soot cake. The substrate wall capacity is therefore minimized by the large Mode 5 particles, as seen in the pressure drop curve. On the contrary, this wall capacity is higher for Mode 3 and 4 since they have better diffusion mechanisms and contain smaller particles (Pe, N_r, and Stk numbers are lower than mode 5). Mode 4 has less filtration velocity and the diffusion effect is stronger than Mode 3 as seen from its lowest Pe, N_r, and Stk numbers among all examined engine modes which could lead to higher wall capacity (this observation was made in [6]).

This study suggests that differences in PM concentration, size distribution, and filtration velocity strongly effect DPF wall capacity. The increase in diffusion and decrease in convection mechanisms suggest increased soot penetration into the pores and higher soot packing densities inside the pores during the wall loading phase which can lead to better catalytic oxidation performance in regeneration [6].

DPF FILLING – TOTAL NUMBER FILTRATION EFFICIENCY

NUMBER-BASED PM BREAKTHROUGH

The changes in PM size distribution downstream of each DPF were measured using a SMPS and EEPS. The EEPS has the benefit of 1 second time resolution, while the minimum SMPS time resolution is 90 seconds. However, the SMPS is well accepted by the aerosol community for measuring the particle size distribution during steady-state operation. Good comparison of using the EEPS and SMPS to measure particle size distribution in raw engine exhaust can be found in previous literature [10]. Both methods are applied to study PM downstream of a DPF while only the SMPS is used for DPF upstream measurement.

An example of changes in particle size distribution shapes measured by the EEPS downstream of a DPF, as a function of time and compared to upstream SMPS measurements, is shown in Figure 16. Time zero is when the exhaust is rerouted to the DPF (Figure 16 and Figure 17). The evolution of the downstream size distribution shapes are similar to the upstream one which corresponds to the literature [11]. Due to the isothermal two-stage dilution process, measurement artifacts from the sampling system at the downstream side are minimized as seen from consistent shapes of particle size distribution between upstream and downstream.

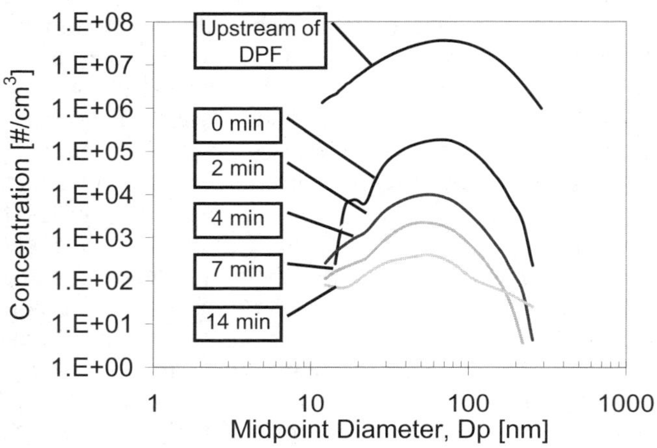

Figure 16. Changes in downstream particle size distribution during DPF filtration (Mode 3, Filter C)

The total number-based filtration efficiency, E (Equation 1), is adopted as a means of measuring the filter performance [11].

$$E = \left(1 - \frac{N_D}{N_U}\right) x100 \qquad \text{Equation 1}$$

N can be either total number concentration (sum of total size spectrum particle concentration) or number concentration at each particle diameter for total filtration efficiency and fractional filtration efficiency, respectively. Subscripts D and U represent the position downstream and upstream of a DPF, respectively.

A comparison of number-based total concentration as measured by the SMPS and EEPS downstream of a DPF is shown in Figure 17. The result shows that the SMPS captured the trends of PM breakthrough well. Discrepancies in the measurements after steady state has been reached can be explained by very low PM concentrations downstream of the DPF. These concentrations are very close to the lower detection limit of the EEPS, while the SMPS can measure much lower concentrations.

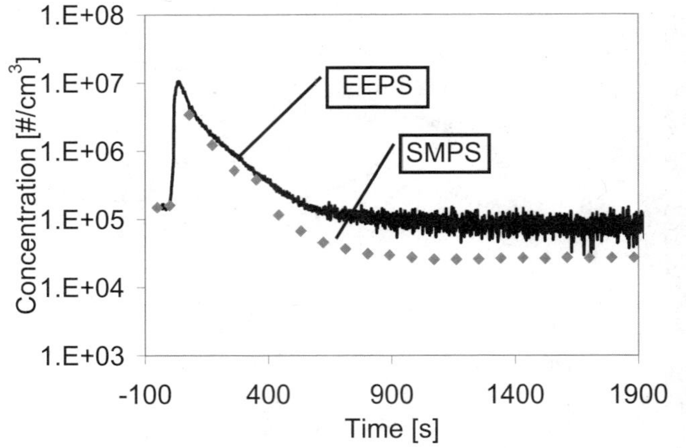

Figure 17. Comparison of total concentration (number-based) between EEPS and SMPS data during filling (Mode 3, Filter C)

INFLUENCE OF SUBSTRATE COATING

Number-based filtration efficiency was measured by the SMPS and calculated by Equation 1 for Filters A and B as shown in Figure 18. Note that the data for Filter C is not included in Figure 18 since its filtration performance is similar to Filter B as seen from the pressure drop curves in Figure 12.

No significant difference can be observed in DPF total number filtration efficiency from washcoat variation. Although the coatings contribute to the amount of the pressure drop across the DPF (Figure 12), it appears to have no influence on the filtration efficiency of the DPF during filling.

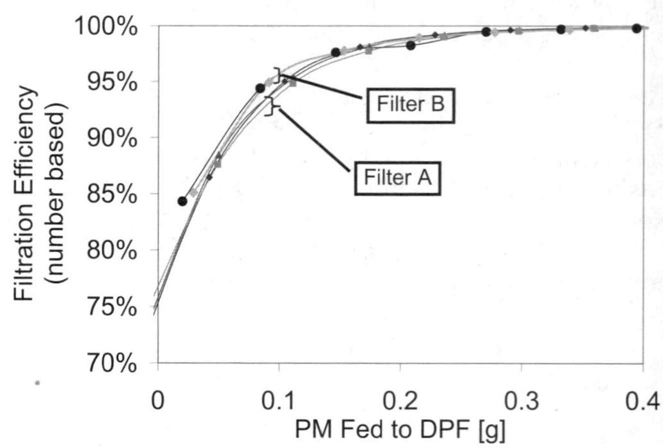

Figure 18. Filtration efficiency (number-based) for different DPF coatings (Mode 3)

INFLUENCE OF SUBSTRATE

A plot of PM breakthrough and the pressure drop for Filter C is shown in Figure 19. The numbers on the pressure drop curve refer to the filling stages, as shown in Figure 11. During DPF filling, the filtration efficiency was shown to be very poor during the initial wall loading phase (stages 1 to 3). Downstream measurements of the real-time total particle concentration lead to significant PM breakthrough until a cake layer (stage 4) is established [20].

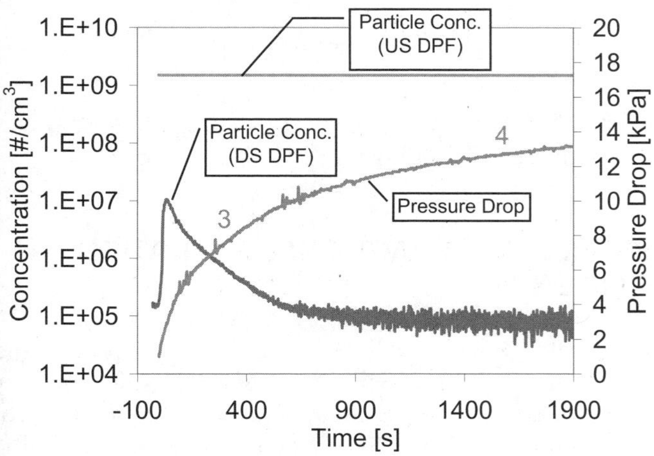

Figure 19. Total concentration upstream (US) and downstream (DS) of the filter with the pressure drop (Mode 3, Filter C)

A plot of PM breakthrough and the pressure drop for Filter D is shown in Figure 20. The increase in the filtration area of Filter D prolonged the filtration stages. Significant decrease in PM breakthrough can be observed during the transition from stages 2 to 3 (Figure 11) where pore bridging occurs.

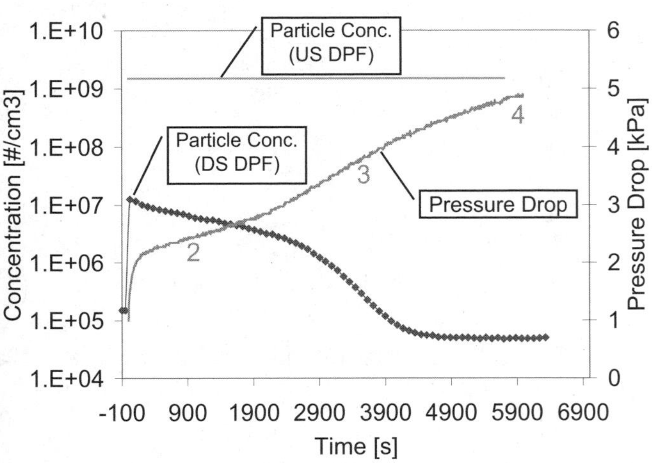

Figure 20. Total concentration upstream (US) and downstream (DS) of the filter with the pressure drop (Mode 3, Filter D)

A comparison between the number-based filtration efficiency of Filters B and D substrates was also made using downstream SMPS data. The filtration efficiency is plotted against the area normalized mass [2] to account for the difference in filtration area between Filters B and D. It is evident from Figure 21 that Filter D has much lower initial filtration efficiency and takes much longer to reach peak filtration efficiency. More than 99% filtration efficiency for Filter B was achieved when PM loading was at 0.17 g/m^2 and for Filter D at 1 g/m^2. The differences in substrate parameters as mentioned in the pressure drop section combined with the reduced filtration velocity of Filter D explain this drop in filtration efficiency.

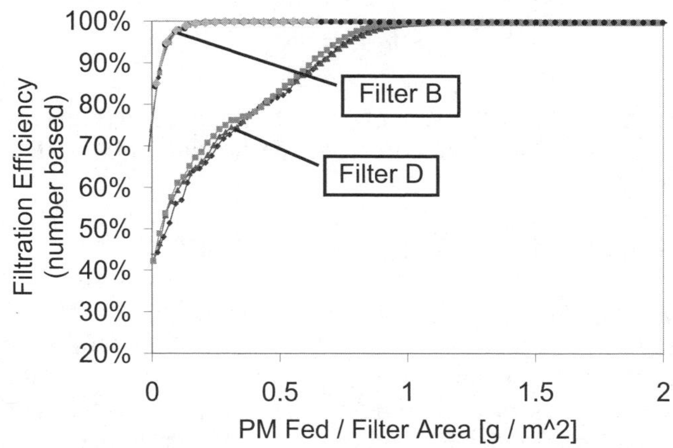

Figure 21. Filtration efficiency (number based) on DPF different substrates (Mode 3)

Based on the results from Figure 18 and Figure 21, it is clear that the progression of the filtration efficiency in the DPF is highly dependent on the properties of the DPF substrate, not the substrate coatings. As was previously shown by varying engine operating conditions on a single DPF substrate, the interaction of the particulate with the substrate (particle diameter, filtration velocity, etc.) strongly affected the DPF pressure drop. This engine-DPF coupling must be taken into consideration when designing an aftertreatment system because initial filtration efficiencies have the potential to drop very low in certain cases.

INFLUENCE OF ENGINE OPERATING CONDITION

Only PM breakthrough from Modes 3 and 4 were studied due to Mode 5 concentrations exceeding the maximum measurement range of the EEPS. The effect of engine operating condition will be carried on in future studies.

Figure 22 shows the comparison of normalized PM breakthrough between Modes 3 and 4 to account for different PM loading rates. As can be seen from the figure, PM breakthroughs for these two engine modes reached a minimum at approximately the same PM loading (about 0.5 g of PM fed to the DPF).

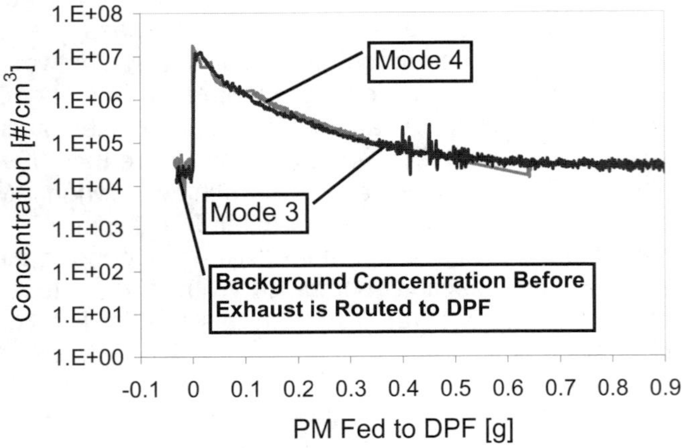

Figure 22. Comparison of soot breakthrough between Modes 3 and 4 by total number of particle concentration of Filter C

MASS BASED FILTRATION EFFICIENCY

PM breakthrough was also detected using the MEXA 1230 PM analyzer which measures real-time PM (soot+SOF) mass concentration. Due to the discrepancies of SOF readings at the downstream side of the examined DPF, only soot concentration data is presented in this section. Figure 23 shows the progression of soot breakthrough while filling with Mode 3 on Filter C. Comparing Figure 23 to Figure 19, the same trends of initial breakthrough were captured by two different instruments. A plot of mass-based (from MEXA 1230 PM) and number-based (from SMPS) filtration efficiencies for Filter C is shown in Figure 24. These two measuring devices give very similar results.

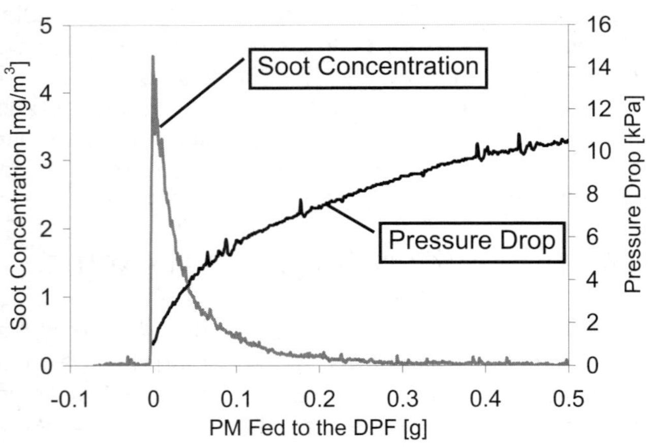

Figure 23. Soot concentration measured by mass downstream of the filter and pressure drop (Mode 3, Filter C)

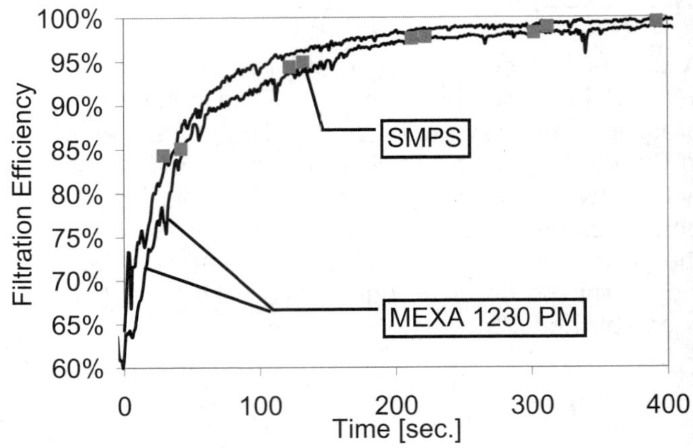

Figure 24. Filtration efficiency (mass-based vs. number-based, Mode 3, Filter C)

DPF FILLING – FRACTIONAL FILTRATION EFFICIENCY

Using the SMPS size distributions upstream and downstream of the DPF, the fractional efficiency by particle diameter was measured for Filter B (Figure 25) and Filter D (Figure 26). Since the results from the previous section suggested that the substrate coating had little effect on the overall filter filtration efficiency, differences in DPF substrate will be examined. Only Mode 3 exhaust was used to investigate the influence of the substrate geometry on the fractional filtration efficiency.

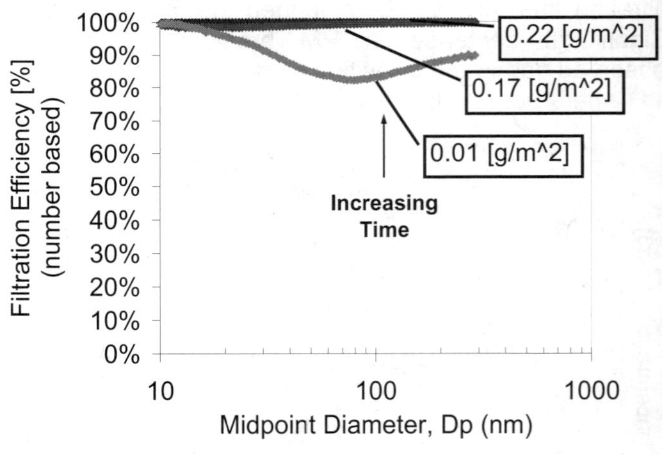

Figure 25. Fractional filtration efficiency downstream of a DPF (Mode 3, Filter B)

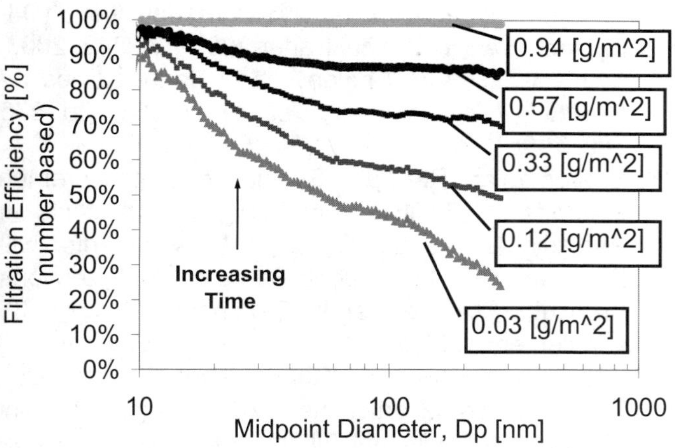

Figure 26. Fractional filtration efficiency downstream of a DPF (Mode 3, Filter D)

The Greenfield gap location [11, 18] of Filter B was around 80 nm while the location of Greenfield gap for Filter D cannot be seen and assumed to be above the size range of the SMPS. The shift in the Greenfield gap to higher diameters for Filter D operation is due to the reduction of filtration velocities and differences in physical filter properties as mentioned in a previous section. For Filter D, the lower filtration velocity, lower tortuosity, and higher porosity decrease the effect of convective filtration and increase the relative effect of the diffusive mechanisms. In addition, the mean pore size of Filter D is larger than Filter C. Because of all these effects, larger particles pass through the DPF and shift the Greenfield gap toward larger particle diameters.

SUMMARY AND CONCLUSIONS

Different DPF types in terms of substrate materials and substrate coatings were tested on a heavy-duty single-cylinder Cummins N14 research engine to study their filtration mechanisms. The filters were loaded at various engine operating conditions, ranging from low to high load and speed. This allowed the DPF to be exposed to PM of different characteristics. The exhaust instrumentation included simultaneous measurements with an FTIR, TEOM, SMPS, MEXA 1230PM real-time particle mass analyzer, real-time Engine Exhaust Particle Sizer (EEPS), and Teflon and quartz filters for detailed characterization of PM in the exhaust. The three different engine operating conditions were shown to result in distinct DPF face velocities and PM physical and chemical properties such as concentration, EC/OC ratio, and particle size distributions. The following conclusions are made.

- A two stage dilution system allowed full control of the dilution process and was implemented into the exhaust system of the current research engine.

- Total PM concentrations for a wide rage of engine operating conditions agreed well among different types of PM measurement equipment.

- The pressure drop regimes occurred essentially at the same time for a DPF substrate with different coatings. This implies that the pressure drop regime progression is less a function of the substrate washcoat variation and more affected by DPF geometry.

- Significant differences in pressure drop curves have been observed during loading of the various filters with the same engine mode. This is due to differences in DPF properties such as filtration area, pore tortuosity, and mean pore diameter.

- Slopes of the pressure drop curves during the cake layer phase were seen to be consistent over a wide range of engine operating conditions for these examined filters and research engine.

- Mode 5 PM showed the fastest transition to cake filtration because it had the largest particle diameters which led to faster pore bridging and lower PM wall capacity.

- Initial breakthrough phenomenon is observed for all clean DPFs. The filtration efficiency increases as a cake layer is formed. A lower pressure drop filter shows longer initial breakthrough than the other ones examined in this study.

- The results suggest that DPF filtration efficiency is more affected by substrate geometry and physical properties (such as porosity, tortuosity, filtration velocity, etc.) than substrate coatings.

- The trends of number-based and mass-based filtration efficiencies were very similar for the examined filter.

- The lower pressure drop filter had lower filtration efficiency and pressure drop during its initial stages of filtration. This is because it had much lower filtration velocities, pore tortuosity and larger mean pore size.

- The Greenfield gap of the higher pressure drop filter was ~80nm, but was too large to measure on the lower pressure drop filter at the same engine condition. This behavior can be explained by a reprioritization of different filtration mechanisms due to differences in substrate geometrical and physical properties.

ACKNOWLEDGMENTS

The authors are sincerely thankful to Honda R&D, Corning, Johnson Matthey, Horiba, Cummins Filtration, TSI, and BP Amoco for their supports in this study. The experimental work and analysis were equally contributed among the first six authors. This work was performed at the University of Wisconsin-Madison Engine Research Center (ERC).

REFERENCES

1. Johnson, J. H. et al. "A Review of Diesel Particulate Control Technology and Emissions Effects – 1992 Horning Memorial Award Lecture" SAE Technical Paper 940233, 1994.
2. Konstandopoulos, A.G. et al., "Fundamental Studies of Diesel Particulate Filters: Transient Loading, Regeneration, and Aging", SAE Technical Paper 2000-01-1016, 2000.
3. Miller, R.K., Haberkamp, W.C., Badeau, K.M., Liu, Z.G., Shirk, R.C., and Wood, T., "Design, Development and Performance of a Composite Diesel Particulate Filter", SAE Technical Paper 2002-01-0323, 2002.
4. Konstandopoulos, A.G. and Kostoglou, M., "Microstructural Aspects of Soot Deposits in Diesel Particulate Traps", SAE Technical Paper 2002-01-1015, 2002.
5. Konstandopoulos, A.G. and Kostoglou, M., "Microstructural Aspects of Soot Oxidation in Diesel Particulate Filters", SAE Technical Paper 2004-01-0693, 2004.
6. Schmidt, N., et al., "Detailed Diesel Exhaust Particulate Characterization and DPF Regeneration Behavior Measurements for Two Different Regeneration Systems", SAE Paper No. 2007-01-1063.
7. Hanamura, K., Suzuki, T., Tanaka, T., and Miyairi, Y., "Visualization of Combustion Phenomena in Regeneration of Diesel Particulate Filter", SAE Technical Paper 2003-01-0836, 2003.
8. Stanmore, B.R., Brilhac, J.F., and Gilot, P., "The Oxidation of Soot: A Review of Experiments, Mechanisms, and Models", Carbon, Vol. 39, pp. 2247-2268, 2001.
9. Maamouri, M. et al. "Real-Time Exhaust PM Mass Measurements for DPF Development Studies" Emission Control Conference, Dresden, 18-19 May 2006.
10. Johnson, T. et al. "A new electrical mobility particle sizer spectrometer for engine exhaust particle measurements" SAE Technical Paper 2004-01-1341, 2004
11. Liu, G.Z. et al. "Transient Performance of Diesel Particulate Filters as Measured by an Engine Exhaust Particle Size Spectrometer", SAE Technical Paper 2005-01-0185, 2005.
12. Federal Register, Vol. 66, No. 12, January 18, 2001, pp 5172-5178.
13. Kusaka, Takeshi "New Real-Time PM Measurement System MEXA 1230 PM", Horiba Inc., Presented Sept 28th 2005.
14. Kweon, C.B., "Effect of Engine Operating Conditions on Particle-Phase Organic Compounds in Engine Exhaust of a Heavy-Duty Direct-Injection (D.I.) Engine", SAE Technical Paper 2003-01-0342, 2003.
15. Kittelson, D.B., "Engine and Nanoparticles: A Review", J. Aerosol Sci., Vol. 29, No. 5/6, pp. 575-588, 1998.
16. Subarmanian, R. et al. "Evaluation of Measurement Methods. Positive and Negative Artifacts in Particulate Organic Carbon Measurements with Denuded and Un-denuded Sampler Configurations" Aerosol Science and Technology, vol. 38, supplement 1, 2004.
17. Xu, Shuhong. et al. "Comparison of Heavy-Duty Truck Diesel Particulate Matter Measurement: TEOM and Traditional Filter" SAE Technical Paper 2005-01-2153, 2005.
18. Hinds, W.C., Aerosol Technology: Properties, Behavior, and Measurement of Airborne Particles, John Wiley & Sons, Inc., USA, 1999.
19. Merkel, G.A. et al. "New Cordierite Diesel Particulate Filters for catalyzed and non-catalyzed Applications' Proceedings of the 9th Diesel Engine Emissions
20. Murtagh, M.J., Sherwood, D.L., and Socha, S.L. Jr., "Development of Diesel Particulate Filter Composition and Its Effects on Thermal Durability and Filtration Performance", SAE Technical Paper No. 940235, 1994.
21. Konstandopoulos, A.G., "Flow Resistance Descriptors for Diesel Particulate Filters: Definitions, Measurements and Testing" SAE Technical Paper 2003-01-0846, 2003.
22. Heck, R.M., Farrauto, R.J., and Gulati, S.T., Catalytic Air Pollution Control: Commercial Technology, John Wiley & Sons, Inc., New York, 2002.
23. Konstandopoulos, A.G. and Johnson, J.H., "Wall-Flow Diesel Particulate Filters-Their Pressure Drop and Collection Efficiency", SAE Technical Paper 890405, 1989.
24. Merkel, G.A., et al., "Effects of Microstructure and Cell Geometry on Performance of Cordierite Diesel Particulate Filters", SAE Technical Paper 2000-01-0193, 2000.
25. Ohno, K., et al., "Characterization of High Porosity SiC-DPF", SAE Technical Paper 2002-01-0325, 2002.

CONTACT

David E. Foster, Phil and Jean Myers Professor

127 Engineering Research Building
1500 Engineering Drive
Madison, WI 53706-1687

E-mail: foster@engr.wisc.edu

2007-01-1063

Detailed Diesel Exhaust Particulate Characterization and DPF Regeneration Behavior Measurements for Two Different Regeneration Systems

Niklas Schmidt and Thatcher Root
University of Wisconsin-Madison

Ekathai Wirojsakunchai, Eric Schroeder, Christopher Kolodziej and David E. Foster
University of Wisconsin – Engine Research Center

Toshiyuki Suga
Honda R&D Corp.

Terunao Kawai
NTSEL

ABSTRACT

Three distinct types of diesel particulate matter (PM) are generated in selected engine operating conditions of a single-cylinder heavy-duty diesel engine. The three types of PM are trapped using typical Cordierite diesel particulate filters (DPF) with different washcoat formulations and a commercial Silicon-Carbide DPF. Two systems, an external electric furnace and an in-situ burner, were used for regeneration. Furnace regeneration experiments allow the collected PM to be classified into two categories depending on oxidation mechanism: PM that is affected by the catalyst and PM that is oxidized by a purely thermal mechanism. The two PM categories prove to contribute differently to pressure drop and transient filtration efficiency during in-situ regeneration.

INTRODUCTION

Continual improvement is being made in the field of diesel particulate filters. The objective is to ensure continuously high filtration efficiency with minimum pressure drop. Initially a clean DPF utilizes deep bed filtration to trap PM in the porous wall. As the wall capacity is reached, soot builds up in a discrete layer (soot cake) on the walls of the inlet channels.

In the soot cake phase, filtration efficiency is close to 100% and filter pressure drop is a linear function of accumulated soot mass.

For the period of deep bed filtration, the relationship between PM mass and pressure drop is more complicated and filtration efficiency can be significantly lower. In this phase the interaction between individual particles and the wall microstructure determines DPF behavior. Depending on particulate composition, flow conditions, and governing transport mechanisms, the amount of soot in the porous wall varies widely.

During regeneration it is of interest what proportion of the PM is deposited with tight contact to the precious metal particles (i.e. in the porous wall) and, as such, is oxidized catalytically.

LABORATORY SETUP

The single cylinder engine used in this study is adapted from an inline six cylinder Cummins N14. The engine is run with Ultra Low Sulfur Diesel (BP, <15 ppm sulfur, Cetane: 53) and low ash oil (Rotella, 0.45 wt% sulfur). Engine specifications are given in Table 1.

Table 1. Research engine specifications

Base Engine	1-cylinder Cummins N14
Compression Ratio	16:1
Displacement	2.3 L
Injection System	Direct Injection
Injector	Electronic Unit Injector

To simulate a turbocharger, the intake pressure, intake temperature, and exhaust backpressure are controlled independently for each engine operating condition.

Particulate measurements are taken upstream and downstream of the DPFs. A Rupprecht & Patashnick Model 6100 partial dilutor is used. With this instrument, dilution is carried out in two stages. The first dilution is conducted in a perforated micro-drilled stainless sample probe just outside of the exhaust system. The incoming dilution air surrounds the raw exhaust gas sample to avoid particle losses on the internal walls of the probe. The second stage is an ejector-type diluter. Here a controlled amount of secondary dilution air is forced through an ejector nozzle at high velocity, causing a pressure drop. The pressure drop at the nozzle draws an air stream flow from the primary diluter. A series of magnetic valves for the primary and secondary dilution air determine sample flow rate and dilution ratio. The instrument is controlled by a real-time PM mass monitoring device (TEOM 1105: Tapered Element Oscillating Microbalance) from the same manufacturer. A detailed description of this instrument can be found in [9].

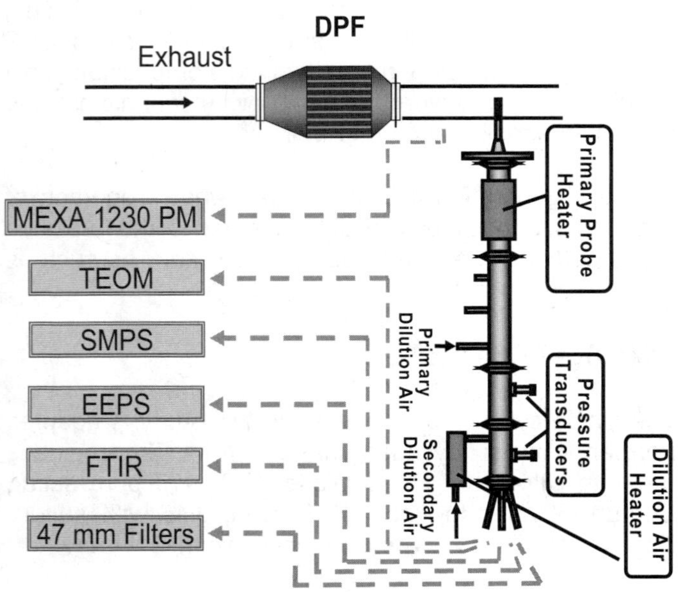

Figure 1. Instrumentation downstream of DPF

DILUTION RATIOS

The dilution ratio is defined as the ratio of the standard volumetric flows of raw and diluted exhaust. In the course of this work, the following dilution conditions were used (Table 2). Downstream of the DPF a low dilution ratio was necessary to detect low PM concentrations.

Table 2. Dilution conditions used

PDAT [°C]	SDAT [°C]	Dilution Ratio
230±7	100±4	21.2±1
230±7	100±4	[a]7.1±0.2

PDAT: Primary Dilution Air Temperature

SDAT: Secondary Dilution Air Temperature

[a]*downstream of DPF*

MEASUREMENT SYSTEMS

An Engine Exhaust Particle Sizer (TSI EEPS Model 3090) Spectrometer was used to measure the size distribution and total concentration of PM with a 10 Hz resolution. Its operating principle is detailed in [4] and it has been previously applied to PM measurements downstream of particulate filters [8].

Gas phase emissions from the exhaust of the engine and the electric furnace are measured by infrared spectroscopy using a Thermo Nicolet Nexus Fourier Transform Infrared Spectrometer (FTIR).

Gravimetric analyses are performed using 47mm Teflon (Pall Teflo 2 μm pore size) filters. EC/OC (elemental carbon/organic carbon) measurements of the particulate matter are performed by collecting soot on 47mm quartz filters and performing a thermal-optical transmittance analysis according to NIOSH 5040 [2].

DPF SPECIFICATIONS

A description of the examined filters is given in Table 3. Filters A, B, and C are made of Cordierite with different washcoat formulations. Filter D is a commercial SiC DPF designed for use in high speed diesel applications. Filters A, B, and C were shown to have higher pressure drop and reach the initial soot cake stage faster than Filter D during filtration. The filtration performances of these filters are studied extensively by Wirojsakunchai, et al. [12].

Table 3. DPF specifications

Designation Number	A	B	C	D
Substrate	Cordierite			Silicon-Carbide
Cell Density	88 CPSI			302 CPSI
Geometric Filter Area	0.62 m²/l			1.08 m²/l
Substrate Porosity ε	48.8 %	45.7 %	45.1 %	49.1 %
Mean Pore Diameter	12.8 μm	11.8 μm	11.8 μm	26.6 μm
Substrate Volume V_s	2.44 l			3.46 l
Washcoat	no	yes	yes	yes
Pt Loading	0 g/l	0 g/l	3 g/l	1 g/l

PM CHARACTERIZATION

A detailed study of the PM properties for the selected engine operating conditions was performed and is reported in [12]. A summary is given in Table 4.

Table 4. PM properties for examined engine operating conditions.

Engine Mode	Mode 3	Mode 4	Mode 5
Speed [rpm]	1800	1200	1200
IMEP [bar]	7.3	5.2	14.2
PM Conc. [mg/m³] (Teflon @ 47°C)	**14.5**	**5.1**	**225**
Filtration Velocity[c] [cm/s]	[a]5.0/[b]2.0	[a]3.8/[b]1.5	[a]3.6/[b]1.4
Geometric Mean Particle Dia. [nm]	**66**	**65**	**127**
EC/OC ratio (quartz @ 47°C)	**0.36**	**0.23**	**0.85**

[a]*Filter A, B, and C*

[b]*Filter D*

[c]*At 230 °C and 1 bar*

Particle size distribution measurements are shown in Figure 2.

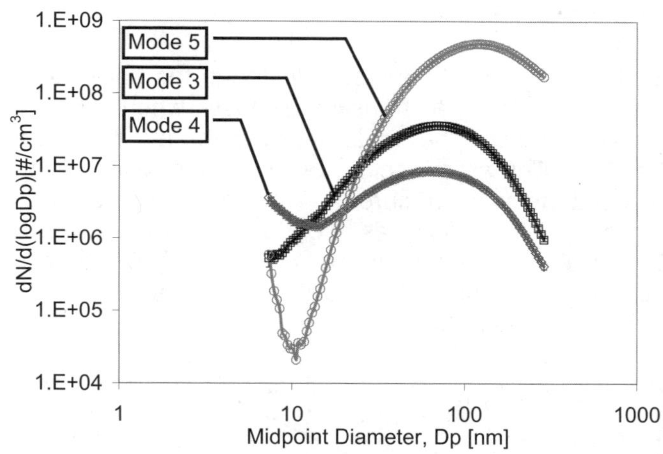

Figure 2. Particle size distribution for raw exhaust

A normalized plot of this data – together with the typical size ranges of nucleation and accumulation modes as proposed by Kittelson [5] — is given in Figure 3.

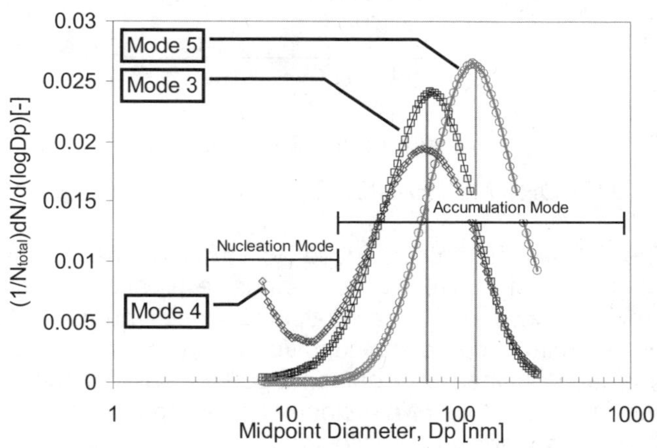

Figure 3. Normalized particle size distribution for raw exhaust

As shown in Figure 3, only Mode 4 displayed a pronounced nucleation mode (*volatile particles*). Accumulation mode particles (*carbonaceous agglomerates*) are present in all three modes. In Modes 3 and 4, most accumulation mode particles are in the 60 nm range. For Mode 5 these particles are much larger (around 120 nm). This is reflected in the mean particle diameters given in Table 4.

DPF FILLING

Filters were loaded with PM from each of the three selected engine operating conditions. The engine exhaust was allowed to reach steady-state conditions in each mode before being routed to the DPF. The exhaust temperature was controlled at 230°C at the DPF inlet for all operating conditions by the use of heat exchanger. Filling times were selected to collect equivalent PM loads for the different engine operating conditions. The pressure drop, DPF internal temperatures, and downstream particle concentration were recorded during filling to ensure the repeatability of DPF performance. A detailed report of this subtask can be found by Wirojsakunchai, et al [12].

As the DPF fills, PM is first collected in the pores of the filter wall. After a transition phase where soot is deposited across the bottle-necks of the surface pores, the PM builds as a discrete layer, or soot cake, on the inlet channel walls (Figure 4). The filter pressure drop is known to rise quickly as the pores become plugged. During the soot cake phase, pressure drop increases linearly with mass of PM as the thickness of the cake layer increases [10].

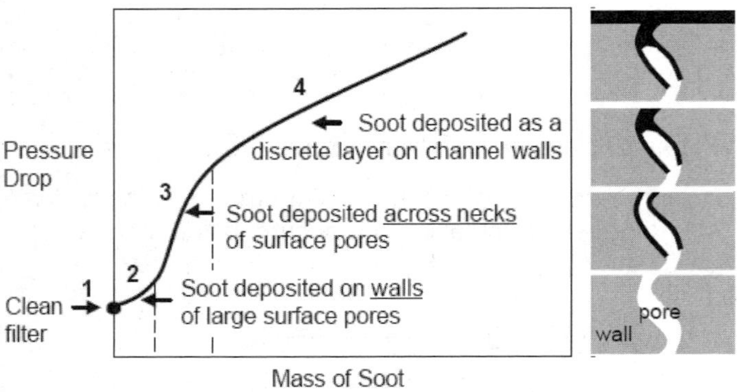

Figure 4. Conceptual model of DPF filling [10]

TRANSPORT MECHANISMS

The Peclet number is used to quantify the relative magnitude of convective and diffusive filtration through Brownian motion. To calculate the Peclet numbers for the examined operating conditions and substrates, the approach outlined by Konstandopoulos was chosen [5]. The microstructure of the clean filter is represented as a matrix of spherical unit collectors.

$$Pe = \frac{U_{pore} \cdot d_c}{D_p}$$ Equation 1

with: U_{pore} pore velocity [m/s]

d_c unit collector diameter (clean filter) [m]

D_p particle diffusion coefficient [m^2/s]

$$U_{pore} = \frac{U_{filt}}{\varepsilon}$$ Equation 2

with: U_{filt} DPF wall filtration velocity [m/s]

ε DPF wall porosity [-]

$$D_p = \frac{k_B T}{3\pi d_p \mu_{exh}} C$$ Equation 3

with: k_B Boltzmann constant, 1.381x10^{-23} [J/K]

d_p mean particle diameter [m]

μ_{exh} dynamic viscosity [kg/m-s]

C Cunningham slip correction factor [-]

The Cunningham slip flow correction factor C is employed to extend the validity of Equation 3 into the free molecule regime [3]

$$C = 1 + Kn\left[1.257 + 0.4e^{\left(\frac{-1.1}{Kn}\right)}\right]$$ Equation 4

with: $Kn = \frac{2\lambda}{d_p}$ particle Knudsen number [-]

$$\lambda = \nu_{exh}\left[\frac{\pi MW_{exh}}{2RT}\right]^{\frac{1}{2}}$$ exhaust gas mean free path [m]

The spherical unit collector diameter of the clean filter wall is calculated from Ergun's equation (Equation 5) [1].

$$d_c = \frac{3}{2}\left[\frac{1-\varepsilon}{\varepsilon}\right] d_{pore}$$ Equation 5

with: d_{pore} DPF wall pore diameter [m]

After a soot layer has been established on the walls of the inlet channels, this layer becomes the governing filtration body. In this stage, the characteristic length scale of the unit collector is the diameter of the primary soot particles that form the soot cake [7].

PECLET NUMBER

Table 5 shows the Peclet numbers for the analyzed engine operating conditions on clean Filter C and D substrates. Since this paper concerns the particle interaction with a clean substrate, the Pe number will be calculated accordingly.

Table 5. Peclet numbers for all investigated engine modes with a clean substrate

Engine Mode	Mode 3	Mode 4	Mode 5
Pe (Filter C)	**1074**	**673**	**2005**
Pe (Filter D)	**767**	**480**	**1431**

For larger mean particle diameters (Mode 5) and higher pore velocities (Filter C substrate) Peclet numbers are higher and mass transport is dominated by **convection**. In cases with smaller mean particle diameters (Mode 4) or lower pore velocities (Filter D substrate) Peclet numbers are lower and **diffusive filtration** becomes more important than with other filters.

EXTERNAL FURNACE REGENERATION

External furnace regenerations were performed to quantify the relative amounts of PM in the DPF wall and soot cake and to estimate PM reactivity. In this experiment, the filled DPF is placed in an electric furnace (shown in Figure 5) and exposed to a prescribed linear temperature ramp of 6.7 °C/min. The temperature is held at 700 °C after the ramp to ensure complete oxidation of the PM. During regeneration, 9 thermocouples measure the internal temperatures.

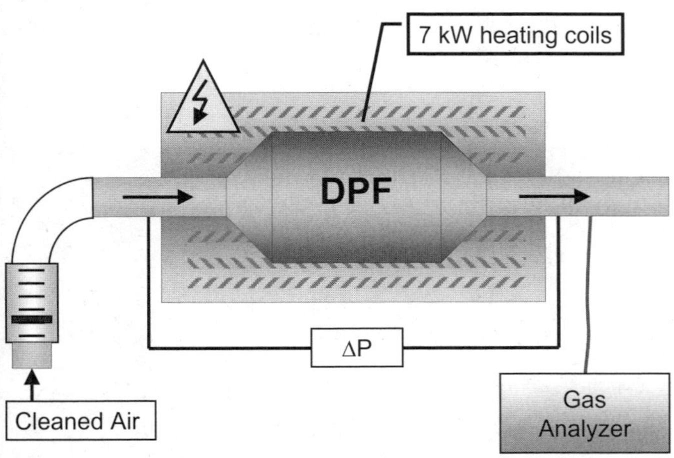

Figure 5. Electric regeneration furnace schematic.

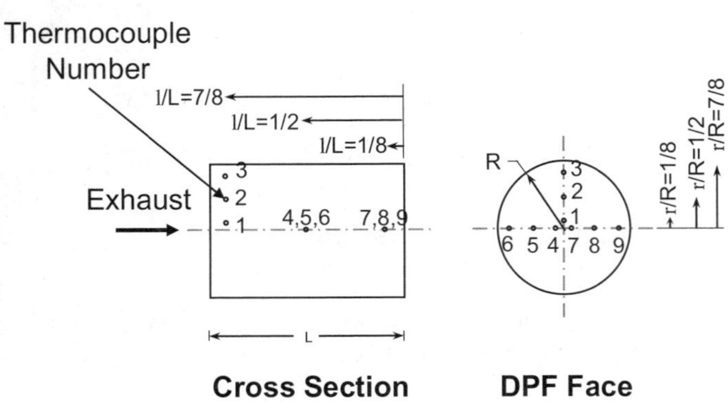

Figure 6. Layout of thermocouple locations

The resulting radial temperature distribution is shown in Figure 7. Axial temperature gradients are less pronounced [1].

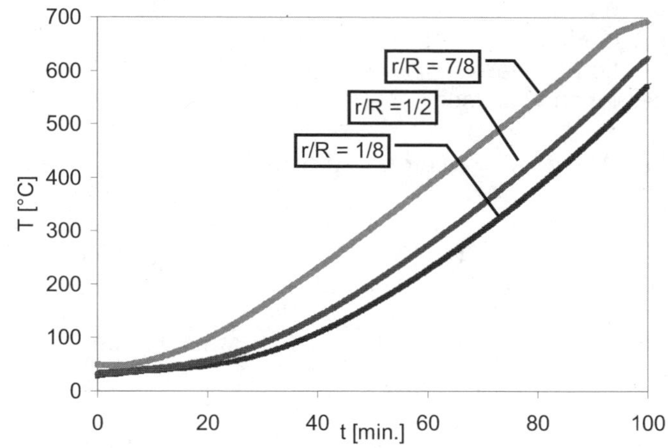

Figure 7. Internal DPF temperatures during furnace regeneration.

Using infrared spectroscopy, the gas-phase concentrations of hydrocarbons, CO, CO_2 and SO_2 evolving from the DPF are detected during the regeneration process. Hydrocarbon emissions from volatilization were not present during any of the experiments.

From this data, a temperature programmed oxidation (TPO) plot, showing the measured molar concentration of evolved CO_x ($y_{cox,meas}$) emissions over the average DPF temperature, is constructed (Figure 8). CO_x is a summation of carbon monoxide and carbon dioxide emissions.

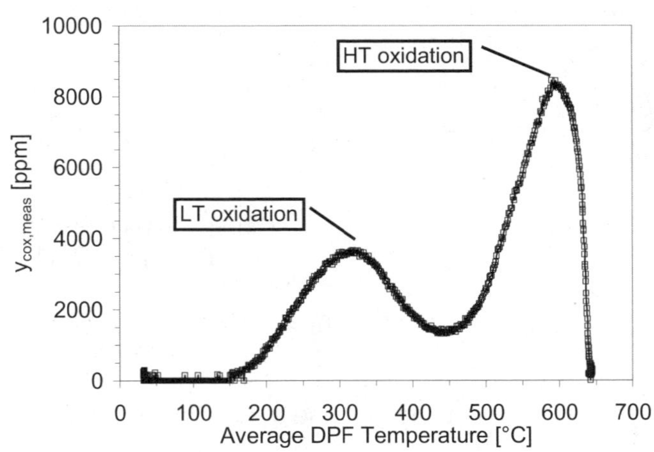

Figure 8. TPO plot of furnace regeneration.

It is evident that PM oxidation occurs in two distinct events in a catalyzed DPF. They are designated low temperature oxidation (LT at ~300°C) and high temperature oxidation (HT at ~575°C). The integral of these emissions gives the relative amount of LT and HT PM (by carbon content) present in the DPF. In all examined cases the LT and HT peaks are convoluted, making an exact partitioning of the PM impossible.

KINETIC PARAMETER ESTIMATION

Mathematical Model

To de-convolute the emissions peaks during furnace regeneration and estimate the kinetic parameters of the oxidation reaction, the following mathematical description was used.

The PM oxidation rate r is expressed as

$$r = \frac{dN_C}{dt} = -kN_C(t)^{n1}[O_2]^{n2} \qquad \text{Equation 6}$$

With: r — reaction rate [mol/s]

k — rate constant

N_C — amount of carbon (PM) [mol]

n1 , n2 — reaction order of carbon , oxygen [-]

$[O_2]$ — oxygen concentration [ppm]

with temperature dependency according to Arrhenius.

$$k = Ae^{\left(\frac{-E_A}{RT}\right)} \qquad \text{Equation 7}$$

With: A — frequency factor

E_A — activation energy [J/mol]

R — ideal gas constant [J/mol·K]

A molar balance over the reactor and sampling system results in Equation 8.

$$\frac{dN_C}{dt} = -\frac{dN_{COX}}{dt} = y_{COX}(t)\frac{dV}{dt}\frac{1}{v_N} \qquad \text{Equation 8}$$

With: N_{COX} — amount of carbon species emissions (CO_2 and CO) [mol]

y_{COX} — concentration of measured carbon species [ppm]

v_N — molar volume at 87 kPa, 165 °C [l/mol] (40.04)

$\frac{dV}{dt}$ — volumetric flow rate at sampling system (FTIR) [l/s]

Oxidation with NO_2 and H_2O is disregarded due to low concentrations of these species.

Assuming that the oxygen concentration $[O_2]$, oxygen reaction order n2, and volumetric flow rate are constant, they can be lumped into the kinetic parameter κ. Equation 9 gives a relationship between DPF temperatures and measured concentrations of evolved oxidation products as a function of the kinetic parameters κ, E_A, n1 and the initial soot amount $N_{C,init}$.

$$y_{CO_x}(t) = \underbrace{\left(\frac{Av_N}{[O_2]^{n2}\frac{dV}{dt}}\right)}_{\kappa} e^{(-E_A/RT)} N_C^{\ n1} \qquad \text{Equation 9}$$

$$\theta = \begin{bmatrix} \kappa \\ E_A \\ n_1 \\ N_{C,init} \end{bmatrix} \qquad \text{Equation 10}$$

With: θ — parameter estimation vector

Approach

Large thermal gradients are present in the DPF during furnace regeneration. Especially for the Cordierite case, the average DPF temperature was found to be a poor choice for the parameter estimation. Therefore the full scale filter is discretized as a system of 9 smaller-volume reactors with experimentally determined temperatures. Areas of the filter with high thermal gradients are more finely discretized.

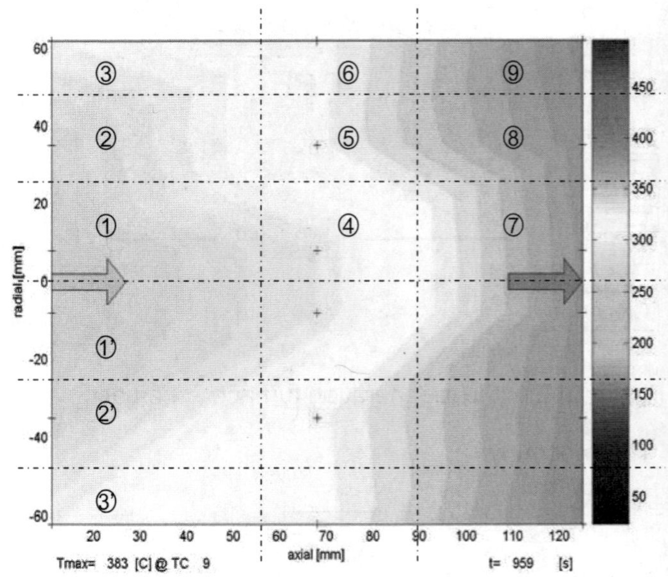

Figure 9. Temperature distribution and discretization of full scale DPF

The experimentally measured emissions are the sum of the emissions evolved from all smaller reaction volumes.

$$y_{CO_x}(\theta, T(t)) = \sum_i y_{CO_x,i}(\theta, T_i(t))$$ Equation 11

With: i reaction volume index [-]

Figure 10 shows Equation 9 evaluated for a set of DPF temperatures T_i and a parameter estimation vector θ. Each curve represents emissions from each of nine small reaction volumes as shown in Figure 9.

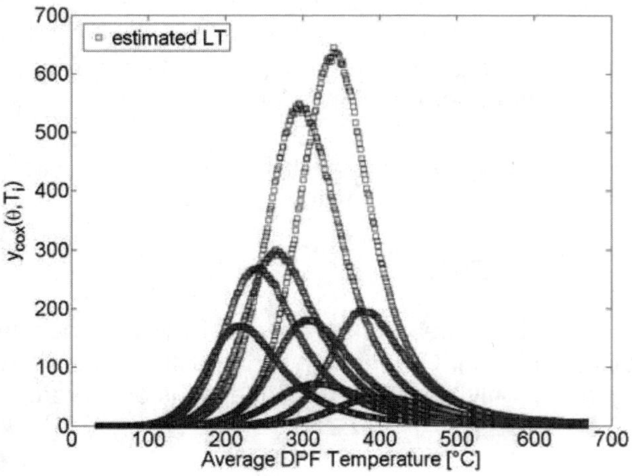

Figure 10. Evolved emissions of the nine individual reaction volumes i (LT)

Very different oxidation rates are seen in the individual reaction volumes depending on their temperature (Figure 10).

With the minimization function Φ in Equation 12, a non-linear least squares method is used to estimate the kinetic parameters for both LT and HT oxidation.

$$\Phi(\theta) = \left(\underline{y_{co_x,est}}(\theta, T_i(t)) - \underline{y_{co_x,meas}}\right)^2$$ Equation 12

With: y_{est} , y_{meas} estimated , measured conc. of carbon species [ppm]

Figure 11 shows the *estimated* emissions $y_{COX,est}$ of the individual reaction volumes for an **initial guess** of the kinetic parameter values θ.

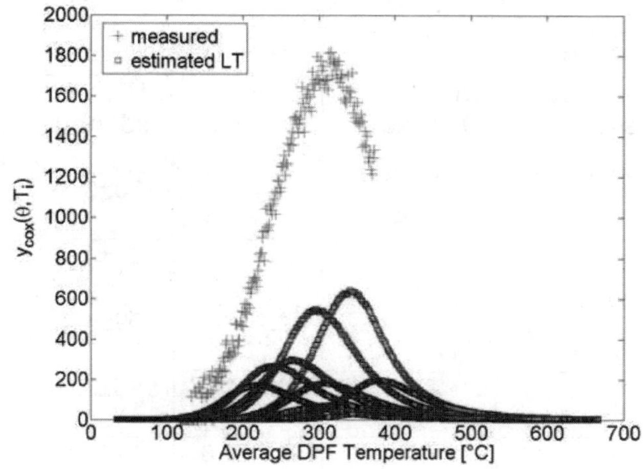

Figure 11. Evolved emissions of individual reaction volumes (LT, initial guess of θ)

In Figure 12, the final result of the iterative parameter estimation is shown and a good fit for the experimental values has been found.

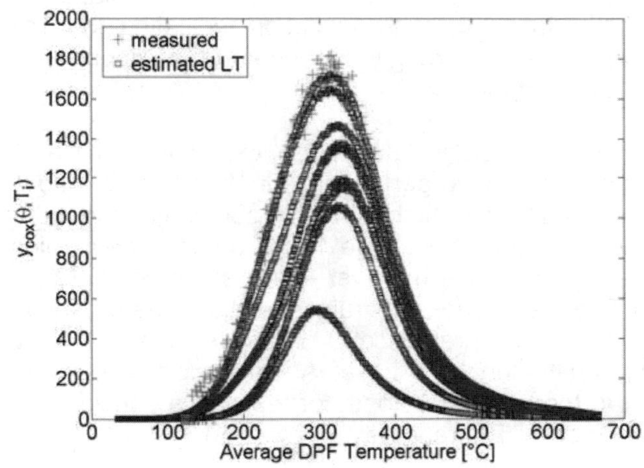

Figure 12. Cumulative sum of evolved emissions of individual reaction volumes (LT, optimal θ)

This approach yields data on PM reactivity (i.e. activation energy) for different PM types and DPF substrates. Additionally it allows for clear separation and quantification of the overlapping LT and HT peaks (Figure 18 to Figure 20).

PM KINETICS

Activation energies derived from this methodology are given in Table 6. Reaction orders with respect to PM were found to be very close to unity for all considered cases.

Table 6. Activation energies E_A (95 % confidence intervals)

Engine Mode ([a]Case)	DPF Designation	**LT E_A [kJ/mol]**	**HT E_A [kJ/mol]**
3	A	n/a	**190±3.3**
3	B	n/a	**128±4.1**
3 (a)	C	**42.2±1.3**	**109±1.4**
3 (b)	D	**44.5±1.3**	**195±3.6**
4	C	**42.9±1.6**	**111±2.1**
5 (c)	C	**46.5±7.2**	**125±2.1**

LT oxidation was not observed in the uncatalyzed cases. Activation energies are in the 40-50 kJ/mol range for catalytic oxidation and between 110-200 kJ/mol for thermal oxidation. Activation energies display little variability during catalytic oxidation indicating that this mechanism is not strongly dependent on composition or morphology of the PM.

Thermal, non-catalytic (HT), oxidation activation energies cover a larger range. On Filter A, B, and C substrates, PM with higher volatile content was found to be more reactive and have lower activation energies. These differences —however— are small compared to the influence of the substrate coating. The washcoat was seen to dramatically increase the reactivity compared to the bare case. Adding a precious group metal to the washcoat does not result in a considerable increase of HT PM reactivity.

On Filter D substrate, thermal oxidation (HT) activation energies were found to be lower for a given PM type (in spite of the presence of a washcoat and precious group metal). This may be explained by differences in filter geometry, pore microstructure and soot deposit properties between Filters A, B, C and Filter D substrates.

For non-catalytic oxidation of PM by O_2, activation energies were found to range from 100 kJ/mol to 190 kJ/mol. These ranges are in agreement with previous work [13, 14].

Most kinetic studies of soot oxidation are performed in flow reactors, single DPF channels or DPF cores. The approach taken in this work —utilizing a full-scale DPF brick— has the advantage that the complexity of the system in terms of microstructure, geometry and size is retained. As such, the results are immediately applicable to a regeneration study as performed in the following.

LT AND HT RATIOS

The same TPO methodology used to derive activation energies also results in the quantification of the relative amount of PM in the two observed oxidation peaks (catalytic and thermal oxidation) of a furnace regeneration (Figure 8). This quantification is of interest for the following reasons:

- The PM fractions contribute differently to filter pressure drop and filtration efficiency
- Differences in energy requirements for oxidation (light-off temperature)
- Ensuring full use of the available catalytically active areas of the filter.

Influence of Catalyst

The LT oxidation peak is the result of PM experiencing catalytic oxidation. On uncatalyzed substrates as shown in Figure 13, only thermal (HT) oxidation is observed.

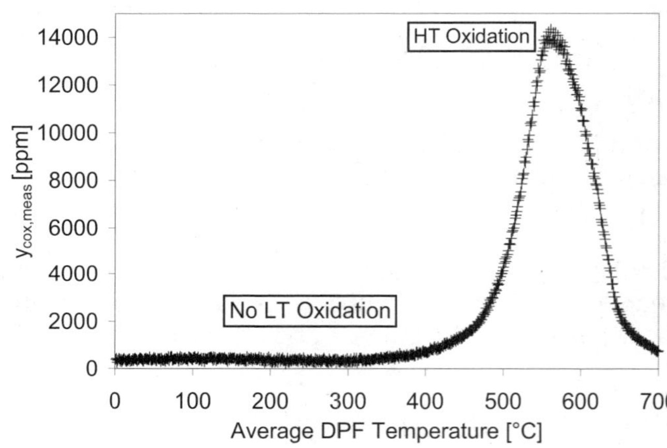

Figure 13. Influence of catalyst (Mode 3, Filter B)

Influence of PM loading

A TPO plot for a catalyzed DPF filled with different amounts of PM from the same engine mode is shown in Figure 14. Using the kinetic approach, as shown in the previous section, and assuming that those activation energies for two oxidation regimes is not a function of PM mass gives relative amount of PM from two oxidation peaks and the result is shown in Figure 15. In Figure 15, a limitation effect is visible as PM mass increases. After about 0.5 g/l of accumulated PM, the available catalytically active surface area becomes extinguished and no further PM deposition in these areas is possible. Accordingly, the PM accumulates in the thermally oxidized areas.

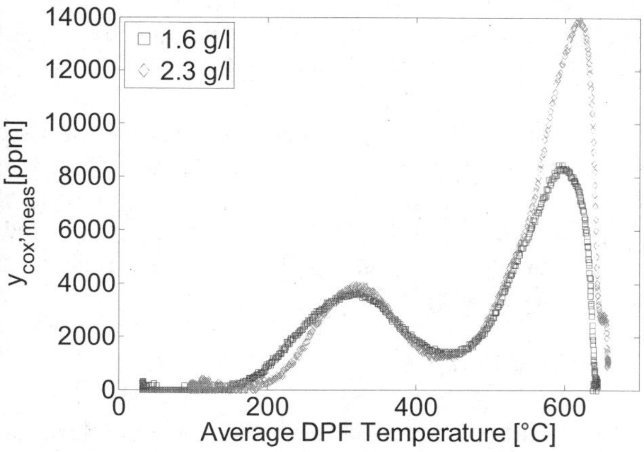

Figure 14. Evolved emissions for different PM loading (Mode 4, Filter C)

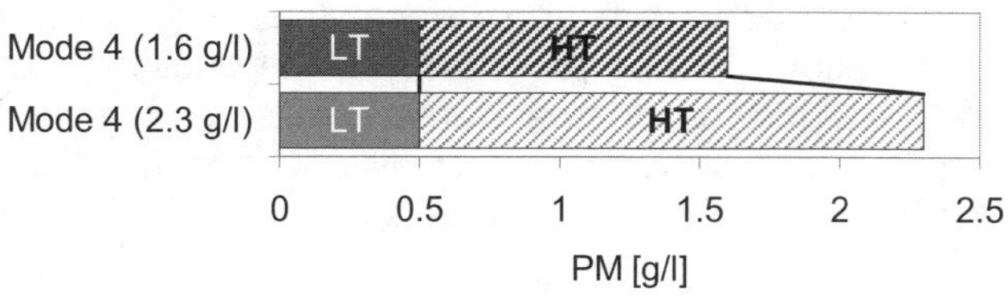

Figure 15. LT/HT ratio for different PM loading (Mode 4, Filter C)

Influence of PM Properties

The influence of transport phenomena is observed in Figure 16 and Figure 17. The dominant filtration mechanisms, i.e. convective vs. diffusive deposition, determine the location of PM deposition in the DPF microstructure. Utilization of the catalyst area varies widely between the operating conditions.

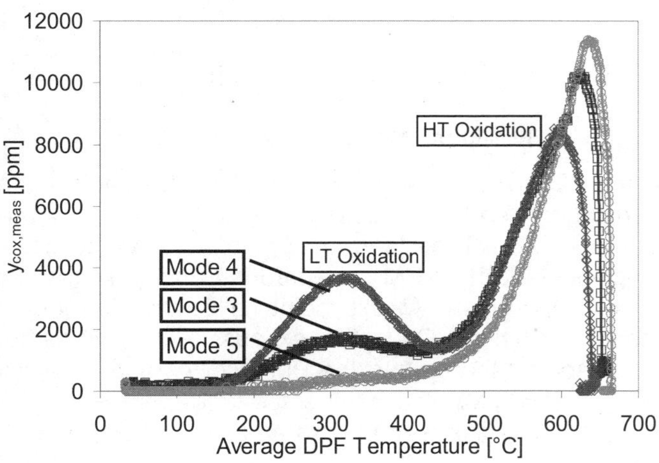

Figure 16. Influence of engine operating condition (Filter C)

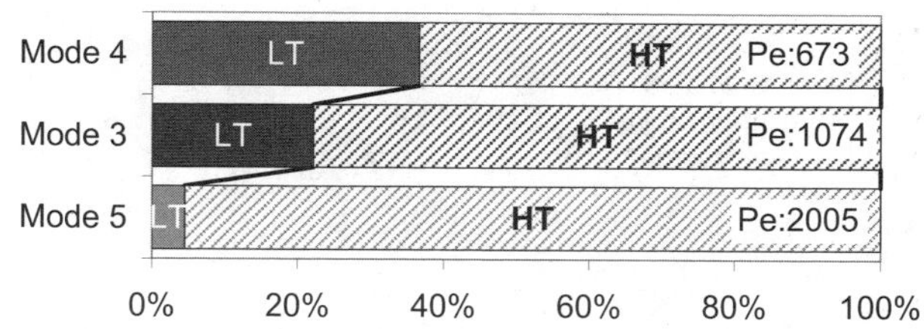

Figure 17. Influence of engine operating condition on LT/HT ratio (Filter C)

In engine modes with small particles and low filtration velocities, the dominant filtration mechanism is diffusion through Brownian motion. This results in a very effective utilization of the catalyst area in the porous wall (Mode 4, Figure 16 and Mode 3 on Filter D, Figure 19). For engine modes with large (accumulation mode) particles (Mode 5, Figure 16) catalyst utilization is poor. The filtration is dominated by the convective mechanism which results in a deterministic deposition of PM and in some cases (Mode 5, Figure 20) can lead to pore blocking and limitations on catalyst utilization. Wall penetration and catalyst utilization correlates with the transport mechanism (i.e. Peclet number) of the system.

Three reference cases are selected to further examine the impact of PM type and substrate properties (Table 4).

a.) Mode 3 on Filter C (high amount of LT soot, high catalyst loading, high thermal gradients)

b.) Mode 3 on Filter D (high amount of LT soot, low catalyst loading, low thermal gradients)

c.) Mode 5 on Filter C (low amount of LT soot, high catalyst loading, high thermal gradients)

The constructed TPO plots are given in Figure 18 to Figure 20.

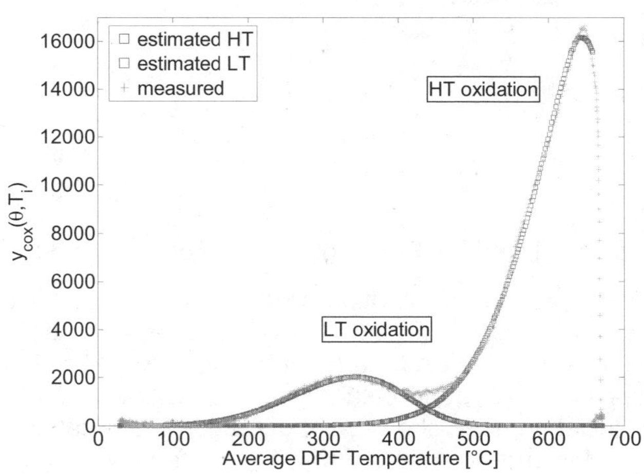

Figure 18. Evolved emissions for case a (Mode 3, Filter C)

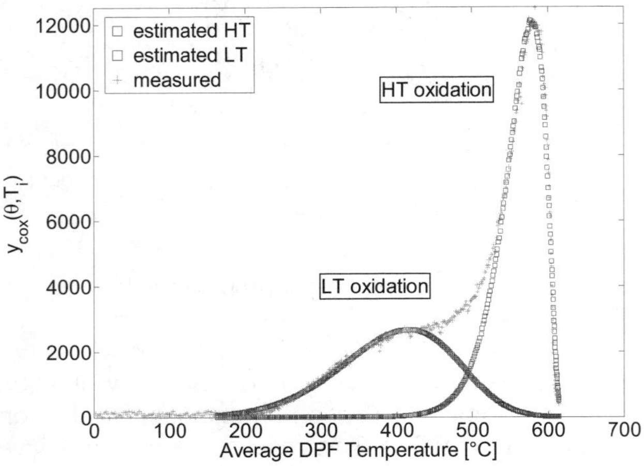

Figure 19. Evolved emissions for case b (Mode 3, Filter D)

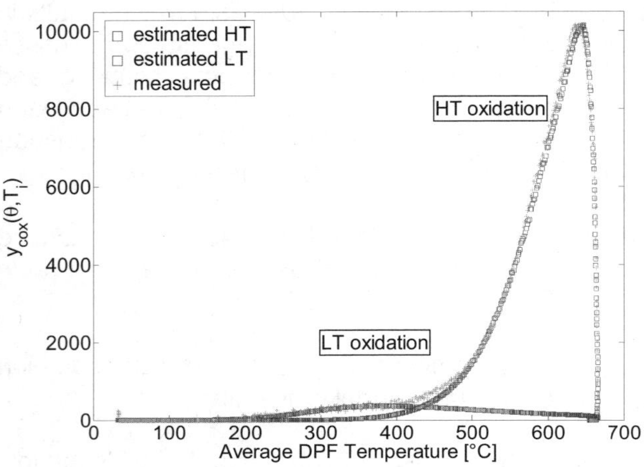

Figure 20. Evolved emissions for case c (Mode 5, Filter C)

In Figure 21, the relative LT and HT PM amounts for cases a, b, and c are quantified.

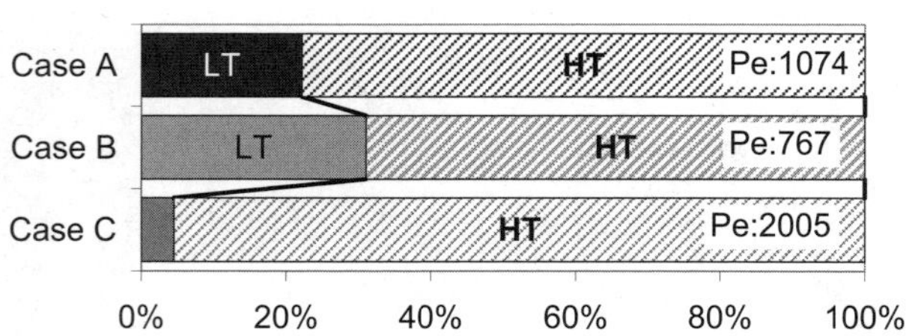

Figure 21: LT and HT PM amount for cases a, b, and c.

In case b, the larger filtration area and higher effective porosity of Filter D (Table 3) results in a low Peclet number. For this case, a large proportion (~30%) of PM is deposited with good catalyst contact. Case c, with a high Peclet number, shows poor utilization of the catalyst area and a large amount of PM in the HT fraction.

In both Figure 17 and Figure 21, the amount of LT PM is seen to be influenced by the governing deposition mechanism. Low Peclet number cases, that favor diffusive deposition, display high amounts of LT PM. High Peclet number cases show the opposite trend and the majority of soot is found in the HT fraction.

This is consistent with the diffusive (low Pe) mechanism depositing PM uniformly on the pore walls (with good catalyst contact) and the convective mechanism resulting in a deterministic deposition at flow obstructions (thereby limiting the catalyst utilization through pore blocking).

IN-SITU REGENERATION

With the kinetic parameters derived from the furnace regeneration, the rate of soot oxidation r (θ,T) in the in-situ case can be calculated. For this experiment (Figure 22), the filters are filled to the same loading as in the furnace case. After filling, the exhaust stream is rapidly heated to 600°C with an oxy-acetylene torch to trigger regeneration. During regeneration, the engine is run in Mode 4. Internal temperatures, pressure drop across the filter, and downstream particle concentrations are measured in real-time with the TSI EEPS during regeneration.

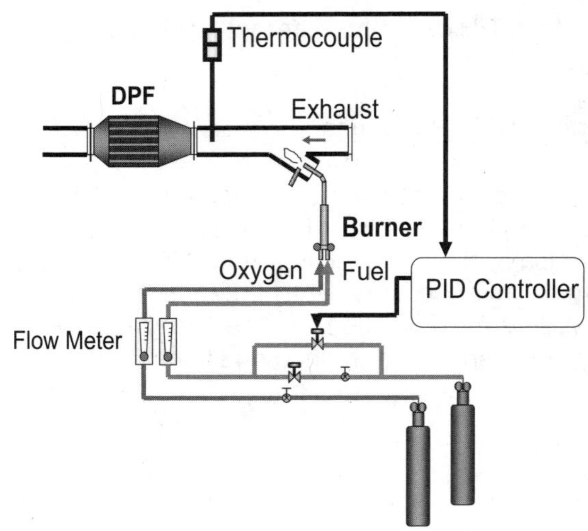

Figure 22. In-situ burner schematic

During filling and in-situ regeneration, the downstream particle concentration passes through the stages shown in Figure 23. The clean DPF displays poor collection efficiency and high PM breakthrough until pore necking occurs (Figure 4). At this point collection efficiency is steady and high. Once the exhaust temperatures are increased by the burner, the PM oxidizes and collection efficiency drops until a final steady state is reached. At this point, the filter is ready to start collecting PM again.

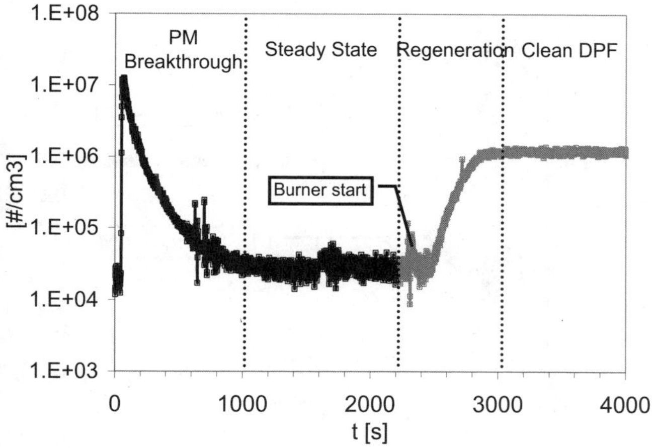

Figure 23. Downstream total particle concentration during fill–regeneration sequence (in-situ)

Figure 24 shows the evolution of the pressure drop across the filter in these stages. The pressure drop rises rapidly during the initial stage of filling. Once a soot cake is established, it displays a linear increase over time. During regeneration, the pressure drop is seen to overshoot as the physical properties of the exhaust, viscosity and density, change due to the change in temperature. It then slowly returns to the level of a clean DPF as the PM is oxidized.

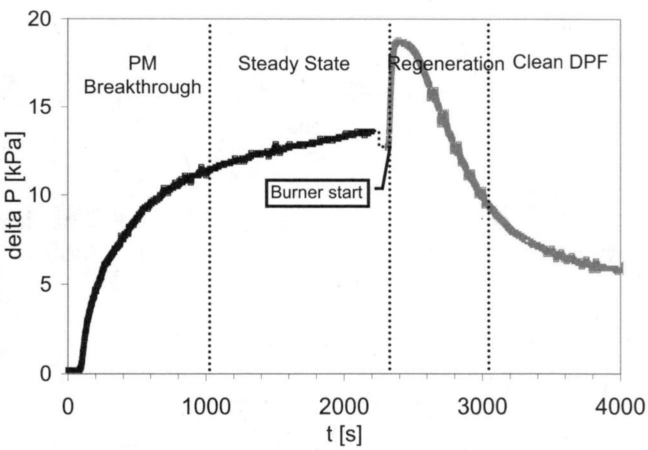

Figure 24. Pressure drop across DPF during fill–regeneration sequence (in-situ)

LT AND HT OXIDATION (IN-SITU CASE)

PM oxidation rates at a given temperature are calculated using the kinetic parameters given in Table 6. Applying this methodology to the experimentally determined internal temperatures results in oxidation rates during in-situ DPF regeneration. With a known initial PM amount, the remaining LT and HT PM in the filter can be tracked over time. Together with measurements of the performance parameters, filtration efficiency and pressure drop, the effect of the PM and substrate selection is examined by assessing the three previously mentioned reference cases.

a.) Mode 3 on Filter C (high amount of LT soot, high catalyst loading, high thermal gradients)

b.) Mode 3 on Filter D (high amount of LT soot, low catalyst loading, low thermal gradients)

c.) Mode 5 on Filter C (low amount of LT soot, high catalyst loading, high thermal gradients)

The average temperature during in-situ regeneration and the calculated PM conversion *(fraction of PM remaining in the filter)* are shown in Figure 25 to Figure 27.

The necessary time for complete regeneration is determined by the reactivity of the HT PM on a given substrate and the thermal diffusivity α of the substrate. The influence of the reactivity is seen when comparing Cases a (Figure 25) and c (Figure 27) with activation energies of 109 kJ/mol and 125 kJ/mol respectively. In Case a, regeneration is completed at ~1000 sec. While in Case c, even at over 1500 sec., unoxidized HT PM remains in the filter. Comparing Filter C (Case a) and D (Case b) substrates, the influence of the thermal diffusivity becomes apparent. Even though it is known to be less reactive, oxidation progresses faster in Case b due to the faster heat up of the substrate and more even temperature distribution.

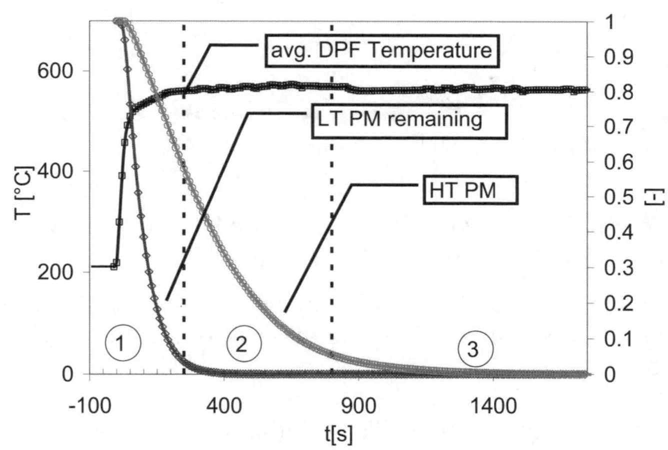

Figure 25. PM oxidation and average DPF temperature during in-situ regeneration (a: Mode 3, Filter C)

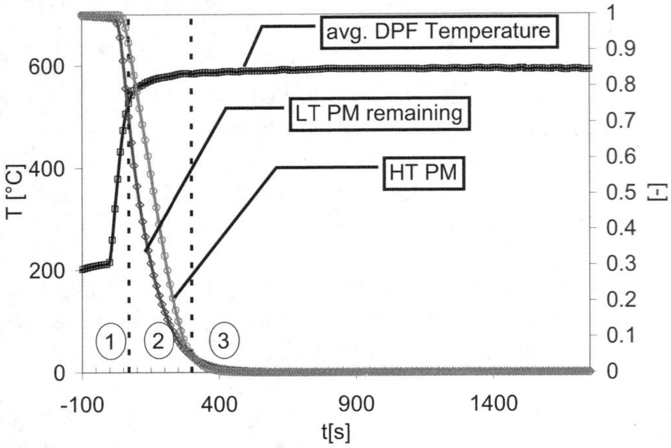

Figure 26. PM oxidation and average DPF temperature during in-situ regeneration (b: Mode 3, Filter D)

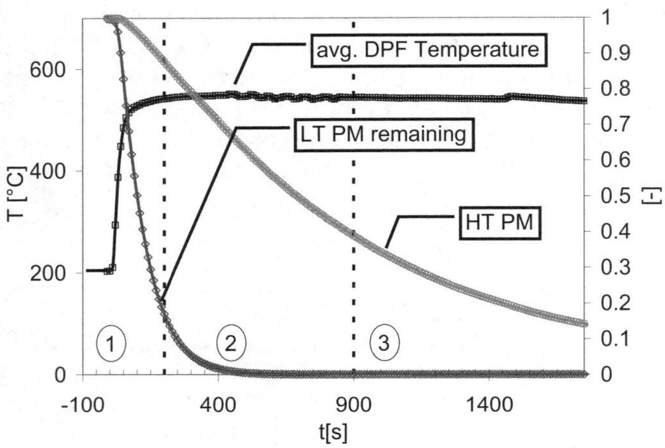

Figure 27. PM oxidation and average DPF temperature during in-situ regeneration (c: Mode 5, Filter C)

In Figure 28 to Figure 30, filtration efficiency is used to define three stages of regeneration. These are described as:

Stage 1: High filtration efficiency (intact soot cake)

Stage 2: Decreasing filtration efficiency (transition)

Stage 3: Low filtration efficiency (clean DPF)

Filtration efficiency and remaining HT PM in the filter are correlated. The transition phase (Stage 2) in filtration efficiency corresponds to the oxidation of the majority of HT PM. This reinforces the notion that HT PM can be assigned to the PM layer on the channel walls which is known from filling experiments to be responsible for high filtration efficiency.

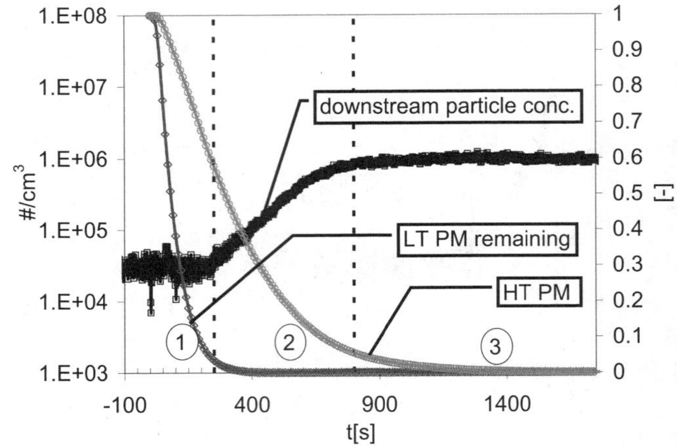

Figure 28. PM oxidation and downstream total particle concentration during in-situ regeneration (a: Mode 3, Filter C)

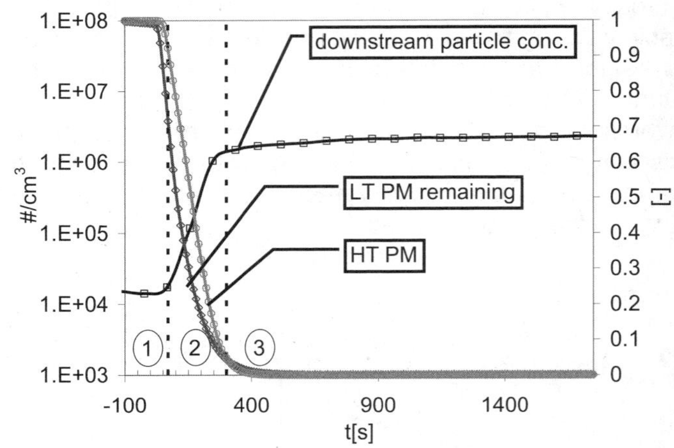

Figure 29. PM oxidation and downstream total particle concentration during in-situ regeneration (b: Mode 3, Filter D)

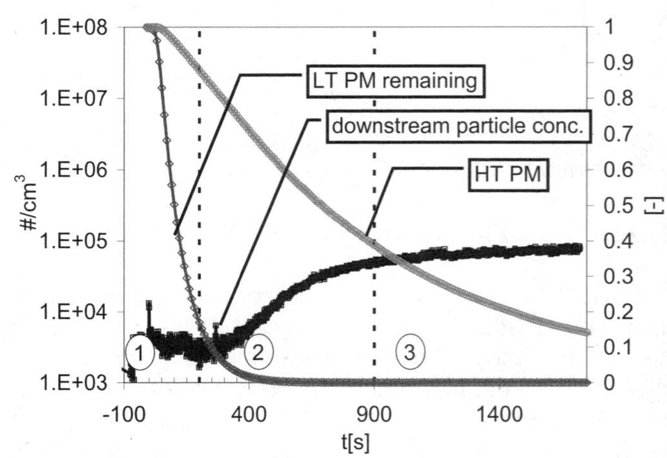

Figure 30. PM oxidation off and downstream total particle concentration during in-situ regeneration (c: Mode 5, Filter C)

The measured pressure drop during regeneration shows the characteristic overshoot and eventual steady state

(Figure 24) in all examined cases (Figure 31 to Figure 33). The overshoot can be attributed to the flow acceleration through the rapid temperature change of the exhaust flow. In the examined cases, LT PM is seen to oxidize during this phase. Once appreciable HT PM oxidation occurs, the pressure drop starts falling and eventually decreases to a steady state when all HT PM is oxidized.

Note, in Case c (Figure 33), with incomplete HT PM oxidation, the pressure drop does not reach the clean filter state.

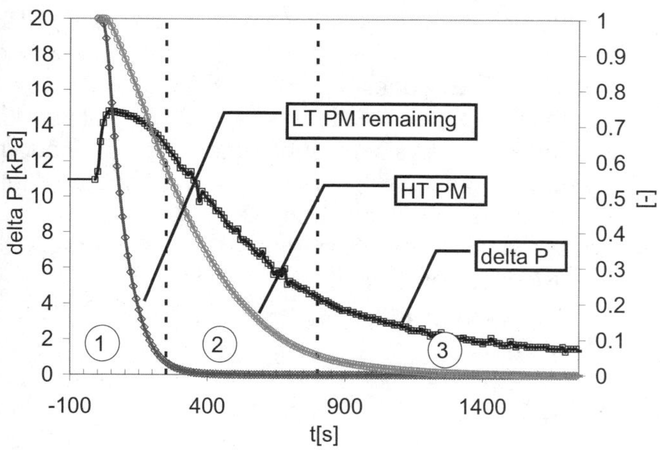

Figure 31. PM oxidation and pressure drop across the filter during in-situ regeneration (a: Mode 3, Filter C)

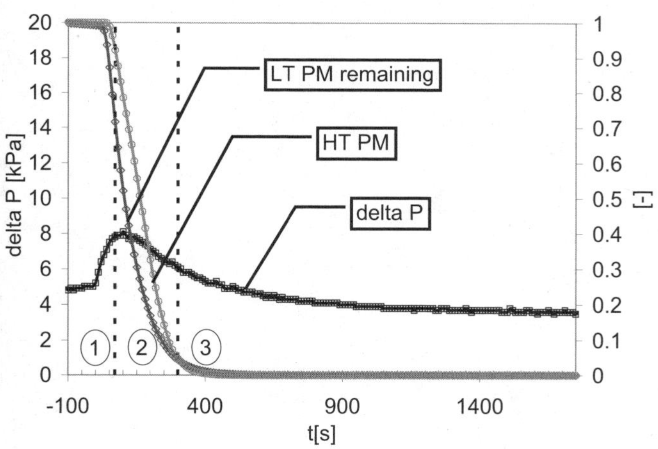

Figure 32. PM oxidation and pressure drop across the filter during in-situ regeneration (b: Mode 3, Filter D)

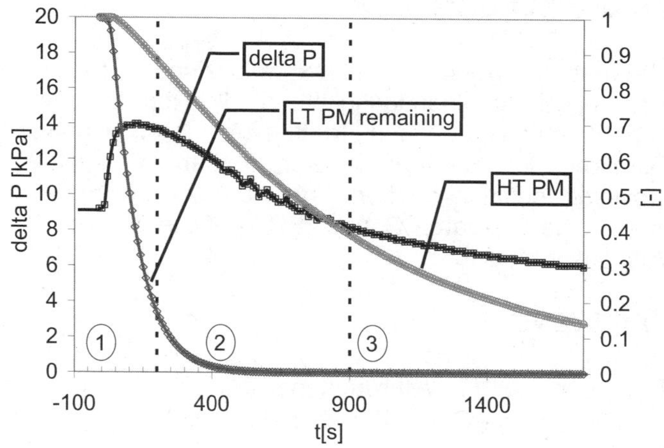

Figure 33. PM oxidation and pressure drop across the filter during in-situ regeneration (c: Mode 5, Filter C)

CONCLUSION

An electrical furnace with precise temperature control was applied to study the PM oxidation process in a DPF. By applying a TPO analysis on the various DPF substrates and substrate coatings, filling with equivalent PM amounts from three selected engine modes, PM oxidation partitioning into HT (soot cake, thermal oxidation) and LT PM (particles in the porous media wall, catalytic oxidation) was quantified. This partitioning was found to vary strongly with the Peclet number of the examined operating conditions. Additionally this method allowed the estimation of kinetic parameters for oxidation of various kinds of PM on different full-scale substrates. Activation energies were found to be in the 40-50 kJ/mol range for catalytic oxidation and between 110-200 kJ/mol for thermal oxidation. Catalytic oxidation does not depend strongly on the composition or morphology of the PM and activation energies display little variability. Thermal oxidation activation energies cover a larger range. On the high pressure drop filter examined in this study, activation energies decrease with increasing volatile content of the PM. These differences are small compared to role of the substrate. Mode 3 PM activation energies are highest on the low pressure drop filter. On the high pressure drop filter, the bare (non-washcoated) case is the least reactive, followed by the washcoated and catalyzed cases. It is concluded that transport phenomena play a major role in determining the location of deposition inside the DPF wall and the utilization of the available catalyst surface area.

Using the kinetic analysis from the ex-situ regeneration, the kinetic parameters of LT and HT PM can be transposed as a model to real-world in-situ regeneration, where the progress of the oxidation is difficult to measure. Good correlation was found between amount of HT (non-catalytically oxidized) PM and the performance parameters of pressure drop and filtration efficiency.

ACKNOWLEDGMENTS

The authors would like to express sincere thanks to Honda R&D for sponsoring this study. Representatives from Corning, Johnson Matthey, Horiba, Cummins Filtration, and BP Amoco were extremely helpful in providing technical support. This work was performed at the Engine Research Center (ERC) of the University of Wisconsin-Madison.

REFERENCES

1. Ergun, S. "Flow through packed columns", Chem. Eng. Progress., v. 48, n. 2, p. 89, 1952.
2. Federal Register, Vol. 66, No. 12, January 18, 2001, pp 5172-5178.
3. Friedlander S.K. "Smoke, dust and haze: Fundamentals of aerosol behavior", John Wiley & Sons, New York, 1977.
4. Johnson, T. et al. "A new electrical mobility particle sizer spectrometer for engine exhaust particle measurements" SAE Paper No. 2004-01-1341, 2004.
5. Kittelson, D.B. "Engines and Nanoparticles: A Review," J. Aerosol Sci., Vol. 29, No. 5/6, pp. 575-588, 1998
6. Konstandopoulos, A.G. et al. "Fundamental Studies of Diesel Particulate Filters: Transient Loading, Regeneration and Aging" SAE Paper No. 2000-01-1016, 2000.
7. Konstandopoulos, A.G. et al. "Microstructural Properties of Soot Deposits in Diesel Particulate Traps" SAE Paper No. 2002-01-1015, 2002.
8. Liu, G.Z. et al. "Transient Performance of Diesel Particulate Filters as Measured by an Engine Exhaust Particle Size Spectrometer", SAE Paper No. 2005-01-0185, 2005.
9. Maamouri, M. et al. "Real-Time Exhaust PM Mass Measurements for DPF Development Studies" Emission Control Conference, Dresden, 18-19 May 2006.
10. Merkel, G.A. et al. "New Cordierite Diesel Particulate Filters for catalyzed and non-catalyzed Applications" Proceedings of the 9th Diesel Engine Emissions Reduction Conference, August 24-28, 2003, Newport, Rhode Island.
11. Neeft J.P.A. et al. "Catalytic Oxidation of Carbon Black – I. Activity of Catalysts and Classification of Oxidation Profiles", Fuel 77(3), 111-121, 1998.
12. Wirojsakunchai, E. et al. "Detailed Diesel Exhaust Particulate Characterization and Real Time DPF Filtration Efficiency Measurements during PM Filling Process", SAE Paper No. 2007-01-0320, 2007.
13. Stratakis, G.A., Stamatelos, A.M. "Thermo-gravimetric analysis of soot emitted by a modern diesel engine run on catalyst-doped fuel" Combustion and Flame 132(1-2), 157-169, 2003.
14. Yezerets, A. et al. "Experimental Determination of the Kinetics of Diesel Soot Oxidation by O_2 – Modeling Consequences" SAE Paper No. 2003-01-0833, 2003.

CONTACT

David E. Foster, Phil and Jean Myers Professor

121 Engineering Research Building
1500 Engineering Drive
Madison, WI 53706-1687

E-mail: foster@engr.wisc.edu

2007-01-0317

Soot Emission Behavior from Diverse Vehicles and Catalytic Technologies Measured by a Solid Particle Counting System

Rahman M. Montajir, Takeshi Kusaka, Inoue Kaori, Nobutaka Kihara and Ichiro Asano
HORIBA Ltd. Kyoto, Japan

Masayuki Adachi and Qiang Wei
HORIBA Instrument Inc., USA

ABSTRACT

A Solid Particle Counting System (SPCS) has been developed according to the ECE draft regulation proposed by the particle measurement program (PMP). In the previous report the basic performance of the SPCS has been mentioned in detail [1-6]. It has been reported that the SPCS demonstrates very stable dilution of sample with air and the error of real time dilution factor is less than 6% up to the total dilution factor of 1000. Penetration of solid particles through the SPCS is over 95% and volatile particles removal efficiency is over 99%.

In this study the SPCS has been used to investigate the soot emission behavior from different vehicles with different after-treatment technologies. Direct injection (DI) diesel vehicles without diesel particulate filter (DPF), and with different DPFs (catalyzed and non-catalyzed) have been tested. Direct injection gasoline (DIG) vehicle with oxidation and NOx reduction catalysts have also been tested. Tests were performed in different chassis dynamometer test cells under New European Driving Cycle (NEDC), for both hot and cold start conditions. Exhaust gas was diluted via full flow dilution tunnels. A cyclonic classifier with 2500 nm (D_{50}) cut point was used to eliminate the coarse mode particles.

It was found that all the tested DI diesel vehicles with DPF emit about 90% of the total soot particles within initial 200 seconds of the driving cycle when driven under NEDC in cold start condition. Similar trend can be seen in some other driving cycles (Japanese JC08). Once the vehicle and the DPF system becomes warmed-up, soot emission decreases drastically. Compared to a DI diesel vehicle fitted with DPF, a DI gasoline vehicle without DPF tested in this study emits higher number of particles. In this case soot particle emission occurs continuously during the entire driving cycle rather than emitting at a particular time independent of the driving cycle.

INTRODUCTION

PMP has proposed a draft regulation to ECE for light-duty diesel (LDD) vehicle exhaust particle number measurement together with gravimetric mass measurement [7]. It is recommended that number concentration of solid particles should be measured after eliminating the volatile particles including sulfate from the exhaust gas. The draft regulation recommends a particle measurement system should consist of a pre-classifier, a heated diluter, an evaporation tube, a cold diluter, and a particle counter. Figure 1 shows the schematic of the PMP recommended particle measurement system.

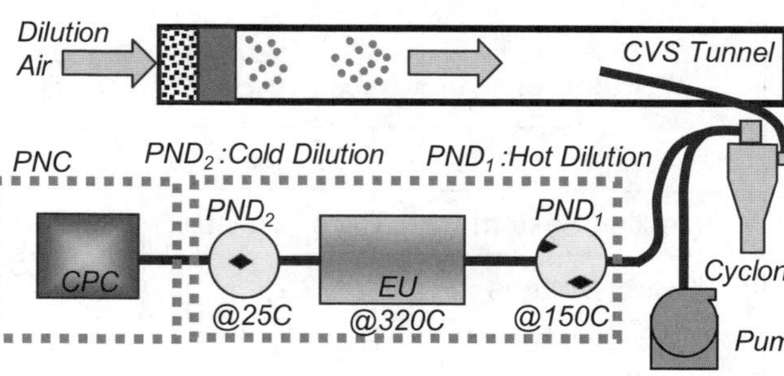

Figure 1: PMP recommended particle measurement system

The pre-classifier provides particle cut off size between 2500 and 10000 nm. The heated diluter dilutes the exhaust gas with particle-free dilution air. The dilution air is heated to above 150 °C. Dilution with high temperature air suppresses the formation of volatile particles. The evaporation tube with the average temperature of 300 to 400 °C is used to evaporate volatile particles including sulfate. By following cold dilution, concentrations of gaseous compounds are decreased to minimum in order to minimize the possibility of re-nucleation and to form particle [8, 9]. By cooling the heated sample to room temperature quickly with the cold diluter, particle loss by thermo-pherotic is minimized as well. Finally, the concentration of the solid particle is counted in the condensation particle counter (CPC), where the cutoff size is 23 nm (D_{50}). The solid particle counting system (SPCS) has been developed according to the above recommendation. The detail design and basic performances have been reported in pervious publications [1~6]. In this study the SPCS has been applied to evaluate soot emission from various vehicles with different after treatment technologies.

DESIGN OF SPCS IN BRIEF

The SPCS integrates a pre-classifier, a hot diluter (PND1), an evaporation tube (EU), a cold diluter (PND2), and a condensation particle counter (CPC) in a single system. A stainless steel cyclone is used as the pre-classifier. In order to make an easy operating instrument, data acquisition and control systems were also developed. Many functions, some of which are recommended by PMP, such as particle penetration, zero checks for CPC and the system, flow check for CPC, and daily linearity check, data saving were automated. Two PID loops were built in the software for controlling the PND1 and PND2 respectively. They control dilution ratios at set points even when there is pressure and temperature fluctuation in the CVS. Thus, the SPCS is insensitive to the sampling condition. Major parameters most of which are recommended by the PMP such as dilution ratio, particle penetration, removal efficiency have been evaluated. Most of them satisfy the PMP recommendation and sometimes showed improved results [1-6].

Therefore, the SPCS is good for measuring engine-emitted solid particle during steady-state and transient conditions. As the system developed is based on the method recommended by PMP, it can be assumed that the particles detected by the SPCS are solid.

TEST VEHICLES AND TEST METHOD

Table 1 shows the detail specification of the test vehicles. There are a few unknown parameters in the table due to respective manufacturer's confidential information. Three DI diesel vehicles with catalyzed DPF named as DDPF-1, DDPF-2, and DDPF-3 were tested. These three vehicles had almost similar engineering features. One DI diesel vehicle without DPF was also considered to observe the emission behavior. In addition, one DI gasoline vehicle was also tested. Tests were performed in various chassis dynamometer test cells. The NEDC test cycle was used both in hot and cold start conditions. All the vehicles were preconditioned well before each test in order to have comparable data. For test under cold start condition the vehicles were soaked at least five hours after preconditioning. In this study there were slight variations in soaking time between the morning test and evening test. Full dilution of exhaust was implied in all tests. Tests are limited to sampling from tunnel rather than direct sampling. All regulated gaseous components were also measured and PM was collected on filter for gravimetric analysis. However in this study only the concentration based particle number was focused.

Table 1: Vehicle Specifications

Parameter	Vehicles				
	Non-DPF	**DDPF-1**	**DDPE-2**	**DDPF-3**	**DIG**
Engine	DI Diesel	DI Diesel	DI Diesel	DI Diesel	DI Gasoline
Injection System	Common Rail	Common Rail	Common Rail	Common Rail	DI-EFI
Swept Volume	2.2 L	2.0 L	2.2 L	2.0 L	3.0 L
Max Power kW/rpm	105/4000	Unknown	105/4000	86/3500	188/6200
Max Torque N-m/rpm	340/2000	Unknown	340/2000	178/2000	314/3600
Transmission	6-Manual	6-Manual	6-Manual	5-Manual	Automatic
Vehicle Weight kg	1637	1590	1637	1430	1855
Fuel Economy km/L	14.5	Unknown	14.5	15.4	11.8
After treatment	Unknown	DPF (SiC)+ FBC	DPF + DOC	DPF + DOC	TWC + NRC
Intake	TCIC	TCIC	TCIC	TCIC	NA
Emission level	Unknown	EURO-IV	EURI-IV	JP (2000)	JP (2000)

RESULTS AND DISCUSSIONS

COMPARISON WITH PMP RECOMMENDED SYSTEM

It is already indicated that the PMP recommended a method for measuring the number of solid particles in exhaust gas. At present there is no benchmark instrument for this purpose. Therefore, to evaluate the performance, the SPCS was compared with an analogous instrument that followed the PMP recommendation and was used in the inter-laboratory correlation exercise for light duty vehicle. The correlation is summarized in Figure 2 for low emission vehicles and for high emission vehicles. The alternative system data is plotted on X-axis while the SPCS data is plotted on Y-axis. Particle emission is expressed in particle/km. The SPCS shows only about 5% higher value due to low particle loss in the system. However the linearity of both systems is very good and the R^2 value is about 0.94 [10]. Confirming excellent correlation was between the two systems.

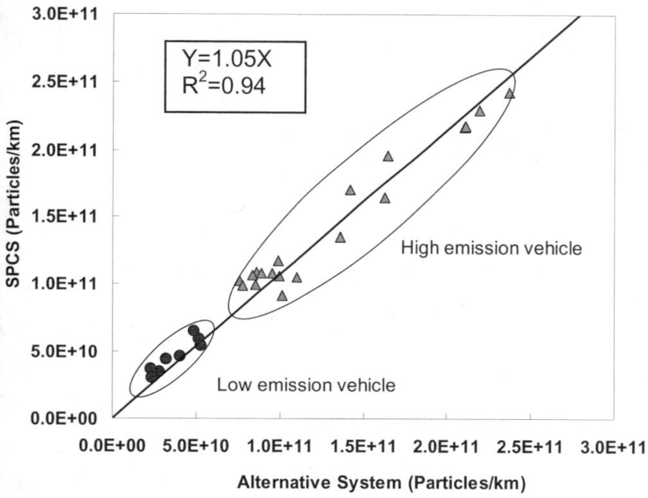

Figure 2: Correlation between SPCS and alternative system (PMP Recommended)

PARTCILE EMISSION FROM VEHICLES WITH AND WITHOUT DPF

Two DI diesel vehicles i.e. non-DPF and DDPF-1 have been considered in this purpose. These two vehicles build with engines having very similar engineering features as mentioned in Table 1. The NEDC driving cycle under hot start condition was used for each vehicle in the same test facility. Exhaust gas was measured after being diluted in a full dilution tunnel at the same dilution condition. Figure 3 represents the results obtained from test done in this case. Graph at the top of figure 3 shows the real-time solid particle emission from both vehicles. Left vertical axis corresponds concentration for vehicle without DPF (DDPF-1) while the right vertical axis corresponds the concentration for vehicle with DPF. It is clear from the figure that vast number of particles is emitted from the non-DPF vehicle. The emission behavior well correlates with the driving pattern. Huge number of particles comes out in the EUDC part of the driving cycle, and especially the emission increases significantly during acceleration. In the EUDC part of the cycle more than 3.0×10^8 particles per cm^3 were detected instantly. On the other hand, the vehicle with DPF shows very clean emission behavior. Particle emission occurs only during the first 200 seconds and during the high speed zone. However the emission level does not exceed 1.0×10^4 particles per cm^3 in diluted exhaust gas.

To understand the change in emission behavior due to DPF the "Particle Emission Rate" was calculated from the real-time particle emission. The term is defined as the ratio of the real-time cumulative particle number emission (N_{cum}) to the total particle number emission (N_{total}) during the driving cycle. It can be expressed by the following equation:

$$\text{Particle Emission Rate} = \frac{\Sigma N_{cum}}{N_{total}} \times 100 \quad \text{------------[1]}$$

Bottom side graph of figure 3 shows the particle emission rate for non-DPF and DDPF-1 vehicles. It shows that the particle emission rate for DDPF-1 vehicle remains almost constant through out the driving cycle under hot start condition. Under this condition a slight portion of fine particles those are difficult to be trapped on the filter surface are assumed to pass through the DPF due to intermittent pressure fluctuation in the exhaust pipe [11, 12]. On the other hand the particle emission rate from the non-DPF vehicle remains low in the ECE part of the driving cycle while it increases very sharply in the EUDC part where the vehicle is accelerated sharply. For non-DPF vehicle the particle emission rate is directly correlated with the engine operating condition. At high load and high speed condition, many particles are emitted and the particle emission rate increases [13]. However it is clear from the figures that the SPCS can successfully detect particle emission from vehicles with and without DPF independent of emission level.

PARTICLE EMISSION FROM VEHICLES FITTED WITH CATALYZED DPF

Emission behavior of solid particles from three vehicles; DDPF-1, DDPF-2 and DDPF-3, has been investigated. These three vehicles have very similar engineering features. Detail specifications are mentioned in Table 1. NEDC driving cycle under cold start condition was used for each vehicle but in different facilities. In addition there was some unavoidable test to test time shifting due to location of test facilities. Sometimes it caused difference in weather which has slight effect on test result even done in strictly controlled environment. Despite this the comparative tests were attempted in this study and the results are represented in Figure 4.

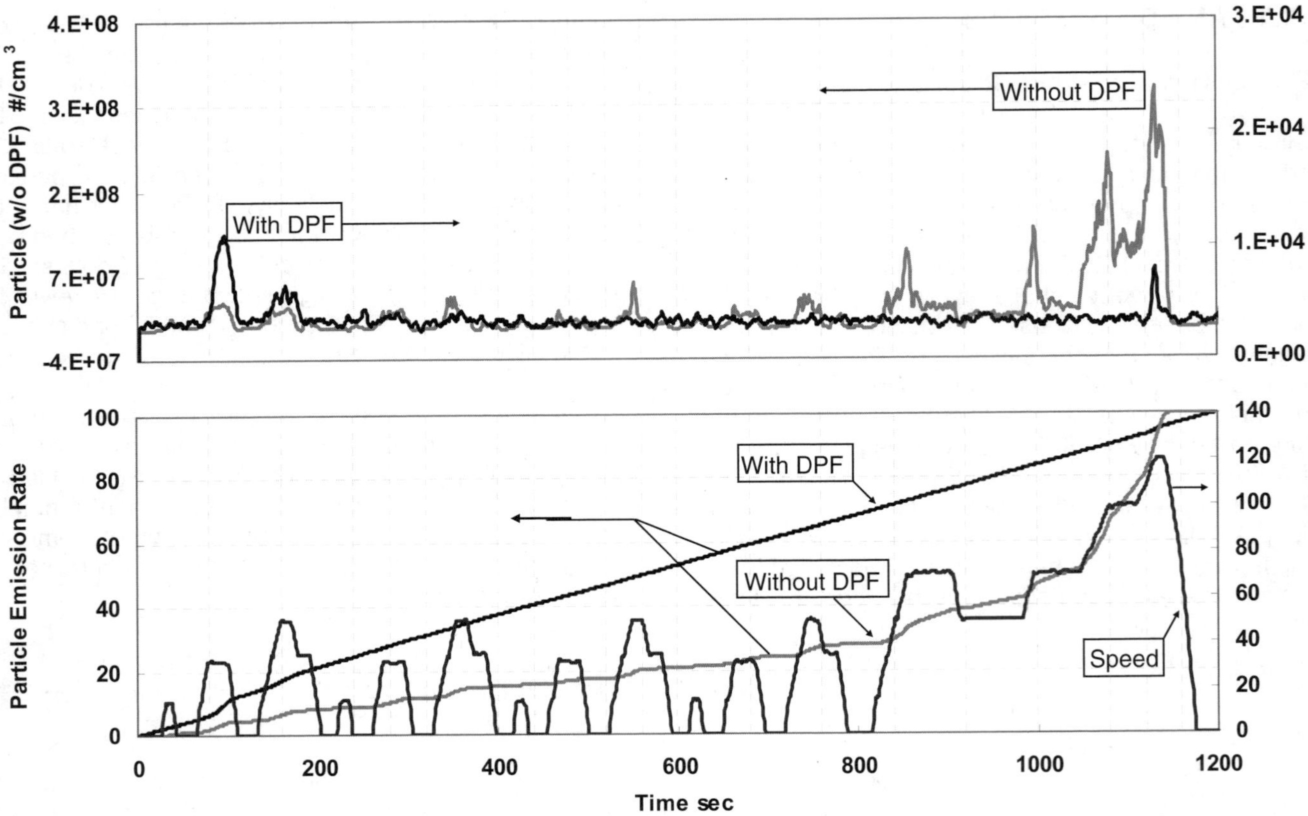

Figure 3: Particle number emission from DI diesel vehicles with and without DPF

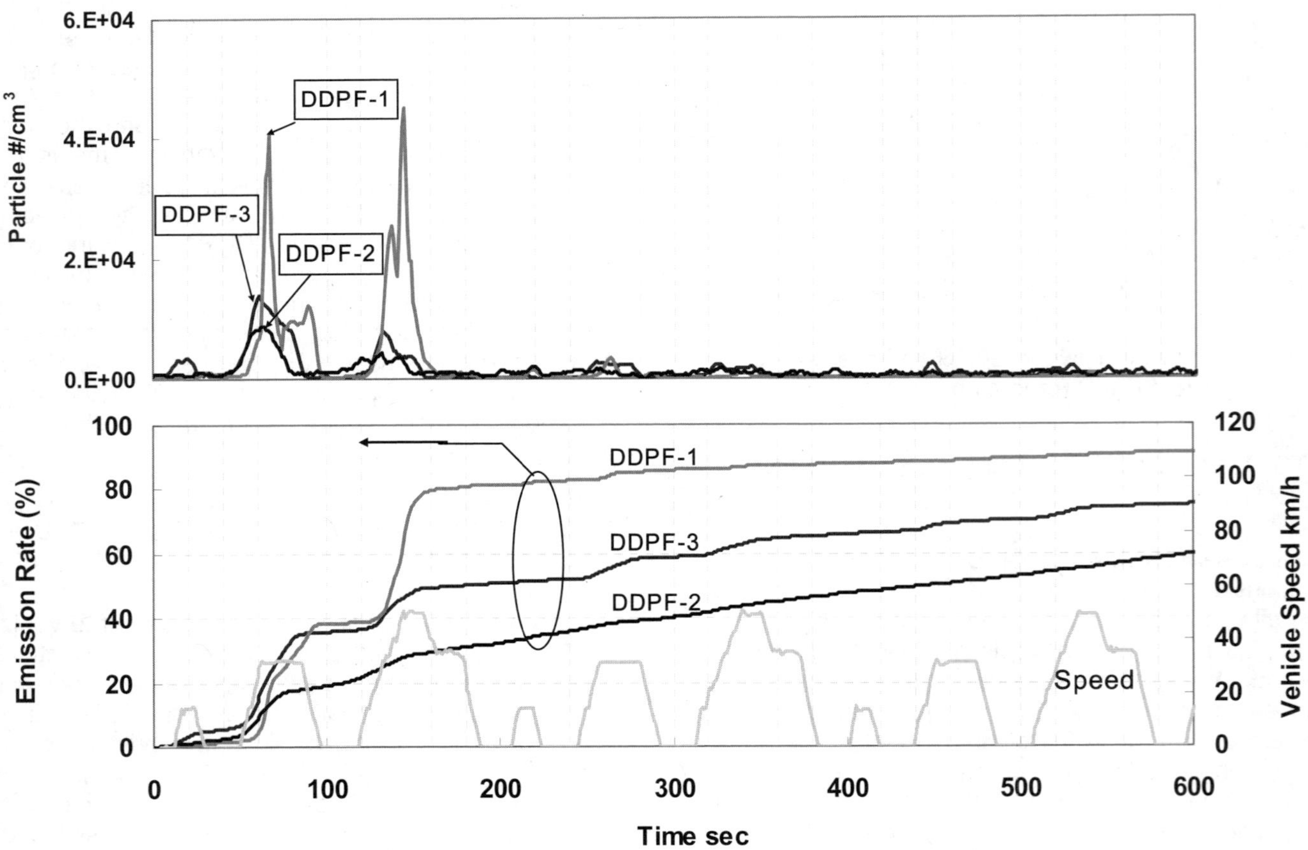

Figure 4: Particle number emission from different DI diesel vehicles with catalyzed DPF

Only first half (600 seconds) of the driving cycle is plotted in this figure rather than whole cycle because of no significant difference in the later part. Graph at the top of figure 4 represents the real-time solid particle emission from three vehicles for NEDC driving cycle under cold start condition. It shows that the DDPF-1 vehicle emits higher number of solid particles during the first 200 seconds of the driving cycle as compared with the other two vehicles. During the (EUDC part) high speed drive, the DDPF-2 vehicle shows the highest number of solid particle emission while the DDPF-1 vehicle shows the lowest. Except these two zones in driving cycles slight variation in emission was also found. There was also slight difference in the background concentration of the dilution air which had negligible effect on the total particle emission and is not discussed in this study. It was found that SPCS detects the slight difference in solid particle emission from three vehicles.

The particle emission rate calculated from the real-time emission graph is shown in the bottom side graph of figure 4. It shows that the particle emission rate for vehicle DDPF-1 is relatively higher than the other two vehicles. Soaking time of the DDPF-1 vehicle might be longer than the other two vehicles. Therefore it emits higher particles during the early stage (ECE part). When the engine as well as the DPF becomes warm-up sufficiently, the particle emission rate decreased significantly and shows similar emission behavior. There may be slight difference in the filtration efficiency of DPFs fitted in each vehicle though it was not confirmed.

PARTICLE EMISSION BEHAVIOR FROM DIESEL VEHICLE DURING AND AFTER REGENERATION OF DPF

Particle emission behavior during regeneration and in successive tests after the regeneration has been investigated by using the DDPF-2 vehicle. The vehicle was driven under NEDC driving cycle in hot start condition. One test was done before regeneration and then regeneration was initiated externally just in the EUDC mode. Just after regeneration several successive tests were also performed under hot start condition. The results are shown in Figure 5.

Graph at the top of figure 5 shows the real-time particle emission behavior during regeneration of the DPF. It shows that huge number of particles is emitted during regeneration in the EUDC part of the driving cycle. Graph at the middle of figure 5 shows a comparison of the real-time particle emission behavior for three different tests; before regeneration, first test just after regeneration and second test after regeneration. Before regeneration very low number of particles was detected. In the first test immediately after the regeneration huge number of particle emission was observed. The particle concentration is about 2 orders higher than normal DPF condition. However the concentration decreases gradually in the successive test cycles after regeneration.

Bottom side graph of figure 5 shows the particle emission rate calculated from the real-time particle emission during regeneration, before regeneration and after regeneration. Higher particle emission rate was observed within the EUDC part of the driving cycle during regeneration while the emission rate gradually shifts to the ECE part of the driving cycle with successive tests after regeneration.

Average particle concentration measured by the SPCS and estimated filtration efficiency are summarized in Figure 6. The filtration efficiency was estimated by assuming the non-DPF emission as the maximum. It shows that from the fourth test after regeneration excellent repeatability in number counting can be achieved. Filtration efficiency reaches to maximum after fourth test and the DPF itself assumed to be stable [16]. Upon regeneration, filter becomes clean and filtration performances decreases. It is reported that PM is first collected in the pores of the filter wall [17]. After a transition phase, the PM builds up as a discrete layer, or soot cake, on the inlet channel walls, therefore, filtration efficiency increases. During regeneration particle number emission is only about 20% of the non-DPF particle number emission. However this emission is about 200 times of the particle number emission in case of a stable DPF operation with sufficient loaded soot.

COMPARISON OF PARTICLE EMISSION FROM DIFFERENT VEHCLES

Comparison of a DPF fitted diesel vehicle with a DI gasoline vehicle has been attempted in this case. The DDPF-1 and DIG vehicles were used in this purpose. The specifications are mentioned in Table 1. Both vehicles were driven under NEDC driving cycle in cold start condition in the same test facility. Results are summarized in Figure 7. Graph at the top of figure 7 shows the real-time particle emission from DDPF-1 and DIG vehicles. Concentration of particles emitting from DDPF-1 vehicle is very high in the early part of the driving cycle while the DIG vehicle emits consistently higher concentration of particles through out the driving cycle. The DDPF-1 vehicle emits significantly lower total number of particles in comparison with the DIG vehicle. However it is clear that the SPCS can be used to measure the very low particle emissions from diesel with catalyzed DPF and relatively high particle emission from gasoline vehicles.

Particle emission rate calculated from this real-time emission is presented in the bottom side graph of figure 7. It shows that DDPF-1 vehicle exhibits higher particle emission rate during the first 200 seconds of the driving cycle. However the DIG vehicle shows relatively constant emission rate. Therefore DDPF-1 vehicle should be considered for first 200 seconds in order to improve it's particle emission quality. On the other hand, DIG vehicle should be considered for overall driving cycle for further improvement.

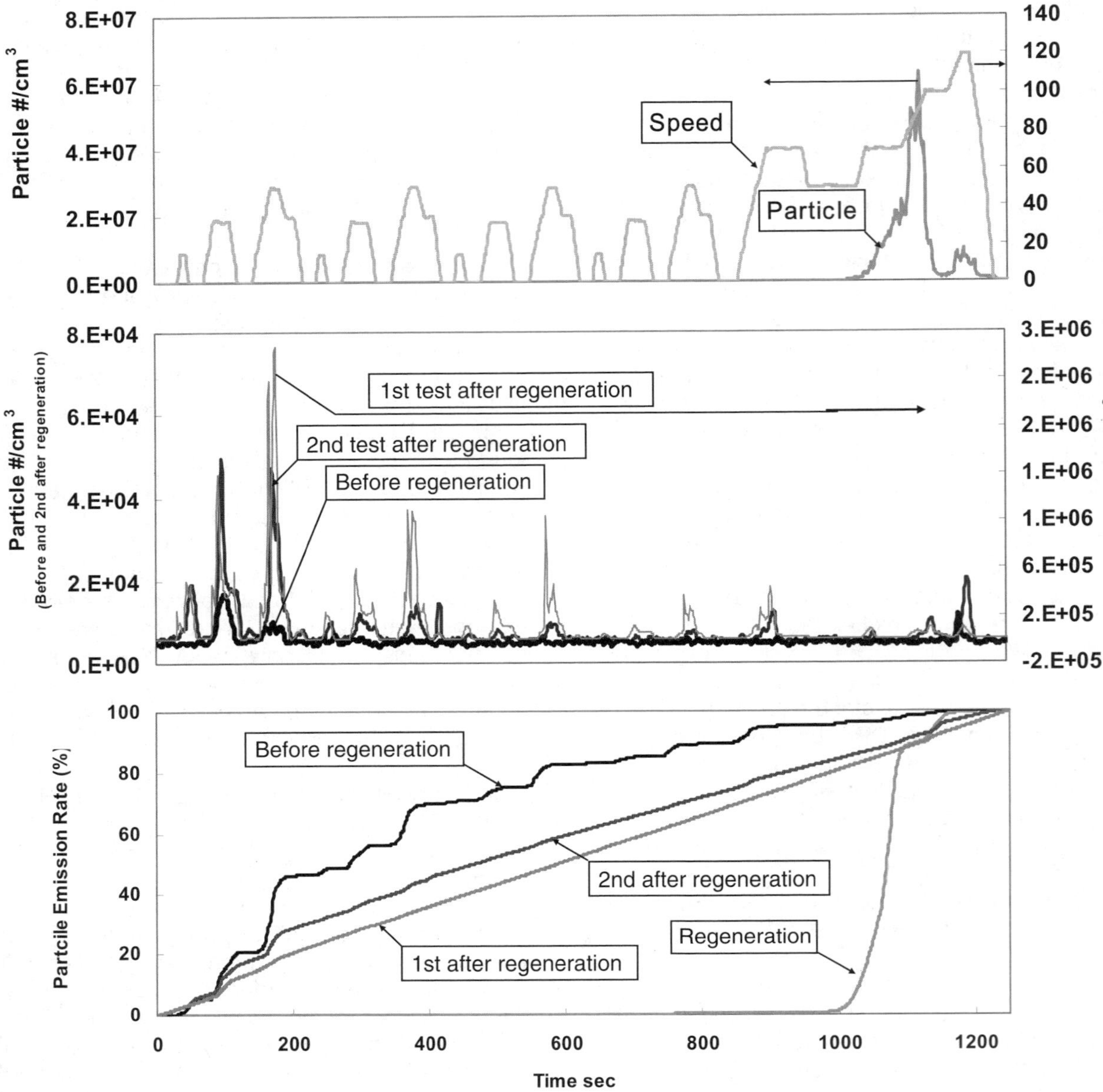

Figure 5: Particle emission behavior from DI diesel vehicle before, during and after regeneration

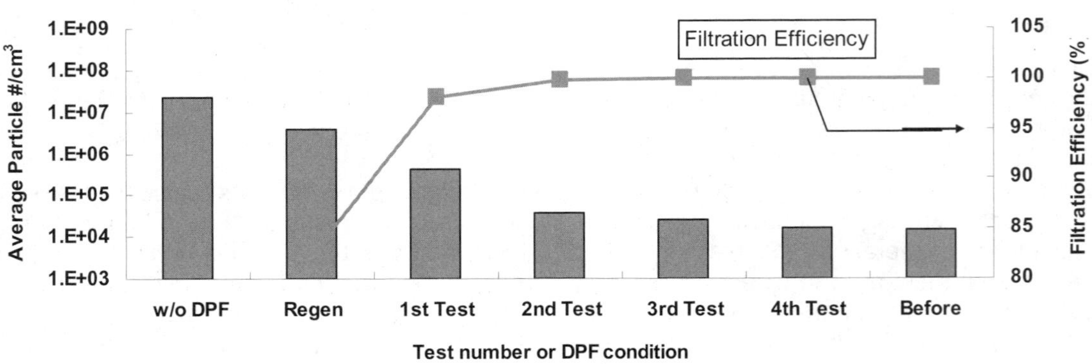

Figure 6: Average particle emission and filtration efficiency

Figure 8 shows the overall comparison of particle emission in number per kilometer from all the test vehicles used in this study. DDPF-1, DDPF-2 and DDPF-3 vehicles show almost similar particle number emission and the value is lower than 10^{11} particles/km; a value that received attraction from many organizations recently. The DIG vehicles emit higher number of particles and may need further consideration. The Non-DPF vehicle emits the highest number of particles as expected.

CONCLUSION

A Solid Particle Counting System has been developed. The system is used for measuring particles emissions from various vehicles with different after treatment technologies. Though some limited number of vehicles was tested the following results can be obtained summarized:

1. Significant particle are emitted from vehicle without DPF. DI diesel vehicle with catalyzed DPF shows very clean emission behavior. In this case most of the particle emission occurs only during the first 200 seconds and slight emission occurs in the high speed zone in cold start condition driven under NEDC cycle.

2. High number of particles is emitted during regeneration of DPF as compared to the normal DPF condition in a DI diesel engine. However this is approximately 20% comparing to the particle number emission from a non DPF diesel vehicle. Just after regeneration particle emission continues but concentration decreases with the successive tests. Excellent repeatability in number counting may be achieved from 4th test after regeneration.

3. The SPCS can measure very low level solid particle emission from diesel vehicles with DPF even the exhaust gas is diluted sufficiently and relatively high level emission from vehicles without DPF as well as from DI gasoline vehicle.

ACKNOWLEDGMENTS

Many of the results presented in this study have been obtained from tests done in different laboratories. The authors would like to acknowledge the kind permission of using test facilities of the Environment Research Division, National Traffic Safety and Environment Laboratory, Tokyo, Japan, Honda R&D Co. Ltd., Tochigi, Japan, and Joint Research Center (JRC), Ispra, Italy.

REFERENCES

1. Qiang Wei, Asano Ichiro, Masayuki Adachi, Rahman M. Montajir, Takeshi Kusaka, and Yuichi Goto, Real-time measuring system for engine exhaust solid particle number emission – Performance and Vehicle tests, SAE Paper No. 2006-01-0864
2. Qiang Wei, Karl Oestergaard, Scott Porter, Asano Ichiro, Masayuki Adachi, Rahman M. Montajir, and Yuichi Goto, Real-time measuring system for engine exhaust solid particle number emission – Design and Performance, SAE Paper No. 2006-01-0865
3. Qiangg Wei and et. al., Real-Time Exhaust Solid Particle Measurement with a Prototype Particle Counting System. ETH 2005
4. Ichiro Asano, and et. al., Development of a Solid Particle Counting System for Engine Exhaust, JSAE paper No. 20065044, Japanese.
5. Rahman Montajir, and et. al., Performance of HORIBA-SPCS in the PMP LDD ILCE, ETH 2006.
6. Rahman Montajir and et. al., Design and Performnace of a Solid Particle Counting System for Real-Time Measurement of Engine Exhaust, Paper No. 2006P175, FISITA.
7. Conclusions on improving particulate mass measurement procedures and new particle number measurement procedures relative to the requirements of the 05 series of amendments to regulation No. 83, GRPE-48-11-Rev. 1, 2004
8. Markus Kasper, The number concentration of non-volatile particles – design study for an instrument according to the PMP recommendations, SAE 2004-01-0960
9. Montajir, R., Kawai, T., Goto, Y., and Odaka, M., Thermal conditioning of exhaust gas: potential for stabilizing diesel nano-particles, SAE 2005-01-0187, 2005
10. PMP Light Duty Inter-Laboratory Correlation Exercise Final Results Summery, Working paper No. GRPE-PMP-17-3, Sept., 2006
11. Johnson, J. H. et al. "A Review of Diesel Particulate Control Technology and Emissions Effects – 1992 Horning Memorial Award Lecture" SAE Technical Paper 940233, 1994.
12. Konstandopoulos, A.G. et al., "Fundamental Studies of Diesel Particulate Filters: Transient Loading, Regeneration, and Aging", SAE Technical Paper 2000-01-1016, 2000.
13. Schmidt, N., et al., "Detailed Diesel Exhaust Particulate Characterization and DPF Regeneration Behavior Measurements for Two Different Regeneration Systems", To be appeared at SAE 2007.
14. Maamouri, M. et al. "Real-Time Exhaust PM Mass Measurements for DPF Development Studies" Emission Control Conference, Dresden, 18-19 May 2006.
15. Konstandopoulos, A.G., "Flow Resistance Descriptors for Diesel Particulate Filters: Definitions, Measurements and Testing" SAE Technical Paper 2003-01-0846.
16. Konstandopoulos, A.G. and Johnson, J.H., "Wall-Flow Diesel Particulate Filters-Their Pressure Drop and Collection Efficiency", SAE Technical Paper 890405, 1989.
17. Ekathai Wirojsakunchai, Eric Schroeder, Chris Kolodziej, Dave Foster, "Detailed Diesel Exhaust

Particulate Characterization and Real-Time DPF Filtration Efficiency Measurements during PM Filling Process" to be appeared in the 2007 SAE World Congress.

CONTACT

Any question and comment should be directed to the following address. Rahman Montajir, Automotive Test Systems Division, HORIBA Ltd., 2-Miyanohigashi, Kisshoin, Minami-ku, Kyoto, 601-8510, Japan. E-mail: montajir.rahman@horiba.com.

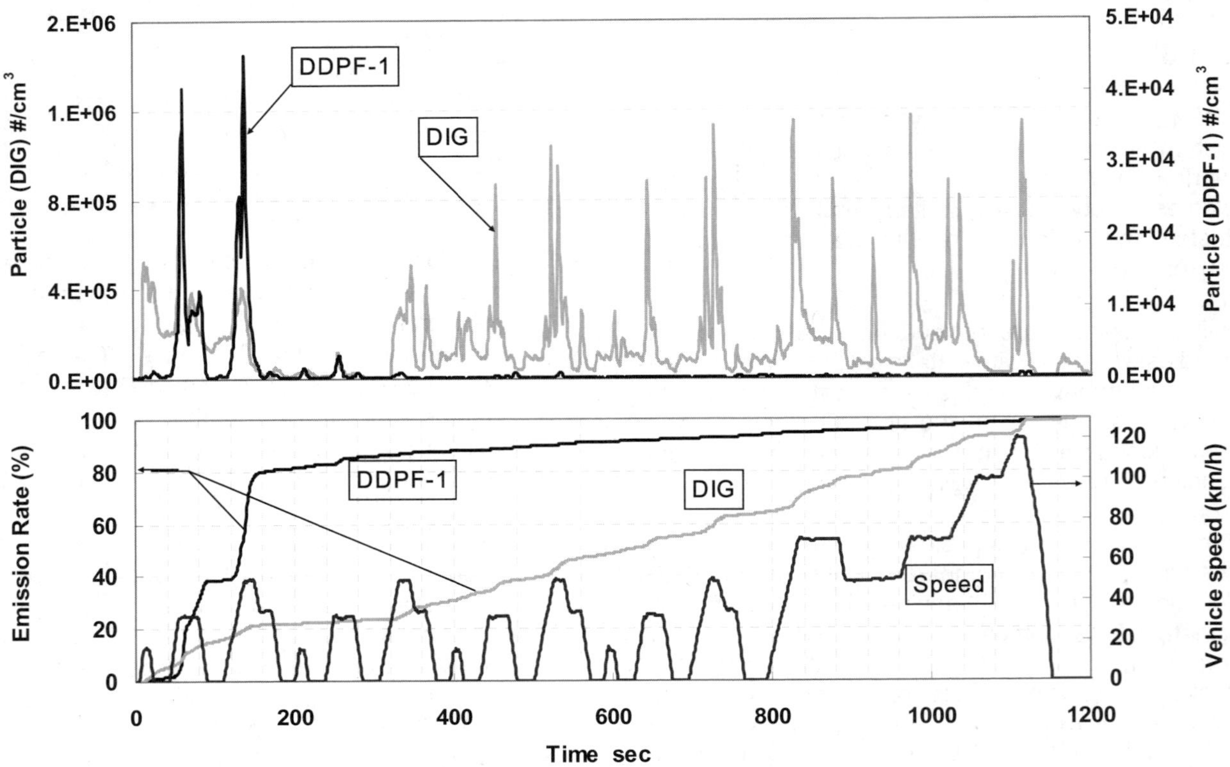

Figure 7: Comparison of particle number emission from different vehicles

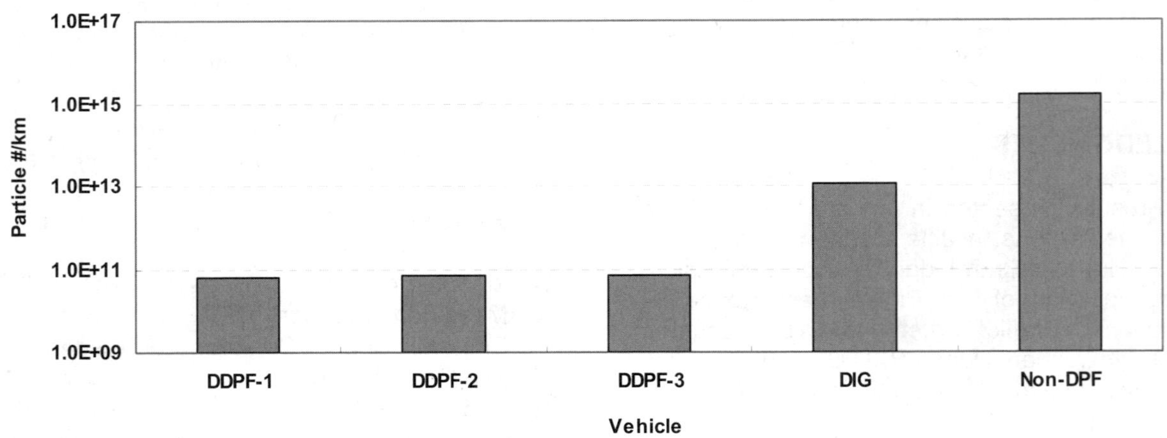

Figure 8: Overall comparison of particle number emission from different vehicles and after treatment technologies

2006-01-0878

DPR with Empirical Formula to Improve Active Regeneration of a PM Filter

Toshikatsu Muramatsu, Tetsuichi Kominami, Jinichi Minamikawa and Munetomo Matsunami
Hino Motors, Ltd.

Junji Komatsu
Hitachi Metals,Ltd.

Jun Ozawa
Tokyo Roki Co., Ltd.

ABSTRACT

Diesel Particulate active Reduction system (DPR) is a system that traps particulate matter in diesel exhaust gas with a particulate filter and actively regenerates the filter when PM accumulates to a specific level. In 2003, DPR was installed on Hino's light-, medium-, and heavy-duty diesel engines, and about 50,000 units of these DPR-equipped diesel engines are currently on the market. This paper reports results of further progress made on optimization of the active regeneration function of DPR.

The goal of successful development of DPR is to optimally control the system under various engine-operating conditions to regenerate the filter without producing abnormal combustion of PM and to minimize the amount of unburned PM to keep the filter from clogging.

To improve the control of DPR, the combustion phenomena of PM collecting on the filter were studied through visualization, and the factors influencing combustion were determined. Based on quantitative information obtained in the captured images of combustion, an empirical formula was developed to calculate the combustion rate of particulate accumulating on the filter. The formula was verified through engine tests and used to optimize the control of DPR.

INTRODUCTION

Compared to a gasoline engine, a diesel engine generates less carbon dioxide (CO_2) emission. As powerplants that contribute to reduced global warming, diesel engines have appeared in recent trends indicating that they are being popularized not only in Europe, but also worldwide. On the other hand, worldwide demand for reduced nitrogen oxides (NOx) and particulate matter (PM) emissions from diesel engines has increased.

In Japan, public perception of diesel engines is still negative because of the persistent impression that diesel exhaust gas is dirty and not environmentally friendly. Thus, it is important for diesel engine manufacturers to continue their efforts to improve diesel exhaust gas emissions regardless of the mandatory legislative requirements.

Hino Motors developed the Diesel Particulate active Reduction system™ (DPR) to enable reduction of both PM mass and number in exhaust gas under various engine-operating conditions. Major functions of DPR are filtration of PM with a highly efficient wall-flow particulate filter and precise control of active regeneration with a common-rail fuel injection system [1].

Since the introduction of DPR-equipped diesel engines in the Japanese market in 2003, both medium- and heavy-duty engines have accumulated 400,000 km. Figure 1 shows the production volume of vehicles equipped with DPR for light-, medium-, and heavy-duty applications.

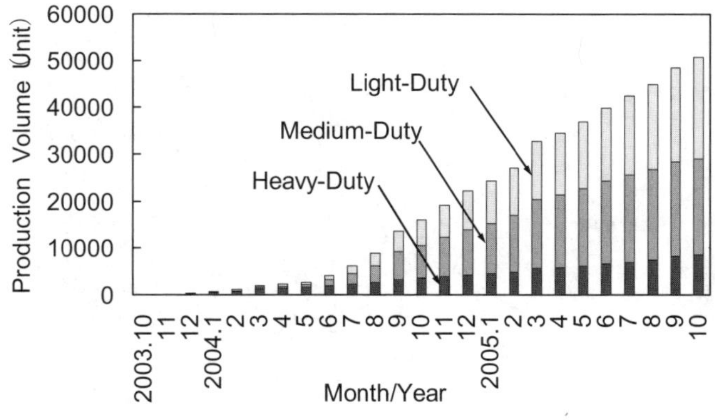

Figure 1. Production Volume of Hino Vehicles Equipped with DPR

Recently, the health effects of PM, particularly in nanometer scale, have been discussed [2]. Therefore, it is believed that PM reduction technology will become the important element of future diesel engine development. Combined with NOx reduction technology, PM reduction technology will need to be more advanced in order to comply with the stringent worldwide emissions regulations scheduled for 2009 and 2010. Under these circumstances, an efficient filter such as one used for DPR that can reduce both PM mass and number has the potential to become a key technology in future diesel engine development.

Yet, methodology for more efficient filtration of PM and more reliable regeneration of the particulate filter has not been established. This paper reports the results of investigations conducted to find ways to achieve optimal combustion of particulate collecting on the filter. The combustion rate of particulate accumulating on the filter was analyzed according to experimental results. Based on these results, an empirical formula was developed to precisely predict the amount of particulate accumulating on the filter.

THE KEY ELEMENTS OF DPR

ENGINE SYSTEM

Figure 2 shows a diesel engine equipped with DPR. A multiple fuel injection system, which is part of DPR, is used for active regeneration of the particulate filter. After the main fuel injection is completed, additional fuel is injected at specific timings before the exhaust valves are opened, as shown in Figure 3, depending on the level of exhaust gas temperature.

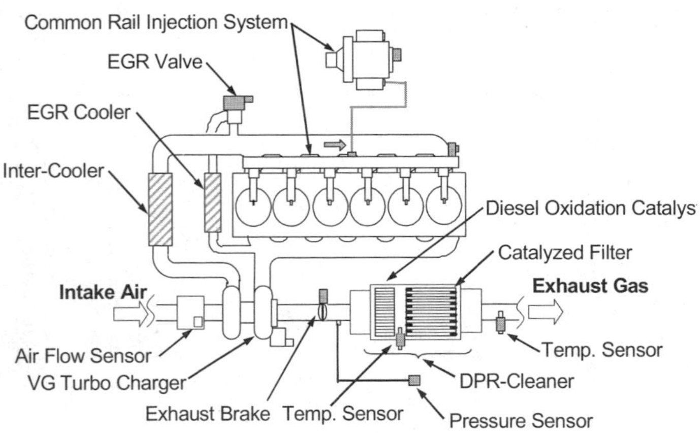

Figure 2. Hino Diesel Engine Equipped with DPR

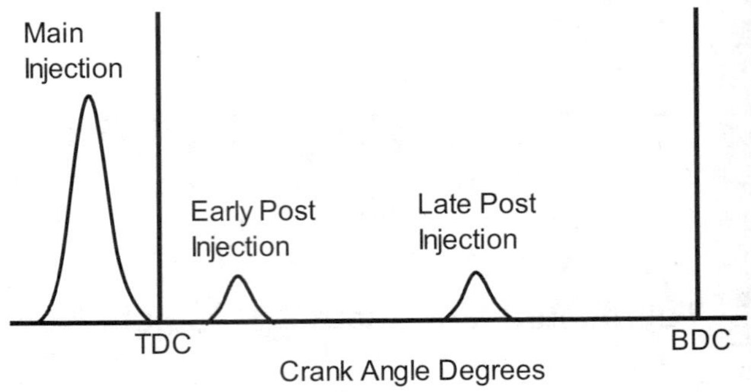

Figure 3. Fuel Injection Schedule Including Early and Late Post Injections

Late post injection takes place immediately before the exhaust valves are opened to supply unburned fuel to an oxidation catalyst installed in front of the particulate filter. However, when filter temperature is low, early post injection is used to inject a small amount of fuel immediately after top dead center. This approach preheats the filter before late post injection can be used. When the exhaust gas temperature needs to be increased in a short time, the amount of late post injection is increased while the filter temperature is rising. The fuel injection is precisely controlled with the feedback information of the filter temperature.

Thus, temperature of the particulate filter can be maintained within the desired range. The system regenerates the filter by actively burning particulate collected on the filter and does not rely on passive regeneration that is based on the chemical reaction of nitrogen dioxide (NO_2) with carbon contained in exhaust gas.

DPR CLEANER

The DPR cleaner shown in Figure 4 consists of a catalyzed wall-flow cordierite particulate filter, an oxidation catalyst, and silencers. The oxidation catalyst installed in front of the particulate filter raises the exhaust gas temperature when DPR performs to actively regenerate the filter.

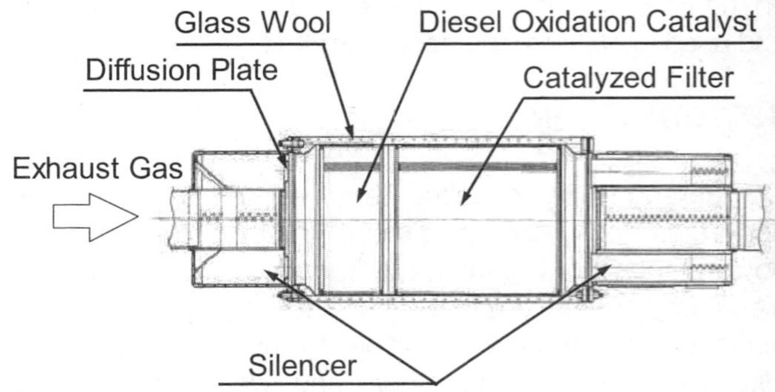

Figure 4. DPR Cleaner

ACTIVE REGENERATION

Figure 5 shows a flow chart of the particulate regeneration process. While the vehicle is driven, an engine electronic control unit (ECU) evaluates the amount of particulate collecting on the particulate filter. As soon as the computer determines that the amount of accumulated particulate has reached the level at which the filter would need regeneration, regeneration is activated to start increasing the filter temperature.

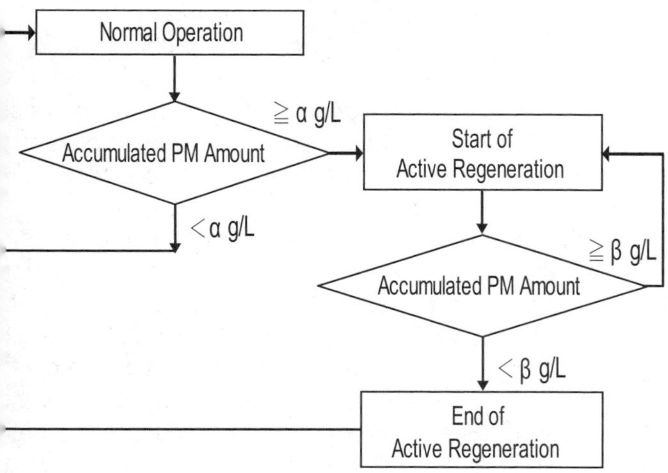

Figure 5. The Particulate Filter Regeneration Process

With the engine ECU, the amount of particulate burned on the filter is calculated based on the measured exhaust gas temperature, and the level of regeneration performed on the filter is evaluated. As soon as the computer has determined complete combustion of particulate on the filter, engine operation returns to normal.

Calculation of the amount of particulate burned on the filter requires the combustion rate of particulate, which is an extremely important element of the regeneration system. If the calculated combustion rate is not correct, all the particulate collected on the filter may not be completely burned, and this remaining particulate may clog the filter. Also, late post fuel injection may be performed unnecessarily, increasing fuel consumption.

To raise the filter temperature in a short time and stabilize it with acceptable precision, late post fuel injection is controlled based on feedback information. Thus, the proper amount of fuel is injected as needed. Figure 6 shows an example of the filter temperature being maintained at the proper level by the fuel injection control.

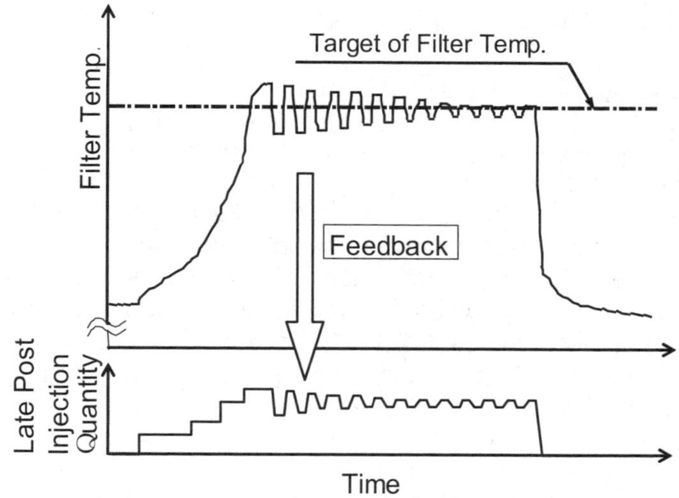

Figure 6. Fuel Injection Control to Raise and Maintain the Particulate Filter Temperature

DEVELOPMENT OF AN EMPIRICAL FORMULA TO OBTAIN THE PM COMBUSTION-RATE CONSTANT

As described earlier, the combustion rate of particulate collected on a particulate filter must be accurately predicted in order to completely burn particulate on the filter with minimal use of late post injection while simultaneously maintaining healthy operation of the particulate filter. The authors developed an empirical formula to increase the accuracy of predicting the combustion rate of particulate collected on the filter. The calculated value is called the "PM combustion rate constant" in this report.

In considering this calculation approach, it is important first to understand the self-ignition characteristics of particulate collected on a particulate filter. Engine tests were conducted under various conditions to measure the self-ignition temperature of particulate collected on the filter. Subsequently, the combustion characteristics of particulate were investigated through visualization of particulate burned on the specimens of the filter used in a test engine, and various effects such as filter temperature, particulate amount, and the oxygen flow rate were evaluated to determine sensitivities. With the information obtained through these experiments, the empirical formula was developed.

SELF-IGNITION CHARACTERISTICS OF PM COLLECTED ON A FILTER

Specifications of the Test Engine

A 13-liter, inline-six, heavy-duty diesel engine equipped with DPR was employed for the engine tests. A specimen of the particulate filter actually used on the test engine was used for the laboratory experiments. Table 1 lists the specifications of the test engine. The engine was operated on Japan Industrial Standard (JIS) No. 2 diesel fuel and SAE 10W30 engine lubricating oil. Table 2 lists the properties of both diesel fuel and engine lubricating oil.

Table 1. Engine Specifications

Displacement	12.913 liters
Bore x Stroke	137 x 146 mm
Cylinder Arrangement	Inline six
Combustion Type	Direct Injection
Aspiration	Turbocharger with an inter-cooler
Fuel Injection	Common-rail system

Table 2. Fuel and Engine Lubricating Oil Properties

JIS No. 2 Diesel Fuel	Density at 15℃	830.7 kg/m^3
	Temp. at 90% distillation	335.0℃
	Cetane Number	58.2
	Sulfur Content	34 wt.ppm
Engine Oil	Grade	SAE 10W30
	Sulfated Ash	0.95%

Table 3 lists specifications of both the oxidation catalyst and the catalyzed particulate filter used for the DPR cleaner of the test engine.

Table 3. Specifications of Both the Oxidation Catalyst and Catalyzed PM Filter of the DPR Cleaner

	Type	Catalyst	Substrate	Volume (Liters)
Oxidation Catalyst	Straight flow	Platinum	Cordierite	6.0
Catalyzed PM Filter	Wall flow	Platinum	Cordierite	17.0

Self-Ignition Temperature of PM Collected on a Filter

According to research results found in the literature [3], the oxidation reaction of carbon is described as follows:

$$C + O_2 \rightarrow CO_2$$

$$m_c = P_\infty (O_2) / (1 / k_d + 1 / k_s)$$

$$k_s = F_s \exp(-E_s / R T_s) \qquad k_d = 4 M_c D / (d R T_m)$$

Where m_c = Reaction rate
k_s = Constant of the surface reaction rate
k_d = Constant of the diffusion rate
$P_\infty (O_2)$ = Partial pressure of oxygen
F_s = Frequency factor of the reaction rate
E_s = Activation energy
R = Gas constant
T_s = Surface temperature of carbon
M_c = Atomic weight of carbon
D = Diffusion constant
d = Diameter of a carbon particle
T_m = Temperature of gas phase

This theoretical expression of the oxidation process involves numerous variables such as the diameter of a carbon particle, and the formulae are quite complex.

To obtain the PM combustion rate constant, the authors believed that it was important to know the self-ignition temperature of particulate collecting on the particulate filter. Using measured self-ignition temperature, it would be possible to determine whether or not particulate could be burned easily. Therefore, particulate collected on the filter tested in the engine was used to measure the self-ignition temperature with a differential thermal analyzer (DTA).

The engine was operated under two different driving patterns as shown in Figure 7: Pattern A represents highway driving conditions and Pattern B represents city driving conditions. The engine tests under these driving conditions were repeated to collect particulate on the filter.

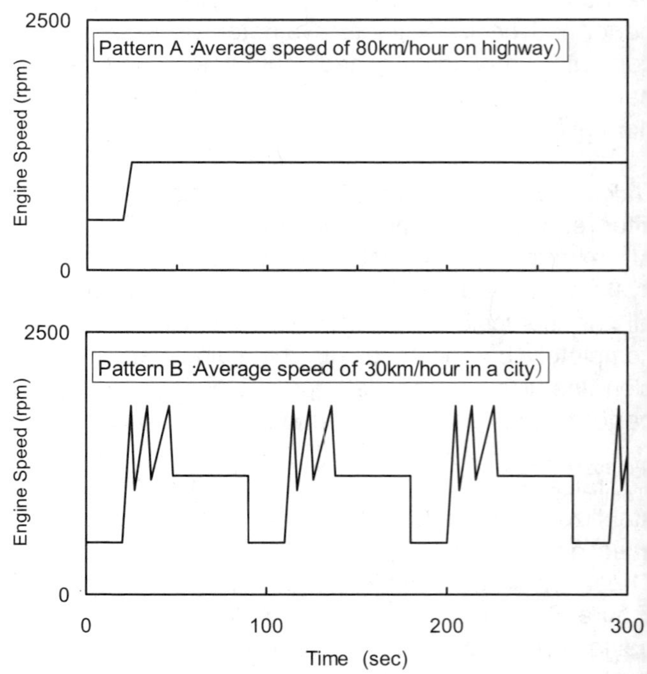

Figure 7. Two Driving Patterns Used for the Engine Tests

Temperature measurements with DTA were performed to determine the self-ignition temperature of particulate collected on the filter under Pattern A driving conditions. As shown in Figure 8, the temperature at which the DTA's signal increased was defined as the self-ignition temperature of the particulate. According to this result, the self-ignition temperature was about 560°C.

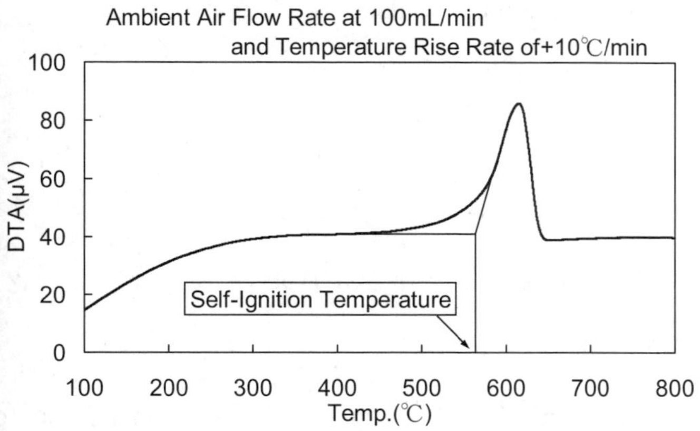

Figure 8. A Method to Obtain the Self-Ignition Temperature of Particulate on a Filter Using the DTA's Signals

For particulate collected on the filter under Pattern B driving conditions, the self-ignition temperature was a little lower than that obtained for Pattern A, as shown in Figure 9. Thus, the self-ignition temperature of particulate collected on the filter can be different depending on the vehicle driving pattern.

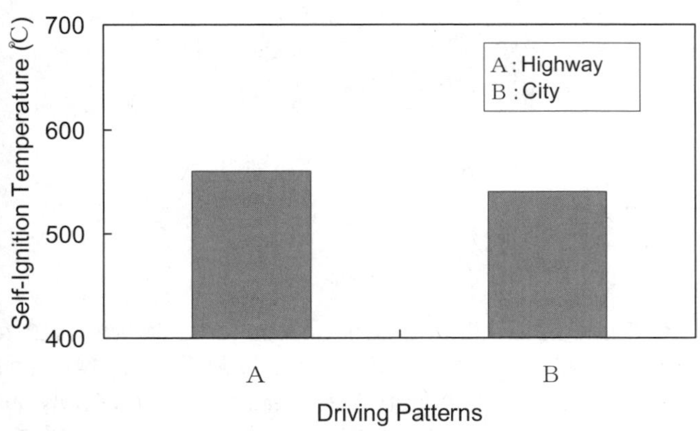

Figure 9. The Effect of Driving Pattern on the Self-Ignition Temperature of Particulate Collected on a Particulate Filter

Parametric tests were conducted to evaluate the sensitivity of the self-ignition temperature of particulate to various parameters. DTA was performed on particulate collected on the filter used for the engine operating under Pattern A driving conditions. The test parameters were (1) increased fuel injection pressure by 20 MPa, (2) fuel injection timing retarded by 4°crank angle (CA), (3) the EGR rate decreased by 8 percent, and (4) doubled space-velocity (SV). Note that the SV value was increased by raising the engine speed.

As shown in Figure 10, the self-ignition temperature of particulate decreased when fuel injection pressure was increased and when the EGR rate was decreased. However, the self-ignition temperature was somewhat insensitive to both fuel injection timing and the SV value. Thus, the self-ignition temperature of particulate varies depending on changes in both fuel injection pressure and the EGR rate. Namely, these parameters affect the combustion rate of particulate collected on the filter.

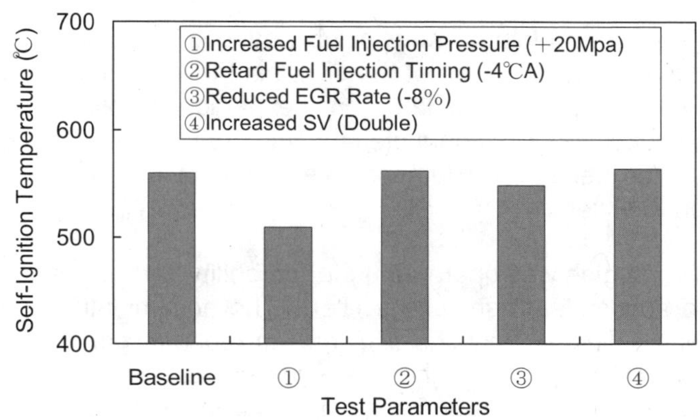

Figure 10 . The Effects of Various Parameters on the Self-Ignition Temperature of Particulate collected on a Particulate Filter

According to Vander Wal et al. [4], the soot oxidation rate varies depending on the nanostructure of soot particles. Therefore, the authors observed soot particles in detail using a transmission electron microscope (TEM) and evaluated the effect of a higher fuel injection pressure. When the fuel injection pressure was increased, the pattern of some lines near the outer skin of the soot particle was different from that of the baseline, as indicated by an arrow in Figure 11, although the pattern of lines observed in the core of the soot particle was similar. It is currently unknown whether or not the above difference made on the soot particle by the higher fuel injection pressure affects the characteristics of the self-ignition temperature. Thus, further investigation is necessary for better understanding of the above phenomena.

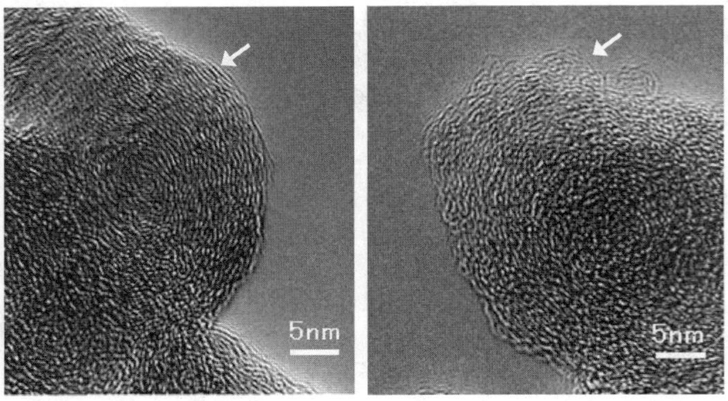

Baseline With a Higher Fuel Injection Pressure

Figure 11. The Effect of a Higher Fuel Injection Pressure on the Nanostructure of Soot Observed in TEM Images

The ash content of engine lubricating oil is emitted in exhaust gas because of the oil consumed in the engine, and this ash accumulates on the particulate filter. An investigation was conducted to evaluate the effect of ash on the self-ignition temperature of particulate collected on the filter.

The engine was operated under durability test conditions to accumulate ash on the filter. The accumulated ash was collected from the filter and mixed with particulate for evaluation of the self-ignition temperature of particulate. The major component of the ash was calcium sulfate ($CaSO_4$). One-half part ash was mixed with one part of particulate for one test, and ash was mixed with an equal part of particulate for the other test. Results of these evaluations demonstrated that the effect of ash on the self-ignition temperature of particulate was very small, as shown in Figure 12.

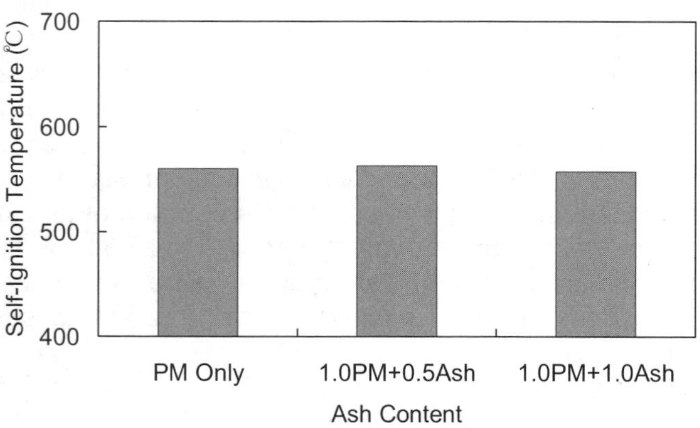

Figure 12. The Effect of Ash Content in Particulate on the Self-Ignition Temperature of Particulate Collected on a Particulate Filter

According to these test results, vehicle driving patterns and engine parameters are very important factors that should be considered for obtaining the PM combustion rate constant. The engine parameters should be those already optimized for the engine and not those still under development. As far as the vehicle-driving pattern is concerned, it is necessary to understand the actual vehicle driving pattern in the field clearly prior to selection of the pattern for developing the PM combustion rate constant. For this project, Pattern A driving conditions were used for the analysis.

VISUALIZATION OF PM COMBUSTION

Visualization Technique

The test rig shown in Figure 13 was constructed to observe the combustion of particulate collected on the particulate filter with a provision for easy alteration of the test condition. The rig was equipped with a heating device that enabled various temperature settings and a digital microscope to visualize the combustion process through a quartz window.

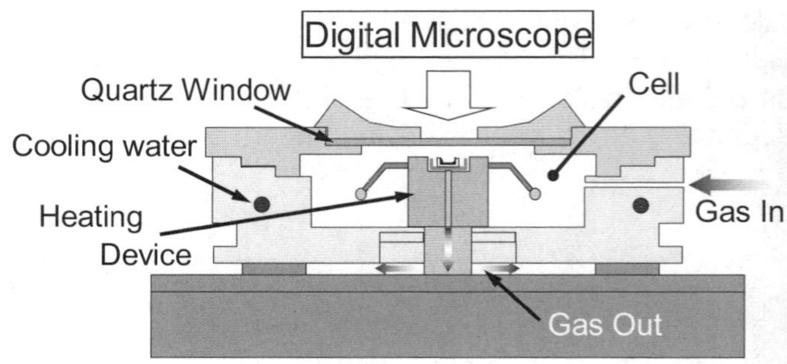

Figure 13. Test Rig to Observe PM Combustion

The particulate filter tested on the engine was cut to make a test specimen, as illustrated in Figure 14. This specimen was placed on the heating device, shown in Figure 15, which was installed in the cell. An ambient gas such as inert gas or oxygen containing gas was run through the cell, and the heating device raised the temperature of the test specimen. The combustion process of particulate on the test specimen was observed with a digital microscope.

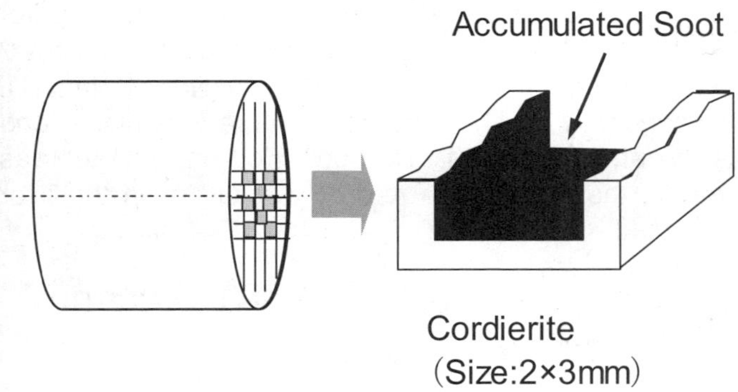

Figure 14. Test Specimen of a Particulate Filter Used in an Engine

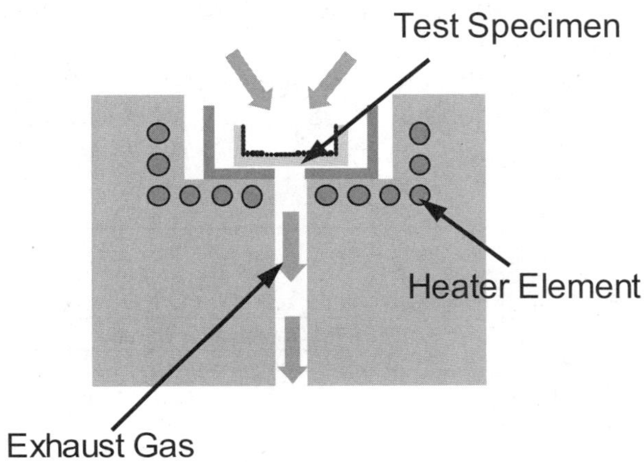

Figure 15. The Electric Heater

Visualization Results of PM Combustion

Figure 16 shows the images of particulate collected on the test specimens (the dark area indicates particulate). Two test specimens are shown in the figure: one that collected particulate at specific particulate accumulation of 3 g/liter (particulate accumulation in weight per unit volume of a filter) and the other that collected particulate at 7 g/liter. The temperatures of the specimens were set at 500, 600, and 700°C. The cell was initially filled with nitrogen so it would not burn particulate while the temperature was set at either of the above levels. As the temperature increased to the target value, ambient gas was switched from nitrogen to gas containing 10 percent oxygen to start combustion of particulate. This gas with added oxygen flowed at 4 milliliters/minute, which assumes the SV value of 80,000 hour^{-1}.

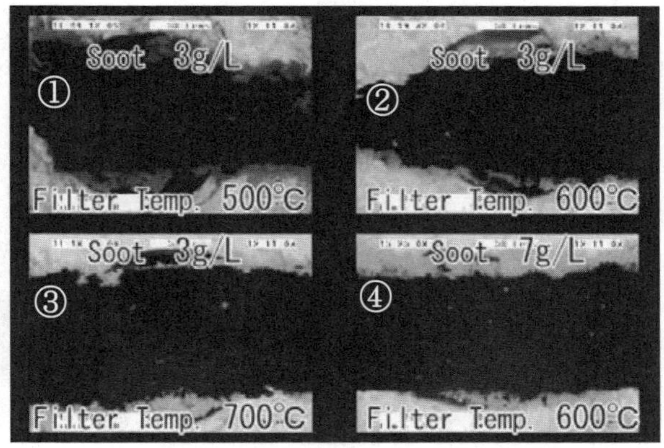

Figure 16. Test Specimens and Test Conditions

Figure 17 shows images similar to those in Figure 16 which were obtained several seconds after the gas containing oxygen was supplied to the cell. According to these images, particulate began to burn on the specimen with particulate collected at 3 g/liter when the temperature was set at 700°C (see No. 3 in the figure). Particulate on the filter material was cracked and broken into small pieces of agglomerated particulate. These pieces were burning as the combustion continued.

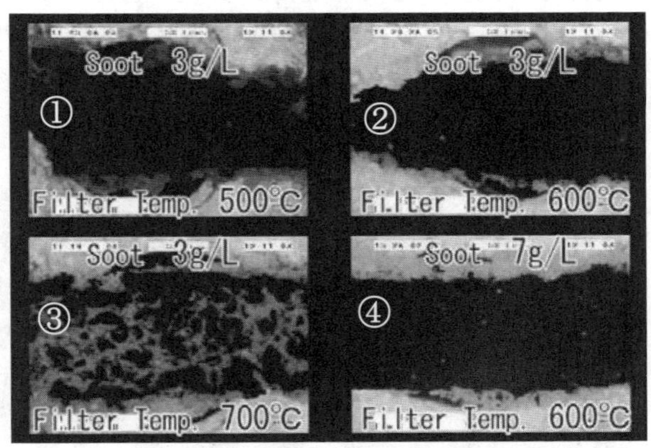

Figure 17. Combustion Process of Particulate on the Test Specimen at Several Seconds in Gas Containing 10 Percent Oxygen

Figure 18 shows images captured after the above combustion process continued for several tens of seconds. According to these images, particulate collected on the filter material at 7 g/liter with the temperature setting of 600°C began to burn (see No. 4 in the figure). No cracks in the particulate on the filter material were observed, yet the large pieces of agglomerated particulate were burning during the combustion process.

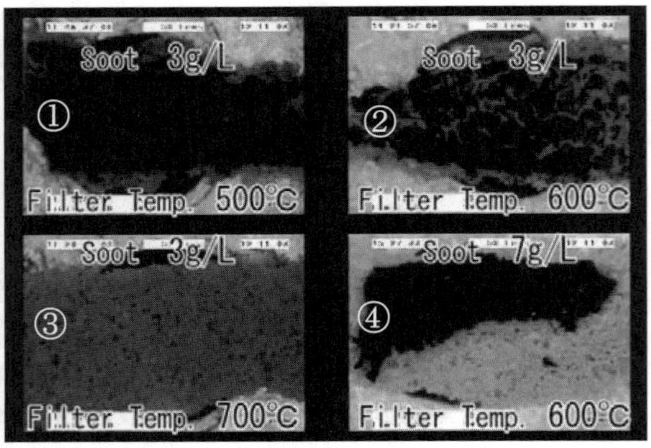

Figure 18. Combustion Process of Particulate on the Test Specimen at Several Tens of Seconds in Gas Containing 10 Percent Oxygen

Thereafter, particulate collected on the filter material at 3 g/liter with a 600°C temperature setting started burning (see No. 2 in the figure). The way this particulate burned was similar to the image of No. 3.

Figure 19 shows images captured several minutes after the ambient gas was switched from nitrogen to gas containing 10 percent oxygen. The images labeled as No. 2, No. 3, and No. 4, indicate the end of particulate combustion. However, image No. 1, namely particulate collected on the filter material at 3 g/liter with the temperature setting of 500°C, did not indicate combustion of particulate.

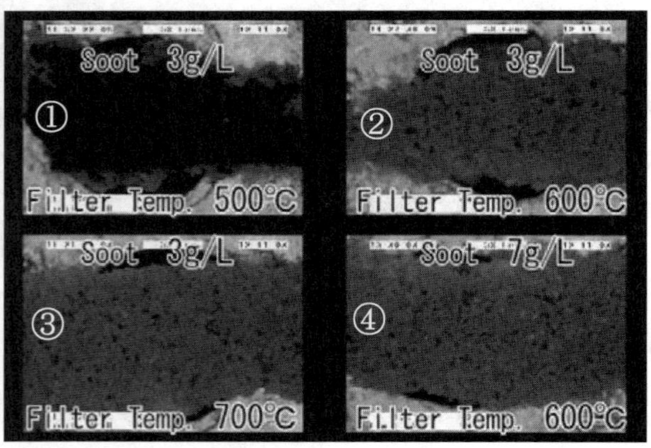

Figure 19. Combustion Process of Particulate on the Test Specimen after Several Minutes in Gas Containing 10 percent Oxygen

According to these observation results, the combustion rate of particulate varied depending on the amount of particulate collected on the filter and the filter temperature. The highest combustion rate was observed in the test with the specimen of particulate collected at 3 g/liter and with the temperature setting at 700°C (No. 3 image) followed by the tests with 7 g/liter and 600°C (No. 4 image), 3 g/liter and 600°C (No. 2 image), and 3 g/liter and 500°C (No. 1 image). Consequently, as the filter temperature was higher and as the amount of particulate collected on the filter was greater, the combustion rate of particulate on the filter increased.

A METHOD TO CALCULATE THE PM COMBUSTION RATE CONSTANT

The still images of particulate combustion captured at every unit time were processed to measure the area of particulate shown in each image. This area decreases as combustion continues. Thus, the change in the area of particulate during combustion was measured to calculate the degree of particulate reduction, as shown in Figure 20. Combustion duration was defined as the time between the replacement of nitrogen with gas containing 10 percent oxygen in the cell and the time when the degree of particulate reduction reached 5 percent.

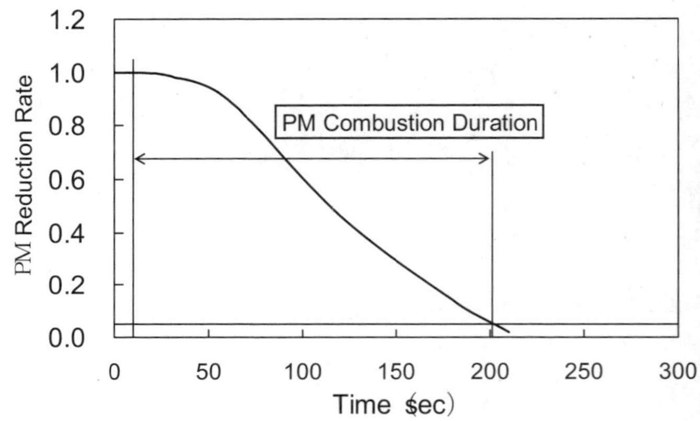

Figure 20. Particulate Combustion Duration Obtained from Combustion Images

The degree of particulate accumulated on a particulate filter was divided by the above combustion duration to obtain the PM combustion rate constant, as explained by the following formula:

$$S = G / t$$

Where S = PM combustion rate constant (g/liter/min)

G = Specific particulate accumulation defined by particulate accumulation in weight per unit volume of a filter (g/liter)

t = PM combustion duration obtained from the images (min)

Various Effects on the PM Combustion Rate Constant

Parametric tests were conducted to evaluate the effects that the parameters of filter temperature, the specific particulate accumulation, and the ambient gas flow rate have on the PM combustion rate constant. Figure 21 shows the effect of filter temperature on the PM combustion rate constant when the following parameters were constant: the specific particulate accumulation at 7 g/liter, the concentration of oxygen at 10 percent, and the SV value of 80,000 $hour^{-1}$. As the filter temperature was increased, the PM combustion rate constant increased.

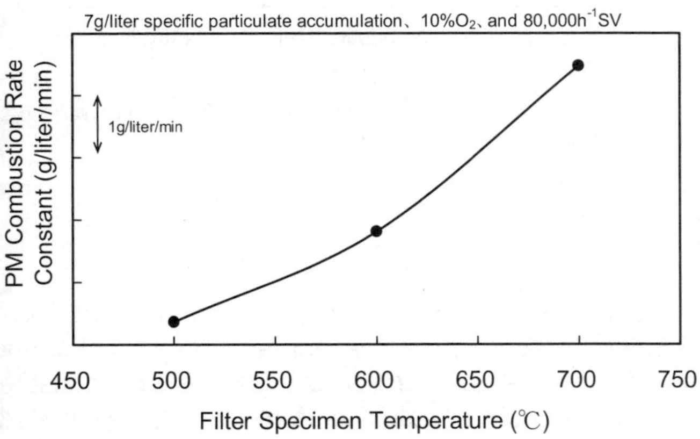

Figure 21. The Effect of Filter Temperature on PM Combustion Rate Constant

Figure 22 shows the effect of the specific particulate accumulation on the PM combustion rate constant. The filter temperature, the concentration of oxygen in the ambient gas, and the SV value were kept constant as follows: 600°C, 10 percent, and 80,000 $hour^{-1}$, respectively. The PM combustion rate constant increased as the specific particulate accumulation on the filter was increased.

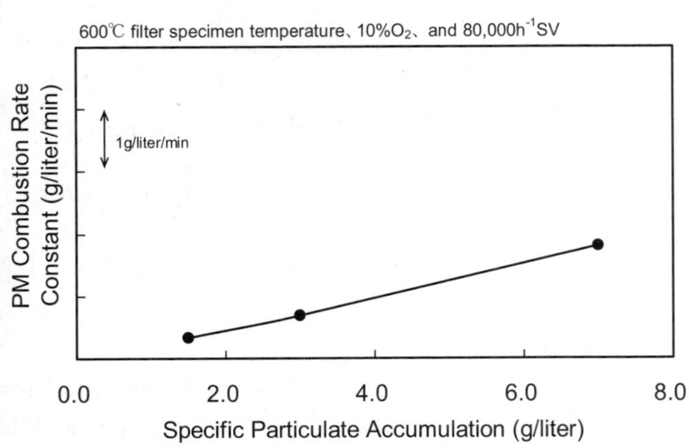

Figure 22. The Effect of the Specific Particulate Accumulation on PM Combustion Rate Constant

Figure 23 shows the effect that the oxygen flow rate had on the PM combustion rate constant as a function of the SV value multiplied by the concentration of oxygen in the ambient gas in the cell. Note that as the ambient gas flow rate was changed, the amount of oxygen supplied to the cell was varied as well as the SV value. Thus, the specific particulate accumulation and the filter temperature were only kept constant at 7 g/liter and 600°C, respectively. As a result, the PM combustion rate constant increased somewhat as the oxygen flow rate was increased.

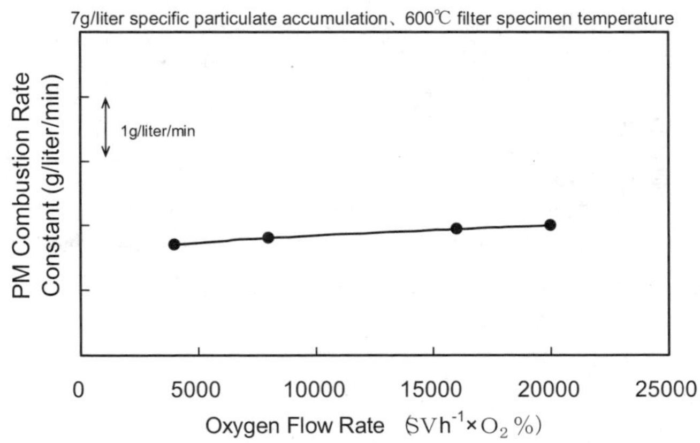

Figure 23. The Effect of the Oxygen Flow Rate on PM Combustion Rate Constant

An Empirical Formula of PM Combustion Rate Constant

Multiple regression analysis was conducted using 70 test data obtained through the visualization of particulate combustion to develop an empirical formula for calculation of the PM combustion rate constant (S). Following is the empirical formula in terms of the regression equation:

$$S = \exp(A_1 \cdot X_1 + A_2 \cdot X_2 + A_3 \cdot X_3 + B)$$

Table 4 lists the standard regression coefficients of three independent variables. According to these results, the PM combustion rate constant is sensitive to the filter temperature followed by the specific particulate accumulation and the oxygen flow rate.

Table 4. The Sensitivity of the PM Combustion Rate Constant

n	Independent Variable (X_n)	Standard Regression Coefficient (β^*)
1	Filter Temperature	0.90706
2	Specific Particulate Accumulation	0.48803
3	Oxygen Flow Rate	0.04632

Figure 24 shows the correlation between the PM combustion rate constant estimated using the above empirical formula and the measured PM combustion rate constant. This result indicates that both estimated and measured values correlated well, and the coefficient of multiple determination (R^2) was 0.9387.

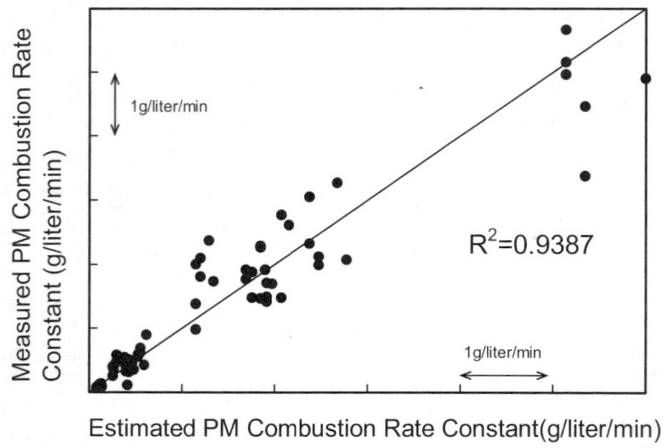

Figure 24. Correlation between Estimated and Measured PM Combustion Rate Constants

VALIDATION OF THE EMPIRICAL FORMULA THROUGH ENGINE TESTS

The empirical formula was tested to validate the accuracy of estimating the PM combustion rate constant on the engine. The particulate filter was weighed after particulate was collected on the filter during the engine test. The weighed filter was reinstalled on the exhaust system, and the engine was operated with late post injection to regenerate the filter. After the filter temperature reached a certain temperature, the filter was weighed. The change in the weight between the above tests with and without regeneration was used to calculate the PM combustion rate constant. Because the particulate filter made of cordierite tends to adsorb water, the filter was heated in an oven to remove water before the filter was weighed.

The PM combustion rate constant was calculated by the following formula:

$$S = (G_0 - G_1) / V / t$$

Where S = PM combustion rate constant (g/liter/min)

G_0 = Weight of a filter before regeneration (g)

G_1 = Weight of the filter after regeneration (g)

V = Volume of the filter (liters)

t = Regeneration time period (min)

The estimated PM combustion rate constant correlated well with the measured value, as shown in Figure 25.

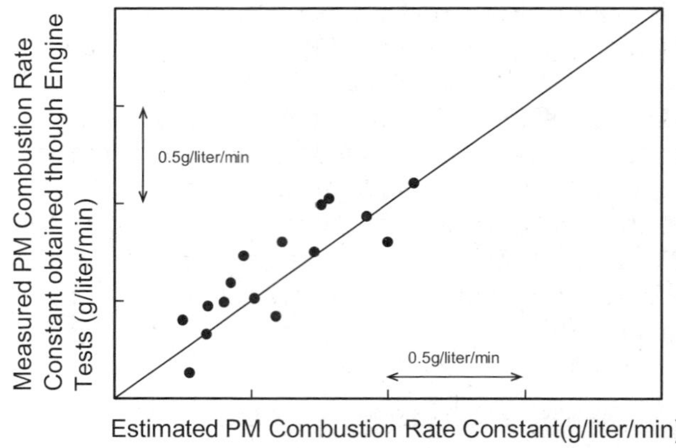

Figure 25. The Estimated PM Combustion Rate Constant correlated with the Measured Value

The engine test was repeated to perform particulate collection followed by active regeneration 20 times, and the effectiveness of the PM combustion rate constant calculated with the empirical formula was evaluated. Figure 26 shows the filter temperature and the pressure drop across the filter while the engine was operated. The pressure drop was stable throughout the engine test.

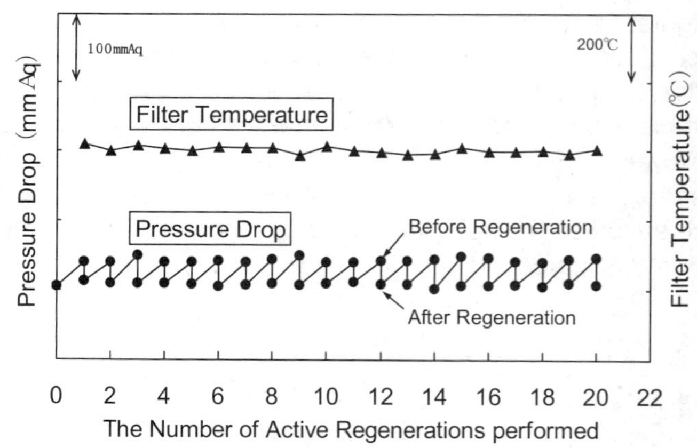

Figure 26. Filter Temperature and Pressure Drop Measured during 20 Active Regenerations

Figure 27 shows the particulate filter sectioned after the engine test was completed. Very little particulate remained on the filter, which indicates that the active regeneration performed well and that particulate collected on the filter was effectively burned.

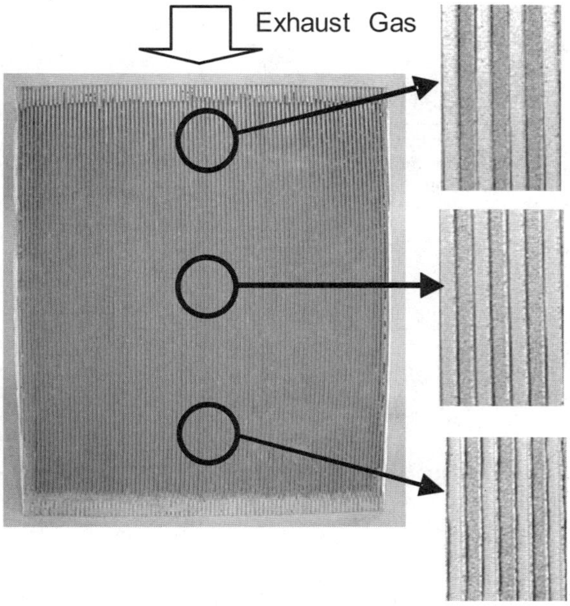

Figure 27. The Sectioned Particulate Filter

OPTIMIZATIONS OF SPECIFIC PARTICULATE ACCUMULATION AND FILTER TEMPERATURE

According to the results of the multiple regression analysis reported in the above section, combustion of particulate collected on the particulate filter was sensitive to both the specific particulate accumulation and the filter temperature [5]. Consequently, active regeneration was optimized with these parameters to efficiently regenerate the filter without creating abnormal combustion.

Engine tests were conducted to perform active regeneration of the particulate filter under various conditions of both the specific particulate accumulation and the filter temperature and to determine conditions that would cause abnormal combustion. Figure 28 shows the locations on the filter at which temperature was measured.

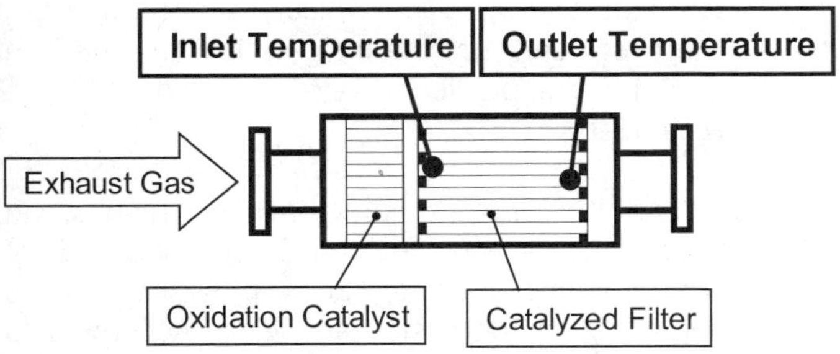

Figure 28. Filter Temperature Measurement Locations

Figure 29 shows the step-transient engine operating conditions with an indication of the late post fuel injection period. The engine was operated under 20-percent load conditions at 80 percent of the rated speed. Late post injection was used to increase the temperature of the particulate filter. Immediately after the temperature reached a certain level, engine speed was reduced to idle.

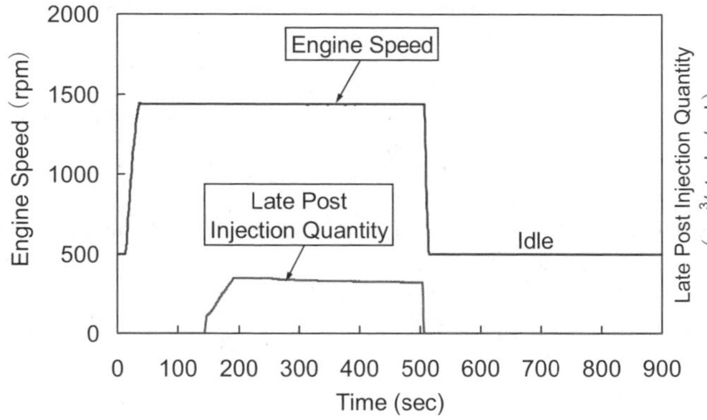

Figure 29. Step-Transient Engine Operating Conditions

Under the above engine operating conditions, abnormal combustion of particulate on the filter and normal combustion were observed. Figure 30 shows the temperatures measured at both the inlet and outlet of the filter as the engine was operated when the combustion of particulate on the filter was normal. The temperature difference between inlet and outlet of the filter was small, and the temperatures were stable during the active regeneration period.

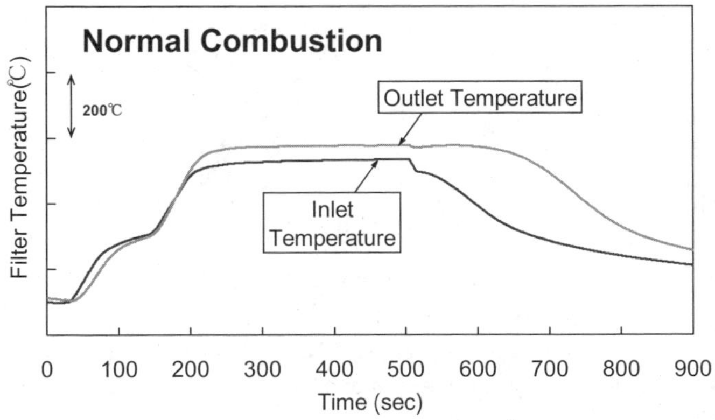

Figure 30. Normal Combustion of Particulate on a Particulate Filter with Active Regeneration

Figure 31 shows similar results obtained when abnormal combustion took place. Temperature at the outlet of the filter significantly increased while temperature at the inlet was stable during active regeneration. The temperature increased high enough to cause deterioration of catalysts.

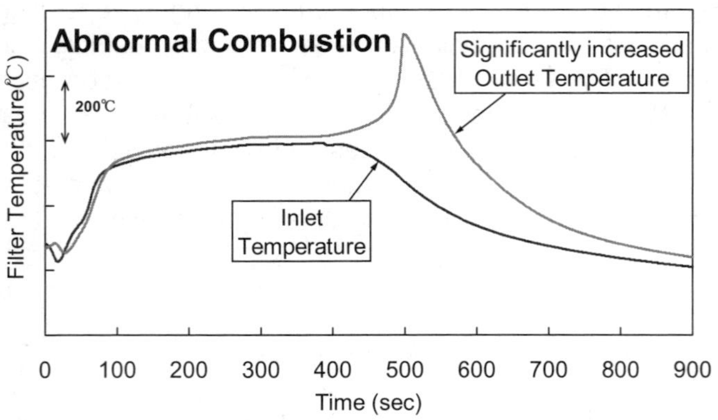

Figure 31. Abnormal Combustion of Particulate on a Particulate Filter with Active Regeneration

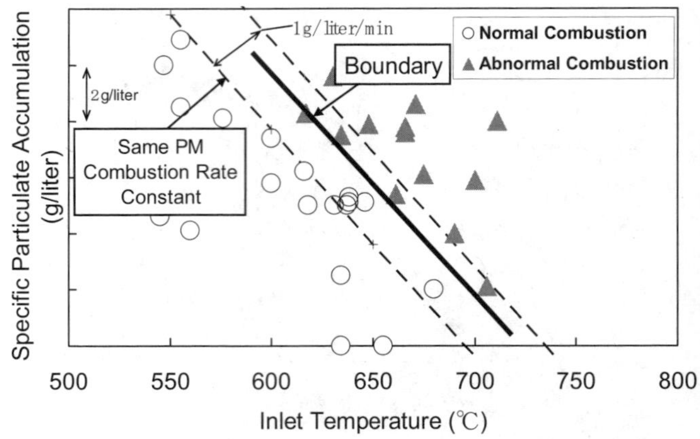

Figure 33. PM Combustion Rate Constant and Abnormal Combustion Conditions

Figure 32 shows the relationship between the specific particulate accumulation on the filter and the temperature at inlet of the filter. When the filter's inlet temperature was near 600°C, abnormal combustion took place in the case of a higher specific particulate accumulation. However, despite the lower specific particulate accumulation, abnormal combustion occurred when the temperature was near 700°C. Therefore, the boundary that separates normal combustion and abnormal combustion exists between the filter's inlet temperatures of 600 and 700°C, as shown in the figure.

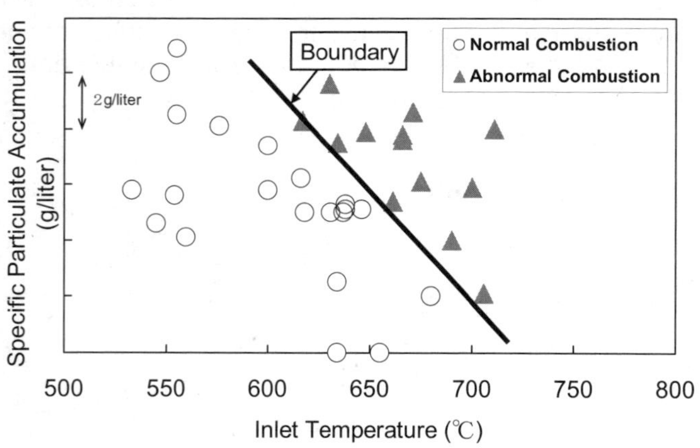

Figure 32. The Relationship between the Specific Particulate Accumulation and Filter's Inlet Temperature

The PM combustion rate constant was calculated with the empirical formula. Figure 33 shows the line that connected the same PM combustion rate constants obtained for various conditions in the relationship between the specific particulate accumulation and the filter's inlet temperature. The slope of this linear line is similar to that of the straight line indicating the boundary between normal combustion and abnormal combustion.

The above results indicate a strong relationship between the PM combustion rate constant and the boundary line. Therefore, abnormal combustion can be eliminated if the calculated PM combustion rate constant is appropriately set while the specific particulate accumulation and the filter's inlet temperature are controlled.

CONCLUSION

1. Particulate collected on a particulate filter self-ignites at about 560°C. However, self-ignition temperature varies depending on vehicle driving patterns and engine parameters such as fuel injection pressure and the EGR rate while the effect of ash accumulated on the filter is small.

2. The combustion process of particulate collected on a filter of DPR was observed using a test rig equipped with a provision to visualize the process in order to understand the combustion characteristics. As a result, the combustion characteristics varied depending on the specific particulate accumulation on a filter and filter temperature.

3. The images captured during visualization of the particulate combustion on a test rig were used to calculate the PM combustion rate constant. The effects that the specific particulate accumulation, filter temperature, and the oxygen flow rate had on the PM combustion rate constant were characterized.

4. Based on the above characterization results, an empirical formula was developed for calculation of the PM combustion rate constant based on the multiple variable regression analysis. According to the regression analysis, the PM combustion rate constant was sensitive to the specific particulate accumulation and filter temperature. Using the empirical formula, the PM combustion rate constant

was estimated when active regeneration was performed on the engine. Both calculated and measured PM combustion rate constant correlated well.

5. While the specific particulate accumulation and the filter's inlet temperature are controlled, appropriately setting the calculated PM combustion rate constant can optimize active regeneration to efficiently burn particulate at high temperature without producing abnormal combustion.

ACKNOWLEDGMENTS

The authors would like to express sincere appreciation to the staff of Hino Motors, Ltd.; Tokyo Roki Co., Ltd.; and Advanced Electronics Research Laboratory and Casting Technology Research Laboratory of Hitachi Metals, Ltd., for their contributions to the project.

REFERENCES

1. H. Toorisaka et al., "DPR Developed for Extremely Low PM Emissions in Production Commercial Vehicles," SAE Paper No. 2004-01-0824.

2. M. Hosoya et al., "The Study of Particle Number Reduction Using After-Treatment Systems for a Heavy-Duty Diesel Engine" SAE Paper 2004-01-1423.

3. Y. Mizutani, Combustion Engineering (3rd Edition), Morikita, October 2002.

4. Randy L. Vander Wal, "Soot Nanostructure: Definition, Quantification and Implications," SAE Paper No. 2005-01-0964.

5. K. Hanamura et al., "Visualization of Combustion Phenomena in Regeneration of Diesel Particulate Filter," SAE Paper No. 2003-01-0836.

II.2 SUBSTRATE GEOMETRY EFFECTS

2004-01-0948

Ash Storage Concept for Diesel Particulate Filters

Dianna M. Young, David L. Hickman, Garima Bhatia and Natarajan Gunasekaran
Corning Incorporated

ABSTRACT

Ash accumulation in heavy duty and light duty diesel filters has become a growing concern due to its negative impact on filter performance over time. Performance issues include increased backpressure and increased fuel penalty. An additional concern is frequency of filter ash cleaning which contributes to overall maintenance and operational costs.

A new ash storage concept filter is discussed in this paper. This concept proposes an exchange between inlet and outlet cells, redistributing surface area and volume resulting in more ash storage and improved pressure drop over traditional filters. Overall filter performance (pressure drop, regeneration, ash/ceramic interactions) was evaluated in the laboratory and results are reported in this paper. This paper will discuss in detail the ash storage concept and its benefits in filter performance.

INTRODUCTION

Diesel particulate wall flow filters are made by extruding high porosity material through a standard die. The product, before plugging, resembles a piece of graph paper with all cells identical. Cells are plugged in a checkerboard fashion with each cell plugged at one end and all the cells that share a wall are plugged at the opposite end, see Figure 1.

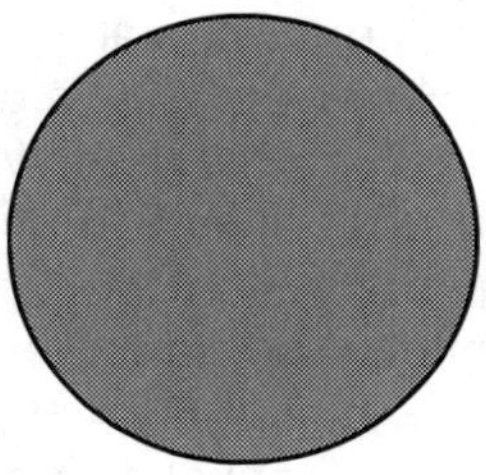

Figure 1. Front face of diesel particulate filter showing checkerboard plug pattern.

The soot and ash accumulates in and on the walls of the inlet channels and gradually narrows the cell openings. As the particulate matter accumulates, the exhaust gas must pass part way down the inlet channel and then through the soot and wall and then out the exit cell, see Figure 2. Therefore, as the filter fills with soot and ash, the pressure drop across the filter increases due to accumulation on the inlet channels.

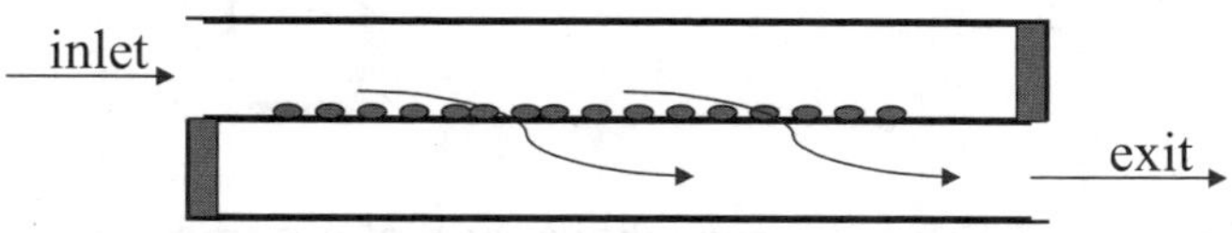

Figure 2. Axial view of an inlet and exit channel; air flows down the inlet channel, across the soot and wall, and exits through the exit channel.

During the life of the filter, the soot will be burned off in regeneration cycles leaving the ash behind. Over time, ash will accumulate to the point where the filter component will likely have an unacceptable pressure drop. Today, the options of addressing ash build up over vehicle lifetime are either periodic removal of ash from, or replacement of, the filter. Both options are unattractive due to cost.

This paper describes a filter design which targets increased ash storage capability. The design encompasses an asymmetric pattern where the inlet channels and outlet channels are different in hydraulic diameter, see Figure 3.

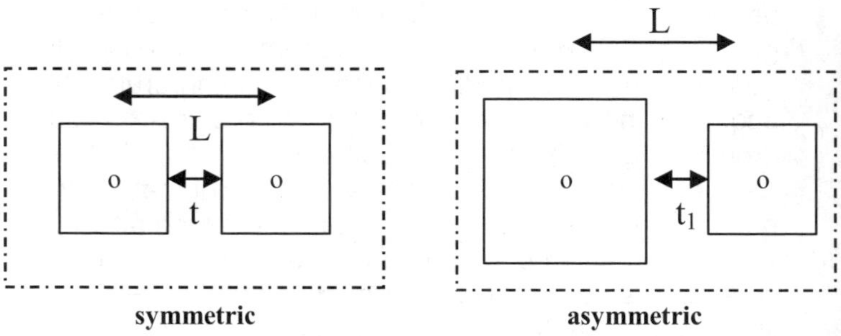

Figure 3. Symmetric and asymmetric designs. Dotted line shows unit cell. Notice that within the unit cell, if CFA and CPSI remain constant, then $t \neq t_1$.

This asymmetric design does offer more upstream filter volume. This means that it can hold more ash. The soot and ash layer would be spread over a greater surface area in an asymmetric design leaving inlet channels more open by comparison to the symmetric design, see Figure 4. Over time, the large inlet small exit design is expected to yield a lower pressure drop due to "down the channel" effects. In this paper modeling and experimental data are discussed.

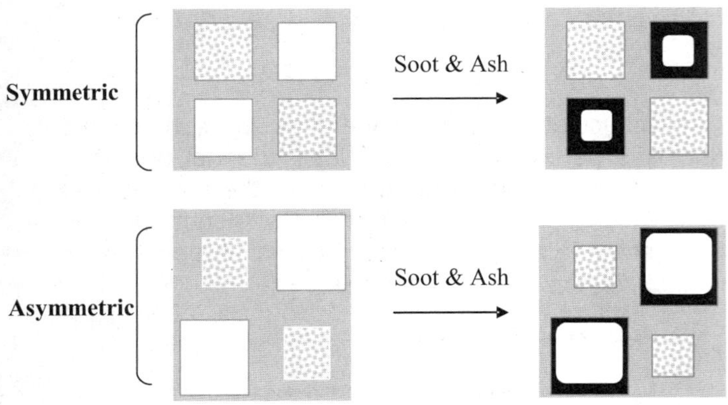

Figure 4. Schematic representation of symmetric and asymmetric designs. Inlet cells are represented by open cells, and exit cells are shaded. Asymmetric designs theoretically hold more soot and ash than symmetric designs.

MODELING

Notice in Figure 3, that as one moves from a symmetric to asymmetric design, it is impossible to keep the open frontal area (OFA), cells per square inch (CPSI), and web thickness constant. This is an important point, in that geometrical comparisons between large inlet small exit designs and symmetric designs are difficult to obtain, and thus comparisons require some use of a pressure drop calculator.

It has been demonstrated in previous work that the pressure drop equation works relatively well in predicting pressure drop of cordierite diesel filters of symmetric geometry.[1,2] However, in an asymmetric case, the equation must be modified. Corning Incorporated has acquired and validated a model which predicts pressure drop of filters which vary in cell density, wall thickness, soot permeability, soot density, and soot-loaded wall permeability. One version of the model equation for symmetric design is shown in Figure 5 and indicates the contributors to the overall pressure drop. Terms are listed in the nomenclature section of this paper.

In the asymmetric case, the inlet channel is larger than the exit channel. The geometrical change also affects the cross terms in the equation, particularly the "down the channel component." Thereby, if one is to predict the pressure drop of asymmetric designs, one will need to modify the equation to reflect this geometrical change. A generalized pressure drop equation has been derived and used to predict optimal asymmetric geometries for diesel filter applications. A simplified version of the generalized pressure drop equation is shown in Figure 6. The terms associated with equation are listed in the nomenclature section.

through the wall component · through the soot component · down the channel compo[nent]: inlet, outlet

$$\Delta P = \frac{\mu Q}{2V_{eff}N}\left[\frac{w_c}{k_o'\alpha} + \frac{1}{2k_{soot}}\ln\left(\frac{D}{D-2w_{soot}}\right) + \frac{4FL_{eff}^{\;2}}{3}\left(\frac{1}{(D-2w_{soot})^4} + \frac{1}{D^4}\right.\right.$$

$$+\frac{\rho Q^2}{V_{eff}^{\;2}N^2D^2}\left(\frac{\beta w_c}{4} + 2\zeta\left(\frac{L_{eff}}{D}\right)^2\right)$$

Forchheimer's term · expansion-contraction term

Figure 5. The pressure drop equation.

$$\Delta P = \frac{\mu Q}{2V_{eff}N}\left[\left\{\frac{2\alpha w_0}{k_0' D_1(1+\alpha)} + \frac{1}{2k_0'}\ln\frac{(1+\alpha)^2}{4\alpha} + \frac{1}{2k_s}\ln\left(\frac{D_1}{D_1'}\right)\right\}\left\{f_1 + f_2(\alpha'^4\right.\right.$$

$$\left. + \frac{4FL_{eff}^2}{D_2^4}\left\{\frac{1}{1+\alpha'^4} + f_3 + f_4\left(\alpha'^4 - 1\right)\right\}\right]$$

$$+\frac{\rho Q^2}{V_{eff}^2N^2}\left[4\frac{L_{eff}^2}{D_2^4} + 2\zeta\frac{L_{eff}^2}{D_1'^4}\right]$$

Figure 6. Generalized pressure drop equation which illustrates fourth power inverse dependence on outlet channel diameter.

In Figure 6, the equation shows two highlighted terms which indicate the fourth power inverse dependence of the total pressure drop on the diameter of the outlet channel. Thus choking down the outlet diameter leads to an increase in the overall pressure drop. This is not obvious if one argues that the increase in the inlet channel diameter should compensate for this effect (for example, the "down the channel" component in Figure 5 which has the sum of the reciprocal fourth powers of the inlet and outlet). However, this argument is easily countered by Figure 7 which shows the effect of increasing the I/E ratio (I = inlet hydraulic diameter, E = exit hydraulic diameter), α, on the average reciprocal inlet and exit channel diameters, raised to the first, second, and fourth powers. At α = 1.7 the average of the inverse fourth power channel diameters is more than double that of the symmetric case. Thus pressure drop advantage would be observed only if soot load is large enough that the positive effect of the reduced soot layer thickness at high I/E ratios can compensate for the negative effect of the choking down the outlet channel diameter.

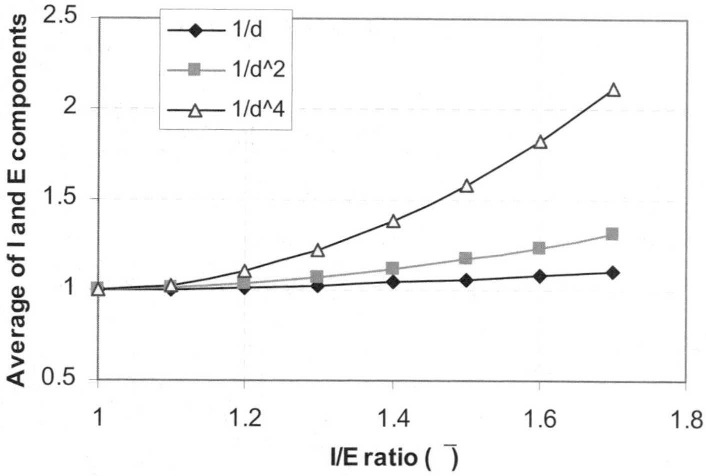

Figure 7. Illustration of reciprocal fourth power dependence on channel diameter.

ASH

Recently there has been much attention drawn to the problem of ash accumulation in diesel particulate filters. Ash accumulation usually results in an increase in pressure drop across the filter, and eventually the ash build-up must be removed. Increased pressure drop does increase fuel penalty and removing ash from the filter raises maintenance cost of the filter; both effects are highly undesirable.

Ash is the result of decomposed lube oil and fuel components during engine use and therefore it is a component of the exhaust gas that reaches the diesel filter. Ash is usually a mixture of metal oxides, sulfates and phosphates. Calcium sulfates and zinc phosphates are compounds often found in ash. Sulfur compounds are caused by sulfur present in the diesel fuel and oil. If fuel additive is used, there are likely iron or cerium oxides present in ash.

Based on internal studies and analysis of various types of ash obtained from the field, the density of ash is approximately 0.4 g/cm^3 (theoretical density of about 2.5 g/cm^3 at 85% porosity). Permeability measurements on ash suggest that it is about ten times more permeable than cordierite used for diesel particulate filters.

ASH STORAGE BENEFIT

The percent increase of inlet channel volume versus I/E ratios can simply be calculated and compared with symmetric designs of the same open frontal area, see Figure 8. An asymmetric design having an I/E = 1.4 results in a 32% increase in inlet volume over a symmetric design. Hydraulic diameter increase of asymmetric design over symmetric are shown in Figure 9. An inlet hydraulic diameter of an asymmetric design having a I/E = 1.4 would be 15% larger than the inlet of a symmetric design. Another example, increasing I/E = 1.6 results in an increase inlet volume of about 44%; the hydraulic diameter increases by 20% respectively, see Figures 8 and 9.

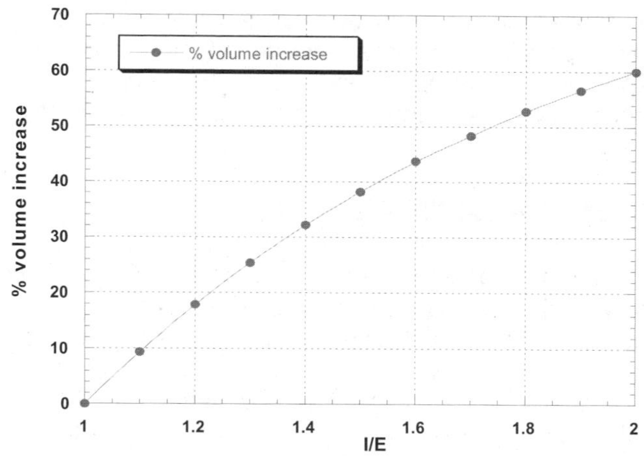

Figure 8. Percent inlet volume increase versus I/E (Inlet/Exit) ratio. I/E = 1 represents a symmetric design.

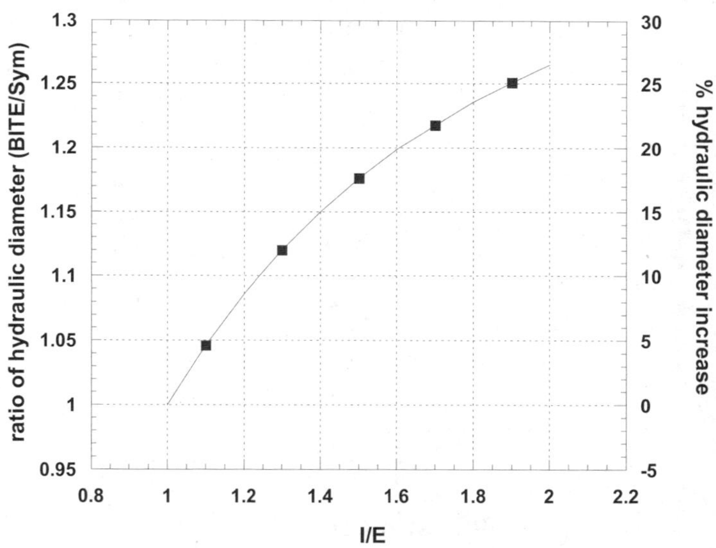

Figure 9. Ratio of hydraulic diameter increase (asymmetric/symmetric) versus I/E ratio. I/E = 1 represents a symmetric design.

The ash storage volume increase (inlet volume increase) can be calculated and one would predict that ash does physically take up space when accumulated in the inlet channel. If ash takes up space in the inlet channel, then there would be less room for a given amount of soot in a symmetric filter than in an asymmetric.

It is difficult to predict the pressure drop effects of ash since it is very permeable, light, and fluffy in comparison to filter walls and soot. One probably needs a significant amount of ash accumulation before seeing any effects.

Furthermore, it takes a long time for large amounts of ash to build up in filters; this means several hundred hours of running time on vehicle. Anecdotal evidence from the field suggests that considerable amounts of ash can be accumulated into filters and that filters still function.

Unfortunately, the pressure drop calculator can not predict pressure drop of ash-loaded filters. Therefore experiments were performed in the laboratory to understand ash-loaded pressure drop of asymmetric designs.

MODEL PREDICTIONS

The modified pressure drop equation (Fig. 6) was used as a guide to predict optimum asymmetric design geometries (favorable inlet to exit ratios), under typical engine running conditions. Figure 10 shows pressure drop versus I/E for 11.25" x 14" cordierite 100, 200, and 300 cpsi geometires for conditions: 250°C inlet, 1000Kg/hr, 3g/L and 8 g/L soot loading. To examine 200 cpsi geometries further, Figure 11 shows pressure drop versus I/E ratio for 200 cpsi geometries at higher flowrates and higher temperatures.

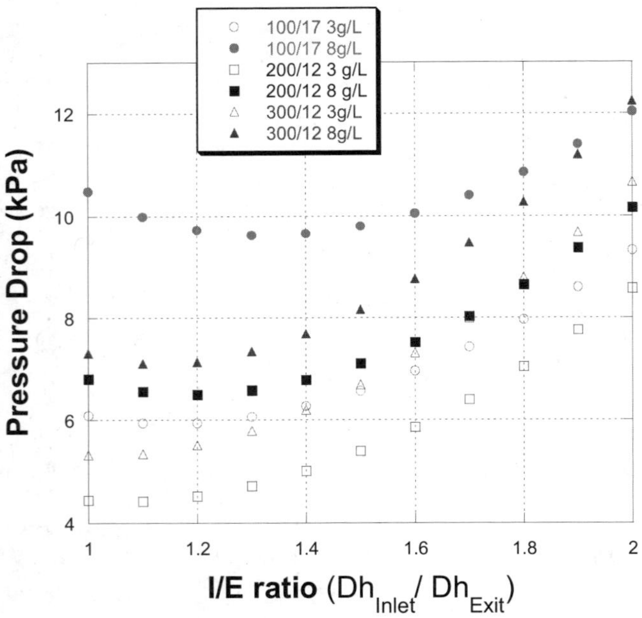

Figure 10. Calculated pressure drop versus I/E for 100, 200, and 300 cpsi designs. Parameters were: 250C inlet temperature, 1000 kg/hr, and 11.25" x 14" sample size.

Based on the modeling, favorable I/E ratios are dependent on soot loading and flowrate. Figures 10 and 11 show that higher pressure drop would be obtained at higher I/E ratios.

To test these pressure drop predictions, small two inch diameter dieletts were fabricated to replicate select I/E geometries. Dieletts were used to make two inch diameter cordierite filter samples such that laboratory testing could be performed.

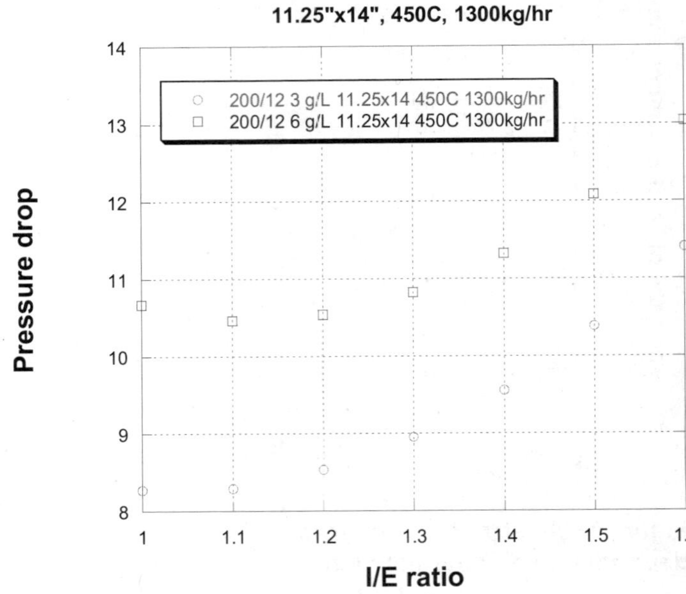

Figure 11. Calculated pressure drop verus I/E for 200 cpsi designs.

EXPERIMENTAL AND RESULTS

PRESSURE DROP

Asymmetric design dieletts were fabricated in 1.3, 1.7, and 2.0 I/E geometries. As mentioned earlier in this paper, it is impossible to keep closed frontal area, cells per square inch and web thickness constant. One of the variables must be changed if the other two are kept constant. In the case of these two inch diameter samples, CFA and CPSI were kept constant and web thickness was allowed to vary for chosen designs.

Pressure drop data versus soot loading data was obtained on 2" x 6" and 2" x 12" cylindrical filter samples, see Figures 12 and 13. Experimental apparatus and data collection procedure has been discussed a previous publications.[3, 4]

The pressure drop data on two inch diameter samples indicate that the model predictions are realistic.

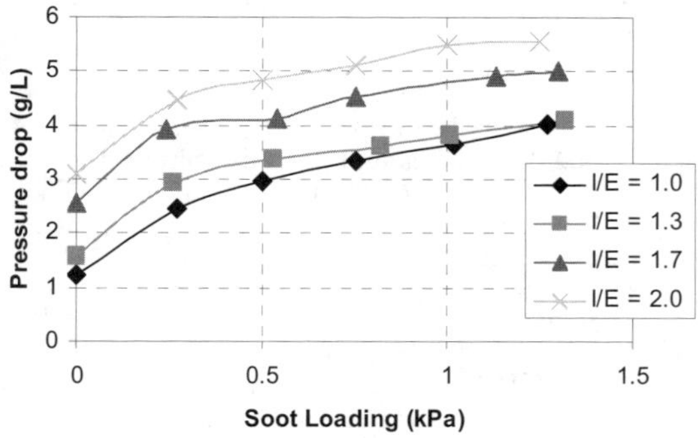

Figure 12. Pressure drop versus soot loading for 2" diameter asymmetric samples. I/E = 1.0 represents symmetric design.

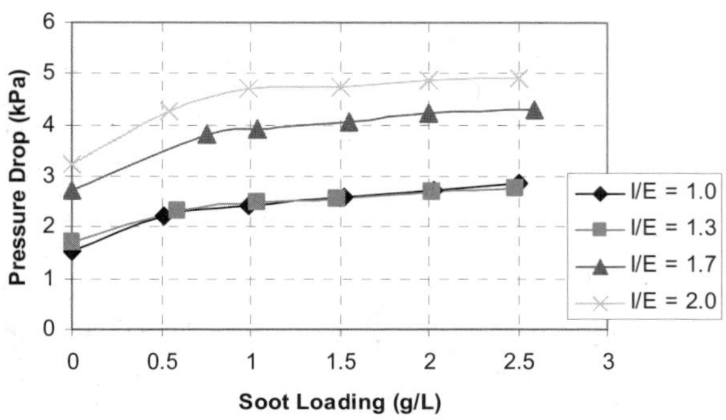

Figure 13. Pressure drop verus soot loading for 2" x 12" asymmetric samples. I/E = 1.0 represents symmetric design.

ASH LOADING

To determine if the asymmetric design had merit for ash storage capacity, ash accumulation into samples was necessary. Large quantities of engine generated ash were obtained from the field. Table 1 summarizes the physical and chemical properties of the ash obtained and used for laboratory experiments.

Particle size distribution (μ)	d10 = 2.38, d50 = 8.5, d90 = 37.61
Surface area ($m^{2/g}$)	45.58
Chemical composition (wt%)	Al_2O_3 1.28%, CaO 27.6%, Cr_2O_3 .14%, CuO 0.22%, Fe_2O_3 1.57, MgO 1.66%, P_2O_5 10.4%, SiO_2 3.81%, ZnO 8.22%
CHN analysis (wt%)	6.58%, 0.31%, 0.14%
Crystalline phases detected (X-ray diffraction)	$CaSO_4$, ZnO

Table 1. Physical and chemical properties of ash used for ash loading.

Several types of laboratory ash loading apparatus were investigated. Set-ups included genset engines, air aspirated nozzles, and vacuum filtration. The techniques that performed most reliably and predictably were the air aspirated nozzle and vacuum filtration. Figure 14 shows the vacuum filtration apparatus. Ash is loaded onto the filter by introducing it at the top end of the delivery tube while vacuum is in operation. This ash load method results in uniform ash deposition down inlet channels at low and moderate loads. Boroscope observations down the channel reveal even distribution throughout. However, at very high ash loads, it becomes difficult to boroscope down the channel due to the thick ash layer closing down of hydraulic diameter.

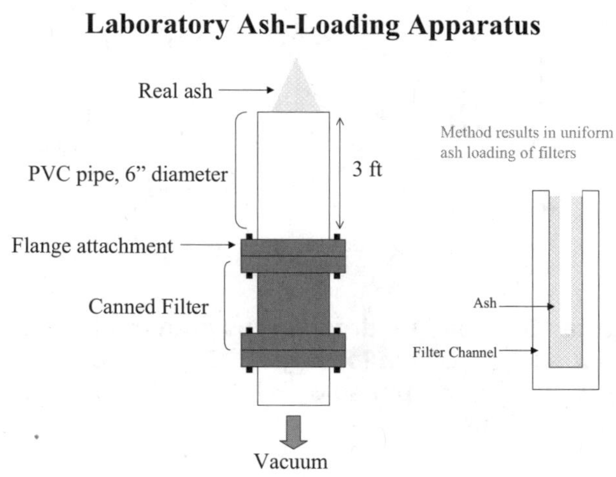

Figure 14. Schematic of laboratory ash loading apparatus and ash loaded inlet channel.

ASH LOADED PRESSURE DROP

Laboratory pressure drop measurements were performed on ash loaded samples. The pressure drop apparatus and experimental conditions have been previously described[5].

Initial experiments were conducted on samples which were 2 in. in diameter. Pressure drop results for diesel particulate filters and are extremely sensitive to the filter diameter. This is a serious problem, especially for small filters. In order to have results in which we can have the greatest confidence it was necessary to work with larger diameter filters. Constructed samples were made from 1.5" x 1.5" x 10" segments assembled to 5.66" diameter and 10" length. In addition to assembled samples, monolithic 5.66" x 6" samples were also loaded with ash to establish ash levels that would dramatically increase pressure drop, see Figure 15. It is apparent that a significant amount of ash must be loaded onto the filter before seeing a substantial increase in pressure drop.

Pressure drop versus ash load data were obtained on assembled asymmetric samples, see Figures 16 and 17.

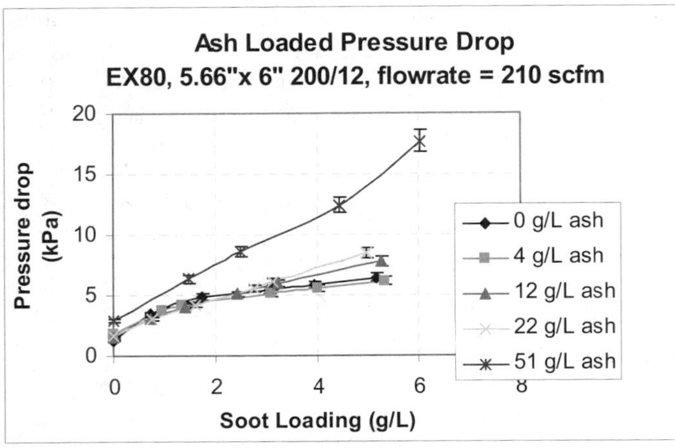

Figure 15. Pressure drop versus soot loading at various levels of ash load (g/L). Graph shows pressure drop response of ash loaded monolithic 5.66" x 6" sample.

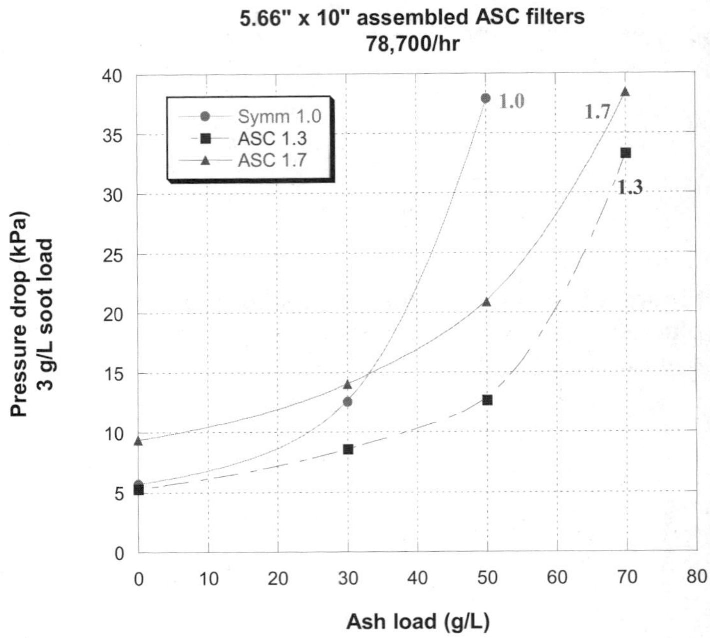

Figure 17. Pressure drop (at 3 g/L soot) versus ash load (g/L).

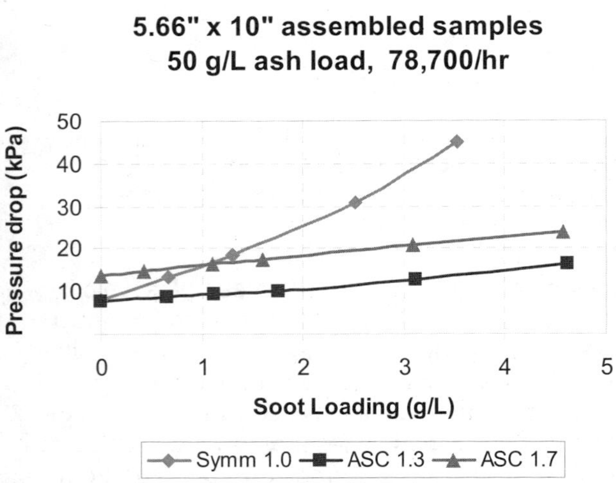

Figure 16. Pressure drop versus soot loading for assembled filters.

REGENERATION

Regeneration experiments on ash containing Cordierite filters were carried out in the laboratory using an artificial carbon soot. The details of the carbon black powder, soot loading and regeneration equipment used in the laboratory can be found elsewhere.[5] A standard monolithic filter of 5.66"D x 6"L was used to determine the influence of ash on the exotherm temperature and regeneration efficiency. 5.66"Dx10"L size filters were used to compare symmetric and asymmetric designs. A soot load of 8g/l was chosen for this ash study based on the data obtained from the soot loading versus exotherm temperature curve. A maximum ash loading of 50g/l was employed for both 6" and 10"long filters. The clean filter was first loaded to 4 g/l of soot and then the required amount of ash was introduced using an apparatus described in Figure 14. After the ash addition, the filter was placed back in the soot loading rig and the remaining 4 g/l of soot was added to it. The regeneration of the soot + ash loaded filter was carried out by following a nitrogen presoak at 650°C inlet temperature at a flow rate of 700L/min. The gas flow was switched after 28 minutes to a mixture containing 18% oxygen and balance nitrogen to trigger the soot combustion.

The filter temperature at desired locations was monitored by placing Type K thermocouples from the exit end of the filter as shown in Figure 18.

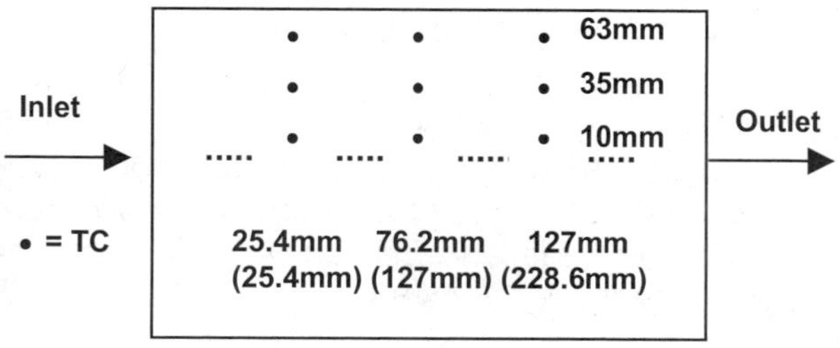

Figure 18. Schematic diagram of thermocouple (TC) pattern used in this study. Values in the brackets represent TC for long filter (10").

The filter temperature profile, pressure change and downstream gas composition were continuously recorded during the soak and regeneration procedure. The percent regeneration efficiency was calculated from the weight changes obtained before and after the regeneration.

Some of the physical properties and regeneration performance parameters of the filters tested in this study are presented in Table 2. The bulk density values were calculated from the effective filter volume, which excludes plug lengths in the filter. The maximum temperature given in the table represents the highest temperature seen at any given location in the filter during soot combustion. Due to the test condition employed to regenerate the soot loaded filter (flow, oxygen, soot and inlet temperature), a 100% regeneration efficiency has been observed on most of the runs.

Filter Type	Filter Size	Bulk Density (g/Cm3)	Soot (g/l)	Ash (g/l)	Max Temp.(°C)	Regeneration Efficiency (%)
Monolith-Standard	5.66" x 6"	0.49	8	0	993	94.2
Monolith-Standard	5.66" x 6"	0.50	8.6	30	1096	100
Monolith-Standard	5.66" x 6"	0.50	8	50	1119	100
Symm. Filter I/E=1.0	5.66" x 10"	0.68	7.94	50	974	100
ASC Filter I/E=1.3	5.66" x 10"	0.66	7.98	50	1055	100

Table 2. Selected physical properties and regeneration performance parameters of various filters used in this study.

Although the temperature profile at various locations in the filter has been monitored continuously, the exotherm temperatures seen along the axial center was further considered here for comparison purposes. Accordingly, Figure 19 shows the axial temperature profile monitored during soot combustion with and without ash on 5.66" x 6" filters. The curves "a, b, c" represent the corresponding exotherm temperature seen at the front, middle and back of the filter respectively for 8 g/l soot with no ash in it. The order of peak temperature values (c>b>a) is very typical during regeneration for any wall flow type diesel particulate filter. Similarly, the exotherm temperatures noticed along the filter axial direction with 30 and 50g/l ash content were shown as "d,e,f" and "g, h, i" respectively. In these cases, the maximum exotherm temperatures are seen at the mid section of the filter due to ash sandwiched between soot layers (~4g/l soot layer). A higher peak temperature values compared to that obtained without ash can be accounted for by the fact that soot per unit volume in the inlet channel may be higher with ash in it. Since the ash was added between soot loadings (4 g/l each to a total of 8 g/l), the soot probably gets uneven distribution in the filter. Due to this soot distribution, the maximum exotherm value was observed at the center of the filter instead of at the back end of the filter. Further, the amount of soot occupied in a given volume can be much higher (g/L) when ash is filled in the inlet channel. The fast rate of soot combustion occurred at the front half of the filter can be interpreted in terms of oxygen availability to the soot layer.

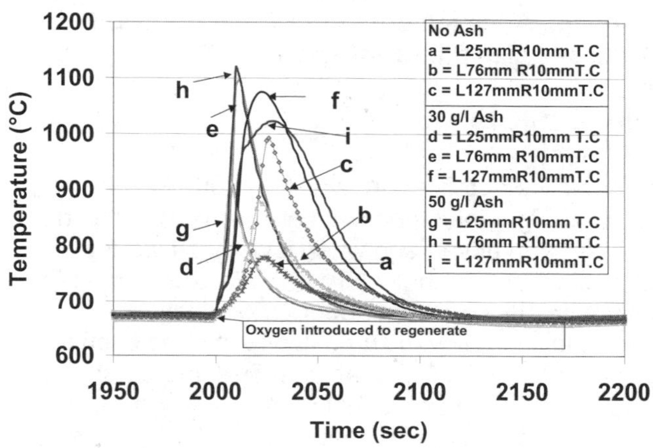

Figure 19. Axial temperature profiles obtained on 5.66"x6" size filters with and without ash.

Figure 20 shows the axial temperature exotherm responses obtained on symmetric and asymmetric design filters at 8g/l soot and 50g/l ash. Although nearly 100% regeneration efficiency has been obtained in these experiments, the broader peak shapes indicate the slow soot combustion process due to reduced oxygen transport down the inlet channel at the flow rate considered here. Further, the presence of ash layer delays the combustion at the back of the filter (L228.6mm or 9" from inlet face). Some of the inconsistency in the soot and ash loading procedure may be attributed to the slight differences in the peak values seen on these two filter designs.

From the above results, it can be clearly seen that the presence of ash in the inlet channel shifts the location of the exotherm temperature during regeneration. The shorter length filter (6" or 150mm) regenerates faster and quicker than the long filter (10" or 254mm). The availability of oxygen to the soot layer is restricted by the presence of ash in the inlet channel and thereby prolonging the time to complete the soot combustion. The thermal gradient or thermal stress experienced by the filter material can be much different with ash than without ash in the channel. Therefore, the location and amount of ash in the channel can be detrimental to the durability over the life time of the filter.

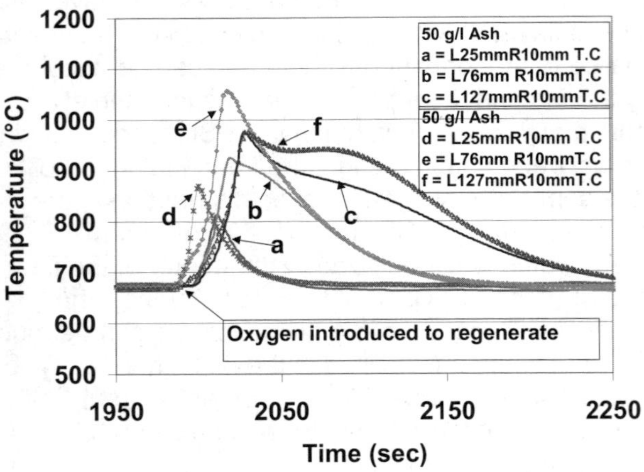

Figure 20. Axial temperature profile obtained on 5.66" x 10" size filters with ash.

CONCLUSION

The objective of this effort, was to determine whether a filter with larger inlet than exit cells would result in a pressure drop advantage when it was ash loaded. Figure eight shows that with a 1.3 inlet to exit ratio the inlet volume is increased about 25% relative to a symmetric filter. It was surprising to find, as shown in figures 15,16 and 17, that at least 50% more ash could be accumulated in an ASC filter than in a symmetric filter. While we really do not understand why the ash storage advantage should be this large, we have come to believe the results after a number of repetitions of the experiments.

Our observations have shown that the ash accumulates both along the sides and at the end of the inlet cells. The regeneration results showed that the peak temperature location moves toward the front of the filter as ash accumulates in the filter. This suggests that the soot is to some extent being pushed toward the front of the filter by the ash accumulating in the rear of the filter. For the same absolute amount of soot, this results, in effect, in a higher specific loading in the front portion of the filter. The result is a very slightly higher peak temperature due to the higher specific soot loading. This result is not substantial enough to outweigh the higher ash holding capacity of the ash storage concept.

ACKNOWLEDGMENTS

The authors would like to acknowledge the help of Mr. D. Madaffari Jr. for preparation of samples and property characterization, and Ms. C. Sawyer, Ms. K. Robbins, Ms. K.. Harris, and Ms. L. Seeley for measurements.

REFERENCES

1) A. G. Konstandopoulos, J. H. Johnson. *Wall-Flow Diesel Particulate Filters-Their Pressure Drop and Collection Efficiency*. SAE Technical paper series No. 890405, 1989.

2) M. Masoudi, A. G. Konstandopoulos. *Validation of a Model and Development of a Simulator for Predicting the Pressure Drop of Diesel Particulate Filters*. SAE Technical paper series No. 2001-01FL-116, 2001.

3) D. M. Young, C. J. Warren, K. P. Gadkaree, L. Johannesen. *Silicon Carbide for Diesel Particulate Filter Applications: Material Development and Thermal Design*. SAE Technical paper series No. 2002-01-0324, 2002.

4) G. A. Merkel, D. M. Beall, D. L. Hickman, M. J. Vernacotola. *Effects of Microstructure and Cell Geometry on Performance of Cordierite Diesel Particulate Filters*. SAE Technical paper series No. 2001-01-0193, 2001.

5) R.J. Locker, N. Gunasekaran and C. Sawyer. *Diesel Particulate Filter Test Methods*. SAE Technical paper series No. 2002-01-1009, 2001.

APPENDIX

D	Channel width (symmetric design)	m
D_1	Inlet channel width (BITE design)	m
D_1'	Soot loaded inlet channel width (BITE design)	m
D_2	Outlet channel width (BITE design)	m
D_f	Overall diameter of DPF	m
f_1, f_2, f_3, f_4	Calculated parameters in generalized pressure drop equation	-
F	Parameter in square channel pressure drop correlation (friction factor)	-
k_0	Clean wall permeability of DPF	m^2
k_0'	Soot-loaded wall permeability of DPF	m^2
k_s	Permeability of soot layer	m^2
L	Overall length of DPF	m
L_{eff}	Effective length of DPF	m
L_p	Plug length	m
M_{sw}	Mass of soot in the wall	g
n	Number of open channels	-
N	Cell density of DPF	cpsi
OFA	Open frontal area of DPF	%
p_{atm}	Atmospheric pressure	N/m^2
Q	Volumetric gas flow rate at the inlet	m^3/s
V_{eff}	Effective filter volume	m^3
w_0	Wall thickness	m
w_s	Soot layer thickness	m

Q	Volumetric gas flow rate at the inlet	m^3/s
V_{eff}	Effective filter volume	m^3
w_0	Wall thickness	m
w_s	Soot layer thickness	m

Greek symbols

α	Ratio of inlet:outlet channel diameters	-
α'	Ratio of soot loaded inlet channel:outlet channel diameters	-
β	Forchheimer's coefficient	/m
μ	Gas phase viscosity	kg/m/s
ρ	Density of gas	kg/m^3
ρ_s	Effective density of the soot layer	kg/m^3
ζ	Channel expansion-contraction coefficient	-

2004-01-0949

Ash Storage Capacity Enhancement of Diesel Particulate Filter

Kazutake Ogyu, Kazushige Ohno, Sungtae Hong and Teruo Komori
IBIDEN Co.,LTD

ABSTRACT

The life of DPF would be decided by ash durability because pressure loss of DPF is increased by accumulation of ash that is produced when soot is burnt in the DPF.
We could succeed to increase maximum ash capacity and extend the life of the DPF by an approach from the filter design, making inlet cell volume larger than outlet cell. Furthermore, when the inlet cell is designed octagon, we found that slanted wall blocking the inlet cells is functional worked for soot filtration. As the results, we could improve to manage both transient pressure loss while the soot is collected and the ash durability by the cell shape with (inlet cell/outlet cell) = (octagon/square).

INTRODUCTION

Global warming is now considered to be a serious problem. he need for the diesel engine will increase more and more in the future from the viewpoint of protection of the global environment because the diesel engine has merits such as a low rate of CO_2 discharge, which is thought to be one of the causes of this global warming. On this account, development of engines and development of technologies for after-treatment of exhaust gas are actively being carried out in parallel with the tightening up of the allowed PM and NO_x discharge [1].

Among the technologies for the after-treatment of diesel exhaust gas, the trump card for PM treatment is considered to be DPF. This is an alternately plugged honeycomb structure, and PM contained in engine exhaust gas flowing into the DPF is trapped as it passes through the wall of the structure. In the year 2000, practical utilization of DPF was started using the method in which recrystallized SiC DPF was combined with post injection in the common rail and a fuel additive and the collected PM were burned periodically [2].

PM contains incombustible substances, although their quantities are small. Therefore, when PM collected in the DPF is burned and regenerated, ash is produced and accumulates inside the DPF. Ash carries the danger of causing a high-pressure loss for the filter due to adhesion, clogging, etc., and therefore, its treatment is very important [3]. In the above-mentioned system that has been put into practical use, the ash is collected in a compact manner behind the inlet cell, and therefore, an increase in sudden pressure loss is avoided. However, the pressure loss increases gradually as the amount of the ash accumulated increases, and thus, the life of the DPF is shorter than the lives of cars. So extending the life of a DPF, in the limited capacity as an automobile part, has become an important challenge.

This time, we dealt with the challenge by approaching it from the filter design. We worked out several cell shapes based on the thought that the allowable amount of ash accumulated can be increased by making the capacity of the cell on the inlet side larger than that on the outlet side. Then, we made prototypes and carried out experiments. As a result, we found that when the shape of the inlet cell is octagonal, soot also accumulates on the slanted wall blocking between the two inlet cells. Then, we found that it would be possible to maintain the pressure loss at a low value and increase the ash accumulation capacity. So we report this.

CONTRIBUTING FACTORS AND LEVELS

The shape of the cell on the inlet side was designed in such a way that the cell capacity on the inlet side is larger than that on the outlet side. Then, we proposed three types and made prototypes. The values of the physical properties of each cell shape when cell density and wall thickness is nearly same are shown in Table .

(Inlet cell / outlet cell)= (Hexagon/Triangle) has the remarkable larger volume of the ash accumulation capacity and smaller filtration area than the conventional cell design. It is because that this cell shape must be had three times larger inlet cell volume than outlet cell. In this paper, we call this cell design "HT cell structure". (Inlet cell/outlet cell)= (Square/Rectangle) is designed by shifting the wall alternately from the conventional cell design and has inlet cell of two different sizes. It can increase the ash capacity without decreasing the filtration area. We call this cell design "SR cell structure". (Inlet cell/ outlet cell)= (Octagon/Square) has the structure of having the slanted wall blocking the inlet cells. We call this cell design "OS cell structure" in this paper.

The filters are made of recrystallized SiC (made by extruding a material of the same composition in a general manner and burning it at the same temperature). The pore diameter is about 9μm and the porosity is about 42% concerning all the cell structure.

The values of the physical properties in Table1 are the one of the example to make the difference of each cell shape clear. They can be changed by changing the wall thickness, cell density and pitch. We can estimate the influence affected by the dependent factor described below for filter performance [4] [5] [6].

1. Inlet volume: To enhance the inlet volume is the most important in this test. Larger inlet volume is expected to keep lower pressure loss while the ash is accumulated in the DPF and enhance the ash durability.
2. Aperture ratio: Higher aperture ratio brings lower pressure loss.
3. Bulk density: Higher bulk density brings lower maximum temperature during filter

Table1 : three types of cell design for enhancement of the inlet cell volume

	conventional	(in/out)= (Hexagon/Triangle)	(in/out)= (Square/Rectangle)	(in/out)= (Octagon/Square)
cell design	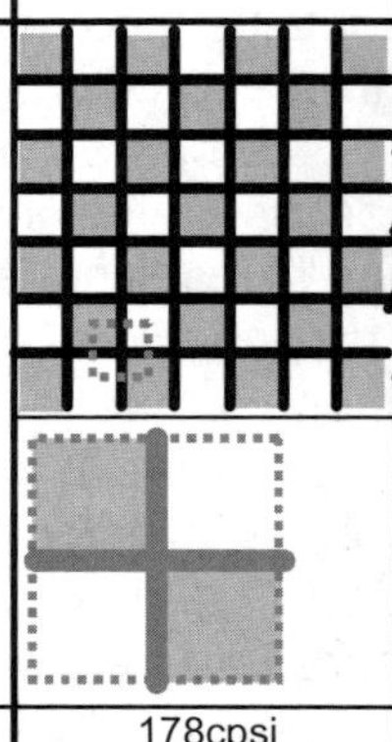	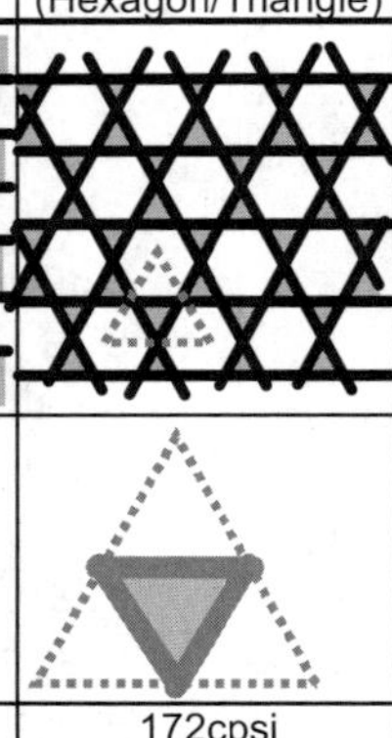	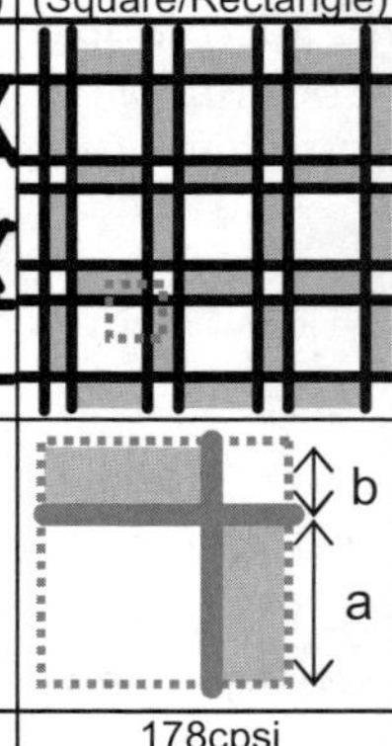	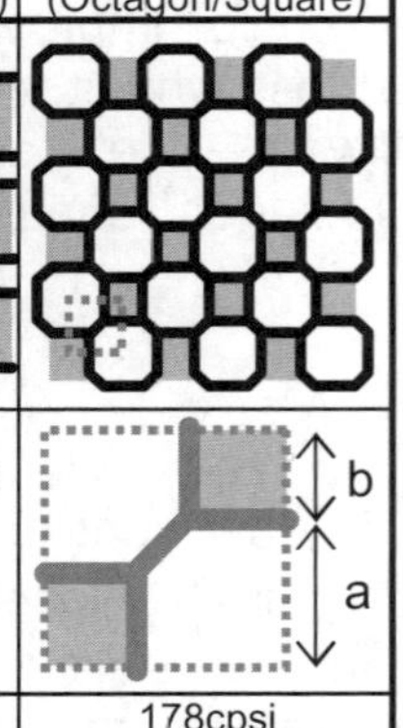
cell density	178cpsi	172cpsi	178cpsi	178cpsi
wall thickness	0.4mm	0.4mm	0.4mm	0.4mm
inlet volume	0.30L/L (1)	0.50L/L (1.67)	0.39L/L (1.30)	0.46L/L (1.53)
honeycomb density	0.72g/cm^3	0.71g/cm^3	0.72g/cm^3	0.70g/cm^3
aperture ratio (inlet)	30.6%	53.3%	39.6%	37.0%
(outlet)	30.6%	7.0%	21.5%	24.2%
filtration area	0.80m^2/L	0.58m^2/L	0.80m^2/L	0.63m^2/L (w/o slanted wall) 0.84m^2/L (with slanted wall)
			e.g: a/b=2.5	e.g: a/b=1.5

* filter length:150mm, plug length:3.5mm was used by this calculation

Table2 : Factors and levels

FACTOR		LEVEL							
cell design		conventional	Hexagon/Triangle			Square/Rectangle		Octagon/Square	
			HT1	HT2	HT3	SR1	SR2	OS1	OS2
pore diameter /um		9							
porosity /%		42							
wall thickness /mm		0.4	0.2	0.3	0.3	0.4	0.3	0.4	
cell density /cpsi		178	124		172	178		158	
(pitch-a)/(pitch-b)		—	—	—	—	2.3	1.7	1.2	1.4
effetive volume /(L/L)		0.30	0.63	0.58	0.55	0.37	0.38	0.38	0.46
aperture ratio /%	(in)	30.6	65.3	60.8	58.4	37.3	38.1	38.0	44.8
	(out)	30.6	16.0	12.3	10.5	21.9	31.2	25.0	19.2
bulk density /(g/cm^3)		0.72	0.35	0.50	0.58	0.76	0.57	0.69	0.67
filtration area /(m^2/L)		0.80	0.90	0.85	0.98	0.80	0.86	0.84	0.88

regeneration (soot combustion). It is because higher density leads higher heat capacity.

4. Filtration area: Larger filtration area bring smaller pressure loss gradient while the soot collecting because the soot layer that has very low gas permeability would be thinner.

We made good use of our experience to decide the level tested this time. And we choose several cell designs to be worth doing experiment for assessment. Factors and levels are shown in Table2.

COMPARISON OF INITIAL PRESSURE LOSSES

Room temperature air was made to flow through a filter with an apparent cross-sectional area of 1180 mm^2 (□34.3mm) and a length of 150 mm, and then, the difference between the pressure at the filter inlet and that at the outlet was measured.

Fig.1 shows the relationship between the air flow rate of inlet and the pressure loss.

It was found that by making the cell capacity on the inlet side larger than that on the outlet side, the pressure loss would increase, and all of the cell design tested this time has higher pressure loss than conventional one when the filter is clean.

Fig.2 shows the arrangement of Fig.1 by the relationship between the cell capacity on the outlet side and the pressure loss at a gas flow speed of 5 m/sec. As the gas passage on the outlet side becomes narrower, the pressure loss increases rapidly. As shown in Table1 and 2, the cell capacity on the outlet side of HT cell structure is one-third smaller than inlet cell in design. Based on this, HT cell structure would cause an increase in the initial pressure loss, although it makes it possible to make the cell capacity on the inlet side large. For this we judged that this HT cell structure would not be worthy of practical utilization.

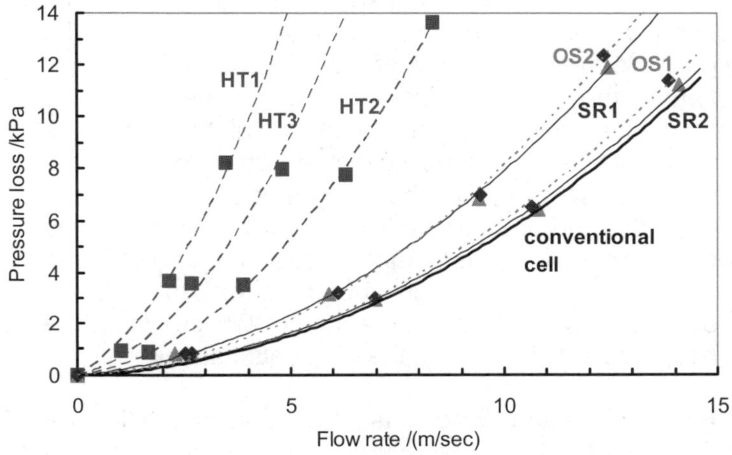

Fig.1 Initial pressure loss of each cell shape

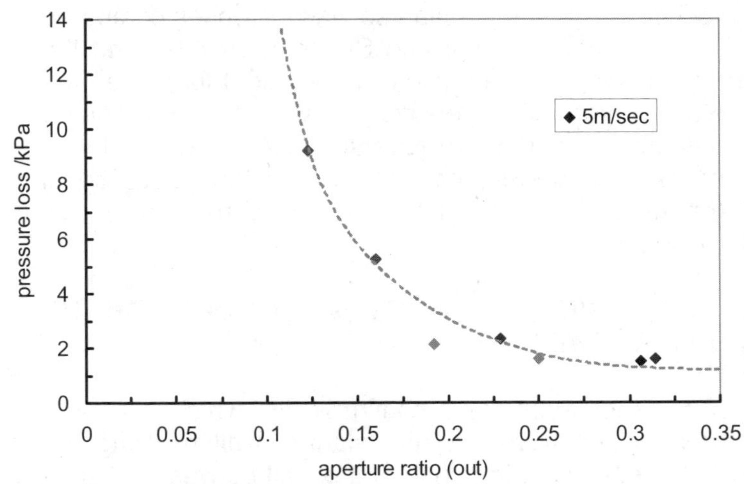

Fig.2 Relationship between aperture ratio of outlet cell and initial pressure loss

COMPARISON OF TRANSIENT PRESSURE LOSSES

The pressure loss of DPF would increase gradually because of the soot layer accumulated on the wall and the ash deposited inside of the DPF.
Subsequently to that, we examined changes in the transient pressure loss while the soot was collected in connection with the cell shapes with SR and OS cell structure. The experimental system and the engine's operating conditions are shown in Fig.3.
Fig.4 shows the experimental results in the form of changes in the pressure loss with respect to the amount of soot collected. The pressure loss is increased in proportion to the amount of soot collected in the DPF. This is because that the soot layer that has very low gas permeability and difficulty to flow the gas is formed on the wall of the DPF. According to this figure, the pressure loss in the SR cell structure increases while maintaining an initial high-pressure loss even after soot is collected compared with the conventional cell. On the other hand, it was made clear that the pressure loss increasing while soot was collected was modest in the case of the OS cell structure. And surprisingly enough, after 2 to 3 g/L has been collected, the pressure loss of OS cell structure becomes smaller than in the conventional cell.

Fig.5 shows the conditions of soot accumulation for the wall of OS2 cell structure when the soot amount is 10g/L. In this way, the soot accumulates even on the slanted wall blocking the inlet cells, and it shows that the slanted wall of OS cell structure is good used for the soot filtration. As a result, total filtration area of OS cell including the slanted wall increases compared with the conventional and SR cell structure, and the soot layer become thinner. It is thought that when the soot layer with relatively low gas permeability becomes thinner, a reverse phenomenon of the transient pressure loss concerning the conventional cell and OS cell structure as mentioned above will occur.

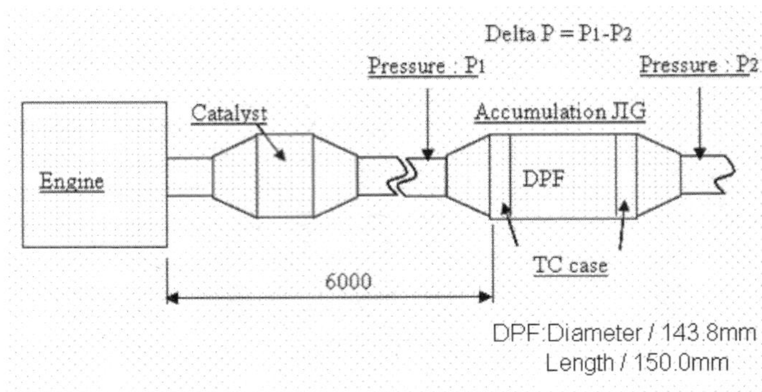

Filter cell structure	Conventional cell
	SR1, SR2, OS1, OS2
engine	DW10
Displacement	1.997Litre
engine condition	3000rpm/50Nm
fuel additive in fuel	CeO_2 25ppm

Fig.3 Experimental system and engine operating condition

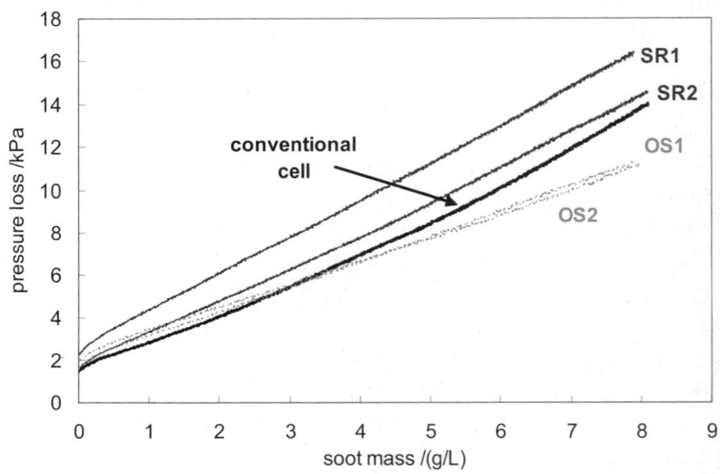

Fig.4 Transient pressure loss with collected soot mass

SIMULATION OF SOOT ACCUMULATION BY FLOW ANALYSIS

A phenomenon was confirmed in which the soot accumulates even on the slanted wall blocking the inlet cells, as in the case of the wall between the inlet cell and outlet cell. Therefore, an analysis by simulation was performed to see how this slanted wall performs the soot filtration function.

The ordinary CFD (Computational Fluid Dynamic) is used, and we did the simulation by the assumption described below;

1. analyzing gas flow during the fresh filter condition(there isn't accumulated soot)
2. assuming that there is certain concentration of soot in the gas and the soot is collected at the surface of the wall
3. increasing the pressure loss in proportion to the integrating mass flow passed through per mesh (unit area) according to equation(1) and (2)

	conventional cell	OS2
pic.		
soot layer thickness	0.135mm	0.130mm

Fig.5 Accumulated soot observation (soot: 10g/L)

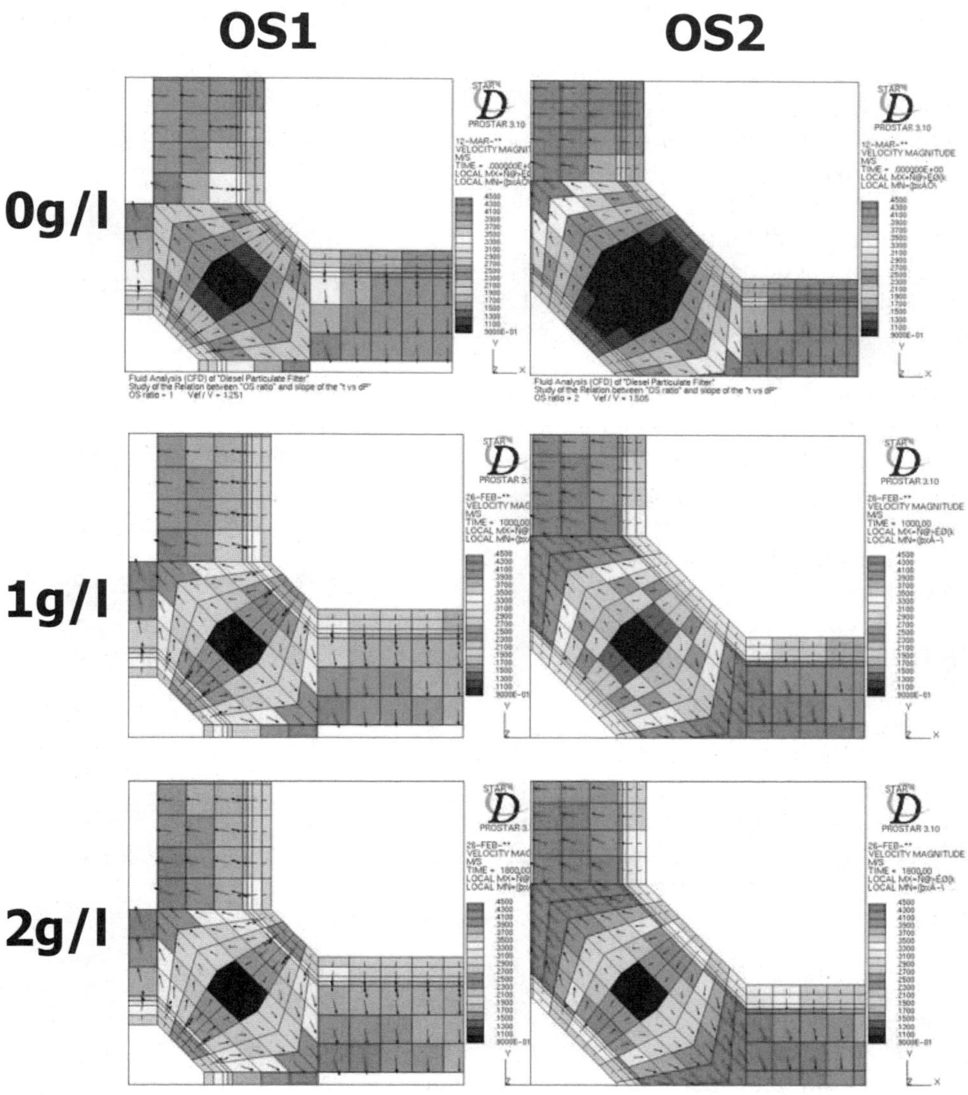

Fig.6 Gas flow simulation of OS cell structure

$$\frac{\Delta P}{t} = \beta * V_{fil} \quad \text{------Eq. (1)}$$

V_{fil}: flow rate that go through the filter [m/s]

$$\beta = \beta_0 + \sum_{iter=1}^{n} \left(\frac{V_t}{V_{fil}} v_f \right) \quad \text{Eq. (2)}$$

β: a coefficient of pressure loss
V_t: flow rate that go through a mesh at every moment [m/s]
v_f: a coefficient of flow rate

4. analyzing gas flow transition with time based on the pressure loss transition with time described 3
5. the fluid is regarded as compressible gas

Fig.6 shows the calculated results. In this figure, arrows express the gas flow vector and colors express the length of flow rate vector. Red expresses the highest flow rate, and yellow, green, blue color expresses lower flow rate in turn.
At first, in the condition in which the soot did not accumulate, the flow did not occur on the slanted wall (slanted wall is painted blue color). However, as soot accumulated on the wall between the inlet and outlet, making it difficult for the gas to flow, passing by the slanted wall and flowing to the outlet cell would make the resistance smaller than passing by the soot layer. It was found that in this way, the gas also flowed to the slanted wall from the inlet cell to outlet cell, like as going around the soot layer. We discovered by this simulation that soot also accumulated on the slanted wall and the slanted wall has the effect on filtration.

VERIFICATION OF DURABILITY AGAINST ASH

For the purpose of improving the ash durability, some performance evaluations were performed on the various cell shapes that would increase the cell capacity on the inlet side. As a result, it was found that the OS cell structure was most advantageous concerning the pressure loss (both the initial case and soot accumulated case).

Therefore, this cell structure was verified by actually using an engine to see if it really had excellent ash durability performance as designed. In the experiment, a cycle of soot collection and regeneration was repeated under the operating conditions shown in Table3 and changes in the pressure loss as ash accumulated were measured. In addition, CT (Computerized Tomography) was used at each 20 cycles to observe the cell capacity closed up by the ash and determine the quantity of ash by measuring the weight. Fig.7 shows a flow chart of this experiment. Here, the regeneration operation timing was set to initiate the regeneration when 6.5 to 7.5 g/L soot has been collected in the effective filter capacity * (Fig.8).

Table3 Engine operating conditions

Filter cell structure		Conventional cell
		(in/out)=(octagon/square) No.2
filter size		D143.8mm*L150mm
engine		DW10
Displacement		1.997Litre
engine condition	accumulation	3000rpm/50Nm
	regeneration	1250rpm/60Nm
fuel additive in fuel		CeO_2 25ppm

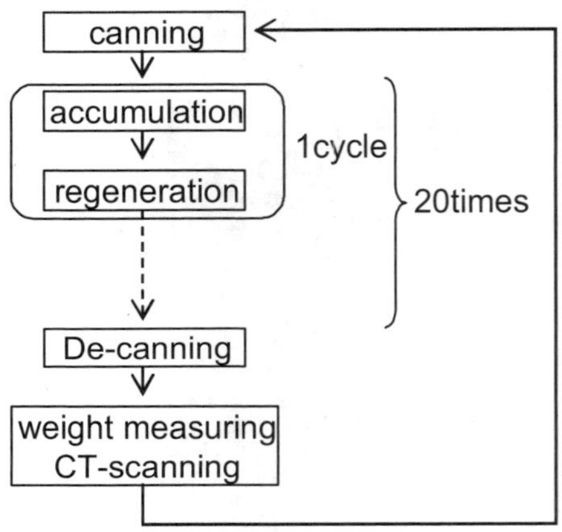

Fig.7 Flow chart of the ash durability test

*(effective filter volume)
=(filter volume)-(volume packed by ash)

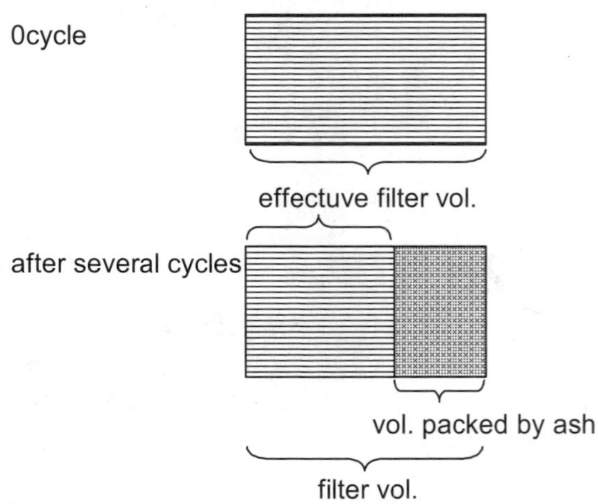

Fig.8 Explanation of Effective Volume

	conventional cell	OS2	conventional cell	OS2
CT photo				
ash length	63mm	44mm	100-101mm	
ash mass	73.3-73.6g/Filter		108.6g/Filter	172.4g/Filter

Fig.9 Deposited ash comparison between conventional and OS cell structure

Fig.9 shows the filter cross section with the same amount of the ash. The ash is reflected for white color in the CT scanning picture. It can be confirmed that by increasing the cell capacity on the inlet side to 1.5 times, the average height of the ash that came out white in the photo became about 2 / 3 times as much.

In addition, Fig.10 shows the changes in the pressure loss with the increase in the amount of ash. Here, the average value of the pressure loss just after the regeneration (amount of soot: 0 g/L) and before the regeneration (amount of ash: 6.5 to 7.5 g/L) is taken for un axis of ordinate. Based on this figure, OS cell structure can accumulate a larger amount of ash before the same pressure loss is reached, compared with the conventional filter. In other words, we were able to confirm that the ash durability increased. It was verified that even if the filter capacity is the same, it is possible to extend the filter life by exercising ingenuity in choosing the cell shape to increase the cell capacity on the inlet side.

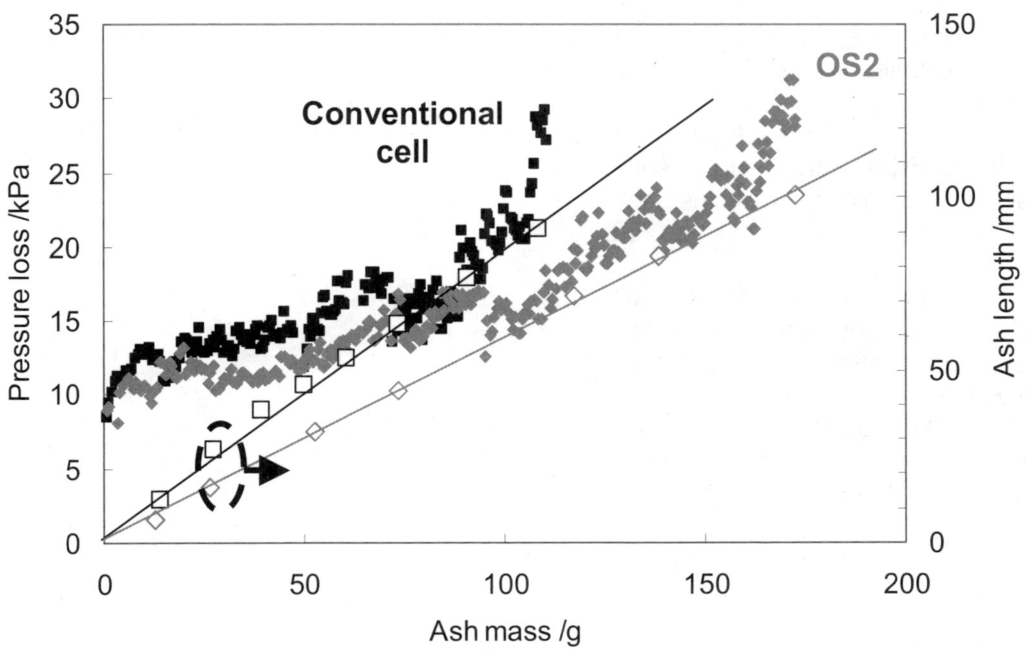

Fig.10 Ash durability effect of OS cell structure

CONCLUSION

1. Filter capacity for the ash accumulation is increased in proportion to the increasing of the inlet cell volume. It brings the improvement of the ash durability and extending the life of DPF

2. In OS cell structure, the soot is accumulated on the slanted wall blocking the inlet cells. The filtration area is become larger than the conventional cell structure because slanted wall is made full use of filtration

3. OS cell structure has relatively thinner accumulated soot thickness because the filtration area is larger. As the result, it makes a pressure loss gradient lower while soot accumulating.

4. In case of devising the cell shape, initial pressure loss has a tendency to depend heavily on the aperture ratio of outlet cell. So there is trade-off relationship between the ash durability and the initial pressure loss. And HT cell structure brings the very high initial pressure loss because it has the small aperture ratio of outlet cell in design.

REFERENCES

1. Timothy V. Johnson, "Diesel Emission Control in Review – The Last 12 Months" SAE paper No.2003-01-0039
2. O.Salvat,et.al., "Passenger Car Serial Application of a Particulate Filter System on a Common Rail Direct Injection Diesel Engine" SAE paper No.2000-01-0473
3. Blanchard Gilbert et al., "Series Application of a Diesel Particulate Filter with Ceria-Based Fuel-Borne Catalyst: Preliminary Conclusions After One Year of Service" SAE paper No.2002-01-2781
4. K.Ohno,et.al., "Characterization of SiC-DPF for Passenger Car" SAE paper No.2000-01-0185
5. N.Taoka,et.al., "Effect of SiC-DPF with High Cell Density for Pressure Loss and Regeneration" SAE paper No.2001-01-0191
6. K.Ogyu ,et.al., "Characterization of Thin Wall SiC-DPF" SAE paper No.2003-01-0377

2004-01-3046

New Strategies for Soot Emission Reduction of HD Vehicles

Dieter Rothe, Frank Ingo Zuther and Eberhard Jacob
MAN Nutzfahrzeuge Gruppe, Business Unit Engines

Armin Messerer, Ulrich Pöschl and Reinhard Niessner
Institute of Hydrochemistry, Technical University Munich

Christian Knab, Matthias Mangold and Christian Mangold
Oberland Mangold

ABSTRACT

The reduction of particulate emissions limits requires new strategies for the tuning of engines and exhaust after treatment systems. A non-blocking exhaust after treatment solution for HD vehicles to achieve the EuroIV emission limits has been developed which consists of a platinum oxidation catalyst and a stainless steel deposition structure with open channels for continuous soot storage and oxidation [1]. The significant advantages compared to standard particulate filters are caused by the open channel structure and characterized by a relatively low level of back pressure, selective separation of soot (EC) and penetrability for engine lubrication oil incineration ash.

Extensive laboratory studies, including a screening of numerous promising deposition structure designs, led to a metal catalyst structure with microsphere coating. The processes dominating particle deposition could be determined, leading to a target-oriented optimization of the deposition structure. Particle deposition caused by diffusion, interception and thermophoresis during instationary test conditions occurs. Measurements on the test bench with an EuroII HD engine show a gravimetric particulate matter reduction of 18 and 55 mg/kWh for the ESC/ETC, which is consistent with the laboratory-studies.

A new photo acoustic instrument (PASS) was applied allowing reliable determination of soot deposition efficiency even during the transient test cycle (ETC) for modern HD engines with a low PM raw emission. Close correlation between number based (SMPS) and mass based (PASS, gravimetric) particle concentration measurements were observed.

The exhaust aftertreatment system investigated here system provides a reliable solution yielding a conversion of up to 70% (ESC/ETC) for elemental carbon at a moderate back pressure.

INTRODUCTION

Particulate matter (PM) in the exhaust of diesel engines is subject to increasing criticism as being a potential risk to public health. Reducing the emission of PM is therefore a very important topic in the development of HD engines and exhaust after treatment systems.

As shown in [2, 3] the EuroIV emission limits could be reached with the combination of a low emission engine with the novel aftertreatment system called PM-KAT.

The primary goal of this study was not to reach the Euro IV PM emission limits what was demonstrated before. Instead, this study comprises a full characterization of a new system.

THE FILTER SYSTEM

The novel system consists of specially designed stainless steel catalyst structures of 50 μm thickness. A corrugated and a flat foil are wound together into a spiral-section cell matrix, called BS-structure.(Fig. 1). The surface of the BS-structure is coated with metal micro spheres, which affect the gas flow over the cell walls due to an significant increase of the surface roughness. Its characteristics comprise both high soot deposition efficiency with an high soot storage capacity, permeability for particulate oil ash compounds and an overall low pressure drop over the system.

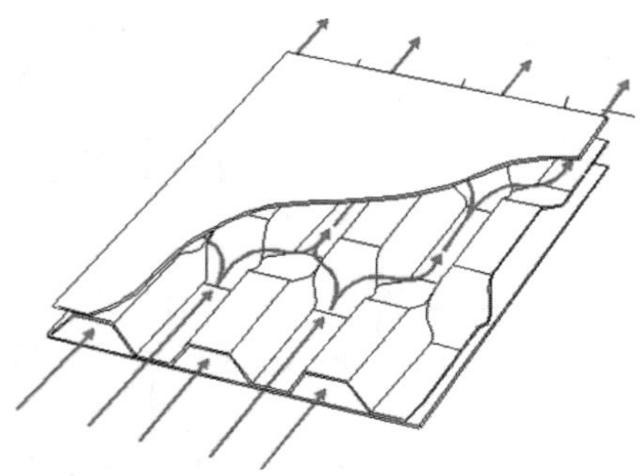

Fig. 1 Schematic drawing of the sandwich structure

Like the standard metal substrates with linear cell structures, which have been used for a considerable time for catalytic converters, a high thermal conductivity as well as high thermal and mechanical stability also characterise the aftertreatment system. Moreover, the deposition structures can be variously coated increasing the particle deposition as well as the conversion of limited gaseous exhaust components like HC (hydrocarbons) and CO. For HDV application the structures are continuously regenerated by soot oxidation processes which are significantly enhanced by catalytically generated NO_2 with an oxidation catalyst.

In contrast to conventional diesel particulate filter (DPF) systems this system allows the continuous reduction of soot particulate emissions at an favorable pressure drop level without complex regeneration techniques and clogging phenomena. It represents a reliable solution for the modern HDV engine technologies. Current research focuses on the further enhancement of soot particle deposition and storage capabilities based on experimental work in laboratory and test bench-scale as well as numerical modeling.

LABORATORY INVESTIGATIONS

MODEL CATALYTIC SYSTEM

The development and optimization of the catalyst structure mainly relied on laboratory investigations, which allowed a fast and economic screening of feasible deposition structures. Moreover, these laboratory investigations under well-defined experimental conditions enable the characterization of the determining deposition mechanisms acting on the fractal-like soot agglomerate particles.

Deposition of typical HDV engine soot particles is determined by diffusion, interception, electrostatics and thermophoresis [4]. The later can contribute to the high deposition efficiency during transient test runs. Particle deposition can be described by the fractional deposition efficiency $T(d_p)$:

$$T(d_p) = \frac{c_{in}(d_p) - c_{out}(d_p)}{c_{in}(d_p)}$$

$c_{in}(d_p)$ and $c_{out}(d_p)$ represent the particle number concentration per size class before and after the catalyst system. As an integral parameter to describe the overall particle removal efficiency the number based total deposition efficiency E_0 is applied:

$$E_0 = \int_{d_{p,\min}}^{d_{p,\max}} T(d_p) c_{in}(d_p) dd_p$$

The mechanical mobility B_0 is the parameter determining the soot particles' deposition behavior under exhaust gas conditions. Therefore the model soot aerosol for the laboratory investigations was conditioned in an extensive experimental setup to yield mobility based particle number distributions similar to the ones observed in HDV exhaust gas (Fig. 2).

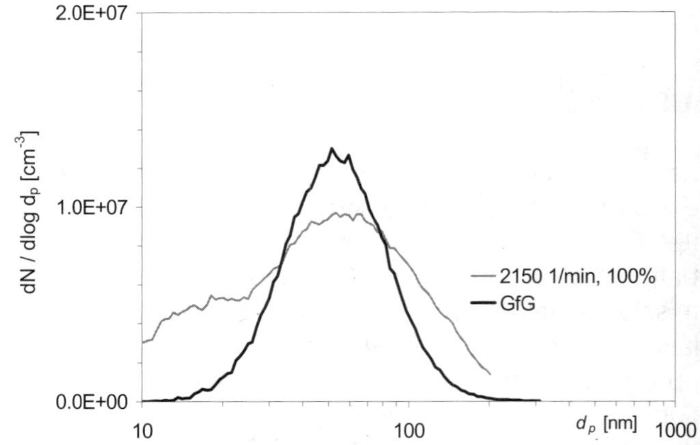

Fig. 2 Mobility based particle size distribution of HD engine and laboratory model soot (GfG).

The comprehensive experimental setup has already been described in a previous paper [5]. The model soot aerosol was generated by spark discharge (Palas GfG 1000, [6]) with an agglomeration, neutralization and mixing conditioning as shown in Fig. 3.

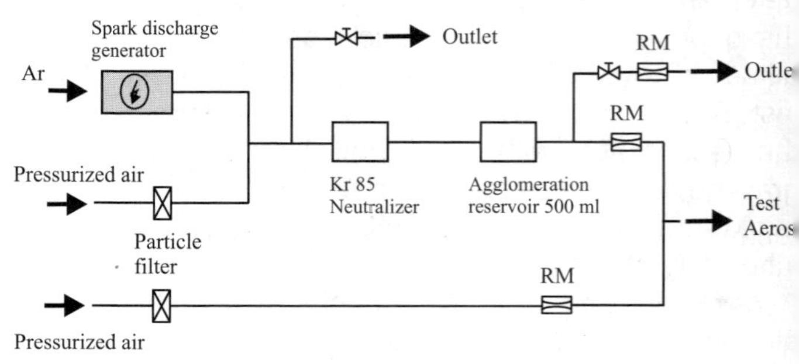

Fig. 3 Setup of the spark discharge soot aerosol generation and conditioning system.

With this setup the particle size distributions observed in HDVs can be modeled for a wide number of test conditions. The particle size distribution was measured with an SMPS-System (TSI EC 3071 and CPC 3025 A) and a dynamic dilution system (TOPAS DDS 560). For data analysis a resolution of 16 particle size classes per decade yields a high signal to noise ratio of SNR > 35.

Therefore the experimental system allows very detailed investigation of particle removal characteristics.

For an efficient screening process of the numerous deposition structures a novel isothermal flat bed reactor (FBR) had been designed which can rapidly be opened to change the catalyst structures and investigate deposition patterns macroscopically and microscopically (SEM). It can be heated to 450°C and allows the investigation of temperature effects [7]. The catalyst structures are fit into the flow channel with a flown through cross sectional area of 10 x 6 mm. Due to the labyrinth style sealing the FBR is leak proof under typical test conditions. The pressure drop in the model system across the different catalyst structures is found to be 5 to 10% higher than in the real exhaust system. However, relative changes in pressure drop resulting from structure variation are within 2% accuracy compared to the real catalyst system. For space velocities (SV) lower than 30000 h^{-1} diffusional losses in the test system were significant. They were measured for an empty FBR ($T(d_p)_{empty}$) and thus the deposition efficiency measurements of the structures for low SVs ($T(d_p)_{tot}$) were corrected applying [8]:

$$T(d_p) = T(d_p)_{tot} - T(d_p)_{empty} + T(d_p)_{tot} \cdot T(d_p)_{empty}$$

The evaluation of the single structures' ability to remove fractal-like soot particles was based on experiments with relatively low soot mass concentration so that the increase of soot removal by soot deposits did not influence the measurements. Therefore the laboratory experiments aimed at the determination of the minimal soot deposition efficiency of the catalyst structures which can be significantly increased by soot deposits [7].

PARTICLE DEPOSITION MEASUREMENTS

Fig. 4 shows $T(d_p)$ determined for the uncoated plain BS structure as well as two different types of metallic micro particle coatings. Particles with mobility diameters d_p < 50 nm are more pronouncedly deposited due to particle movement induced by Brownian diffusion. Coatings with about 350 g/l stainless steel micro spheres (A) or irregularly shaped steel particles (B) significantly enhance particle deposition.

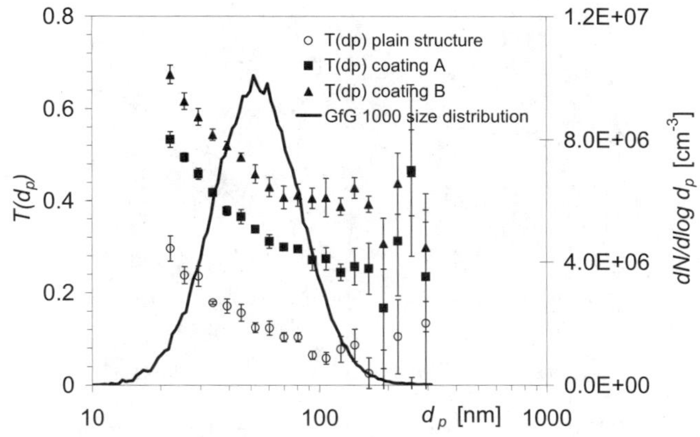

Fig. 4 Size resolved particle deposition efficiency *$T(d_p)$* for the corrugated stainless steel structure and two different coatings.

The number based total deposition efficiency E_0 decreases with increasing flow velocity for the uncoated structure and increases for all varieties of coating applied (Fig. 5). An alternative approach based on partial filtration through a metal fleece also yields an initial deposition efficiency, which decreases from 0.5 to 0.44 for the experimental conditions.

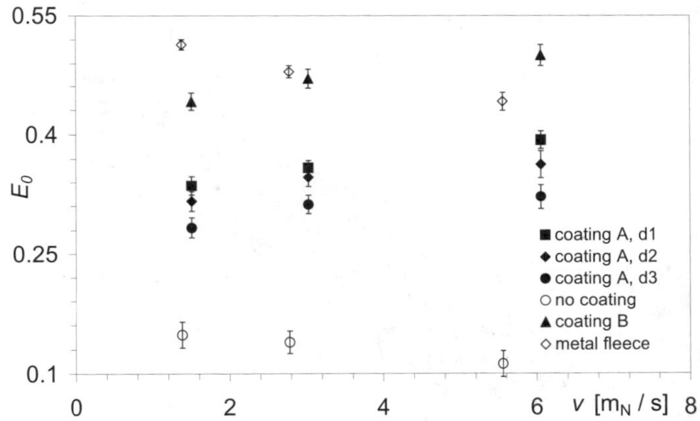

Fig. 5 Number based total deposition efficiency E_0 for the structure with different coatings and an alternative approach based on metal fleece as a function of flow velocity.

Decreasing the diameter of the coating spheres ($d_1<d_2<d_3$) increases E_0 due to the higher surface/volume ratio. The effect of the micro sphere coating on the particle deposition is shown in Fig. 6.

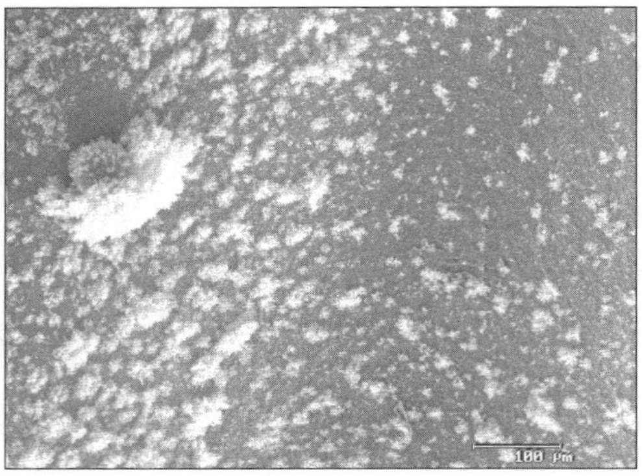

Fig. 6 SEM photograph illustrating the influence of the micro sphere coating on the soot particle deposition.

CFD-MODELLING AND DEPOSITION MECHANISM

The investigation of the particle deposition mechanisms is supported by CFD calculations of the flow field. The model comprises three elementary cells of one layer of the sandwich structure (Fig. 7). It has been finally validated by a pressure drop accuracy of 15% compared to the experimental results. Grid independency of the numerical solution was obtained for 1,125,000 computational cells.

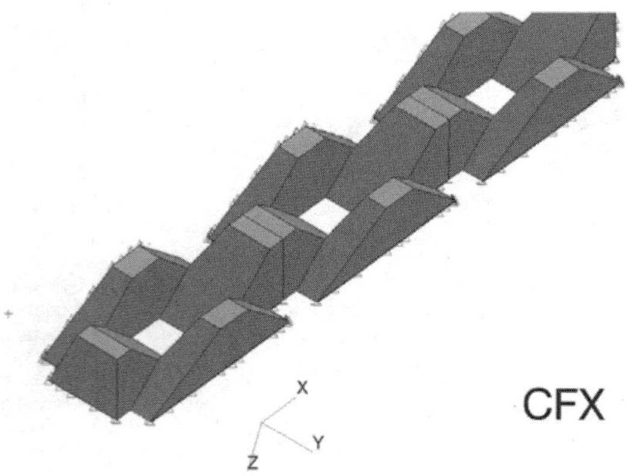

Fig. 7 3D rendering of the simulated geometric part

The flow field is found to be viscously dominated with only a limited number of vortices behind the major corrugations at high flow velocities. Therefore no isotropic turbulence is generated in the deposition structure. Particle removal is therefore a consequence of subsequent diffusional transport towards the wall where the concentration profile is homogenized at the narrow flown through diameters at each corrugation (Fig. 8).

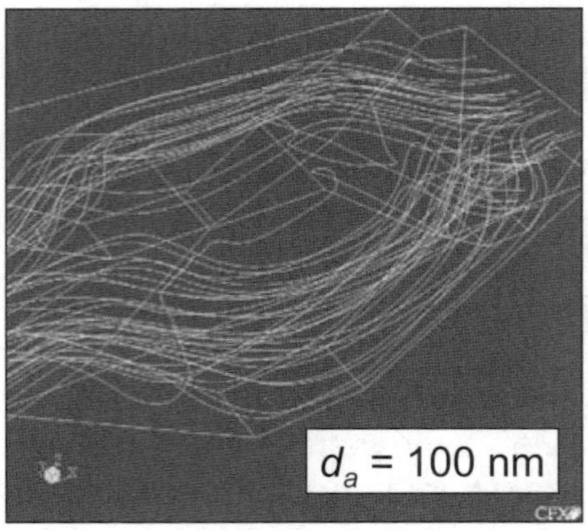

Fig. 8 Particle trajectories indicating the homogenization of the concentration profile at each corrugation

Therefore the flow design of the structures enable continuous transport of particles towards the walls where then specially coatings such as micro spheres capture the particles due to diffusion and interception.

ENGINE TEST BENCH MEASUREMENTS

EXPERIMENTAL PROCEDURE AND TEST CONDITIONS

A MAN D2866 LF30/31 (EuroII) engine is used for the experiments. Maximum exhaust gas volume flow 1400 Nm^3/h at nominal load The gravimetric PM results are compared with the results obtained with a MAN D2066 LF(EuroIII). The experiments are carried out on a hydraulic, electronically controlled test bench. Replacing the motoring phases with zero-load phases approximates the transient ETC cycle. However, from calibration experiments it is known that primarily the measured particle values tend to be better on the highly dynamic test bench. The test procedures ESC and ETC are described elsewhere [9].

A "sulfate free" diesel fuel (9ppm S) and a engine oil (Shell Rimula) with a sulfate ash content of 1.9 % was used. Oil consumption measured at full power 0.19 g/kWh.

The aftertreatment system is built in a muffler (Fig. 9). A 3.2l oxidation catalyst (200 cpsi, 40 g Pt/ft^3, Umicore) and 4 modules (2x 75 mm long) with an OM(Oberland Mangold) BS structure coated with microspheres.

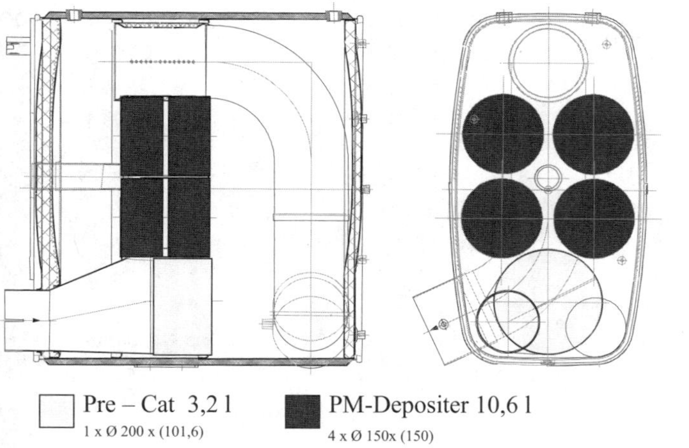

Fig. 9 Afterttreatment system in a TGA truck muffler (RTA)

PARTICLE SAMPLING AND ANALYSIS

The gaseous compounds before and after the system were measured with a Horiba MEXA 7100-exhaust gas analyzer and a Siemens SESAM 2 FTIR spectrometer.

The determination of particle mass was carried out in accordance with the legally prescribed procedure [ESC, 10] using a micro tunnel (NOVA Mikrotrol 4). The micro tunnel is based on the partial flow dilution procedure. Here, part of the engine exhaust gas is abstracted, diluted to a defined degree and fed through the measurement filter. The total particle mass on this filter is determined gravimetrically. The micro tunnel is also used as dilution device for the SMPS and PASS measurements (dilution ratio 5-50).

The soluble organic fraction was determined with a modified extraction method (dichloromethane/toluene) with accelerated solvent extraction ASE instead of an ultrasonic bath [11, 12].

For particle size distribution measurements a SMPS system (TSI 3936 with components DMA 3081 and CPC 3025) was used.

For direct measurement of the soot concentration in the exhaust a PASS (photo acoustic soot sensor, prototype TU Munich) system is used. This method of measurement involves the specific heating of soot particles by absorption of light produced by a modulated laser. The resultant acoustic wave is detected. The PASS signal detected is direct proportional to the concentration of soot [13, 14].

RESULTS

NO$_2$ CONVERSION AND EXHAUST GAS TEMPERATURE

The conversion efficiency of NO to NO$_2$ by the oxidation catalyst is shown in Fig. 10. Because of the temperature dependent equilibrium between NO and NO$_2$ the concentration of NO$_2$ decreases at temperatures above 300°C.

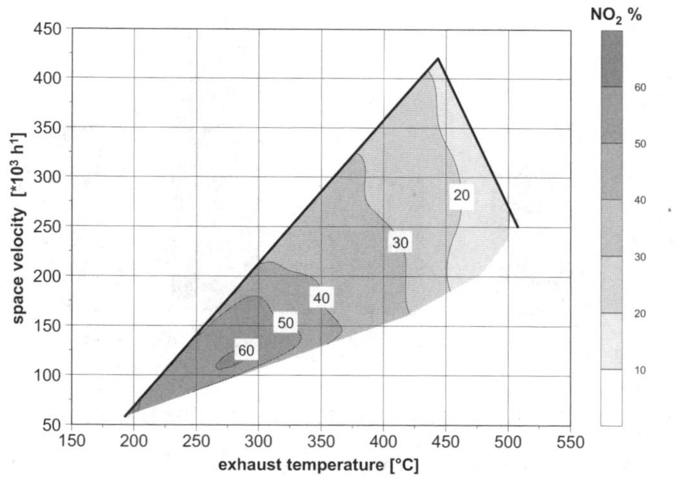

Fig. 10 Percentage of NO$_2$ converted by the oxidation catalyst

In addition to a preferably high NO$_2$ content, a high exhaust temperature is necessary for the oxidation of the deposited soot. The exhaust gas temperatures in the ESC and ETC test measured before the system are shown in Fig. 11.

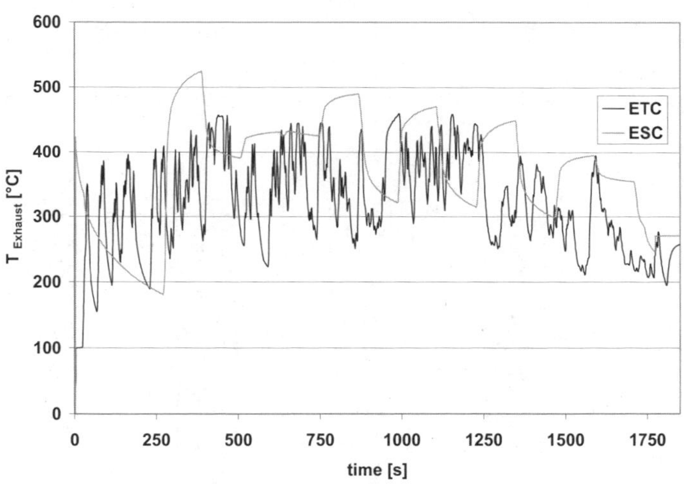

Fig. 11 Exhaust gas temperatures (before the system) in the ETC and ESC tests

GRAVIMETICAL RESULTS IN ESC AND ETC

Fig. 12 shows the PM mass result (average of 3 samples). For the D2066 LF measurements we observed a reduction of 45% in the ESC/ETC (exhaust back pressure 90mbar at speed C 100% load). For the D2866 LF 30/31 engine the reduction was 30/50% in the ESC/ETC (exhaust back pressure 104mbar at speed C 100% load).

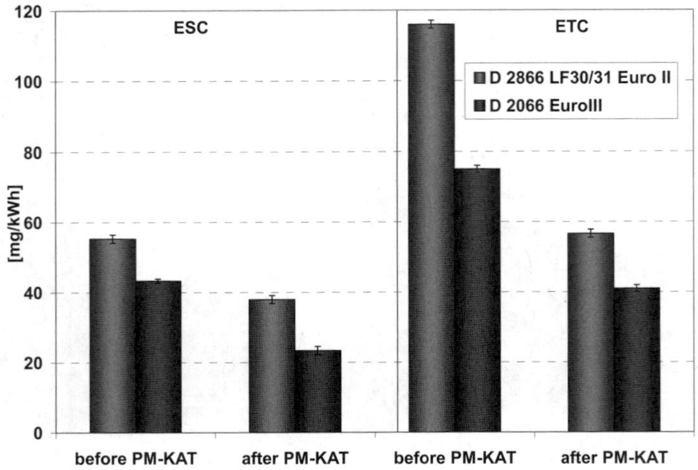

Fig. 12 PM mass results before and after the system

If we consider the PM reduction potential of the system it will be possible to match the combustion parameters as fuel injection and turbo charging to accomplish the demands of EuroIV in the ESC as well as in the ETC [2].

PARTICLE ANALYSIS

The measured SOF content before and after the system is shown in Fig. 13. We had a conversion of 60/53 % in the ESC/ETC. Due to a lack of experimental data the sulfate content was calculated estimating conversion factors of 2% before and 80 % after the system. From earlier measurements it is well known that there is a good correlation between the calculated sulfate values with measurements[12] .

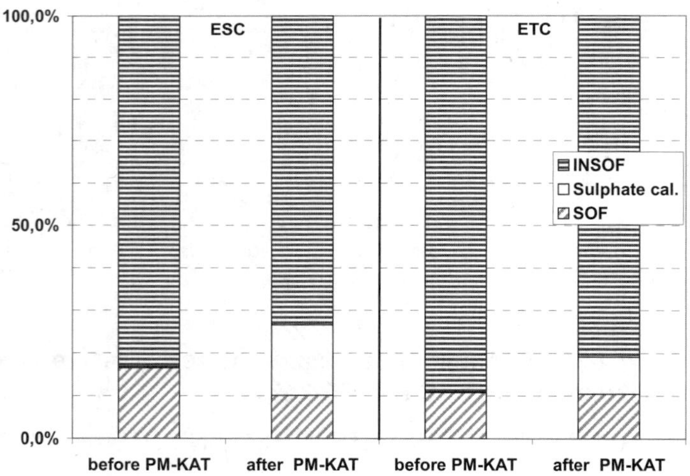

Fig. 13 Measured / calculated particle composition before/after the system

PASS SOOT CONCENTRATION

Fig. 14 shows the results of the PASS soot concentration measurements for the ESC. The shown PASS signal is direct proportional to the soot concentration. The measurement before the system was repeated and showed a good reproducibility. The soot conversion rates are between 40 and 65% except for the idle mode.

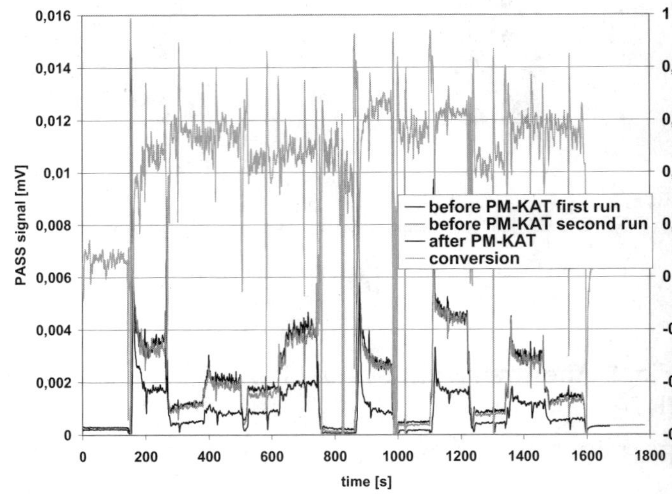

Fig. 14 Soot emission before and after the system in the ESC measured with the PASS system

The conversion measured with the PASS system is higher than the gravimetrical results because of the formed sulfates after the system that are not measured with the soot specific PASS system.

The soot emission before and after the system in the ETC shows Fig. 15. Here the micro tunnel was set to a constant value dilution factor of 1:10.

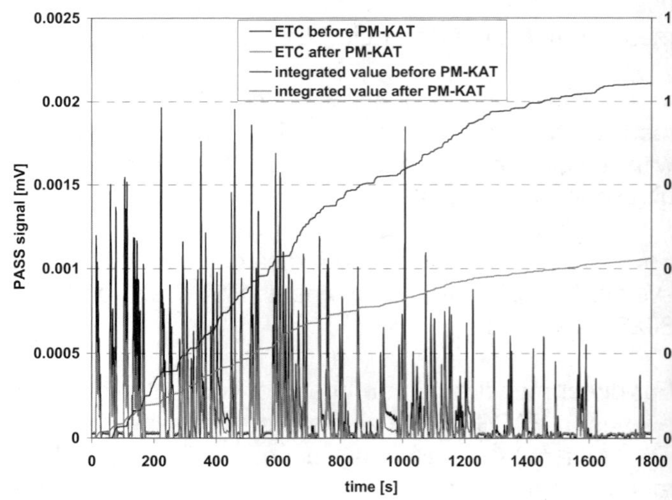

Fig. 15 Soot emission in the ETC before and after the system measured with the PASS system

In Fig. 15 and Fig. 16 we can clearly see the good soot deposition efficiency at high emission peaks with soot conversion of up to 70%. The soot conversion (integrated value) over the whole ETC is 50% which correlates well with the PM mass value.

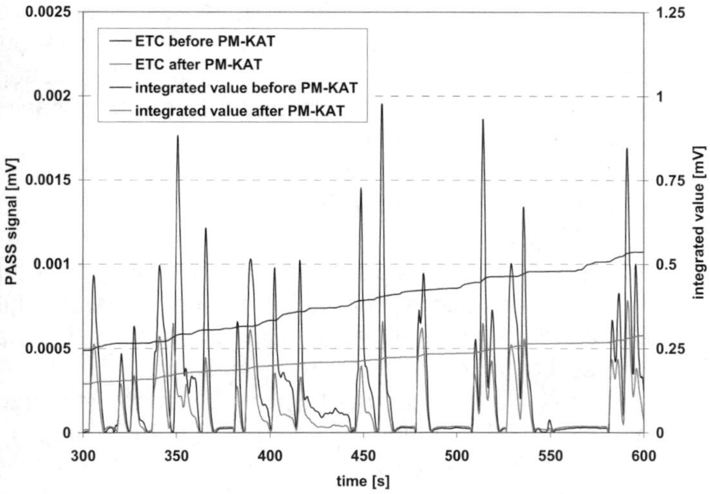

Fig. 16 Cut out from the city part of the ETC

Based on the measurements after the system no significant resuspension ("blow off") of deposited agglomerated particulate matter could be observed in the transient ETC test.

PARTICLE SIZE DISTRIBUTION

The agglomerate particle size distribution was determined with the SMPS system. To achieve certain persistence the duration of every mode of the ESC was prolonged from 2 to 8 min. The measurement started after 5min. The y-axis value in Fig. 17 shows the calculated weighted cumulative value over the whole test. The measured single mode results in Particle/cm^3 were recalculated in Particle/kg.

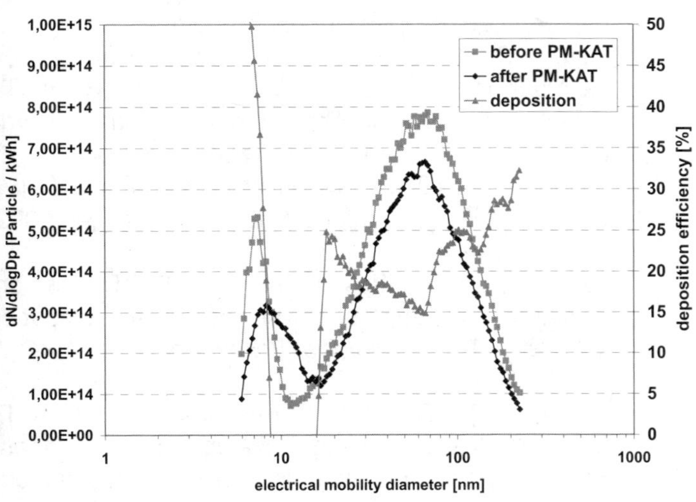

Fig. 17 Particle size distribution ESC before and after the system and deposition efficiency

The obtained values where then step weighted and power corrected accumulated for the whole test. In Fig. 17 we can see a bimodal size distribution. We can see the nucleation mode with a maximum at diameters smaller than 10 nm and the accumulation mode with a maximum at 60 nm. It is supposed that the nucleation mode particles consist mainly out of hydrocarbon or sulfuric acid droplets. The core of these particles is believed to consist out of oil ash particles, which serve as condensation particles [15]. This mode is difficult to reproduce and strongly influenced by the sampling parameters (dilution, temperature, humidity, time, etc.). The accumulation mode describes the actual soot particles. This mode is highly reproducible. The soot particles can also contain condensed hydrocarbons, sulfuric acid and oil ash particles.

Due to the formation of sulfuric acid the deposition efficiency for the nucleation mode particles is negative. Because of their small size these particles have only a small mass and contribute only little to the total mass. The deposition efficiency in the accumulation mode starts at 20 % and increases to 40% for the larger particles. This is in agreement with the mass based value because it is supposed that some sulfate particles have already been formed on the soot particle surface.

LONG TIME MEASUREMENTS (SMPS)

In Fig. 18 the long time behavior (2h) of the particle size distribution at a constant stage point (ESC step 7 1424min^{-1}, 320Nm) measured before the system is shown. The good reproducibility of the accumulation mode (10 % variation) showed the constant emission of the engine. Compared to that the nucleation mode showed a decrease of the particle concentration over 50 min and after that a constant behavior. This shows the strong influence of the nucleation as described before.

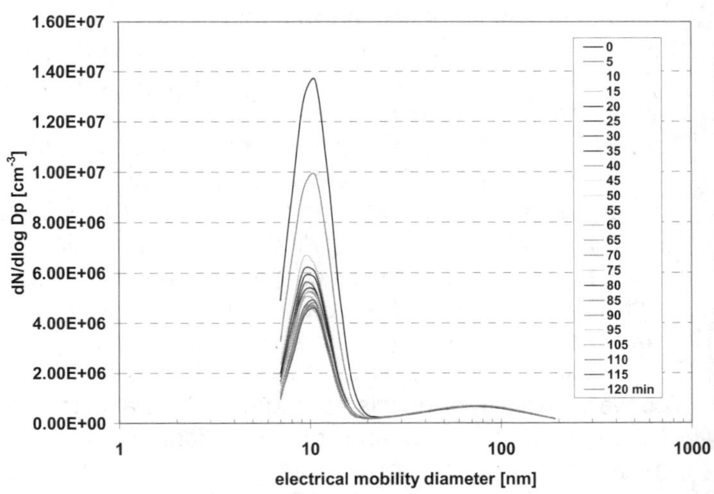

Fig. 18 Long time measurements with the SMPS system ESC stage 7 (speed A 25% load) before the system

In Fig. 19 the long time behavior (1h) of the particle size distribution after the system is shown. There is also the variability of the nucleation mode, which declines over the time.

The accumulation mode increases slightly during the first 15 min up to 10% to stay then stable. This behavior shows clearly the good soot storage and oxidation capacity of the investigated aftertreatment system. The measurements were repeated for a high load point (speed C 100%) and show similar results.

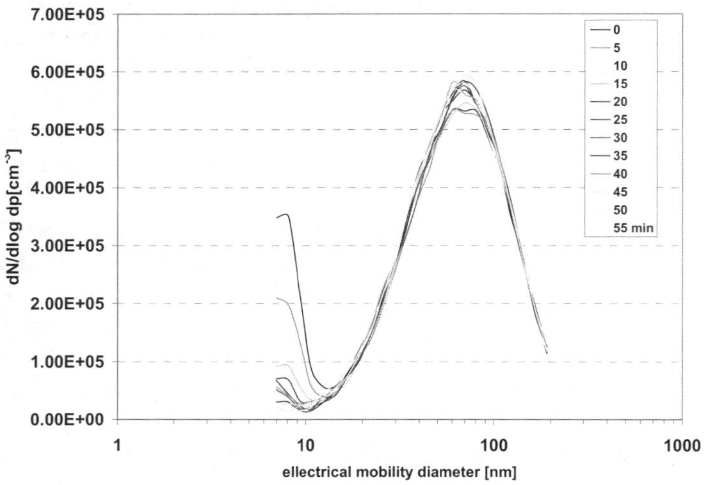

Fig. 19 Long time measurements with the SMPS system ESC stage 7 (speed A 25% load) after the system

CONCLUSION

An extensive laboratory-screening program on a specially designed model diesel exhaust gas test bench supported the development of novel stainless steel structures. With model soot experiments it was found that about 350 g/l of coating with micro particles led to a significantly increased deposition efficiency of up to 60%. Smaller particles with d_p< 50 nm are more pronouncedly removed due to diffusion driven deposition. With the novel aftertreatment system applied on engines matched for EuroIV the conversion rates of the system measured here are sufficient to fulfil the Euro IV emission limits. When applied for Euro IV optimized engines the here measured conversion rates are highly sufficient to undergo the Euro IV emission limits. With optimized structures there is a potential to obtain even higher soot conversion rates. It is favorable that the soot conversion values for this structure in the transient test (ETC) are higher compared to the stationary test (ESC) because the ESC is much easier controllable through engine modification than the ETC. Care has to be taken with the fuel sulfur content. Even with a quite low fuel sulfur value of 9 ppm about 30 % of the Euro IV PM limit is reached only through sulfate formation. Therefore with the soot specific PASS measurements we obtain higher conversion rates compared to the gravimetric values. The often discussed problem of particle reentrainment ("blow off") in open PM structures like this system could not be found. The deposition efficiency determined with the SMPS system is in agreement with the gravimetric values. The "negative" conversion rates for the nucleation mode are most likely caused by sulfuric acid formation. The long time SMPS measurements show clearly the good reproducibility of the accumulation mode and the good soot storage and oxidation capacity of the investigated system

ACKNOWLEDGEMENTS

This work is part of the project "Katalytisches System zur filterlosen kontinuierlichen Rußpartikelverminderung für Fahrzeugdieselmotoren"(engl. „Catalytic system for the filterless continuous soot particle reduction for diesel engine vehicles") supported by the Bavarian Research Foundation (BFS), Munich. The support from Dr.-Ing. H.-J. Schmid during the CFD simulations is gratefully acknowledged. The authors thank Dr. Haisch and Dr. Beck for their assistance in setting up the PASS-System. We would also like to express our gratitude to Dipl.-Ing. Stephan Reisch for the D2066 LF results and Tobias Fleischmann for technical assistance at the engine test bench.

REFERENCES

1. Döring, A.; Hübner, A.; Jacob, E, „Verfahren zur Abscheidung von Feinstpartikel aus dem Abgas von Brennkraftmaschinen"199 34 932 (1999)
2. Jacob, E.; D'Alfonso, N.;Döring, A.; Reisch, S.; Rothe, D.; Brück, R.; Treiber, P. (2002) " PM-KAT®: a Non-Blocking Solution to Reduce Carbon Particulate Emissions from EuroIV Heavy Duty Engines" 23. Internationales Wiener Motoren-symposium, Band 2: Fortschritts-Berichte VDI-Reihe 12 No. 490 (VDI Verlag, Düsseldorf) http://emitec.de/pdf/wien_paper2002man.pdf
3. Rothe, D.; D'Alfonso, N.; Reisch, S.; Jacob, E.; Heimgärtner, C.; Leipertz, A.; Kremer, H.; PM-KAT®: Erfahrungen und Möglichkeiten zur Minderung von Dieselruß für Euro IV Nutzfahrzeugmotoren, in Leipertz, A. (Edit.) Motorische Verbrennung (VI. Congress) BEV Heft 2002.1, Esytec, Erlangen 2002, P. 487-496
4. Johnson J.E., Kittelson D.B. (1996) "Deposition, diffusion and adsorption in the diesel oxidation catalyst". *Appl. Catalysis B:Environmental* 10:117-137.
5. Messerer A., Rothe D., Poeschl U., Niessner R. (2004) "Advances in the Development of Filterless Soot Deposition Systems for the Continuous Removal of Diesel Particulate Matter". *Topics in Catalysis* 30/31*:* 247-250.
6. Evans D.E., Harrison R.M., Ayres J.G. (2003) "The generation and characterisation of elemental carbon aerosols for human challenge studies". *J. Aerosol Sci.* 34:1023–1041.
7. Messerer A., Schmid H.-J., Knab C., Poeschl U., Niessner R. (2004) „Erhöhung der Abscheidung ultrafeiner Dieselrußpartikeln durch Mikrokugelbe-schichtung auf metallträgerbasierten Katalysator-

strukturen". *Chemie Ingenieur Technik 76*: 1092-1096.
8. Brockmann J.E. (1993) „Sampling and transport of aerosols". In Aerosol Measurement: Principles, Techniques and Applications (Edited by Willeke K. and Baron P.A.). Van Nostrand Reinhold, New York.
9. http://www.dieselnet.com/standards
10. EU Guideline 1999/96/EG
11. Institute of Petroleum, „Extraction of the soluable organic fraction (SOF) of diesel particulates colleted of filters – Soxhlet extraction method", IP Method 443/99
12. Rothe, D., Kamm, S.; Baumann, H.; Jacob, E.; Niessner, R.; Pöschl, U. (2001) „Determination of the Sulfate Content of Diesel Exhaust Particulate Matter", *Journal of Aerosol Science*, 32: S75-S76,
13. Schindler, W.; Haisch, C.; Beck, H. A.; Niessner, R.; Jacob, E.; Rothe, D.;"A Photoacoustic Sensor System for Time Resolved Quantification of Diesel Soot Emissions", *SAE Technical Paper Series* 2004, 2004-01-0968
14. Beck, H. A.; Niessner, R.; Haisch, C. (2003) "Development and characterisation of a mobile photoacoustic sensor for on-line soot emission monitoring in diesel exhaust gas." *Anal. Bioanal. Chem.* 375: 1136-1143.
15. Abdul-Khalek, I. S.; Kittelson, D. B.; Graskow, B. R.; Wie, Q, Brear, F.; „Diesel Exhaust Particle Size: Measurement Issuses and Trends", *SAE Technical Paper Series* 1998, 980525

CONTACTS

Dipl.-Chem. Dieter Rothe
MAN Nutzfahrzeuge Gruppe
Business Unit Engines
Nürnberg, Germany
Dieter_Rothe@mn.man.de

Dipl.-Ing. Frank Ingo Zuther
MAN Nutzfahrzeuge Gruppe
Business Unit Engines
Nürnberg, Germany
Frank_Zuther@mn.man.de

Dr. Eberhard Jacob
MAN Nutzfahrzeuge Gruppe
Business Unit Engines
Nürnberg, Germany
Eberhard_Jacob@mn.man.de

Dipl.-Ing. Armin Messerer
Institute of Hydrochemistry
Technical University of Munich, Germany
Armin.Messerer@ch.tum.de

Dr. Ulrich Pöschl
Institute of Hydrochemistry
Technical University of Munich, Germany
Ulrich.Poeschl@ch.tum.de

Prof. Dr. Reinhard Niessner
Institute of Hydrochemistry
Technical University of Munich, Germany
Reinhard.Niessner@ch.tum.de

Dipl.-Ing. Christian Knab
Oberland Mangold
Eschenlohe, Germany
knab@oberland-mangold.de

Matthias Mangold
Oberland Mangold
Eschenlohe, Germany
m.mangold@oberland-mangold.de

Christian Mangold
Oberland Mangold
Eschenlohe, Germany
c.mangold@oberland-mangold.de

DEFINITIONS, ACRONYMS, ABBREVIATIONS

ASE:	Accelerated Solvent extraction
CFD:	Computational Fluid Dynamics
DPF:	Diesel Particulate Filter
OM:	Oberland Mangold
EC :	Elemental Carbon
ESC:	European Stationary Cycle
ETC:	European Transient Cycle
FBR:	Flat Bed Reactor
HDV:	Heavy Duty Vehicle
PASS:	Photoacoustic Soot Sensor
PM:	Particulate Matter
PM-KAT:	germ.: **PM-Kat**alysator (**P**articulate **M**atter-**Cat**alyst)
RTA:	Roth Technik Austria (Muffler manufactory)
SEM:	Scanning Electron Microscopy
SMPS:	Scanning Mobility Particle Sizer
SNR:	Signal to Noise Ratio
SOF:	Soluble Organic Fraction
TGA:	Trucknology Generation A (New MAN Truck series)

2007-01-0232

New Flow-Through Trap System Targeting 50% PM Removal for Diesel Emission Control

Yongtaek Choi, Zhongyuan Dang, Ron Stone and Marty Morrill
Sud-Chemie Inc.

Donald Floyd
Porvair

ABSTRACT

A diesel particulate Flow Through Trap (FTT) system, composed of either silicon carbide (SiC) ceramic foam or stainless steel foam, has been developed and evaluated. The basic concept of the system is to configure several different pore sizes of the foam filters gradually, from large pore sizes to small pore sizes, in order to achieve high efficiency of diesel PM reduction and simultaneously low pressure drop. Each component of the FTT system was coated with Sud-Chemie's proprietary catalyst support and further catalyzed with platinum for better performance. Multiple configurations of the FTT system were evaluated by varying pore sizes (20~50 PPI), substrate sizes, and the number of beds. Tests were conducted using 0.74 Liter and 1.86 Liter diesel engine / generator sets.

Based on the results of these engine tests, the FTT is found to be an effective device to remove more than 50% of PM from the diesel engine exhaust. The Balance Point Temperature of the FTT systems was around 260°C, and the FTT prototypes can be regenerated successfully without any further active regeneration control when the exhaust gas temperature is maintained at 280°C and higher. For the system tested in this paper, the size should be bigger than 1.3 times of engine displacement to meet the target, 50% of PM removal efficiency.

INTRODUCTION

For the removal of PM (Particular Matter) from the diesel engine, it is not sufficient for a Diesel Oxidation Catalyst (DOC) to remove 30% of total PM as in most applications. The DOC catalyst mainly works on the Soluble Organic Fraction (SOF) portion of the PM and this portion varies with the engines and operating conditions. A Diesel Particulate Filter (DPF) is a more effective tool for removing greater than 85% of PM, however, it also has various limitations related to continuous regeneration, system reliability, and economics of the system. A device which gives PM removal efficiency in-between that of a DOC and DPF is required in many applications, like retrofitting in-use diesel vehicle. [1~3]

As the interest in emission control devices has been growing recently, the California Air Resources Board (CARB) classifies various diesel emission control systems, based on their verification procedure, into three categories as outlined in Table1 [4]. Many strategies such as diesel particulate filters, diesel oxidation catalysts, fuel additives, selective, catalytic reduction systems, exhaust gas recirculation systems, and alternative diesel fuels meet the verification requirements.

Table 1. Verification Classifications for Diesel Emission Control Strategies

Pollutant	Reduction	Classification
PM	< 25%	Not verified
	> 25%	Level 1
	> 50%	Level 2
	> 85% or < 0.01 g/bhp-hr	Level 3

For the Level 2 device several types of filter substrate, such as wire mesh, metal flow-through with notches, and ceramic or metal foam can be used. The CARB has verified the Environmental Solutions Worldwide (ESW) Particulate Reactor™ system which employs a catalyzed wire mesh substrate to achieve at least a 50 percent reduction in particulate matter emissions, qualifying it for Level 2 verification. Also, the CARB has verified the Donaldson Company's DFM diesel multi-stage filter (DMF) system and Johnson Matthey's PCRT system which utilizes a metallic foil/metal fleece substrate by Emitech. Emitech substrate is composed of porous flat foils and porous layer made of sintered metal fiber fleece which allow trapping particles flowing through mixing foil into fleece material [3].

Operating as a "partial DPF" system, Sud-Chemie's newly developed FTT unit is composed of several bricks of metal foam filters. This paper describes the basic concept of the FTT, the characterization of the components, and the performance of the system through engine testing.

CONCEPT OF FTT SYSTEM

The main design concept of the FTT system involves uniform distribution of PM through the entire length of the foam filter assembly to minimize the backpressure of the device. A typical FTT system uses 4~8 foam filter segments, which are configured to create a porosity gradient by strategically arranging segments of different pore size and densities.

In order to meet the target PM reduction of 50%, foam substrate filters are chosen for their effectiveness at trapping PM, especially with high pore density foams. However, the back pressure increase due to the formation of a PM cake-type build-up on the inlet surface of the filter is critical. To overcome the formation of this build-up, many diesel PM filter systems have been developed by incorporating different filter material, geometric configuration, regeneration technologies and control/monitoring options.

It is well known that the pressure drop of porous media can be described by the following equation derived from Darcy's law.

$$\Delta P = \frac{\mu \cdot L}{k} \cdot \frac{Q}{A} \qquad (1)$$

In diesel exhaust systems, Q stands for Diesel exhaust gas flow rate, μ is for Diesel exhaust gas viscosity, A is Inlet front area of the foam filter exposed to flow, *L* is the Thickness of the foam filter, and k is the Permeability factor. The k value is related to the pore density of the foam substrates. Increasing inlet front area (A) and decreasing thickness (*L*) could reduce the back pressure build-up.

Dupuit was the first scientist to use theoretical principles to interpret Darcy's equation. A consequence of Dupuit's study is the inclusion of a quadratic velocity term added to the above equation thus modeling the resistive form drag imposed on the fluid by the solid permeable medium where the coefficient C is the form coefficient of the porous layer. Replacing (Q/A) of the equation (1) with the linear velocity (U), the Hazen-Dupuit-Darcy equation (2) can be derived.

$$\frac{\Delta P}{L} = \frac{\mu}{K} \cdot U + C \cdot \rho \cdot U^2 \qquad (2)$$

Even though many system designs can be considered to increase the inlet front area of foam filters to reduce the pressure drop, Sud-Chemie adopted a new concept of PM filter device (FTT, or Flow Through Trap) to achieve high efficiency of diesel PM reduction as well as low pressure drop. The approach of the Sud-Chemie FTT system is to enhance Permeability Factor (k) of the foam filter material, instead of changing the geometric parameters of the foam filter.

With deep bed filters, it is found that the distribution of PM accumulation across the filter element is not uniform. Usually the frontal section traps more particulate than rear section of the filter element. A preferred way of designing the system is to stack the filter elements from large pore size filter elements to small size pore filter elements, as shown in Figure1. The advantage of this design is that the PM distribution through the system is more uniform because the PM accumulation on the front large pore size filter element is lower than that on small pore size filter element. Only partial PM is trapped and the rest of the flow goes to the next filter elements.

With a combination of the graduated pore density (from low to high) of foam substrates, the Permeability Factor (k) of foam substrates will increase in comparison with the single high pore density foam substrate as shown in Figure1. The filtration modeling calculation indicates that the configuration of the graduated pore density of foam substrate will result in a more uniform distribution of diesel PM throughout the entire foam substrate trap as opposed to a single high pore density foam substrate. This uniform distribution of diesel PM will keep PM filtration as a deep-bed filtration mechanism and also avoid developing the cake-type filtration and flow channel being blocked, leading to the FTT device having low pressure drop and high PM reduction efficiency.

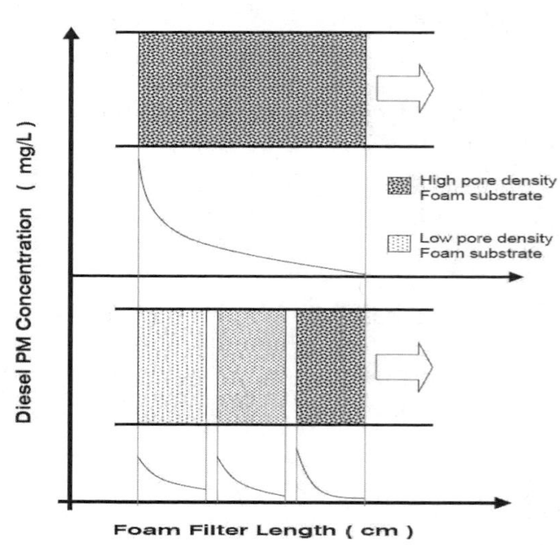

Figure 1. The PM accumulation pattern through an FTT in one single foam and graduated configured foams

Deep bed filter element foams can be made of SiC, alumina, zirconia, silica aluminate, as well as wire-mesh filter elements made of stainless steel and ceramic fibers. These foam and wire-mesh elements are commercially available. The filter elements can be coated with catalyst to assist the combustion of particulate and gaseous emissions, with Pt catalysts preferred for coating. Over Pt catalysts, NO from the engine exhaust is converted into NO_2 which then helps oxidize soot.

$$NO + O2 \rightarrow NO_2$$

$$NO_2 + C \rightarrow NO + CO$$

Coating of Pt catalyst on the filter media lowers the combustion temperature of the soot and helps continuous passive regeneration of the system. Pt catalyst is also active for the removal of gaseous CO and hydrocarbons. Catalyst coating on the filter media is done by conventional methods.

CERAMIC AND METAL FOAM FILTERS

In general, SiC Ceramic Foam Filter is composed of Silicon Carbide(SiC) and several different ceramic materials such as aluminum oxide, zirconia (ZrO_2). It has many advantages such as high porosity, large surface area, high temperature mechanical properties, and thermal shock resistance. Figure 2 illustrates the photos of Foseco's commercial SiC Ceramic Foam (a), and the porosity of this ceramic foam filter (b). This filter has been widely used in the casting industry for a long time as a foundry filter. The effectiveness of Foseco foundry filters is known as a result of their foam structure, some 90% of the volume consists of open, reticulated, interconnected pores. [5]

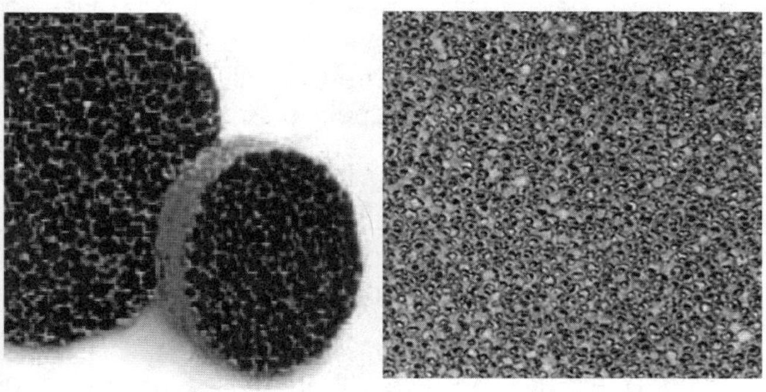

(a) (b)

Figure 2. FOSECO SiC Foam Filters (a) and the porosity of 30 ppi filter (b)

Conversely, metal media and cellular construction provide a more robust substrate than ceramic substrate, while providing a tortuous path and deep bed filtration. It can be produced in a range of metals, pore sizes, densities and shapes.

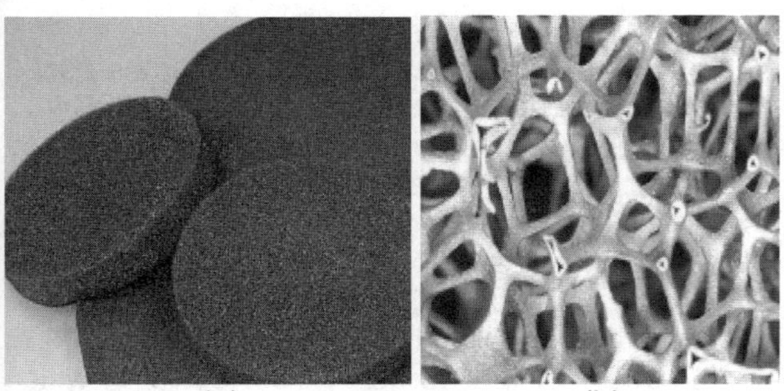

(a) (b)

Figure 3. Photos of (a) the metal foams and (b) the SEM images of the cell and the connecting struts

Therefore, the density and porosity can be independently modified to optimize filtration and back pressure. Figure 3 shows the shapes of the metal foam filter (Metpore® EFCS supplied by PORVAIR) and its pore structure. The SiC ceramic foam filters have more geometrical surface area and economic merits while metal foam filters have benefits of mechanical strength, fast thermal response, and high internal surface area.

EXPERIMENTAL

FOAM FILTERS AS A CATALYST SUPPORT

Two types of foam filters, SiC Ceramic foam and stainless steel foam, have been used for the FTT system and engine tests. For SiC Ceramic Filters, three different pore densities from 20 ppi to 40 ppi of FOSECO filters were used. 4.66" D by 1" thick filter was used for the small engine (0.74L) tests and 5.66" OD by 1.38" thick filters were used for large engine (1.86L) tests. For metal foam filters, stainless steel metal foam supplied by PORVAIR was used [6]. Four different pore densities (20 ~ 50 ppi) with 7% material density were used. The size of the filter for 1.86 L engine was 5.6" OD by 1.0" thick. Detailed specifications of the metal foams are summarized in the Table 2.

Table 2. PORVAIR's Metpore® EFCS Material Property Data

Properties	Units	Specifications
Pore Density	Pores Per Inch	20ppi, 30ppi, 40ppi, 50ppi
Maximum Operating Temperature	°C	1600
Density Range	% Bulk	3~15
Coefficient of Thermal Expansion	x 10^{-6} in/in/°C	11
Thermal Shock Resistamce	n/a	Excellent
Modulus of Rupture at 25 C	Psi	210
Compressive Strength @ 5% density	N/mm^2	0.86
Tolerance Specifications are +-2% for Length and Width, +-0.06" Thickness for an as fired part.		
Tolerance Specifications are +-0.10" for Length and Width, +-0.06" Thickness for a post fired cut part.		

WASHCOATING AND PRECIOUS METAL COATING

A proprietary washcoat developed by Sud-Chemie was used to coat the FTT foam segments [7]. Emphasis was placed on having uniform coating while meeting requirements to provide enough surface area for proper catalytic function and avoiding channel plugging during the wash-coating process. Catalyst loading was done by conventional means followed by curing, drying, and calcination.

FTT SYSTEM DESCRIPTION

Various FTT configurations were used by varying several different pore sizes of SiC ceramic foam and metal foam substrates. The thickness of individual foam is limited to the length of 1.0~1.38 inch for SiC ceramic foam and 1.0 inch for metal foam due to the intrinsic property of the foam production procedure. The diameter of the foam filter was 4.66 inch for 0.74L engine and 5.66 inch for 1.86 L engine. Figure4 shows the schematic diagram of FTT system composed of 4-segments of foam filters configured for a large pore to small pore gradient.

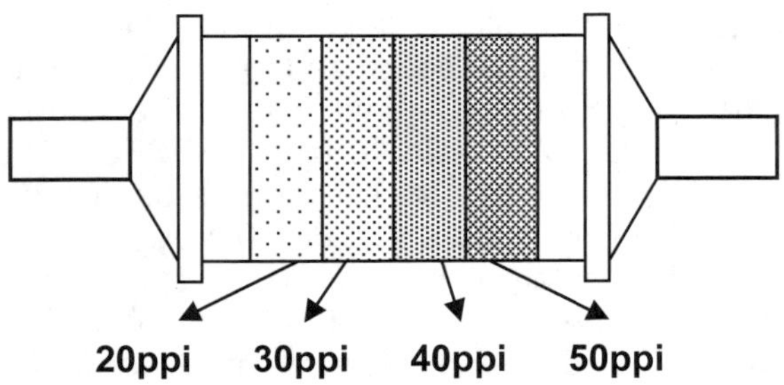

Figure 4. The Configuration of 4 Bed FTT System

Seven different types of FTT prototypes were prepared for engine testing. The detailed specifications and configuration of the system are summarized in Table 3. Volume ratio of FTT unit to engine displacement is important factor for FTT sizing and system design. The ratio of different FTT type varies 0.87~1.30 which is in-between size of DOC and DPF. For example, Type F FTT is composed of six 5.66" RD by 1" thick metal foams in series and total volume is 2.42 Liter which is 1.3 times of tested engine. All prototype samples are wash-coated and catalyzed before they are tested. Type G FTT used catalyzed wire-mesh in place of the first foam filter.

Table 3. The Description of FTT Sample Prototypes

Type	Substrate	Foam Layout (ppi)	Avg ppi	FTT Vol. (Liter)	Engine Size (Liter)	Vol. Ratio
A	SiC Foam	30/40/50	40	0.84	0.74	1.13
B	SiC Foam	30/40/50	40	1.67	1.86	0.90
C	Metal Foam	20/30/40 /50	35	1.61	1.86	0.87
D	Metal Foam	20/30/40 /40/50	36	2.02	1.86	1.08
E	Metal Foam	20/30/30 /40/40/50	35	2.42	1.86	1.30
F	Metal Foam	30/30/40 /40/50/50	40	2.42	1.86	1.30
G	Metal Foam	WM/30/40 /40/50/50	42	3.43	1.86	1.84

TEST ENGINES AND FUEL

The test engine was a Lister-Petter LPA2 two-cylinder diesel engine and LPW4 four-cylinder engine. The engine displacement was 0.74 Liter and 1.86 Liter respectively. The maximum power output was 6.3kw and 17kw respectively. The engines were each integrated with a generator set and had the engine speed fixed at 1800 rpm. An Avtron K490 10 kw and 20kw Load Bank were used to apply varying loads to the generator output. The Test cell Engine specification is shown in the Table 4 Two diesel fuels were used, a street diesel fuel (or LSD, low sulfur diesel) with approximately 300~400 ppm sulfur, and an ultra low sulfur diesel (ULSD) with a sulfur level less than 15ppm.

Table 4. Test Cell Engines Specifications

Model	Lister-Petter LPA2	Lister-Petter LPW4
Displacements	0.73 L	1.86 L
Type	4 stroke	4 stroke
Max. Power	6.3 kW	17 kW
Cylinders	2	4
RPM	1800 fixed	1800 fixed

CONDITIONS AND TEST CYCLES

For the laboratory engine tests, three different types of test modes were performed. The first test mode was a Constant Load Test to meet specific target temperatures of engine exhaust. Two or three specific engine loads were selected to achieve the specific target temperatures ranging from 200°C to 280°C. For exhaust temperatures of 200~230°C, it is very unlikely for PM removal to occur without any back pressure build-up; while the range of 270~280°C is good for continuous regeneration of PM without much back pressure increase.

The second test mode is a Standard Test Cycle used to evaluate performance of the FTT system by measuring gaseous and PM emission as well as monitoring back pressure increase through the trap. Every engine load and corresponding exhaust temperature of this test mode is shown in the Figure 5. The load test in a standard test cycle composed of 6 different load steps and very similar to the ISO 8178 test cycle which is currently used as the regulation test mode for off-road diesel engines.

The third test mode is a Durability Test Cycle composed of four different phases of preconditioning, PM build-up at low load, regeneration, and maintaining at balance point temperature. It includes 15~20 hours of a soot build-up phase and another 14 hours of maintaining phase; totalling more than 35 ~ 40 hours. Therefore this test mode is very useful for long time durability of the FTT system. This durability test cycle is summarized in Table 5. Phase 3 of the Durability Test Cycle, in which the engine load is increased stepwise for regeneration, is very similar to the Standard Test Cycle.

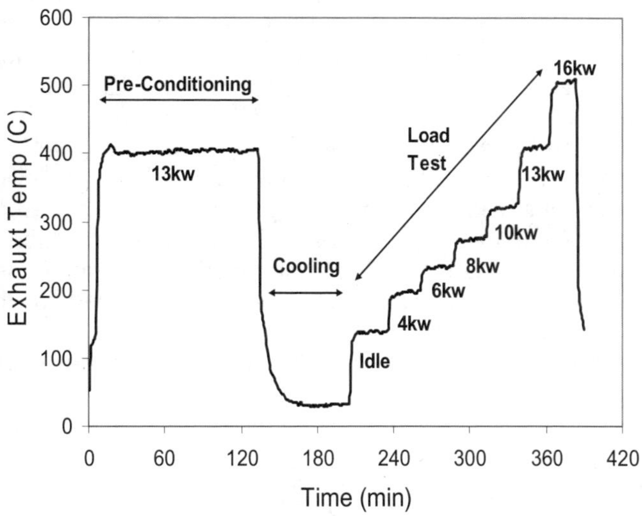

Figure 5. The Diagram of Standard Test Cycle to evaluate the performance of FTT

Table 5. Durability Test Cycle for long term durability of FTT system

	Description	Engine Load	Temp. (deg C)	Time
1	Preconditioning at high load	13kw	400	4 hrs
2	PM Build-up at low load	5kw	220 ~230	15~20 hrs
3	Regeneration with ramping load	5kw~6kw 7kw~8kw 9kw~10kw 11kw~12kw 13kw~14kw	240 ~500	20 min. each load
4	Maintain at Balance Temp.	8kw	285 ~295	14 hours

EMISSION MEASUREMENTS

For gaseous emission, three analyzers supplied by California Analytical Instruments, (Model ZRJ-1 NDIR for CO, Model 600 HCLD for NOx, Model 300 HFID for HC) were used. For PM measurements, two methods are used. The first method collects a PM sample directly from the exhaust stream without any dilution using an in-house sampling device. In this case the temperature of the sampler was maintained at 110°C to avoid water condensation. The PM weight was determined by micro balance using a conventional gravimetric method.

To measure PM weight and perform a more detailed analysis of the SOF, soot, and sulfate, the conventional gravimetric method is widely used. In the EPA regulation and many diesel emission applications, it is required to use this gravimetric method. However, this method is very complicated and time consuming because it involves multiple chemical analysis, hardware and specific conditions.

In our test, Horiba PM Analyzer, MEXA-1370PM, has been used an alternative method for PM measurement. This analyzer provides separate accurate measurements of soot, SOF, sulfate, and total PM measures in units of micro gram through the use of vaporization and oxidation techniques [8, 9]. In this method a filter with a PM sample is placed in the furnace and heated to 980°C firstly in a flow of N_2 gas and secondly in a flow of O_2; CO_2 and SO_2 is then measured by individual detectors. For accurate PM sampling a AVL mini-dilution system, SPC- 472, is used to collect PM samples on 47 mm quartz filters. With this dilution sampling method, the temperature of sample filter holder was maintained at 50±3°C.

RESULTS AND DISCUSSION

PRESSURE DROP OF METAL FOAMS

Controlling material properties of the foam filter is very important for many different reasons. For instance, a very small pore size has a larger surface area for catalytic applications than a large pore size, but comes with the cost of a higher pressure drop. A foam structure with a high relative density has better conductive and convective heat transfer properties as well as greater strength than foam with a low relative density. But an increase in relative density means a heavier component with a slower time response in thermal applications. In order to use metal foams effectively, the interaction between the physical characteristics (relative density, pore size, material type) and the material properties (strength, pressure drop, surface area) must be understood.

Nearly 180 runs were conducted with stainless steel samples of different diameter, thickness, inlet velocity, ppi, and relative density. The values of μ/K and ρC were computed at various densities for each ppi number. The values are tabulated in Table 6 and show an increase in the two constants as the relative density is increased and as the ppi number is increased. The constants were then fit to the relative density at each ppi number. Pressure drop is found to be highly dependent upon material pore size, and less dependent upon material density.

Table 6. Coefficients for Velocity Terms in Equation (2)

PPI Number	Relative Density (%)	μ/K (N s/m^4)	ρC (kg/m^4)
20	4.7	286.23	392.62
20	7.7	739.84	619.62
20	9.5	442.16	75.754
30	4.1	342.94	96.617
30	4.4	431.82	180.24
40	5.5	508.61	715.18
50	4.5	557.02	870.11
60	4.4	585.67	641.57
60	5.4	720.62	2257.4
60	7.9	565.38	5287.9
60	8.9	1969.2	243738

CONSTANT LOAD TEST

In order to confirm the concept of FTT design in Figure 1, continuous operation was performed for 20 hours on the 0.74L engine using uncoated ceramic foam filter. An FTT system using three disks measuring 4.66 inch diameter by 1 inch thick composed of SiC ceramic was used for this engine test. To keep the engine exhaust temperature at 270°C ~ 280°C, the load of the engine was controlled between 3.5 kw ~ 3.75 kw. As shown in Figure 6, the back pressure build up through the 50 ppi-uncoated SiC foam filters was so fast that it has reached 14 kPa in 10 hours of operation. In contrast, the 30 ppi foam showed no significant pressure build-up after almost 20 hours, an indication of no filtering function through the trap.

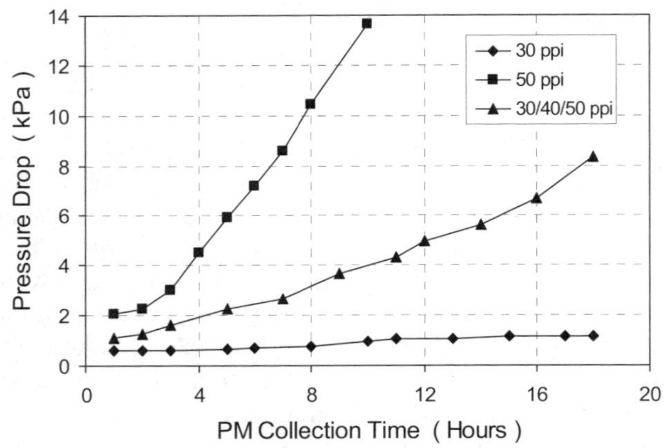

Figure 6. Constant Mode Test for FTT Back Pressure Build-up over the 0.73\4L LP Diesel Engine for various SiC Ceramic Foams without catalyst coating

A SiC Ceramic Foam FTT device coated with Sud-Chemie's proprietary washcoat and catalyst formulation was tested using 0.74L LP diesel engine. The first step was at low engine load and exhaust temperature of 200ºC. The catalytic FTT device ran for 40 hours to observe the pressure drop trend and PM collection time. The second step was at middle engine load with the exhaust temperature at 275ºC. The catalytic FTT device ran for 30 hours and pressure drop trends were recorded over the time. As shown in Figure 7, the back pressure of the catalyst coated FTT system (30-40-50 ppi) did not increase for up to 50 hours of operation at 280°C and showed continuous regeneration; unlike the uncoated FTT which showed very fast back pressure build-up and no observable regeneration. At a lower load, it was not sufficient for the coated DPF to continuously regenerate at a 230°C exhaust temperature.

Similar Constant Load Tests were confirmed using the 1.86 Liter engine with SiC Ceramic Foam FTT and metal foam FTT (TYPE B and C as described in Table 3). The engine load was maintained at 5kw and 8kw to keep exhaust temperatures at 230°C and 280°C. Back pressure build-up, regeneration pattern, FTT inlet/outlet temperatures, PM, and gaseous emissions were measured and monitored.

Figure 8 compares pressure trends for both uncoated and wash coated metal FTT filters. According to the pressure build-up pattern of the coated metal FTT, it is evident that the particulate matter in the FTT system is continuously regenerated at 280°C~ 300°C. Lack of pressure build up in the coated system at 8kw (300°C or higher) is an indication of that the FTT is achieving continuous regeneration at high load by NO_2/Carbon reaction. At low load, there is only partial regeneration of PM on the coated FTT as indicated by the increased back pressure, which is still less than observed for the uncoated FTT. Most of these results (pressure build up, PM removal) for the metal FTT are similar to the results for the SiC FTT system.

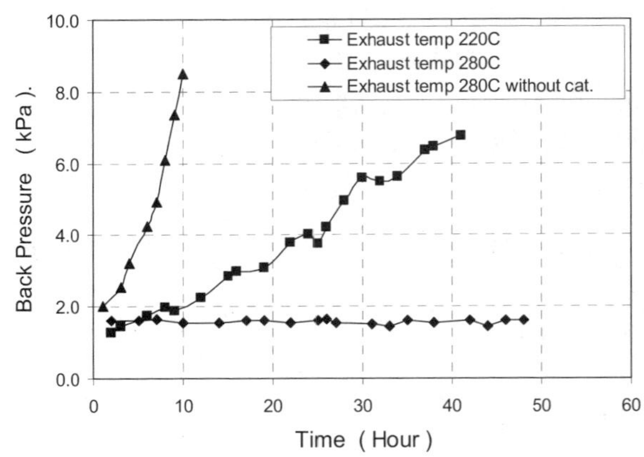

Figure 7. Constant Mode Test, 0.74L Engine: The trends of back pressure build-up with SiC Ceramic Foam FTT

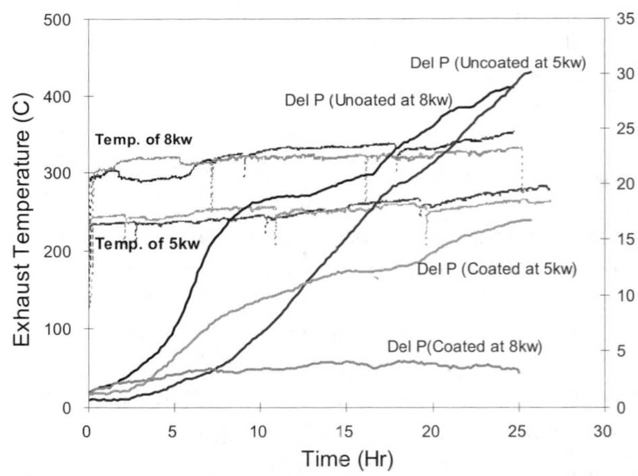

Figure 8. The Exhaust temperature and back pressure build-up trends with coated and uncoated metal FTT at 5kw and 8kw of 1.86L Engine

Visual inspection of the FTT foam filters after engine testing shows that the residual soot is accumulated mostly in the front area of all filters rather than the rear side. Figure 9 is the picture showing the state of the front area and back side of one selected SiC filer. All FTT filters are weighed before and after each individual engine test in both Constant Load Test and Durability Test Cycle. The soot amount remaining on individual FTT segments is increasing with the pore density of the filter. For example, in case of TYPE C FTT (20/30/40/50 ppi), the weight of the first bed is 14~16% of the total and the last fourth bed takes 37~38% of the total weight of the soot. The basic concept of the FTT system, a "graduated pore density of foam substrate will result in a more uniform distribution of diesel PM throughout the entire foam substrate trap as opposed to a single high pore density foam substrate" was confirmed again through the above analysis based on the regeneration pattern, visual inspection, and soot weight analysis.

Figure 9. The Pictures of SiC FTT after the 50 hours of Engine Test, Left: Inlet Surface Right: Outlet Area of 50 ppi SiC

DURABILITY TEST CYCLE AND BALANCE POINT TEMPERATURE (BPT)

The Durability Test Cycle (Table 5) is designed for two main purposes, to determine the Balance Point Temperature (BPT) of the unit and to see the long time reliability of the FTT system. To determine the BPT, a certain period of soot accumulation is necessary and can be done by operating the engine at low load and low temperature, conditions at which continuous regeneration will not occur, for an extended period of time. It should be noted that this operation window at the low load is representative of the exhaust gas conditions often found in light duty diesel vehicles that operate in the city. This phase 2 test period was maintained until the back pressure reached to 12 ~ 15kPa, usually taking 14~16 hours at 5kw in the 1.86Liter Engine. After this phase 2 test, the load was increased step by step to determine the BPT. The BPT is the temperature where the rate of PM removal (or regeneration) over the catalyst is at equilibrium with the rate of PM emission from the engine. Most practical and widely used method to determine BPT is the step wise raise in exhaust temperature by 20~30 minutes interval and measuring the slope (or dP/dt) of the pressure drop. The value is positive when the back pressure is built-up through the unit. The BPT is the point when the value becomes negative as the regeneration rate exceeds the PM accumulation rate.

Figure 10 shows the results of SiC ceramic foam FTT (TYPE B) and metal foam FTT (TYPE C) using one cycle of the Durability Test Cycle. At phase 1, when the exhaust temperature is maintained at 400°C for four hours, the back pressure of both systems was kept constantly at 3kPa, similar to what was observed in the Constant Load Test at 280°C. After reaching 13~14 kPa of maximum pressure drop in the PM build-up phase, the FTT was regenerated almost completely when the exhaust temperature was increased step-wise according to the phase 3 procedure. The detailed temperature and pressure profile of phase 3 is shown in Figure 11 from which the BPT can be determined. According to the figure, the BPT of the FTT was 259°C for SiC ceramic foam FTT and 264°C for metal foam FTT. Also, it was observed that the pressure drop falls significantly when the exhaust temperature exceeds 300°C. The backpressures of TYPE D~G were comparably low as 2 ~ 4 kPa with a couple of Standard Test Cycles.

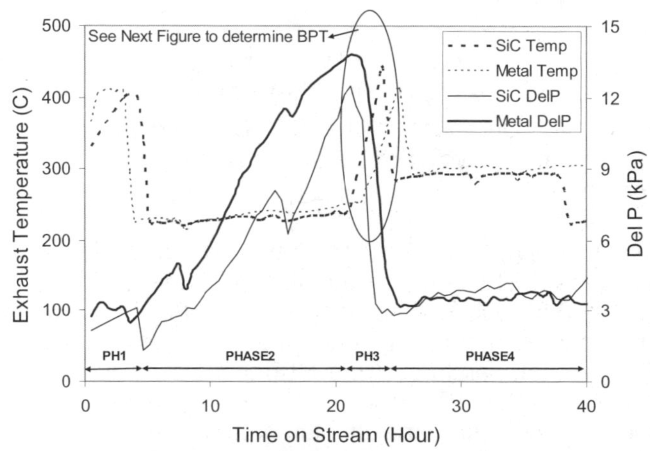

Figure 10. The Exhaust temperature and back pressure trends on the SiC ceramic foam FTT(TYPE B) and metal foam FTT(TYPE C) during the Durability Test Mode

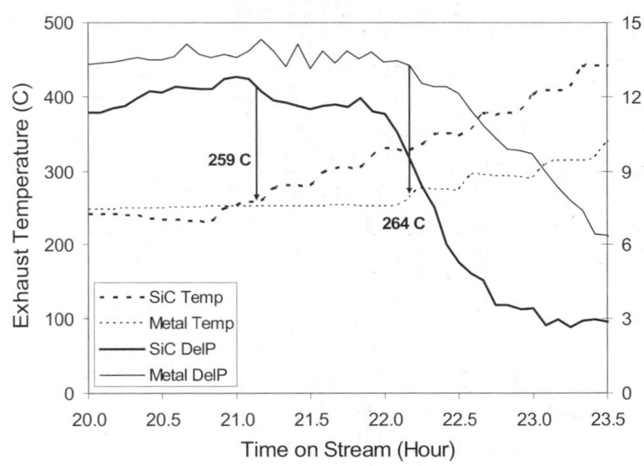

Figure 11. Determination of Balance Point Temperature of FTT system

To evaluate the reliability of the FTT system for a longer time of period, 3 cycles of Durability Test Mode were performed using SiC ceramic foam, TYPE B, using the 1.86 Liter Engine. Figure 12 shows the pressure drop trends for more than 100 hours in this test mode. During all three test cycles, it was found that 2~3 hours at phase 3 conditions, increasing the engine load and temperature step-wise, is enough to regenerate all the particulate accumulated during 14~15 hours at low load (phase 2). From these findings it can be inferred that 20~30% of the high temperature window 280°C ~ 450°C could be sufficient to apply the FTT system in a real vehicle.

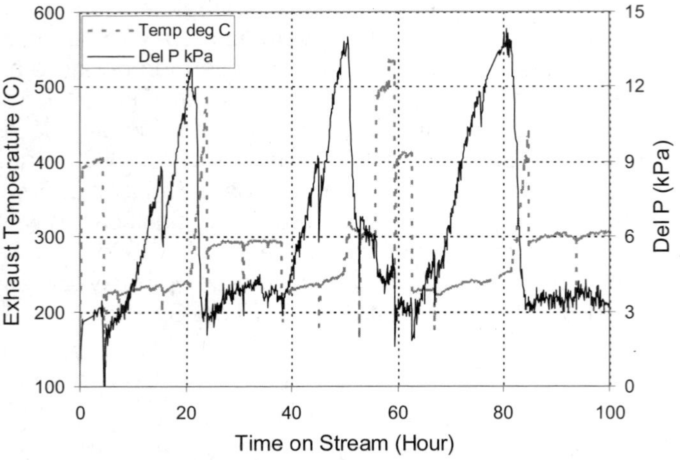

Figure 12. Durability of the SiC Ceramic Foam Filter, TYPE B, three cycles of Durability Test Cycle, 100 hours operation.

Another long term durability test was performed using metal foam FTT. A catalyzed metal FTT unit (TYPE D, 20-30-40-40-50 ppi, 2.06 L) was tested on the 1.86L Engine for a long term durability and low temperature application. The inlet temperature of the FTT unit was maintained at low temperature (fixed load at 2kw) for more than 100 hours. During this time the temperature increased from 200°C to 250°C and higher likely due to a back pressure increase. The FTT system was successful in making continuous regeneration at 280 °C or higher; however, it will likely eventually fail to operate due to high back pressure if the engine is kept at the low temperature range of 200°C ~250 °C. In this temperature range of 250°C ~300°C, some regeneration can occur, but the back pressure of the system is so high that it can exceed the maximum allowable level (~2400 mmH2O) for most diesel engine operations. For this reason it should be required to set a typical applicable or guarantee condition for the FTT unit in which it is specified that a certain percentage of the temperature window should be higher than 280°C.

Back pressure or pressure build up looks acceptable in this FTT configuration and current engine. But considering all FTT performance tests in this paper were performed on fresh catalyst, the durability test should be revisited in the future to further evaluate the high back pressure effect.

GAS EMISSION PERFORMANCE

The emission performance of five different FTT prototypes, TYPE C, D, E, F, and G, have been evaluated by the Standard Test Mode using the 1.86L engine. Figures 13 and 14 show the performance curves of CO and THC. Most FTT prototypes, except TYPE C, have similar activity to remove the gaseous emission. As shown in Table 7, the 50% light-off temperature was 210~216°C for CO and 230~234°C for HC.

The total amount of NOx was unchanged in all test ranges regardless of FTT types. In most cases the NOx conversion through the FTT was usually in the range of 2~3%. However, the NO to NO_2 conversion is critical because NO_2 is directly related to the regeneration of the FTT system for removing the soot trapped in the FTT foam filter. In gaseous emission measurement, the NO_2/NOx ratio is an important parameter to evaluate NO to NO_2 activity of the catalyst. According to the Figure17, most FTT units have the maximum NO2/NOx value around 40% at 350°C.

Table 7. The Performance Summary of Gas Emission

Sample ID	CO 50 % Light-Off Temp. (°C)	THC 50 % Light-Off Temp. (°C)	MAX NO_2/NO_x Ratio (%)
TYPE C	237	257	25
TYPE D	210	234	34
TYPE E	211	231	43
TYPE F	216	233	38
TYPE G	212	230	38

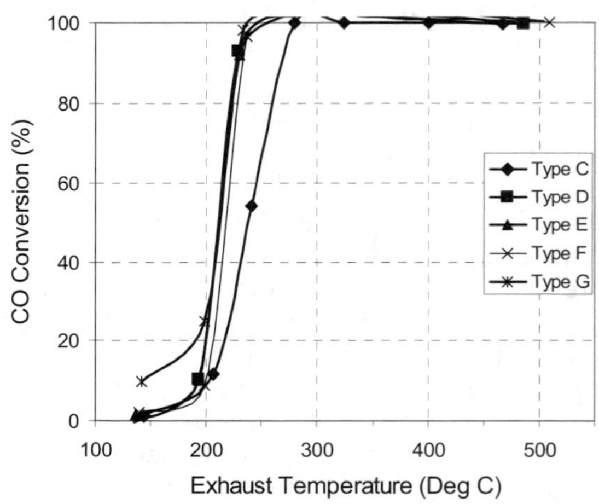

Figure 13. CO Conversion with Various FTT Prototypes

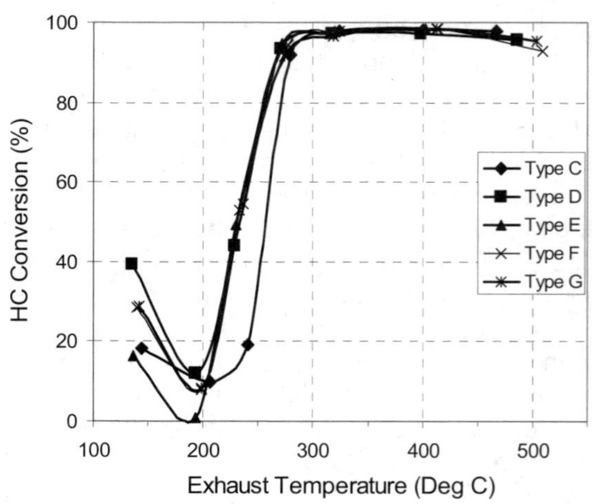

Figure 14. THC Conversion with Various FTT Prototypes

PM REMOVAL EFFICIENCY

It is no doubt that accurate PM measurement is very important in the development and verification of the FTT system. Even though PM emission characteristics of the engine can be affected by several conditions and environmental factors, a separate baseline test is required in most verified engine testing labs. To calculate the PM removal efficiency over the FTT system, a baseline engine test was performed before installation of the FTT unit.

Figure 15 shows the PM emission characteristics of the 1.86 L engine without any FTT system (empty can) over the Standard Test Cycle. With the HORIBA PM analyzer the collected PM can be analyzed into three components, SOF, Soot, and Sulfate. It shows the amount of each component according to the exhaust temperature. At the idle and low load, SOF weight takes 60% of the total PM while the soot portion takes more than 90% at the high load. The sulfate only takes 2~4% of the total PM amounts found during the entire operating window of the test cycle.

The results of PM measurement with various FTT prototypes are summarized in the Table 8 and Table 9. The PM measurement in Table 8 is based on the first method using direct sampling from the raw exhaust. In this method, only the weight of PM was determined by gravimetric method. In the case of 0.74 L engine test, the PM removal efficiency was very high, 69%~78% at low loads and maintained 45% at the high load. In the 1.86L engine test at two different constant loads (5kw and 8kw), the PM removal efficiency was similar in level between SiC metal foam based FTT(Type B) and metal foam based FTT (Type C). The size of these two FTT prototypes was 1.67 and 1.61 Liter and the ratio of FTT to engine displacements was 0.90 and 0.87 respectively. Considering the PM removal efficiency of 45~48%, this FTT size looks a little small to meet the target performance. The next size FTT, Type D had better results which showed 50~54% PM removal.

More detailed and extensive analysis of the PM was performed using the HORIBA PM Analyzer. For these PM measurements, five different types of FTT units were tested on the 1.86L Engine using the Standard Test Cycle. In this engine test, the PM samples were each collected at 6 different load points and analyzed into SOF, soot, and sulfate.

Table 9 shows the summary of the PM measurements with all types of FTT. As shown in this table the total PM removal efficiency increases with FTT volume shown in the Table 3. Type F and G FTT have good performance for target PM removal efficiency (50% or higher). It shows that the FTT size needs to be 1.3 times of engine displacement or bigger.

Figure 16 shows the amounts of PM, SOF, and soot before and after the FTT (Type F) unit at various load conditions. <PM in>, <SOF in>, and <Soot in> in this diagram represents the amount determined from the baseline engine test before the FTT is installed.

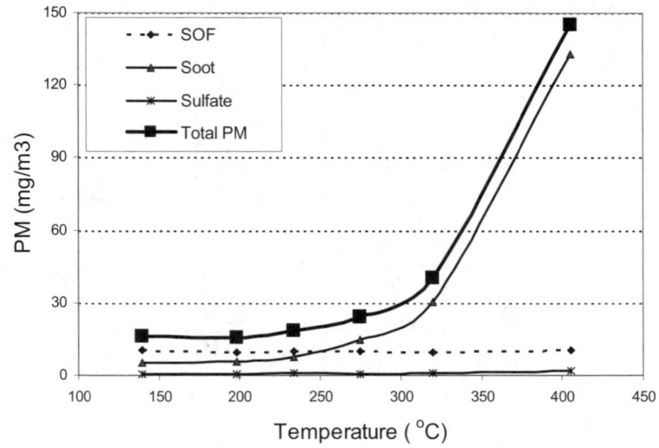

Figure 15. The composition of the PM with the engine load and the exhaust temperature in the baseline test

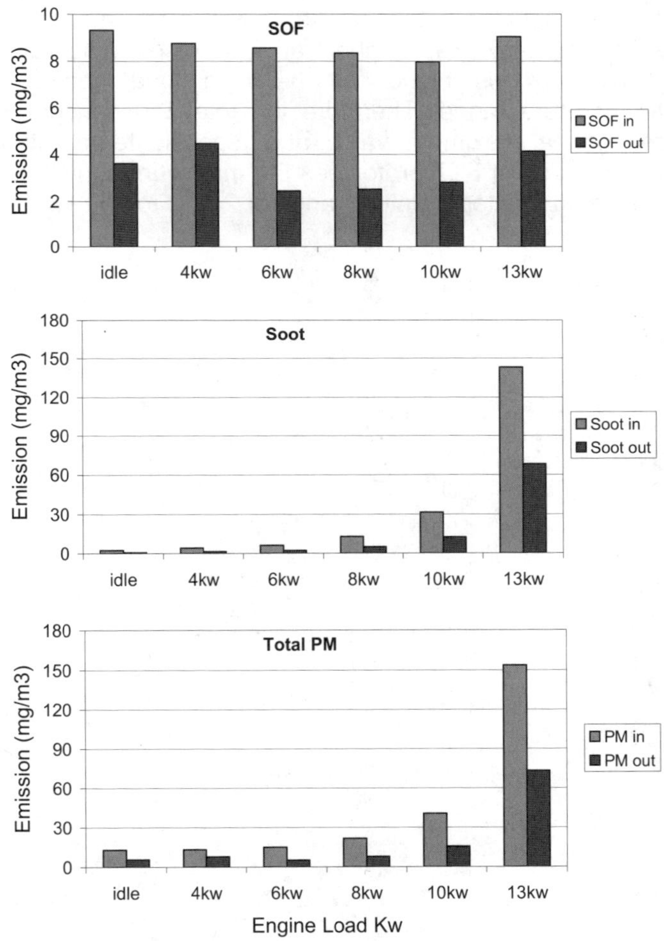

Figure 16. PM Emission before and after FTT (Type F) with various engine loads

Because the absolute amount of PM increases rapidly at high load (13kw) and high exhaust temperature, the PM amount in this step greatly affects the total PM removal efficiency of the system. But the performance of the FTT at the low load and low temperature is also important.

Table 8. The summary of PM removal efficiency measured by raw exhaust sampling without dilution and conventional gravimetric method

Engine Size	Type	Load (Kw)	Temp (°C)	Conv. (%)
0.74L	A	1.5	200	69%
		3.0	270	78%
		5.0	370	45%
1.86L	B	5.0	250	45%
		8.0	310	66%
	C	5.0	250	48%
		8.0	310	45%
	D	5.0	250	54%
		8.0	310	50%

Figure 17 shows the conversion of PM, SOF, and Soot as a function of the exhaust temperature with various Types of FTT. Since the sulfate portion of the PM is not significant, the summation of SOF and Soot is very close to the amount of total PM. As shown in this figure, the soot conversion is higher at low load while SOF conversion wis higher at the high load and high temperature range. Therefore the total PM conversion reaches a maximum value at a specific temperature range. In most FTT prototypes the maximum conversion is located in a temperature range of 230°C to 280°C.

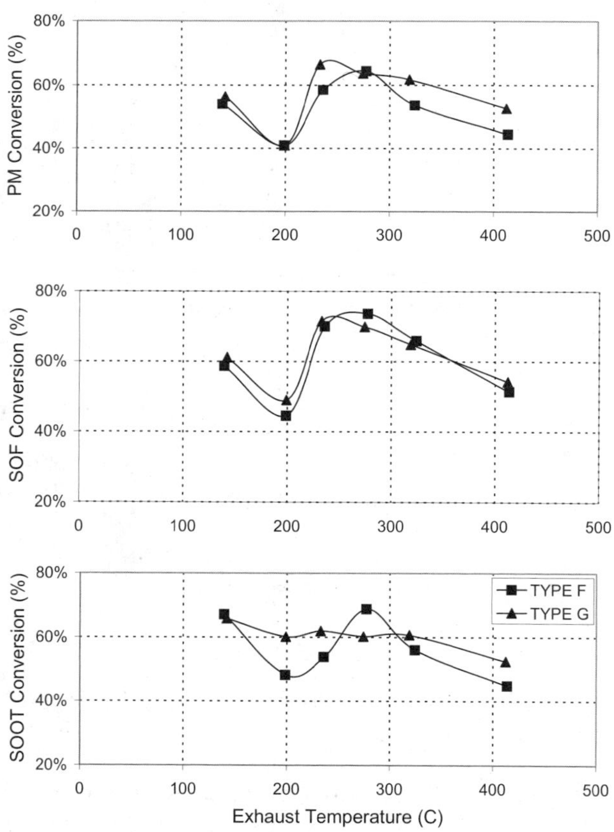

Figure 17. The conversion of PM, SOF, and Soot as the function of the exhaust temperature with various Types of FTT

Table 9. The summary of PM removal efficiency with 1.86L engine measured by HORIBA MEXA-1370 PM Analyzer with AVL Mini Dilution Tunnel sampling

Emission in mg/m^3		Type C*	Type D	Type E	Type F	Type G
SOF	Base	69.7	60.3	60.3	55.2	52.0
	FTT	42.4	30.0	26.4	21.6	20.0
	Conv(%)	39.2	50.3	56.1	60.8	61.6
Soot	Base	217	185	185	196	200
	FTT	119	128	118	98.4	90.4
	Conv(%)	45.3	30.8	36.2	49.8	54.9
Sulfate	Base	5.02	4.94	4.94	4.57	5.01
	FTT	46.40	6.54	2.36	8.04	4.68
	Conv(%)	-823	-32.4	52.1	-75.9	6.6
Total PM	Base	292	251	251	252	257
	FTT	207	165	147	128	115
	Conv(%)	28.9	34.3	41.3	49.2	55.3
Fuel Used		LSD	ULSD	ULSD	ULSD	ULSD

CONCLUSION

The various FTT systems using either SiC ceramic foam filter or metal foam filter were tested on two diesel engines. Based on the results of these engine tests, the FTT is found to be an effective device to remove 50% of PM from the diesel engine exhaust. Also, the basic concept of the FTT system, that a "graduated pore density of foam substrate will result in a more uniform distribution of diesel PM throughout the entire foam substrate trap as opposed to a single high pore density foam substrate" was confirmed.

Catalyst coating on the foam filters is greatly beneficial for lowering back pressure build-up and continuous regeneration of the trap. The Balance Point Temperature of the FTT systems was around 260°C, and the FTT prototypes can be regenerated successfully without any further active regeneration control when the exhaust gas temperature is maintained at 280°C and higher for a minimum specified percentage of time.

Based on the durability tests of 100~150 hours, it was found that the system is very reliable and the trap can be regenerated repeatedly. In order to apply the FTT concept for in-use diesel vehicles, it is expected that 20~30% of the temperature window might have to be at 280°C or higher. Based on the PM removal efficiency with various FTT Types, the FTT size is should be 1.3 times of engine displacement or bigger to meet the target PM removal efficiency (50% or higher).

Since PM and SOF removal efficiency shows pretty high level over the wide range of exhaust temperatures, the FTT system has the potential to be a very promising solution for various emission control needs.

FUTURE WORK

As the basic feasibility study and performance test for the FTT system has been completed in the engine lab, a longer durability test and field demonstration using real diesel vehicle is necessary. Such validation programs with in-use vehicles are now on-going with several potential customers and test institutes.

REFERENCES

1. Guido Saracco, et al. " Diesel particulate abatement via catalytic traps ", Catalysis Today, 60, 33(2000),

2. P. Russo, etal, " Simultaneous filtration and catalytic oxidation of carbonaceous particulates " Topics in catalysis, 22, 123 (2003).

3. Todd Jacobs, Sougato Chatterjee, Ray Conway, Andy Walker, Jan Kramer and Klaus Mueller-Haas, "Development of Partial Filter Technology for HDD Retrofit", SAE Technical Paper 2006-01-0213

4. Final Regulation Order, California Code of Regulation, Title 13, Chapter 14, Section 2700~2704: http://www.arb.ca.gov/diesel/verdev/vt/vt.htm

5. The Ceramic Foundry Filter with Foam Structure, Foseco Brochure, http://www.foseco-at-gifa.com/gifa/download/brochures_en/SEDEXbrochure.pdf

6. Metpore® EFCS, Diesel Particulate Filter Media, Porvair Advanced Materials, http://www.porvairmetalfoam.com/documents/MetporeEFCS.pdf

7. Zhongyuan Dang, Yinyan Huang, and Amiram Bar-Ilan, "Oxidation catalyst on a substrate utilized for the purification of exhaust gases", United States Patent Application 20060211569

8. "Comparison of an Alternative Particulate Mass Measurement with Advanced Microbalance Analysis", SAE Technical Paper 2004-01-0589

9. "An Alternative Technique for Low Particulate Measurement", SAE Technical Paper 2001-01-0218

CONTACT

Yongtaek Choi
Sud-Chemie Inc.
Yongtaek.Choi@sud-chemie.com
781-444-5188 (ext.37)

Zhongyuan Dang
Sud-Chemie Inc.
Zhongyuan.Dang@sud-chemie.com
781-444-5188 (ext.30)

II.3 TECHNOLOGIES

2000-01-0473

Passenger Car Serial Application of a Particulate Filter System on a Common Rail Direct Injection Diesel Engine

O. Salvat, P. Marez and G. Belot
PSA PEUGEOT CITROEN

1 - ABSTRACT

Although combustion system improvements allow extremely low level of particulate (EURO 4 particulate level is already achieved) the particulate filter appears to be the only solution to reduce solid particulate emission by two to three order of magnitude over the whole size range starting from 10 nm. This is the reason why an active particulate filter system has been developed.

The major issue was to fill in the gap between particulate regeneration temperature (550°C) and the naturally low exhaust gas temperature of modern DI diesel engine (150°C to 200°C in city driving conditions). This low exhaust gas temperature is the result of overall efficiency improvement to reduce fuel consumption. The active system created is based on three main groups of elements: first a specific software dedicated for regeneration control and system diagnosis, second a filtration medium made of porous Silicon Carbide (SiC) located downstream an oxidation catalyst plus a differential pressure sensor and two temperature sensors and finally a fuel additive in a specific tank and a dosing system in the main fuel tank.

The active system continuously monitors the mass of soot in the filter from the filter pressure loss to start or stop the regeneration process. This process begins with an in-cylinder gas temperature increase by specific management of multiple injection and air boost pressure to keep the torque at the same level for driveability. As soon as the oxidation catalyst in front of the filter has lit off the amount of hydrocarbon in the exhaust gas is increased by multiple injection tuning to create catalytic post combustion. The temperature rise generated by these two actions (+300°C) completed by the effect of the additive in the particulate insures the rapid and complete regeneration of the filter.

This paper presents the features and the performances of the first active particulate filter system available on a serial passenger car. This system developed in just eighteen months and whose operating modes are unnoticeable to the driver makes the most of the flexibility and the performance of the Common Rail fuel injection system. It also illustrates the continuously growing potential of modern direct injection diesel engines.

2 - INTRODUCTION

2.1 - DIESEL ENGINES – Compared to gasoline engines diesel engines are well known for their natural durability and robustness but also and mainly for their low fuel consumption that has continuously been improving over the past years. This last advantage has become all the most important, as countries around the world are committed to reduce CO_2 emissions. In addition to inherent high efficiency leading to low emission level of CO2, diesel engines possess low CO and HC exhaust level but emit relatively high NOx and particulate as well.

For more than forty years, engine manufacturers have been committed to reduce particulate emissions from diesel exhaust as shown in figure 1. As a result the level has been divided by four since 1980. This reduction partly comes from the use of oxidation catalyst since the end 1996 that deal with the Soluble Organic Fraction (SOF) of the soot as well as from the progressive improvement of diesel fuel formulation. However the major part of the reduction is essentially due to improvement in the control of the combustion process which resulted on the latest HDi engine in a further 60% particulate reduction compared to IDI technology. This gain is the conjunction of:

- High fuel injection pressures all over the engine operating range by Common Rail technology.
- Fully optimized combustion chamber design by 3D modeling.
- Elaborated liner machining to significantly reduce oil consumption

Though, diesel exhaust gas still contains a higher amount of particulate than exhaust gas of gasoline engine. To date, the diesel particulate trap is the only system that can curtail the IOF particulate count to the limit of the measurement techniques all over the particulate range from 10 nm to several microns as it has been demonstrated in several publications [1-7].

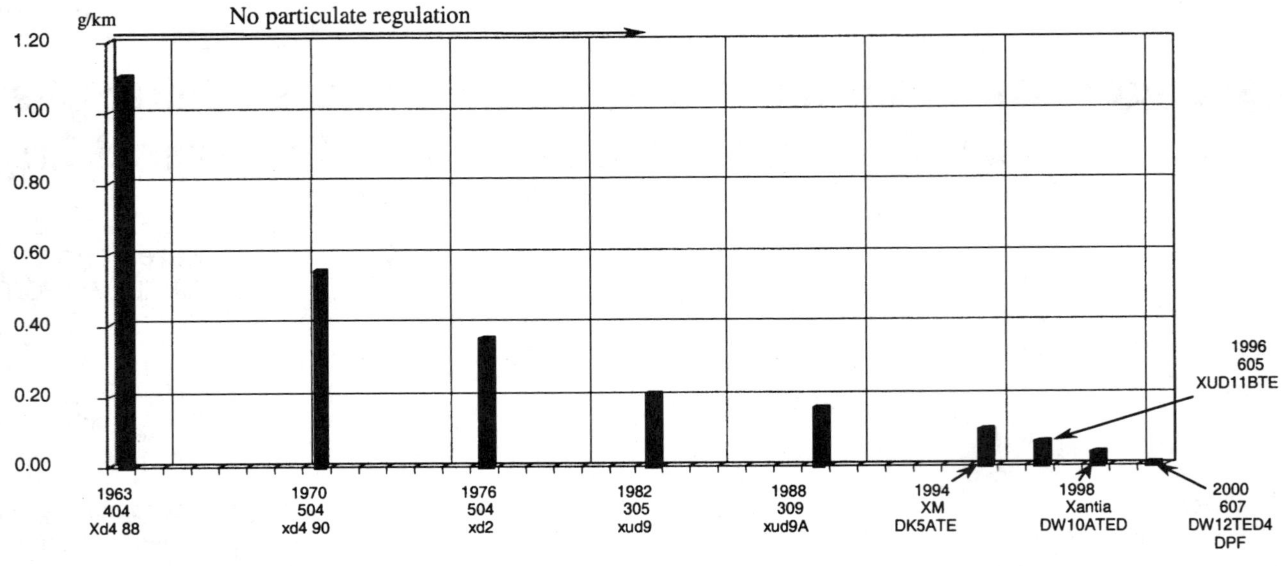

Figure 1. Particulate reduction of PSA vehicles

2.2 - TRAP SYSTEMS – For more than twenty years the diesel particulate filter is being considered on production vehicles [8 – 23]. However its wide introduction on passenger car applications was limited by the resolution of the following difficulties:

- Safe, reliable and periodic regeneration under all driving conditions to prevent filter melting or cracking and to insure system durability.
- Minimum fuel penalty which implies low back pressure systems and management of filter loading by ash and inorganic residues
- Development of a low cost system. The regeneration systems investigated so far were generally based on complex and costly regeneration techniques.

To tackle this challenge a new technology has been developed which is the basis of this paper. The regeneration principle employed in the presented system is based on the conjunction of two techniques:

- A full use of the flexibility of the Common Rail system as the only system that can manage multiple injections with different timing and quantity in an engine cycle to give:
    - direct exhaust gas temperature increase in the combustion chamber plus,
    - catalytic post oxidation by management of exhaust gas composition (hydrocarbon, CO),
- The use of RHODIA "EOLYS" additive (cerium based) in the fuel to decrease the ignition temperature of particulate.

Before detailing the system it is however necessary to briefly describe particulate matter.

3 - PARTICULATE MATTER

3.1 - PARTICULATE FORMATION – In diesel engines the fuel is injected at high velocity and vaporizes and mixes with the high-temperature high-pressure cylinder air. Then spontaneous ignition of parts of the already-mixed fuel and air takes place after a short delay period, which increases the cylinder pressure and temperature. While fuel injection proceeds, fuel is atomized, vaporized, air-mixed and burnt. Moreover, mixing of the air with burning and already burned gases both from the combustion and from EGR occurs throughout the combustion process. Therefore despite constant efforts to improve the quality of the combustion, zones in the chamber can be locally too rich which leads to incomplete combustion due to a lack of oxygen and to formation of particulate precursors.

Although complex, the mechanisms of particulate formation can be describe according to a chain of successive chemical and physical phenomena from the combustion chamber to the atmosphere [24 – 25].

At the end of the combustion the precursors of the particulate coexist in the chamber. They are composed mainly of:

- Solid carbon micro spheres (about 0.01 to 0.08 µm diameter)
- Gaseous hydrocarbons from diesel fuel and lube oil
- Sulfur dioxide (SO_2)
- Water vapor.

During the expansion stroke and the cooling in the exhaust line these precursors will progressively aggregate to form agglomerates (roughly 0.05 – 1.0 µm diameter when collected on palflex filter and about 10 nm to 800 nm when measured in exhaust gas by SMPS).

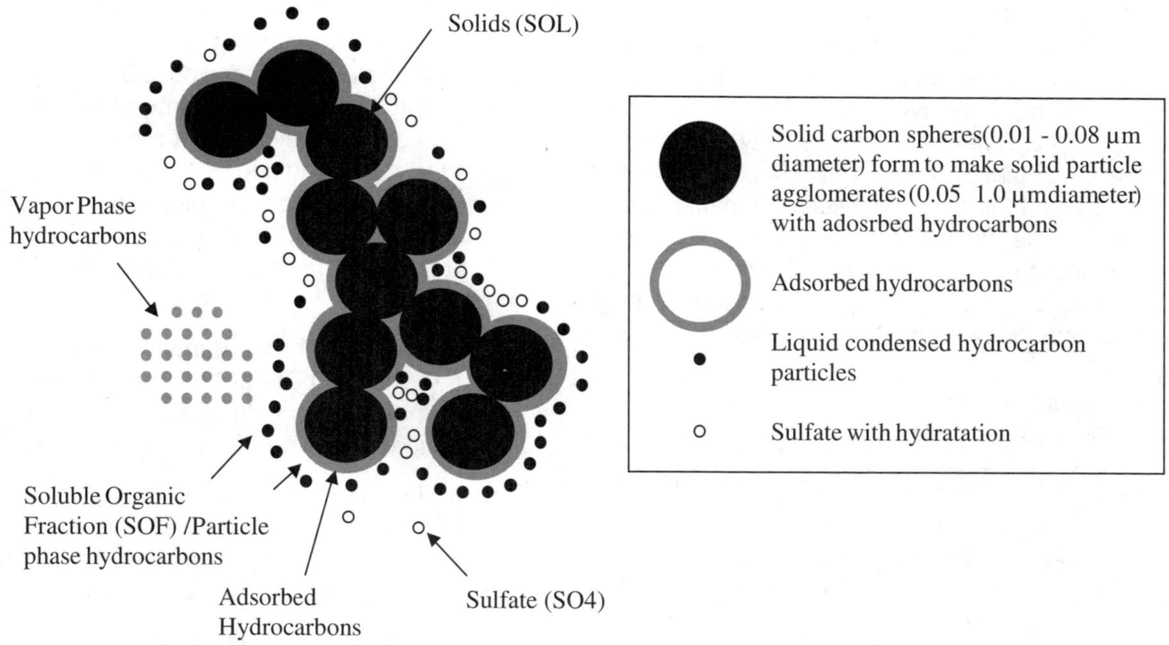

Figure 2. Schematic of Diesel Particulates and vaporphase compounds

Reprinted with permission from SAE paper No 940233 c 1994
Society of Automotive Engineers, Inc.

3.2 - PARTICULATE COMPOSITION – Diesel particulate matter is therefore a complex mixture of organic and inorganic compounds in solid phase: soot component (spherical carbon particles) and in condensed phase: organic or hydrocarbon component and sulfate (SO_4 – primarily H_2SO_4 and some metal ion sulfates) as illustrated on figure 2. The amount of particulate matter formed and the ratio of its compounds is closely linked to:

- Combustion process (Combustion chamber design, swirl ratio, injection pressure and holes diameter, injection rate…)
- Diesel fuel quality: high natural Cetane number and low polyaromatic compounds reduce the number of carbon particulate precursors thus the particulate matter,
- Sulfur content in the fuel
- After treatment devices: oxidation catalysts reduce the SOF fraction of the TPM but generate sulfate at high exhaust gas temperature.

The combustion of particulate matter as well as particulate size distribution will be discussed in chapter 4.1.1 and 5.1 respectively.

4 - FILTER SYSTEM

4.1 - REGENERATION PRINCIPLES – To insure durability and prevent filter cracking due to soot overloading it is necessary to completely burn the particulate matter in the filter at each regeneration. This implies to create and maintain regeneration conditions long enough for regeneration propagation.

4.1.1 - Particulate combustion – Particulate regeneration temperature has been extensively studied. For example evaluation by Thermal Gravimetric Analysis (TGA) indicates that particulate matter burns at about 550°C / 600°C. Various metal additive have been investigated to lower the oxidation temperature of the soot. The additive selected is based on Ceria It reduces soot regeneration temperature to about 450°C (100°C gain) and also significantly increases combustion rate as shown on figure 3.

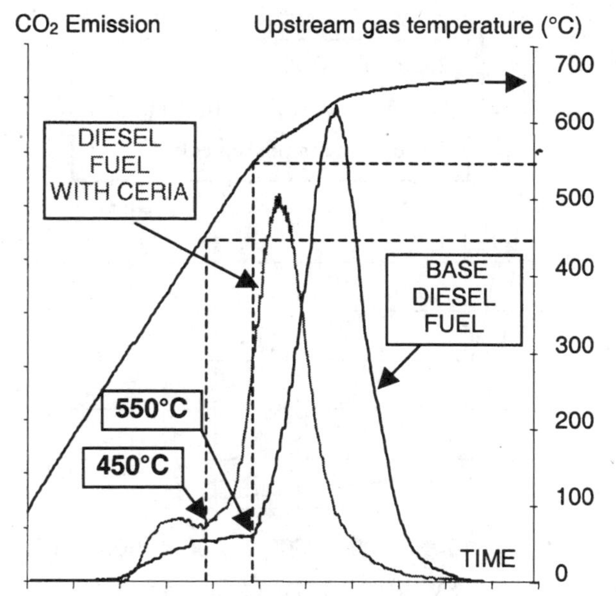

SGB Conditions : O_2 = 18 to 4%, H_2O= 3%, N_2=balance, SV=25000h^{-1}, temperature rampe=100°C to 630°C in 1320s and stabilisation at 630°C,
Filter loading : 2580 rpm, 97 Nm

Figure 3. Thermal Gravimetric comparison between Ceria additized and non additized particulate

4.1.2 - Exhaust gas temperature generation – The soot regeneration temperature obtained with the additive is however still far above usual exhaust gas temperature of light duty diesel engine. Moreover the fuel consumption and CO_2 reduction achieved over the last few years on modern DI diesel engines by a significant efficiency improvement has lead to further reduction of exhaust gas temperature all over the engine operating range. Table 1 shows typical exhaust gas temperatures upstream DPF on a 406 equipped by a 2l HDi engine. Consequently DPF regeneration has become even more difficult.

Table 1. Exhaust gas temperature in °C upstream DPF on a 406 Vs engine type.

Vehicle speed (kph)	20	50	70	100	120	150
2l HDi	120	150	175	230	270	340
2.1l IDI	140	175	200	280	320	400

To increase exhaust gas temperature, the approach was to modify the thermodynamic cycle as described in figure 4 in order to:

1. Obtain maximum exhaust gas temperature at exhaust valve opening
2. Minimize the amount of fuel required to create the temperature level requested
3. Keep same BMEP and cycle to cycle variability for driveability and to render the regeneration tuning unnoticeable to the driver
4. Keep the same low noise level for the same reason

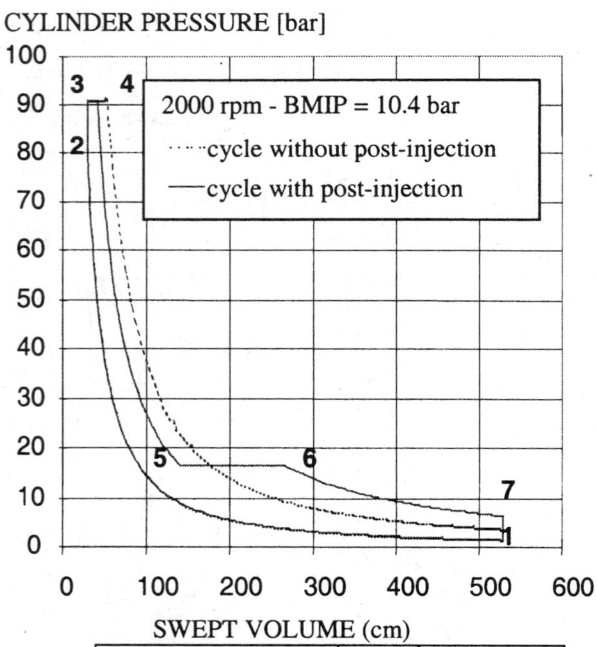

Fuel quantity (mg/St)	Base	Multiple inj
Pilot (Q_{23})	2	2
Second inj (Q_{34})	22	12
Third inj (Q_{56})	0	20

Figure 4. thermodynamic cycle with multiple injection compared with base cycle

To satisfy the first and second criteria it is necessary to completely burn the maximum amount of fuel as late in the expansion stroke as possible without by-product formation (CO, unburned hydrocarbons and soot). To do so while keeping the BMEP constant it is also necessary to reduce the amount of work done by the combustion close to top dead center while controlling the thermodynamic conditions in the cylinder since they are a key parameter to insure stable and complete late combustion.

This theoretical thermodynamic cycle was obtained thanks to the flexibility of the Common Rail fuel injection system that allows multiple fuel injections with different timing and quantity during an engine cycle.

During the usual engine operating mode the fuel is injected in two phases: a pilot and a main injection to tune the engine according to best emission, noise, fuel consumption and performance trade off.

During the regeneration mode three injections are performed:

- a pilot injection to control the start of the heat release and therefore the combustion noise
- a second injection to participate in BMEP formation and to insure stable and late combustion
- a third injection to create the exhaust gas temperature and to complete BMEP.

Depending on the engine operating conditions the second injection can be significantly smaller than the main injection of the usual mode (low load, low engine speed) while the third injection can be significantly more important as this is illustrated on figure 5. On the contrary the change in the thermodynamic cycle doesn't need to be important at high load high engine speed since exhaust gas temperature is closer to soot regeneration temperature.

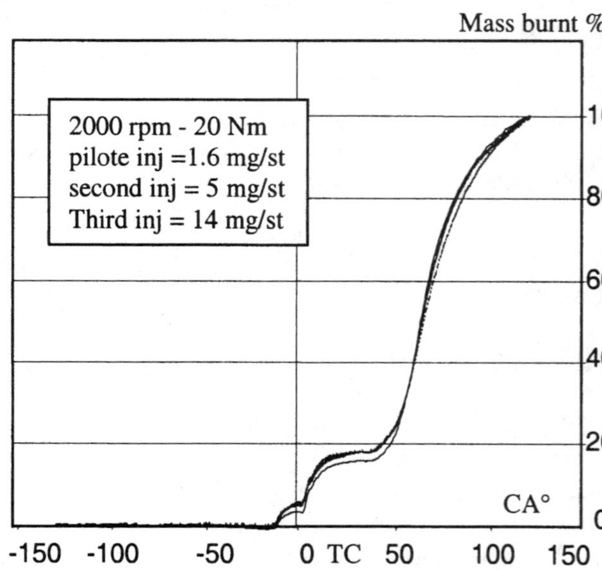

Figure 5. Calculated mass burnt fraction on 2000 rpm – 20 Nm engine operating conditions (50 kph on 406 – 2l HDi).

To completely control the whole process it is moreover necessary to regulate the turbocharger on specific values due to the enthalpy increase at the turbine and to also have a good control of the intake air temperature. Figure 6 illustrates the different phases of engine tuning.

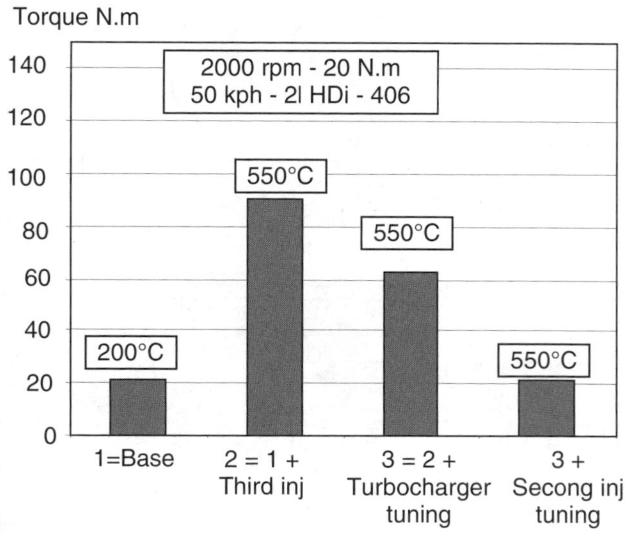

Figure 6. Illustration of multiple injection and turbocharger tuning to increase temperature and maintain torque

4.1.3 – Catalytic combustion – Under very low engine load and speed, generation of exhaust gas temperature up to 450°C without a BMEP increase is difficult. Therefore catalytic combustion in front of the filter as a complementary solution was developed to reach soot regeneration temperature.

Since natural hydrocarbon emission of diesel engine is extremely low, generation of sufficient hydrocarbon quantity was necessary to create the exotherm desired. This very fine control of the exhaust gas chemical composition by the Common Rail system is achieved by slightly retarding the third injection timing while adjusting its quantity. The quantity of the second injection is also modified to maintain BMEP and control cylinder temperature. This strategy is only possible when catalyst efficiency is sufficient to prevent any hydrocarbon slip. Therefore a two-mode strategy has been defined based on average catalyst temperature monitoring from two temperature sensors as illustrated on figure 7. When T_{inlet} or T_{outlet} are below the threshold defined on figure X multiple injection tuning is set to only generate exhaust gas temperature. As soon as the catalyst has lit off as a result of this action multiple injection is tuned to a different setting to also increase hydrocarbon concentration in the exhaust gas and generate heat by catalyst oxidation.

The result of all the complementary actions to regenerate is shown on figure 8.

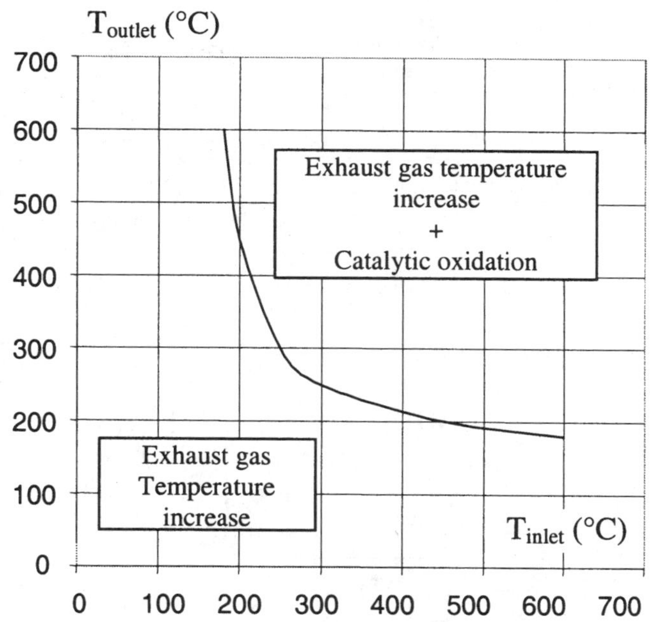

Figure 7. Zones of Multiple injection setting as a function of catalyst temperature

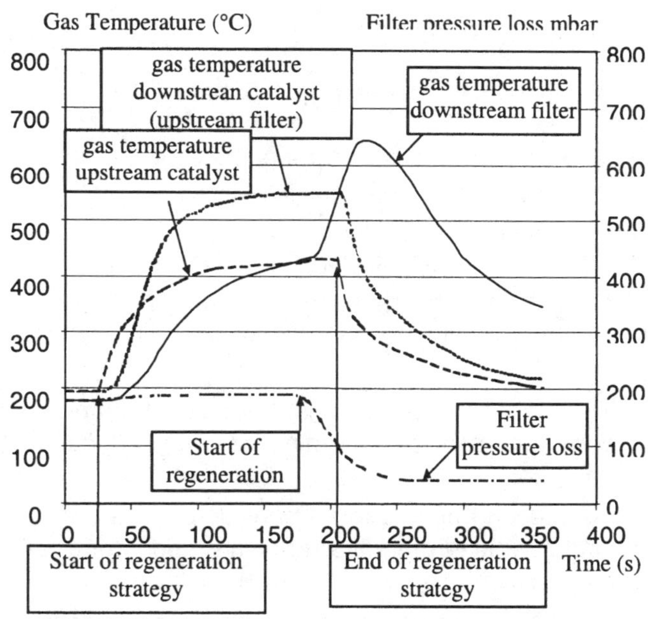

Figure 8. Example of exhaust gas temperature increase and particulate filter regeneration under steady state conditions

4.2 - SYSTEM COMPONENTS – As illustrated on figure 9 the system is composed of:

- a specific software and sensors (filter differential pressure and exhaust gas temperature upstream and downstream the catalyst) to:
  - determine the optimal time to regenerate and to stop a regeneration procedure
  - regenerate the filter by multiple injections and turbocharger management
  - insure system diagnosis.

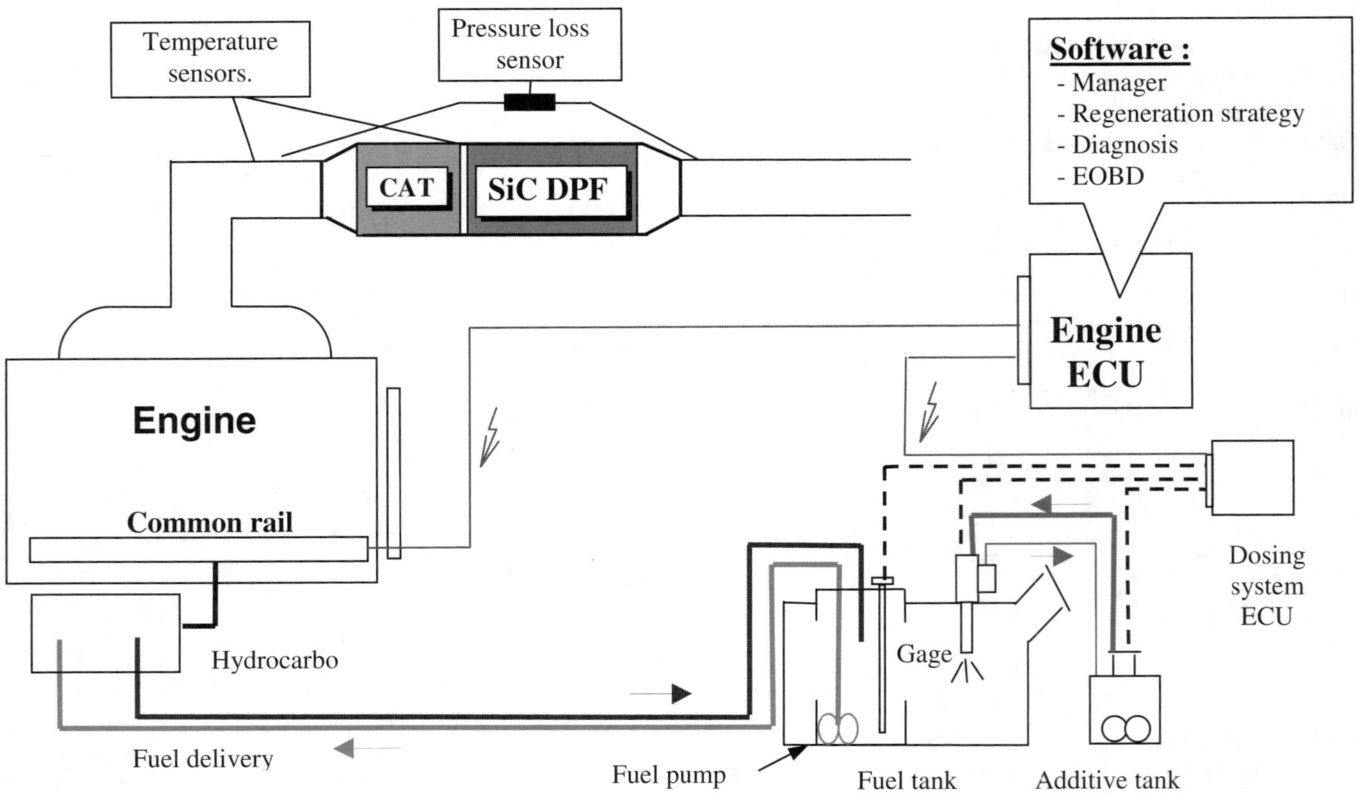

Figure 9. Schematic of the active diesel particulate filter system developed

- an oxidation catalyst in front of the filter
- a wall flow silicon carbide filter
- a dosing system for the fuel additive

4.2.1 - Software description – Different aspects have to be taken into account to determine when a filter has to be regenerated.

1. The most important one is the maximum amount of soot that can burn within the filter without filter damage.
2. The second aspect to be considered is a minimum soot quantity in the filter to insure good combustion propagation
3. Finally it is also necessary to optimize fuel penalty by managing the distance between two consecutive regenerations.

First of all it is necessary to determine the amount of carbon matter in the filter. The strategy developed is based on the relationship between filter pressure loss and mass and composition of the material trapped. This material is composed of:

1. particulate
2. residue from the additive
3. ashes from lube oil
4. metal particulate from engine wear
5. inorganic material passing through the air filter

The effect of material 4) and 5) on filter differential pressure can be neglected due to their very limited quantity even over 80000km.

Oil consumption, which has been significantly reduced by elaborated liner machining to limit particulate formation, is now well controlled. Typical value for HDi engine is 0.1 l / 1000 km and the ash content in the oil is roughly 8g/l. to 10 g/l of oil. Therefore pressure build up in the filter from lube oil is relatively slow and linear but could further be reduced by low ash or free ash lube oil.

The amount of additive residue in the filter depends directly on treat rate as well as on fuel consumption.

Thus the amount of additive residue and the quantity of ashes from lube oil is calculated at every moment by the engine control unit (ECU) to estimate the corresponding filter pressure build up.

The last step is to estimate the mass of soot trapped in the filter. Results from experiments show that the shape of particulate aggregates in the filter depends strongly on the engine operating condition as illustrated on figure 10 and 11. The mass of soot is thus estimated from the determined filter pressure loss due to particulate matter and a map inside the ECU function of engine parameters.

In theory it would have been simpler to directly calculate and integrate soot emission of the engine. However this method cannot estimate the amount of soot left in the filter in case of partial regeneration while this aspect is taken into account when filter pressure loss is measured.

To minimize fuel penalty it is necessary to take two aspects into account:

- fuel consumption due to filter pressure build up
- amount of fuel needed to regenerate the filter.

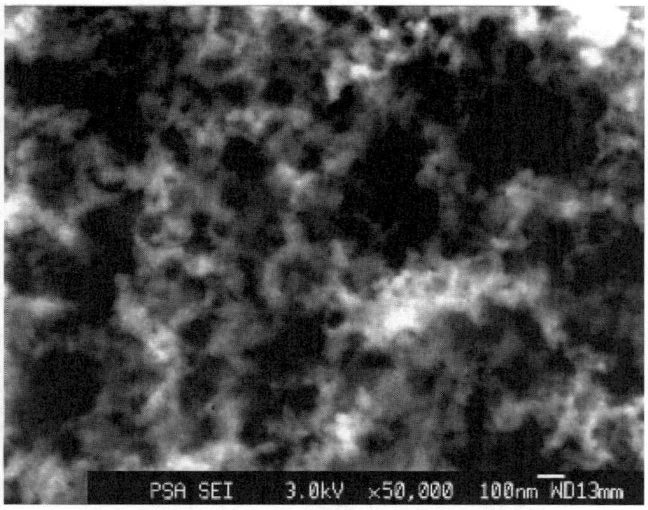

Figure 10. MEB measurements of particulate aggregates – engine-operating condition: 2030 rpm – 21.4 N.m

Figure 11. MEB measurements of particulate aggregates – engine-operating condition: 2580 rpm – 97 N.m

For a given vehicle operating condition fuel consumption increase linearly with the distance between two regenerations while fuel penalty due to temperature increase to regenerate is a constant. Thus fuel penalty for a given vehicle operating condition is minimum in a band of kilometric regeneration period as illustrated on figure 12.

As soon as the system defines that the filter needs to be regenerated the software changes fuel injection and turbocharger parameters to increase exhaust gas temperature as described in chapter 4.1. Then fuel injection and turbocharger parameters are switch back to normal, some time after filter regeneration is detected from a reduction of filter pressure loss, to insure the complete combustion of the soot in the filter.

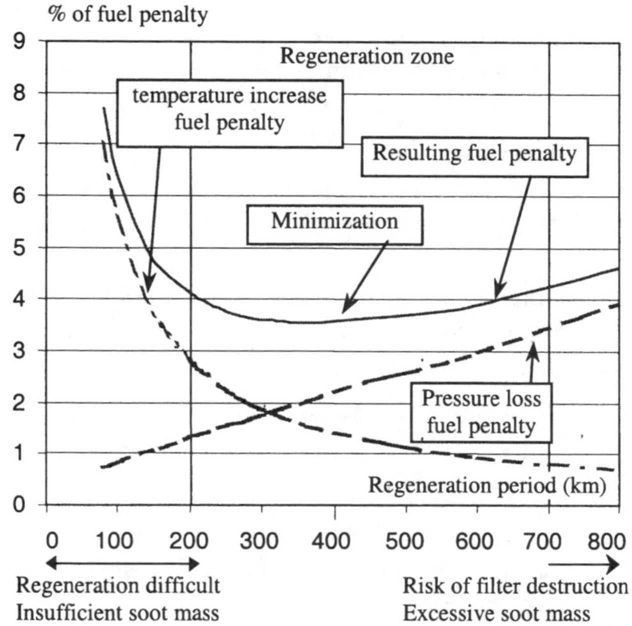

Figure 12. Minimization of fuel penalty depending on vehicle operating conditions and DPF system characteristics

Finally the software also insures filter and system diagnosis by:

- monitoring the filter pressure loss and exhaust gas temperature to determine, as illustrated on figure 13:
    - a cracked filter or any filter tampering (the filter pressure loss is below a defined threshold)
    - a clogged filter (the pressure loss is above a defined threshold)
    - a decrease of the oxidation function of the catalyst.
- monitoring the level of additive left in its tank.

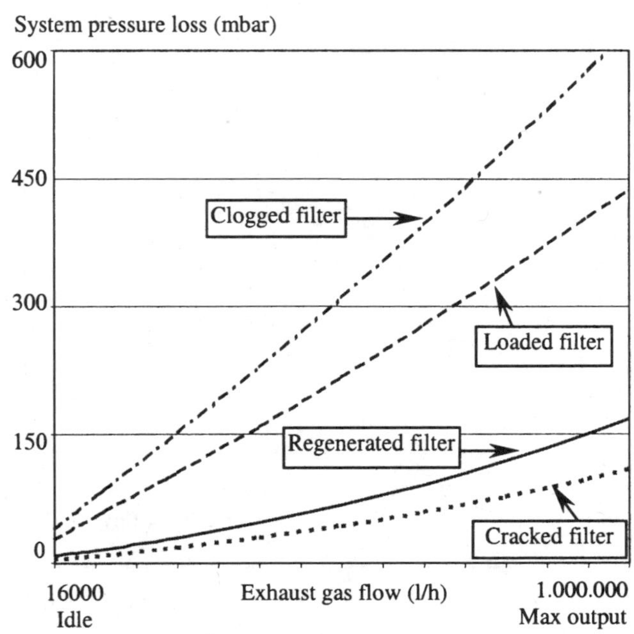

Figure 13. System monitoring and diagnosis from pressure loss measurement and exhaust flow calculation

4.2.2 – Filter – In addition to the control of fuel injection and turbocharger parameters excellent filter characteristics are essential to the system [26 – 32]. Filter type and material were carefully chosen according to:

- filtration efficiency
- soot storage capacity and thermal shock durability (parameter controlling the regeneration activation frequency)
- filter pressure loss (with and without soot, with and without residues)
- pressure loss scattering
- heat capacity and mass
- thermal conductivity
- performance to volume ratio
- Chemical resistance to:
  - diesel exhaust gas composition (sulfur based compounds)
  - ashes from lube oil
  - Ceria residues
- Potential of further development

The filter selected is a wall flow type made of porous SiC by IBIDEN. This already very high performance medium was then further co-developed to reach among the required characteristics:

- superior filtration efficiency as this is detailed in chapter 5
- three to five times soot storage capacity of equivalent cordierite type filter while its weight remains low (SiC filter density is close to 1).
- relatively low pressure loss and limited pressure loss scattering (less than ±5%) as this is illustrated on figure 14.

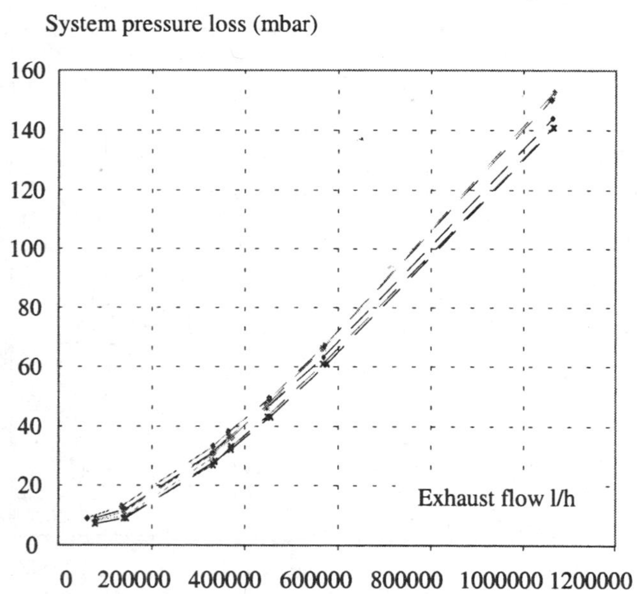

Figure 14. Illustration of system pressure loss level and scattering (Oxi cat + SiC filter)

4.2.3 – Additivation unit – As illustrated on figure 9 this unit is composed of:

- a small specific tank equipped with a sensor to detect min additive level. The overall autonomy is 80000 km
- an injector and a regulator in the main fuel tank
- a small ECU connected to the vehicle and engine ECU's to:
  - detect an addition of fuel in the main tank and inject the additive according to the treat rate selected.
  - transmit to the engine ECU the amount of additive injected for calculation of filter pressure loss due to Ceria residues
  - inform the driver of the need to refill the additive tank
  - insure the system diagnosis

4.2.4 – Additive – Beside its performance on soot regeneration the fuel additive was also selected on the following criteria:

- Health and environmental innocuousness. This additive and its treat rate combined with the selected SiC filter was accepted by the French and German environmental authorities.
- Compatibility with filter and vehicle components (filter material, fuel injection system, fuel tank material...)
- Temperature and time stability
- Solubility in the fuel
- Viscosity Vs temperature

4.3 – MAINTENANCE – Every 80000 km it is necessary to refill the additive tank and to clean the filter of Ceria residues, ashes from lube oil and any other inorganic compounds.

A cleaning procedure based on alternative filter back blowing with water and air cycles at high pressure as been developed. This procedure that will be applied by the after sales network also integrates the recycling of the residues.

5 – PERFORMANCE

5.1 – PARTICULATE REDUCTION

5.1.1 – Number of particulate – SMPS measurements indicate that the SiC filter selected reduces particulate matter emission by three to four orders of magnitude all over the particulate range as this is illustrated on figures 15 and 16.

Moreover SMPS measurement shows no difference in particulate concentration between particulate matter emission downstream SiC filter and particulate count in air dilution all over the particulate range as indicated on figures 17 and 18.

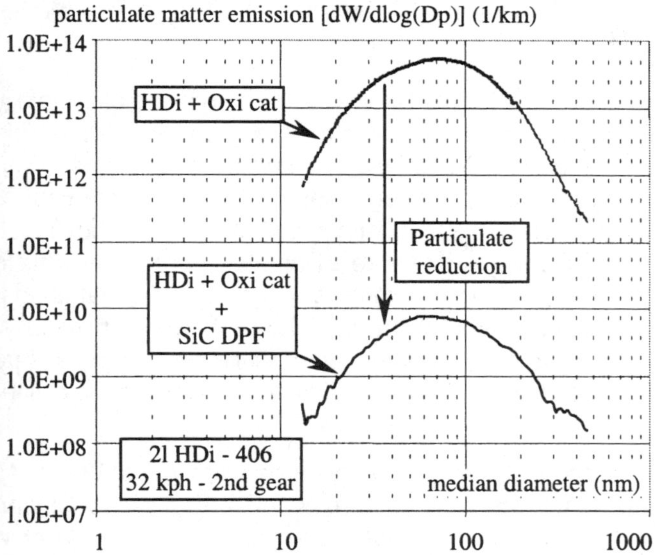

Figure 15. SiC filter reduction of particulate matter emission on a 2l HDi on a 406 (32 kph)

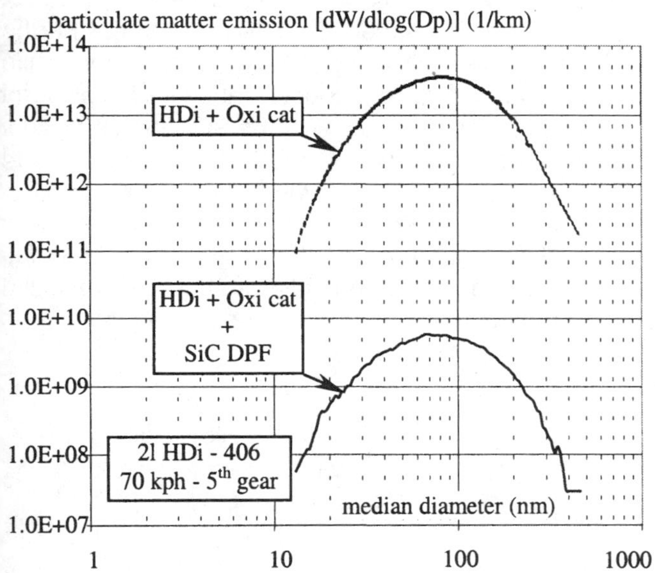

Figure 16. SiC filter reduction of particulate matter emission on a 2l HDi on a 406 (70 kph)

Based on the measurements filtration efficiency of the SiC filter selected is very close to 100% all over the particulate range. However since filter particulate matter emission is equivalent to air dilution level very accurate calculation of effective filter efficiency is difficult.

5.1.2 – Mass of particulate – In agreement with SMPS measurements particulate matter emission per kilometer on MVEG cycle of HDi vehicles with SiC filter is equivalent to reference tests made with air dilution only (no vehicle) and is significantly lower than the generation of vehicle equipped with IDI engines, as shown on figure 19. Results of reference tests made with air dilution only to check out the quality of the installation are generally in between 0.002 g/km and 0.004 g/km.

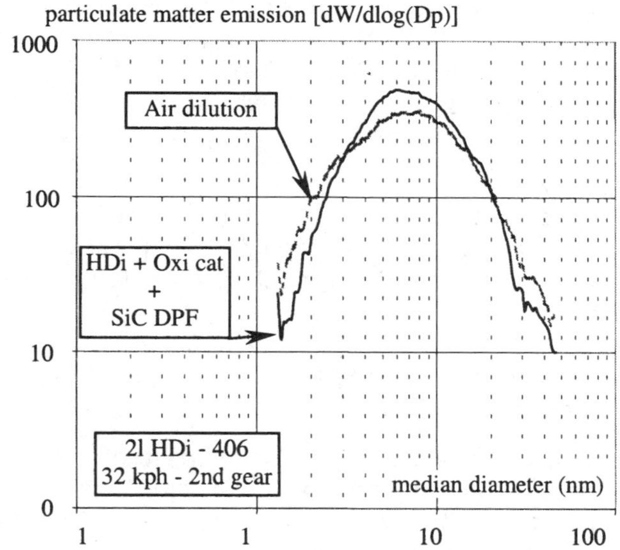

Figure 17. Comparison of SiC filter particulate matter emission with air dilution on a 406 - 2l HDi (32kph)

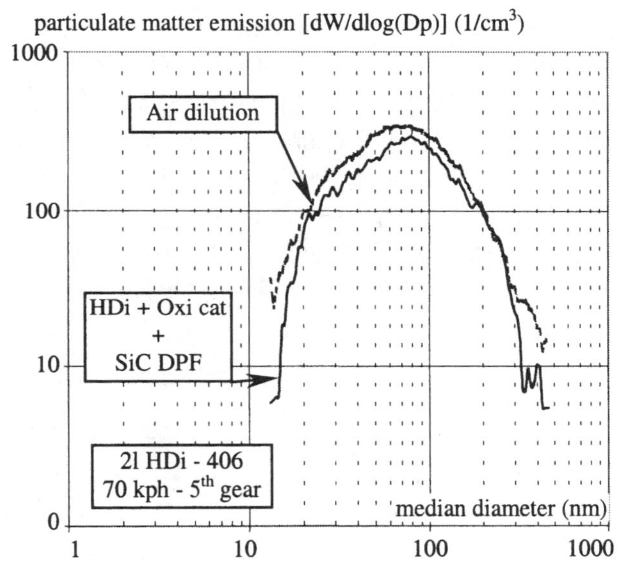

Figure 18. Comparison of SiC filter particulate matter emission with air dilution on a 406 - 2l HDi (70kph)

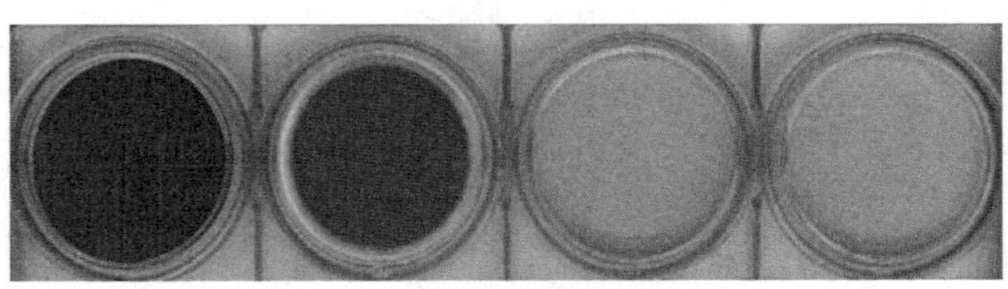

605-DK5	605-HDi	605-HDi+DPF	Air Dilution
Part 0,1 g/km	**Part 0,035 g/km**	**Part 0,004 g/km**	**Part 0,004 g/km**

Figure 19. Particulate matter emission in mass (MVEG cycle)

According to MVEG cycle results SiC filter efficiency is also close to 100% but a more accurate efficiency estimation is difficult due to the limits of the mass measurement technique. For the same reason direct comparison between mass reduction and number reduction is ticklish as well.

The most noticeable effect of this excellent filtration efficiency is the complete suppression of black smoke emission of diesel powered vehicles as illustrated on figure 20.

5.3 – VEHICLE PERFORMANCE – Current system definition lead to:

- an average fuel consumption increase of about 5% for city driving condition. This fuel consumption increase is however reduced when more favorable operating conditions are used: road, speedway, mountain...essentially because natural exhaust gas temperature is closer to soot combustion temperature.
- A 2% reduction of maximum engine output due to system pressure loss at maximum air mass flow. However excellent low speed driveability due to the natural high torque of HDi engines is not affected.

Complementary work is under way to significantly improve both fuel penalty and max output.

The vehicle which is equipped by this system reaches EURO 3 legislation level with a "0" particulate level (0.004 g/km – equivalent to air dilution level) and further developments are already engaged in order to propos an EURO 4 definition.

5.4 – FUEL QUALITY – Results of complementar SMPS and steady state cycle experiments on HDi vehicl equipped with oxidation catalyst plus SiC filter showe higher particulate matter emission than the same vehicl without catalyst and filter for vehicle speed above 10 kph as illustrated on figure 21 and 22. For steady vehicl speed of 120 kph composition analysis indicates tha particulate matter of the vehicle equipped with catalys plus filter is mainly composed of while virtually no sulfat was detected in the particulate when catalyst and filte were removed.

Thus fuel sulfur content effect was investigated [33 – 34] The same experiments were conducted with fuel sulfu content of 500 ppm (base), 300 ppm, 50 ppm and belo 10 ppm. Results are shown on figure 24 and 25. The clearly indicate that with fuel sulfur content below 50 pp particulate matter emission of vehicles equipped wit oxidation catalyst and SiC filter is equivalent to air dilutio level for vehicle speeds above 100 kph. Furthe experiments will have to be done to confirm fuel sulfu content for higher vehicle speeds (above 150 kph). Th results also confirm the ability of the SiC filter selected t suppress carbonaceous and metallic particulate matte of diesel exhaust gas all over the particulate range for a engine operating conditions. Particulate matter emissio measured downstream the filter is the result of sulfat formation on the oxidation catalyst due to fuel sulfu content. To suppress this phenomenon fuel sulfur conten below 50 ppm is necessary.

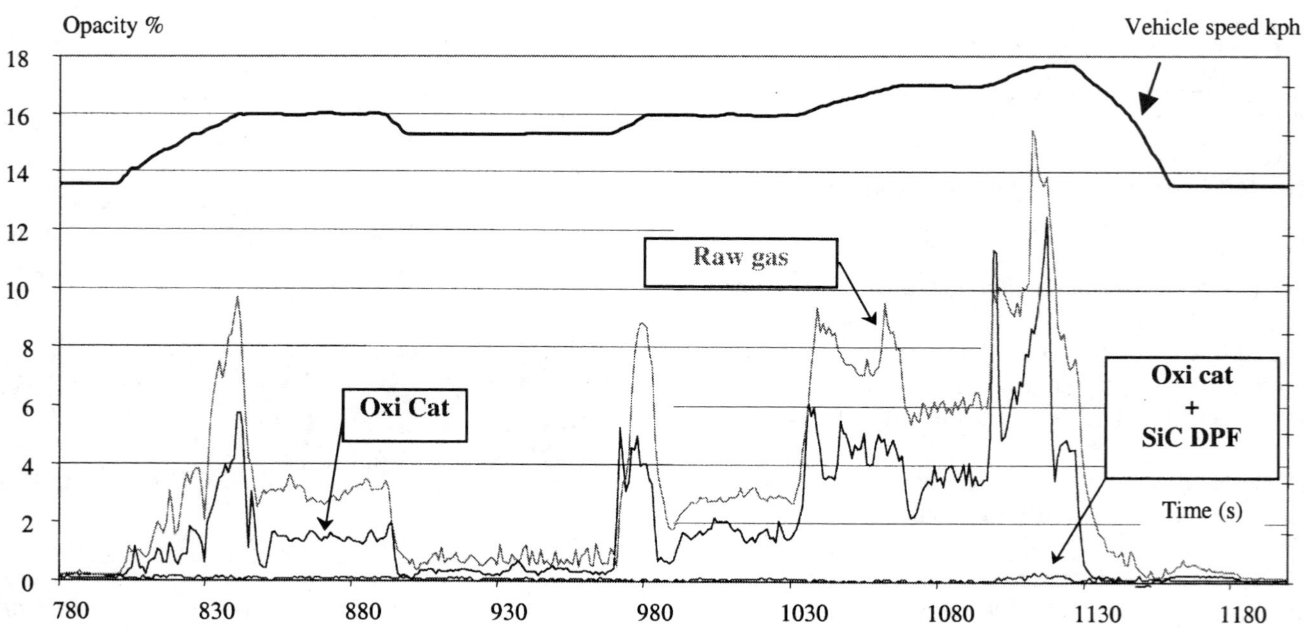

Figure 20. DPF system efficiency on black smoke emission on a 605 – 2.11 IDI Engine (EUDC cycle)

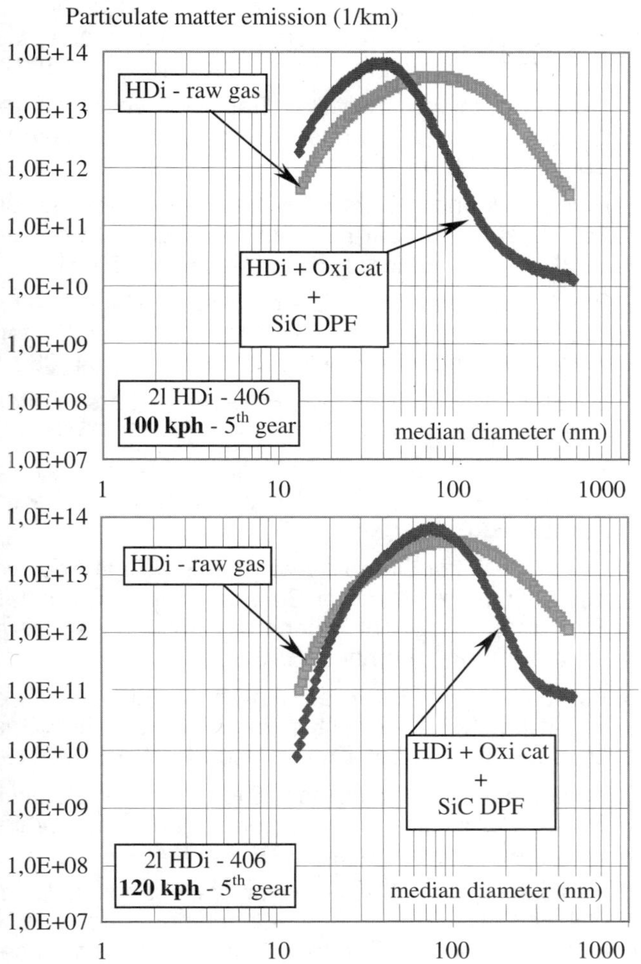

Figure 21. particulate matter emission at 100 kph and 120 kph (SMPS measurements)

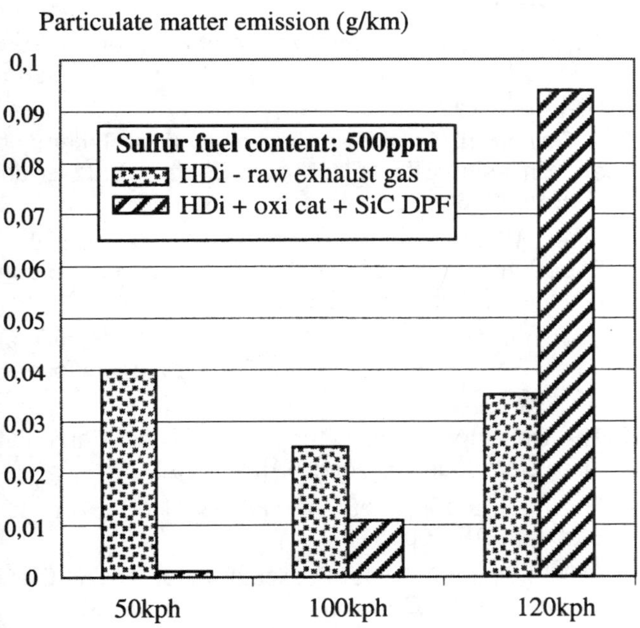

Figure 22. Steady state particulate matter emission versus vehicle speed

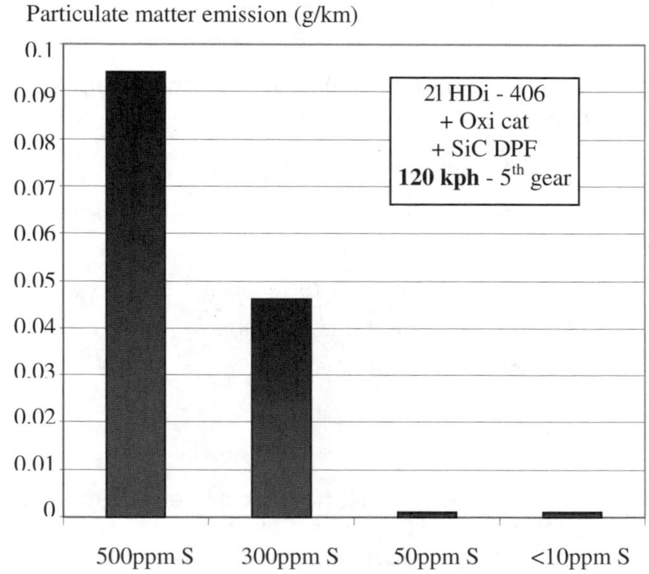

Figure 24. Influence of fuel Sulfur content on particulate matter emission in mass downstream HDi engine + Oxi cat + SiC DPF (vehicle speed 120 kph)

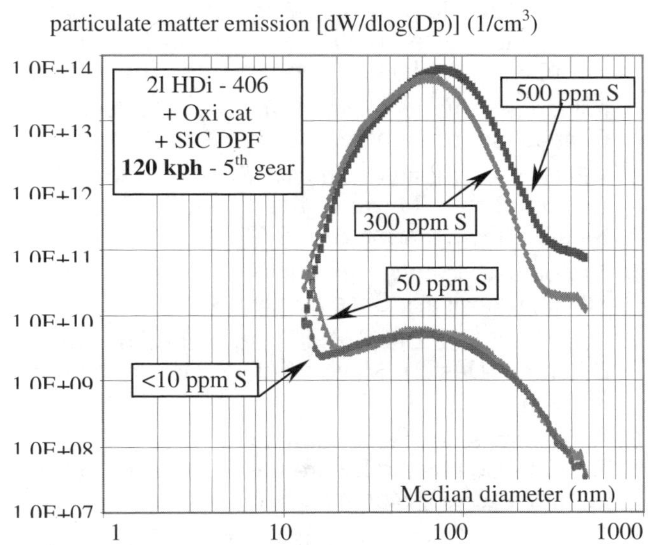

Figure 25. Effect of fuel sulfur content on particulate matter emission in median diameter downstream HDi engine + Oxi cat + SiC DPF (vehicle speed 120 kph)

6 – CONCLUSION

The first active particulate filter system which has been described represents a significant step forward for the environment since it permits to reach the “0” particulate level aimed and it also reinforces the natural potential of modern direct injection diesel engines:

- Low fuel consumption and CO_2 emission
- Durability and robustness

which also provide when equipped with the SiC filter selected:

- particulate matter emission of diesel powered vehicle reduced to air dilution level
- black smoke puffs totally suppressed

Finally the system presented can be further improved to reduce mainly:

- maintenance (frequency reduction)
- vehicle implementation constrains
- fuel consumption penalty

7 – REFERENCE

1. VERT: Diesel Nano-Particulate Emissions: Properties and Reduction Strategies, Mayer, Matter, Scheidegger, Czerwinski, Wyser, Kieser, Weidhofer, SAE 980539
2. Effect of DPF on Particulate Size Distribution Using an Electrical Low Pressure Impactor, Pattas, Kiriakis, Samaras, SAE 980544
3. Effect of a Continuously Regenerating Diesel Particulate Filter on Non-Regulated Emissions and Particulate Size Distribution, Hawker, Hüthwohl, Henn, Koch, Lüders, Lüers, Stommel, SAE 980189
4. Influence of Various Diesel Traps on Particulate Size Distribution, Dementhon, Martin, SAE 972999
5. VERT – Clean Diesel Engines for Tunnel Construction, Mayer, Czerwinski, Scheidegger, Kieser, Bigga, Wyser, SAE 970478
6. Trapping Efficiency Depending on particulate Size, Mayer, Czewinski, Scheidegger, SAE 960472
7. Particle Size Distribution Downstream Traps of Differents Design, Mayer, Egli, Burtcher et al., SAE 950373
8. Applications for Regeneration of Diesel Particulate Traps by Combining Different Regeneration Systems, Lüders, Backes, Stommel, SAE 970470
9. The effect of Size and Positioning of Ceramic DPFs on the Regeneration Induced by a Cerium Based Additive, Pattas, Kyriakis, Samaras, Mustel, Rouveirolles, SAE 970184
10. Experience with a New Particulate Trap Technology in Europe, Hawker, Myers, Hüthwohl, Vogel, Bates, Magnusson, Bronnenberg, SAE 970182
11. Regeneration of DPF at Low Temperature with the Use of a Cerium Based Fuel Additive, Pattas, Samaras, Roumbos, Lemaire, Mustel, Rouveirolles, SAE 960135
12. Sequential Regeneration System for Diesel Particulate Traps with an Intelligent, Self-Learning Control Management, Chung, Wi, Backes, Braun, Stommel, Lüders, SIAT Symposium on International Automotive Technology, Pune, India (1996)
13. An Aero dynamically Regenerated Diesel Particulate Trap Coupled to an Electric Soot Incinerator with Dual Wall Flow Filters, Caceres, Yannis, SAE 950371
14. Quasi-Continous Particle Trap Regeneration by Cerium Additives, Lepperhoff, Lüders, Barthe, Lemaire, SAE 950369
15. Fiber Wound Electrically Regenerable Diesel Particulate Filter Cartridge for Small Diesel Engines, Shirk, Bloom, Kitahara, Shinzawa, SAE 950153
16. Fuel Additive Supported Trap Regeneration Possibilities by Engine Management System Measures, Lemaire, Mustel, Zelenka, SAE 942069
17. Ceramic Trap System for Passenger Cars, Pattas, Lee, Cho, Papadopoulos, Kyriakis, SAE 940543
18. A Diesel Particulate Regeneration System Using a Copper Fuel Additive, Daly, McKinnon, Martin, Pavlich, SAE 930131
19. Cordierite Filter Durability with Cerium Fuel Additive: 100,000 km of Revenue Service in Athens, Pattas, Samaras, Sherwood, Umehara, Cantianni, Auguerre Chariol, Barthe, Lemaire, SAE 920363
20. Particulate Trap Regeneration Induced by Means of Oxidizing Agents Injected Into the Exhaust Gas, Hardenberg, Daudel, Erdmannsdörfer, SAE 870016
21. Advanced Techniques for Thermal and Catalytic Diesel Particulate Trap Regeneration, Rao, White, Wade, Aimme, Cikannec, SAE 850014
22. Particulate Control Systems for Diesel Engines Using Catalytically Coated and Uncoated Traps with Consideration of Regeneration Techniques, Öser, Thoms, SAE 830087
23. Thermal and Catalytical Regeneration of Diesel Particulate Traps, Wade, White, Florek, Cikanek, SAE 830083
24. Formation and Characterization of Particules, Health Effects Institute, Report of the 1996 HEI Workshop, HEI No.5, 1997
25. A Review of Diesel Particulate Control Technology and Emission Effects, Johnson, Bagley, Gratz, Leddy, SAE 940233
26. Fiber Wound Diesel Particulate Filter Durability Experience with Metal Based Additives, Bloom, Brunner, Schroeer, SAE 970180
27. Development of Porous Metal for Diesel Particulate Filter, Ban, Ihara, Okamoto, Shinozaki, Shinoyama, SAE 950739
28. The Development of Fiber Wound Diesel Particulate Filter Cardridges, Bloom, SAE 950152
29. Thermal Loading of SiC Filters, Hoi, Sorenson, Stobbe, SAE 950151
30. SHW's Sintered Metal Particulate Trap for All Diesel Engines, SHW, Commercial Brochure, 1995

31. St udy of SiC Application to Diesel Particulate Filter (Part 1): Material Development, Itoh, Shimato, Komori (Ibiden Co., Ltd.), Okazoe, Yamada, Niimura, Watanabe (Nissan Diesel Motor Co., Ltd), SAE 930360

32. Study of SiC Application to Diesel Particulate Filter (part 2): Engine Tests Results, Okazoe, Yamada, Niimura, Watanabe (Nissan Diesel Motor Co., Ltd), Itoh, Shimato, Komori (Ibiden Co., Ltd.), SAE 930361

33. The Effect of Low Sulfur Fuel and a Ceramic Particle Filter on Diesel Exhaust Size Distribution, Baumgard, Johnson, SAE 920566

34. Exhaust Particulate Size Distribution: Vehicle and Fuel Influences in Light Duty Vehicles, Rickeard, Bateman, Kwon, McAughey, Dickens,

2003-01-0378

Ceria-Based Fuel-Borne Catalysts for Series Diesel Particulate Filter Regeneration

G. Blanchard and T. Seguelong
RHODIA Electronics & Catalysis

J. Michelin, S. Schuerholz and F. Terres
Tenneco Automotive - Heinrich Gillet GmbH & Co. KG

ABSTRACT

Ceria-based fuel-borne catalysts have been used to promote regeneration of the DPF, by lowering the soot-combustion temperature and favoring propagation of the combustion process throughout the soot layer, and thus to ensure durability of DPF efficiency over the EURO 4 standard limits.

Since the market introduction of the PSA Peugeot Citroën DPF System for series applications, extensive works have been done to improve and simplify the DPF technology. A new enhanced ceria-based fuel-borne catalyst has been designed and developed to limit the DPF system maintenance.

This paper proposes comparative studies of two different Ceria-based fuel-borne catalysts for DPF regeneration, particularly focussing on the enhanced catalytic activity of soot combustion : correlation between soot combustion and oxygen buffering is proposed to explain difference in catalytic activity, while catalytic activity on NOx species is not affected.

The improvement of catalytic activity leads to decrease of global dosing rate in the diesel fuel that significantly reduces the ash-build up effect.

From field evaluations, characterizations of several DPFs show that ash arising from the fuel-borne catalyst strongly depends on the lubricant oil consumption, the average diesel fuel consumption and the raw engine-out PM emissions.

In relationship with those field evaluations, a second section of the paper presents mathematical models to design and optimise the DPF size in order to limit its maintenance, taking into account the global vehicle ash production.

Basically, the use of enhanced Ceria-based fuel-borne catalyst significant extends the length of the interval between DPF maintenance (ash removal by washing process) and leads to a simplification of the global DPF system offer.

That improvement places the use of Ceria-based fuel-borne catalyst to promote DPF regeneration as an attractive technical and economical approach for DPF regeneration in series passenger car applications

INTRODUCTION

Diesel particulate emissions have been subject to intense research in recent years. It has been shown that the Diesel Particulate Filter (DPF) is the only efficient technology to reduce emissions of particulate matter, including the ultra-fine solid particulate, to a level similar to that of ambient air.

The DPF system consists of a filter positioned in the exhaust stream designed to physically collect the PM emissions, preventing thus their release to the atmosphere, while allowing the exhaust gases to pass through the system /**1**/. The most common DPF design is the wall-flow monolith, which consists of many small parallel ceramic channels running axially through the part. Adjacent channels are alternatively plugged at each end in order to force the diesel exhaust gases through the porous substrate walls which act as a mechanical filter. Since the volume of PM generated by a diesel engine is sufficient to fill up and plug a reasonably sized filter in a matter of few hundred kilometers, combustion of the accumulated soot in the DPF must occur frequently in order to regenerate the trap for the next accumulation step.

If filtration efficiency can be achieved using ceramic-based wall-flow filtration, such as silicon carbide or cordierite, durability of filtration performance is considered to be the main issue for serial applications.

In order to have long-term and reliable operation, the choice of the filter material and careful adaptation of the regeneration strategy is of crucial importance.

Figure 1 gives an overview of the main possibilities which exist for oxidation (burning off) of the soot particulate.

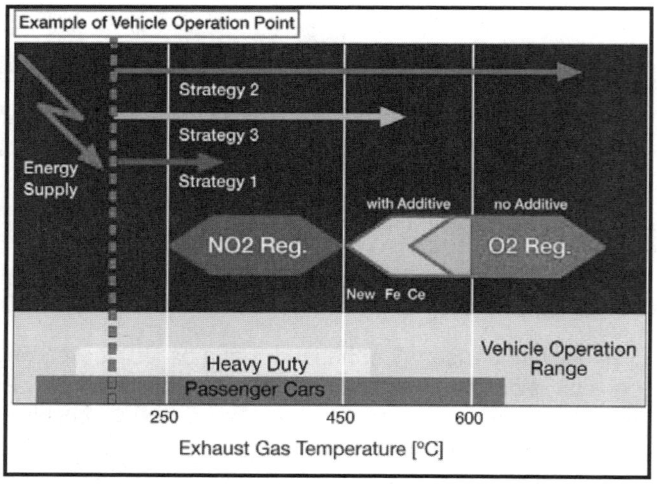

Figure 1 : Current DPF regeneration strategies

In order to effectively burn a Carbon-particle with oxygen (O_2), high temperatures (> 600°C) are needed. In diesel exhaust systems, these temperatures can hardly be met even under full load operation.

Through the use of fuel-borne catalyst based on Ceria (Ce) or Iron (Fe), these ignition limits can be reduced, and soot burns at approx. 400/450°C /**2**/.

The advantage of fuel-borne catalyst is that the ignition temperature is reached during regular driving more frequently or that, during low-load operation, the expenditure of energy needed to reach regeneration temperature is lower. Apart from this, regeneration is also considerably accelerated by the fuel-borne catalyst resulting in a shorter regeneration phase. On the other hand, fuel-borne catalysts produce residual ashes that, in addition to the unavoidable oil ashes, may clog the filter sooner.

In principle, it is also possible to burn off the soot with nitrogen dioxide (NO_2). Here much lower temperatures (approx. 250°C) are needed than when using O_2. This oxidation is very slow and nearly not exothermal. The NO_2 is produced by oxidation of NO, either in an upstream Oxicat and/or by the catalytic coating of the filter itself /**3**/. Unfortunately, the possibility of NO_2-regeneration is limited to 450°C because NO_2 is not stable at higher temperature levels. Furthermore, it should be noted, that an appropriate NO_2 to soot ratio is crucial for acceptable efficiency, which might become an issue for future NOx-optimised vehicles. In addition, the Sulfur level in diesel fuel is crucial for NOx catalysis, and when Sulfur level decreases, as we can get with Ultra-Low Sulfur Level fuel, NO_2 release can be observed what is considered as more toxic than NO alone /**4**/.

Whether O_2 or NO_2, both procedures have in common that they do not work under certain normal driving conditions (e.g. low-load operation in town traffic). Additional energy, therefore, is always needed in order to achieve regeneration conditions. This results in a discontinuous accumulation and regeneration process. There are various possibilities for providing the energy required:

- Post-injection of fuel and exothermal reactions in the Diesel Oxidation Catalyst (DOC)
- Electric heater or ignition device (available electric power in the 12V system is max.1.0 – 1.5kW)
- Fuel Burner
- Non Thermal Plasma (NTP), whereby the soot is regenerated by oxidation with O-radicals, which can be created by the plasma even at the lowest temperatures. The electrical power required for today's systems is under 1 kW.

Looking again at **Figure 1**, different strategies for the regeneration are indicated:

- Strategy 1 - Increase of the temperature into the NO_2 regeneration range

- Strategy 2 - Increase of the temperature into the O_2 regeneration range

- Strategy 3 - Increase of the temperature into the fuel-borne catalyst supported O_2 regeneration range

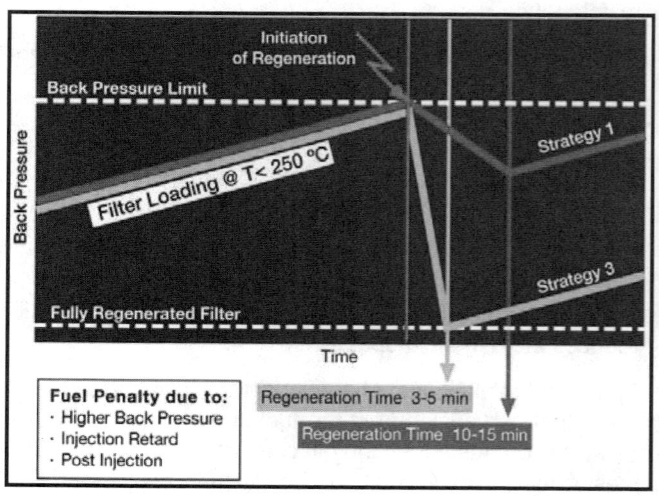

Figure 2 : Back pressure for different regeneration strategies

Generally, the supply of energy to initiate the regeneration process increases the fuel consumption. On the other hand, soot accumulation in the particulate filter increases the back pressure and thus also increases fuel consumption. **Figure 2** shows the regeneration characteristics for two regeneration strategies (1 and 3).

In the first strategy when critical back-pressure level has been reached only little energy needs to be provided in order to reach an exhaust gas temperature of approx. 300°C. Regeneration takes place very slowly and is not complete. Although less energy needs to be provided, it has to be provided over a longer period. As the filter is only partly regenerated it still has a high back-pressure level (→fuel consumption) and needs to be regenerated again after only a short driving period.

As opposed to this, in the third strategy more energy has to be provided to achieve the ignition temperature of approx. 500°C but this only for a short time because as a result of the fuel-borne catalyst supported reactions the filter is already completely regenerated after 3 – 5 minutes. As the filter is completely regenerated, the back-pressure level is low again and the time till the next regeneration is comparatively long.

It can be clearly seen that the fuel-borne catalyst supported strategy 3 is energy-wise the more favourable.

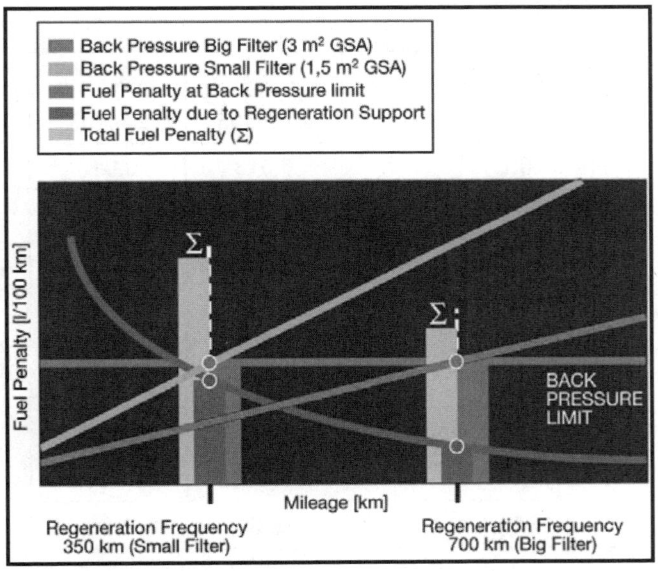

Figure 3 : Fuel penalty overview

The design of a DPF must always be made in direct relation to the chosen regeneration strategy. **Figure 3** shows the energy balance for two filter sizes and the respective amount of regeneration energy required. On the assumption that there is a maximum permissible back pressure level, a small filter has to be more frequently regenerated than a large filter. The bars on the chart showing the sums demonstrate that it is more advantageous from an energy point of view to have a larger filter and therefore less frequent regeneration. Furthermore, due to higher ash storage capacity, a larger filter increases much slower in back pressure with ash accumulation which again has a beneficial effect on fuel consumption. Opposed to this are the higher costs which arise for larger filters, the heavier weight and, possibly, space limitations for vehicle integration.

The issue is to develop appropriate regeneration strategies which clean the filter of the trapped particulate. In view of the extreme thermal conditions which could occur during this regeneration, the filter system needs to be adapted properly. Another remaining issue is the handling of the oil ashes which are trapped in the filter as well, by cleaning strategies or appropriate filter design.

CERIA-BASED FUEL-BORNE CATALYSTS FOR DPF SYSTEM IMPROVEMENT

The attractive method to meet the regeneration challenge is to dose diesel fuel with a Ceria-based fuel-borne catalyst (called "Ce-FBC reference") at low concentration. PSA Peugeot Citroën has marketed the Strategy 3 for series vehicle applications equipped with High Pressure Diesel Direct Injection engines /**5**/. Since market introduction in May 2000, more than 400000 vehicles, DPF-equipped, have been sold throughout Europe.

In order to limit the ash build-up effect and extend the DPF interval service, a new Iron-doped Ceria-based fuel-borne catalyst (called "Fe/Ce-FBC") has been designed and developed by Rhodia.

Improvement mainly focussed on enhancement of the catalytic activity to decrease the dosing rate, using a new chemical synthesis route and Iron as doping element, to tune the Ceria catalytic activity for enhanced soot combustion.

The catalytic performance were extensively presented in previous paper /**6**/ and are summarized in the **Table 1**.

Fuel-Borne Catalysts	**Ce-FBC reference**	**Fe/Ce-FBC**
Active Ceria	**Ceria**	**Iron-doped Ceria**
Dosing rate (ppm)	**25**	**10**
Soot burn-off (°C)	**460**	**410**
Regeneration time (second) **(at 500°C - 1600rpm/42Nm)**	**1600**	**400**
Max. temperature peak (°C) **(with 8g/l soot loading)**	**700**	**750**
Thermal stress (°C/cm) **(with 8g/l soot loading**	**55**	**55**

Table 1 : Comparison of the Ceria-based fuel-borne catalysts

Basically, as opposed to the Ce-FBC reference, regeneration with Fe/Ce-FBC occurs at about 410°C, even at very low dosing rate of 10ppm, while keeping the DPF functionality. This temperature level is reached considerably more often under regular driving conditions which significantly reduces the frequency of forced regeneration and thus the expenditure of energy.

1 – KINETIC OF DPF REGENERATION USING POST-INJECTION

A 2,2 liters Peugeot DW12 TED4 direct diesel injection engine, turbo-charged with inter-cooler air, fitted with the original Peugeot 607 HDi model exhaust line has been used for DPF regeneration evaluations. All tests were carried out according to the loading and regeneration procedures (at 450°C and 500°C, using post-injection strategy) already reported in /**6**/.

Diesel fuel quality is a European standard diesel fuel, with 280ppm sulfur content.

Figure 4 shows the kinetics of DPF regeneration (in percentage) versus time, determined from the back-pressure evolution during regeneration.

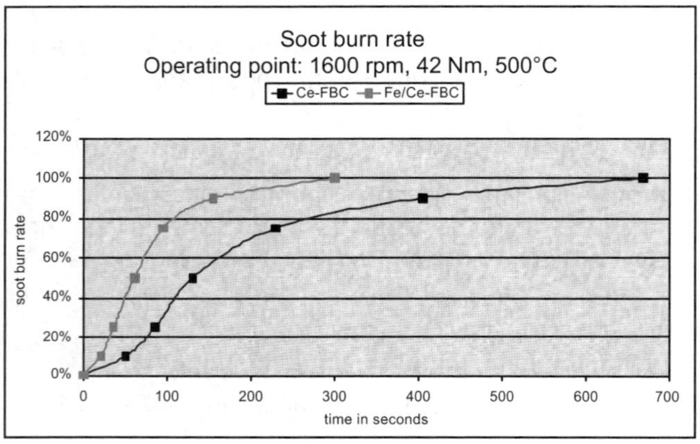

Figure 4 : Soot Burn Rate vs. time at 500°C

The Soot Burn Rate evolution (SBR) versus time can be fit with the equation :

$$SBR = [1 - \exp(-at)] \times [1 - \exp(-bt)]$$

where :

SBR is the Soot Burn Rate in %

t is the time, expressed in seconds

the coefficient ***a*** stands for the burning rate speed in s^{-1}

the coefficient ***b*** describes the ignition speed in s^{-1}

and example of fitting curves is reported in **Figure 5**.

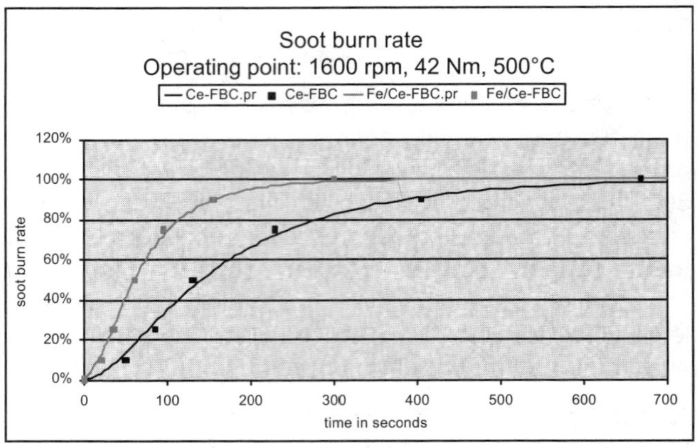

Figure 5 : Modeling of Soot Burn Rate regeneration at 500°C

The coefficients are reported in **Table 2**, for different conditions.

FBC	Ce-FBC reference	Fe/Ce-FBC
Dosing rate (ppm)	**25**	**7**

a (at 450°C) (in second^{-1})	**0,0034** ± **0,0005**	**0,0115** ± **0,003**
b (at 450°C) (in second^{-1})	**0,013** ± **0,0025**	**0,015** ± **0,000**
a/b (at 450°C)	**0,26**	**0,76**

a (at 500°C) (in second^{-1})	**0,006** ± **0,0006**	**0,0168** ± **0,0012**
b (at 500°C) (in second^{-1})	**0,015** ± **0,0025**	**0,028** ± **0,002**
a/b (at 500°C)	**0,40**	**0,60**

Table 2 : Mathematical model of kinetic of DPF regeneration

Fe/Ce-FBC mainly affects the burning rate and DPF regeneration (as higher ***a***, as faster the burning rate and DPF regeneration), while keeping the ignition of DPF regeneration in the same range (as higher ***b***, as faster the ignition of DPF regeneration).

Those results could demonstrate that the ignition of soot burn-off is related to the soot composition, local temperature and gas phase, and not to the catalyst composition.

On the other hand, the Ceria-based catalyst composition changes significantly the kinetic of DPF regeneration process : Iron-doped Ceria-based Catalyst shows faster

DPF regeneration (3 to 4 times faster), with providing locally active Oxygen species in the soot bed.

In combination with the local exothermic phenomenon related to the soot combustion itself, the Oxygen released from the catalyst can accelerate the whole soot combustion process.

2 – OXYGEN STORAGE CAPACITY (OSC)

Oxygen Storage Capacity measurements were performed to verify the impact of Oxygen and catalyst composition effect on the soot combustion process.

The Ceria-based fuel-borne catalysts were impregnated on inert Alumina support at loading 6,3% by weight in order to evaluated the Oxygen Storage Capacity of Ceria and Iron-doped Ceria catalysts.

The Oxygen Storage Capacity (OSC) measurement procedure was previously described in /**7**/. OSC measurements of catalysts as powder were carried out using alternate pulses of gaseous O_2 and CO passed through the oxide bed at different temperature in order to simulate the redox conditions existing in automotive exhaust gases. O_2 was diluted to 2.5% in He and CO to 5%. The flow rate of He was 10 $l.h^{-1}$. A catalyst volume of 0.1 cm^3 was packed in a glass reactor placed in a vertical furnace. Gas chromatography was used to determine the O_2, CO and CO_2 content in the exhaust gases.

Results are expressed in % of CO conversion versus temperature of measurement and reported in **Figure 6**.

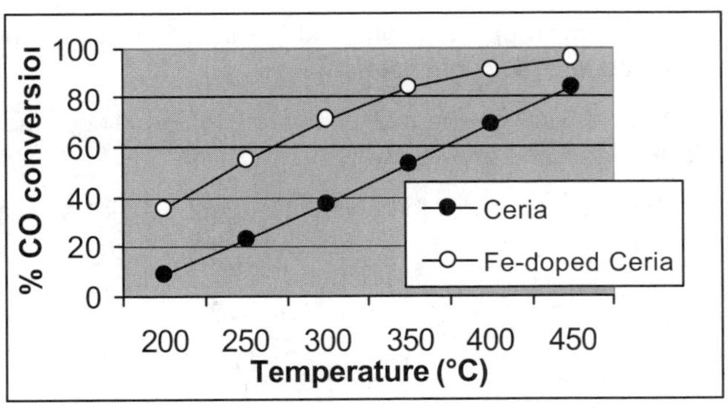

Figure 6 : Oxygen Storage Capacity

As we can see, Iron-doped Ceria leads to an Oxygen availability at lower temperature, compared with pure Ceria : at 350°C, about 82% of Oxygen is available for oxidation reaction, compared with 50% of Ceria.

If we take into account the Oxygen availability at 400°C, Iron-doped Ceria is nearly fully reduced, while it occurs between 450 and 500°C with Ceria : that could explain the main difference between Ce-FBC reference and Fe/Ce-FBC.

Then, the exothermal reaction from the soot combustion process is used for the diffusion process in the DPF, according to the Semenov's theory, proposed several years ago /**8**/.

CERIA-BASED FUEL-BORNE CATALYSTS : IMPACT ON THE ASH ACCUMULATION

1 – GENERAL TREND IN ASH ACCUMULATION

: Ash deposits in the DPF during accelerated ash loading on the engine test bench differ in general from the deposits which have been found after normal road driving conditions /**10**/. **Figure 7a** shows an example of the wall deposits, which is typical for tests carried out on the engine bench.

In view of chemical composition, accelerated ash loading procedure produces a very huge unbalance between compound coming from engine lubricant which cannot be simply increased and Ceria-based oxide coming from extra high fuel-borne catalyst treat rate. Moreover the main difference comes surely from the way to accumulate ashes in the filter. Actually, accelerated cycles create ash produced at high temperature without going through the normal process : soot accumulation and periodic regeneration. Therefore the ash morphology and deposit pattern are definitely not comparable.

The consequences in term of back pressure behavior and service interval will be develop later.

2 – PRELIMINARY RESULTS ON PEUGEOT 406 HDI

: The X-ray picture in **Figure 7b** shows that under vehicle operating conditions on road, the ashes fill the filter from the back to the front. Here, a Peugeot 406 HDI 2 liters, fitted with a 5,66"x9" SiC-DPF was tested over a field evaluation with Fe/Ce-FBC. After approximately 70000km, about 15% of its ash storage capacity was filled up.

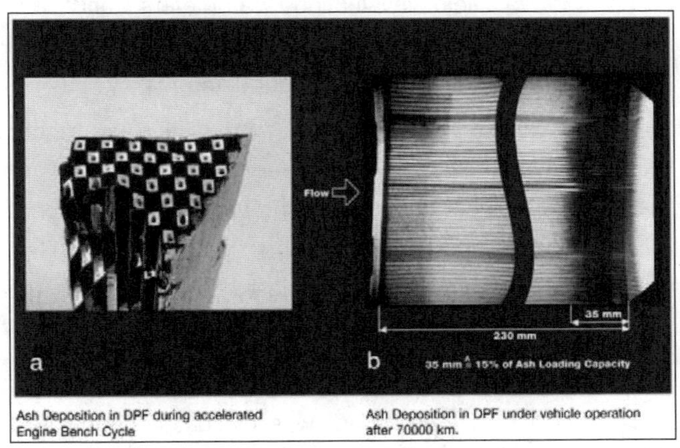

Figure 7 : Typical ash deposits during accelerated test bench (a) and field evaluation (b)

Optimistically it could be concluded that, with a slightly bigger filter, even with a fuel-borne catalyst, the useful life of the engine (approx. 240000 km) could be covered without cleaning the filter.

3 – PEUGEOT 607 HDI : A Peugeot 607 HDi 2,2 liters diesel engine, DPF-equipped, was evaluated over the similar mixed drive cycle. The length of evaluation was 81700km with following parameters :

- SiC-DPF : 5,66" x 10" (4,12 liters) / 200cpsi;
- European standard diesel fuel (300mg/kg of sulfur);
- average fuel consumption of 9 liters/100km;
- Ce-FBC dosing rate of 25ppm;

After 81700km, the canned SiC-DPF was directly characterized using X-Ray analysis, to check ash situation and package (**Figure 8**) : as previously, ash package is located at the rear of the DPF, and the ash package volume corresponds to 50% of the total DPF volume.

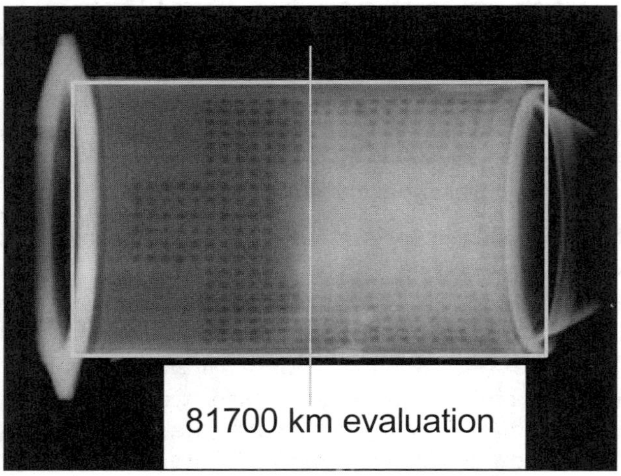

Figure 8 : X-Ray analysis of DPF after 81700km operation (using Ce-FBC Reference)

According to field evaluation parameters and ash characteristics, we can specify that :

- the global ash density is about 0,45 g/cm^3;
- the average Ce-FBC ash production is about 2,32g/1000km (80% of total ash);
- the average lubricant ash production is about 0,52g/1000km (18% of total ash);

the rest of ash is about 2% related to the Diesel fuel additives part and engine wear.

4 – PEUGEOT 406 HDI : A Peugeot 406 HDi 2,0 liters diesel engine, DPF-equipped, was evaluated during field operation, over a mixed drive cycle included urban, extra-urban and highway. The length of evaluation was 90000km with following parameters :

- SiC-DPF : 5,66"x6" (2,47 liter) / 200cpsi;
- European standard diesel fuel (300mg/kg of sulfur);
- average fuel consumption of 7 liters/100km;
- Fe/Ce-FBC dosing rate of 10ppm;

After 90000km, the canned SiC-DPF was directly characterized using X-Ray analysis, to check ash situation and package (**Figure 9**).

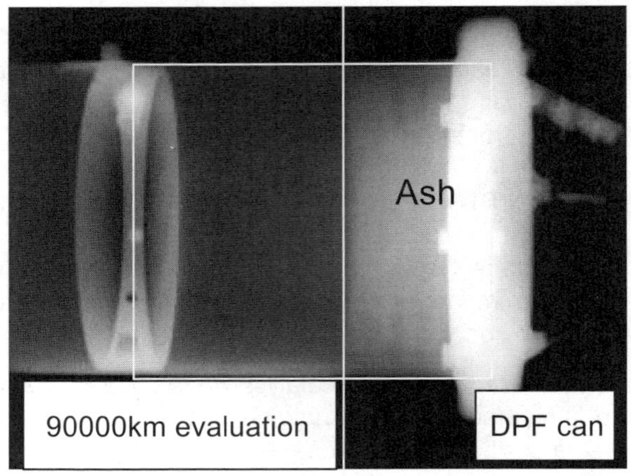

Figure 9 : X-Ray analysis of DPF after 90000km operation (using Fe/Ce-FBC)

As we can see, ash package is still located at the rear of the DPF. According to **Figure 9**, the ash package volume corresponds to 40% of the total DPF volume.

According to the average diesel fuel consumption of 7 liters/100km, the 10ppm dosing rate, the chemical compositions, the lubricant composition and the volume of ash trapped in the DPF, we can specify that :

- the global ash density is about 0,43 g/cm^3;
- the average Fe/Ce-FBC ash production is about 0,76g/1000km (57% of total ash);
- the average lubricant ash production is about 0,55g/1000km (41% of total ash);

the rest of ash is about 2%, related to the diesel fuel additives part and engine wear.

All the results recorded with Peugeot vehicles are summarized in **Table 3** :

Vehicle	**607 HDi**	**406 HDi**
FBC class	**Ce-FBC**	**Fe/Ce-FBC**
Dosing rate (ppm)	**25**	**10**
Fuel consumption (l/100km)	**9**	**7**
Evaluation duration (km)	**81700**	**90000**
Global ash density (g/cm^3)	**0,45**	**0,43**
FBC part (%)	**80**	**57**
Lube part (%)	**18**	**41**
Rest of ash (%)	**2**	**2**
FBC ash production (g/1000km)	**2,32**	**0,76**
Lubricant ash production (g/1000km)	**0,52**	**0,55**
Global ash production (g/1000km)	**2,90**	**1,33**

Table 3. Summary of ash evaluations

Basically, with HDi engines, the Fe/Ce-FBC ash contribution is normally in the range of 40% to 60%, depending on lubricant oil consumption, the Fe/Ce-FBC dosing rate (related to the engine-out soot emissions) and diesel fuel consumption.

5 – VOLKSWAGEN PASSAT 1,9 TDI : A

Volkswagen Passat TDI 1,9 liters diesel engine, DPF-equipped, was evaluated during field operation, over a mixed drive cycle included urban, extra-urban and highway. The length of evaluation was about 55000km with following parameters :

- SiC-DPF : 5,66"x6" (2,47 liter) / 200cpsi;
- European standard diesel fuel (300mg/kg of sulfur);
- average fuel consumption of 6,4 liters/100km;
- high Fe/Ce-FBC dosing rate of **28**ppm;

At 55000km, the evolution of DPF back-pressure with the exhaust flow rate for regenerated DPF is not really affected by the presence of accumulated ash package, as showed in **Figure 10** : it follows the initial back-pressure evolution of the original clean DPF.

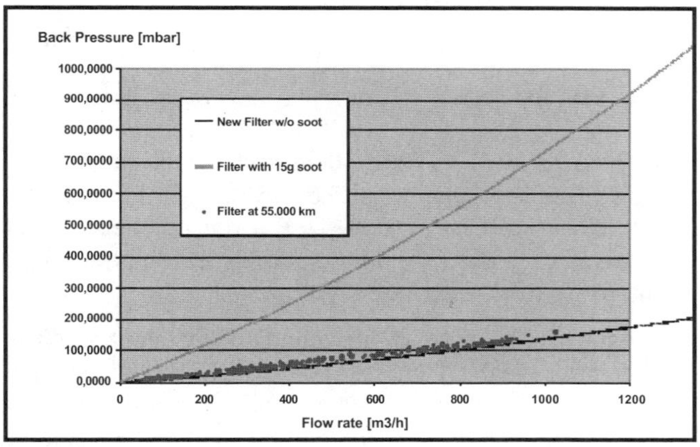

Figure 10 : Impact of ash accumulated in DPF on the back pressure

On the other hand, at 55000km, the back-pressure of DPF loaded with amount of soot of 6 g/l trap (approx. 15g of soot loading), increases sharply with exhaust flow rate, indicating that the ash accumulation strongly limits the DPF volume available for soot accumulation (**Figure 10**).

In addition, the influence of soot loading on the back-pressure is not similar with and without ash accumulation. **Figure 11** clearly shows that the DPF regeneration at similar back-pressure threshold has to be triggered at smaller total soot loading with mileage and ash accumulation.

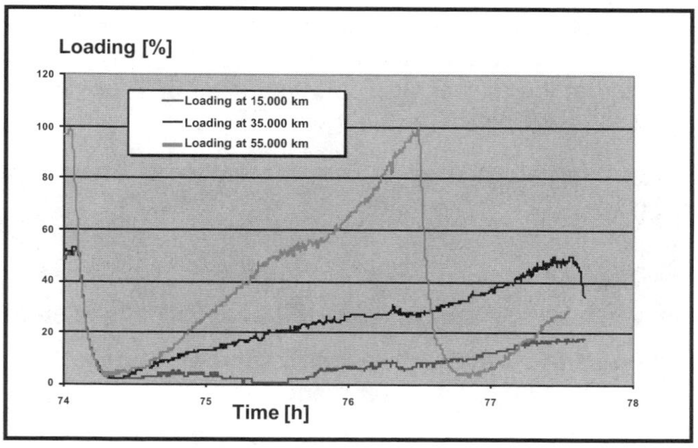

Figure 11 : Evolution of interval for DPF regeneration

Figure 12 reports the evolution of the length of the interval between DPF regeneration at different mileage (15000, 35000 and 55000km). That clearly shows a decrease of volume available for soot accumulation during the loading step, with mileage and ash build-up phenomenon.

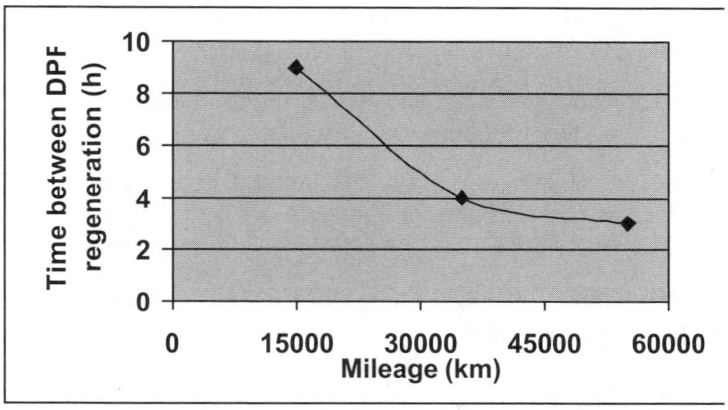

Figure 12 : Average frequency of DPF regeneration vs. mileage

After 55000km, the DPF was characterized using X-Ray analysis, to check ash situation and package (**Figure 13**). As previously observed, ash package is located at the rear of the DPF, and the ash package volume corresponds to about 50% of the total DPF volume.

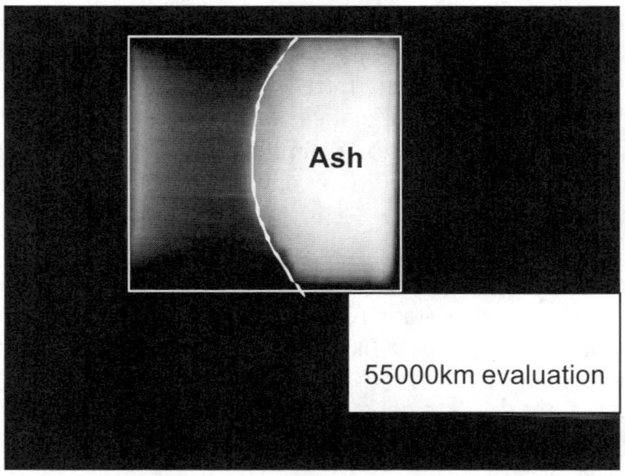

Figure 13 : X-Ray picture of ash accumulation

Optical picture of one modular SiC segment of DPF (from middle core) confirms that the ash accumulation is located at the rear of the DPF, and about 50% of volume is filled up with global ash (**Figure 14**).

Figure 14 : Optical characterization of module SiC segment

In that case, the ash arising from Fe/Ce-FBC is about 49% of total ash. The part of diesel fuel additives is negligible due to the low diesel fuel consumption and the high Fe/Ce-FBC dosing rate.

According to field evaluation parameters and ash characteristics, we can specify that :

- the global ash density is about 0,47 g/cm^3;
- the average ash production is about 1,94g/1000km (58% of total ash);
- the average engine ash production is about 1,35g/1000km (40% of total ash);

the rest of ash is approx. 2% related to the Diesel fuel additives part and engine wear.

DISCUSSIONS ON DPF INTERVAL SERVICES IMPROVEMENT

The four previous real cases show clearly, in spite of their different origins, different lube-oil types as well as different fuel-borne catalysts and treat rates, that ash can be treated as a single component characterized by a incremental mass over driving period and its density.

Based on field operation results, we have developed a simple mathematical model for ash accumulation evolutions, taking into account: the average diesel fuel consumption (l/100km); the diesel fuel density (usually 0,84g/cm^3); the average lubricant ash production (g/1000km); the engine wear and fuel ash production (g/1000km); the dosing rate of fuel-borne catalyst (whatever Ce-FBC Reference or Fe/Ce-FBC); and the global ash density to estimate the maximum DPF operation length before maintenance, according the maximum of 60g/liter-DPF /**9**/.

1 – PEUGEOT 607 : with the average diesel fuel consumption of 9 liters/100km, the average lubricant ash production of 0,60g/1000km, the engine wear and fuel ash production of 0,06g/1000km and the dosing rate of Fe/Ce-FBC of 10ppm, a 5,66"x10" SiC-DPF can be used over 150000km before maintenance (about 247grams of global ash).

The results are reported in the **Figure 15**, and that estimation is really consistent with references /**9**/ and /**10**/.

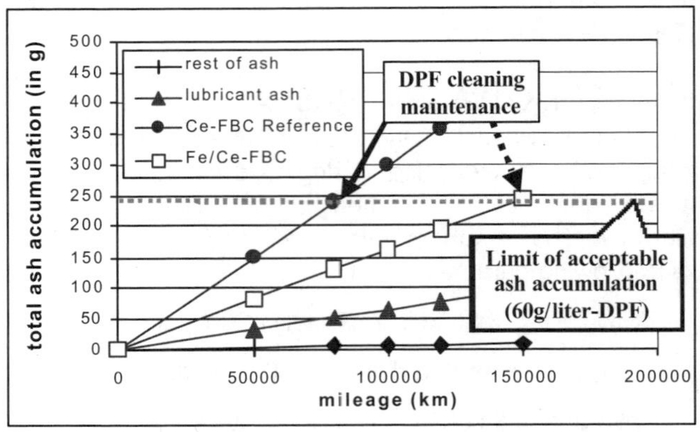

Figure 15 : Model for ash accumulation evolution

2 – PEUGEOT 406 : with the average diesel fuel consumption of 7 liters/100km, the average lubricant ash production of 0,60g/1000km, the engine wear and fuel ash production of 0,03g/1000km and the dosing rate of Fe/Ce-FBC of 10ppm, the 5,66"x9" SiC-DPF could be used over 160000km before DPF cleaning step (about 223grams of global ash).

3 – VW PASSAT : with the average diesel fuel consumption of 6,5 liters/100km, the average lubricant ash production of 1,35g/1000km and the dosing rate of Fe/Ce-FBC of 10ppm, the 5,66"x6" SiC-DPF could be used over 72000km before DPF cleaning step (about 148grams of global ash). In that case, the ash part arising from Fe/Ce-FBC is about 34% by weight.

The length of the interval between DPF maintenance (ash removal by washing process) is directly limited by DPF size : to achieve a 100000km maintenance interval (to meet EURO 4 limits), our mathematical model indicates a DPF size of 5.66"x9".

4 – CONCLUSIONS : that mathematical model is really flexible, and can be adapted to the specific lubricant consumption and ash production, and SiC-DPF size, that it strongly depends on the diesel engine type and engine maker strategy.

In that first approach, the model can help us in DPF design, but it does not take into account the impact on the DPF back pressure during soot loading what is considered as the major part of fuel-penalty after engine operation.

However, that simple model has been set up with Ibiden's 200 cpsi-based SiC-DPF and small diesel engine sizes (< 2,5 Liters).

In addition, neither changes in DPF material, design and cpsi characteristics nor diesel engine sizes larger than 2,2 Liters are considered. For further developments, those parameters have to be included in a predictive model for DPF System.

DPF SYSTEM PREDICTION TOOL

1 – PRESSURE DROP IN HONEYCOMB FILTERS : Honeycomb filters always work according to the same principle, (**Figure 16**). The inlet and outlet channels are alternatively closed and as a result the exhaust gas entering the inlet channel has to flow through the filter wall in order to reach the outlet channel. In doing so the particulate is trapped in the filter wall in the inlet channel causing three major sources for the loss of pressure:

- friction in the inlet and outlet channel
- pressure drop through the permeable filter wall
- contraction and expansion losses

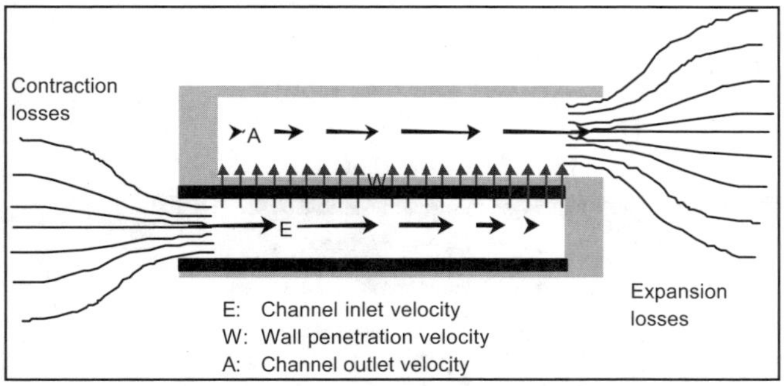

Figure 16: Diagram of a honeycomb filter

There is a characteristic length for every type of honeycomb filter at which the total pressure drop is the lowest. This is due to the fact that with increasing filter length the friction losses in the channels increase and through the filter walls decrease. Contraction and expansion losses have only a secondary effect.

2 – MATHEMATICAL MODEL OF AN UNLOADED CHANNEL FILTER : our improved calculation model is based on previous developments /**11**/. Using momentum and mass conservation laws and isothermal conditions a non linear differential equation of 2nd order can be found for the local channel outlet velocity valid for low and high mass flow rates. Based on this equation the pressure drop of an unloaded filter can be calculated.

Through the non-linear term the equation can not be analytically solved. With the help of mathematical software, such as Maple, Mathematica or Matlab, the solution for the channel outlet velocity can be numerically found. Widespread use of such expert tools and a simple connection of the gained results to other software is not always possible. A differential procedure was therefore developed in order to solve the boundary value problem which could be integrated into various software /**12**/.

3 – MODELLING OF A LOADED FILTER : The approach introduced in the previous paragraphs for the calculation of the pressure drop, without any further transformation, applies to unloaded filters only. In order to allow for the consideration of additional loading in the model an exact knowledge of the soot and ash distribution is necessary.

At the Engine Symposium 2002 in Vienna /**13**/, the AVL presented detailed investigations which have been made into the deposit behaviour of soot. Using CFD-calculations and technical measurements taken from tests an almost even distribution of soot has been able to be proven length-wise along the channel. This fact enables the formulation of the variable wall permeability at increasing soot loading. The hydraulic diameter of the inlet channel which is also variable can be calculated using the thickness of the soot layer.

X-rays analyses taken of the filters from driving operation in the field show no ash deposits on the filter wall but rather a filling up from back to front (**Figure 8, 9 & 13**). This fact can be taken into consideration by reducing the active filter length by the amount of the filled channel. The pressure drop of the opposite outlet channel can then be separately calculated using the so-called Hagen-Poiseuille Law (**Figure 17**).

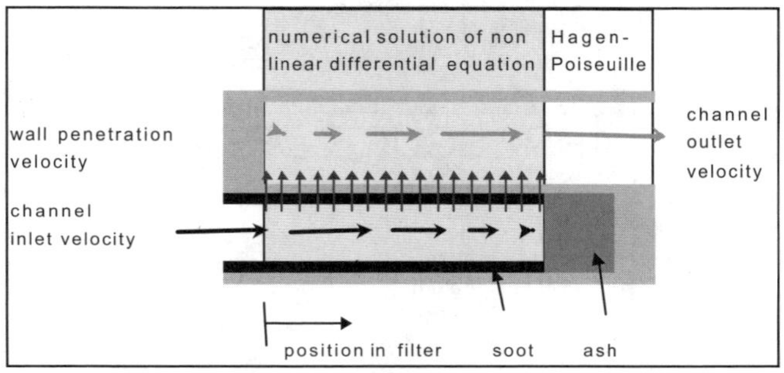

Figure 17 : Schematic presentation of a filter model

4 – EVALUATION OF THE PHYSICAL VALUES : The physical values must be specially evaluated for the substrate, soot and ash. The following values must be defined :

Substrate

- Permeability
- Forchheimer coefficient (turbulent behaviour)

Ash

- Density

Soot

- Permeability
- Forchheimer coefficient (turbulent behaviour)
- Density

Detailed parameter studies are needed to evaluate the different physical values. Pressure drop graphs for the various lengths have to be measured on a flow test bench for the manufacturer's particular filter type. The permeability can then be defined from the pressure drops at very low mass flow rates as in this case the non-linear part of the differential equation tends towards 0.

Using the known permeability the Forchheimer coefficient can then be calculated. Measurements have led here to the result that the coefficient must be defined for different lengths.

Definition of the physical soot values are much more difficult because permeability and density have to be defined simultaneously. This requires program aided algorithm which calculates the values being sought from the pressure drop graphs measured on the flow test bench.

The Forchheimer coefficient is calculated using the same principle as was used for the substrate.

The pressure loss coefficients of the expansion and contraction losses are taken from the perforated walls. In this case the losses are approximated depending on the hydraulic channel diameter and the open filter cross section area.

4 – THE PREDICTION TOOL P-CAT AND ITS FUNCTIONS: The procedure for the solution of the calculation of the pressure drop is programmed in Visual Basic. This enables the use of Microsoft EXCEL and in parameter studies provides a printout of the results without any marked time delay. The development engineer has two different possibilities available here for carrying out calculations. The pressure drop for one complete system consisting of the inlet cone, catalyst, DPF and outlet cone can be defined in one document, (**Figure 18**). Besides the pressure drop of the individual components, the α□ and β□ values needed for the CFD-calculations are also given. In the sheet "Velocities", the channel and wall velocities are shown.

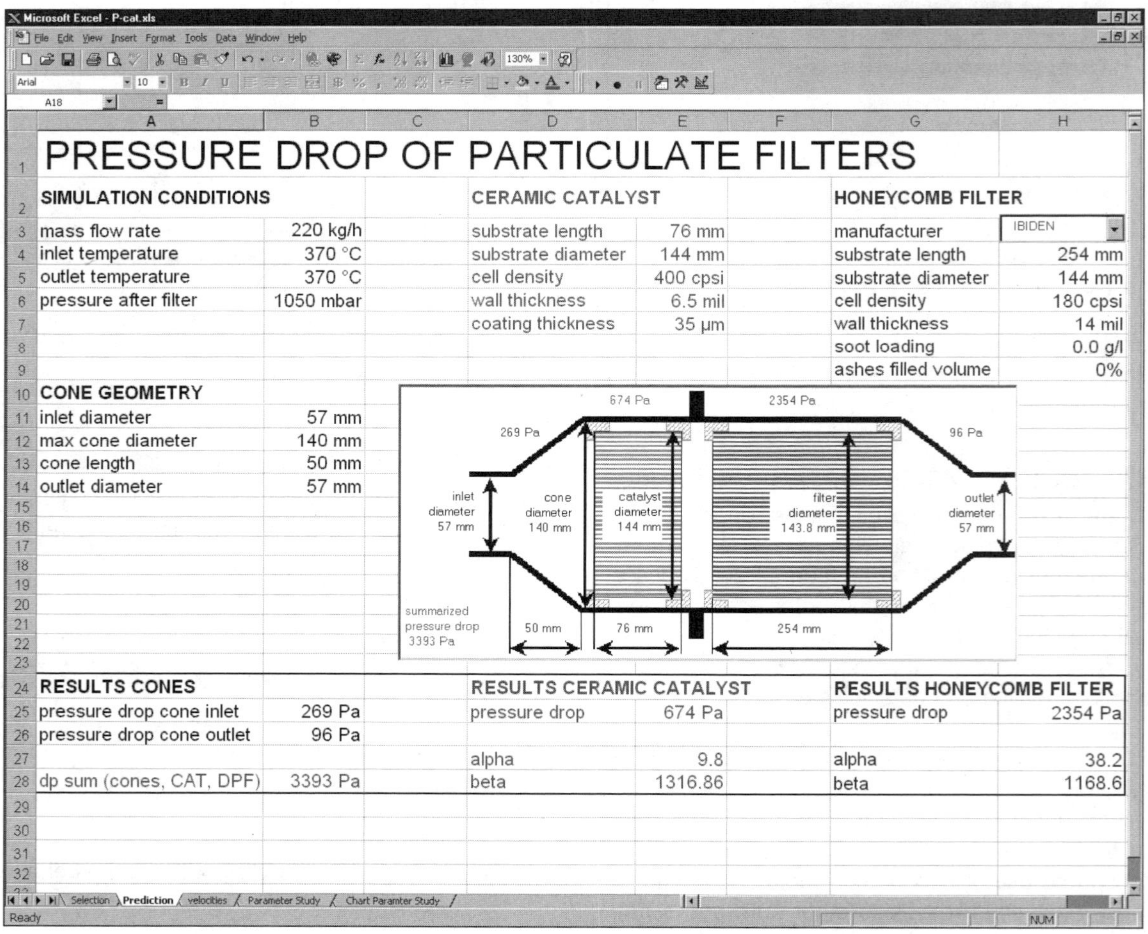

Figure 18: Input mask P-CAT

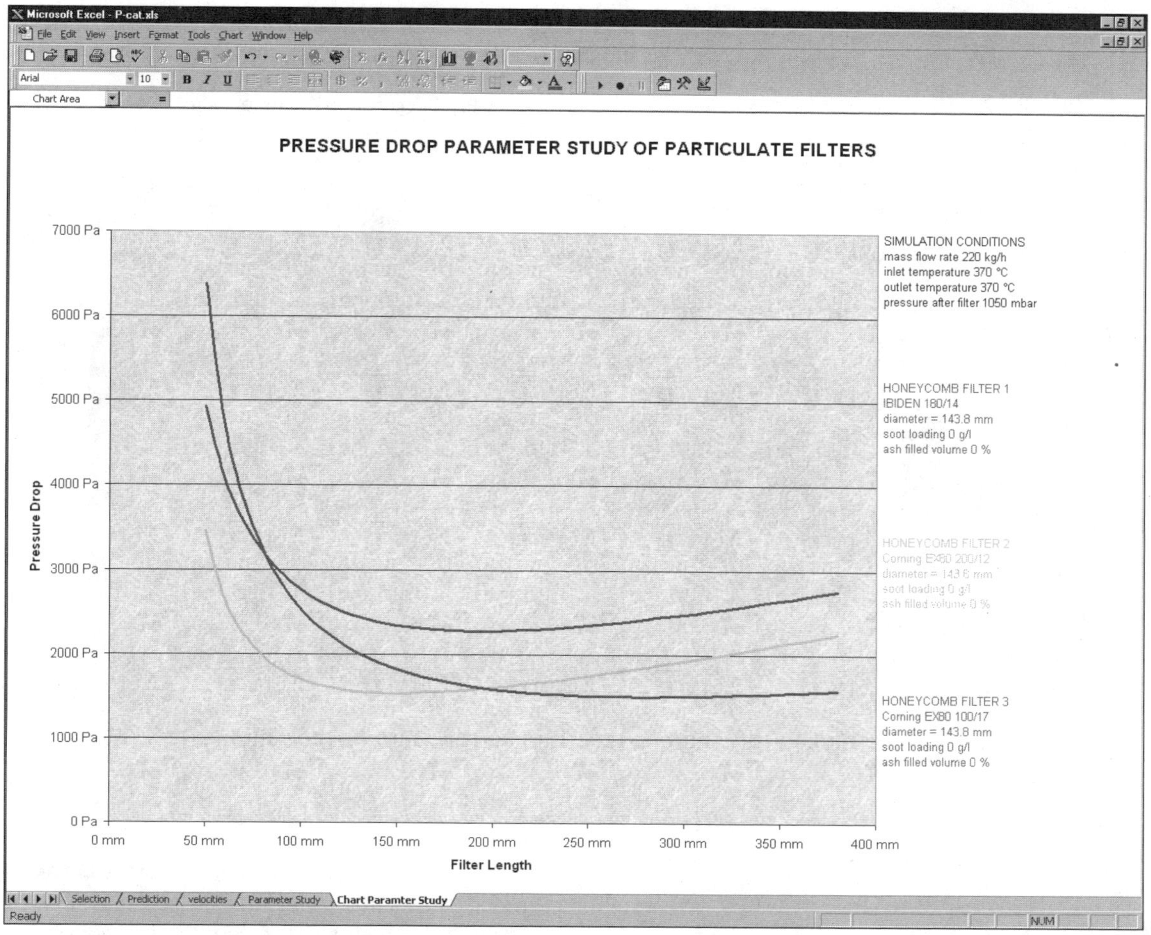

Figure 19: Diagram parameter study

The worksheet "Parameter Study" permits (under constant simulation conditions, **Figure 19**) the direct comparison of the various substrates with different soot and ash loading. The pressure drop is shown here over the substrate length.

DESIGN OF A DPF SYSTEM USING P-CAT

In the following system design, to give an example, an Ibiden's DPF system is optimised for various driving conditions and fuel borne catalysts (Ce-FBC Reference and Fe/Ce-FBC) for a 2-litres diesel engine.

A 5.66"x10" (180/14) DPF after 80000 km of driving is used as the basis for an optimisation example. It is proceeded on the assumption here that the effective filter volume is loaded with 55% ash (with Ce-FBC Reference). In further scenarios, the influence on the filter of the better performing Fe/Ce-FBC is more closely examined.

In order to evaluate the pressure drop two types of engine operation are considered:

- full load
- constant driving on the motorway

Driving at full load is of interest to those engineers constructing the engine in order for them to calculate maximum power, whereas a cruising speed of approx. 130 km/h gives an estimation of the higher fuel consumption which is caused by the soot load.

All results are reported in **Figure 20**.

Basis and reference : an ash load of 55% causes at full load an increase of pressure drop of less than 50%.

A soot load of 8g/l, based on the free filter area, results in a doubling of the pressure drop (at 130 km/h).

The ash is deposited in the inlet channel contrary to the flow direction and through this causes a shortening of the effective filter channel length. If the full load pressure drop of an unloaded filter for differing lengths is considered, it is determined that the pressure drop for filter lengths between approx. 150 mm – 250 mm is almost constant, (**Figure 21**). This means that for a soot-free filter (full load) the ash in large areas is of secondary importance although the average wall penetration velocity is increased.

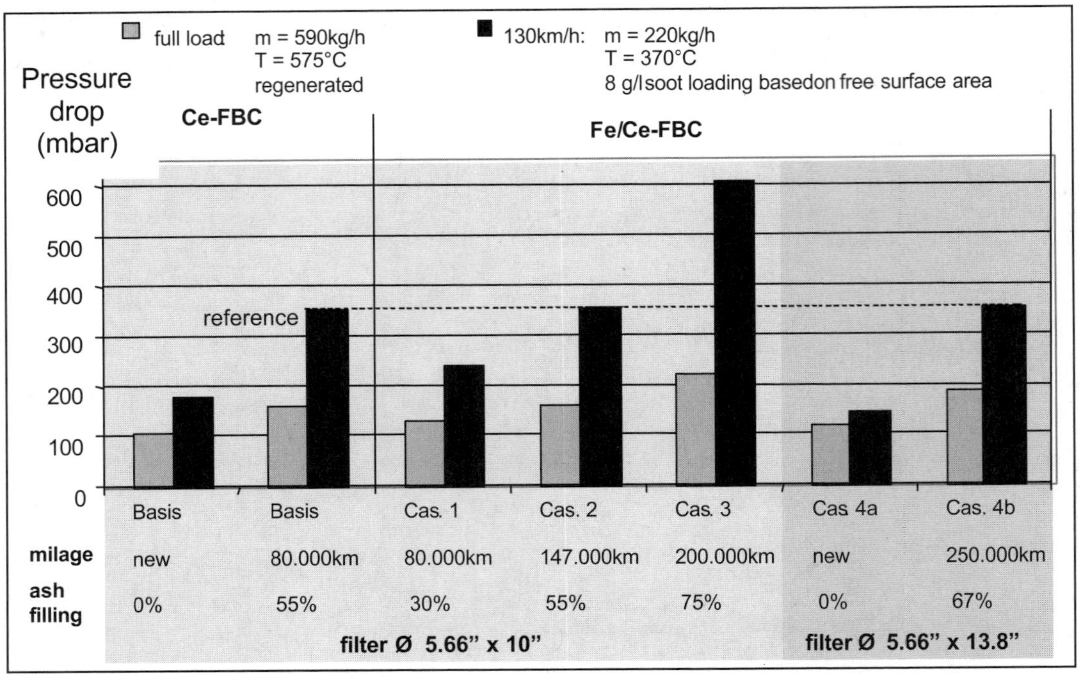

Figure 20: DPF pressure losses for various marginal conditions

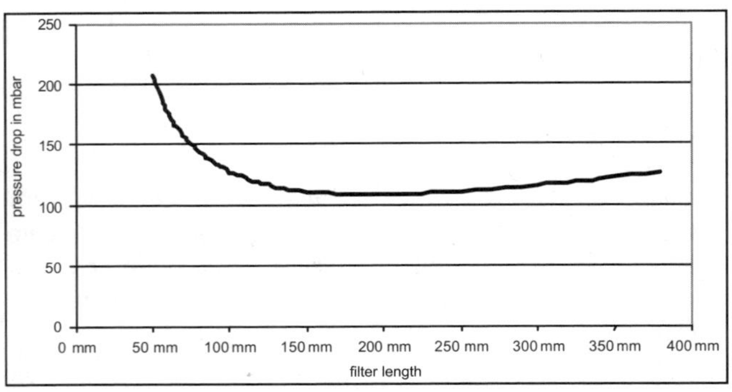

Figure 21: Full load pressure drop for various unburdened Ibiden – filter lengths at full load (m=590kg/h, T=575°C)

Only with increasing soot loading (8g/l) does the increase in velocity have an effect (caused by the very low permeability of the soot in comparison to the substrate) and the pressure drop in the channel is proportionally small, so that halving the filter area causes a doubling of the pressure drop.

Cases 1 and 2 : through the use of a more efficient diesel fuel borne catalyst (with limited ash production), the pressure loss falls in comparison to the basic case because the degree of ash deposits after a running performance of 80000 km is reduced from 55% to 30%. The pressure drop at 130km/h increases by approx. 35% compared to a new filter.

If the filter is then also filled to 55% with ash in order to reach the same pressure drop as in the basic case, through the new additive the running performance can be increased by approx. 67000 km to 147000 km.

Cases 3 and 4 : the question to be answered is whether the filter can be designed for the lifetime of the vehicle through the use of a more efficient additive. In case 3 it is seen that the filter is already 75% filled with ash at 200000 km. As a result of the 25% remaining filter area the pressure drop at 130km/h is more than three times that of a new filter.

In order to reach a running performance of 250000 km the filter has to be lengthened by nearly 40% to 13.8", so that at 130 km/h it is equal to the basic case. At full load the pressure drop then increases by 16%.

CONCLUSIONS

The combination of a Diesel Particulate Filter (DPF) with a Ceria-based fuel-borne catalyst has been demonstrated to be an highly efficient, reliable and durable technology for the abatement of particulate matter (PM) emissions from diesel passenger cars.

To limit maintenance constraints, a new Ceria-based fuel-borne catalyst (called "Fe/Ce-FBC") has been designed with an enhanced catalytic activity. Evaluations on modern diesel engine have shown optimal dosing rate of 10ppm (related to engine-out PM emissions), that could limit DPF maintenance, compared with the Ce-FBC reference.

Real field tests showed a huge extension on the DPF interval service, and most of them could be used over 150000km. However, that trend is really dependent on the diesel fuel consumption, the Fe/Ce-FBC dosing rate (related to the engine-out PM emissions), the lubricant oil consumption and the DPF size and design.

On the other hand, mathematical models, based on field operations, have been developed to optimize the DPF size, taking into account different parameters. Different solutions can be foreseen, depending on the global car maker strategy for DPF service interval :

- Increase of DPF size could lead to a zero DPF maintenance,

- Decrease of DPF size limits the DPF service interval, and an optimum should be specified to fulfill EURO 4 requirements

- finally, limiting the dosing rate (directly related to the engine-out PM emissions), and improvement in lubricant oil could also give an opportunity to extend the DPF interval service.

In conclusion, due to the enhanced catalytic activity, the new Ceria-based fuel-borne catalyst, in conjunction with SiC-DPF could decrease the total cost of the DPF system, especially in the maintenance issue.

Further new developments are in hand, notably to reduce engine-out PM emissions and to minimize maintenance constraints.

ACKNOWLEDGMENTS

The authors would like to thank :

- PSA Peugeot Citroën
- Volkswagen AG
- HTW Dresden
- and our colleagues

for making available the detailed results, graphs and pictures.

REFERENCES

[1] - Diesel Particulate Traps, DieselNet Technology Guide, www.dieselnet.com, (Ecopoint Inc.).

[2] - F. Novel-Cattin et al. Evaluation Method for Diesel Particulate Trap Regeneration Additives : Application to Five Additives, SAE Paper 2000-01-1914

[3] - Ronny Allansson et al., SAE Paper 2002-01-0428

[4] - MSHA's letter, Program Information Bulletin N° P02-4 / May 31, 2002

[5] - Salvat Olivier et al., Passenger Car Serial Application of a Particulate Filter System on a Common Rail Direct Injection Diesel Engine, SAE paper 2000-01-0473.

[6] - Blanchard Gilbert and al., Series Application of a Diesel Particulate Filter with a Ceria-Based Fuel-Borne Catalyst: Preliminary Conclusions After One Year of Service, SAE paper 2002-01-2781.

[7] - Jean Pierre Cuif et al. (Ce,Zr)O2 solid Solutions for Three-Way Catalysts, SAE Paper 970463

[8] - G. De Soete, Aspects chimiques de la combustion du charbon pulvérisé in Revue de l'Institut Français du Pétrole, vol.37, 514-519 (1982)

[9] - Vonarb Regis, Faurecia's Experience in DPF Serial Application, 6th ETH Conference on Nanoparticle-Measurement (Zurich, 19th-21st August 2002).

[10] - J.Michelin et al., Partikelfilter für Diesel-Pkw beladungs- und Regenerationsverhalten, MTZ 7-8/2002 Jahrgang 63 (p. 568-577).

[11] - A. G. Konstandopoulos et al. SAE Paper 1999-01-0468

[12] - S. Schuerholz et al., Auslegung und Druckverlustberechnung von Diesel-Partikel-Filter-Systemen, AVL Forum Partikel Emissionen (24th – 25th September 2002)

[13] - Cartus, Th.; Diewald, R.; Herzog, P.; Strigl, Th.; Wanker, R.: "Dieselpartikelfilter-Systemintegration – Von der 3D –Simulation bis zur Serie", Fortschrittsberichte VDI, Series 12, No. 490, Volume 2

2004-01-0694

The Diesel Exhaust Aftertreatment (DEXA) Cluster: A Systematic Approach to Diesel Particulate Emission Control in Europe

A. G. Konstandopoulos, D. Zarvalis, E. Papaioannou and N. D. Vlachos
Aerosol & Particle Technology Laboratory, CERTH/CPERI

G. Boretto, M. F. Pidria, P. Faraldi and O. Piacenza
Centro Ricerche Fiat S.c.p.A.

P. Prenninger, T. Cartus and H. Schreier
AVL List GmbH

W. Brandstätter and C. Wassermayr
Christian Doppler Laboratory for Applied Computational Thermofluiddynamics

G. Lepperhof, V. Scholz, B. Luers and J. Schnitzler
FEV Motorentechnik GmbH

M. Claussen, A. Wollmann, and M. Maly
Clausthaler-Umwelttechnik Institut GmbH

G. Tsotridis
European Commission-Joint Research Centre, Institute for Energy

B. M. Vaglieco and S. S. Merola
Istituto Motori, National Research Council of Italy

D. Webster, D. Bergeal and C. Görsmann
Johnson Matthey PLC

H. Obernosterer
Obernosterer Strickstoffe GmbH

D. Fino, N. Russo, G. Saracco and V. Specchia
Politecnico di Torino

N. Moral
Renault Recherche Innovation

A. D'Anna and A. D'Alessio
Universita degli Studi di Napoli Federico II

R. Zahoransky and E. Laile
Wizard Zahoransky KG

S. Schmidt and M. Ranalli
Zeuna-Staerker GmbH. & Co. KG

ABSTRACT

The DEXA Cluster consisted of three closely interlinked projects. In 2003 the DEXA Cluster concluded by demonstrating the successful development of critical technologies for Diesel exhaust particulate after-treatment, without adverse effects on NO_x emissions and maintaining the fuel economy advantages of the Diesel engine well beyond the EURO IV (2000) emission standards horizon.

In the present paper the most important results of the DEXA Cluster projects in the demonstration of advanced particulate control technologies, the development of a simulation toolkit for the design of diesel exhaust after-treatment systems and the development of novel particulate characterization methodologies, are presented.

The motivation for the DEXA Cluster research was to increase the market competitiveness of diesel engine powertrains for passenger cars worldwide, and to accelerate the adoption of particulate control technology.

INTRODUCTION

The European diesel engine industry and its suppliers, represents a vital sector across the continent, with more than two million direct work positions and a turnover of over 400 billion Euro. Meeting proposed emission levels (Euro IV in 2005 and beyond, "Euro V") for NO_x and PM requires the development of a number of critical technologies, an important element of which is represented by advanced Diesel Exhaust After-treatment (DEXA) technology.

In 2000 a group of passenger car manufacturers, exhaust after-treatment system suppliers, engineering research and development industries, instrumentation manufacturers, software vendors and research institutes/universities formed a collaborative consortium (DEXA Cluster) to address in an integrated manner diesel emission control, with emphasis on particulate.

The DEXA Cluster consists of the following three projects:

- the component technology integration aspect (Project ART-DEXA, managed by Centro Ricerche Fiat)
- the system design aspect (Project SYLOC-DEXA, managed by AVL)
- the quality assessment/measurements aspect (Project PSICO-DEXA, managed by the Aerosol & Particle Technology Laboratory of CERTH/CPERI, who also served as the Cluster coordinator)

The interactions and synergies between the projects are shown in Figure 1. The present paper presents the most important results of the DEXA Cluster projects which include: 1) the development of advanced catalytic particulate control technologies and their integration on two demonstrator vehicles, 2) the development of a user-friendly simulation toolkit for the computer-aided design and engineering of integrated diesel exhaust after-treatment systems and 3) the development of novel particulate size and composition measurement methodologies.

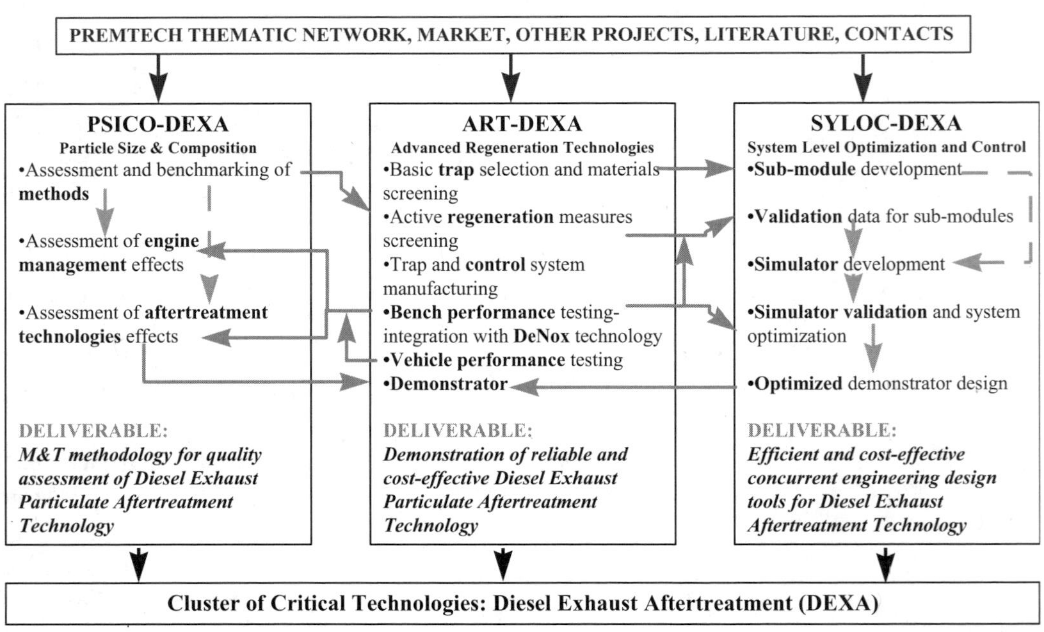

Figure 1. Structure of the DEXA Cluster.

COMPONENT TECHNOLOGIES

FILTER MEDIA

Filter configurations considered were wall-flow monoliths, ceramic foams and fibrous cartridges. In this sub-task, testing & procurement of commercially available filter media and filters, as well as examination of several structures of knitted ceramic fibrous filters developed within the DEXA Cluster was carried out. Screening and testing of different systems was done in two steps. The first step included testing of different filter media in small scale with respect to their permeability, filtration efficiency and soot loading behavior. The second step included testing of different prototype filter systems developed by the project partners as well as acquired from external suppliers.

Filter media screening

For the purpose of filter media screening, several filter disks were tested with respect to permeability, filtration efficiency and loading behavior. Permeability and loading tests were done by the use of a small-scale filter disk testing unit (for the loading test a part of the exhaust flow of a 1.9 L TDI engine, which was running at 2400 rpm and 91.5 Nm was used). Filtration efficiency tests were performed with the use of a Scanning Mobility Particle Sizer (SMPS) and a mini dilution tunnel system. Table 1 shows the permeability and initial filtration efficiency tests results for SiC, metal felt and fiber filter materials.

Table 1. Filtration efficiency and permeability of three different filter materials.

	Metal Felt	Fiber Filter	SiC
Permeability (m^2)	$7.1 \cdot 10^{-11}$	n.a.	$4.4 \cdot 10^{-11}$
Filtration Efficiency (%) based on particle number	64.7	89.5	96.6
Filtration Efficiency (%) based on particle area	43.5	82.6	95.6
Filtration Efficiency (%) based on particle volume	36.7	79.1	95.2

Subsequently, larger scale systems were evaluated with respect to their permeability, filtration efficiency and loading behavior. A full-scale wound fiber filter assembly was built in order for the fiber filter concept to be tested in conjunction with two other reference particulate trap concepts. The first one was a SiC 5.66"x6" wall-flow filter while the other was a 5.66"x6" wall flow cordierite monolith. Permeability tests were conducted in a testing facility specifically manufactured for this reason. The results of the permeability tests, together with the other characteristics of the filter specimens, are shown in Table 2.

Table 2. Filtration efficiency and permeability of three different filter materials.

Material	Prototype SiC 5.66"x6"	Cordierite 5.66"x6"	Fiber filter cartridge assembly
Pore size (μm)	11.14	12	N/A
Cell density (cpsi)	92	88	N/A
Porosity %	36	48	67.4
Filtration area (m^2)	1.19	1.46	0.2694
Fiber diameter (μm)	N/A	N/A	8.2
Permeability (m^2)	$2.7 \cdot 10^{-12}$	$6.3 \cdot 10^{-13}$	$1.8 \cdot 10^{-10}$

After the permeability tests the three particulate traps were tested for their loading behavior together with their filtration efficiency. The traps were placed in the engine exhaust and were exposed at a flow rate of 182 kg/hr, at a temperature of 350 C. The SiC filter showed the best loading behavior (Figure 2).

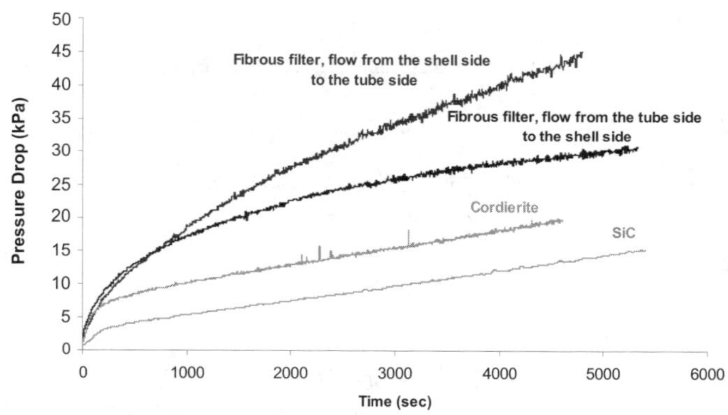

Figure 2. Pressure drop measured during identical loading conditions of four different particulate trap systems.

Collection efficiency tests were performed with the use of a SMPS together with a mini dilution tunnel system. The SMPS unit was able to take measurements before and after the installed particulate trap system. The collection efficiency of the best performer, the SiC monolith, was excellent with an initial efficiency of 99.8 %. The filtration efficiency of the cordierite monolith was also very good (96.4 %) while the filtration efficiency measured for the fiber filter cartridge was 73.2 %.

The filtration technology selected to fulfill the requirements of low pressure drop (<15 kPa on a NEDC cycle) and high collection efficiency (>90 %), was the SiC filter and all subsequent work focused mainly on SiC filters.

CATALYSTS

Powders

Studies of soot oxidation catalysts were initially performed at the powder scale at participating research laboratories. In one laboratory, a number of alternative catalyst-formulations were synthesized by wet-chemistry [1] as well as aerosol processing techniques [2] and they were evaluated using TGA analysis. The most active (designated as E-37 in Figure 3) was applied and tested on a small-scale monolith. A good soot conversion above 450 C has been observed. Hence, the synthesis procedure of the catalyst powder was delivered to industrial partners in the consortium for scale-up.

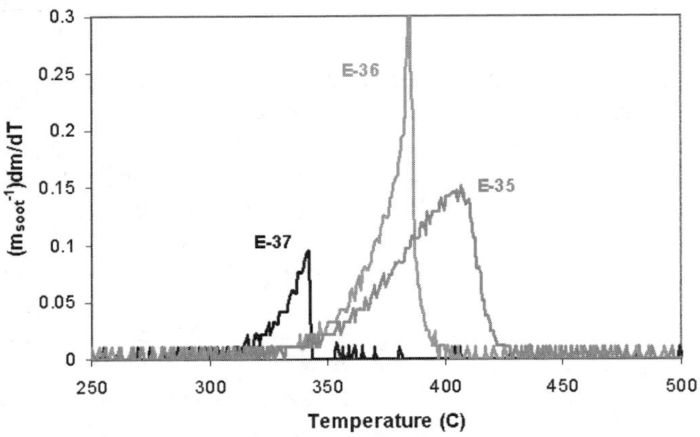

Figure 3. TGA Analysis of three alternative catalyst formulations: E-35, E-36 and E-37.

Firstly in another laboratory, the $Cs_4V_2O_7$ [3] and the eutectic mixture of Cs_2O and V_2O_5 [4] catalysts were at first developed, but found to be effective only in ceramic foam traps but not in wall-flow monoliths. For this reason, a developmental campaign started at this lab aimed at the synthesis of perovskite-type catalysts, combining good catalytic activity with a good stability [5]: several perovskite and mixed oxide based catalysts were produced by means of the combustion synthesis technique and then tested by Temperature Programmed Oxidation (TPO) analysis [6] (Figure 4). The synthesis of the most active catalyst based on perovskites was also transferred to industrial partners for scale-up.

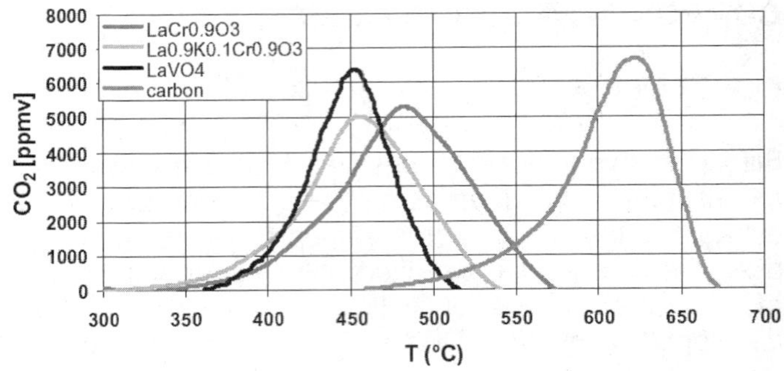

Figure 4. TPO tests of both perovskite and mixed oxide based catalysts versus non catalytic combustion of carbon (initial condition = 9:1 catalyst to soot mass ratio).

Catalytic Filters

A number of catalytic filters using catalyst formulations developed within the project, as well as obtained from external suppliers, were screened by several partners in the project. After the screening, the work has focused on two parallel directions: (i) on a "Mainstream" catalyst technology (based on an industrial-scale produced catalyst formulation), and (ii) on "Sidestream" catalyst technologies (perovskites and the E-37 catalyst family) which were also coated on full scale samples.

Overall twelve different DPF systems as given in Table 3 have been tested, in terms of their behavior under steady state loading and regeneration. Eleven of them were based on a Silicon Carbide 5.66"x6" monolith and one was based on catalyzed fibrous filters, produced within the project.

Table 3. Characteristics of the systems under evaluation.

S	DPF	Catalyst
1	SiC 5.66"x6", 200 cpsi	BMO
2	SiC 5.66"x6", 200 cpsi	BMO/ NM
3	SiC 5.66"x6", 200 cpsi	E-37 (2x)
4	SiC 5.66"x6", 200 cpsi	-
5	SiC 5.66"x6", 200 cpsi	BMO/NM
6	SiC 5.66"x6", 200 cpsi	BMO/Low NM
7	SiC 5.66"x6", 200 cpsi	BMO/High NM
8	Fibrous filter	BMO/NM
9	SiC 5.66"x6", 200 cpsi	E-37 (1x) / NM
10	SiC 5.66"x6", 200 cpsi	E-37 (2x) / NM
11	SiC 5.66"x6", 200 cpsi	Perovskite
12	SiC 5.66"x6", 200 cpsi	E-37 (2x)/ NM2
BMO: Base Metal Oxide NM: Noble Metal NM2: Noble Metal coating both in inlet and outlet channels		

As shown in Figure 5 a broad dispersion of the pressure drop behavior of the systems is evident attributable to the different catalyst coatings of each DPF.

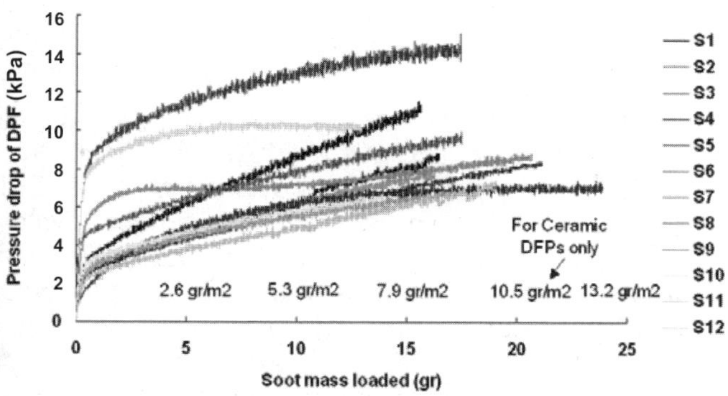

Figure 5. Soot loading behavior of all filters tested. Engine at 2400 rpm and 91.5 Nm, DPF inlet temperature: 385 C, Soot emission rate: 3.4 mg/s, Exhaust flow 180 kg/h, NO_2 concentration: 150 ppm.

REGENERATION STRATEGIES

Starting from fuel post injection by common rail and EGR rate variation, a complete and systematic investigation of possible other engine measures, like boost pressure variation or throttling was performed. The goal was to achieve synergistic integration with the catalytic filter technologies by raising the temperature of the exhaust, without adversely affecting least fuel penalty and driveability conditions.

Engine tuning modification offers great opportunities to increase exhaust temperature. Nevertheless, none of the independently screened internal engine measures could offer sufficient potential to reach the required exhaust temperature, and it can be noted that all of them can be associated to varying degree with an increase of fuel consumption and/or an vehicle performance decrease. Thus, a combination of several measures (for example, throttle + EGR rate + post-injection) is required to reach the objectives. Specific strategies therefore, focusing on combination of boost pressure, EGR rate increase, and fuel injection modification have been developed.

The potential of the multiple injection concept (Figure 6) has been also assessed, with extremely positive results.

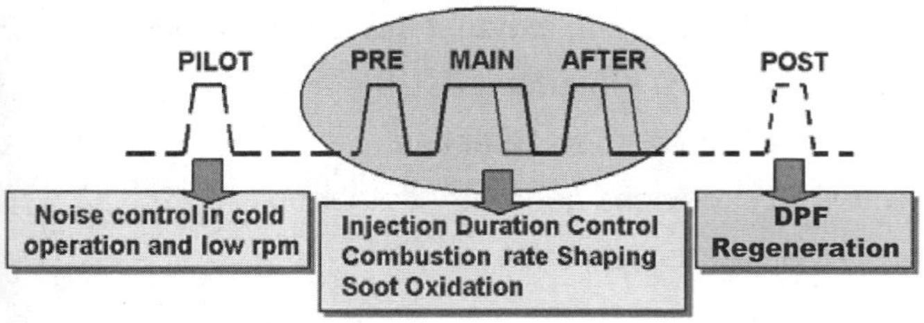

Figure 6. Multiple injection concept.

It was thus possible to increase the temperature of the exhaust gas upstream the DPF, under all engine conditions of practical interest, up to values that would insure fast ignition and regeneration of the aftertreatment systems that would be installed on demonstrator vehicles, without any effect on driver's feeling.

REGENERATION CONTROL

Regeneration control algorithms were developed by the engine manufacturers and implemented in the ECU. Emergency procedures in case of a system (sensors) failure have been taken into account (the regeneration is initiated always after a certain time period). In case that the sensors work correctly again, the program returns to the normal operation mode.

The control algorithms, developed according to the constrains of the ECU include:

- data input and system check
- calculation of the soot load
- regeneration decision

After the start of the program, the ECU tests whether the engine is running or not. If the engine is running, the measured variables are read (Figure 7). These are:

- air mass flow
- fuel mass flow
- pressure drop over the filter
- temperature upstream the filter

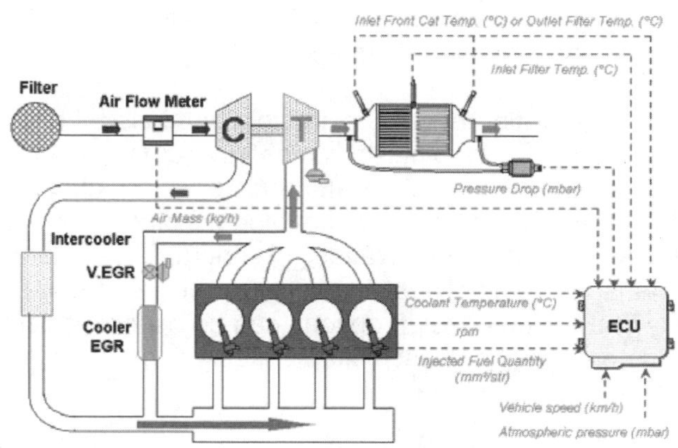

Figure 7. Controllable parameters for DPF particulate load diagnostic.

After that the exhaust mass flow is calculated and the plausibility of the values is checked. If one of the values is out of the system limits an emergency program is started. Then the soot content calculation of the filter is performed exploiting a sub-procedure based on published models [7, 8].

Figure 8 shows the pressure drop evolution of a catalytic DPF for a transient loading experiment compared with the soot load model algorithm. This experiment was performed in order to demonstrate the applicability of the developed soot mass estimation model for on-board vehicle use. The cycle chosen for this application includes four engine points that represent the 25, 50, 75 and 100 % load of a Euro 3 common-rail engine.

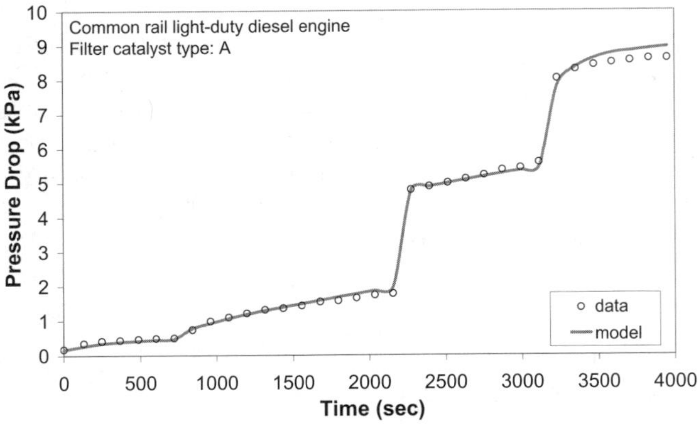

Figure 8. Comparison of experimental data vs. soot load model prediction for a silicon carbide filter, at transient loading.

SIMULATION TOOLS

Engine development work so far has proven that it is not possible to optimize the behavior of the individual components of a Diesel engine exhaust system independently: the optimization has to consider a synergistic interaction of individual elements of the exhaust system and the engine.

The identification of successful combinations by building and testing entire after-treatment systems or vehicles requires often prohibitive amounts of effort and costs. The alternative to this is to develop predictive computer models for the optimization of entire Diesel engine exhaust systems. These models can be used in such a way that various design options can be assessed rapidly to determine the most suitable design for a particular engine-prototype or vehicle.

Research work focused on laboratory experiments and steady-state engine tests for collecting measurement data as input to the simulation software. The consortium partners have investigated many commercial particulate filter types and conversion systems in synergy with the ART-DEXA project described in preceding sections.

Physical models for each of the candidate after-treatment devices have been implemented in the form of interfaceable sub-programs, which can be coupled to the CFD model of the entire exhaust system for the simulation of component behavior, such as loading and regeneration of the particulate traps. Also, a database containing information on after-treatment device materials and geometric characteristics has been compiled with continuous updates/extensions during the run-time of the project.

SIMULATION METHODOLOGY

The basis for the simulation toolkit developed in the DEXA Cluster (project SYLOC-DEXA) is the coupling of sub-models for exhaust system devices to the 3-D CFD simulation of the flow inside the entire exhaust system [9, 10]. The exhaust gas model is that of a non-isothermal compressible gas with variable properties (conductivity, heat capacity, viscosity). The Reynolds-averaged Navier-Stokes equations are solved with closure for the Reynolds stresses provided by the k-ε model. The main outline of the simulation methodology was summarized in [11]. The sub-models [8, 12, 13] included in the simulation toolkit were for the following exhaust system aspects:

- wall-flow Diesel particulate filter pressure drop
- wall-flow Diesel particulate filter regeneration behavior
- hydrocarbon oxidation and NO $\rightarrow$ NO_2 conversion by Diesel oxidation catalyst
- fibrous cartridge and ceramic foam filter pressure drop
- silencer back-pressure
- heat losses to the exterior (forced/natural convection, radiation)
- heat transfer within ceramic honeycomb devices
- exhaust gas properties dependence on temperature and pressure by polynomial or power-law relations

The simulation can provide steady-state and transient (time-stepping) solutions. Furthermore, the sub-module execution control adaptively switches to efficient quasi-transient calculations when the response times of the devices modeled are much longer than those of the other system dynamics.

EXHAUST SYSTEM SIMULATION TOOLKIT

The simulation toolkit consists of the following components:

- 1-D/3-D simulation interface
- 3-D flow solver
- subroutines of the exhaust device sub-models
- graphical user interface (GUI)
- material properties database
- exhaust devices geometry database
- experiments database

The flow solver used was Fire v.7.3 (AVL List) for which an interface to solver internal data and access to the execution-control code were provided. An interfacing code was produced to allow the component subroutines to be coupled to the flow solution. The graphical user interface (GUI) was custom-made for the purposes of the project and handles all the necessary pre-processing (Figure 9). The GUI system includes links to the material property and device geometry database, allowing the user to easily select characteristic physical and geometrical properties of the materials and device components, (monolith types, catalysts, etc.) for which reliable (manufacturer or project-generated) data is available. The user interface also permits the use of experimental test results, which have been compiled into a database of an agreed standard format, to be used as input conditions to the simulation. Finally, the GUI provides a means for very rapidly modeling an exhaust system geometry with all the required exhaust piping and after-treatment components, freeing the user of the tedious task of manual 3-D CFD mesh generation via a general purpose pre-processor (which also requires specialist experience).

Based on the positive experience with the GUI developed in the DEXA Cluster, the programme package FIRE 8 already incorporates an extended version of a specific GUI for the simulation of aftertreatment systems.

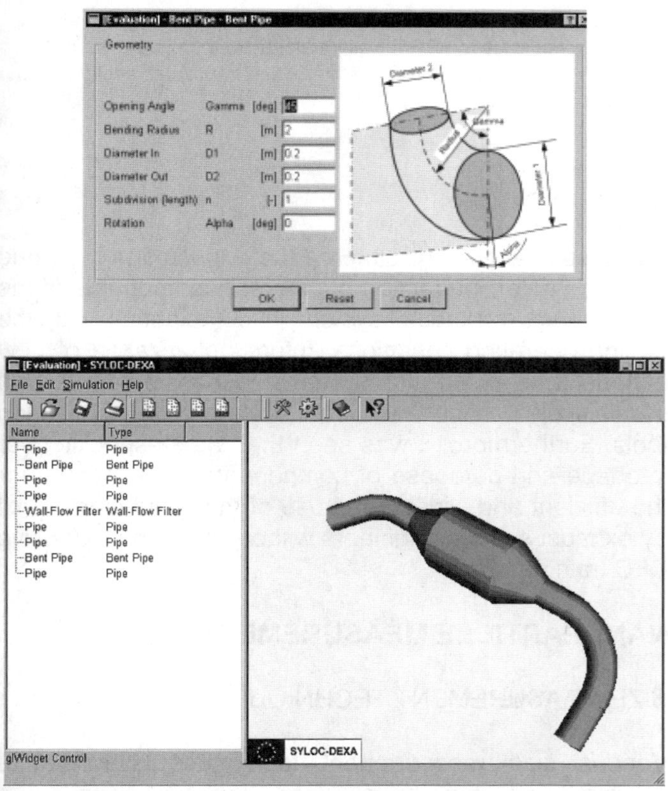

Figure 9. Screenshots of elements of the Graphical User Interface of the SYLOC-DEXA toolkit. The upper image depicts a menu for adding a segment to the exhaust system, in this case, a variable radius pipe bend. The lower figure shows the main GUI panel with the sequence of the exhaust system elements and a 3-D rendering of the resulting system geometry.

VALIDATION

Validation of the simulation toolkit was one of the critical objectives of the SYLOC-DEXA project. Validation was performed at two levels: for the sub-models of the after-treatment components and the entire system simulation. The former is a check of the physical validity of the individual device models whereas the second (system-level) validation is a check of the correctness of the numerical coupling. Out of the exhaust after-treatment elements investigated, the most mission critical (and scientifically challenging) is the model of the wall-flow DPF regeneration process, which was based on work previously described in [14, 15, 16].

Numerous engine-bench experiments with the candidate DPF devices for the DEXA demonstrators were used for comparison with the model, which showed good agreement. Example results are shown in Figure 10. Very good agreement was generally obtained (as seen in Figure 11) in similar comparisons performed with results from simulations of entire exhaust systems (layout shown in Figure 12). This not only supports the validity of the SYLOC-DEXA simulation toolkit, but also indicates that the physical component models used have much to benefit from the more realistic and spatially detailed boundary conditions provided by the CFD model of the exhaust system (Figure 13).

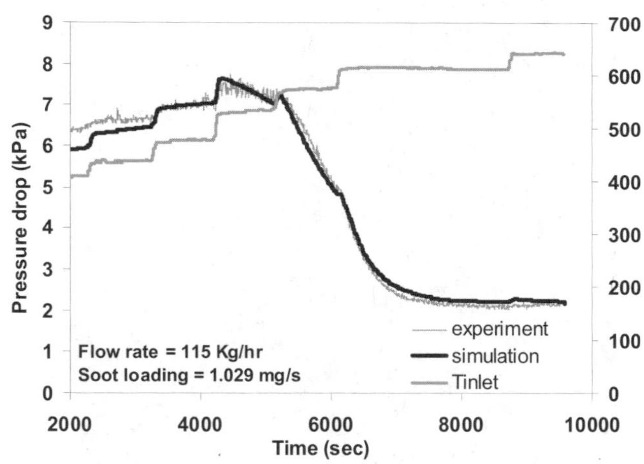

Figure 10. Comparison of the DPF simulation module solution against experimental results for regeneration by exhaust temperature control. The DPF consisted of a SiC 5.66"×6" wall-flow monolith and was tested on the exhaust of a 1.9 liter direct injection (Euro III) Diesel engine.

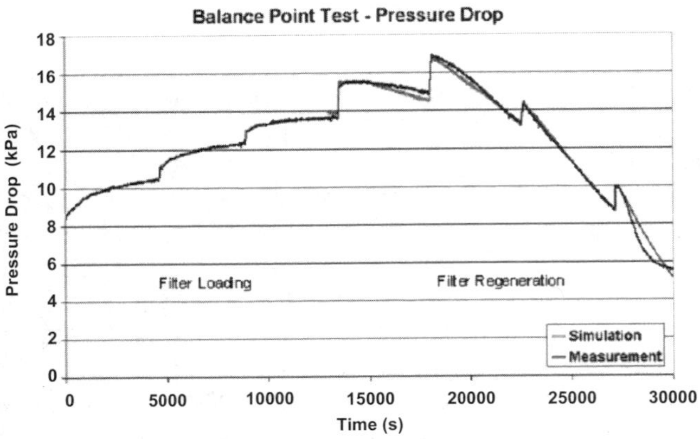

Figure 11. Comparison of measured and calculated exhaust system pressure drop during a (DPF) balance point test sequence.

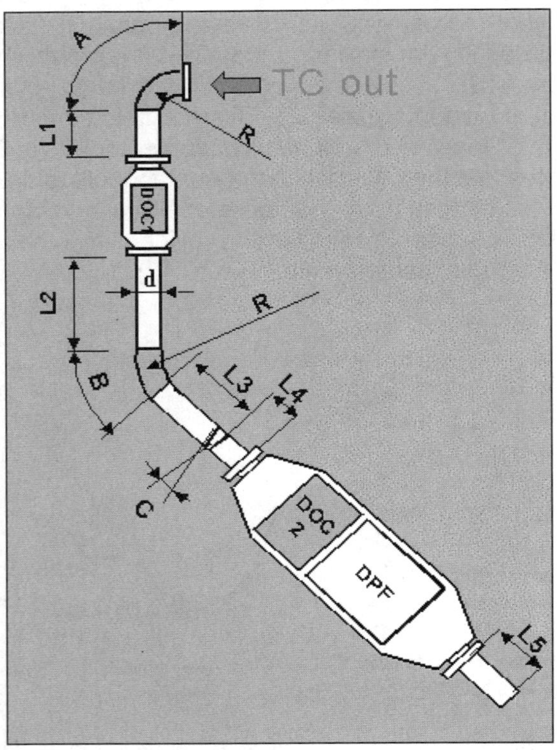

Figure 12. Schematic layout of complete exhaust system and the parameters that could be varied in the investigation of exhaust system performance.

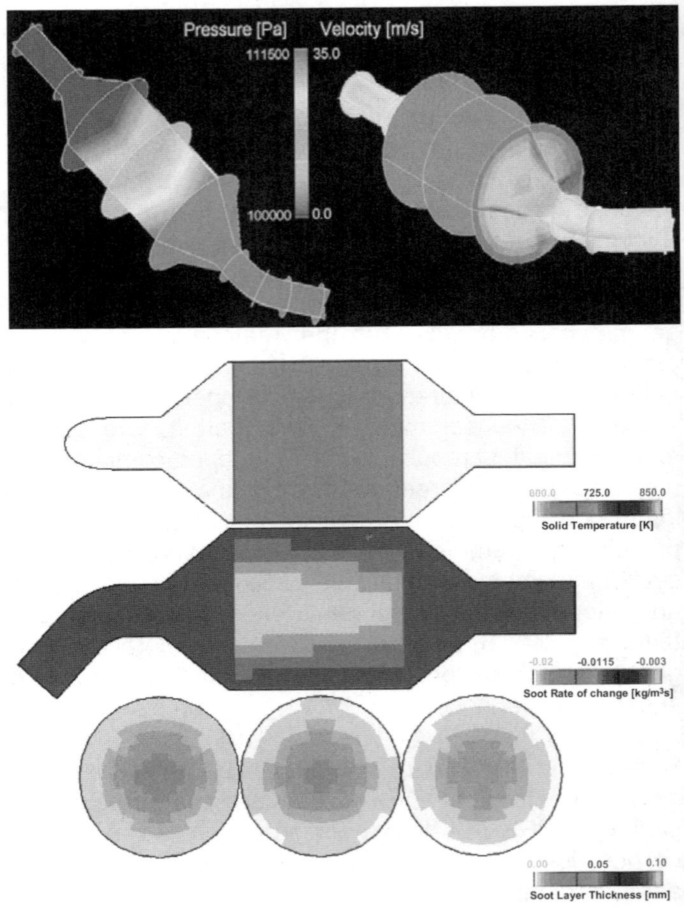

Figure 13. Detailed information is provided within components of interest, in this case the DPF. Solid material temperature, soot mass loading/removal rate are given on lateral and transverse sections of the device.

The results achieved reinforce the expectation that, due to the complex interaction of different components of the Diesel engine exhaust system in combination with the engine operating conditions, future optimization of new exhaust after-treatment systems in passenger cars is prohibitively difficult without system-level simulation tools. Furthermore, it was seen that a task-specific user interface and database of components are essential for the efficient and widespread use of the simulation toolkit by exhaust system engineers without requiring extensive CFD training.

NANOPARTICLE MEASUREMENT

SIZE MEASUREMENT TECHNIQUES

Various size measurement techniques available at participating laboratories were deployed within the DEXA Cluster. An assessment of each technique employed for measuring diesel particle size in the cluster is shown in Table 4 classified according to their measurement principle.

Table 4. Assessment of particle size measuring techniques.

Technique	Comments
Electrical Mobility based techniques	
SMPS 7-1,000 nm	*Scanning Mobility Particle Spectrometer* Method of choice for steady-state conditions
DDMPS 2-1,000 nm	*Double Differential Mobility Particle Spectrometer* Method of choice for steady-state conditions
TrDMPS 10-500 nm	*Transient Differential Mobility Particle Spectrometer* Highest measurement resolution under transient conditions. Expensive and demanding
Aerodynamic based techniques	
ELPI 30-5,000 nm	*Electric Low Pressure Impactor* Good for following transient trends. Needs care to avoid particle losses and additional data analysis to compare to SMPS. Useful in conjunction with SMPS for morphological characterization
BLPI 25-11,400 nm	*Berner Low Pressure Impactor* Allows chemical characterization of particles but it is time consuming
TOFPS 200-700,000 nm	*Time-of-Flight Particle Spectrometer* Not applicable for diesel particles, only for re-entrained large aggregates
Optical based techniques	
UV-vis extinction/scattering > 2 nm	*Ultraviolet to visible extinction and scattering* Measures primary particle size. Applicable under transient conditions
LPME > 10 nm	*Long Path Multiwavelength Extinction* Applicable under transient conditions. Available as prototype from the project for volume fraction and primary particle size determination. Needs improvement with respect to transient response [17, 18].
Thermophoretic Sampling/AEM	Provides absolute measure of particle size and morphology

The available measurements indicate that the majority of particles measured with the electrical mobility and aerodynamic techniques lie in the range of 30-200 nm. The optical techniques on the other hand, so far result in measurements representative of the primary particle diameters, which are found to lie in the range of 8 to 40 nm. Transmission Electron Microscopy (TEM) pictures of mobility segregated particles demonstrate the big difference that geometric and "operational" particle size can have for soot aggregates (Figure 14).

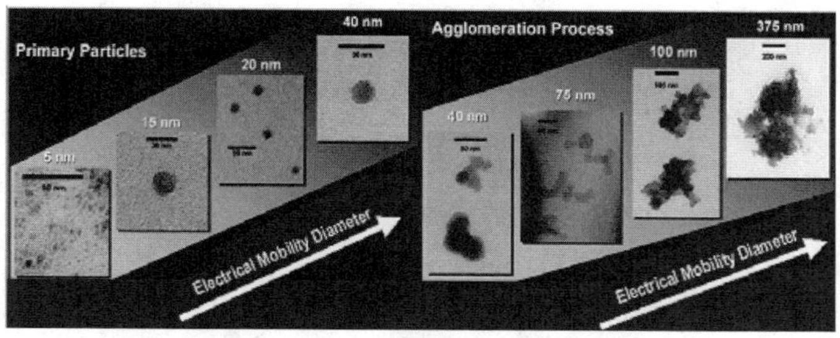

Figure 14. Particle morphology vs. mobility diameter.

Correlations of the aerodynamic (ELPI) to mobility (SMPS) determined diameters were obtained enabling thus the characterization of the fractal morphology the of diesel particles (Figure 15).

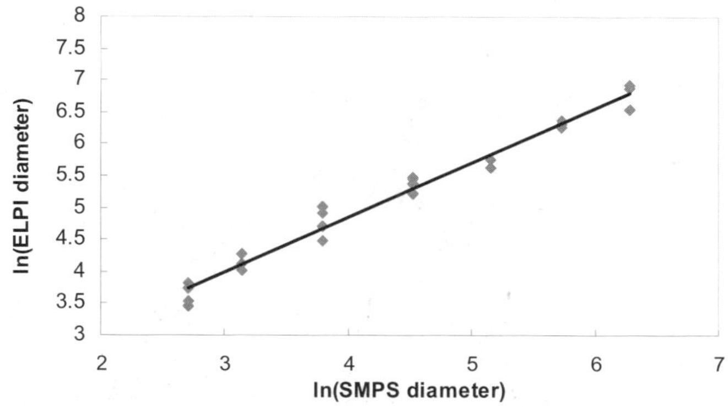

Figure 15. Correlation of ELPI to SMPS diameter.

Measurements under transient conditions were possible with the TrDMPS system, which traces very well the burst of particles produced during acceleration as it can be seen in Figure 16.

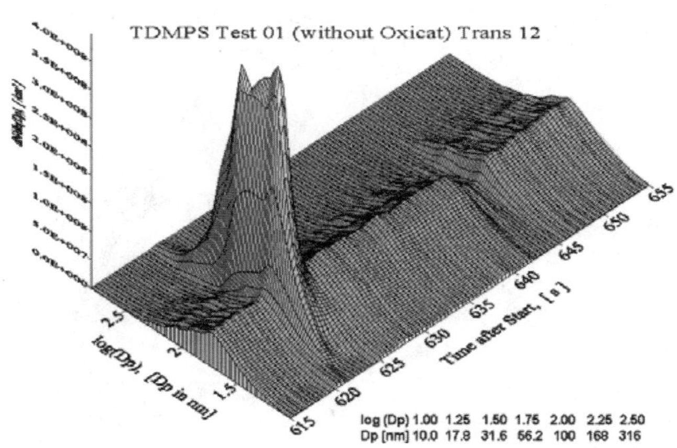

Figure 16. Transient Particle Size Distributions measured with the TrDMPS.

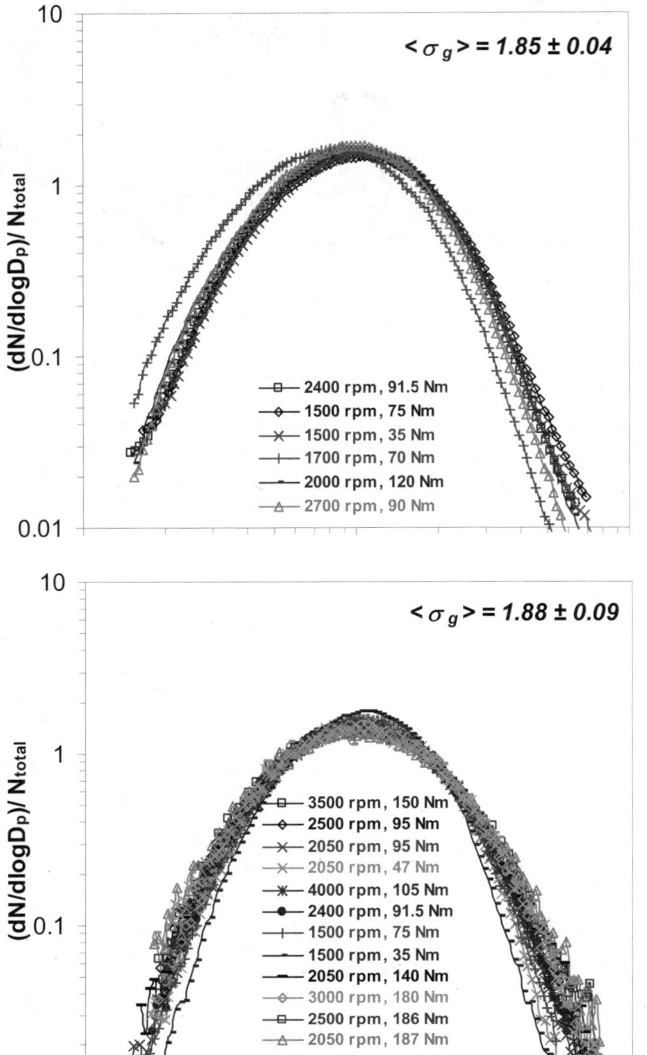

Figure 17. Universal size distributions obtained for three combustion engine technologies (Top: common rail, Middle: rotary pump, Bottom: pump unit injector).

An important result of the DEXA Cluster has been the confirmation of the existence of a universal size distribution [19] for the emitted soot particles (irrespective of combustion technology, Figure 17) and its interpretation using a population balance description of the random continuous oxidative fragmentation and coagulation of fractal aggregates [20].

COMPOSITION MEASUREMENT TECHNIQUES

Particulate composition measurement techniques available at partners' laboratories were set-up for measurements in the exhaust at the selected engine operating points. In addition fuel and lube oil samples were submitted to analysis. Trace elements in lube oil, sulphur content in fuel and in lube oil, and inorganic analysis on diesel exhaust filter samples were performed. The significant trace elements detected on the particulate matter sampled were Ba, Cr, Cu, Fe and Zn. The analyses have been carry out by Inductively Coupled Plasma – Optical Emission Spectrometry (ICP-OES).

Ex-situ composition measurement techniques are based on filter collection from the raw and dilute exhaust to determine total particulate matter breakdown to soluble organic fraction/volatiles, solids and sulphate, according to standard practices.

A strong decrease of the measured emission with the increase of the dilution ratio has been observed attributed mainly to the soluble organic fraction. When a backup filter was employed, the amount of particles (volatiles) collected at the backup filter exhibited the same decrease with the dilution ratio as the soluble fraction collected at the main filter. The non - soluble fraction exhibited a slight but significant decrease with the increase of the dilution ratio.

Several particle - loaded filters as well as oil samples were also analyzed, by neutron radiation spectroscopy for trace elements. Ca, K, Na, Sn and Zn are the major trace metals identified in the particles.

In-situ/on-line techniques for particulate composition (distinguished into visible-transparent organic and soot-like carbonaceous components) based on UV absorption and UV laser excited fluorescence [21, 22, 23, 24] have been implemented and calibrated with combustion aerosols from laminar flames. These have been also adapted to engine exhaust (Figure 18).

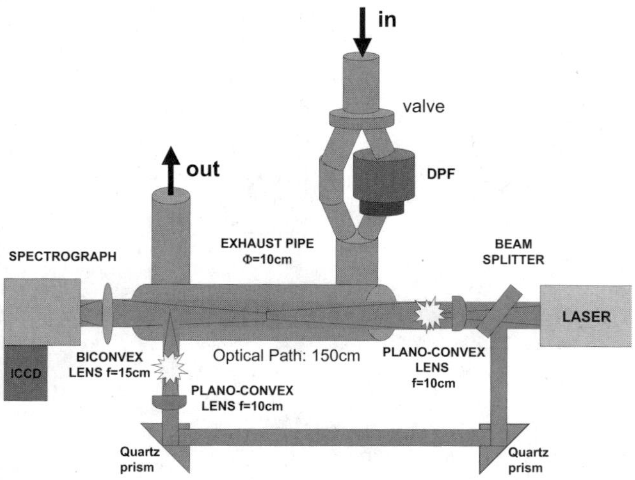

Figure 18. In-situ/on-line optical techniques set-up in engine exhaust.

A Thermophoretic sampler [25] was used to collect the emitted particles selected by electrical mobility size class by both a DDMPS and an SMPS to provide insight into the nature and morphology of the particles emitted. For samples recorded during filter regeneration, particles emitted from the filter were the primary interest. Particles in the range from 16nm to 50 nm were captured as agglomerates during the regeneration events. Inspection using the EDX showed them all to be carbon based although high vacuum conditions during EDX examination may cause evaporation of any volatile components captured on the TPS grid.

COMBINED MEASUREMENT TECHNIQUES

By combining a range of size measurement techniques (SMPS, LPME, ELPI) and a surface composition sensitive technique (the Nanomet DC/PAS) important information for the nature of the particles has been obtained [26, 27, 28, 29, 30, 31], including their density and composition differences. It is possible to combine the SMPS and LPME with simultaneous Raw Exhaust gravimetric Sampling (RES) to obtain information on the fractal morphology and primary particle density of diesel aggregates, as follows:

The SMPS measures the diameter and concentration of soot fractal aggregates, which consist of a number of primary particles obeying the relation:

$$\widetilde{N} = k_f \left(\frac{2R_g}{d_p}\right)^{D_f} \qquad (1)$$

where $\widetilde{N}$ is the number of primary particles per aggregate, k_f is the fractal prefactor, R_g is the radius of gyration of the aggregate (assumed to scale with the SMPS mobility diameter), d_p is the primary particle diameter and D_f the fractal dimension. Therefore, aggregate number concentration can be transformed to volume fraction summing contributions from each size class of the SMPS according to:

$$V_f = \sum_{i=1}^{n} N_i * \widetilde{N}_i * \frac{\pi}{6} d_p^{\,3} \qquad (2)$$

where Ni (i=1,..n) is the soot aggregate concentration at each of the n size classes measured by the SMPS. Given the primary particle diameter (average about 32 nm as determined from independent measurements) and the fractal dimension equal to 1.82 (limit of cluster-cluster diffusion limited aggregation) or equal to 2.4 (as communicated in the 2002 ETH Nanoparticle workshop by Prof. D. Kittelson of the U. of Minnesota), the values of k_f can be calculated by equating the value obtained from Equation (2) to the value of the volume fraction acquired from an independent technique, such as the LPME.

This can be defended with a high degree of confidence since the concentration of soot particles expressed as volume fraction is a primitive quantity of the LPME determined by the following equation, valid in the limit of particle size much smaller than the laser wavelength (Rayleigh limit):

$$V_f = \ln\left(\frac{I_o}{I}\right) \frac{\lambda}{36\pi F(\lambda) L} \qquad (3)$$

where $F(\lambda)$ is a function of the real (n) and imaginary (k) parts of the refractive index and L is the optimal path length. I_o and I is the incident and transmitted light intensity respectively and l is the laser wavelength.

Following this procedure a value for k_f based on the available data sets of 3.9 was calculated if D_f =1.82 and a value of 1.76 if D_f =2.4. A comparison of the SMPS volume fractions calculated using Equation (2) against the LPME is given in Figure 19.

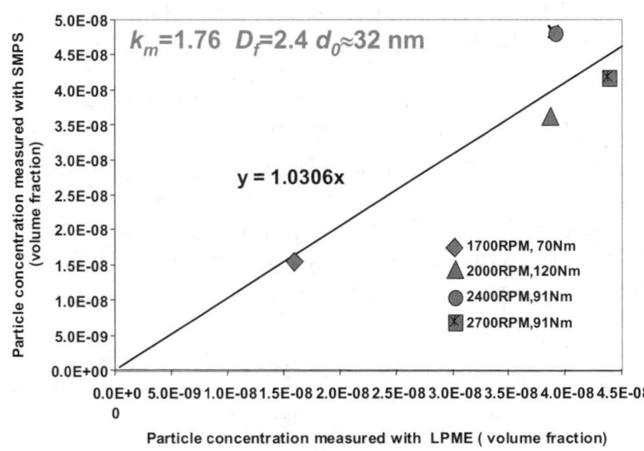

Figure 19. Comparison of SMPS and LPME volume fraction measurements.

The LPME has been also calibrated against the gravimetric technique of Raw Exhaust Sampling (RES), which provides an accurate method for determining soot mass emission rates. Special highly efficient glass fiber filters are used to sample in triplicate soot particles for each mode of engine operation simultaneously with the SMPS and the multiwavelength sensor. The slope of the curve in Figure 20, which is equal to 2339 kg/m^3, represents the soot primary particle solid density based on the very good assumption that these particles have spherical shape.

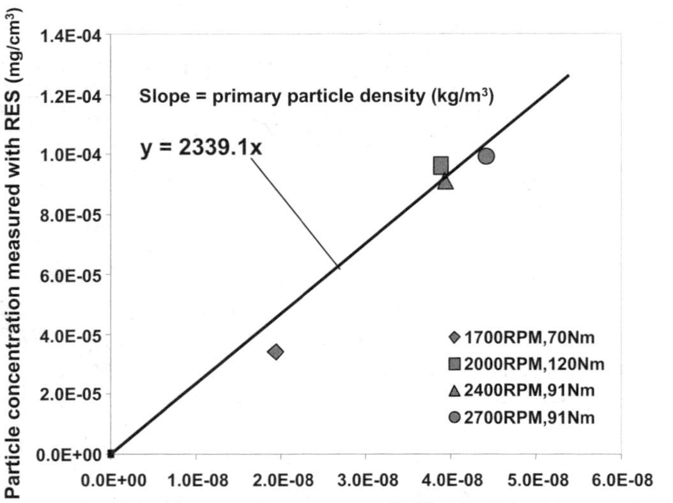

Figure 20. Determination of primary particle density from LPME and RES measurements.

The joint size-composition distribution has been investigated employing a tandem arrangement of a Differential Mobility analyzer (DMA) with a Diffusion Charger (DC) and a Photoelectric Aerosol Sensor (PAS) from the Nanomet instrument as well as a Condensation Particle Counter (CPC) and measuring with the above instruments for each discrete Particle Diameter selected by the DMA. The experimental arrangement of the Tandem DMA-(PAS-DC-CPC) system is shown in Figure 21. The DMA was set at a certain particle size (18, 30, 45, 60, 80, 100, 130, 170 and 225 and 240 nm) and the PAS, DC and CPC signals were recorded. Then a new size was selected in the DMA and the procedure was repeated. The dilution ratio in the first step is 17 and in the second step 3.3, resulting in a total dilution ratio of 57. The temperature of the rotary diluter was 80 C. The dilution is accomplished with dry, particle – free air.

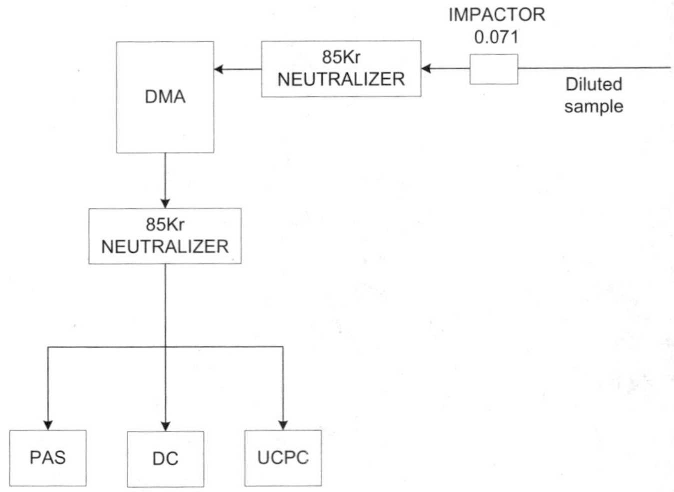

Figure 21. Tandem DMA - (PAS/DC/CPC) set-up.

Very good agreement is seen between the size specific DC sensor signal, and the total Fuchs surface area of the soot (Figure 22), lending support to the claim that the DC signal is proportional to the accessible surface of the aerosol.

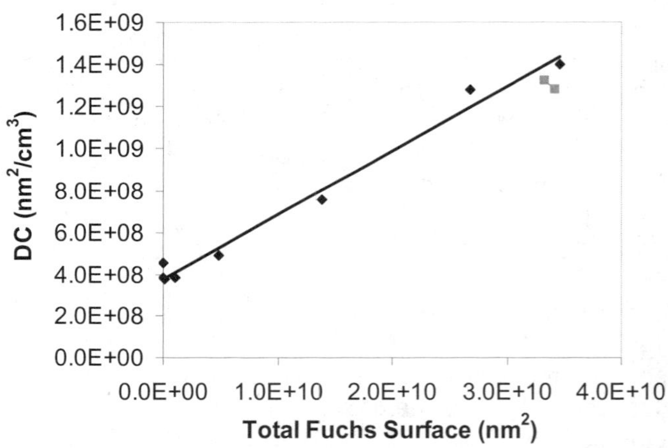

Figure 22. DC signal vs. Total Fuchs Surface for different particle diameters measured with the tandem DMA/PAS-DC-CPC system at 1500 rpm, 5 bar.

In addition, it is shown (Figure 23) that there is a correlation between the PAS sensor signal, and the total Fuchs surface area of the soot particles.

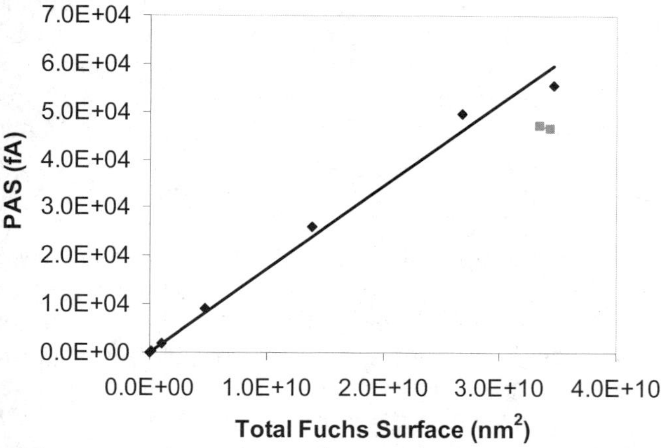

Figure 23. PAS signal vs. Total Fuchs Surface for different particle diameters measured with the tandem DMA/PAS-DC-CPC system at 1500 rpm, 5 bar.

The size dependence of the ratio of PAS/Total Fuchs Area shown in Figure 24 illustrates that the developed technique can follow size-specific shifts with engine load with respect to the amount and the composition of the soot aggregates.

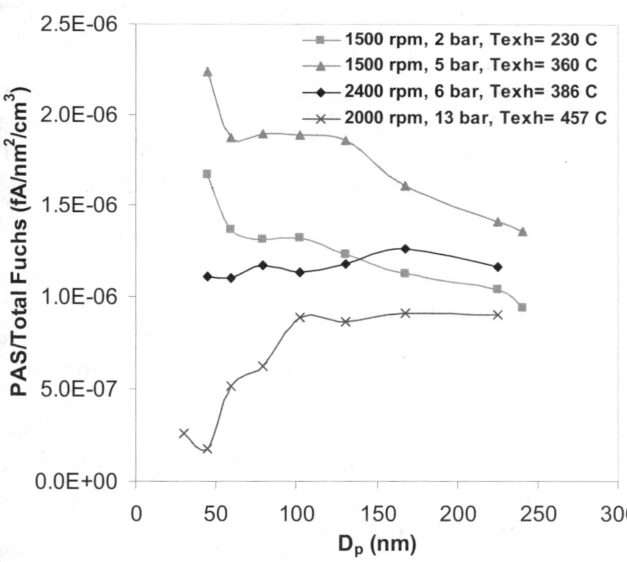

Figure 24. PAS/Total Fuchs Surface ratio vs. particle diameter measured with the Tandem DMA/PAS-DC-CPC at all engine points.

In addition the data in Figure 24 demonstrate that the PAS/Fuchs Surface Area metric (which is related to the elemental carbon content of the particles) increases as the solid fraction of the particulate (Table 5) lending further support to the use of the technique for on-line size specific composition characterization.

Table 5. SOF content at Diluted Exhaust at different engine operation conditions.

Engine Conditions (rpm, bar)	1500, 2	1500, 5	2400, 6	2000, 13
SOF fraction (%)	0.189	0.1185	0.2083	0.3973
SOL fraction (%)	0.811	0.8815	0.7917	0.6027

A Low Pressure Impactor was employed to collect on filter samples particles of a given aerodynamic size, which were subsequently analyzed for trace species with neutron activation gamma spectroscopy.

The trace element analysis on size segregated samples typically shows (Figure 25) that within the uncertainties of the measurements the trace elements do not seem to occupy a preferential size, a fact that can be rationalized based on the composition "mixing" that particle coagulation brings.

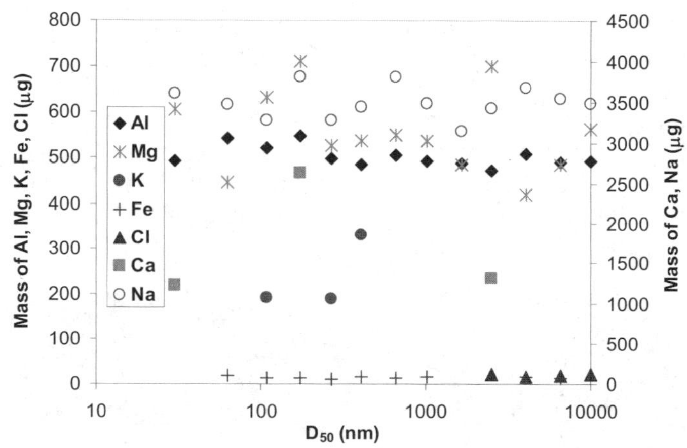

Figure 25. Size specific distribution of trace elements on particles at (1500 rpm, 5 bar) as determined by neutron activation spectroscopy by JRC on transmitted samples.

ASSESSMENT OF EMISSION CONTROL SYSTEM

The particulate size and composition measurement techniques were applied for the characterization of the complete emission control system. The typical findings shown in Figure 26, demonstrate the exceptional collection efficiency of the developed and selected DPF system.

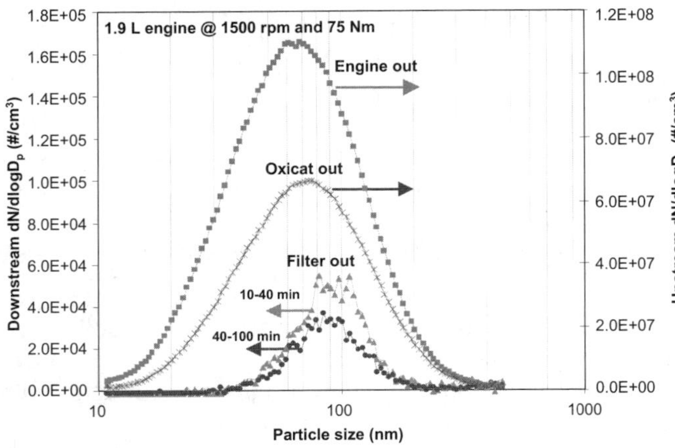

Figure 26. Effect of DEXA emission control system (DOC and DPF combination) on particle size distributions.

DEMONSTRATORS

Based on the results of the component technology investigations, the use of simulation and the possibilities offered by the regeneration control algorithms investigated, two emission control system configurations based on a combination of a Diesel Oxidation Catalyst and a Catalyzed Diesel Particulate Filter were selected for demonstration of DEXA technology on vehicles (Figure 27) as explained below.

Figure 27. DEXA-Cluster vehicle demonstrators.

Demonstrator A had the following characteristics: Engine displacement 1.9 L, common rail fuel injection, with VGT and cooled EGR. The exhaust system consisted of a close coupled catalyst (volume of 0.6 L), a cordierite oxidation catalyst (5.66"x3.54", 400 cpsi / 6.5 mil) and a silicon carbide DPF (200 cpsi / 14 mil) with a catalyst coating based on the "Mainstream" catalyst technology of a base metal oxide in combination with a noble metal.

Demonstrator B had the following characteristics: Engine displacement 2.2 L, common rail fuel injection, with VGT and cooled EGR. The exhaust system consisted of a cordierite oxidation catalyst (5.66"x3.54", 400 cpsi / 6.5 mil) and a silicon carbide DPF (200 cpsi / 14 mil) with a catalyst coating based on the "Sidestream" catalyst technology of the E37 catalyst formulation.

Both demonstrator vehicles were found to comply with Euro IV NO_x and particulate emissions. Notably particulate emissions lower than 0.0025 g/km on the NEDC cycle have been assessed, i.e. –90 % compared to Euro IV limits. Ageing the emission control system for 26000 km showed no deterioration of the DOC and DPF system.

Cost estimates for the entire emission control system were assessed to be competitive compared to a fuel-additive based solution.

CONCLUSION

1. Two demonstrator vehicles, characterized by Euro IV compliant NO_x and particulate emissions (particulate emission lower than 0.0025 g/km on NEDC cycle have been assessed, i.e. –90 % compared to Euro IV limits) have been built up, combining the most promising particulate filter solution (catalyzed silicon carbide wall-flow filter) with the most effective active regeneration strategy (capable to promote the particulate filter regeneration in all driving conditions) developed within the DEXA Cluster. The average increase of fuel consumption is lower than 2 %. The cost and the environmental impact (LCA) estimated for the whole system appears to be competitive compared to a fuel-additive based solution.

2. A simulation toolkit was developed based on the coupling of sub-models for exhaust system devices to a 3-D CFD solver for the flow inside the entire exhaust system. The simulation tool has been validated against an experimental database containing information on after-treatment device materials and geometric characteristics as well as numerous engine-bench experiments with the candidate DPF devices for the DEXA demonstrator vehicles.

3. Measurement technologies, analysis methods, and protocols for size and composition analysis of particles have been developed and assessed. Mobility based techniques are the method of choice and real time performance has been demonstrated with a new system (TrDMPS). An analysis to connect number and mass based distributions for fractal particles has been also developed. Optical techniques deployed in the raw exhaust are a useful research tool and look promising but require further development for routine use. The deployment of different instrumentation in tandem operation (DMA-PAS/DC/CPC) or in simultaneous operation

(SMPS/ELPI) was shown to provide useful in-situ size specific composition and morphological information of the particles. The aftertreatment technologies chosen for the DEXA demonstrators reduce particle numbers by ~ 2 orders of magnitude (close to ambient) during regular engine operation. AEM/EDX analysis of particles emitted from the filter during regeneration showed them to be carbon based (but the evaporation of any volatile material cannot be precluded in the high vacuum environment of the instrument).

The positive results of the DEXA Cluster research are expected to increase the market competitiveness of Diesel engine powertrains for passenger cars in more stringent emission regulations and will accelerate the adoption of particulate control technology.

ACKNOWLEDGMENTS

The DEXA Cluster of projects - ART-DEXA (Contract number: G3RD-CT-1999-00016), SYLOC-DEXA (Contract number: G3RD-CT-1999-00014), PSICO-DEXA (Contract number: G6RD-CT-1999-00038) - has been funded in part by the European Commission GROWTH program within the 5th Framework Program. We would like to thank the European Commission officers Daniel Chiron, Richard Gilmore, Maurizio Maggiore and Patrick Mercier-Handisyde for their excellent collaboration. In addition, we are grateful to Evdoxia Kladopoulou of APTL CERTH/CPERI for her assistance in the editing and preparation of the paper.

REFERENCES

1. Papaioannou E., Stabouli A., Konstandopoulos A. G., (1997), "A Laboratory Screening Study of Soot Oxidation Catalysts", Proc. 1st Mtg Greek Section of Comb. Inst., pp. 371.
2. Karadimitra K., Papaioannou E., Macheridou G., Konstandopoulos A. G., (2001), "Ceria Nanoparticle Coated Filters for Soot Emission Control", PARTEC 2001, International Congress for Particle Technology, 27-29 March 2001, Nuremberg, Germany.
3. Fino P., Saracco G., Specchia V., (2002), "Filtration and Catalytic Abatement of Diesel Particulate from Stationary Sources", *Chem. Eng. Sci.*, **57**(22-23), pp. 4955-4966.
4. Fino D., Russo N., Badini C., Saracco G., Specchia V., (2003), "Effect of Active Species Mobility on Soot-Combustion over Cs-V Catalysts", *AIChE Journal*, **49**(8), pp. 2173-2180.
5. Fino D., Russo N., Saracco G., Specchia V., (2003), "The Role of Suprafacial Oxygen in some Perovskites for the Catalytic Combustion of Soot", *Journal of Catalysis*, **217**(2), pp. 367-375.
6. Fino D., Fino P., Saracco G., Specchia V., (2003), Innovative Means for the Catalytic Regeneration of Particulate Traps for Diesel Exhaust Cleaning", *Chem. Eng. Sci.*, **58**(3-6), pp. 951-958.
7. Skaperdas E., Konstandopoulos A. G., (2001), "Prediction of Diesel Particulate Filter Loading Behavior for System Control Applications", CD AUTO 01, 3rd International Conference on Control and Diagnostics in Automotive Applications, 4-6 July 2001, Sestri-Levante, Italy.
8. Konstandopoulos A. G., Skaperdas E., Masoudi M., (2002), "Microstructural Properties of Soot Deposits in Diesel Particulate Filters", *SAE Tech. Paper No.* 2002-01-1015 (SP-1673).
9. Wassermayr C., Brandstätter W., Schreier H., Prenninger P., (2003), "Virtual Engineering of Diesel Exhaust Systems to Meet Future Emission Legislation", *JSAE Tech. Paper No.* 20035426.
10. Brandstätter W., Wassermayr C., Prenninger P., (2002), "An Integrated Approach for the Design of Diesel Engine Exhaust Systems to meet Euro 4 and beyond Emissions Legislations", THIESEL 2002 Thermodynamic Processes in Diesel Engines, 11-13 September 2002, Valencia, Spain.
11. Wassermayr C., 'Numerical Modelling and Simulation of Diesel Exhaust Gas Aftertreatment Systems', Dissertation, University of Leoben, Austria (2003).
12. Konstandopoulos A. G., and Kostoglou M., (2000), "Reciprocating Flow Regeneration of Soot Filters", *Comb. and Flame*, **121**(3), pp. 488-500.
13. Konstandopoulos A. G., Kostoglou M., Housiada P., Vlachos N., Zarvalis D., (2003), "Multichannel Simulation of Soot Oxidation in Diesel Particulate Filters", *SAE Tech. Paper No.* 2003-01-0839 (SP-1755).
14. Konstandopoulos A. G., Kostoglou M., Skaperdas E., Papaioannou E., Zarvalis D., Kladopoulou E., (2000), "Fundamental Studies of Diesel Particulate Filters: Transient Loading, Regeneration and Aging" *SAE Tech. Paper No.* 2000-01-1016 (SP-1497).
15. Konstandopoulos A. G., Kostoglou M., Housiada P., (2001), "Spatial Non-Uniformities in Diesel Particulate Trap Regeneration" *SAE Tech. Paper No.* 2001-01-0908 (SP-1582).
16. Konstandopoulos A. G., Skaperdas E., Masoudi M., (2001), "Inertial Contributions to the Pressure Drop of Diesel Particulate Filters," *SAE Tech. Paper No.* 2001-01-0909 (SP-1582).
17. Laile E., Zahoransky R. A., Claussen M., (2003), "Time Resolved Particle Measurements in the Exhaust of Diesel Engines During Different Cycles", European Aerosol Conference EAC 2003, 31 August - 5 September 2003, Madrid, Spain.

18. Fisenko S. P., Laile E., Zahoransky R. A., (2003), "Dynamic Behavior of a Long Path Optical Cell For On-Line Soot Particle Measurements", European Aerosol Conference EAC 2003, 31 August - 5 September 2003, Madrid, Spain.
19. Harris S. J., and Maricq M. M., (2002), "The Role of Fragmentation in Defining the Signature Size Distribution of Diesel Soot", *J. Aerosol Science*, **33**, pp. 935–942.
20. Kostoglou M. and Konstandopoulos A. G., (2003), "Oxidative Fragmentation and Coagulation of Diesel Soot Aggregates", European Aerosol Conference EAC 2003, 31 August - 5 September 2003, Madrid, Spain.
21. Sgro L. A., Minutolo P., Basile G., D'Alessio A., (2001), "UV-Visible Spectroscopy of Organic Carbon Particulate Sampled from Ethylene/Air Flames", *Chemosphere*, **42**, pp. 671-680.
22. Vaglieco B. M., Beretta F., D'Alessio A., (1990), "In Situ Evaluation of the Soot Refractive Index in the UV – Visible from the Measurement of the Scattering and Extinction Coefficients in Rich Flames", *Comb. and Flame*, **79**, pp. 259-271.
23. Vaglieco B. M., Monda O., Corcione, F. E., Menguc, M. P., (1994), "Optical and Radiative Properties of Particulate at Diesel Engine Exhaust", *Comb. Sci. Tech.*, **102**, pp. 283-299.
24. Merola S.S., Vaglieco B. M., Consales S. (2001), "Characterization of CR Diesel Exhaust by UV-Visible Extinction and Scattering Spectroscopy", 5th International Conference of Internal Combustion Engines, 23-27 September 2001, Capri, Italy.
25. Cartus T. & Wewerka A., (2001), "New Approaches in Particle Size and Morphology Measurements", 5th ETH Conference on Nanoparticle Measurements, 6-8 August 2001, Zurich, Switzerland.
26. Konstandopoulos A. G., et al., (2001), "Optical, Electrical Mobility and Gravimetric Measurements of Soot Particles Before and After Filter Media in Diesel Exhaust", 5th International Conference of Internal Combustion Engines, 23-27 September 2001, Capri, Italy.
27. Nikitidis M. S., Konstandopoulos A. G., Zahoransky R. A., Laile E., (2001), "Correlation of Measurements of a new Long Optical Path Particle Sensor Against Gravimetric and Electrical Mobility Based Particle Measurements in Diesel Exhaust", 5th International Conference of Internal Combustion Engines, 23-27 September 2001, Capri, Italy.
28. Zahoransky R. A., Laile E., Saier T., Konstandopoulos A. G., Nikitidis M. S., (2001), "Optical Multi-wavelength Technique Applied to the Online Measurement of Particle Emissions of Engines", 5th International Conference of Internal Combustion Engines, 23-27 September 2001, Capri, Italy.
29. Zahoransky R. A., Laile E., Claussen M., Wollmann A., Konstandopoulos A. G., (2002), "Time Resolved Measurements of Soot Concentrations and Mean Particle Sizes During EUDC and ECE Cycles", 6th ETH Conference on Nanoparticle Measurement, 19-21 August 2002, Zurich, Switzerland.
30. Claussen M., Benker B. and Wollmann A., (2003), "Comparison of Diesel Soot Particle Size Distributions Measured with a Cascade Impactor and a Mobility Spectrometer", European Aerosol Conference EAC 2003, 31 August - 5 September 2003, Madrid, Spain.
31. Baltzopoulou P., Papaioannou E. and Konstandopoulos A. G., (2003), "Size Specific Characterization of Diesel Particulates", European Aerosol Conference EAC 2003, 31 August - 5 September 2003, Madrid, Spain.

CONTACT

Dr. Athanasios G. Konstandopoulos, Head of Aerosol & Particle Technology Laboratory, CERTH/CPERI, P.O. Box 361, Thermi, Thessaloniki 57001, Greece.
e-mail: agk@cperi.certh.gr, web: http://apt.cperi.certh.gr

DEFINITIONS, ACRONYMS, ABBREVIATIONS

DPF Diesel particulate filter

PM Particulate matter

TC Turbo-charger

2005-01-1756

The Second Generation of Catalyzed Diesel Particulate Filter Systems for Passenger Cars

- Particulate Filters with Integrated Oxidation Catalyst Function -

M. Pfeifer, M. Votsmeier, M. Kögel, P. C. Spurk and E. S. Lox
Umicore AG & Co. KG, Hanau, Germany

J. F. Knoth
Darmstadt University of Technology, Institute of Technical Chemistry, Darmstadt, Germany

ABSTRACT

Several diesel passenger car manufacturers in the European Union recently announced the future use of catalyzed diesel particulate filter systems on their vehicles.

The major technical challenge is the periodical regeneration of the filters loaded with the retained diesel particulates.

In order to promote filter regeneration, catalytic activation of the accumulated soot is advantageous. Therefore, the first serial application of diesel particulate filter systems uses catalytically active fuel additives. These systems were introduced about four years ago.

Since that time, other systems, using a dedicated catalytically activated diesel particulate filter combined with an upstream diesel oxidation catalyst, have been introduced as well. This allows filter regeneration without extra fuel additives.

In the past, adding catalytic coating to a filter substrate has often resulted in increasing the pressure drop over the filter to an unacceptable level. Using a newly developed, dedicated processing technology, coatings with outstanding catalytic functions even on standard filter substrates can be applied without a severe backpressure increase. This so-called "macrowashcoating" technology enables the possibility to integrate the diesel oxidation catalyst's functions directly in the filter substrate.

The HC and CO oxidation performance of a filter with integrated oxidation catalyst functionality is compared to a conventional flow-through monolith by computer simulation and experiment. Simulation and experiment both show a systematic advantage of the catalyst filter over the conventional monolith configuration. Computer simulation provides some explanation for this observation.

Finally, the technology's suitability for application to passenger cars is proven by intensive engine bench and vehicle tests.

INTRODUCTION

The particulate emissions of diesel-engine vehicles have been reduced by over ninety percent within the last thirty years and introduction of Euro IV legislation in 2005 will prescribe a further reduction by one-half of the particulate threshold. Nevertheless many vehicles without particulate filters are already capable of meeting the Euro IV particulate standards today due to further developments in the combustion process, the combustion chamber design, as well as the injection pressures and injection profile [1].

While particulate filters are expected to become an integral component of an exhaust aftertreatment system for reaching US07 limits, these are by no means a mandatory requirement for the particulate levels currently being discussed for the Euro V legislation. In any case, the particulate filter represents an effective measure to reduce or even eliminate diesel particulate emission. For this reason, most of the automobile manufacturers are following the lead of PSA Peugeot Citroën and are advancing or already implementing the use of particulate filters in Europe [2].

However, the positive effect of an over 95% reduction of particulate emissions with the use of currently available cordierite or silicon carbide based filter substrates has some negative side effects, such as an increase in the exhaust gas backpressure, and thus also in fuel consumption. Other materials based on the principle of deep bed filtration do not increase the exhaust gas backpressure in the same way, but demonstrate significantly lower filtration efficiencies. [3]

The regeneration of the loaded filter under all operating conditions represents a major challenge in the development of filter systems. The regeneration should take place if possible without additional energy, yet quickly and reliably. Uncoated, non-catalytic filters cannot adequately meet this requirement.

For this reason, some of the passenger car applications currently on the market employ thermal regeneration supported by additives. Reliable regeneration of the soot-loaded filter is ensured by a combination of engine (post-injection, post-combustion, throttling) and catalytic measures (oxidation catalyst, fuel additives). The engine post-injection increases the temperature of the exhaust gas to approx. 450 °C. A portion of the unburned post-injected fuel mixes with the exhaust gas and is oxidized by the diesel oxidation catalyst in front of the particulate filter. This additional increase in temperature raises the exhaust gas temperature in front of the filter to between 500 °C and 550 °C [4]. In this temperature range, the soot activated by the fuel additive spontaneously combusts, ensuring on the one hand a fast regeneration of the filter while making the control of the combustion process more difficult on the other hand. In addition to the higher technical complexity of the system, consisting of an additional tank and metering system, the use of the additive results in a substantial increase in the amount of ash introduced into the filter.

This ash loading is associated with a corresponding increase in the exhaust gas backpressure over the service life and makes filter servicing or even replacement necessary after about 100,000 – 150,000 km for current system architectures (additive type and filter design) [5, 6]. The further development of this design aims to increase the service life of the filter to approx. 240,000 km.

Alternatively, a catalyzed filter can be used. This design is used currently at DaimlerChrysler, among others [5]. Similar to the fuel additive, the purpose of the coating is to reduce the activation energy for the oxygen-based oxidation of the soot. Another advantage of the catalyzed soot filter compared to the additive-based systems is a complete suppression of undesirable secondary emissions of carbon monoxide and various hydrocarbons that occur during regeneration [7]. Good HC/CO light-off behavior for supporting the function of the upstream oxidation catalyst is also desired. And finally, the catalytic particulate filter can support the upstream diesel oxidation catalyst in the generation of NO_2, which is capable of oxidizing diesel soot already at temperatures above approximately 300°C. This reduces the so-called balance temperature. The balance temperature is hereby defined as the temperature at which the same amount of soot in the particulate filter is oxidized as is emitted by the engine in the same unit of time [8]. Therefore, in many applications the time interval between two active oxygen-based regenerations can be increased depending on the driving profile.

All these requirements should be accompanied by a high thermal stability of the catalytic filter coating and preferably no negative impact on backpressure.

A further optimization can be achieved through the integration of the diesel oxidation catalyst in the filter; i.e., the filter coating assumes the functions of the diesel catalyst. The advantages of such a combined system are obvious: simplicity of the system, optimization of costs and design space. During the regeneration of the filter, the exothermic energy released in the oxidation of post-injected hydrocarbons is generated there where it is needed; namely, directly in the filter.

Currently no such application is in use in a passenger car; however, Volkswagen has announced the intent to introduce these systems to the market in 2004/2005 [9].

The goal of the studies presented here is to examine the potential and functionality of catalytic filters with integrated oxidation catalyst functions and to make initial statements on aging behavior and durability of these systems.

INTEGRATION OF THE DIESEL OXIDATION CATALYST ON THE PARTICULATE FILTER (STAND-ALONE FILTER)

A stand-alone filter must fulfill the function of the diesel oxidation catalyst. The advantages are simplicity of the system and optimization of cost and design space. The precious metals needed for functioning of the diesel catalyst can be used in synergy also for the oxidation of the soot.

However, a critical question remains, whether a particulate filter on silicon carbide basis with significantly reduced cell density and high thermal capacity can be a suitable substrate for an oxidation catalyst coating. Additional issues to be clarified are the durability within the scope of filter ash loading and the stability of the oxidation coating in light of typical filter aging mechanisms.

CATALYTIC COATING OF DIESEL PARTICULATE FILTERS

Today European automobile manufacturers use silicon carbide-based particulate filters exclusively. These filters are made of recrystallized or silicon-bonded silicon carbide. In addition to conventional filter matrices with a porosity of 40 – 45% and an average pore diameter of approx. 10 µm, substrate manufacturers have developed filters specifically for coated applications with a porosity of up to 60% and an average pore diameter in the range of approx. 20 µm [10]. The advantage of these new filter substrates is a higher tolerance regarding washcoat loadings with respect to the backpressure behavior. However, lower mechanical stability and reduced maximum soot loading represent disadvantages [10].

In contrast to standard applications such as three-way or oxidation catalysts, in which the substrate serves exclusively as the carrier for the catalyst, the diesel particulate filter has its own functionality which can be changed by the catalyst coating. For example, the coating can increase the filtration efficiency of the filter and influence its maximum soot loading. Decisive, however, is the effect of the coating on the backpressure behavior of the filter.

These interactions between filter substrate and coating must be considered in the development of suitable catalysts. Taking these factors into account, Umicore has developed and optimized a coating process for particulate filters. This process is referred to as "macrocoating."

In the coating of standard catalysts in the automotive sector, first the precious metals are fixed to support oxides. These support oxides are suspended in water, and finally the substrate is coated with this suspension. If this method is used on particulate filters, the catalyst lies as a layer on the surface of the inlet-side channels of the substrate. Only a very small portion of the catalyst penetrates into the first pore layers of the substrate and plugs most of the pore necks.

With the newly developed macrocoating technology, the catalytic components are deposited homogeneously into the pore system of the substrate. Figure 1 shows a comparison of the distribution of catalytically active components in a standard coating (right) and in a macrocoating (left).

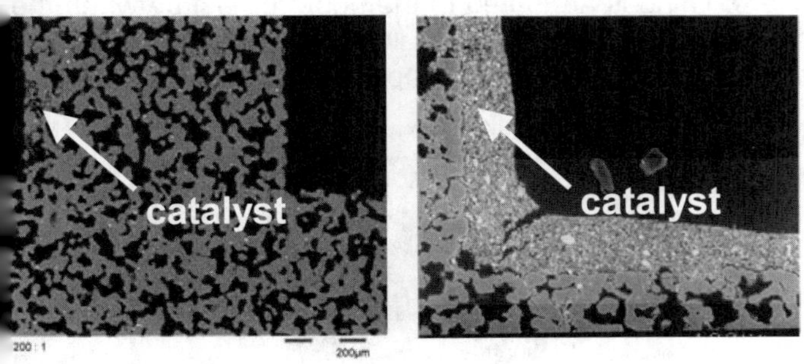

Figure 1: Scanning electron microscope images of a SiC particulate filter (42 % porosity) with standard coating process (right) and macrocoated (left)

The different distribution of the catalyst in the filter has a decisive influence on its backpressure behavior. With macrocoating the backpressure rises compared to the uncoated substrate at a given flow rate depending on the permeability of the filter substrate in the range of 5 – 30 %; while for the standard coating process, on the other hand, it rises by a factor of two or three.

REGENERATION BEHAVIOR OF CATALYTIC FILTERS WITH AND WITHOUT PRE-CATALYST

One critical question in connection with stand-alone filters (CC-DPF) is to which extent the regeneration behavior of such a system differs compared to the standard systems consisting of upstream oxidation catalyst, uncatalyzed filter and fuel additive or upstream oxidation catalyst and coated particulate filter.

Test set-up

The study was performed on a direct-injected 2.2 l common-rail diesel engine with exhaust gas turbo-charging, charge cooling and 100 kW output.

Diesel fuel with a sulfur content of 30 ppm was used. An AMA 2000 Exhaust Gas Measuring System from Pierburg was applied for detection of gaseous components.

In an initial evaluation the regeneration behavior of the three following systems at two different filter inlet temperatures was compared:

System 1: stand-alone macrocoated filter (90 g/ft³ Pt)

System 2a: diesel oxidation catalyst (120 g/ft³ Pt) upstream of the macrocoated filter (90 g/ft³ Pt)

System 2b: diesel oxidation catalyst (120 g/ft³ Pt) upstream of an uncatalyzed filter using a fuel additive

In system 2a a diesel oxidation catalyst (∅ 5.66" x 3.5"; 400 cpsi, 6.5 mil; 120 g/ft^3 of Pt) was mounted upstream of a macrocoated SiC filter (∅ 5.66" x 6"; 300 cpsi, high porosity) with a platinum content of 90 g/ft^3. For system 1 this macrocoated filter was positioned instead of the diesel catalyst in system 2a. To investigate the fuel additive system (system 2b) the oxidation catalyst was placed in front of an uncoated 5.66" x 6" SiC filter substrate (∅ 5.66" x 6"; 200 cpsi, 42 % porosity). A cerium additive (Eolys DPX-9 from Rhodia) with a mass concentration of 25 ppm was used. Figure 2 shows a schematic drawing of the exhaust gas system.

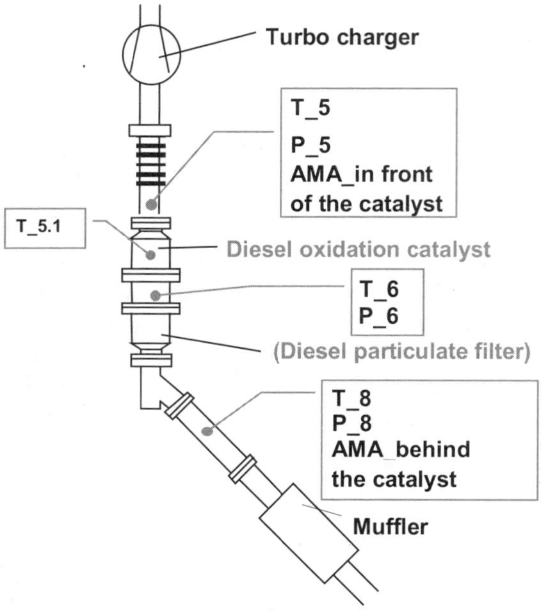

Figure 2: Schematic diagram of the exhaust gas system with the corresponding measuring points

Before each regeneration the filters were loaded at a speed of 2300 rpm and a torque of 40 Nm within 12 hours with around 20 g (8 g/l) of soot (absolute). Subsequently the regeneration was introduced through a post-injection. Two different regeneration modes were used: In the first mode the target temperature in front of the filter was 450°C for system configuration 2a and 2b. The duration of the post-injection was 15 minutes. In the second mode the post-injection amount was increased in such a way that a temperature of 600°C in front of the filter was reached, again based on system configuration 2a and 2b. Due to the higher temperature the duration was reduced to 10 minutes. For system 1 an identical post-injection strategy was used. After completion the filters were cooled down to 80°C and weighed. The amount of soot oxidized was determined by the difference in the weight before and after regeneration. Table 1 is listing the soot quantities for the different systems which were combusted during the post-injection mode.

Table 1: Soot amount for the three systems combusted during the post-injection mode. The initial soot loading was around 20 g.

System	Combusted soot amount @ 450°C [g]	Combusted soot amount @ 600°C [g]
System 1: stand-alone CDPF	10,2	17,0
System 2a: DOC / CDPF	1,7	13,5
System 2b: DOC / DPF / FBC	5,5	20,8

At an inlet temperature of 450°C system 1 performs best. This is probably due to the combustion of post-injected hydrocarbons directly on the filter leading to locally exothermic reactions with much higher temperatures than the overall gas temperature and to a partially soot combustion. Thus this is more a thermal than a catalytic based form of regeneration. Nevertheless in no case could the whole amount of soot trapped inside the filter be oxidized within the time frame.

Using the second post-injection mode with 600°C in front of the filter the additive based system (system 2b) shows the highest regeneration efficiency. According to the run of the pressure-drop curve (not shown here) the filter was regenerated totally only within 4 minutes of post-injection. This is the result of the tight contact between the soot particulates and the catalyst particulates originating from the additive. The reduced activity of the filter coating is attributed to the poor contact between the solid soot and the solid catalyst. As shown in Figure 3, the soot is located as a compact layer on the surface of the channel.

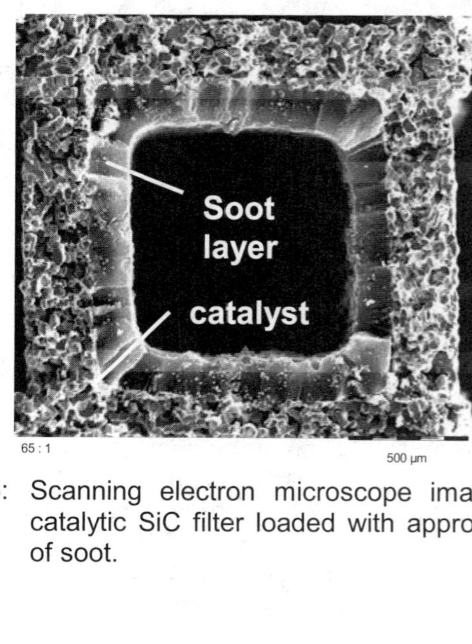

Figure 3: Scanning electron microscope image of a catalytic SiC filter loaded with approx. 10 g/l of soot.

Anyway the stand-alone filter shows a better regeneration performance than in combination with a heat-up catalyst.

To evaluate this in more detail a further investigation was performed. The filters were again loaded with 8 g/l soot (20 g absolute) according to the loading procedure described above. With the regeneration strategy leading to a filter inlet temperature of 600°C (see above) the duration of post-injection was varied between 100 s and 600 s in increments of 50 or 100 s. Again after completion of the post-injection the remaining soot in the filter was detected by filter weighing.

Figure 4 shows the relative amount of soot which was combusted as a function of post-injection time for system 1 and 2a.

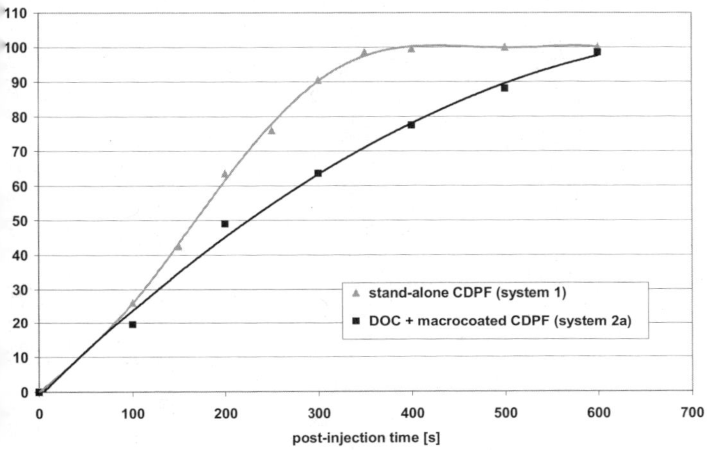

Figure 4: Relative soot amount that was oxidized during the regeneration with various post-injection times for the macrocoated filter alone (system 1) and in combination with a pre-catalyst (system 2a)

Using the macrocoated filter application without the oxidation catalyst upstream a post-injection time of only about 350 s was required for complete filter regeneration, whereas with the combined system it took around 600 s for a complete regeneration. This is apparently due to the direct combustion of post-injected hydrocarbons on the DPF leading locally to temperatures inside the filter much higher than the overall gas temperature.

SUITABILITY OF WALL-FLOW FILTERS AS SUBSTRATE FOR OXIDATION COATINGS

Development of first generation catalyzed particulate filters focused on the coating's ability to support filter regeneration. With the emergence of the more advanced macrocoating technology substantial degrees of freedom have been gained for the design of wall integrated washcoats. This means that now the wall integrated washcoats can be optimized for their substrate - precious metal interactions in much the same way as known for conventional flow-through catalysts. As a consequence, today most of the catalyst functionality that can be implemented in a conventional catalyst can also be implemented in a particulate filter. The advantages of implementing a diesel oxidation catalyst functionality directly on the filter have been outlined in a previous section of this contribution.

Besides regeneration behavior, the second obvious characteristic of a catalyzed stand-alone filter (CC-DPF) is its ability to perform the oxidation catalyst function. The main question here is how the catalyst filter performance compares to the performance of a conventional flow-through catalyst.

From a reaction engineering point of view, a wall flow filter is a fundamentally different reactor concept than the conventional flow-through catalyst. In the following section these differences and their potential impact on an overall system design are analysed by means of detailed flow simulation. The simulation results demonstrate that with an assumed identical activity of the catalytic sites, the wall-flow configuration can show higher conversion than the conventional flow-through filter. This theoretical result is confirmed experimentally in synthetic exhaust bench tests. Finally, the performance of catalytically coated filter is demonstrated under transient conditions in a vehicle test.

Modeling

In this section the differences between a filter with catalyst integrated in the filter wall and a conventional monolith reactor are studied from a reaction engineering point of view. Intuitively, the main differences can be summarized as follows: In a wall-flow filter the exhaust gas (including the reactants) is forced through the catalytic wall by the pressure gradient. This means very short contact times on one hand but absence of transport limitation on the other hand. In a conventional monolith reactor the exhaust including the reactants moves along the channel which means longer contact times but potential diffusion resistance effects. In this section a mathematical model of the wall-flow filter is presented that allows development of a more detailed understanding of these intuitively obvious differences. In order to quantitatively cover the differences between the two configurations it is essential that the flow field in the gas phase and in the porous wall as well as the coupling of diffusion and reaction are explicitly taken into account. Diffusion effects in the filter wall are neglected in commercially available filter simulation software and to the authors' knowledge have not been covered in the open literature. For this reason a new reactor model has been set up in this work. The model setup will be briefly described and the results obtained with this model are presented in the following sections.

Model Setup

A two dimensional numerical model has been set up that describes the reactive flow in one pair of input/output channels and the connecting filter wall. Depending on whether a wall flow filter or a conventional flow-through catalyst is modeled either both channels are open or the inlet and outlet channels are appropriately blocked to ensure flow-through filter operation. Due to symmetry, only half of each of the inlet and outlet channel needs to be considered so that the computed domain consists of half an inlet channel, half an outlet channel, the filter wall and an entrance region upstream of the inlet channel (table 2).

Table 2: Characteristics of the simulated wall-flow filter

Channel length [m]	$50 \cdot 10^{-3}$
Channel half width [m]	$7{,}7 \cdot 10^{-4}$
Porous wall length [m]	$51 \cdot 10^{-3}$
Porous wall width [m]	$3{,}6 \cdot 10^{-4}$
Plug length [m]	$1 \cdot 10^{-3}$
Plug width [m]	$7{,}7 \cdot 10^{-4}$
Complete geometry length [m]	$71 \cdot 10^{-3}$
Complete geometry width [m]	$1{,}9 \cdot 10^{-3}$
Inlet velocity [m s^{-1}]	1

The pressure and velocity field in the entrance region, the inlet and the outlet channel are described by the Navier-Stokes equations coupled to the flow field in the porous wall that is described by Darcy's equation. The velocity and pressure fields computed in this way well correspond to the analytical solutions known from the literature [11]. Based on the flow field the diffusion/reaction equation is solved in the gas phase and in the porous medium. A first order reaction has been assumed in the wall for the CO-oxidation and zero reaction rate was assumed in the gas phase. The temperature dependence of the reaction rate is described by an Arrhenius equation with the preexponential factor adapted for 50% conversion at a temperature of 200°C in the wall flow filter configuration. A value of 104 $kJmol^{-1}$ [12] is used for the activation energy. Isothermal conditions are assumed for all simulations presented. Except for the simulations presented in Fig. 8 a porous wall effective diffusion coefficient of $1 \cdot 10^{-6}$ m^2 s^{-1} is used for all simulations. The model has been set up and is solved in the commercial software package FEMLAB 3.0 a. The simulations of the flow-through monolith are performed with an identical model setup as the wall-flow filter simulations except that the plugs in the channels are removed.

Modeling results

A) General qualitative differences between wall-flow configuration and conventional flow-through monolith

Fig. 5 compares CO concentration profiles in the wall-flow configuration and in the conventional flow-through configuration.

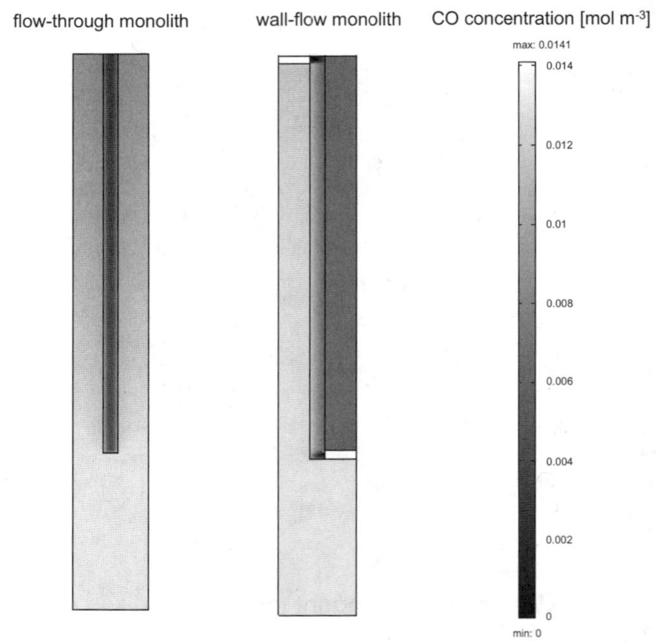

Figure 5: Reactant (CO) concentration distribution in both catalyst geometries

In these simulations the reaction rate for the first order reaction in the wall has been adapted for 50% conversion of the wall flow case. The main difference in the concentration profiles is that in the wall-flow configuration the concentration gradient is mainly inside the wall and the reaction coordinate is in the direction perpendicular to the channel axis. In the flow- through configuration the reaction coordinate is mainly along the channel axis. These effects are further illustrated in Fig. 6 that shows concentration profiles parallel and perpendicular to the channel axes for both reactor configurations.

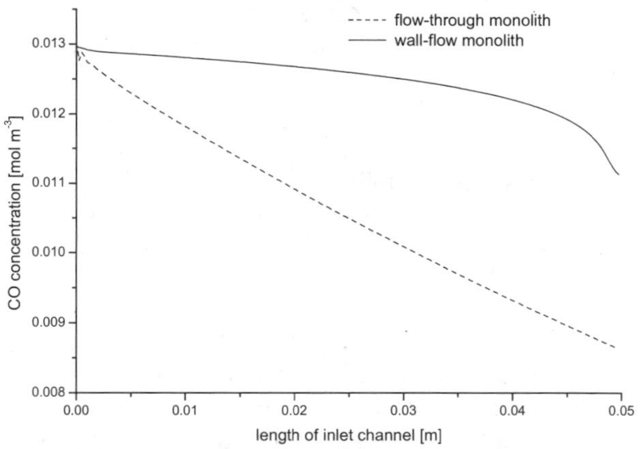

Figure 6a: Reactant (CO) concentration distribution along the inlet channel.

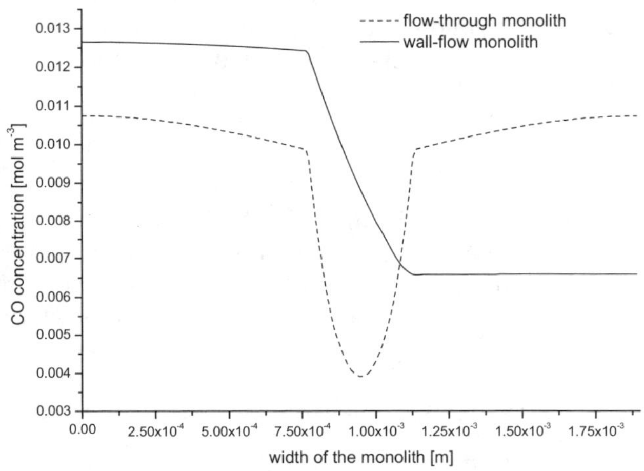

Figure 6b: Horizontal reactant (CO) concentration distribution

As apparent from Figs. 5 and 6 the CO concentration in the inlet channel remains nearly constant at the level of the inlet concentration. This means that different wall segments along the channel see an identical reactant concentration. As a consequence in first approximation the overall reactor can be represented by one plug flow reactor with the reaction coordinate oriented in radial direction along the flow in porous wall. On the other hand the conventional flow-through reactor is best approximated by a plug flow reactor oriented along the channel axis. Some concentration gradient perpendicular to the flow direction is observed inside the wall of the flow-through catalyst. This gradient causes an internal mass transfer limitation that is well described by the well-known effectiveness factor equation based on the Thiele modulus.

The different orientation of the reaction coordinate in the two reactor configurations has important implications for reactor operation and performance, especially under transient conditions. For example, during reactor heat-up for the conventional flow-through reactor the heat-front moves parallel to the reaction coordinate. In case of the wall-flow filter the heat front moves perpendicular to the reaction coordinate.

The orientation of the reaction coordinate perpendicular to the channel axis is characteristic of wall-flow catalyst operation in the range of low overall conversion responsible for light-off behavior. At higher temperatures back-diffusion from the wall into the inlet channel becomes more important and significant concentration gradients can be observed in the direction of the channel axis.

B) Performance comparison between wall-flow configuration and conventional flow-through monolith

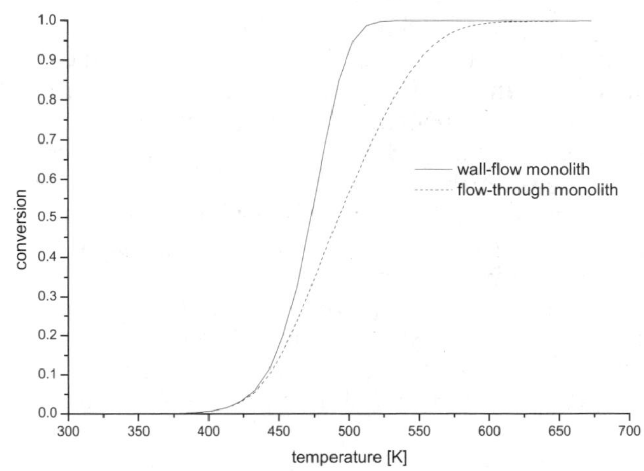

Figure 7: Computed light-off curves for the CO-oxidation

Fig. 7 compares computed light-off curves for the wall-flow configuration and the conventional flow-through monolith. The two simulations have been obtained from an identical simulation setup including identical meshes with the only difference, that for the simulation of the flow-through monolith the plugs in the inlet- and outlet channel have been removed as described in the previous section. A significantly lower light-off temperature is found for the wall-flow filter. In Fig. 8 for a fixed reaction rate the obtained conversion for the two reactor configurations is compared as function of assumed diffusion coefficients for the porous wall.

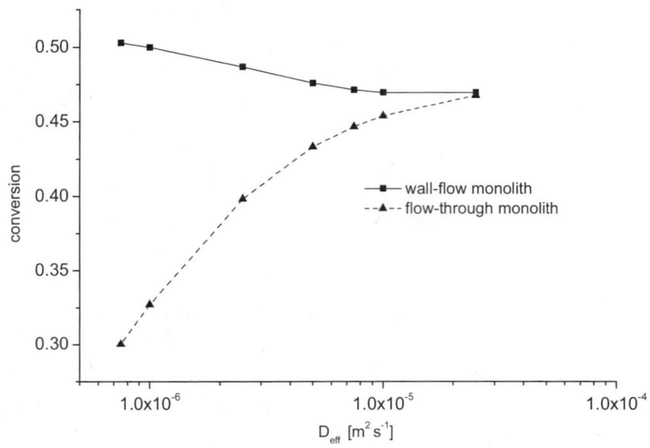

Figure 8: Computed CO-conversion as a function of the porous wall diffusion coefficient

Again, significant differences are observed . The differences increase with decreasing diffusion coefficient. As expected the conversion for the conventional flow-through monolith increases with increasing diffusion coefficient. This increase of the conversion with diffusion coefficient can be quantitatively understood based on the standard effectiveness factor equation for first order reaction. The conversion of the wall-flow filter decreases with increasing diffusion coefficient. This behavior can be understood if one treats the wall-flow filter as a plug flow reactor with the reaction coordinate perpendicular to the channel axis as described above. In the plug flow approximation, diffusion along the radial direction of the filter introduces back mixing along the reaction coordinate. Back mixing in a plug flow reactor with a first order reaction reduces the overall reactor conversion [13].

C) Conclusion from the modeling study

It has been shown that the wall-flow reactor shows some fundamentally different behavior than the conventional flow-through monolith. These differences are best illustrated by the different orientation of the reaction coordinate in the two reactor configurations.

At identical activity of the catalytic sites and identical reactor geometry the wall-flow filter shows systematically higher conversion compared to the conventional flow-through monolith. The differences between the two reactor concepts is shown to decrease with increasing diffusion coefficient for the porous wall. This indicates that mass transfer resistance effects play an important role as an explanation for the observed effects.

The purpose of the preliminary simulation results presented in this publication is to study the fundamental differences between the two reactor configurations. A first order rate law is assumed since it best allows focussing on the fundamental differences between the two reactor concepts without any further complication by more complex effects specific to any more complex rate laws.

It is well known in the literature that catalytic conversion of CO and other pollutants is not well described by a first order rate law. First simulations with Langmuir-Hinshelwood-type kinetics show qualitatively similar results.

The trends found in the simulation study using first order kinetics qualitatively confirm the experimental results presented in the following section. Simulations with more realistic Langmuir-Hinshelwood kinetics will allow a more quantitative comparison between experiments and simulation.

Synthetic gas bench tests

Light-off tests were performed on a synthetic gas bench to investigate the influence of different flow conditions in a wall-flow filter on the hydrocarbon and CO oxidation activity. To eliminate the impact of different material properties only SiC based substrates were used.

Test set-up

The light-off curves were measured on a synthetic gas line with a nominal heat ramp of 15°C/min and a flow rate of 32,2±0.5 l · min $^{-1}$. All samples were conditioned for 20 minutes at 500 °C. The temperatures varied from 50°C up to 450°C. The standard gas composition is given in table 3.

Table 3: Gas composition used for the light-off tests

Component	Concentration
CO	350 vol.-ppm
H_2	116 vol.-ppm
C_3H_6 (rel. To C1)	270 vol.-ppm
NO	270 vol.-ppm
CO_2	10.7 vol.-%
H_2O	10 vol.-%
O_2	6 vol.-%
N_2	Balance

A ∅1" x 6" filter core made of silicon carbide with a cell density of 178 cpsi, a wall thickness of 16mil, a porosity of 42 % and a medium pore size of 9 µm was coated with a DOC based catalyst using a standard coating process so that the catalyst was deposited as a layer on the surface of the inlet channels. A relatively low platinum loading of only 13 g/ft³ was chosen to observe higher differences in activity. The light-off performance was investigated using the conditions described above with a corresponding space-velocity of 25 000 h^{-1}. After the measurement the plugged channels on the downstream side of the filter core were cut off (~ 4 mm in length) so that a kind of flow-through substrate was formed with the same physical properties as the wall-flow filter core. The light-off measurement was repeated. Figure 9 shows a comparison of the different light-off curves for hydrocarbons and CO.

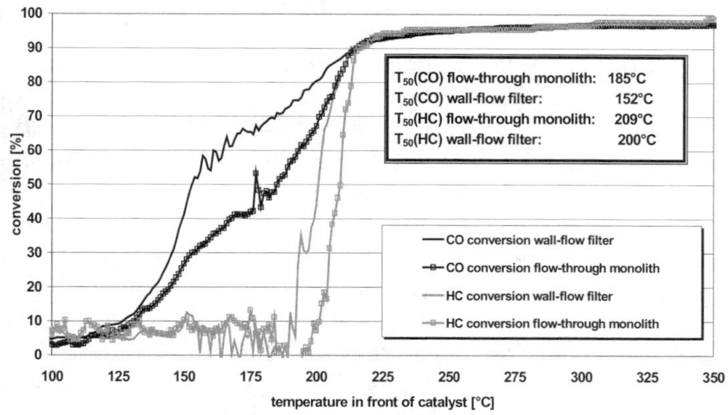

Figure 9: HC and CO light-off curves of the filter core before and after cutting the plugged outlet channels

The test results are in good accordance with the observations described in the modeling section. They clearly reveal that the changed flow conditions in a wall-flow filter compared to a flow-through substrate can give some benefits in the conversion of gaseous components like CO or HC. This is strongly dependent on the test conditions such as space-velocities and on physico-chemical properties like the reaction rate for the given reaction.

Vehicle Tests

As noted in the previous chapter, good HC and CO oxidation activities can be achieved with catalytic filters under stationary conditions. However, the performance in transient operating states must be studied.

For this reason two SiC filters (∅ 5.66" x 6"; 300 cpsi) were catalytically coated. The platinum loading of the macrocoated filters was 20 and 90 g/ft^3 respectively. These filters were tested – without additional upstream oxidation catalyst (CC-DPF) – on an Euro III certified, automatic transmission vehicle with a 2.7 l diesel engine in the NEDC. All substrates were installed a few centimeters behind the turbo charger, in correspondence to the serial system. The HC and CO emissions in the NEDC are shown in Figure 11.

To examine the influence of aging, an appropriate aging cycle was developed on a special aging unit with a range of temperatures that reliably covers the temperatures occurring during controlled filter regenerations. Therefore, the temperature in the filter inlet fluctuates periodically between 200°C and 820°C. Each cycle lasts eight minutes. The filters used for the test were aged 100 cycles, which, depending on the distance between two active regenerations in the vehicle, should correspond to 50,000 km and 100,000 km driven. A section of the aging temperature profile is shown in Figure 10.

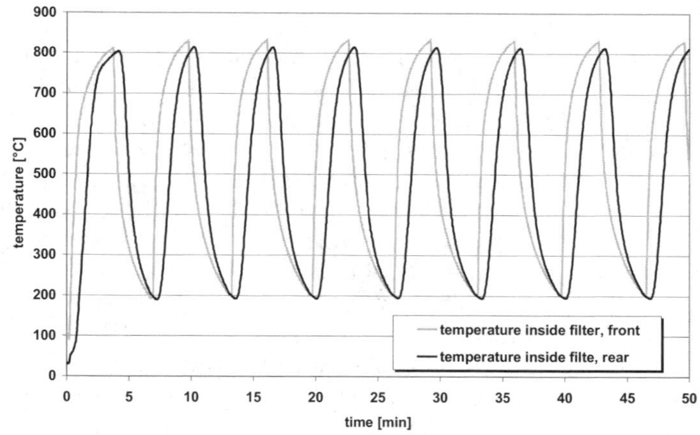

Figure 10: Section of the aging temperature profile for the filters on the specially developed unit for accelerated aging of catalysts

The emissions after aging were shown in Figure 11 as well. In addition to the tests using the coated particulate filters, a volume-production oxidation catalyst with 90 g/ft^3 of platinum was tested (cordierite carrier: ∅ 5.66" x 6", 400 cpsi, 6.5 mil). The corresponding results are shown also in Figure 11, but only in fresh state.

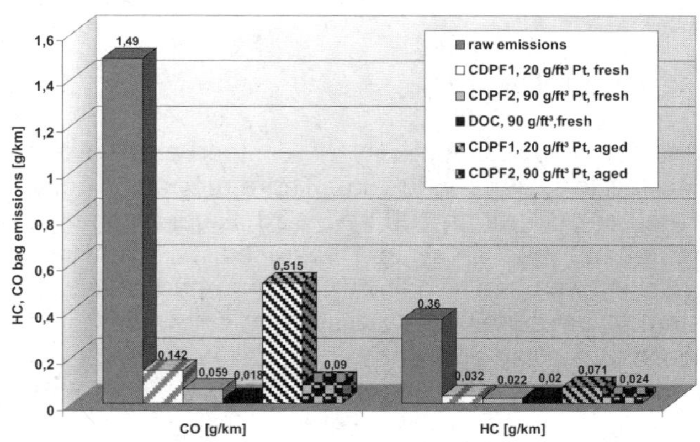

Figure 11: Bag emissions in the NEDC of macrocoated SiC filters with 20 and 90 g/ft3 of platinum, fresh and after aging, and of a fresh diesel oxidation catalyst with 90 g/ft3 of platinum as reference

In fresh state, conversion rates of 90% of CO and 91% of HC are achieved, even with a minimally loaded particulate filter (20 g/ft^3 Pt). Compared to the corresponding oxidation catalyst on a cordierite flow-through substrate, the fresh CO conversion rate is reduced for the filter with 90 g/ft^3 platinum concentration by 3% (absolute), while for HC it is almost identical. Even after hydrothermal aging HC and CO emissions remain well below the Euro IV thresholds with this filter.

Despite of the fact that the oxidation catalyst has a conversion rate slightly higher than the stand-alone catalyzed filter this do not contravene the observations and statements described in the modeling section and the synthetic light-off tests. The NEDC as a dynamic test cycle is depending on much more parameters than the simple test matrix used for the modeling work and is much more complex than a stationary light-off test. Furthermore neither the catalytic coating of the DOC and of the filter nor the substrate type was the same: a cordierite substrate with a cell density of 400 cpsi and a wall thickness of 6.5 mil was used for the DOC and a SiC based filter with 300 cpsi and 12 mil for the CDPF in the NEDC tests. In particular the heat capacity of the catalysts has a very big influence on the emissions and conversion rates and this was much higher for the catalyzed filter than for the oxidation catalyst. In the modeling studies and the light-off tests the same substrate material with the same cell geometry was postulated / used for the wall-flow filter and for the flow-through monolith as well.

Nevertheless these results clearly demonstrate the basic potential of the stand-alone filter; however, the system must be evaluated for each application and analyzed in detail.

DURABILITY TESTING OF THE STAND-ALONE FILTER

Conducting of the test

To investigate the durability of a stand-alone filter an aging cycle was conducted. Therefore an Euro III certified vehicle with a 100 kW, 2.2 l diesel engine with a common rail injection system was used. This vehicle was equipped with an additive-assisted particulate filter system. The serial catalyst system consisting of pre-oxidation catalyst and uncoated filter was replaced for the test series with a macrocoated SiC filter (∅ 5.66" x 6") in an exchangeable canning (Figure 12). The platinum loading of the filter, at 7.3 g (absolute), was comparable to that of the serial system.

The fuel additive tank of the vehicle was emptied, rinsed, and filled with pure diesel fuel for the tests to avoid additive entrainment during testing.

Figure 12: Exchangeable canning used with stand-alone filter in comparison to volume-produced canning. The canning is in the underfloor position, approx. 60 cm behind the turbo-charger.

The vehicle was operated in AMA cycle for a total of 25,000 km with Euro IV certified fuel. During this period, the exhaust gas temperatures before and after the filter, as well as the backpressure over the filter, were recorded with a data logger. The regeneration strategy of the serial application was used. The catalytic performance of the filter was determined in new condition as well as every 5000 km in the NEDC.

Test results

The AMA cycle is a test run with relatively low exhaust gas temperatures. Despite the low level of the exhaust gas temperatures, filter blockage did not occur during the 25,000 km test run. Due to the regeneration strategy of the serial application used in this durability test an active filter regeneration was initiated by the engine control unit every 400 to 500 kilometers. The regeneration always led to a complete soot combustion with moderate peak-temperatures of about 800 – 850 °C inside the filter. No uncontrolled regeneration event was observed during the 25,000 km of operation. One example is shown in Figure 13.

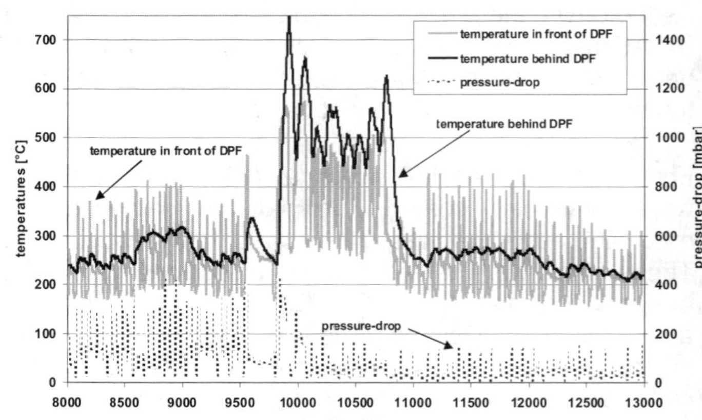

Figure 13: Filter regeneration in the AMA cycle. During the regeneration the gas temperature behind the filter is higher than in front of the filter due to the oxidation of post-injected hydrocarbons; after the soot burn-off is completed, the backpressure level returns to a substantially lower level.

Figure 14 shows the influence of the aging on the HC and CO emissions downstream of the filter. The figure shows the mean of two individual measurements.

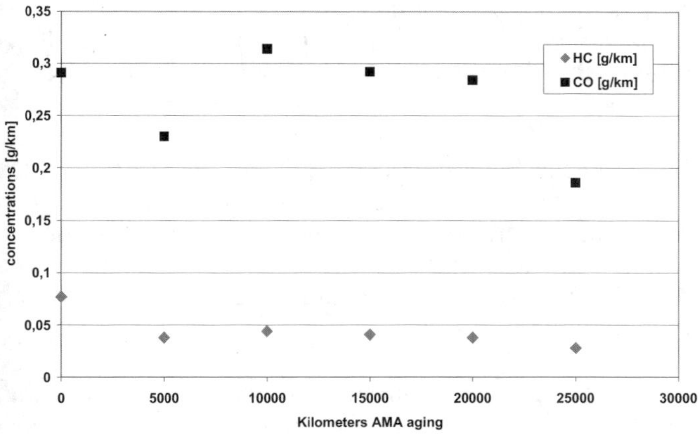

Figure 14: Change in HC and CO emissions behind the macrocoated filter over the run time

The average raw emissions of the vehicle were approx. 1.3 g/km for CO and 0.25 g/km for HC. Therefore, average conversion rates of about 80% for CO and 85% for HC were achieved. The results were thus substantially below the Euro IV limits.

Whereas the HC bag results after a first conditioning phase stabilized with values ranging from 0.028 and 0.044 g/km, the fluctuations in the CO emissions were significantly greater. However, no real aging behavior of the catalytic coating was actually observed. Instead, it can be presumed that the differences in the CO bag values resulted from different soot loadings on the filter.

To assess the influence of the soot loading on the performance of a catalytic filter, therefore, a second identically constructed system was measured on the same vehicle in new condition and after soot loading in the NEDC. The soot loading was performed on an engine test bench on the 2.2 l engine, whereby the macro-catalyzed filter was loaded with a defined amount of 10.3 g/l of soot and subsequently measured on the vehicle.

As shown in the results in Table 4, the soot loading had a significant influence on the emissions, particularly on CO. Consequently, the CO emissions increased after soot loading from 0.154 g/km to 0.257 g/km. The influence of the soot on the HC emissions, however, was much lower.

Table 4: Influence of the soot loading on HC and CO bag results of a macrocoated porous SiC filter with 84 g/ft^3 of platinum

	CO [g/km]	HC [g/km]
Raw emissions	1.27	0.23
CDPF, soot-free	0.154	0.024
CDPF with 10.3 g/l of soot	0.257	0.029

INFLUENCE OF ASH LOADING ON THE CATALYTIC PERFORMANCE OF A STAND-ALONE FILTER

The oil ash loading and distribution of the macrocoated filter aged for 25.000 km in the AMA cycle (see chapter above) was determined by using three-dimensional computer tomography with a 225 kV micro focus X-ray tube. The voltage was 220 kV, the current 150 µA. The pictures revealed very low amounts of ash only located on the rear side of the inlet channels. This kind of ash distribution is well known for active regenerating systems. In applications with only passive regenerating systems the ash is distributed more or less homogeneously over the filter length. Figure 15 shows a tomography of the AMA aged filter.

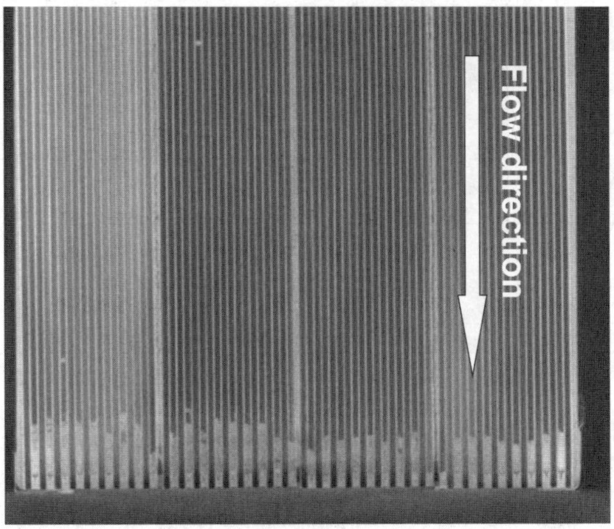

Figure 15: Oil ash distribution inside the macrocoated filter aged for 25,000 km in an AMA cycle

The ash amount in this filter was so low that a possible influence on the catalytic performance is negligible.

To investigate the impact of higher ash loadings a macrocoated filter (∅ 5.66" x 6", 300 cpsi, 42 % porosity, 40 g/ft³ platinum content) was loaded stepwise with a synthetic ash based on silica balls with a diameter of 90 – 150 µm.

The ash-free filter was first degreened for two hours at 625°C in an oven. Afterwards the HC and CO oxidation activity was determined in a stationary light-off test. For this test series, the 2.2 l engine described in a previous chapter was used. In the light-off test, the temperature in front of the filter was increased through appropriate adjustments of the engine load at a constant speed of 1200 rpm in increments of 10°C, beginning at 120°C; in the temperature range between 200°C and 350°C, the temperature was increased in 25°C steps. After the target temperature was reached, the system was conditioned respectively for 6 minutes before the exhaust gas concentrations were measured over 4 minutes.

The filter was filled-up progressively up to 3" with the silica balls in increments of 1". After every loading step the filter was calcined half an hour at 625°C to sinter the ash. Afterwards pressure-drop at 235 Nm and 3000 min^{-1} and light-off performance were measured on the engine bench. The results are reported in table 5.

Table 5: Influence of ash loading on pressure-drop and HC and CO light-off temperatures of a macro-coated SiC filter with 40 g/ft3 of platinum

Ash fill level ["]	$T_{50}(CO)$ [°C]	$T_{50}(HC)$ [°C]	Δp @ 360 kg/h, 520°C [mbar]
0	140	167	35
1	140	168	40
2	144	166	51
3	144	172	69

Despite the fact that the pressure-drop of the filter was increasing tremendously, the impact of the ash on the catalytic activity was rather small. Since with increasing ash loadings the mean residence time of exhaust gas in the outlet channels is rising as well there is no significant difference in the conversion of hydrocarbon and CO molecules flowing along the inlet or the outlet channels of the filter.

It can be expected that the influence of ash loading under dynamic test conditions is more critical as there can exist a temperature gradient between filter inlet and outlet channels. A similar test as described above will be performed soon on a vehicle measured in the New European Driving Cycle.

CONCLUSION

The regeneration of loaded soot filters presents a great challenge to the development of soot filter systems. The proposed solution described here – the integration of the oxidation catalyst on the filter (stand-alone CDPF) – offers the potential of ensuring reliable system functionality.

The catalytic requirements of a filter-only system, which combines the functions of oxidation catalyst and particulate filter, are high. The use of a "macrocoat" technology developed especially for the filter allows superior catalytic function and thus fulfills these high requirements.

Simulation and experimental studies show that with identical catalytic activity and reactor geometry a wall-flow filter systematically achieves higher conversion than the corresponding flow-through configuration. This improved performance of the catalyzed can be explained by some fundamental differences in reactor operation that are revealed by the simulation study.

It was shown that a particulate filter can represent a substrate that is basically suitable for a diesel catalyst coating. The regeneration capability of a stand-alone filter system is significantly improved in comparison to the combination of diesel catalyst and coated filter.

An aging study over a 25,000 km AMA cycle shows no loss of activity of the filter used.

The influence of higher oil ash loadings on the catalytic filter performance is rather small.

Volkswagen has announced the intent to introduce such a system on the market end of 2004 / beginning of 2005.

ACKNOWLEDGMENTS

The authors would like to thank Mr. Roger Staab, Mr. Christian Kühn, Mr. Pascal Adolph and Mr. Oliver Scherg for their support in the manufacture of the catalytic filter and in the analysis and evaluation of the measuring results. J.F.Knoth thanks Dr. A. Drochner and Prof. Dr.-Ing. Vogel from the Technical University of Darmstadt for their support.

REFERENCES

[1] Duvinage, F., Nolte, A., Paule, M., DaimlerChrysler AG, "Diesel Particulate Filters for Passenger Cars – Yesterday, Today, and Tomorrow," *Emission Control, Dresden 2002,* 271 – 284

[2] http://222.vcd.org/, "Diesel with Filter"

[3] E. Jacob et al, 23 *Vienna Intern. Engine Symposium*, VDI Progress Reports Series 12, No. 490, Volume 2, 2002, 196 – 216

[4] Salvat, O. Marez, P., Belot, G., "Passenger car serial application of a particulate filter system on a common rail direct injection diesel engine," SAE Paper 2001-01-0473

[5] Schommers, J., Enderle, C., Binz, R., Duvinage, F., Ruzicka, N., DaimlerChrysler AG, "Dieselpartikelfilterkonzepte für Pkw in Verbindung mit der Abgasstufe Euro IV", 25th *Vienna Intern. Engine Symposium* 2004, Edition 1500, M. L. Gaukler Werbeagentur Stuttgart

[6] Vogt, C. D., Wiehl, J., Hashimoto, S., Kordesch, V., Waras, H., „Regeneration Behaviour of SiC DPF", *Forum Partikel Emissionen AVL*, 24. – 25.09.2002 Nürnberg, 266

[7] Gieshoff, J., Pfeifer, M., Schäfer-Sindlinger, A., Hackbarth, U., Teysset, O., Colignon, C., Rigaudeau, C., Salvat, O., Krieg, H., Wenclawiak, B. W., "Regeneration of Catalytic Diesel Particulate Filters," SAE 2001-01-0907

[8] Spurk, P. C., Pfeifer, M., van Setten, B., Hohenburg, G., Gietzelt, C., "Examination of Engine Measuresfor the Regeneration of Catalytic Activated Diesel Particulate Filters in Commercial Vehicles" 24th *Vienna Engine Symposium* May 15-16, 2003, VDI Series 12, No. 539 Volume 1; Page 337 – 358

[9] Dorenkamp, R., Volkswagen AG, "Massnahmen zur Emissionserfüllung bei Diesel-Pkw," *FAD Dresden 2003*

[10] Schäfer-Sindlinger, A., Vogt, C. D., Hashimoto, S., Hamanaka, T., Matsubara, R., *Auto Technology*, Volume 5, 2003, 64 - 67

[11] Opris, C., Dissertation, Michigan Technological University, 1997

[12] Voltz, S., Morgan, C., Liederman, D., Jacob, S., *nd. Eng. Chem. Prod. Res. Develop.,* Volume 12, 1973, 294-301

[13] Fitzer, E., Fritz, W., Emig, G., "Technische Chemie", 4.edition, Springer-Verlag, 1995

CONTACT

Dr. Marcus Pfeifer
Umicore AG & Co. KG

e-mail: marcus.pfeifer@eu.umicore.com

Rodenbacher Chaussee 4
P.O. Box 1351
D-60453 Hanau-Wolfgang

Germany

2002-01-0428

Optimising the Low Temperature Performance and Regeneration Efficiency of the Continuously Regenerating Diesel Particulate Filter (CR-DPF) System

Ronny Allansson, Philip G. Blakeman, Barry J. Cooper, Howard Hess, Peter J. Silcock and Andrew P Walker
Johnson Matthey, Catalytic Systems Division

ABSTRACT

As legislation tightens in the Heavy Duty Diesel (HDD) area it is essential to develop systems with high activity and excellent durability for both Particulate Matter (PM) and NOx control. The Continuously Regenerating Trap (CRT™) system controls hydrocarbon (HC), CO and PM emissions from HDD vehicles with efficiencies of over 90%, and has demonstrated very good field durability over distances exceeding 700,000 km. The system is widely used in Europe, and is demonstrating the same high performance and excellent durability within field applications in North America. The Continuously Regenerating Trap (CRT™) system has been developed and patented by Johnson Matthey [1]. Throughout this paper this system will be referred to as the Continuously Regenerating Diesel Particulate Filter, CR-DPF.

The CR-DPF comprises an oxidation catalyst, optimised for NO_2 generation from the engine-out NOx, and a downstream DPF. The DPF collects the PM, and the NO_2 generated over the catalyst combusts this PM (carbon) at temperatures much lower than is the case for combustion with oxygen (250°C for the NO_2-carbon reaction, and 550°C for the O_2-carbon reaction). The oxidation catalyst is also very efficient at converting the engine-out CO (into CO_2) and HC (into CO_2 and H_2O).

The current CR-DPF system uses a DPF without a catalytic coating. Within this work we have compared the respective performance of the CR-DPF, a Catalysed Soot Filter (CSF, comprising a Pt-based catalyst coated onto a DPF), and an Oxicat + CSF system. The work presented shows that the low temperature performance of the CR-DPF system is significantly better than that of the CSF, and that the best low temperature regeneration is obtained by using the Oxicat + CSF system. In addition, it is shown that the efficiency of the reaction between NOx and carbon in the filter can be significantly increased by applying a catalytic coating to the filter.

These results demonstrate that it is possible to widen the operating window of the CR-DPF system, and clearly reveal that the preferred PM control system is the Oxicat + CSF system, in which the CSF has an optimised catalytic coating.

INTRODUCTION

The progressive tightening of the emissions standards for Heavy Duty Diesel vehicles throughout the world presents challenges for the engine development and emission control communities. In Europe, the major challenges are in PM control for 2005 (Euro IV) and in both NOx and PM control for 2008 (Euro V), while in North America the 2007 limits present major challenges in both NOx and PM control. Table 1 summarises the evolution of the emissions legislation in Europe, and Table 2 outlines the North American regulations.

Table 1: Emissions Legislation Limits in Europe (ESC Test Cycle; g $kW^{-1}hr^{-1}$)

Year	HC	CO	NOx	PM
2000 (Euro III)	0.66	2.1	5.0	0.10
2005 (Euro IV)	0.46	1.5	3.5	0.02
2008 (Euro V)	0.25	1.5	2.0	0.02

Table 2: Emissions Legislation Limits in North America (ESC Test Cycle; g $bhp^{-1}hr^{-1}$)

Year	HC	CO	NOx	PM
1998	1.3	15.5	4.0	0.10
2004	0.5	15.5	2.5	0.10
2007	0.14	15.5	0.2	0.01

The HC and CO emission limits do not present any significant problem, because the engine-out emissions of these pollutants are already very low. However, major reductions in PM and NOx emissions are required in the near future. In Europe in 2005, the legislative limits will tighten by a factor of 5, from 0.10 g $kW^{-1}hr^{-1}$ to 0.02 g $kW^{-1}hr^{-1}$ over the European Steady state Cycle (ESC). In addition, the NOx limit will decrease from 5.0 to 3.5 g $kW^{-1}hr^{-1}$. It is likely that for many systems this reduction in NOx emissions will be achieved using Exhaust Gas Recirculation (EGR), but, while the application of EGR will decrease the engine-out NOx emissions, it will increase the PM emissions of the engine. Therefore, the major challenge is to control the PM emissions of the system. In North America in 2007 the emissions limits will require the implementation of both PM and NOx control technologies. This paper focuses on PM control, which will be central to emission compliance in both Europe and North America.

One system which has shown great promise in controlling PM emissions is the CR-DPF. There are currently over 15,000 CR-DPF systems in operation throughout the world, and the long-term field durability of the system has been clearly demonstrated. Indeed, systems have been shown to provide very high levels of CO, HC and PM reduction even after over 500,000 km (and six years) of field operation [2].

The CR-DPF system comprises an oxidation catalyst followed by a wall-flow Diesel Particulate Filter (DPF), as shown schematically in Figure 1.

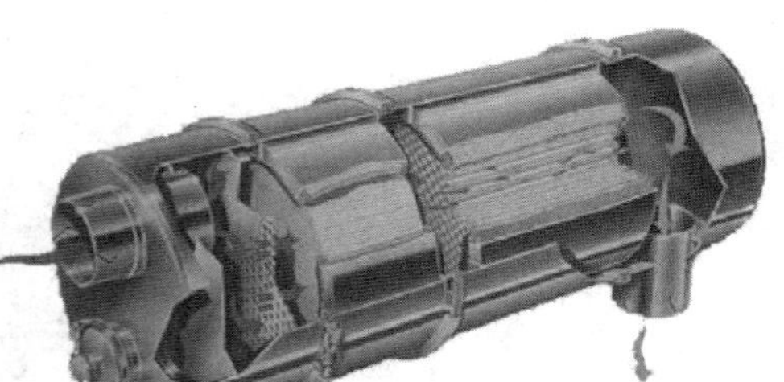

Figure 1: Schematic Representation of the CR-DPF System

The DPF traps the PM and the oxidation catalyst oxidises a portion of the engine-out NO to generate NO_2. This NO_2 combusts the PM at a much lower temperature than does oxygen, as shown in Figure 2.

The data shown in Figure 2 were collected using a laboratory reactor into which a controlled amount of PM (0.02 g) was loaded. The PM (soot) was obtained from a Euro II heavy duty diesel engine. The temperature of the sample was increased gradually while the oxidising gas feed was passed over the system at 50 ml/min and the amount of CO_2 generated from the combustion of the PM was monitored using a mass spectrometer. It can be seen that the onset temperature for PM combustion using NO_2 is approximately 200°C - much lower than the 450-500°C required for the onset of combustion using O_2.

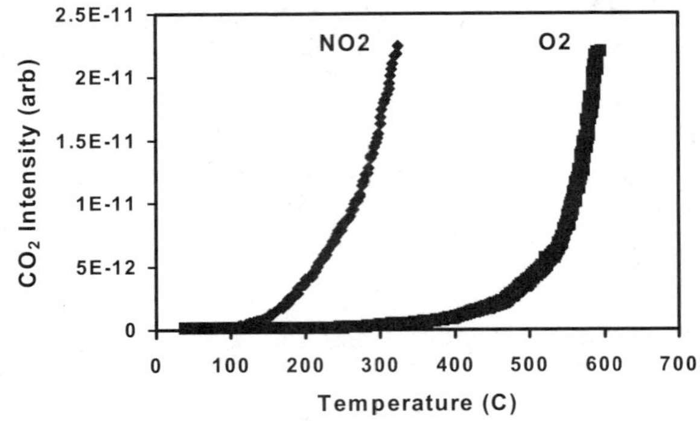

Figure 2: Comparison of the Combustion Temperature of PM in NO_2 and O_2

The CSF also uses NO_2, but less effectively than the CR-DPF (see below). In this system, a Pt-based oxidation catalyst is coated onto the walls of the DPF. This Pt generates NO_2, which can react with the carbon accumulated around it in the filter. So this system also operates by using the low temperature NO_2-carbon combustion reaction.

Within the CR-DPF system, the oxidation catalyst, which is optimised for the NO oxidation reaction, also has very high activity for both CO and HC oxidation. Table 3 shows the typical performance of the CR-DPF system over the European Stationary Cycle (ESC) on a Euro I engine using Swedish MK1 fuel containing 2-3 ppm S.

Table 3: Performance of the CR-DPF System on a Euro I Engine (ESC Test Cycle; g $kW^{-1}hr^{-1}$)

	HC	CO	NOx	PM
Engine-Out	0.162	0.989	7.018	0.163
Tailpipe-Out	0.003	0.002	6.874	0.008
Euro IV Limits	*0.460*	*1.500*	*3.500*	*0.020*

The HC and CO are reduced to levels which are barely detectable, and the PM is reduced to a level which is comfortably below the 2005 Euro IV legislated level, even on this engine with relatively high emissions. The NO_2 generated by the oxidation catalyst is converted back into NO when it reacts with carbon, so very little NOx conversion is observed. The small amount of NOx conversion observed is believed to be due to the reaction of a small proportion of the NO_2 with carbon in the filter to produce nitrogen, since no N_2O was observed downstream of the DPF in any of these experiments.

The temperature dependence of the NO oxidation activity of a Pt-based oxidation catalyst optimised for NO_2 generation is shown in Figure 3 [3].

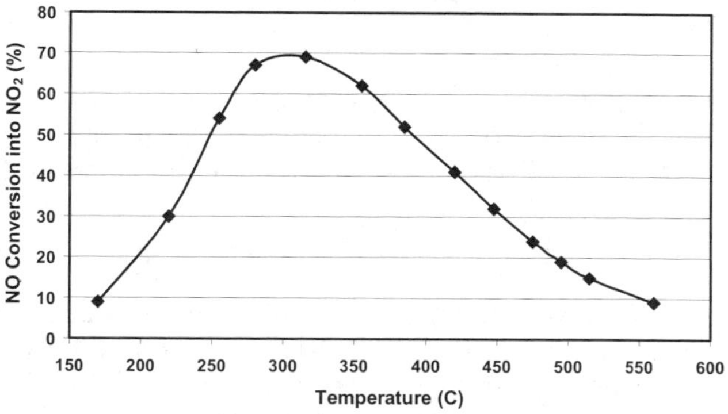

Figure 3: Temperature Dependence of the NO Oxidation Activity of a Pt-based Catalyst

It can be seen that the temperature for the onset of significant NO oxidation activity is approximately 200°C. The activity then rises rapidly before the NO to NO_2 conversion passes through a maximum at approximately 300°C. The NO conversion then decreases as a result of thermodynamic constraints. Note, however, that while the NO oxidation activity falls below 20% at 500°C, this is the temperature at which the oxygen can start to combust the PM. So, at the point at which the NO_2-PM reaction becomes less effective because relatively little NO_2 can be generated by the system, the O_2-PM reaction takes over. In addition, when the engine is generating temperatures of the order of 500°C in the exhaust, the engine-out NOx level is very high, so even a 20% conversion of this NO into NO_2 will provide a high absolute level of NO_2 to the filter. In practice, therefore, the CR-DPF system can provide effective PM control at all temperatures from approximately 270°C upwards. (This lower temperature limit is defined by the capability of the oxidation catalyst to generate NO_2 at low temperatures and by the reaction rate between NO_2 and carbon at low temperature.)

As outlined above, the CR-DPF has been used very successfully in the retrofit market over a wide range of operating duty cycles, since most heavy duty applications have duty cycles which are sufficiently warm to guarantee continuous regeneration of the CR-DPF system. However, there are some applications (eg garbage trucks) in which the operating duty cycle is sometimes too cold to ensure continuous regeneration of the CR-DPF system. Therefore, it is desirable to extend the effective operating window of the CR-DPF system to lower temperatures, such that the system will work passively over such duty cycles. Within OE applications of the CR-DPF the system will be integrated into the engine management system of the vehicle, such that if the duty cycle is sufficiently cold that the CR-DPF does not regenerate effectively (which will be assessed by an observed increase in back pressure) an active regeneration strategy will be triggered (see eg [4]).

In this paper the performance of the CR-DPF has been compared with that of a Catalysed Soot Filter (CSF; in which a Pt-based catalyst is coated onto a wall-flow particulate filter) and of a system comprising a CR-DPF incorporating a CSF (ie Oxicat + CSF). The low temperature regeneration characteristics of the three systems have been investigated, as have their respective regeneration rates at higher temperatures. Based upon these results, a new CR-DPF system with improved low and high temperature operating characteristics has been developed.

EXPERIMENTAL

CATALYST DETAILS

The catalysts used were coated onto ceramic monolith substrates with a cell density of 62 cells.cm^{-2} (400 cells.in^{-2}) using standard processing techniques. The geometric dimensions of the catalysts used for the engine evaluations, given as diameter x length, were 266.7 x 152.4 mm (10.5 x 6 in) with a wall thickness of 0.15 mm (0.006 in). The cordierite particulate filters used were production standard types, with 15 cells.cm^{-2} (100 cells.in-2) and with geometric dimensions (again given as diameter x length) of 266.7 x 304.8 mm (10.5 x 12 in), with a wall thickness of 0.43 mm (0.017 in). The Catalysed Soot Filter (CSF) was made by coating a Pt-based catalyst onto the cordierite particulate filter, using standard processing techniques. The Pt loading on the CSF was higher than that within the CR-DPF. The combined system, comprising an Oxicat + CSF in the CR-DPF configuration, was generated by using the oxidation catalyst from the CR-DPF system in combination with the CSF.

BENCH ENGINE EVALUATION

The ESC evaluations were carried out on a 12 litre, 310 kW turbocharged aftercooled engine using standard procedures. The engine was operated for the prescribed time in each mode, and completed the engine speed and load changes in the first 20 seconds at each mode. The sampling times were at least 4 seconds per 0.01 weighting factor. One PM sample was taken over the complete test cycle by means of a dilution tunnel. The filter papers were stabilised under controlled temperature and humidity conditions for 24 hours prior to the testing and then installed for the test. After the test cycle had been completed, the filter papers, with the collected PM, were again stabilised before being weighed.

The Balance Point Temperature (BPT) evaluations were carried out on a 9.0 litre, 250hp engine using fuel with a sulfur content of 3ppm. The systems were pre-loaded overnight with PM at low temperature (225°C) until the back pressure was approximately 2" Hg above that of the clean system. The engine was then operated at a constant speed of 1580 rpm and the load of the engine was increased to raise the temperature of the system. The experiments were started at 210°C, and the temperature was subsequently increased in 20°C increments until a temperature of around 370°C was reached. The system was held at each temperature for 15 minutes, and the back pressure of the system was recorded continuously. The BPT is defined as the temperature at which the rate of change of back pressure with respect to time is zero. That is, the temperature at which the rate of soot combustion in the filter is equivalent to the rate of soot accumulation.

The Low Temperature Cycle evaluations were carried out on a 10 litre, 210 kW turbocharged European bus engine in a cell with multiple gas stream analysis. The high temperature (350°C) regeneration experiments were performed on a 12 litre, 310 kW turbocharged aftercooled engine in a cell with the same analytical facilities. Swedish MK1 fuel (fuel sulfur level < 10 ppm) was used during these tests. The fuel was purchased from a standard refinery source. The engine oil used during the evaluations was Shell Myrina 15W 40 (S level ~ 0.5%).

Gaseous emissions were measured raw before and after the catalyst system using a range of analysers: chemiluminescence for NO and NO_2, heated FID (Flame Ionisation Detector) for HC, NDIR (Non-Dispersive Infra Red) analysers for CO, CO_2 and N_2O, and a paramagnetic detector for O_2.

In some experiments the gas phase composition was measured between the catalyst and DPF, as well as after the DPF. In every such experiment it was observed that the vast majority (~95%) of the NO_2 consumed by reaction with carbon in the filter was converted back into NO. No N_2O was observed downstream of the DPF. Therefore, it is inferred that the small amount (up to 5%) of NO_2 which was not reduced to NO, must have been reduced to N_2.

RESULTS AND DISCUSSION

BALANCE POINT TEMPERATURE MEASUREMENTS

The Balance Point Temperature (BPT) test provides an indication of the performance of filter-based systems, particularly at low temperatures. Specifically, the BPT is defined as the temperature at which the rate of soot combustion within the filter system is equivalent to the rate of accumulation of soot in the filter. Therefore, a low BPT is desireable, since this indicates good low temperature system performance. The BPT test was performed on each of the three systems: CR-DPF (ie oxidation catalyst followed by bare DPF), Catalysed Soot Filter (CSF) alone, and Oxicat + CSF. Figure 4 shows a typical balance point temperature trace; this one was obtained over the CR-DPF.

As expected, at low temperatures the back pressure increases slightly since the temperature is not high enough to enable the catalyst to convert a significant proportion of the NO into NO_2; nor is the filter hot enough for the NO_2-carbon reaction to proceed at an appreciable rate.

As the temperature is increased the rates of both these reactions increase, so the catalyst makes more NO_2. This NO_2 reacts effectively with the carbon stored in the filter. The overall consequence is that the rate of change of back pressure of the system first becomes zero as a function of time; at this point the rate of soot accumulation in the filter is equivalent to the rate of soot combustion due to the NO_2-carbon reaction in the filter. This occurs at a temperature of around 265°C, and is defined as the balance point temperature of the CR-DPF under these conditions.

As the temperature is further increased the back pressure of the system decreases, as the rate of carbon consumption in the filter exceeds the rate of carbon accumulation (ie the carbon accumulated during both the initial low temperature pre-loading and during the early (cold) part of the experiment is burned out). Further increases in temperature result in a substantial increase in the rate of soot combustion, which leads to the high rate of back pressure decrease observed at temperatures of around 325°C and above.

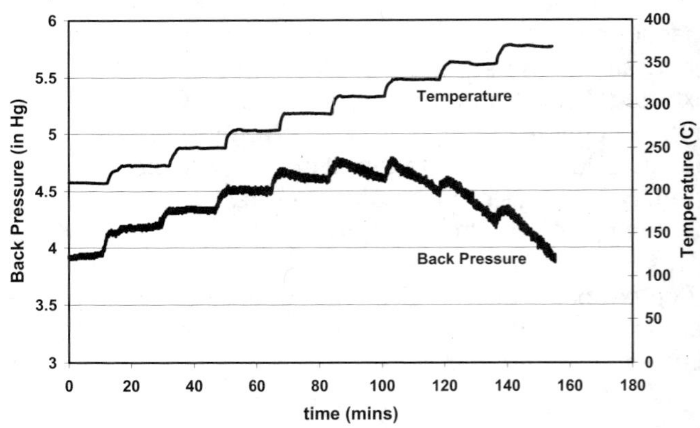

Figure 4: Balance Point Temperature Data Obtained Over the CR-DPF

The BPT test was also carried out for the CSF alone and for the Oxicat + CSF system. For each system the gradient of the back pressure profile was calculated at each temperature, to provide an indication of the rate of back pressure change as a function of inlet temperature. These data are plotted in Figure 5.

The data shown in Figure 5 can be used to extract a formal Balance Point Temperature, since the temperature at which the profile crosses the temperature axis is the BPT. Table 4 summarises the Balance Point Temperatures of the three systems studied.

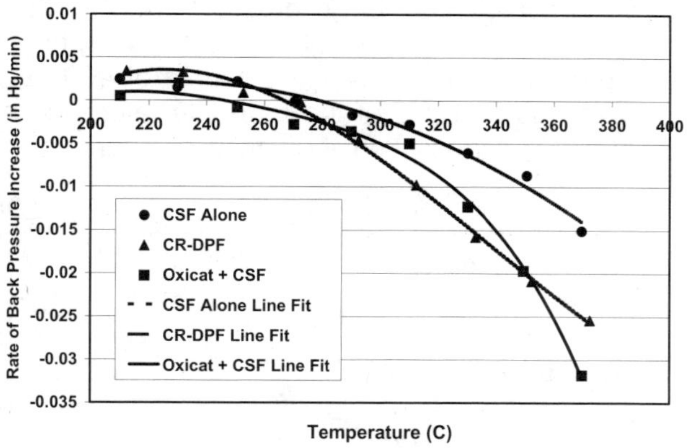

Figure 5: Rate of Increase of Back Pressure as a Function of Temperature for the Three Systems

Table 4: Balance Point Temperatures of the CR-DPF, the CSF Alone and the Combined System Comprising Oxicat + CSF

System	Balance Point Temperature
CR-DPF	265°C
CSF Alone	280°C
Oxicat + CSF	250°C

It can be seen that the ranking of the systems (with the best-performing system first) is:

Oxicat + CSF > CR-DPF > CSF alone

The superior performance of the CR-DPF over the CSF alone is consistent with the results obtained in the DECSE program [5], and with those obtained during Swiss studies of different PM control technologies [6].

This superior low temperature performance can be explained by considering the reaction probability of the NO_2 with the carbon in the different systems. Within the CR-DPF the NO_2 is generated upstream of the filter, and this gas phase NO_2 then passes through the whole of the carbon layer before being released into the atmosphere. This passage through the carbon layer provides numerous NO_2-carbon collisions, resulting in a relatively high reaction probability. In contrast, within the CSF, the NO_2 is generated by the Pt coated on the walls of the filter, ie adjacent to the carbon on the walls of the filter, and underneath the carbon that has built up within the filter. This presents the possibility of far fewer collisions between the NO_2 molecule and the carbon species, which leads to a correspondingly lower reaction probability.

The combined system, ie Oxicat + CSF, has the best performance of the three systems. This is because this system combines the good low temperature performance of the CR-DPF system (since again the NO_2 passes through the whole of the carbon layer) with the ability to maximise the usage of the available NO_2. Within the CR-DPF system the NO_2 molecule generated by the upstream oxidation catalyst can only be used once to burn carbon, since the reaction proceeds according to:

$$2\,NO_2 + C \rightarrow 2\,NO + CO_2 \qquad (1)$$

That is, the NO_2 reacts with the carbon and is converted back to NO. However, within the Oxicat + CSF system there is the possibility that the NO generated by this reaction can be re-converted into NO_2 by the Pt on the walls of the filter, and that this NO_2 can react with carbon adjacent to it on the walls of the filter. Therefore, the NO_2 generated upstream on the oxidation catalyst reacts with the carbon towards the top of the accumulated carbon layer within the filter, and the NO_2 is converted back into NO. This NO is subsequently re-oxidised into NO_2 by the Pt on the walls of the filter, where it can react with carbon adjacent to it. In this way, each NO molecule produced by the engine can potentially be used more than once by the Oxicat + CSF system. This leads to the observed superior low temperature performance with respect to the other systems studied here.

Note that the reaction given in Equation (1) represents the dominant stoichiometry occurring within the CR-DPF system. This stoichiometry is inferred because the alternative reaction (Equation (2) leads to CO formation:

$$NO_2 + C \rightarrow NO + CO \qquad (2)$$

Only very low levels of CO were observed downstream of the DPF in the CR-DPF system, so the dominant reaction stoichiometry within this system is that defined by Equation (1).

LOW TEMPERATURE CYCLE PERFORMANCE

The Balance Point Temperature measurements indicate that the best performing system at low temperature is the Oxicat + CSF system. In order to further probe the low temperature regeneration performance of the three systems a Low Temperature Cycle was developed to simulate the transient low temperature conditions which might occur on eg a slow moving city centre bus. This engine bench cycle comprises a number of engine speed-load points, and is shown schematically in Figure 6.

The temperature profiles observed at the inlet to the catalyst, the outlet of the catalyst and the outlet of the filter during the Low Temperature Cycle are shown in Figure 7. Note that the average temperatures at each location in the exhaust stream are very similar (before catalyst = 227°C, after catalyst = 231°C and after filter = 231°C).

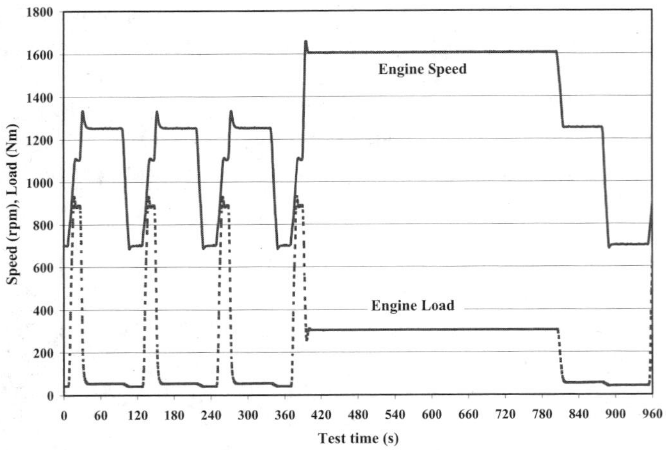

Figure 6: Engine Speed and Load Profiles During the Low Temperature Cycle

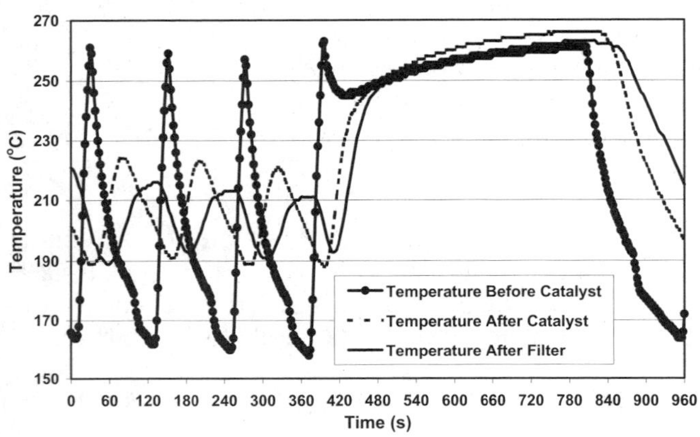

Figure 7: Temperature Profiles During the Low Temperature Cycle

The cycle comprises a transient phase followed by a steady state phase. The maximum temperature observed during the Low Temperature Cycle (LTC) is 260-270°C. The temperature profiles demonstrate the thermal inertia characteristics of the system. The catalyst inlet temperature profile is highly transient during the first part of the cycle, but most of these thermal variations are damped out by the thermal mass of the catalyst and the filter, such that the amplitude of temperature variation is much lower at the filter outlet. During the second, steady state phase of the cycle (ie between about 400 and 800s), all of the temperatures approach the same level, as expected.

The key feature, however, is that the temperatures throughout the cycle are low, which makes it a suitable cycle over which to test the low temperature performance characteristics of the three filter-based systems.

Figure 8 compares the back pressure of the three systems (CR-DPF, CSF alone, and Oxicat + CSF) during repeated runs of the LTC, over the course of 40 hours. The back pressure profile of the bare filter alone (ie no catalyst in the system) is also shown, for comparison.

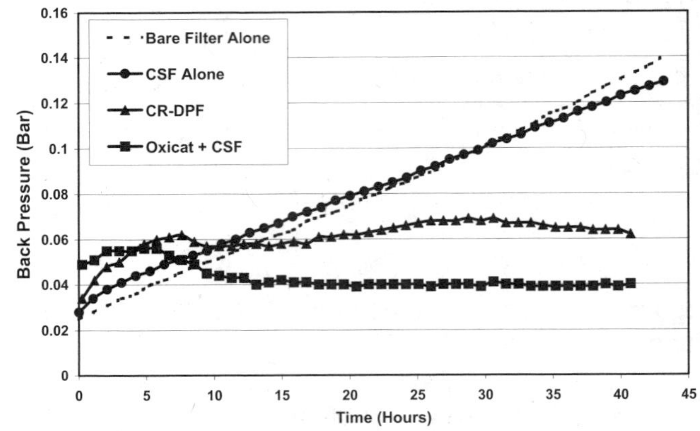

Figure 8: Back Pressure of the Bare Filter Alone, CR-DPF, CSF Alone and Oxicat + CSF Systems During Repeated Runs of the LTC Over 40 Hours

When no catalyst is present in the system, ie over the bare filter alone, the back pressure increases linearly with time as expected, since the soot is being collected in the filter and no regeneration is occurring. When the CSF is tested alone over this Low Temperature Cycle the back pressure also increases rapidly, with no sign of stabilisation. The rate of increase of back pressure is slightly lower than that observed over the bare filter alone, revealing that a very small amount of regeneration is occurring over the CSF. This observation is consistent with the balance point temperature measurements discussed above, where it was found that the balance point of the CSF alone is 280°C. Therefore, over such a low temperature cycle it is clear that the CSF is not capable of operating effectively.

The CR-DPF shows an initial increase in back pressure before stabilising at an operating back pressure of around 60 mbar. Therefore, it is clear that the CR-DPF operates effectively over this low temperature duty cycle, in marked contrast to the performance of the CSF.

The best performance, however, is observed over the combined system, comprising the Oxicat + CSF system in the CR-DPF configuration. The back pressure of this system starts off at around 50 mbar, before increasing slightly and then decreasing to a stable level of 40 mbar. The CSF used in this experiment was the same as was used previously in the "CSF alone" experiment, so it is most likely that the back pressure is initially a little higher than expected due to the incomplete cleaning of the CSF between the two experiments. (The CSF was cleaned by running the engine at a condition giving rise to a temperature of 550°C for around 20 minutes.) This combined Oxicat + CSF system results in a lower stabilised back pressure than that of the CR-DPF, revealing that this is the most effective PM control system studied here. Again, this result is consistent with the balance point temperature measurements, which showed that the Oxicat + CSF system gave rise to the lowest balance point, ie the best low temperature performance. The performance of the systems over the Low Temperature Cycle follows the ranking below, with the best-performing system first:

Oxicat + CSF > CR-DPF >> CSF Alone > Bare Filter

This ranking is identical to the one established for the balance point temperatures of the three filter-based systems above.

HIGH TEMPERATURE REGENERATION PERFORMANCE

It is also important that the PM control device can regenerate rapidly at elevated temperatures. This will be important for duty cycles in which the system spends long periods of time operating at relatively low temperatures, and then operates for a short period at high temperatures via eg highway driving. Within such a duty cycle the filter will accumulate soot under the low temperature conditions, and it will be beneficial if the system can remove this stored soot rapidly during the short period of time at high temperature. Rapid high temperature regeneration capability will also be particularly important under active regeneration conditions, eg following prolonged low temperature operation resulting in significant levels of soot build-up in the filter. Under such conditions the active regeneration strategy will be triggered, and the temperature of the system will be raised to promote the regeneration of the system. However, there will be a fuel economy penalty associated with such active regeneration measures, so it is clearly beneficial if the system regenerates rapidly under such conditions, to minimise the amount of time for which the active regeneration system needs to be deployed.

The regeneration characteristics of the three systems (CR-DPF, CSF alone and Oxicat + CSF) have been studied. Each of the filters was loaded with the same level of soot (corresponding to around 6 g $litre^{-1}$ of filter), by running the systems for long periods of time at low temperature (ie at temperatures at which significant filter regeneration did not occur). Note that the back pressure of the CSF only system is slightly lower than that of the CR-DPF and Oxicat + CSF systems. It is believed that this is due to the absence of a pre-oxidation catalyst in this system, since this oxidation catalyst will contribute to the overall back pressure of the system. The temperature was then rapidly increased to 350°C, and the evolution of the back pressure was monitored as a function of time. This temperature was chosen because it is regarded as a temperature which will be relatively easy to access within active regeneration strategies. The engine-out NOx level at this engine condition was approximately 620 ppm. Figure 9 shows the back pressure profiles observed.

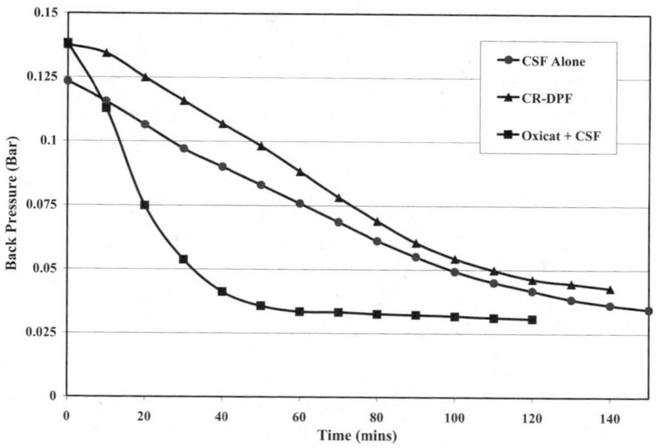

Figure 9: Back Pressure Reduction as a Function of Time at 350°C Over the Loaded CR-DPF, CSF Alone and Oxicat + CSF Systems

At these relatively high temperatures the CSF has equivalent regeneration performance to the CR-DPF, in contrast with the situation at lower temperatures, where the CR-DPF performs significantly better than the CSF alone. However, the rate of back pressure reduction observed over the Oxicat + CSF system is substantially higher than the other two systems. Under these conditions (which correspond to very high soot loadings in the filters), the CR-DPF and CSF alone take around 140 minutes to achieve their baseline back pressure, while the Oxicat + CSF system reaches the baseline back pressure (corresponding to a clean filter) after just 40 minutes. Therefore, the rate of regeneration at such a temperature is approximately 3 times faster over the combined Oxicat + CSF system than it is over the CR-DPF alone or CSF alone.

It is believed that the combined system regenerates at a much greater rate because of the possibility to re-use the NOx within such a system. That is, the oxidation catalyst generates NO_2 upstream of the filter and, at a temperature of 350°C, this NO_2 reacts rapidly with the accumulated carbon at the top of the carbon layer within the filter. This NO_2 is then converted back into NO. When the NO reaches the filter wall (after passing through the carbon layer) it can be oxidised by the Pt coated onto the wall of the filter, to re-form NO_2. This NO_2 can then react with carbon adjacent to it on the filter wall. In this way, the carbon is rapidly removed from both the top and the bottom of the accumulated carbon layer.

This increase in regeneration rate at high temperature is another significant advantage obtained by combining the Oxicat + CSF within the CR-DPF configuration.

CONCLUSIONS

Three filter-based PM control devices have been tested under three different operating conditions. The systems comprised: Continuously Regenerating Diesel Particulate Filter (CR-DPF) – oxidation catalyst followed by uncoated Diesel Particulate Filter (DPF); Catalysed

Soot Filter (CSF) – Pt-based catalyst coated onto the walls of a DPF; Oxicat + CSF - combined system comprising oxidation catalyst followed by CSF, in the CR-DPF configuration. The results can be summarised as follows:

- The Balance Point Temperatures (BPT) of the systems are ranked as follows (with the lowest BPT, ie best performing system, first):

  Oxicat + CSF < CR-DPF < CSF Alone

  That is, the Oxicat + CSF system has the best low temperature performance within the BPT test, followed by the CR-DPF, followed by the CSF alone. The superior low temperature performance of the CR-DPF with respect to the CSF is consistent with previous reports in the literature (see above).

- The performance of the three systems was studied over a Low Temperature Cycle, developed to simulate difficult (ie cold) operating conditions in the field. The back pressure of the systems was monitored, and performance was assessed based upon the stabilised back pressure observed after repeated running of the cycle. Over this cycle the performance of the systems ranked as follows, again with the best-performing system first:

  Oxicat + CSF > CR-DPF >> CSF Alone

  The stabilised back pressure observed over the Oxicat + CSF system was around 40 mbar, while that of the CR-DPF was around 60 mbar. The back pressure of the CSF alone increased rapidly over the Low Temperature Cycle, and showed no sign of stabilising. The experiment was terminated when the back pressure reached 130 mbar. Indeed, the rate of back pressure increase observed over the CSF was only slightly lower than that observed over a bare filter, indicating that the system was hardly regenerating at all over this low temperature cycle. These results are consistent with the ones obtained during the Balance Point Temperature measurements.

- The high temperature regeneration characteristics of the systems were studied by loading the filters with a high level of soot, and then monitoring the rate of decrease in back pressure under regeneration conditions at 350°C. Under these conditions the systems were ranked as follows (with the best performing system first):

  Oxicat + CSF > CR-DPF = CSF Alone

Overall, therefore, it is clear that the system comprising the Oxicat + CSF in the CR-DPF configuration is the best performing system under all conditions studied in this work. It is also clear that the low temperature performance of the CR-DPF is significantly better than that of the CSF; this difference manifested itself dramatically during the Low Temperature Cycle experiments. The improvements in low temperature performance offered by the Oxicat + CSF system in the CR-DPF configuration are particularly significant, since the developments reported in this paper should allow the application of filter-based PM control devices to vehicles that would normally be considered too cold even for CR-DPF application.

ACKNOWLEDGMENTS

The technical assistance of Mr Jim Harris and Mr Ari Uusimaki are gratefully acknowledged for their invaluable help during the engine bench measurements discussed in this paper. We are very grateful to the Johnson Matthey heavy duty diesel team for its input, as well as to Johnson Matthey plc for permission to publish this paper.

REFERENCES

[1] BJ Cooper, HJ Jung and JE Thoss, US Patent 4,902,487 (1990)

[2] R Allansson, BJ Cooper, JE Thoss, AJ Uusimaki, AP Walker and JP Warren, SAE 2000-01-0480, Detroit 2000

[3] JP Warren, R Allansson, PN Hawker and AJJ Wilkins, Proc. Diesel Engines - Particulate Control, IMechE 1998 S491/006 p.45

[4] R Verbeek, M van Aken and M Verkiel, SAE 2001-01-1947, Orlando 2001

[5] DECSE program, Final Report (January 2000), sponsored by the US Department of Energy, Engine Manufacturers Association and Manufacturers of Emissions Controls Association

[6] A Mayer, P Nothiger, R Zbinden and R Evequoz, SAE 2001-01-0187, Detroit 2001

CONTACT

Dr Andy Walker

Johnson Matthey, Catalytic Systems Division

walkeap@matthey.com

III. NO_x CONTROL

III.1 SCR SYSTEMS

2000-01-0189

Improved SCR Systems for Heavy Duty Applications

J. Gieshoff, A. Schäfer-Sindlinger, P. C. Spurk and J. A. A. van den Tillaart
Degussa-Hüls AG

G. Garr
ICT Inc.

ABSTRACT

This paper describes the function and application of the preoxidation, hydrolysis and SCR catalysts individually and as a combined system for urea SCR both in model gas and engine bench tests. Using the basic system and a non-optimized urea injection strategy 45% NO_x conversion was achieved in the ESC engine test. Adding a preoxidation catalyst significantly improved the NO_x conversion in the low temperature region of the engine mapping. NO_x conversions over 75% can be achieved in the ESC test using this improved system. With a 50% reduced SCR catalyst volume still a NO_x conversion of over 65% could be achieved. Tests after 200 hours engine aging show that the activity of the system is stable.

INTRODUCTION

In view of the US 2002/2004 and EU IV/V legislation, the major concern of exhaust gas aftertreatment on the future heavy duty diesel vehicles is reducing the NO_x and/or the particulate emissions (see figure 1). Several solutions for this problem seem to be technical feasible, e.g. various diesel particulate traps, cooled EGR. Most of these solutions, however, make a clear trade off between emissions and fuel consumption. Some of these are even associated with a distinct fuel penalty. Especially in the heavy duty market fuel efficiency is a major engineering target as the end users will see a clear effect on their operation costs. A fuel efficient mode typically gives high NO_x raw emissions and low particulate emissions as is visible from the NO_x-particulate trade-off curve. A SCR (Selective Catalytic Reduction) exhaust gas aftertreatment system using urea as reductant has high NO_x reduction potential. Therefor the engine can be operated fuel efficient, and the SCR system can reduce the emitted NO_x in most cases enough to meet legislation. Depending on the engine design, however, it can be expected that additionally also some form of particulate aftertreatment is needed.

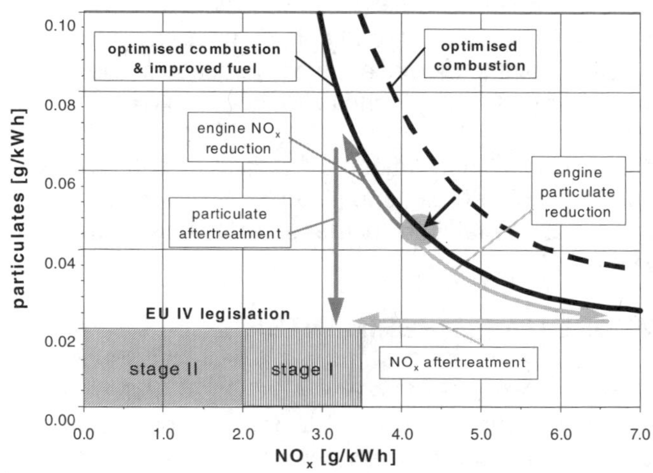

Figure 1. Trade–off curve between NO_x and particulate emissions for a typical Heavy Duty application together with different reduction routes and the future European legislative limits.

In this paper the technical feasibility of the SCR catalyst system will be shown and discussed. Other problems which have to addressed before the SCR system could be successfully applied, like e.g. injection strategy, injection system, urea availability and OBD (On Board Diagnosis) will not be discussed in this paper.

DESCRIPTION OF THE SCR SYSTEM

The basic SCR system (see figure 2), the HSO system, consists of three different catalysts in series after the urea injection point; a hydrolysis catalyst (H), a SCR catalyst (S) and a guard oxidation catalyst (O). The urea is sprayed evenly onto the hydrolysis catalysts which converts the urea with water selectively to ammonia (NH_3) and carbon dioxide (CO_2). Then the ammonia reacts on the SCR catalyst with the NO_x present in the exhaust gas to form nitrogen (N_2). Finally an oxidation catalyst is added as guard catalyst to avoid any secondary emissions of ammonia during dynamic operation.

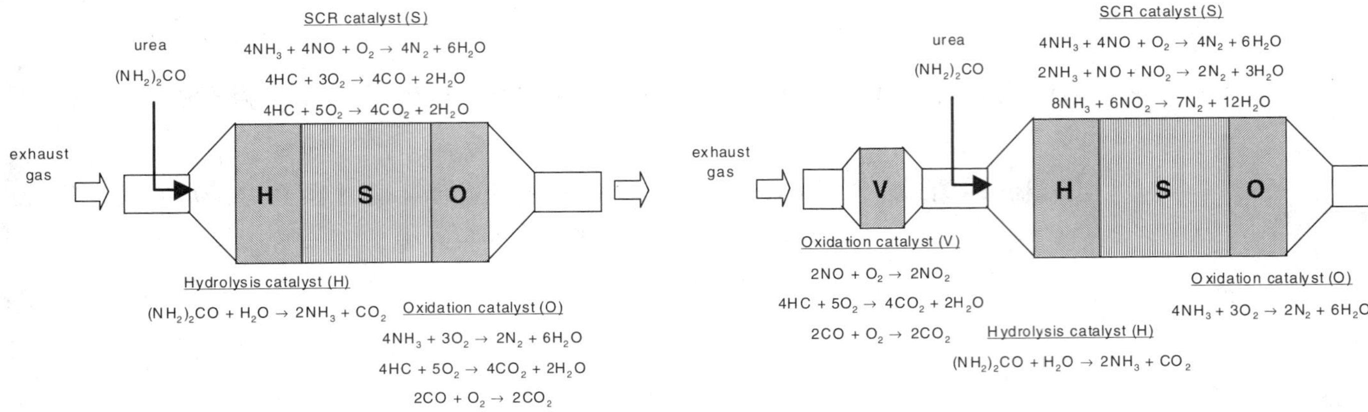

Figure 2. Schematic description of the proposed basic, HSO, and the more advanced, VHSO, SCR system together with the most important occurring reactions.

On the improved VHSO SCR system a preoxidation catalyst (V) is added (see figure 2) upstream of the urea injection point to increase the NO_2 fraction in the exhaust gas. Most of the NO_x in the exhaust gas of diesel engines is present in the form of NO as the NO_2 fraction in NO_x in diesel exhaust gas amounts typically to only 10 vol.-%. Due to a higher NO_2 content, the NO_x conversion activity on the SCR catalyst is increased at temperatures below ca. 300° C.

MODEL GAS EXPERIMENTS

SET-UP AND EXPERIMENTAL CONDITIONS – The equipment which was used to test the catalysts in the laboratory is schematically shown in figure 3. The set-up consists of a set of mass flow controllers for the various gasses used, a HPLC pump to control the feed of water, an evaporator in which a nitrogen stream can be saturated with a hydrocarbon, a quartz reactor with inset in which the catalyst core (∅:1" L:3") is mounted, a furnace to heat the reactor and a FTIR spectrometer to analyze both the inlet and outlet gas. The FTIR spectrometer is connected to the rest of the set-up by lines heated to 180° C. The pressure in the FTIR cell is kept constant. This set-up was used to determine the steady state conversion at a standard space velocity of 30000 hr^{-1} at different temperatures between 150° and 520° C using, unless referred otherwise, the standard gas composition given in table 1.

Table 1. Standard gas composition for model gas experiments.

component	concentration
NO	500 v.-ppm
NH_3	450 v.-ppm
H_2O	1.3 vol.-%
O_2	5 vol.-%
N_2	balance

In the model gas experiments ammonia was used to determine directly the activity of the SCR catalyst. The NH_3/NO_x molar ratio, alpha, was for the standard conditions 0.9, which essentially limits the expected maximum NO_x conversion to 90%. It was checked that the absence of CO_2 and CO did not influence the NO_x conversion over the SCR catalyst used in this study. The water concentration was kept relatively low (typically diesel exhaust gas contains ca. 6 vol.-% water) to minimize the problems in the quantitative FTIR gas analysis. In the absence of water the activity of the SCR catalyst is completely different from the activity in the presence of water. Above water concentrations of typically 0.5 vol.-% no effect of the water concentration is observed other than a dilution effect.

The SCR catalyst used in the model gas investigations reported here was a ∅:1", L:3" 400 cpsi/6.5 mil ceramic substrate coated with a $V_2O_5/WO_3/TiO_2$ formulation.

SCR CATALYST

Activity map – By varying the molar ratio of NH_3 to NO_x a map of SCR activity as function of temperature and alpha ratio can acquired. Figure 4 shows this map for the SCR catalyst at the standard conditions defined in the previous section. At temperatures above 300° C the NO_x conversion increases linearly with the alpha ratio as is expected for the stoichiometric reaction between NO and NH_3. At lower temperatures, however, the conversion increases only linearly with the alpha ratio up to a certain critical alpha ratio above which an increase in alpha ratio does not result in a significant increase in NO_x conversion. Moreover, increasing the alpha ratio above the critical alpha ratio results in a stationary NH_3 slip. Measured N_2O concentrations after the catalyst were always below 10 v.-ppm and resulted partially from impurities in the gases used and also from oxidation of NH_3 in the equipment itself.

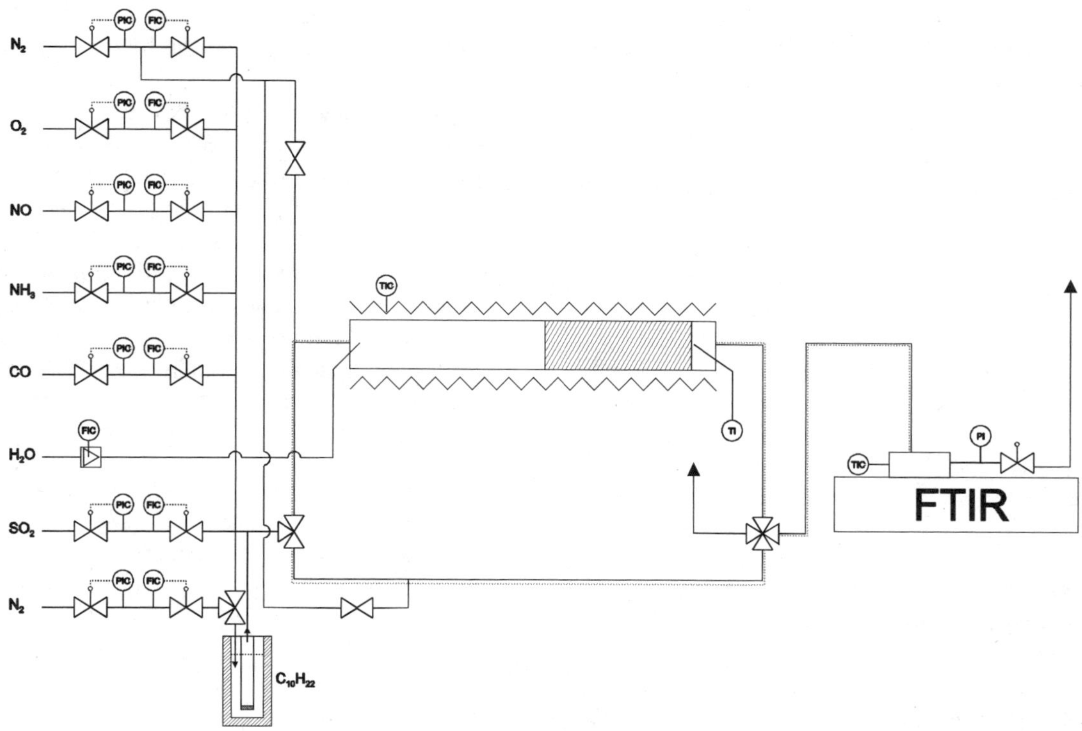

Figure 3. Schematic description of the model gas test set-up.

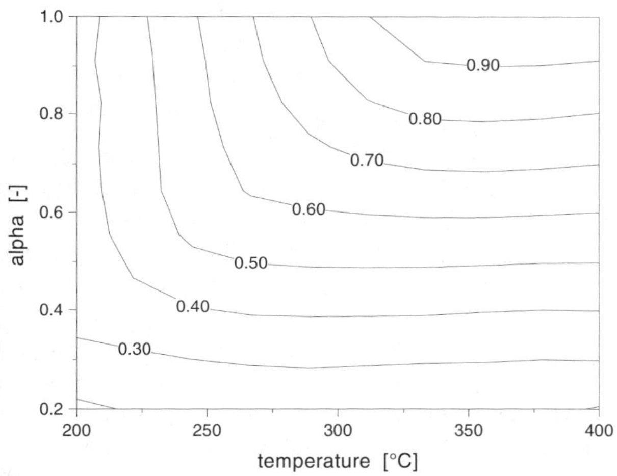

Figure 4. NO_x conversion as function of NH_3/NO_x ratio, alpha, and temperature for the SCR catalyst at standard conditions.

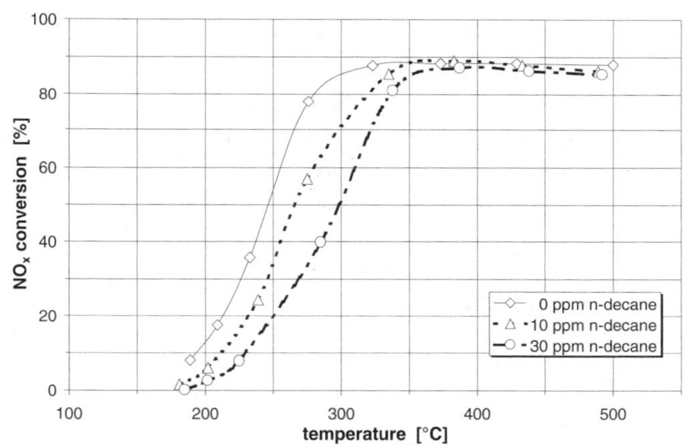

Figure 5. NO_x conversion as function of temperature at various n-decane concentrations for the SCR catalyst at standard conditions.

Effect of hydrocarbons – Figure 5 shows the effect of hydrocarbons on the NO_x conversion over the catalyst at standard conditions. In this case n-decane ($C_{10}H_{22}$) was taken as a model component. With other hydrocarbons similar effects were observed. It is clear that the curve for NO_x conversion versus temperature is shifted to higher temperatures with higher hydrocarbon concentrations. The n-decane is oxidized to CO and CO_2 over the catalyst between 250° and 400° C. However when no NH_3 is added no significant conversion of NO_x is observed, which leads to the conclusion that the hydrocarbons are not involved in the reduction of NO_x on the SCR catalyst.

Effect of NO_2 fraction in NO_x – The effect of increasing the NO_2 fraction of NO_x at a constant NO_x concentration of 500 v.-ppm was studied in the model gas setup at an increased space velocity of 60000 hr^{-1}. The results of these tests are visualized in figure 6. Upon increasing the NO_2 content up to 50 vol.-% a clear increase in the low temperature activity is observed. At higher contents, however, the high temperature activity and the maximum NO_x conversion are reduced significantly and stationary NH_3 slip is observed. It is clear that at NO_2 contents of up to 50%, NO_x can be converted more efficiently at lower temperatures and higher space velocities than without NO_2. At higher contents however the higher stoichiometry needed to convert NO_2, i.e. 1.33 compared to 1 for NO, and the slower reaction of NO_2 with NH_3 which results in NH_3 slip, limits the maximum conversion without secondary emissions.

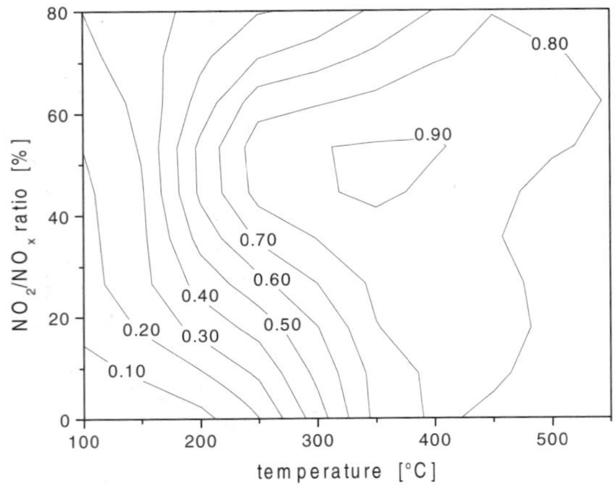

Figure 6. NO_x conversion as function of NO_2/NO_x ratio and temperature for the SCR catalyst at a space velocity of 60000 hr^{-1}.

Stability – To determine the thermal stability of the SCR formulation, cores were aged for 100 hours in a furnace in air containing 10 vol.-% water and 20 v.-ppm SO_2 at temperatures of 650°, 700° and 750° C. Figure 7 shows clearly that this aging procedure up to aging temperatures of 700° C has no significant effect on the activity. At higher temperatures the activity of the catalyst is dramatically reduced. At these temperatures the V_2O_5 catalyzes the phase transition of the high surface area anatase TiO_2 to the low surface area rutile TiO_2.

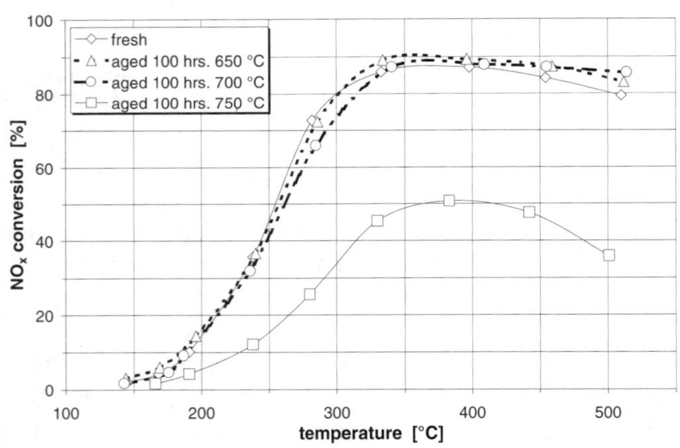

Figure 7. NO_x conversion as function of temperature after furnace aging (in flowing air containing 10 vol.-% water and 20 v.-ppm SO_2) for the SCR catalyst tested at standard conditions.

PREOXIDATION CATALYST – Using a preoxidation catalyst it is possible to increase the NO_2 fraction in the NO_x up to the maximum given by the thermodynamic equilibrium between NO and NO_2 in O_2. Figure 8 gives the thermodynamic equilibrium for different O_2 concentrations together with the composition after an oxidation catalyst (containing 50 g/ft^3 Pt) as function of temperature. The gas at the inlet contained 6 vol.-% O_2, 270 v.-ppm NO, 10 vol.-% H_2O and N_2 as balance. The space velocity during this test was 50000 hr^{-1}. At temperatures above 280° C this catalyst completely equilibrates the gas phase and hence the NO_2 content is thermodynamically controlled. At lower temperatures the activity of the catalyst is not sufficient to convert enough NO to reach the thermodynamic equilibrium and the NO_2 content is kinetically controlled. At temperatures below 150 °C Pt based oxidation catalysts are not active enough to give any significant oxidation of NO to NO_2 at space velocities typical for automotive exhaust aftertreatment.

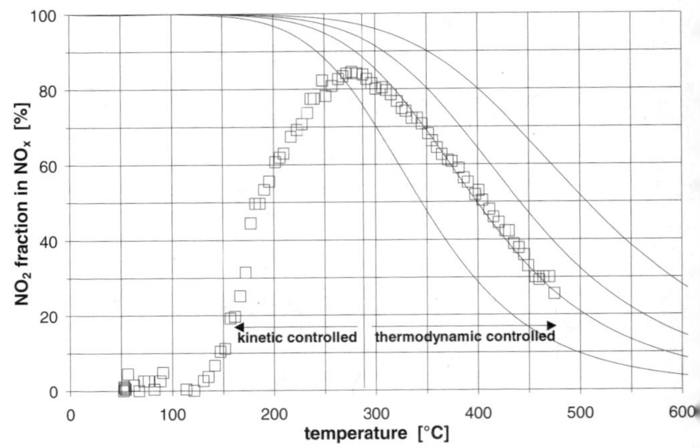

Figure 8. NO_2 fraction in NO_x after an 50 g/ft^3 Pt oxidation catalyst with at inlet 270 v.-ppm NO, 6 vol.-% O_2, 10 vol.-% H_2O in N_2. In full lines the thermodynamic NO/NO_2 equilibria are depicted in 1, 6, 20 and 100% O_2 respectively at atmospheric pressure.

If hydrocarbons and CO are present, NO_2 formation will be inhibited and start to occur only at higher temperatures. Moreover, as NO_2 will react preferentially with hydrocarbons and CO, the NO_2 fraction in the exhaust gas could even be decreased when the catalyst is not correctly dimensioned.

COMBINATION OF PREOXIDATION CATALYST WITH SCR CATALYST – A few tests have been run on the model gas setup with a combination of preoxidation catalyst (90 g/ft^3 Pt) and a SCR catalyst. The preoxidation catalyst was heated to a temperature 30° C higher than the SCR catalyst in a separate furnace upstream of the furnace with the SCR catalyst at a space velocity of 60000 hr^{-1} to simulate typical application conditions. The NH_3 was injected after this preoxidation catalyst. The SCR catalyst was tested at a space velocity of 30000 hr^{-1} at different temperatures and alpha ratios. The results in figure 9 show that now, even at temperatures below 300° C, the NO_x conversion increases linearly with the NH_3/NO_x ratio.

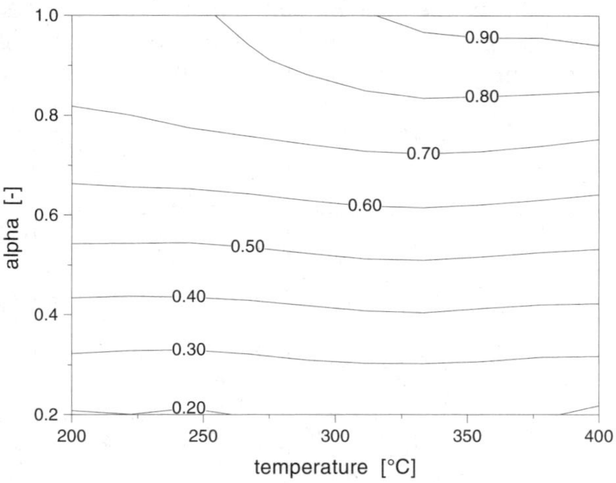

Figure 9. NO_x conversion as function of NH_3/NO_x ratio, alpha, and temperature for the SCR catalyst at standard conditions with a NO_2 content in the gas determined by a preoxidation catalyst (90 g/ft^3 Pt) at a temperature 30° C higher than the SCR catalyst and at a space velocity of 60000 hr^{-1}.

Compared to the measurements were only NO was present (see figure 4) the maximum conversion at low temperatures was increased by typically 30-40%.

DISCUSSION OF MODEL GAS RESULTS – The model gas test results show that significant improvement in the low temperature activity of the SCR system can be obtained by increasing the NO_2 fraction in the NO_x. The optimum NO_2 fraction is 50%. A Pt based oxidation catalyst can increase the NO_2 fraction in NO_x when the catalyst is correctly dimensioned to oxidize the hydrocarbons, CO and NO. The oxidation catalyst must also not be too large as higher levels than the optimum level of 50% NO_2 will be detrimental for the NH_3 efficiency but will also yield secondary emissions of NH_3. At temperatures below 150° C no significant amounts of NO_2 can be produced at the oxidation catalyst and the activity of the system, even with direct NH_3 injection is low. Also for other reasons injection of urea at these low temperatures is not advisable.

ENGINE BENCH EXPERIMENTS

SET-UP AND EXPERIMENTAL CONDITIONS – For the engine bench experiments reported here, a dynamic engine test bench (see figure 10) equipped with a state of the art 4.0 L heavy duty turbo-charged direct injection diesel engine was used. The standard exhaust components were analyzed with the Horiba MEXA exhaust analyzer system. Additionally a laser diode spectrometer (Altoptronics LDS3000) was used to detect NH_3.

The catalysts used consisted of coated metallic substrates (400 cpsi). These could be combined in a modular fashion to change catalyst arrangement or space velocities. The volume of the SCR catalyst was varied between 4.6 and 9.2 L. The volume of the hydrolysis catalyst was in all experiments 4 L. In the experiments with preoxidation catalyst 2 L of a diesel oxidation catalyst (90 g/ft^3 Pt) was added upstream of the urea injection point. The catalyst volumes have not been optimized and are probably oversized. The potential of SCR as a NO_x reduction technology on heavy duty engines can however be demonstrated by this non-optimized system.

Both gaseous NH_3 and a urea solution (32.5%) could be injected into the system. The urea solution injection system and the inlet cone of the hydrolysis catalyst were optimized to guarantee a homogeneous distribution of urea on the hydrolysis catalyst. The urea injection system consists of a pump, a pulse controlled valve and an injection nozzle which enables injection rates between 1 and 30 g/min. It was verified by comparing on one side experiments without hydrolysis catalyst and with direct gaseous NH_3 injection to, on the other side, experiments with hydrolysis catalyst and with urea solution injection,

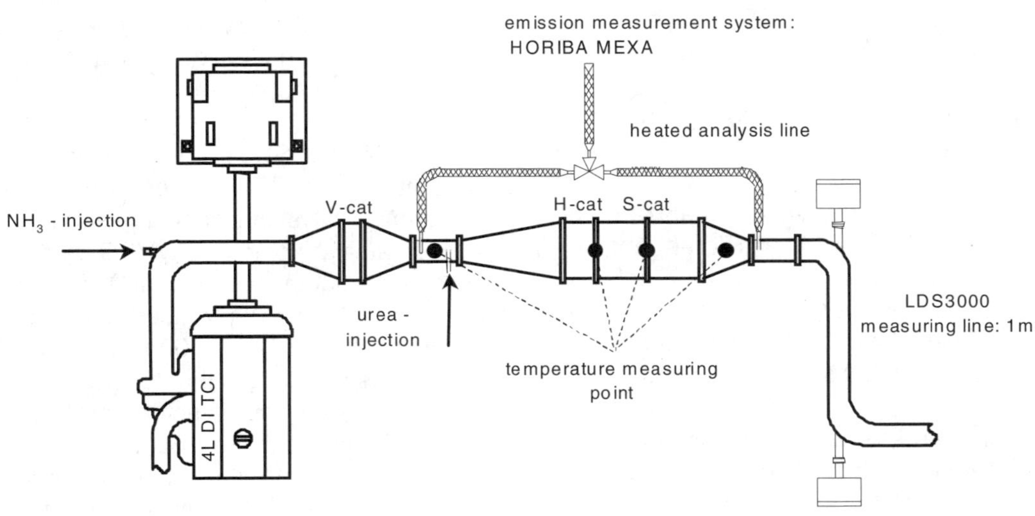

Figure 10. Schematic description of engine bench set-up.

that the distribution of urea on the hydrolysis catalyst and the activity of the hydrolysis catalyst were not influencing the results for the SCR system. The results reported here have all been obtained with the urea solution injection system.

Before every experiment the catalyst system was preconditioned to eliminate any adsorbed NH_3 by running the engine 10 minutes at a high load and intermediate speed point giving high temperatures in the hydrolysis and SCR catalysts. Low sulfur fuel was used in the experiments described here.

SYSTEM ACTIVITY MAPS – By varying the injected amount of urea solution (the NH_3/NO_x ratio, alpha) and the engine load at a constant speed of 1400 rpm (European Stationary Cycle, ESC, test cycle base speed), curves for the NO_x conversion and NH_3 slip versus alpha ratio for different temperatures could be obtained (see figure 11). The data show clearly the effects discussed extensively in the model gas experiments section. The conversion increases linearly with the alpha ratio up to a critical alpha ratio. Above this alpha ratio the NO_x conversion does not increase significantly and NH_3 slip occurs. Above temperatures of 300° C NH_3 slip free conversions above 80% can be achieved.

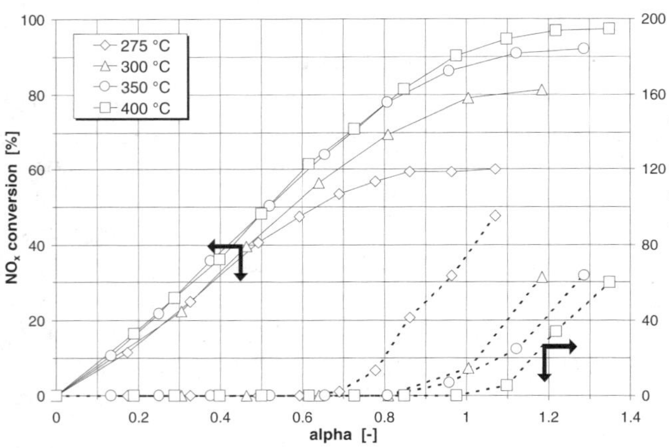

Figure 11. NO_x conversion and ammonia slip for basic HSO system directly after SCR catalyst on engine bench as function of NH_3/NO_x ratio, alpha, at various engine operation points characterized by the corresponding exhaust gas temperature.

Using these curves, the NO_x conversion and alpha ratio at starting NH_3 slip were determined for different engine operation points. Using different volumes of the SCR catalyst, a map of NO_x conversion as function of temperature and space velocity was obtained for both the basic HSO system and the VHSO system (see figures 12 and 13). The basic HSO system results show a clear effect of temperature and space velocity on NO_x conversion. Below temperatures of ca. 300° C the conversion is limited to 55%. This behavior changes dramatically when using a preoxidation catalyst (VHSO system). Now the conversion below 300° C varies between 50 and 85%. Moreover the effect of the space velocity is far less pronounced than in the basic system.

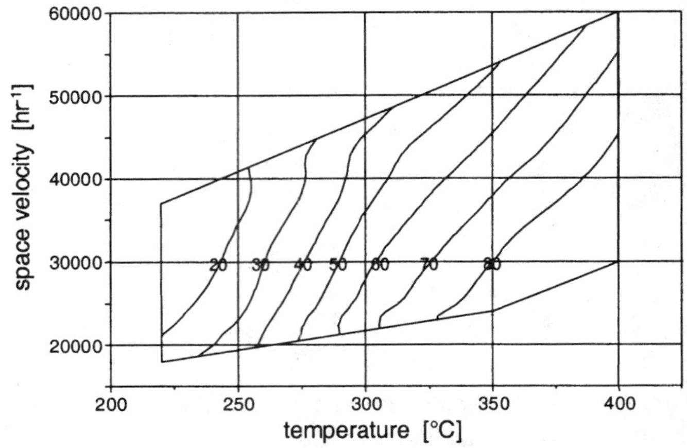

Figure 12. NO_x conversion at maximum NH_3/NO_x ratio, alpha, without NH_3 slip for basic HSO system directly after the SCR catalyst for different stationary engine operation points characterized by space velocity and temperature.

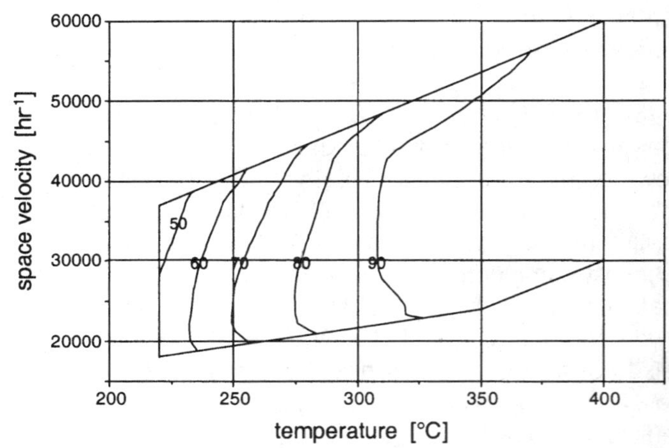

Figure 13. NO_x conversion at maximum NH_3/NO_x ratio, alpha, without NH_3 slip for the improved VHSO system with preoxidation catalyst directly after the SCR catalyst for different stationary engine operation points characterized by space velocity and temperature.

ESC TEST RESULTS – To show the NO_x reduction potential of a SCR system ESC tests were run with different catalyst systems: a basic HSO system with 9.2 L SCR catalyst volume, a VHSO system with 9.2 L SCR catalyst volume and a VHSO system with 4.6 L SCR catalyst volume. The temperature at the inlet of the SCR catalyst varied between 300° and 450° C. Only during the first phase ("idling") the temperature was 200°C. The urea solution was injected starting from the second phase (mode 2) of the ESC with a constant injection rate. This injection rate is sufficient to convert in every phase of the ESC test all NO_x stoichiometrically. The results are shown in figure 14 and table 2.

Table 2. ESC test cycle NO_x conversions for the three different systems tested.

phase	relative weight	flow	NO_x-conc.	NO_x-conversion HSO 9.2 L SCR	VHSO 9.2 L SCR	VHSO 4.6 L SCR
	[-]	[kg/h]	[vppm]	[%]	[%]	[%]
1	0.15	160	370	0.0	0.0	0.0
2	0.08	445	610	23.2	69.5	53.3
3	0.10	505	340	64.6	86.2	77.0
4	0.10	590	380	48.6	74.8	69.9
5	0.05	300	450	70.9	93.4	92.7
6	0.05	380	520	53.6	86.0	81.3
7	0.05	230	280	93.3	99.5	97.5
8	0.09	610	550	31.7	74.6	59.2
9	0.10	365	250	85.7	98.3	92.6
10	0.08	695	450	30.5	69.0	50.2
11	0.05	490	200	71.8	98.5	93.9
12	0.05	670	320	55.0	84.3	71.2
13	0.05	620	260	61.2	87.0	77.1
total				**45**	**75**	**65**

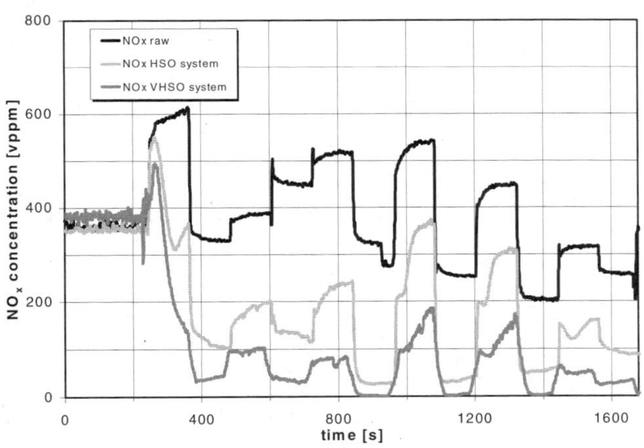

Figure 14. Raw NO_x concentration and NO_x concentrations after the SCR catalyst in the HSO and VHSO system in the ESC test cycle with a non-optimized stationary urea injection rate.

The data shows that significant NO_x conversions can be reached: 45 % for the basic HSO system with 9.2 L SCR catalyst volume. Conversions for the individual phases vary between 25% and 95%. Adding a preoxidation catalyst (VHSO system) the SCR system converts 75% with conversions in the individual phases varying between 70% and 100%. Even when the SCR catalyst volume is reduced to 4.6 L the total ESC conversion is significantly higher than in the basic HSO system: 65%.

The conversions obtained here could be increased by injection of urea during the first phase as this part contributes for almost 5% to the total ESC test result. Assuming 40% for the VHSO system in the first phase, the total ESC conversion could amount to 77%.

ENGINE AGING – To determine the aging behavior of the SCR catalyst system, a VHSO system with respectively a 1.6 pre-oxidation catalyst, a 1.25 hydrolysis catalyst, a 4.9 SCR catalyst and a 1.25 L guard oxidation catalyst was mounted on a 2.5 L DI TCI passenger vehicle engine on a dynamic engine bench. There the catalyst system was aged for 200 hours using the aging cycle depicted in figure 15. This cycle consist of different phases which all expose the system to conditions which are considered to be extreme for aging. The maximum temperature observed during this aging cycle was 500° C. During all the phases urea solution was injected in amounts determined by the NO_x raw emissions and the temperature for that phase. At temperatures below 250° C an alpha ratio of 1 was used. At temperatures above 250° C the alpha was kept at 0.7. The fuel used for aging was a standard diesel fuel containing 230 wt.-ppm sulfur. Every 50 hours the activity of the system was determined by running two times a simulated MVEG-A test on the dynamic engine bench with an inertia mass of 1900 kg. The fuel used in this activity test was a low sulfur diesel fuel. The NO_x conversion in the MVEG-A test is quite sensitive to small changes in the activity of the complete system because of the relatively cold ECE part. The results of this aging test are shown in figure 16.

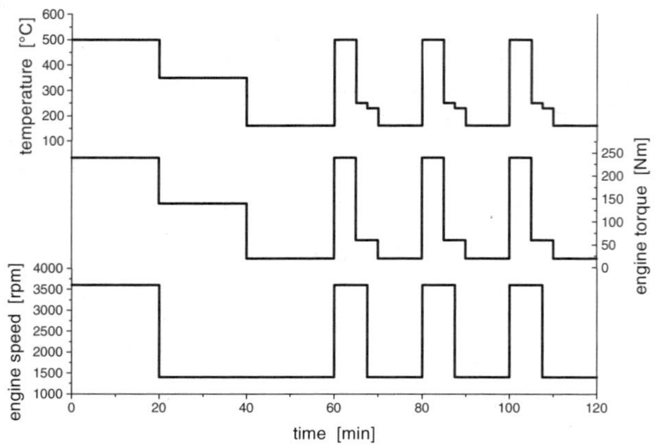

Figure 15. Engine aging cycle for the SCR system.

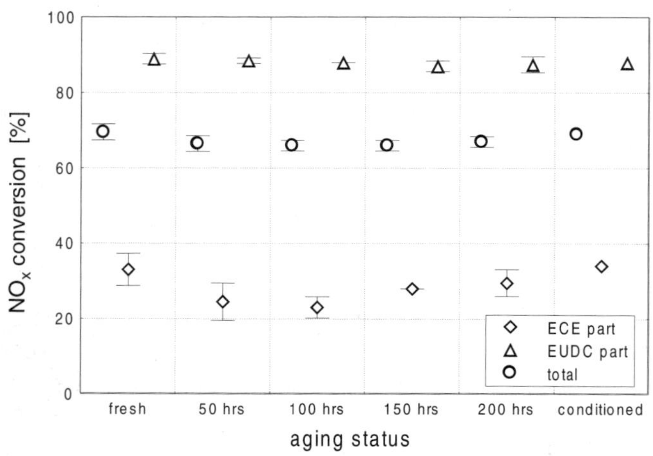

Figure 16. NO_x conversion (with standard deviation) in the MVEG-A cycle as function of engine aging for the ECE part, EUDC part and the total cycle.

This test shows that almost no activity loss can be observed during the 200 hours of intensive aging. After aging no significant activity change is observed in the EUDC part of the cycle. In the ECE part a decreasing activity can be observed as trend. In parallel a sternly decreasing CO conversion is observed, so the activity loss is most probably due to a deactivation of the preoxidation catalyst, which also lowers the NO_2 fraction in NO_x in the exhaust gas at low temperatures. After conditioning at a high speed and load operation point for 10 hours using the low sulfur diesel, this activity is restored completely. The deactivation of the preoxidation catalyst is most probably due to sulfate deposition. These results give confidence for the applicability of the system on a heavy duty engine, however, prolonged aging tests have to be run to confirm the aging stability.

CONCLUSIONS

The results reported here show that the proposed SCR systems can give significant NO_x reductions in the ESC test. Already with a rude constant urea injection rate conversions of 77% can be achieved by injecting during all phases of the ESC test. By optimizing the injection strategy, e.g. by taking into account the actual level of NH_3 stored in the system, it is expected to get to conversions higher than 75% based on the results reported here. The engine aging results are promising with respect to the long term stability of the SCR catalyst system. The catalyst system in itself is clearly technically feasible.

However introduction of the SCR system on heavy duty vehicles still depends on the solution of many other problems like e.g. injection strategy, injection system, urea availability, OBD and the acceptability of the system to the end users. All these problems can only be solved by a close cooperation of all the people involved. This way the many hurdles for a successful introduction of the SCR system on heavy duty vehicles can be overcome.

2001-01-0514

Advanced Urea SCR Catalysts for Automotive Applications

J. Gieshoff, M. Pfeifer, A. Schäfer-Sindlinger and P.C. Spurk
dmc^2 AG

G. Garr, T. Leprince
ICT, Inc.

M. Crocker
dmc^2

ABSTRACT

The LEV II and EURO V legislation in 2007/2008 require a high conversion level for nitrogen oxides to meet the emission levels for diesel SUVs and trucks. Therefore, U.S. and European truck manufacturers are considering the introduction of urea SCR systems no later than model year 2005. The current SCR catalysts are based mainly on systems derived from stationary power plant applications. Therefore, improved washcoat based monolith catalysts were developed using standard types of formulations. These catalysts achieved high conversion levels similar to extruded systems in passenger car and truck test cycles. However, to meet further tightening of standards, a new class of catalysts was developed. These advanced type of catalytic coatings proved to be equivalent or even better than standard washcoat formulations. Results will be shown from ESC, MVEG and US-FTP 75 tests to illustrate the progress in catalyst design for urea SCR. The new catalyst systems show a very good aging performance and will be the next generation of urea SCR catalysts for mobile applications.

INTRODUCTION

Nitrogen oxide emissions contribute significantly to photochemical smog and also to acid rain. Therefore many attempts are currently being made to further reduce these emissions especially from large diesel engines used for non-stationary applications.

Iwamato and co-workers showed that Cu-zeolites are suitable for the decomposition of NO [1] under lean exhaust gas conditions. Copper ion exchanged ZSM-5, for example, is known to decompose NO into N_2 and O_2, but the kinetics are unfavourable and temperatures above 500 °C are required in order to achieve significant rates. Usually hydrocarbons had been used as reductants in the case of the Cu ZSM-5 catalysts. According to Hall et al. byproducts as N_2O are also formed [2]. Furthermore, high quantities of hydrocarbons are needed to achieve a good NO_x conversion [3].

Use of ammonia as reductant leads to the SCR reaction with a high selectivity for nitrogen and a low reductant consumption. In this case, primarily V_2O_5 based catalysts are used [4]. Zeolite-containing catalysts were also tested in the past for stationary power plant applications. Durability was always an issue in these tests. The well-known $V_2O_5/WO_3/TiO_2$ catalyst systems are commonly applied to stationary engines [5].

First attempts to transfer these types of catalysts on mobile applications showed very promising results [6, 7]. However, the temperature stability of these catalysts are limited to 650°C. Centi et al. found that the activity of Cu ZSM-5 was higher than V_2O_5 on TiO_2 [8] and Anpo and co-workers demonstrated that certain natural zeolites might also be applicable for NO_x reduction with ammonia on diesel engines [9]. Ion exchanged zeolites currently achieve considerable attention as catalysts for NO_x reduction by ammonia for instationary applications as well.

This paper: demonstrates the NO_x reduction potential of ion exchanged synthetic zeolites, describes the impact of NO_2 on the NO_x conversion as well as their aging behavior and shows the applicability for passenger cars and SUVs. Furthermore a view on truck engine test results is given in this paper.

MODEL GAS TESTS

The test equipment used for the model gas tests is described earlier [7]. The system consisted of: a gas dosing system with a set of mass flow controllers for the various gases listed in table 1, a HPLC pump to feed water to the gas mixture, an evaporator to add hydrocarbons, if necessary, and a quartz reactor with inset in which the catalyst cores were fixed. The reactor tube was mounted in a tube furnace and connected via heated lines to a FTIR spectrometer to analyze both inlet and outlet gases.

TEST CONDITIONS – The NO_x conversion of the SCR catalyst samples was determined under steady state conditions normally at space velocities of 30,000 hr^{-1} and at temperatures varying between 150 and 600°C. The standard gas composition for each test is given in table 1 (Gas Mixture 1). The molar ratio between NH_3 and NO (alpha ratio) was kept at 0.9 under standard conditions. This limited the expected NO conversion to 90%. In some tests 50% of the NO was replaced by NO_2 (Table 1: Gas Mixture 2) to investigate the NO_2 impact on the NO_x conversion.

Table 1: Model gas composition for sample evaluation

Gas Component	Gas Mixture 1	Gas Mixture 2
NO	500 v.-ppm	250 v.- ppm [a]
NO_2	-	250 v.- ppm [a]
NH_3	450 v.-ppm	450 v.-ppm
H_2O	1.3 vol.-%	1.3 vol.-%
O_2	5.0 vol.-%	5.0 vol.-%
N_2	Balance	Balance

[a] NO_2 impact study

Diesel exhaust gas typically contains 6.0 vol.-% of water. A water concentration above 0.5 vol.-% was found to have no further effect other than dilution of the model gas. Therefore, 1.3 vol.-% of water was selected to simulate the water impact on the catalyst performance and to minimize problems with quantitative FTIR gas analysis. The absence of CO and CO_2 does not influence the result of the model gas test.

The SCR catalysts prepared for the model gas tests consisted of different ion exchanged zeolite containing formulations. The various transition metal containing zeolite samples were prepared according to new confidential pocedures. For comparison reasons, a standard $V_2O_5/WO_3/TiO_2$ SCR catalyst described in [7] was added as a reference to the model gas experiments. All samples were coated on ceramic honeycomb substrates (dia. 1", length 3", 400 cpsi, 6.5 mil).

MODEL GAS TESTS WITHOUT NO_2 – To examine the NO conversion of the ion exchanged zeolite samples the coated substrates were mounted in the model gas reactor mentioned above. Figure 1 shows an example for the Cu ZSM-5 catalyst. The NO conversion started at around 150°C and lead to a maximum between 250 to 300°C. It is known from hydrocarbon DeNOx experiments that at higher temperatures the NO_x conversion of Cu ZSM-5 is affected by water [10]. This might also be the explanation for the significant drop of the conversion curve above 325°C. In addition figure 1 shows the NO conversion of the Cu ZSM-5 reference catalyst under conditions listed in table 2 after aging at 650 °C (Aging 1 @ 650 C).

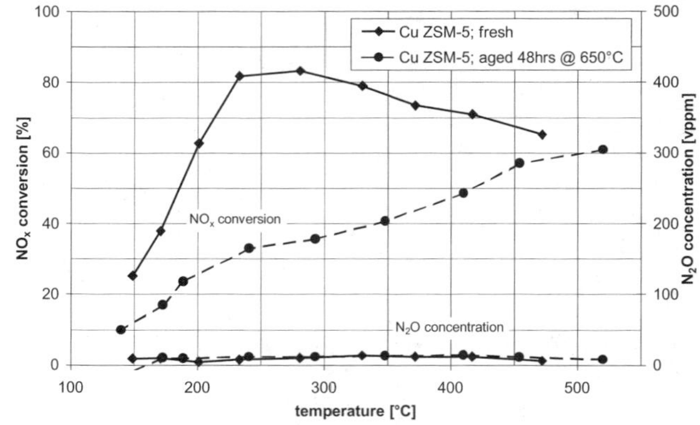

Figure 1. NO_x conversion and N_2O concentration as a function of temperature for the fresh Cu ZSM-5 catalyst and after aging measured without NO_2.

The data shows a significant loss in NO conversion efficiency. Details of the aging results will be discussed later.

Table 2: Model gas composition for aging tests

Gas Component	Aging 1 @ 650 °C	Aging 2 @ 800 °C
SO_2	20 v.-ppm	-
H_2O	10.0 vol.-%	10.0 vol.-%
O_2	10.0 vol.-%	10.0 vol.-%
N_2	Balance	Balance

The aging properties of Cu based zeolite catalysts proved to be insufficient for mobile applications and are one of the reasons why SCR catalysts based on the $V_2O_5/WO_3/TiO_2$ oxide system were the first choice in current heavy duty applications.

Besides Cu ZSM-5 a number of other zeolite containing catalyst samples were evaluated. An overview is given in figure 2.

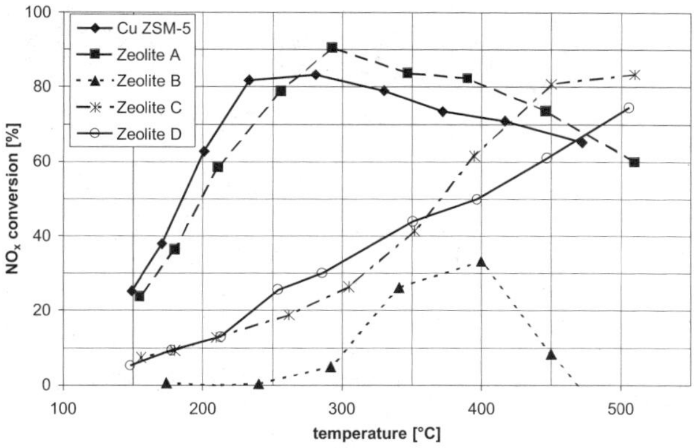

Figure 2 Catalyst screening with different zeolites, fresh, measured without NO_2.

With the exception of catalyst sample Zeolite A, none of the samples showed a high potential for NO reduction. Since it is known from the $V_2O_5/WO_3/TiO_2$ catalysts that NO_2 has a strong impact on the low temperature activity of ammonia / urea SCR catalysts [7], all screening tests were repeated with NO_2 in the feed gas.

MODEL GAS TESTS WITH NO_2 - In the case of our sample Cu ZSM-5 the overall activity of the fresh and the aged sample was significantly improved (Figure 3). At temperatures between 200 and 450°C the catalyzed decomposition of NO_x is observable lifting the NO_x conversion over 90 %. In addition, the N_2O emissions also rose significantly.

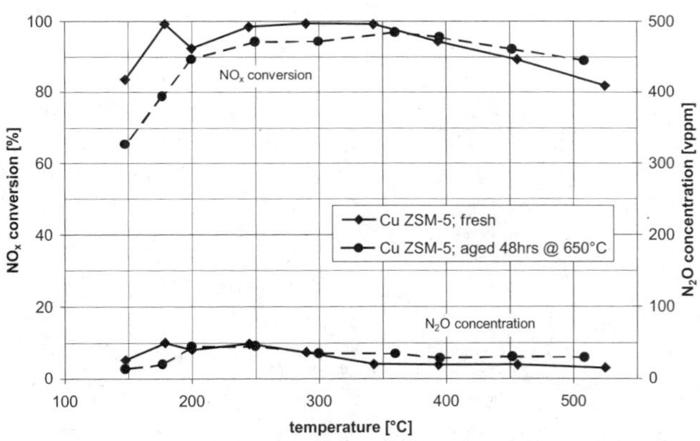

Figure 3 NO_x conversion and N_2O emissions of Cu ZSM-5 fresh and after aging for 48 hours at 650°C with NO_2 (gas mixture 2).

Compared to Cu ZSM-5 other zeolite containing samples showed less low temperature activity but a slightly higher NO_x conversion efficiency at temperatures above 400°C (Figure 4). Catalyst sample B exhibited significantly less N_2O formation than Cu ZSM-5.

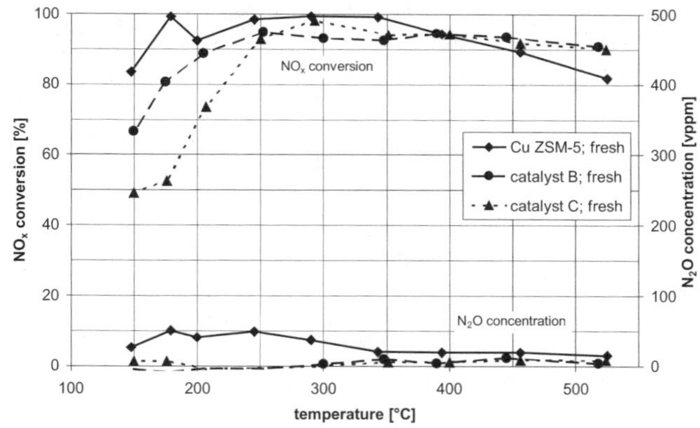

Figure 4 NO_x conversion and N_2O emissions for different types of ion exchanged ZSM-5 catalysts under fresh conditions.

Because of the good low-temperature behavior and better selectivity to N_2 of catalyst sample B, this material was selected for all further tests.

SYNTHETIC GAS AGING – In the next phase of the catalyst evaluation all samples were aged for 48 hours at 650°C in a synthetic gas atmosphere given in table 2 (Aging 1 @ 650°C). These temperature conditions were selected according to standard aging procedures for SCR catalysts and represent the maximum operation temperature for $V_2O_5/WO_3/TiO_2$ containing systems. The results of these aging experiments for catalyst sample B are shown in figure 5.

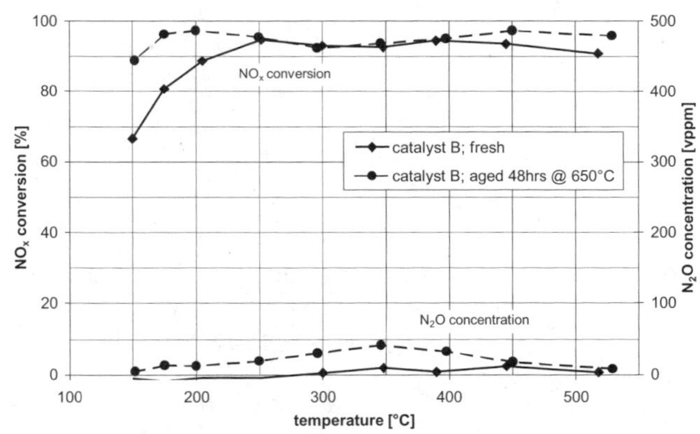

Figure 5 NO_x conversion as a function of temperature fresh and after aging for catalyst B with NO_2 (gas mixture 2).

After 650°C aging for 48 hours no loss in activity was observed. SO_2 did not deactivate the catalyst and at lower temperatures (< 250°C) the performance was even better compared to the fresh results. More importantly than the conversion is the selectivity of this catalyst. Significantly less N_2O formation was observed at low and high temperatures in the model gas tests after aging than in the case Cu ZSM-5 (Figures 1, 3).

The space velocity has no impact on the overall performance of catalyst sample B within the range examined (Figure 6).

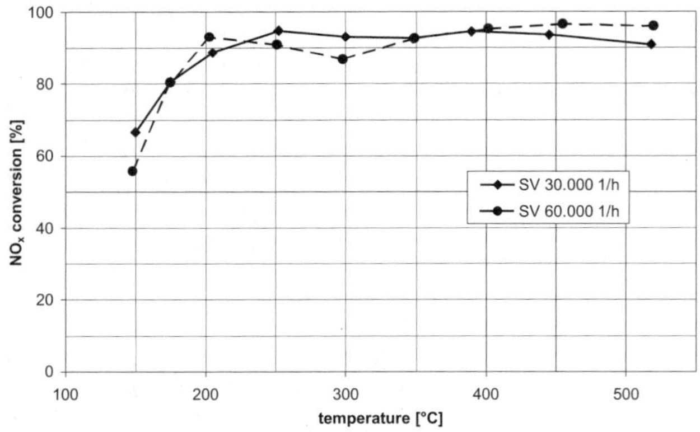

Figure 6 NO_x conversion as a function of space velocity for catalyst B measured with NO_2 (gas mixture 2).

Compared to the standard $V_2O_5/WO_3/TiO_2$ catalyst described in [7] catalyst sample B shows a slight improvement at very low and very high temperatures (Figure 7).

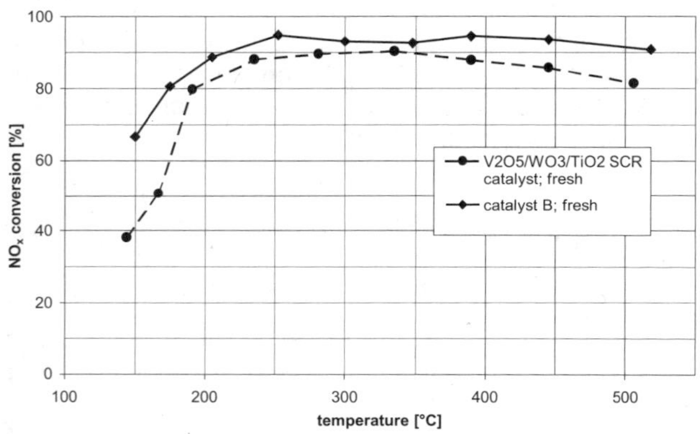

Figure 7 NO_x conversion as a function of temperature for catalyst B and the standard $V_2O_5/WO_3/TiO_2$ SCR catalyst measured with NO_2.

V_2O_5 containing SCR catalysts are irreversibly deactivated at temperatures above 700°C. This is due to a phase change of the TiO_2 support oxide from anatase to rutile, loss surface area and evaporation of V_2O_5 and WO_3. Catalyst sample B was aged for 48 hours at 800°C. The gas composition for the aging test is given in table 2 (Aging 2 @ 800°C). The impact of these severe aging conditions on the sample is shown in figure 8 in comparison to the fresh catalyst.

The overall NO_x conversion of the aged sample is slightly lower compared to the fresh and N_2O emissions were sufficiently low.

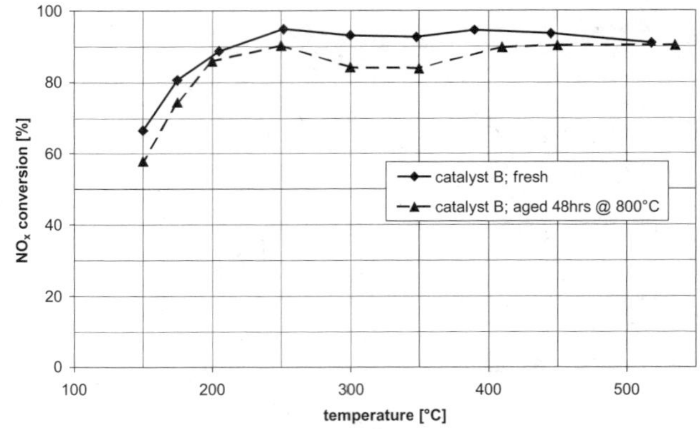

Figure 8: NO_x conversion as a function of temperature fresh and after aging for 48 hours at 800°C for catalyst B

Since the model gas experiments gave sufficiently promising results, especially with catalyst sample B, engine bench tests were performed to elucidate its NO_x reduction potential under real exhaust gas conditions and its long term durability.

ENGINE BENCH EXPERIMENTS

SET UP AND EXPERIMENTAL CONDITIONS – For the experiments reported here a dynamic engine test bench (see figure 9) was used equipped with a state of the art 4.0 L DI TCI engine. Besides the standard exhaust gas measurement system a laser diode spectrometer (ALTOPTRONICS LDS3000) was used to detect ammonia [7].

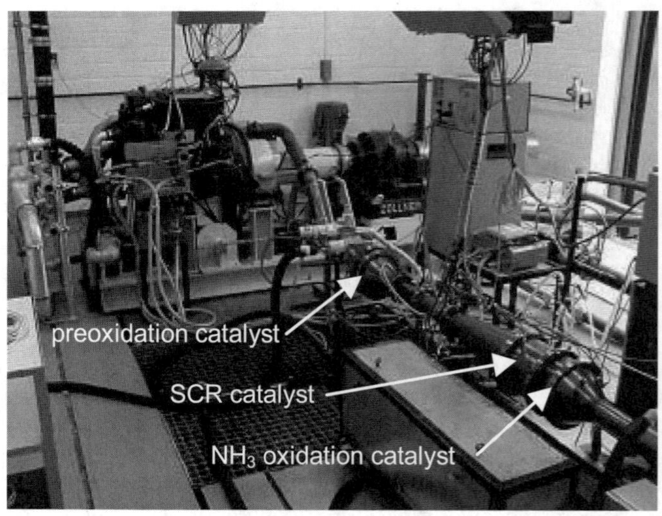

Figure 9. Dynamic engine test bench.

The VSO system, described in [7], used for these tests consisted of coated metallic substrates (400 cpsi). The pre-oxidation (V-) catalyst had a volume of 2.0 L with a loading of 90 g/ft³ Platinum and was added upstream of the urea injection point (see figure 9). The volume of the SCR catalyst B was 4.6 L. Additionally a 2.0 L NH_3

oxidation catalyst (O-catalyst) loaded with 10 g/ft³ Pt was used. The volumes and precious metal and washcoat loading of the oxidation catalysts were not optimized and are probably oversized.

For all tests a 32.5% urea solution was used as reductant. The urea injection system consisted of a pump, a pulse controlled valve and an injection nozzle which enables a wide range of injection rates. The injection strategy was not optimized for this application; nevertheless, the potential of the new developed SCR catalyst can be demonstrated.

Before every experiment the catalyst system was preconditioned by running the engine 10 minutes at an engine operating point giving high temperatures in the SCR catalyst. This procedure eliminates any adsorbed NH_3 in the SCR catalyst. A low sulfur fuel was used for the testing.

ESC RESULTS – To show the NO_x reduction potential of the developed SCR catalyst ESC tests were run comparable to those tests described in [7]. The temperature at the inlet of the SCR catalyst is shown in figure 10 together with the space velocity in the SCR catalyst.

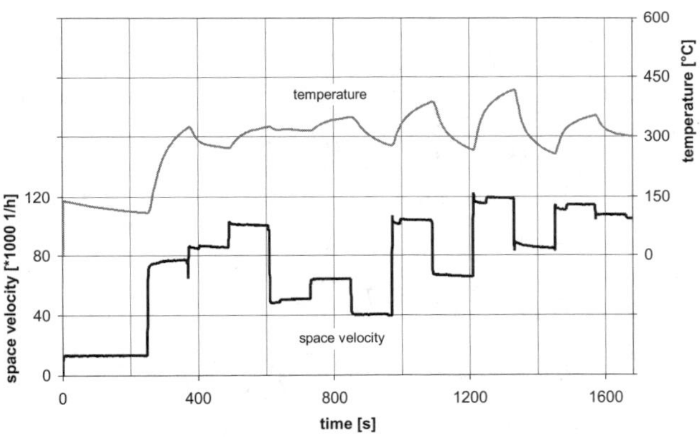

Figure 10. Space velocity and SCR catalyst inlet temperature in the ESC.

The temperature at the inlet varied between 275° and 450°C. Only during mode 1 did the temperature not exceed 150°C. For space velocity a maximum of 120,000 hr^{-1} was observed in mode 10. The urea injection started at mode 2 of the ESC with a constant injection rate. The selected injection rate is sufficient to convert all NO_x in every phase of the ESC cycle. The result of the test is shown in figure 11.

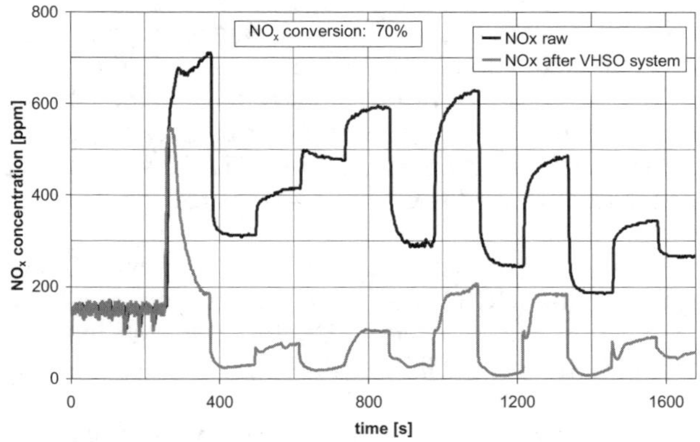

Figure 11. Raw NO_x concentration and NO_x concentration after the VSO system in the ESC test cycle.

Using the SCR catalyst type B an overall NO_x reduction in the ESC cycle of 70% was achieved with a fresh catalyst. This is a slight improvement compared to the standard SCR catalyst described in [7]. Optimizing the urea injection strategy can increase the NO_x performance. Injecting urea while in the idling phase would also improve the overall efficiency.

ENGINE AGING – To determine the aging behavior of the SCR catalyst, a VSO system was aged for 200 hours using the aging cycle shown in figure 12 [7]. The cycle consists of different phases which were considered to be extreme for aging. The maximum temperature reached in this aging cycle was 500°C.

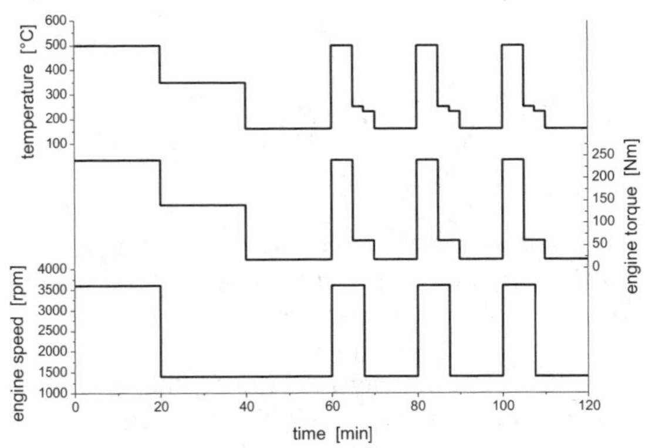

Figure 12: Engine aging cycle for the VSO system.

During the whole cycle urea was injected. At temperatures below 250°C an alpha ratio above 1.0 was used. The fuel used for the aging was standard diesel fuel with a sulfur content of 230 ppm-wt.. The catalyst activity was determined every 50 hours by running the NEDC test twice. The fuel used for this NEDC tests was a low sulfur diesel.

The catalyst system used for aging consisted of a 2.5 L pre-oxidation catalyst, a 2.5 L SCR catalyst type B and a 1.25 L NH_3 oxidation catalyst and was mounted on a 2 L DI TCI engine. Besides a standard exhaust gas measurement system used to determine CO, HC and NO_x, a mass spectrometer and a FTIR were used to detect NH_3, NO_2 and N_2O secondary emissions.

The urea injection strategy used for the NEDC was based on the SCR inlet temperature and stationary conversion maps. For the NEDC test the urea injection started when the SCR inlet temperature reached 120°C.

Figure 12 shows the NO_x in the raw gas and after the VSO system depending on the cycle time. Beginning with urea injection, a NO_x conversion is observed even in the low temperature part ECE. For the whole ECE part of the NEDC cycle a NO_x conversion of 49% was achieved. In the EUDC part of the cycle, the NO_x conversion reaches 73%. Over the whole NEDC a NO_x reduction of 63% was achieved using a fresh SCR catalyst.

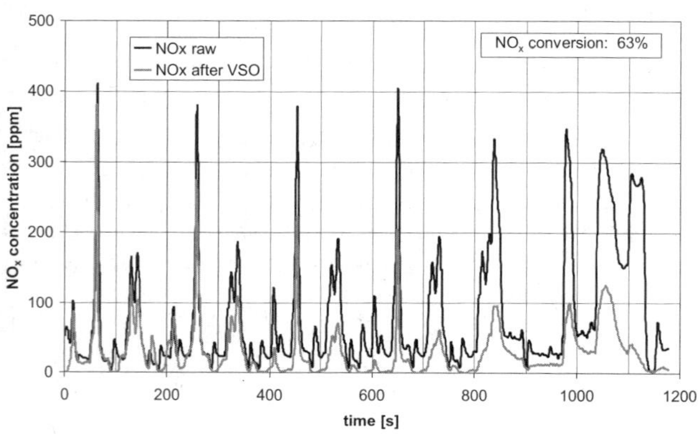

Figure 12: Raw NO_x concentration and NO_x concentration after VSO system in the NEDC test cycle

Figure 13 shows the NO_x conversion in the ECE part, the EUDC part and in the total cycle as a function of aging.

This test shows no activity loss during the 200 hours of intensive aging. Although a slight decrease in the ECE activity is observed, the conversion in the EUDC part of the cycle increases from 73% up to 80% at the end of the aging cycle. As described in [7] the activity loss in the ECE part is most probably due to a partial deactivation of the pre-oxidation catalyst because of sulfate deposition.

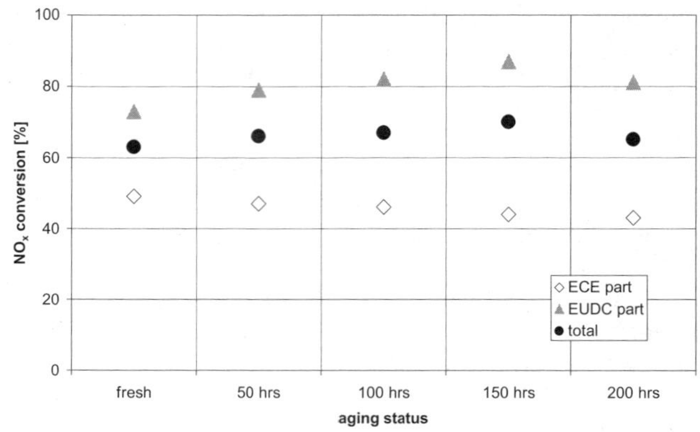

Figure 13: NO_x conversion in the NEDC cycle as function of engine aging for the ECE part, the EUDC part and the total cycle.

LOW TEMPERATURE AGING – To determine the effect of HC adsorption on the activity of the SCR catalyst the VSO system was again aged for eight hours under idling conditions. Figure 14 shows the HC and CO concentration in the outlet of the VSO system for this aging. The temperature during this aging did not exceed 130°C in the pre-oxidation catalyst.

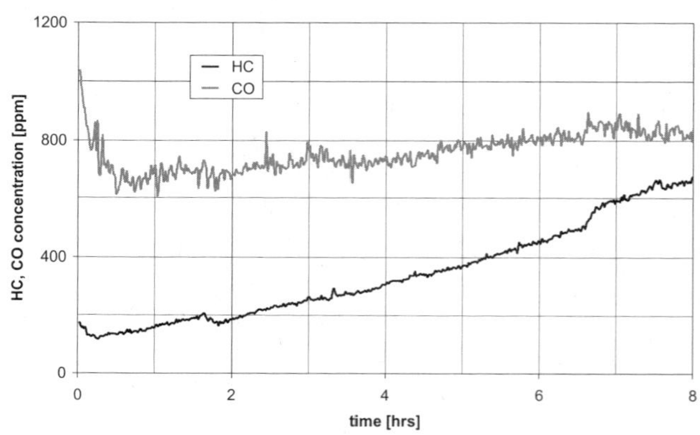

Figure 14: HC and CO concentration during eight hours of low temperature aging under idling conditions.

After eight hours of aging the systems activity was determined by running a NEDC cycle. A NO_x conversion of 6% was achieved in this cycle. As can be seen in figure 15, the NO_x concentration after the VSO system exceeded the NO_x raw concentration. The increased NO_x concentration is most likely due to ammonia oxidation on the NH_3 oxidation catalyst. Because the SCR catalyst was deactivated by adsorbed hydrocarbons the injected amount of urea was oxidized on the oxidation catalyst.

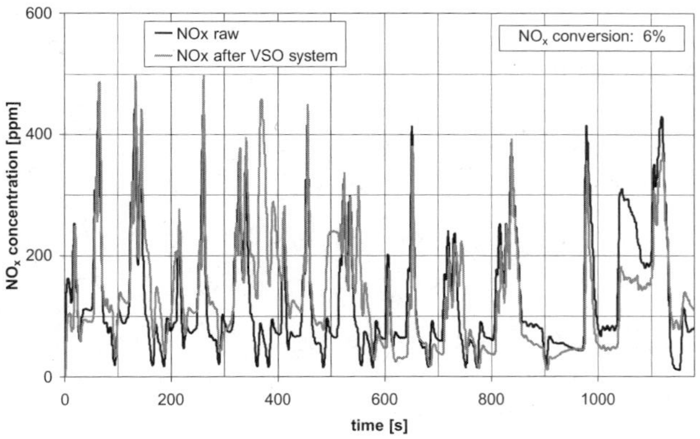

Figure 15: Raw NO_x concentration and concentration after VSO system in the NEDC test cycle after eight hours of low temperature aging at idling conditions.

After conditioning the VSO system by running the NEDC cycle two more times an activity of 66 % was achieved (see figure 16) which corresponds with the results shown in figure 13.

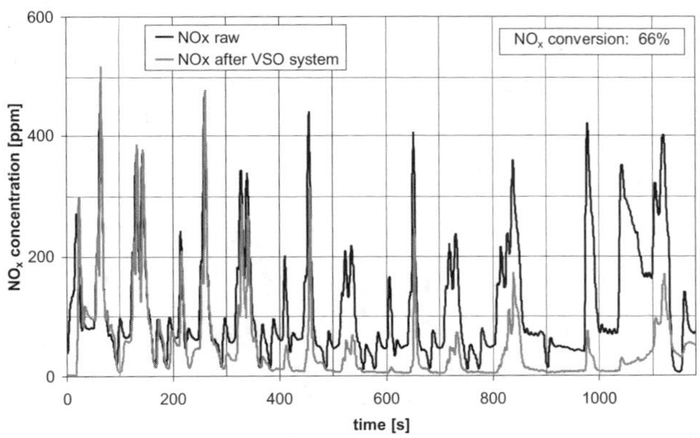

Figure 16: Raw NO_x concentration and concentration after VSO system in the NEDC test cycle after subsequent conditioning the VSO system.

US-FTP 75 ENGINE RESULTS – The activity of the aged VSO system was also measured in the US-FTP 75 cycle using the 2 L DI TCI engine. Figure 17 shows the NO_x concentration in the raw exhaust gas and after the VSO system. In this test an overall NO_x conversion of 80% was achieved using the VSO system with the new developed SCR catalyst.

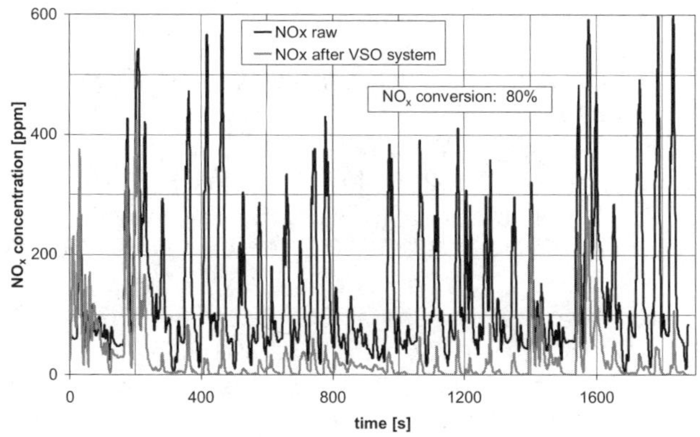

Figure 17: Raw NO_x concentration and concentration after VSO system in the FTP75 test cycle.

CONCLUSION

The results reported here show the high performance of ion exchanged zeolite catalysts when used in a catalyst system with a pre-oxidation catalyst. The crucial component to achieve high NO_x conversion levels is NO_2. If a sufficiently high NO_2 concentration level can be generated during the operation of the catalyst system obviously similar conversion levels are achieved as with the standard $V_2O_5/WO_3/TiO_2$ SCR catalysts. Once more NO_2 proved to be the key component for NO_x reduction. Work is in progress to determine the optimum NO_2 concentration for the process. Levels above 50% of the total NO_x concentration obviously lead to a decreased selectivity i.e. increasing N_2O formation.

Also in the case of ion exchanged zeolites the re-oxidation of the active center at low temperatures plays an important role as shown by the NO_2 impact.

The model gas and engine aging results are promising. At least for medium duty applications it might be possible to apply vanadia-free SCR catalysts due to long term stability. However the performance of a system does not depend on the catalyst performance alone. An optimized urea injection strategy is necessary to achieve higher conversion rates.

ACKNOWLEDGMENTS

The authors like to thank Mrs. Y. Demel, M. Muller, and A. Harth for preparation of samples and for performance of the model gas experiments as well as Prof. G. Hohenberg, University of Technology Darmstadt and Prof. W. Müller, University of Kaiserslautern for engine testing and fruitful discussions of the results.

REFERENCES

1. M. Iwamoto, S. Yokoo, K. Sakai, S. Kagawa, J. Chem. Soc., Faraday Trans., 77(1981)1629-1638; M. Iwamoto, H. Yahiro, Y. Mine, S. Kagawa, Chem. Lett., (1989)213-216; M. Iwamoto, H. Yahiro, K. Tanda, N. Mizuno, Y. Mine, S. Kagawa, J. Phys. Chem., 95(1991)3727-3730; M. Iwamoto, H. Yohiro, N. Mizuno, W. Zang, Y. Mine, H. Furukawa, S. Kagawa, J. Phys. Chem., 96(1992)9360-9366;
2. Y. Li, W.K. Hall, J. Catal., 129(1991)202-215;
3. M. Kawanami, A. Okumura, M. Horiuchi, A. Schäfer-Sindlinger, K. Zerafa, SAE Technical Paper No. 961129;
4. M. Baacke, R. Brand, B. Engler, A. Kiss, P. Kleine-Möllhoff, P. Kleinschmidt, E. Koberstein, M. Siray, DE 3841990, (14.12.1988);
5. J.J. Janssen: "Environmental Catalysis – Stationary Sources" in Handbook of Catalysis, Volume 4, 1636-1644, Edit. G. Ertl, H. Knözinger, J. Weitkamp, Wiley-VCh, Weinheim 1997;
6. E. Jacob, G. Emmerling, A. Döring, U. Graf, M. Harris, J.A.A. van den Tillaart and B. Hupfeld, 19. Vienna Engine Symposium, 1998;
7. J. Gieshoff, A. Schäfer-Sindlinger, P.C. Spurk and J.A.A. van den Tillaart, SAE Technical Paper No. 2000-01-0189;
8. G. Centi, S. Perathoner, Y. Shioya, M. Anpo, Res. Chem. Intermed., 17(1992) 125-135;
9. H. Mishima, K. Hashimoto, T. Ono, M. Anpo, Applied Catalysis B: Environmental, 19(1998)119-129;
10. M. Iwamoto and H. Hamada, Catal. Today, 17(1991)94; M. Shelef, C.N. Montreuil and H.W. Jen, Catalysis Letters, 26(1994)277-284.

CONTACT

Dr.-Ing. Paul C. Spurk
d**mc**^{2} AG, AC-RT-AT

e-mail: paul.spurk@dmc-2.de

Rodenbacher Chaussee 4
P.O. Box 1351
D-60453 Hanau-Wolfgang

Germany

or

Dr. Juergen Gieshoff
d**mc**^{2} AG, AC-RT-RD

e-mail: juergen.gieshoff@dmc-2.de

2001-01-3625

Recent Advances in the Development of Urea-SCR for Automotive Applications

Manfred Koebel, Martin Elsener and Giuseppe Madia
Paul Scherrer Institute

ABSTRACT

In the recent years, considerable progress has been achieved in the development of urea-SCR for nonstationary, mobile applications. Main challenges are the reduction of the catalyst volumes and the optimization of the dosing strategy to minimize transient ammonia emissions during load changes. Catalysts with increased cell density and enhanced intrinsic activity are one way to obtain smaller catalyst volumes. Another approach is the introduction of additional catalysts upstream and downstream of the SCR catalyst itself. Most effective is an oxidizing precatalyst that converts a part of NO to NO_2, thus allowing a faster SCR reaction to occur. However, such a system requires the use of fuel with reduced sulfur content.

Work performed at the Paul Scherrer Institute comprises of fundamental experiments in a laboratory scale reactor with synthetic exhaust gas and applied experiments on a diesel test stand. In addition to the evaluation of various commercial and proprietary catalysts, important problems of the process have been investigated as, for example, the fast SCR reaction, ammonia adsorption on the catalysts, the decomposition of urea, and catalyst aging.

INTRODUCTION

Forthcoming European and U.S. emission legislation for diesel vehicles calls for a significant reduction of both particles and NO_x. However, an important difference is the fact that in Europe an increasing number of passenger cars are equipped with diesel engines. Their maximum allowable emissions of particles and NO_x will be also subjected to stringent regulations. The present work focuses mainly on HD diesel vehicles.

Figure 1 shows the large gap between feasible raw emissions of HD diesel engines for trucks and forthcoming emission standards. It is now generally assumed that the EURO 4 emission standards proposed for the year 2005 will be no longer feasible by primary measures alone, but will require additional exhaust gas aftertreatment techniques. There exist two basic strategies to attain these future standards:

1. Optimize the combustion with respect to low NO_x emissions, leading to a high emission of unburned material (soot, CO, hydrocarbons). Then use a particulate filter in the aftertreatment to clean up the particulate matter.

2. Optimize the combustion with respect to low emissions of unburned material, leading to a high emission of NO_x. Use a DeNOx process in the aftertreatment.

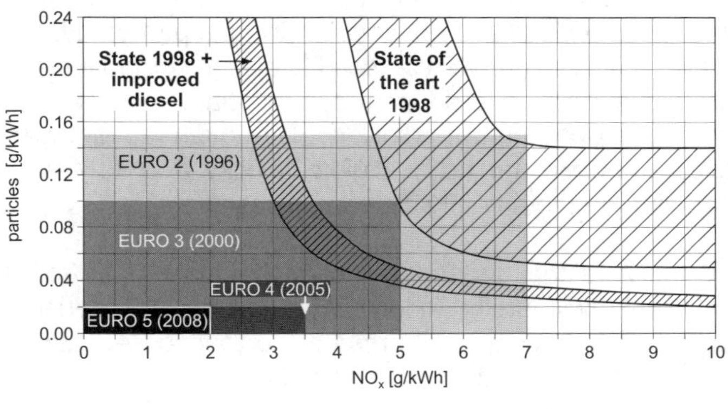

Figure 1: Trade-off between particulate matter and NO_x for HD truck engines and prospected EURO emission standards

The first strategy does not require a complicated chemistry, but leads to an increase in fuel consumption. If optimized fuel economy is an additional target, then the second strategy is the better choice. This calls for the selective catalytic reduction of NO_x in the presence of oxygen.

Various selective DeNOx techniques have been proposed and studied in the last years. These include SCR (Selective Catalytic Reduction) with N-containing reducing agents like ammonia and urea [1, 2], the NOx storage-reduction (NSR) catalyst [3, 4] and the so-called HC-SCR [5, 6]. In the latter process hydrocarbons are used to reduce nitrogen oxides. However, although much effort has been directed into the development of these catalysts, selectivity is still poor resulting in the preferential oxidation of oxygen that, in turn, increases the fuel consumption and the CO_2 emissions. The NSR catalyst utilizes the ability of basic oxides to store NO_x as nitrates, but it is very prone to deactivation by sulfur oxides due to the higher stability of the corresponding sulfates.

In our opinion the SCR process using urea as selective reducing agent has the highest chances of maturing into a commercial technique for heavy duty diesel applications. Urea-SCR has been studied by various groups in the last ten years [2, 7-9]. The present paper resumes the challenges of this development and further reports some recent advances.

CHEMISTRY OF THE SCR PROCESS

It is well known that diesel exhaust contains the nitrogen oxides mainly in the form of NO (nitrogen monoxide, typically over 90 %) and only a minor fraction in the form of NO_2 (nitrogen dioxide). Therefore, the SCR process must mainly remove NO by reduction.

The selectivity of the process refers to the selectivity of the reducing agent, which should not reduce the excess oxygen typical to a lean exhaust. Experience has shown that N-containing reducing agents like ammonia, urea and cyanuric acid are outstanding for this purpose. Urea may be considered as a solid storage compound for ammonia and is produced commercially by reacting ammonia with carbon dioxide. The main advantages over ammonia are its negligible toxicity and its easier handling due to a negligible vapor pressure.

To facilitate the understanding of the main SCR reactions, we will firstly focus on ammonia as reducing agent. Only later we will present the reaction equations for the generation of ammonia from urea.

The main reaction between ammonia and NO on the usual SCR catalysts is:

$$4\,NH_3 + 4\,NO + O_2 \rightarrow 4\,N_2 + 6\,H_2O \qquad (1)$$

We will call this reaction subsequently "standard SCR". According to this reaction 1 mole of NO will consume 1 mole of NH_3 and ¼ mole of O_2. Definitely faster than reaction (1), is the reaction of a 1:1-mixture of NO and NO_2:

$$4\,NH_3 + 2\,NO + 2\,NO_2 \rightarrow 4\,N_2 + 6\,H_2O \qquad (2)$$

This reaction will be subsequently denoted "fast SCR". The increase of the reaction rate has been long known [10, 11] and it is now proposed as a practical measure to increase the performance of automotive DeNOx systems. To increase the fraction of NO_2 in the exhaust, a strong oxidation catalyst (Pt-based) is placed upstream of the SCR catalyst. However, the fraction of NO_2 should not exceed 50% of NO_x because the reaction involving pure NO_2 is again slower than both reactions (1) and (2):

$$4\,NH_3 + 3\,NO_2 \rightarrow 3.5\,N_2 + 6\,H_2O \qquad (3)$$

At high temperatures (>400°C) the commonly used SCR catalysts based on TiO_2-WO_3-V_2O_5 tend to form nitrous oxide (N_2O). One of the possible reactions leading to its formation is:

$$4\,NH_3 + 4\,NO + 3\,O_2 \rightarrow 4\,N_2O + 6\,H_2O \qquad (4)$$

At still higher temperatures, the oxidizing properties of the catalyst get even more pronounced. This causes direct oxidation of ammonia to NO, thus limiting the maximum possible NO_x conversion:

$$4\,NH_3 + 5\,O_2 \rightarrow 4\,NO + 6\,H_2O \qquad (5)$$

In all these reactions ammonia may be substituted by urea and the above equations will change accordingly. The overall decomposition reaction of urea in the presence of water is:

$$NH_2\text{-}CO\text{-}NH_2 + H_2O \rightarrow 2\,NH_3 + CO_2 \qquad (6)$$

This reaction involves both a thermal and a hydrolytic decomposition step. We will discuss the respective details further below.

IMPROVING SCR CATALYSTS

PROBLEM

Gas hourly space velocities (GHSV) typical to stationary SCR catalysts are around 10,000 h^{-1} (this means 10,000 volumes of gas at standard conditions per volume of catalyst and hour). On the other hand, the space velocities of three-way catalysts (spark ignited engines) are typically of the order of 100,000 h^{-1}, i.e. about ten times higher. The latter range of space velocities results in catalyst volumes similar to the cylinder volume of the gasoline engine.

Due to the limited space available onboard a vehicle, higher space velocities in the SCR catalyst are necessary, i.e. the volumetric activity of the SCR catalysts must be drastically increased. This requirement becomes even more important given the fact that the diesel engine has a widely varying range of exhaust gas temperatures. However, the catalysts based on TiO_2-WO_3-V_2O_5 have their maximum activity between 300 and 400°C, the activity decreasing rapidly with falling temperature. The following measures are useful to reduce the catalyst volume:

1. Increase the intrinsic activity of the SCR catalyst. If a high selectivity for N_2-formation is to be kept, i.e. little formation of N_2O and NO_x at higher temperatures, the standard SCR catalyst systems based on TiO_2-WO_3-V_2O_5 or TiO_2-MoO_3-V_2O_5 are still the systems of choice. Their drawback is a limited activity, which is

particularly evident at the lowest loads/exhaust temperatures of the engine. The concentration of vanadia may be increased to boost the intrinsic activity, but concentrations above ≈3% will lead to a loss of selectivity of these catalysts.

2. Increase of the cell density of the monolith, i.e. the geometric surface-to-volume ratio of the catalysts. Increasing the cell density from 36 cpsi to 300 cpsi is accompanied by an increase of the volumetric surface from ≈770 m^2/m^3 to ≈2200 m^2/m^3. SCR extruded catalysts based on TiO_2-WO_3-V_2O_5 with cell densities up to 300 cpsi have been already developed by some manufacturers. Still higher cell densities can be realized by coating techniques using cordierite or metallic substrates. This is due to the fact that cordierite or metal foils have much better mechanical properties and, therefore, allow thinner walls of the monolith. In extruded catalysts the entire wall consists of TiO_2-WO_3-V_2O_5 plus admixtures of binders and glassfibers to increase the mechanical properties. Higher cell densities will be especially effective at high temperatures, where the SCR reaction proceeds mainly in the top layer of the catalyst.

RESULTS

We started to investigate coated SCR catalysts in 1996 [12]. The reason for this engagement was the fact that such catalysts were not available to us at that time. We expected superior performance in nonstationary applications due to the prospect of using higher cell densities and the lower mass of active material per volume. The latter reduces the problem of ammonia spikes following sudden load increases.

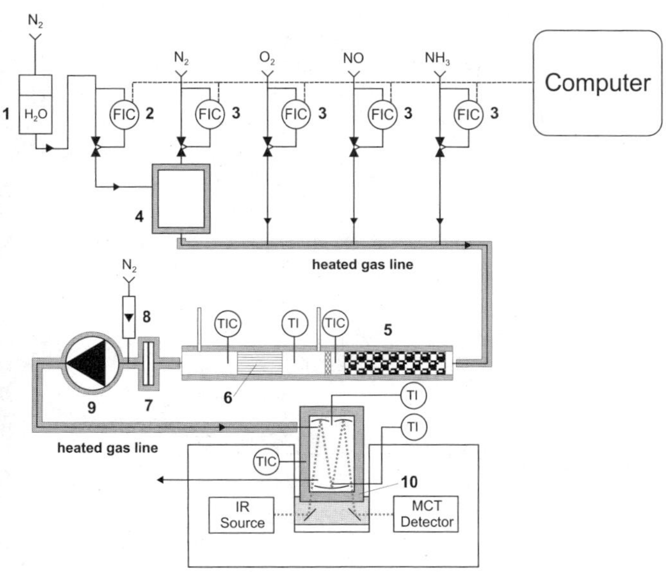

Figure 2: Laboratory test stand: 1, water reservoir; 2, liquid mass flow controller; 3, mass flow controller; 4, water evaporator; 5, reactor; 6, catalyst sample; 7, filter; 8, flow meter; 9, diaphragm pump; 10, gas cell

In the standard test, the catalysts are first tested in the laboratory test stand at a GHSV of 52,000 h^{-1} (Fig. 2). The basic gas composition is adapted to a typical diesel exhaust containing 10% oxygen and 5% water with balance nitrogen. Standard tests are made with 1000 ppm NO and varying the concentration of NH_3 in the feed, i.e. varying the stoichiometric ratio NH_3:NO. Gas analysis is performed by FTIR spectroscopy (Nicolet Magna 560) allowing to measure simultaneously NO, NO_2, NH_3, N_2O.

In order to asses the performance of catalysts we opt to plot NH_3-slip = f(DeNOx), where DeNOx denotes the conversion of NO_x in percent. For the practical use of a catalyst, the maximum possible DeNOx at high ammonia slip is of no interest, the important issue being the lower DeNOx at a limited ammonia slip (typically 10 ppm). Figures 3 to 5 compare the performance of three catalysts at various temperatures. The closer a curve is to the right lower edge of the graph, the better the catalyst is at this temperature. The strong improvement over the last years is evident from these figures. The basic properties of the three catalysts are resumed in Table 1.

Table 1: Properties of the three catalysts

property	units	D31	D42	K51
type	--	extruded	extruded	coated
production year	--	1995	1999	2000
cell density	cpsi	200	300	400
V_2O_5-content	%	≈ 3	≈ 3	≈ 3

According to Fig. 3 catalyst D31 could hardly reach 30 % DeNOx @ 10 ppm NH_3. This is a commercial extruded catalyst with 200 cpsi. This cell density was already a strong improvement over the usual commercial catalysts available in 1995. Figure 4 shows the performance of D42, an extruded catalyst with 300 cpsi. This catalyst reaches almost 90 % DeNOx at temperatures above 350°C. Still somewhat better results are obtained with the coated catalyst K51 from our own development (Fig. 5). Both the increased cell density and the higher intrinsic catalytic activity of K51 contribute to the better performance of this catalyst compared to D42.

However, an initial high activity is not the unique quality criterion for such a catalyst. We have found that thermal stability should also be tested. Figure 6 compares the activity of several coated catalysts with 400 cpsi cell density. For simplicity, only the DeNOx at 10 ppm NH_3-slip is given. Thermal aging was performed in this case during 100 hours at 550°C. More critical aging tests are performed at temperatures of 600°C and even 650°C. From these results it is evident that K99 is a much improved version of the earlier catalyst (K51). We have also found thermal stability to be a problem for some commercial catalysts.

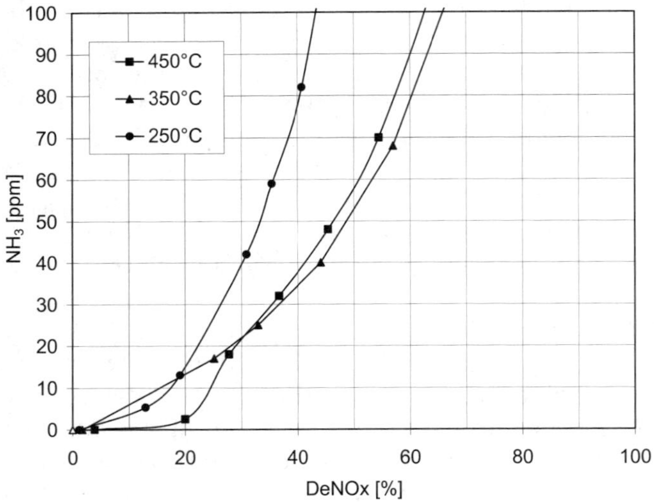

Figure 3: Lab test of catalyst D31 (200 cpsi, extruded)

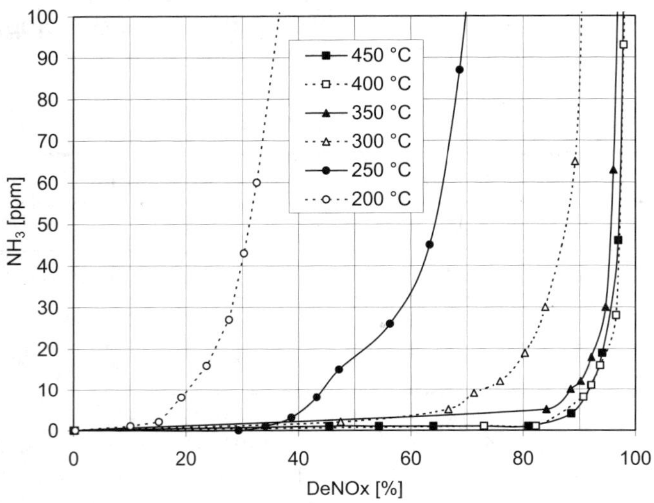

Figure 4: Lab test of catalyst D42 (300 cpsi, extruded)

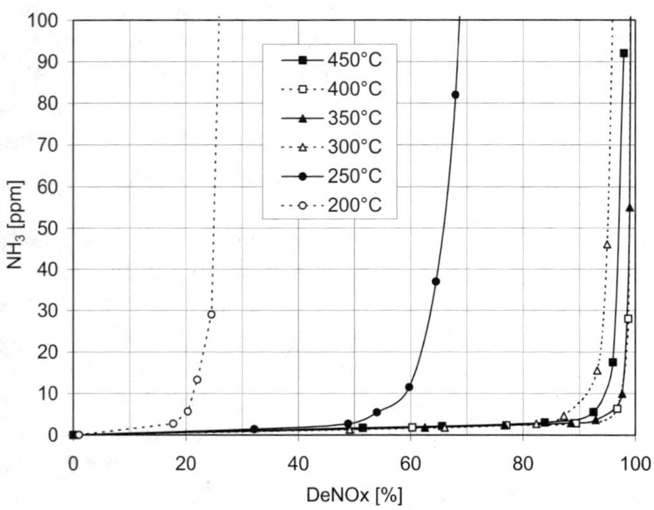

Figure 5: Lab test of catalyst K51 (400 cpsi, coated)

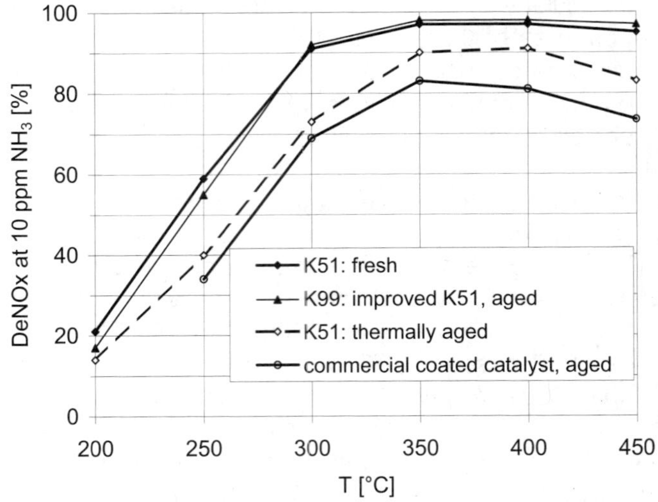

Figure 6: DeNOx @ 10 ppm NH_3 after thermal aging (100 h @ 550°C) for three coated catalysts with 400 cpsi

THE FAST SCR REACTION

As already mentioned in the chemistry section, the fast SCR reaction is another means to increase the NO_x conversion for a given catalyst which is particularly effective at low temperatures (below 300°C).

Figure 7 shows the improvement in performance that can be obtained by using a 1:1 mixture of NO + NO_2 instead of NO only. These results refer again to catalyst K51 and should be thus compared with Fig. 5 (NO only). Comparing the results for 10 ppm NH_3 slip at 200°C, only ≈21 % DeNOx are obtained with pure NO but ≈93 % with the NO-NO_2 mixture. The mixture works even at 150°C where still a DeNOx of ≈43 % is obtained.

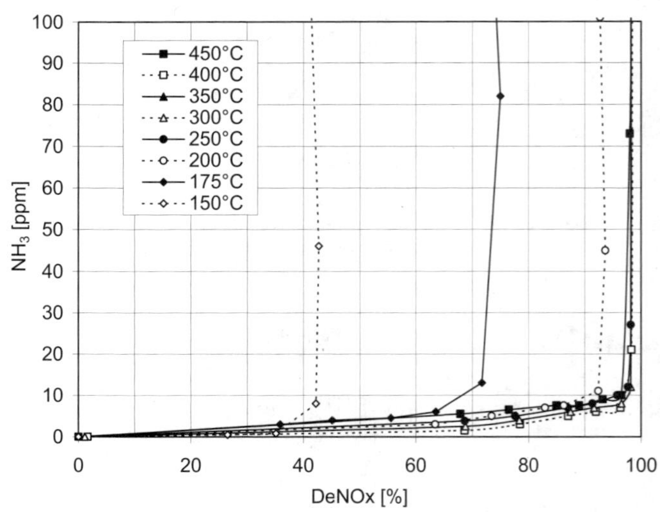

Figure 7: Lab test of catalyst K51 in a 1:1 mixture NO+NO_2. GHSV = 52,000 h^{-1}

The conversion of NO to NO_2 is best performed by a strong oxidation catalyst (Pt-based) upstream of the SCR catalyst. Such a catalyst has two additional effects:

1. It oxidizes also SO_2 to SO_3 in the exhaust. This will increase the sulfate fraction in the particulates and may also cause other undesired effects. It is therefore generally assumed that an SCR system with oxidation-precatalyst should be operated only with low sulfur fuel (S<30ppm).

2. A positive effect is the oxidation of hydrocarbons at low temperatures (T<300°C) that results in a better DeNOx performance of the SCR catalyst at these temperatures. However, experiments on the diesel test stand have shown also a decrease of NO_2 over the oxidation catalyst [13]. This means that the oxidation of hydrocarbons is at least partially due to the reduction of NO_2 to NO (confer the CRT trap, where NO_2 reduces soot). Taking both effects together, the removal of hydrocarbons overrides the removal of NO_2, the net effect being an improved DeNOx at low temperatures.

However, even with an oxidation precatalyst, the exhaust temperatures cannot be lowered to any desired value. This is because at temperatures below 200°C the fast SCR reaction (2) is competed by another reaction that leads to the formation of ammonium nitrate:

$$2\,NO_2 + 2\,NH_3 \rightarrow NH_4NO_3 + N_2 + H_2O \qquad (7)$$

Ammonium nitrate is a solid with a melting point of 170°C. It may thus deposit in the pores of the catalyst in solid or liquid form and cause its temporary deactivation. The detailed behavior of SCR catalysts in NO-NO_2 mixtures at temperatures below 200°C has been investigated in detail [13, 14]. The results from these experiments suggest that for typical SCR catalysts the lower threshold temperature, at which no degradation of catalyst activity is observed, is around 180°C.

MINIMIZING AMMONIA SLIP

In the aforementioned discussion on improving SCR catalysts it was stated that the ammonia slip should not be too high and we have set this limit to ≈10 ppm. In the nonstationary operation of the catalysts, fast increases of the load will cause additional spikes in the emission of ammonia. The reason is that SCR catalysts adsorb considerable amounts of ammonia and this adsorption decreases with temperature.

Figure 8 compares the total amount of ammonia adsorbed in an extruded and a coated catalyst sample with 7 cm^3 volume. It can be seen that the quantity of adsorbed ammonia decreases strongly with temperature. The extruded catalyst adsorbs much higher quantities than the coated catalyst as it has about five times more active mass. Due to the method of determination, i.e. under SCR conditions with NO present, this expected difference is only visible at low temperatures.

Further details on the subject of ammonia adsorption on these catalysts have been published elsewhere [15].

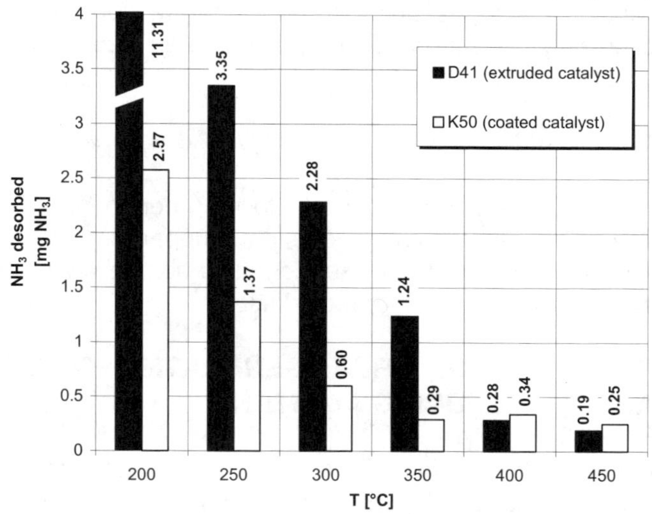

Figure 8: Total ammonia adsorption of coated catalyst K50 and extruded catalyst D41 under SCR conditions. NH_3(in) = NO(in) = 1000 ppm. Catalyst volume = 7 cm^3

Figure 9 shows the effect of a fast load increase in a diesel test stand experiment. A considerable ammonia peak is obtained if the dosage of urea is increased abruptly with load. Much better results are obtained if the actual catalyst activity is taken into account, i.e. the actual catalyst temperature. Still better results are possible by a closed feedback control. In the present example, the slip of NH_3 was measured and taken as the controlling parameter. It is evident that an optimized dosing strategy can also help in minimizing ammonia peaks associated with load changes.

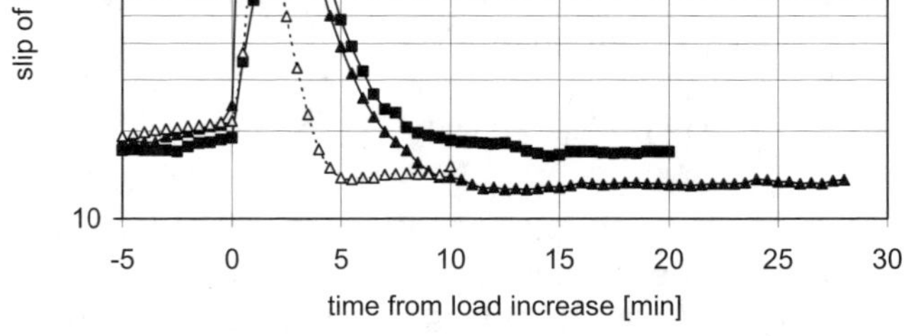

Figure 9: Load increase 25→75 kW, corresponding to a temperature increase 230→400°C

Another possibility to diminish ammonia slip is the addition of an oxidation catalyst downstream of the SCR catalyst. The use of such an oxidation catalyst is also proposed in the "VHRO-system" [8, 20] using a structured catalyst composed of four single catalysts. However, oxidation catalysts that convert NH_3 fully into nitrogen and water are not known, their lacking selectivity causing the formation of considerable fractions of N_2O especially at intermediate temperatures (250-300°C) according to:

$$4\,NH_3 + 4\,NO + 3\,O_2 \rightarrow 4\,N_2O + 6\,H_2O \qquad [8]$$

N_2O is highly undesirable due to its greenhouse effect and its ability to destroy ozone in the upper atmosphere. The additional cost and the increased complexity of the system represent further problems.

REACTIONS OF UREA: GENERATION OF THE ACTUAL REDUCING AGENT NH_3

For reasons of safety and toxicity, urea is the preferred selective reducing agent for mobile SCR. However, using urea instead of ammonia will complicate the DeNOx chemistry by the appearance of isocyanic acid as an important intermediate.

Urea is usually applied in the form of an aqueous solution of 30-40 % which is atomized into the exhaust. The decomposition of urea involves the following steps:

1. Heating in the hot exhaust will first cause the evaporation of water from the droplet, resulting in a droplet consisting of concentrated urea solution or molten urea:

$$NH_2\text{-}CO\text{-}NH_2(aq) \rightarrow NH_2\text{-}CO\text{-}NH_2(melt) + x\,H_2O \qquad (9)$$

2. Further heating will cause decomposition of urea by thermolysis yielding ammonia and isocyanic acid. In this way equimolar amounts of ammonia and isocyanic acid are formed:

$$NH_2\text{-}CO\text{-}NH_2(melt) \rightarrow NH_3 + HNCO \qquad (10)$$

3. Isocyanic acid may react further with water contained in the exhaust. This is a hydrolysis reaction yielding ammonia and carbon dioxide:

$$HNCO + H_2O \rightarrow NH_3 + CO_2 \qquad (11)$$

This reaction does not occur in the gas phase but it occurs on the typical SCR catalysts based on TiO_2-WO_3-V_2O_5. This reaction is exothermic whereas the two previous ones are endothermic.

We have shown that the rate of HNCO hydrolysis is much faster than the rate of the SCR reaction, at least at low-to-medium temperatures [16]. Therefore, we believe that NH_3 is still the intrinsic reducing agent even if urea is the applied reducing agent.

In order to speed up the thermohydrolytic decomposition of urea, the use of a hydrolyzing catalyst in front of the SCR catalyst has been proposed [8, 17]. However, the hydrolysis of HNCO is not the only reaction contributing to the delayed liberation of ammonia, but all three reactions require additional time. A mobile SCR-system will have only a short distance between the engine exhaust (where urea may be injected) and the catalyst entrance. This leads to a short residence time of urea in this section and therefore steps (9) and (10) will not be completed at the catalyst entrance. As a result, a mixture of NH_3, HNCO and urea will be fed to the catalyst. Figure 10 shows the relative amounts of these three components at the catalyst entrance measured in a typical experiment on the diesel test stand at PSI.

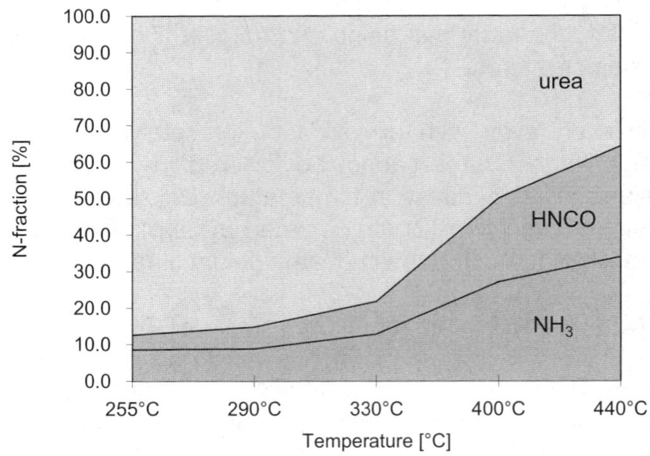

Figure 10: Ammonia, isocyanic acid and urea at various temperatures at the catalyst entrance. Residence time urea injection-catalyst entrance = .09s at 440°C

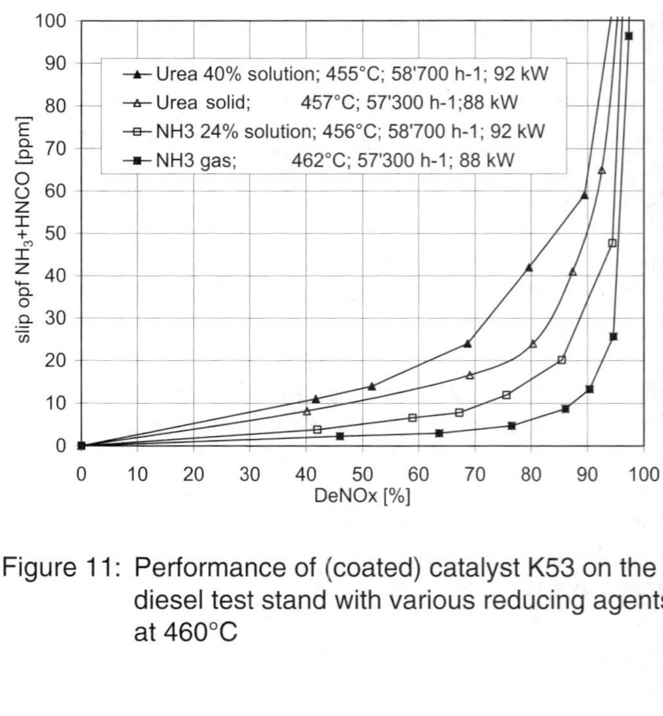

Figure 11: Performance of (coated) catalyst K53 on the diesel test stand with various reducing agents at 460°C

The delayed liberation of ammonia will cause a loss of SCR-performance compared to the use of ammonia. This effect is particularly pronounced at high space velocities and at high temperatures as shown in Fig. 11 for various reducing agents at 460°C: the best performance is obtained with gaseous ammonia, followed by ammonia solution, solid urea, and finally urea solution. It is evident that ammonia outperforms urea.

Aiming at compact DeNOx systems will lead to short residence times ahead of the SCR catalyst. This causes considerable mixing problems for NO_x with the reducing agent. The mixing problem is further aggravated if an additional precatalyst for oxidizing NO is used. In this case urea can be injected only after the oxidation catalyst leading to even shorter mixing and residence times.

FURTHER PROBLEMS WITH UREA

Due to its low toxicity and its ease of handling, urea is presently the most favored candidate for automotive applications. We have just outlined the difficulties due to its thermal decomposition. In a recent study we have reevaluated four possible SCR agents: urea solution, ammonia, ammonia solution and ammonium carbamate [18]. According to this study, urea has some further disadvantages:

1. Emission of isocyanic acid. The isocyanic acid formed during the thermolysis of urea is not fully decomposed to ammonia; hence, in addition to ammonia isocyanic acid may be also found in the treated exhaust ("slip of isocyanic acid"). We have measured HNCO routinely in experiments on the diesel test stand. Our experience is that the emission is usually below 5 ppm; however, at very low and at very high temperatures concentrations up to 50 ppm have been observed with underdimensioned catalysts. It is presently not yet clear how the toxicity of HNCO should be classified because organic isocyanates have a recent bad reputation of inducing hyperreactive respiratory diseases. On the other hand, there are claims that the toxicity of HNCO should be low due to its fast hydrolysis in the lungs yielding ammonia and carbon dioxide.

2. The high freezing point of aqueous urea solutions. The preferred concentration is 32.5% which is the eutectic with a freezing point of -11°C. This will pose problems in winter use. Therefore, it has been suggested to use solid urea. The use of solid urea would also greatly simplify the infrastructure for distribution of the reducing gent. Some efforts are already directed toward the development of dosing systems for solid urea, but we expect this dosing unit to be an expensive and critical component.

3. Possible formation of high molecular weight compounds. Due to the high reactivity of isocyanic acid other compounds may be formed. These include cyanuric acid (trimerized form of HNCO, a solid), biuret and melamine etc. We had studied this problem already in the early years of research on stationary SCR, as it seemed quite probable that nozzles might be clogged and deposits formed in the flue duct. Today, we think that this can be handled technically and the emission of secondary high molecular weight products is very small [2].

DEGRADATION OF CATALYSTS

In the section "improving SCR catalysts" we have already discussed the problem of high temperature stability. We have seen that temperatures above 500°C may become critical for some catalysts. Optimized catalysts should stand temperatures of 600-650°C for a short time. Although exhaust temperatures up to 700°C are within the range of modern diesel engines, such operating conditions are rare, of short duration and the heat capacity of the exhaust system will further reduce possible peak temperatures in the catalyst itself. Therefore, we do not expect thermal aging to be a serious problem for the application of vanadia-based catalysts.

The ternary catalyst system TiO_2-WO_3-V_2O_5 is resistant to sulfur. As a rule the presence of sulfur oxides even increases the activity at high temperatures. However, at temperatures below ≈250°C a temporary deactivation may be observed which is due to the deposition of ammonium sulfates in the pores. This deactivation will disappear when the catalyst reaches again higher temperatures. We have already mentioned above other temporary deactivations at low temperatures due to hydrocarbons and ammonium sulfate (at very low temperatures).

A serious problem is the sensitivity of these catalysts to compounds with alkaline properties and phosphorous. Both are found in typical motor oil formulations. ZDDP, the very common antiwear booster, contains zinc and phosphorous. An alkalinity reserve in the oil is usually obtained by the addition of calcium or magnesium compounds. It is thus no surprise to find a degradation of the catalysts over extended test periods. Figure 12 illustrates this degradation for a coated catalyst during a 1000 hours test at ≈400°C. The plots show the performance after both 100 and 1000 hours of test, i.e. the slip of NH_3 alone and also the sum (NH_3 + HNCO) is plotted as a function of DeNOx for various temperatures. Due to the fact that the slip of (NH_3 + HNCO) increases much more than the slip of NH_3 alone, we conclude that the hydrolyzing activity has degraded much more than the SCR activity. The loss of both activities is especially pronounced at the higher temperatures. Due to the fact that the catalysts do not show thermal aging at the test temperature of 400°C we attribute the loss of activity to the effect of lubricating oil additives. We suggest that this point needs further investigation: the ash content of lubricating oils (especially phosphorous and basic metal compounds) should be reduced and the oil consumption of the engine should be made as low as possible.

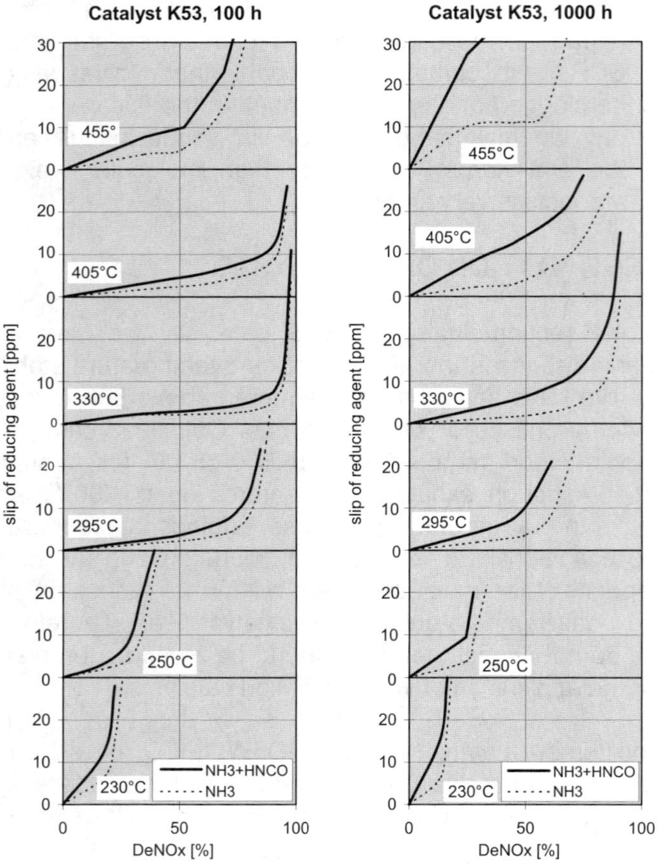

Figure 12: Aging of catalyst K53 (coated) on the diesel test stand - performance at various temperatures

CONCLUSIONS AND DISCUSSION

The development of urea-SCR systems for automotive applications requires a deeper understanding of the underlying reactions and the development of catalysts of higher activity. Considerable progress has been obtained in both areas during the last years. The volumetric activity of the catalysts could be strongly improved. By the addition of other catalysts upstream and downstream of the SCR catalyst the performance could be further enhanced. However, the question still remains if such a complex system will be the only solution for every application or if there exist situations where a simple system containing just the SCR catalyst would perform equally well. Especially the costs of a sophisticated system with a multi-structured catalyst are still an open question.

The SINOx system for trucks uses an SCR catalyst only. It was proposed by Siemens as early as 1996 [7] and has been tested by a number of trucking companies in the following years. The original system was quite voluminous and has been developed further to a more compact shape [19]. Whereas the original system needed 63 liters of an extruded catalyst with 100 cpsi, the newer system uses 30 liters of a 200 cpsi extruded catalyst (for an engine with 12 liters of cylinder volume). The commercial introduction is planned in this year.

More refined systems composed of structured catalysts have been described by MAN [8, 17] and Degussa [20]. Up to four catalysts are used in these systems and the maximum version is given in Fig. 13. The functions of the individual catalysts are:

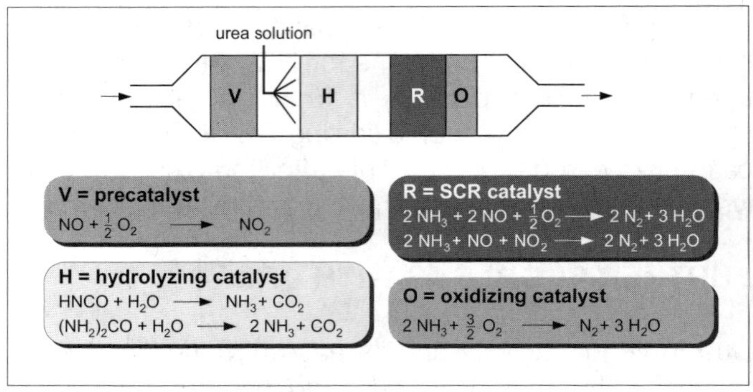

Figure 13: Scheme of the "VHRO-system" according to MAN [8, 17]

1. **V** is a strongly oxidizing precatalyst for the conversion of NO to NO_2 permitting utilization of the fast SCR reaction (2).

2. **H** is a hydrolyzing catalyst for the conversion of urea into ammonia and carbon dioxide. In order to obtain also a high trapping effect for urea aerosol particles, optically dense structures are the preferred geometry.

3. **R** (MAN) or **S** (Degussa) is the actual SCR catalyst on which both the fast and the standard SCR reaction take place.

4. **O** is another oxidation catalyst for the elimination of ammonia slip.

For all systems high cell density coated catalysts are used. Originally, cell densities up to 800 cpsi were suggested. However, due to pressure drop considerations there is probably a useful limit at 300 to 400 cpsi. In a more recent paper [17] Jacob et al. even observed a significant reduction of the carbon particle mass, which is attributed to a temporary trapping of soot particles on the H-, R- and O-sections of the catalyst and oxidation by NO_2. This is another interesting observation as it allows the reduction of soot particles without the use of soot filters and their well-known regeneration problems.

Degussa has also studied a simpler design, HSO, where the preoxidation catalyst has been omitted [20]. In the ESC test, the HSO system reached only 45 % NO_x reduction compared to 75 % with the VHSO system. Other structured catalyst systems are certainly possible. The question as to how complex a urea-SCR system of the future will be, cannot yet be answered definitely. In trying to find the best answer, the following points should be taken into account:

- A system containing V- or O-oxidation catalysts (platinum metal based) requires the use of low sulfur fuel. It is now sure that the sulfur content of diesel fuel will be limited to 50 ppm by EU regulations from 2005 on, as this is a basic requirement for practically all aftertreatment methods. ACEA now even asks for a sulfur limit of 10 ppm, as some techniques require such levels, especially the NO_x-storage-reduction catalyst.

- The primary effect of the preoxidation catalyst is to increase the NO_2 fraction of the exhaust thus permitting the fast SCR reaction. The second effect is a considerable oxidation of hydrocarbons which inhibit the SCR reaction at low temperatures. This will also help to increase the NO_x conversion at low temperatures, especially in the case of engines having a substantial primary emission of hydrocarbons.

- The decision for installing a preoxidation catalyst will strongly depend on the minimum exhaust temperatures, at which a substantial NO_x conversion should be obtained. Trucks or busses for long distance driving will probably not require such a catalyst, because driving periods at low load (= low exhaust temperatures) are rare. On the other hand, busses and trucks that run mainly in city areas rarely reach exhaust temperatures above 250°C and thus might profit considerably from a preoxidation catalyst. The same is true for passenger cars: due to the generous sizing of the engine exhaust temperatures will hardly reach 200°C in a city test cycle thus requiring at least the use of a preoxidation catalyst.

- The true value of the hydrolyzing or H-catalyst is not generally embraced. This is due to the fact that the SCR catalyst itself possesses excellent hydrolyzing properties. The real value of an additional H-catalyst might be more its additional contribution in trapping urea aerosol particles thus preventing delayed thermolysis to happen also on the SCR catalyst.

- The performance of a complex system comprising various catalysts should be compared with a system of the same volume with just an SCR catalyst. The latter system would be much simpler, as it contains only an SCR catalyst of an increased volume.

- Finally, the much higher costs of a complex system will have to be also considered.

ACKNOWLEDGMENTS

The financial support of the Swiss Federal Office of Energy (BfE) is gratefully acknowledged and we are indebted to Liebherr Machines Bulle SA for providing us the diesel engine.

REFERENCES

1. Bosch, H.; Janssen, F. Catalytic Reduction of Nitrogen Oxides. *Catal. Today* **1988**, *2,* 369.
2. Koebel, M.; Elsener, M.; Marti, T.. NO_x-Reduction in Diesel Exhaust Gas with Urea and Selective Catalytic Reduction. *Combust. Sci. and Tech.* **1996**, *Vol. 121*, 85.
3. Takahashi, N. et al. the new concept 3-way catalyst for automotive lean-burn engine: NO_x storage and reduction catalyst. Proc. 1st World Congress on Environmental Catalysis, Pisa, May 1-5, 1995, p 45-48.
4. Brandt, S. et al. Entwicklungsfortschritte bei NO_x-Adsorber-Katalysatoren für magerbetriebene Ottomotoren. Kraftfahrwesen und Verbrennungsmotoren - 3. Stuttgarter Symposium February 23-25, 1999, 83.
5. Shelef, M. Selective Catalytic Reduction of NO_x with N-Free Reductants. *Chem. Rev.* **1995,** *95*, 209.
6. Misono, M. Catalytic reduction of nitrogen oxides by bifunctional catalysts. *Cattech* **1998**, *3,* 53.
7. Fränkle, G. et al. SINOx, The Exhaust Gas Purification System for Trucks. 18th Int. Vienna Motor Symposium,, April 24 and 25, 1996.
8. Jacob, E.; Emmerling, G. et al. NOx-Verminderung für Nutzfahrzeugmotoren mit Harnstoff-SCR-Kompaktsystemen (GD-KAT). *Fortschritt-Berichte VDI* Reihe 12, Nr. 348, Bd. 1, p 366. 19th Int. Vienna Motor Symposium,. May 7-8, 1998.
9. Koebel, M.; Elsener, M.; Kleemann, M. Urea-SCR: A promising technique to reduce NOx emissions from automotive diesel engines. *Catalysis Today* **2000**, *59*, 335.
10. Kato, A.; Matsuda, S.; Nakajima, F.; Kuroda, H.; Narita,T. *J. Phys. Chem.* **1981**, *85*, 4099.
11. Tuenter, G.; Leeuwen, W.; Snepvangers, L. Kinetics and mechanism of the NO_x reduction with NH_3 on V_2O_5-WO_3-TiO_2 catalyst. *Ind. Eng. Chem. Prod. Res. Dev.* **1986**, *25*, 633.
12. M. Kleemann: Beschichtung von Cordierit-Wabenkörpern für die selektive katalytische Reduktion von Stickoxiden. Thesis ETH Zurich Nr. 13401 (1999)
13. Koebel, M.; Elsener, M.; Madia, G. NO_x Reduction in Diesel Exhaust by Urea SCR at Low Temperatures. *MTZ wordwide* **2001**, *62*, 35
14. Koebel, M.; Elsener, M.; Madia, G. Reaction Pathways in the Selective Catalytic Reduction Process with NO and NO_2 at Low Temperatures. *Ind. Eng. Chem. Res.* **2001,** *40*, 52
15. Kleemann, M.; Elsener, M.; Koebel, M.; Wokaun, A. Investigation of the Ammonia Adsorption on Monolithic SCR Catalysts by Transient Response Analysis. *Appl. Catal. B* **2000**, *27,* 231

16. Kleemann, M.; Elsener, M.; Koebel, M.; Wokaun, A. Hydrolysis of Isocyanic Acid on SCR Catalysts. *Ind. Eng. Chem Res.* **2000,** *39,* 4120
17. Jacob, E.; Döring, A. GD-Kat: Abgasnachbehandlungssystem zur simultanen Kohlenstoffpartikel-Oxidation und NO_x-Reduktion für EURO 4/5-Nfz-Dieselmotoren. 21st Int. Vienna Motor Symposium, May 4-5, 2000
18. Koebel, M. Vor- und Nachteile stickstoffhaltiger Reduktionsmittel beim Einsatz in mobilen SCR-Systemen. 3. Dresdener Motorenkolloquium, Mai 20-21, 1999, 76-83
19. Hofmann, L.; Mathes, W.; Fischer, S. Die Entwicklung des SINOx-Systems für Nutzfahrzeuge zur Serienreife. 20th Int. Vienna Motor Symposium, May 6-7, 1999
20. Gieshoff, J.; Schäfer-Sindlinger, A.; Spurk, P.C.; van den Tillaart, J.A.A.; Garr, G. Improved SCR Systems for Heavy Duty Applications. SAE Nr. 2000-01-0189

CONTACT

Dr. Manfred Koebel
Exhaust Gas Aftertreatment
Paul Scherrer Institute
CH-5232 Villigen PSI
Tel: ++41-56-310 26 04
Fax: ++41-56-310 21 99
e-mail: manfred.koebel@psi.ch

2003-01-3362

Control Strategy for NO_x - Emission Reduction with SCR

Dieter H. E. Seher, Michael Reichelt and Stefan Wickert
Robert Bosch GmbH

ABSTRACT

Future emission standards for heavy-duty vehicles like Euro 4, Euro 5, US '07 require advanced engine functionality. One contribution to achieve this target is the catalytic reduction of nitrogen oxides by injection of urea water solution to the exhaust gas. An overview on a urea dosing system, also called DENOXTRONIC, is given and a dosing strategy is described.

INTRODUCTION

The current emission limits like Euro 3 and US '02/'04 in general are realized by measures on combustion optimization along with exhaust gas recirculation. While exhaust gas recirculation in of Euro 3 engines in combination with advanced timing can be used to improve the fuel consumption, exhaust gas recirculation is necessary for achieving US '02/'04, [1]. Future emission levels like Euro 5 require additional measures on exhaust gas.

Euro 4 can be met by different strategies [2], see fig. 1: One possibility is a cooled exhaust gas recirculation possibly in combination with a particulate trap.

An alternative approach is the selective catalytic reduction technology (SCR). By using SCR systems the

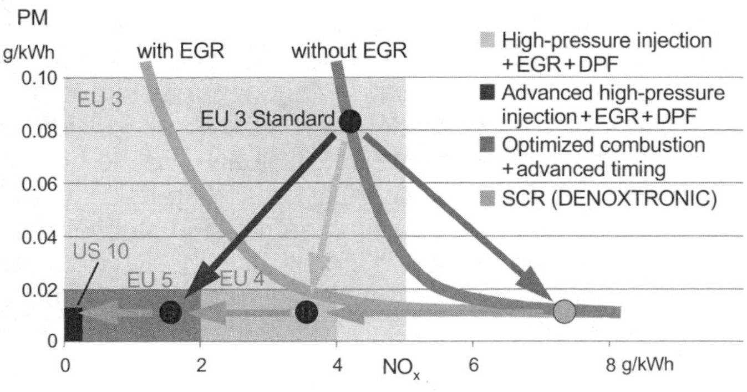

Fig. 1: Strategies for emission reduction

fuel economy can be improved compared to alternative aftertreatment systems because the engine can be operated at higher NO_x raw emissions due to a more advanced begin of injection. Furthermore the particulate emission is being reduced.

In a basic study in 1993 the potential of the SCR-technology [4] was evaluated. It was shown that the NO_x emissions can be significantly reduced in stationary and transient engine operation. A comparison of different technologies [5] demonstrates that SCR systems using ammonia obtained from urea as reducing agent are the preferable solution for commercial vehicle engines. Further studies [6, 7, 8] analyzed SCR systems with urea as reducing agent deeper. They concluded that this technology is an important approach to meet future emission limits.

Thus the usage of the SCR-system is the primary path in order to fulfill Euro 5. For meeting the emission limits US´07/´10 the SCR system is also a reasonable technical solution.

SYSTEM OVERVIEW

The basic principle of the SCR process is the reduction of NO_x by ammonia. The ammonia is formed from a urea water solution with 32.5 % urea content, which is injected into the exhaust gas. This solution is also called AdBlue. The major advantage of urea is its availability on the market in various fields of applications and that it is nontoxic. Disadvantage is the comparably high freezing point of the urea water solution (– 11 ^{0}C). However this can be resolved by an appropriate heating strategy of the urea dosing system (DENOXTRONIC).

An overview of the SCR-system is shown in fig. 2. Exhaust gas from the engine is first led through an oxidation catalyst. Target is that about 50 % of the exhaust gas nitrogen oxides are oxidized to NO_2. $NO:NO_2$ = 1:1 is the best ratio for NO_x reduction. Furthermore hydrocarbons are converted to carbon dioxide and water. This pre-catalyst helps to improve conversion rate of the SCR catalyst. However this process requires fuel with a low sulfur content in order to avoid increased particulate emission by forming of sulfates. Downstream of the oxidation catalyst the urea water solution is injected through the spray tip. For this purpose the required urea quantity corresponding to the NOx emission of the engine is dispersed in a mixing chamber by using compressed air. The result is sprayed into the exhaust gas stream upstream the SCR-catalyst.

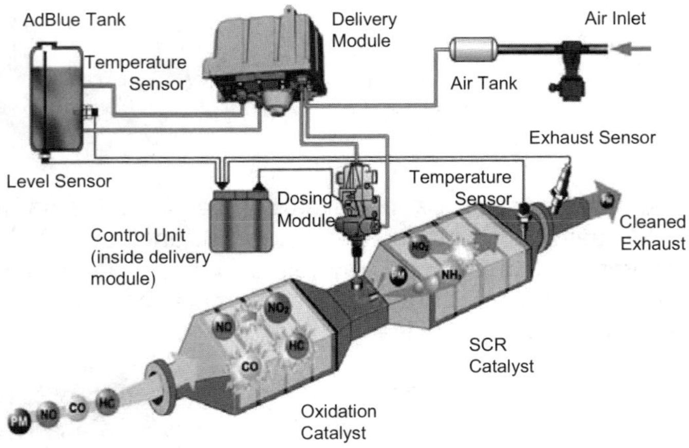

Fig. 2: SCR system with reduction agent urea (AdBlue)

In the first section of the SCR - catalyst the urea is decomposed and ammonia is formed (hydrolysis). In the following section of this catalyst the ammonia reduces the NO_x to nitrogen and water. The main reaction is:

$$2\,NH_3 + NO + NO_2 \rightarrow 2\,N_2 + 3\,H_2O$$

In order to avoid the emission of excess ammonia an oxidation catalyst can be placed downstream to the SCR catalyst. The dosing strategy is designed to avoid this.

The DENOXTRONIC system has the following major tasks: At first the necessary amount of urea according to the NO_x emission of the engine is being injected into the exhaust gas stream. OBD requirements like detecting "major functional failures" have to be fulfilled and diagnostic functions are checking the system all the time. Finally all pipes of the system must be kept at temperatures above the melting point of the urea water solution in order to satisfy operation at low temperatures.

The DENOXTRONIC system works as follows, fig. 2: The urea water solution is pumped from a tank to the dosing valve. In order to maintain an always-defined dosing quantity, the pressure of the solution is monitored by the ECU. The pressure of the air stream is also monitored and kept to the required level during the dosing operation. Both circuits are joined together in the mixing chamber. The necessary urea amount is transported in the dosing pipe by the compressed air and is atomized in the exhaust gas tube by a spray tip.

The electronic control unit (ECU) controls all operations. It contains the required functions, evaluates the sensor signals and activates the actuators. The necessary engine data like engine speed, injected fuel quantity, etc. are obtained from the engine control unit via CAN-interface.

Fig. 3 shows the DENOXTRONIC system. The components are subdivided into a delivery module and a dosing module. The delivery module contains all components processing the urea water solution and compressed air including the ECU, sensors for pressures and temperatures and a filter. It can be located somewhere at the chassis of the vehicle. The dosing module containing the mixing chamber, the dosing valve and a temperature sensor is usually placed near the exhaust gas pipe. Advantage of this modular realization is that the delivery module containing the majority of the system components can be located more flexibly at the chassis.

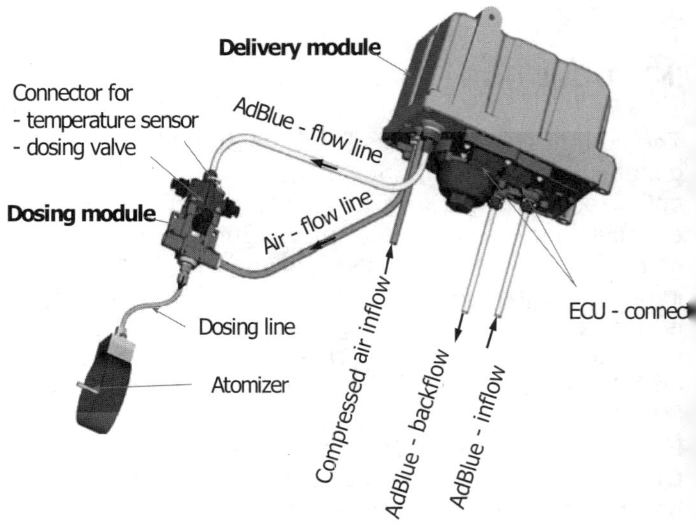

Fig. 3: DENOXTRONIC system

DOSING STRATEGY

OVERVIEW

The dosing strategy takes into account different processes. The required amount of urea depends on the NO_x raw emission of the engine as functions of engine speed, injected fuel quantity and engine temperature. Additionally the conversion rate of the catalyst and the accumulation of ammonia in the catalyst are to be taken into account. The system must be able to react to larger changes of the catalyst temperature due to changes of the engine load. Target is minimizing the NO_x emission and avoiding slip of ammonia under all boundary conditions.

An overview is depicted in fig. 4. First a set value of the urea water solution (UWS) is calculated on a stationary basis. A dynamic model corrects this quantity. It represents the dynamic behavior of the following processes:

- the amount of NO_x produced by the engine,
- the temperature and efficiency of the catalyst,
- the amount of ammonia stored in the catalyst

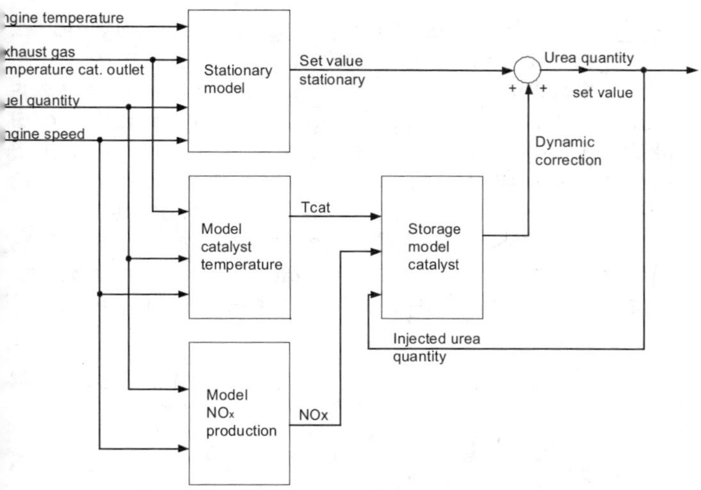

Fig. 4: Overview on dosing strategy

The temperature inside the catalyst, the NO_x produced by the engine and the injected urea quantity are the input variables for the storage model of the catalyst.

STATIONARY MODEL

Fig. 5 shows the algorithm of the stationary model.

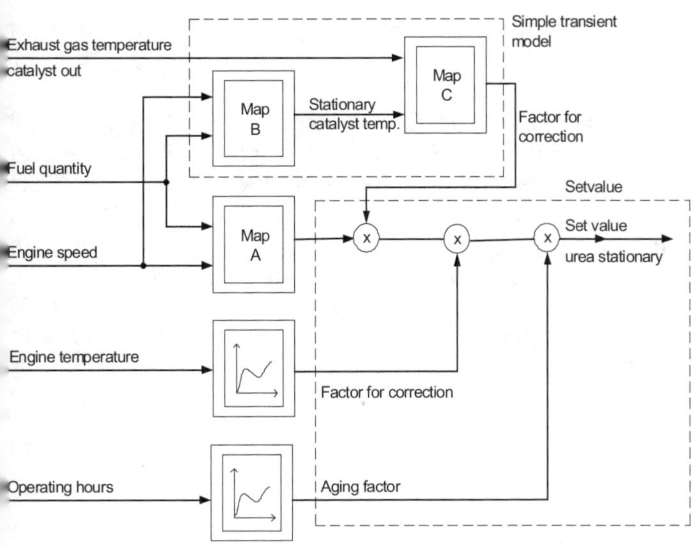

Fig. 5: Stationary model

The map A, a function of engine speed and injected fuel quantity, calculates the basic urea quantity set-value. It represents roughly the amount of NO_x to be converted. This basic set-value is corrected by a function depending on the engine temperature. Thus the effect of the engine temperature to the NO_x emission is taken into account. An aging of the catalyst can be taken into account by a second correction depending on the operating hours of the engine.

In case the catalyst has only a small capacity for accumulating ammonia, an approximation using two maps can be applied. The basic idea of this approach assumes that the catalyst temperature under stationary conditions and its actual temperature are the main effects to the conversion rate and the ammonia accumulation. The exhaust gas temperature measured at the outlet of the catalyst obtains the latter. With map B the stationary temperature is estimated depending on engine speed and injected fuel quantity. Using map C, a function of stationary and actual catalyst temperature, a quasi-dynamic correction of the basic urea quantity is performed.

DYNAMIC MODEL

If catalysts with a higher storage capacity of ammonia are used a more detailed dynamic model can be applied. The temperature inside the catalyst, the NO_x produced by the engine and the injected urea quantity are the input variables for the storage model of the catalyst.

An overview of the dynamic model is shown in fig. 6.

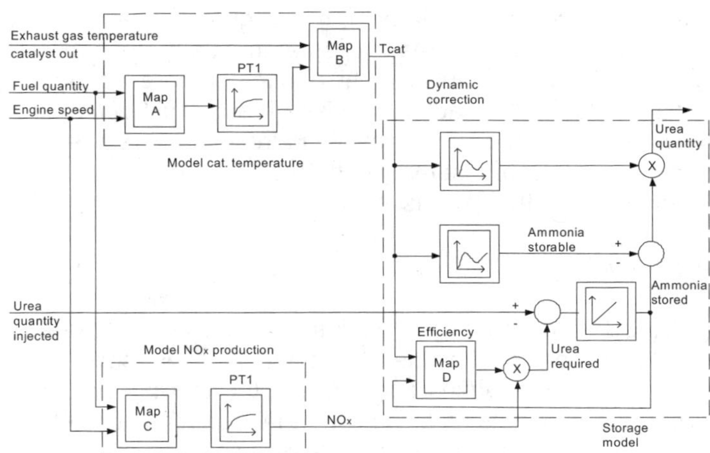

Fig. 6: Dynamic model

One major state variable to be represented by the model is the effective temperature inside the catalyst T_{cat}. It has a major effect on capacity storing ammonia and to the conversion rate of NO_x. It is calculated in the following steps: a map A as a function of engine speed and injected fuel quantity is followed by a first order delay function (PT1). This approximates the temperature of the exhaust gas entering the catalyst. The output of the PT1 and the measured exhaust gas temperature at the outlet of the catalyst result in the estimated temperature of the catalyst T_{cat} calculated by map B

The production of NO_x is estimated by a map C followed by a first order delay function (PT1). The PT1 approximates the delays of the NO_x-production and the transport to the catalyst.

The accumulation process of ammonia is represented by the integrator I. Its Input variable is the difference of the urea quantity dosed to the exhaust gas minus the urea quantity required due to NO_x-reduction. The required urea quantity is calculated from the NO_x production model. It is multiplied afterwards by the efficiency of the catalyst resulting from map D as a function of the catalyst temperature and the amount of ammonia stored in the catalyst (output variable of the integrator I).

In order to calculate the dynamic portion of the urea quantity to be dosed to the exhaust gas, the amount of ammonia accumulated in the catalyst must be compared with a nominal amount of ammonia the catalyst should store for optimal NO_x conversion with minimal risk of ammonia slip. This nominal amount of ammonia as a property of the catalyst is calculated by a function depending on the temperature of the catalyst. Afterwards the difference of this amount minus the stored amount of ammonia is calculated. This difference represents the need of additional ammonia, if it is greater than zero. A negative value indicates excess ammonia in the catalyst. The difference is multiplied by a function representing the catalyst's ability to adsorb or desorb ammonia depending on its temperature. This results in the dynamic correction of the stationary set-value of the urea quantity to be sprayed into the exhaust gas.

SUMMARY

The dosing strategy contains a stationary approach, which is corrected by the output of a dynamic model. The dynamic model takes into account the storage behavior of the catalyst, its efficiency and the production of NO_x. A major influence on the processes in the catalyst is its temperature inside which is obtained by a dynamic model too.

The algorithms described are realized in the electronic control unit (ECU). Apart from that the ECU controls the dosing process itself like operating the valves and heating of the pipes at low temperatures. Additional functions are monitoring and diagnostics of the DENOXTRONIC system and onboard diagnosis (OBD) and offboard diagnosis.

ENGINE RESULTS

The SCR system was tested on a six-cylinder truck engine. The Diesel fuel injection system used is a Bosch in-line pump system. The catalyst used is a first prototype with a maximum NO_x conversion of 65 % at steady state conditions. Modern catalysts are able to operate at conversion rates of 80 %. With this setup European transient cycle tests (ETC) were performed. A first application gave the following results.

Fig. 7 shows ETC-test result using the stationary model. The dynamic behavior is taken into account with the simple dynamic correction by using maps. The engine is operated with a comparably high NO_x raw emission of 9 g/kWh. Using the SCR system results in a reduction of NO_x to 3.9 g/kWh. The ammonia slip is 2.0 ppm on an average. For a short time 46 ppm ammonia slipped through. An average NO_x conversion rate of 57 % is obtained. The SCR system required around a mass of urea of 4 % of the diesel fuel consumed during the test.

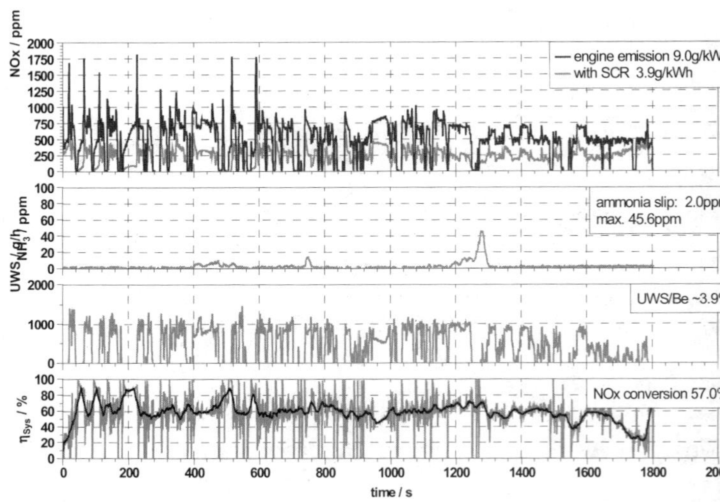

Fig. 7: European Transient Cycle (ETC) results with stationary model

The results of the ETC with the dynamic storage model are shown in fig. 8. The raw emission of the engine and other boundary conditions were the same. The SCR system reduces the NO_x emission to 3.8 g/kWh. The ammonia slip is significantly lower: 1.1 ppm on an average, 25.6 ppm maximum for a short time. NO_x conversion rate and urea consumption are in the same order of magnitude as by using the stationary model with three maps.

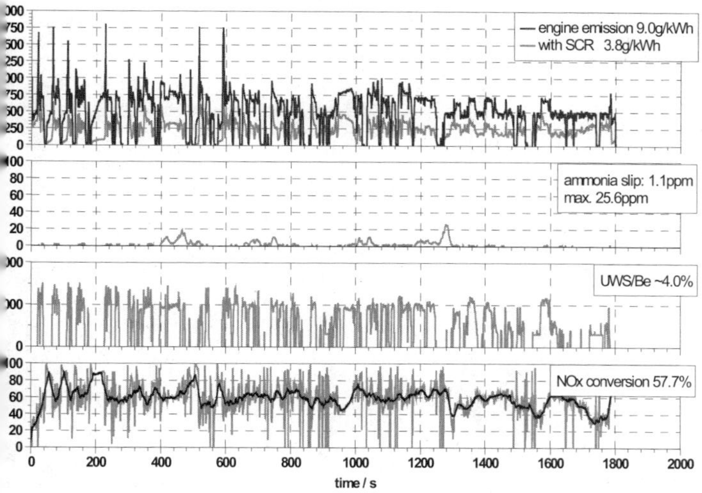

Fig. 8: European Transient Cycle (ETC) results with dynamic model

CONCLUSION

In order to fulfill future emission limits for heavy-duty vehicles like Euro 4, Euro 5 and US'07/'10 measures on the exhaust gas are required. The selective catalyst reduction system (SCR system) with urea as reducing agent is a promising approach for reduction of NO_x emissions. Along with that the fuel consumption can be reduced by upto 5 % - 7 % compared to alternative measures of NO_x reduction. This takes into account environmental aspects like CO_2-emission as well as economical aspects.

The DENOXTRONIC system provides the reducing agent urea water solution, also called AdBlue, according to the raw emission of the engine. The properties and the behavior of the catalyst are taken into account by mathematical models.

Engine tests based on the European transient cycle show with first prototypes of catalysts a NO_x conversion rate of approximately 60 % with an average ammonia slip of 2 ppm. Using a dynamic model for the dosing strategy the NH_3 slip can be reduced to an average value of approximately 1 ppm compared to the stationary model. With modern catalysts a NO_x conversion rate of 80 % is achievable without the usage of an exhaust gas sensor.

Thus an SCR-catalyst system with urea as reducing agent provides an important contribution for achieving future emission limits.

Due to the market demand Bosch is developing the DENOXTRONIC system. It is currently in the field test phase in different trucks. Start of serial of production will be in 2004.

The next step in the ongoing development is a urea dosing system, which does not need a medium like compressed air for spraying urea water solution into the exhaust gas.

REFERENCES

1. S. Becher, S. Forthmann, B. Tichy: Abgasrückführregelung beim NKW-Dieselmotor im dynamischen Betrieb; Tagung: emission control, Dresden
2. R. Maier, B. Mahr, U. Projahn: Development of Fuel Injection Systems for Future Demands. 2nd AVL International Commercial Powertrain Conference, Graz, Austria, 2003.
3. B. Mahr: Future and Potential of Diesel Injection Systems. THIESEL 2002 Conference on Thermo- and Fluiddynamic Processes in Diesel Engines.
4. G. Hüthwohl, Q. Li, G. Lepperhoff: Untersuchung der NO_x- Reduzierung im Abgas von Dieselmotoren durch SCR-Katalysatoren. MTZ Motortechnische Zeitschrift 54 (1993)
5. W. Weisweiler: Entfernung von Stickstoffoxiden aus Sauerstoff enthaltenden Automobil-Abgasen. Chemie Ingenieur Technik 72 (2003)
6. W. Kind, E. Jacob, W. Müller: NO_x-Verminderung bei Dieselmotoren - Motoruntersuchungen mit Harnstoff als Reduktionsmittel für die katalytische NO_x-Reduktion. MTZ Motortechnische Zeitschrift 62 (2001) 1
7. G. Fränkle et. al.: SINOx Ein Abgasreinigungs--system für Nutzfahrzeuge. 18 Wiener Motoren-symposium 1997
8. E. Jacob et. al.: NOx-Verminderung für Nutzfahrzeugmotoren mit Harnstoff-SCR-Kompaktsystemen. 19. Wiener Motorensymposium 1998

2004-01-1291

NO_x Control Development with Urea SCR on a Diesel Passenger Car

Paul Tennison, Christine Lambert and Michael Levin
Ford Research & Advanced Engineering

ABSTRACT

Diesel vehicles have significant advantages over their gasoline counterparts including a more efficient engine, higher fuel economy, and lower emissions of HC, CO, and CO_2. However, NO_x control is more difficult on a diesel because of the high O_2 concentration in the exhaust, making conventional three-way catalysts ineffective. Two current available technologies for continuous NO_x reduction onboard diesel vehicles are Selective Catalytic Reduction (SCR) using aqueous urea and lean NO_x trap (LNT) catalysts. This paper discusses an application with SCR. SCR with ammonia has been used for many years at stationary sources. Aqueous urea is a convenient way to deliver ammonia onboard a vehicle and high NO_x efficiencies have been shown in past work by Ford and others using urea.

Tailpipe NO_x emissions from a modified European production level 1.8L diesel Ford Focus TDCi were reduced to the range of ULEVII levels (0.05 g/mi NO_x) with a green catalyst system. The green system also had results within the Tier 2 SFTP US06 standard (0.14 g/mi $NMHC+NO_x$). An oxidation catalyst was used to convert engine-out HC and CO upstream of the urea SCR system. Aqueous urea was added to the exhaust using a Ford-developed air-assisted injection system. A base metal/zeolite SCR catalyst utilized the added reductant to convert NOx to N_2 under lean conditions.

INTRODUCTION

Emissions of NOx and particulate matter (PM) are of primary concern for diesel vehicles to meet future emissions standards. Diesel vehicles potentially have a 40% higher fuel economy than current gasoline vehicles with approximately 20% lower CO_2 emissions. However, removal of NOx from lean (high O_2) exhaust gas is not trivial due to a lack of reducing compounds in the exhaust gas. Selective Catalytic Reduction (SCR) with ammonia as the reductant has been used extensively for stationary source NOx control [1]. The high selectivity of ammonia for reaction with NOx in high O_2 environments makes SCR attractive for use on light-duty diesel vehicles. Compared to ammonia, aqueous urea is more convenient for use onboard a vehicle. Feasibility of using urea onboard a vehicle has been proven by past work at Ford [2], Volkswagen [3] and Mack Truck [4], and Daimler-Chrysler [5].

For a successful implementation of Urea SCR in a personal use vehicle the following is required: (1) a supply and distribution infrastructure to make aqueous urea available at motor vehicle refueling and service facilities, (2) aqueous urea delivery to the vehicle in a manner that does not inconvenience or burden the driver.

The viability of aqueous urea supply and distribution infrastructure has been studied [6,7,8]. The price of aqueous urea has been estimated to be between $0.50 and $3.00 per gallon depending upon sales volume. Based on Ford estimates, urea consumption for a light-duty vehicle could vary from 4000 miles/gallon for a small car to 1000 miles/gallon for a heavy SUV. Consequently, the lifetime (120k mi) cost of urea for a light-duty vehicle would be approximately $15-360, or < 5% of the diesel fuel cost.

The preferred delivery method would be to add urea during vehicle refueling through a common fuel nozzle and fill pipe interface: urea / diesel co-fueling. Aqueous urea is well suited for delivery in this fashion. The person fueling the vehicle needs no new skills or knowledge, and may not even be aware of the second fluid, depending on choices made for display and pricing on the dispenser [9]. The co-fueling method of onboard delivery of urea has been demonstrated on both light and heavy-duty vehicles [10].

In this study, conversion efficiencies of an exhaust system containing an oxidation catalyst, SCR catalyst with aqueous urea injection, and catalyzed diesel particulate filter (CDPF) were evaluated on a modified European 1.8L TDCi diesel Ford Focus. The main goal was to demonstrate emissions comparable to ULEVII standards (0.05 g/mi NO_x, 0.01 g/mi PM, 0.040 g/mi NMOG and 1.7 g/mi CO). Another goal was to show feasibility of reaching the Tier 2 SFTP US06 standard (0.14 g/mi NMHC+NO_x). Low sulfur diesel fuel of about 5 ppm sulfur was used. All vehicle testing was performed at the Ford Research & Advanced Engineering Vehicle Emissions Research Lab, Dearborn, MI.

EXPERIMENTAL

CHEMISTRY

When urea is exposed to warm exhaust gas, it readily decomposes to ammonia (NH_3) [11]. A molar ratio of NH_3/NO_x = 1 (or slightly less than stoichiometry) is typically used to avoid "slip" of NH_3 past the catalyst and out the tailpipe. Additionally the catalyst can store NH_3 and therefore acts as a buffer to rapid changes in speed and load. Decomposition of urea and subsequent reduction of NOx may occur by the following scheme:

urea decomposition:

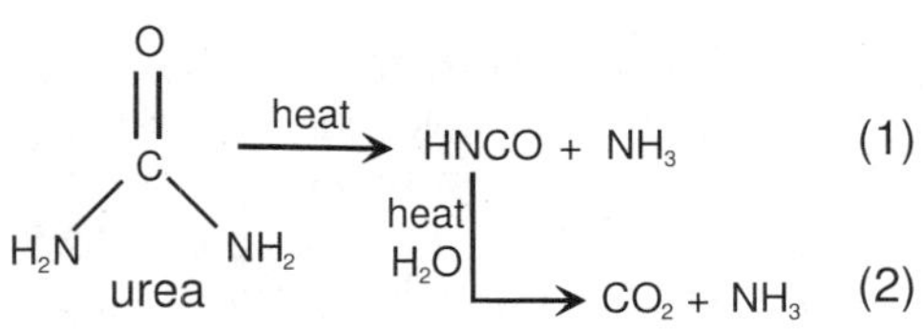

NOx reduction:

$$4NO + 4NH_3 + O_2 \longrightarrow 4N_2 + 6H_2O \quad (3)$$

$$6NO_2 + 8NH_3 \longrightarrow 7N_2 + 12H_2O \quad (4)$$

$$2NH_3 + NO + NO_2 \longrightarrow 2N_2 + 3H_2O \quad (5)$$

Undesired reactions include formation of N_2O instead of N_2 and oxidation of NH_3. Also urea may not be completely decomposed before reaching the catalyst surface [2,12].

The SCR catalyst used in this study was a base metal/zeolite formulation that did not contain vanadium. It was chosen for its wide temperature window for very high NOx conversion as shown in Figure 1. This window covered the underbody temperatures typically experienced on light-duty vehicle test cycles. A mixture of NO and NO_2 in the catalyst feedgas enhanced low temperature NOx conversion.

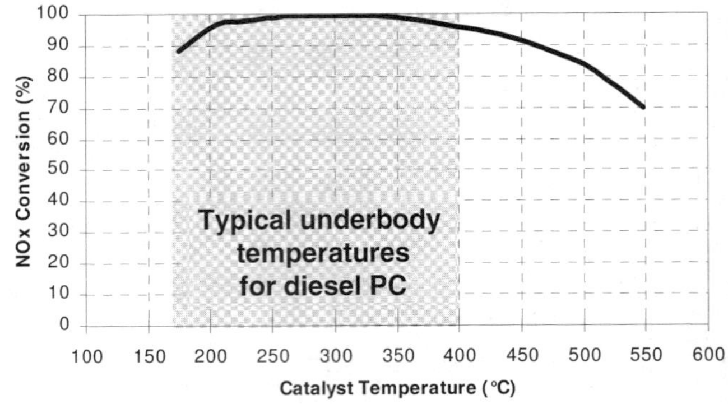

Figure 1. Typical SCR catalyst performance in diesel exhaust gas at 30k h^{-1} assuming a blend of NO and NO_2 in the feed. Data were collected in a laboratory flow reactor with a monolith core. The NH_3/NOx molar ratio was set to 1.

Base metal/zeolite catalysts are well-suited for use on light-duty diesel vehicles but are not without issues. It is well known that zeolites adsorb NH_3 and the adsorption capacity decreases with increasing temperature, possibly resulting in NH_3 slip past the SCR catalyst. Zeolites also adsorb hydrocarbons (HC), and this has a negative effect on NO_x conversion with NH_3 below about 400°C, impacting LD FTP performance [13]. Other poisons include sulfur that is present in the diesel fuel and engine oil [14], so low-sulfur formulations should be used to minimize its impact. In addition, zeolites are limited to temperatures below 800°C, so thermal management is critical to durable SCR function.

EXHAUST SYSTEM, UREA INJECTION SYSTEM AND INSTRUMENTATION

The exhaust system configuration used in this study is shown in Figure 2. All catalyst bricks were used in a de-greened but non-aged condition ("fresh"). A diesel oxidation catalyst (DOC) was located close-coupled to the engine and had an overall volume of 1.6 L. The DOC converted engine-out HC and CO while also converting a portion of the NO to NO_2.

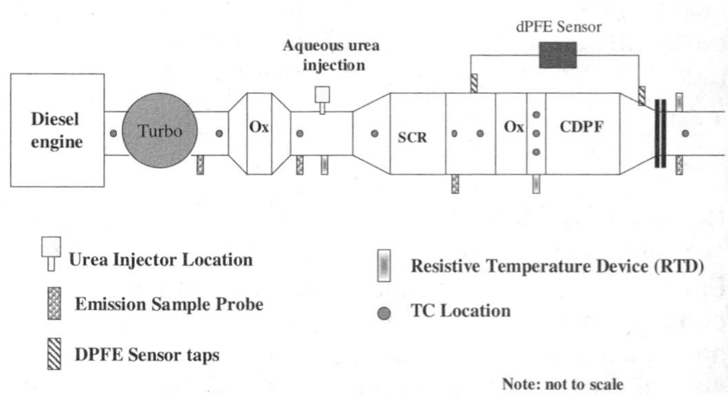

Figure 2. Exhaust system configuration onboard the Focus vehicle (DOC = diesel oxidation catalyst, SCR = Selective Catalytic Reduction, CDPF= Catalyzed Diesel Particulate Filter).

Aqueous urea was introduced approximately 55 cm upstream of a 3.7 L SCR catalyst using an air-assisted injection system developed at Ford. As a part of the development process the injection distance and cone angle entering the catalyst were optimized for best urea mixing and utilization of the catalytic surface area. A solution of 30 wt% urea in water was used with atomizing air for good dispersion of the reductant in the exhaust gas. This concentration was close to the eutectic mixture of 32.5 wt% urea with a slight margin for evaporation of water. The aqueous urea was added to the exhaust such that the calculated NH_3/NOx molar ratio entering the SCR catalyst was kept approximately equal to one during the engine operation, assuming complete urea decomposition. The flow rate of urea was controlled based on information from the average SCR catalyst temperature (from a resistive temperature device (RTD) sensor before the SCR and one after the SCR), an engine out NO_x sensor (NGK Insulators, Ltd., Japan; 1000 ppm NO_x range) and base engine parameters. An SCR catalyst inlet temperature of 150°C was chosen as the criteria to begin urea injection based on the melting point of urea (135°C) and catalyst activity data. The relevant aqueous urea properties are shown below in Table 1. Additionally, a differential pressure sensor was used to monitor the soot loading on the catalyzed diesel particulate filter (CDPF).

Table 1
Aqueous urea properties.

fluid property	value
urea content	30 wt%
water content	70 wt%
freeze point	15°F
urea melt point	135°C

TEST FUEL / CONDITIONS

Diesel fuel containing 5 ppm S was used throughout the vehicle testing. The fuel was obtained from a fuel blender and was meant to simulate European diesel fuel with 2007 US sulfur. The cetane level was higher and aromatics were lower than typical US diesel fuel. The fuel properties are shown below in Table 2.

Tests were performed at the Vehicle Emissions Research Laboratory (VERL) at Ford Research & Advanced Engineering (Dearborn, MI) using a 48" single-roll electric dynamometer under controlled temperature and humidity conditions. The VERL site utilizes 3 full emissions benches (measuring HC, CO, NO_x, CO_2 and O_2 at engine-out, post-catalyst(s) and tailpipe positions), a full dilution tunnel (for PM measurements), raw FTIR capability and dilute FTIR capability (to measure NH_3, NO_2, N_20, etc.).

Table 2
Diesel fuel properties.

fuel property	value
cetane	58.7
cetane index (calculated)	56.9
flash point	53°C
pour point	-23°C
cloud point	-18°C
sulfur	5 ppm
aromatics	2 wt%

SCOPE OF WORK

Cold-start Federal Test Procedures (FTP-75) and also US06 cycles were performed using a 1.8L diesel Focus. The placement and sizing of the catalysts were designed with the goal of demonstrating ULEVII emission levels (0.05 g/mi NOx, 0.01 g/mi PM, 0.040 g/mi NMOG and 1.7 g/mi CO). Another goal was to show feasibility of reaching the Tier 2 SFTP US06 standard (0.14 g/mi NMHC+NOx). The amount of un-reacted NH_3 ("NH_3 slip") was monitored downstream of the SCR catalyst and at the tailpipe. Additionally, the restriction of exhaust gas flowing through the particulate filter was monitored to provide an indication of when filter regeneration was needed. The main focus of this work was to test proof of concept for NOx reduction and not to judge long-term durability of the exhaust system.

RESULTS AND DISCUSSION

FAST WARM-UP ROUTINE

The key to achieving the high NO_x conversion required for the ULEVII standards is to bring the exhaust system up to operating temperature as quickly as possible. A fast warm-up routine was developed to result in an SCR (NO_x) catalyst inlet temperature of 150°C within the first 100 s after a cold start. To maximize catalyst performance the routine was also designed to maintain a minimum SCR catalyst temperature of 200 °C regardless of operating conditions (i.e. not cycle dependent). The fast warm up strategy utilized the various features available on the 1.8L DuraTorq TDCi common-rail diesel engine A resistive temperature device (RTD) was used as the control point for the strategy. Figure 3 shows the resulting inlet SCR temperatures from Phase 1 of the FTP (cold start) with and without the fast warm-up routine utilized. With fast warm-up the front face of the SCR reaches operating temperature (150°C) approximately 100 s earlier. Driver variations and other noise factors can affect this result. The resulting tailpipe NOx value was approximately 50% lower for the test with the fast warm-up routine than for the test without it.

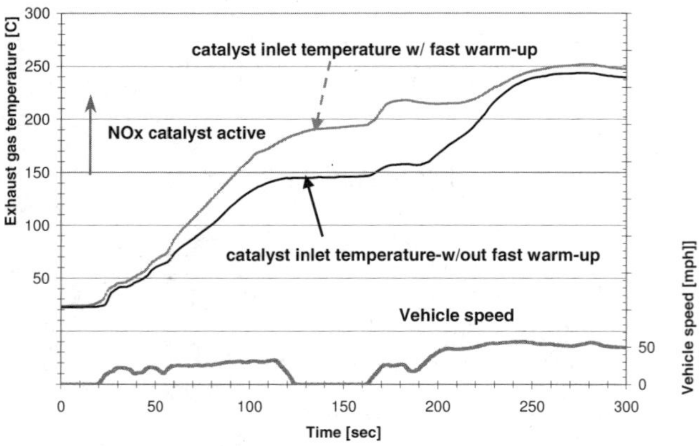

Figure 3. SCR catalyst inlet temperatures during the first 300 s of the FTP-75 (cold start) with and without fast warm-up.

FTP PERFORMANCE (HC, CO, NOx, PM)

Average data from five cold-start FTP-75 tests are shown in Table 3. Emission levels comparable to the ULEVII standards were achieved at low mileage. The FTP weighted average fuel economy of this vehicle was approximately 38.5 mpg. The average metro-highway fuel economy (M-H FE) was 45 mpg.

Table 3
Cold-start FTP-75 weighted average emission levels and efficiencies from five vehicle tests.

Species	Engine -out [g/mi]	Tailpipe [g/mi]	ULEV II 50k mi PC Standard [g/mi]	Efficiency [%]
NMHC	0.45	0.012	0.04[a]	82%
CO	0.80	0.143	1.7	60%
NO_x	0.48	0.037	0.05	92%
Phase 1	*0.50*	*0.082*	---	*83%*
Phase 2	*0.47*	*0.017*	---	*96%*
Phase 3	*0.48*	*0.040*	---	*92%*
PM	---	0.001	0.01[b]	

[a] NMOG standard.
[b] PM standard at 120k mi.

The SCR catalyst inlet temperature, vehicle speed (for reference), SCR catalyst inlet NO_x and tailpipe NO_x are shown in Figure 4. The initial conversion of NO_x was low until the SCR catalyst reaches its operating temperature window at about 100 s. Note that by the time of the second acceleration the tailpipe NO_x is minimized to slight peaks under heavy acceleration. For this test, conversion was over 85% in phase 1, nearly 100% in phase 2 and over 95% after the soak and restart in phase 3. The overall cycle conversion rate on this test was approximately 95%. The SCR catalyst inlet temperature (via thermocouple), cumulative tailpipe NO_x (as a percent of total engine out NO_x) and vehicle speed over the FTP-75 are shown in Figure 5. Conversion was over 95% in the second phase of the test cycle. The overall NO_x system efficiency for this test was approximately 95%.

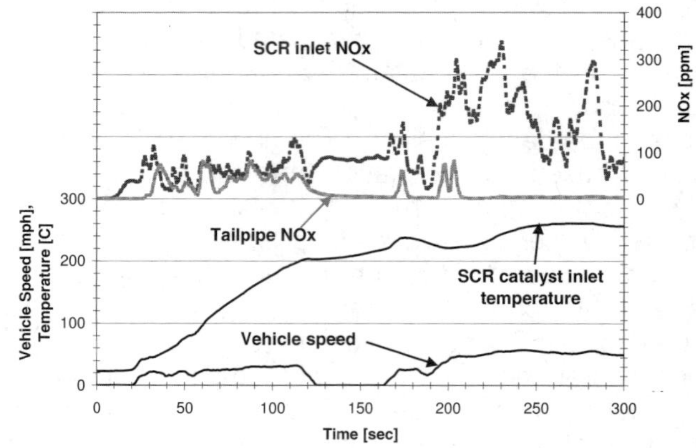

Figure 4. SCR catalyst inlet temperature, vehicle speed, SCR catalyst inlet NO_x and tailpipe NO_x during the first 300 seconds of Phase 1 in an FTP-75 (cold start).

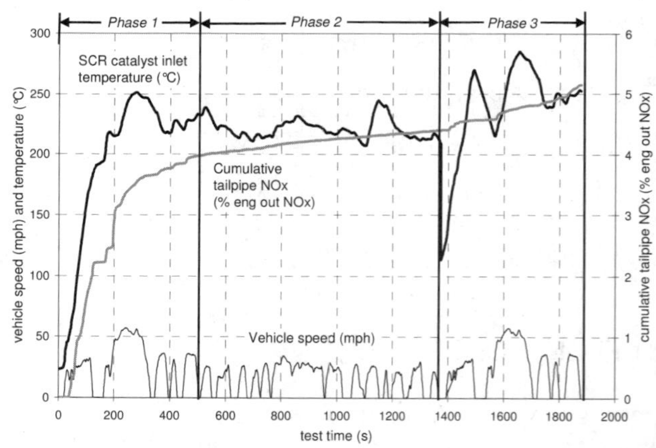

Figure 5. SCR catalyst inlet temperature, cumulative tailpipe NO_x, and vehicle speed over the FTP-75 (cold start).

HC POISONING OF THE SCR CATALYST

During the development of the emission control system, it was found that un-reacted hydrocarbons (HC) during cold-start were adsorbed by the zeolite-based SCR catalyst and had a detrimental effect on NO_x reduction. HC slip past the oxidation catalyst was investigated by utilizing several different calibrations of the fast warm-up routine. Figure 6 shows the cumulative total hydrocarbon results (THC) during Phase 1 of the FTP-75 (cold start) downstream of the oxidation catalyst. A test without fast warm-up and three tests with different warm-up calibrations are shown. Depending upon the calibration, the total hydrocarbons entering the SCR catalyst were almost three times the amount than with no fast warm-up. A variation on Calibration #1 was adopted to minimize the amount of HC to which the SCR catalyst was exposed.

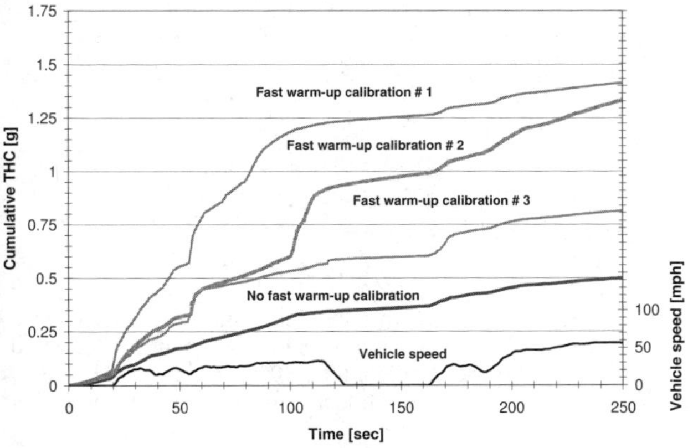

Figure 6. Cumulative total hydrocarbon at SCR inlet using fast warm-up during FTP Phase 1 cold starts.

The reduction in NO_x efficiency can be significant depending upon the amount of hydrocarbons stored on the SCR. The amount of time it takes to achieve a high enough level of HC storage to affect NO_x performance will vary with driving style, conditions and calibration. With the chosen, optimized fast warm-up calibration, over ten (consecutive) cold-start FTP-75s were completed without any significant reduction in NO_x efficiency. Additionally, the stored hydrocarbons may be effectively removed from the catalyst by raising its temperature to > 450°C. A strategy is currently under development at Ford to achieve this (both independently as conditions warrant and simultaneously with a DPF regeneration).

EFFECT OF FAST WARM-UP ON PM GENERATION

In addition to hydrocarbon poisoning of the SCR catalyst, the effect of the fast warm-up routine on the quantity of engine-out particulate matter (PM) was investigated. Several different calibrations were utilized to study these effects, as was done with the HC slip investigation. The PM results from Phase 1 of the FTP with and without fast warm-up are shown in Figure 7. Two tests were performed at each condition. As can be seen from the figure, the fast warm-up can significantly increase the PM over the base calibration, but this is not significantly larger than other widely performed engine calibration changes (such as retarding timing and / or increased EGR rates). Additionally, the effect of fast warm-up on the amount of soluble organic fraction (SOF) was studied and the change in SOF was minimal. As mentioned previously, a variation of Calibration #1 was used for the majority of later testing to minimize engine-out PM and HC slippage into the SCR.

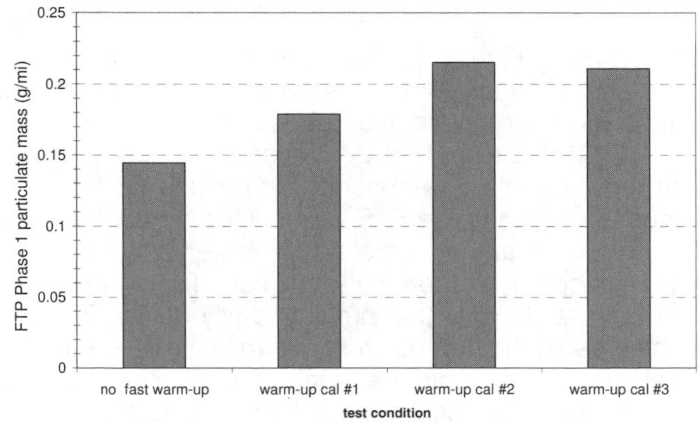

Figure 7. Engine-out particulate matter (PM) results with and without fast warm-up during Phase 1 (cold start) of the FTP-75 cycle.

US06 TEST PERFORMANCE

It was desirable to test the performance of the SCR system at higher temperatures than those experienced during the FTP-75 cycle. During the US06 cycle, exhaust gas temperatures were significantly warmer with higher exhaust flow rates due to the overall higher vehicle speed and rapid accelerations. The inlet temperature to the SCR catalyst approached 400°C. For this investigation, US06 tests were run in sets of three: one warm-up cycle followed by two hot cycles.

Average data from nine US06 cycles (1/3 cold, 2/3 hot) are shown in Table 4. Emission levels comparable to the Tier 2 SFTP US06 standards were achieved at low mileage. Average NO_x conversion was over 80%.

Table 4
US06 average emission levels and efficiencies from nine vehicle tests including warm up cycles.

Species	Engine-out (modal) [g/mi]	Tailpipe (bag) [g/mi]	4K Tier 2 SFTP US06 Standard [g/mi]	Efficiency [%]
NMHC	0.15	0.015	N/A	90%
CO	3.56	0.196	8.0	93%
NO_x	0.71	0.122	N/A	83%
NMHC + NO_x	0.86	0.137	0.14	83%

There is a potential to improve US06 cycle NO_x conversion by further optimizing the urea injection process and having better understanding of the NH_3 storage/release functions of the SCR catalyst as is shown with the following data. During the first (cold) US06 cycle of each set, a significant amount of ammonia was usually (depending upon prior driving conditions / NH_3 storage) detected downstream of the SCR catalyst. As mentioned previously, the SCR catalyst can store ammonia and this storage capacity decreases with increasing temperature. Three US06 cycles were performed with varying starting conditions and the results are shown in Figure 8. Test A was a cold start US06 after ten consecutive FTP-75 cycles. Test B was a cold start US06 after several US06 cycles. Test C was a hot start US06 following Test B. Therefore, Test A began with the most stored ammonia (due to the lack of high temperature during the FTP's) and Test C had the least. As the SCR catalyst temperature increased, ammonia was released from the catalyst in both Test A and Test B. This data implies that prolonged city driving followed by high-speed highway driving can result in NH_3 release. However, no ammonia was emitted from the tailpipe in any test due to the oxidizing function of the CDPF.

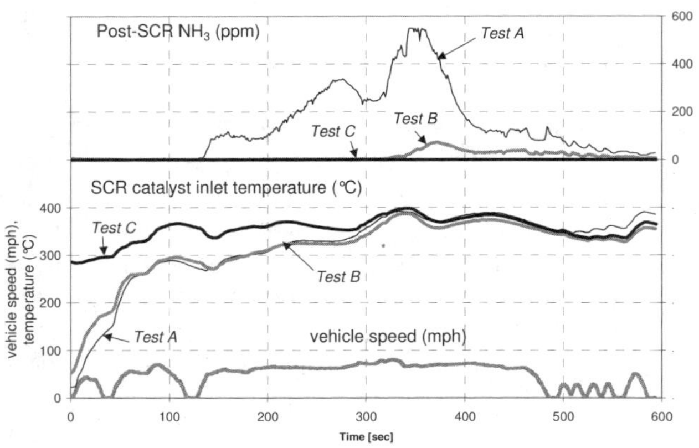

Figure 8. Post-SCR ammonia emission, SCR catalyst inlet temperature, and vehicle speed during US06 cycles. Test A was a cold start US06 after ten repeated FTP-75 cycles. Test B was a cold start US06 after previous US06 cycles. Test C was a hot US06 conducted immediately following Test B.

Nearly all of the released ammonia was converted to NO_x in the downstream CDPF. This had a negative effect on overall system NOx conversion efficiency. As shown in Figure 9, Test A had over five times the cumulative tailpipe NO_x than Test C. The increase in tailpipe NO_x emission corresponded directly with post-SCR ammonia emission. Excess injected urea (i.e. NH_3) is stored on the SCR at low temperatures and released at high temperatures (creating NO_x remake across the CDPF). With improved, finer control of the urea injection system the amount of excess injected urea can be reduced, thus reducing NO_x remake. NO_x remake from NH_3 slip can be controlled further by developing / utilizing advanced NH_3 storage strategies and exhaust temperature management. Thus, even though the NH3 storage buffer is reduced at high temperature, SCR efficiency can be maintained at a high level with a urea dosing system that is closely aligned with engine-out NO_x.

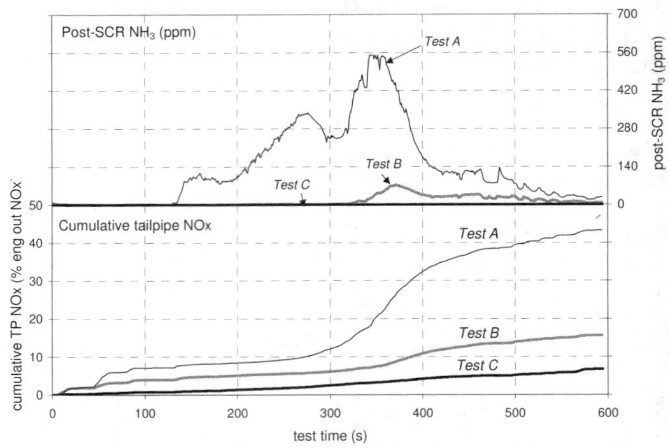

Figure 9. Post-SCR ammonia emission and cumulative tailpipe NOx during US06 cycles. Test A was a cold-start US06 after ten repeated FTP-75 cycles. Test B was a cold-start US06 after previous US06 cycles. Test C was a hot US06 conducted immediately following Test B.

CONCLUSIONS

A diesel Ford Focus demonstrated emissions comparable to ULEVII levels through the use of an oxidation catalyst, aqueous urea SCR system, and CDPF. The Tier 2 SFTP US06 standards were also shown to be feasible with this vehicle. However, this work was performed with a fresh system and the durability of such a system remains unproven. The conversion level for NO_x was over 90% for the cold start FTP-75 and over 80% for the US06 cycle. A fast warm-up strategy, very active SCR catalyst, and excellent control of urea injection quantities made these high NO_x conversions possible. Slip of ammonia from the SCR catalyst was converted to NO_x and was included in the net NO_x efficiency. No measurable ammonia escaped into the atmosphere.

Areas for future work include demonstration of a similar emission control system on a light duty truck, catalyst durability improvements, and emissions testing during a DPF regeneration event. Additionally, the storage and release functions of the SCR catalyst need to be further understood and predictive models established to further the control capability of the urea injection system.

ACKNOWLEDGMENTS

The authors acknowledge the Vehicle Emissions Research Laboratory team at Ford Research & Advanced Engineering (Dearborn, MI) for their expertise in vehicle testing. Other Ford colleagues to thank include Michiel van Nieuwstadt and Devesh Upadhyay for development of the urea injection and CDPF control strategy, Paul Laing for modeling support, and Rick Soltis for providing exhaust gas sensors. Special thanks go to Dick Baker for his enduring support of this project.

CONTACT INFORMATION

Paul Tennison
Ford Motor Company
20901 Oakwood Blvd
Dearborn, MI 48124
e-mail: ptenniso@ford.com

REFERENCES

1. R.M. Heck and R.J. Farrauto, "Catalytic Air Pollution Control," Van Nostrand Reinhold, New York, 1995.
2. H. Luders, R. Backes, G. Huthwohl, D.A. Ketcher, R.W. Horrocks, R.G. Hurley, and R.H. Hammerle, "An urea lean NOx catalyst system for light duty diesel vehicles," SAE 952493.
3. W. Held, A. Konig, T. Richter, L. Puppe, "Catalytic NOx reduction in net oxidizing exhaust gas," SAE 900496.
4. W.R. Miller, J.T. Klein, R. Mueller, W. Doelling, J. Zuerbig, "The development of urea-SCR technology for US heavy duty trucks," SAE 2000-01-0190.
5. W. Mueller, H. Ölschlegel, A. Schäfer, N. Hakim and K. Binder, "Selective Catalytic Reduction - Europe's NOx Reduction Technology", SAE 2003-01-2304.
6. "Selective Catalytic Reduction (SCR) Infrastructure Study", Project Review Meeting, Arthur D. Little, October 1, 2001, NREL Subcontract No. ACL-1-31038-01, ADL Case No. 72457.
7. M.D. Jackson, S. Fable, S. Venkatesh, "Viability of Urea Infrastructure for SCR Systems", TIAX, LLC, U.S. Environmental Protection Agency, Clean Diesel Engine Implementation Workshop, Chicago, Illinois, August 6-7, 2003
8. S. Fable, S. Venkatesh, "Development of a Urea Infrastructure to Support MY2007+ Heavy-duty Vehicle Emission Requirements", Final Report, NREL Contract No. ADZ-3-33020-01.
9. D. Baker, B. Hammerle, R. Backes, "Harnstofflogistik zur NOx-Reduzierung von Fahrzeugen (Urea infrastructure for control of vehicular NOx emissions)", Aachener Kolloquium Fahrzeug- und Motorentechnik 2000, Aachen, Ger., October 2000.
10. M.B. Levin, R.E. Baker, "Co-fueling of Urea for Diesel Cars and Trucks", SAE 2002-01-0290.
11. H.T. Hug, A. Mayer, A. Hartenstein, "Off-highway exhaust gas after-treatment combining urea-SCR, oxidation catalysis and traps," SAE 930363.
12. M. Koebel, M. Elsener, T. Marti, "NOx-Reduction in diesel exhaust gas with urea and selective catalytic reduction," Combust. Sci. and Tech. 121 (1996) 85-102.
13. C. Lambert, C. Montreuil, J. Vanderslice, "Application of Organic Freeze-point Depressants in Aqueous Urea Solutions: Effect on NOx Reduction," SAE 2003-01-0775.
14. R. Hughes, "Deactivation of Catalysts," Academic Press, New York, 1984.

DEFINITIONS, ACRONYMS, ABBREVIATIONS

CDPF: Catalyzed Diesel Particulate Filter.
CO: Carbon monoxide.
CO_2: Carbon dioxide.
FTP: The FTP-75 (Federal Test Procedure) has been used for emission certification of light duty vehicles in the U.S. The entire FTP-75 cycle consists of the following segments: cold start phase (Phase 1), transient phase (Phase 2), and hot start phase (Phase 3). The distance traveled is 11.04 miles, duration is 1874s, and the average speed is 21.2 mph.
HC: Hydrocarbon
HD: Heavy-duty
LD: Light-duty
O_2: Oxygen.
NO_x: Oxides of nitrogen, including NO and NO_2.
N_2: Nitrogen.
N_2O: Nitrous oxide.
NH_3: Ammonia.
NMHC: Non-methane hydrocarbons
NMOG: Non-methane organic gases
PC: Passenger car
PM: Particulate matter.
SCR: Selective Catalytic Reduction. Refers to the conversion of NO_x to nitrogen in a high oxygen environment through the use of a reducing species such as hydrocarbon or ammonia.
SFTP: Supplemental Federal Test Procedure.
TC: thermocouple.
TP: tailpipe.
ULEV: Ultra Low Emission Vehicles.
US06: Part of the Supplemental federal Test Procedure (SFTP) introduced by the EPA in January of 1995. The US06 is an aggressive, high speed and high acceleration emissions test for vehicles operating on a single-roll electric dynamometer. Total distance traveled is 8.0 miles with a test duration of 600 seconds.

2004-01-1292

Technical Advantages of Urea SCR for Light-Duty and Heavy-Duty Diesel Vehicle Applications

Christine Lambert and Robert Hammerle
Ford Research & Advanced Engineering

Ralph McGill
Oak Ridge National Laboratory

Magdi Khair and Christopher Sharp
Southwest Research Institute

ABSTRACT

The 2007 emission standards for both light-duty and heavy-duty diesel vehicles remain a challenge. A level of about 90% NOx conversion is required to meet the standards. Technologies that have the most potential to achieve very high NOx conversion at low temperatures of diesel exhaust are lean NOx traps (LNTs) and Selective Catalytic Reduction (SCR) of NOx using aqueous urea, typically known as Urea SCR. The LNT has the advantage of requiring no new infrastructure, and does not pose any new customer compliance issues. However, Urea SCR has high and durable NOx conversion in a wider temperature window, a lower equivalent fuel penalty, and lower system cost. On a technical basis, Urea SCR has the best chance of meeting the 2007 NOx targets. This paper reviews the results of some demonstration programs for both light- and heavy-duty applications.

INTRODUCTION

Reduction of NOx in the lean exhaust gas of diesel engines is not trivial. The 2007 NOx standards represent a 90+% reduction from previous standards as shown in Table 1. Also shown are NMOG (Non-Methane Organic Gases), NMHC (Non-Methane Hydrocarbon), CO (Carbon Monoxide) and PM (Particulate Matter) standards. In general, the two leading technologies for very high NOx conversion are Lean NOx Traps (LNTs) and Selective Catalytic Reduction of NOx using aqueous urea (Urea SCR).

LNTs contain an oxidative component such as Pt that oxidizes engine-out NO to NO_2 and a storage component (typically an alkali metal salt) that forms a nitrate compound during lean (excess oxygen) conditions. During rich (low oxygen) conditions, the nitrate compound releases NOx that is then reduced to N_2.

The reduction of NOx with either ammonia (NH_3) or urea has been used extensively for stationary source emission control. NOx reduction is possible due to the high selectivity of the NH_3 and NOx reaction to form elemental N_2. The typical reaction scheme using urea is as follows:

urea decomposition:

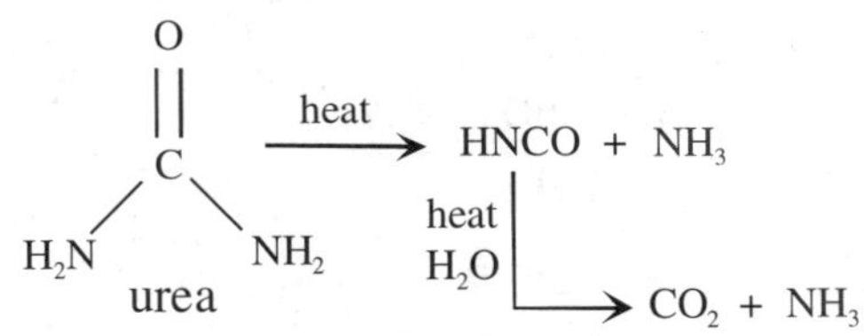

NO_x reduction:

$$4NO + 4NH_3 + O_2 \longrightarrow 4N_2 + 6H_2O$$
$$6NO_2 + 8NH_3 \longrightarrow 7N_2 + 12H_2O$$
$$2NH_3 + NO + NO_2 \longrightarrow 2N_2 + 3H_2O$$

Urea is the preferred means of delivering ammonia onboard a vehicle because it is safely transported and can be easily injected as an aqueous solution. Urea has a very low toxicity and is widely used as a fertilizer. The mixture with the lowest freeze point (eutectic) with about 32.5 wt% urea in water is preferred.

There have been successful light-duty (LD) and heavy-duty (HD) testing of both LNT and Urea SCR on US emission cycles. Table 2 summarizes the most recent results.

Table 1

2007 standards for LD and HD diesel vehicles.

	NMOG/NMHC	CO	NOx	PM
LD Tier 1 diesel (100k mi)[a]	0.31-0.56 g/mi	4.2-7.3 g/mi	0.97-1.53 g/mi	0.10-0.12 g/mi
LD Tier 2 Bin 5 (120k mi)	0.090 g/mi	4.2 g/mi	0.07 g/mi	0.01 g/mi
HD 2004 MY	0.5 g/hp-hr	15.5 g/hp-hr[b]	2.0 g/hp-hr	0.10 g/hp-hr
HD 2007 MY	0.14 g/hp-hr[c]	15.5 g/hp-hr[b]	0.20 g/hp-hr[c]	0.01 g/hp-hr

[a] Tier 1 standards vary according to vehicle weight.
[b] HD CO standard is carried over from 1998 MY.
[c] 2007 NMHC and NOx standards to be phased in from 2007-2010.

Table 2

Summary of reported results from recent diesel technology testing with fresh catalysts.

Application	Technology	Cycle	NOx Conv.	Tailpipe NOx	Ref.
PC	Urea SCR	FTP-75	90+%	0.05 g/mi	1
PC	Urea SCR	US06	90+%	0.05 g/mi	1
PC	LNT (DPNR)	FTP-75	80+% [2]	0.05 g/mi	3
PC	LNT (DPNR)	US06	80+% [2]	0.14 g/mi	3
LDT	Urea SCR	FTP-75	82%	0.17 g/mi	4
LDT	LNT	FTP-75	72%	not reported	5
HDT	Urea SCR	HD FTP	85%	0.86 g/hp-hr	6
HDT	Urea SCR	SET	86%	0.85 g/hp-hr	6
HDT	Urea SCR	HD FTP	90%	0.22 g/hp-hr[a]	7
HDT	Urea SCR	SET	90%	0.18 g/hp-hr	7
HDT	LNT	HD FTP	95%	0.13 g/hp-hr	8[b]
HDT	LNT	SET	94%	0.12 g/hp-hr	8[b]

Notes:
All results obtained with very low sulfur diesel fuel (< 30 ppm).
PC = Passenger Car
LDT = Light-Duty Truck
HDT = Heavy-Duty Truck
DPNR = Diesel Particulate - NOx Reduction
FTP-75: Federal Test Procedure for LD vehicles; cold-start three-bag cycle.
US06: High speed, high load cycle for LD vehicles; part of Supplemental FTP.
HD FTP: Heavy-Duty Federal Test Procedure; transient dynamometer cycle.
SET: Supplemental Emission Test for HD; steady-state points.
[a] Composite results for HD FTP (1/7 cold, 6/7 hot start).
[b] Results obtained with catalyst preconditioning.

The case against using Urea SCR for vehicle applications is twofold: (1) an infrastructure is required to deliver the reductant onboard the vehicle and (2) customer compliance is required to maintain adequate reductant for continuous, high NOx conversion. These issues make Urea SCR difficult to implement, but not impossible. In Europe, Urea SCR is the technology chosen by HD manufacturers to meet Euro IV and Euro V standards [9, 10]. Urea SCR does not have a direct fuel penalty because the engine can be tuned for optimum fuel consumption rather than minimum engine-out NOx, and durability is confirmed for over 500,000 km (>300,000 mi) [9]. Development of an aqueous urea infrastructure for LD diesels is more uncertain, but also not impossible, especially if an approach like co-fueling of diesel and urea simultaneously is adopted [11,12]. Such an approach requires no extra action by the customer other than normal refueling. In fact, refueling a vehicle equipped with either SCR or LNT emission control technologies can be transparent to the end user.

This paper will focus on the technical advantages of Urea SCR over LNT that include wider temperature window for very high NOx conversion, higher resistance to sulfur poisoning, better thermal durability, lower fuel economy penalty, lower system HC emissions, lower greenhouse gas emissions, and lower system cost.

EXPERIMENTAL

LABORATORY TESTING & AGING CONDITIONS: UREA SCR

Fundamental catalyst activity data were obtained at Ford using a laboratory-scale flow reactor system. A round sample core with dimensions of 1" diameter and 1.5" length was taken from a washcoated cordierite monolith obtained from a supplier. The catalyst was a base metal/zeolite type that did not contain vanadium. Simulated diesel exhaust gas flowed through the sample core at a space velocity of 30k h^{-1}, measured at standard conditions. The composition of the feedgas is shown in Table 3. The composition of the inlet NOx was varied from 0 to 80% NO_2 (balance NO).

Table 3

Composition of simulated diesel exhaust gas for SCR activity measurement.

Component	Concentration
O_2	14%
H_2O	4.5%
CO_2	5%
NOx	350 ppm
NH_3	350 ppm
N_2	Balance

Catalyst temperature was maintained with a tube furnace. An FTIR (infrared) spectrometer was used at the reactor outlet to measure NOx and NH_3. Data were taken at catalyst temperatures from approximately 170 to 600°C in order to cover the normal operating range expected on a diesel vehicle. NOx conversion was allowed to equilibrate before raising the catalyst temperature to the next level.

SCR catalysts were aged to represent 120k miles via two modes: low temperature sulfur exposure and hydrothermal aging. For aging with sulfur, the monolith core was installed into a tube furnace that was set at 350°C. N_2, O_2, H_2O and CO_2 at the appropriate levels for diesel exhaust were flowed through the catalyst at a space velocity of 30k h^{-1}. SO_3 was generated from SO_2 over a Pt catalyst and added to the gas stream at a level to represent 120k miles of exposure over 24 hours for a range of diesel vehicles from PC to LDT. A future fuel sulfur level of 10 ppm was assumed for this calculation. After aging, NOx performance was measured without any attempts to remove sulfur from the catalyst. Hydrothermal aging was performed at 670°C and 30k h^{-1} with N_2, O_2, H_2O, CO_2 and SO_2 at the appropriate levels. An aging time of 64 hours was chosen to represent well over 120k miles based on Ford's projection of accumulated time at high temperature if a diesel particulate filter is periodically regenerating in the system.

LABORATORY TESTING & AGING CONDITIONS: LNT

Monolith cores of LNT samples were tested at Ford in a similar manner to the SCR catalysts except that the reactor feed gas was operated in a cyclic fashion of lean and rich durations. The feed gas concentration and the duration of lean and rich stages were controlled. A typical cycle for the LNT performance measurement contained 25 s of lean condition and 5 s of rich condition. The lean and rich feedgas conditions are shown in Table 4. Total gas flow was 30k h^{-1}, measured at standard conditions.

Table 4

Composition of simulated diesel exhaust gas for LNT activity measurement.

Component	Concentration (lean)	Concentration (rich)
CO	500 ppm	4%
H_2	167 ppm	1.3%
C_3H_6	300 ppm C_1	5000 ppm C_1
NO	500 ppm	500 ppm
O_2	10%	1%
CO_2	5%	5%
H_2O	5%	5%
N_2	Balance	Balance
Lambda	1.96	0.90

LNTs were aged under the conditions that may occur onboard a vehicle when removing sulfur from the catalyst (deSOx). The deSOx operation is typically done at high temperature (> 600°C) and predominantly rich conditions. For laboratory aging of LNTs, an inlet temperature between 600-700°C was chosen.

MODELING TOOLS

The Ford proprietary computer simulation tool, SIMTWC, is used to evaluate and compare the expected performance of aftertreatment configurations, saving experimental time and effort. A detailed description of this model has already been published and is readily available [13]. SIMTWC is integrated into the Matlab®/Simulink® environment to provide a flexible means to build and configure a wide variety of simulated exhaust aftertreatment systems. It includes a library of exhaust system components that range from thermocouples and flanges to pipes (single and dual wall) and catalysts. Inputs to the model include geometric specifications of the exhaust system components, performance characteristics of the catalysts, exhaust flow rate, temperature and emission species levels.

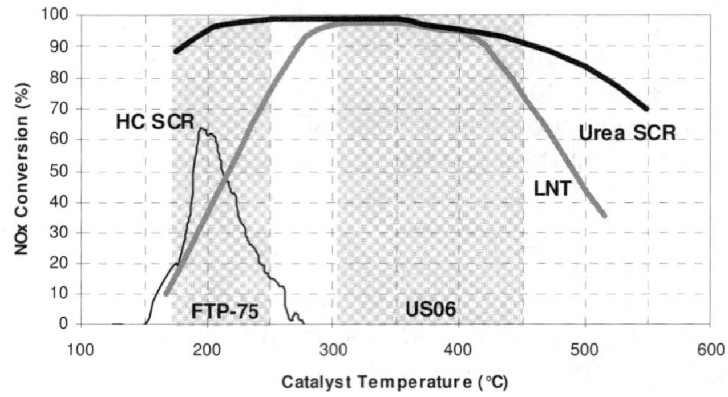

Fig. 1. Typical performance curves for slightly aged catalysts on LD test cycles. Typical underfloor temperature ranges for PC and LDT are shown for reference (see shaded areas). Space velocity was about 30k h-1, as measured at standard conditions, corresponding to a catalyst volume of about equal to the engine swept volume.

RESULTS

It has been found that Urea SCR has a wider temperature operating window, greater resistance to sulfur poisoning, greater thermal durability, lower fuel penalty, lower system HC emissions, lower greenhouse gas emissions, lower system cost and lower usage cost than LNT.

TEMPERATURE OPERATING WINDOW

The temperature operating windows at 30k h^{-1} for the best available LNT and Urea SCR catalysts as tested by Ford are shown in Figure 1. These performance data are for slightly aged catalysts. A typical HC SCR curve for diesel fuel and a Pt catalyst is shown for reference. Urea SCR has the widest operating window for >90% NOx conversion. Ammonia was used as the reductant here, but similar performance has been verified at Ford on engine exhaust systems using aqueous urea as the reductant. LNT has a narrower window that is shifted to temperatures higher than those that generally occur on the LD FTP-75 test. This means that the LNT must be heated by at least 100°C to give high efficiency on that test cycle. HD diesels typically operate with slightly higher temperature windows (250-500°C). LNT efficiency falls off drastically above 400°C.

RESISTANCE TO SULFUR POISONING

Sulfur is found in both diesel fuel and lube oils and generally has a negative effect on catalyst functionality. An LNT adsorbs sulfur in a similar way as it adsorbs nitrates. Sulfates are more stable and more difficult to remove than the nitrates. Sulfur must be removed periodically from the LNT (deSOx) in order to maintain high NOx conversion, and the time beween deSOx operations has been estimated to be about 3000 miles for a LD application. Urea SCR catalysts are inherently more resistant to sulfur since they do not possess the same type of storage component as an LNT. In Figure 2, the laboratory NOx reduction performance of an SCR catalyst core that was exposed to 120k miles of flowing SO_3 is shown. The exact amount of sulfur exposure (g/l catalyst) was calculated based on 10 ppm fuel sulfur and a catalyst volume of twice engine displacement. By using the most durable formulation available and a combination of NO and NO_2 in the feedgas, the effect of sulfur on performance is minimized.

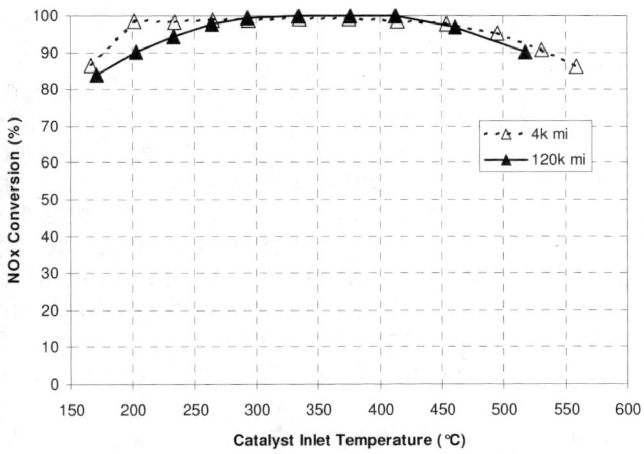

Fig. 2. Resistance of SCR catalyst to poisoning by sulfur. The catalyst was exposed to SO_3 at 350°C to simulate 120k mi exposure from diesel fuel containing 10 ppm sulfur. NOx conversion was measured with no attempt to remove the sulfur. Feedgas NOx contained equimolar NO and NO_2 with total flow at 30k h-1, measured at standard conditions.

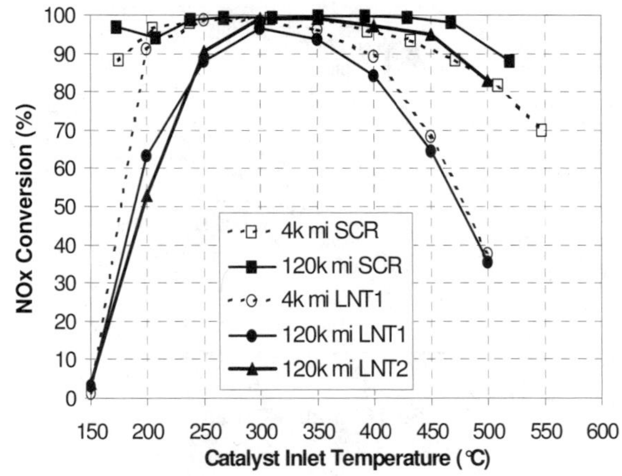

Fig. 3. Degreened (4k) and aged (120k) performance comparison of urea SCR and LNT in steady state flow reactors. The SCR catalyst was hydrothermally aged under lean conditions at 670°C to represent 120k LD vehicle miles. LNTs were aged under deSOx conditions. Total flow was 30k h-1, measured at standard conditions.

THERMAL DURABILITY

The deSOx operation for an LNT is done typically at 600-700°C under predominantly rich conditions, leading to the issue of thermal durability. A recent study looked at LNT thermal degradation at 510°C, not the most demanding of conditions [14]. In another study, repeated deSOx operations were performed at 700°C, leading to a steady decline in LNT performance [15].

A comparison of Urea SCR and LNT thermal durability is shown in Figure 3. SCR performance is relatively unaffected by the hydrothermal treatment and in fact shows a slight improvement in the high temperature region. Again a combination of NO and NO_2 is used in the SCR reactor feedgas. Thermal aging of a low temperature LNT formulation (LNT1) causes a dramatic decrease in performance at 200°C from about 90% to 65% NOx conversion. Another aged LNT formulation (LNT2) shows similar low temperature performance to aged LNT1 with better high temperature performance. However, the performance of aged LNT2 is inferior to the aged SCR.

FUEL ECONOMY PENALTY

LNT technology is expected to have a higher fuel economy penalty because rich conditions are required to remove and reduce NOx and SOx stored on the trap, and raise the temperature of the LNT during low temperature (< 300°C) operation. Urea SCR has no significant fuel penalty because it can provide high NOx conversion at low temperatures and does not use fuel to reduce NOx. An equivalent fuel penalty may be calculated to account for the energy used in the manufacture of urea.

The expected performance curves for LNT and Urea SCR vs fuel penalty are shown in Figure 4 for the LD FTP-75. A similar curve for HC SCR using diesel fuel over a Pt catalyst is shown for reference. The NOx conversions are optimistic projections for the best formulations available without any type of cold start strategy. It is assumed that the NOx catalyst is located upstream of the PM filter, and similar feedgas conditions were used for each technology. A penalty of 3% is included for all three systems to account for the increased backpressure and periodic regeneration of a diesel PM filter. An equivalent fuel penalty for use of urea based on the energy required for manufacture is estimated to be approximately 0.2%. This is insignificant compared to the 7% penalty projected for LNT to achieve 80% NOx conversion. Similarly, a recent study at Cummins found a 7% fuel penalty on a hot FTP-75 simulation for 84% NOx conversion with an LNT, and 11.6% fuel penalty for 89% NOx conversion on an FTP-72 cycle without LNT preconditioning [5].

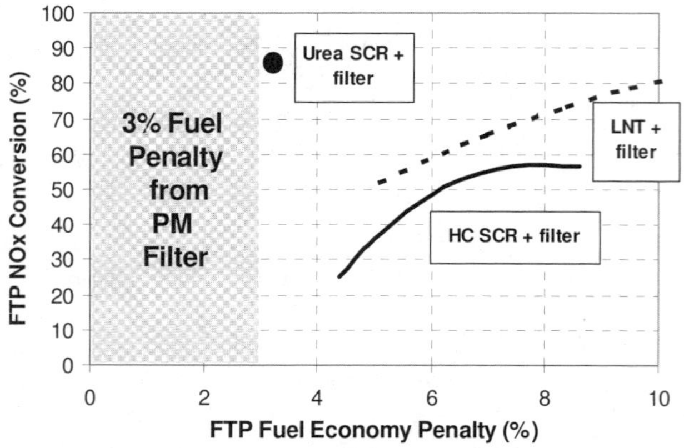

Fig. 4. Estimated fuel penalties of lean NOx reduction technologies vs NOx conversion for the LD FTP-75.

The expected performance curves for LNT and Urea SCR vs fuel penalty are shown in Figure 5 for the HD FTP. Here it is assumed that the NOx catalyst is located downstream of the PM filter since cold-start is not heavily weighted for HD emission testing. An equivalent fuel penalty for use of urea is estimated to be approximately 0.7%. Estimated fuel penalties for HD truck in Europe for an LNT system are 5% for Euro IV and 7% for Euro V [10]. Fuel consumption penalties for high NOx efficiency (>90%) on US HD cycles are around 7% [16].

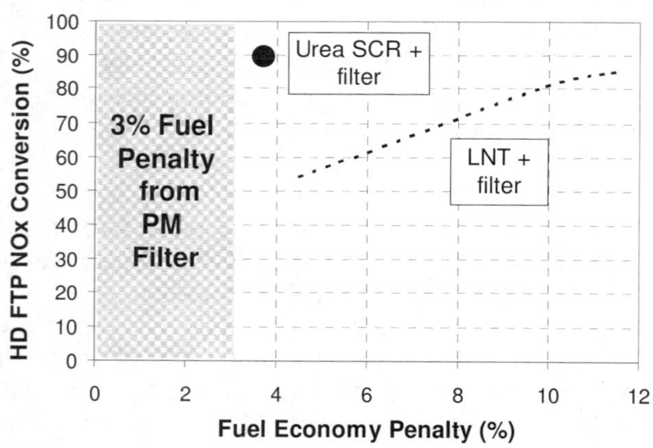

Fig 5. Calculated fuel penalties of lean NOx reduction technologies for HD engines vs NOx conversion.

SYSTEM HC EMISSIONS

For 2007, the LD NMHC (Non-Methane Hydrocarbon) emission standards must be met in addition to the NOx standards. For LNTs, this can be a challenge due to the frequent rich conditions and excess fuel used to heat the catalyst. On the other hand, Urea SCR catalysts for diesel vehicles do not contain precious metal, so an oxidation function must be added to the system to oxidize HC. In Table 5, data are shown from the recent LD diesel technology demonstrations described previously in Table 2 that reported NMHC.

It is apparent that, at least with fresh catalysts, both systems can provide enough HC oxidation to achieve the FTP-75 standard. More challenging is the US06 cycle with its high speeds and loads. The DOC (Diesel Oxidation Catalyst) + Urea SCR system has essentially no NMHC emissions and 0.05 g/mi NOx, well below the 0.14 g/mi standard. The DPNR system delivers 0.14 g/mi NOx and 0.19 g/mi NMHC.

GREENHOUSE GAS (GHG) EMISSIONS

Diesel vehicles emit CO_2 at a rate that is about 20-30% lower than comparable gasoline vehicles based on European vehicle data. Increased CO_2 emissions can result from excess fuel injection to enhance the operation of the aftertreatment system. This is directly related to a decrease in fuel economy as explained above. Also lean NOx aftertreatment devices often do not completely reduce NOx to N_2 and can form nitrous oxide (N_2O) as a byproduct. It is well known that N_2O is a GHG that has about 270 times the global warming potential of CO_2. Therefore, even if N_2O is formed in small amounts, it can have a potentially large impact. Both LNTs and Urea SCR catalysts can form N_2O in varying amounts depending upon operating conditions and catalyst formulation. The total CO_2 penalty may be calculated using an "equivalent CO_2 penalty" from N_2O formed on a driving cycle as shown below, where the base emission is without excess fuel injection:

Total CO_2 penalty = CO_2 (g/mi) / base CO_2 (g/mi) + N_2O (g/mi) x 270 / base CO_2 (g/mi)

A breakdown of CO_2 penalties for LNT and Urea SCR on a LD vehicle (FTP-75) is shown in Table 6. The LNT produces more GHG than a Urea SCR system operating at equivalent high NOx efficiency.

Table 5

Typical HC emissions on various test cycles.

Application	NOx Technology	Cycle	NMHC	NMHC + NOx	Ref.
		FTP-75	*0.090 g/mi**	---	
LD PC	DOC + Urea SCR	FTP-75	0.012 g/mi	---	1
LDT	DOC + Urea SCR	FTP-75	0.066 g/mi	---	4
LD PC	LNT (DPNR)	FTP-75	0.07 g/mi	---	3
		US06	---	*0.14 g/mi**	
LD PC	DOC+Urea SCR	US06	---	0.05 g/mi	1
LD PC	LNT (DPNR)	US06	---	0.33 g/mi	3

* 2007 LD Tier 2 Bin 5 and SFTP emission standards.

Table 6

CO_2 penalties for LNT and Urea SCR operating at high NOx efficiency on the LD FTP-75.

GHG source	CO_2 penalty (%)	GHG source	CO_2 penalty (%)
Urea manufacture	0.2	LNT deNOx	2.2
Urea SCR N_2O make	2.0	LNT deSOx	0.2
		LNT heating	4.5
		LNT N_2O make	0.9
Urea SCR Total	2.2	LNT Total	7.8

SYSTEM COST

System costs for Urea SCR and LNT systems can be estimated using historical cost information for washcoated catalysts and recent precious metal prices. Details for exact costs are proprietary to each manufacturer and are not discussed here; however, general comparisons can be made. Those knowledgeable can do similar calculations based on their own assumptions.

In Figure 6, catalyst configurations assumed for LD applications are shown. The cost estimate for Urea SCR system includes: diesel oxidation catalyst, aqueous urea injection system and SCR catalyst. The cost estimate for LNT includes: diesel oxidation catalyst, secondary fuel injection system and LNT catalyst. The PM filter is not included for simplicity and is assumed to be identical for both systems. Geometric volumes of the Urea SCR catalyst and the LNT are assumed to be equal and twice the engine displacement. Estimated cost of the LD LNT system is about twice that of the Urea SCR system.

Catalyst configurations assumed for HD applications are shown in Figure 7. The cost estimate for Urea SCR includes: aqueous urea injection system and SCR catalyst. The cost estimate for LNT includes: secondary fuel injection system and LNT catalyst. Again the PM filter is not included in the cost estimates and is assumed to be catalyzed and identical for both NOx systems. Geometric volumes of the Urea SCR catalyst and the LNT are assumed to be equal and twice the engine displacement. A means of heating the LNT may be required to achieve the needed cold-start performance on the HD FTP and is not included here. The estimated cost of the HD LNT system is about three times that of the Urea SCR system.

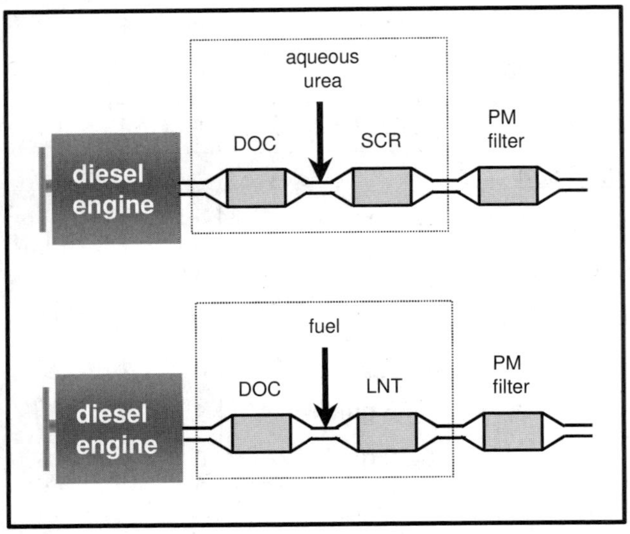

Fig. 6. Configurations used for LD system cost comparison.

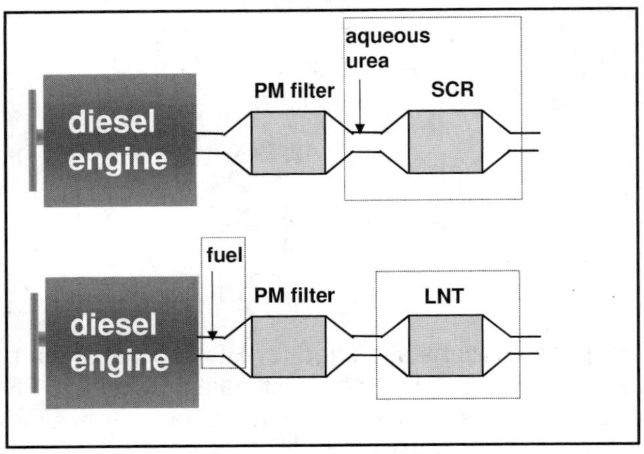

Fig. 7. Configurations used for HD system cost comparison.

The reasons why LNT systems are significantly more expensive than Urea SCR systems are:

(1) The LNT contains precious metal while the Urea SCR catalyst contains base metal. The ratio of average precious metal prices to average base metal prices over the last year is roughly 2000:1.

(2) The LNT precious metal cost is much greater than the urea injection system cost. A simple, air-assisted, reductant injection system has been developed by Ford that has a lower projected cost than previous systems [17]. The point at which the injection system equals the cost of the LNT occurs at an engine displacement much smaller than current PC diesel engines.

(3) The LNT system cost / Urea system cost ratio is higher for HD than for LD because while the LNT and Urea SCR catalysts costs scale with engine size, the urea injection system cost stays largely fixed. Since the LNT contains precious metal and the SCR catalyst contains base metal, the cost differential between the two technologies becomes more disparate as engine size increases.

USAGE COST

Urea SCR – Light-Duty Usage Costs

For a Tier 2 Bin 5 LD diesel vehicle, it is assumed that 90% NOx conversion is required to meet the 0.07 g/mi NOx standard at 120k miles. This allows one to estimate the consumption of a 32.5 wt% aqueous urea solution at approximately 2700 mpg. An average cost of $1.50/gal for diesel fuel may be assumed, but the projected cost of aqueous urea for vehicle use is more uncertain. A recent study on potential HD urea consumption estimates that average future urea price could be as low as $1/gal [18]. For the LD case, no such studies currently exist because of the relative uncertainty in the future US LD diesel market. If a slightly higher and more conservative urea cost is assumed ($1.50/gal), then the total usage cost over 120k miles can be estimated for a range of fuel consumption rates as shown below in Table 7. The usage cost for an LNT system is also shown, assuming a 7% fuel economy penalty for high NOx conversion as discussed in a previous section.

Table 7

Lifetime usage costs for a LD diesel vehicle with high efficiency NOx emission control.

diesel fuel consumption (mpg)	Urea SCR Usage Cost ($)	LNT Usage Cost ($)	Cost Differential ($)
25	7267	7704	437
30	6067	6420	353
35	5210	5503	293
40	4567	4815	248
45	4067	4280	213
50	3667	3852	185

It is interesting to estimate the "break-even points" where Urea SCR and LNT may cost the customer the same amount to operate. Assuming aqueous urea costs $1.50/gal, the LNT has to have a fuel penalty of about 1.5% to have the same usage cost as the Urea SCR system. Conversely, if the LNT fuel penalty is 7%, aqueous urea would have to cost $7-10/gal to have the same usage costs.

The real-world cost differential between use of Urea SCR and LNT is variable according to how the customer actually drives the vehicle. The estimates made here are based on 0.7 g/mi NOx engine-out over the FTP-75 that would require 90% NOx conversion to meet the Tier 2

Bin 5 standard. Other driving cycles would result in different engine-out NOx levels requiring different conversion levels, but our conclusion is that the Urea SCR system will in general cost less to operate than an LNT system at 90% NOx conversion.

Urea SCR – Heavy-Duty Usage Costs

For HD applications, a ratio of aqueous urea consumption relative to fuel consumption of around 5% is quite common [6]. A similar result is reported for a NOx conversion efficiency of 71% [19]. Recent work with more advanced SCR formulations, as experienced in the APBF-DEC program, disclose that as high as 90% NOx efficiency may be achieved in HD applications using Urea SCR technology. This reduction is realized with a more efficient urea consumption of approximately 3% of the fuel consumption [20].

For an average urea consumption over the lifetime of a HD vehicle, we may assume a ratio of urea/fuel consumption of 4%. Additionally, we assume that the vehicle lifetime is the EPA mandated durability requirement of 435k miles and that the vehicle fuel mileage is 6 mpg. Based on these assumptions, the fuel used over the lifetime of the vehicle is 72500 gallons and the urea consumed is 2900 gallons. Using the most likely scenario/pathway for urea distribution published in a recent study [18], the following table is suggested:

Table 8

Retail cost of aqueous urea [18].

Aq. Urea Costs	$/gal	Remarks
Production	0.21	Average of 0.12 – 0.30 reported in reference*
Distribution	0.70	Most likely distribution pathway
Dealer Mark-up	0.08	Average of 0.05 – 0.10 reported in reference*
Total Estimated	**0.99**	

Based on the above table and other assumptions made here, the cost of a lifetime supply of urea for a HD vehicle is about $3000. In comparison, an LNT system with a 7% fuel penalty results in approximately $7600 in extra fuel costs. Additionally, the use of Urea SCR may allow for additional cost savings, because fuel economy is maximized through injection timing advance. A 6% fuel consumption advantage for a MY2007 HD Urea SCR and CDPF system over a MY2004 HD engine is noted in a recent publication [21]. A similar study reports a few percent improvement in BSFC (Brake Specific Fuel Consumption) after engine recalibration [6]. If a conservative 3% advantage in fuel consumption is assumed for Urea SCR, a lifetime fuel savings of 2175 gal is realized, or about $3250. This savings in fuel not only more than pays for the additional price of the urea, it also brings the total lifetime usage cost differential between LNT and Urea SCR to $7350.

CONCLUSIONS

Urea SCR can provide the 90+% NOx conversion required for 2007 LD and HD NOx standards with the following technical advantages over LNTs:

- wider operating temperature window
- greater durability
- lower fuel economy penalty
- lower HC emissions
- lower greenhouse gas emissions
- lower system cost
- lower usage cost

ACKNOWLEDGMENTS

This paper was prepared in part with the support of the U.S. Department of Energy, under Award No. DE-FC26-01NT41103. However, any opinions, findings, conclusions, or recommendations expressed herein are those of the authors and do not necessarily reflect the views of DOE.

Urea SCR work at Southwest Research Institute (see Ref. 7) was sponsored by the APBF-DEC (Advanced Petroleum Based Fuels - Diesel Emissions Control) program, a partnership between DOE, EPA, CARB/SCAQMD and industry.

We thank Yisun Cheng and Will Belanger for obtaining the LNT performance results in the laboratory at Ford. We also thank Paul Laing, also at Ford, for providing predictions of catalyst performance and fuel economy penalties.

REFERENCES

1. P. Tennison, M. Levin, and C. Lambert, "NOx Control Development with Urea SCR on a Diesel Passenger Car", SAE 2004-01-1291.
2. K. Nakatani, S. Hirota, S. Takeshima, K. Itoh, T. Tanaka, K. Dohmae, "Simultaneous PM and NOx Reduction System for Diesel Engines", SAE 2002-01-0957.
3. J. McDonald and B. Bunker, "Testing of the Toyota Avensis DPNR at U.S. EPA-NVFEL", SAE 2002-01-2877.
4. R. Hammerle, "Urea SCR and DPF System for Diesel Sport Utility Vehicle Meeting Tier II Bin 5", presented at 8th Diesel Engine Emissions Reduction (DEER) Conference, 2002.

5. R. Mital, J. Li, S.C. Huang, B.J. Stroia, R.C. Yu, J.A. Anderson, K. Howden, "Diesel Exhaust Emissions Control for Light Duty Vehicles", SAE 2003-01-0041.
6. B. Scarnegie, W.R. Miller, B. Ballmert, W. Doelling, S. Fischer, "Recent DPF/SCR Results Targeting US2007 and Euro 4/5 HD Emissions", SAE 2003-01-0774.
7. Magdi Khair, "Demonstration of Potential for Selective Catalytic Reduction and Diesel Particulate Filters", presented at World Fuels Conference, Washington, DC, October 10, 2002.
8. C. Schenk, J. McDonald, C. Laroo, "High-Efficiency NOx and PM Exhaust Emission Control for Heavy-Duty On-Highway Diesel Engines - Part Two", 2001-01-3619.
9. W. Muller, H. Ölschlegel, A. Schäfer, N. Hakim and K. Binder, "Selective Catalytic Reduction - Europe's NOx Reduction Technology", SAE 2003-01-2304.
10. "Selective Catalytic Reduction (Final Report): The Most Promising Technology to Comply with the Imminent Euro IV and Euro V Emission Standards for HD Engines", presentation by the ACEA (Verband der europäischen Automobilindustrie), Brussels, June 23, 2003.
11. D. Baker, B. Hammerle, R. Backes, "Harnstofflogistik zur NOx-Reduzierung von Fahrzeugen (Urea Infrastructure for Control of Vehicular NOx Emissions)", Aachener Kolloquium Fahrzeug-und Motorentechnik 2000.
12. M. Levin and R. Baker, "Co-fueling of Urea for Diesel Cars and Trucks", SAE 2002-01-0290.
13. P.M. Laing, M.D. Shane, S. Son, A. A. Adamczyk, and P. Li, "A Simplified Approach to Modeling Exhaust System Emissions: SIMTWC," SAE 1999-01-3476.
14. C. Schenk and C. Laroo, "NOx Adsorber Aging on a Heavy-Duty On-Highway Diesel Engine - Part One", SAE 2003-01-0042.
15. S. Geckler, D. Tomazic, V. Scholz, M.V. Whalen, D. McKinnon, J. Orban, R.A. Gorse, O. Bailey, J.C. Hoelzer, "Development of a Desulfurization Strategy for a NOx Adsorber Catalyst System", SAE 2001-01-0510.
16. M. May, "APBF-DEC Heavy Duty NOx Adsorber/DPF Project: Heavy Duty Linehaul Platform Project Update", presented at 9th DEER Conference, 2003.
17. C-H. Wu and R. Hammerle, "An On-board Reductant Delivery System for Diesel Aftertreatment", SAE 2001-01-3622.
18. S. Fable, F. Kamakate, S. Venkatesh, "Selective Catalytic Reduction Urea Infrastructure Study", subcontractor report for the National Renewable Energy Laboratory (NREL/SR-540-32689), July 2002, http://www.ott.doe.gov/pdfs/adl_urea.pdf.
19. M. Khair, D. McKinnon, "Performance Evaluation of Advanced Emission Control Technologies for Diesel Heavy-Duty Engines", SAE 1999-01-3564.
20. M. Khair, "Low Emissions Potential of EGR-SCR-DPF and Advanced Fuel Formulations – A Progress Report", presented at 9th DEER Conference, 2003.
21. M. Schittler, "State-of-the-art and Emerging Truck Engine Technologies", presented at 9th DEER Conference, 2003.

CONTACT

Christine K. Lambert
PO Box 2053, MD3179, SRL
Dearborn, MI 48121
clamber9@ford.com

DEFINITIONS, ACRONYMS, ABBREVIATIONS

CO: Carbon Monoxide

CO_2: Carbon Dioxide

deSOx: Removal of S from LNT

DOC: Diesel Oxidation Catalyst

DPNR: Diesel Particulate - NOx Reduction

FTIR: Fourier Transform Infrared

FTP: Federal Test Procedure

GHG: Greenhouse Gas

HC: Hydrocarbons

HD: Heavy-Duty

HDT: Heavy-Duty Truck

LD: Light-Duty

LDT: Light-Duty Truck

LNT: Lean NOx Trap (NOx adsorber)

NH_3: Ammonia

NMHC: Non-Methane Hydrocarbons

NMOG: Non-Methane Organic Gases

N_2O: Nitrous Oxide

NOx: Oxides of Nitrogen (NO, NO_2)

O_2: Oxygen

PC: Passenger Car

PM: Particulate Matter

SCR: Selective Catalytic Reduction

SET: Supplemental Emission Test

SIMTWC: Ford proprietary aftertreatment model.

SOx: Oxides of Sulfur (SO_2, SO_3)

Urea SCR: SCR using aqueous urea.

US06: High speed, high load cycle for LD vehicles

2005-01-1860

Development of Urea-SCR System for Heavy-Duty Commercial Vehicles

Kiminobu Hirata, Nobuhiko Masaki, Hiroki Ueno and Hisashi Akagawa
Nissan Diesel Motor Co., Ltd.

ABSTRACT

In Japan there is currently a strong social demand for exhaust emissions reduction from heavy-duty diesel engines. Therefore, new Long-Term Regulation will come into effect in October 2005, setting the NOx standard at 2.0 g/kWh and the PM standard at 0.027 g/kWh. At the same time, customers always demand exceptional fuel economy from heavy-duty commercial vehicles. A urea-based Selective Catalytic Reduction (SCR) system was developed to satisfy both these demands, and will be introduced in the fall of 2004.

The operating conditions of vehicles in Japan are different from those in the US and Europe. Basically, average vehicle speeds are significantly lower. To improve the low temperature SCR performance, an oxidation catalyst was located upstream of the SCR, and an additional oxidation catalyst was located downstream of the SCR for emergency NH3 slip. The muffler size with all three catalysts was similar to a conventional muffler.

The fuel and urea water solution consumption of the system was studied and compared with the fuel consumption of a similar system equipped with EGR and a Diesel Particulate Filter (DPF) unit.

The properties of the urea water solution used here were not well known at the time of this project. Therefore, the specifications and corrosion characteristics of relevant materials were investigated and suitable materials were selected. Urea water solutions freeze at around –11 deg C, so a thawing system was also developed.

Vehicle tests were carried out at various conditions, during which no major problems arose.

INTRODUCTION

A large proportion of the expenses of a transportation business is for fuel. So commercial vehicles are exclusively powered by diesel engines due to their high efficiency. On the environmental side, carbon dioxide emissions (a greenhouse gas) are low, but further reduction of oxides of nitrogen (NOx) and particulate matter (PM) are strongly required. New long-term exhaust emissions regulations will be introduced in Japan in October of 2005. The new maximum values will be NOx = 2.0g/kWh and PM = 0.027g/kWh, which will be the lowest values in the world at that time.

It is impossible to meet the 2005 NOx and PM levels by engine modification only. Notably, NOx levels have a trade-off relationship with fuel economy, so if an after-treatment system can be used, which can substantially reduce NOx levels, good fuel economy can be maintained. Therefore we combined engine modification with SCR (Selective Catalytic Reduction), a popular de-NOx catalyst technique, and chose urea solution as the reducing agent [1-3]. This combination made it possible to achieve the new long-term exhaust emission levels and get improved fuel economy as well, without using a DPF (Diesel Particulate Filter).

On the other hand, use of a urea-water solution with an SCR is still somewhat new, and there are many issues that must be clarified, an example of which is handling the urea-water solution. Urea-water solutions are not used in the automotive field, so the characteristics and influence of many SCR system components had to be carefully planned out. Therefore, we investigated the influence of various materials, and the fundamental characteristics and stability of the urea-water solution. Since it freezes at –11 deg C, low-temperature countermeasures were applied. Development of the optimum urea-water supply system was a major part of this study.

The other important characteristic is the overall system durability, including the catalyst. We evaluated this parameter by on-road running tests.

The vehicle which used this system and meets the new long-term emissions regulations was released to the market in the Fall of 2004, about one year before the new regulations take effect.

DEVELOPMENT OF UREA SCR SYSTEM

Figure 1 shows a schematic of the Urea-SCR system developed in this project. The main components are the catalysts and the plumbing to add the reducing agent to the exhaust stream. The reducing agent is a 32.5% urea water solution, and it is injected into the exhaust gas stream in front of SCR catalyst. Once injected, the urea undergoes hydrolysis reactions to form ammonia (NH_3), and the NH_3 reacts with NOx, resulting in N_2.

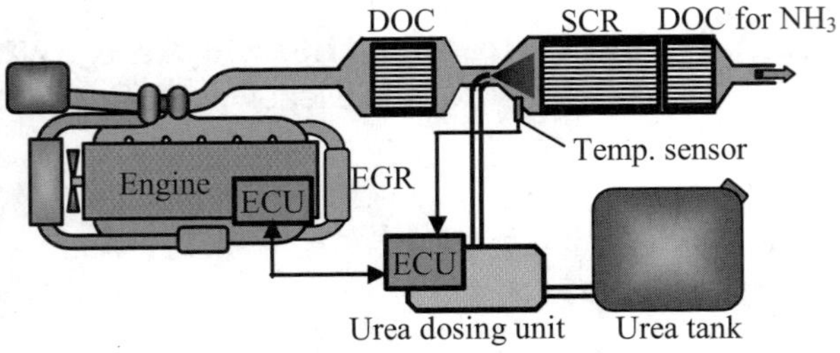

Fig. 1. Schematic of the Urea-SCR System

CATALYST DEVELOPMENT

Improvement in catalyst performance at low-temperatures

The Japanese transient cycle (JE05) adopted for the new long-term regulations has a relatively higher emphasis on low speed and low load operation, compared with the U.S. (FTP) and European (ETC) test cycles (Figure 2). As a result, the exhaust gas temperatures are generally lower than those of the European and US test cycles.

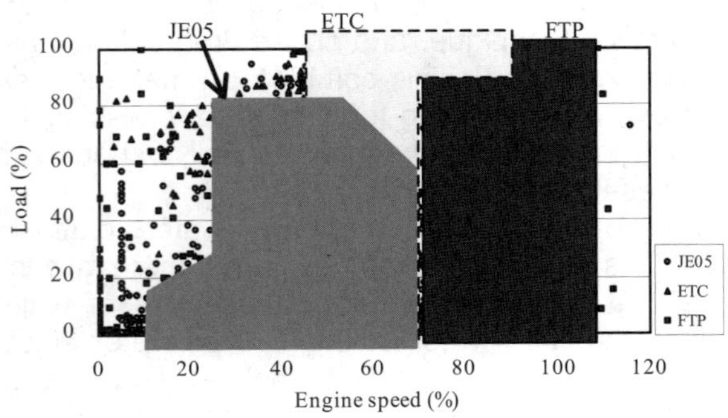

Fig. 2. Comparison of engine speed and load during JE05, ETC, FTP test cycles

It was thus decided to focus upon improving NOx reduction reactions of the general low-temperature for the catalyst used here. The urea hydrolysis and NOx reduction reactions are shown by reactions (1) - (4). There is a 1:1 ratio of $NO:NO_2$ in reaction (4), and since this reaction is most active, it is effective to focus here for improvement in the rate of conversion at low temperatures.

$$CO(NH_2)_2+H_2O \rightarrow 2NH_3+CO_2 \quad (1)$$

$$4NO+4NH_3+O_2 \rightarrow 4\,N_2+6H_2O \quad (2)$$

$$6NO_2+8NH_3 \rightarrow 7N_2+12H_2O \quad (3)$$

$$NO+NO_2+2NH_3 \rightarrow 2N_2+3H_2O \quad (4)$$

Note that the DOC (Diesel Oxidation Catalyst) was located upstream of the SCR catalyst and optimized the activity. Table 1 shows the catalyst specifications and Figure 3 shows NOx to N_2 conversion with and without the DOC. DOC "A" (in table 1) was selected for use in this project, mainly due to its high NO to NO_2 activity. The SCR with the DOC can sharply improve low-temperature NOx conversion during many low-speed low-load portions of the JE05 test cycle .

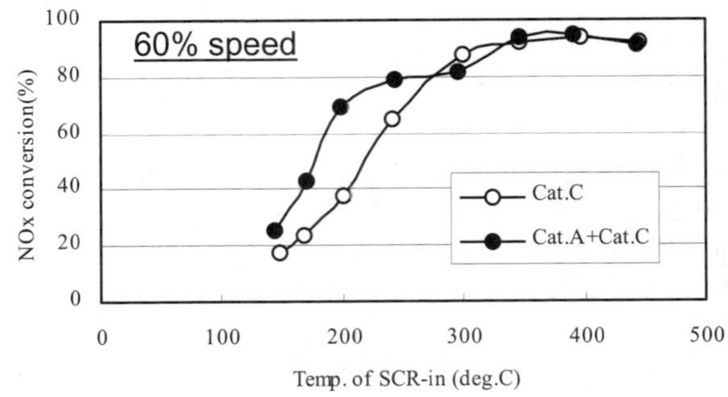

Fig. 3 Effects of DOC on NOx conversion

Selection of an SCR catalyst

There are vanadium and zeolite type SCR catalyst systems. The vanadium type has already been adopted as the SCR system of choice for stationary engines, and it is coming into more widespread use in vehicular applications. But the vanadium type catalyst is formed by squeeze out, and is considered to be large in size. It is thus difficult to integrate this unit into a vehicle at this time. On the other hand, the zeolite type catalyst is coated on a ceramic substrate, and can easily be small in size compared with the vanadium type.

There is not a significant difference in NOx conversion between the vanadium and zeolite type catalysts, as shown in Figure 4. Furthermore, the catalyst volume was considered, as shown in Figure 5. This plot shows a comparison of NOx conversion with the vanadium and zeolite type catalysts against catalyst volume. The vanadium type shows a tendency for the NOx conversion to decrease with reduction in volume, while for the zeolite type a higher conversion efficiency is indicated for a given volume.

As mentioned above, the zeolite type catalyst was chosen for the present system, in consideration of the vehicle mounting situation and integration circumstances.

Table1 Urea-SCR catalyst specifications

		DOC		SCR				NH3-SLP
		A	B	C	D	E	F	G
Diameter	mm(inch)	266.7(10.5)	←	241.3(9.5)	266.7(10.5)	←	←	←
Length	mm(inch)	152.4(6.0)	101.6(4.0)	254(10)	152.4(6.0)	←	←	101.6(4.0)
Volume	L	8.5	5.7	11.6	8.5	←	←	5.7
Cell Density	cell/inch2	400	←	300	400	←	←	←
Composition		Pt-type	←	Vanadium-type	Zeolite-type I	Zeolite-type II	Zeolite-type III	Pt-type

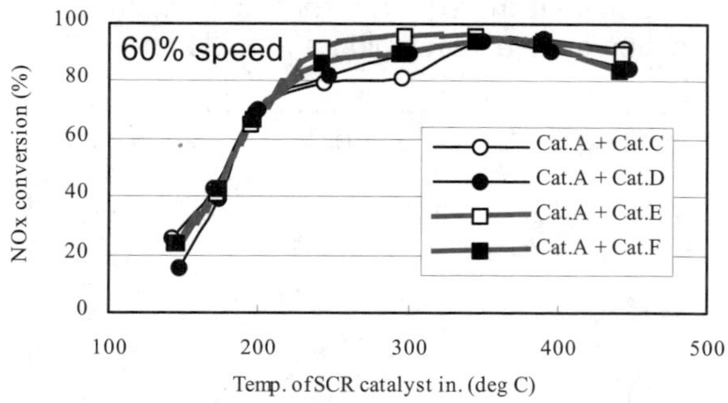

Fig. 4 Iufluence of catalyst on NOx conversion

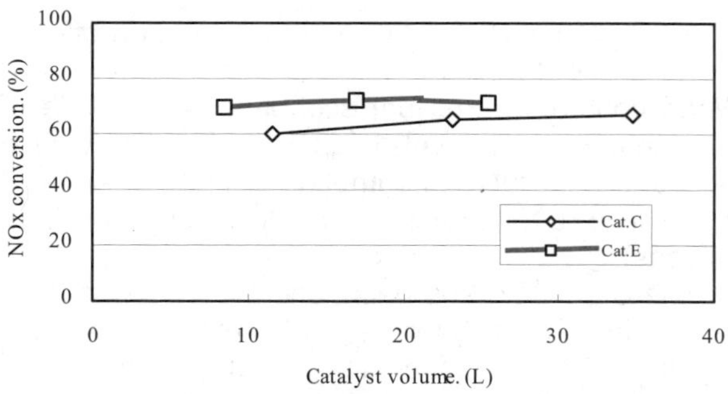

Fig. 5 Influence of catalyst volume on NOx conversion (JE05)

Prevention of NH_3 slip

If more-than-sufficient NH_3 reducing agent is supplied to the SCR for the amount of NOx present, the excess NH_3 (which could not react with NOx) will slip. In general, an equivalent amount of urea solution is injected for the NOx present, but the downstream DOC is included for such cases of emergency NH_3 slip.

The oxidation performance of the NH_3 slip catalyst is shown in Figure 6. By use of this catalyst, even when there is NH_3 slip, the tailpipe levels can be reduced sharply.

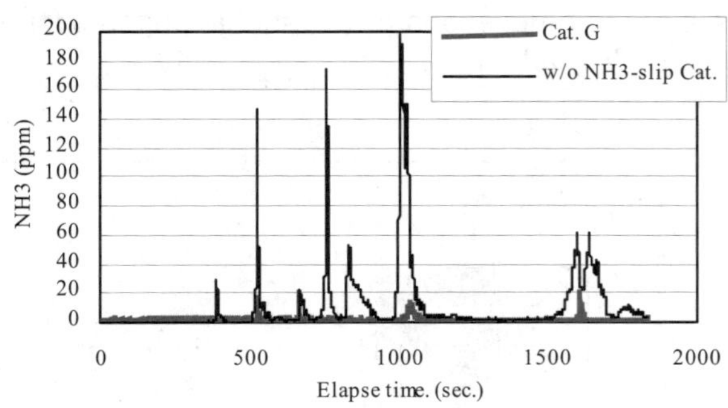

Fig. 6 Performance of NH_3-Slip catalyst (JE05)

DEVELOPMENT OF UREA WATER-INJECTION EQUIPMENT

A Urea-SCR system has to inject the solution at the proper location and time for optimum NOx reduction, to minimize NH3 slip, and to efficiently utilize the urea solution. Figure 7 shows the urea solution addition equipment, made by Bosch.

An air assist type nozzle was used to inject the urea solution. This type of injector has good atomization and dispersion of the urea solution and resists plugging. Furthermore, an electric heater is built into the injector body and piping to prevent freezing, and for thawing frozen solution. Further details are provided in a later section.

The urea water-Ingection equipment has ECU and it took on the task of urea solution injection control. Input variables included activity temperature, adsorption and desorption of NH_3, NOx discharge level, etc. With this arrangement, system optimization was carried out. Results indicated high levels of NOx conversion without NH_3 slip. (Figure 8)

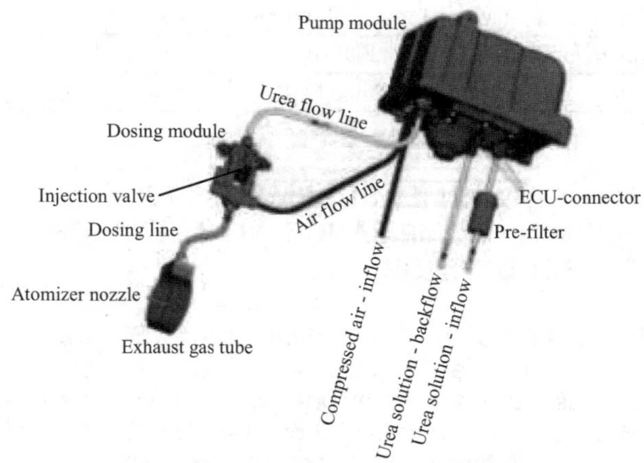

Fig. 7 Schematic of urea metering unit

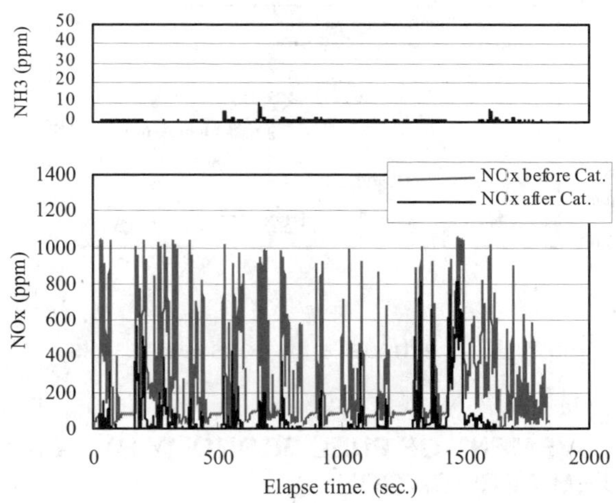

Fig. 8 NOx and maximum NH_3 concentration on JE05 mode.

ENGINE DEVELOPMENT FOR OPTIMIZATION WITH UREA SCR

Engine development was carried out on the assumption that NOx can be reduced with a urea-SCR system. The fuel-injection pressure was increased to reduce PM emissions. For the 13L engine used here, we adopted a new miniaturized unit injector. Then the fuel injection system and combustion system were re-optimized, which allowed us to achieve PM levels below the new long-term regulated value - without a DPF. Furthermore, a cooled EGR system was used to reduce NOx emissions in the low exhaust temperature region, in which the catalyst does not perform well. Table 2 shows the specifications of the two engine types.

Table2 Engine specifications

Engine Type	DI, L6,TCI,4-Stroke	DI, L6,TCI,4-Stroke
Displacement	13.074 L	9.203 L
Bore × Stroke	ϕ 136mm×150mm	ϕ 125mm×125mm
Compression ratio	17.5	16.5
Injection system	Unit Injector	Common rail
Maximum power	280kW/1800rpm	250kW/2200rpm
Maximum torque	1648Nm/1400rpm	1446Nm/1400rpm
Maximum Inj. Press.	200MPa	160MPa
EGR	Cooled EGR	Cooled EGR

EXHAUST-GAS MEASUREMENTS TO DETERMINE COMPLIANCE WITH NEW LONG-TERM REGULATIONS

Test results of the 13L engine for the JE05 cycle are shown in Figure 9. Values are shown comparing emissions levels for the engine equipped with the Urea-SCR system to an engine without the Urea-SCR system. The results show that NOx and PM were reduced sharply and did not exceed the new long-term regulated values.

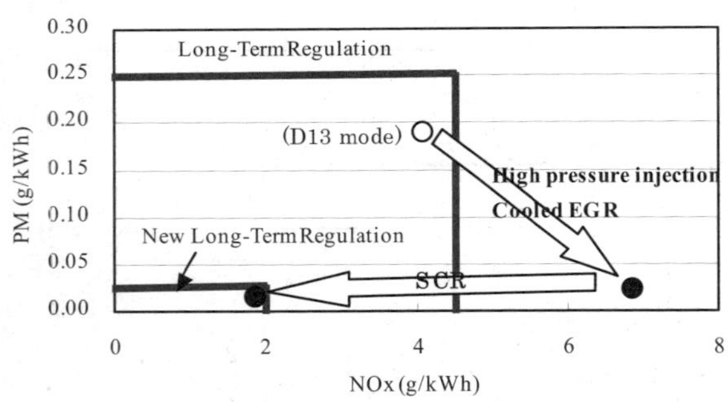

Fig. 9 JE05 exhaust emissions test results

IMPROVEMENT OF FUEL ECONOMY BY SCR SYSTEM APPLICATION

There are two approaches to achieving the new long term emissions regulations [4]. One is to reduce PM by in-cylinder combustion improvement and reduce NOx with a Urea-SCR catalyst. The other is to reduce NOx by high levels of EGR and reduce PM by using a DPF. Fuel economy may be modeled and compared when using these two approaches to determine which is more promising. Generally, fuel economy is related to NOx emissions levels. Figure 10 shows fuel economy as a function of NOx level at 60% speed and 60% load, with the NOx level varied by injection timing and EGR rate. It can be seen that there is a strong dependence of fuel consumption on injection timing and EGR rate.

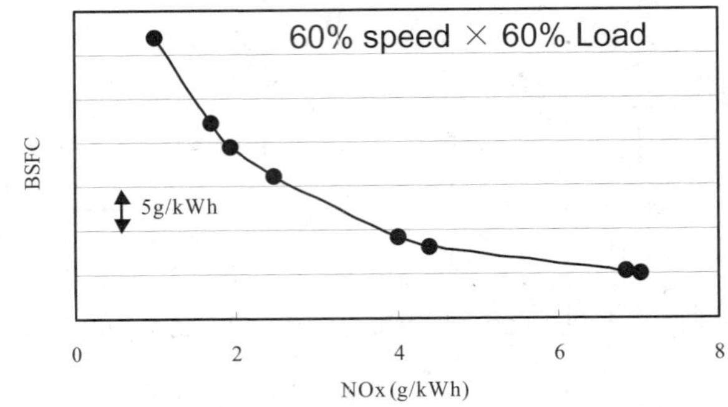

Fig. 10 NOx Fuel Consumption Trade-off

When comparing the fuel economy of a vehicle with an engine using high levels of EGR and a DPF, we need to account for a fuel consumption penalty of about 3% due to active regeneration of the DPF (for some driving conditions).

For the Urea-SCR approach, one must account for the urea solution's cost, after determining the consumption and price. The consumption changes with driving conditions, but it is at most about 5% of fuel consumption. If the urea solution price is half the fuel price, the cost penalty is 2.5% of fuel cost.

As mentioned above, we simulated the fuel economy of vehicles utilizing the two above approaches to achieve the new long-term regulations. Conditions were 90km/h constant speed, which is a typical running condition for heavy-duty commercial vehicles in Japan. In this case, DPF does not need active regeneration, so the resulting fuel consumption penalty is not observed. The simulation results are shown Figure 11. For the Urea-SCR system, the fuel consumption is 6.5% better than the high EGR + DPF case. Accounting for the urea solution's cost, it is 4% better. For the city driving mode, which has more low speed conditions, urea consumption is lower, and active regeneration is needed for the DPF, so a greater fuel economy advantage appears for the Urea-SCR system.

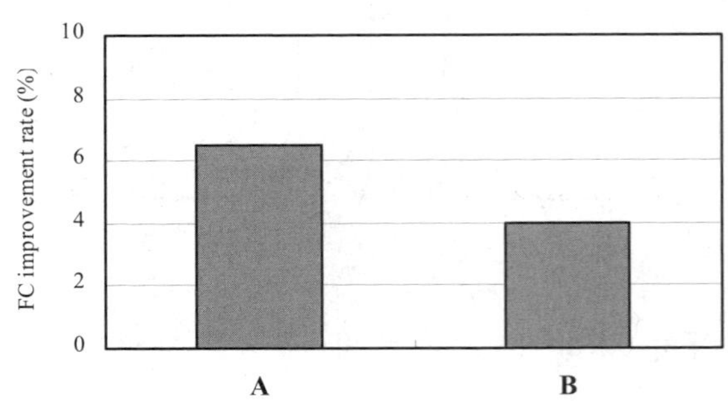

Fig. 11 Fuel consumption Improvement for the Urea-SCR system compared to high EGR+DPF system at 90km/h speed
A: Not including urea solution cost
B: Including urea solution cost

UREA WATER SOLUTION

CHARACTERISTICS OF UREA WATER SOLUTION

The SCR system uses ammonia (NH_3) as the actual NOx reducing agent. NH_3, however, requires high-pressure gaseous storage and handling. This makes ammonia unsuitable for vehicular use. Urea is easier to handle and safer to store and use. Urea is a white odorless crystal, is used for cosmetics (moisturizer) and medical supplies. Additional information is available on the product safety data sheet.

Urea, which is in crystalline form, dissolves in water and is used at a concentration of 32.5%. The main characteristics of the 32.5% urea water solution are shown in Table 3. Its specific gravity is about 1.1, it is a colorless and odorless liquid. Its freezing point is -11 deg C.

Table 3 Properties of 32.5% urea solution

Concentration (wt%)	32.5
Density at 20℃	1.090
Freezing point (℃)	-11
pH	8.8
Color	colorless
Odor	odorless

Considering typical vehicular operating conditions, a low freezing point was desired. The minimum freezing temperature is achieved at the 32.5% urea concentration. Figure 12 shows the relation between urea concentration and freezing point. The 32.5% concentration results in the lowest freezing point of -11 deg C. Stronger or weaker solutions have a higher freezing point.

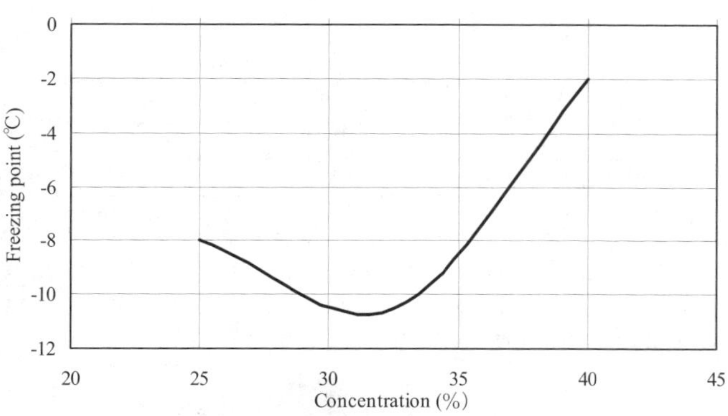

Fig. 12 Influence of concentration on freezing point

From this point we will always assume a 32.5% urea concentration in this work. For other work using different concentrations, one should remember that the NOx conversion rate, and NH_3 slip amount may be different. Since the solution is injected through relatively small tubing, plugging due to impurities must be avoided.

Japan and Europe are establishing standards for urea water solutions on the basis of "Refined Urea" and ion exchange water. Automotive Standard JASO E502 was established in Japan, and this will become a JIS standard in the near future. Similarly, the German federal standard DIN70070 was enacted by the European commission. Work is being done to create an ISO standard based on the JASO and DIN standards.

Table 4 Highlights of the JASO standards

Characteristics	Unit	Limits	
		Min	Max
Concentration	weight-%	31.8	33.3
Density at 20℃	g/cm3	1.087	1.092
Refractive index at 20℃	—	1.3817	1.384
Alkalinity as NH3	%	—	0.2
Carbonate as CO2	%	—	0.2
Biuret	%	—	0.3
Aldehyde	mg/kg	—	10
Insolubles	mg/kg	—	20
Phosphate(PO4)	mg/kg	—	0.5
Calcium	mg/kg	—	0.5
Iron	mg/kg	—	0.5
Copper	mg/kg	—	0.2
Zinc	mg/kg	—	0.2
Chromium	mg/kg	—	0.2
Nickel	mg/kg	—	0.2
Magnesium	mg/kg	—	0.5
Sodium	mg/kg	—	0.5
Potassium	mg/kg	—	0.5

STABILITY OF UREA WATER SOLUTION

Urea water solutions may be left in storage for long periods during the distribution process and when the vehicle is not regularly operated. We investigated the long-term stability of urea water solution in terms of the solution's pH, density, and sediment formation.

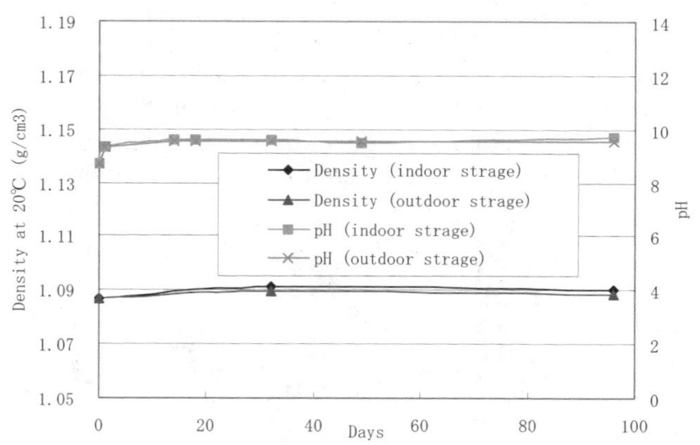

Fig. 13 Urea water stability in storage

Figure 13 shows analysis results for a urea water solution left outdoors for about 14 weeks (96 days). The pH and density were measured. In spite of the sunlight, the concentration was unchanged, and there was also no sediment formation.

INFLUENCE OF UREA WATER SOLUTION ON WETTED MATERIALS

In order to investigate the influence of urea water solution on wetted materials in the Urea-SCR system, such as the tank and piping, soak testing of metal, resin, and rubber materials was performed. A sample of test material was placed in the urea water solution and another in distilled water for comparison. From here, tests were performed with immersion at 80 deg C test temperature for a maximum 1,000hr. We describe here the test results of 80 deg C x 120hr, which were the initial material selection conditions.

(1) Tests of Metallic Materials

The amount of corrosion was measured for stainless steel, aluminum, cast iron, steel (galvanized), brass, and copper. These materials may be used for storage tank or piping. The test results are shown in Figure 14.

Although notable corrosion was found on the copper and brass, others were unchanged. Therefore the stainless steel was adopted for the storage tank.

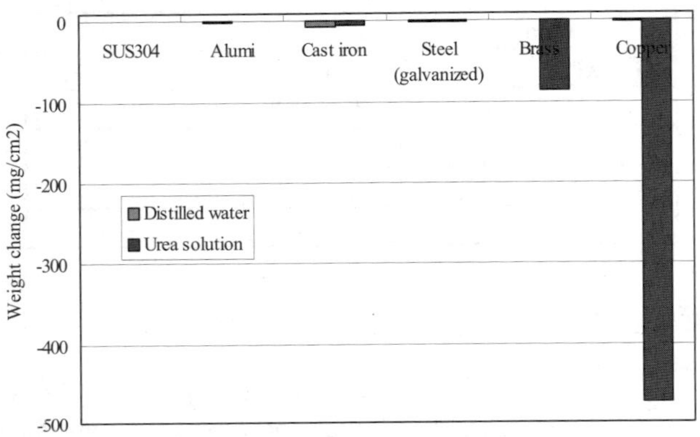

Fig. 14 Corrosion of metal samples

(2) Tests of Resin Materials

Volume, Dimensional and weight changes were measured for PFA, PA, PVC, and PVC+PTFE. Volumetric change of these test results are shown in Figure 15. PFA and PA12 were adopted for piping, since these samples exhibited little volumetric, dimensional or weight changes compared with the other resin materials.

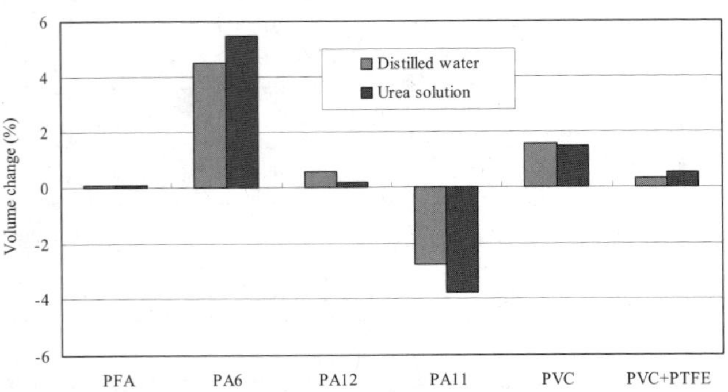

Fig. 15 Volumetric change of plastics

(3) Tests of Rubber Materials

The change in volume and hardness were measured for EPDM, NBR, and NBR+PVC. Figure 16 shows volumetric change of these samples. It can be seen that EPDM shows the least volume and hardness change compared with the other materials. EPDM was therefore used for seals in the Urea-SCR system.

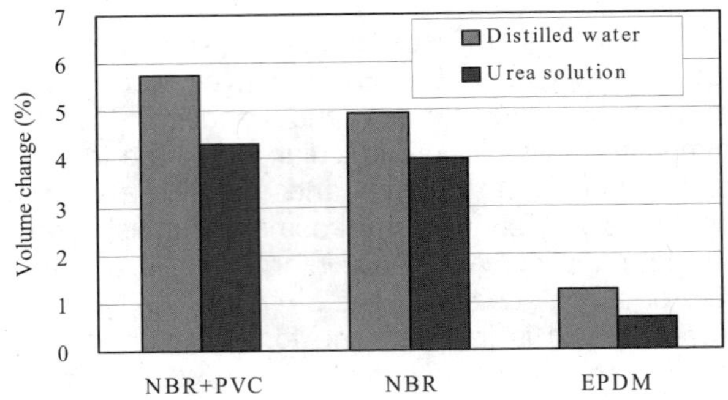

Fig. 16 Volumetric change of rubber materials

UREA WATER SUPPLY SYSTEM

It is convenient for a truck driver to fill the urea water solution simultaneously with a fuel. About 80% of the domestic medium and heavy-duty truck drivers use own fuel tank or truck station, as shown in Table 5. So, for drivers who use own fuel tank, they can supply urea water solution for they own its tank too. There are seven main in-tank bases in Japan, which are nearly ready to supply urea water solution to vehicles. The wholesale supply organization to about 80 percent of the roughly 1,000 truck stations nationwide is also ready to begin supplying urea water solution. Further, the supply organization to all NISSAN DIESEL dealers nationwide is also ready to begin supplying urea water solution. A container for storage of emergency backup urea water solution was also developed. When our trucks equipped with the Urea-SCR system were released to the market, adequate supply infrastructure was already attained.

Table 5 Share of fuel filling stations (Heavy-duty truck users)

	number	share
Truck station	≈1,000	20～25%
Freight own tanks	≈10,000	50～60%
Gas station	≈44,000	20～25%

The driver can monitor a urea water solution indicator in the instrument panel, which tells at a glance the quantity remaining in the tank (Figure 17).

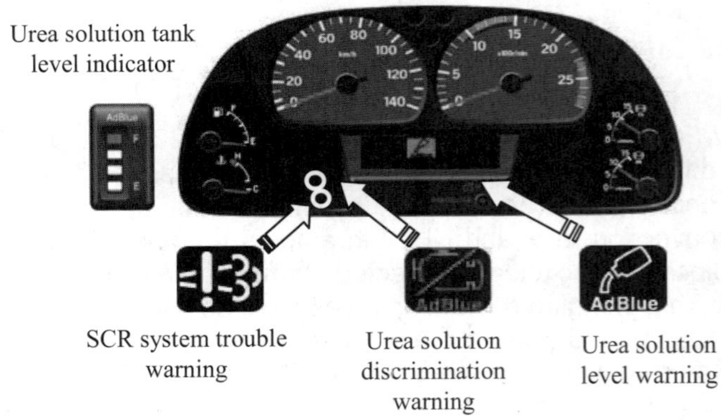

Fig. 17 Urea solution level indicator

If the level drops below a minimum amount, a warning lamp indicates this. In order to prevent refilling the urea-water tank with fuel, the urea-water tank cap is color-coded and the mouth is smaller than fuel filler nozzle.

Moreover, if the urea water solution is sub-standard, or a different type of liquid is added, a urea water solution recognition sensor detects this and a warning is given to the driver.

COLD TEMPERATURE CHARACTERISTICS

UREA THAWING SYSTEM

As mentioned above, the urea-water solution freezes at -11 deg C. For cold weather operation, the urea water tank has thawing and freezing prevention systems. First, the urea water tank has passages through which engine coolant flows, to provide quick thawing and warm-up (Figure 18). Also, the urea solution piping and injection unit have an electric heater. These controls use input from air and urea solution temperature sensors.

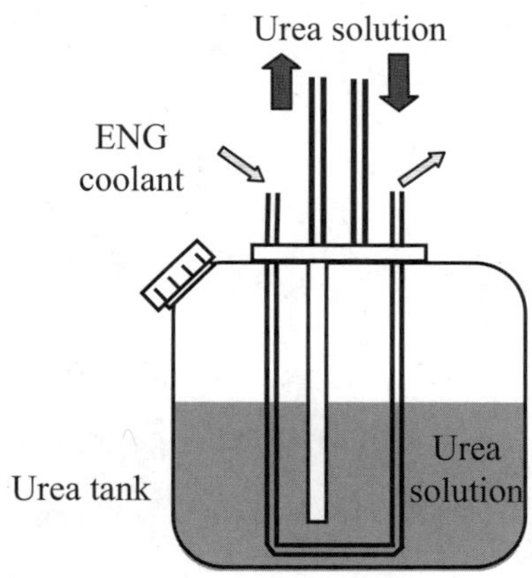

Fig. 18 Schematic of tank heating system

LOW TEMPERATURE EMISSIONS PERFORMANCE

The rate of NOx conversion falls as the ambient temperature drops, due to the decrease in catalyst inlet temperature. Thus, the NOx conversion rate was investigated during vehicle operation at about 0 deg C ambient temperature. The test results are shown in Figure 19. An approximate 70% NOx conversion rate is obtained at this operating temperature.

DURABILITY

The practical durability of a Urea-SCR system is the most important. Several test vehicles were operated at highway, suburban, and urban conditions. Test vehicles were equipped with the best catalysts, as determined by the bench tests. The total distance covered by these durability tests thus far is about 700,000km.

An example of results revealed to date follows. The change in NOx conversion rate from 36,000km driven to 50,000km driven is shown in Figure 20. Table 6 shows the specifics of this test vehicle and engine.

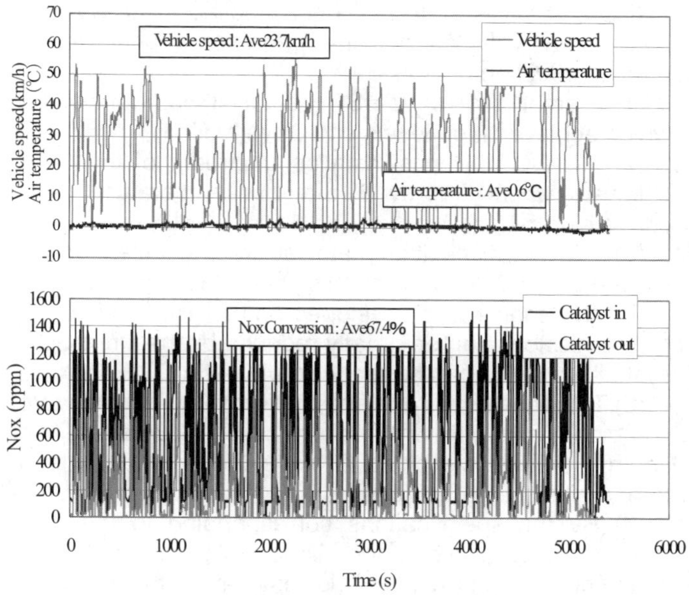

Fig. 19 Low-temperature NOx emissions

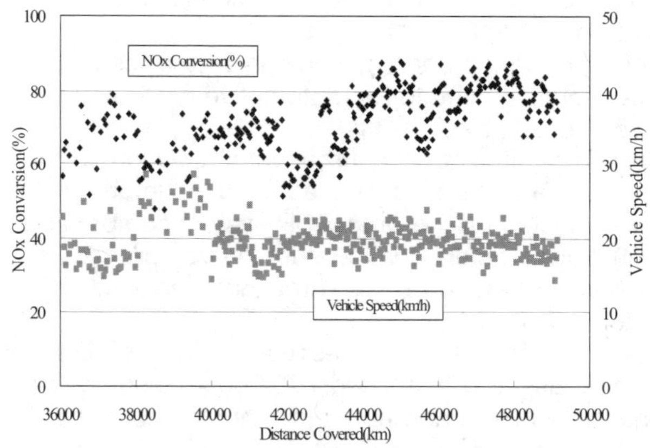

Fig. 20 Durability test results

Although test result show a slight decrease in performance at low operating temperature, it was possible to maintain approximately 70% NOx conversion rate on average. No SCR system trouble or plugging problems were observed in the durability tests to date.

It may be pointed out that the durability tests are still ongoing.

Table 6 Specifications of test vehicle and engin

Vehicle	Type	Heavy Duty Truck
	Gross vehicle weight	24,840kg
	Vehicle weight	8,730kg
Engine	Type	DI , TCI
	Numbers of cylinder	6
	Bore×Stroke	ϕ 136×150mm
	Displacement	13.074L
	Rated power	272kW/1900rpm

CONCLUSION

A Urea-SCR system has been applied to heavy-duty commercial vehicles, for the purpose of achieving a drastic reduction of emissions.

(1) A small zeolite type SCR catalyst was developed, which achieved high levels of NOx reduction.

(2) A DOC unit was installed upstream of the SCR catalyst, which improved the low-temperature NOx conversion rate.

(3) A NH_3 oxidization catalyst was installed downstream of the SCR catalyst; this prevented NH_3 slip.

(4) The engine was developed on the assumption that NOx emissions can be greatly reduced by the Urea-SCR system, thus PM can be reduced without a DPF.

(5) The above catalysts and the improved engine were combined, and with further optimization of urea addition control, the vehicle could achieve the new 2005 long term emissions regulations.

(6) By the specifications corresponding to the new long term regulation, the Urea-SCR system's fuel consumption improvement is estimated to be 6.5% compared with a similar unit using the high-EGR + DPF technique for high-speed operation. Even if we take the consumption of urea solution into consideration, there is still a 4% advantage.

(7) Important properties and characteristics of the urea water solution used for an automotive urea SCR catalyst have been clarified.

(8) The effect of the urea-water solution on wetted system materials was studied, and optimum materials were selected for the Urea-SCR system components.

(9) Low ambient temperature freeze prevention and thawing systems for the urea storage and injection systems were developed.

(10) Several test vehicles using Urea-SCR systems have amassed a combined 700,000km distance, and to date there have been no major troubles nor a decline in the NOx conversion rate.

Many vehicles equipped with this Urea-SCR systems will arrive on the market in the Fall of 2004, about one year before the new long-term emissions regulations take effect. We plan to continually make improvements to these emissions control systems and the vehicles, as many customers are relying on us.

ACKNOWLEDGMENTS

The authors are deeply grateful to Tokyo-Roki Co., Ltd. and Bosch Automotive Systems Corp. for their cooperation on this project.

REFERENCES

1. Havenith C., Verbeek R. P. (1997) Transient Performance of a Urea deNOx Catalyst System for Low Emissions Heavy-Duty Diesel Engines. SAE Paper 970185
2. Fritz N., Mathes W., Zuerbig J., Mueller R. (1999) On-Road Demonstration of NOx Emission Control for Diesel Trucks with SINOx Urea SCR System. SAE Paper 1999-01-0111
3. Miller W. R., Klein J. T., Mueller R., Doelling W., Zuerbig J. (2000) The Development of Urea-SCR Technology for US Heavy Duty Trucks. SAE Paper 2000-01-0190
4. Moser F. X., Sams T., Cartellieri W. (2001) Impact of Future Exhaust Gas Emission Legislation on the Heavy Duty Truck Engine. SAE Paper 2001-01-0186

2007-01-1577

Innovative Substrate Technology for High Performance Heavy Duty Truck SCR Catalyst Systems

Michael Rice, Jan Kramer, Klaus Mueller-Haas, Raimund Mueller
Emitec, Inc.

ABSTRACT

Diesel engine emissions regulations throughout the world have changed dramatically over the last 20 years, with the most stringent phase for medium/heavy duty being implemented starting 2007 – 2010. The general perception is, that combination systems including PM reduction devices and NOx aftertreatment, likely SCR will be applied, hence significant additional emissions equipment will need to be added to a diesel engine. The consequences of this will be increased space for packaging and increased cost, to meet the demanding regulations. Traditional substrate technology, while developing extensively for gasoline engines by use of higher cell densities has limitations when applied to CI diesel engines. New developments in substrate and coating technology now enable considerable improvement in the efficiency and packaging space needed to meet certain limit values. A novel approach combining a continuously operating PM-reduction with advanced "turbulent" SCR-catalyst in a special system configuration has been investigated in an engine cell program with a HDDE, demonstrating the capability to approach US 2010 emission limits with an extremely compact and cost effective design.

INTRODUCTION

The U.S. 2007 and EU 4/5 emission limits for HD-trucks can be met with the on-going engine development successes and application of one exhaust emission control technology, either for PM-reduction (DPF, PM-Metalit) or for NOx reduction (SCR systems). For US 2010 and the announced but not yet finalized EU 6, it is considered common understanding in the industry that a combination of both PM and NOx reduction exhaust systems will be required, together with further advanced engine technology. It is generally expected that the engine will apply high–pressure EGR systems, dual stage turbo charging, high pressure common rail injection, and as well additional advanced features. While these developments will bring the engine-out emissions to a much lower level, they will also add cost. It will be necessary to reconsider the traditional approaches of emission control systems, to maintain the diesel engines perception as the highly efficient, durable, cost-effective prime mover for trucks. Future emission control systems shall be designed in view of the control and emission performance possibilities of future engine technology, while minimizing space allocation in the truck and to keep the demands on maintenance and operator's interaction to a minimum.

The future engines will emit NOx and PM at significantly lower levels as could be expected some years ago. Taking this into account, this paper will describe the status of an on-going investigation to apply the continuously operating PM-Metalit to a low emitting engine, minimizing the demand for active regeneration equipment and filter maintenance. Further, advanced catalyst and coating technology has shown the potential to offer a drastic reduction of diesel oxidation catalyst volumes. Similar advantages are demonstrated regarding the required SCR catalyst volumes when applying "turbulent" catalyst substrates and appropriate catalyst coating. In addition, for low load/low temperature operation, a system will demonstrate the application of "turbulent" substrate technology to aid in the hydrolysis of urea and to influence the flow within the system itself to enhance performance.

Based on the application of these advanced substrate technologies, a high-performance, yet extremely compact emission control system to approach the US 2010/EU 6 emission limits is introduced.

To understand how each new product can be used in an exhaust system, a "pre-development" section discusses different technological "pieces" and the relevant test work, and finally this is demonstrated by the presentation of results from different emission control systems tested on a Heavy Duty Diesel Engine at AVL.

PRE-DEVELOPMENT

BUILDING BLOCKS FOR DIESEL EXHAUST SYSTEMS – STRUCTURED FOIL SUBSTRATES

For both ceramic and metal substrates, the inlet flow can be characterized as "turbulent" while as the exhaust gas passes down the straight channel, very quickly a "laminar" flow pattern is established.

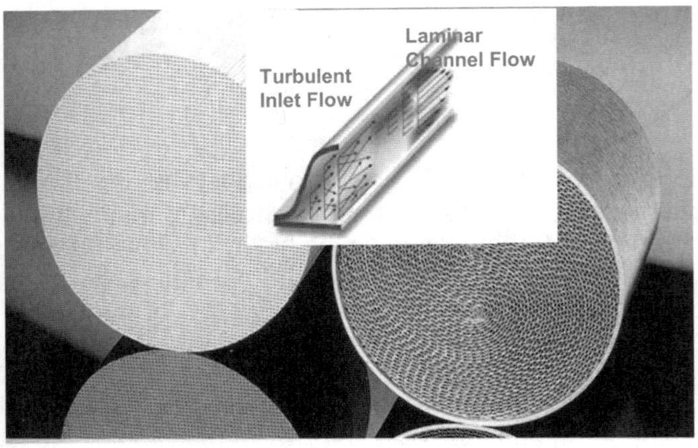

Figure 1. Traditional substrate technology, ceramic and metal, has historically been dominated by straight channels producing a laminar flow profile.

In the following paragraphs, different metallic substrate technologies are described along with a brief statement on benefits for each. In all structured foil substrates, the laminar flow regime is influenced to enhance and re-condition gas flow or to improve flow distribution.

Standard Substrates: with parallel channels, (metallic or ceramic), considered "standard" technology. Metal substrates use thin foils for backpressure optimization. This class of substrates is characterized by straight channels where laminar flow in the exhaust gas is the dominant factor controlling overall performance.

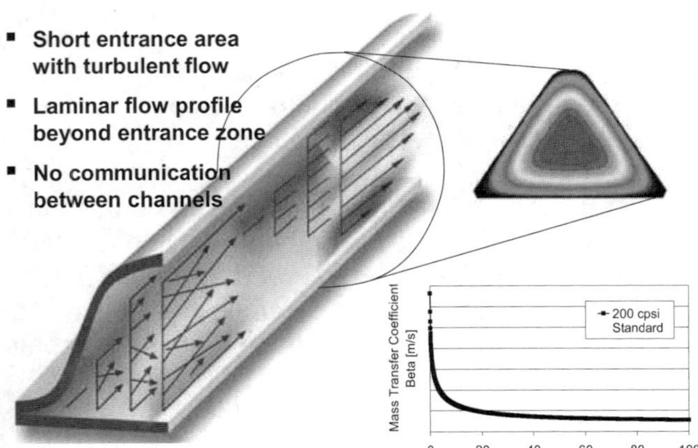

Figure 2: Standard foil design for metallic substrates.

LS Metalit®: Substrates with a "counter-corrugation" which is repeated along the length of each channel, with the objective to improve mass transfer.

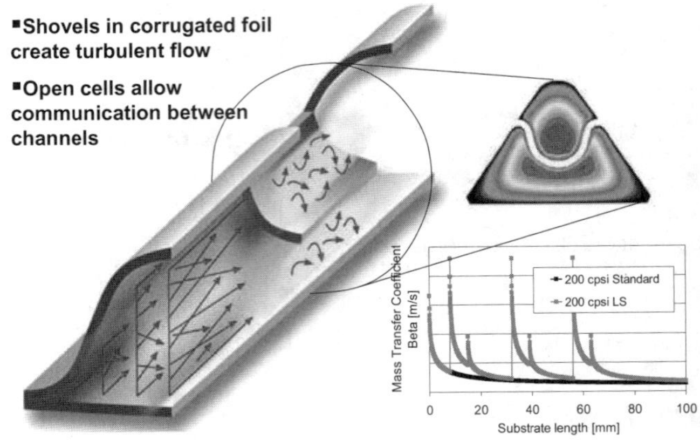

Figure 3: LS foil with counter-corrugation to improve masstransfer

LS/PE Metalit®: Substrates using a corrugated layer with the "counter-corrugation" technique, but adding a flat layer using perforated foil to increase the mass transfer and allow mixing of the exhaust gas from channel to channel within the substrate itself. Exhaust gas which can mix from each channel improves the potential that NO_x and NH_3 will come into contact in a reduction catalyst.

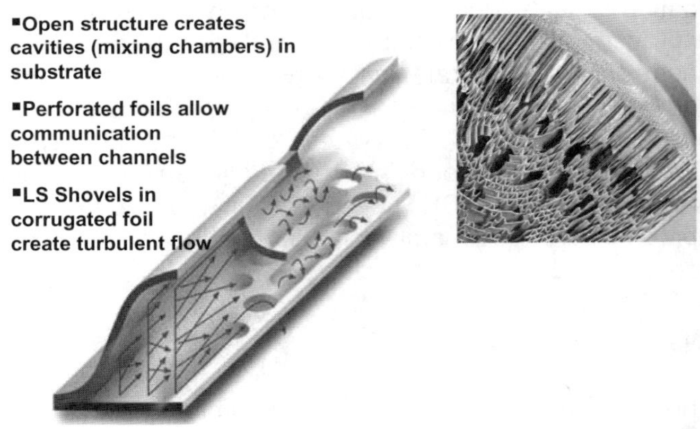

Figure 4: LS/PE foil with counter-corrugation and perforated flat layer

MX/PE Metalit®: Metal substrate using "shovels" on the corrugated layer combined with a flat layer with perforations. The objective of this substrate is to actively force the exhaust gas from one neighboring channel into the next channel and provide for a "micro-mixing" of the exhaust gas. This is useful to improve the flow distribution, destroy large droplets of remaining urea, and improve the distribution of urea/ammonia over the catalyst cross section. Reference Figure 5.

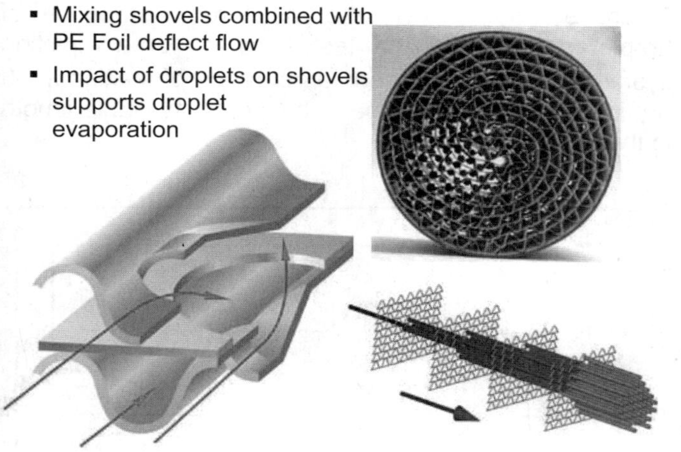

Figure 5: Mixing foil with corrugated "shovel" layer and perforated flat layer

PM Metalit®: Metal substrate using a corrugated layer of foil to divert the exhaust gas from each channel into the flat layer, constructed of stainless steel "fleece" material used to trap and reduce particulate matter [9, 10].

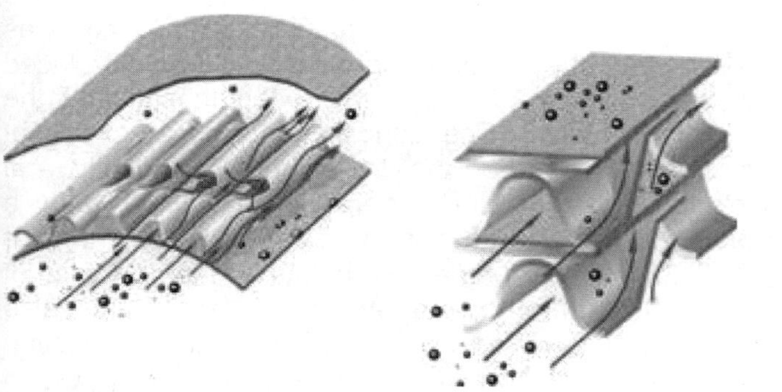

Figure 6. PM Metalit® deflects particulate matter into flat fleece layer for reduction of particles.

SCR FUNDAMENTALS

A typical layout of SCR systems is shown in Figure 7 below:

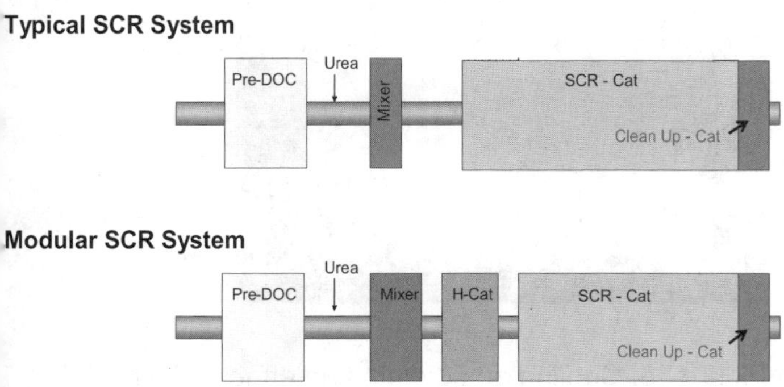

Figure 7. General layout of SCR aftertreatment systems.

DOC Function: A DOC in front of the SCR system is used to generate NO_2 with

$$2NO + O_2 \rightarrow 2NO_2 \qquad (1)$$

for facilitation of fast SCR reaction, which is especially important at low temperatures [1, 2]. Depending on system configuration, the DOC also plays an important role in passive and active regeneration of the downstream PM Metalit or DPF.

Hydrolysis Catalyst: The hydrolysis function is to complete the formation of NH_3 from the injected urea, and this is comprised of three steps as shown below:

Evaporation
$$\{(NH_2)_2CO{\bullet}7H_2O\} \rightarrow (NH_2)_2CO + 7H_2O \qquad (2)$$

Thermolysis
$$(NH_2)_2CO \rightarrow HNCO + NH_3 \qquad (3)$$

Hydrolysis
$$HNCO + H_2O \rightarrow CO_2 + NH_3 \qquad (4)$$

The hydrolysis catalyst can be integrated into the reduction catalyst or placed as a separate module between DOC and SCR substrate.

Reduction Catalyst: The function of this component is the selective reduction of NO_x with NH_3.

Standard reaction:
$$4NO + O_2 + 4NH_3 \rightarrow 4N_2 + 6H_2O\ () \qquad (5)$$

Fast reaction:
$$2NO_2 + 2NO + 4NH_3 \rightarrow 4N_2 + 6H_2O\ (Fast) \qquad (6)$$

NH_3 Clean Up Catalyst: Typically an additional catalyst is placed behind the SCR substrate to oxidize excess NH_3 with O_2 to avoid objectionable odors.

$$4NH_3 + 3O_2 \rightarrow 2N_2 + 6H_2O \qquad (7)$$

SCR CHALLENGES – OVERVIEW

Successful SCR systems have numerous challenges, and several innovative substrate technologies have been developed with the ability to enhance performance as compared to current standard substrate technologies. In particular the following challenges are addressed:

1) NO_2 production
2) Urea evaporation and decomposition
3) Flow distribution

1) NO_2 PRODUCTION

NO_2 production is important during the "fast-SCR" reaction process occuring in the downstream reduction catalyst and for passive reduction of PM. Although for this set of experiments standard 300 cpsi substrates are used as pre-catalysts for NO_2 production, later experiments have shown the potential of structured foil (LS/PE) for improved volume specific NO_2 production. Figure 8 below shows a comparison of 200 cpsi, LS/PE 100/200 cpsi, and LS/PE 200/400 cpsi. Over a wide range of space velocities, the LS/PE 200/400 cpsi shows a significant advantage of 25-50% better NO_2 production.

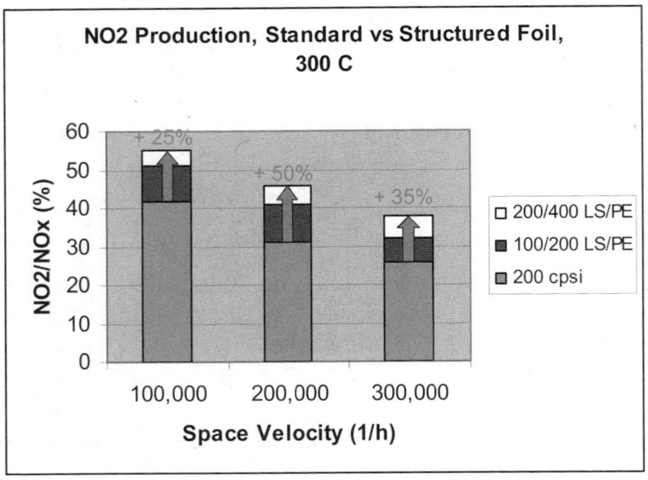

Figure 8: Comparison of NO2 production for different foil technologies and cell densities.

The trend for NO_2 production of structured foil substrates is being studied further in additional programs as this is important not only for SCR systems, but also for passive and active regeneration of DPF systems, i.e. the higher the passive regeneration component the lower the fuel penalty due to "forced" regeneration events. In addition, structured foil substrates are being intensively studied for downsizing due the higher volumetric efficiency and subsequent cost benefits [3].

2) UREA EVAPORATION AND DECOMPOSITION

Urea spray consists of a distribution of droplets, some of which can be large. The first process in the formation of NH_3 from urea is the evaporation of the water. Larger droplets of urea can take much more "distance" in the exhaust pipe for evaporation, and in some cases beyond the available length of the exhaust system. Values have been calculated as a reference, using a mass flow of 200 kg/hr and 430 C a droplet of urea with diameter of 30 μm would evaporate in a length of 1.1m, while a droplet having diameter of 70 μm would take approximately 6.1 m for evaporation of the water.

Figure 9 below shows the measured distribution of droplet diameters for a tested airless urea injection system. Typically, an ideal target range of maximum 30 μm is expressed for diameter based on available length in the exhaust system.

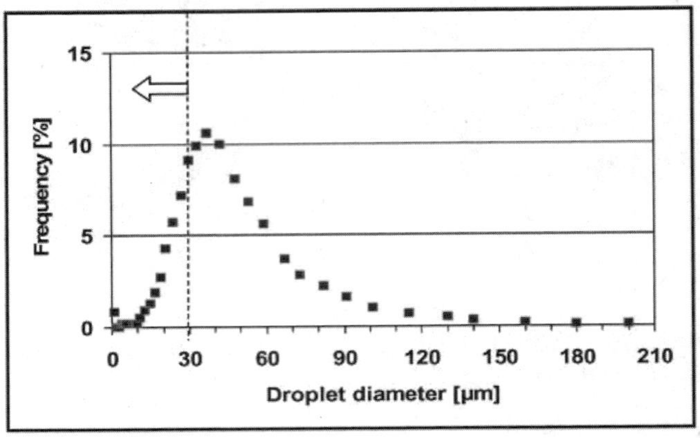

Figure 9: Typical distribution of urea droplets with an airless dosing system.

In this example, a significant amount of urea would reach the reduction catalyst still as aqueous urea and with a large droplet diameter. To demonstrate this effect, an experiment on a flow bench rig was constructed where a thermographic camera was placed on the exit side of a substrate used to simulate the reduction catalyst. The conditions for the experiment were limited based on the test equipment capacity, and included an exhaust temperature of 135 C with a volumetric flow of 250 m^3. These green areas in Figure 10 are showing where large droplets hit the front face of the catalyst and have a "cooling" effect. This also coincides with the spray pattern of the particular injection system used. Although in practice, urea injection would not be advisable at 135 degree C, the experiment demonstrates the effect on the reduction catalyst if, under any operating conditions the injected urea is not fully evaporated within the given exhaust pipe length.

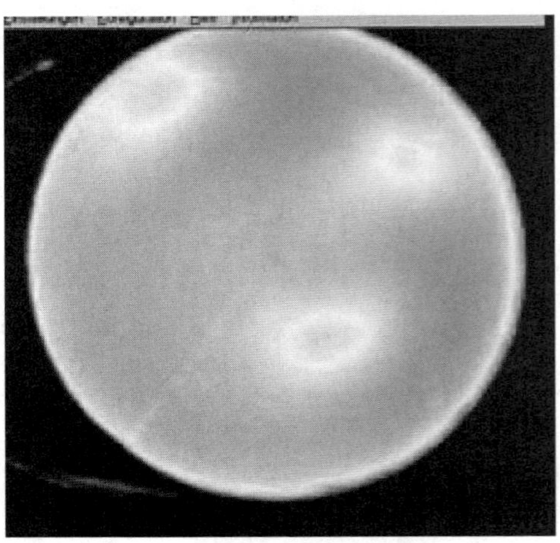

Figure 10: Cooling and distribution effects of large droplets and spray pattern of urea injection system.

For the present study, the challenge of evaporation and decomposition of urea was addressed by designing a special hydrolysis catalyst substrate with two stages. The first stage was specifically designed to destroy the large remaining droplets of aqueous urea through a coarse cell density "mixing" device, and it is followed by second stage which is tuned for optimized hydrolysis performance.

To find the optimum foil types for best evaporation and hydrolysis performance, two experiments were conducted on a model gas test bench. First, three different foil types were compared for their efficiency to destroy the large urea droplets, evaporate the water, and to improve the distribution across the cross-section.

This experiment was conducted with CO_2 used as a tracer gas injected upstream of the catalyst – in the middle of the exhaust pipe. Downstream of the catalyst a probe was positioned which moved along the pipe radius and extracted gas samples for FTIR analysis. The sampling probe could be positioned in 0.5 mm increments along the radius starting from the center. Values for the CO_2 concentration were then recorded and fitted with a Gaussian distribution to calculate mixing efficiency as follows:

$$\eta = 1 - 0,5 \cdot \sigma \tag{8}$$

Sigma (σ) in this case is a measure of deviation from perfect mixing and can vary between 0 and 2, with the larger values representing a lower mixing performance:

$$\sigma = \sum_i \left(\left| \frac{c_i - \bar{c}}{\bar{c}} \right| \cdot \frac{\dot{V}_i}{\dot{V}} \right) \tag{9}$$

ci: measured local concentration, volume specific [ppm]
c: bulk-concentration, volume specific [ppm]
Vi: Volume flow at measuring position [m3/h],
V: Total-Volume flow [m3/h]

A Mixing efficiency of η=1 would represent perfect mixing, while η=0 means no mixing.

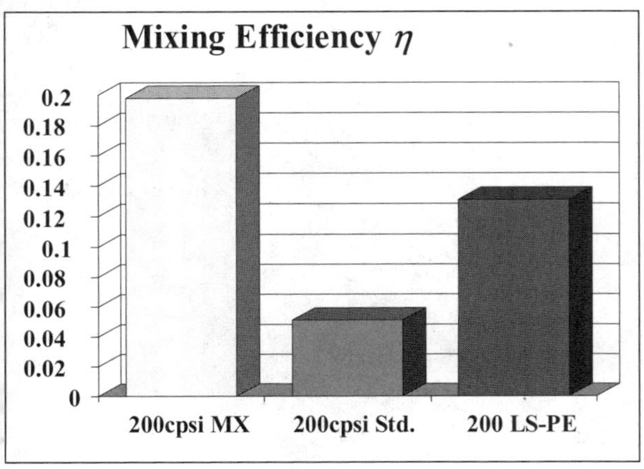

Figure 11: Mixing efficiency comparison and example equations.

As can be seen from Figure 11, MX or mixing structured foil attained the highest mixing performance. Based on this result, MX foil was choosen for the first stage of the hydrolysis catalyst [4,6].

A second set of experiments, again conducted on a model gas test bench, was designed to compare different foil types for their ability to break down the injected Adblue mixture into Ammonia (NH_3) and Isocyanic Acid (HNCO). To characterize this performance, a "urea decomposition coefficient" is calculated:

$$Y_{(NH_2)_2CO} = 0,5 \cdot \frac{c_{NH_3} + c_{HNCO}}{c_{(NH_2)_2CO}} \tag{10}$$

which describes the quantity of urea that is transformed into its gaseous components. Figure 12 below shows the measured urea decomposition coefficient for the tested foil types.

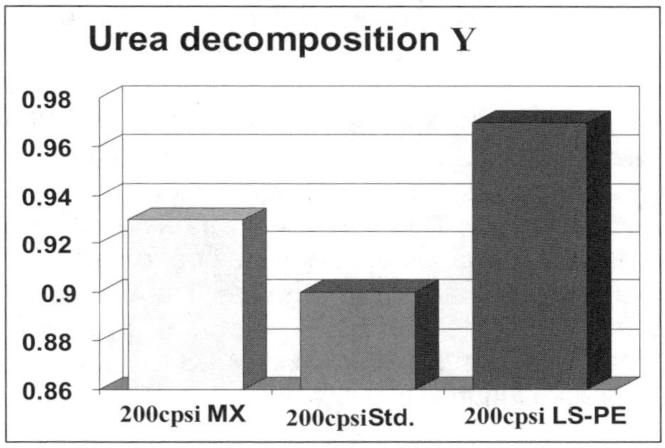

Figure 12: Urea Decomposition for different foil types.

Finally a second characteristic, the "Hydrolysis Coefficient", was calculated:

$$\eta_{Hydrolyse} = \left(\frac{c_{NH_3}}{c_{NH_3} + c_{HNCO}} - 0,5 \right) \tag{11}$$

which describes the further conversion of the intermediate product Isocyanic acid (HNCO) into the desired Ammonia (NH_3).

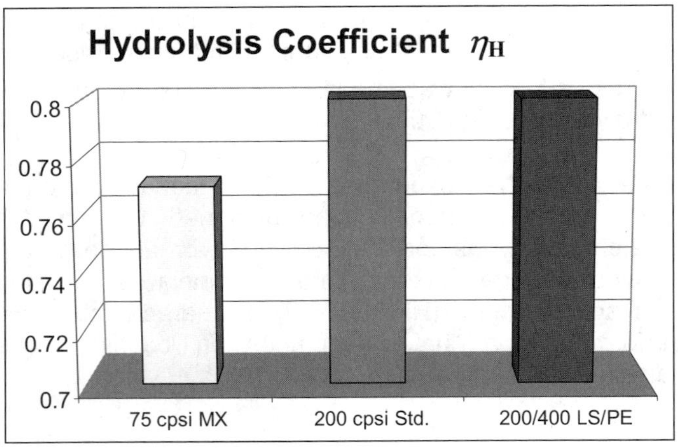

Figure 13: Hydrolysis Coefficient of different foil types.

The experiments to determine these two coefficients were run with a mass flow containing an AdBlue aqueous urea mixture so that NH_3 concentrations were in the range of 2,500 ppm downstream of the catalyst. Figures 12 and 13 above show the urea decomposition coefficient and the hydrolysis coefficient for the tested systems at a space velocity of 175,000/h.

Referencing Figure 12 and 13, the 200/400 cpsi LS/PE foil showed superior urea decomposition and hydrolysis efficiency compared to standard straight channel substrate technology. The results will vary as a function of temperature, but based on the overall performance of urea decomposition and hydrolysis, the final choice for the second substrate in the "two-stage" hydrolysis catalyst was to utilize the 200/400 cpsi LS/PE structure [4, 6]. Reference Figure 14 below.

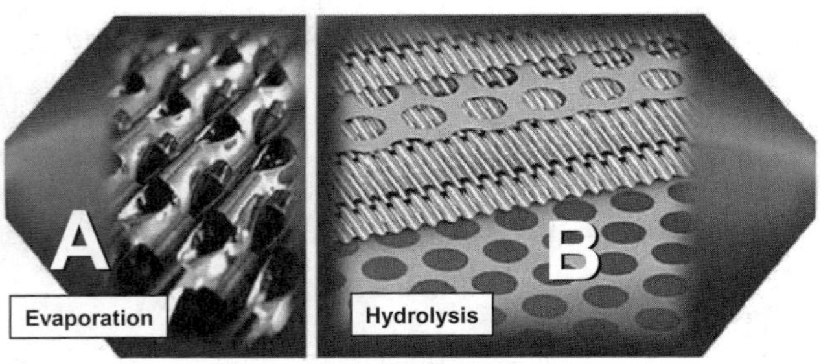

Figure 14: Two stage hydrolysis catalyst, first stage for destruction of large droplets of urea followed by 200/400 LS/PE substrate.

3) FLOW DISTRIBUTION

Flow distribution is an important element of all catalyst designs. For SCR catalysts, not only is it important to achieve a uniform distribution of urea/NH_3 across the face of the reduction catalyst, its also needed that the flow distribution of the exhaust gas across the front face is optimized to avoid poor utilization.

An experiment on a cold flow bench was designed to demonstrate exhaust gas flow velocity distribution in a large diameter SCR substrate. For this purpose, the flow velocity at different points behind the SCR substrate is measured with a sensor. Measurements were conducted in increments of 5 mm in X and Y direction. Figure 15 depicts the test setup with a straight inlet cone in front of a SCR substrate.

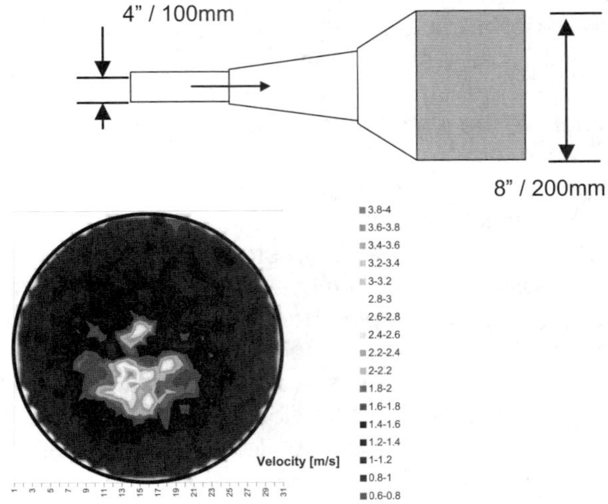

Figure 15: Flow velocity distribution behind SCR catalyst with standard empty cone.

In this case the flow is dominated by a pattern indicating "jet flow" where a large portion of the gas flow is in the center of the SCR substrate, while the gas flow near the cone walls is stalled.

A unique type of substrate, called the Conicat® for conical converter (Figure 16) is available to fill the space used by the empty cone. The Conicat® has been described in previous papers [7], originally as a close-coupled converter for fast light-off and improved flow distribution.

The system of Figure 15 was then re-tested with a Conicat® installed in the formerly empty cone. The results shown in Figure 17 show a much improved flow distribution over the front face of the large diameter SCR substrate.

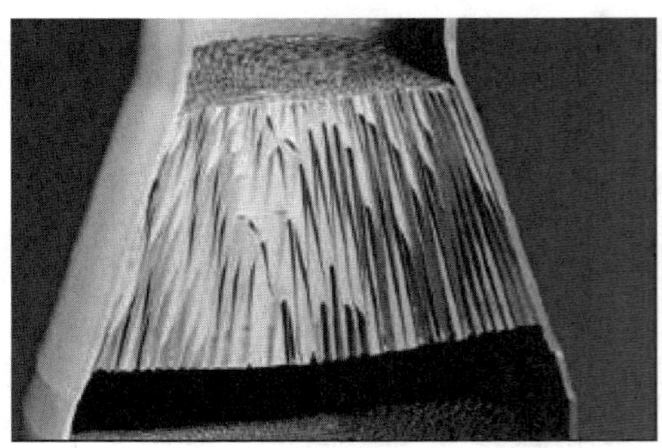

Figure 16: Example of Emitec Conicat®.

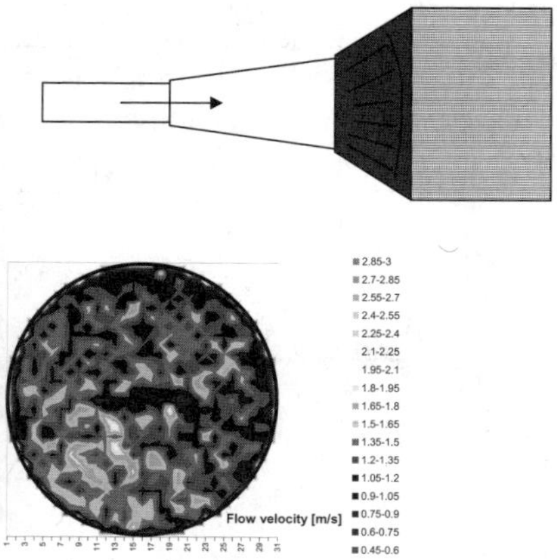

Figure 17: Flow velocity distribution with Emitec Conicat®

Both systems, with the empty inlet cone and with the Emitec Conicat®, were then tested on a cold flow bench for pressure drop comparison. The results show a lower pressure drop over the system with Conicat® at high exhaust flow rates. In high flow conditions, the flow stall near the walls of the empty cone causes a pressure drop which is higher than the pressure drop over channels of the Conicat®. At lower flow rates, the pressure drop from flow stall is slightly lower than the channel pressure drop in the Conicat®. Reference Figure 18 below.

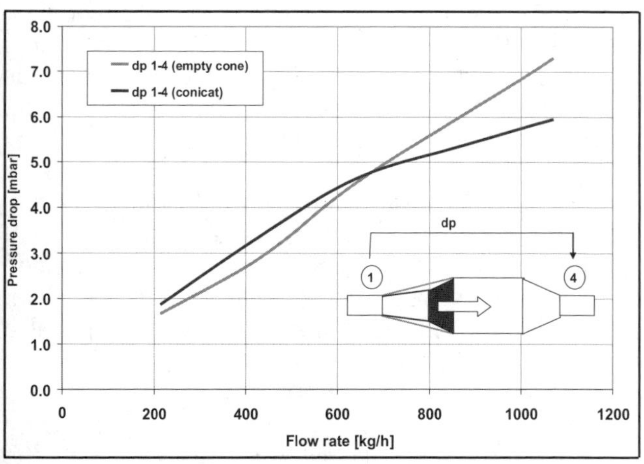

Figure 18: Pressure drop test results with Emitec Conicat®.

WASHCOAT PROCESS FOR TURBULENT CATALYSTS

With both the 40 cpsi Mixer substrate and the 200/400 cpsi LS/PE substrate, a coating of Titanium Dioxide is used to provide for a porous surface for effective evaporation (limit the effect of a "steam" insulation around the droplet), also having in the coating a high mechanical stability against erosion, and very importantly a high catalytic activity for HNCO- hydrolysis.

To apply this washcoat technology, a new production process was developed. Coating of standard substrates with parallel channels is a process that is well established and in high production volumes. Different methods exist, either by applying exact quantities in a "precision" process, or by a waterfall of washcoat and subsequent blowing out excess material to clear the channels. Substrates which utilize structured foil present new challenges due to their interrupted channel wall and internal cavaties, as present with LS, LS/PE, PE, and MX type foils. For these technologies, blowing out the excess washcoat material with compressed air for example may not work as the air follows the path of least resistance. To eliminate the excess material, a process which relies on centrifugal forces has been developed and shown to be effective in providing a uniform and consistent washcoat layer.

EXPERIMENTAL TEST SET-UP AND RESULTS

EXPERIMENTAL SETUP

The following section describes the test set-up comparing SCR systems using Structured Foil substrates and a baseline SCR system currently available in European Trucks (fully extruded reduction catalyst). A heavy duty truck engine developed to meet Euro 5 emissions limits (NO_x = 2 g/kWh and PM = 0.02 g/kWh (ESC) and 0.03 g/kWh (ETC) was utilized. The baseline SCR system (Vanadium type, fully extruded) was designed to meet the Euro V emissions requirements. The testing was designed to demonstrate the benefits of new foil technologies for significant reduction in space and system cost. The building blocks for this test program included: a) a DOC for low temperature performance, b) structured metal substrates for both urea mixing and hydrolysis, and c) Vanadium type coated structured metal substrates for the NOx reduction catalyst to investigate enhanced mass transfer and internal mixing, and the PM Metalit® in an extended step of the program for continuous reduction of particulate matter.

The investigation was carried out with a heavy duty engine in the range of 2 ltr/cylinder. Due to the use of an SCR system the engine was calibrated for lowest particulate matter emissions and optimized fuel consumption, and did not have an EGR system. The engine was installed in a fully dynamic and transient operation test cell at AVL. The test rig set-up is shown below in Figure 19.

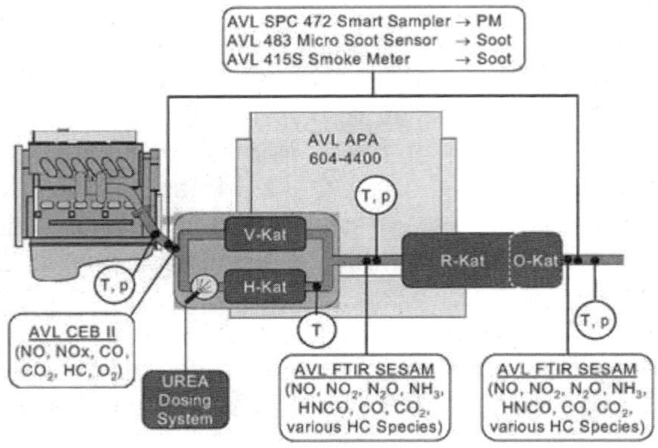

Figure 19: Test rig and exhaust set-up at AVL.

Points where temperature and pressure were monitored are marked as T and p. At the turbine outlet location gaseous engine-out emissions were measured (NO, NO_x, CO, CO_2, HC, O_2) with standard analyzers (ref AVL Combustion Emissions Bench II). The exhaust gas composition downstream of the V (oxidation) and H (hydrolysis) Catalysts and as well in the "tailpipe" location was measured simultaneously to the engine out emissions by FTIR SESAM technology.

Table 1 below reviews the specific system configuration, system 0 is the baseline, systems 1, 2, and 3 all integrate different elements for comparative testing. As a final step, the PM Metalit® is added before the reduction catalyst and is described in a later section.

System #	Hydrolysis Cat	Pre-DOC	Reduction Cat	Clean-up
0 (baseline)	Serial Production	No	300cpsi V- Extruded 100% Volume	Zone Coated
1	40cpsi MX + 200/400 LS/PE	No	300cpsi V-Extruded 100% Volume	Zone Coated
2	40cpsi MX + 200/400 LS/PE	300 cpsi	300/600 LS/PE V-coated 70% Volume	Zone Coated
3	40cpsi MX + 200/400 LS/PE	300 cpsi	300cpsi V-Extruded 100% Volume	Zone Coated

Table 1: System description for comparative testing

All testing was carried out at AVL's laboratories in Graz, Austria. The tested systems are defined in Figure 20.

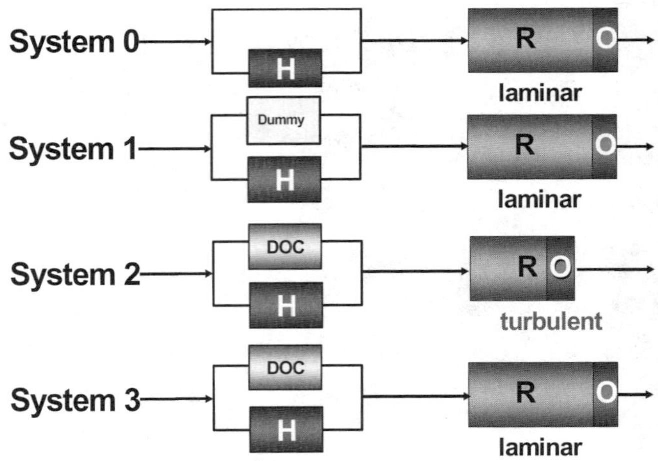

Figure 20: System layout and description.

The baseline system, System 0 in the test matrix, consisted of a hydrolysis catalyst followed by a large volume of fully extruded reduction catalysts. This system did not use a pre-oxidation catalyst. The outlet side of the reduction catalysts was zone coated to integrate the slip cat function.

In system 1, the hydrolysis catalyst was substituted with an advanced two-stage hydrolysis catalyst (reference Figure 14) with approx. 10% larger volume. At the same time, a dummy catalyst was placed in a parallel pipe to allow a direct comparison to a system with pre-oxidation catalyst later. The benefits in system performance with a parallel arrangement of pre-DOC and hydrolysis cat has been demonstrated in previous studies [6,8,10]. The reduction catalysts remained the same.

System 2 used the same hydrolysis catalyst as system 1, and a pre-oxidation catalyst replaced the dummy cat in the parallel exhaust pipe. Second, the reduction catalysts were removed and substituted with a smaller volume of 300/600 LS/PE substrates, which were also zone coated for integrated slip cat function.

Finally, for System 3 the 300/600 LS/PE reduction catalysts were removed again and replaced with the original fully extruded reduction catalysts. This system used the same hydrolysis catalyst and pre-oxidation catalyst as in system 2. This setup allowed a direct comparison of the fully extruded and the turbulent reduction catalysts.

All systems were tested for backpressure and showed similar overall system backpressure within 5%.

The urea dosing system was not modified and represented the production equipment and calibration. All catalysts were degreened on the test engine with a specific proprietary test cycle.

TEST RESULTS

INFLUENCE OF HYDROLYSIS CATALYST

For the first part of the test program, the influence of the hydrolysis catalyst was analyzed by comparing systems 0 and 1, as shown in Figure 21 below.

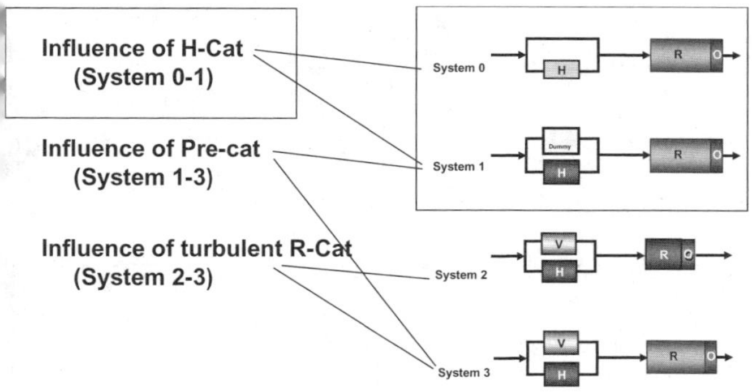

Figure 21: Influence of hydrolysis catalyst is compared between systems 0 and 1.

System 0, the baseline production system, included a hydrolysis catalyst, with the urea injection located approximately 150mm upstream. System 1 used an advanced hydrolysis catalyst in the two stage design (40 cpsi mixer foil followed by 200/400 cpsi LS/PE). While in the model gas bench a 200cpsi mixer was used, a lower cell density of 40cpsi was chosen for the engine tests to keep backpressure optimized.

Figure 22 below shows the total decomposition of the injected urea into NH_3 and HNCO during the ESC test. Both systems have a similar rate of urea decomposition.

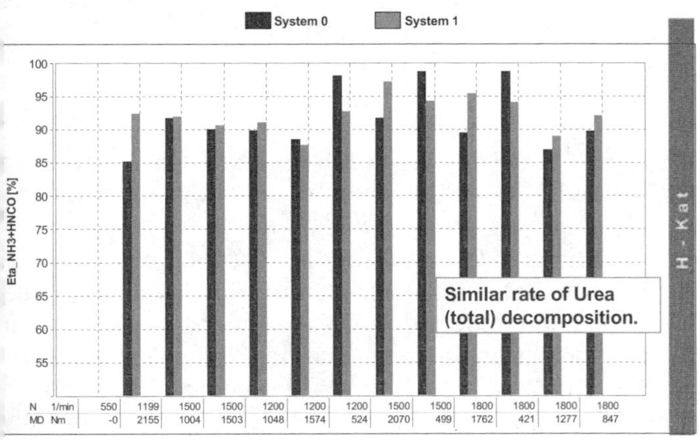

Figure 22: Rate of Urea decomposition for Systems 0 and 1

While the total decomposition of urea was similar, it was found that system 1 with the two stage hydrolysis catalyst shifted the concentration of the decomposition products more towards NH_3.

As more NH_3 was formed in the advanced hydrolysis catalyst of system 1, the amount of intermediate HNCO was reduced. Figure 23 shows the NH_3 Yield for both tested systems.

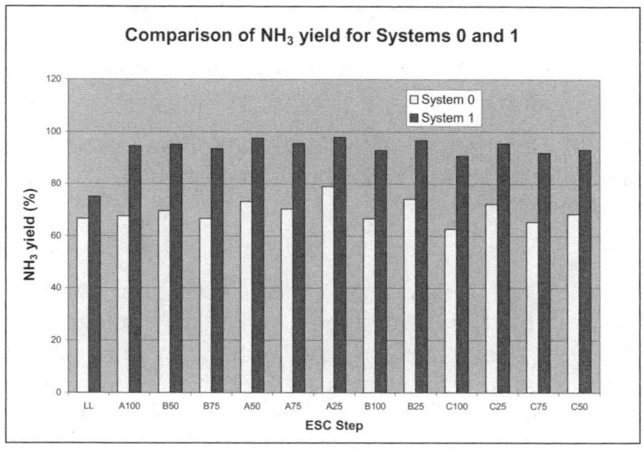

Figure 23: Comparison of NH_3 yield for systems 0 and 1.

This result could be confirmed also for transient operation during following ETC tests. It can be concluded that the substitution of the baseline hydrolysis catalyst in system 0 with the proposed structured foil design (1st stage mixer foil + second stage LS/PE foil) in system 1 resulted overall in similar total urea conversion (NH_3 + HNCO), but the break down yielded a more complete conversion of urea to NH_3. This test result was consistent with the experience discussed earlier on the model gas test bench.

With the modifications in system 1, the urea injection nozzle was moved to a location 700mm upstream of the hydrolysis catalyst. In the baseline system, the nozzle was located only 150mm upstream of the hydrolysis catalyst. To understand the influence of the greater distance in system 1, a second set of experiments was carried out comparing nozzle positions of 150mm and 700mm upstream of the hydrolysis catalyst in system 1.

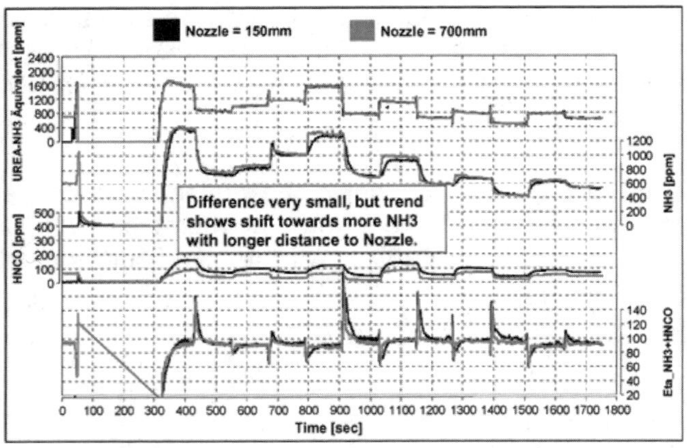

Figure 24: Comparison of system 1 with urea injector located at both 150 and 700 mm from front face of hydrolysis catalyst.

Figure 24 illustrates the urea decomposition and hydrolysis of the two-stage hydrolysis catalyst at different nozzle positions. As expected, the longer distance from injector nozzle to front face of hydrolysis catalyst slightly improved urea conversion towards more NH_3. It is notable, that the use of the two stage hydrolysis catalyst allowed for a wide range in placement of the urea injector.

INFLUENCE OF PRE-OXIDATION CATALYST

In the second phase, the influence of the pre-catalyst was investigated by comparison of systems 1 and 3, reference Figure 25 below.

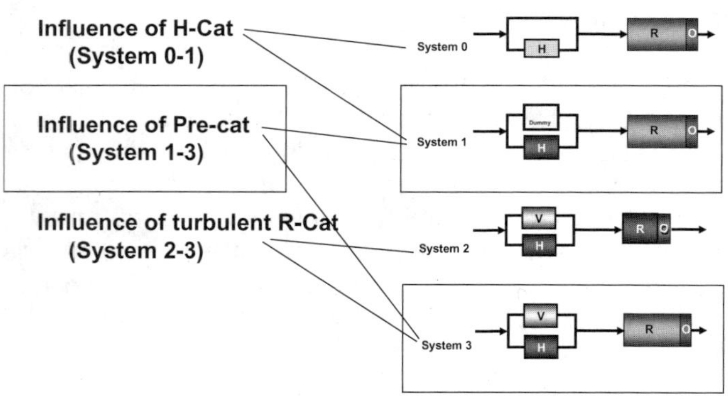

Figure 25: Comparison of systems 1 and 3 to understand the effect of a pre-catalyst

For this experiment, the dummy catalyst of system 1 was replaced with a 300cpsi oxidation catalyst.

As it was expected, the pre-catalyst had the effect of increasing the NO_2 concentration in the exhaust stream, The NO_2 production during both the ESC and ETC testing are presented in Figure 26 and Figure 27.

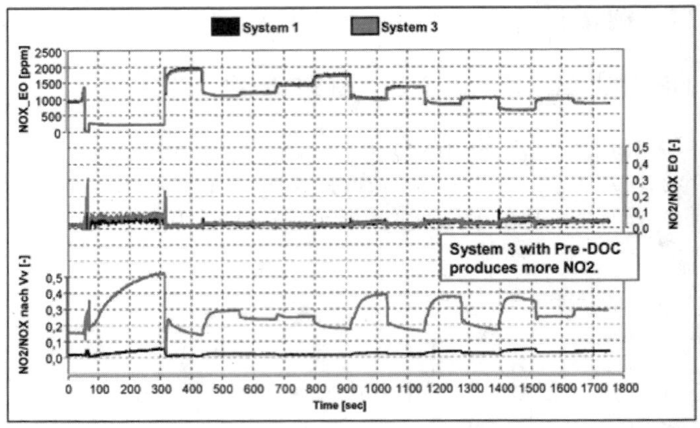

Figure 26: NO_2 production with pre-catalyst, ESC test.

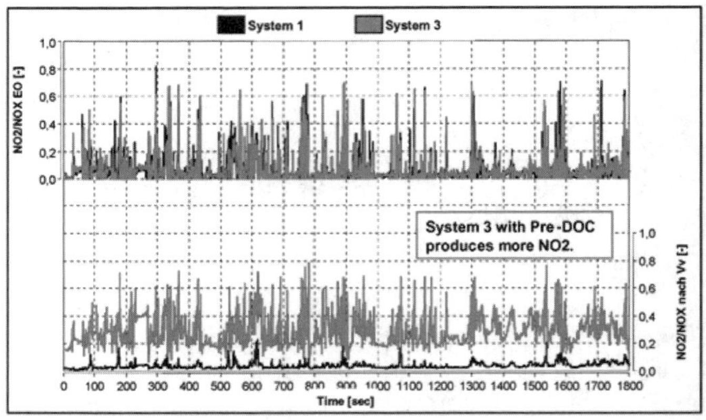

Figure 27: NO_2 production with pre-catalyst, ETC test.

Previous studies have shown that increased NO_2 level have the effect of improving the low temperature performance of the reduction catalyst system by using the so called "fast SCR reaction" [1, 2]. The effect of increased NO_2 production for overall system efficiency in this experiment is discussed later on. NO_2 production is also important for continuous PM reduction and this is also discussed in a later section.

INFLUENCE OF TURBULENT REDUCTION CATALYSTS

The 3rd phase of the test program was designed to study if turbulent substrates, such as LS/PE foils, can be used to reduce the required SCR volume because of their better mass transfer and internal flow mixing. Thus, system 2 utilized a structured foil reduction catalyst with 300/600 LS/PE foil that was downsized by approximately 30% compared to the baseline which utilized a fully extruded 300 cpsi reduction catalyst. In each case the last 20 mm of each reduction catalyst was coated with a DOC washcoat for function as a slip catalyst. For this phase, System 2 was compared to system 3. Both systems used the identical pre-oxidation and hydrolysis catalyst, reference Figure 28.

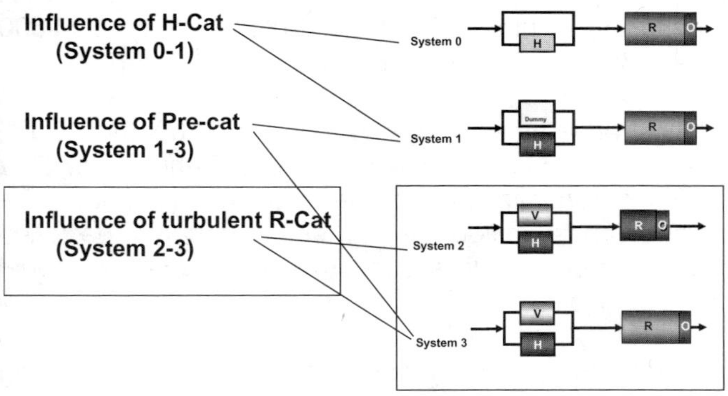

Figure 28: Comparison of Systems 2 and 3, reduction catalyst of System 2 approximately 30% smaller than baseline System 3.

The smaller volume of the turbulent reduction catalysts in system 2 lead to a higher space velocity and reduced resident times in the channel, reference Figure 29. The goal was now to investigate, if this higher space velocity leads to reduced SCR performance, or if the use of turbulent 300/600 LS/PE substrates allows for volume reduction at equal or similar performance.

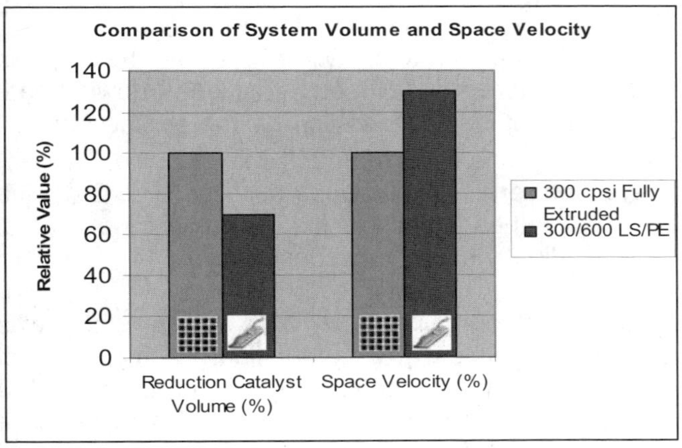

Figure 29: Space velocities in reduction catalyst for systems 2 and 3

First, tailpipe NH_3, HNCO and N_2O emissions were compared in an ESC test for system 2 and system 3, reference Figure 30 below.

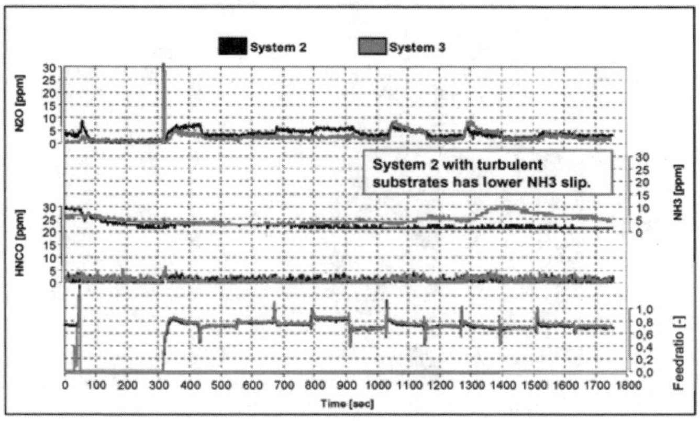

Figure 30: Comparison of System 2 and 3, ESC test.

It was found that at equal urea feed rate the ammonia (NH_3) slip was reduced with the turbulent reduction catalysts of System 2. At the same time, the HNCO concentration was nearly identical for both systems. These results could be confirmed also during transient ETC tests.

The N_2O emission of both systems was with maximum 7 ppm very low. However, it was notable, that system 2 had a slightly increased N_2O tailpipe concentration. Because of its higher volumetric efficiency, it is possible to reduce the Pt loading of the NH_3 slip catalyst, which would further reduce the N_2O production of system 2.

DISCUSSION OF SYSTEM PERFORMANCE

The comparison of overall system performance shows a similar NOx reduction for all tested systems reference Figure 31 and Figure 32 below. Each system was tested 4 times. For all the testing, the dosing strategy was the same as the baseline production system.

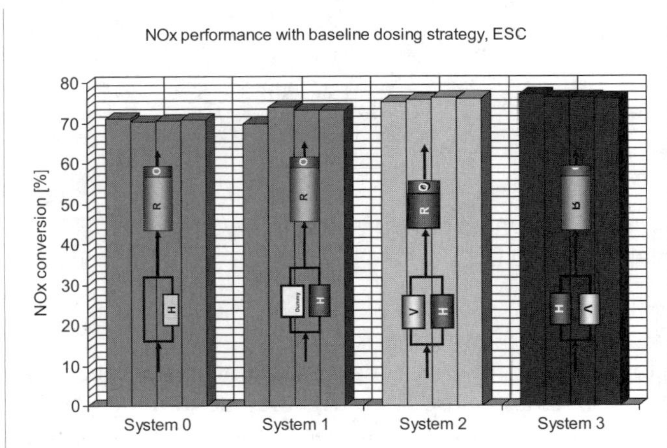

Figure 31: NOx efficiency over different systems for ESC test.

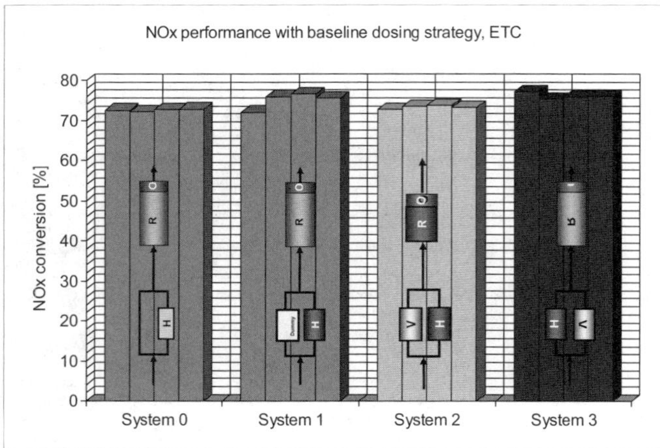

Figure 32: NOx efficiency over different systems for ETC tests.

Comparing system 0 with system 1, it is apparent that the better NH_3 production of the two-stage hydrolysis catalyst did not lead to a significant improvement of NOx reduction.

Also, a comparison of system 1 with system 3 does not yield a significant increase in NOx performance with the elevated level of NO_2 as it was expected. The reason is that in the exhaust gas temperatures of this engine during european HD test cycles are already in the optimum range for a SCR reaction, see Figure 33.

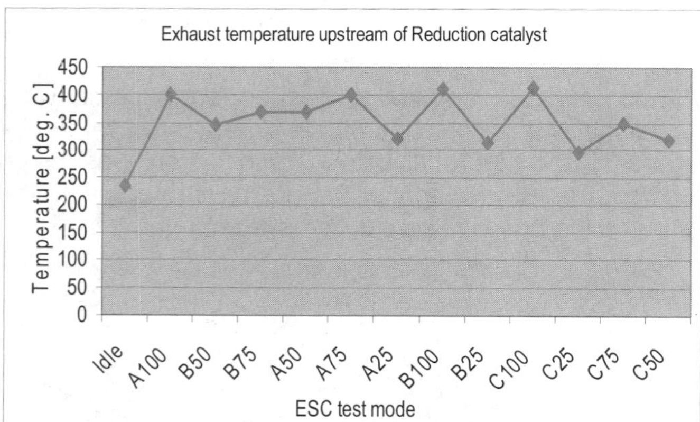

Figure 33: Exhaust temperatures upstream of SCR catalyst during ESC test

In test cycles and under real world driving operation with lower exhaust temperatures, the advantage of the system with pre-oxidation and two-stage hydrolysis catalyst is expected to be greater.

A comparison of system 2 with system 3 shows also a similar NOx performance during ESC and ETC test runs. Here, system 2 utilizes the turbulent reduction catalyst with approximately 30% lower volume compared to the fully extruded reduction catalyst of system 3. The result confirms that it is possible to reduce the SCR catalyst volume using turbulent foil substrates such as 300/600 LS/PE, while maintaining similar NOx performance.

At the same time, system 2 with the 300/600 LS/PE substrates had lower tailpipe NH_3 levels which allows for additional optimization of the system calibration. For instance, a modified dosing strategy could be used which can be optimized for higher dosing rate at equal NH_3 slip levels.

A further study to investigate the potential of downsizing the baseline reduction catalyst was considered, however the higher NH_3 slip of system 3 indicates that it may not be possible to downsize the reduction catalyst as it was done with the 300/600LS/PE.

INTEGRATED PM REDUCTION SYSTEM

Many studies have been conducted to investigate the potential integration of particulate reduction devices into SCR systems. Most of these systems include a combination of DOC and DPF positioned upstream of the SCR catalysts to use available NO_2 for continuous regeneration of the DPF. Such layout requires special high temperature SCR technology and protection measures to avoid damage to the SCR catalyst during active DPF regeneration. Other systems have been introduced where the DPF was positioned downstream of the SCR catalyst. Depending on the engine out and tailpipe emissions this approach may still be feasible, however for certain applications such as U.S. 2010 the DPF would then need to be actively regenerated, leading to slightly higher fuel consumption.

As part of an on-going study, this test program was extended and system 2 was modified. Each module of the SCR catalysts was combined with an upstream PM Metalit as shown in Figure 34 below. A "Photo Acoustic Soot Sensor" (PASS) system was used to record PM emissions behind each catalyst system.

Figure 34: Combination System with PM Metalit and SCR catalysts

Previous laboratory tests have reported that PM reduction can be improved when NH_3 is present [8]. It was shown that it is necessary to completely convert the injected urea into NH_3 to avoid deposits from intermediate products such as HNCO in the filter.

The results of this experiment confirmed the previous experiences. Over a wide range of the ESC, a PM reduction of 50% could be achieved. Figure 35 shows the PM reduction, of the combined system compared to system 2 in the ESC test.

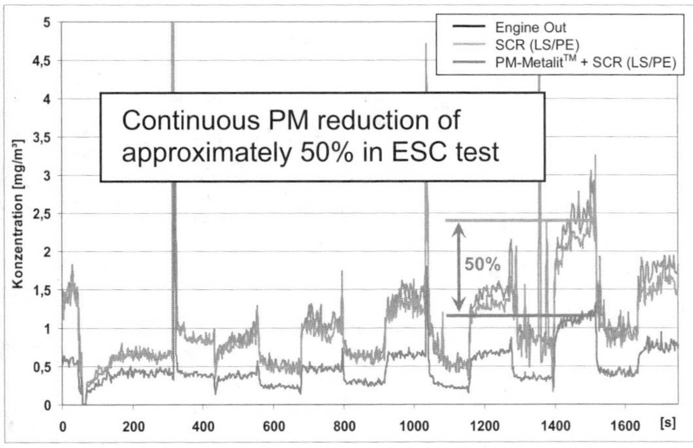

Figure 35: PM reduction in ESC test with System 2 and combined system

This result was repeated also in the transient test using gravimetric PM measurement. To determine a degradation of PM reduction from possible deposit of urea on the PM Metalit ® filter, the ETC tests were conducted first without urea injection, then with the baseline injection calibration, and finally with a modified higher dosing rate. Figure 36 below shows the achieved PM reduction for all test runs.

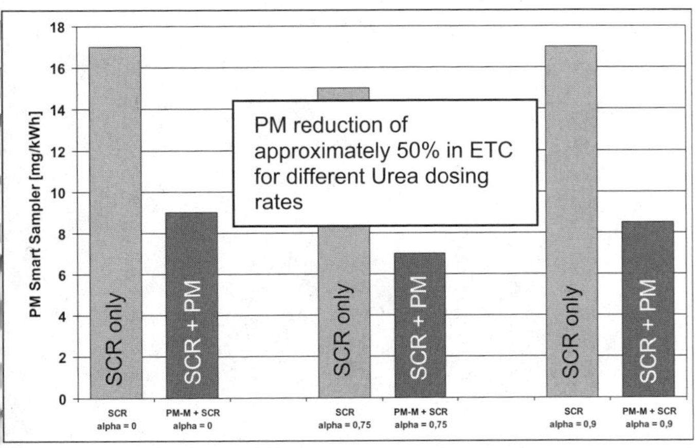

Figure 36: PM reduction in ETC test

Even with increased urea dosing rate, no negative effect from deposits in the PM Metalit ® filter could be found. Further studies are planned to demonstrate long term durability and robustness against degradation of this ultra compact PM Metalit ® SCR combined system.

CONCLUSIONS

Different modular SCR systems were designed and tested on a Heavy Duty Euro V diesel engine. The systems utilized a pre-oxidation catalyst, a two-stage hydrolysis catalyst, and either a fully extruded or turbulent 300/600 cpsi LS/PE reduction catalyst. An additional system utilized a PM Metalit® in combination with the optimized SCR system.

- The two-stage hydrolysis catalyst using MX + LS/PE foil and a newly developed washcoat process produced a more complete conversion of urea to NH_3.
- The SCR system using the turbulent 300/600 LS/PE substrate achieved similar NOx performance with 30% less catalyst volume while also having a lower NH_3 slip at the tailpipe.
- A significant space savings could be achieved with reduced volumes, enabling the packaging of SCR systems in areas with tight space requirements.
- Further system optimization is possible through changes to the urea dosing strategy and using the internal mixing benefits of LS/PE foil in the reduction catalyst.
- An additional test was conducted to understand the potential of combining the advanced SCR system with the PM Metalit®. The results yielded an additional 50% reduction in PM in a very compact combination SCR/PM aftertreatment system.
- Future testing is planned to demonstrate long term robustness of a system which can meet US2010 and EU VI requirements.
- Development of combustion and fuel economy optimized engines allows the application of the PM Metalit®, a device which continuously reduces the PM in the exhaust stream, without the need for active regeneration or the ash service cleaning needed for the majority of today's systems.

ACKNOWLEDGMENTS

Grateful acknowledgments are given to:

Dr. Stefan de Buhr, Mr. Holger Herrmuth, Mr. Thomas Cartus of AVL LIST GMBH.

Dr. Joerg Spengler of Sud-Chemie.

REFERENCES

1. "Devlopment of urea SCR system for HD Commercial vehicles", K.Hirata et al, SAE 2005-01-1860
2. "Optimization of urea SCR deNOx Systems for HD Diesel Engines.", R.v.Helden et al, SAE 2004-01-0154
3. "Faster Is Better: The Effect of Internal Turbulence on DOC Efficiency," K Dawson, J Kramer, SAE 2006-01-1525
4. "Urea-SCR-Technology – Performance of Structured Metallic Substrates," S. Steinbach, J. Grunwald, T. Sattelmayer, TU Munich, SAE 2006-01-3505
5. "Innovative Metallic Substrates for Exhaust Emission Challenges for Gasoline and Diesel Substrates", K.Mueller-Haas, M.Rice, SAE 2005-01-3851
6. "High Performance SCR Catalyst System: Elements to Guarantee the Lowest Emissions of NOx," Dr. Jacob, Dr. Scheeder, T. Cartus, H.P. Mai, Dr. Paulus, Dr. Spengler, 27th International Vienna Motor Symposium, 2006.
7. "The Conical Converter and its potential for future Close-Coupled converter Concepts," Wolfgang Maus, Rolf Brueck, Emitec, GmbH, SAE 980414.
8. "Downsizing of Future Exhaust Aftertreatment Systems for HD vehicles," Dr. Scheeder, Dr. Mueller, Emitec, Dr. Paulus, Dr. Spengler, Sud Chemie AG, FAD Conference, Emissions Systems for Diesel Motores, 2006.
9. "Application of New Diseel Aftertreatment Strategies on a Production 1.9 L Common-Rail Turbocharded Engine," Saroglia, Basso, Fiat-GM Powertrain Italia, Presti, Reizig, Stock, Emitec GmbH, SAE 2002-01-1313.
10. "Emission Limits of Future HD Truck Engines," Dr. Jacob, MAN, 26th International Engine Conference Vienna, 2005.

2007-01-1574

Is Closed-Loop SCR Control Required to Meet Future Emission Targets?

Frank Willems, Robert Cloudt, Edwin van den Eijnden, Marcel van Genderen and Ruud Verbeek
TNO Automotive

Bram de Jager and Wiebe Boomsma
Eindhoven University of Technology

Ignace van den Heuvel
DAF Trucks

ABSTRACT

To meet 2010 emission targets, optimal SCR system performance is required. In addition, attention has to be paid to in-use compliance requirements. Closed-loop control seems an attractive option to meet the formulated goals. This study deals with the potential and limitations of closed-loop SCR control.

High NO_x conversion in combination with acceptable NH_3 slip can be realized with an open-loop control strategy. However, closed-loop control is needed to make the SCR system robust for urea dosage inaccuracy, catalyst ageing and NO_x engine-out variations. Then, the system meets conformity of production and in-use compliance norms.

To demonstrate the potential of closed-loop SCR control, a NO_x sensor based control strategy with cross-sensitivity compensation is compared with an adaptive surface coverage/NH_3 slip control strategy and an open-loop strategy. The adaptive surface coverage/NH_3 slip control strategy shows best performance over simulated ESC and ETC cycles.

SCR catalyst dynamics, time delay in the urea injection and maximum NH_3 slip targets limit the performance of closed-loop SCR control. If new reagent dosage systems and future catalyst technology are able to relieve these limitations, closed-loop control has the potential to reduce the calibration effort and to improve the transient control performance.

INTRODUCTION

Urea SCR technology is widely accepted to meet Euro-4 emission targets for heavy-duty diesel applications. With the realization of the urea infrastructure in Europe, this technology becomes an attractive alternative for light-duty diesel applications. Moreover, urea SCR is considered as a promising technology to meet US2010 emission standards. A summary of the proposed European and US heavy-duty emission targets is given in Table 1.

Table 1: Proposed European and US heavy-duty diesel emission targets.

	Euro-4 (2005)		Euro-5 (2008)		US2010
Test cycle	**ESC**	**ETC**	**ESC**	**ETC**	**FTP**
NO_x (g/kWh)	3.5	3.5	2.0	2.0	0.27
PM (g/kWh)	0.02	0.03	0.02	0.03	0.013
CO (g/kWh)	1.5	4.0	1.5	4.0	20.8
HC (g/kWh)	0.46	0.55[1)]	0.46	0.55[1)]	0.19[1)]

To meet (post-)2010 emission targets, the main challenges for urea SCR systems are:

- ***Maximal NO_x conversion*** Compared to current Euro-4 levels, tailpipe NO_x emissions have to be further reduced by 90% to meet 2010 targets, see Table 1. Depending on the engine-out NO_x calibration, the desired NO_x conversion is in the order of 70-80% for Euro-5 limits and 80-90% for US2010 limits. By optimizing the NO_x conversion, safety margins associated with ageing issues and

[1] NMHC

climate effects have to be reduced and the transient performance becomes important. Then, the risk of NH_3 slip increases; although NH_3 emission is not regulated, a limit of 10 ppm average NH_3 and 25 ppm peak NH_3 is often applied, see *e.g.,* [1].

- ***Enhanced low-temperature performance*** Currently applied SCR catalysts are characterized by reduced NO_x conversion for low temperatures. To optimize the NO_x conversion over an FTP test cycle, performance has to be increased, especially for the relatively cold part of the cycle. Low-temperature performance is also an issue in light-duty applications.

- ***Meet in-use compliance requirements*** Besides emission targets for specified test cycles, future legislation for heavy-duty application will also include limits on performance degradation in the field. This introduces the need for SCR systems that are more robust for system variations.

As a result, optimal SCR system performance has to be realized to meet future emission legislation. This requires high NO_x conversion, NH_3 slip suppression and improved control quality. Open-loop SCR control strategies are essentially able to achieve optimal performance over a test cycle, but this requires a considerable calibration effort. Closed-loop SCR control can relieve this calibration effort and improve dynamic control quality. More important, it offers robustness for system variations. This is essential to meet in-use compliance and conformity of production (COP) requirements.

This paper discusses the potential and limitations of closed-loop SCR control to optimize the SCR system performance. First, an overview of the current status of closed-loop SCR control is presented. From simulations with two control strategies, the achievable performance over an ESC and ETC test cycle is demonstrated. Finally, possible developments are discussed which will further improve the closed-loop SCR performance.

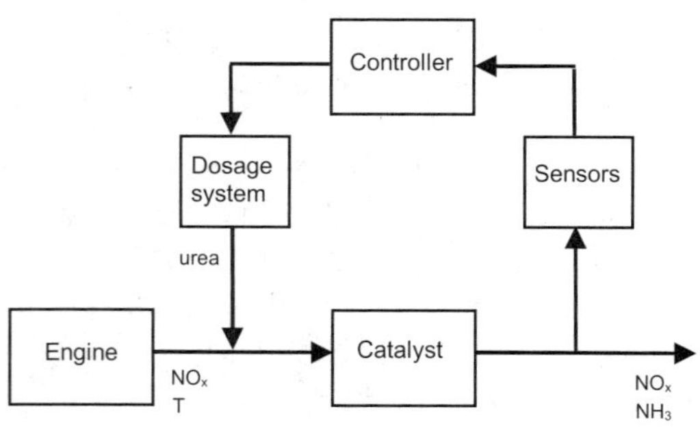

Figure 1: Automotive urea SCR system lay-out

SCR SYSTEM PERFORMANCE LIMITATIONS

Figure 1 shows a typical lay-out of an engine equipped with a urea SCR system. In this system, three subsystems can be distinguished: NH_3 formation, catalyst, and control system. In this section, the subsystems are discussed in more detail. Focus is on performance limiting aspects.

NH_3 FORMATION

To form the required reducing agent for NO_x reduction in the SCR catalyst, an aqueous urea solution (AdBlue) is injected through a nozzle, such that it is atomized in the exhaust pipe. The following steps can be distinguished in the NH_3 formation process [2]. Evaporation of water from the injected droplets ($T \geq 100$ °C). As urea starts melting at 133 °C, it stays solid in this stage. Subsequently, the resulting urea ($H_2N\text{-}CO\text{-}NH_2$) will be decomposed in two steps:

$$H_2N\text{-}CO\text{-}NH_2 \rightarrow NH_3 + HNCO \quad (1)$$

$$HNCO + H_2O \rightarrow NH_3 + CO_2 \quad (2)$$

Thermal decomposition (1) can take place upstream of the SCR catalyst. However, the amount of formed NH_3 depends on temperature and space velocity. This is illustrated in [3].

From measurements in a flow reactor, it is concluded that the contribution of the hydrolysis (2) to NH_3 formation upstream of the SCR catalyst is negligible (Figure 2). As a result, maximal 50% of NH_3 is available at the SCR catalyst inlet. The hydrolysis needs to be catalyzed; this reaction occurs inside the SCR catalyst.

For optimal performance, it is important that the injected urea is well distributed over the pipe diameter; inhomogeneous distribution can lead to reduced NO_x conversion and increased NH_3 slip. A mixing device can be applied to improve the distribution.

Possible ways to increase the formed NH_3 upstream of the SCR catalyst are: i) application of a hydrolysis catalyst [4]; ii) (off-line) formation of gaseous NH_3; iii) thermal management measures to increase the exhaust temperature.

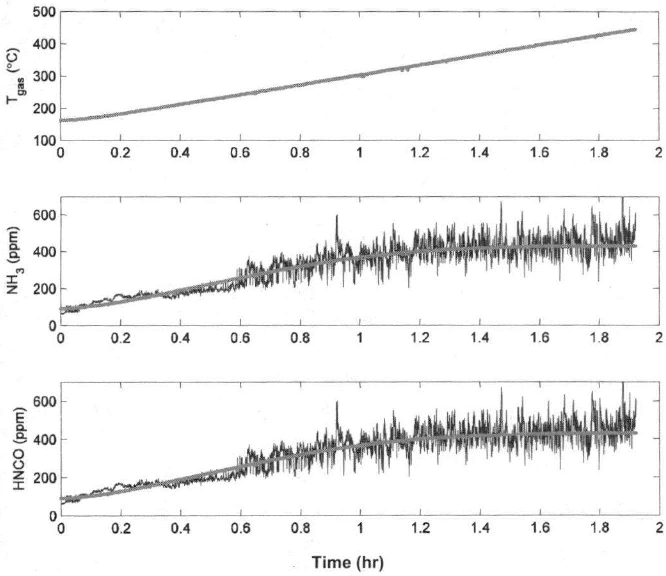

Figure 2: Flow reactor measurements of urea decomposition upstream of SCR catalyst.

CATALYST SYSTEM

Using the formed NH_3, the nitrogen oxides (NO_x) emitted by the engine are reduced over an SCR catalyst according to the following reaction mechanisms:

$$4NH_3 + 4NO + O_2 \rightarrow 4N_2 + 6H_2O \quad (3)$$

$$4NH_3 + 2NO + 2NO_2 \rightarrow 4N_2 + 6H_2O \quad (4)$$

$$4NH_3 + 2NO_2 + O_2 \rightarrow 3N_2 + 6H_2O \quad (5)$$

The most desired pathway is the "fast-SCR" reaction (4), which is considerably faster than the "standard SCR" reaction (3) and reaction (5). However, this path requires an NO:NO_2 ratio of 1:1. As the engine out NO:NO_2 ratio is 9:1, a pre-oxidation catalyst is applied. At temperatures down to 100 °C, NO_x conversion can be increased up to 70% [5]. In that case, NH_3 formation is an issue when AdBlue is used as reagent.

NO_x conversion depends on exhaust gas temperature and space velocity, see *e.g.* [1]. For low temperature applications, NO_x conversion can be enhanced by measures that increase the exhaust gas temperature; *e.g.*, throttling, post-injection in combination with an oxidation catalyst or variable valve timing.

Achievable NO_x conversion also relies on available NH_3. For high temperatures (T > 450 °C), maximal achievable steady state NO_x conversion is mainly constrained due to NH_3 oxidation:

$$4NH_3 + 5O_2 \rightarrow 4NO + 6H_2O \quad (6)$$

Other side reactions that can play a role are the formation of ammonium nitrate (NH_4NO_3) and the formation of nitrous oxide (N_2O), at respectively temperatures below 200 °C and above 450 °C.

For a system in which the urea decomposition is not completed before the catalyst, part of the thermolysis (1) and the entire hydrolysis (2) will take place inside the SCR catalyst. By increasing the available NH_3 on the catalyst surface, system performance is enhanced and the required catalyst dimensions can be reduced [4].

CONTROL SYSTEM

The SCR control system consists of sensors that monitor the system, a controller that computes the desired actions and a urea dosage system (see Figure 1).

With the introduction of On-Board Monitoring, the application of a post-SCR catalyst NO_x sensor is foreseen. This sensor can be applied in closed-loop control strategies. However, current available sensors are cross-sensitive for NH_3, as illustrated in [1].

Urea dosage systems can introduce a significant time delay in the system. Achievable transient performance is strongly affected by this aspect. Therefore, it will be dealt with in this study. With the introduction of airless systems this time delay is expected to be reduced significantly.

SCR control strategies will be discussed in the next section.

SCR CONTROL STRATEGIES

The main focus in SCR control system development is to realize the desired NO_x emission over a test cycle: *nominal performance*. More precisely, the urea dosage is manipulated such that a desired amount of NO_x reduction is realized. In case a tailpipe NO_x target is made explicit, this is often called *reference tracking*. With the introduction of On-Board Diagnostics (OBD) regulations, controllers that make the system robust against system variations [6] gain attention. These controllers have to deal with *e.g.* urea injection inaccuracy, reduced NO_x conversion due to catalyst ageing and NO_x engine-out variations associated with varying ambient conditions. This is often referred to as *disturbance rejection*. Then, compliance with real-life emission targets can be demonstrated.

This section briefly describes an open-loop strategy and gives a rough overview of the status of research into closed-loop strategies. Finally, aspects that limit the applicability of closed-loop SCR control are discussed.

OPEN-LOOP SCR CONTROL

Open-loop urea dosage strategies have proven to be sufficient to meet Euro-4 and Euro-5 emission standards.

This requires about 50-60% and 70-80% NO_x reduction respectively, in a typical heavy-duty application. This open-loop or feedforward control strategy generally comprises an engine-out NO_x prediction functionality, a post-catalyst NO_x target and a so-called NSR map (see Figure 3).

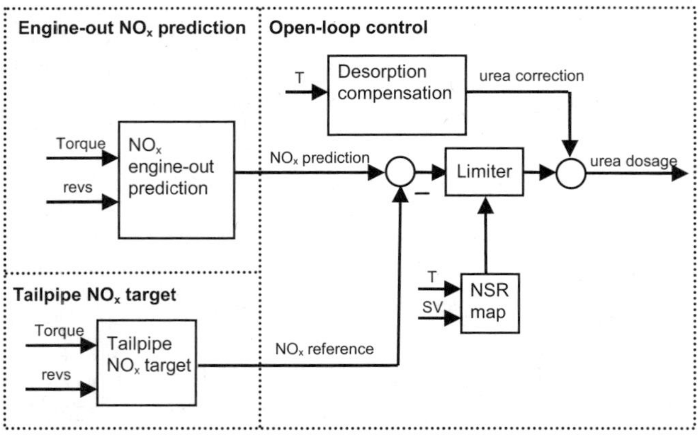

Figure 3: Block scheme of an open-loop SCR control strategy

The desired NO_x reduction is calculated as the difference between engine-out emissions and the desired post-catalyst NO_x concentration. Stoichiometry gives the theoretical amount of urea needed to accomplish the desired NO_x reduction. This amount is limited by a catalyst temperature and space velocity dependent NSR (Nominal Stoichiometric Ratio) map, in order to prevent NH_3 slip.

Nominal control performance of the presented open-loop strategy over a test cycle is determined by the tailpipe NO_x target and NSR map calibrations. In achieving maximal NO_x conversion using this open-loop strategy, it becomes increasingly difficult to find calibrations which achieve high NO_x reduction in combination with low NH_3 slip; especially for transient conditions (including step changes in an ESC cycle). A dynamic compensation, like a functionality which compensates for NH_3 adsorption and desorption, can be added to the open-loop strategy to improve control performance during transients. This introduces even more calibration effort.

Note that a pre-catalyst NO_x sensor can be introduced to the presented open-loop control strategy to enhance robustness in face of varying engine-out NO_x emissions.

CLOSED-LOOP SCR CONTROL

Recently, closed-loop SCR control attracts considerable attention, see, *e.g.*, [1], [3], [7], [8]. The presented control strategies mainly rely on ammonia surface coverage control and NO_x sensor feedback for Vanadium catalysts. [1] deals with the limitations corresponding to the NO_x sensor's cross-sensitivity for NH_3.

Control performance improvement through the application of a closed-loop SCR control strategy is not evident. Several aspects limit the success of a closed-loop SCR control strategy:

- Slow catalyst dynamics;
- NH_3 slip prevention;
- Time delay in the urea dosage system.

For closed-loop control strategies that use a post-catalyst NO_x sensor, the sensor cross-sensitivity can limit the performance.

Catalyst dynamics

The catalyst dynamics are linked to the extent to which a varying urea injection affects the post-catalyst NO_x concentrations. NH_3 buffering and slow reaction kinetics filter the effect of urea dosage variations on the post-catalyst NO_x concentration. If the post-catalyst NO_x concentrations are insensitive to the urea dosage variations, then the SCR catalyst is said to have slow dynamics. Due to this insensitivity, high performance closed-loop control has to apply a large feedback gain. This high feedback gain can potentially destabilize the SCR system and causes an aggressive actuator signal which potentially leads to NH_3 slip.

If the catalyst dynamics are fast, a (high gain) NO_x feedback controller is able to track a tailpipe NO_x reference signal regardless of engine-out NO_x dynamics and open-loop calibration. In this case, the calibration effort can be greatly reduced.

Cross-sensitivity of the NO_x sensor

The cross-sensitivity of the NO_x sensor to NH_3 can also limit the closed-loop control performance. Ignoring this cross-sensitivity causes the control loop to become potentially unstable. A control strategy that copes with the NO_x sensor's cross-sensitivity has been developed and is described in the next section. It relies on perturbing the urea injection and observing its effect on the cross-sensitive sensor signal. It is only able to compensate for slowly changing phenomena during stationary conditions.

Time delay

In an SCR system with fast dynamics, the time delay of the urea dosage system limits the performance of a closed-loop SCR control. In control engineering, delays are well known to limit control performance.

The potential of NH_3 slip feedback control is also limited by the time delay of the urea dosing system, but it is more heavily affected by NH_3 storage in the SCR catalyst.

NH$_3$ storage

When the catalyst shows a considerable NH$_3$ storage capacity, the closed-loop NH$_3$ controller tends to load the catalyst with NH$_3$, while it is pursuing a certain level of NH$_3$ slip. The adsorbed amount of NH$_3$ will desorb during the next temperature increase, probably causing NH$_3$ slip.

Figure 4 shows the steady-state NH$_3$ storage as a function of temperature for two catalyst types. From this figure, it is seen that the Vanadium catalyst can store less NH$_3$ than the Zeolite catalyst for a given temperature. The steady-state NH$_3$ storage exponentially decreases with increasing temperature. At high catalyst temperatures, the NH$_3$ buffering is limited and adsorption and desorption mechanisms are fast. This makes closed-loop NH$_3$ slip control feasible for high temperatures.

For low temperatures, the NH$_3$ storage has to be limited to an amount that does not cause unacceptable high NH$_3$ slip peaks during a sudden temperature rise. At low temperatures, these levels are considerably lower than the steady-state NH$_3$ storage. This kind of NH$_3$ slip prevention can limit the NO$_x$ conversion at low temperatures.

Controlling the NH$_3$ surface coverage of an SCR catalyst provides an ideal combination of achieving maximum NO$_x$ conversion and preventing NH$_3$ slip. Furthermore, it requires less calibration effort than SCR control strategies containing NSR maps and tailpipe NO$_x$ target maps.

As the surface coverage of an SCR catalyst can not be measured directly, it has to be estimated. In the next section, an SCR control strategy is presented which uses an on-line storage model to estimate the NH$_3$ surface coverage of the SCR catalyst. Note that as no information from feedback sensors is used in the NH$_3$ storage estimation, surface coverage control is essentially an open-loop strategy.

Due to the requirements associated with future emission targets, the remaining part of this study focuses on maximal NO$_x$ conversion. In the following sections, the potential of NO$_x$ and NH$_3$ sensor based closed-loop control strategies will be studied. These strategies will be applied to a SCR system with the following characteristics:

- Nonlinearities (including non-negative actuator constraint);

- Time delay in the dosage system;

- NH$_3$ emission constraints;

- Cross-sensitivity of NO$_x$ sensor for NH$_3$

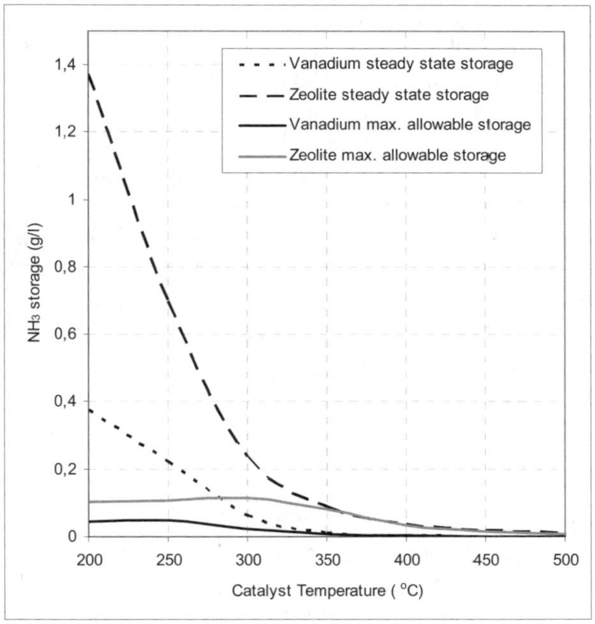

Figure 4. Steady state and maximum allowable NH$_3$ storage as a function of catalyst temperature, for a 34 l SCR catalyst, SV = 25,000 h^{-1}, NSR = 1.2 and engine-out NO$_x$ = 250 g/h.

STUDIED CONTROL STRATEGIES

An adaptive control strategy which combines NH$_3$ slip feedback with surface coverage control is presented. Its performance is compared with a NO$_x$-sensor based control with cross-sensitivity compensation and an open-loop map based strategy.

OPEN-LOOP MAP BASED CONTROL

The open-loop strategy is shown in Figure 5. The applied NSR map gives for each combination of catalyst temperature and space velocity the maximal allowable nominal NH$_3$ to NO$_x$ stoichiometric ratio which results in a steady-state NH$_3$ slip of 10 ppm. The urea dosage is deduced from this map and estimates of the catalyst temperature and space velocity. As the objective is maximal NO$_x$ conversion, no tailpipe NO$_x$ target is included and the NSR map is directly applied to the predicted engine-out NO$_x$ concentration.

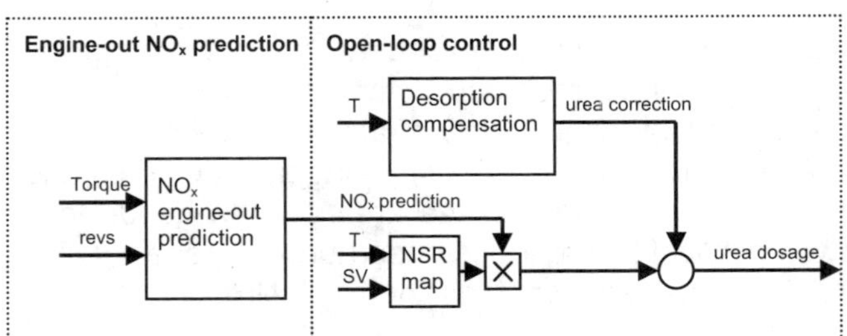

Figure 5: Block scheme of open-loop control strategy

The engine-out NO_x prediction is based on an engine specific map of nominal NO_x emissions. These steady-state values are corrected for dynamic temperature effects through first order filters.

The desorption compensation block is included to compensate for NH_3 desorption during temperature transients. It is based on the steady state NH_3 storage data presented in Figure 4. During a temperature increase a certain amount of NH_3 will desorb from the catalyst surface. This amount is subtracted from the urea or NH_3 injection amount dictated by the NSR map, in order to prevent NH_3 slip.

NO_x SENSOR BASED CONTROL

The structure of the proposed NO_x-sensor based controller is presented in Figure 6. The strategy modulates a block signal on the nominal urea injection, and observes the signal from the cross-sensitive NO_x sensor. Urea is alternately overdosed and under dosed with respect to the average urea dosage. The chosen amplitude and period of the block signal are dependent on the SCR catalyst dynamics and operating point. The effect of the urea dosage variations have to be clearly observable in the NO_x sensor signal. The urea perturbation is stopped during low temperature regimes. Under these conditions, the SCR catalyst dynamics prevent the urea injection variations to be observable in the NO_x sensor signal.

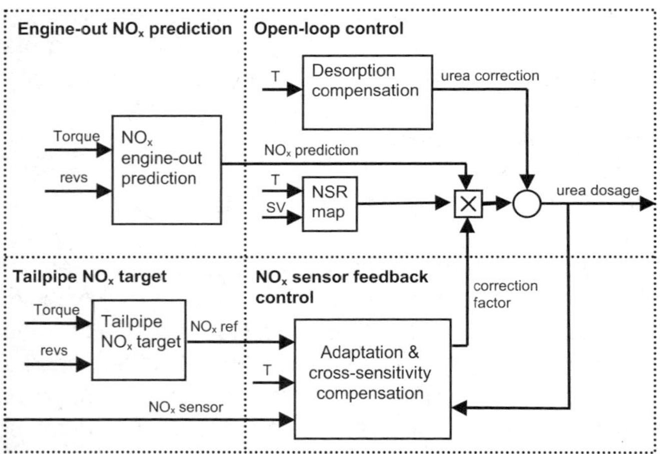

Figure 6: Block scheme of NO_x sensor based control strategy

If the urea injection is raised and a decreasing NO_x sensor signal is recorded, then it is safe to further increase the average injection rate. In case an increased urea dosage results in an increasing sensor signal, then NH_3 slip is detected and the urea dosage needs to be lowered. Vice versa, a decreased urea dosage resulting in a decreased sensor signal indicates NH_3 slip.

For the algorithm to function properly, stationary SCR operating conditions are required; *i.e.*, stationary NO_x engine-out emissions and exhaust flow. Transient conditions make it impossible to judge whether a change in NO_x sensor signal has been caused by the varying urea injection, or by variation of another quantity (e.g. engine out NO_x or temperature variations). This property limits the performance of the proposed control strategy. It is only able to compensate for the effects of slowly changing phenomena like ambient conditions.

The NO_x sensor based control strategy relies on the map based open-loop strategy for the nominal urea dosage. The tailpipe NO_x target map is consequently calibrated for NO_x levels which correspond to the steady state post-catalyst NO_x concentrations which are achieved by the map based open-loop strategy.

ADAPTIVE SURFACE COVERAGE / NH_3 SLIP CONTROL

In closed-loop NH_3 control, maximum NO_x conversion is pursued under a given NH_3 slip constraint. However, depending on the NH_3 buffering on the SCR catalyst, NH_3 slip feedback control is only feasible at high and increasing catalyst temperatures. Under 'low' or decreasing temperature conditions, an NH_3 slip feedback controller tends to load the SCR catalyst with NH_3.

In order to prevent uncontrollable NH_3 slip, the NH_3 storage on the SCR catalyst needs to be limited. Surface coverage control limits this amount to a level that maintains a considerable NO_x conversion, but prevents NH_3 slip in case of an increasing temperature transient. The surface coverage reference map (θ map) in Figure 7 is calibrated to prevent undesired NH_3 slip peaks during a severe temperature rise.

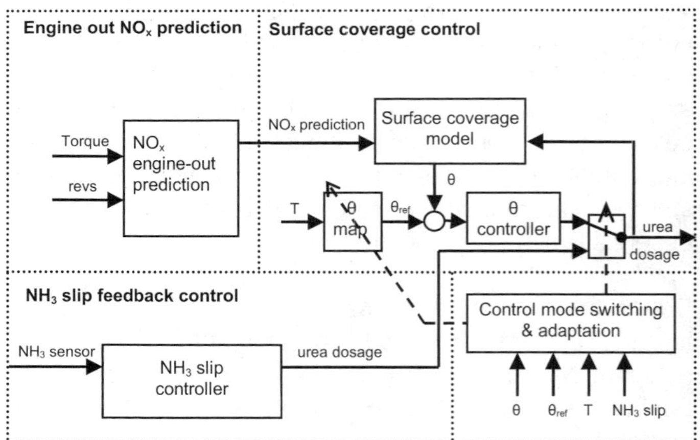

Figure 7: Block scheme of adaptive surface coverage / NH3 based control strategy

The NH_3 storage or surface coverage of an SCR catalyst cannot be measured directly and has to be estimated. Besides NH_3 adsorption/desorption mechanisms, the actual NH_3 storage depends on NO_x conversion chemistry, NH_3 oxidation and possible side reactions with NH_3. For current simulations, the same SCR model is used for surface coverage prediction and SCR catalyst simulation. A sensitivity study has pointed out that 20% discrepancy in the surface coverage prediction causes less than 2% increase of total NO_x emissions during steady state and an ETC driving cycle. This observation suggests that a simplified, less accurate model can be used for surface coverage prediction.

The open-loop surface coverage control is combined with the closed-loop NH_3 slip control to provide robustness and adaptation capabilities. The control mode switching & adaptation block in Figure 7 adapts the θ map such that NH_3 slip feedback control mode is active during high and increasing catalyst temperatures and surface coverage control is active during low and decreasing temperatures. The developed controller is capable of adapting to changing conditions.

SIMULATION RESULTS

The three presented control strategies are evaluated using TNO's simulation tool SimCat. More information on this simulation tool is presented in the appendix. The simulated SCR system contains two parallel Vanadium SCR bricks with a length of 12 inch and a diameter of 10.5 inch, making the total catalyst volume about 34 liters. In the model, complete urea decomposition is assumed upstream of the SCR catalyst. The injection system is assumed to impose a time delay of 0.5 s. Engine-out data has been gathered from a Euro-2 315 kW heavy-duty diesel engine with a swept volume of 12 liter.

The three control strategies are evaluated for maximal NO_x conversion using ESC simulations for a nominal case, and a case in which engine-out NO_x emissions are increased by 30%. This second case simulates the effects of climate and engine variation on engine-out NO_x emissions [9], and serves to evaluate the robustness of the closed-loop controls. The simulated ESC emission results are presented in Table 3. Figure 8 shows the ESC simulation traces for the 30% increased engine-out NO_x case.

As the control strategy based on NO_x feedback is based on the map based open-loop strategy, it applies little correction and achieves comparable NO_x and NH_3 emissions for the nominal case, as can be seen in Table 3. The adaptive surface coverage/NH_3 slip control strategy shows a NO_x cycle result of 0.87 g/kWh and excels in avoiding NH_3 slip by limiting the surface coverage. It demonstrates a maximum slip level of 24 ppm, whereas the open-loop and NO_x sensor based control strategies show a maximum NH_3 slip peak of about 90 ppm over the ESC cycle.

Figure 8 shows how the NO_x sensor based control strategy and the adaptive surface coverage/NH_3 control slowly increase the reagent dosage in order to compensate for the increased engine-out emissions. The map based open-loop strategy is not able to compensate for the increased engine-out NO_x emissions as it uses a prediction of nominal engine-out NO_x concentrations instead of an engine-out NO_x sensor. The post-catalyst NO_x concentrations for the NO_x sensor based strategy slowly converge to the nominal open-loop NO_x levels which were calibrated through the tailpipe NO_x target map. Adaptive surface coverage/NH_3 control again avoids excessive NH_3 slip. It increases the reference NH_3 storage and consequently the reagent injection, until it is able to control the NH_3 slip at a 10 ppm level during the relatively hot ESC modes.

The emission results for the 30% increased engine-out NO_x case presented in Table 3 correspond to the traces in Figure 8. The cross-sensitive NO_x feedback and adaptive surface coverage/NH_3 control strategies show an improved NO_x conversion in comparison to the open-loop strategy, but the cycle NO_x emissions are significantly higher than the NO_x emissions for the nominal engine-out NO_x case. The cycle results improve when the strategies are allowed to adapt to the 30% increased NO_x situation before the cycle result are recorded. In this case, the NO_x sensor based strategy accomplishes a NO_x cycle result of 0.82 g/kWh, but shows an unacceptable high NH_3 slip peak. For the adaptive surface coverage/NH_3 slip control strategy, the NO_x emissions over the ESC cycle drop to 1.36 g/kWh, while the average and maximum NH_3 slip are 5 ppm and 16 ppm respectively.

The map based open-loop strategy and the adaptive surface coverage/NH_3 control strategy are also compared over the ETC cycle. The ETC performance of the NO_x sensor based feedback strategy is not considered, because this strategy is only applicable in stationary conditions. The NO_x sensor based control strategy is not able to apply any corrections during the ETC cycle and will achieve a control performance similar to the open-loop performance.

Table 3. ESC simulation results

Engine-out NO_x	Control strategy	g NO_x / kWh	avg. NH_3 slip (ppm)	max. NH_3 slip (ppm)
Nominal	Map based open-loop	0.78	9	91
	Cross-sensitive NO_x feedback	0.83	8	89
	Adaptive θ/NH_3 control	0.87	6	24
30% increased	Map based open-loop	2.09	2	12
	Cross-sensitive NO_x feedback	1.25	6	45
	Adaptive θ/NH_3 control	1.63	3	15

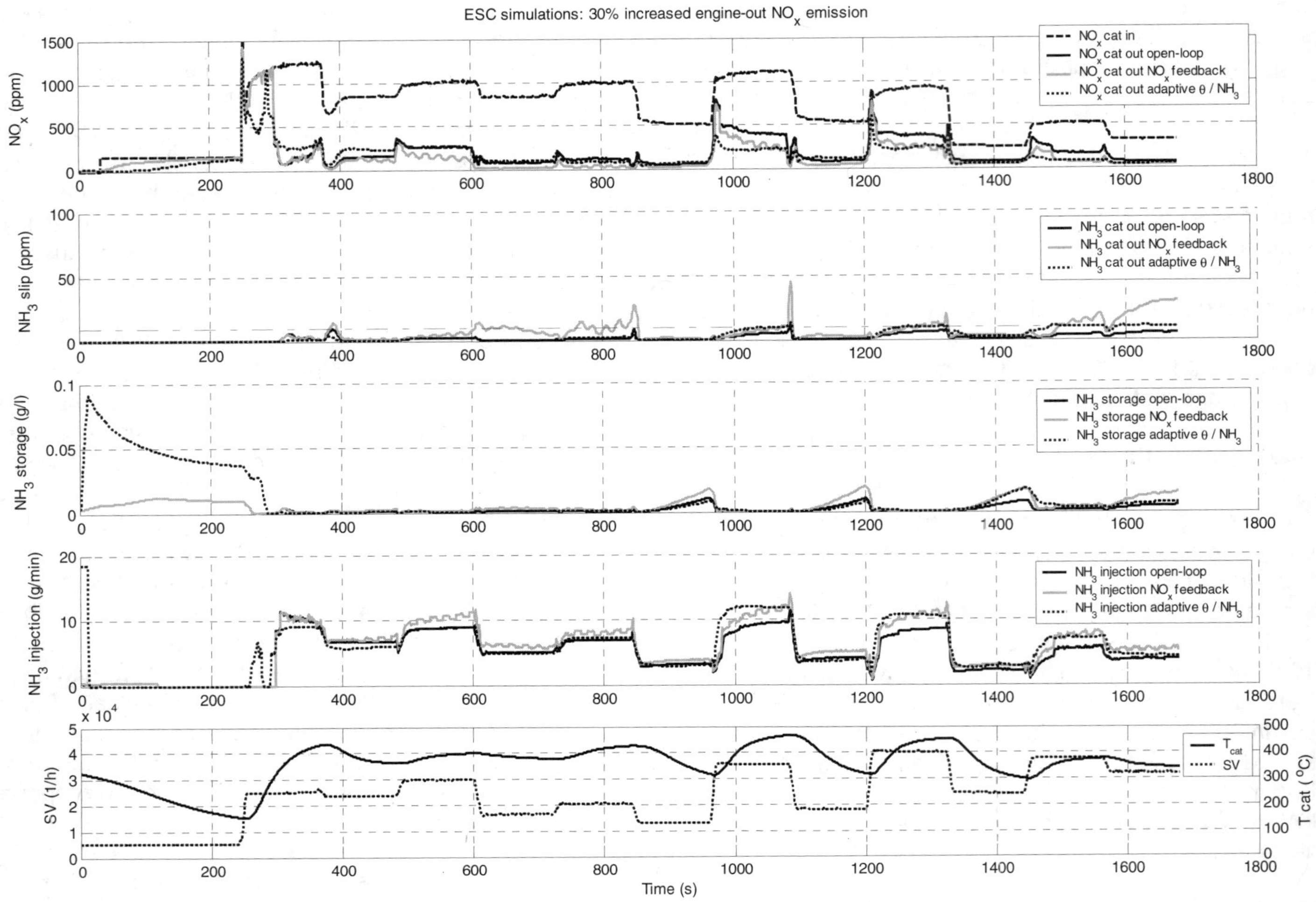

Figure 8. ESC simulation results for 30% increased engine-out NO_x emissions

Table 4. ETC simulation results

Engine-out NO_x	Control strategy	g NO_x / kWh	avg. NH_3 slip (ppm)	max. NH_3 slip (ppm)
Nominal	Map based open-loop	2.67	3	48
	Adaptive θ/NH_3 control	1.66	3	34
30% increased	Map based open-loop	4.22	1	29
	Adaptive θ/NH_3 control	2.88	2	33

Table 4 demonstrates that the adaptive surface coverage/NH_3 slip control strategy shows a control performance superior to the map based open-loop strategy when it comes to both NO_x conversion and NH_3 slip prevention. For this dynamic cycle, its surface coverage control is better suited to achieve high NO_x conversion than the map-based open-loop calibration The NH_3 feedback mechanism allows compensation for the increased engine-out NO_x emissions; the tailpipe NO_x emissions have been lowered by 32% relative to the open-loop performance in case of the 30% increased engine-out NO_x emissions.

For maximal NO_x conversion, The maximum NH_3 slip constraint of 25 ppm seems hard to satisfy according to the presented simulation results. The open-loop map based strategy and the NO_x sensor based strategy show large NH_3 slip peaks, especially during ESC simulations. Surface coverage control offers the best solution of preventing NH_3 slip peaks.

The demonstrated emission levels are not yet near the levels required by the US2010 standard: 0.27 g NO_x/kWh. SCR system improvements as well as measures to decrease engine-out NO_x emissions will be necessary to further lower tailpipe NO_x emissions.

DISCUSSION

Besides the selected control strategy, the achievable performance is strongly affected by the characteristics of the studied system, To realize optimal SCR system performance, an outlook is given on which developments are desirable from a control point of view.

Catalyst

The SCR system dynamics are dominated by the relatively slow surface coverage dynamics. As illustrated in [10], fast dynamics are favorable to reduce calibration time; in that case, an open-loop calibration is no longer required. However, it will serve as a fall-back strategy in case the feedback sensor fails.

To optimize the performance of the SCR catalyst, a pre-oxidation catalyst will be applied. This does not fundamentally change the closed-loop SCR strategy; generally speaking, low temperature NO_x conversion will be enhanced.

Zeolite SCR catalysts are supposed to be applied to meet future emission targets. These catalysts are characterized by high steady-state NO_x conversion and relatively high NH_3 storage capacity (see Figure 4), which makes the NH_3 slip control more challenging.

To guarantee negligible NH_3 slip, an NH_3 oxidation catalyst can be placed downstream of the SCR catalyst. To make this an attractive option for future systems, the NH_3 conversion and selectivity of state-of-the-art NH_3 oxidation catalysts has to be improved.

Urea dosage system

Due to the relatively slow catalyst dynamics, the time delay in the dosing system seems currently not the most limiting factor on performance. However, it constraints the closed-loop system dynamics. With the use of airless dosage systems, the time delay can significantly be reduced. Then, the nominal performance can be enhanced for systems with fast catalyst dynamics. This will result in less calibration effort and increased NO_x conversion over a dynamic test cycle.

Sensors

With the introduction of future OBD requirements, measurement of tailpipe NO_x emissions will be required. Therefore, the application of a post-SCR catalyst NO_x sensor is foreseen. Its cross-sensitivity complicates the use in closed-loop SCR control strategies. In the presented strategy, the adaptation speed is slow and tailpipe NO_x can not be reduced below a specific value (minimum of NO_x + NH_3 concentrations added) due to the cross sensitivity characteristic of the sensor. An SCR catalyst model could help in estimating the fractions NO_x and NH_3 in the cross-sensitive sensor signal, but this requires an accurate model.

NH_3 slip feedback is limited to high temperature operating conditions, whereas NO_x feedback control can be applied in all operating conditions. Realization of a specified tailpipe NO_x level is only possible if a model can predict this level. Furthermore, production type NH_3 sensors are expected not to be available within a few years. This makes NO_x sensor based closed-loop SCR control solutions most promising. Especially, when new sensor technology is able to cancel the cross-sensitivity problem.

With the introduction of fast SCR system dynamics, the need for high bandwidth sensors will grow. Moreover, sensor accuracy and durability becomes more important when maximal NO_x conversion is required.

Interaction with engine and DPF

Finally, it is foreseen that SCR systems will be combined with DPF systems to meet future emission targets. In case a DPF is placed upstream of the SCR system, integration aspects have to be assessed, see *e.g.* [11]; the DPF can influence the NO/NO_2 ratio upstream of the SCR catalyst and the exhaust gas temperature due to thermal inertia of the DPF or during active regeneration. In the end, the performance of the combined system of engine, DPF and SCR system is of interest. This combined system has to achieve the desired tailpipe emission with minimal fuel consumption and acceptable drivability. This requires an integrated approach: *Integrated Emission Management*. In that case, the engine is controlled such that the required conditions (emissions, exhaust gas temperature) are realized for the DPF and SCR system; for instance, using engine measures to increase the exhaust gas temperature during a specific period in time.

CONCLUSIONS

To meet future emission targets, SCR systems have to achieve maximal NO_x conversion in combination with minimal NH_3 slip. In this study, the potential and limitations of closed-loop SCR systems are discussed to realize optimal performance.

From the reported results, the following can be concluded:

- Closed-loop SCR control is essential to meet in-use compliance requirements. Main benefit is that it adds adaptation capabilities and robustness for system variations. Closed-loop control is not required to enhance nominal performance over a test cycle for the current SCR systems.

- Performance of closed-loop SCR control is limited due to slow catalyst dynamics, a time delay in the urea dosage and limits on NH_3 emission. Large NH_3

buffering capacity of an SCR catalyst limits the applicability of NH_3 slip feedback control and exposes the need for surface coverage control.

- The presented adaptive surface coverage/NH_3-slip control strategy achieved highest NO_x conversion with minimal NH_3 slip over an ESC and ETC cycle, for the studied SCR system with time delay in urea injection. The NO_x sensor based strategy with cross-sensitivity compensation is only able to compensate for slowly changing phenomena during stationary conditions.

- If reagent dosage systems and future sensor and catalyst technology relieve the specified limitations, closed-loop SCR control has the potential to reduce the calibration effort and to improve transient control performance.

Future work will focus on the demonstration of control strategies on an experimental set-up and on the further development of the SCR system model. Control developments also have to concentrate on the interaction of the SCR system with the engine and DPF system: *Integrated Emission Management.*

ACKNOWLEDGMENTS

Part of this research was done in the framework of the PREDUCE research program (Program for the Reduction of the Emissions from Diesels towards Ultra Clean Engines). This program is partially funded by the Dutch Ministry of Economical Affairs.

REFERENCES

1. Schär, Onder, Elsener and Geering, Model-based control of an SCR system for a mobile application, SAE 2004-05-0412
2. Koebel, Elsener and Kleemann, Urea-SCR: a promising technique to reduce NOx emissions from automotive diesel engines, Catalyst Today, 59:335-345, 2000
3. Chi and DaCosta, Modeling and control of a urea-SCR aftertreatment system, SAE 2005-01-0966
4. Jacob, Müller, Scheeder, Cartus, Dreisbach, Gotre, Mai, Paulus and Spengler, High performance SCR catalyst system: Elements to guarantee the lowest emission of NOx, proc. of 27th International Vienna Motor Symposium, pp. 240-264, 2006
5. Hammer and Bröer, 1998, Plasma enhanced selective catalytic reduction of NOx for diesel cars, SAE 982428
6. Nebergall, Hagen and Owen, Selective catalytic reduction on-board diagnostics: past and future challenges, SAE 2005-01-3603
7. Upadhyay and Van Nieuwstad, Control design of an automotive urea SCR catalyst, Proc. of ASME: IMECHE, pp. 1-8. 2002
8. Qingwen Song and George Zhu, Model-based closed-loop control of urea SCR exhaust aftertreatment system for diesel engine, SAE 2002-01-0287
9. Van Helden, Verbeek, Willems, Van der Welle, Optimization of urea SCR DeNOx systems for HD Diesel engines, SAE 2004-01-0154
10. Boomsma, Design of a controller for a closed-loop SCR exhaust aftertreatment system, Master's thesis, Eindhoven University of Technology, 2006.
11. Acharya, Alam and Boehman, Fuel and system interaction effects on urea-SCR control of NOx in diesel exhaust aftertreatment, SAE 2006-01-0638.

APPENDIX

TNO's SimCat is a tool for exhaust gas after treatment system modeling in MATLAB/Simulink®. With this modular tool, various SCR system configurations can be modeled. It consists of one dimensional models for urea decomposition in the exhaust pipe, pre-oxidation catalyst, SCR catalyst and an NH_3 oxidation catalyst. These models are based on first principle modeling, including mass and energy balances.

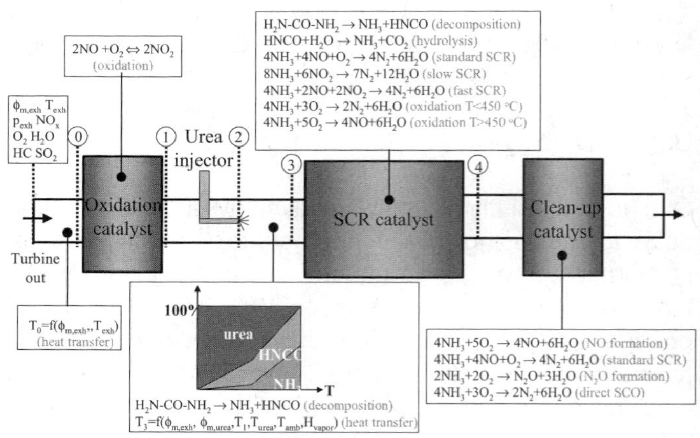

Figure 9. Illustration of SCR system model

The SCR catalyst model takes exhaust flow, upstream exhaust gas temperatures and upstream concentrations of several species as its inputs. The model predicts quantities like NO_x conversion, NH_3 slip and temperatures as a function of the location in the catalyst and as a function of time. The modeled chemical reactions are shown in Figure 9. A description of the reactions can be found in [2] and [9].

The applied models are validated using flow reactor data and engine dynamometer data. As shown in Figure 10, the model shows good agreement with flow reactor measurements.

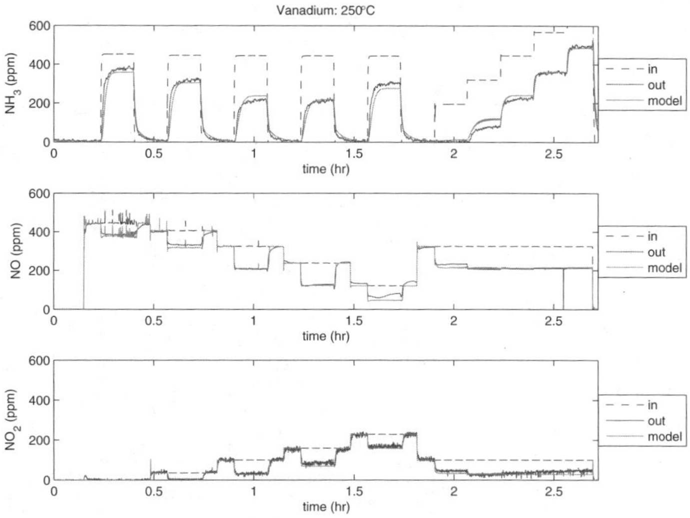

Figure 10. Flow reactor experiments and model fit of a Vanadium catalyst using NH_3 reagent at 250 °C and space velocity of 100,000 h^{-}

CONTACT

Dr. Frank Willems
TNO Automotive
Schoemakerstraat 97
P.O. Box 6033
2600 JA Delft
The Netherlands
Email: Frank.Willems@tno.nl
Tel: +31 15 2697471

III.2 LNT, LNC SYSTEMS

2001-01-0510

Development of a Desulfurization Strategy for a NO_x Adsorber Catalyst System

Sam Geckler and Dean Tomazic
FEV Engine Technology, Inc.

Volker Scholz
FEV Motorentechnik GmbH

Margaret V. Whalen
National Renewable Energy Laboratory

Dale McKinnon
MECA

John Orban
Battelle

Robert A. Gorse
Ford Motor Co.

Owen Bailey
ASEC

James C. Hoelzer
Detroit Diesel

ABSTRACT

The aggressive reduction of future diesel engine NO_x emission limits forces the heavy- and light-duty diesel engine manufacturers to develop means to comply with stringent legislation. As a result, different exhaust emission control technologies applicable to NO_x have been the subject of many investigations. One of these systems is the NO_x adsorber catalyst, which has shown high NO_x conversion rates during previous investigations with acceptable fuel consumption penalties. In addition, the NO_x adsorber catalyst does not require a secondary on-board reductant.

However, the NO_x adsorber catalyst also represents the most sulfur sensitive emissions control device currently under investigation for advanced NO_x control. To remove the sulfur introduced into the system through the diesel fuel and stored on the catalyst sites during operation, specific regeneration strategies and boundary conditions were investigated and developed. To achieve the required exhaust temperature under slightly rich conditions, a pre-catalyst may be required. Under constant engine operating conditions, the duration of the desulfurization process was varied to determine the required desulfurization period. In addition, the catalysts were aged and periodically desulfurized to determine the influence of multiple desulfurization events on long-term catalyst performance.

The test results show that using a pre-catalyst, the NO_x conversion capability of a NO_x adsorber catalyst can be restored even after multiple desulfurizations. However, a decline in conversion efficiency can be observed over time. In anticipation of future catalyst improvements, it can be concluded, that the adsorber catalyst is a promising technology for future vehicle applications.

INTRODUCTION

NO_x adsorber catalysts can exhibit very high NO_x conversion efficiencies over a wide-temperature range

[1, 2] in application to diesel engines, and therefore, because this technology does not require a secondary on-board reductant, it is attractive as an emission control device. However, current fuel sulfur levels, and even the U.S. EPA proposed cap level, would result in deactivation of the NO_x adsorber catalyst from sulfur contamination without an appropriately applied sulfur maintenance strategy.

EXPERIMENTAL RESULTS

The test results presented in this paper are divided in two sections. The first describes general NO_x adsorber demonstration tests conducted to demonstrate the potential of the technology specifically related to rich diesel engine operation. The second focuses on the development of a desulfurization strategy used to regenerate the NO_x adsorber storage capacity from sulfur contamination.

NO_X ADSORBER FUNDAMENTALS – The basic operation of NO_x adsorber catalysts has been described in detail in several publications [1, 2, 3]. Therefore, it is only briefly covered within this paper. A typical NO_x adsorber catalyst is applied as a washcoat to a conventional automotive catalyst substrate such as a monolithic ceramic. The coating includes basic oxides that trap acidic oxides (NO_x) in lean environments. The NO_x adsorber catalyst includes these trapping compounds and also precious metals to facilitate the lean mode trapping and enable the rich mode reduction of NO_x. In a rich environment, the desorbed NO_x is reduced to diatomic nitrogen over precious metal sites in the same way that a three-way catalyst operates. The lean mode storage and rich mode desorption and reduction process are schematically shown in Figure 1.

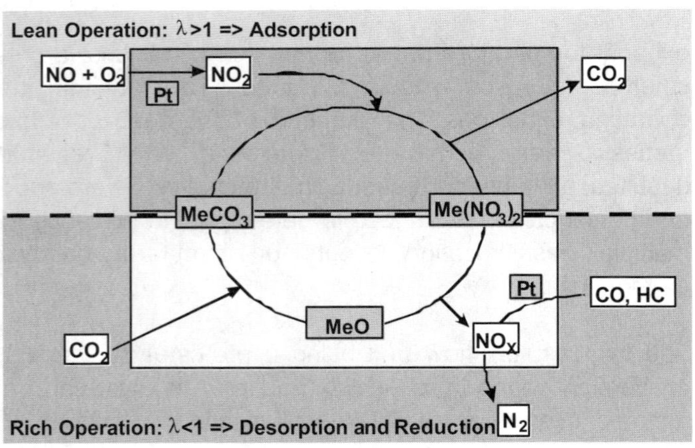

Figure 1: Schematic of NO_x adsorber operation

<u>NO_x Adsorber Configurations</u> - At least two configurations are under development by various organizations. In the first, the engine is operated normally and the exhaust gas flow is switched between two duplicate legs of the exhaust system with a NO_x trap in each leg. The catalyst traps NO_x from the exhaust in one leg, while in the other, the stored NO_x is desorbed and reduced. This is much simpler from the engine point of view; however, it requires double the catalysts and a separate fuel injection system (for each leg) to deliver the reductant (fuel) to create the reducing environment required for desorption and catalytic reduction.

The second configuration requires only a single exhaust system and trap. To accomplish regeneration of the NO_x adsorbed, the engine must be used to supply the reducing environment to regenerate the trap for NO_x. *Results obtained using the second configuration are herein described.*

In contrast with the normal lean engine operation mode ($\lambda > 1$), the diesel engine must be operated for a short period of time under rich conditions ($\lambda < 1$) in order to regenerate a NO_x adsorber catalyst. Two methods are available to achieve this relative air/fuel ratio condition for the diesel engine: (1) by reducing the air mass or (2) by increasing the injected fuel quantity (or a combination of both). Figure 2 provides an overview of these possibilities with respect to stoichiometric engine operation.

Due to the relatively slow response time, considering a targeted regeneration period of one to two seconds, methods to achieve rich conditions on the intake air side are less useful. The fuel injection system, on the other hand, allows switching from lean to rich operating conditions within one single engine cycle. In this context, post-injection must be considered as a very effective tool. This, however, requires a flexible high-pressure fuel injection system such as a common rail system. In order to minimize the torque and speed fluctuation during steady-state engine operation when switching from lean to rich mode, additional fuel injection parameters including quantity and beginning of pilot- and main- injection must be adjusted accordingly.

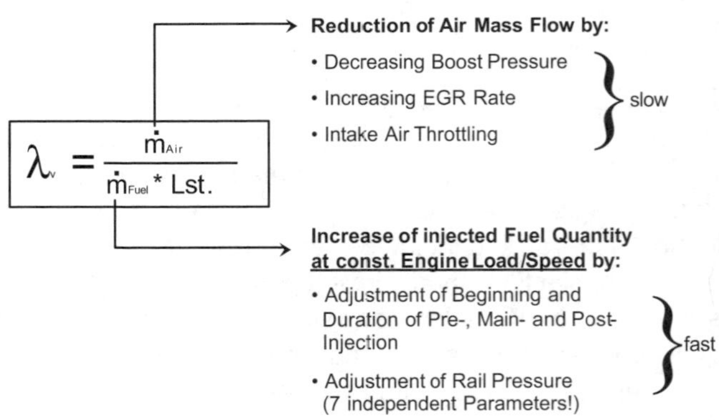

Figure 2: Methods for reducing air/fuel ratio

Due to the inhomogeneous mixture formation in a diesel engine, rich operation leads to locally enriched zones, which usually results in a significant increase in black smoke emissions. However, an optimized calibration of all engine operating parameters can reduce the increase of the black smoke to an acceptable level.

Typically, a successful NO_x adsorber catalyst regeneration strategy is also characterized by a high content of carbon monoxide (CO > 3%) and a simultaneously low oxygen content (O_2 < 2%) in the exhaust gas. During rich operation, the post-injection quantity cannot be completely oxidized for some operating points; therefore, an oxidation catalyst located upstream of the NO_x adsorber catalyst can be used to oxidize the unburned fuel utilizing the remaining oxygen present in the exhaust gas.

Figure 3 shows a comparison between lean and rich engine operation for a steady-state operating point of 2000 rpm and 2 bar BMEP. For the part load operating point shown, the smoke emission during rich operation reaches an acceptable value of 1.1 Bosch units despite a relatively high EGR rate of about 60-65%, while the oxygen content in the exhaust is at a favorably low level of 1%. The brake specific fuel consumption (BSFC) is increased by about 57% as a result of the rich operation compared to the lean engine operation mode. Assuming, that the adsorption period for the NO_x adsorber catalyst during lean conditions is 60 seconds and the regeneration (rich) period lasts for 2 seconds, an average increase in BSFC below 2% can be achieved.

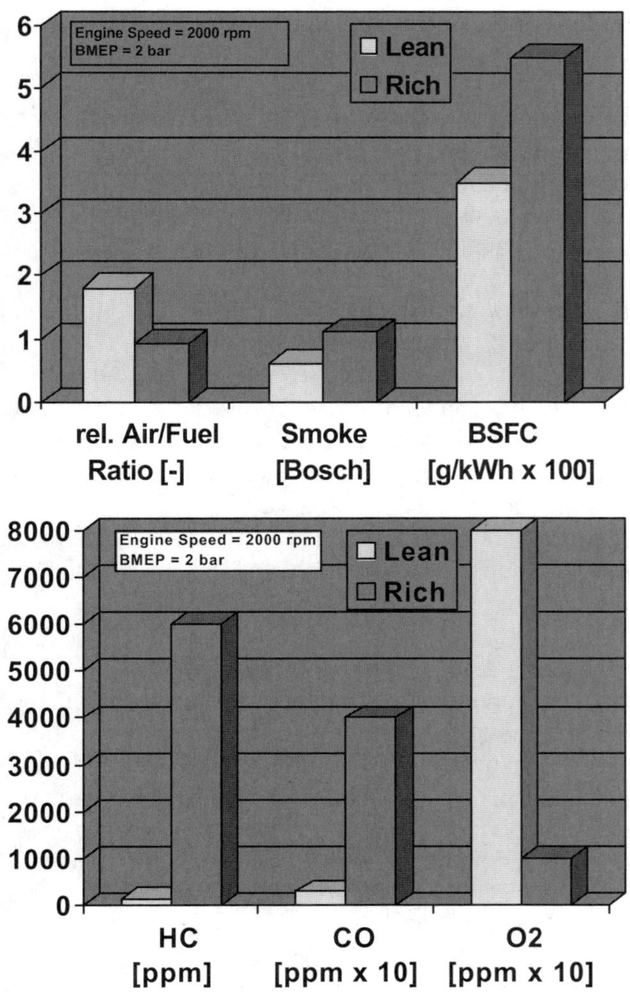

Figure 3: Comparison between lean and rich engine operation of a 2 liter common rail diesel engine at 2000 rpm and 2 bar BMEP

Under steady-state conditions, an adjustment of all engine parameters required for rich operation in the entire engine map results in a significant effort. The transition to transient engine operation requires even further effort by optimizing the transient response between lean and rich operation to improve the vehicle behavior (acoustics, dynamics, etc.).

NO_X ADSORBER DESULFURIZATION – NO_x adsorber catalysts become contaminated with sulfur as the sulfur is trapped as a sulfate on the same storage sites intended to trap NO_x. As this occurs, an increase in NO_x slip will begin to appear downstream of the catalyst. The rate of sulfur contamination is proportional to the sulfur species concentrations upstream of the catalyst (i.e., fuel sulfur level). In the presence of sulfur, regardless of the source, the ability of the catalyst to adsorb nitrates will be reduced over time.

To counteract the effect of sulfur contamination, it would be useful if a procedure existed to drive stored sulfate from the catalyst similar to the way NO_x is regenerated. The conditions necessary for sulfur release have been generally reported by other researchers, although the specific conditions depend on the catalyst formulation. It is necessary to bring the catalyst to high operating temperature and to provide exhaust conditions corresponding to rich operation. Several researchers have published information related to desulfurization of NO_x adsorbers [4, 5], although not using a state-of-the-art diesel engine to provide those conditions. The following sections describe results obtained in an effort to answer the following questions:

1. Can a diesel engine operating strategy be developed to remove sulfur trapped in the NO_x adsorber?
2. How does repeated aging and desulfurization affect NO_x adsorber durability in the presence of various fuel sulfur levels?

Experimental System – To conduct the experimental investigations, a state-of-the-art HSDI diesel engine was used equipped with a common rail fuel injection system. The engine contained the following characteristics and features:

• Total Engine Displacement:	1943 cc
• Bore/Stroke:	82 /92 mm
• Compression Ratio:	19:1
• Maximum Cylinder Pressure:	150 bar
• Maximum Injection Pressure:	1350 bar
• Charging System:	WG Turbocharger
• Maximum Boost Pressure:	2.05 bar
• Rated Power:	81 kW @ 4200 rpm
• Additional Features:	Intake Air Throttling EGR

The engine was installed into the test facility with a dual exhaust system allowing two catalysts, each sized for half the exhaust flow, to be installed simultaneously. In this configuration two prototype catalysts could be subjected to identical conditions thereby increasing confidence levels in the results. A schematic of the test hardware installation is shown in Figure 4.

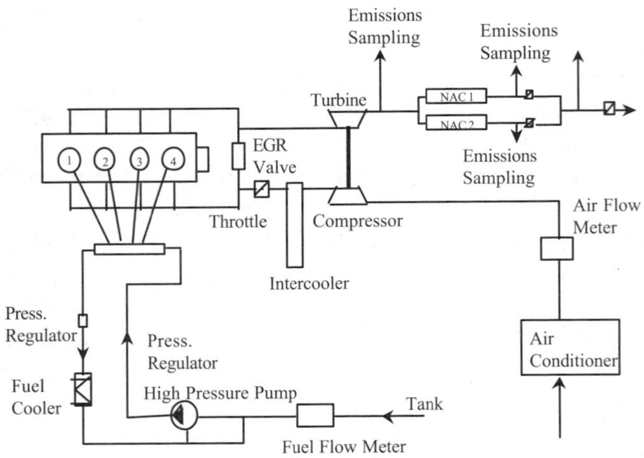

Figure 4: Test Hardware Schematic

Experimental Catalysts – The NO_x adsorber catalysts utilized to conduct the testing were of a type previously tested under the Diesel Emission Control – Sulfur Effects (DECSE) program [6]. For comparison purposes catalysts with identical formulation were used within this program although the technology represents an older technology being state-of-the-art in 1998.

The general NO_x adsorber catalyst characteristics that can be described are:

- 2.5 liter
- 400 cpsi, 6.5 ml wall thickness ceramic substrates
- 5.66" by 6" round geometry
- 1998 Experimental formulation at 164 g/ft3
- Pt:Pd:Rh at 10:3.9:1 ratio

Desulfurization Conditions - The primary conditions controlling the desulfurization process include the relative air/fuel ratio and the catalyst inlet temperature. Based upon information received from the catalyst supplier, a temperature of 700 °C was set as the target for sulfur release. Temperatures above 800 °C were undesirable due to an increase in performance degradation due to exposure to higher temperature. Therefore, the ideal catalyst inlet temperature for desulfurization of the catalysts for this program would be at least 700 °C without exceeding 800 °C. Richer gas mixtures facilitate more rapid desulfurization, whereas leaner (although still rich) mixtures affect slower desulfurization. Relative air/fuel ratios from 0.7 to 1.0 have been reported as effective for sulfur release [4]. As a compromise and to eliminate extended development beyond the scope of work, a value of 0.9 was set as the target.

Desulfurization conditions were only established at an engine load level corresponding to a catalyst inlet temperature of 400 °C. This point was representative of the NO_x regeneration calibration in the middle region of the temperature window investigated considering fuel economy penalty, turbine outlet temperature, and maximum rich mode CO level.

Defining the target operating point and desired desulfurization conditions was only part of the challenge in removing stored sulfur from the NO_x trap; achieving those conditions using only the engine hardware was another. The front face of the NO_x adsorber catalysts was about 1.5 meters downstream of the turbocharger outlet in this test cell configuration causing a significant temperature drop. Therefore, it would have been difficult to create engine out temperature levels sufficiently high to ensure greater than 700 °C at the NO_x adsorber inlet.

The approach for generating the high temperature required at the NO_x adsorber inlet was to utilize a "warm-up" catalyst "close-coupled" to the turbocharger outlet. This catalyst would provide complete or partial oxidation of unburned hydrocarbons thereby developing a large exotherm over the catalyst. This exotherm would be transferred through the exhaust system to the NO_x adsorber inlet providing the necessary temperature conditions to release sulfur from the NO_x trap.

The results of the desulfurization procedure development are separated into two sections. First described is the process of demonstrating the exotherm over the "warm-up" catalyst to achieve the high catalyst inlet temperatures and relative air/fuel ratio necessary to release sulfur. Tests corresponding to these results were conducted with only the "warm-up" catalyst installed utilizing straight pipes to replace the NO_x adsorber catalysts.

The second part of developing the desulfurization procedure concerns identifying the time length required to complete the desulfurization event under constant engine operating conditions. Several tests were conducted to determine a satisfactory duration for the desulfurization event that could be carried forward into the final tasks of the project.

The initial tests conducted to identify a suitable desulfurization strategy, included the evaluation of different EGR levels and different post-injection quantities intended for oxidation over the "warm-up" catalyst. Figure 5 shows the influence of a variation in post-injection quantity on several important parameters including relative air/fuel ratio and temperature at the adsorber catalyst inlet. At this level of EGR, about 30 %, there is sufficient fuel in the exhaust to consume all the oxygen present in the exhaust downstream of the turbocharger. Therefore, no combination of this EGR rate and the post-injection quantity levels shown in Figure 5 achieve both the target temperature level and relative air/fuel ratio at the catalyst inlet position.

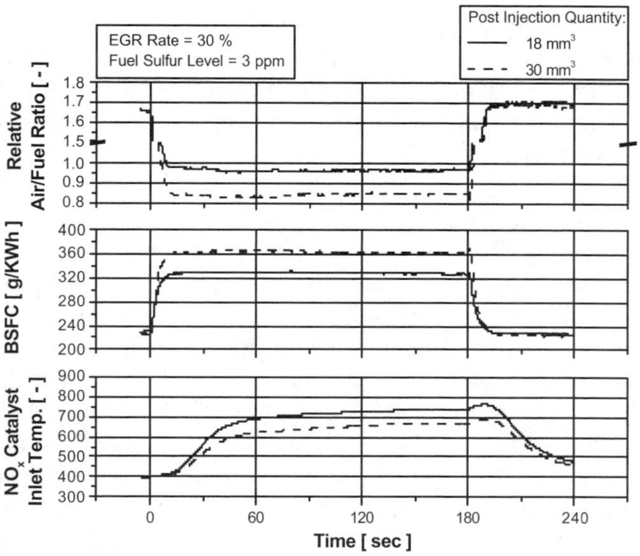

Figure 5: Desulfurization conditions corresponding to an EGR rate of 30%

The temperature increase is higher by using a slightly lower EGR rate as shown in Figure 6. The time required to reach 700 °C at the inlet position of the NO_x adsorber catalyst is about 1½ minutes. This time period was longer than desired; however, with the catalyst installed, the temperature at the position of the NO_x adsorber catalyst inlet is expected to increase more quickly due to the presence of the higher pressure and the hot brick.

An EGR rate for the desulfurization procedure of about 27 % was selected. Although lower EGR rates may further reduce the time to get up to temperature, it would also increase the relative air/fuel ratio requiring a higher post-injection quantity. Figure 6 shows, that to achieve the desired relative air/fuel ratio, a post-injection quantity of 30 mm^3/stroke is necessary, so this value was selected for the desulfurization strategy. Because 30 mm^3/stroke is already rather high and to limit the development effort, lower EGR rates were not investigated.

In summary, the desulfurization conditions were:

Engine Speed:	3000 rpm
Engine Load (BMEP):	167 Nm (10.8 bar)
Catalyst Inlet Temperature:	>700 °C
Rel. Air/Fuel Ratio:	~0.9
EGR Rate:	27 %
Post-injection Quantity:	30 mm^3

It should be noted that the value of the exotherm over the "warm-up" catalyst is rather high as shown in Figure 7. By the time stable temperatures are achieved, the temperature increase over the catalyst has reached approximately 350 °C. This large value is partially necessary due to the test cell installation of the catalysts and the length of the exhaust system.

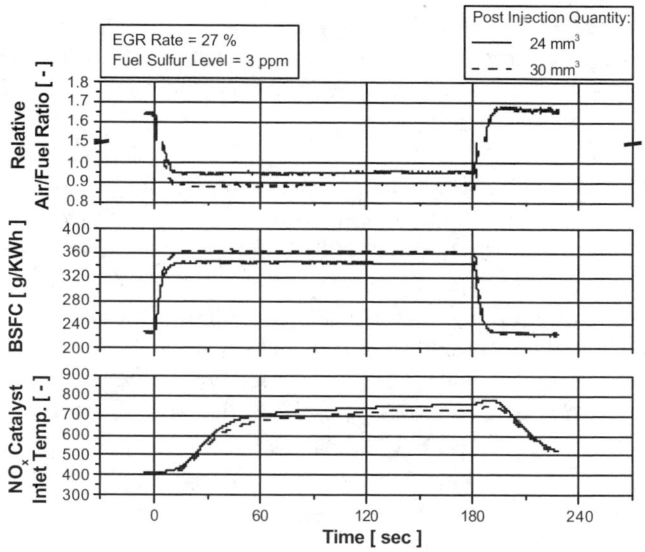

Figure 6: Desulfurization conditions corresponding to an EGR rate of 27%

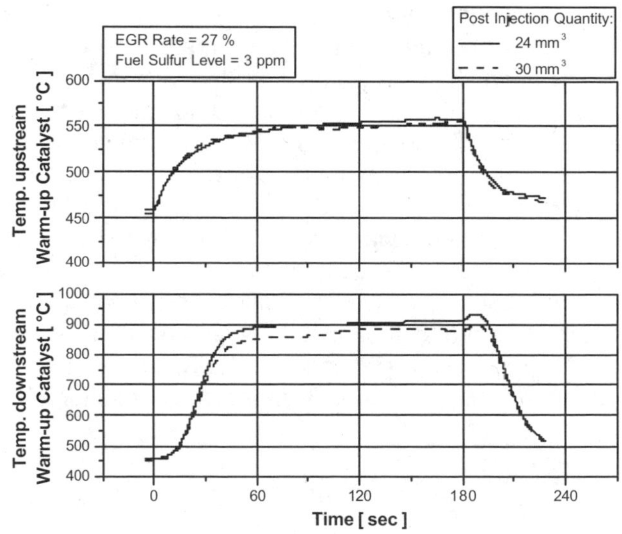

Figure 7: Exotherm over the "warm-up" catalyst for an EGR rate of 27 % and several post-injection quantities

The simplified approach described provides a tool to study the possibility of desulfurizing a NO_x adsorber catalyst. It is probable that increasing the complexity of the hardware and software utilized to achieve these conditions will allow them to be achieved more quickly and precisely with lower stress on the components involved.

Desulfurization Duration - With the conditions for sulfur regeneration defined, it was necessary to determine what desulfurization time length was required to return these NO_x adsorber catalysts to a previously high level of NO_x conversion efficiency following a defined sulfur contamination event.

A catalyst was installed and mapped to measure the actual NO_x conversion efficiency at the catalyst inlet temperature of 400 °C. This initial conversion efficiency, 80 %, was used to define the target contamination level with which to test different desulfurization event lengths.

The catalyst was then systematically contaminated through operation on a high sulfur level fuel. A commercially available fuel with a mass percent sulfur of 0.033 % was utilized.

The conversion efficiency of the catalyst at the 400 °C point was frequently monitored until the conversion efficiency had dropped by 25 % compared to the base value. In this case, that target absolute NO_x conversion efficiency was 60 %. Because the effectiveness of the desulfurization procedure did not always result in a return to the initial 80 % conversion efficiency, the catalyst was contaminated to as near the 60 % absolute conversion efficiency as possible between desulfurization attempts. Beginning from this level of performance and applying various desulfurization time lengths, the most effective desulfurization duration was identified corresponding to the exhaust conditions previously developed for the specific catalyst tested.

Figure 8 shows the influence of the various desulfurization event lengths operated during these tests. The process used to create these results consisted of the following steps: (1) the catalyst was contaminated, as described; (2) the fuel was changed to the DECSE 3 ppm sulfur level fuel to confirm the conversion efficiency level; (3) the desulfurization was conducted; (4) and finally, the conversion efficiency was then re-measured using the DECSE 3ppm fuel. The contamination event control was good with absolute conversion efficiencies varying in a small tolerance band around the target value. The time period for each contamination event was approximately two (2) hours using the 0.033 % mass percent sulfur fuel.

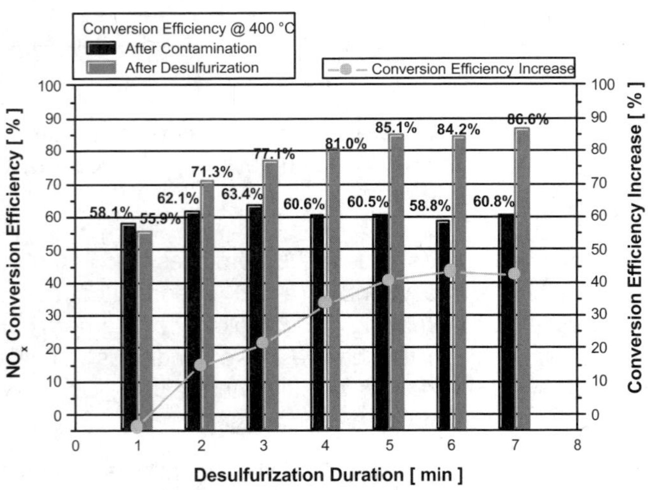

Figure 8: The influence of desulfurization duration on conversion efficiency

The desulfurization event must be several minutes in length using the conditions developed to achieve high levels of sulfur removal and NO_x conversion efficiency recovery. For example, if a one (1) minute event is utilized, because of the delay in the temperature increase, the catalyst does not achieve a sufficient temperature level for sulfur release. Actually, slightly lower conversion efficiency was measured after the one (1) minute desulfurization event compared to before desulfurization, although the difference is within typical measurement-to-measurement variability for NO_x conversion efficiency that is estimated at ± 2.5 %.

Increasing the length of the desulfurization event tended to increase the absolute NO_x conversion efficiency. The point of diminishing return with respect to NO_x conversion efficiency occurred around the five (5) or six (6) minute length. The event length selected for the desulfurization strategy was six (6) minutes. This event length should always provide a high degree of desulfurization even though the fuel economy penalty would be higher than for the five (5) minute event. This is considered an acceptable compromise due to the limited number of times the desulfurization event will be necessary throughout a vehicle life.

<u>Conversion Efficiency Recovery</u> - After developing the procedure for desulfurization, the effectiveness of this strategy was demonstrated using the catalysts aged with various sulfur levels during the DECSE I program [6]. In that program three (3) catalyst pairs were aged using three (3) different sulfur level fuels. The following table lists the designation, aging time, and fuel sulfur level used for aging in that program.

Catalyst Designation	Aging Duration (hours)	Fuel Sulfur Level (ppm)
3a/b	250	3
16a/b	200	16
30a/b	150	30

Each catalyst pair was installed into the exhaust system and the NO_x conversion efficiencies were measured. Figure 9 includes these results and results obtained from re-mapping the catalyst performance after the six minute desulfurization.

Clearly, the desulfurization procedure was successful in returning the catalyst performance to a very high level. The catalyst pairs achieved greater than 80 % NO_x conversion from a low temperature of 300 °C to an upper limit of 450 °C. Even at 500 °C, the catalysts still exhibited more than 60 % NO_x conversion. The desulfurization was particularly effective in restoring activity in the mid-temperature range.

The efficacy of the desulfurization procedure in removing sulfur is consistently good when examining catalysts having been aged with various sulfur level fuels. Figure 9 shows that the NO_x conversion efficiency performance of catalyst pairs 16a/b and 30a/b is largely similar to the catalyst pair aged with an ultra-low sulfur level of 3 ppm after desulfurization.

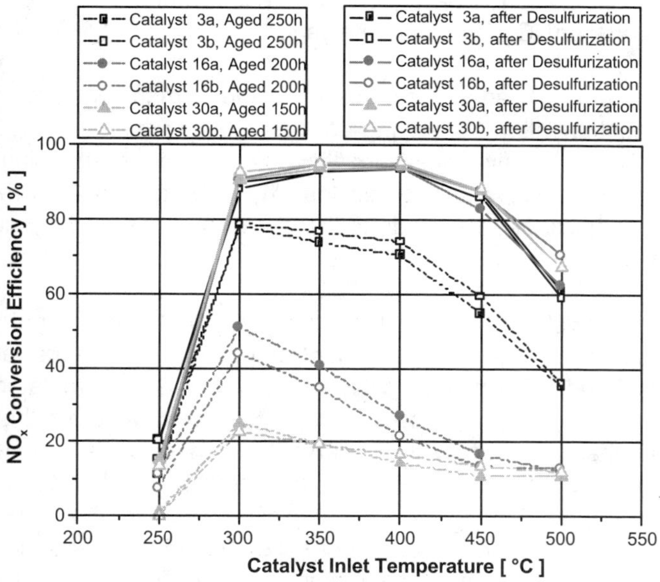

Figure 9: Conversion efficiencies for contaminated and desulfurized NO_x adsorber catalysts

Durability Investigation – The desulfurization demonstration provides encouraging results that sulfur can be released from NO_x adsorber catalysts using the engine and catalyst installation approach described. However, the results in the previous section provide no indication if this process can be repeated to recover NO_x conversion efficiency for a catalyst repeatedly exposed to high sulfur concentration exhaust.

The following tests were therefore conducted to make an initial assessment of the catalyst's ability to survive repeated aging and desulfurization. The catalysts were aged and then mapped for NO_x conversion efficiency over the desired temperature window, and then desulfurized followed by mapping again. This process was repeated using separate catalyst pairs and different sulfur level fuels for the aging process. The schematic in Figure 10 represents the testing conducted and includes the catalyst designations used for these tests. Prior to beginning the repeated aging schedule shown in Figure 10 each catalyst pair was de-greened and aged for 10 hours, mapped for its conversion efficiency, desulfurized, and re-mapped all with the 3 ppm sulfur level fuel.

Summary data from the mapping events in Part I are shown in Figure 11. The first mapping data from catalysts S5/6 showed a very wide range (300 to 450 °C) of temperature over which the catalysts operate at very high NO_x conversion efficiency (> 90 %). As time progressed, the conversion efficiency dropped after desulfurization.

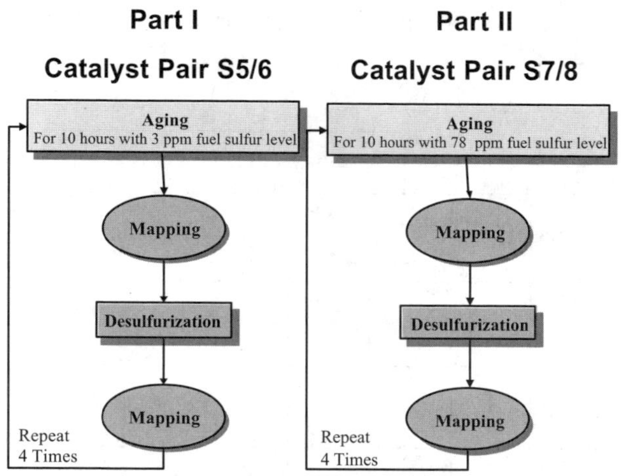

Figure 10: Test program to evaluate the long-term effects of repeated aging and desulfurization

The rate of reduction in the NO_x conversion efficiency varies with catalyst inlet temperature. At 500 °C, the rate is the fastest; at the end of the cycle, the mapping after the sixth desulfurization event provided a conversion efficiency of less than 60 %. At the other extreme, 400 °C, the rate of reduction was low enough to provide a final conversion efficiency of approximately 90 % at the end of the cycle. The data after the sixth desulfurization represent the performance of a catalyst operated for 100 hours under the defined cycle.

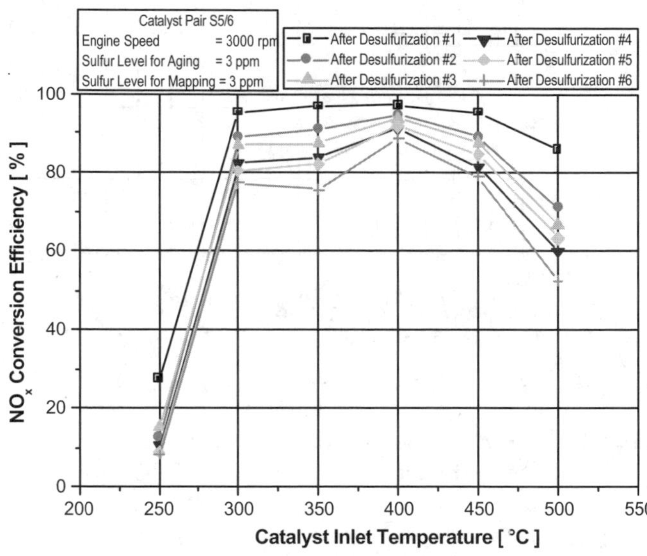

Figure 11: Conversion Efficiency Results After Desulfurization in Part I

The data from Part I must be considered in conjunction with the companion data set from Part II where the aging was conducted with the higher fuel sulfur level. These results are shown in Figure 12. Although, higher contamination rates could be expected using the higher sulfur level fuel for aging, the data in Figure 12 are nearly identical to the after desulfurization conversion efficiencies shown in Figure 11. To visualize the differences between the test cycles the data must be presented in another form, as shown in Figure 13, that includes both the after aging and after desulfurization NO_x conversion efficiencies for both catalyst pairs at a constant catalyst inlet temperature of 350 °C.

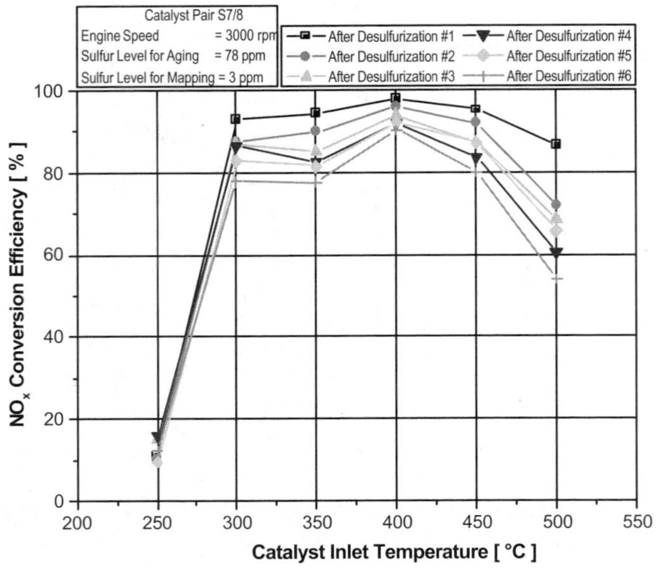

Figure 12: Conversion Efficiency Results After Desulfurization in Part II

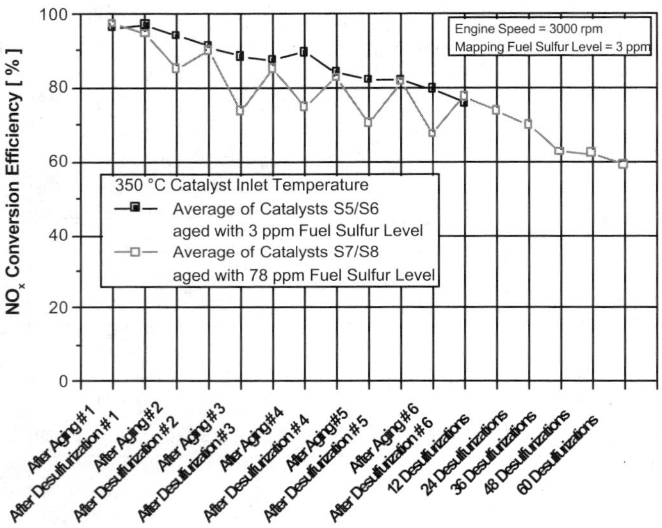

Figure 13: Conversion efficiencies for Parts I and II at a catalyst inlet temperature of 350 °C

In Figure 13, a "saw-tooth" shape resulting from the consecutive aging and desulfurization events can be observed due to the sulfur level used during the aging processes for catalysts S7/8. The drops in efficiency observed after the aging periods indicate that sufficient sulfur is being adsorbed onto the trap to decrease the NO_x conversion efficiency, and subsequently, the increase in conversion efficiency after desulfurization demonstrates to what extent the desulfurization strategy developed is able to restore the NO_x regeneration effectiveness.

Testing under Part II was extended at the end of the project relating to the decrease in conversion efficiency as a function of time. The intention of the tests conducted under the extension was to evaluate the influence of many desulfurization events on the effectiveness of these particular NO_x adsorber catalysts.

Therefore, after completing the mapping after Desulfurization #6, testing continued by conducting twelve (12) consecutive desulfurization events and re-mapping. This process was repeated four (4) times to add the five (5) additional data points to the curve for catalysts S7/8 in Figure 13.

Each twelve (12) desulfurization cycle represents about one hour at the desulfurization temperature. Therefore, at the end of Part II, catalysts S7/8 had accumulated approximately 5½ hours at high temperature. Over this number of desulfurization events, the conversion efficiency does not reach a stable value, but continues to decline.

There are some conclusions to be made regarding comparison of the data obtained under Parts I and II to investigate long-term desulfurization effects. Most importantly, the NO_x conversion efficiencies measured in each part return to similar levels following desulfurization, independent of the sulfur level used to conduct the aging. This phenomenon is shown in Figure 13 for a catalyst inlet temperature of 350 °C. Each catalyst pair begins with NO_x conversion efficiency >95 % and exhibits slow reduction with time. After desulfurization #6, the average absolute conversion efficiency remains within a few percent for each catalyst pair at just under 80 %.

Unfortunately, the desulfurization strategy developed, in application to the specific catalyst formulations tested (1998 status technology), continually reduced the NO_x conversion efficiency. It remains for future investigations to determine if an alternative desulfurization approach would prevent this or if additional catalyst development will provide catalysts with enhanced durability or sulfur release mechanisms under less severe conditions.

CONCLUSION

A desulfurization procedure was developed and demonstrated for a single engine operating point corresponding to a catalyst inlet temperature of 400 °C. NO_x adsorber catalysts previously aged for various lengths with fuel sulfur levels of 3, 16, and 30 ppm, exhibited excellent NO_x conversion efficiency after being exposed to a single desulfurization event. The procedure utilized the features of the engine in combination with tailored heterogeneous catalysis to prepare the conditions necessary for sulfur release.

The primary conclusions stemming from these tests concerning desulfurization of NO_x adsorber catalysts on diesel engines include:

- A desulfurization procedure can be developed that regenerates sulfur from a NO_x adsorber catalyst during operation on an HSDI diesel engine

- Catalysts aged with fuel sulfur levels up to 30 ppm and for up to 250 hours can be returned to NO_x conversion efficiency levels similar to unpoisoned catalyst using only one desulfurization event
- While some drivability issues remain, particularly transient response, the general boundary conditions of the desulfurization event indicate that the strategy is promising for vehicle integration

The long-term effects of repeated aging and desulfurization were also investigated. Although, the testing was of a limited nature, important conclusions could be drawn. The most important is that with repeated aging and desulfurization, the NO_x conversion efficiency continually decreased. Observation of the post-desulfurization conversion efficiency indicates that this decrease is independent of the fuel sulfur level.

ACKNOWLEDGMENTS

The authors gratefully acknowledge the contributions of the Diesel Emission Control – Sulfur Effects Program NO_x adsorber technical team including:

Wendy Clark	National Renewable Energy Lab
Brian West	Oak Ridge National Lab
Yi Liu	Detroit Diesel

This work was supported by the Department of Energy with the sponsoring representatives being Steve Goguen and John Garbak and also by the Manufacturers of Emission Controls Association.

The authors would also like to thank Andreas Ervens, Philipp Loebbert, and Sven Einhert who operated the engine and collected the data summarized herein. Their commitment and attention to detail improved the program and allowed the testing to be completed on schedule.

REFERENCES

1. Parks, J.E., G.J. Wagner, W.E. Epling, M.W. Sanders, and L.E. Campbell, "NO_x Sorbate Catalyst System with Sulfur Catalyst Protection for Aftertreatment of No. 2 Diesel Exhaust", SAE Paper 1999-01-3557, 1999.
2. Luders, H., P. Stommel, and S. Geckler, "Diesel Exhaust Aftertreatment – New Approaches to Ultra Low Emission Diesel Vehicles", SAE Paper 1999-01-0108, 1999.
3. Asanuma, T., S. Takeshima, T. Yamashita, T. Tanaka, T. Murai, and S. Iguchi, "Influence of Sulfur Concentration in Gasoline on NO_x Storage – Reduction Catalyst", SAE Paper 1999-01-3501, 1999.
4. Guyon, M., F. Blejean, C. Bert, and Ph. Le Faou, "Impact of Sulfur on NO_x Trap Catalyst Activity – Study of Regeneration Conditions", SAE Paper 982607, 1998.
5. Muller, W., W. Strehlau, J. Hoehne, A. Okumura, U. Gobel, W. Lox, and M. Hori, "Durability Aspects of NO_x Adsorption Catalysts for Direct Injection Gasoline Vehicles with Regard to European Application", SAE Paper 1999-01-1285, 1999.
6. Diesel Emission Control – Sulfur Effects (DECSE) Program, "Phase I Interim Data Report No. 2: NO_x Adsorber Catalysts", U. S. Department of Energy, 1999.

2006-01-0211

Study of 2-LEG NOx Storage-Reduction Catalyst System for HD Diesel Engine

Ichiro Tsumagari, Hiroshi Hirabayashi, Yoshihide Takenaka, Mitsuru Hosoya and Masatoshi Shimoda
Hino Motors, Ltd.

ABSTRACT

A 2-LEG NOx Storage-Reduction (NSR) catalyst system is one of potential after-treatment technology to meet stringent NOx and PM emissions standards as Post New Long Term (Japanese 2009 regulation) and US'10.

Concerning NOx reduction using NSR catalyst, a secondary fuel injection is necessary to make fuel-rich exhaust condition during the NOx reduction, and causes its fuel penalty. Since fuel injected in the high-temperature (~250 degrees Celsius) exhaust instantly reacts with oxygen in common diesel exhaust, the proportion of fuel consumption to reduce the NOx stored on NSR catalyst is relatively small. A 2-LEG NSR catalyst system has the decreasing exhaust flow mechanism during NOx reduction, and the potential to improve the NOx reduction and fuel penalty. Therefore, this paper studies the 2-LEG NSR catalyst system. The after-treatment system consists of NSR catalysts, a secondary fuel injection system, flow controlled valves and a Catalyzed Diesel Particulate Filter (CDPF).

The relationship between NOx reductions and fuel penalties is investigated on several steady-state engine operating points using a 7.685 liter medium-heavy-duty diesel engine. The 2-LEG system achieves high efficiency NOx reduction compared with the single flow NSR system, and has the significant advantage of NOx reduction in high-exhaust flow conditions.

INTRODUCTION

Recently, drastic PM and NOx reductions in diesel exhaust gas are desired in the world. So, emissions levels have been strictly regulated in Japan. To decrease these emissions, new technologies such as cooled Exhaust Gas Recirculation (EGR), high pressure common-rail injection system and Diesel Particulate active Reduction system (DPR)™ using Catalyzed Diesel Particulate filter (CDPF)[1, 2] have been developed. However, the regulations are tightened in future. NOx reduction more than 65% is necessary to achieve Japanese Post New Long Term (2009~) emissions standard; NOx<0.7(g/kWh), PM<0.01 (g/kWh), as shown in Figure 1. Further NOx reduction is required for US'10 emission standard; NOx<0.3(g/kWh). A NSR catalyst which is one of a NOx reduction after-treatment device has been studied [3-6]. In particular, Multi-Path (2-LEG) systems using NSR catalysts have been studied and achieved good performance of NOx reduction [7-13].

This paper studies the potential of a 2-LEG NSR catalyst after-treatment system from a viewpoint of the oxygen. The differences of the performance on NOx reduction and fuel penalty between 2-LEG and single systems are investigated.

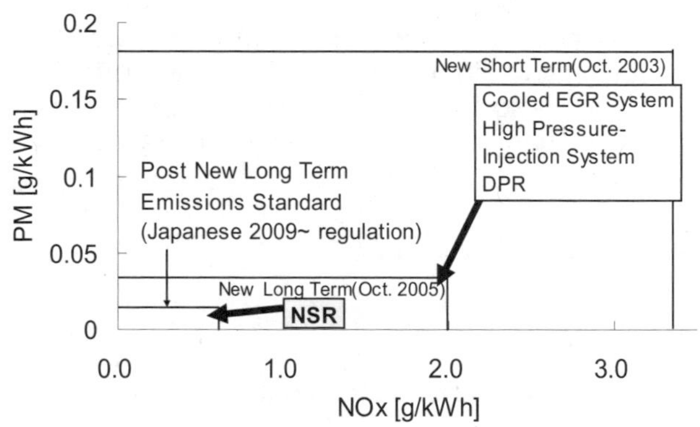

Figure 1: Japanese NOx and PM Emissions Standards and Emissions Reduction Technologies

NOx STORAGE-REDUCTION CATALYST

Figure 2 shows the reaction mechanism in the NSR catalyst [14-16]. Under fuel-lean conditions, NOx is oxidized on the precious metals and reacts on the NOx storage components to form mainly nitrates. Thereafter, under fuel-rich of stoichiometric conditions, nitrates on the catalyst surface are reduced to nitrogen. Generally, the diesel exhaust condition is fuel-lean, so it is necessary that fuel is injected into exhaust in the pipe before the NSR catalyst to achieve fuel-rich state for reducing NOx. Therefore, fuel is consumed much to reduce NOx stored on a NSR catalyst especially for diesel engine.

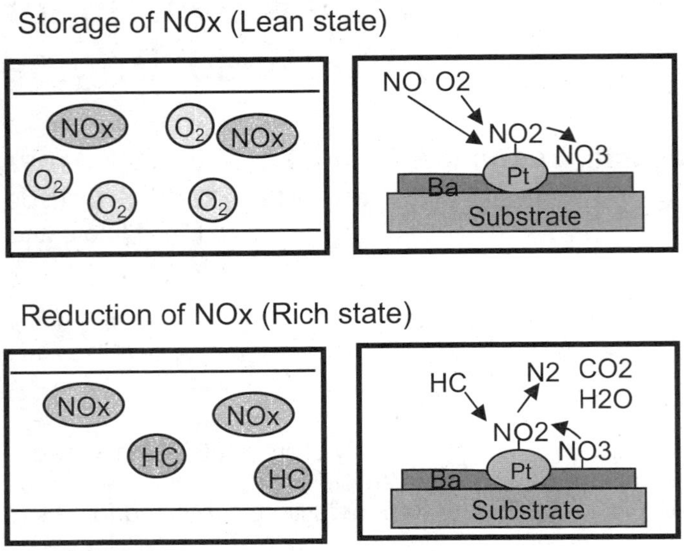

Figure 2: Reaction mechanism in NSR

SUBJECT OF NSR SYSTEM

Figure 3 shows the result of NOx reduction and fuel penalty with our single flow architecture of NSR catalyst system over a Japanese Emission 2005 transient cycle (JE05). In particular, NOx reduction efficiency is not enough to achieve our target for Post New Long Term emissions standard. And the fuel penalty must be improved for fuel economy. So, the two improved points of NOx reduction and fuel penalty must be studied to achieve the future standards of Post New Long Term emissions and also US'10. A NSR catalyst system for diesel engine has about 4 improved points described below.

1) NOx reduction efficiency

2) Fuel penalty

3) Durability (with desulfurization control)

4) Cost

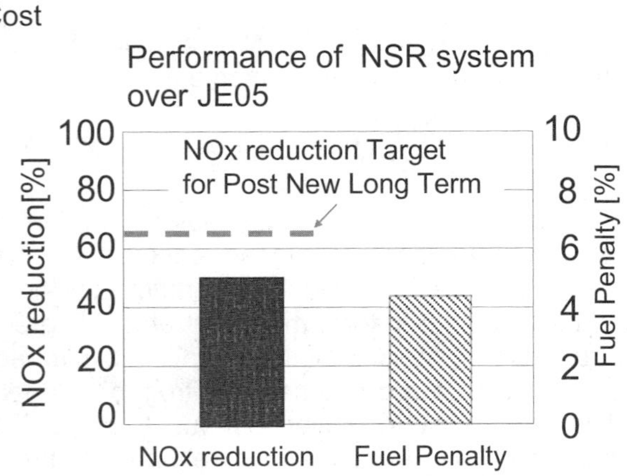

Figure 3: NOx reduction and fuel penalty with single NSR system over JE05.

Figure 4 shows the proportion of fuel injected consumption measured with a single NSR catalyst system with approximately 90% NOx reduction and about 300 degrees Celsius on catalyst surface. Since NOx reduction occurs in fuel-rich state, amount of fuel injected into exhaust is necessary to react with a large quantity of oxygen in diesel exhaust. Then, if fuel is injected effectively, it is expected that the improvement of fuel penalty is possible.

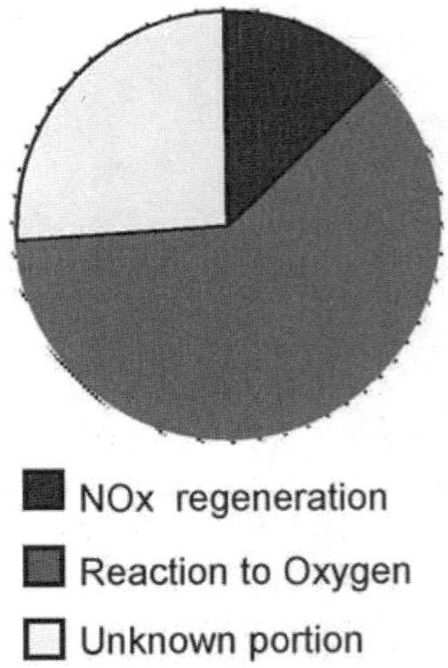

Figure 4: Proportion of fuel consumption

2-LEG NOx STORAGE-REDUCTION CATALYST SYSTEM

CONCEPT OF 2-LEG NSR SYSTEM

1. A quantity of oxygen should be decreased to improve fuel penalty.

A quantity of fuel injected into exhaust to be fuel-rich state is due to oxygen quantity at a NSR catalyst. So exhaust flow limited using the valve at one of 2-LEG exhaust pipes deceases a quantity of oxygen.

2. Exhaust flow should be decreased to increase a reaction-time of NOx reduction.

The improvements of NOx reduction and HydroCarbons (HC) slippage are expected under long reaction-time which provided the sufficient time of the NOx reduction and the fuel oxidation. Figure 5 shows a time-series illustration of the oxygen concentration profile as ideal reaction downstream of NSR catalyst during fuel injected into exhaust. These effects can be confirmed by considering the length of reaction-time and minimum of oxygen concentration.

Investigations of 2-LEG system have been carried out [7-13], and have found that high-efficiency of NOx reduction is obtained under less fuel. However, in these previous works, it has not been investigated enough the effects of oxygen flow and concentration compared with single flow NSR catalyst system as the NOx reductions and fuel penalties.

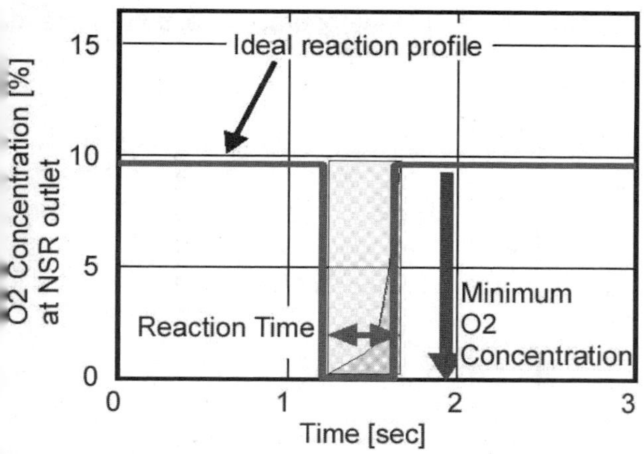

Figure 5: O2 concentration during fuel injection

DEPENDENCY OF MINIMUM OXYGEN CONCENTRATION AND R-TIME FOR NOX REDUCTION

Figure 6 shows relationship between NOx reduction and minimum O2 concentration. Oxygen concentration falls as increase of fuel injected in exhaust. It found that the concentration less than 3.8% condition is an active point of NOx reduction. Continuous time during O2 concentration of less than about 3.8% is defined as R-time. Figure 7 shows relationship between NOx reduction and R-time. It is considered that R-time is the effective NOx reduction time. So the efficiency improvement of NOx reduction needs both decrease of minimum of O2 concentration and extension of R-time. Therefore, it is important to reduce exhaust O2 flow in NSR catalyst. Consequently, 2-LEG NSR catalyst system using alternative flow control to reduce oxygen flow is adopted to improve NOx reduction efficiency and fuel penalty.

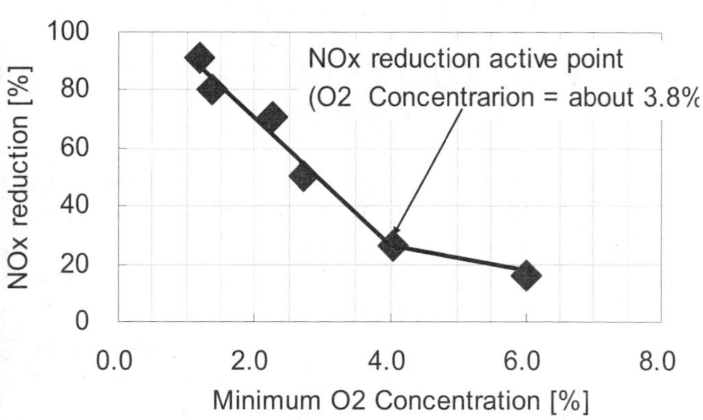

Figure 6: Relation between reached O2 concentration and NOx reduction

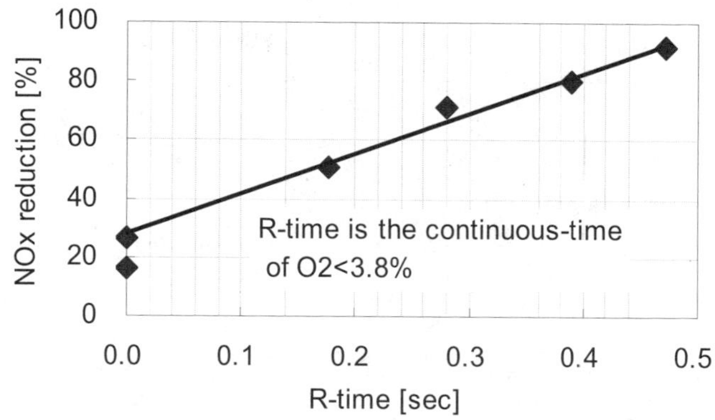

Figure 7: Relation between R-time and NOx reduction

EXPERIMENT

2-LEG NSR SYSTEM

Table 1 shows the engine specifications used in this study. Exchange valves with DC motor that could adjust an angle by controller voltage were installed in front of each 2-LEG NSR catalyst. Fuel injectors were installed between the exchange valves and NSR catalysts (Figure 8). Fuel Injection is driven at 0.7MPa of fuel supplies pressure to reduce NOx stored on NSR catalyst during exchange valve closed. Fuel injectors have 18 holes nozzle for a good spray distribution and 720 m liter / minute of injection flow rate to manufacture for NSR catalyst systems.

Table 1: Engine specifications

Engine Model	2005 HINO J08E
Displacement (Liters)	7.685
Bore x Stroke (mm)	112 x 130
Cylinder Arrangement	Inline six
Combustion Type	Direct Injection
Aspiration	Turbocharger with an intercooler
Fuel Injection	Common-rail system
EGR	High pressure Cooled system

Table 2 shows the NSR catalysts and CDPF specifications used in this study.

Diesel fuel for engine and injection into exhaust contains less than 10 ppm sulfur. NSR catalyst aged for 20hr using rated power of the engine (450 degrees Celsius).

Table 2: NSR & CDPF specifications

	Type	Catalyst	substrate	Vol. (Liters)
NSR	Flow through	NOx absorption material, Pt	Cordierite	10.4 X 2
CDPF	Wall flow	Pt	Cordierite	8.5

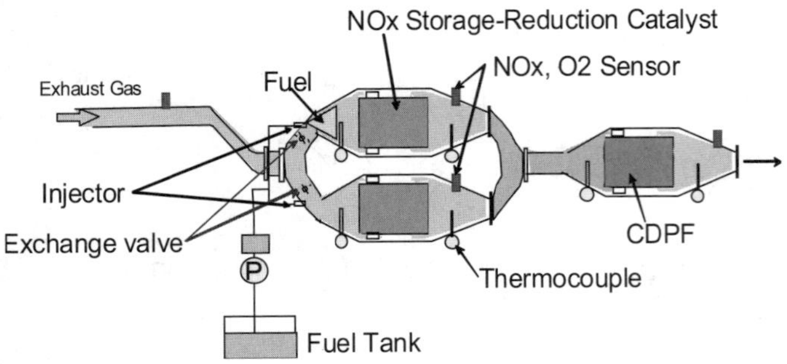

Figure 8: Schematic of 2-LEG NSR catalyst system

Temperature is measured in front and behind of NSR catalysts, and O2 concentration is measured at downstream of catalysts using zirconia- NOx and O2 concentration sensor.

CONTROL OF 2-LEG NSR SYSTEM

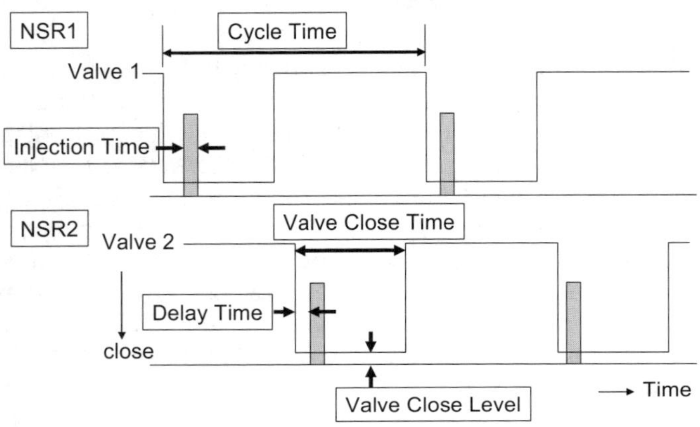

Figure 9: Parameters of 2-LEG NSR system control

Figure 9 shows the exchange valves open/close and fuel injection schedules. The parameters (cycle time, valve closed level, valve closed time, fuel injection time / timing (delay time)) were optimized at each steady state engine operation point. Restrict exhaust flow at one of 2-LEGs is optimized 5~15 % of whole exhaust flow controlled exchange valves. The reduction of NOx stored on NSR catalyst is performed alternately. The injected fuel consumption is mainly determined by cycle time.

RESUTLS AND DISCUSSION

Performance of 2-LEG and Single NSR system are compared for several engine operating points.

MINIMUM O2 CONCENTRATION AND R-TIME

Figure 10 shows time-series plots of the oxygen concentration downstream of the NSR catalyst during fuel injection into exhaust using single NSR system. Figure 11 shows one's plots using 2-LEG NSR system. Experimental conditions are followings, both fuel penalties are same 1.3%, catalyst temperature is 350 (degrees Celsius), and at the operation point 60% speed - 60% load.

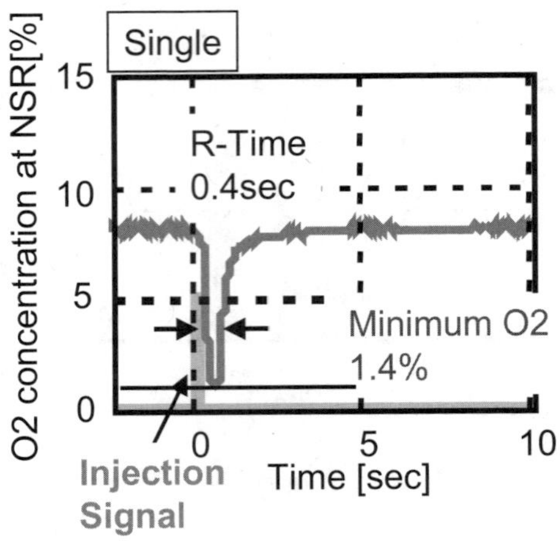

Figure 10: O2 concentration using single flow NSR system

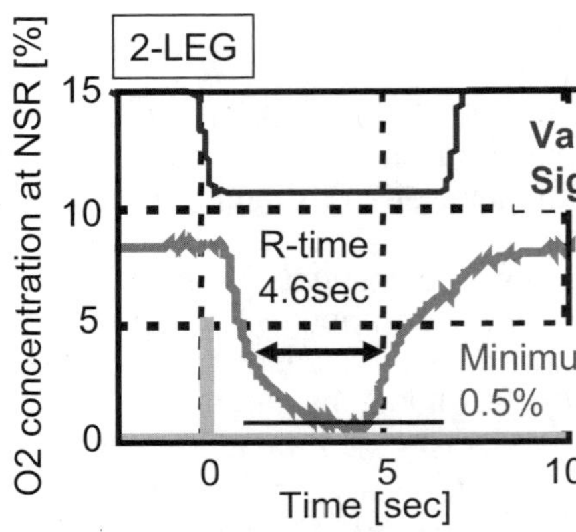

Figure 11: O2 concentration using 2-LEG NSR system

Apparently, R-time of 0.4 seconds using single NSR system is shorter than R-time of 4.6 second using 2-LEG NSR system. By reducing exhaust flow quantity on half of 2-LEG pipes, it is confirmed to decrease minimum O2 concentration and to keep R-time using 2-LEG system.

MINIMUM O2 CONCENTRATION AND R-TIME ON FUEL INJECTION QUANTITIES

Figure 12 shows minimum oxygen concentration between single system and 2-LEG system depends on fuel injection time that is continuous time during power supplying the injector. So, fuel injection time corresponds to amount of fuel injected quantity during fuel injection. Figure 13 shows R-time between of single system and 2-LEG system depends on fuel injection time. Even if quantities of fuel injection increase, 2-LEG NSR system keeps advantages of lower minimum O2 concentration and longer R-time compared with single NSR system. As these results indicate, it is expected to improve efficiency of NOx reduction by using 2-LEG system.

Data of single system were measured under the condition of full exhaust flow using 2-LEG NSR system with two valves opened. The other experimental conditions (i.e., state of engine, durability level of catalyst, spray-distribution and quantity of fuel injection.) between single and 2-LEG NSR system are same on measuring continually

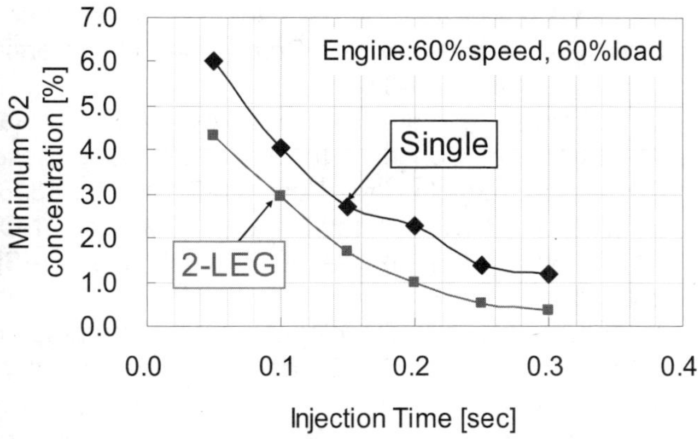

Figure 12: Minimum O2 concentration on fuel injection times

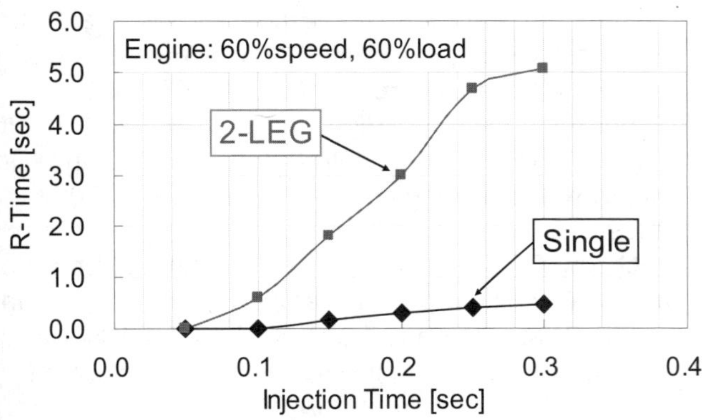

Figure 13: R-time on fuel injection times

IMPROVEMENT OF NOx REDUCTION

Improved NOx reduction on medium engine load

NOx reduction is measured for previous experimental conditions, shown in Figure 14. Oxygen flow quantity is about 45 kg/h on 60%speed-60%load engine operating point. The fuel penalty is defined as the fuel injected in exhaust divided by the fuel consumption of each engine operating point. NOx reduction is calculated as the NOx concentration at CDPF outlet divided by NOx concentration at engine outlet. The data are measured for 10 minutes after stabilizing the temperature of NSR catalysts. Comparing with single system, NOx reduction of 2-LEG NSR system is improved 15.8% in same fuel penalty. Fuel penalty is improved 0.4% in same NOx reduction (approximately 90%). It is confirmed that 2-LEG NSR system has better efficiency of NOx reduction than one of single NSR system for medium engine load.

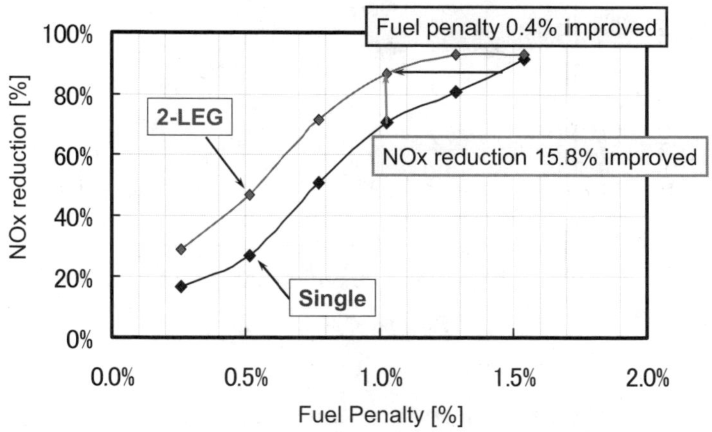

Figure 14: NOx reduction on 60%speed-60%load

Improved NOx reduction on rated power of engine operating point

Under the condition of high exhaust flow (100%speed-100%load of engine operating point); experiment is performed under the condition above exhaust O2 flow quantity of 105 kg/h at catalyst temperature of 425 degrees Celsius. Comparing with single NSR system, NOx reduction of 2-LEG system can be improved about 69% (Figure 15) on 1.35% fuel penalty. In addition, it is difficult to achieve 90% NOx reduction using single NSR system on any fuel penalty.

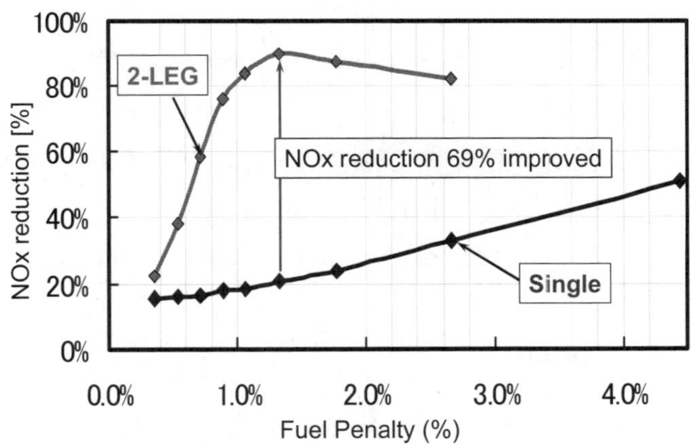

Figure 15: NOx reduction on 100%speed-100%load engine operating point.

NOx reduction for oxygen flow quantity

As above results indicate, it is expected that oxygen flow quantity, that is exhaust flow quantity, affects NOx reduction efficiency. Figure 16 shows NOx reduction on several O2 flow quantities. Plots are about 90% NOx reduction of 2-LEG NSR system and NOx reduction of single NSR system are plotted for same fuel penalty with 2-LEG NSR system. As a result, especially in high-exhaust flow, 2-LEG NSR system has good-advantage of NOx reduction efficiency.

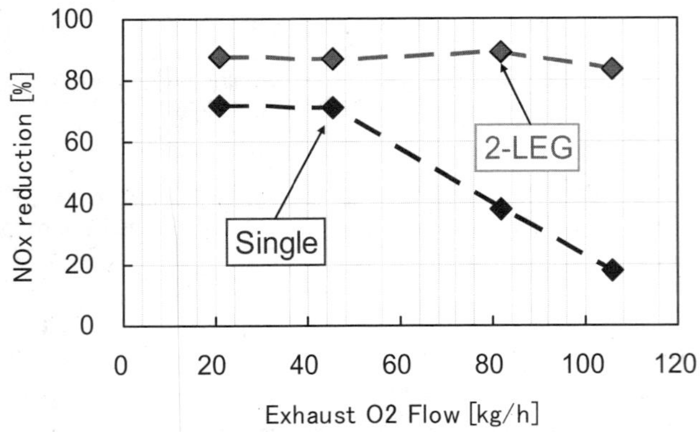

Figure 16: NOx reduction and exhaust O2 flow

Catalyst Volume Reduction

On high exhaust flow condition, 2-LEG NSR system has enough NOx reduction. So, catalyst volume was changed half and measured NOx reduction efficiency. Figure 17 shows 2-LEG NSR system can keep high NOx reduction using 10.4 liter of catalyst.

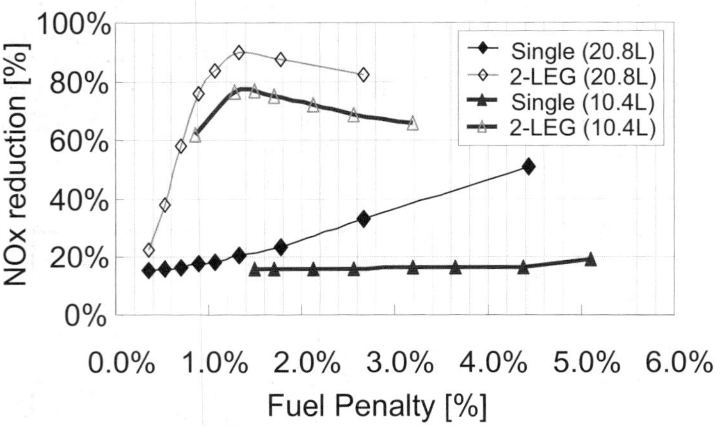

Figure 17: Fuel penalty and NOx reduction on 100%speed-100%load

CONCLUSION

1) It becomes possible to improve NOx reduction and fuel penalty by reducing oxygen flow quantity (gas flow quantity) using 2-LEG NSR catalyst system

2) Especially in high-exhaust flow, 2-LEG NSR system has a good-advantage of NOx reduction efficiency.

3) 2-LEG NSR catalyst system can reduce NSR catalyst volume keeping high NOx reduction efficiency for high load operation condition.

ACKNOWLEDGMENTS

Authors sincerely appreciate the staffs of Hino Motors Ltd. for development of 2-LEG NSR catalyst system.

REFERENCES

1. J. Minamikawa et al. ,"Development of Diesel Particulate active Reduction system", JASE 20035600
2. H. Toorisaka et al. , "DPR Developed for Extremely Low PM Emissions in Production Commercial Vehicles" SAE 2004-01-0824
3. M. Kramer et al. , "Possible Exhaust Gas Aftertreatment Concepts for Passenger Car Diesel Engines with Sulphur-free Fuel", SAE 1999-01-1328
4. S. Hirota et al. ,"Simultaneous PM and NOx Reduction System in Diesel Exhaust Gas", JASE 20015186
5. N. Kobayashi et al. ,"Development of a Simultaneous Reduction System of NOx and Particulate Matter", JSAE 20035129
6. T. Asanuma et al. , "Effect of Sulfur-free and Aromatics-free Diesel Fuel on Vehicle Exhaust Emissions using Simultaneous PM and NOx Reduction System", SAE 2003-01-186
7. Schenk, C., et al, "High-Efficiency NOx and PM Exhaust Emission Control for Heavy-Duty On-Highway Diesel Engines", SAE 2001-01-1351
8. Schenk, C., et al, "High-Efficiency NOx and PM Exhaust Emission Control for Heavy-Duty On-Highway Diesel Engines –Part Two", SAE 2001-01-3619
9. Jackson, C., et al, "Comparison of Exhaust Emission from Application of the Ramped Modal Cycle and Steady-State Nonroad Test", SAE 2005-01-1615
10. Webb, C., et al, "Achieving Tier 2 Bin 5 Emission Levels with a Medium Duty Diesel Pick-Up and a NOx Absorber, Diesel Particulate Filter Emissions System-Exhaust Gas Temperature Management", SAE 2004-01-0584
11. Whitacre, S, et al, "Systems Approach to meeting EPA 2010 Heavy-Duty Emission Standards Using a NOx Absorber Catalyst and Diesel Particle Filter on a 15L Engine", SAE 2004-01-0587
12. Kong, Y., et al, "NOx Trap Regeneration with an ON-Board Hydrogen Generation Device", SAE 2004-01-0582
13. S. Watanabe, et al, "An Improvement of Diesel PM and NOx Reduction System", 2005 Diesel Engine Emissions Reduction Conference (DEER), session8, http://www.eere.energy.gov/vehiclesandfuels/pdfs/deer_2005/session8/2005_deer_watanabe.pdf
14. N. Miyoshi et al. ,"Development of New Concept Three-Way Catalyst for automotive Lean-burn Engines ",SAE Paper NO. 950809
15. K. Katou, et al. "Development of NOx Storage-Reduction Three-way Catalyst 1st Report ", Jidousya Gijyutukai Ronbunsyu 9630705
16. T. Tanaka et al. "Development of NOx Storage – Reduction Three-way Catalyst 2nd Report ", Jidousya Gijyutukai Ronbunsyu 9540958

2007-01-0239

Study on Low NO_x Emission Control Using Newly Developed Lean NO_x Catalyst for Diesel Engines

Tomoko Morita, Norio Suzuki, Naohiro Satoh, Katsuji Wada and Hiroshi Ohno
Honda R&D Co., Ltd.

ABSTRACT

In recent years, emission regulations have become more stringent as a result of increased environmental awareness in each region of the world. For lean-burn diesel engines, since it is not possible to use three-way catalytic converters, reducing NO_x emissions is a difficult technical challenge.

To respond to these strict regulations, an exhaust gas aftertreatment system was developed, featuring a lean NO_x catalyst (LNC) that uses a new chemical reaction mechanism to reduce NO_x.

The feature of the new LNC is the way it reduces NO_x through an NH_3-selective catalytic reduction (SCR), in which NO_x adsorbed in the lean mixture condition is converted to NH_3 in the rich mixture condition and reduced in the following lean mixture condition. Thus, the new system allows more efficient reduction of NO_x than its conventional counterparts. However, an appropriate switching control between lean and rich mixture conditions along with compensation for catalyst deterioration was necessary.

As a result, this study established a method of regulating the supply of reductant to the catalyst by monitoring the air-fuel ratio of exhaust gases and the amount of intake air, and simultaneously realized a lower fuel penalty and emission cleanliness.

1. INTRODUCTION

Although the modern diesel engine has gained a significant market share, particularly in the EU, due to its excellent fuel economy and high torque generation, there remain the issues of reducing particulate matter (PM) and decreasing NO_x in oxygen-rich conditions, issues which are incompatible. Considering BIN5 emission regulation in the US and EURO VI, due to be seriously discussed, a new technology is called for in order to minimize fuel penalty and further reduce emissions.

As shown in Figure 1, several new methods for reducing NO_x have been investigated and put into practical use, such as Urea-SCR, HC-SCR and NO_x storage-reduction. The Urea-SCR is spreading mainly among heavy duty diesels, but application to passenger vehicles has yet to progress due to equipping and infrastructure problems. On the other hand, for HC-SCR, the issue lies in the restricted temperature range where high efficiency is attainable. With the NO_x storage-reduction method, though considered an effective method for passenger vehicles, technical issues are low NO_x reduction efficiency in the low temperature range, development of a method of ensuring a rich air-fuel ratio, and fuel penalty.

This study focused on a new exhaust gas aftertreatment system using a newly developed LNC catalytic reaction mechanism, which should resolve the above issues.

	System configuration	Issues
Urea-SCR	Urea 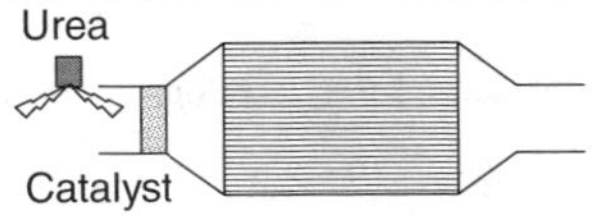 Catalyst $4NO + 4NH_3 + O_2 \rightarrow 4N_2 + 6H_2O$ $6NO_2 + 8NH_3 \rightarrow 7N_2 + 12H_2O$ $4NH_3 + 2NO + 2NO_2 \rightarrow 4N_2 + 6H_2O$	· Establishment of new infrastructure · Equipping · Freezing of urea-water
HC-SCR	HC (Fuel) 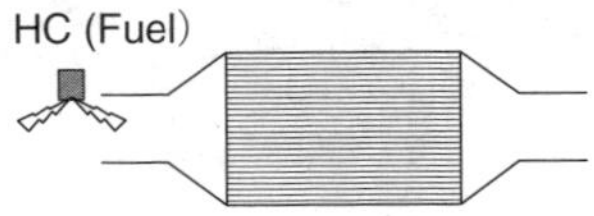 $4NO + C_3H_8 + 2O_2$ $\rightarrow 2N_2 + 3CO_2 + 4H_2O$	· Restricted temperature range where high NO_x reduction efficiency can be achieved
NO_x Storage-reduction	Rich air-fuel ratio 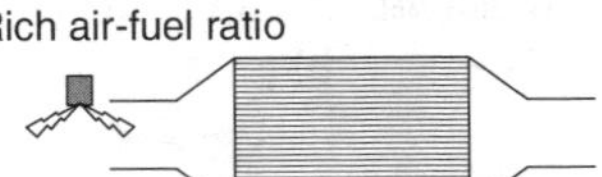 Lean: $NO + 1/2O_2 + MO \rightarrow MNO_3$ Rich: $MNO_3 + 2CO \rightarrow MO + 2CO_2 + 1/2N_2$ $MNO_3 + 2H_2 \rightarrow MO + 2H_2O + 1/2N_2$ $9/2MNO_3 + C_3H_6$ $\rightarrow 3CO_2 + 3H_2O + 9/2MO + 9/4N_2$	· NO_x reduction efficiency at low temperatures · Technology ensuring rich mixture · Fuel penalty

Figure 1 Conventional NO_x reduction technology[1], [2], [3]

2. NEW LNC DEVELOPMENT CONCEPT

Figure 2 shows overviews of the conventional methods and the chemical reaction mechanisms for the new concept. In the conventional NOx storage-reduction catalyst, NO is oxidized to NO_x in the lean mixture condition and stored by NO_x storage material. In the ensuing rich mixture condition, NO_x reacts with reductants such as HC and CO to form N_2. In Urea-SCR, urea dosed by a urea dosing unit is hydrolized to NH_3. The NH_3 then reacts with NO_x on an SCR catalyst to form N_2. In contrast with these methods, in developing the new LNC, attention was paid to self-generating NH_3 through NO_x adsorption during the rich mixture condition and making effective use of this.

More specifically, NO_x undergoes selective catalytic reduction with NH_3 to generate N_2 in the lean mixture condition, a reaction used in urea-SCR. Thus, catalysts that would adsorb NH_3 in the rich mixture condition and then emit it in the lean mixture condition to promote the NH_3-SCR reaction were examined, along with a control system for the process.

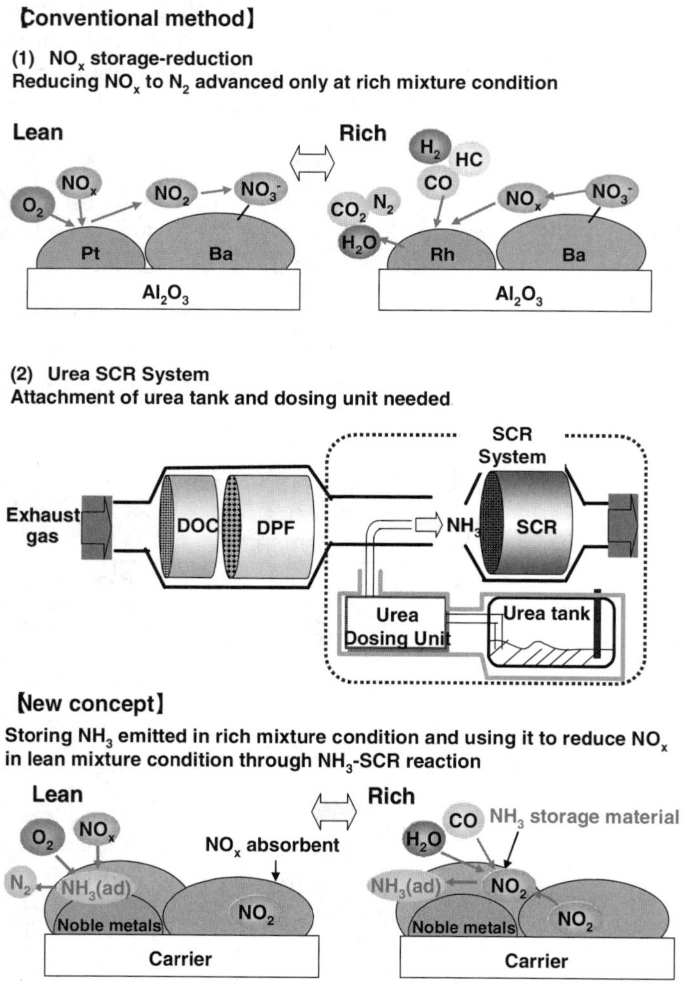

Figure 2 Comparison between the new concept and conventional methods[4], [5]

3. LNC SYSTEM CONFIGURATION

Figure 3 shows the catalyst structure adopted for the new LNC concept. The catalyst consists of two layers; the lower features platinum supported on CeO_2 material, while the upper layer has a transition metal ion-exchanged zeolite as a solid acid.

The two layers each play significant roles in promoting the reaction in the new concept in both lean and rich mixture conditions. The lower layer promotes adsorption of NO_x in the lean mixture condition and the conversion of NO_x to NH_3 in the rich. The upper layer ensures NH_3 storage in the rich mixture condition and the NH_3-SCR reaction in the lean.

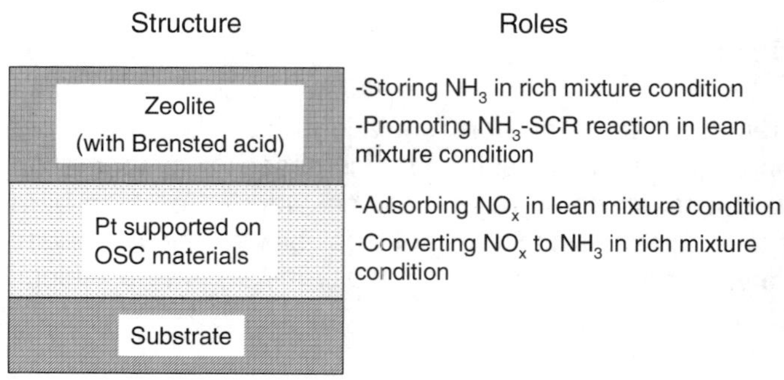

Figure 3 LNC system structure[6]

3.1. UPPER LAYER MATERIAL PROPERTIES

As the material for the upper layer, the well-known NH_3 storage material zeolite was selected[7]. Zeolite features the two acid sites, Brønsted acid and Lewis acid, and efficiently stores the alkaline NH_3. However, the material has the shortcoming of poor hydrothermal durability, so thermal durability was improved through metal modification, resulting in a material capable of sufficiently storing NH_3 in a temperature range of 150 to 300 °C.

3.2. LOWER LAYER MATERIAL PROPERTIES

In the NO_x storage catalyst featured in conventional vehicles, mainly alkaline and alkaline-earth metals have been used as the NO_x storage materials. Development was performed focusing on CeO_2-based materials of rare earth oxides, as a possible material for NO_x adsorption at low temperatures [8]. Platinum supported on the material promotes oxidation of NO to NO_2, resulting in increased NO_x adsorption.

As described above, the lower layer also functions as an NH_3 generator. NH_3 generation occurs only when H_2 is generated during the water gas shift reaction between CO and H_2O. With an alkaline or alkaline-earth metal, the reaction takes place at a high temperature due to degraded activation of the noble metals, making it impossible to ensure effective generation of NH_3. However, the Pt supported on CeO_2 developed in this study ensures excellent water gas shift reaction performance even at low temperatures. Pt supported on CeO_2 can be said to feature the double functions of NO_x adsorption and NH_3 generation.

4. REACTION MECHANISM

The behavior of N_2 generation was investigated using a mass spectrometer in order to better understand the NO_x reduction mechanism. The composition of the model gas used for the experiment is shown in Table 1, while the pattern of N_2 generation is depicted in Figure 4. A large amount of N_2 was found to be generated, especially when the air-fuel ratio was lean. The graph provides evidence that the NH_3-SCR reaction takes place in the lean mixture condition, as aimed for in the new LNC system concept.

Table 1 Model gas for N_2 generation pattern

	Lean	Rich
O_2	6%	-
H_2	-	2.5%
NO	500ppm	-
He	Bal.	Bal.
Time	55 sec	5 sec

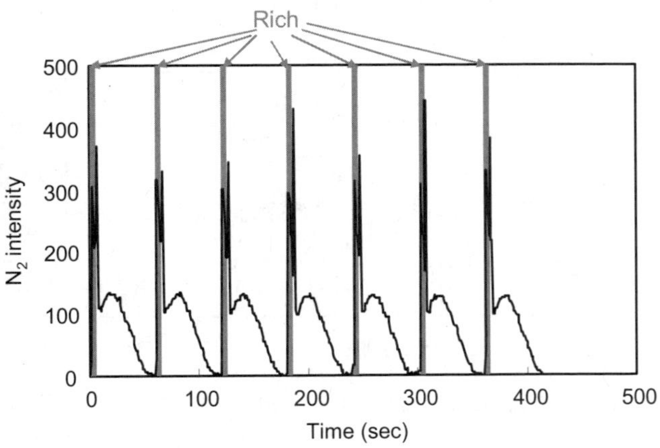

Figure 4 N_2 generation pattern in lean/rich air-fuel ratio conditions (Temp.: 200 °C)

The reaction mechanism of the new LNC is shown in detail in Figure 5.

Step One: In the lean mixture condition, the CeO_2 material adsorbs NO_x. In this case, the material also adsorbs NO, but adsorption becomes more efficient when NO is oxidized to NO_2 by Pt (a).

Step Two: In the rich mixture condition, the NO_x adsorbed in the previous step is desorbed and forms NH_3 through reactions with H_2 on the Pt. At this time, the H_2 used as a reductant is derived from the H_2 in the exhaust gas and that generated by the water gas shift reaction between CO and H_2O. The NH_3 is then stored by the solid acid sites on the zeolite (b).

Step Three: When the condition inside the catalytic converter becomes lean again, the NH_3 stored by the zeolite reacts with gaseous NO_x and O_2 in an NH_3-SCR reaction and, as shown in Figure 4, N_2 is generated and emitted (c). At this time, reactions (a) and (c) occur simultaneously in the lean mixture condition.

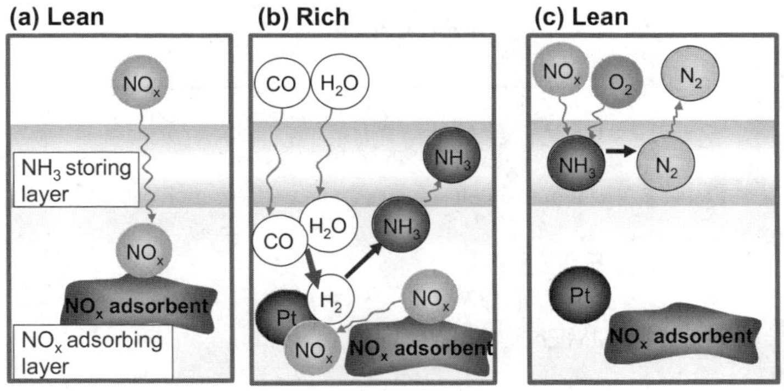

Figure 5 Reaction mechanism

5. LABORATORY TEST RESULTS

A laboratory test was conducted using a test piece with a simulated age of 120 kmiles of usage. The aging condition was set up in consideration of the time exposed to high temperatures due to such process as DPF regeneration and sulfur purge. The test conditions are shown in Table 2. The space velocity (GHSV) was $50000h^{-1}$ and one lean/rich cycle was set at 55 sec/5 sec.

Table 2 Model gas components

	Lean	Rich
O_2	6%	0%
CO_2	6%	6%
C_3H_6	500 ppmC	500 ppmC
CO	900 ppm	20000 ppm
NO	110 ppm	110 ppm
N_2	Bal.	Bal.
H_2O	7%	7%
GHSV(h^{-1})	50000	50000

Figure 6 compares NO_x reduction efficiency in the new LNC and the conventional NO_x storage catalyst after ten lean/rich cycles. The conventional system performance was not very high at 300 °C or less since NO_x adsorption is limited at low temperatures. In contrast, in the new LNC, the use of CeO_2 material improves NO_x adsorption and NH_3 generation at low temperatures, resulting in greater NO_x reduction efficiency at 250 °C or less. Thus, the new LNC system is advantageous in diesel vehicles, which have relatively low emission temperatures.

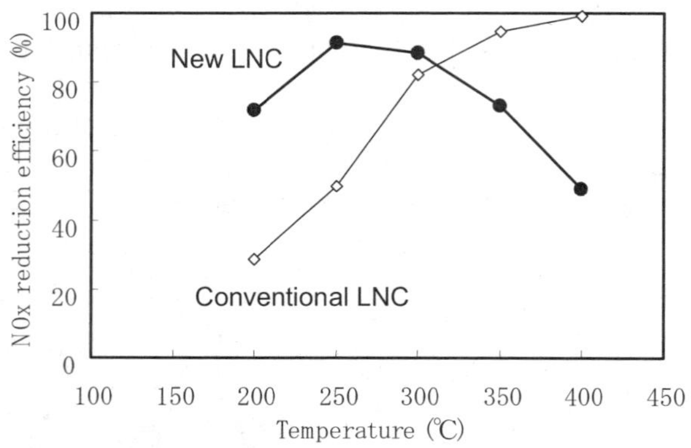

Figure 6 NO_x reduction efficiency in repeated lean/rich cycle testing

6. VEHICLE TEST

6.1. EXPERIMENTAL SYSTEM

The engine used for the test was based on the 2.2-liter direct-injection diesel engine in the European market Honda Accord with modifications made to the exhaust gas aftertreatment system. Table 3 lists the major specifications of the engine. The compression ratio was set at 16.7:1 and fuel was supplied through the common-rail system. Additionally, fuel with a cetane index of 59 was used.

Table 3 Test engine major specifications

Cylinder configuration	Inline 4
Displacement volume [cm^3]	2204
Bore × Stroke [mm]	85 × 97.1
Compression ratio	16.7
Valve mechanism	DOHC 4 valve
Fuel injection system	Common rail system
Turbocharger	Variable nozzle turbo
Swirl ratio	Continuously variable
EGR system	With EGR cooler
Cetane index	59

6.2. EXHAUST GAS AFTERTREATMENT SYSTEM CONFIGURATION

Figure 7 illustrates the exhaust gas aftertreatment system using the new LNC. A diesel particulate filter (DPF) was placed close to the engine out with the aim of removing PM, and immediately in front of that, a diesel oxidation catalyst (DOC) was attached to ensure quick temperature rise at DPF regeneration and reduce CO and HC. The LNC features an underfloor configuration to avoid thermal deterioration.

Three sensors were attached to the test unit. An airflow sensor was placed upstream of the intake manifold to assume the exhaust gas amount, while both upstream and downstream of the DOC, universal exhaust gas oxygen (UEGO) sensors were positioned to monitor the air-fuel ratio.

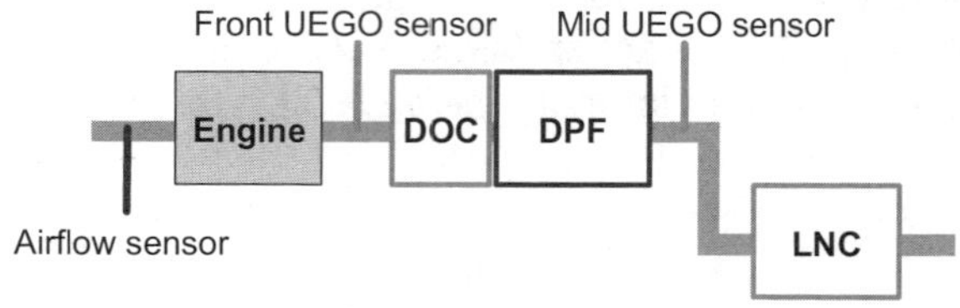

Figure 7 Exhaust gas aftertreatment system configuration

Figure 8 shows the LNC temperature pattern when such a diesel emissions system was used. The drive mode was FTP-75. The LNC catalyst temperature remained 250 °C or lower, and as noted previously in the laboratory test, the new LNC system ensured a relatively high NOX reduction efficiency in this temperature range, suggesting that the new system is appropriate for reducing diesel NOX emissions.

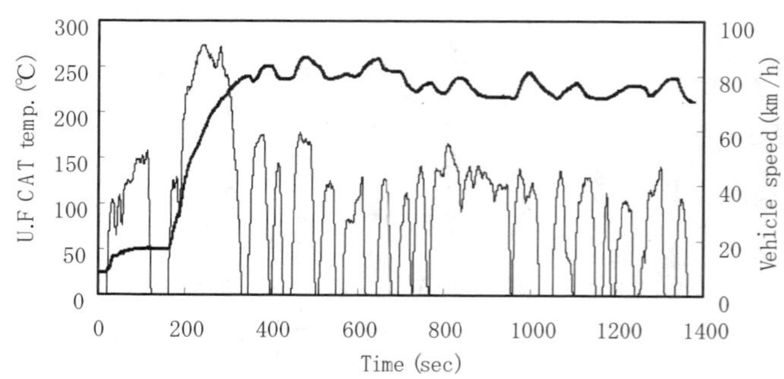

Figure 8 Underfloor catalyst temperature in FTP-75 mode

7. CONTROL METHOD

7.1. INVESTIGATION OF REDUCTANT SUPPLY METHOD

Table 4 shows a method of creating a rich air-fuel ratio. General methods are fuel injection to the exhaust ports, post injection supplying fuel to cylinders after combustion, and rich combustion achieved by controlling the combustion air-fuel ratio. With the first option, a problem is the possibility of thermal deterioration of the catalyst due to sudden temperature increases caused by the direct supply of fuel to the catalytic converter. In post injection, fuel injected may adhere to the cylinder wall, diluting the lubricant and leading to increased fuel consumption and decreased reliability. Rich combustion offers the advantage of greater CO generation, which contributes to NO_x reduction by the LNC, although it raises the technical challenges of increased PM generation and noise and vibration (NV).

Table 4 method ensuring rich air-fuel ratio

method	Issues
Exhaust port fuel injection	Increased catalyst temperature
Post injection	Oil dilution (increased fuel consumption and decreased reliability) Increased HC emissions
Rich combustion	Increased PM and NV

Figure 9 compares the ratios of CO and HC flowing into the LNC during one rich air-fuel ratio control. Assuming that the weight of CO supplied by post injection is 1, during rich combustion, 2.7 times more CO can be supplied to LNC. In contrast, the ratio of HC is lower in rich combustion than in post injection, so that even from the viewpoint of HC emissions, rich combustion can be said to more efficient.

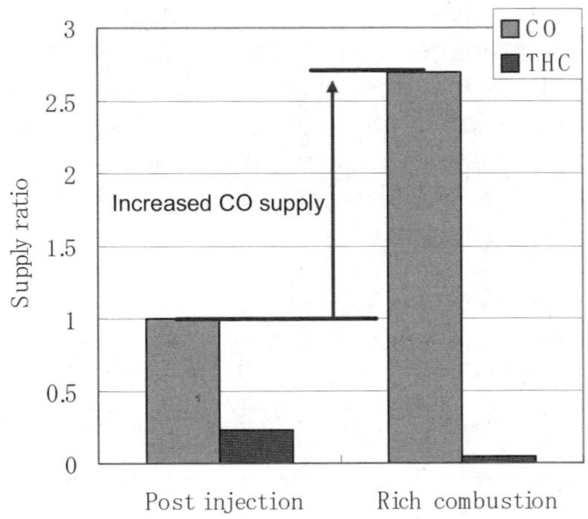

Figure 9 Comparison of CO supply

7.2. INVESTIGATION OF REDUCTANT AMOUNT

Reductant supply amount was investigated in order to ensure NO_x is efficiently reduced using the new LNC.

Figure 10 shows changes in the amounts of NO_x and NH_3 adsorbed by the LNC during lean and rich mixture conditions. When lean combustion began with no substances adsorbed, only the adsorption of NO_x proceeded (a).

When the NO_x adsorption amount reached a certain level, the system switched to the rich mixture condition and the NO_x adsorbed was converted to NH_3, resulting in an increase in the amount of NH_3 stored (b).

In the following lean mixture condition, since NO_x adsorption and the NH_3-SCR reaction between NO_x and the NH_3 stored took place simultaneously, the pace of NO_x adsorption was slower and thus the lean mixture condition period was able to be extended (c).

This extendable duration depends on the amount of stored NH_3 generated during the rich mixture condition. Therefore, a sufficient amount of reductant needs to be delivered at this time. If this amount is insufficient, as shown at the bottom of Figure 10, the amount of NH_3 stored will be reduced, resulting in accelerated NO_x adsorption in the following lean mixture condition, shortening the duration of the lean mixture condition.

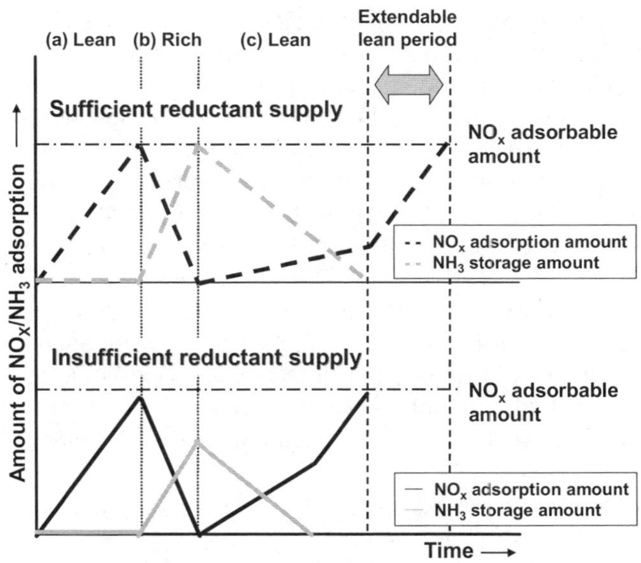

Figure 10 Changes in NO_x and NH_3 adsorption amounts

Figure 11 shows the test results for the optimized amount of inlet reductant according to NO_x emissions. For the test, LNC temperature was 280 °C, NO_x flow in LNC was constant and lean/rich switching operation was conducted for 10 cycles. The amount of NO_x emitted (NO_x emissions after/before LNC) against the amount of reductant was recorded and plotted on the graph. In this case, since the duration of the lean mixture condition was unchanged, the amount of adsorbable NO_x during the lean mixture condition decreased if the amount of reductant supplied was insufficient and as a result, NO_x emissions increased. When the amount of reductant was less than the optimal value, NO_x emissions increased linearly and became constant above the optimal value. The amount of reductant at this threshold was determined to be the required amount.

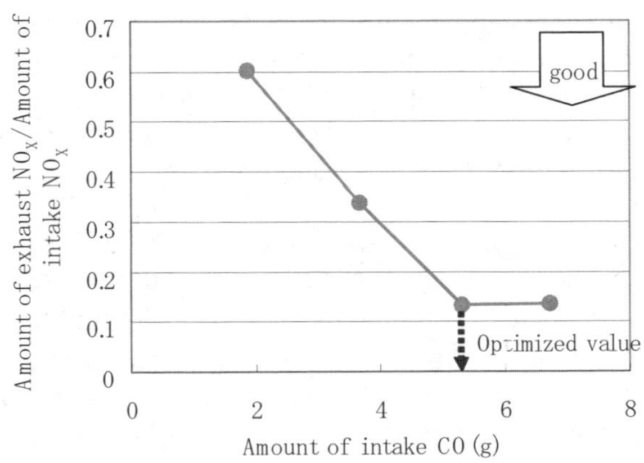

Figure 11 Determination of required reductant amount

7.3. DETERMINATION OF RICH DURATION

The rich mixture condition should have finished when the actual amount of reductant exceeds the required amount. Therefore, a method of estimating the actual amount to suit driving conditions was considered.

The estimate method based on the input from the UEGO and airflow sensors is shown in Figure 12. First, the current air-fuel ratio is deducted from a predetermined air-fuel ratio close to the stoichiometric air-fuel ratio. The result is regarded as an index of the richness of emissions. If the ratio is greater than the stoichiometric ratio of 14.7:1, it is proportional to the amount of reductant. These results are then multiplied by the airflow sensor input and integrated from the beginning of rich operation to estimate the amount of reductant supplied.

Measurements conducted on the actual system are plotted in the graph in Figure 12. The estimated parameter of the reductant amount is proportional to the total actual amount of actual CO and HC. From this relationship, it is possible to estimate the actual amount.

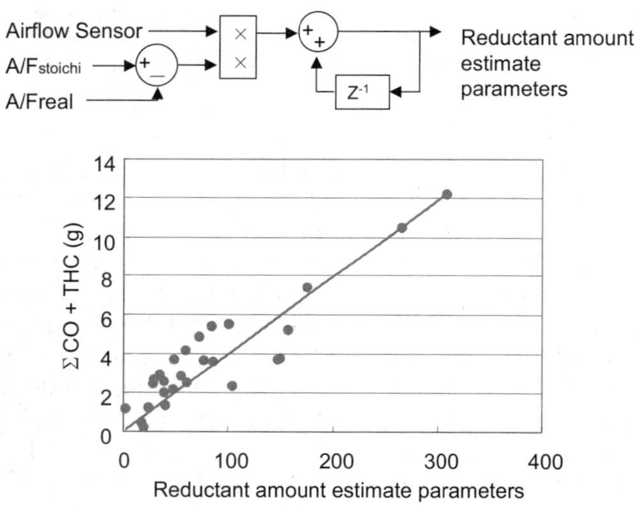

Figure 12 Methods of estimating reductant amount

7.4. CONTROL LOGIC INCLUDING DETERIORATION COMPENSATION

As mentioned above, the rich duration was determined by the reductant at the front of the DOC. However the oxidation efficiency of the DOC changed as a result of deterioration; thus, it is necessary to adjust the required reductant to cope with DOC deterioration.

Figure 13 shows an overview of the control logic determining the lean/rich cycle. The amount of reductant after the DOC (Σmid_reductant) is estimated by the same method as the amount of supply reductant (Σfront reductant) computed in Figure12. Comparing Σfront_reductant with the required amount of reductant, commands are given to change to the lean mixture condition from the rich condition. Additional compensation to cope with DOC deterioration is determined based upon the deviation of Σmid_reductant against Σfront_reductant, which is continuously monitored.

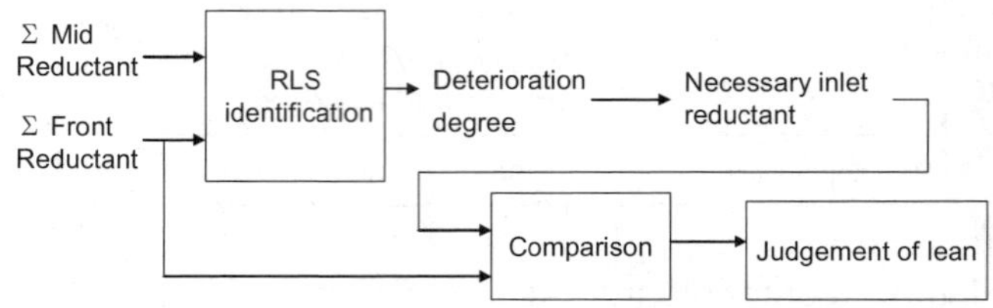

Figure 13 Lean/Rich cycle control logic

Figure 14 shows catalyst thermal aging of 4, 50 and 120 kmiles. Plotting Σmid_reductant / Σfront_reductant over the mileage, the ratio was found to increase with deterioration, which can be attributed to the amount of reductant consumed decreasing as a result of DOC deterioration, resulting in an increased flow of the reductant downstream. The deterioration is continuously monitored and the determination for switching to the lean mixture condition is made at an early stage accordingly, ensuring an approximately 20% reduction in the fuel penalty with a catalyst aged 120 kmiles as compared with one aged 4 kmiles.

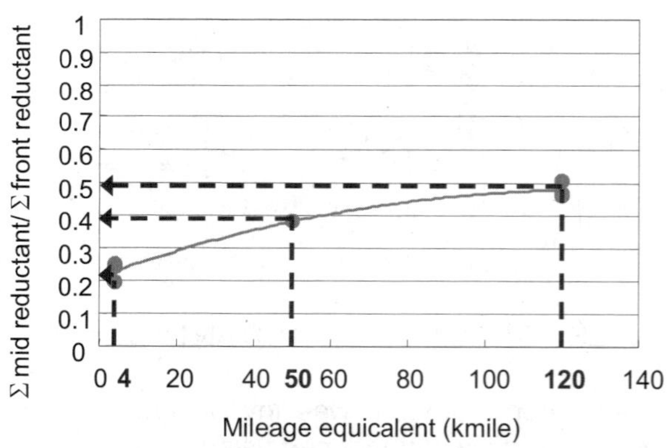

Figure 14 Simulated catalyst deterioration after operation for 4 k, 50 k and 120kmiles

8. VEHICLE TEST RESULTS

Figure 15 shows the NOx reduction efficiency with the new LNC system adopted, in FTP-75 mode. The margin of efficiency was sufficient to achieve emission regulations for CO and HC since the DOC had a strong reducing ability. This was also achieved for PM due to the ability of DPF to remove PM. The emissions target was achieved by optimizing the lean/rich timing in conditions corresponding to 120 kmiles of use. An identical system was applied to a test catalyst simulating 4 kmiles of use. Achieving the target was more difficult with the less-aged catalyst, mainly due to the greater amount of reductant being consumed by the DOC. To cope with this, the timing of switching to the lean mixture condition was compensated for in accordance with the DOC deterioration, as shown in Figure 13, making it possible to reach the target.

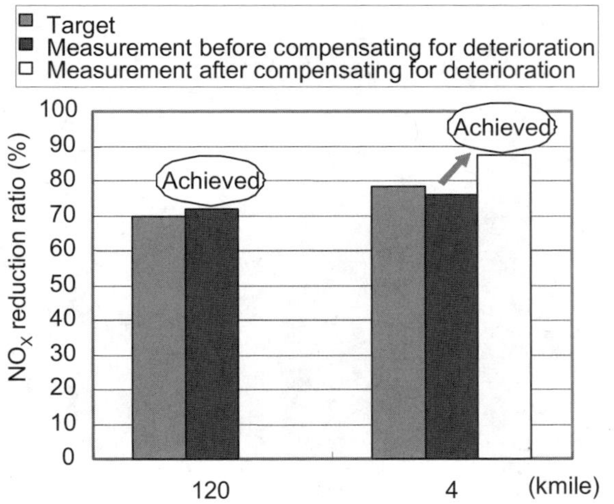

Figure 15 NO_x reduction efficiency in FTP-75 mode

9. CONCLUSION

A new chemical reaction concept was established to create a new LNC designed for diesel engines. In this study, special attention was paid to self-generating ammonia (NH_3) and lean/rich switching control was developed in order to make use of this catalyst. As a result, the target NO_x reduction efficiency was achieved in FTP-75 mode. The results of this research are described below.

1. The new LNC NO_x reduction mechanism was as follows. NO_x was adsorbed during the lean mixture condition and in the ensuing rich mixture condition, it was converted to NH_3 and stored. The stored NH_3 reduced NO_x to N_2 through the SCR reaction occurring in the next lean condition. The effective use of NH_3 minimizes the fuel penalty.
2. The new LNC features a double-layer structure. The upper layer consists of a transition metal ion-exchanged zeolite that stores NH_3, whereas the lower layer consists of Pt supported on CeO_2, infrequently used as an NO_x adsorbent. Thus, the catalyst ensures sufficient NO_x and NH_3 adsorption even at low temperatures and offers high NO_x reduction efficiency in frequently used high temperature regions in FTP-75 mode in a diesel passenger car..
3. In order to make full use of the reaction characteristics of the new LNC, determining the appropriate amount of reductant supplied for a single rich mixture cycle was required. With the duration of the lean mixture condition constant and the amount of reductant varied, the point at which NO_x emissions were lowest and most stable was selected.
4. The target NO_x reduction efficiency was achieved in FTP-75 mode using the control logic mentioned below. For the test, catalysts with thermal aging corresponding to 4 and 120 kmiles of use were used.

- The amount of reductant was estimated from the integration of the results of multiplying the intake air amount with the value achieved by deducting the current air-fuel ratio from the predetermined ratio, which was close to the stoichiometric air-fuel ratio.

- The DOC deterioration was estimated by UEGO sensors upstream and downstream of the system. In accordance with this deterioration estimate, the amount of reductant required for a single rich cycle was compensated for to obtain the optimal value.

In order to fully use the NO_x reduction mechanism offered by the new LNC, ensuring rich mixture conditions through rich combustion is considered optimal, but there is the need to address such issues as reducing PM emissions and sufficiently controlling NV.

REFERENCES

[1]Shimazaki, N. 2003. Practical Application of Premixed Compression Ignition Diesel Engines. *Jidosha Gijutsu Kaishi* (Journal of Society of Automotive Engineers of Japan) Vol. 57, No. 9: 65–70.

[2]Hara, S. 2003. Performance of Urea-SCR System for Heavy-Duty Diesel Engines. *Jidosha Gijutsu Kaishi* (Journal of Society of Automotive Engineers of Japan) Vol. 57, No. 9: 88–93.

[3] Yoshihara, Y. 2005. Diesel Combustion: Formation of Nitrogen Oxides. *Jidosha Gijutsu Kaishi* (Journal of Society of Automotive Engineers of Japan) Vol. 59, No. 4: 70-74.

[4]Collier, T. Brogan, M. Retman, P. and Bye, R. 2003. Development of a Rapid Sulfation Technique and Fundamental Investigations into Desulfation Process. *SAE Paper.* 01-1162

[5]Shimoda, M., Hosoya, M and Takenaka, Y. 2006. Technologies of Aftertreatment System for Heavy-Duty Clean Diesel Engine in the Future. *Jidosha Gijutsu Kaishi* (Journal of Society of Automotive Engineers of Japan) Vol. 60, No. 9: 64-69.

[6]Satoh, N., Ohno, H., and Nakatsuji, T. 2006. A NO_x Reduction System Using Ammonia Storage-Selective Catalytic Reduction in Rich and Lean Operations. *Aachen Colloquium "Automobile and Engine Technology"* 15: 259-270.

[7]Gongshin, Q., and Ralph, Y. 1998. *Appl. Catal. B: Environmental* 15: 107.

[8]Nakatsuji, T., Shimizu, H., and Yasukawa, R. 1994. *Shokubai* 36: 100.

2007-01-1241

Ceria Doped Ba-Alumina Oxides for Durable NOx-Storage Catalysts Efficient at Low Temperature

V. Bellière-Baca, V. Harlé, C. Pitois and E. Rohart
RHODIA Research & Technologies

M. Allain
Electronics and Catalysis

ABSTRACT

The reduction of NOx from Diesel Engines or Lean-burn Gasoline Engines is a major issue in automotive catalysis. Over the last several years, many solutions to remove NOx under lean operating conditions have been considered. Attention is now focused on two main technologies: (i) Selective Catalytic Reduction using ammonia as reductant (urea SCR) and (ii) NOx-Storage Catalyst (NSC).

This paper deals with materials for NOx storage catalysts. Even if the fresh NSC shows high efficiency, its durability is low due to a fast deactivation of the active sites [1,3]. This problem relates to a drastic sintering of the materials after ageing, especially during the regeneration and the desulfation [3-5]. Moreover, the current materials are sensitive to SOx poisoning [1,3,4].

In this paper, modified Ba-alumina oxides used as NOx-Storage materials have been investigated. These oxides contain high Ba loading (such as 20 wt% Ba) and are thermally stable. They present high surface area (more than 100 m^2/g) and available Ba sites for the NOx-Storage process after ageing at 900°C.

The corresponding aged Pt model catalysts show high NOx-Storage capacity from 200°C to 450°C, even after hydrothermal ageing at 700°C and sulfation treatments. The low temperature NSC efficiency can further be increased by adding a third compound, such as ceria, without decreasing the high temperature NSC performances.

1 - INTRODUCTION

It is commonly known that fresh NOx-trap catalysts show high efficiency but low durability due to a fast deactivation of the active sites. The widely used reference catalyst consists of Pt-BaO/Al_2O_3 with a Ba loading in the range of 4 to 20 wt.% [6,7] associated with a TWC catalyst based on Rh formulation. During the NOx-storage process in lean conditions, Pt oxidizes NO molecules into NO_2 which are further trapped as nitrites or nitrates by the 'Ba species' [3]. Under lean conditions, this type of combination is highly temperature sensitive. After ageing at 700 to 800°C it shows drastic sintering of both Pt and Ba. More precisely $BaAl_2O_4$ crystallization is one of the degradation pathways for a typical NSC catalyst. It leads to fast catalytic deactivation.

In addition to this thermal deterioration, Ba is strongly poisoned by SO_2 from the fuel, even for low sulfur loaded fuels, and this yields to poor durability. It is reported that complex desulfation strategies at temperatures higher than 750°C under rich conditions only recover partly the NOx-trap sites. This high desulfation treatment increases the thermal stress of the catalyst and accelerates the sintering and the deactivation.

Consequently, it is important to develop a catalyst that can absorb NOx quickly and has a sufficiently large storage capacity. Suitable materials have to be thermally stable and have to be less sensitive to SO_2 poisoning meaning more sulfate resistant or more easily desulfated.

Ceria-zirconia-based compounds have become key components in the formulation of several commercial materials, especially for their high stability and for their high redox behaviour [8-12].Many studies claim the interest to use CeO_2-ZrO_2 mixed oxides in intimate contact with precious metals to boost the CeO_2-ZrO_2 redox properties. Consequently, CeO_2-ZrO_2 are tested to enhance the Pt activity toward the NO oxidation to NO_2.

To increase the NOx-Trap efficiency in the whole temperature range (200-450°C), Rhodia has developed rare earth mixed oxides and modified Ba-alumina materials with rare earth oxides that show high thermal stability and high sulfation resistance. In this paper, Pt powder model catalysts prepared with given carriers, i.e. CeO_2-ZrO_2, CeO_2-ZrO_2-Pr_6O_{11}, CeO_2-ZrO_2-La_2O_3, BaO-Al_2O_3, CeO_2-BaO-Al_2O_3 and CeO_2-ZrO_2-BaO-Al_2O_3, are evaluated in NOx-storage under model fuel-lean conditions as well as a usual Pt-BaO/Al_2O_3 powder model catalyst prepared by successive impregnations of respectively Barium and Pt onto alumina.

The following sections deal with the experimental conditions – catalysts preparation and activation, BET surface analysis, XRD characterization, NOx-storage capacity – key results, discussions and conclusion.

2 - EXPERIMENTAL

Materials. In the following sections, the composition of the materials is expressed as weight oxide percent. These compositions are given in table 1. CeO_2-ZrO_2-based samples were obtained through a wet chemistry process, starting from aqueous solutions and patented by Rhodia [13] leading to pure solid solutions. The sample referred as BaAl-Im was prepared by incipient-wetness impregnation method. The Al_2O_3 support was obtained after calcination of SB3 Condea boehmite at 800°C for 4h. The barium precursor, barium nitrate [$Ba(NO_3)_2$], was purchased from Prolabo Rectapur 99 %. Alumina impregnation of barium nitrate solution was realized in two steps due to the limited solubility of Ba nitrate precursor. After each alumina impregnation (10wt% of BaO each time) at 80°C, the powder was dried at 120°C for 2h and calcined at 500°C for 2h. The modified BaO-Al_2O_3 sample referred as BaAl, CeBaAl and CeZrBaAl were obtained by a Rhodia proprietary process.

Surface area is measured by the Brunauer-Emmett-Teller (BET) method. N_2 sorption isotherms are measured at -197°C using a Micrometrics ASAP 2010 instrument. The thermal stability is the surface area measured by the BET method after calcining under air the oxides in a muffle furnace respectively at 900°C / 4h and 1000°C/4h.

X-ray powder diffraction (XRD) data are collected for all samples at room temperature with a X'Pert PRO MPD PANalytical powder diffractometer system using Cu $K\alpha 1+\alpha 2$ radiation (λ=1.54182 Å) and a detector with variable slits. XRD patterns are shown after correction of the baseline drift induced by the variable slits detector.

Table 1. Material composition expressed as weight oxide percent and labelling of the samples.

Composition (wt oxide %)	Sample Label	Type
CeO_2/ZrO_2 70/30	CeZr	Rhodia Process
$CeO_2/ZrO_2/Pr_6O_{11}$ 65/27/8	CeZrPr	Rhodia process
$CeO_2/ZrO_2/La_2O_3$ 20/70/10	CeZrLa	Rhodia process
BaO/Al_2O_3 20/80	BaAl-Im	Impregnation
BaO/Al_2O_3 20/80	BaAl	Rhodia Process
$CeO_2/BaO/Al_2O_3$ 5/20/75	CeBaAl	Rhodia Process
$CeO_2/ZrO_2/BaO/Al_2O_3$ 3.5/1.5/20/75	CeZrBaAl	Rhodia Process

Catalyst preparations. All these supports were then impregnated with a solution of tetraamine platinum hydroxide (1wt% Pt in solution), evaporated, dried at 120°C for 2h, calcined at 500°C for 2h, and then aged at 700°C under a wet atmosphere (air-water volume ratio of 9) for 4h. The platinum salt was purchased from Prolabo. The Pt loading is 1 weight percent for all the catalysts.

Sulfation treatments are done on hydrothermally aged catalysts in 20 vpm SO_2 - 10 vol%O_2 – 10 vol%H_2O – balance N_2 at 300°C/12h. Elemental S loadings of the sulfated model catalysts are determined by chemical analysis on a LECO CS 444 based on catalyst combustion at 1300°C under oxygen and detection of SOx formed on an infra-red spectrometer. The maximum

S loading in our experimental sulfation conditions is 1.28 wt% S considering that SO_2 is fully oxidized in SO_3 and further fully trapped as sulfates by the catalyst.

Testing of powder model catalysts. Before testing, sulfated model catalysts are activated in-situ on a synthetic gas bench (see Figure 1). Model catalysts are treated at a space velocity of 200 000 h^{-1} under reducing conditions (4.9% O_2 + 10% CO + 5% CO_2 + 250vpm of C_3H_6 + 250vpm of C_3H_8 + 5% H_2O in N_2) before testing. To do this, the powder model catalyst is put in a quartz U-shaped down-flow reactor (Length 255 mm; internal diameter 5 mm), the temperature is increased at the rate of 10°C/min from 150°C to 600°C, kept at 600°C for 20 min, and then cooled down. Note that the oven temperature is set by the reactor temperature avoiding thus high exothermal phenomenon within the catalyst bed during the activation of the model catalysts. After this treatment, catalysts should be at least partially desulfated. After activation, the catalysts are evaluated on the same synthetic gas bench at a space velocity of 200 000 h^{-1}. The gas composition, generated by independent mass flow controllers, is lean (10% O_2 + 5% H_2O + 10% CO_2 + 300 vpm NO in N_2). Prior to the catalytic performance estimation, the powder model catalysts are set at the desired testing temperature while the gas flow was stabilized. The conversions are then measured through the FTIR analysis (MAGNA 560 Nicolet) of NO and NO_2 outlet gasses. Performances are expressed as percentage of NOx trapped after 1 minute running and as the conversion of NO into NO_2 after 1 minute running. In other words, after 1 minute running at the reaction temperature, the difference between the NO conversion and the NO_2 detected in the gas phase corresponds to the NOx trapped.

This testing only gives indications on the NOx storage capacity of the materials but in any case on the global NOx conversion during lean/rich cycles.

At the end the remaining elemental S loadings of the desulfated catalysts are determined as described above and compared to the S loading after sulfation treatment.

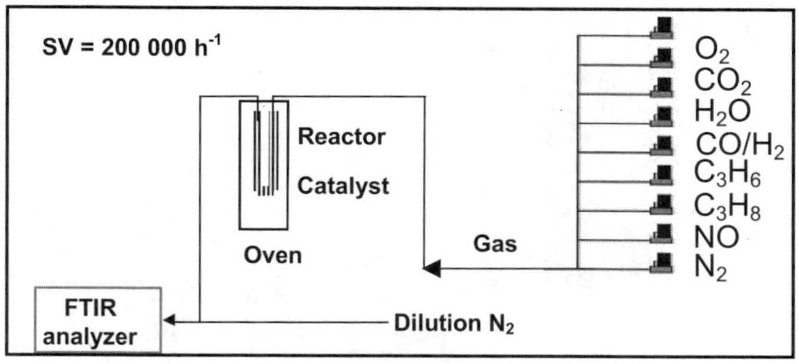

Figure 1. Schematic view of the synthetic gas bench.

3 – THERMAL STABILITY OF THE MATERIALS

Figure 2 shows the surface area of the aged materials at 900°C and 1000°C under air during 4 hours. The CeO_2-ZrO_2-based mixed oxides show a surface area in the range of 55m²/g – 65m²/g after ageing at 900°C/4h. The addition of Pr or La oxides increases the surface area. Figure 3 shows the X-Ray Diffraction patterns of the 4 hours-air-aged at 900°C CeO_2-ZrO_2-based oxides. They present pure solid solution with either a cubic structure for Ce-rich samples such as CeZr and CeZrPr either a tetragonal structure for Zr-rich sample, i.e. CeZrLa.

Moreover, as illustrated in figure 2, the BaAl sample prepared by a Rhodia process shows a higher surface area at 900°C and at 1000°C than the material prepared by impregnation BaAl-Im. Furthermore, CeO_2 and CeO_2-ZrO_2 doped BaO-Al_2O_3 show the same stability.

The BaO-Al_2O_3-based solids were characterized by XRD after 4 hours-air-ageing at 900°C. The XRD patterns are shown in figure 4. Whatever the material, Barium carbonate ($BaCO_3$) and Alumina peaks are present. $BaAl_2O_4$ phase is detected on BaAl-Im prepared by impregnation, whereas no $BaAl_2O_4$ phase is detected on BaAl, CeBaAl and CeZrBaAl prepared by Rhodia process. It is assumed that the missing $BaAl_2O_4$ phase does not crystallize due to the homogeneous dispersion of Ba in the matrix.

CeO_2 XRD peaks are detected on CeBaAl and CeO_2-ZrO_2 on CeZrBaAl. The estimated average CeO_2 particle size is 7nm, indicating well dispersed CeO_2 particles on BaO-Al_2O_3 even after a thermal treatment.

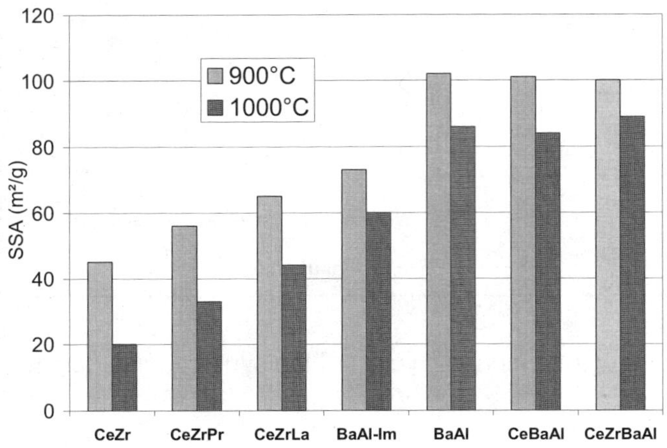

Figure 2. Surface area of the 4 hours-air-aged materials at 900°C and 1000°C.

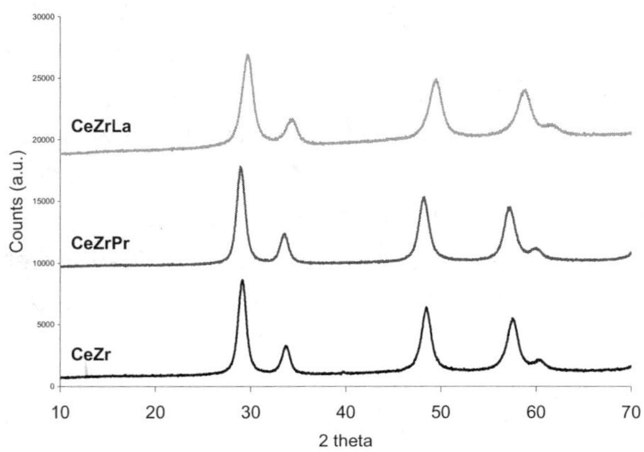

Figure 3. XRD patterns on air-aged CeO_2-ZrO_2 matrix at 900°C.

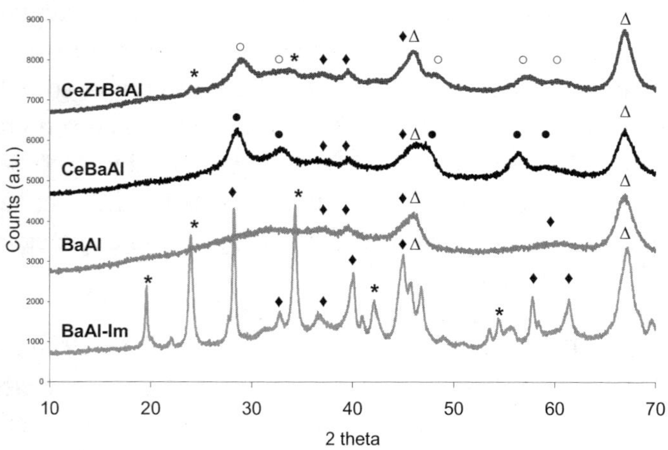

Figure 4. XRD patterns on air-aged BaO-Al_2O_3 modified materials at 900°C (Δ:Al_2O_3, ♦:$BaCO_3$, *:$BaAl_2O_4$, ●:CeO_2, ○:CeO_2-ZrO_2).

4 – IMPACT OF SULFATION-DESULFATION ON HYDROTHERMALLY AGED Pt POWDER MODEL CATALYSTS.

NOx-storage catalysts must be thermally stable and less sensitive to SO_2 poisoning. The catalysts must be either sulfate resistant or more easily desulfated. The sulfur resistance of hydrothermally aged Pt powder model catalysts is estimated by comparing their sulfur loadings after sulfation at 300°C and after desulfation at 600°C under rich conditions (λ=0.93). The results are presented in table 2.

1wt%Pt supported on	S loading (wt%)		Desulfation Rate (%)
	after sulfation 300°C	after desulfation 600°C	
CeZr	0.52	< detection limit	100
CeZrPr	0.92	< detection limit	100
CeZrLa	0.51	< detection limit	100
BaAl-Im	1.10	0.20	81
BaAl	1.28	0.30	77
CeBaAl	0.70	0.15	79
CeZrBaAl	1.10	0.25	77

Table 2. Sulfur loading on aged catalysts after sulfation at 300°C and after desulfation at 600°C (=0.93).

Column 2 in table 2 shows the total amount of S stored onto the model catalysts after the sulfation treatment. For CeO_2-ZrO_2-based samples, the sulfur loading varies between 0.51wt% and 0.92 wt%. It is much less than the expected value of 1.28 wt% that would correspond to 100% of SO_2 trapped by the catalyst. The addition of Pr_6O_{11} to CeO_2-ZrO_2 matrix increases the sulfur loading. It is assumed that this high sulfur loading is related to the basicity of ceria and praseodymium oxide as already mentioned by Trovarelli et al [8]. It is assumed that the higher sulfatation level when Pr_6O_{11} is added to CeO_2-ZrO_2 matrix is related to the higher concentration of basic sites – ie Ce and Pr - at the surface of the mixed oxide. For CeZrLa, the addition of La does not increase the sulfur loading due to the zirconia-rich composition which shows less basic sites (Ce and La).

The two BaO-Al_2O_3 solids, BaAl-Im and BaAl, show a total sulfur loading close to the theoretical value (1.28%) as well as CeZrBaAl. So, doping BaAl with 5wt% CeO_2-ZrO_2 (70/30) does not affect BaAl sulfation resistance. CeBaAl has stored less sulfates than BaAl and (CeZr)BaAl.

The column 3 in table 2 shows the remaining sulfur amount on the catalyst after the desulfation treatment (at $\lambda = 0.93$). The column 4 in table 2 indicates the desulfation rate, the amount of stored sulfur that is removed after the desulfation treatment.

The CeO_2-ZrO_2-based materials do not show any remaining sulfur after the desulfation treatment at 600°C whereas BaAl based materials keep 20% of the adsorbed sulfur after desulfation at the same temperature.

As a conclusion, these results point out :

- the high sulfur resistance of CeO_2-ZrO_2-based oxides that can be totally desulfated at 600°C under reducing conditions.

- Modifying BaO-Al_2O_3 with well dispersed CeO_2 (5wt%) enables to reduce the amount of stored sulfur to the level of CeO_2-ZrO_2-based oxides.

- 80% of the sulfates adsorbed onto the AlBaCeOx based oxides are removed after desulfation at 600°C

5 – NOX STORAGE CAPACITY OF THE HYDROTHERMALLY AGED Pt POWDER MODEL CATALYSTS

Figure 5 shows the percentage of NOx stored on CeO_2-ZrO_2-based oxides after 1 minute running with experimental temperatures varying from 200°C to 450°C. The Pt/CeZr model catalysts cannot trap more than 50% of converted NOx in the whole temperature range, i.e. 200°C to 450°C. However, whatever the experimental temperature, doping CeO_2-ZrO_2 with Pr_6O_{11} or with La_2O_3 increases the amount of NOx-trapped after 1 minute running.

Pt/CeZrPr and Pt/CeZrLa model catalysts show high NOx-storage efficiency at low temperature, with a 90% maximum efficiency at 250°C and 300°C. These very high efficiencies are related to the very high NO conversion into NOx species as illustrated in Figure 6. This figure highlights the percentage of NO molecules transformed into NO_2 (in the gas phase) and NO_2 (adsorbed on the material as nitrites or nitrates). For these 2 materials the NO conversion between 250 and 300°C exceeds 95% while the level of NOx trapped is above 90%, meaning that the NO_2 produced is trapped as 'absorbed-nitrates' without any limiting step.

Moreover, as indicated in Figures 5 and 6, CeZrPr is the most promising carrier to trap NOx at very low temperature, i.e. 200°C (60% of NO_2 being trapped) while CeZrLa is efficient in a broader temperature range (up to 70% of NOx trapped from 250°C to 400°C). In addition, CeZrLa shows a good trade-off between high redox properties on one hand and a number of NOx sorption sites high enough to broaden the NSC activity window to high temperature. Note that the undoped CeO_2-ZrO_2 material shows low NO conversion and low NOx-storage capacity in the same range of temperature. Therefore, rare earth doped CeO_2-ZrO_2 materials improve strongly the NOx storage properties by enhancing the NO conversion i.e. enhancing NO to NO_2 oxidation while keeping high efficiency in NOx trapping.

Figure 7 shows the percentage of NOx stored on BaO-Al_2O_3-based solids after 1 minute running at temperatures varying from 200°C to 450°C and figure 8, the respective NO conversion into NO_x (NO_2(g) + NO_2(ads)).

As illustrated in figures 7 and 8, Pt/BaAl and Pt/BaAl-Im model catalysts, are efficient only at high temperatures, above 350°C. In that range of temperature, Pt/BaAl shows the highest efficiency with more than 80% NOx stored when temperature varies from 350°C to 450°C. This high storage efficiency is related to a fast storage of NO2 as nitrates by the surface with no or few release of NO2 in the gaz phase. This material presents significantly higher efficiency than the Pt/BaAl-Im model catalyst prepared by impregnation route. This result is related to a poor NO conversion i.e; poor NO oxidation of the Pt/BaAl-Im catalyst

As shown in figure 7, Pt/CeZrBaAl and Pt/CeBaAl show high NOx-storage efficiency in the whole temperature range, i.e. from 200°C to 450°C. These very high efficiencies are related to the very high NO conversion, on the whole temperature range (Figure 8). At low temperatures (200°C and 250°C), Pt/CeBaAl and Pt/CeZrBaAl NO conversion exceeds 90%.

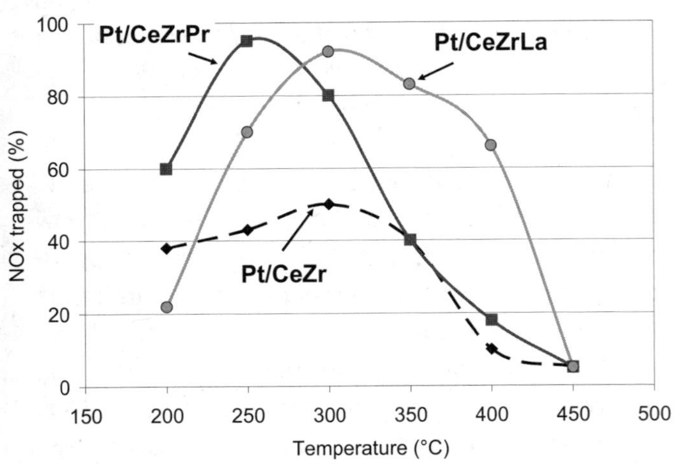

Figure 5. Amount of NOx stored on CeO_2-ZrO_2-based oxides after 1 minute running at different temperatures.

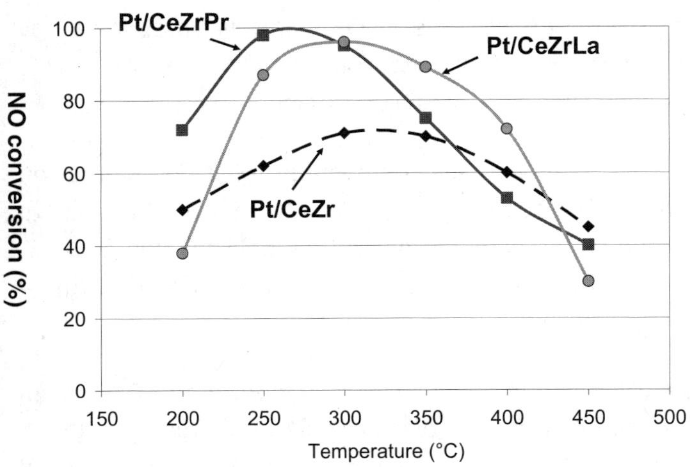

Figure 6. CeO_2-ZrO_2-based oxide performances in NO conversion after 1 minute running at different temperatures.

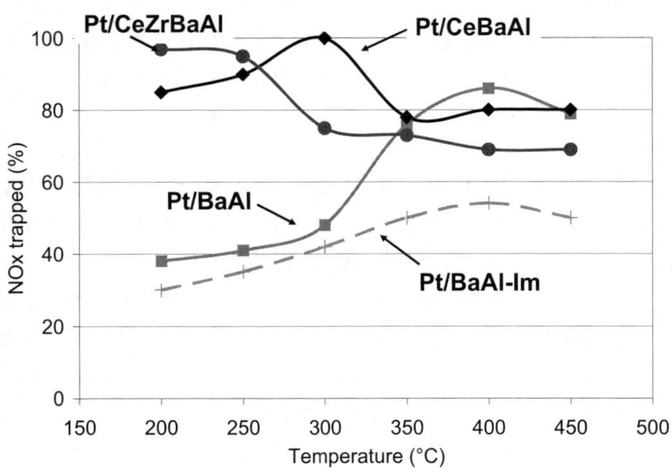

Figure 7. Amount of NOx stored on BaO-Al_2O_3-based solids after 1 minute running at different temperatures.

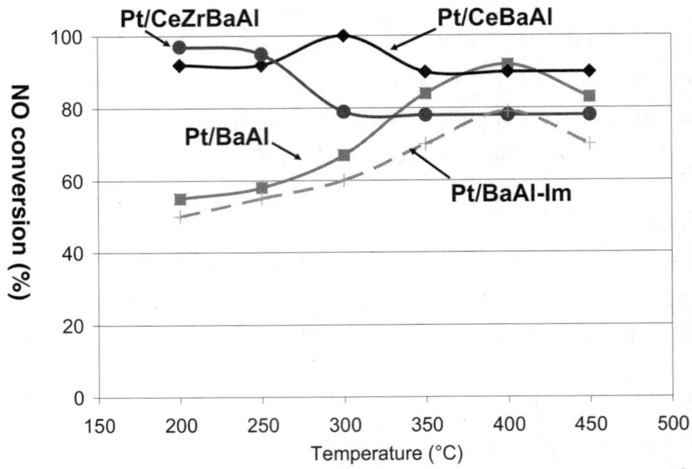

Figure 8. BaO-Al_2O_3-based material performances in NO conversion after 1 minute running at different temperatures.

6 – CONCLUSION

- CeZr based mixed oxides with high thermal stability, high NOx storage capacity at low temperature – with more than 90% of the NOx stored at 250°C - and 100% sulphur removal at 600°C in rich conditions

- AlBaCe and AlBa(CeZr) based oxides with very high surface area, and high NOx storage capacity on the whole temperature range – from 200°C to 450°C.

- their use in the formulation of full size catalysts would be of interest to consider the implications of these new materials on global NOx conversion in lean/rich cycles conditions and on fuel penalty too.

REFERENCES

[1] L.J. Gill, P.G. Blakeman, M.V. Twigg, A.P. Walker, Top. Catal. 28 (2004)157.

[2] J. Theis, J. Lupescu, J. Ura, R. McCabe, SAE Technical Paper no. 2006-01-1069.

[3] W.S. Epling, L.E. Campbell, A. Yezerets, N.W. Currier, J.E. Parks II, Catal. Rev. 42 (2004) 163.

[4] M. Takeuchi, S. Matsumoto, Top. Catal. 28 (2004) 151.

[5] T. Armaroli, A. Lambert, J. Lavy, S. Raux, F. Tihay, S. Dorge, H. Mahzoul, P. Gilot, P. Climaud, A. Sassi, T. LeTallec, F. Rohr, SAE Technical Paper no. 2006-01-1068.

[6] M. Piacentini, M. Maciejewski, A Baiker, Appl. Catal. B. 60 (2005) 265.

[7] S. Pishinger, Top. Catal. 30 (2004) 5.

[8] M. Shelef, G.W. Graham and R. Mc Cabe in 'Ceria and other oxygen storage components in automotive catalysts' – Catalysis by Ceria and Related Materials edited by A. Trovarelli (Imperial College Press 2001).

[9] E. Rohart, O. Larcher, S. Deutsch, C. Hédouin, A. Huang, F. Fajardie, M. Allain, P. Macaudière. – Topics in Catalysis vols 30./31 July 2004 p.417 - From Zr-rich to Ce-rich : thermal stability of OSC materials on the whole composition range.

[10] E. Rohart, O. Larcher, C. Hédouin, M. Allain, P. Macaudière, SAE Technical Paper no. 2004-01-1274.

[11] E. Rohart, O. Larcher, E. Ottaviani, S. Pelissard, M. Allain, SAE Technical Paper no. 2005-01-1107.

[12] E. Rohart, S. Verdier, B. Pacaud, A. Demourgues, V. Harlé, M. Allain, A. Baylet, H. Takemori, E. Suda, SAE Technical Paper no. 2006-01-0852.

[13] M. Aubert, T. Birchem, G. Blanchard, O. Touret, WO 97/02213.

III.3 COMBINED SYSTEMS

2006-01-3552

Advanced NO_x Aftertreatment System And Controls For On-Highway Heavy Duty Diesel Engines

Haoran Hu, Johannes Reuter, John Yan and James McCarthy, Jr.
Eaton Corporation

ABSTRACT

This paper describes a NO_x aftertreatment system and control strategy for heavy-duty diesel engines to achieve US EPA 2010 emissions regulations. The NO_x aftertreatment system comprises of a fuel reformer catalyst, a LNT catalyst, and a SCR catalyst. The only reductant required to operate this system is diesel fuel; hence, no urea infrastructure is required to support this approach. The fuel reformer is used to generate reformate which is a combination of hydrogen, carbon monoxide and unburned hydrocarbons. This reformate provides a more efficient feedstock to improve LNT NO_x regeneration efficiency. Engine out NO_x is reduced using a two-step process. First, NO_x is stored in the LNT catalyst during lean operation. During rich operation, portions of the stored NO_x are converted to nitrogen and ammonia. Next, the ammonia released from the LNT is captured by the downstream SCR catalyst. The stored ammonia is further used to reduce the NO_x that slips past the LNT catalyst.

System hardware was developed and evaluated in dynamometer test cells using 9L and 14L commercial-available heavy-duty diesel engines. System control strategies were developed to optimize the NO_x conversion efficiency and fuel penalty.

INTRODUCTION

NO_x emissions from diesel-powered commercial vehicles are an environmental problem recognized worldwide. Several countries, including the United States, have long had pending regulations that will limit NO_x emissions from vehicles. Manufacturers and researchers have put considerable efforts toward meeting those regulations. In conventional gasoline powered vehicles that use stoichiometric fuel-air mixtures, three-way catalysts have been shown to control NO_x emissions. In diesel-powered vehicles, however, the exhaust is too oxygen-rich for three-way catalysts to be effective.

Several solutions have been proposed for controlling NO_x emissions from diesel-powered vehicles. One set of approaches focus on the engine. Techniques such as exhaust gas recirculation and homogeneous charge compression ignition (HCCI) can reduce NO_x emissions by reducing the adiabatic flame temperature [1]. These techniques alone, however, will not be sufficient to reduce NO_x emissions to meet stringent emissions regulations, such as the US EPA 2010 emissions regulations for on-highway heavy duty vehicles. Another set of approaches is to remove NO_x from the vehicle exhaust. These include the use of lean-burn NO_x catalysts, lean NO_x traps (LNT), and selective catalytic reduction (SCR).

Lean-burn NO_x catalysts [2] promote the reduction of NOx under oxygen-rich conditions. It has proved challenging to find a lean-burn NO_x catalyst that has the required activity, durability, and operating temperature range. Lean-burn NO_x catalysts also tend to be hydrothermally unstable. A noticeable loss of activity occurs after relatively little use. Lean burn NO_x catalysts typically employ a zeolite washcoat, which is thought to provide a reducing microenvironment. The introduction of a reductant, such as diesel fuel, into the exhaust is generally required and introduces a fuel economy penalty of 3% or more. Currently, peak NO_x conversion efficiency with lean-burn catalysts is in the ranges of 15% to 30%.

LNT or NO_x adsorber catalysts offer more attractive NOx reduction efficiency and relatively low fuel penalty in comparison with lean NO_x catalysts [3,4]. The concept of the NO_x adsorber has been developed based on acid-base washcoat chemistry. It involves adsorption and storage of NO_x in the catalyst washcoat during lean conditions and releasing it under rich operations. The released NO_x is catalytically converted to nitrogen and other species. However, sulfur, primarily as sulfur dioxide, in the diesel exhaust causes the LNT performance gradually to decline. Periodic desulfation is required to maintain the NO_x reduction performance of the LNT catalysts. The frequent desulfation will cause durability concerns of the LNT catalysts.

Typical configurations of the LNT catalysts include dual leg systems or single leg systems. The dual leg LNT systems allow the entire exhaust gas to pass through a single LNT leg while the other leg is either regenerating (transforming NO_x to nitrogen and other species) or desulfating (removing sulfur from the LNT). A valve can be used to switch the flow between LNT legs such that the freshly cleaned LNT stores NO_x while the other leg is cleaned. The downside to this approach is high system cost due to the increased catalyst volume of dual legs along with the high cost of precious metals associated with LNT catalysts. Additionally, the added complexity of a valving system adds cost and poses reliability and durability concerns. The single leg LNT systems eliminate the complex valving system and reduce system volume. This system is forced to regenerate the LNT while maintaining high NO_x conversion efficiencies since the LNT is always in contact with the exhaust gas. However, this system is prone to lower NO_x conversion efficiencies than a dual leg system, higher fuel penalties due to removing high amounts of oxygen to force rich operation (unlike dual leg approach), NOx spikes during regeneration can be up to twice the normal NO_x level if not controlled properly and ammonia slip during regeneration.

SCR involves the reduction of NO_x by ammonia [5]. The reaction takes place even in an oxidizing environment. Ammonia can be fed continuously into the exhaust. SCR can achieve NO_x reductions in excess of 90%, however, there are concerns over the lack of infrastructure for distributing ammonia or a suitable precursor, urea, in the North American market. SCR also raises concerns relating to the possible release of ammonia into the environment.

Other efforts have been made to enhance the LNT catalyst NO_x reduction performance and to overcome the SCR infrastructure constraints. Dalla Betta et al.[6] report that an in-line fuel reformer/LNT system will improve LNT NO_x conversion efficiency by increasing the hydrogen concentration in the exhaust stream. Lambert et al. [7] describe a combined LNT/SCR system for automotive diesel NO_x reductions. The system uses LNT and SCR catalysts in series to utilize the ammonia generated by the LNT during regeneration to further reduce NO_x emissions by the SCR catalysts.

This paper will report an advanced exhaust aftertreatment system with fuel reformer/LNT/SCR catalysts in series for on-highway heavy duty diesel engines to meet the US EPA 2010 emissions regulations. This system utilizes a fuel reformer to generate H_2 and CO to improve LNT NO_x reduction efficiency and the conversion efficiency of NO_x to NH_3. The NH_3 species will be stored and be utilized to reduce NO_x further by the subsequent SCR catalysts. This approach converts the drawbacks of the single leg LNT approach (low conversion during regeneration and NH_3 slip) into an advantage while capitalizing on the fact that a urea infrastructure is not required since diesel fuel is the only reductant.

AFTERTREATMENT SYSTEM LAYOUT

Figure 1 shows a schematic of the aftertreatment system. The system is comprised of a diesel fuel reformer followed by a LNT catalyst and an NH_3-SCR catalyst. The fuel reformer generates reformate to improve the efficiency of the regeneration cycle as compared to injecting fuel directly into the exhaust system. The functions of LNT are three-fold: 1) adsorb NO_x when engine is running under normal (lean) conditions (Eqn 1); 2) reduce a portion of stored NO_x to N_2 under regeneration (rich) conditions (Eqn 2); and 3) reduce the rest stored NO_x to ammonia under regeneration conditions (Eqn 3).

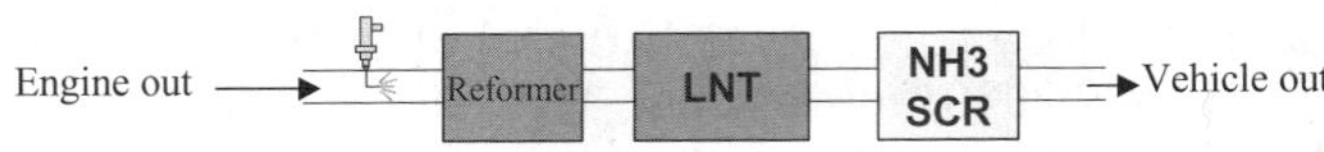

Figure 1: Aftertreatment System Schematics

$$BaCO_3 + 2NO_2 + \tfrac{1}{2} O_2 \rightarrow Ba(NO_3)_2 + CO_2 \qquad (1)$$
$$Ba(NO_3)_2 + 5CO \rightarrow BaCO_3 + 4\, CO_2 + N_2 \qquad (2)$$
$$Ba(NO_3)_2 + 8\, H_2 + CO_2 \rightarrow BaCO_3 + 2\, NH_3 + 5\, H_2O \qquad (3)$$

The functions of NH_3-SCR catalyst are two-fold: 1) capture the NH_3 released from LNT regeneration by adsorption; and 2) use the stored NH_3 to reduce NO_x that have slipped out of LNT (Eqn 4). More detailed chemistry of LNT and SCR reactions can be found elsewhere [8].

$$4\, NH_3 + 4\, NO + O_2 \rightarrow 4\, N_2 + 6\, H_2O \qquad (4)$$

The chemistry of diesel fuel reformer can be generally summarized into three reactions: combustion (Eqn 5) to consume oxygen and generate heat to reach reaction temperature for reforming, steam reforming (Eqn 6) to produce carbon monoxide and hydrogen and a water gas shift reaction (Eqn 7) to form hydrogen. The resultant product H_2 promotes higher efficiency LNT regeneration and desulfation (i.e., high reaction rates even at low temperatures) and promotes the formation of NH_3.

combustion

$$C_mH_n + (m+n/4)\, O_2 \rightarrow mCO_2 + n/2\, H_2O, \qquad (5)$$

steam reforming

$$C_mH_n + mH_2O \leftrightarrow mCO + (m+n/2)H_2 \qquad (6)$$

water gas shift

$$CO + H_2O \leftrightarrow CO_2 + H_2 \quad (7)$$

TEST SETUP AND DESCRIPTION

The aftertreatment system is being developed using two dynamometer test cells and a vehicle. Commercially available engines that met the 2002-2004 emission standards were used as development tools. A 14L Detroit Diesel Series 60 engine was used in one of the test cells and also in the vehicle. The engine was rated at 500 hp @ 2100 rpm, 1650 ft-lb at 1200 rpm. A Class 8 Freightliner vehicle was used for this test. The second test cell used a Caterpillar C9 (9L) engine that was rated at 350 hp @ 2300 rpm.

Figure 2 shows a typical layout of emission instrumentation for one of the test setups. A Horiba MEXA-120 portable sensor monitors NO_x/O_2 concentrations at engine-out position. A Horiba chemi-luminescence (MEXA-7000) analyzer $NO_x/CO/CO_2$/HC is used at the reformer-out to monitor the feed gas to the LNT catalyst. In order to monitor NH_3 trace, two FTIRs (Model IGS by Thermo Electron) are placed at the LNT-out and SCR-out. These monitor NO_x, NH_3, and CO, though total HC measurement is not very accurate.

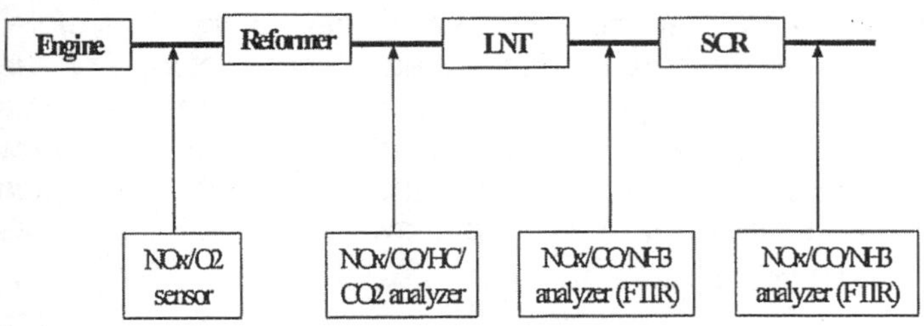

Figure 2: Test Setup Schematics

Figure 3 shows a photograph of the setup of using the DDC 14L Series 60 engine. In order to match the available catalyst volume, the exhaust flow was cut in half for the 14L Series 60 engine such that the catalysts were exposed to 7L equivalent engine displacement. A special valving mechanism was installed in the test cell to direct half of the exhaust flow through a catalyst setup. The valve position was controlled using exhaust flow meters in each exhaust pipe to maintain half the engine exhaust flow through the catalyst setup. Three separate catalysts are shown, which include the fuel reformer (bottom) followed by the LNT catalysts (vertically) and SCR catalysts (top). The fuel reformer was 9"ODx3" long (3.1L), the LNT included two 9.5"ODx6"long bricks (14L) and the SCR included two 9.5"ODx6"long bricks (14L) for this prototype evaluation. Emission equipment included a full species Horiba analyzer and two Thermo Electron FTIRs. Correlation was established between all NO_x measurements by placing each analyzer at the same position in the exhaust pipe. The Horiba bench was placed downstream of the fuel reformer to measure NO_x, CO and HC. The FTIR's were commonly placed at both LNT and SCR out to monitor NO_x, NH_3 and CO. A NO_x sensor was placed downstream of the engine to monitor engine out NO_x. An intake throttle valve was installed upstream of the intake manifold to reduce the oxygen concentration in the exhaust to required levels during a regeneration event. The oxygen content was measured using the Horiba bench (slow response) and oxygen sensors (fast response) to monitor and effectively operate the fuel reformer. It is not Eaton's intention to require an intake throttle valve for production; however, this was an effective means to demonstrate the aftertreatment concept.

Figure 3: Test Setup Using the 14L Series 60 Engine

Figure 4 shows a photograph of the test setup using the CAT C9 engine. The exhaust flow did not require splitting since the catalyst volume for the reformer and LNT catalysts were appropriately sized for a 9L engine. The focus of this test setup was LNT desulfation. As a result, early testing did not include a SCR catalyst downstream of the LNT. The fuel reformer was 9.0"ODx3"long (3.1L) while the LNT included two bricks 9.5"ODx6"long (14L). Current testing includes the SCR catalyst downstream. Emissions equipment included dual full species Horiba benches, V&F hydrogen analyzer, V&F mass spectrometer and a Thermo Electron FTIR. The dual Horiba benches were commonly placed at reformer out (monitor NO_x, CO and HC) and LNT out (NO_x, CO and HC) to quantify the NO_x reduction, CO and HC breakthrough. The hydrogen analyzer was placed downstream of the fuel reformer to measure hydrogen production. The mass spectrometer was placed downstream of the LNT to measure sulfur release. The FTIR was also placed downstream of the LNT to measure ammonia production. Similar to the Series 60 test setup, a NO_x sensor was placed at engine out to measure NO_x which equated to the Horiba bench downstream of the reformer during lean operation. An intake throttle valve was used to control exhaust oxygen, which was monitored by oxygen sensors at engine out, reformer out and LNT out.

Figure 4: Test Setup Using CAT C9 Engine

CONTROL STRATEGY

The objective of the control algorithms that need to be developed for operating the system can be stated in the following optimization problem: Minimize the additional amount of fuel needed to operate the aftertreatment system subject to the constraints that the NO_x level at tail pipe out meets the 2010 emission standard together with the other regulated emissions.

In order to achieve such a goal, several control and estimation strategies have been developed to perform the necessary underlying tasks during regeneration and desulfation events. These can be summarized in the following way:

- Evaluation of the state of the LNT. This includes the estimate of the amount of trapped NO_x, as well as the analysis of aging condition.
- Control of the Partial Oxidation (POX) process during regeneration and desulfation.
- Supervisory control of the relevant conditions of the engine and LNT to schedule the required regeneration and desulfation cycles.
- Diagnostics of the complete system including sensors and actuator, reformer monitoring, etc.

The control architecture used for accomplishing these tasks is depicted in Figure 5.

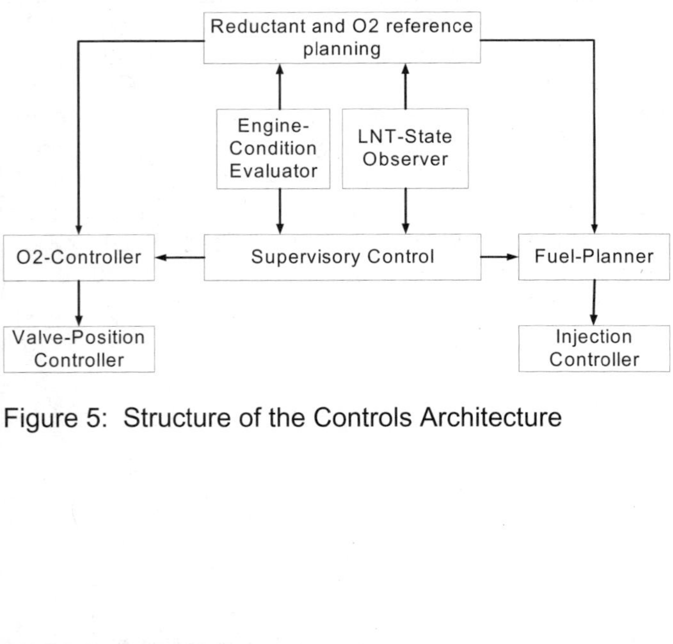

Figure 5: Structure of the Controls Architecture

The central part is a supervisory control module that uses engine information and a LNT state observer. The LNT state is used to determine when regeneration is necessary and the engine data is used to evaluate if the conditions are suitable to actually allow and command a regeneration event. The upper block determines the optimal way to command the oxygen and reductant flow during regeneration. NO_x storage, exhaust O_2 content, LNT temperature, and space velocity are evaluated in order to calculate fuel and oxygen flow profiles to regenerate the LNT in the most efficient way. The O_2 controller sets the intake throttle position and the fuel planner sets the fueling profile. Subsequently, the underlying valve-position and injection controller ensure an accurate fuel dosing and oxygen flow.

Evaluation of the LNT State

The LNT NOx trapping capacity is constantly monitored to assure a suitable NOx reduction efficiency. Degradation in NOx trapping efficiency due to sulfation is monitored and a desulfation event is scheduled to regain trapping capacity.

A variety of controls oriented models for LNT catalysts can be found in the literatures [9,10]. Adaptive features to account for the aging of the LNT have been published [11,12]. It has been found that these models are not sufficient to cover all aspects involved during lean operation, regeneration and desulfation. Therefore, ongoing research in this area is conducted. For the results presented in this paper, the trapped NO_x has been estimated by using concentration and flow measurements in and out of the LNT. The regeneration was triggered when a certain level of NO_x in the LNT has been stored.

Reformer Control

The early development control algorithms for the fuel reformer were focused on reducing oxygen to low levels (3-6%) using air intake throttling and injecting a prescribed fuel profile over a fixed duration. This approach is referred to as the 'initial' control strategy.

Figure 6 shows an example of a typical reformer control event. The engine out oxygen is reduced via intake throttling from approximately 10% to 5%. Fuel is injected into the reformer to consume the remaining oxygen as depicted by the reformer out oxygen dropping to zero. The fuel rate is increased to create a rich atmosphere in the reformer where reformate is produced as depicted by the carbon monoxide profile. Hydrogen has a similar profile as carbon monoxide. Upon completion of fuel injection, the air intake throttle returns to normal operation and the engine out oxygen returns to its normal value of 10%.

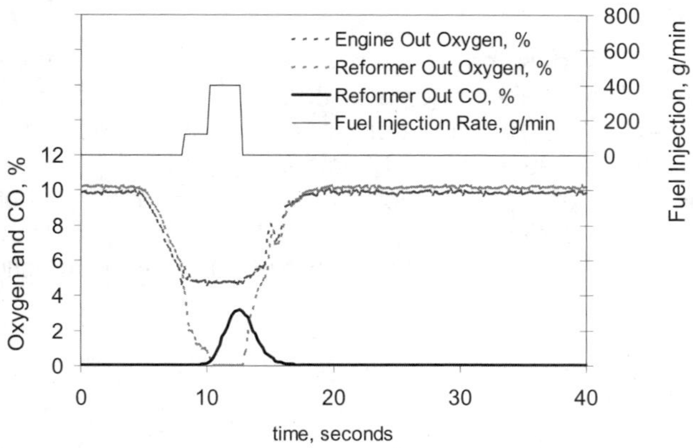

Figure 6: Example of Oxygen Control, Fueling and Reformate Production.

Later development efforts have focused on a more advanced strategy to better reflect the physical properties of the process as illustrated in Figure 7.

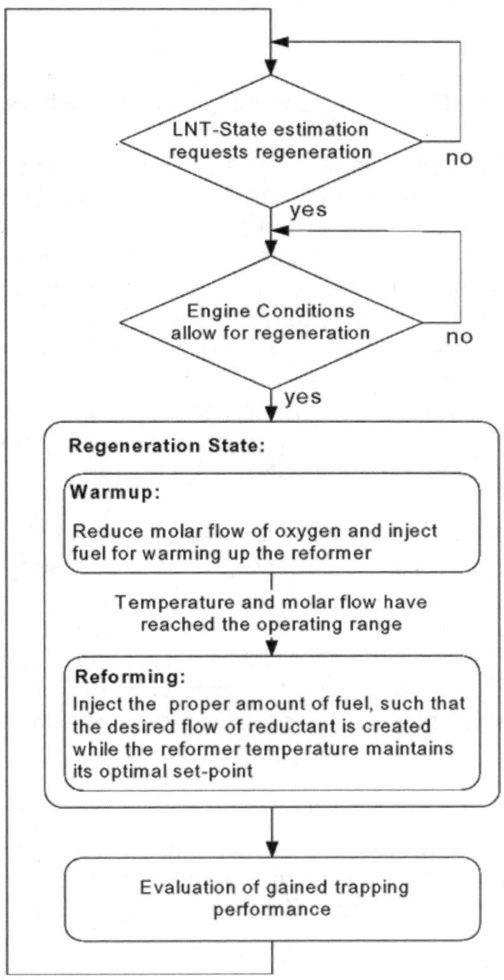

Figure 7: Two Phase Reformer Control

The task of creating the desired amounts of reformate has been divided into two phases. During the first phase, oxygen is depleted and the reformer catalyst is brought up to the preferred operating temperature by injecting fuel below stoichiometry. A Kalman-Filter based state observer is used to estimate the catalyst temperature. The second phase begins once the desired catalyst temperature is reached. In this phase, the actual reforming process is controlled. It has been found that best results can be achieved if the oxygen during this phase can be controlled in such a way that the excessive heat from the exothermic oxidation and water gas shift (Eqns 5 and 7) minus the heat consumed by the endothermic steam reforming process (Eqn 6) is sufficient to keep the catalyst temperature at the set point for optimal reforming efficiency. The partial oxidation process requires the concurrent control of oxygen in the exhaust gas stream and fuel injected upstream of the reformer catalyst.

It is important to consider the NO_x release dynamics during the rich phase to further improve the interaction between reformer and LNT. Operating the reforming process in a suitable way can control the formation of ammonia. These requirements can be implemented as additional elements in the cost function for the optimization problem stated above.

In the chemical process industry, model predictive control (MPC) is commonly used to control processes involving chemical reactions. The MPC strategy simplifies the operating states, (e.g. temperatures and concentrations) as well as limitations in the input quantities. The reformer can be operated in a switched mode, which basically leads to a hybrid (continuous-discrete) control system. In order to include this phenomenon in the control algorithm, applying a predictive control strategy has been found to be beneficial.

Supervisory Control

While the above-mentioned strategies are reasonably easy to operate under steady state or quasi-transient conditions, it might not be feasible to operate the reformer in a suitable fashion under certain transient conditions. In particular, during fast load changes it is difficult to maintain the oxygen flow within the optimal region. Therefore, a supervisory control module continuously monitors the engine conditions and depending on the results of the evaluation permits or terminates a regeneration event.

Diagnostics

Safety aspects are considered in the controller by implementing a scaled diagnostics layer. On the lowest level, the electrical properties of all connections from the sensors and actuators are checked for open input/output and short circuits. On the next level signal range plausibility tests are performed. On the highest level, model-based onboard diagnostics are performed such as comparing the evolvement of the reformer temperature during the fueling event with the temperatures predicted by a thermodynamic model. This approach provides better temperature control to protect against catalyst thermal degradation.

The strategies outlined here have been implemented in Matlab®/Simulink®. A dSPACE MicroAutobox has been used as a Rapid Prototyping Control Platform. For driving the fuel dosing system, a dedicated electronic powerstage module with enhanced diagnostics capabilities has been developed.

RESULTS AND DISCUSSION

LNT/SCR Efficiency Improvement

Figure 8 shows a system efficiency plot at a high speed, high load operating point commonly referred to as C100. The steady state engine out NO_x is approximately 250 ppm while the NO_x out of the LNT is around 50 ppm. The cycle time between LNT regenerations is 100 seconds. As a result, the engine out NO_x is 12.9 grams per cycle while the LNT out NO_x is 3.4 grams per cycle. This accounts for a 74% reduction in NO_x using the LNT catalyst. The regeneration shown by the fuel reformer pulse creates a rich atmosphere in the LNT catalyst that converts stored NO_x to nitrogen, ammonia and other species. For this case, a large amount of the stored NO_x is converted to ammonia as depicted by the high ammonia peak (approximately 1500 ppm). This ammonia is stored in the SCR catalyst and later used to further reduce NO_x in the SCR catalyst such that the NO_x at tailpipe out is 1.9 grams per cycle or approximately 25 ppm. Although the NO_x efficiency in the SCR is only 44% since a large amount of the NO_x was mitigated in the LNT catalyst, the SCR contribution increases the system NO_x efficiency to 85%. This system efficiency is in the proper ballpark to convert a 2007 engine at approximately 1.2 g/hp-hr to the 2010 emission level of 0.2 g/hp-hr NO_x. Note that this experiment was run using a 2004 engine around 2.5 g/hp-hr NO_x that converted 85% of the NOx to around 0.2 g/hp-hr NO_x. These are encouraging results since neither the LNT nor the SCR catalysts are required to mitigate all of the NO_x while the combination shows encouraging results to meet the NO_x target.

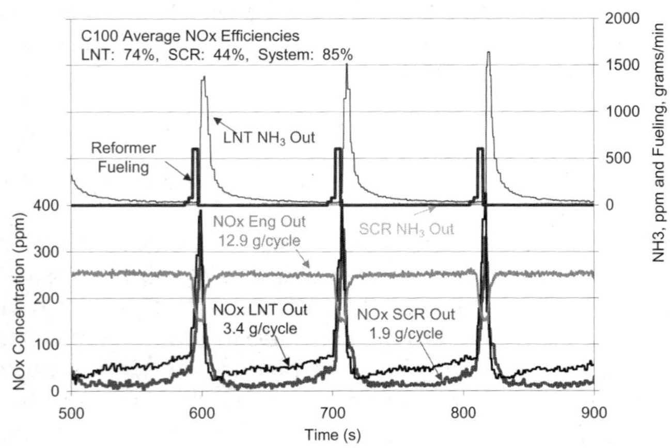

Figure 8: System Efficiency at C100 Operating Point

Figure 9 shows that NH_3 yield increases as more fuel is injected into the reformer during LNT regeneration. However, this apparent correlation with fuel penalty does not show whether the NH_3 yield is controlled by richness (lambda) of the reducing gas, reductant quantity, reductant composition (e.g. H_2/CO/HC ratio) or other factors affected by changing reformer fuel injection quantity.

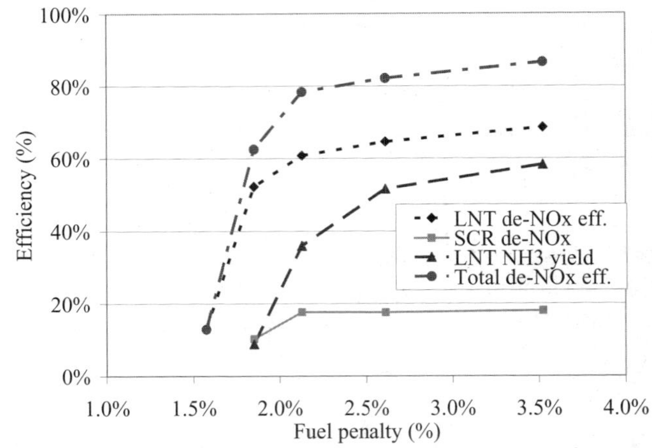

Figure 9: NH_3 Generation and NO_x Reduction Efficiency

Under steady-state (single engine mode) operation, LNT NO_x conversion increases with the fuel penalty (see Figure 9). However, when fuel penalty is over a certain threshold (e.g. @ 2% in this case), further increase LNT NO_x conversion is less efficient by simply increase fuel penalty. This may be due to the nature of the LNT, and thus, catalyst formulation may need to be tuned in order to further improve the performance. From a system integration standpoint, reformer control to optimize the gas composition entering into LNT may be another approach to further improve LNT NO_x conversion.

The SCR NO_x reduction efficiency also levels off when fuel penalty reaches a threshold. This result suggests that increasing NH_3 input to the SCR does not lead to improvement of SCR NO_x reduction efficiency. In other words, the NH_3 utilization for NO_x reduction is reduced at higher NH_3 yields. Further improvement of NH3 utilization efficiency will be addressed in future studies.

Desulfation

The major drawback to LNT systems is the need of desulfation, or sulfur removal, which blocks the NO_x storage sites. Proper desulfation is required to ensure reliability and durability of the LNT to maintain high NOx reduction efficiencies over 435,000 miles for heavy duty vehicles. A worst-case analysis of the number of desulfation events required over the life of the catalyst system is conducted. This analysis assumes a 14L engine displacement having LNT catalysts at twice the engine swept volume (28L). The sulfur level in the exhaust is assumed to be 30 ppm, which is on the high side. This is a combination of the sulfur from the fuel and the oil. An average vehicle fuel economy is assumed to be 6 miles per gallon. Initial estimates of the useful life and sulfur loading are 435,000 miles and 0.5 g/L, respectively. Using these assumptions, the required number of desulfation events is 482 as shown in the equation below.

$$\frac{(435{,}000\ \text{mi})*(3.785\ \text{l/gal})*(0.82\ \text{kg/l})*(1000\ \text{g/kg})*(30\ \text{g S}/10^6\ \text{g fuel})}{(6\ \text{mi/gal})*(0.5\ \text{g S/ L DeSOx trigger})*(28\ \text{L catalyst / vehicle})} = 482$$

Variations in useful life and sulfur loading are shown in Table 1. This analysis suggests that for a Class 8 truck application, the worst case number of desulfation events over 435,000 miles will range between 200 and 500 for LNT sulfur loadings between 0.5 and 1.0 g/L. Likewise, LNT catalysts used for 1 million miles may experience up to 1100 desulfation events.

Table 1: Desulfation Frequency Analysis

DeSOx Frequency	Sulfur Loading, g/L	Useful Life, miles
241	1.0	435,000
482	0.5	435,000
554	1.0	1,000,000
1108	0.5	1,000,000

Figure 10 shows the sulfur release for a typical desulfation event after it was hydrothermally aged for approximately 100 equivalent desulfation events and was subjected to 157 desulfation events. As a result, this LNT catalyst was subjected to approximately 250 desulfation events which puts this catalyst aging on the lower end limit of the worst case desulfation limit for 435,000 mile useful life. The figure shows that the majority of the sulfur is released as H_2S followed by COS and SO_2 over a 500 second desulfation event.

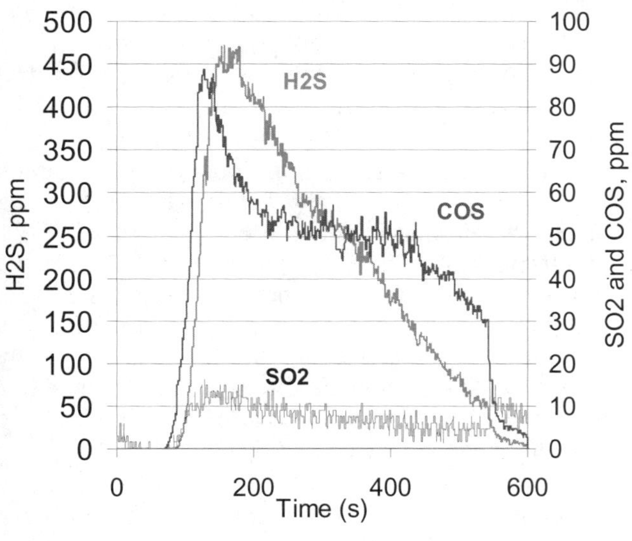

Figure 10: LNT Sulfur Release

FUTURE DEVELOPMENT

System Durability

Figure 11 shows a projected reduction of the desulfation event of the Reformer/LNT/SCR system in comparison with a LNT only system. The performance curve (dashed) of a LNT only system is based on the engine test conditions in B50 operating mode. The horizontal axis shows the hours of operation of the catalyst system, and the vertical axis is the engine out NO_x. The EPA 2010 NTE (not to exceed) emission limit is shown as a horizontal dashed line. The LNT catalyst efficiency will decrease due to the sulfur poisoning from the exhaust stream. Desulfation of the LNT catalyst has to be performed to recover the LNT NO_x reduction performance after approximately 40 to 50 hours of operation. The solid curve is a simulated performance of the Reformer/LNT/SCR system. The ammonia generated during the LNT regeneration process will be stored on the SCR catalyst and will further perform NO_x reduction up to 20% of the engine out NO_x. Therefore, the requirement of NO_x reduction efficiency of LNT catalyst is lowered from 85% of LNT only system to 65% of the Reformer/LNT/SCR system. The time between desulfation events (i.e., desulfation interval) can be increased by up to 30% for the Reformer/LNT/SCR system as compared to a LNT only system.

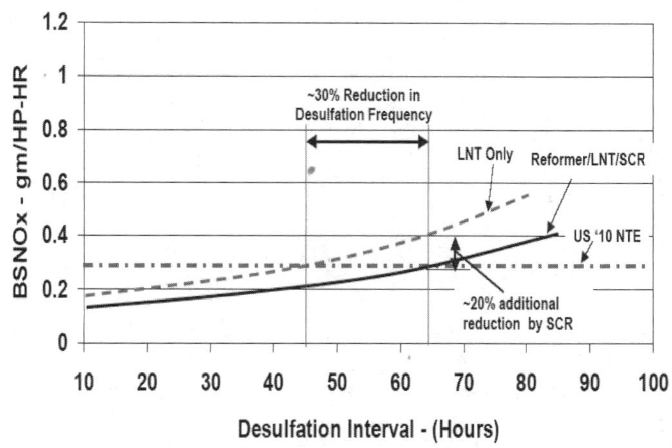

Figure 11: Expected Longer Life with LNT-SCR Combination

Systems Optimization

The results presented in Figure 12 and 13 show a comparison of the results achieved with the initial simple strategy and a more advance approach. It has been found that applying feedback strategies mentioned previously yields superior performance over simple open loop strategies.

Figure 12 shows the production of ammonia during regeneration events using two control strategies. The ammonia yield is significantly increased and can be used to improve the downstream SCR performance. The strategy provides a significant reduction of the typical and undesired NO_x spike during regeneration is shown in Figure 13.

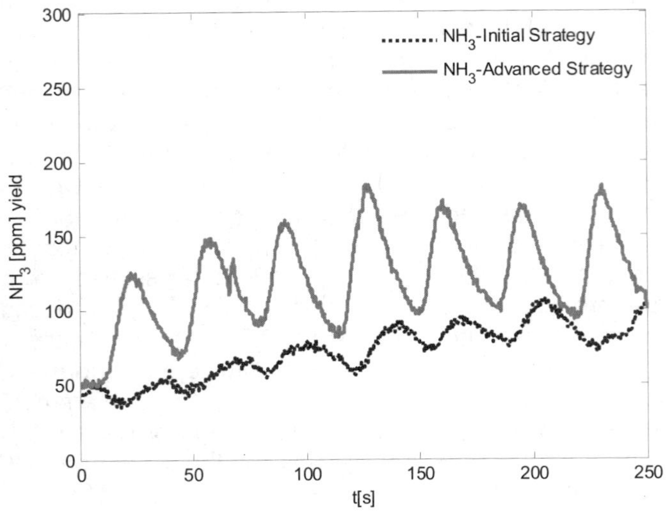

Figure 12: Ammonia Yield: Initial (Dashed) vs. Advanced Control Strategy

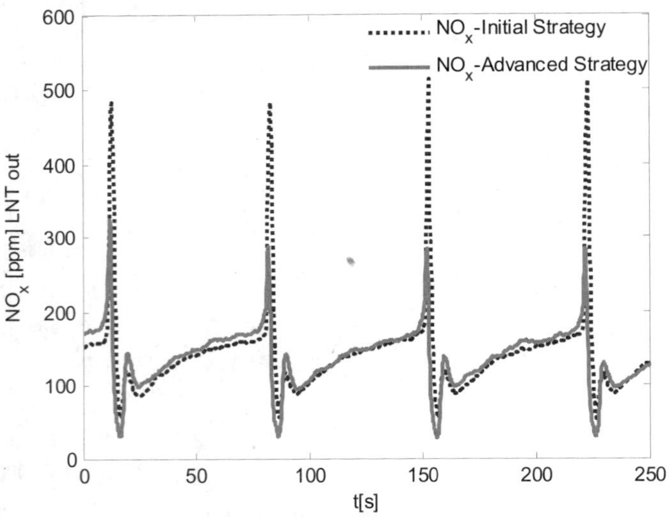

Figure 13: NO_x Spike With Initial (Dashed) And Advanced Control Strategy

Figure 14 shows further refinement to the advanced control strategy that reduces the required fuel injection quantity. A model-based approach was used for controlling the reformer that reduced the calibration effort to optimize the fuel injection quantities.

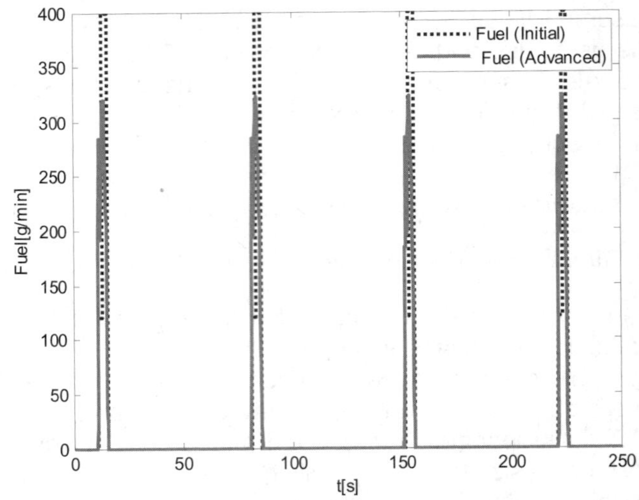

Figure 14: Reduced Fuel Consumption By Applying Advanced Control Strategies

CONCLUSION AND RECOMMENDATIONS

A NO_x aftertreatment system with an advanced control strategy has been proposed for heavy-duty diesel engine emission reductions without requiring additional reductant supply infrastructure. The proposed system uses an in-line fuel reformer followed by a LNT catalyst and a SCR catalyst. The proposed system utilizes the fuel reformer to generate H_2 and CO to improve the subsequent LNT catalyst NO_x reduction efficiency and NH_3 yield. The LNT generates onboard ammonia during LNT regeneration that is used in a downstream SCR for further NOx reduction. The benefits of the proposed aftertreatment have been demonstrated through engine dynamometer testing and can be summarized as follows:

- Both LNT and SCR catalysts contribute to the overall NO_x conversion efficiency. Over 50% of the NO_x stored by the LNT can be converted into ammonia, which is subsequently used by the downstream SCR to further reduce engine out NO_x by up to 20%.

- The LNT catalyst of the proposed aftertreatment system has a relatively lower NO_x conversion efficiency requirement than the LNT only aftertreatment system. For equal overall system NO_x conversion efficiency, the time between desulfation events (i.e., desulfation interval) of the proposed system can be increased up to 30% under the current testing conditions. Less frequent desulfation events will contribute to durability improvement of the LNT catalyst as well as fuel saving.

- A control system has been developed and implemented that allows regeneration of the LNT during steady state and transient engine conditions. Onboard diagnostics enable the safe operation of the system. The control system has been used under lab conditions and in a vehicle during various test trips. Control strategies have been implemented to optimize the aftertreatment system. The optimized control strategies maximize ammonia generation, provide significant reduction of the NO_x spikes and allow for a lower fuel penalty. Future development will focus on further improvement of the strategies in particular for desulfation.

ACKNOWLEDGMENTS

The authors would like to thank Eaton Corporation for its permission to publish this work. The authors would like to acknowledge the support of Karen Bevan, Jeff Boris, Dave Ginter, Marty Nadeau, Mike Rosbury, Dmitry Shamis, Vishal Singh, Bill Taylor, Tom Stover and Kevin McGovern.

REFERENCES

1. Duffy K. et al.," Heavy duty Diesel Engine HCCI Development", US DOE DEER Conference, Coronado, CA, August 2004.

2. Burch, R., et al., "A Review of the Selective Reduction of NO_x with Hydrocarbons under Lean-burn Conditions with Non-zeolite Oxide and Platinum Group Metal Catalysts," Applied Catalysis B: Environmental 39 (2002) 282-303.

3. Laroo, C. et al., "NOx Adsorber Desulfation Techniques for Heavy-Duty On-Highway Diesel Engines" SAE 2002-01-2871

4. Mizuno T. et al., "Development of a New DPNR Catalyst" SAE 2004-01-0578

5. Hirata, K. et al." Development of Urea SCR System for HDD Commercial Vehicles", SAE 2005-01-1860.

6. Dalla Betta R. el al." Transient Dynamometer Testing of a Single leg NOx Adsorber Combined with a Fuel Processor for Enhanced NOx Control", US DOE DEER Conference, Chicago, Sept. 2005.

7. Lambert, C. et al., "Urea SCR and DPF System for diesel LDT/SUV meeting Tier 2 Bin 5", US DOE DEER Conference, Chicago, September 2005

8. Acharya, R. et al., "Fuel and System Interaction Effects on Urea-SCR Control of in Diesel Exhaust Aftertreatment", SAE 2006-01-0638.

9. Ketfi-Cherif, A. et al.,"Modeling and Control of a NOx Trap Catalyst", SAE 2000-01-1199

10. Sunohara, D. et al. ,"Control of Rich Spike with Adaptive Estimation of Stored NOx in Lean NOx Trap for a Diesel Engine", SAE 2005-01-1090

11. Kim Y.-W. et al., "A Phenomenological Control Oriented Lean NOx Trap Model", SAE 2003-01-1164

12. Asik, J. et al., "Closed Loop Temperature Control Of A Lean NOx Trap During Desulfation", in proceedings of the American Control Conference, San Diego, California, 1999

13. Koebel, M., Elsener, M. and Kleemann, M., "Urea-SCR: a promising technique to reduce NOx emission from automotive diesel engines," Catalysis today, Volume 59, pgs. 335-345, 2000.

CONTACT

Haoranhu@eaton.com

2007-01-1244

Calibration of a LNT-SCR Diesel Aftertreatment System

Rachel Snow, Giovanni Cavatatio, Doug Dobson, Cliff Montreuil and Robert Hammerle
Ford Motor Company Research and Advanced Engineering

ABSTRACT

An investigation into the post-LNT NH_3 generation and NO_x slip over an LNT was completed for the purposes of determining the proper calibration for NO_x reduction over a downstream SCR. Experiments were conducted in an engine dynamometer on the DW12 2.2L diesel engine that was calibrated for rich operation.

The catalyst system was evaluated at 325°C for three different engine speed/load points. Engine speeds of 1750, 2000, and 2500 rpm were chosen, resulting in a LNT space velocities of 35,000 – 68,000 hr^{-1}. Increased lean times were tested, which resulted in NOx loadings on the LNT that ranged from 0.10 - 0.50 grams NO_x/L catalyst. In addition, varied degrees of richness were tested, corresponding to lambda sensor readings of 0.90, 0.92, and 0.94.

LNT performance was evaluated based on it's ability to provide a stoichiometric ratio of reactants to the SCR for NO_x conversion. Gross and net NO_x conversion was evaluated for a LNT-SCR system. Optimal NO_x conversion for this type of system required a balance between providing sufficient reductant for NH_3 generation without poisoning the SCR from HC. System performance was controlled by calibration of NO_x loading and degree of richness.

INTRODUCTION

Diesel engines are a proven technology for providing the customer with fuel economy superior to stoichiometric gasoline engine operation. However, NO_x emissions are significantly higher than that of stoichiometric gasoline operation because the lean combustion of diesel engines precludes the use of traditional three-way catalysts. To meet 2007 US and future European NO_x emissions standards, a reduction of tailpipe NOx is required.

Two leading contenders for diesel NOx aftertreatment are the Lean NOx Trap (LNT) and Selective Catalytic Reduction (SCR) catalysts. Each technology has pros and cons associated with implementation. NH_3-SCR catalysts have the benefit of containing no precious metal, a significant cost saving relative to LNTs, in addition to a wider range of operating temperatures where NO_x conversion can be achieved [1]. However, the NH_3-SCR systems traditionally use urea as the feed reductant, thus requiring a currently non-existent infrastructure to supply vehicles with the reductant necessary to maintain catalyst function.

LNTs have the benefit of using diesel fuel that is already present on the vehicle as a NO_x reductant. However, large amounts of precious metal are required to achieve future emissions targets. In addition, high temperature events are required to clean sulfur that has accumulated on the LNT [2,3]. High temperature exposure is known to sinter precious metal, which can result in loss of its ability to reduce NO_x to N_2.

LNTs use periodic rich purges with diesel fuel to reduce NO_x stored on the trap. Careful control of the purge event is required to prevent NO_x reduction to NH_3 rather than the desired N_2 [4]. A synergy of the two systems is possible if one takes advantage of the ability of the LNT to produce NH_3 to provide the reductant to a NH_3-SCR catalyst, a system referred to as LNT-SCR.

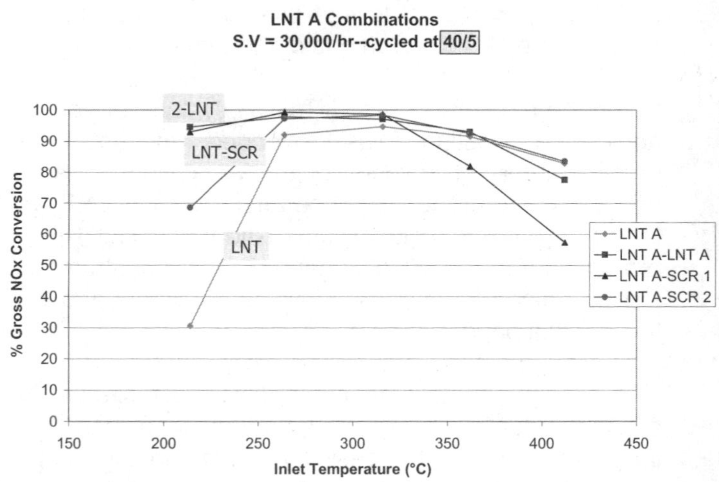

Figure 1. LNT-SCR system performance compared to an LNT system with equivalent precious metal

Experiments conducted at Ford Research Laboratories in Dearborn, Michigan in 2001 led to a new concept of combining an LNT upstream of an SCR to obtain improved NOx conversion and reduced ammonia slip, relative to a traditional LNT only aftertreatment system. The patent application describing the embodiments of such a system was submitted in 2002 [5]. Further enhancements were noted by positioning the LNT and SCR formulations in alternating zones and ultimately a same brick double-layer LNT/SCR configuration.

Figure 1 shows the results from flow reactor studies comparing this new LNT-SCR system with a traditional LNT [6]. The catalyst cores had been hydrothermal aged 64 hrs in a oven at 670°C prior to evaluation in a flow reactor. Evaluations were performed at a space velocity of 30,000 hr^{-1} for each catalyst and at temperatures ranging from 200-400°C. The reactor was programmed for lean/rich cycling at 40 sec lean/ 5 sec rich. The gas mixture used for the lean and rich conditions are shown in Table 1.

GAS	LEAN	RICH
CO	0 ppm	4.00%
H_2	0 ppm	1.30%
C_3H_6	0 ppm	5000 ppm C1
O_2	10%	1%
NO	500 ppm	500 ppm
CO_2	5%	5%
H_2O	5%	5%

Table 1. Gas mixture used in flow reactor evaluation of hydrothermal aged LNT and LNT-SCR aftertreatment systems

The LNT-SCR system had improved performance over a traditional LNT with no additional precious metal content. Furthermore, the LNT-SCR system achieved NOx conversion equivalent to that of a LNT with 2 times the precious metal under portions of the temperature regime.

Given the potential of the LNT-SCR system for vehicular application, testing was extended to an engine dynamometer. The focus of the work was to gain an understanding of the NH_3 generation over the LNT using HC provided from a diesel combustion process. This is a prerequisite to the development of a control strategy for the operation of a combined LNT-SCR system and is the topic of this paper.

The NH_3 at the tailpipe of an LNT-SCR system was examined for a variety of operating conditions. It is important to prevent excess NH_3 from slipping past the catalysts of lean NOx aftertreatment systems. If a Catalyzed Diesel Particulate Filter (CDPF) containing precious metal is downstream, the NH_3 may be oxidized to NO_x, resulting in reduced aftertreatment system performance [7]. There is also a risk of the formation of N_2O, which is a potent greenhouse gas with 270 times the warming potential of CO_2.

EXPERIMENTAL CONDITIONS

A series of experiments were conducted on an engine dynamometer using a PSA DW12 2.2L diesel engine at Ford Research Laboratories in Dearborn, Michigan. An LNT-SCR catalyst system underwent accelerated aging on the dynamometer to 50,000 equivalent miles. A schematic of the system is shown in Figure 2.

The system included a 1.25L Diesel Oxidation Catalyst (DOC) upstream of the LNT. The LNT and SCR bricks were each 5.66-inch (144 mm) diameter by 6 inch (152 mm) long, resulting in catalyst volumes of 2.5L (1.12 x ESV) each.

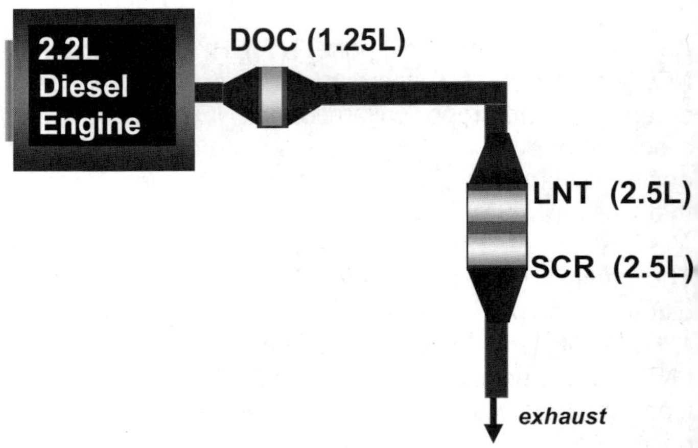

Figure 2. Aftertreatment system for the control of diesel NO_x emissions by using a DOC followed by an LNT and SCR catalyst

The 50,000 mile accelerated aging included three phases - sulfur exposure, desulfation (deSOx), and DPF regeneration. Sulfur exposure was achieved by SO_2 injection upstream of the LNT. The engine was operated using a NO_x regeneration mode of 55 sec lean/ 5 sec rich at a steady operating point of 2000 rpm/ 68 N-m during the sulfur exposure. The average LNT inlet temperature was 320°C.

Desulfation was achieved by raising the catalyst bed temperature by means of lean post-injection. Once the desired deSOx temperature was reached, the engine was cycled lean/rich (2 sec lean/ 15 sec rich) providing an average catalyst midbed temperature of 700°C. DPF regeneration conditions were achieved by raising the catalyst midbed temperature to 580 °C using lean post-injection.

Desulfation and DPF regeneration times were based an assumed vehicle fuel economy of 40 mpg (5.88 L/100 km) using 10 ppm sulfur fuel and oil consumption of 1 quart (0.946 L) every 5000 miles (8000 km). DPF regeneration time was based on an estimate of engine particulate level and a soot limit of 6 g soot/L filter.

After aging, the catalyst system was evaluated at three different engine speed/load points. Engine speeds of 1750, 2000, and 2500 rpm were chosen to provide space velocities on the LNT ranging from 35,000 – 68,000 hr^{-1}. Engine throttle was set such that the average catalyst midbed temperature was held constant at ~325°C. For each point, the lean time between rich regeneration pulses was varied to provide NO_x loading on the LNT ranging from 0.10 - 0.50 grams NO_x/L catalyst. The 5 second rich regeneration pulse used to generate ammonia was tested at lambda sensor values of 0.90, 0.92, and 0.94. A summary of the test conditions is shown in Table 2.

T = ~325 °C	CASE #1	CASE #2	CASE #3
RPM	1750 rpm	2000 rpm	2500 rpm
Lean time	60-100 sec	60-85 sec	60-85 sec
LNT NOx Loading	0.1 – 0.5 (g NOx/L LNT)	0.2 – 0.5 (g NOx/L LNT)	0.3 – 0.5 (g NOx/L LNT)
Lambda	0.94 - 0.90	0.94 - 0.90	0.94 - 0.90
HC	3000-5000 ppm	5000-8000 ppm	1-2%
CO	1.3- 2.0%	1.4- 1.8%	1.5-1.8%

Table 2. Test Condition for Calibration Experiments on a 50,000 mile Equivalent Aged LNT-SCR systems

A Midac M-series analyzer was used to measure nitrogen species including NO, NO2, and NH_3 by Fourier Transform Infrared (FT-IR) spectroscopy. A Horiba MEXA 7100 Gas Analyzer with a heated Flame Ionization Detector (FID) was used to measure CO and total hydrocarbons (THC).

RESULTS AND DISCUSSION

LNT-SCR SYSTEM NET NOX CONVERSION

Figure 3 shows the net NO_x conversion versus NO_x loading for a LNT-SCR system over the experimental matrix described in Table 2. Net NO_x conversion does not count NH_3 as a converted product and is the best measure of system performance for a system that may have precious metal components downstream. As discussed in the introduction, NH_3 has the potential to be oxidized into NO_x or N_2O, either of which is undesirable. The results in Figure 2 show there was a strong decline in net NO_x conversion as the NO_x loading between purge events was increased.

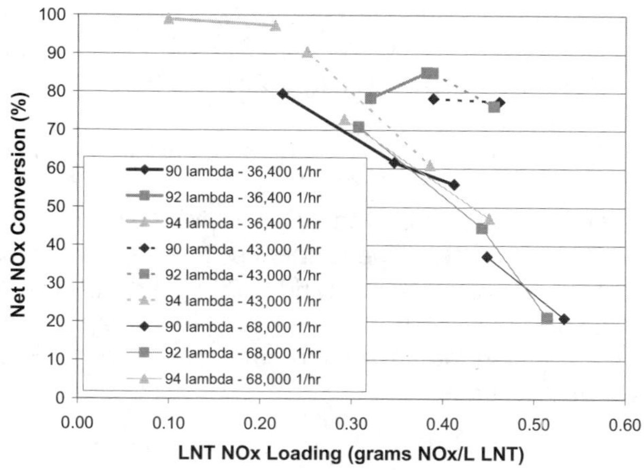

Figure 3. Net NOx conversion for a LNT-SCR system as a function of NOx loading on the LNT for varied space velocity and purge richness

The first portion of a rich purge event is used to deplete excess oxygen from the exhaust stream and only the later portion of the rich purge contributes to the reduction of NO_x to NH_3. It is therefore desirable to maximize the interval between purge events in order to minimize the fuel penalty associated with the depletion of excess oxygen. Typical O_2 concentrations in diesel exhaust range from 10-14%, thus a significant amount of fuel is required to consume the O_2 present in the exhaust prior to NH_3 generation. Thus, an objective of LNT-SCR system calibration should be to reduce purge frequency in an effort to maintain the strong fuel economy benefit of a diesel engine.

Increased NOx loading between regeneration events corresponded to reduced purge frequency, and therefore a reduced fuel penalty. Despite the general trend of decreased performance with increased NOx loading, there were a few points where ~80% NOx conversion was achieved at high levels of NOx loading. The remainder of this paper examines the factors that contributed to achieving this high system performance while minimizing fuel penalty; including space velocity, LNT NOx loading, and richness of the purge event.

NH3/NOX RATIO AND NET NOX CONVERSION

Key to successful operation of an LNT-SCR system is the generation of the stoichiometric proportions of NO_x and NH_3 at the exit of the LNT for conversion over the downstream SCR. NH_3/NO_x = 1 is the stoichiometric ratio for two of the three reduction reactions equations (Eq.1-3) that take place on an SCR catalyst.

$$(1)\quad 4\,NO + 4\,NH_3 + O_2 \rightarrow 4\,N_2 + 6\,H_2O$$

$$(2)\quad NO + NO_2 + 2\,NH_3 \rightarrow 2\,N_2 + 3\,H_2O$$

$$(3)\quad 6\,NO_2 + 8\,NH_3 \rightarrow 7\,N_2 + 12\,H_2O$$

In this section, the response of post-LNT NH_3/NO_x ratio and LNT-SCR system performance to the calibration parameters is examined at 1750, 2000, and 2500 rpm. In a later section, laboratory data is used to explain the effect of changing the calibration parameters on the SCR catalyst performance.

Case 1 – Low Space Velocity (1750 rpm)

Figure 4 shows the NH_3/NO_x ratio as a function of post-DOC HC (increased richness of purge) for LNT NO_x loadings of 0.22, 0.33, and 0.40 grams NO_x/L LNT. A broad range of NH_3/NO_x ratios was generated as the degree of richness was increased from 0.90 to 0.94 lambda. For instance, when the trap was lightly loaded (0.22 g/L LNT), the NH_3/NO_x ratios ranged from 1 to 35.

Figure 5 shows the gross and net NO_x conversion at the outlet of the SCR for the 1750 rpm case. High NH_3/NO_x ratios at the inlet to the SCR lead to low net NOx conversion because most of the excess NH_3 went unreacted through the SCR. Only the leanest purge (lambda = 0.94) provided net NO_x conversion above 90%.

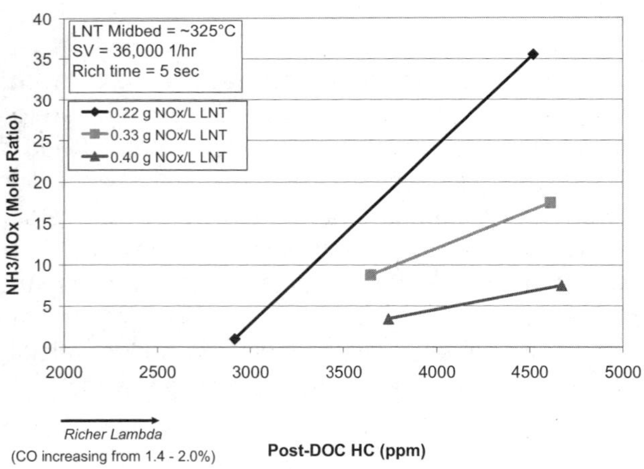

Figure 4. Effect of rich pulse characteristics on post-LNT NH_3/NO_x Ratio for 1750 rpm case

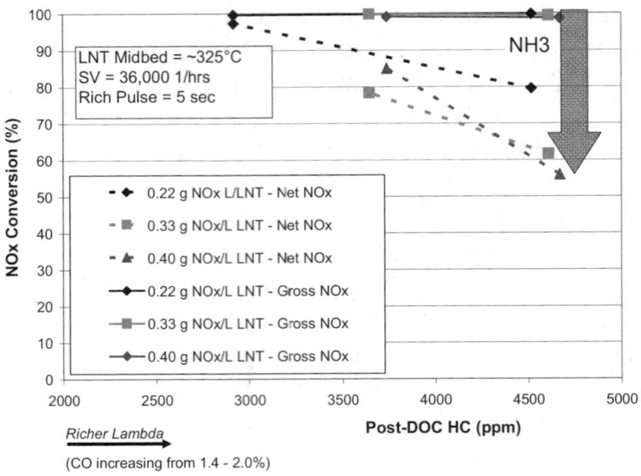

Figure 5. Net NOx conversion as a function of richness of purge for the 1750 rpm case

Case 2 – Moderate Space Velocity (2000 rpm)

Figure 6 and 7 shows the same graphs of NH_3/NO_x ratio and NO_x conversion for the 2000 rpm case. Figure 6 shows near stoichiometric NH_3/NO_x ratios for SCR operation were achieved at both 0.38 and 0.45 grams NO_x/L LNT. Interestingly, the higher loading was less sensitive to the degree of richness. This may be an important consideration for system robustness, as aging of the upstream precious metal components will result in changing purge characteristics over its in-vehicle life [8].

Figure 7 shows that while increased richness provided better gross NO_x conversion, the net NO_x conversion was in fact worse. Thus the improvement in NO_x gross conversion was due to additional NO_x going to NH_3, rather than N_2. For the conditions tested, the highest net NO_x conversion, 85%, was found at the moderate purge (lambda = 0.92) with a NO_x loading of 0.38 g NO_x/ L LNT.

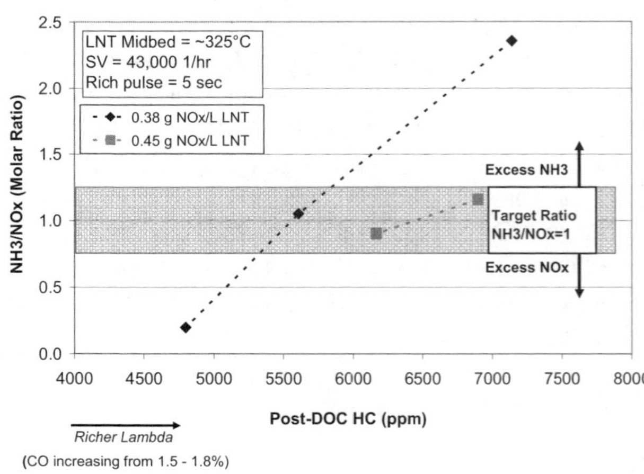

Figure 6. Effect of rich pulse characteristics on post-LNT NH_3/NO_x ratio for 2000 rpm case

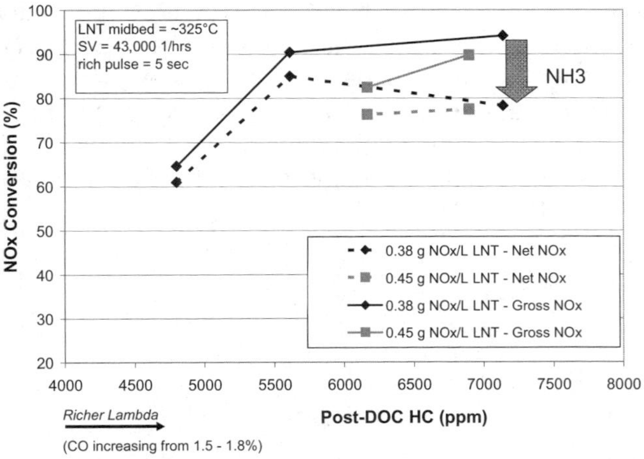

Figure 7. Net NOx conversion as a function of richness of purge for the 2000 rpm case

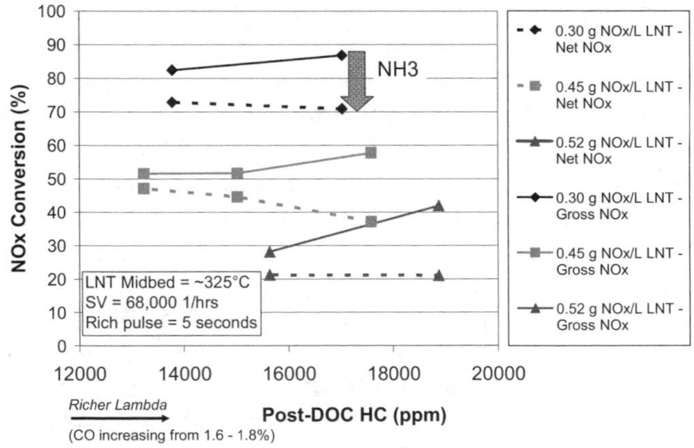

Figure 9. Net NOx conversion as a function of richness of purge for the 2500 rpm case

Case 3 – High Space Velocity (2500 rpm)

Figure 8 shows stoichiometric ratios of NH_3/NO_x were achieved for all three of the NO_x loadings tested. As NO_x loading increased, more reductant was required to achieve the proper NH_3/NO_x. Furthermore, the amount of reductant required to achieve the proper ratio was 2-3 times greater at this engine speed than for the lower engine speeds.

Figure 9 reveals system NOx conversion fell dramatically as the LNT NOx loading was increased, despite the fact the proper mix of NH_3 and NO_x was produced by the LNT. One explanation for this behavior is that at high space velocity (68,000 hr^{-1}), conversion is limited by mass transfer considerations on both the LNT and SCR. Later in this paper, the data will show that HC breakthrough was also an important factor in this regime.

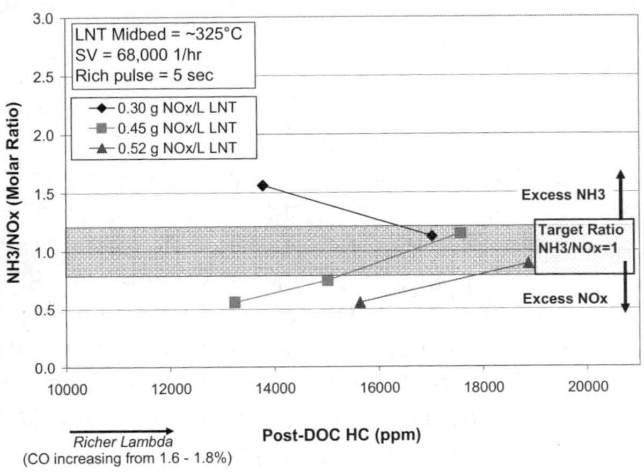

Figure 8. NH3/NOx ratio as a function of richness of purge for the 2500 rpm case

POST-LNT GAS COMPOSITION

In the previous section, the affect purge richness had on post-LNT NH_3/NO_x ratio and NO_x conversion over the LNT-SCR system was discussed. In this section, the impact LNT NOx loading and space velocity had on post-LNT gas composition is examined.

Figure 10 shows the mass of NO_x that slipped past the LNT during a single lean/rich cycle. NO_x slip increased with the quantity loaded on the LNT, both during the lean storage period as the trap neared saturation and during the rich purge event where a large spike of NO_x was desorbed and remained unconverted [9]. NO_x slip increased with NO_x loading more rapidly at the high space velocity (2500 rpm) because a greater proportion of slip occurred during the lean storage period.

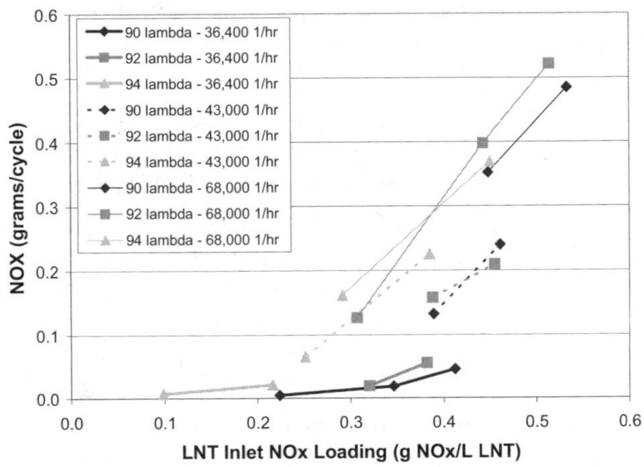

Figure 10. Post-LNT NOx emissions as a function of NOx loading, space velocity (engine rpm), and purge richness

At low space velocity (1750 rpm) there was very little NOx slip during the lean storage period. The NO oxidation to NO_2 and subsequent storage was very efficient. Therefore, the measured NOx slip was primarily from the desorption spike during the rich purge. As a result there was a clear trend of lower NOx slip as purge richness increased and the purge event became more effective.

The low levels of NOx at the outlet of the LNT at 1750 rpm explains why there was such a large range of NH_3/NO_x ratios, shown previously in Figure 4. The LNT was very effective at the low space velocity conditions; thus, it was easy to generate too much ammonia relative to the very small quantities of NO_x.

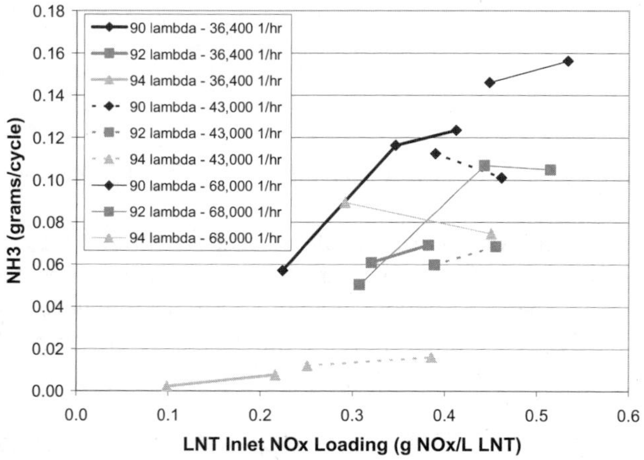

Figure 11. Post-LNT NH_3 emissions as a function of NOx loading, space velocity, and purge richness

Post-LNT NH_3 generation is shown in Figure 11. NH_3 generated over the LNT initially increased with NO_x loading. This is consistent with the literature, which has documented increased NH_3 emissions from a LNT as the NOx storage sites became saturated [2]. However, Figure 11 shows that further increasing NOx loading failed to generate more NH_3. In this regime, NH_3 generation was limited by the amount of available reductant.

The reductant limited regime can be observed clearly for the experiments run at 1750 rpm and 0.90 lambda. As the NO_x loading was increased from 0.21 to 0.35 gram/L, the NH_3 generated more than doubled from 0.055 to 0.115 grams/cycle. Yet when the NO_x loading was further increased to 0.41 gram/L, the NH3 generated increased only slightly to 0.125 grams/cycle.

As NOx loading was increased, the NO_x slip increased at a greater rate than the NH_3 generated, which resulted in a decrease in NH_3/NO_x, shown in Figure 12. One strategy to find the proper NH_3/NO_x ratio may be to start from an NH_3 rich point and increase the NO_x loading until the molar ratios are equal, or conversely start from a NH_3 lean point and decrease NO_x loading.

For example, the best NH_3/NO_x ratio at 1750 rpm was at 0.94 lambda and less than 0.25 grams NOx/L LNT. However, the fuel penalty was greater than 10% because of the low LNT NOx loading (high regeneration frequency). A reduced fuel penalty may be achieved by increasing the NOx loading to match the NH_3 generated at 0.92 or 0.90 lambda.

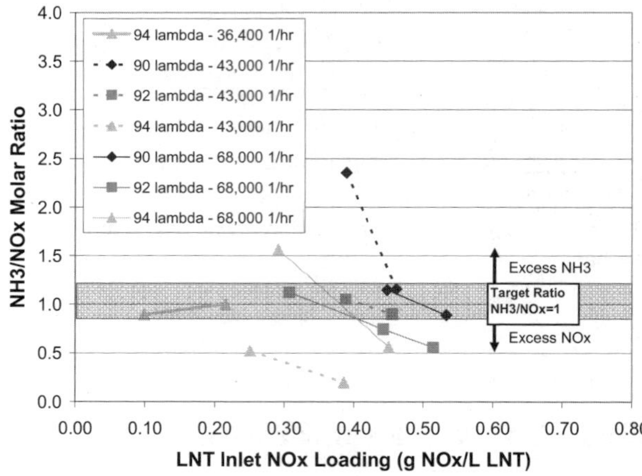

Figure 12. Effect of NOx loading, space velocity, and richness of purge on post-LNT NH3/NOx Ratio Figure 10 shows the molar NH_3/NO_x ratio that resulted from the combined NOx slip and NH_3 generation. The shaded band indicates the target area for operation, where the NH_3/NO_x ratio is 1

Unfortunately, no conclusion can be drawn regarding the trend of NH_3 generation with space velocity because the data is confounded by changes in the composition of the rich purge, which play a role in the chemistry of NH_3 generation on the LNT. Space velocity was altered by changing engine speed and pedal command. The result was significant differences in rich pulse composition due to the rich calibration and the ability of the DOC to oxidize HC at the different speed/load points. For example, as engine speed increased from 1750 to 2500 rpm the HC concentrations increased from 3,000 ppm to 10,000 ppm, despite having the same lambda sensor reading of 0.94.

SCR NOX CONVERSION

In the last section, the impact LNT NO_x loading had on both NO_x slip and NH_3 generation and a method for arriving at a stoichiometric ratio of NH_3/NO_x at the outlet of the LNT for subsequent conversion over an SCR catalyst was discussed. In addition, the effect of space velocity and purge richness was examined. In this section, the SCR performance is analyzed and the factors that are important for achieving high LNT-SCR system NO_x conversion are discussed.

Gross NO_x conversion over the SCR versus inlet NH_3/NO_x ratio is shown in Figure 13. The target area for system calibration near stoichiometric NH_3/NO_x ratios is represented by the shaded region on the graph. The area to the left of the shaded target area had low NO_x conversion due to insufficient ammonia generation. The highest level of conversion was found to the right of the target area, where there was an abundance of NH_3 available for NO_x reduction. Operation in this region is not desirable however, because there was a great deal of NH_3 slip past the LNT-SCR system, which could lead to the formation of undesirable products if a catalyzed DPF were downstream of the system.

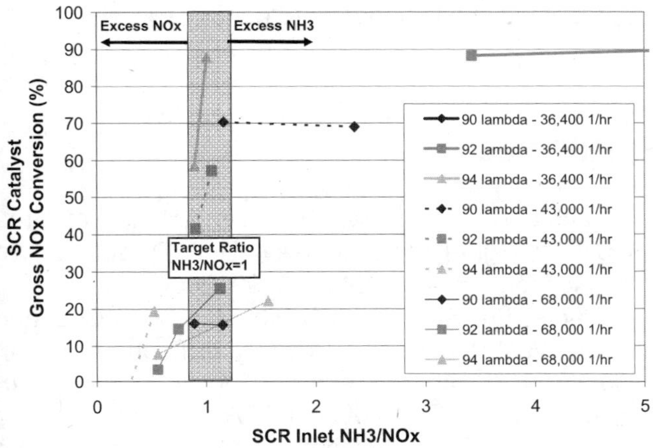

Figure 13. Gross NOx conversion over the SCR catalyst in an LNT-SCR diesel aftertreatment system

SCR NO_x conversion was much lower than expected for several points, despite the fact the post-LNT NH_3/NO_x ratios were near stoichiometric. These points corresponded to the high space velocity condition of ~68,000 hr^{-1} (2500 rpm).

Figure 14 shows the results of a space velocity study conducted in a laboratory flow reactor using a similar SCR catalyst formulation. The catalyst core had been hydrothermally aged in a oven for 64 hrs at 670°C, a thermal exposure equivalent to 120,000 miles in a vehicle equipped with a DPF. NO conversion over the SCR catalyst was evaluated over a range of temperatures at both 30,000 and 80,000 hr^{-1}. SCR feedgas was comprised of 350 ppm NH_3, 350 ppm NO, 5% CO_2, 5% H_2O, 14% O_2, with a balance N_2. At 325°C, the conversion fell from 87% to 75% as space velocity increased from 30,000 and 80,000 hr^{-1}.

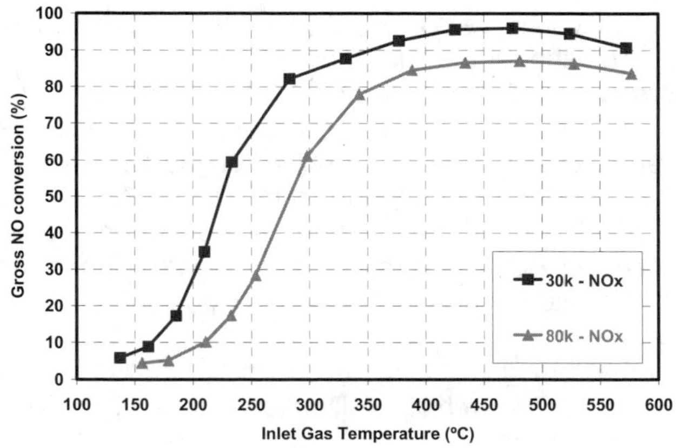

Figure 14. Gross NOx Conversion of a SCR catalyst that underwent hydro-thermal aged at 670°C for 64 hrs at both 30,000 and 80,000 hr^{-1} space velocity

The results from the dynamometer showed SCR NO_x conversion fell from 88% at 36,400 hr^{-1} (1750 rpm) to less than 25% at 68,000 hr^{-1} (2500 rpm). While the laboratory data shows increased space velocity can account for a portion of the loss in performance, there must be other factors present that caused the degradation in performance. Another factor that is known to affect SCR performance is HC poisoning [10]. Figure 15 shows the result of the addition of HC to the laboratory evaluation conditions described previously. Upon exposure to 50 ppm n-Decane (500 ppm C_1 HC), there was a rapid deterioration in SCR NO_x conversion from 82% to 65%.

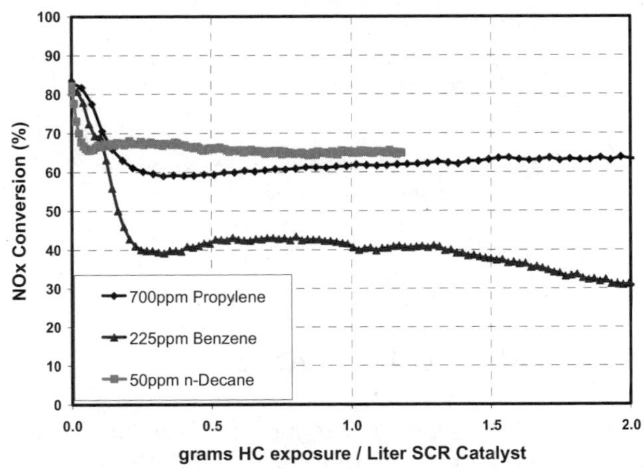

Figure 15. Effect of exposure to Propylene, Benzene, and n-Decane on NO_x conversion for a SCR catalyst that was hydro-thermal aged at 670°C for 64 hrs

It was noted earlier in this paper that the HC concentrations increased significantly with engine speed. Figure 16 shows the mass of HC reaching the SCR catalyst was 3-8 times greater for the 2500 rpm case than for the lower engine speeds. The average concentration during the rich pulse ranged from 4000 - 9000 ppm C_1 HC at 2500 rpm, compared to 750 - 2000 ppm C_1 HC at 1750 rpm. There is a direct correlation between the HC exposure shown in Figure 16 and the SCR gross NOx conversion shown in Figure 13.

The breakthrough of such high levels of HC is not acceptable for a vehicle calibration. Ideally, the rich purge would have minimal HC and be comprised primarily of H_2 and CO. Refinement to the rich calibration may improve the rich pulse composition and prevent HC poisoning of the SCR. Rich engine operation is achieved by reducing mass airflow through intake throttling and EGR, and increasing fuel quantity while adjusting injection timing to provide the desired torque. Experiments conducted at Oak Ridge National Laboratory has shown significant differences in the relative amounts of HC and CO during rich operation depending on the calibration approach [11].

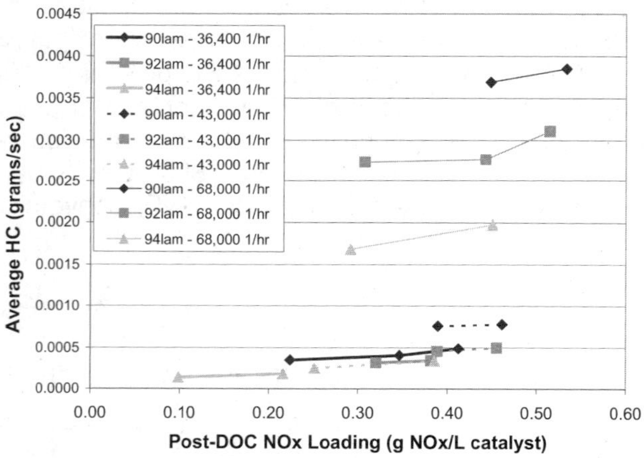

Figure 16. Mass of HC at Outlet of LNT during dynamometer evaluations of a LNT-SCR aftertreatment system

SUMMARY AND CONCLUSIONS

In this paper, the effect of varying space velocity, NOx loading, and richness of purge had on LNT-SCR system performance was discussed in the context of maximizing NOx conversion while minimizing fuel penalty. The result showed,

- Increasing LNT NOx loading typically decreased the NH_3 to NOx ratio.
- NH_3 generation increased with NOx loading, and eventually levels off due to limited reductant availability.
- NO_x loading and purge richness values can be chosen to minimize fuel penalty

In addition, the results showed gross NO_x conversion over the SCR remained high as long as a sufficient amount of NH_3 was available to reduce the NO_x and HC slip was limited so that poisoning did not occur.

ACKNOWLEDGMENTS

Mike Odell and Gar Flint provided for the maintenance and smooth operation of the dynamometer where the data presented in this paper were generated.

REFERENCES

1. C. Lambert, et al., "Technical Advantages of Urea SCR for Light-Duty and Heavy-Duty Diesel Vehicle Applications", SAE 2004-01-1292, 2004
2. J. Parks, Shean Huff, Josh Pihl, Jae-Soon Choi and Brian West, "Nitrogen Selectivity in Lean NOx Trap Catalysis with Diesel Engine In-Cylinder Regeneration", SAE 2005-01-3876, 2005
3. J. Li, et al., "Sulfur Poisoning and Desulfation of the Lean NOx Trap", SAE 2001-01-2503, 2001
4. Y. Cheng., et al., "Factors Affecting Diesel LNT Durability in Lab Reactor Studies", SAE 2004-01-0156, 2004
5. H.S. Gandhi, J.V. Cavataio, R.H. Hammerle, Y. Cheng, U.S. Patent No. US2004/ 0076565, April 22, 2004.
6. C. Lambert, et al., "Urea SCR and DOF System for Diesel LDT/SUV Meeting Tier 2 Bin 5", US Department of Energy Diesel Engine Emissions Reductions (DEER) Conference, Chicago, September 2005
7. J.Girard et al., "The Influence of Ammonia Slip Catalysts on Ammonia, N2O and NOx Emissions for Diesel Engines", 07PFL-103, September 25, 2006
8. R.Snow et al., "Robustness of a LNT-SCR System to Aging Protocol", SAE 07PFL-677, 2007
9. Joseph R. Theis, et al., "NOx Release Characteristics of Lean NOx Traps During Rich Purges", SAE 2003-01-1159, 2003
10. Steven J. Schmieg and Jong-Hwan Lee, "Evaluation of Supplier Catalyst Formulations for the Selective Catalytic Reduction of NOx with Ammonia", SAE 2005-01-3881, 2005
11. Brian West, et al., "Assessing Reductant Chemistry During In-Cylinder Regeneration of Diesel Lean NOx Traps", SAE 2004-01-3023, 2004

CONTACT

Rachel Snow

Ford Research and Advanced Engineering

2101 Village Rd

Rm 3215

MD 3179

Dearborn, MI, 48121

rsnow5@ford.com

DEFINITIONS, ACRONYMS, ABBREVIATIONS

CDPF: Catalyzed Diesel Particulate Filter

deSOx: Desulfation or sulfur removal

DOC: Diesel oxidation catalyst

FID: Flame Ionization Detector

FTIR: Fourier Transform Infrared

LNT: Lean NOx Trap, sometimes called NSR for NOx Storage Reduction catalyst

NOx: Oxides of nitrogen; generally taken as NO plus NO_2

SCR: Selective Catalytic Reduction

THC: Total Hydrocarbons

IV. INTEGRATED AND COMBINED SYSTEMS FOR SIMULTANEOUS PM AND NO_x CONTROL

IV.1 HEAVY DUTY

2000-01-0187

Achieving Heavy-Duty Diesel NO_x/PM Levels Below the EPA 2002 Standards--An Integrated Solution

Magdi Khair
Southwest Research Institute

Jacques Lemaire
Rhodia

Stefan Fischer
Siemens

ABSTRACT

The diesel engine has long been the most energy efficient powerplant for transportation. Moreover, diesels emit extremely low levels of hydrocarbon and carbon monoxide that do not require post-combustion treatment to comply with current and projected standards. It is admittedly, however, difficult for diesel engines to simultaneously meet projected nitrogen oxides and particulate matter standards. Traditionally, measures aimed at reducing one of these two exhaust species have led to increasing the other. This physical characteristic, which is known as NO_x/PM tradeoff, remains the subject of an intense research effort.

Despite this challenge, there is significant evidence that heavy-duty highway engine manufacturers can achieve substantial emission reductions. Many development programs carried out over the last five years have yielded remarkable results in laboratory demonstrations. Yet, even with the progress achieved thus far it is evident that engine design measures alone will fall short of meeting future NO_x/PM emissions limits.

In the meantime, many advances were achieved in diesel exhaust post-combustion emission control devices (after-treatment) technology. Selective catalytic reduction (SCR), used in the past to reduce NO_x from large industrial installations, is now being developed for heavy-duty on-highway diesel engines. Diesel particulate filters (DPF), which are known for their ability in reducing PM, have previously suffered from regeneration, reliability, and durability problems. Now they are being included in plans for serial production. In their advanced development and planned production configurations, both technologies depend heavily on engine speed, load, exhaust temperature, and other characteristics. In fact, the move to integrate engine and post-combustion emission control devices (PC-ECD) into one overall emission control system is gaining popularity.

This paper describes the application of SCR and DPF technology to a heavy-duty on-highway diesel engine. Regeneration of the DPF was further enhanced with a cerium-based fuel-borne catalyst, which reduced the ignition temperature of the collected particulate. Results of this development effort not only met the 2002 EPA heavy-duty diesel NO_x and PM standards, but went beyond these requirements to achieve less than 1.5 and 0.05 g/bhp-hr respectively, for these pollutants. Detailed accounting of engine- and DPF-out cerium is also provided.

INTRODUCTION

Developing the diesel engine to meet mandated emissions standards can be traced back to the early 1980s where composite NO_x emissions for heavy-duty engines in EPA FTP were greater than 11 g/bhp-hr. Particulate matter (PM) was not even defined, let alone regulated at that time. In 1988, the diesel engine industry was first required to meet standards for the particulate emission as newly defined. At the same time, composite NO_x was first limited to 10.7 g/bhp-hr, and later to 6.0 g/bhp-hr, with particulate at 0.6 g/bhp-hr. Soon researchers discovered the difficulty in meeting both NO_x and PM standards simultaneously, since actions to reduce NO_x seemed to increase PM, and vice versa. The challenge was met mainly by injection timing retard and minor design changes to the fuel injection system.

Tough as it may have been, the 1988 challenge paled in comparison with that of 1991 when the NO_x standard was reduced a modest 17 percent from 6.0 to 5.0 g/bhp-hr, but the particulate emission standard was reduced by about 58 percent, from 0.6 to 0.25 g/bhp-hr. While seen by some as a formidable obstacle, many felt that the 1991 heavy-duty standards would force diesels out of existence. Predictions of the diesel engine's demise were proven wrong, and diesels not only survived, they appeared to even thrive. With higher injection pressure, and in some cases electronic fuel injection control, cooler intake charge temperature, and a new breed of combustion chamber design, diesel engines became cleaner-burning and sported excellent performance characteristics. In the years following 1991, the diesel engine continued its steady march toward cleaner exhaust with superior performance, while maintaining its superior fuel economy and durability. Figure 1 illustrates the history of diesel exhaust emissions since particulate matter was first regulated as a pollutant.

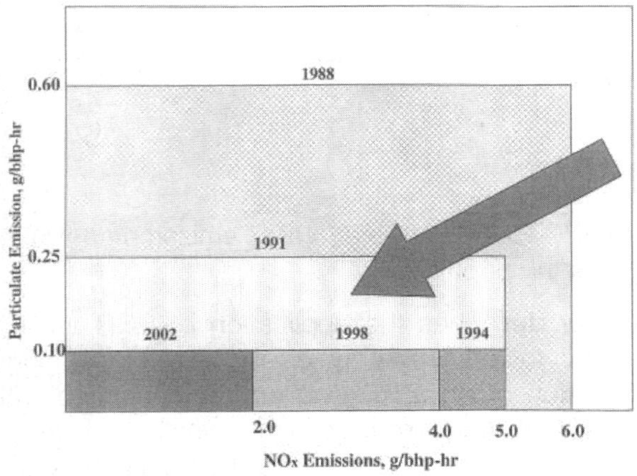

Figure 1. History of NO_x and PM Standards for Heavy-Duty Diesels

Beyond 1991, NO_x standards were further restricted to 4.0 g/bhp-hr in 1998, while PM was reduced to and remained at 0.10 g/bhp-hr in 1994. Thanks to wider use of electronic engine controls, better engine breathing with careful intake port designs, better mixture preparation, and increased air utilization, the diesel engine improved its standing as a premier powerplant for commercial transportation.

Now the industry is faced with the formidable task of meeting a newly mandated NO_x level of 2.0 g/bhp-hr (2.5 g/bhp-hr for NO_x+NMHC) for 2002. Early indications show that heavy-duty diesels will also meet this latest standard. The enabling technology this time is exhaust gas recirculation. While this technology in certainly not new, its application to the heavy-duty diesel engine is a novel development.

As further improvements are made in the engine and its systems there appears to be no end in sight for stricter standards. In the meantime, many advances were achieved in diesel post-combustion emission reduction technology. Many of these devices function in intimate relation to engine performance and share input parameters with the engine electronic control module (ECM). The SCR technology is a prime example of PC-ECDs whose performance is integrated with that of the engine. The DPF is another PC-ECD that allows the engine to achieve extremely low PM emissions while promising to be a good tool in limiting fine particles emanating from the combustion process.

Both SCR and DPF technologies were applied to a 1998 heavy-duty diesel engine. Extensive engine mapping was performed to achieve optimum NO_x control via SCR. A non-catalyzed DPF aided by fuel-borne catalyst was also used to maintain excellent particulate control. This work is described in the balance of this paper.

TEST ENGINE AND FUEL SULFUR SPECIFICATIONS

The test engine was an in-line six-cylinder DDC Series 60 equipped with EUI and calibrated for the 1998 heavy-duty diesel emission standards. The engine displacement was 12.7 L (775 cu.in), and it was turbocharged , wastegated, and aftercooled. Power output was 295 kW (400 hp) at 1800 rpm, and maximum torque was 2237 N$m (1650 lb-ft) at 1200 rpm.

Emission grade No. 2 diesel fuel was obtained from Phillips Petroleum for this test program. The actual sulfur content of this fuel was 368 ppm by weight. This fuel had a Cetane Number of 47.5 and 28.8 percent aromatics.

TEST PROCEDURES

Emissions were measured over the US heavy-duty transient Federal Test Procedure (FTP). A 13-mode test designed to investigate emissions outside the FTP conditions and during steady-state operation was also part of the test protocol. Table 1 outlines the conditions of the 13-mode steady-state test and shown graphically in Figure 2. Additional off-cycle testing was performed using the European Stationary Cycle test procedure (ESC) toward the end of the demonstration program. Each of the 13 modes was run for 30 minutes to collect adequate loadings on the particulate filters.

Table 1. Conditions for the Steady-State 13-Mode Test

Mode	Speed, rpm	Load, %
1	600	0
2	840	50
3	840	100
4	1080	50
5	1080	100
6	1380	50
7	1380	100
8	1560	50
9	1560	75
10	1560	100
11	1800	50
12	1800	75
13	1800	100

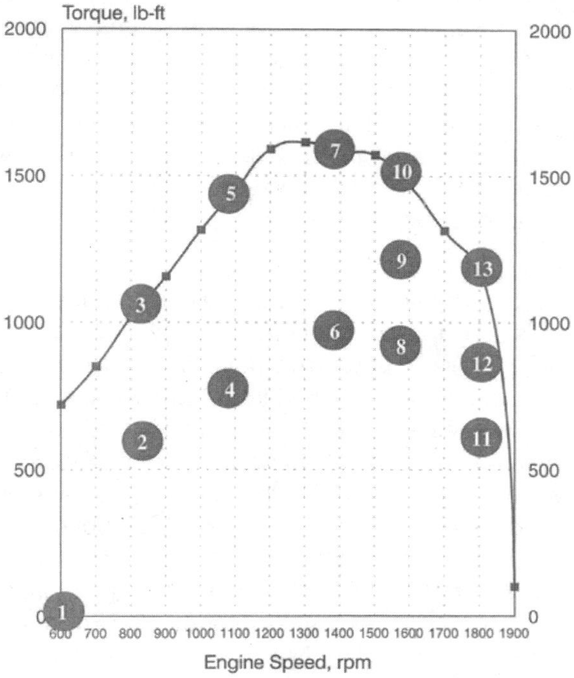

Figure 2. Graphical Representation of HMT Modes

Gaseous emissions of hydrocarbon (HC), carbon monoxide (CO), and oxides of nitrogen (NO_X), as well as particulate matter (PM) were sampled and analyzed in this program. Hydrocarbons were measured using continuous sampling techniques employing a heated flame ionization detector. Carbon monoxide was measured using non-dispersive infrared instruments. Oxides of nitrogen were measured continuously using a chemiluminescence detector. Total PM was determined by collecting particulate matter on a set of 90mm Pallflex filters that were weighed before and after the test cycle.

TEST RESULTS AND DISCUSSION

SELECTIVE CATALYTIC REDUCTION – Several tests were conducted with the engine in its baseline configuration to map NO_X emission concentrations typical of an EPA transient cycle. Shown in Figure 3 is a topographical map of NO_X emission in g/h for the EPA transient cycle as a function of engine speed and load conditions. This map shows that the highest NO_X mass emission rates (above 3500 g/h) were mostly at part-load conditions (700 to 1250 lb-ft) and speeds of 1200 to 1800 rpm. The specific NO_X emission map is given in Figure 4, and the corresponding exhaust temperature map is given in Figure 5. The highest exhaust temperatures were experienced at engine speeds below peak-torque speed, and at full-load conditions (see Figure 5). These speed, torque, NO_X, and exhaust temperature maps established the baseline characteristics for the DDC Series 60 engine. Initial testing used a homogeneous extruded SCR catalyst (TiO2-V2O5 type) and a 32 percent urea solution. The electronic control system was programmed on that basis, resulting in NO_X reduction of about 50 percent. The target, however, was for a NO_X calibration of less than 1.5 g/bhp-hr and therefore, the urea concentration was raised to 46 percent while the control strategy remained unchanged. In addition, the initial goal was to limit ammonia slip to an absolute value of about 20 ppm, which was later changed to 10 ppm. The initial thrust was directed toward developing a control strategy for the EPA transient cycle. Later, another strategy was developed for the 13-Mode steady-state HMT.

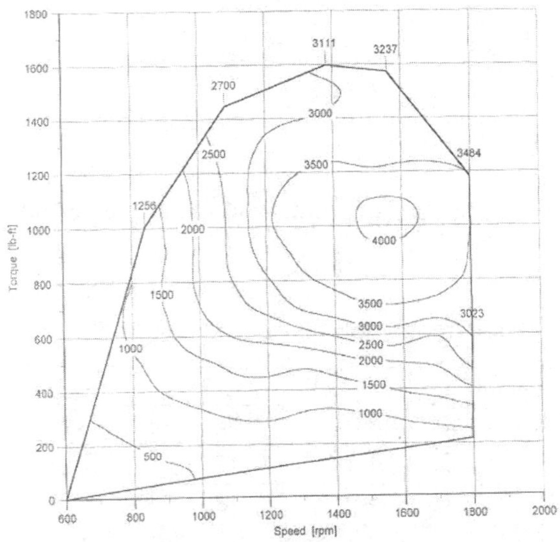

Figure 3. NO_X Emission Map for the Baseline DDC Series 60 Engine

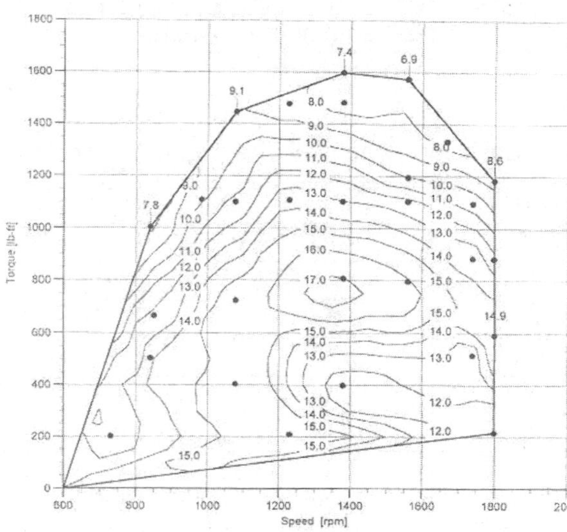

Figure 4. Specific NO_X Emissions for DDC Series 60 Baseline

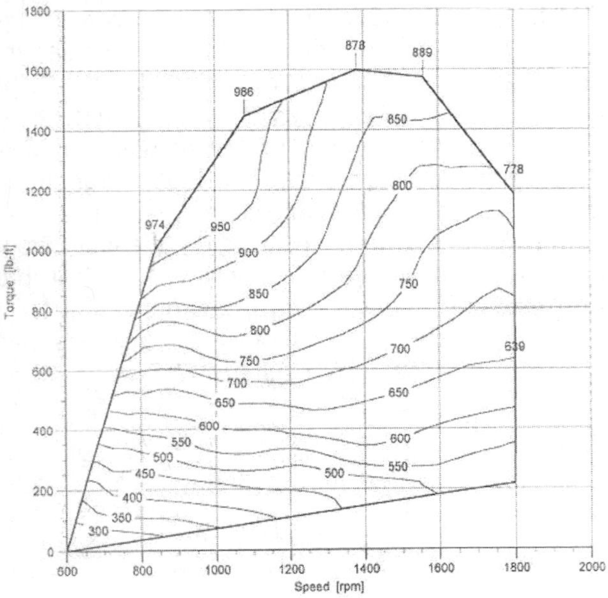

Figure 5. Exhaust Temperature Map for DDC Series 60 Baseline

Several experimental runs were made to verify the operation of the urea injection system. These runs were made while observing raw NO_x emission traces rather than processing the data, in order to expedite the development process. Results of the final calibration are given in Table 2, where the engine baseline results are also given. Reduction of NO_x emission was about 70 percent, while PM was also reduced by about 15 percent.

Off-cycle emission testing was also performed with the SCR system installed. The SCR system reduced modal NO_x emissions from 65 percent to 99+ percent, as shown in Table 3. It is important to point out that the steady-state urea calibration was generated with 32 percent urea solution, whereas the testing was conducted with a 46 percent solution. Therefore, while some of these modes have shown impressive NO_x reductions, another calibration would have to be generated that would optimize NO_x reduction while maintaining ammonia slip at a minimum. In the HMT most of the 13 modes had ammonia slip greater than 10 ppm except for Modes 13. Further work is anticipated for reducing NO_x while maintaining ammonia slip to below 10 ppm.

Table 2. FTP Transient Emissions--SCR And 46 Percent Urea--368 ppm Sulfur-Fuel

Test No.	Test Condition	Emissions, g/bhp-hr				Cycle Work, hp-hr	Cycle BSFC, lb/ bhp-hr	NH_3 Slip, ppm
		HC	CO	NO_X	PM			
120298C1	Cold-Start	0.010	1.863	1.936	0.073	30.82	0.398	40
120298H1	Hot-Start	0.005	1.158	1.094	0.060	30.36	0.381	43
SCR Composite[a]		0.006	1.260	1.216	0.062	30.43	0.384	43
070898C1	Cold-Start	0.215	1.671	4.168	0.084	30.81	0.413	-
070998H1	Hot-Start	0.098	1.017	3.961	0.071	30.55	0.398	-
Baseline Composite[a]		0.115	1.111	3.991	0.073	30.59	0.401	-

[a] The sum of one-seventh of the cold-start and six-seventh of the hot-start tests

Table 3. Emission Results of The Highway Map Test--with SCR and 46 Percent Urea--368 ppm Sulfur-Fuel

Mode No.	Emissions, g/bhp-hr				NOx Reduction, %	NH_3 Slip, ppm
	HC	CO	NO_X	PM		
1	- a	- a	- a	- a	- a	- a
2	0.00	0.27	1.98	0.032	86	16
3	0.00	2.45	0.03	0.108	100	14
4	0.00	0.28	1.67	0.030	89	119
5	0.00	1.65	1.27	0.118	87	20
6	0.00	0.20	4.37	0.030	76	16
7	0.00	0.21	1.36	0.025	83	21
8	0.00	0.15	4.48	0.033	74	56
9	0.00	0.16	2.53	0.021	77	45
10	0.00	0.18	1.55	0.032	79	25
11	0.00	0.20	3.26	0.095	79	71
12	0.00	0.13	4.82	0.053	65	11
13	0.00	0.14	2.68	0.030	71	7

[a] Specific emissions left out due to questionable accuracy because of wide speed/torque variability at idle.

DIESEL PARTICULATE FILTER WITH FUEL-BORNE CATALYST – Exhaust system modifications were made to accommodate testing the SCR system with a non-catalyzed DPF, installed downstream from the SCR (see Figure 6). Testing was carried out with 368ppm sulfur-fuel and 46 percent urea concentration. Cerium-based fuel-borne catalyst (FBC) was mixed with the 368ppm sulfur-fuel according to instructions obtained from the supplier. A 20-hour conditioning period followed, the engine being operated with the FBC. During the conditioning period the engine was run for 30-minute intervals interchangeably between peak-torque speed and maximum power speed, both points being run at full-load conditions. A non-catalyzed, ceramic cordierite, wallflow EX-80, 200 cpsi, diesel particulate filter was canned and installed in the exhaust system as shown in Figure 6. Results of the EPA FTP and the HMT using this setup are given in Tables 4 and 5, respectively.

Results of the SCR, DPF, and FBC technology integration on the 1998 DDC Series 60 heavy-duty diesel engine indicate that NO_x + NMHC emissions far less than 1.5 g/bhp-hr can be achieved, with extremely low PM (0.002 g/bhp-hr), in the EPA transient emission test. This represents about 70, and 97 percent NO_x and particulate emission reduction from the engine baseline values, respectively. This remarkable result was achieved without apparent loss in fuel economy.

Figure 6. Photograph of the Test Cell Installation Showing the SCR and DPF in the Exhaust System

Table 4. FTP Test of SCR With 46 Percent Urea, DPF, And FBC--368 ppm Sulfur-Fuel

Test No.	Test Condition	Cycle BSFC, lb/bhp-hr	Cycle Work, bhp-hr	Emissions, g/bhp-hr			
				HC	CO	NO_x	PM
031899C1	Cold-Start	0.401	30.42	0.056	2.014	1.561	0.005
031799H1	Hot-Start	0.381	29.98	0.009	1.100	1.101	0.002
Composite[a]		0.384	30.04	0.016	1.232	1.168	0.002

[a] One sixth of cold-start plus six-sevenths of hot-start

Table 5. HMT with SCR#3--46 Percent Urea-DPF--368 ppm Sulfur-Fuel

Mode No.	Emissions, g/bhp-hr			
	HC	CO	NO_x	PM
1	a	a	a	a
2	0.00	0.22	1.65	0.007
3	0.00	1.93	0.04	0.006
4	0.00	0.19	1.25	0.003
5	0.00	1.64	1.37	0.045
6	0.00	0.15	5.39	0.002
7	0.00	0.20	2.47	0.008
8	0.00	0.13	6.98	0.002
9	0.00	0.18	4.17	0.001
10	0.00	0.17	2.39	0.011
11	0.00	0.22	3.41	0.003
12	0.00	0.15	4.54	0.000
13	0.00	0.18	3.14	0.000

[a] Specific emissions left out due to questionable accuracy because of wide wide speed-torque variability at idle.

FBC RESIDUE COLLECTION EFFECTIVENESS OF DPFS – Recent efforts in DPF development have concentrated on improving regeneration quality. A method to achieve this objective is using fuel-borne catalysts such as the cerium-based FBC used in this work. However, regulatory agencies have expressed some concern regarding possible emission of the fuel-borne catalyst and its oxidation products into the environment. In December 1997, the Umwelt Bundes Amt (German directorate of the environment) issued a statement indicating that the use of some selected FBCs must be associated with placing an approved DPF in the exhaust of a diesel engine. Motivated by this concern, a short test was conducted to account for cerium with and without a DPF. The engine was operated for one hour at a steady-state condition selected from a 13-mode HMT. Mode 10 of the HMT was used, since it had the highest particulate emission of the steady-state test in the engine as-received configuration. The FBC concentration was specified by the supplier to be representative of production intent. Three conditions were tested:

- Engine-out (without diesel particulate filter) particulate collected for one hour.

- DPF installed in the exhaust and accumulating particulate for one hour.
- Sampled particulate for 15 minutes while inducing DPF regeneration. This process followed Step 2 without performing any DPF conditioning. Engine speed was brought down to Mode 5 of the HMT where exhaust temperature had reached 519°C to assist in forcing DPF regeneration.

Particulate samples were collected on four 47 mm Pallflex filters which were analyzed for trace metals with emphasis on cerium. Measured air flow during the test was used to calculate trace metal concentrations per cubic meter for the three tests described earlier.

Results in Table 6 indicate that the DPF's cerium retention efficiency while in the particulate accumulation mode was 99.8 percent, and 97.3 percent in the regeneration mode.

Table 6. Cerium Test Results

Test Condition	Sample, μg/filter
Engine-Out (w/o DPF)	257
DPF-Out (Accumulation)	0.595
DPF-Out (Regeneration)	1.72

CONCLUSION

- A heavy-duty DDC Series 60 diesel engine calibrated to meet 1998 emission standards and equipped with an SCR gave NO_x + NMHC emissions of less than 1.5 g/bhp-hr in the EPA FTP.
- The above result was achieved while maintaining PM emissions below 0.01 g/bhp-hr by using a non-catalyzed DPF assisted with a FBC, downstream of the SCR system.
- The use of a DPF in cases where a FBC is used to assist in its regeneration is quite effective in retaining cerium oxides from escaping into the environment.
- Integrating PC-ECDs and engine control adds a new dimension to the potential of diesel engines for meeting extremely low emission levels.
- More work is needed to optimize urea injection rates and minimize ammonia slip.

ACKNOWLEDGMENTS

The Department of Emissions Research at the Southwest Research Institute expresses its appreciation to Rhodia and Siemens for sponsoring this work.

REFERENCES

1. Gill, A. P., "Design Choices for 1990's Low Emission Diesel Engines," SAE Paper 880350, February 29-March 4, 1988.
2. Clerc, J. C., Miller, R. K., McDonald, A. C., and Schlamadinger, H., "A Diesel Engine/Catalyst System for Pick-up and Medium-Duty Trucks" SAE Paper 932982, November 1-4, 1993.
3. Ball, D. J. and Stack, R. G., "Catalyst Considerations for Diesel Converters," SAE Paper 902110, October 22-25, 1990.
4. Kojetin, P., Janezich, F., Sura, L., and Tuma, D., "Production Experience of a Ceramic Wall Flow Electric Regeneration Diesel Particulate Trap," SAE Paper 930129, 1993.
5. Wyatt, M., Manning, W. A., Roth, S. A., D'Aniello, Jr., M. J., Andersson, E. S., and Fredholm, S. C. G., "The Design of Flow-Through Diesel Oxidation Catalysts," SAE Paper 930130, 1993.
6. Khair, M. K., "Diesel Particulate Analysis for Effective Exhaust Aftertreatment," SAE Paper 932954, November 1-4, 1993.
7. Krutzsch B. and Wenninger, G., "Effect of Sodium- and Lithium-Based Fuel Additives on the Regeneration Efficiency of Diesel Particulate Filters," SAE Paper 922188m October 19-22, 1992.
8. Rao, C. D. N., Cikanek, H. A., and Horrocks, R. W., "Diesel Particulate Control System for Ford 1.8L Sierra Turbo-Diesel to Meet 1997-2003 Particulate Standards," SAE Paper 940458, 1994.
9. Stiglic, P., Bagelman, B., and Hundemer, R., "Emission Testing of Two Heavy-Duty Diesel Engines Equipped with Exhaust Aftertreatment," SAE Paper 900919, April 3-5, 1990.
10. Newkirk, M. S., "Application of Methods for Determining Total Organic Contribution to Diesel Particulates," ASME Paper 92-ICE-16, January 26-30, 1992.
11. This paper is a condensation of a final report summarizing activities and findings of SwRI Project 08-1854.
12. Fritz N., Mathes W., Zuerbig J., and Mueller R., "On-Road Demonstration of NO_x Emission Control for Diesel Trucks with $SINO_x$ Urea SCR System," SAE Paper 1999-01-0111, March 1-4, 1999

2000-01-0188

An Integrated SCR and Continuously Regenerating Trap System to Meet Future NOx and PM Legislation

Guy R. Chandler, Barry J. Cooper, James P. Harris, James E. Thoss, Ari Uusimäki, Andrew P. Walker and James P. Warren
Johnson Matthey, Catalytic Systems Division

ABSTRACT

The tightening NOx and particulate matter (PM) emission standards for heavy duty diesel powered vehicles are stimulating the development of aftertreatment systems to reduce NOx and PM emissions from such vehicles. Here we present data on a new system which combines NO_2-based continuously regenerative trap particulate removal technology with urea-based Selective Catalytic Reduction (SCR) NOx removal technology. There are a number of beneficial synergistic effects associated with combining these two technologies, including a significant improvement in the low temperature NOx removal performance of the SCR system. The development of this PM/NOx control system is described, and the main features of this novel strategy are outlined.

The PM/NOx control system has been evaluated on a number of different engines and over a number of different drive cycles. During testing we have shown that this technology enables simultaneous NOx conversions of 75-90% and PM conversions of 75-90% to be obtained on current engines using both US and European test procedures. These results are presented and discussed.

1. INTRODUCTION

The proposed emissions standards for heavy duty diesel engines in North America and in Europe will require significant reductions in both NOx and particulate matter (PM) emissions from such vehicles. There are a number of possible approaches currently being considered to achieve these objectives. Many of these include the combination of engine modifications to control one pollutant with an aftertreatment system to control the other. So, for example, the engine can be modified to give low NOx and high PM, and this PM can be controlled using filter technology, such as a Continuously Regenerating Diesel Particulate Filter (CR-DPF) [1]. Alternatively, the engine can be calibrated to give low PM and high NOx, and this NOx can be controlled using Selective Catalytic Reduction (SCR) technology, most probably using ammonia (derived from urea) as the reductant [2].

However, a third alternative is now becoming available, which allows the engine to be calibrated for optimum performance/fuel economy etc, and which provides very high simultaneous conversion of NOx and PM. This technology combines advanced SCR NOx control catalyst technology with the PM control of a CR-DPF to deliver high conversions of HC, CO, NOx and PM. This paper describes the processes involved in the development of this combined PM/NOx control technology, and demonstrates the potential of this technology to deliver very low emissions over different test cycles and on different engines.

2. EXPERIMENTAL DETAILS

2.1 CATALYST DETAILS – The Continuously Regenerating Diesel Particulate Filter (CR-DPF) used throughout this work was the Johnson Matthey CRT™ (Continouusly Regenerating Trap). Throughout this paper the CRT™ is referred to as the "CR-DPF". The CR-DPF comprises an oxidation catalyst upstream of a DPF; this oxidation catalyst oxidises a proportion of the engine-out NO into NO_2, and this NO_2 combusts the particulate (trapped on the DPF) at low temperature.

The catalysts used were coated onto ceramic monolith substrates with a cell density of 62 cells.cm^{-2} (400 cells.in^{-2}) using standard processing techniques. The geometric dimensions of the catalysts used for the engine evaluations, given as diameter x length, were 266.7 x 152.4 mm (10.5 x 6 in) with a wall thickness of 0.15 mm (0.006 in). The cordierite particulate filters used were production standard types, with 15 cells.cm^{-2} (100 cells.in^{-2}) and with geometric dimensions (again given as diameter x length) of 266.7 x 304.8 mm (10.5 x 12 in), with a wall thickness of 0.43 mm (0.017 in).

2.2 BENCH ENGINE EVALUATION ON THE US ENGINE – Heavy Duty Transient (HDT) and European Steady state Cycle (ESC, OICA) evaluations were performed using a 7.2 liter, 224 kW (300 hp) turbocharged aftercooled engine in a test cell with motoring dynometer capability.

Emissions on the HDT and ESC (OICA) tests were measured using a full flow dilution tunnel with a secondary dilution tunnel and filter holder to measure particulate. The NOx emissions were measured using a heated chemiluminescence analyzer, but the NOx was measured dry. The HC emissions were measured using a heated Flame Ionisation Detector. Both of these systems sampled continuously. CO and CO_2 emissions were measured using continuous bag samples read post test by NDIR analyzers. N_2O emissions were measured by gas chromatograph from bag samples and NH_3 was measured using wet chemistry techniques. All regulated emissions during the OICA and HDT tests were measured in accordance with US EPA guidelines.

An airless injector with a separate automatic control system that varied injector volumes based on engine speed and torque was used to inject urea. The urea used was a commercially available formulation of 32.5 % urea in water. The engine out NOx was mapped over the transient and ESC (OICA) cycles and separate urea injection maps were developed for these cycles providing a 1:1 NOx: NH_3 ratio.

Testing was performed using test fuel with a cetane number of 41.6, aromatics content of 15.5 % and sulfur content of 15 ppm.

2.3 BENCH ENGINE EVALUATION ON THE EUROPEAN ENGINE – The ESC evaluations were carried out on a 12 litre, 310 kW turbocharged aftercooled engine in a cell with multiple gas stream analysis. The temperature window experiments were performed on a 10 litre, 210 kW turbocharged engine with the same analytical facilities. Two types of fuel were used during the evaluations: Swedish MK1 and UK pump diesel fuel with sulphur levels of <10 ppm and <50 ppm respectively. The MK1 fuel has a cetane number of 53 and an aromatics content of <4 %, while the UK pump diesel fuel has a cetane number of 52 and an aromatics content of ~30%. The fuels were purchased from existing refinery sources. The engine oil used during the evaluations was Shell Myrina 15W 40.

The engine temperature window experiments were carried out at fixed engine speed, and the temperature was increased by increasing the load on the engine. The ESC evaluations were carried out using standard procedures. The catalysts had all been run on the engine for at least 10 hours before ESC evaluations were carried out. Prior to each ESC run the system was conditioned by running one ESC cycle. The engine was operated for the prescribed time in each mode, and completed the engine speed and load changes in the first 20 seconds at each mode. The sampling times were at least 4 seconds per 0.01 weighting factor. One PM sample was taken over the complete test cycle by means of a dilution tunnel. The filter papers were stabilised under controlled temperature and humidity conditions for 24 hours prior to the testing and then installed for the test. After the test cycle had been completed, the filter papers, with the collected PM, were again stabilised before being weighed.

Gaseous emissions were measured raw before and after the catalyst system using a range of analysers: chemiluminescence for NO and NO_2, heated FID (Flame Ionisation Detector) for HC, NDIR (Non-Dispersive Infra Red) analysers for CO, CO_2 and N_2O, and a paramagnetic detector for O_2. A 32.5% aqueous urea solution was used as the source of urea; this solution was injected into the exhaust directly using a commercially available urea injector system. The required urea dosage rate was calculated from the engine-out NOx levels using dosage rates of NO:NH_3 = 1:1 and NO_2:NH_3 = 1:1.3. The volumetric flow rate of the urea solution to the injector nozzle was measured to check that the urea dosing rate was correct. Ammonia was detected using a laser-based AltOptronic LD 3000 analyser which measures the ammonia sample in situ.

2.4 MICROREACTOR EVALUATION – The microreactor (SCAT) experiments were carried out using powder catalyst samples at a gas hourly space velocity of 50,000 hr^{-1}. Two different gas feeds were used, with the following compositions: Gas Feed 1: 200ppm NO, 200ppm NH_3, 200ppm CO, 100ppm C_3H_6, 12% O_2, 4.5% CO_2, 4.5% H_2O and 20ppm SO_2. The second gas feed was identical to the first, except that 100ppm of the NO was replaced by 100ppm of NO_2 (ie the NOx composition of this second gas feed was 100ppm NO and 100ppm NO_2). The experiments were carried out by increasing the temperature at a rate of 5^oC/min. NH_3 and NOx were monitored using a chemiluminescence analyser, while the CO and CO_2 were monitored using NDIR detectors. The O_2 and C_3H_6 were monitored by gas chromatography.

3. RESULTS AND DISCUSSION

3.1 DEVELOPMENT OF THE COMBINED PM/NO_X CONTROL SYSTEM – It is clearly desirable to be able to attain high simultaneous conversion of both PM and NOx from heavy duty vehicles. The CR-DPF is well established as a particulate control strategy, and also offers very high conversions of HC and CO [3]. In addition, SCR is becoming widely regarded as a promising technology to control NOx emissions [4]. SCR enables high NOx conversion over a wide temperature window. However, the low temperature performance of SCR catalysts tends to be very dependent upon the space velocity, as illustrated in Figure 1. The Figure shows that as the number of SCR catalysts is increased from 1 to 2 to 3 the low temperature NOx conversion increases significantly. These light-off experiments were conducted on a 10 litre, 210 kW turbocharged engine; the volume of each of the SCR catalysts was 8.5 litres. It can be seen that the 3 catalyst system gives high NOx conversion over a very wide temperature window.

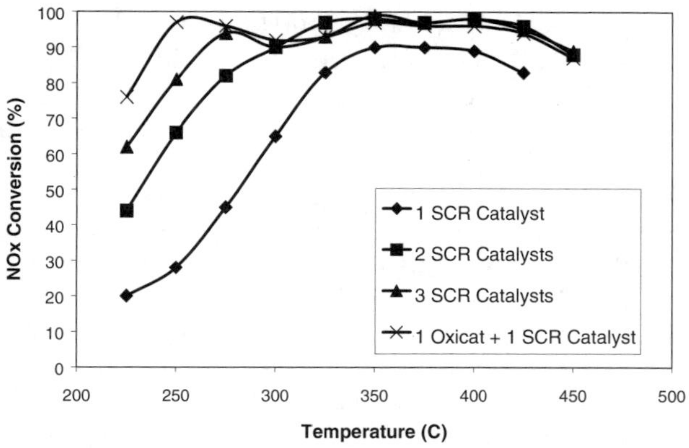

Figure 1. Effect of Increasing the Number of SCR Catalysts on the Activity of the System, Compared with the Effect of Placing a Pt-based Oxidation Catalyst Upstream of a Single SCR Catalyst

However, this 3 catalyst system corresponds to a catalyst volume of 25.5 litres on a 10 litre engine, which is a rather high catalyst volume. We have found that an alternative way to boost the low temperature NOx activity of SCR catalyst systems is to place an oxidation catalyst upstream of the SCR catalysts. Figure 1 also shows the impact of placing a Pt-based oxidation catalyst in front of a single SCR catalyst, with urea injection between the oxidation catalyst and the SCR catalyst (so that the urea/ammonia is not oxidised to NOx over the oxidation catalyst). It is clear that the resulting enhancement in low temperature activity is significantly greater than that obtained by increasing the number of SCR catalysts to 3. Therefore, this approach facilitates good low temperature activity with a relatively low catalyst volume.

A number of possible explanations can be put forward for this low temperature activity enhancement, including:

1. the removal of HC and CO by the Pt-based oxidation catalyst improves the low temperature SCR performance,
2. the presence of NO_2 in the gas stream after the Pt-based oxidation catalyst enhances the low temperature SCR activity.

Microreactor (SCAT) experiments have been carried out to test these hypotheses, and the experiments showed that there is a definite small enhancement in low temperature activity associated with the removal of HC and CO from the gas feed. This is because these species can inhibit the ammonia-NO reaction at low temperatures. However, these experiments revealed that the major promoting effect is associated with the generation of NO_2 over the oxidation catalyst. This NO_2 strongly promotes the low temperature activity of the SCR catalyst, as shown in Figure 2, which compares the NOx conversion efficiency of an SCR catalyst using an inlet feed containing 200ppm NO and 200ppm NH_3 with the efficiency using 100ppm NO, 100ppm NO_2 and 200ppm NH_3. The low temperature activity enhancement in the presence of the NO_2 can be clearly seen.

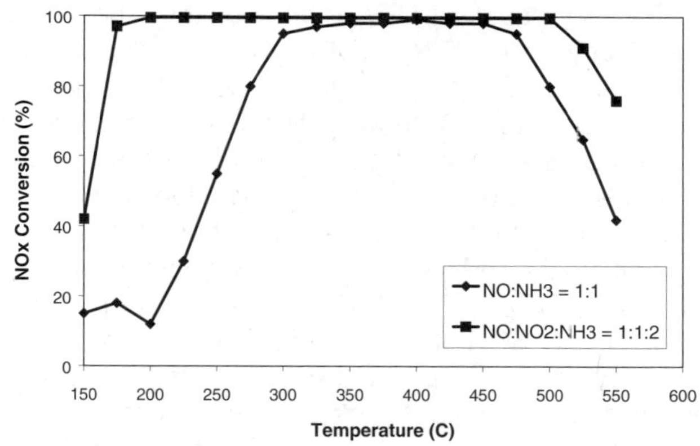

Figure 2. A Comparison of the NOx Conversion Activity of an SCR Catalyst Using a NO:NH_3 feed (200:200ppm) and a NO:NO_2:NH_3 feed (100:100:200ppm)

The equations representing the reactions between ammonia and NO and NO_2 are as follows:

$$4\ NO + 4\ NH_3 + O_2 \Rightarrow 4\ N_2 + 6\ H_2O$$

$$6\ NO_2 + 8\ NH_3 \Rightarrow 7\ N_2 + 12\ H_2O$$

Therefore, the CR-DPF delivers high PM, HC and CO control, and an SCR system can provide high NOx conversion over a wide temperature window. These two emission control strategies have now been combined to assess the potential of such systems to deliver high simultaneous conversions of PM and NOx (as well as HC and CO) [5]. The results presented above suggest that there may be a significant NOx conversion enhancement associated with placing the CR-DPF upstream of the SCR system, since the CR-DPF operates by producing NO_2 which subsequently combusts the PM. If some of this NO_2 reaches the SCR catalyst it will lead to an improvement in the low temperature NOx conversion of the SCR system. The configuration of the combined HC, CO, NOx and PM control system is shown schematically in Figure 3.

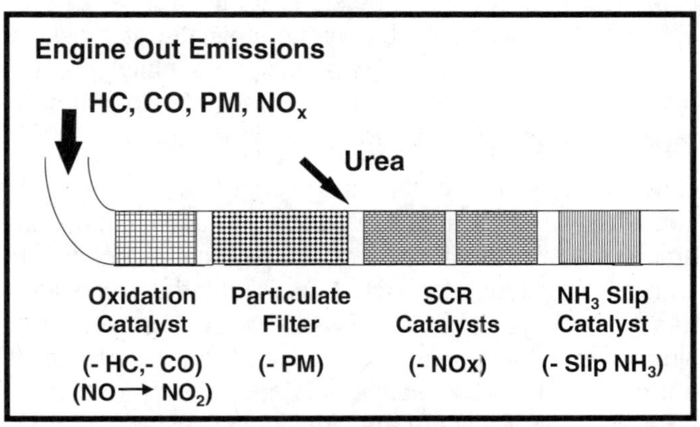

Figure 3. Schematic Representation of the Combined HC, CO, NOx and PM Control System

The engine emissions are initially treated by the CR-DPF, which operates by converting (oxidising) a proportion of the engine-out NO into NO_2 over an oxidation catalyst; this NO_2 then reacts with PM trapped by the downstream Diesel Particulate Filter (DPF) which cleans and regenerates the filter. Numerous studies have demonstrated that NO_2 reacts with PM at much lower temperatures than does oxygen (eg [6]). The oxidation catalyst also provides very high conversion of HC (into CO_2 and H_2O) and of CO (into CO_2). Therefore, the CR-DPF removes the PM, HC and CO, and converts some of the NO into NO_2. The resulting gas stream is then treated by the SCR system, which comprises a urea injection system followed by the SCR catalyst, followed by an oxidation catalyst to minimise ammonia slip from the system. This system has been tested on both US and European engines, using three different SCR catalyst formulations. The results of these investigations are reported below.

3.2 TEST RESULTS ON A NORTH AMERICAN ENGINE – A number of tests have been performed on a US engine to evaluate the performance of the CR-DPF and SCR technologies in isolation, as well as the combined CR-DPF + SCR system. The engine volume was 7.2 litres, the CR-DPF volume was 25.5 litres, the SCR catalyst volume used was 17 litres and the clean-up catalyst had a volume of 8.5 litres. The remainder of the exhaust system was 101.6 mm (4 in) diameter tubing; all tubing and catalyst cans after the urea injector were made of stainless steel. The CR-DPF was tested alone, and then two different SCR catalyst formulations (designated SCR1 and SCR2) were tested alone. Following these experiments, each of the SCR catalysts was tested in combination with the CR-DPF. The systems were tested over the ESC cycle, using 15 ppm S fuel and the weighted emission results are presented in Table 1.

The HC and CO limits are comfortably met by all of the systems. The CR-DPF substantially reduces the PM emissions (88% conversion), and, as expected, has no significant effect on the NOx emissions. The SCR catalysts when tested alone significantly reduce the NOx emissions to levels well below the legislated values, and appear to increase the PM emissions slightly (perhaps due to the presence of small levels of slip ammonia reaching and being trapped by the filter paper).

However, the most outstanding results are those of the combined CR-DPF + SCR systems. These systems give emission levels which are substantially below the legislated values for all four regulated emissions, demonstrating that it is possible to obtain very high simultaneous conversions of NOx and PM (as well as HC and CO). The NOx results obtained with the combined CR-DPF + SCR system are significantly lower than those observed with the SCR catalysts alone, which reflects the low temperature enhancement of NOx conversion associated with the production of NO_2 by the CR-DPF, as discussed above.

Table 1. CR-DPF, SCR and Combined CR-DPF + SCR Emissions on a US Engine over the ESC (OICA) (g/bhp-hr).

Technology	**HC**	**CO**	**NOx**	**PM**
Engine Out	0.01	0.79	5.39	0.050
CR-DPF	0.00	0.03	5.46	0.006
SCR1	0.00	0.04	0.91	0.061
SCR2	0.00	0.12	0.84	0.069
CR-DPF + SCR1	0.00	0.02	0.60	0.015
CR-DPF + SCR2	0.00	0.16	0.41	0.025
2002 ¾ Limits*	***0.50***	***15.50***	***2.50***	***0.10***

1 g/bhp-hr = 1.341 g/kW-hr
(* for those manufacturers subject to the Consent Decree. Other manufacturers will need to meet these standards in 2004
** if NMHC > 0.5g/bhp-hr the NMHC + NOx standard is 2.4 g/bhp-hr NOx; but if NMHC < 0.5 g/bhp-hr the NMHC + NOx standard becomes 2.5 g/bhp-hr)

The two CR-DPF + SCR systems tested above have also been tested on the same engine over the US Heavy Duty Transient (HDT) cycle, again using 15 ppm S fuel. The results of these tests are summarised in Table 2.

Table 2. CR-DPF +SCR Emissions on a US Engine over the US HDT Cycle (g/bhp-hr)

Technology	**HC**	**CO**	**NOx**	**PM**
Engine Out	0.077	1.519	5.633	0.116
CR-DPF + SCR1	0.000	0.161	0.835	0.014
CR-DPF + SCR2	0.000	0.498	0.840	0.007
2002 ¾ Limits*	***0.500***	***15.500***	***2.500***	***0.100***

1 g/bhp-hr = 1.341 g/kW-hr
(* for those manufacturers subject to the Consent Decree. Other manufacturers will need to meet these standards in 2004)

Once again, the emissions from the combined CR-DPF + SCR systems are substantially below the proposed legislated standards.

The PM and NOx emissions of the two CR-DPF + SCR systems over the two test cycles are summarised in Figure 4, which also shows the 2002 3/4 US standards. Note that since the HC emissions of the systems are approximately zero, the NOx emission standard becomes 2.5 g/bhp-hr.

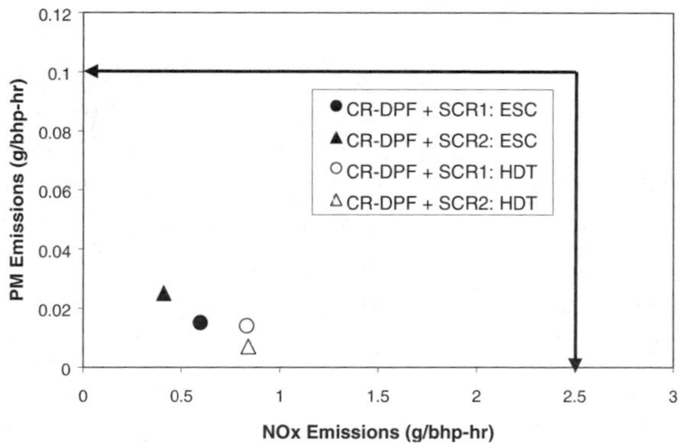

Figure 4. PM and NOx Emissions of the Combined CR-DPF + SCR Systems over the ESC (OICA) and HDT Cycles With Respect to the US Standards.

3.3 TEST RESULTS ON A EUROPEAN ENGINE – Two CR-DPF + SCR systems have also been tested over the ESC cycle on a European engine. A system based on SCR2 catalyst technology was tested, along with a new formulation, SCR3. The engine capacity was 12 litres, the CR-DPF volume was 25.5 litres, the SCR volume was 8.5 litres unless stated otherwise and the volume of the NH_3 slip catalyst was 8.5 litres. During these tests the sulphur level in the fuel was approximately 3 ppm (Swedish MK1 fuel). The results of the tests are summarised in Table 3.

Table 3. CR-DPF + SCR Emissions on a European Engine over the ESC (g/kW-hr).

Technology	HC	CO	NOx	PM
Engine Out	0.162	0.989	7.018	0.163
CR-DPF + SCR2	0.005	0.008	1.592	0.008
CR-DPF + SCR3*	0.003	0.000	1.061	0.007
2005 Limits**	***0.460***	***1.500***	***3.500***	***0.020***
2008 Limits**	***0.250***	***1.500***	***2.000***	***0.020***

1 g/kW-hr = 0.7457 g/bhp-hr
(* SCR3 catalyst volume = 17 litres
** These limits are the *proposed* limits)

Once again, it can be seen that the combined PM/NOx control systems give emissions which are substantially below the proposed standards for 2005, and also comfortably below the proposed 2008 standards. This is the case even with a single SCR catalyst. This is further demonstrated in Figure 5, which plots the NOx and PM results against the proposed future European standards.

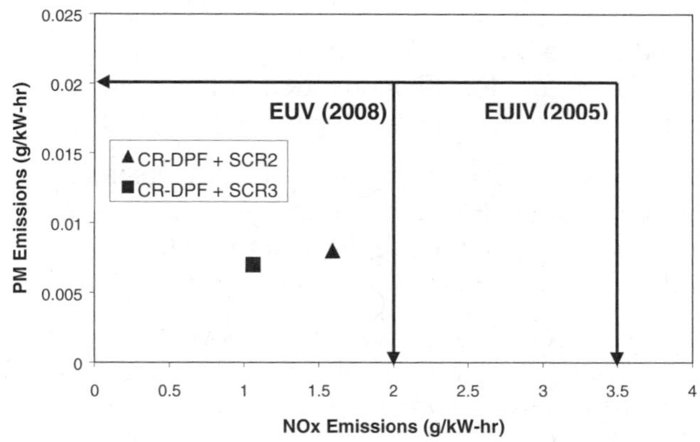

Figure 5. PM and NOx Emissions of two Combined CR-DPF + SCR Systems over the ESC with Respect to the Proposed European Heavy Duty Emission Standards.

One of the issues associated with ammonia SCR-based NOx control systems is the possibility of ammonia slip from the system. As discussed above, the systems used in this work contained an oxidation catalyst as the last catalyst in the system, to oxidise any slip ammonia and therefore minimise ammonia release from the system. During the tests the ammonia slip of the system was monitored. At the outlet of the SCR catalysts the ammonia concentration in the exhaust gas was between 10 and 45 ppm. The ammonia oxidation catalyst removed most of this ammonia, so that the ammonia slip from the whole system was only between 5 and 15 ppm. It has previously been reported that a fraction of the slip ammonia undergoes partial oxidation to N_2O, as well as undergoing more complete oxidation to NO and NO_2 over an oxidation catalyst [4]. This was also observed here, where the level of N_2O after the SCR catalysts was between 0 and 7 ppm. This level was increased slightly over the ammonia oxidation catalyst, but the tailpipe emission was still very low at between 5 and 10 ppm. These levels of ammonia and N_2O emissions compare very favourably with the corresponding levels from gasoline-powered passenger cars fitted with three-way catalysts, where the levels of ammonia and N_2O release are around 60-360 and 12-35 ppm respectively [4]. It is expected that the low levels observed here could be further reduced by better optimising the urea injection control strategy.

4. CONCLUSIONS

By combining the CR-DPF with SCR technology it is possible to reach the NOx, PM, HC and CO emissions standards proposed to be introduced in Europe in 2008 and in North America in 2002 3/4, when using conditioned catalyst systems. This is the case over both the ESC (OICA) and US HDT test cycles. This combined PM/NOx control system potentially enables vehicle manufacturers to calibrate their vehicles for optimum performance and fuel economy and still be able to reach future emissions standards of both PM and NOx.

Low levels of ammonia slip and N_2O release were observed over the PM/NOx control system. It is expected that these low levels could be even further reduced by developing more advanced urea dosing strategies.

Extended durability trials are now required to demonstrate that the excellent performance of the combined PM/NOx control system is maintained during system ageing.

ACKNOWLEDGEMENTS

We would like to thank Dr Philip Blakeman and Mr Colin Maloney for their help in preparing the catalysts. We would also like to thank Dr Raj Rajaram and Dr Isabel Jones for carrying out the microreactor experiments. The technical assistance of Mr Borje Johansson and Mr Anders Andreasson are gratefully acknowledged. We are grateful to Johnson Matthey plc for permission to publish this paper.

REFERENCES

1. BJ Cooper, HJ Jung and JE Thoss, US Patent 4,902,487 (1990).
2. C Havenith, RP Verbeek, DM Heaton and P van Sloten, SAE 952652.
3. JP Warren, R Allansson, PN Hawker and AJJ Wilkins, Proc. Diesel Engines - Particulate Control, IMechE 1998 S491/006 p.45.
4. C Havenith and RP Verbeek, SAE 970185.
5. A Andreasson, G R Chandler, C F Goersmann and JP Warren, Patent Application PCT Appln No WO99/39809.
6. BJ Cooper and JE Thoss, SAE 890404

2005-01-3548

NOx and PM Reduction Using Combined SCR and DPF Technology in Heavy Duty Diesel Applications

Ray Conway, Sougato Chatterjee, Alex Beavan, Claus Goersmann and Andy Walker
Johnson Matthey Environmental Catalysts & Technologies

ABSTRACT

The application of oxidation catalyst and particulate filter technology for the reduction of particulate matter (PM), hydrocarbons (HC) and carbon monoxide (CO) emissions from heavy duty diesel engines has become an established practice. The design and performance of such systems have been commercially proven to the point that the application of these technologies is cost effective and durable. The application of an effective NOx reduction technology in heavy duty diesel applications is more complicated since there are no passive NOx reduction technologies that can be fit onto HDD vehicles. However, Selective Catalytic Reduction (SCR) systems using Urea injection to achieve NOx reduction have become the technology of choice in Europe and have been applied to achieve Euro IV emissions levels on new HDD vehicles. In addition, retrofit SCR emission control systems have also been developed that can provide high NOx reduction when applied on existing HDD vehicles.

This paper will discuss the development of a commercially available four-way (NOx, PM, CO and HC) emission reduction product for HDD vehicles. The system combines the Johnson Matthey CRT® diesel particulate filter system with a urea SCR system and is known commercially as SCRT® system. This paper will discuss the development and application of such a system for OE and retrofit applications. SCRT product development included optimization of the SCR catalyst performance, integration of the urea injection system hardware, development of a control algorithm and detailed alarm systems.

Transient and steady state test cell data demonstrating 70 - 90% NOx reduction with this system are presented. In addition, chassis dyno emissions results and field operational data on SCRT equipped vehicles is also reported.

INTRODUCTION

The tightening of the emissions standards for Heavy Duty Diesel vehicles throughout the world presents challenges for the engine development and emission control communities. In both North America and Europe, the major challenges are in PM and NOx control for the legislation to be introduced in 2005 (Euro IV), 2007 (US '07), 2008 (Euro V), and 2010 (US '10). Table 1 summarizes the evolution of the emissions standards in the US, and Table 2 outlines the European regulations.

Year	HC	CO	NOx	PM
1998	1.3	15.5	4.0	0.10
2004	0.5	15.5	2.5[1]	0.10
2007	0.14	15.5	1.2	0.01
2010	0.14	15.5	0.2	0.01

1. NOx + NMHC limit with a HC limit of 0.5

Table 1: Emissions Legislation Limits for the US (SET Test Cycle and US HDT Cycle; g/bhp-hr)

Year	HC	CO	NOx	PM
2000 (Euro III)	0.66	2.1	5.0	0.10
2005 (Euro IV)	0.46	1.5	3.5	0.02
2008 (Euro V)	0.25	1.5	2.0	0.02

Table 2: Emissions Legislation Limits in Europe (ESC & ETC Test Cycle; g/ kW-hr)

Achieving the HC and CO emission limits is not expected to be difficult, because the engine-out emissions of these pollutants are already very low. However, major reductions in both PM and NOx emissions are required in the near future to meet these regulations

One system that has been widely used to control PM, HC and CO very effectively is the Continuously Regenerating Technology (CRT®) diesel particulate filter, which was developed and patented by Johnson Matthey [1, 2]. Throughout the remainder of this paper, this system will be referred to as the Continuously Regenerating Diesel Particulate Filter, CRDPF. There are currently over 100,000 CRDPF systems in operation throughout the world.

Within Europe, the principal NOx control strategy starting in 2005, is Selective Catalytic Reduction (SCR) using ammonia, derived from urea, as the reductant. Significant NOx reductions have been demonstrated with Urea based SCR systems on Heavy Duty Diesel applications [3].

By combining the CRDPF and SCR technologies it is possible to get significant reductions of all four of the major pollutants

The CRDPF system comprises an oxidation catalyst followed by a wall-flow Diesel Particulate Filter (DPF). The DPF traps the PM and the oxidation catalyst oxidizes a portion of the engine-out NO to generate NO_2. This NO_2 combusts the PM at a much lower temperature (around 250°C) than does oxygen (around 550°C). This low temperature combustion of PM by NO_2 enables the passive operation (with continuous PM removal) of the CRDPF system on a wide range of HDD applications (e.g. buses, trucks, garbage trucks) and is the basis of the CRDPF. The CRDPF has been used very successfully in the retrofit market over a wide range of operating duty cycles [4, 5, 6], since most HDD applications have duty cycles which are sufficiently warm to guarantee continuous regeneration of the CRDPF system. As well as removing PM, the CRDPF virtually eliminates CO and HC, since the oxidation catalyst converts these species into CO_2 and H_2O respectively.

Selective Catalytic Reduction (SCR) for NOx control using ammonia, derived from urea, as the reductant has been demonstrated as highly effective in heavy duty applications [3]. Ammonia is carried on board in the non-hazardous form of urea solution.

Ammonia is generated on board the vehicle via the rapid hydrolysis of urea. One of the keys to the successful implementation of SCR technology is the ability to match the amount of urea injected to the amount of NOx emitted from the engine. Two methods are used to achieve accurate dosing of urea; a) detailed urea injection map based on engine information and b) urea injection based on real time NOx input (NOx sensor based) and calculation logic. Both methods are discussed and demonstrated here.

Vanadium-based SCR systems are used extensively to control NOx emissions in the stationary source market. These catalysts are very well-suited to the demands of this market sector, since vanadium has very high selectivity to nitrogen, and has a wide operating temperature window. Vanadium-based SCR catalysts also have very high sulfur tolerance. In fact, the major deactivation mechanisms pertaining to such catalysts are thermal deactivation and (potentially) poisoning due to oil-derived species, such as phosphorus. Within the stationary source market the temperature of vanadium-based applications is controlled to avoid the high temperatures that could induce catalyst deactivation and selectivity loss. In view of the greater thermal demands of the on-road market, in which operating temperatures in the 500-600°C range are not uncommon, new vanadium-based catalysts have been developed to ensure the required thermal durability. This was one of the major targets of the development work described in previous work [3].

This paper describes the development of the four way reduction system for OE and Retrofit applications. Both development paths involved laboratory and field testing. The CRDPF and SCR catalysts used in both types of systems were of the same formulation but different urea dosing strategies and equipment were used. The OE systems use Urea map based injection and the retrofit system uses a NOx sensor based Urea injection algorithm.

EXPERIMENTAL

The development of the SCR-DPF system consisted of engine testing at test facilities in Europe and North America, combined with field trials in London, Paris and the United States. Details about the SCR-DPF system, test engines, test facilities and test procedures are discussed in this section.

SYSTEM DESCRIPTION

Figure 1 shows the schematic of the SCR-DPF system. The system consists of the following key components:

1. CRDPF
2. SCR catalyst system
3. Urea dosing & injection system

As shown in Figure 1, exhaust gas from the engine passes through the CRDPF system first, followed by the SCR system. The CRDPF eliminates CO, HC and PM. Urea injection takes place immediately after the CRDPF. The atomized urea generates NH_3 through hydrolysis and this NH_3 reduces the NOx over the SCR catalyst. An NH_3 slip control catalyst is also used at the end of the SCR catalyst system to oxidize any NH_3 that is not used during the reaction.

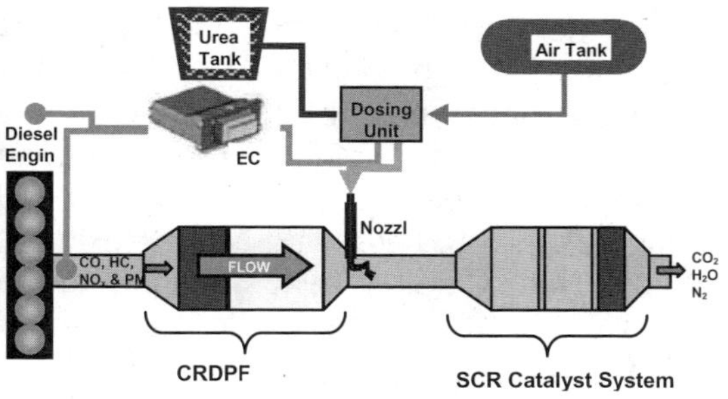

Figure 1: Schematic of SCR-DPF System.

As shown in Figure 1, many of the SCR-DPF systems are configured in the linear fashion, where an SCR + Slip catalyst system follows a CRDPF system. Figure 2 shows a different design of the SCR-DPF system, called the Compact design.

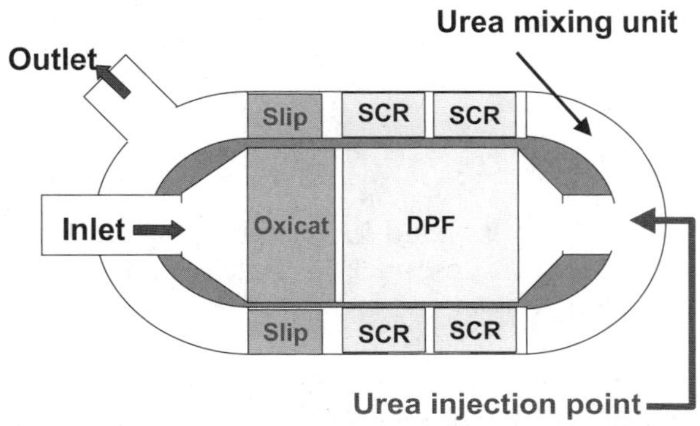

Figure 2: Schematic Representation of the Compact SCR-DPF system

Within the Compact design, the gas first passes through the CR-DPF system, and is then turned through 180° and flows through the SCR catalysts, which are coated onto metallic, annular substrates, fitted around the CR-DPF system. This presents a wider but much shorter packaging envelope for the combined system. The Compact design was developed for one of the OE projects and was was necessary to meet the space constraints of the vehicle.

Details about the primary components of the SCR-DPF system are discussed below.

Continuously Regenerating Diesel Particulate Filter (CRDPF)

The CRDPF device is composed of two primary sections, where the oxidation step in the first is followed by the soot collection/combustion process in the second. The CRDPF comprises four modules which include an inlet head, catalyst section, filter section and an outlet head. The catalyst section contains an oxidation catalyst consisting of a ceramic honeycomb substrate coated with a proprietary highly active platinum group metal. Aside from oxidizing a portion of the NO for soot combustion, the catalyst also oxidizes CO, HC and the SOF portion of the PM. In the filter section, the exhaust flows through a bare ceramic wall-flow filter. The filter is also a honeycomb design with alternate channels blocked at each end, forcing exhaust to flow through the walls of the filter where gaseous components pass through and the soot is trapped. The trapped soot is then combusted by the NO_2 generated by the catalyst. Further details are provided elsewhere [2].

The CRDPF catalyst was applied using standard coating processes onto ceramic substrates with a cell density of 400 cells/in^2. The filter sections of the CRDPF consisted of bare ceramic wall flow filter with a cell density of either 100 or 200 cells/in^2. The size of the CRDPF is specific to a given application. The catalyst section is sized to produce sufficient NO_2 to combust the soot collected in the filter. The filter is sized to provide low operating back pressure.

The CRDPF in the compact design consisted of a PGM catalyst coated onto a 8.44 liter metallic monolith substrate with a cell density of 350 cells/in^2 and a wall thickness of 0.006 in. using standard processing techniques. A cordierite particulate filter is used in the CR-DPF section of the system. The filter is a standard production type of 100 cells/in^2 and a wall thickness of 0.017 in. The volume of the filter was 34.5 liters.

The CRDPF systems used in this project, on various engines, are listed in Table 1.

Engine Model	System Type	CRDPF		SCR System	
		Cat (liter)	Filter (liter)	SCR Cat (liter)	Slip Cat (liter)
Volvo DH10 (10L)	In-line	4.25	17	16.6	3.7
Volvo D12C (12L)	Compact	8.44	34.5	19	9.52
Cat 3126 (7.2L)	In-line	8.5	17	17	4.2
Cum ISX	In-line	11	27.5	27.8	9.2

Table 3: CRDPF and SCR catalyst sizes in different applications

SCR catalyst system

The SCR catalyst in the in-line system consists of vanadium based catalyst mounted on a ceramic monolith. The SCR catalyst is followed by an ammonia slip catalyst also coated on ceramic substrates. The size of the SCR catalyst is based on the engine exhaust flow rates. Typically the volume of catalyst is 1.5 to 2 times the engine displacement. Table 3 also shows the amount of SCR and NH_3 slip catalysts used in each of the test applications. The SCR catalyst was applied on ceramic 400 cells/in^2 substrates as was the ammonia slip catalyst. In the linear system, the ammonia slip catalyst was located downstream of the SCR catalyst and typically the volume of the catalyst is about 0.5 times the engine displacement.

The compact SCR catalysts were coated onto metallic, annular substrates, with a cell density of 350 cells/in^2 using standard processing techniques. Each annular substrate had a volume of 9.52 liters. For some experiments an ammonia slip (clean-up) catalyst coated onto a single annular substrate was used. In other experiments, the ammonia slip catalyst was coated onto a 266.7 x 152.4 mm ceramic substrate with a cell density of 400 cells/in^2 and was located downstream of the compact system.

Urea dosing and injection system

The urea dosing systems used are different but contain similar functions. A metered amount of urea is delivered into a pressurized air stream. The air and urea mixture is then transported to a nozzle that atomizes and distributes the urea in the exhaust flow. The mechanical functions of the system consist of air pressure regulation, pumping urea from the tank to the dosing system and metering the urea into the air flow.

Retrofit Systems

Figure 3 is a schematic of the urea dosing system used in retrofit applications with the various components and control inputs. The amount of urea to be injected is calculated from input signals received from the following:

- Engine Out NOx (Conc.) Sensor
- Engine Parameters via installed sensors or CAN J-1939 data bus
- Exhaust gas temperatures at CRDPF inlet, SCR inlet and SCR outlet
- Urea Temp
- Urea and Air system pressures

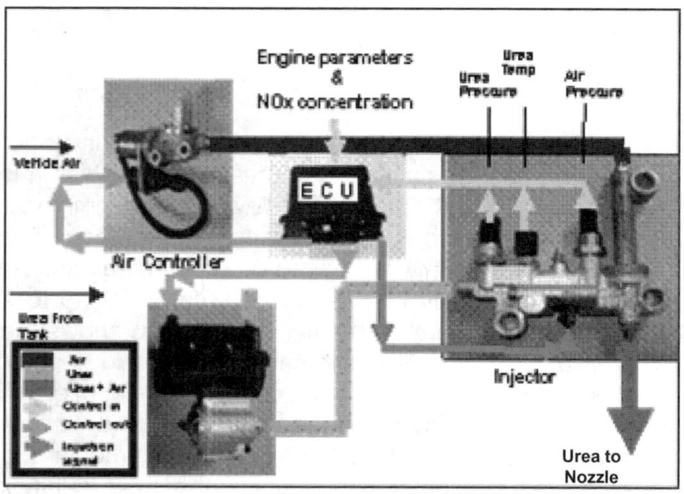

Figure 3: The components in the urea dosing system

This information is used along with application specific data entered into the ECU to calculate the amount of urea needed to get the maximum possible NOx reduction, under that operating condition. A NOx sensor is installed in the exhaust pipe at the outlet of the turbocharger. The retrofit system ECU uses an algorithm that calculates the amount of urea needed based on the engine outlet NOx reading and the exhaust flow of the engine. The ECU then sends a signal to the Urea dosing manifold to deliver the required amount of urea to the SCR catalyst.

Figure 3 also shows the components of the urea injection system. These consist of a urea pump, an air regulator and a dosing manifold. The injection components are manufactured by Hilite International. Urea is pumped from the tank to the urea dosing manifold via a 24 volt accumulator pump capable of delivering up to 157 ml/min of urea. An air regulator is used to deliver pressurized air maintained below 5 bar, to the urea dosing manifold. Either the test cell or the vehicle air is used to supply this air to the regulator. The air regulator is specific to the system.

OE systems

For the systems targeted for OE applications, urea was injected using a Bosch Urea Dosing System (UDS), shown in Figure 4.

The dosing strategy is matched to the NOx emissions of the engine using a pre-programmed engine map. The engine is operated in a test cell at all possible speeds and loads and the engine out NOx is measured. This engine map is programmed into the UDS and used to determine the proper amount of urea to inject. The amount of urea injected can be increased or decreased by adjusting the ammonia NOx ratio (ANR) parameter in the system. An ANR of 1 corresponds to the stoichiometric ratio by moles, which potentially enables 100% NOx conversion. The urea is then injected into a predefined air mass flow delivered from the vehicle air reservoir or compressor in a subsequent mixing chamber. A back-flow valve is integrated into the mixing chamber, to avoid any intrusion of urea into the air supply. The air flow transports the urea through piping and then injects it into the exhaust gas flow through a spray tip. Under these conditions, a fine atomization and even urea distribution in the exhaust flow are pre-requisites for high NOx conversions.

The self-contained control unit is mounted inside the pump. Communication between the dosing control unit and the engine control unit is maintained through the standardized J1939 CAN-bus.

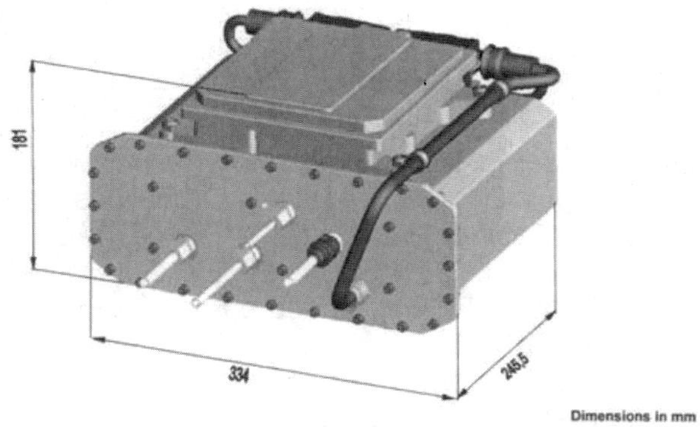

Figure 4: Bosch Urea Dosing System (UDS)

TEST ENGINES AND FACILITES

Development testing

During system development as well as to accurately document the effectiveness of the system, engine dynamometer tests were carried out with the SCR-DPF system on different engines. In addition, following vehicle installations, chassis dynamometer emission tests were also conducted with the SCR-DPF systems.

The engine used for the test cell work in North America was a 1998, 7.2 liter Caterpillar 3126 engine rated at 250 hp. In Europe two Volvo engines were used; a 1997 Euro II 10 liter DH10 rated at 285 horsepower and a 12 liter D12C rated at 465 horsepower and certified to US 2000 emission levels.

For the North American work, steady state development and FTP transient testing were conducted on the Caterpillar engine at Environment Canada's Environmental Technology Centre in Ottawa, Ontario. The emission testing was conducted in the Heavy-Duty engine emissions laboratory of the Emissions Research and Measurement Division. Figure 5 is a picture of the system installed in the test Cell.

The exhaust gas sampling system in this laboratory uses a large double-dilution critical flow venturi (CFV) continuous volume sampler (CVS). Coupled to the dilution tunnel is a secondary dilution tunnel. A continuously integrated analytical system incorporating heated sample lines, filters, heated Flame Ionization Detector (FID) and heated Chemiluminescence instrument are used for THC and NOx measurements, respectively. Similarly, continuous measurements of CO and CO_2 concentrations were taken throughout the engine duty cycle with two separate Non-Dispersive Infrared (NDIR) detection analyzers. NH_3 was measured using a laser based analyzer, down stream of the SCR catalyst, prior to the dilution tunnel. Further details were provided elsewhere [4].

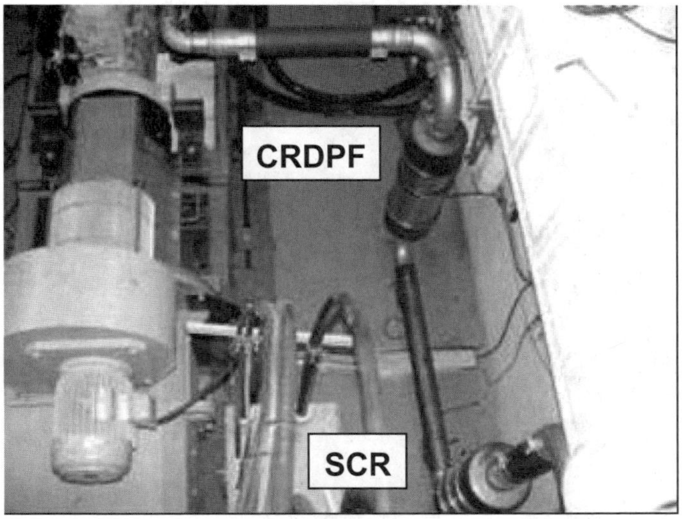

Figure 5: SCR-DPF system in the test cell at Environment Canada

Similar steady state and transient cycle testing were also carried out with a Volvo engine at the Johnson Matthey test facility in Gothenburg Sweden (Figure 6). Both the ESC (European Steady State Cycle) and the ETC (European Transient Cycle) test cycles were used for this validation work.

This facility was also equipped with a total dilution CVS system. Particulate matter was collected using the diluted exhaust on Teflon coated filters and analyzed by gravimetric methods. Gaseous emissions were measured raw before and after the catalyst system using a range of analyzers: chemiluminescence for NO and NO_2, heated FID (Flame Ionisation Detector) for HC, NDIR (Non-Dispersive Infra Red) analyzers for CO, CO_2 and N_2O, and a paramagnetic detector for O_2. Again a laser detrion device was used to measure NH_3 after the SCR catalyst.

The SCR-DPF systems were tested using the US Heavy duty transient cycle (FTP), the European Steady State Cycle (ESC) and European Transient (ETC) test cycles. Regulated emissions (except PM) were measured before and after the Compact SCR-DPF System. Second-by-second measurements of gaseous emissions and engine parameters were made during the test cycles

Various steady state cycles were used to age and de-green the catalysts prior to testing.

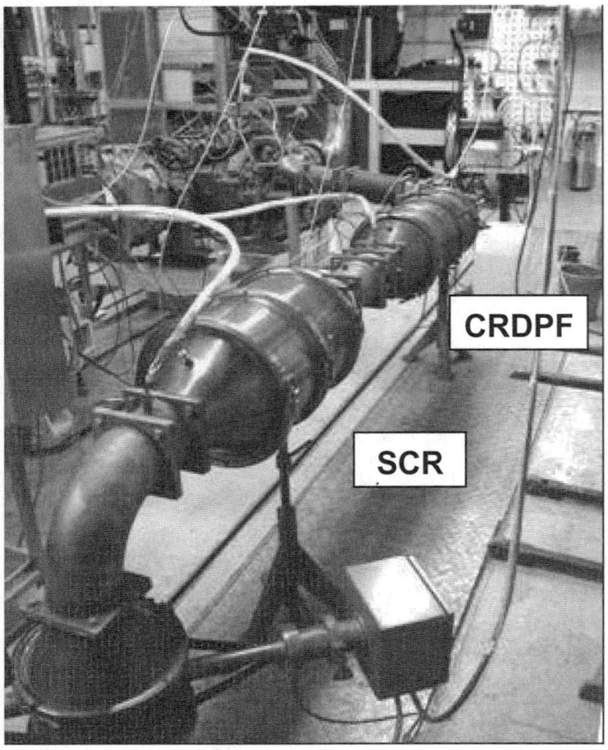

Figure 6: SCR-DPF system installed in the test cell at Johnson Matthey Gothenburg.

Chassis dynamometer emission testing was also conducted with SCR-DPF equipped vehicles at the Millbrook test and development center in England and the Utac center in Montilhery France. A SCR-DPF equipped Dennis DART bus was tested at Millbrook. A complete description of the Millbrook facility is available elsewhere [7]. The emissions are measured using a full CVS system with standard bag sampling as well as using second by second measurement. Ammonia slip is measured via FTIR. The test cycle used was the Millbrook London Transit Bus Cycle (Figure 7)

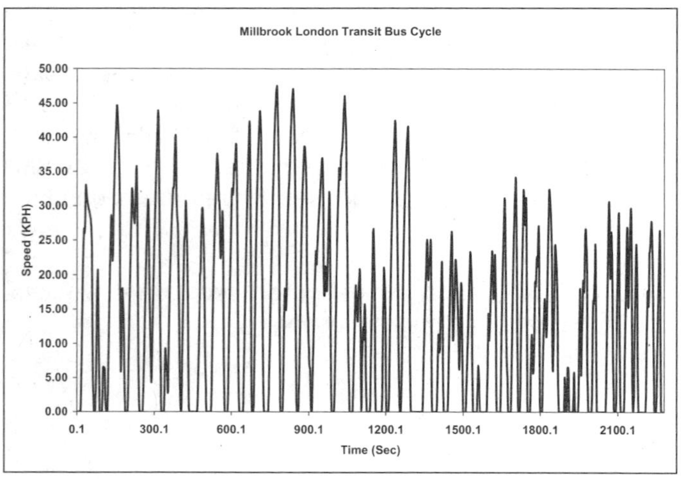

Figure 7: Millbrook London Transit Bus cycle

Two SCR-DPF equipped Paris buses were tested at the Heavy-vehicle test cell at UTAC. Emissions were sampled into bags through a primary dilution tunnel via heated lines. The sampling system used critical flow venturi constant volume sampler (CVS). PM samples were collected on standard filter media following dilution in the CVS system.

On the UTAC chassis dynamometer, the bus was tested over the ADEME/RATP developed urban bus transient duty cycle (Figure 8). This cycle simulates the operation of a bus over Parisian routes. First, 6 test cycles were run with the SCR-DPF equipped bus, but with the Urea injection system turned off. This provided the engine out NOx emissions. Then 6 repeat cycles were run on the bus with the urea injection system turned on, to obtain the NOx reduction.

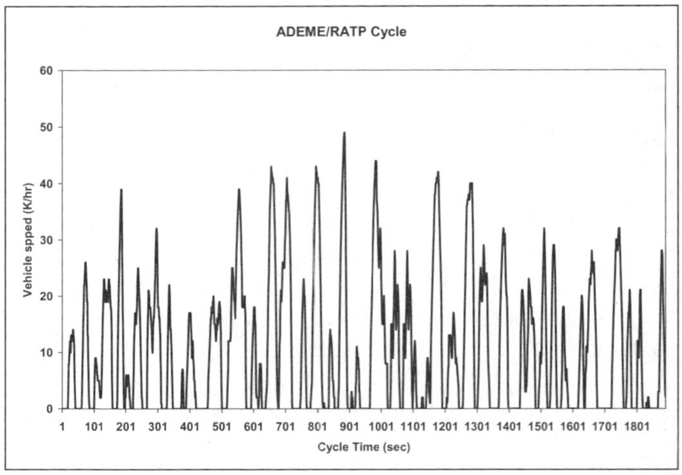

Figure 8: ADEME/RATP transient test cycle

FUEL

Ultra low sulfur diesel (ULSD) fuels from various suppliers were used during the development testing. All of the fuels had a sulfur content of less them 30 ppm.

FIELD TRIALS

Following initial system development and test cell work, SCR-DPF systems were installed in several different vehicles to gather on-road performance information. One system using a map based injection strategy was installed in a Class 8 truck containing a Cummins ISX, 15 liter, 425 hp engine (Figure 9). This truck is used to haul freight from the mid-west to the west coast of the United States.

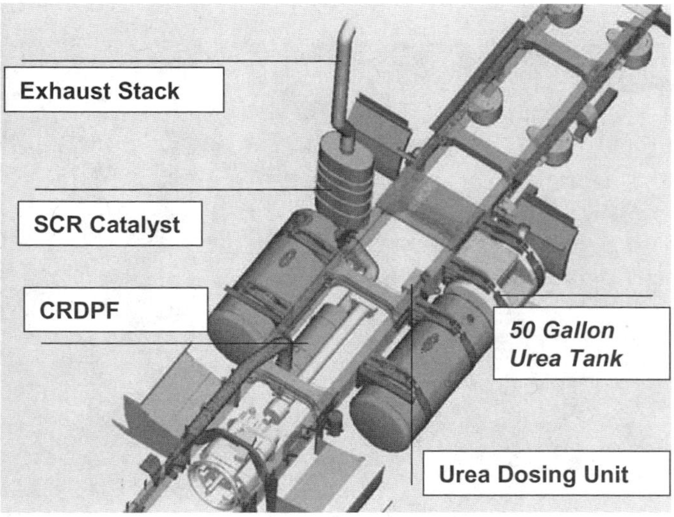

Figure 9: Schematic Representation of the Installation of the SCR-DPF System

The sensor based injection strategy version of the SCR-DPF system was installed on three buses in Europe. One was on a Dennis Dart bus operating in London (Figure 10, 11, and 12) and the other two were on a Renault Agora bus and a IrisBus operating in Paris. The Denis Dart bus was a 30 ft bus powered by a Euro III Cummins ISB 3.9 liter engine rated at 135 horsepower. Both of the Agora and the Iris buses were 40 Ft long. The Renault Agora bus was powered by a 9.8 liter Euro II certified Renault engine rated at 152 kW (206 hp). The Irisbus was powered by a 7.8 liter Euro III (Figures 13 and 14) certified Iveco Engine rated at 180 kW (244 hp). Both the Dart and the IrisBus with the Euro III engines were electronically controlled so the CAN bus data could be used by the SCR-DPF ECU to calculate the exhaust flow. The Agora bus with the Euro II engine was mechanically controlled. This bus was fitted with an air mass flow sensor on the air intake to calculate the exhaust flow.

After the three buses were retrofitted with the SCR-DPF system, they were put back into normal service. These are city buses which spend much of their operating time in heavy stop and go traffic in large cities.

Figure 10: Dennis Dart bus where SCR-DPF was installed

Figure 11: View from rear of the Dennis bus with SCR-DPF installed

Figure 12: SCR-DPF system installed on the Dennis bus

Figure 13: IrisBus Euro III bus with SCR-DPF

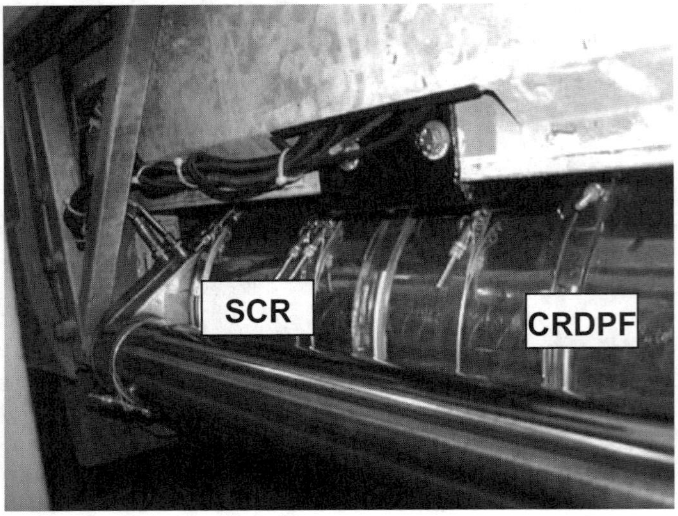

Figure 14: SCR-DPF system installed on IrisBus Euro III bus with

Another sensor based SCR-DPF system was installed on a refuse transfer truck in Los Angeles County, CA, USA. The truck is powered by a 1997 Cummins M11 engine rated at 330 hp (Figure 15). This Los Angeles County Sanitation Department truck transfers recycling from a collection point to a landfill for processing.

Figure 15: LASD truck with SCR-DPF installed

Finally, the Compact SCR-DPF system using map based injection strategy has been fitted to a number of trucks which are currently operating in revenue service in North America. These trucks have been delivered as OE units from Volvo, factory fitted with the SCR-DPF systems. The engine used is a Volvo D12C. Figure 16 shows a picture of the compact SCR-DPF system installed on one such truck.

Figure 16: Compact SCR-Trap System Installed on a Heavy Duty Truck

Table 4 provides the catalyst sizes used for all of the field trial systems.

Application	Engine Model	CRDPF		SCR System	
		Cat (liter)	Filter (liter)	SCR Cat (liter)	Slip Cat (liter)
Class 8 Truck	CUM ISX	11.0	27.5	37.8	9.2
Paris IrisBus E3 Bus	Iveco	8.5	17	17	4.2
Dennis Dart	CUM ISB	4.25	17	16.6	3.7
Volvo Truck	Volvo D12C	8.44	34.5	19.2	9.5
LACSD Truck	CUM M11	8.5	17	25.5	4.25

Table 4: CRDPF and SCR catalyst sizes in field trial applications

All of the vehicles were equipped with NGK NOx sensors to measure engine out and tailpipe NOx levels while in operation. The three buses and one truck equipped with the sensor based control algorithm have an ECU that has datalogging capability so the NOx reduction data was stored in the system ECU. The other two truck applications had separate dataloggers that recorded the engine out and tailpipe NOx data. The field trial systems were evaluated for system durability, performance and maintenance impact.

RESULTS AND DISCUSSIONS

In this section results from the system development engine testing, chassis dyno testing and field evaluation are presented.

TEST CELL DEVELOPMENT RESULTS

The specific development tasks for this project were:

1. Evaluate the combined CRDPF-SCR technology for reduction of PM, HC, CO and NOx
2. Evaluate and optimize system component performance
3. Design, develop and optimize the control software and hardware for retrofit applications
4. Evaluate the performance and durability of the of the SCR-DPF catalyst in a mobile Heavy Duty application

SCR catalysts have been used successfully in stationary applications for twenty years. These applications tend to operate under steady state conditions and use both diesel and natural gas fueled engines. In order to apply the SCR catalyst to on-road applications with their transient nature and heavy fuel and oil, new durability requirements of the catalyst had to be evaluated. Previously, detailed test work [3] was carried out to evaluate the performance of an advanced SCR catalyst. The catalyst evaluated was an advanced version of Vanadium on Titania formulation. The catalyst was exposed to high temperature, sulfur and potential poisoning by lube oil additives. Those tests demonstrated the robust performance of the catalyst under conditions expected in mobile HDD applications. In the present paper, complete SCR-DPF system development will be discussed.

SYSTEM DEVELOPMENT

Figure 17 shows some results from the system development testing with the Caterpillar 3126 engine in Canada. In this chart, NOx conversion and NH_3 slip are reported as a function of urea injection (PWM). The data is from steady state engine operation at 1500 rpm speed and different engine loads. The engine loads basically represent the exhaust temperature, which increased with increasing engine load at constant speed. Therefore, the minimum catalyst temperature was at 220 ft-lb load while the highest temperature was at 666 ft-lb load. Lowest NOx conversion was observed with 220 ft-lb load condition, the lowest exhaust temperature. Highest NOx conversion was observed with 370 ft-lb load and then NOx conversion again decreased with increasing load, but always stayed above 220 ft-lb condition. NH_3 slip was the highest at the lowest load and then decreased continuously with increasing load. With increasing urea injection, NOx conversion at the different loads generally increased almost linearly up to 90%. Beyond 90%, the rate of increase was much slower. The parabolic nature of NOx conversion with increasing load, probably indicated effects of both NO_2:NO ratio and exhaust temperature on the SCR reaction.

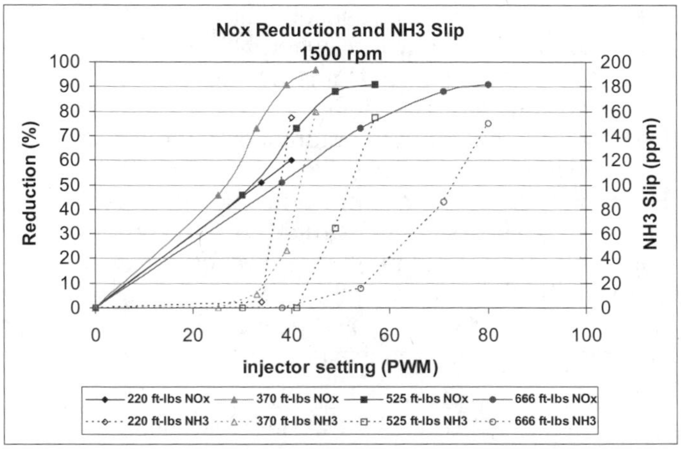

Figure 17: NOx reduction obtained with different urea injection at different engine loads.

The control algorithm and SCR-DPF system were also tested at the JM facility in Sweden on the Volvo engine. A series of NOx conversion curves were generated with the system. For these curves, the engine speed was held constant and the load was changed to achieve particular SCR inlet temperatures. Figure 18 is an example of this conversion curve. Note that the X-axis shows the CRDPF inlet temperature or the engine out exhaust.

In Figure 18 it can be seen that NOx conversion with the system increased from CRDPF inlet temps of 150 – 250°C, but then went through a slight reduction before increasing and stabilizing again from 350°C. The reason for this behaviour is again the percentage of NO_2 in the NOx

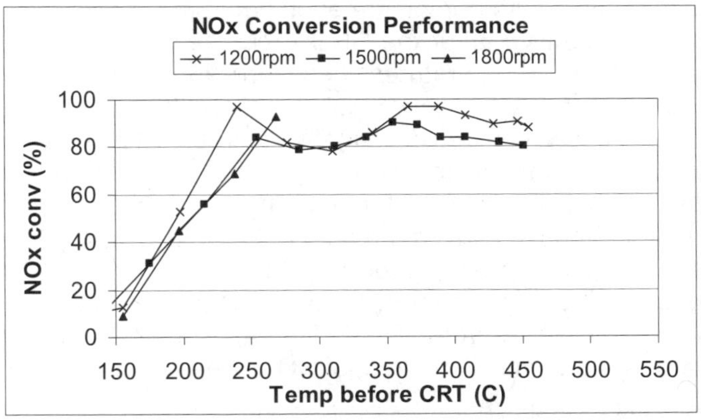

Figure 18: NOx Conversion over steady state conditions

There are three major reactions between NOx species and ammonia which can contribute to the overall reaction process over vanadium-based SCR catalysts:

$$4\,NH_3 + 4\,NO + O_2 \rightarrow 4\,N_2 + 6\,H_2O \qquad (1)$$

$$4\,NH_3 + 2\,NO + 2\,NO_2 \rightarrow 4\,N_2 + 6\,H_2O \qquad (2)$$

$$8\,NH_3 + 6\,NO_2 \rightarrow 7\,N_2 + 12\,H_2O \qquad (3)$$

Reaction (1) is a fast reaction and dominates in the absence of NO_2. However, when NO_2 is present in the NOx feed, a significant improvement in activity is observed due to Reaction (2), which is very fast. However, the extent of the promotion depends upon the ratio of NO_2:NO in the gas feed. At ratios of up to 1:1, Reaction (2) is dominant, and excellent activity is observed. This is demonstrated in the 200 to 250°C, operating regime in Figure 18. However, if the NO_2 level increases further, that is, if the NO_2:NO ratio goes above 1:1, the activity falls sharply because Reaction (3) then becomes important, and this is a very slow reaction.

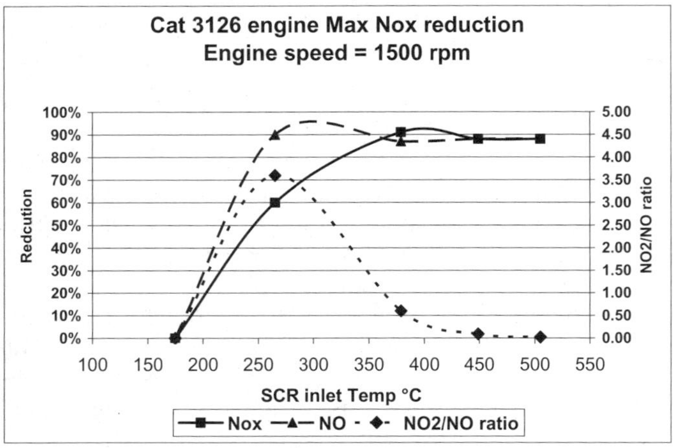

Figure 19: Effect of NO_2/NO ratio on NOx conversion

Therefore, overall NOx reduction could decrease under conditions where NO_2:NO ratio is over 1. Figure 19 clearly demonstrates this effect observed during the engine testing. The ratio of NO_2:NO is a function of the oxidation catalyst in the CRDPF and the exhaust temperature. As demonstrated in figure 19, the NO_2:NO ratio of the NOx after the CRDPF increases over 1:1 at engine out temperatures between 250 and 350°C. Referring back to Figure 18, it can be seen that the overall NOx conversion slightly decreases at CRDPF inlet temperatures between 250 and 350°C. However, this effect is minimized at temperatures above 350°C, as thermodynamics drives the NO_2:NO ratio below 1. This shows that it is very important to carefully design the overall SCR-DPF system, to take into account both the low temperature promotional effect as well as the higher temperature inhibition effect of NO_2 in the combined unit.

ENGINE DYNOMOMETER TESTING

<u>OE system</u>

The compact SCR-DPF system for OE applications was tested over The European Steady-state Cycle (ESC) and the US Heavy Duty Transient (FTP) cycles. The engine used for these studies was a Volvo D12C 12 litre, 347 kW (465 hp) turbocharged, aftercooled engine, originally certified for US'00 but the engine calibration was modified to optimize engine performance and compliment the pollution control system.

Ammonia to NOx ratios (ANRs) of between 0.85 and 1.00 (by moles) were used in this work; the ANR was used to control the urea injection rate. The volumetric flow rate of the urea solution to the injector nozzle was measured periodically to check that the urea dosing rate was correct.

The performance of the system was initially measured over the ESC test cycle with an ANR ratio of 0.95, so the maximum possible NOx conversion in these experiments was 95%. Two Compact SCR-DPF Systems were tested: one containing two annular SCR catalysts followed by an annular ammonia slip catalyst, and the other containing three annular SCR catalysts followed by an ammonia slip catalyst located downstream of the compact unit. In both cases the emissions measurements were made after the ammonia slip catalyst. The ESC emissions of these two compact systems are summarized in Table 5.

System	HC	CO	NOx	PM
Engine Out	0.123	0.324	6.926	0.022
2 SCR Catalysts	0.002	0.000	1.097	0.006
3 SCR Catalysts	0.002	0.000	0.562	0.008
Euro V Limits	*0.460*	*1.500*	*2.000*	*0.020*

Table 5: Emissions of the Compact SCR-DPF System over the ESC (g/kW-hr)

It can be seen that the system with the lower catalyst volume (the one containing only 2 annular SCR catalysts) is very comfortably inside the Euro V emission legislation for all four pollutants, which is the goal of the engine OEs. Indeed, the NOx emissions of the system are approximately half the proposed Euro V NOx standard of 2 g/kW-hr. When the additional SCR catalyst is added to the system, the NOx emissions are reduced to approximately one quarter of the proposed Euro V NOx level. As expected, the addition of the third SCR catalyst does not lead to any significant changes in the other emissions, which all remain far below the Euro V legislated levels.

The HC and CO conversions obtained using the compact systems are extremely high, with both systems providing 98% HC and 100% CO conversion. The NOx conversion obtained using two annular SCR catalysts is 84%, rising to 92% when the third SCR catalyst is added. (Since urea is injected to provide an ANR of 0.95, 95% NOx conversion is the maximum possible in these experiments, so this 92% is a remarkably high conversion figure). The PM conversion from the system is around 70%, but this relatively low figure is a consequence of the very low engine-out PM emissions, which, at 0.022 g/kW-hr, almost meet the Euro V PM standard of 0.020 g/kW-hr without the CR-DPF filter system. Note that the PM emissions from the compact systems are less than half of the Euro V legislated level.

Figure 20 shows the NOx conversion and ammonia slip level from the compact system containing three annular SCR catalysts at each mode of the ESC (except Mode 1, which is the idle mode). Note that the NOx conversion is measured over the complete system (ie *downstream* of the ammonia slip catalyst) while the ammonia slip is recorded in front of the ammonia slip catalyst. The ammonia slip was measured here to show how much ammonia breaks through the SCR catalysts, and therefore to provide an accurate assessment of the precision of the urea injection calibration.

Figure 20 shows that the NOx conversion remains extremely high in each ESC mode, never falling significantly below 90%. The ammonia slip levels measured *before* the slip catalyst are very low, never rising above 10 ppm, which testifies to the excellent urea injection calibration developed in this work. As expected, the ammonia slip level measured downstream of the slip catalyst was almost undetectable.

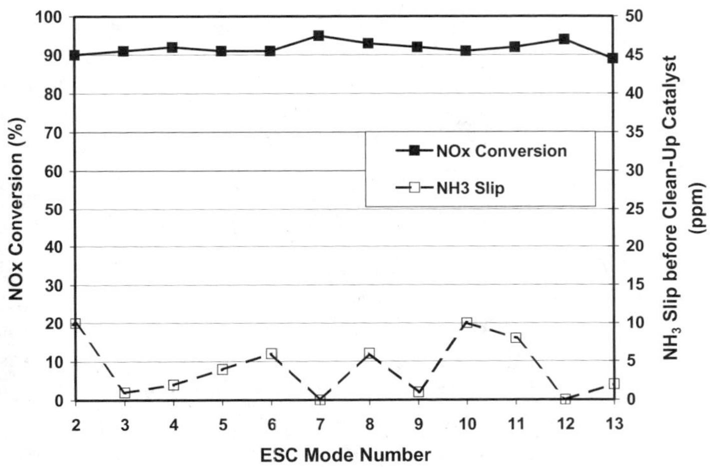

Figure 20: Mode-by-Mode NOx Conversion and Ammonia Slip Levels using 3 SCR Catalysts over the ESC Test Cycle

For the US Heavy Duty FTP Transient test the compact SCR-DPF system comprised two annular SCR catalysts followed by an annular ammonia slip catalyst.

Figure 21 shows the engine-out and tailpipe NOx emissions, as well as the SCR catalyst temperature, during a hot start US Heavy Duty Transient test.

Since the SCR inlet temperature is above 200°C permitting urea to be injected for almost the entire hot cycle, the NOx conversion seen is high. Table 6 summarizes the emissions and conversions during the hot start US FTP test.

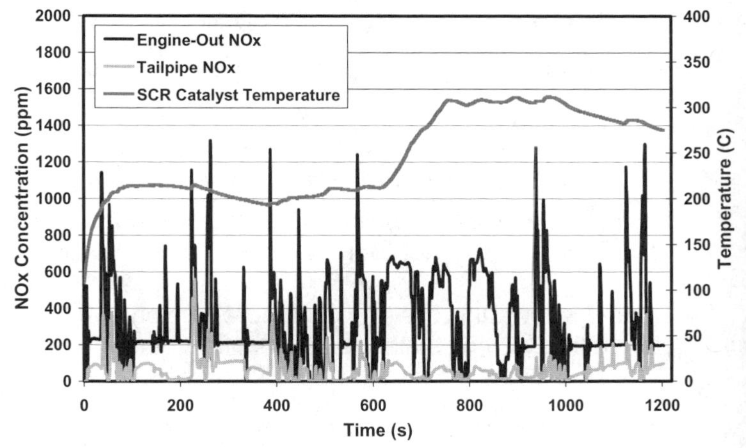

Figure 21: NOx Emissions and SCR Catalyst Temperature During a Hot Start US Heavy Duty Transient Test

System	HC	CO	NOx
Engine Out	0.202	0.832	4.975
Tailpipe	0.007	0.000	0.866
Conversion	97%	100%	83%

Table 6: Gaseous Emissions From the Compact SCR-Trap System in the Hot Start US FTP Test (g/bhp-hr)

Table 7 shows the overall weighted emissions (1/7 cold start plus 6/7 hot start emissions) from the Compact SCR-Trap System over the US FTP test cycle.

System	HC	CO	NOx
Engine Out	0.199	0.899	5.040
Tailpipe	0.008	0.050	1.079
Conversion	96%	94%	79%

Table 7: Weighted Gaseous Emissions From the Compact SCR-Trap System in the US Heavy Duty Transient Test (g/bhp-hr)

The weighted NOx conversion of 79% brings this 5 g/bhp-hr engine inside the US 2007 NOx limit of 1.2 g/bhp-hr. Unfortunately it was not possible to measure the transient PM emissions during these tests. However, it is expected that with the CRDPF, the PM emission will be below 0.01 g/bhp-hr. The HC and CO emissions are significantly beneath the US 2010 limits of 0.14 and 15.5 g/bhp-hr respectively.

Therefore, the combination of the state-of-the-art Volvo engine and the Compact SCR-DPF System is capable of meeting the Euro V and US 2007 emission regulations for all emissions, and can meet the US 2010 limits for HC, CO and PM

Retrofit system

After the system development and integration, the complete retrofit SCR-DPF system was tested on the Caterpillar 3126 and the Volvo DH10A engines using the US HDD FTP cycle and European Transient Cycle (ETC).

Figure 22 shows the results of a hot start US FTP test on the CAT 3126 engine with the SCR-DPF. First, the engine was emissions tested without the SCR-DPF system. This represented the baseline emissions data. Following the baseline testing, the engine was tested with the SCR-DPF system installed. Both the engine out NOx emissions and the SCR-DPF out NOx emissions are plotted in Figure 22 as a function of test time. Please note that this NOx concentration was measured in the dilution tunnel and not the raw numbers. In addition, the SCR inlet temperature and the NH_3 slip during the FTP cycle are also shown in this graph.

The SCR inlet temperature with the CAT 3126 engine under the hot start FTP cycle was relatively low. For nearly 50% of the cycle, the SCR inlet temperature was below 240°C and it was below 200°C for considerable portion of the first 400 seconds. During this time, urea injection was minimized to avoid urea deposit formation and consequently, NOx reduction was low. After the first 400 seconds, urea injection increased significantly and nearly 90% NOx reduction was observed in this last 800 seconds of the FTP cycle. Overall, the cycle average NOx reduction was 76%. The NH_3 slip never exceeded 10 ppm during the entire test cycle. The cycle average NH_3 slip was 1 ppm. These results clearly demonstrated the transient urea injection and NOx reduction capability of the retrofit SCR-DPF system. It was also evident that with higher SCR inlet temperature, >80% NOx reduction is possible with this system under transient operation.

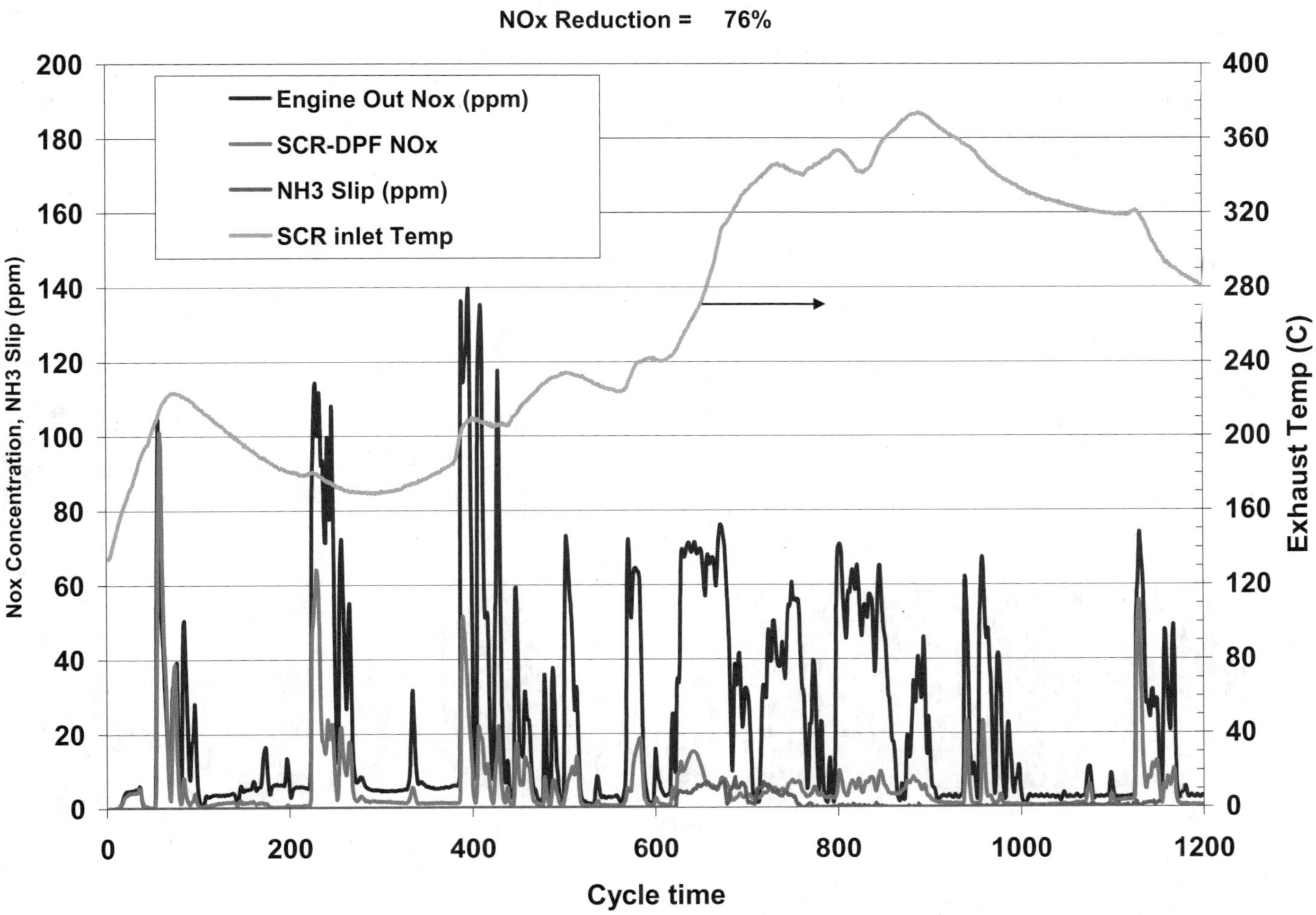

Figure 22: Performance of SCR-DPF system over US FTP cycle on the Caterpillar 3126 engine

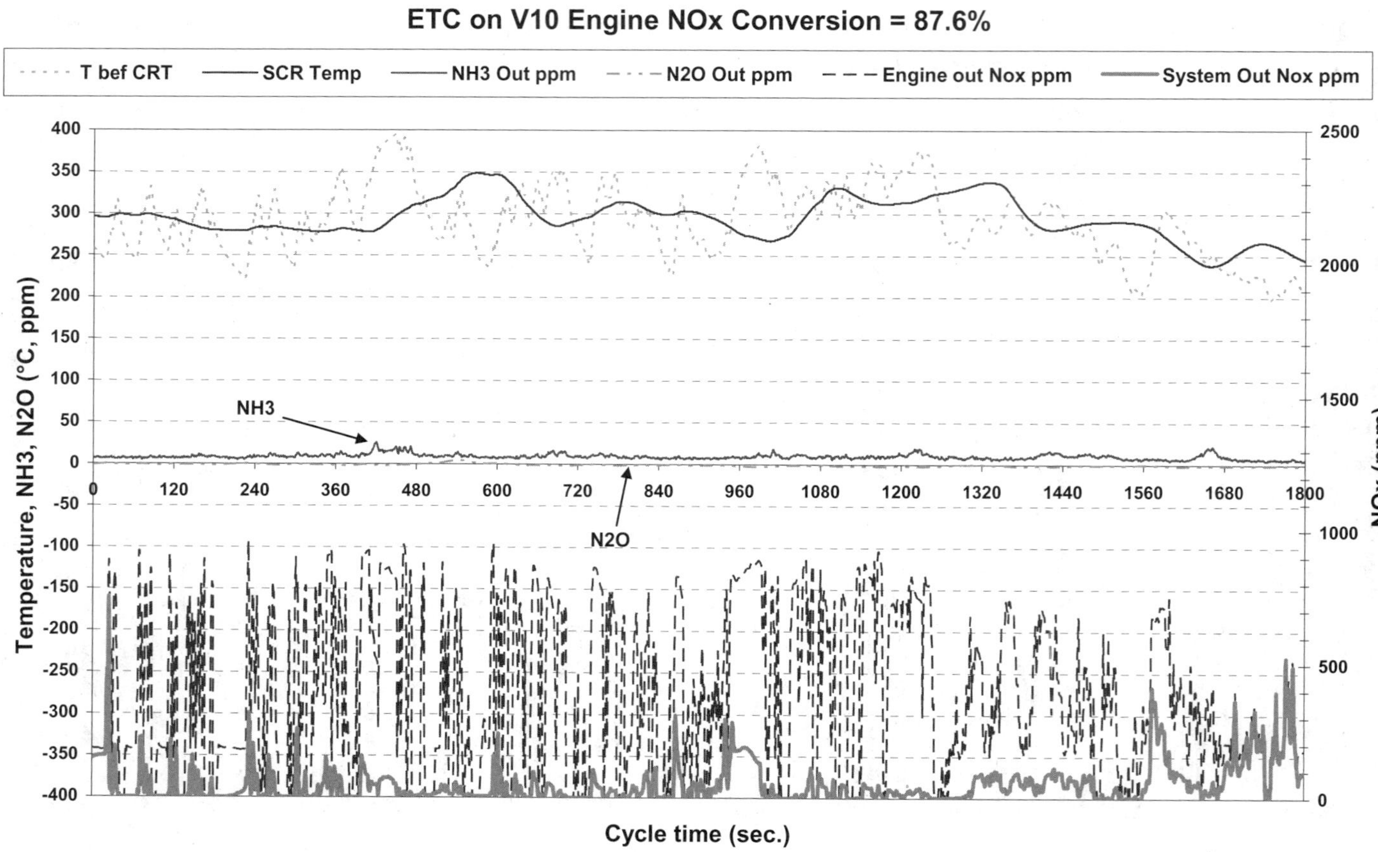

Figure 23: European Transient Cycle test results with SCR-DPF system on the Volvo engine

The SCR-DPF system was also tested under the ETC cycle on the 10 liter Volvo engine. Figure 23 shows the results from the ETC testing. The top part of the graph shows the CRDPF and SCR inlet temperatures during the testing. The SCR inlet temperature with this test was considerably higher and it remained at or above 250°C for nearly the entire test cycle. Consequently, significant NOx reduction was achieved from the beginning of the cycle. The bottom half of the chart shows the engine out and tail pipe NOx emissions. Please note that the NOx concentration reported here is the raw number, measured directly upstream and down stream of the SCR system (Figure 5). The cycle average NOx reduction observed during this test was 87.6%. The ammonia slip (shown in the middle of the graph) during this test was 25 ppm peak and 9 ppm average. These results again demonstrated the excellent transient capability of the system.

ESC test results with the system on the Volvo engine are plotted in Figure 24. Tests were carried out for 13 modes and NOx conversion and SCR inlet temperatures for each mode are shown in the figure. In general the SCR temperature during ESC was warm and it resulted in high overall NOx reduction. The cycle average NOx reduction was 87%.

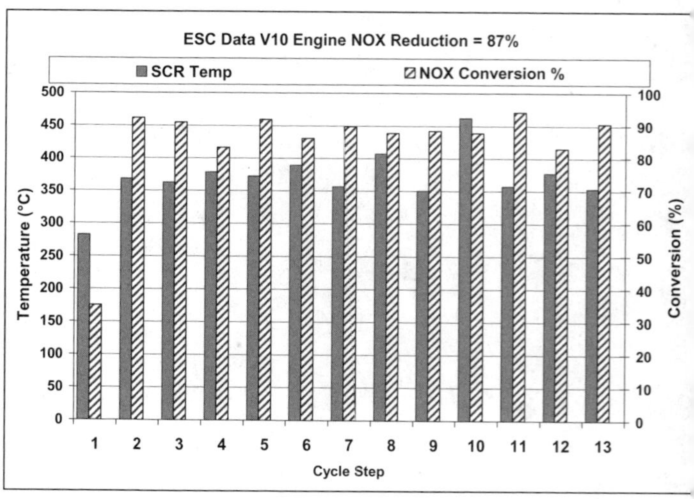

Figure 24: European Steady State Cycle test results

CHASSIS DYNO TESTS

The SCR-DPF equipped London and Paris buses were tested on chassis dynamometers under simulated transient test cycles to evaluate the performance of the system in real vehicle application.

The London Dennis Dart bus was tested at the Millbrook testing facility. The objectives were to optimize the injection calibration as well as to evaluate the system performance under real world vehicle operation. The vehicle was run over repeated MLTB cycles (a real world inner city cycle) with small calibration changes. Figure 25 shows one of the MLTB test cycle results with the SCR-DPF on the bus. The vehicle speed, instantaneous NOx conversion and SCR inlet temperature are plotted in this figure. The Cummins engine on this vehicle exhibited good operating temperature, which stayed above 250°C during the entire test cycle. Therefore, urea injection in this system was never limited by operating temperature and 70-94% NOx conversions were observed during the transient operation. The cycle average NOx reduction for this test was an outstanding 89% with no measurable NH_3 slip.

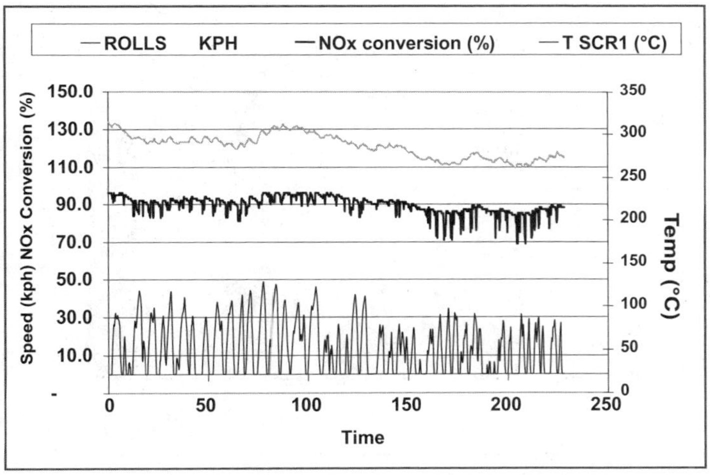

Figure 25: Data from Millbrook test cycle with SCR-DPF

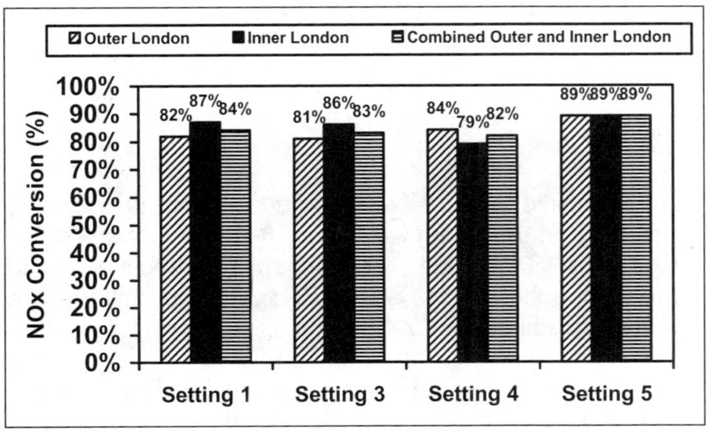

Figure 26: Overall Millbrook test results with SCR-DPF on Dennis Dart bus

The overall MLTB test results are presented in Figure 26. Average NOx conversions for each test cycle are presented as a function of the different system calibration settings. Calibration settings 1, 3, 4 and 5 represent minor modifications in the injection strategy to optimize the system performance under this transient operation. Calibration setting 2 was used to collect engine out NOx emission, without any urea injection. The test results are also broken up to show system performance during the outer London part of the cycle, the inner London part of the cycle and the combined cycle. The test results showed cycle average NOx reductions of 79 to 89% with this system. The tests indicated that calibration setting 5 produced the best performance. That optimized setting was retained in the system for on-road operation. No measurable NH_3 slip was noticed during these tests, attesting to the accuracy of the injection strategy.

Two Paris buses were fitted with prototype SCR-DPF systems and tested on a chassis dynamometer at UTAC. Then the systems were tested in actual transit operation to demonstrate their durability. Besides the primary test cell data, some emissions and temperature data were also collected during these chassis dyno tests, using the ECU in the SCR-DPF system. As previously stated, a second tailpipe NOx sensor was added to these prototypes so that both the engine out NOx and tailpipe NOx could be recorded during testing. This data is presented here.

One of the buses tested was an IrisBus with an Iveco Euro III engine. The SCR-DPF equipped vehicle was first tested over six repeated ADEME/RATP cycles with no urea injection. These results were utilized as engine out NOx baseline. Following the baseline testing, the bus was again tested over six repeat cycles with SCR-DPF operating and urea injection turned on. Temperature and system data from one of those cycles is presented in Figure 27. In this figure, the CRDPF inlet and SCR inlet temperatures are shown along with the vehicle speed and the rolling average NOx conversion. The NOx conversion shown is a 60 sec average of instantaneous NOx conversions calculate by comparing the readings from the engine out and tailpipe NOx sensors. As can be seen, the SCR inlet temperature always remained above 225°C and beyond the first 700 seconds, it remained at or above 250°C.

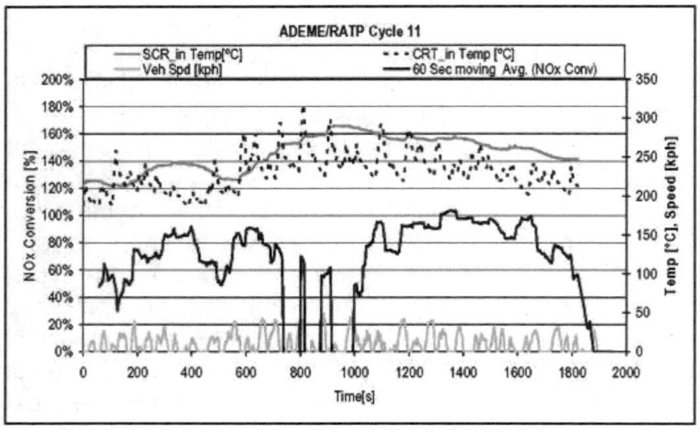

Figure 27: System temperatures and NOx Reduction on the Euro III IrisBus bus over ADEME/RATP cycle

As shown in Figure 27, the average NOx reduction during this cycle was over 80% for most of the cycle. The overall SCR-DPF system performance results are presented in Figure 28. This data was collected by UTAC. The data presented in Figure 28 is an average of the six test cycles. The results compare the emissions of the Euro III bus in engine only, with CRDPF, with SCR-DPF but without urea and with SCR-DPF with urea injection configurations. Please note that the baseline engine emissions and engine with CRDPF only data were collected from separate buses, not the current SCR-DPF equipped bus.

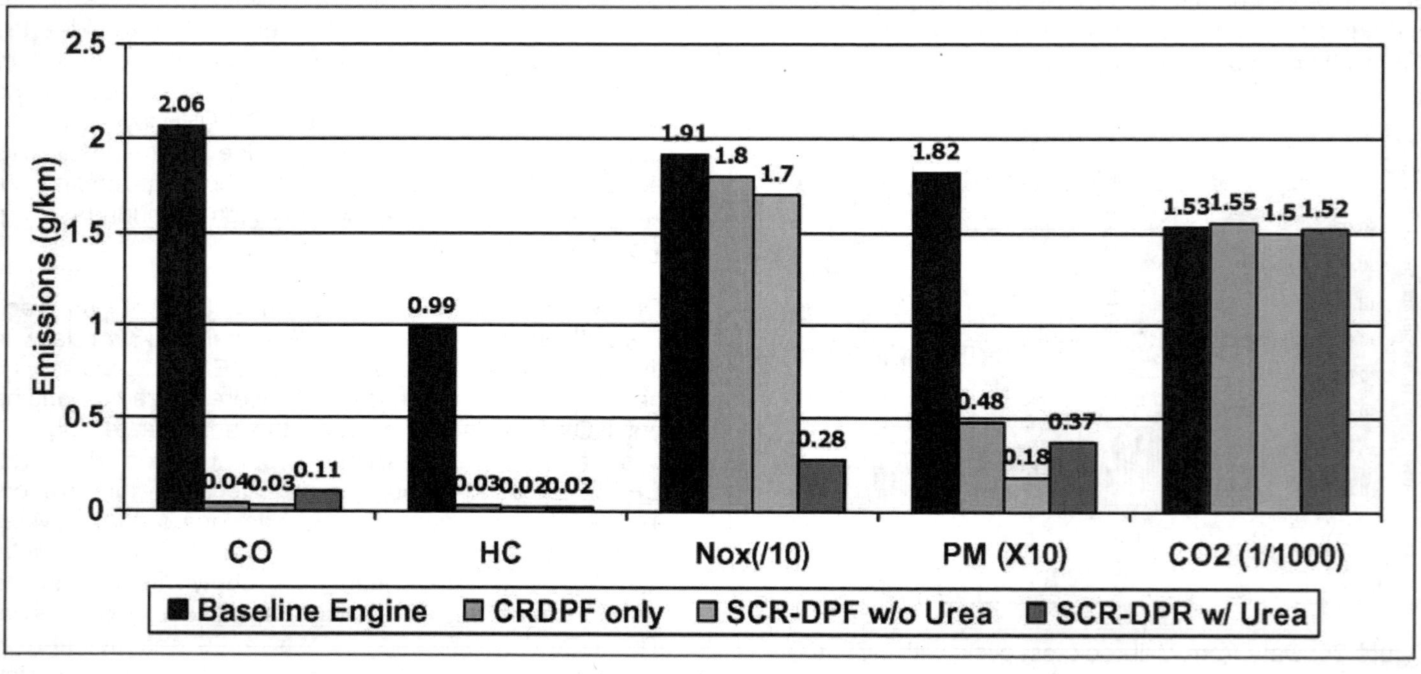

Figure 28: Emission testing results of Euro 3 IrisBus bus under UTAC cycle

The results show that the SCR-DPF system can provide 83-85% NOx reduction on this bus under the ADEME/RATP cycle.

FIELD TRIALS

Besides test cell work, field testing of the SCR-DPF system was carried out on several vehicles. These included different vehicles as well as different configurations of the SCR-DPF system. Both OE type and retrofit systems were evaluated.

OE system

An extensive field trial was carried out with an OE map based system using a Class 8 truck containing a Cummins ISX engine. The dimensions and volume of the components of the SCR-DPF system for this trial are summarized in Table 3.

The system configuration as used in the truck for this field trial is shown in Figure 9. A vertical SCR catalyst system followed a horizontal CRDPF system with Urea injection in between. Urea was injected using a Bosch Urea Dosing System (UDS), and the required quantity of urea at each engine operating point was determined by mapping the engine prior to the field trial. A very conservative urea injection strategy was used, such that the Ammonia-to-NOx Ratio (ANR) was only 0.85, which means that the maximum possible NOx conversion was 85%. In addition, urea was not injected until the temperature reached 250°C, to be absolutely sure that deposits due to the partial decomposition of urea would not be formed during this field trial. (At low temperatures the partial decomposition of urea can produce undesirable species such as cyanuric acid.)

Again, NOx sensors were installed on the vehicle not as part of the control system but just to monitor both the engine-out and tailpipe NOx levels. This was done in order to provide an on-road assessment of the NOx conversion obtained by the system.

This truck is used to haul freight from the mid-west to the west coast. Each round trip comprises 5,000 miles and takes approximately 110 hours. In general, the CRDPF system is only recommended for use with Ultra Low Sulfur Diesel (ULSD), ie fuel with a sulfur level of 50 ppm or lower. This is principally because sulfur present in the exhaust from higher sulfur fuel can reduce the efficiency of the passive DPF regeneration, especially at lower temperatures. In addition, the oxidation catalyst within the CRDPF system can produce sulfate from the fuel sulfur at higher operating temperatures. However, within this field trial it was not practical to develop an infrastructure to supply ULSD, so it was decided that the truck would operate using regular US diesel fuel, with a sulfur level of around 350 ppm. The plan was to monitor the CRDPF very carefully during this testing with higher sulfur fuel.

The temperature profile over the course of a typical freight trip is shown in Figure 29. This temperature was measured downstream of the SCR unit.

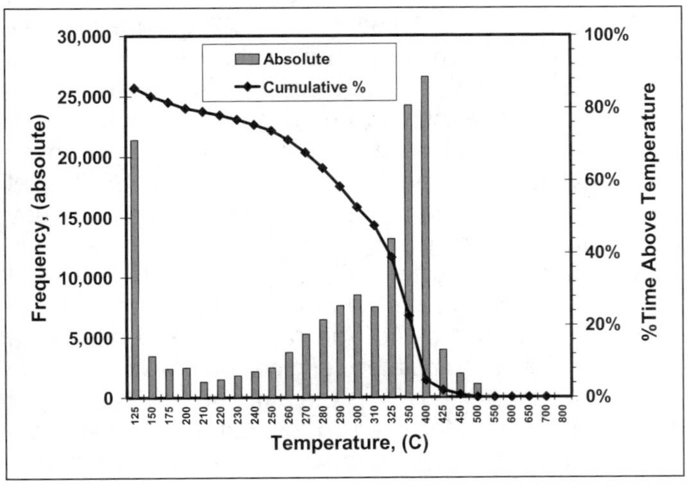

Figure 29: Histogram Showing the Temperature Distribution During a Typical Freight Trip

It can be seen that the temperature of the system is between 325 and 425°C for the majority of the time, with excursions into the 500-550°C range. Note also that these temperatures were measured downstream of the SCR system; it is clear that the temperature at the inlet to the SCR unit will be significantly higher. The truck also spends periods of time at idle, as shown by the time at temperatures below 200°C. This high overall temperature profile suggested that the CRDPF would have minimal issues with passive regeneration due to higher sulfur fuel.

The truck was used in normal revenue service, but was periodically removed from service for brief periods of time to assess the performance of the SCR-DPF system. A route comprising a round-trip of 850 km was developed, and measurements made over the course of these runs were used to assess the efficiency of the system. After each evaluation trip the truck was placed back into revenue service and continued to accumulate mileage at a high rate.

Figure 30 shows the NOx conversion obtained during an on-road system evaluation after approximately one month of normal service.

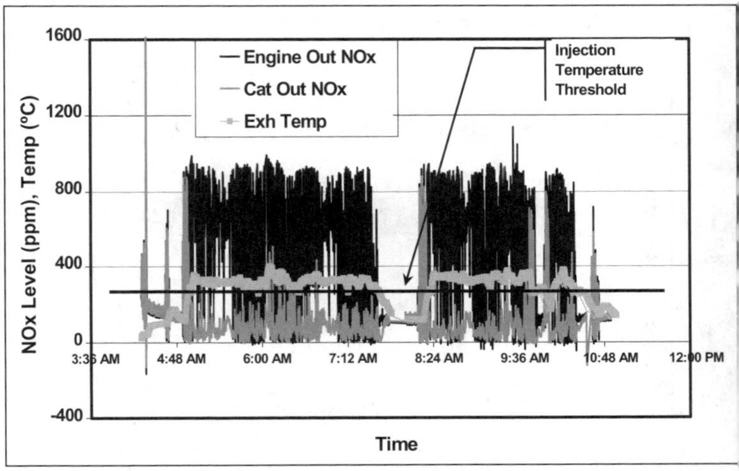

Figure 30: Engine-Out and Tailpipe NOx Measurements Made During an 850 km Evaluation Trip Shortly After the Start of the Field Trial

It can be seen that for the majority of the time the temperature at the inlet to the SCR system was above the critical level of 250°C at which urea injection was enabled. It is clear that a very high level of NOx conversion is observed over the course of this evaluation trip. Indeed, the average on-road NOx conversion measured over this trip was 80%.

Figure 31 shows the engine-out and tailpipe NOx levels over the evaluation route after 6 more months of mileage accumulation on the truck. The test results showed that the SCR-DPF system attained 80% NOx conversion level. Therefore, the system did not appear to lose any NOx conversion activity over the initial 6 month period of the field trial.

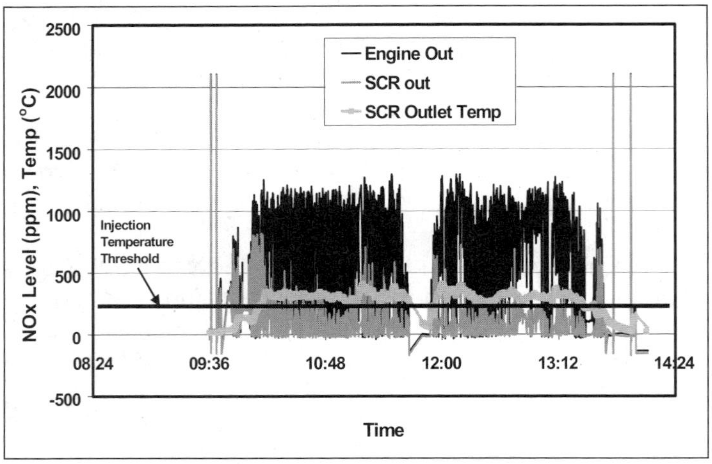

Figure 31: Engine-Out and Tailpipe NOx Measurements Made During an 850 km Evaluation Trip After 6 Months of the Field Trial

Figure 32 shows the average on-road NOx conversion over this 850 km evaluation cycle as a function of the mileage accumulated by the truck.

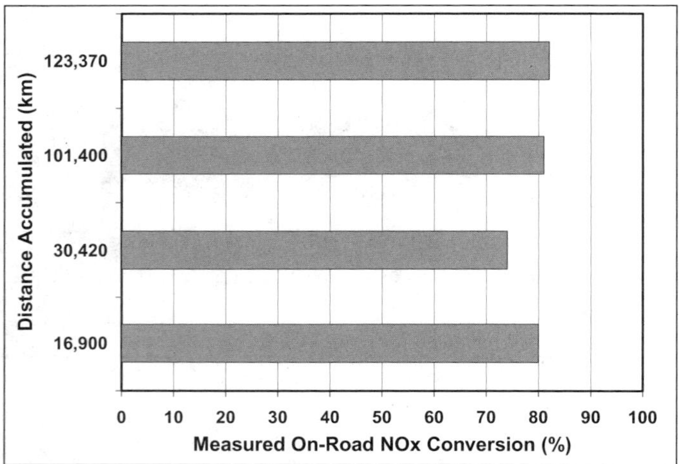

Figure 32: Measured On-Road NOx Conversion Over the 850 km Evaluation Route as a Function of Mileage Accumulated

It is clear that throughout the 125,000 km of this field trial, very high on-road NOx conversion levels were attained and maintained. At the 125,000 km point, the measured on-road NOx conversion was 82%, which corresponds to a true system efficiency of 96% (considering that the ANR was set at 0.85). In addition, no reduction in system NOx conversion performance was observed over the course of the 125,000 km, which attests to the excellent durability and poison resistance of the SCR catalyst system.

Retrofit system

Three of the SCR-DPF retrofitted test vehicles were monitored for on-road operation using the data logger built in the SCR-DPF ECU as well as external data loggers. As discussed earlier, all these systems operated with a NOx sensor input based Urea dosing strategy. Besides engine and SCR-DPF system information, NOx conversion data was also collected via NOx sensors in the exhaust, located upstream and downstream of the SCR-DPF system.

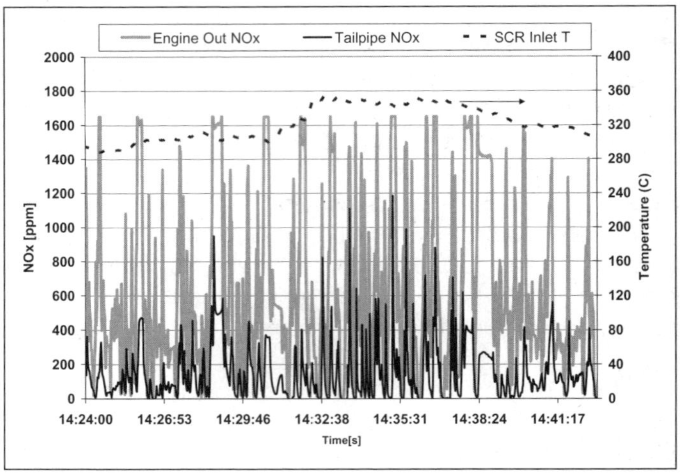

Figure 33: NOx reduction during on-road operation of Dennis Dart bus with SCR-DPF

Figure 33 presents 20 minutes of 1 sec data logged in the system ECU during on-road system operation on the Dennis Dart bus. The bus was operated over a combined route that included in-town and rural route operation. The route was a simulated operation but there were no passengers on the bus. The results show that during this on-road operation, the SCR inlet temperature was at and above 300°C. With such a warm temperature profile, the SCR-DPF system provided 70 - >90% NOx reductions during this on-road operation. This is excellent performance from a retrofit system over a real world duty cycle.

14 Dennis Dart buses have been retrofitted with the SCR-DPF system for a 6 month trial in London. The systems were all installed by May 2005. The trial will officially start in August 2005 after the systems have all been evaluated to see if there is any installation or mechanical problems. Figure 34 shows summary data from one of these Dennis Dart buses in normal service over a two week period. The average SCR inlet temperature and average NOx reduction for each day are shown on the graph. The daily NOx reduction is determined by using the NOx sensors and the exhaust flow rate to accumulate the amount of NOx coming out of the engine and out of the tailpipe. The difference between the accumulated engine out NOx and the accumulated tailpipe NOx is the amount of NOx reduced that day. Figure 34 demonstrates that the high average SCR inlet temperature on this application results in very high NOx reductions. The results demonstrated 80 – 95% NOx reduction during on-road operation of the SCR-DPF system on these buses.

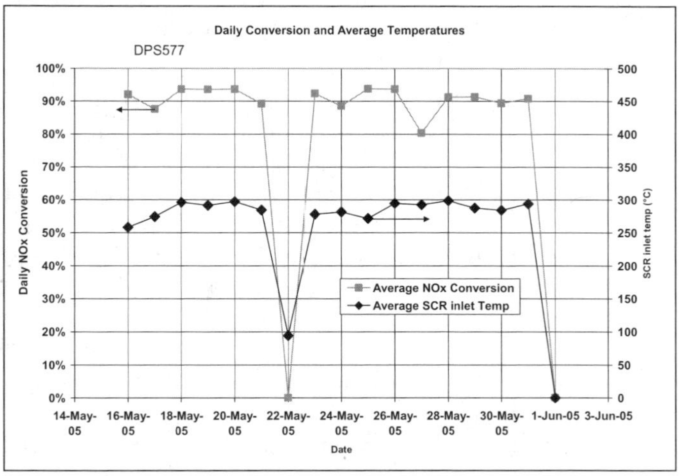

Figure 34: On-road summary data for Dennis Dart Bus in normal operation.

Similar on-road operational data was collected from the SCR-DPF equipped buses operating in Paris. Second by second data from the Euro III bus operating a test route is presented in Figure 35. These graphs show the temperatures and NOx data collected in the SCR-DPF system ECU.

Figure 35 presents data recorded while the Euro III bus left the garage after a brief warm-up and went to the center of Paris (referred to as "out run"). From the center of Paris the bus went to a route (IP) that runs around Paris (return run). The IP portion of the route refers to operation on the Internal Peripherique and back to the garage. During the out run there was road work going on that slowed traffic down to approximately 5 kph. This slow operation resulted in very low exhaust temps and therefore low NOx conversion. More realistic inner city driving happened during the out run from the city center to the IP where the SCR inlet temperture reached 250 °C and NOx conversion was approximately 80%. Once the bus was on the suburban IP portion of the route, the SCR inlet temperature increased above 250 °C and the NOx conversion increased into the 90% range. These results show the importance of temperature to the performance of the system.

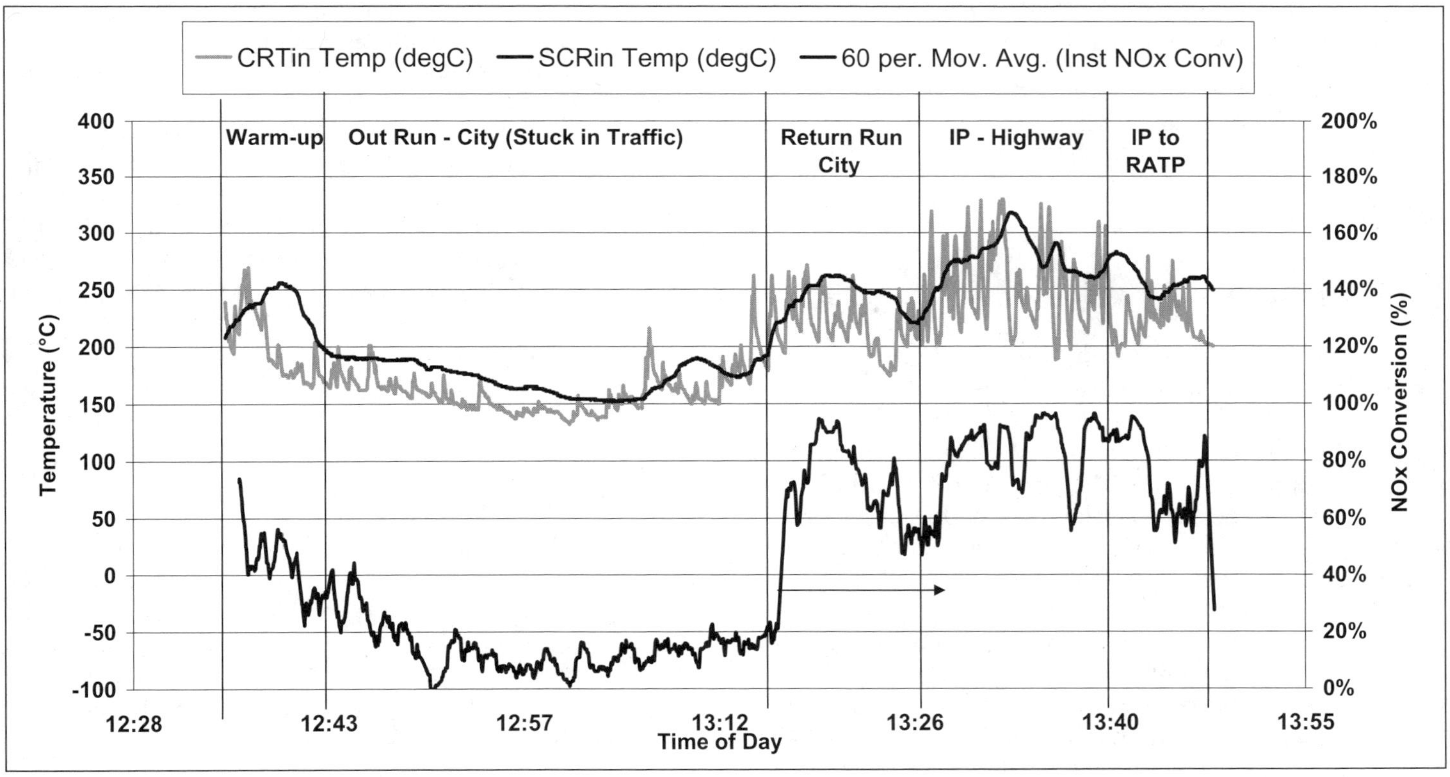

Figure 35: On-road operational data from Euro III IrisBus bus with SCR-DPF system

After the initial testing the Euro III bus was put into normal service. Data was logged at one minute intervals so that, `up to one month of data could be recorded in the system ECU.

Figure 36 shows on-road data obtained during the post installation road test on the SCR-DPF system installed on the Los Angeles County Sanitation Department truck (Figure 15). As can be seen, during on-road operation, the SCR inlet temperature in this application remained well above 200°C, in fact mostly around 300°C. With such exhaust temperature, high NOx reduction is expected. Based on the NOx sensor measurements, the accumulated NOx reduction over this 10 minute on-road test was 78%. This vehicle is being monitored for further on-road performance and durability.

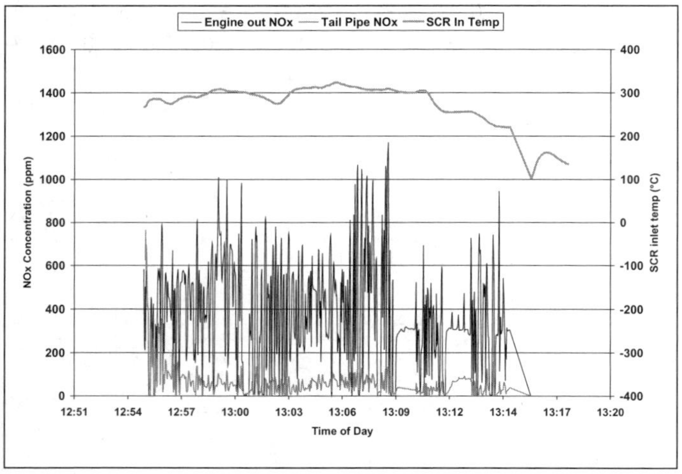

Figure 36 – On road data from LACSD Truck 4134 with the SCR-DPF system installed.

CONCLUSIONS

The following conclusions can be drawn from the work carried out so far with the SCR-DPF system:

- Overall, the combination of Selective Catalytic Reduction and CRDPF has been found to be a highly effective emission control device for NOx, PM, CO and HC emissions reduction from existing diesel vehicles via retrofit installation.

- In such a combination, the NO_2 produced by the CRDPF can benefit as well as possibly hinder NOx reduction, based on the NO_2/NO ratio. Careful attention needs to be paid to the overall system design to maintain NO_2/NO ratio at or below 1:1, in order to enhance NOx reduction.

- Engine testing results showed high NOx reduction capabilities of the SCR-DPF system. It produced 76-88% NOx reduction on the Caterpillar and the Volvo engines under FTP and ETC testing, respectively. The presence of the CRDPF also enables this system to produce 85-95% CO, HC and PM reductions.

- Vehicle testing on chassis dynamometer confirmed the effectiveness of the SCR-DPF system again by exhibiting > 80% NOx reductions on the Cummins Truck, Dennis Dart and the IrisBus Euro III engine buses. These high NOx conversions were obtained using the two different inner city cycles (London and Paris)

- Long term field trial with the SCR-DPF system on OE equipped Cummins ISX engine truck also exhibited >80% NOx reduction during on-road operation. These results attest once more to the very high specific activity of the SCR catalyst used in this system. Furthermore, no change in the NOx conversion of the system was observed over the course of 125,000 km of in-field revenue service within a relatively high temperature duty cycle. This indicated the excellent in-field durability of the SCR catalyst nad the overall SCR-DPF system.

- Additional field trials with retrofitted SCR-DPF systems on London Bus and LACSD truck have demonstrated the high emissions reductions capability of the SCR-DPF system. Durability data from the system will be presented in later publications.

- Finally, the field trial tests also showed the critical nature of the SCR temperature in determining the overall NOx reduction. The average SCR temperature will have to be over 200°C in order to obtain high NOx conversions since effective urea hydrolysis cannot be achieved below this temperature

ACKNOWLEDGMENTS

The following organizations and individuals are acknowledged for their assistance in the development and testing of the SCR-DPF systems:

Eminox
STT Emtec AB
Hilite International
Bosch
The staff of Environment Canada, ERMD
ADEME (Gabriel Plassat)
Cummins Inc.
RATP, Paris, France
London United Busways Limited

REFERENCES

1. BJ Cooper, HJ Jung and JE Thoss, US Patent 4,902,487 (1990)
2. Cooper, B., Thoss, J., "Role of NO in Diesel Particulate Control",SAE Technical Paper 890404. 1989
3. AP Walker, PG Blakeman, T. Ilkenhans, B Magnusson and AC McDonald "The Development and In-Field Demonstration of Highly Durable SCR Catalyst systems
4. Chatterjee, S., Conway, R., Lanni, T, Frank, B., Tang, S., Rosenblatt, D., Bush, C, Lowell, D, Evans, J, McLean, R. and Levy, SJ., "Performance and Durability Evaluation of Continuously Regenerating Particulate Filters on Diesel Powered Urban Buses at NY City Transit – Part II", SAE technical Paper 2002-01-0430, 2002.
5. LeTavec, C, Uihlein, J, et al., "Year-Long Evaluation of Trucks and Buses Equipped with Passive Diesel Particulate Filters", SAE Technical Paper 2002-01-0433, 2002.
6. Allansson, R., Goersmann, C., Lavenius, M., Phillips, PR., Uusimaki, AJ. and Walker, AP., "The Development and In-Field Performance of Highly Durable Particulate Control Systems ", SAE Technical Paper 2004-01-0072, 2004.
7. Andrew Eastlake, "The Latest Developments in Heavy-Duty Vehicle Aftertreatment Testing for Real-World Emissions and Fuel Economy", SAE Technical Paper 1999-01-0470, Detroit

CONTACT

Dr Andy Walker,
Johnson Matthey,
Environmental Catalysts and Technologies
380 Lapp Road, Malvern PA 19355
walkeap@matthey.com

Sougato Chatterjee,
Johnson Matthey Catalysts,
380 Lapp Road, Malvern, PA 19355;
chatterjee@JMUSA.com

IV.2 LIGHT DUTY

2002-01-0957

Simultaneous PM and NOx Reduction System for Diesel Engines

Koichiro Nakatani, Shinya Hirota, Shinichi Takeshima, Kazuhiro Itoh and Toshiaki Tanaka
Toyota Motor Corporation

Kazuhiko Dohmae
Toyota Central R&D Labs., INC.

ABSTRACT

A new after-treatment system called DPNR (Diesel Particulate - NOx Reduction System) has been developed for simultaneous and continuous reduction of particulate matter (PM) and nitrogen oxides (NOx) in diesel exhaust gas. This system consists of both a new catalytic technology and a new diesel combustion technology which enables rich operating conditions in diesel engines. The catalytic converter for the DPNR has a newly developed porous ceramic structure coated with a NOx storage reduction catalyst. A fresh DPNR catalyst reduced more than 80 % of both PM and NOx. This paper describes the concept and performance of the system in detail. Especially, the details of the PM oxidation mechanism in DPNR are described.

INTRODUCTION

Diesel engines with relatively good fuel economy are known as an effective means of reducing CO_2 emissions and preventing the problem of global warming. On the other hand, they emit relatively high NOx and particulate (PM) emissions. Recently, however, remarkable reductions of NOx and PM emissions have been accomplished by engine modifications. For example, electronically controlled fuel injection technology and cooled EGR systems have been introduced. However, the emission level is still high in comparison with the strict future emission standards.

In order to accomplish more reductions in NOx and PM, it is indispensable to develop a new, revolutionary after-treatment technology which has high conversion efficiency such as the three-way catalyst or the NOx storage reduction catalytic system which is available for gasoline engines. Consequently, many after-treatment systems have been developed.

The development of the DPF (Diesel Particulate Filter) began more than 20 years ago. Many systems have been proposed and they can be classified into two types of PM oxidation. In one of them, the forced regenerating type DPF, PM is accumulated temporarily in the filter and then burnt off at proper time by increasing DPF temperature. In the other type, the continuous regenerating type DPF, PM is trapped and oxidized continuously without increasing DPF temperature.

It is impossible under 500 °C to burn PM which is exposed in the exhaust gas temporarily. However, the exhaust gas temperature of modern DI diesel engines is low (150 °C to 250 °C in city driving conditions). In order to burn PM trapped in the DPF, it is necessary to keep the exhaust gas temperature over 500 °C or to decrease the PM ignition temperature. So, the forced PM regeneration type of DPF requires a proper heating device [1] or a device to decrease PM ignition temperature such as a fuel additive [2]. The control system becomes more complex and the filter clogging occurs due to the ash remaining after burning PM. On the other hand, the continuous PM oxidation type traps such as Continuous Regeneration Trap (CRT) [3] and Catalyzed Soot Filter (CSF) [4] oxidizes PM by the function of the catalyst. Both types of DPF have high trapping efficiency, but they are incapable of reducing NOx.

As shown in Figure 1, two types of NOx reduction technologies have been developed. The one is a NOx storage-reduction catalytic system (NSR) [5][6][7] which stores NOx in the catalyst under lean operating conditions and then reduces the stored NOx under momentary rich operating conditions. The other is the selective catalytic reduction of NOx with a reductant originating from urea compound [8] or supplemental HC [9].

However, NSR has not been practical for diesel engines because it is difficult to attain a rich operating condition. On the other hand, in the selective type with a urea compound, it is necessary to install a urea solution injection system and suppress ammonia emissions. The

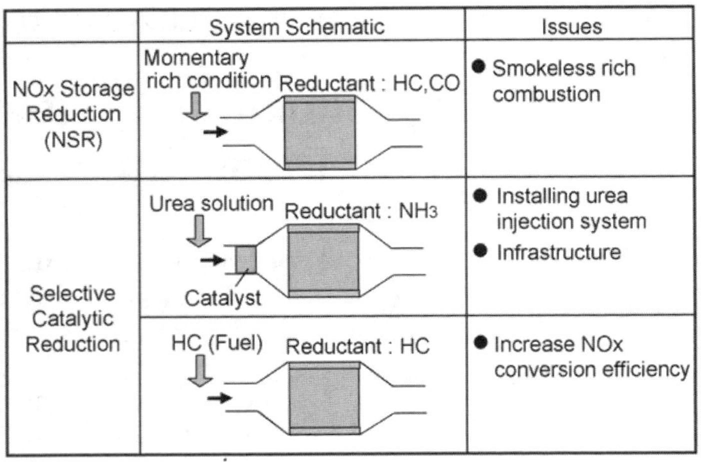

Figure 1 Examples of NOx reduction system

selective type with HC does not have high conversion efficiency.

As mentioned above, many after-treatment systems have been studied, but most of them cannot decrease PM and NOx simultaneously.

In this study, a DPNR (Diesel Particulate-NOx Reduction System) has been developed. This system consists of both a new catalytic technology and a new diesel combustion technology which enables rich operating conditions in diesel engines. As a result, this system can reduce more than 80% of both PM and NOx. Some of the results of this study were published [10]. In this paper we add some new results and elaborate more on the key points.

DIESEL PARTICULATE-NOx REDUCTION SYSTEM (DPNR)

DPNR has been developed to accomplish drastic reductions in NOx and PM. The principal attribute of DPNR is the simultaneous and continuous reduction of PM and NOx with a simple and compact catalytic converter system.

CATALYTIC CONVERTER FOR DPNR

The new catalytic converter for DPNR is a newly developed, fine porous ceramic, coated with NOx storage reduction catalyst which was designed for lean-burn gasoline engines. Figure 2 shows the structure of DPNR catalyst. This catalytic converter consists of a porous monolithic honeycomb structure. Filter channels are alternatively plugged to force the gas flow through the porous walls where PM is trapped temporally. The structure of this substrate is optimized for PM to go into the substrate pores and the catalyst component is coated not only on the substrate wall but also inside the substrate pores.

DPNR catalyst samples was prepared on cordierate filter (300 cpsi, wall thickness : 0.3 mm, porosity : 55 %, mean pore size : 25 μm) for this basic study.

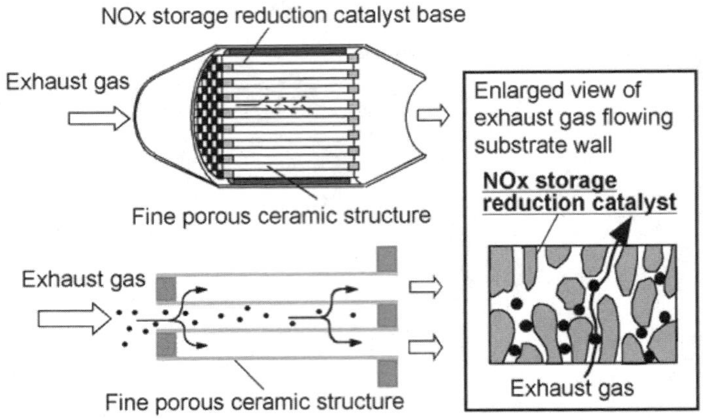

Figure 2 Structure of DPNR catalyst

REACTION MECHANISM IN DPNR CATALYST

Figure 3 shows the reaction mechanisms in the DPNR catalyst. The NOx reduction mechanism in the DPNR is similar to that in a NOx storage reduction catalyst as already reported [6]. Under lean conditions, NOx is oxidized on the precious metals and reacts with the NOx storage components to form mainly nitrates. Thereafter, under rich conditions, nitrates on the catalyst surface are reduced to nitrogen.

With regard to PM, under lean conditions, PM is oxidized by active oxygen released in the NOx storage process and by excess oxygen in the exhaust gas. Under rich conditions, PM is also oxidized by the active oxygen that is released in the reduction process of stored NOx. Consequently, PM and NOx are reduced by controlling the air fuel ratio from lean to rich alternately.

RICH AIR FUEL RATIO IN DIESEL ENGINES

DPNR needs rich or stoichiometric operation periodically to reduce NOx. It was proposed that a rich air fuel ratio in a diesel engine was realized by multi-timing fuel injection including post injection, or supplemental fuel injection into the exhaust gas or smokeless low temperature combustion [11] with a massive EGR.

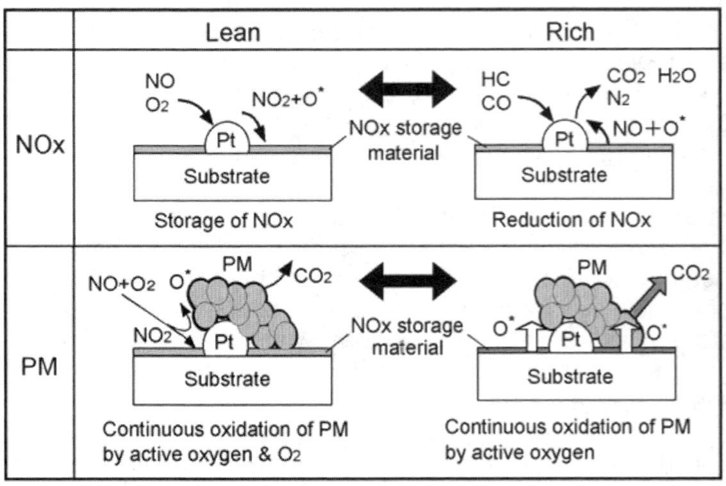

Figure 3 NOx-PM reaction mechanism in DPNR

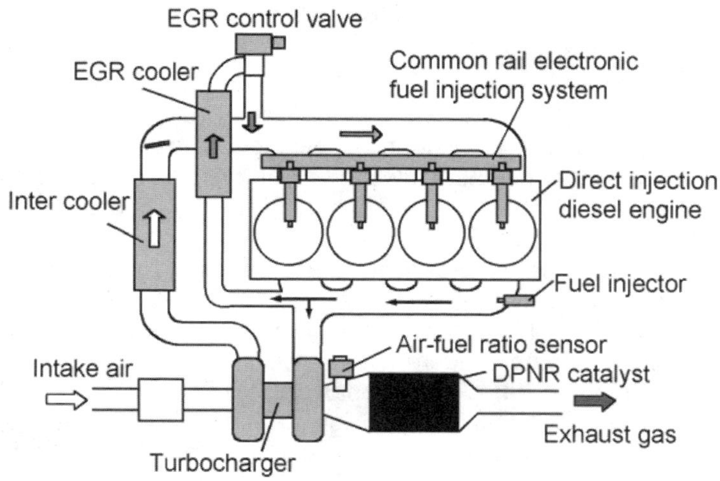

Figure 4 Diagram of DPNR

Figure 4 shows one example of the engine system in which a rich air fuel ratio is realized by low temperature combustion and supplemental fuel injection into the exhaust gas. This system has a direct-injection diesel engine featuring the latest common rail electrically controlled fuel injection system and the high efficiency EGR cooler. The DPNR catalyst is located after the turbocharger and a fuel injector is also added to the exhaust manifold.

In this study, basic studies for DPNR were performed with a prototype, in-line, 4-cylinder, 2.0 L, water-cooled Turbocharged Direct Injection (TDI) diesel engine. Volume of DPNR catalyst used in this study is 1 – 1.5 times larger than engine displacement. Diesel fuel with a sulfur content of 30 ppm was used for all testing.

NOx CONVERSION IN DPNR

Figure 5 shows the NOx regeneration behavior in the DPNR. The test was set up in which the storage phase was set at normal diesel combustion for 60 seconds and the regeneration phase at low temperature combustion (air fuel ratio 12.5) for 3 seconds.

As shown in this figure, NOx in diesel exhaust gas was purified by the DPNR catalyst. The NOx regeneration

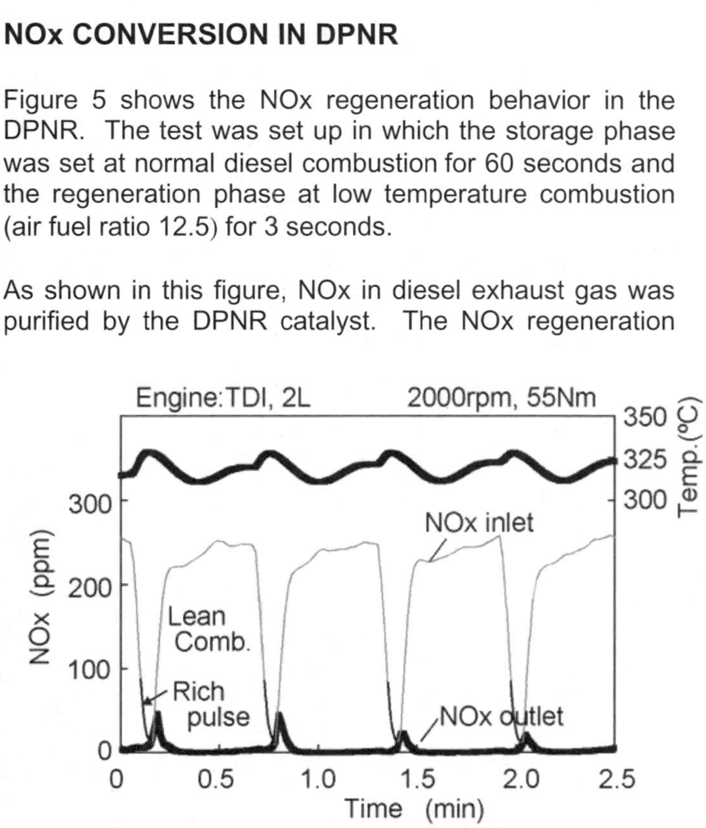

Figure 5 NOx regeneration behavior in DPNR

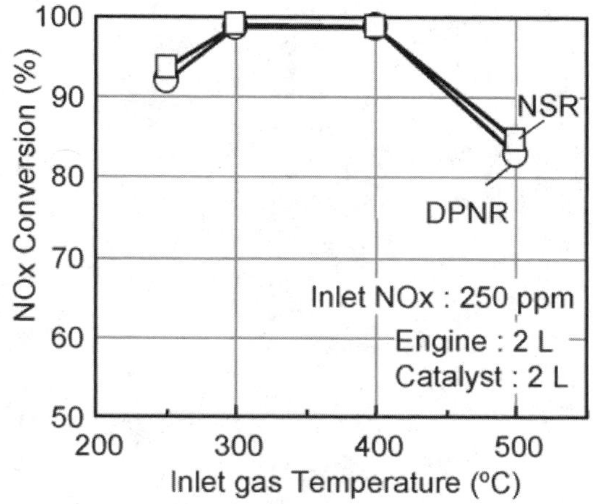

Figure 6 NOx conversion of DPNR

behavior in DPNR is almost similar to that in NOx storage reduction catalyst.

Figure 6 shows NOx conversion efficiency of DPNR catalyst and NOx storage reduction catalyst. NOx conversion efficiency was averaged from cycles running 60 seconds lean and 1 second rich to air fuel ratio 11.5. The NOx conversion efficiency of DPNR catalyst is same as that of NOx storage reduction catalyst.

DPNR with a fresh catalyst has a good performance for NOx conversion. However, because sulfur poisoning influences the DPNR catalyst performance as well as NOx storage-reduction catalyst, it is necessary that low sulfur fuel is introduced in order to educe the potential of DPNR.

PM OXIDATION IN DPNR CATALYST

As mentioned above, the DPNR continuously oxidizes PM by controlling the air fuel ratio. In this section, the PM oxidation mechanism details and the requirements for continuous PM oxidation are described.

TEM OBSERVATION OF PM OXIDATION

The change of PM in the oxidation process was observed using transmission electron microscopy (TEM). The photographs are shown in Figure 7. PM was collected on a thin Al_2O_3 plate, which was coated with platinum and NOx storage material, just after being exhausted from engine. Then, PM was oxidized under the condition of 10 kPa absolute, 350 °C,10% O_2 and 1000ppm NO.

As known well, PM consists of collections of primary particles, which are about 20 nm in diameter. Each of the primary particles becomes smaller as the oxidation process proceeds. As a result, most secondary particles diminish within 30~50 minutes under the pressure of 10 kPa absolute.

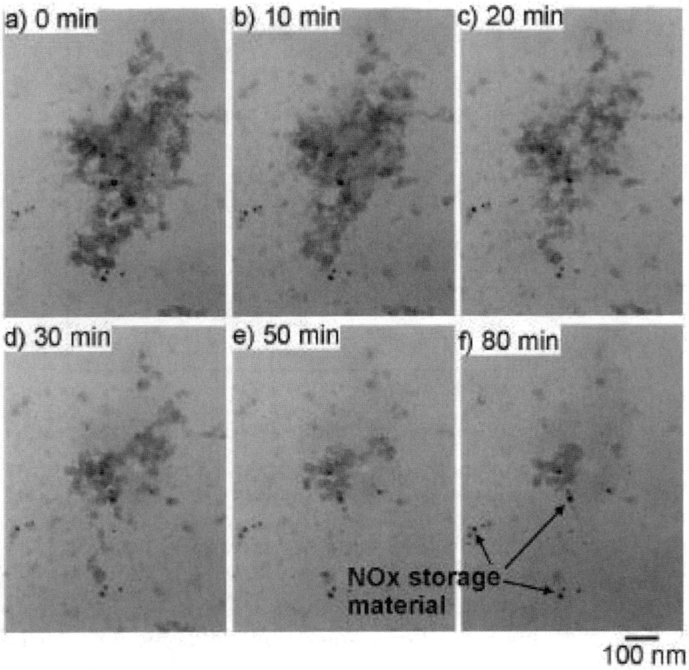

Figure 7 Series of TEM images for diesel soot oxidation on DPNR catalyst (ambient conditions : 350 ºC, 10 kPa, 10 % O_2)

It means that, in the actual exhaust condition, where ambient pressure is ten times larger than under the test conditions, most secondary particles will disappear within a few minutes.

Because PM is a solid carbon, the oxidation rate of PM is not as fast as the gas-phase reactions. Therefore, large space volume and surface area are needed in the DPNR to prevent particulates from combining with each other and forming larger particles which are hard to oxidize.

EFFECT OF PM LOADING ON PM OXIDATION

Figure 8 shows the relationship between the amount of PM trapped in the DPNR catalyst and the PM oxidation rate per amount of PM. PM mass loading in the DPNR catalyst was estimated by the pressure drop across the catalytic converter. The PM oxidation rate was determined by the decreasing rate of pressure drop. It was found that the smaller the PM mass loading in the DPNR catalyst, the faster the PM oxidation rate. The oxidation rate decreased gradually as the loading became greater.

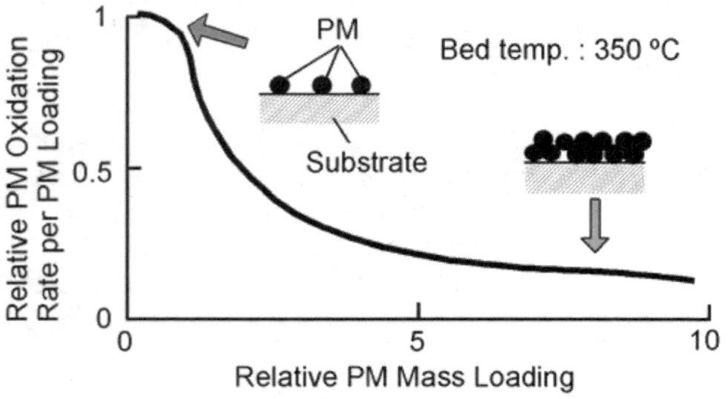

Figure 8 Relationship between the PM mass loading and the PM oxidation rate per PM loading

Next, it is studied how the intensity of contact between PM and catalyst influences on the PM oxidation. The intensity of contact between PM and catalyst is changed by means of mixing. One sample named “tight contact” was made by milling a mixture of PM and catalyst by the use of a ball mill and the other named “loose contact” was made by shaking a mixture of PM and catalyst in a vessel. PM was gathered just after emitted from an IDI diesel engine. Figure 9 shows CO_2 emission indicated by thermal garavimetric analysis. PM, which is in tight contact with catalyst, is oxidized at lower temperatures. It reveals that the physical contact between PM and catalyst is very important to PM oxidation.

From this result, the reason for the result in Figure 8 might be that the number of PM which is trapped far from the catalyst surface increases as PM loading increases and it is more difficult for PM to react with the active oxygen.

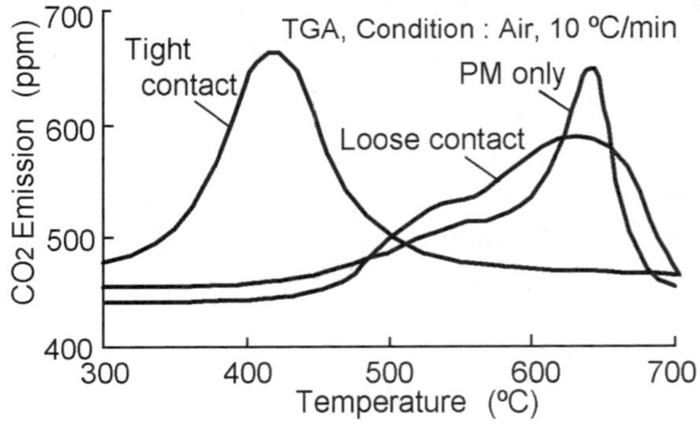

Figure 9 Effect of contact between PM and catalyst on the oxidation of PM

PM OXIDATION RATE IN DPNR CATALYST

Figure 10 shows the relationship between the bed temperature in the DPNR catalyst and the PM oxidation rate at low PM loading. PM oxidation rate means the maximum PM mass which the DPNR can oxidize with keeping pressure drop across the catalyst low. PM can

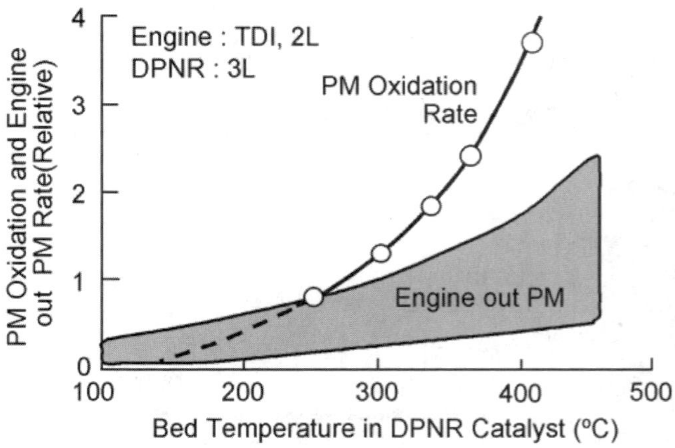

Figure 10 Relationship between the PM oxidation rate and the bed temperature in the DPNR catalyst

be oxidized even at low gas temperatures such as 250 °C. Figure 10 also shows relative engine out PM rate from a latest TDI engine. In most of operating conditions, PM can be oxidized continuously in the DPNR catalyst with keeping the low pressure drop. In the case of low catalyst bed temperatures, under 250 °C, engine out PM exceeds the DPNR oxidation capability. In such cases, the exhaust gas temperature must be increased in order to prevent PM accumulation in the DPNR.

CHANGE OF PM MORPHOLOGY

As mentioned above, PM can be oxidized at lower bed temperatures in the DPNR catalyst. The promotion of PM oxidation with the DPNR might be not only the function of catalyst but also the change of PM morphology. PM morphology might change in exhaust gas, that is, PM just after being emitted from the engine is different from PM exposed to exhaust gas for a long time.

Figure 11 shows the distribution of micro pore diameter in PM measured by N_2 adsorption and the Korvath-Kawazoe method. PM just after being emitted from engine has many micro pores at 0.5 to 1 nm in diameter, but the micro pores decrease remarkably in PM which is exposed to air at 300 °C for 2 hours. The decrease of micro pores indicates that the disorder of PM surface decreases in exhaust gas and the PM oxidation rate lowers. It suggests that the change of structure causes the lowering of the PM oxidation rate.

Figure 12 shows the TEM images of PM annealed in air at 300 °C. PM was captured on glass wool at room temperature. The photograph (a) is the image of PM as received and (b) is the image of what are received and then exposed in air at 300 °C for 20 minutes. In photograph (a), PM consists of collections of primary particles forming aggregates, but the boundary of each particle is distinct. On the other hand, in (b), the primary particles are sintered with each other and to grow into large particles in 50 to 70 nm in diameter. This might be the reason why micro pores decrease remarkably.

It was deduced from the foregoing results that PM just after engine emission are characterized by their

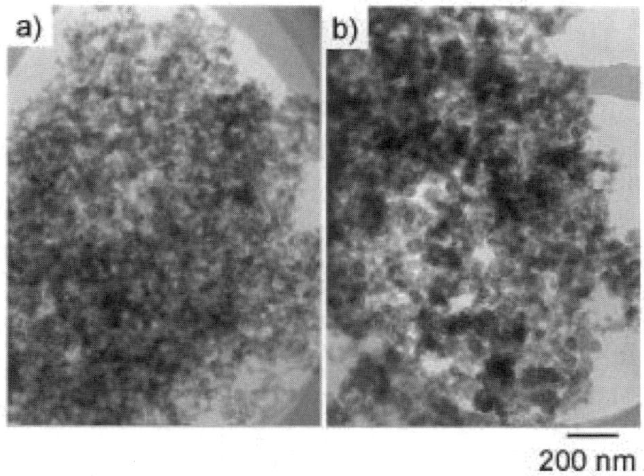

Figure 12 TEM images of PM annealed in air : (a) as received and (b) 300 °C, 20 min

susceptibility to the oxidation reaction because PM consists of primary particles with many micro pores. If PM is exposed to exhaust gas, the primary particles combine with each other and become large particles.

In other words, it is estimated that because by oxidizing frequently, as done in DPNR, PM doesn't have an opportunity to consolidate and therefore oxidation rate is enhanced, this is much better than oxidizing infrequently as in traditional filter systems.

PROMOTION OF PM OXIDATION WITH RICH PULSE

Figure 13 shows the transition of pressure drops across the DPNR catalyst measured with or without a rich pulse. It is obvious that the increase of pressure drop was smaller in the case with the modulation. It indicated that PM oxidation is promoted by the use of a rich pulse. It is thought that there are some effects of rich pulse on PM oxidation. For example, rise of local temperature, catalytic activity improvement of precious metal, generation of the active oxygen, increase of PM oxidation characteristics.

Next, the effect of modulation of the air-fuel ratio on the PM oxidation characteristics was investigated by means of temperature programmed oxidation. Figure 14 shows

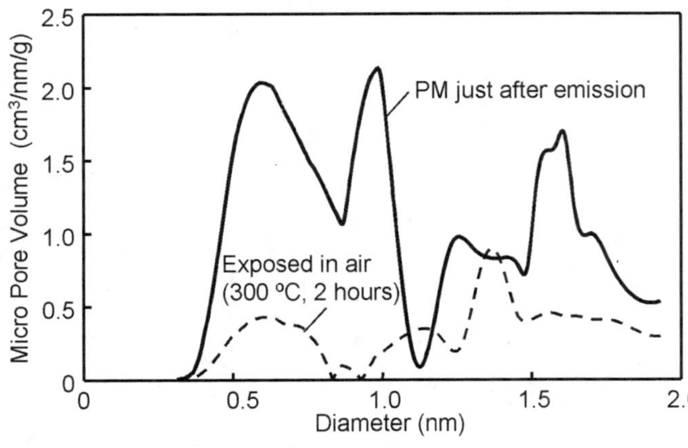

Figure 11 Change of micro pore diameter distribution on PM

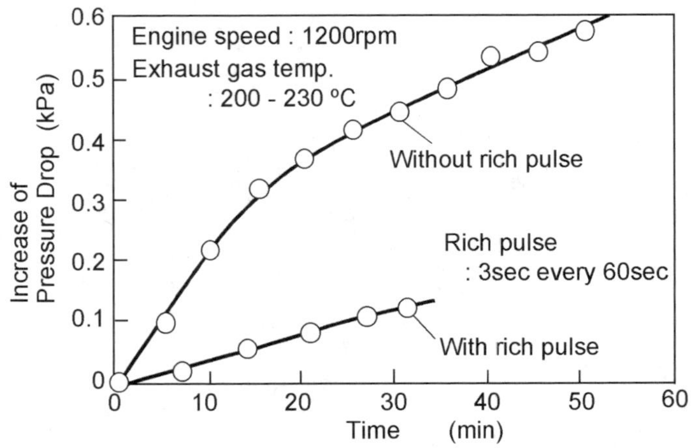

Figure 13 Effect of the rich pulse on the transition of pressure drop

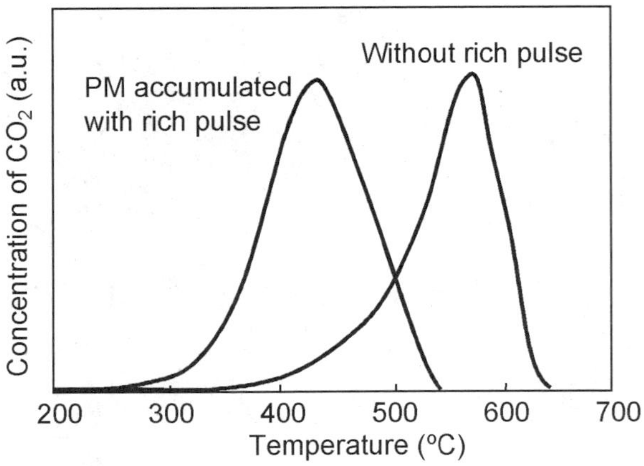

Figure 14 Oxidation characteristics of PM trapped both with and without rich pulse

the comparison of oxidation characteristics of PM trapped in the DPNR catalyst, both with and without rich pulse. In this result, the oxidation temperature of PM trapped during the modulation of the air-fuel ratio was at least 100 degrees lower than that of PM trapped during the steady lean operating condition. As mentioned above, the disorder of PM surface decreases in lean exhaust gas and it causes the lowering PM oxidation activity. However, air fuel ratio modulation keeps PM oxidation activity high. This phenomenon can't be explained as the effect of local temperature rise and catalytic activity improvement. Then, the effect of rich pulse was studied with keeping the influence of temperature little.

Figure 15 shows the PM oxidation rate with keeping the bed temperature as constant as possible. In this test, DPNR catalyst was set in an electric heating furnace and its bed temperature was set at 300 °C. 4 seconds rich pulse is performed every 120 seconds lean conditions. The actual exhaust gas was led in DPNR catalyst for lean condition and 1 vol% H_2 in N_2 was led for rich pulse. In order to lower temperature rise by the oxidation of rich gas, N_2 gas was led before and after rich pulse. It is shown that the PM oxidation rate becomes higher by rich pulse.

From these results, it is estimated that there are some reasons, why rich pulse promotes PM oxidation, except

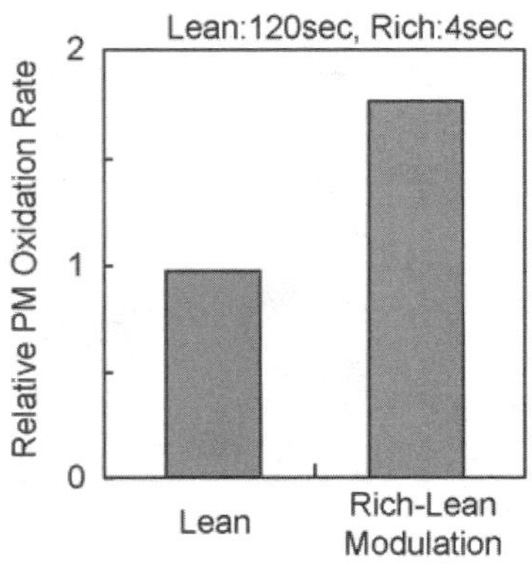

Figure 15 Effect of rich pulse on PM oxidation rate (Bed temp.:300°C, Lean : diesel exhaust gas, Rich : 1 vol% H_2 in N_2)

the effect of local temperature rise. Thereafter, the basic experiments in relation to the active oxygen were made to make clear the effect on PM oxidation.

GENERATION OF ACTIVE OXYGEN BY RICH PULSE

ESR (Electron Spin Resonance) was used to analyze the generation of superoxide as the representation of active oxygen which includes another species, O^-, OH^-, O radical etc. Figure 16 shows the transition of the ESR signal under a steady lean condition and under lean condition just after the rich pulse. In a steady lean condition, no active oxygen was detected, although NOx existed. The ESR signal originating from superoxide emerged just after the rich pulse and weakened gradually under lean conditions. In other words, the active oxygen is generated vigorously just after switching from rich condition to lean condition. When lean condition continues for long, little active oxygen is generated.

Next, the effect of the active oxygen on PM oxidation rate was investigated using the two types of DPNR catalysts which have different amount of superoxide generated under lean-rich modulation. The amount of active oxygen was changed experimentally by the modification of supports. The bed temperature was kept at 300 °C by setting the catalyst in the electric heating furnace. Rich pulse was performed by same methods as figure 15. As

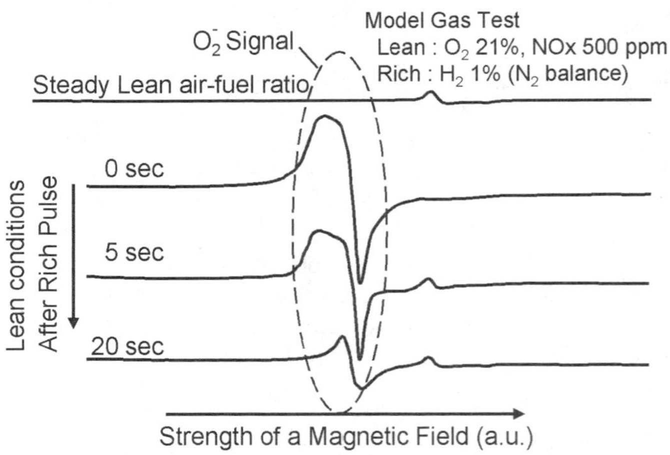

Figure 16 Effect of air-fuel ratio modulation on ESR spectra in DPNR catalyst

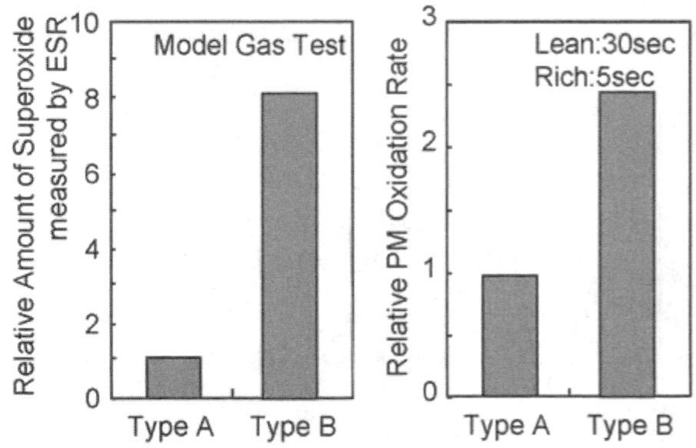

Figure 17 Effect of the amount of active oxygen on PM oxidation rate

shown in figure 17, superoxide generation of type B catalyst, just after rich pulse, is 8 times larger than that of type A catalyst. PM oxidation rate increased as the amount of active oxygen increased and this result indicated that PM oxidation in DPNR catalyst was promoted by the active oxygen generated during lean rich modulation.

Because of this result, it is estimated that the active oxygen which is generated by the catalyst, attacks the PM surface and promotes oxidation. This phenomenon is the outstanding feature of DPNR, it combines PM oxidation with NOx storage reduction.

EFFECTS OF DPNR ON REDUCTIONS OF PM AND NOx

Figure 18 shows the capability of the DPNR to reduce PM and NOx emissions under transient conditions on a vehicle equipped with a TDI diesel engine. The inertia weight of the test vehicle was 1.25t and the specifications of engine were mentioned above. The test mode is the Japanese 10-15 mode. In this case, low temperature diesel combustion was used at idling and steady speed. A fresh DPNR catalyst reduced more than 80% of both PM and NOx.

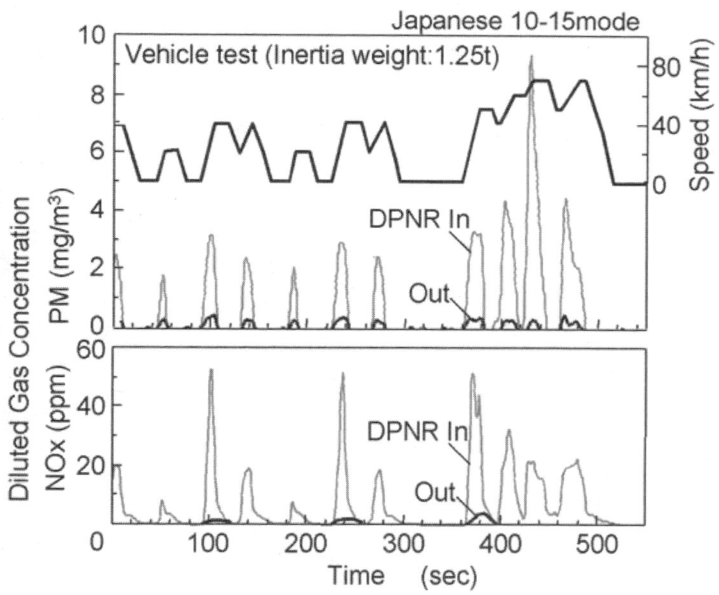

Figure 18 Effect of DPNR on reductions of PM and NOx emissions

CONCLUSION

(1) DPNR (Diesel Particulate-NOx Reduction system) is capable of simultaneous reductions of PM and NOx in diesel exhaust. It has been developed as a new exhaust after-treatment system.

(2) DPNR consists of a porous ceramic structure which is coated with a NOx storage-reduction catalyst. This system is capable of continuous reductions of NOx and PM, using a repetitive control of lean and rich operating conditions.

(3) The PM oxidation rate becomes higher as the amount of PM trapped in the DPNR is smaller. It is thus indicated that the appropriate control of the PM emission level or exhaust temperature is necessary to keep the PM oxidation rate high.

(4) The modulation of the air-fuel ratio (lean and rich) facilitates PM oxidation. This is presumably attributed to the oxidation effect of active oxygen released at the same time upon the storage and reduction of NOx.

ACKNOWLEDGMENTS

The authors gratefully acknowledge the support of Cataler Corporation for this basic study. The provision of DPF substrate through Denso Corporation is acknowledged.

REFERENCES

1. K. Kobashi, K. Hayashi, H. Aoki and S. Takeshima, "Regeneration Capability of Diesel Particulate System with Wall-Flow Monolith Filter", JSAE Review, Vol. 10, No. 4 (1989).
2. O. Salvat, P. Marez and G. Belot, "Passenger Car Serial Application of a Particulate Filter System on a Common Rail Direct Injection Diesel Engine", SAE Paper No.2000-01-0473 (2000).
3. P. Hawker and N. Myers, "Experience with a New Particulate Trap Technology in Europe", SAE Paper No.970182 (1997).
4. U. Dahle, T. Beutel and A. Punke, "Euro 4 - Exhaust After-treatment for Lean Burn Engines (Gasoline/ Diesel)", VDA Technical Congress 1999 (1999), 63-74.
5. N. Miyoshi, et al., "Development of New Concept Three-Way Catalyst for Automotive Lean-Burn Engines", SAE Paper No.950809 (1995).
6. T. Tanaka, K. Katou, S. Takeshima, S. Matsumoto, K. Yokota, K. Kasahara, and S. Iguchi, "Development of NOx Storage-Reduction 3-Way Catalyst -2nd Report", JSAE (in Japanese) 9540958 (1995).
7. T. Asanuma, S. Takeshima, T. Yamashita, T. Tanaka, T. Murai and S. Iguchi, "Influence of Sulfur Concentration in Gasoline on NOx Storage-Reduction Catalyst", SAE Paper No.1999-01-3501 (1999).
8. H. Luders, et al. "An Urea Lean NOx Catalyst System for Light Duty Diesel Vehicles", SAE Paper No.952493 (1995).
9. S. Matsumoto, K. Yokota, H. Doi, M. Kimura, K. Sekizawa and S. Kasahara, "Research on new DeNOx catalysts for automotive engines", Catalysis Today, 22 (1994), 127-146.
10. T. Tanaka, "Simultaneous Reduction of PM and NOx – A New After-Treatment System –", 22nd

International Vienna Motor Symposium (2001), 216-228.

11. S. Sasaki, T. Ito and S. Iguchi, "Smoke-less Rich Combustion by Low Temperature Oxidation in Diesel Engines", 9th Aachen Colloquium Automobile and Engine Technology 2000.

2003-01-0043

Passenger Car Investigations of NO_x-Adsorber and DPF Combination to Fulfill Future Diesel Emission Limits

Ulrich Pfahl, Werner Hirtler and Thomas Cartus
AVL List GmbH, Austria

ABSTRACT

The current study focuses on the operating behavior of a NO_x-adsorber catalyst upstream of a Diesel particulate filter as two separate exhaust gas aftertreatment components. New European drive cycle (NEDC), US FTP 75, and US06 tests with a NO_x-adsorber have been performed with and without a DPF in two passenger cars with initial test weights (ITW) between 1550 and 1750 kg. Transient particle size distributions have also been measured during FTP 75 and US06 tests. Also on road driving with a NO_x-adsorber and DPF has been carried out. During these tests, the focus has been on the NO_x-adsorber catalyst and on the interaction between the NO_x-adsorber and the DPF. The purge events to regenerate the NO_x-adsorber (rich exhaust gas composition, $\lambda < 1$) have been controlled automatically via new engine control unit functionalities. These algorithms have been developed using the rapid prototype tool ETAS ASCET. Exhaust gas back pressure behavior as well as timing and efficiency of $DeNO_x$-purge events over driving distance and during different driving conditions have been investigated and the oil dilution as a consequence of regeneration of NO_x-adsorber ($DeNO_x$-purge) has been evaluated. NO_x conversion efficiency depends on driving patterns and optimization of the corresponding control functions. Overall, a potential for NO_x reduction above 50% during real world driving has been observed with unexpected low DPF back pressure increase.

INTRODUCTION

The share of passenger cars equipped with Diesel engines is still growing significantly in Europe. Manufacturers in North America have also recently renewed their interest in producing light trucks and passenger cars equipped with Diesel engines to significantly improve fleet fuel economy. Lean burning Diesel engines equipped with Diesel oxidation catalysts are very low on carbon monoxide (CO) and unburned hydrocarbon (HC) emissions (1), whereas particulate matter (PM) and nitrogen oxides (NO_x) emission reduction still remains in the focus of intensive investigations. Due to the low exhaust gas temperatures and the high oxygen content in the exhaust gas, which characterizes lean burning engines, the reduction of particulate matter (PM) or nitrogen oxides (NO_x) from the exhaust gas is very difficult.

Beside the reduction of engine out emissions due to improved mixture formation and combustion (2,3), advanced exhaust gas aftertreatment technologies are currently under development to reduce particulate matter and nitrogen oxides emissions of Diesel engines (4). Different exhaust gas aftertreatment components such as the Diesel particulate filter (DPF) and NO_x catalyst systems have been investigated both separately and together as an integrated solution. Diesel particulate filters are already in production to reduce PM emissions of Diesel passenger cars (5,6,7) and are also the subject of further development activities for optimization (8,9). For Diesel passenger cars and light commercial vehicles, NO_x catalyst systems such as the NO_x-adsorber or SCR-catalysts (Selective Catalytic Reduction) are under development (10,11), whereas for passenger cars with direct injection gasoline engines NO_x-adsorber catalysts are already in production.

To meet future Diesel emission legislation limits a combination of both aftertreatment systems – a DPF and NO_x catalyst - seems to be mandatory from today's point of view to reduce particulate matter and nitrogen oxides emissions simultaneously (12,13). Such a combination is also expected to offer advantages with regard to the package situation, especially in passenger cars. The approach to calibrating such a 4-way-catalyst concept as well as possible synergy effects of a combination of both components are to a great extent still unknown. Therefore, this study focuses on the simultaneous reduction of particulate matter and nitrogen oxides emissions in a Diesel passenger car by utilizing a Diesel particulate filter and a NO_x-adsorber catalyst. The operating behavior of the NO_x-adsorber in combination with the Diesel particulate filter as two separate exhaust gas aftertreatment components has been tested and the interaction between NO_x-adsorber and DPF has been evaluated.

EXPERIMENTAL SET-UP AND TEST CYCLES

VEHICLE AND ENGINE

For this study two different production vehicles have been used. The inertia test weight of these vehicles are 1750 (vehicle I) and 1550 kg (vehicle II). Both are equipped with a manual 5 speed transmission. The water-cooled 4 cylinder engines both have displacements of approximately 500 cm^3 per cylinder. They are equipped with air/air intercoolers as well as EGR systems.

FUEL INJECTION AND ENGINE MANAGEMENT SYSTEM

The 4 cylinder direct injection Diesel engines which have been used for these investigations are both equipped with a common rail fuel injection system with a maximum pressure of approximately 1350 bar. The fuel injection systems use solenoid CR-injectors. The purge events to regenerate the NO_x-adsorber have been controlled via new engine control unit functionalities. These algorithms have been developed using the rapid prototype tool ETAS ASCET. For vehicle I the purge events have been controlled automatically considering engine, vehicle and NO_x-adsorber conditions. For vehicle II the purge control during the different test cycles has been time based.

FUEL SPECIFICATIONS

For the current tests a low sulfur diesel fuel with a sulfur content < 10 ppm has been used. For vehicle I, Cerium additive EOLYS DPX9 has been added to the low sulfur fuel to decrease the soot oxidation temperature and increase soot oxidation rate during Diesel particulate regeneration. The concentration of Cerium used in the fuel has been 25 ppm. The lubricant used for these test has been standard synthetic lubrication oil corresponding to the vehicle manufacturers specification.

EXHAUST GAS AFTERTREATMENT SYSTEM

The production aftertreatment systems of both vehicles consist solely of an oxidation catalyst. The oxidation catalyst has been exchanged with a NO_x-adsorber catalyst and a Diesel particulate filter has been added. For both vehicles a DPF with a diameter of 5,66 inch, and a cell density of 200 cpsi has been used. The DPF length has been 10 inch for vehicle I. For vehicle II a catalyzed DPF with 6 inch length has been used. The NO_x-adsorber catalyst has been utilized not only as a NO_x storage catalyst, but also as Diesel oxidation catalyst to convert carbon monoxide (CO) and unburned hydrocarbons (HC).

In comparison to the production configuration, the exhaust gas aftertreatment system has been equipped with two additional sensors. One wide band λ-probe has been used to control rich combustion for NO_x-adsorber regeneration. The measuring range of this λ-probe is from 0.7 up to 8. To monitor the NO_x conversion efficiency, one NGK NO_x-sensor has been used with a NO_x measuring range from 0 up to 1500 ppm.

The configuration of the different components of the exhaust gas aftertreatment system and of the additional sensors is shown in Fig. 1.

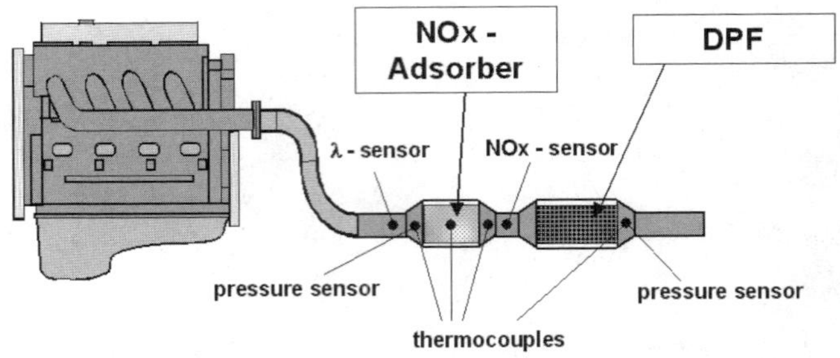

Figure 1. Exhaust gas aftertreatment system.

CHASSIS DYNAMOMETER

For the roller chassis dynamometer testing a Cosine Twin Roller chassis dynamometer with a maximum brake power of 62 kW and a total moment of inertia of 4900 kgm^2 has been used. The data acquisition system for the roller chassis dynamometer testing is an AVL INDIMASTER 670 and INCA PC data recorder.

EMISSION MEASUREMENT

For the roller chassis dynamometer testing a full-flow constant volume dilution tunnel has been used to collect particulate and bag emissions. Furthermore, two AVL combustion emission benches have been used to measure gaseous emissions at two positions in the exhaust gas aftertreatment system, e.g. upstream and downstream of the NO_x-adsorber. The analyzed gaseous emissions are O_2, CO_2, NO_x, CO and HC. Particle size distribution measurements have been carried out with a transient differential mobility particle spectrometer (TDMPS). The measuring range is from 10 to 600 nm with 20 particle size classes. The oil dilution has been evaluated using gas chromatographic investigations.

TEST CYCLES

The passenger car emissions have been evaluated during three different drive cycles. The first test cycle is the New European Drive Cycle (NEDC) with a total duration of 1180 seconds. The first 780 seconds are four similar repetitions of urban driving, whereas the last 400 seconds represent extra urban driving. The total length of the cycle is 11.007 km. Maximum and average speeds are 120 and 33.6 km/h, respectively. The second test cycle is the US Federal Test Procedure 75 (FTP 75) with a total duration of 1877 seconds. The first 505 seconds represent a cold transient phase, the second 867 seconds represent a cold stabilized period and the third 505 seconds are a hot transient phase. The total length of the cycle is 11.04 miles. Maximum and average speeds are 56.68 and 21.29 mph, respectively. The third test cycle is the US06 high speed/high load cycle with a total duration of 600 seconds. Maximum speed and max. acceleration are 80.3 mph and 8 mph/second. The preconditioning of both vehicles and consequently of the aftertreatment systems before testing in one of these three cycles has been two consecutive extra-urban cycle parts of the NEDC.

PASSENGER CAR INVESTIGATIONS

The focus of this study has been the implementation of a NO_x-adsorber catalyst in combination with a Diesel particulate filter into a passenger car exhaust gas aftertreatment system. The NO_x-adsorber upstream of the DPF has also been used for conversion of CO and unburned HC. The upstream position of the NO_x-adsorber relative to the DPF (see Fig. 1) has the main advantage that overheating of the NO_x-adsorber during DPF regeneration can be avoided. Further advantages of this configuration are the minimizing of the thermal inertia upstream of the NO_x-adsorber, DPF heating during NO_x-adsorber regeneration and HC reduction upstream of the DPF. Disadvantages are reduced C oxidation by NO_2 and possible ash poisoning of the NO_x-adsorber.

For the implementation of a NO_x-adsorber into the exhaust gas aftertreatment system, additional ECU functionalities are required. A modular design has been used to integrate and combine the required functionalities. This modular approach offers several advantages such as the possibility to improve or to exchange each module independent from other modules. The NO_x-adsorber catalyst performance has been evaluated based on the up- and downstream NO_x concentrations. The upstream NO_x concentrations have been estimated with the help of engine out raw emission maps. The downstream NO_x concentrations have been measured with a NO_x sensor. This information has been used to calculate an on-line NO_x conversion efficiency. The calculated conversion rate has been compared with a pre-defined reference NO_x conversion which is a function of temperature and is shown in Fig. 2.

The evaluation of NO_x conversion efficiency over the NO_x-adsorber catalyst during vehicle driving has been one of several input signals to the ECU supervisor functionality controlling the purge events of the NO_x-adsorber. The second main input signal for the supervisor has been a vehicle evaluation functionality module (gear, speed, acceleration, water temperature, etc.). This vehicle evaluation module compares several vehicle parameters with reference parameters to decide whether or not the NO_x-adsorber purge can be carried out under the current driving conditions. These new functionalities have been utilized during cycle testing and real world driving with vehicle I, whereas for vehicle II the purge events have been controlled time based.

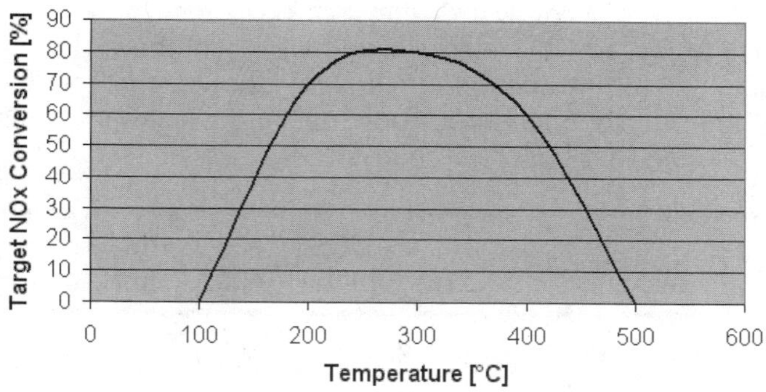

Figure 2. Reference NO_x conversion as a function of temperature.

Based on a present NO_x-adsorber purge demand, the engine has been operated in various modes. Depending on the temperature of the NO_x-adsorber catalyst, the engine out temperature needed to be increased. This increase has been necessary to avoid CO or HC breakthroughs and could be carried out by re-calibration of the engine maps for such heat mode. Above the NO_x-adsorber catalyst light-off temperature, the engine has been operated under rich conditions for desorption and reduction of the stored nitrogen oxides. Therefore, rich combustion calibrations have been developed for the relevant engine map areas (NEDC, FTP 75 and US06). The additional NO_x-adsorber functionalities have been tested and optimized during both on road and test track driving as well as during emission evaluation on the roller chassis dynamometer for vehicle I. Finally, with these very complex functionalities and engine calibrations as well as with the additional Diesel particulate filter, real world driving has been performed to evaluate the efficiency of the complete exhaust gas aftertreatment system.

REFERENCE TESTS VEHICLE I

Several cold and hot start NEDC tests were performed as a base for vehicle I for the current passenger car investigations. The first configuration has been without any catalyst. The second configuration is similar to the production configuration, which is oxidation catalyst only. The third configuration has been oxidation catalyst and DPF. The main results of these tests are given in Table 1.

After the base tests without NO_x-adsorber, a fourth configuration has been tested with NO_x-adsorber only, but with the additional functionalities deactivated (no $DeNO_x$-purges). As a consequence, the NO_x-adsorber catalyst has performed similar to an oxidation catalyst, but without specific NO_x conversion. Additionally, the above described NO_x-adsorber functionalities have been activated and another set of base tests has been performed with NO_x-adsorber but without a DPF. Finally, the last base tests with NO_x-adsorber and DPF have been carried out. The results of the hot start NEDC tests are given in Table 2. The histories during NEDC of the NO_x concentrations upstream and downstream of the

NO_x-adsorber catalyst of the final base test of vehicle I with activated NO_x-adsorber and DPF are shown in Fig. 3. For the same test, Fig. 4 shows the on-line temperatures measured upstream, downstream, and within the NO_x-adsorber catalyst.

Configuration	HC g/km	CO g/km	NO_x g/km	PM g/km	CO_2 g/km
No Catalyst hot start	0.18	0.74	0.60	0.05	162
Oxi-Cat only hot start	0.02	0.01	0.58	0.04	160
Oxi-Cat only cold start	0.04	0.20	0.56	0.04	174
Oxi-Cat & DPF hot start	0.03	0.02	0.62	<0.01	163

Table 1. Reference NEDC results without NO_x-adsorber for vehicle I.

Configuration	HC g/km	CO g/km	NO_x g/km	PM g/km	CO_2 g/km
passive NO_x-adsorber only	0.02	0.01	0.58	0.03	158
active NO_x-adsorber only	0.08	0.02	0.29	0.05	170
active NO_x-adsorber & DPF	0.07	0.02	0.28	<0.01	168

Table 2. Reference NEDC hot start results with NO_x-adsorber for vehicle I.

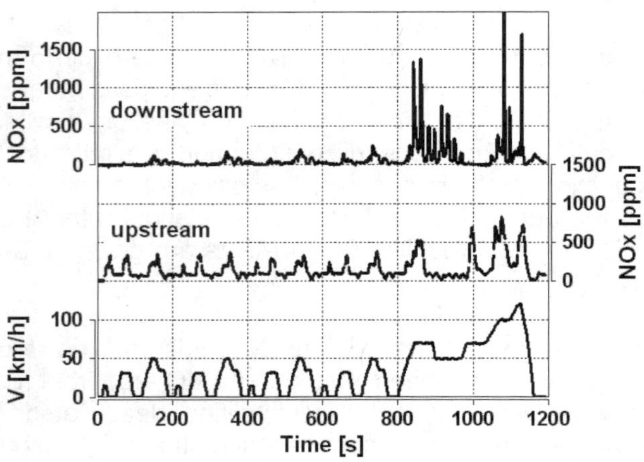

Figure 3. NO_x concentrations up- and downstream of NO_x-adsorber during hot start NEDC (vehicle I).

The base test results of vehicle I show a NO_x conversion efficiency during NEDC above 50%. Due to the rich combustion calibration during $DeNO_x$ purges of the NO_x-adsorber, an increase of HC emissions can be observed. Also the PM emissions increase due to the modified calibration during rich combustion, which can only be measured without DPF. For the operation with DPF, higher engine out PM emission will lead to accelerated soot loading of the particulate filter and consequently will lead to higher required frequency of DPF regeneration.

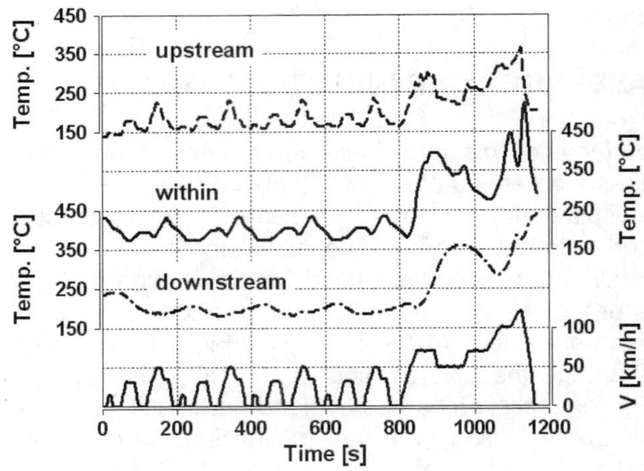

Figure 4. On-line temperatures measured upstream, downstream, and within the NO_x-adsorber catalyst (vehicle I).

REFERENCE TESTS VEHICLE II

Several hot start NEDC tests were also performed as a base for vehicle II. The first configuration tested has been without any catalyst. The second configuration is similar to the production configuration, which is an oxidation catalyst only. Similar to vehicle I, additional base tests with the NO_x-adsorber have been carried out. The first test has been with NO_x-adsorber catalyst only, but with $DeNO_x$-purges deactivated, whereas for the next test the time based controlled NO_x-adsorber purge events have been activated. Finally, a DPF has been added as shown in Fig. 1. The results of these tests are given in Table 3.

After performing the NEDC tests also FTP 75 and US06 tests have been carried out with vehicle II. Hardware configurations have been oxidation catalyst only (production configuration), active NO_x-adsorber catalyst only, and active NO_x-adsorber catalyst in combination with DPF. The results of these tests are given in Table 4 for the FTP 75 tests and in Table 5 for the US06 tests.

The reference results of vehicle II show that the NO_x conversion efficiency during NEDC and FTP 75 are in the range of 60 to 65%. These NO_x conversion efficiencies obtained for vehicle II are even more promising compared to the NO_x conversion efficiency obtained for vehicle I. Nevertheless, the HC and CO emission increase due to NO_x-adsorber purging is significantly higher for vehicle II. Obviously, the advanced control functionalities utilized for vehicle I allows a compromise along the trade-off between NO_x-emission reduction and HC / CO emission increase. The catalytic coating of the DPF in vehicle II leads to a slight decrease of HC and CO emissions during FTP 75 tests. Due to the higher exhaust gas temperatures this effect is even more obvious during the US06 high speed/high load cycle. During the US06 tests with generally higher NO_x emissions a conversion efficiency of approx. 35% could be achieved with vehicle II. Even though the DPF leads to a PM reduction above 90% during US06 tests, the PM emissions with DPF are still clearly higher than during NEDC and FTP 75 tests.

Configuration	HC g/km	CO g/km	NO_x g/km	PM g/km	CO_2 g/km
No Catalyst	0.15	0.76	0.72	0.05	148
Oxi-Cat only	0.03	0.01	0.71	0.04	150
passive NO_x-adsorber only	0.03	0.01	0.64	0.04	149
active NO_x-adsorber only	0.16	0.02	0.26	0.05	156
active NO_x-adsorber & DPF	0.14	0.03	0.28	<0.01	157

Table 3. Reference NEDC hot start results with different exhaust gas aftertreatment configurations for vehicle II.

Configuration	HC g/km	CO g/km	NO_x g/km	PM g/km	CO_2 g/km
Oxi-Cat only	0.03	0.05	0.67	0.04	182
active NO_x-adsorber only	0.62	0.23	0.23	0.05	189
active NO_x-adsorber & DPF	0.59	0.20	0.24	<0.01	190

Table 4. Reference FTP 75 results with different exhaust gas aftertreatment configurations for vehicle II.

The histories of NO_x concentrations upstream and downstream of the NO_x-adsorber catalyst during FTP 75 testing of vehicle II with activated NO_x-adsorber functionalities are shown in Fig. 5. Although NO_x concentration peaks can be higher downstream compared to upstream of the NO_x-adsorber, the NO_x concentrations generally show lower levels downstream of the NO_x-adsorber leading to 60 to 65% reduction of cumulative NO_x emissions during FTP 75 as given in Table 4. While the cumulative NO_x emissions during the three phases of the test are very similar without the NO_x-adsorber, the reductions during phase 2 (Fig. 5: approx. 500 s – 1374 s) and phase 3 (approx. 1950 s – 2455 s) are slightly higher compared to phase 1 (approx. 0 s – 500 s) of the complete test.

Configuration	HC g/km	CO g/km	NO_x g/km	PM g/km	CO_2 g/km
Oxi-Cat only	0.01	0.01	0.94	0.21	187
active NO_x-adsorber only	0.39	0.26	0.58	0.23	200
active NO_x-adsorber & DPF	0.27	0.19	0.60	0.02	198

Table 5. Reference US06 results with different exhaust gas aftertreatment configurations for vehicle II.

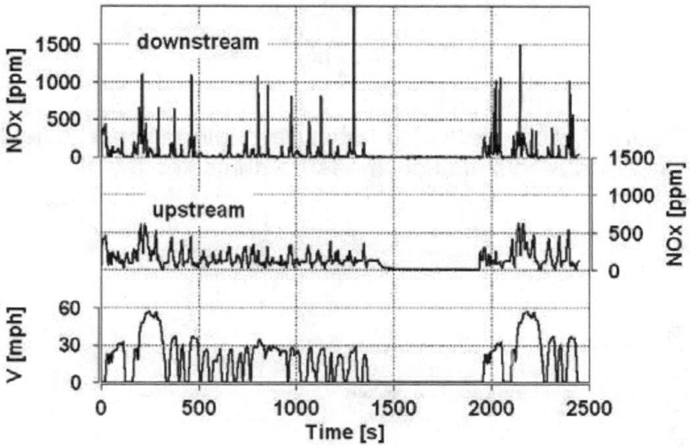

Figure 5. NO_x concentrations up- and downstream of NO_x-adsorber during FTP 75 (vehicle II).

During the FTP 75 tests, transient particle size distribution measurements (measuring range 10 - 600 nm, 20 particle size classes) have been carried out with a transient differential mobility particle spectrometer (TDMPS). The results of total particle number up- and downstream of the DPF are plotted in Fig. 6. As expected, the particle number history shows a very dynamic behavior upstream of the DPF with average particle numbers of approx. 1e07 $1/cm^3$ with maximum peaks up to 1e08 $1/cm^3$. The particle number history downstream of the DPF shows an average value slightly higher than 1e06 $1/cm^3$ in the logarithmic scale.

Corresponding to Fig. 6 the results of the mean particle diameters up- and downstream of the DPF are plotted in Fig. 7. Again, the mean particle diameter history during FTP 75 shows a very dynamic behavior upstream of the DPF. During acceleration phases the mean particle diameter increases whereas the lowest values are obtained during vehicle deceleration phases with fuel cut-off. The mean particle diameter exceeds 50 nm during most acceleration phases upstream of the DPF, whereas downstream of the DPF the mean particle diameter shows a very constant behavior with an average value of 35 to 40 nm. Reduced particle numbers (approximately one order of magnitude) in combination with reduced mean particle diameters lead to a PM emission reduction (mass-based) during the FTP 75 of approx. 85%.

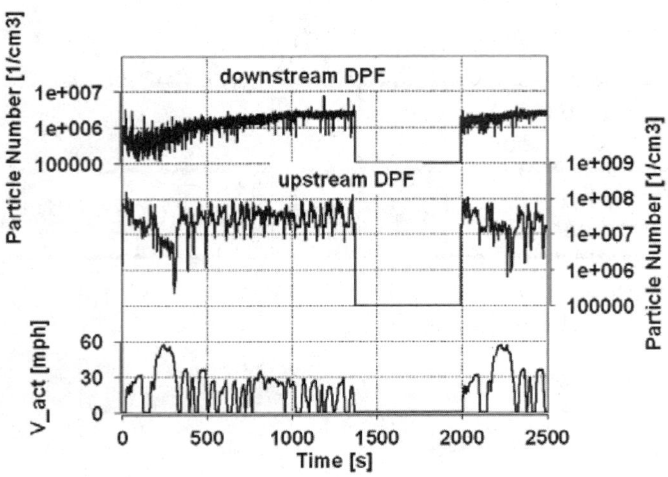

Figure 6. Total particle number calculated from transient particle size distribution measurements for the FTP 75 (vehicle II).

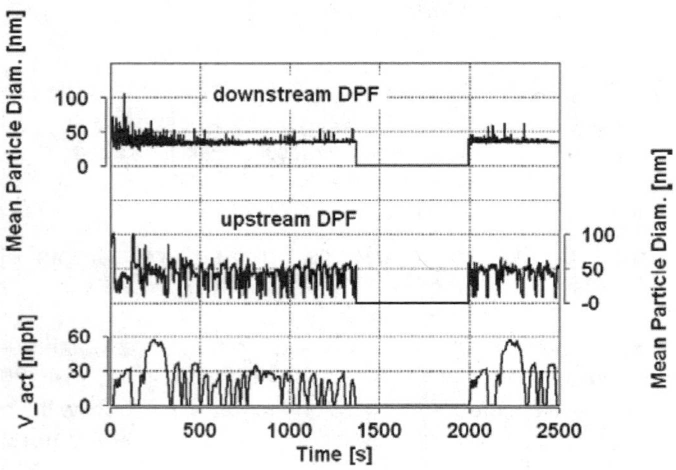

Figure 7. Mean particle diameter calculated from transient particle size distribution measurements for the FTP 75 (vehicle II).

Fig. 8 shows on-line temperatures and NO_x concentrations upstream and downstream of the NO_x-adsorber catalyst during US06 testing of vehicle II with active NO_x-adsorber purge events. The NO_x concentrations show lower reduction during the first 140 seconds and the last 90 seconds of the test with the higher share of transient driving, whereas from 140 seconds to 510 seconds with a lower share of transient driving, the NO_x reduction is higher. As expected, the temperature histories show much higher dynamics upstream of the NO_x-adsorber. The temperatures downstream of the NO_x-adsorber show an average value of approx. 400°C, which is much higher than the average temperature during the NEDC as shown in Fig. 4. Overall, the generally higher NO_x emissions due to higher speed and load as well as the higher temperatures lead to reduced NO_x emission reduction during the US06 compared to the NEDC and FTP 75.

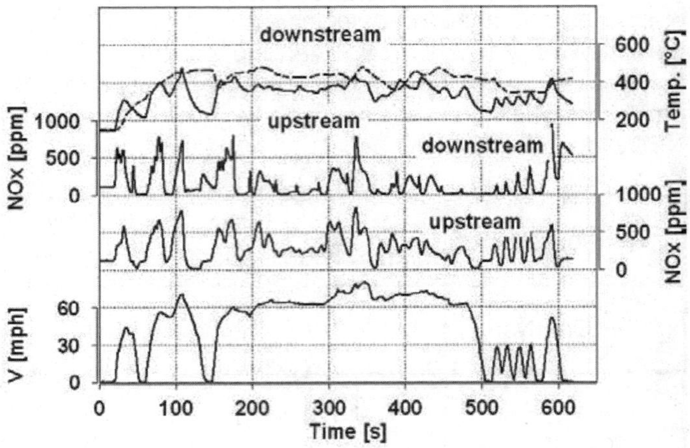

Figure 8. NO_x concentrations and on-line temperatures up- and downstream of NO_x-adsorber during the US06 (vehicle II).

Similar to the FTP 75 tests, transient particle size distribution measurements (measuring range 10 - 600 nm, 20 particle size classes) have also been carried out during the US06 tests. The results of total particle number up- and downstream of the DPF are plotted in Fig. 9. Again, the particle number histories show a more dynamic behavior upstream compared to downstream DPF. Upstream DPF maximum particle numbers exceed 1e08 $1/cm^3$ during acceleration and fall below 1e06 $1/cm^3$ during fuel cut-off.

During US06, the particle number history downstream of the DPF shows an average value of approx. 1e06 $1/cm^3$. Unlike the results obtained for the FTP 75, the particle number during US06 shows a continuous increase during the test duration of 600 seconds. As shown in Fig. 8, the temperatures downstream of the NO_x-adsorber are in the range of 400°C. These are appropriate temperatures for soot oxidation with NO_2 (CRT-effect) within the DPF. The decreasing soot loading leads to decreasing soot filtration efficiency of the DPF and consequently to increasing particle numbers.

The results of the mean particle diameters up- and downstream of the DPF during US06 testing are plotted in Fig. 10. During acceleration phases the mean particle diameter increases whereas lowest values can be obtained during longer vehicle deceleration phases with fuel cut-off. The mean particle diameter mostly exceeds 40 nm upstream of the DPF, whereas downstream of the DPF the mean particle diameter mostly remains under 40 nm with an average value of approx. 35 nm. Reduced particle numbers in combination with reduced mean particle diameters lead to a PM emission reduction (mass-based) during US06 above 90% .

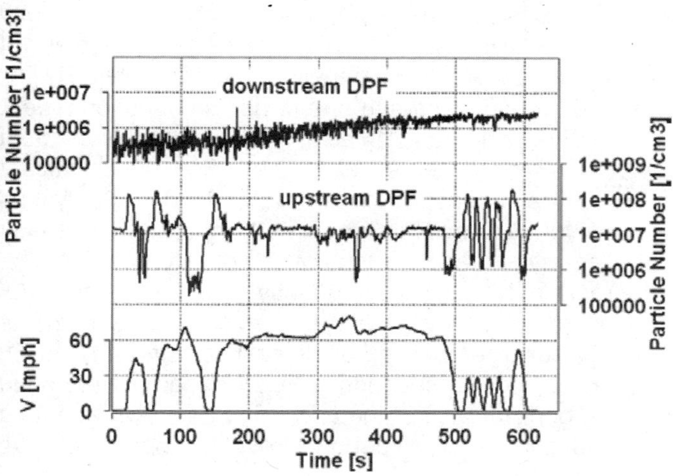

Figure 9. Total particle number calculated from transient particle size distribution measurements for the US06 (vehicle II).

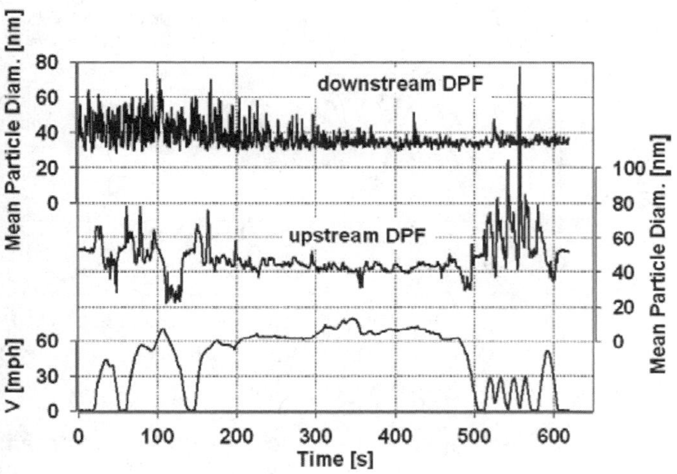

Figure 10. Mean particle diameter calculated from transient particle size distribution measurements for the US06 (vehicle II).

REAL WORLD DRIVING

After implementation of the NO_x-adsorber catalyst and the Diesel particulate filter into the exhaust gas aftertreatment system of vehicle I, the base measurements have been carried out as described above. Based on the successful reduction of PM emissions by over 90% and of NO_x emissions above 50% in the NEDC, "real world driving" has been performed to evaluate the efficiency of the modified exhaust gas system in combination with the additional functionalities.

The vehicle has been operated in different driving conditions. Beside specific urban and extra urban driving, highway testing has also been performed. The split between different driving patterns is shown in Fig. 11. The total distance of evaluated driving has been 617 km with an average vehicle speed of 45 km/h. This corresponds to a total time of 820 minutes. Table 6 shows the detailed distances, average speeds and time duration of the passenger car real world driving tests.

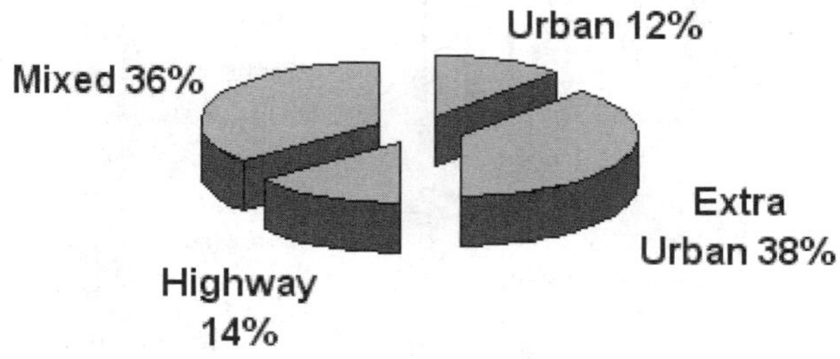

Figure 11. Split of driving patterns

During the complete driving evaluation, the NO_x concentrations upstream of the NO_x-adsorber have been calculated based on the engine out raw emission maps, whereas the downstream NO_x concentrations have been measured with the NO_x sensor as described above. The average CO_2 emission has been 192 g/km.

	Distance km	Average Speed km/h	Time minutes
Urban	76	26	175
Extra Urban	234	46	305
Highway	86	82	63
Mixed	221	48	276
Total	617	45	820

Table 6. Real world driving evaluation – split into different driving patterns.

Highway Driving

The total distance of highway driving has been 86 km as shown in Table 6. In detail, two uninterrupted 25 km highway driving and a third uninterrupted 36 km highway driving have been carried out. The third sequence of highway driving has been analyzed in detail. The average speed of this sequence has been 74 km/h which leads to an evaluation time of approximately 1800 seconds. The vehicle speed and the NO_x concentrations upstream and downstream of the NO_x-adsorber catalyst as well as the calculated cumulative NO_x emissions are shown in Fig. 12. The NO_x-adsorber purge times and the temperatures upstream, downstream, and within the NO_x-adsorber are shown in Fig. 13. Additionally, Fig. 14 summarizes engine speed, air/fuel-ratio λ and pedal sensor with the corresponding injected fuel quantity. Finally, the exhaust gas back pressure difference between upstream of the NO_x-adsorber and downstream of the Diesel particulate filter is also plotted in Fig. 14.

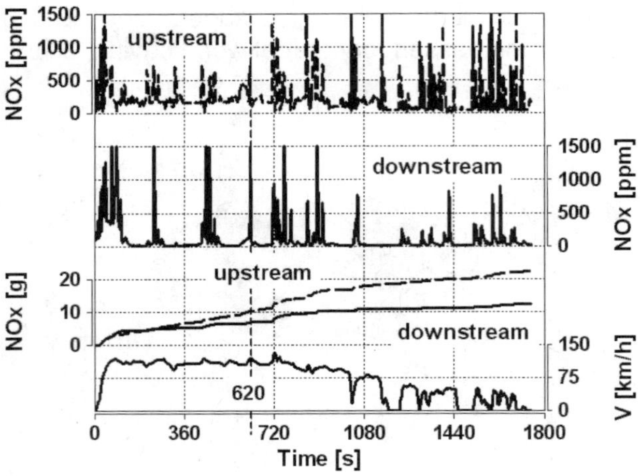

Figure 12. NO_x concentrations and cumulative NO_x emissions upstream and downstream of the NO_x-adsorber catalyst and vehicle speed during highway driving.

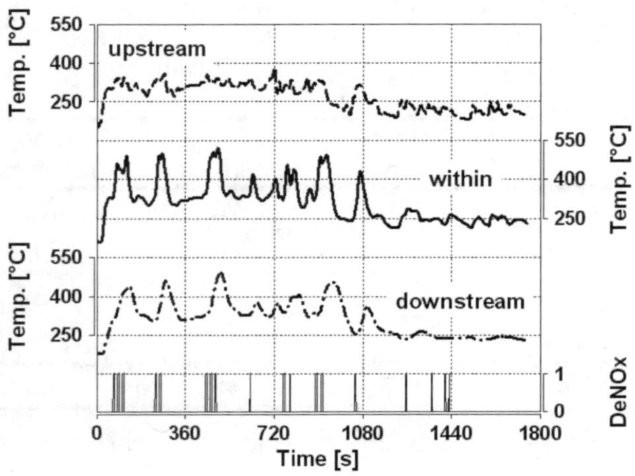

Figure 13. Temperatures upstream, downstream, and within the NO_x-adsorber catalyst and NO_x-adsorber purge events during highway driving.

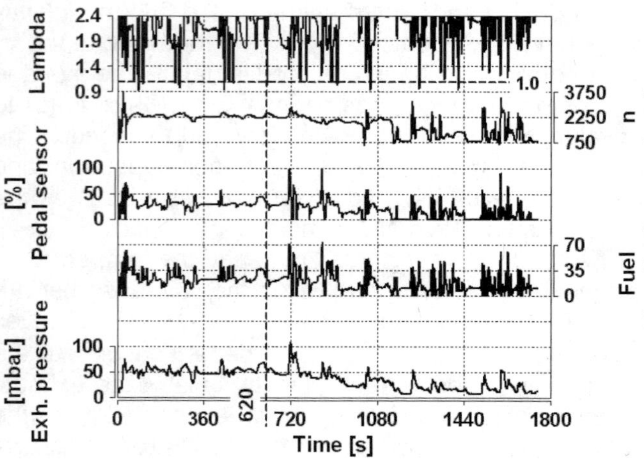

Figure 14. Engine speed, air/fuel-ratio λ, pedal sensor with corresponding injected fuel quantity and exhaust gas back pressure (Δp across the NO_x-adsorber and DPF) during highway driving.

Fig. 13 marks the 18 $DeNO_x$ purge events during the 30 minute driving period. The average of one purge event per 100 seconds (approximately) leads to a relatively low fuel consumption increase due to rich combustion (see λ < 1 in Fig. 14). Fig. 13 also shows, that subsequent $DeNO_x$ purge events can lead to a significant temperature increase within the NO_x-adsorber catalyst as well as further downstream. This temperature increase can lead to thermal desorption of NO_x within the adsorber and can consequently decrease the NO_x conversion efficiency. This effect must be considered for the NO_x-adsorber control functionalities. Efficient $DeNO_x$ purges, e.g. at 620 seconds in Fig. 12, lead to very good storage performance of the NO_x-adsorber catalyst during the subsequent time period provided that excessive temperature peaks can be avoided. Fig. 14 shows the corresponding injected fuel quantity peak of up to 50 mm^3/stroke at 620 seconds, which clearly indicates the rich combustion calibration, since the vehicle speed is almost constant and no acceleration can be observed.

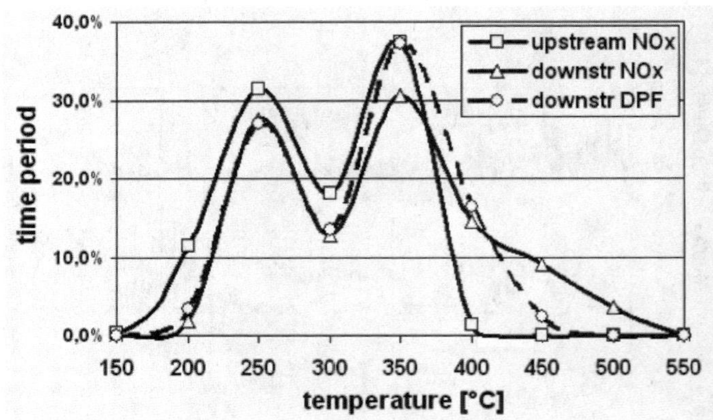

Figure 15. Temperature frequency up- and downstream of the NO_x-adsorber and downstream of the DPF during highway driving.

In addition to the temperature histories in Fig. 13, one further thermocouple within the NO_x-adsorber catalyst inlet indicates a maximum temperature of 555°C. The average temperature within the NO_x-adsorber has been 313°C during highway driving. The temperatures up-stream and downstream of the NO_x-adsorber as well as the temperature downstream of the DPF are plotted by frequency of occurrence in Fig. 15. Two local maximums can be observed at approximately 250°C and 350°C. The temperature downstream of the NO_x-adsorber, but upstream of the DPF, is between 9% and 10% of the time above 450°C. Without activated NO_x-adsorber purge functionalities, meaning without frequent operation of the Diesel engine under rich combustion conditions (λ < 1), the exhaust gas temperature upstream of the DPF would have been always below 360-400°C. This significant shift towards higher exhaust gas temperatures seems to be influencing the soot accumulation within the DPF, especially since the low sulfur fuel has been premixed with 25 ppm Cerium additive to decrease soot oxidation temperature. An increased NO_2/NO-ratio due to the temperature shift could also promote soot oxidation within the DPF.

The total time period of rich combustion to purge the NO_x-adsorber during highway driving has been approximately 3.7%. With a total duration of 1800 seconds and 18 $DeNO_x$ purge events, the average duration of each rich combustion calibration event has been 3.7 seconds. The calculated average engine out NO_x emissions have been 0.62 g/km, whereas the tailpipe emissions have been 0.34 g/km. This corresponds to a NO_x conversion efficiency of 45%.

Urban Driving

The total distance of urban driving during these passenger car investigations has been 76 km at an average speed of 26 km/h, which corresponds to a driving time of 175 minutes. One uninterrupted urban driving period of 16 km has been evaluated in detail. The average speed of this sequence has been 30 km/h which leads to an analyzing time of approximately 2000 seconds. The vehicle speed and the NO_x concentrations upstream and downstream of the NO_x-adsorber catalyst are shown in Fig. 16. NO_x-adsorber purge times, temperatures upstream, downstream, and within the NO_x-adsorber as well as the exhaust gas back pressure difference across the NO_x-adsorber and Diesel particulate filter are shown in Fig. 17.

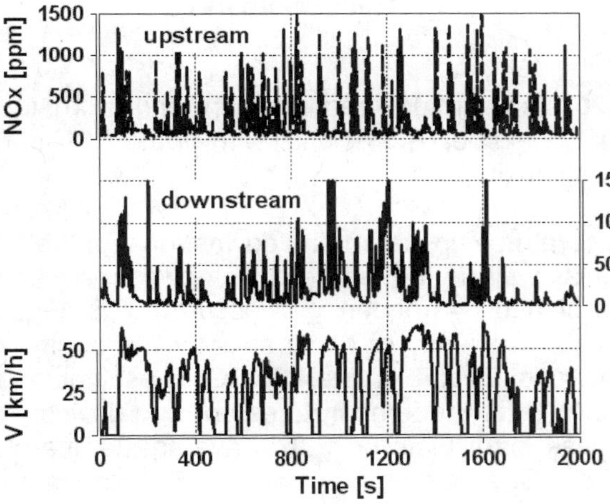

Figure 16. NO_x concentrations upstream and downstream of the NO_x-adsorber catalyst and vehicle speed during urban driving.

Fig. 17 shows 20 $DeNO_x$ purge events over a period of 2000 seconds. Consequently, the average time period between two $DeNO_x$ purges is approx. 100 seconds, which is similar to the results obtained during highway driving. Nevertheless, the total time period of rich combustion to purge the NO_x-adsorber during urban driving has only been approximately 2.2%, which is significantly lower than during highway driving. This leads to an average duration of each rich combustion calibration event of 2.2 seconds. These shorter purge events are less effective compared to the purge events during highway driving. Furthermore, Fig. 16 shows the relatively low vehicle speeds during urban driving. Since the additional NO_x-adsorber functionalities have been optimized for vehicle speeds above 40 km/h, the total NO_x conversion efficiency during urban driving is much smaller compared to highway driving.

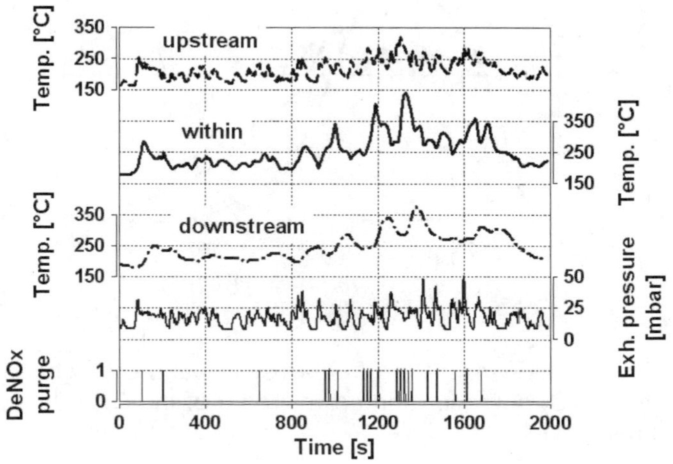

Figure 17. Temperatures upstream, downstream, and within the NO_x-adsorber catalyst, exhaust gas back pressure (Δp across the NO_x-adsorber and DPF) and NO_x-adsorber purge events during urban driving.

Fig. 17 shows the temperatures upstream, downstream and within the NO_x-adsorber catalyst. They are significantly lower during urban driving compared to highway driving. The maximum temperature within the center of the catalyst is approximately 400°C, whereas one further thermocouple within the catalyst inlet indicates a temperature maximum of 440°C, which is 115°C lower than during highway driving. The average temperature within the center of the NO_x-adsorber has been 253°C during urban driving, which is 60°C lower compared to highway driving. Fig. 17 clearly shows, that subsequent $DeNO_x$ purge events around 1200 and 1300 seconds lead to temperature increases within the NO_x-adsorber catalyst above 350°C. This leads to increased NO_x emissions as indicated in Fig. 16.

The temperatures upstream and downstream of the NO_x-adsorber as well as the temperature downstream of the DPF are plotted by frequency of occurrence in Fig. 18. Only one local maximum can be observed of approximately 250°C. The temperature downstream of the NO_x-adsorber, but upstream of the DPF, does not exceed 450°C.

The calculated average engine out NO_x emissions during urban driving have been 0.84 g/km. Due to a lot of acceleration phases, this is 32% higher compared to highway driving. Due to the lower NO_x conversion efficiency as described above, NO_x conversion efficiency is distinctively lower than for highway driving.

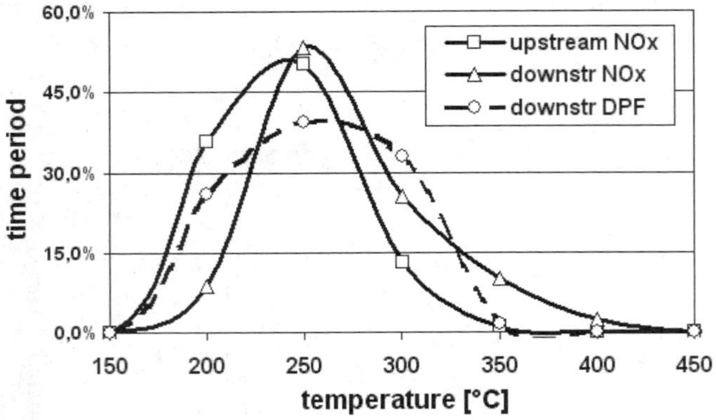

Figure 18. Temperature frequency up- and downstream of the NO_x-adsorber and downstream of the DPF during urban driving.

Extra Urban Driving

The total distance of extra urban driving during the current investigations has been 234 km at an average speed of 46 km/h with a driving time of 305 minutes. One uninterrupted extra urban driving period of 37 km has been analyzed in detail. The average speed of this sequence has been 49 km/h which leads to an analyzing time of approximately 2700 seconds. The vehicle speed, NO_x concentrations upstream and downstream of the NO_x-adsorber catalyst and exhaust gas back pressure difference across the NO_x-adsorber and Diesel particulate filter are shown in Fig. 19.

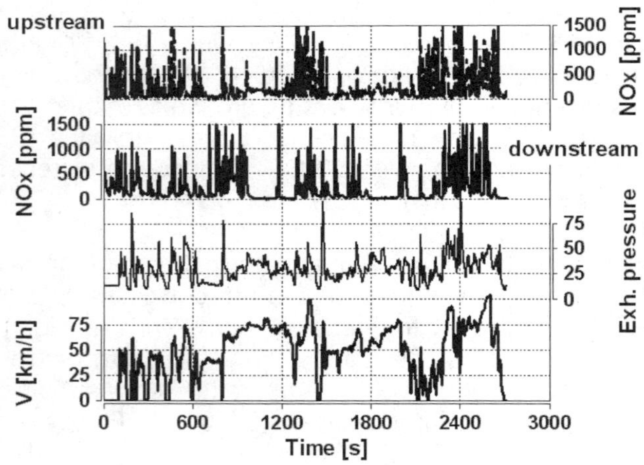

Figure 19. NO_x concentrations upstream and downstream of the NO_x-adsorber catalyst, exhaust back pressure (Δp across the NO_x-adsorber and DPF) and vehicle speed during extra urban driving.

As expected, Fig. 19 shows that the vehicle speeds are lower compared to highway driving and higher compared to urban driving. Especially during the near constant driving speed period from around 900 seconds to 1200 seconds, the NO_x emissions downstream of the NO_x-adsorber catalyst are very low.

The total time period of rich combustion to purge the NO_x-adsorber during extra urban driving has been approximately 2.8%. Again, this is between highway and urban driving as expected. The calculated average engine out NO_x emissions have been 0.82 g/km, whereas the tailpipe emissions have been 0.66 g/km. This corresponds to a NO_x conversion efficiency of 20%.

The maximum exhaust back pressure difference across the NO_x-adsorber and Diesel particulate filter is approximately 100 mbar. This value has been 110 mbar for highway driving and only 51 mbar during urban driving. In all three cases, the lowest back pressure difference during engine idle is around 10 mbar.

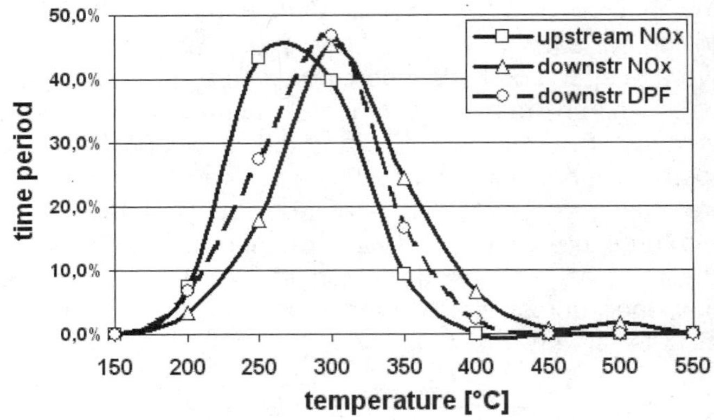

Figure 20. Temperature frequency up- and downstream of the NO_x-adsorber and downstream of the DPF during extra urban driving.

The temperatures upstream and downstream of the NO_x-adsorber as well as the temperature downstream of the DPF are plotted by frequency of occurrence in Fig. 20. The highest frequency of occurrence of the engine out temperature is at approximately 250°C. These values are around 300°C for the temperatures up- and downstream of the Diesel particulate filter. The average temperature within the NO_x-adsorber has been 295°C during extra urban driving, which is only 18°C lower compared to highway driving. The maximum temperature at the NO_x-adsorber inlet has been 545°C during extra urban driving, which is also only 10°C lower compared to highway driving. These overall surprisingly high temperatures during extra urban driving compared to urban and highway driving are caused by too many $DeNO_x$ purge events in too short a time period, which leads to temperature increase and consequently to NO_x desorption without subsequent NO_x reduction. This decreases the NO_x conversion efficiency, which is much lower compared to highway driving.

Mixed Driving and Cold Start

The total distance of mixed driving during this study has been 221 km as shown in Table 6. One uninterrupted sequence of 21 km mixed driving has been analyzed in detail. The average speed of this sequence has been 33 km/h which is only slightly above the average speed of urban driving. Reasons for the low average speed are the delayed vehicle start after cold engine start and the frequent vehicle stops during the second half of this evaluation as well as very low speeds during the last 15% of driving. The evaluated time of this sequence is approximately 2200 seconds. The vehicle speed and the NO_x concentrations upstream and downstream of the NO_x-adsorber catalyst as well as the calculated cumulative NO_x emissions are shown in Fig. 21. The NO_x-adsorber purge times, the temperatures upstream, downstream, and within the NO_x-adsorber and exhaust gas back pressure difference across the NO_x-adsorber and Diesel particulate filter are shown in Fig. 22.

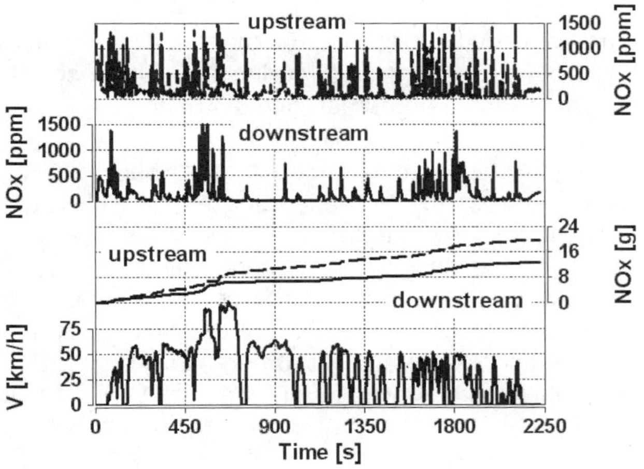

Figure 21. NO_x concentrations and cumulative NO_x emissions upstream and downstream of the NO_x-adsorber catalyst and vehicle speed during mixed driving.

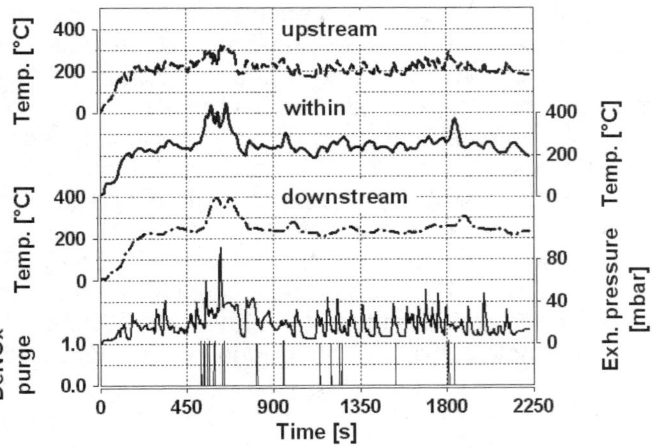

Figure 22. Temperatures upstream, downstream, and within the NO_x-adsorber catalyst and NO_x-adsorber purge events during mixed driving.

The temperatures plotted in Fig. 22 show that the vehicle has been started under cold engine conditions. Since the NO_x-adsorber catalyst has not been conditioned before, the storage performance is insufficient during the first 450 seconds. Only 33%, 2 out of 6 grams, are stored during that period as shown in Fig. 21. It takes approximately 120 seconds for the temperature within the center of the NO_x-adsorber catalyst to exceed 200°C. During the first 500 seconds no $DeNO_x$ purge events occur due to too low engine water temperature.

The total time period of rich combustion to purge the NO_x-adsorber during this evaluation has been only slightly above 1%. With total duration of 2200 seconds and 13 $DeNO_x$ purge events, the average duration of each rich combustion calibration event has been only 1.7 seconds. The calculated average engine out NO_x emissions have been 0.98 g/km, which is very high due to the many acceleration phases. The tailpipe emissions have been calculated to 0.6 g/km. This corresponds to a NO_x conversion efficiency of approx. 40%.

Five subsequent $DeNO_x$ purge events between 500 and 600 seconds lead to a significant temperature increase inside the NO_x-adsorber catalyst up to 445°C as shown in Fig. 22. In addition to the temperature histories in Fig. 22, another thermocouple within the NO_x-adsorber catalyst inlet gives a temperature maximum of 475°C. These high temperatures promote NO_x desorption within the catalyst and lower NO_x conversion efficiency. The average temperature within the NO_x-adsorber has been 244°C during mixed driving, which is even 9°C lower compared to city driving.

The temperatures upstream and downstream of the NO_x-adsorber as well as the temperature downstream of the DPF are plotted by frequency of occurrence in Fig. 23. The highest frequency of occurrence is at approximately 250°C. Temperature distribution is very similar to city driving below 250°C, whereas the frequency of occurrence is lower at higher temperatures.

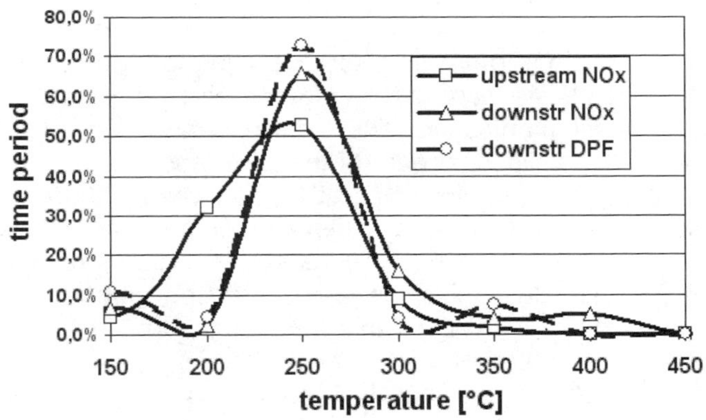

Figure 23. Temperature frequency up- and downstream of the NO_x-adsorber and downstream of the DPF during mixed driving.

Summary Real World Driving

During the complete distance of 617 km real world driving with vehicle I at an average speed of 45 km/h and a total driving time of 820 minutes, as shown in Table 6, the complete vehicle including the additional NO_x-adsorber catalyst functionalities have been very stable in operation. The effect on driveability has been minimal, especially the vehicle acoustics during transition phases between lean and rich combustion as well as during the rich combustion phases themselves. Since only a smaller part of the driving has been highway driving with rich combustion times above 3%, the complete rich combustion time during these tests has been below 2% which leads to a NO_x-adsorber catalyst efficiency noticeably below the efficiency for highway driving. Since the additional NO_x-adsorber functionalities have been optimized for vehicle driving above 40 km/h and driving patterns similar to the New European Drive Cycle, the highest potential for further improvements is optimization during slower driving in lower gears.

After completion of the real world driving, another NEDC test has been carried out without any conditioning of the aftertreatment system. The results are shown in Table 7.

The results for HC and CO emissions are slightly better compared to the NEDC test prior to the real world driving. On the other hand, the NO_x emissions of 0.33 g/km are slightly higher and represent a NO_x conversion of 43% based on the NO_x emissions without active NO_x-adsorber as shown in Table 2.

Configuration	HC g/km	CO g/km	NO_x g/km	PM g/km	CO_2 g/km
active NO_x-adsorber & DPF	0.06	0.01	0.33	<0.01	160

Table 7. Final NEDC hot start results with active NO_x-adsorber and DPF for vehicle I.

INTERACTION BETWEEN THE NO_X-ADSORBER CATALYST AND DIESEL PARTICULATE FILTER

As shown in Fig. 1, the exhaust gas aftertreatment system consists of a NO_x-adsorber catalyst upstream and Diesel particulate filter downstream. The NO_x storage capability upstream of the DPF leads to reduced NO_x concentrations within the DPF. Therefore, the soot oxidation with NO_2 in the relevant temperature range from 250°C up to 450°C decreases. Furthermore, the engine out PM emissions are increased during rich combustion as shown in Table 2. Since the total rich time is only in the range of a few percent, this effect increases the average engine out PM emissions only slightly. Nevertheless, both effects have the tendency to lead to a faster soot loading in the DPF and consequently lead to a quicker exhaust gas back pressure increase over driving distance. On the other hand, an increased NO_2/NO-ratio due to increased temperatures (towards 350°C) could promote continuous soot oxidation within the DPF. Also, the general exhaust gas temperature increase due to frequent rich combustion calibration to purge the NO_x-adsorber catalyst could lead to slower soot loading and slower exhaust gas back pressure increase over the driving distance.

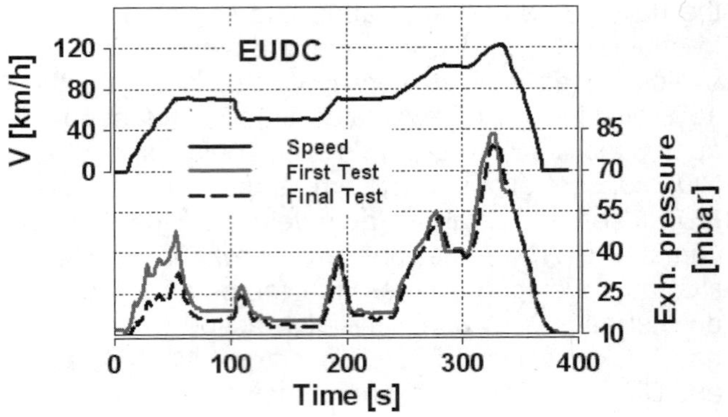

Figure 24. Exhaust gas back pressure difference across the NO_x-adsorber and DPF during the 400 seconds of testing in the extra urban part of the NEDC.

Figure 24 shows the exhaust gas back pressure difference across the NO_x-adsorber and DPF during 400 seconds testing in the extra urban part of the NEDC. The solid bright line shows the corresponding exhaust gas back pressure during EUDC prior to the real world driving, whereas the dashed dark line shows the corresponding exhaust gas back pressure during EUDC after the real world driving of 617 km. The DPF had been regenerated before the start of these investigations, but during the complete driving distance of 617 km no DPF regeneration had been carried out.

The results in Fig. 24 show almost the same pressure histories during EUDC before and after the real world driving tests. The pressure drop is slightly lower at the end of the tests. These results indicate that the filter loading during the driving distance of 617 km must have been at least partly compensated by soot oxidation during this period. The engine out temperatures have been below 400°C during highway driving and even lower during urban and extra urban driving, as shown in several Figures above. Although 25 ppm Cerium additive has been added to the low sulfur fuel (vehicle I) to decrease the soot oxidation temperature and increase soot oxidation rate, temperatures up to 400°C seem to be too low to support the effect of non increasing back pressure over 617 km driving distance. Whereas an increased NO_2/NO-ratio due to the temperature shift could promote continuous soot oxidation within the DPF.

Fig. 25 shows the different positions of temperature measurement within the exhaust gas aftertreatment system used during these investigations. Detailed analyzes of the temperatures show that due to subsequent engine operation under rich combustion the temperature at the outlet of the NO_x-adsorber can significantly exceed 400°C. Fig. 26 shows temperature histories up- and downstream, as well as at inlet, center and outlet of the NO_x-adsorber catalyst. Also temperature history downstream of the DPF and vehicle speed are shown in Fig. 26. Rich engine combustion starts at around 1884 seconds at a vehicle speed of approximately 110 km/h. The duration of the $DeNO_x$ purge event is 5 seconds. All measured temperatures are in the order of 350°C. During rich combustion and NO_x-adsorber purge, the temperature upstream of the NO_x-adsorber does not increase, rather a temperature decrease with decreasing vehicle speed and engine load can be observed. Opposite to this behavior, the temperatures inside the NO_x-adsorber catalyst increase. A temperature wave from front to rear can be observed with temperatures above 480°C at the inlet and around 450°C at the outlet of the NO_x-adsorber catalyst. Further downstream, the temperature increase occurs later. Fig. 26 shows the temperature increase above 400°C downstream of the NO_x-adsorber at approx. 1908 seconds. This effect, as described above, can lead to even higher temperatures if several purge events occur within a short period of time.

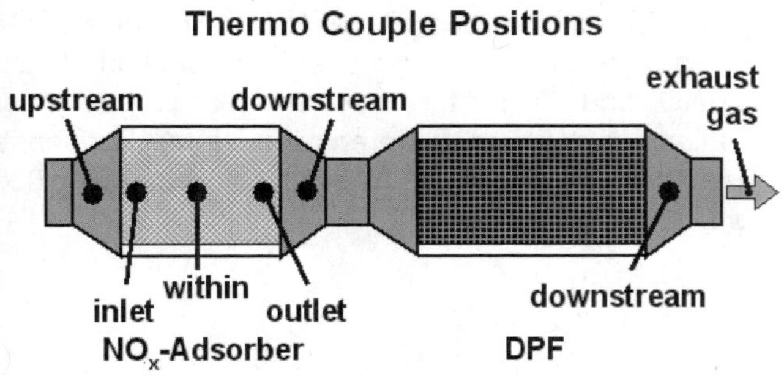

Figure 25. Thermocouple positions during temperature measurements of passenger car investigations.

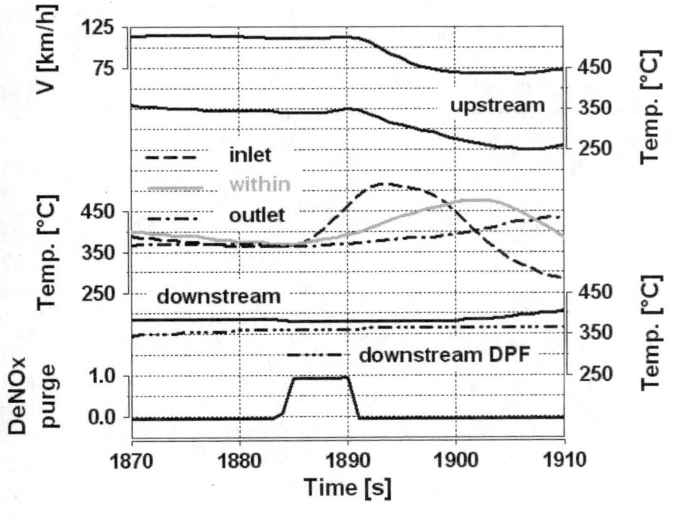

Figure 26. Temperature histories and vehicle speed during a rich combustion event.

OIL DILUTION

During the 617 km of total passenger car driving with vehicle I, approximately 0.5 ltr. of fuel has been injected during post injection. The post injection has been one of several measures to operate the engine under rich conditions to purge the NO_x-adsorber. This fuel quantity has been injected with various timings and amounts depending on the engine operating conditions. Before the vehicle driving, the 5 - 5.5 ltr. of engine oil has been replaced with fresh oil. After the 617 km, the engine oil has been analyzed and approximately 0.3% oil dilution from fuel could be detected. More than 3% of the injected post quantity contributed to the oil dilution. Based on a service interval of about 15000 km this corresponds to an oil dilution by fuel of approximately 7%. Regarding engine durability this value is not acceptable and needs to be reduced.

CONCLUSION

This study focuses on the operating behavior of a NO_x-adsorber catalyst upstream of a Diesel particulate filter as two separate exhaust gas aftertreatment components. New European Drive Cycle, US FTP 75, and US06 tests on roller chassis dynamometer and real world driving with different driving patterns with NO_x-adsorber and DPF have been carried out with two different vehicles. During these tests, the focus has been on NO_x-adsorber catalyst and on the interaction between the NO_x-adsorber and DPF. Transient particle size distribution measurements have been carried out during the FTP 75 and US06 tests. Exhaust gas back pressure behavior as well as timing and efficiency of $DeNO_x$-purge events over the driving distance and during different driving conditions have been investigated. The oil dilution as a consequence of regeneration of the NO_x-adsorber ($DeNO_x$-purge) has also been evaluated in this study.

During cycle testing on chassis dynamometer as well as during highway and mixed vehicle driving the PM emission reduction remains high due to the utilization of a DPF, whereas the actual NO_x reduction depends on driving conditions in connection with optimization of NO_x-adsorber control functionalities. For lower vehicle velocities and driving in lower gears the NO_x-adsorber control functionalities need to be further optimized.

Exhaust gas back pressure histories have been evaluated at the start and end of the real world driving. During this period of more than 600 km without a DPF regeneration, no exhaust gas back pressure increase could be observed. Increased exhaust gas temperatures due to frequent rich engine combustion and $DeNO_x$ purge in combination with 25 ppm Cerium additive in the used fuel seem to increase soot oxidation. Also an increased NO_2/NO-ratio due to the temperature shift could promote continuous soot oxidation. Transient particle size distribution measurements indicate soot oxidation with NO_2 (CRT-effect) during the US06 tests.

The oil dilution during real world passenger car driving due to post injection under rich conditions to purge the NO_x-adsorber has been evaluated. The post injection has been one of several measures used to achieve rich conditions. The oil dilution needs to be included in the further optimization of the NO_x-adsorber control functionalities.

Overall, the results of the passenger car investigations with a lean NO_x-adsorber and Diesel particulate filter show a simultaneous reduction of PM of approx. 90% and above 50% for NO_x. These conversion rates give a high potential for achievement of future emission legislation limits.

REFERENCES

1. Howitt J.S., and Brown K.F., "Diesel Exhaust After-treatment 1999", SAE 1999 SP-1414
2. SAE, "Homogeneous Charge Compression Ignition (HCCI) Combustion 2002", SAE 2002 SP-1688
3. Akihama K., Takatori Y., Inagaki K., Sasaki S., and Dean A.M., "Mechanism of the Smokeless Rich Diesel Combustion by Reducing Temperature", SAE 2001–01–0655
4. Johnson T.V., "Diesel Emission Control – Last 12 Months in Review", SAE 2000–01–2817
5. Salvat O., Marez P., and Belot G., "Passenger Car Serial Application of a Particulate Filter System on a Common Rail Direct Injection Diesel Engine", SAE 2000–01–0473
6. Bouchez M., et. al., "Strategies for the Control of Particulate Trap Regeneration", SAE 2000–01–0472
7. Cartus T., Diewald R., Herzog P., Strigl T., and Wanker R., "Diesel Particulate Filter System Integration – From 3D-Simulation to Production", 23. Internationales Wiener Motorensymposium, Fortschritt-Berichte VDI, Reihe 12, Nr. 490, Band 2, April 2002, pp. 234–256
8. Jacobs E., D'Alfonso N., Doering A., Reisch S., Rothe D., Brueck R., and Treiber P., "PM-KAT: a Non-Blocking Solution to Reduce Carbon Particle Emissions of Euro IV Engines", 23. Internationales Wiener Motorensymposium, Fortschritt-Berichte VDI, Reihe 12, Nr. 490, Band 2, April 2002, pp. 196–216
9. Brown K.F., and Howitt J.S., "Diesel Exhaust Emission Control: Diesel Particulate Filters", SAE 2001 SP-1582
10. Brown K.F., and Howitt J.S., "Developments in Regulation and Catalystic Systems", SAE 2001 SP-1581
11. Klein H., Boehnke H., Hoffmann M., Lox E., Cartus T., Neunteufl K., Buergler L., and Herzog P., "NO_x-Aftertreatment for Diesel Passenger Cars solved? Desulfation of NO_x-Storage-Catalysts", 22. Internationales Wiener Motorensymposium, Fortschritt-Berichte VDI, Reihe 12, Nr. 455, Band 2, April 2001, pp. 192–215
12. Tanaka T., "Simultaneous Reduction of PM and NO_x – A New After-Treatment System", 22. Internationales Wiener Motorensymposium, Fortschritt-Berichte VDI, Reihe 12, Nr. 455, Band 2, April 2001, pp. 216–228
13. Fujimura T., Matushita S., Tanaka T., and Kojima K., "Development towards Serial Production of a Diesel Passenger Car with Simultaneous Reduction System of NO_x and PM for the European Market", 23. Internationales Wiener Motorensymposium, Fortschritt-Berichte VDI, Reihe 12, Nr. 490, Band 2, April 2002, pp. 156–174

CONTACT

Ulrich J. Pfahl
Project Manager
Exhaust Aftertreatment Systems
Powertrain Engineering

ulrich.pfahl@avl.com

AVL LIST GMBH
Hans-List-Platz 1
8020 Graz
Austria

2004-01-1425

Combined NO_x and PM Exhaust Gas Aftertreatment Approaches for HSDI Diesel Engines

Holger Herrmuth, Thomas Cartus, Robert Ducellari and Otfried Derschmidt
AVL List GmbH, Austria

ABSTRACT

While particulate filters are already being used in increasing numbers with production HSDI (High Speed Direct Injection) Diesel powertrains, combined NO_x and PM reducing devices are still in the development phase. In the past, various lean NO_x trap (LNT) applications using internal and external measures to enable enrichment of stoichiometric exhaust gas have been investigated. Some initial results on the integration of LNT and catalytic Diesel Particulate Filter (cDPF) functionality into HSDI powertrains have already been published, and systems consisting of a cDPF and LNT on separate bricks, as well as one brick solutions have been described without being directly compared.

In the present paper a system comparison of various exhaust gas aftertreatment system architectures in combination with a prototype engine designed to fulfil future emissions legislation at lowest CO2 levels performed at AVL is described.

Location of the LNT both up- and downstream of the cDPF and the performance of a single brick 4-Way Catalyst (4-WC) were investigated. With selected exhaust gas aftertreatment devices, fuel enrichment has been assessed using three different approaches: Internal rich combustion, external fuel injection (exhaust port) and fuel evaporation upstream of the exhaust aftertreatment device. Prototype fuel injection devices have been set up to demonstrate the advantages and disadvantages for production application.

The main focus of the investigations of the different systems was their impact on NO_x conversion efficiency, desulfurization and cDPF regeneration capability. Furthermore, the interaction of the cDPF and LNT and their performance under lean and lean/rich conditions were investigated. Finally, a detailed assessment of the compared systems was performed and a possible combination of NO_x and PM reduction device with alternative combustion was considered.

INTRODUCTION

Lowest fuel consumption in combination with pleasurable driveability is leading to a steadily growing market share of the Diesel passenger car in Europe. Studies are showing a tremendous increase in market penetration of Diesel powered cars, most likely reaching a market share of approximately 50 % in Europe in 2007 /1/. The interest in the use of the HSDI Diesel engine as a powertrain for light commercial vehicles, SUVs and PCs is also growing in the USA.

Besides the superior fuel economy of the Diesel powertrain, the steady increase in the specific power and specific torque, currently heading for 65 kW/l and 200 Nm/l, are further reasons for its success /2/. This increase in power and torque density can be utilized in engine downsizing concepts to fulfil future emissions legislation requirements for the simultaneous reduction of toxic exhaust emissions and fuel consumption. In-depth analysis of various CO_2-reduction technologies proves downsizing of Diesel powertrains to be the most efficient way to meet future legislative limits /3/. The investigations described in the present paper were carried out with a prototype engine designed and equipped to be part of such a downsizing concept.

However, shifting the engine operation to higher loads improves efficiency but has a negative impact on the NO_x and particulate trade-off due to lower air excess ratio and hence deteriorated EGR conditions. In order to achieve the necessary reduction of toxic exhaust emissions, two areas have to be considered: raw emissions reduction and exhaust aftertreatment devices.

Regarding raw emission reduction, the application of alternative combustion, based on homogeneous or partly homogeneous mixture preparation /4/, will be mentioned but not explained here.

A promising approach to further reduce NO_x and PM emission downstream of the engine is the combination of a lean NO_x trap (LNT) and a catalytic Diesel Particulate Filter (cDPF). Such a combination can be realized as a two brick system, in which the LNT is located upstream /5/ or downstream of the cDPF, or as a single brick solution /6, 7, 8/.

Each of these system variants require additional measures to realize the different necessary operation modes, for example, NO_x regeneration and desulfurization of the LNT and PM regeneration of the cDPF. Various strategies have been pursued to create rich of stoichiometric exhaust gas composition as well as an increase in exhaust gas temperature. Tuning of the combustion itself /9/, injection or evaporation of Diesel fuel /7, 10/ as well as direct introduction of CO, H_2, C_2H_4 /11/ in the exhaust line were investigated.

However, to date a detailed investigation into the integration of these different system variants has not been performed. Base information to allow the best application specific solution to be found is not yet available.

As a consequence this study focuses on the comparison and assessment of three different exhaust gas aftertreatment systems consisting of LNT as well as cDPF functionality, in combination with three different enrichment strategies.

EXPERIMENTAL SET-UP AND TEST CYCLES

ENGINE

AVL´s in-house developed ECO TARGET engine was utilized as the test engine. The ECO TARGET engine is a 1.2 l inline 3-cylinder TCI-HSDI Diesel with 2 valves per cylinder, variable swirl concept, high efficiency EGR-system and a new 3rd generation Bosch Piezo CR injection system. The engine has been optimized for mid-class passenger cars. Assisted by a 10 kW electric motor the engine is intended to replace current 1.9-2.0 l engines. The powerful electric motor enables the whole range of mild hybrid potentials, idle shut off, regenerative braking, electrically assisted acceleration and cruising (e-boost and e-drive). The drive train concept presented demonstrates that CO_2-emissions of 90-120 g/km can be achieved for vehicles in the 1300-1400 kg ITW-class and engine out emissions will simultaneously be cut to approximately 40% of the Euro 4 limits /3/.

FUEL INJECTION AND ENGINE MANAGEMENT SYSTEM

As engine management system, Bosch´s EDC 16 was applied without any exhaust aftertreatment functionality. To enable simple application of heat combustion for both catalyst heating and cDPF regeneration but also simple switching between these different engine operating modes an ASCET™ (ETAS) bypass was implemented. This bypass enabled the storage of three additional set of engine maps as absolute calibration (not as offsets of the lean engine calibration) without any ambient corrections. Switching between these calibrations was enabled on a time basis, so that various lean/rich ratios could be investigated. Also time dependent switching to heat 1 calibration for catalyst light-off and afterwards to heat 2 as cDPF regeneration mode could be realized. Furthermore, both actuation of the external fuel injection and the fuel evaporator was implemented into the bypass controller. The bypass including the control of the external fuel injection device was programmed by AVL. The software to control the fuel evaporation unit was supported by BOSCH. The external fuel injection system and the fuel evaporator will be described later.

FUEL SPECIFICATIONS

To avoid any effect of aging due to sulfur, sulfur free Diesel fuel was used (sulfur content < 1ppm). More detailed fuel data can be found below:

Cetane Index	[-]	52.6
Density	[g/cm3 at 15°C]	0.822
Net Caloric Value	[MJ/kg]	43.1
H/C Ratio	[-]	1.9
Stoichiometric Ratio	[-]	14.65
Total Sulfur	[ppm]	< 1
Distillation Range		[°C]
	IBP	166
	10%	192
	20%	208
	30%	218
	50%	249
	70%	278
	80%	300
	90%	324
	FBP	354

EMISSIONS MEASUREMENT

As an emissions measurement system, two latest generation AVL CEB II with simultaneous measurement of NO and NO_2 were used. For CO two calibration levels were defined. When exceeding the lower measurement range the CEB switched automatically to the higher calibration range. The exhaust pipe was equipped with various measurement positions so that exhaust gas could be sampled up- and downstream of each exhaust aftertreatment component. The measurement range of the analyzers are listed below:

NO	ppm	0 - 4 000
NO_2	ppm	0 - 4 000
THC	ppm	0 - 50 000
O_2	%	0 – 21
CO – low	ppm	0 – 1 000
CO – high	ppm	0 – 50 000
CO_2	%	0 - 16

Table 1: Measurement range of the CEB II

EXHAUST GAS AFTERTREATMENT SYSTEMS

For passenger car applications, the soot filter is currently being widely introduced into series production, the preferred version being the cDPF, which does not require the use of any additional fuel additives. For NO_x reducing exhaust aftertreatment systems the LNT is currently seen as most feasible exhaust aftertreatment system for the near future. Significant improvements regarding packaging, costs and also calibration effort are seen with the combination of a soot filter with a LNT coating, a so called 4-Way-Catalyst (4-WC). Therefore possible combination of these three exhaust aftertreatment systems in combination with an oxidation catalyst have been investigated.

System 1

System 1 consists of a LNT and a catalyzed soot filter with the LNT located upstream of the cDPF.

System 2

With System 2 the LNT is located downstream of the cDPF. To enable soot regeneration utilizing exothermic reaction upstream of the cDPF, an oxidation catalyst was positioned ahead of the cDPF.

System 3

System 3 consists of the above mentioned 4-Way-Catalyst (4-WC) with an oxidation catalyst located upstream, again to improve soot regeneration.

The three investigated exhaust aftertreatment systems are shown in Fig. 1. Specifications of the exhaust aftertreatment components are given in Table 2. The exhaust aftertreatment catalysts were kindly provided by Johnson Matthey.

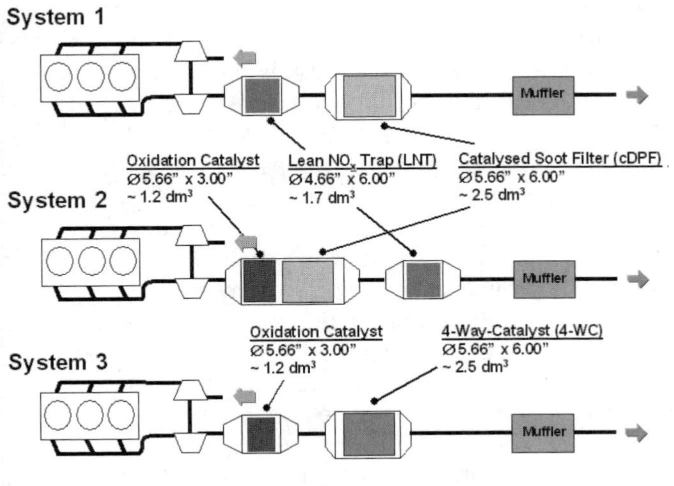

Figure 1: Layout of the exhaust aftertreatment systems

	Oxi-Cat	NO_x adsorber	CDPF	4-WC
Dimension	∅ 5,66" x 3"	∅ 4,66" x 6"	∅ 5,66" x 6"	∅ 5,66" x 6"
Volume	~ 1,2 dm^3	~ 1,7 dm^3	~ 2,5 dm^3	~ 2,5 dm^3
Cell Density	400 cpsi	400 cpsi	200 cpsi	300 cpsi
Wall Thickness	4 mil	4 mil	16 mil	12 mil
Substrate	Ceramic	Ceramic	SiC	Cordierite

Table 2: Specification of exhaust aftertreatment components

TEST CYCLES

For these investigations constant speed points were chosen as representative points of the NEDC to determine NO_x conversion and particulate matter filtration efficiency. Desulfurization capability was investigated at low and high load represented by 35 km/h and 100 km/h vehicle speed. Since active soot regeneration at low exhaust gas temperatures is an essential requirement, the representative point of 35 km/h, 3rd gear was chosen. The impact of lean/rich cycles on LNT, cDPF and 4-WC was investigated with a 10 hrs cycle. Within this cycle all five engine operating points were run for 24 min in the order as listed in Table 3 with the exception of the 32 km/h operating point which run as the last point to enable cooling down of the engine and exhaust aftertreatment system before completion of the test run. This cycle was repeated five times.

	Engine Speed	BMEP	Vehicle Velocity	Gear
1	1/min	bar	km/h	Number
1	1500	2.5	35/50	3/4
2	2100	1.8	32	2
3	1900	4.3	70	5
4	2400	6.5	100	5
5	2900	8.5	120	5

Table 3: Engine operating points

EXHAUST AFTERTREATMENT SYSTEM COMPARISON

NO_X CONVERSION

A series of baseline conversion tests with a lean/rich ratio of 60/2, 60/4, 100/2 and 100/4 s were performed. With a lean/rich ratio of 60/4 s, the highest conversion rates were achieved. Compared to expected lean phases during real vehicle operation the 60s lean phase is short. A rich duration of 3-5 s is realistic, enabling high conversion rates but also with the avoidance of hydrocarbon breakthrough during real vehicle operation. For the following investigations internal rich combustion was applied.

One of the most influential parameters for NO_x conversion is the temperature within either the LNT or 4-WC. To allow a better comparison of the different systems, the average temperature in the center of the LNT and 4-WC was taken from the minimum catalyst temperature during lean operation and the maximum catalyst temperature during or shortly after the rich peak. Fig. 2 shows that at both low load engine operating points comparable temperature within the LNT, and 4-WC in the range of 220-250°C were achieved for all three systems. At mid and high load it was found that the 4-WC has lower temperatures during lean/rich cycling compared to the LNT of System 1 and 2. Generally, it seems, that at mid and high load for both System 1 and 2 the temperature level inside the LNT is in the range of 400 – 450°C.

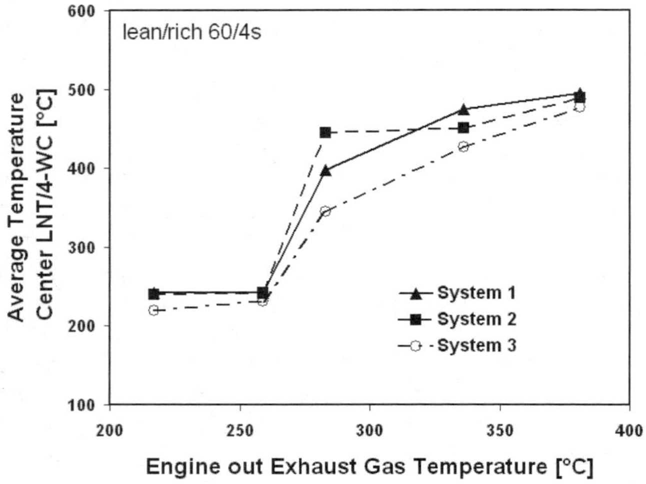

Figure 2: Catalyst temperature center LNT and 4-WC during lean/rich ratio 60/4s

According to the temperature distribution mentioned above NO_x conversion rates of more than 90% can be found in the range of 220-400°C catalyst temperature, Fig. 3. However, it has to be considered, that the "light-off" temperature (50% NO_x conversion) of the LNT is in the range of 220 – 250°C. Especially at 220°C, not all of the stored NO_x is released, resulting in a higher apparent NO_x conversion until a stabilized NO_x loading is obtained within the LNT.

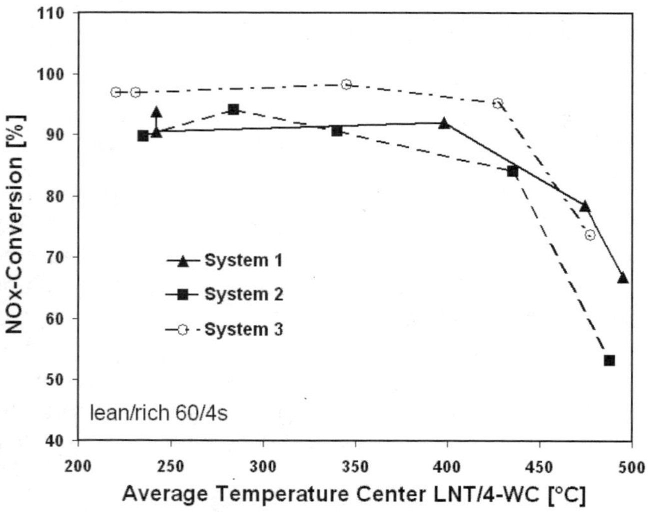

Figure 3: NO_x conversion of System 1, 2 and 3 versus average temperature center LNT/4-WC, lean/rich ratio 60/4s

Due to very low NO_x emissions at the representative low load engine operating points, the lean loading time up to this specific loading of the LNT is long. Therefore, a determination of a real NO_x conversion is rather difficult at very low load. Generally, it can be seen, that System 1 and 2 have comparable NO_x conversion rates in the temperature range of 220 – 400°C. System 3 shows some advantages regarding NO_x conversion in this temperature range.

In the temperature range above 400°C the NO_x conversion rate drops for all three investigated systems. One reason for this decrease is the engine out NO_x emissions, which are high compared to the mid and low load engine operating points so that during a loading time of 60s the LNT is quickly saturated with NO_x. Also at temperatures above around 450°C the nitrate becomes unstable, resulting in reduced NOx conversion. At this high temperature engine operating point, System 2 shows several disadvantages regarding NO_x conversion. This effect is based on oxidization of hydrocarbons and carbon monoxide inside the oxidation catalyst and the catalyzed soot filter resulting in fewer reducing agents remaining in the exhaust gas for NO_x conversion. This tendency can also be seen at lower temperatures starting at 340°C.

The NO_x conversion efficiency is comparable to the conversion efficiency as found in /5/, so that similar NO_x conversion rates derived from these investigations in the range of 55-60% in the NEDC can be expected for all three systems.

HC-, CO- AND O_2 CONVERSION

Besides NO_x conversion, also hydrocarbon, carbon monoxide conversion and oxygen levels have to be considered for a combined NO_x and PM reduction device especially during rich combustion. In Fig. 4-6 hydrocarbon, carbon monoxide and oxygen concentrations up- and downstream of each exhaust aftertreatment component for each system are shown at one engine operating point (1500 1/min; 2.5 bar BMEP). It can be seen that for all three systems, hydrocarbon emissions are significantly reduced after each exhaust aftertreatment component, but they are still at critical levels. The tailpipe hydrocarbon peak of System 1 during rich combustion is in the range of 5000 ppm. Slight advantages regarding peak hydrocarbons were found for System 2 and 3 with approximately 3500 ppm. Even though improvements regarding hydrocarbon breakthrough with optimization of rich combustion and lambda control are seen, hydrocarbon breakthrough has to be considered as extremely critical and remains as one of the major issues for further development.

Tailpipe carbon monoxide levels were not critical for all three systems, with peaks not higher than 35 ppm. However, in System 2 a significant carbon monoxide reduction of 75% by the oxidation catalyst and cDPF were found. The reduction of carbon monoxide, which is the preferred reducing agent for NO_x conversion, and also the reduction of hydrocarbons, result in reduced NO_x conversion at high temperatures as mentioned above.

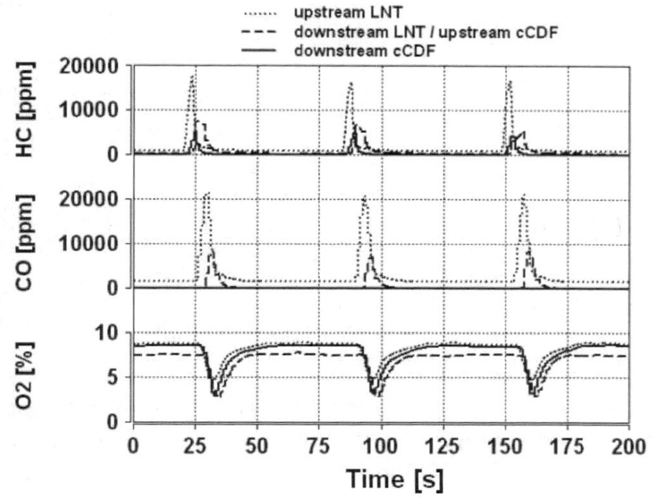

Figure 4: HC-, CO- and O_2-Conversion at 1500 1/min, 2.5 bar BMEP for System 1

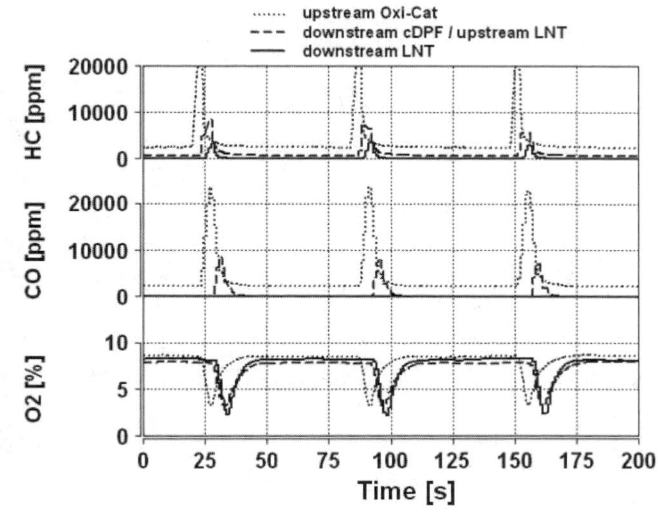

Figure 5: HC-, CO- and O_2-Conversion at 1500 1/min, 2.5 bar BMEP for System 2

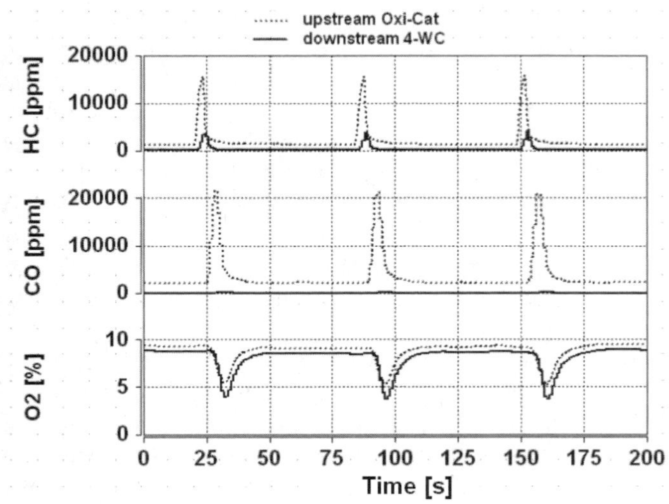

Figure 6: HC-, CO- and O_2-Conversion at 1500 1/min, 2.5 bar BMEP for System 3

For all three systems residual oxygen concentrations in the range of 2-3% have been found. In System 1 with the oxidation catalyst arranged upstream of the LNT and cDPF, the residual oxygen concentration is only reduced by approximately 1%, so that with oxidizing components upstream of the LNT, no optimum reducing atmosphere with no residual oxygen can be achieved.

DESULFURIZATION CAPABILITY

Even though low sulfur fuel is being introduced widely into the market, the sulfur content is still high enough to cause poisoning of the LNT coating. During lean phases sulfur is stored as SO_2 based on the same chemical processes as NO_2. However, the binding energy of SO_2 is higher compared to NO_2, which means, the temperature required to release SO_2 is higher. Therefore, a calibration or a calibration sequence has to be found to achieve not only a reducing atmosphere but also 600-650°C within and over the entire length of the catalyst.

System 1

In two representative engine operating points (1500 1/min, 2,5 bar BMEP; 2400 1/min, 6,5 bar BMEP) a sequence to enable boundary conditions for desulfurization could be found. If rich operation only is applied for a sufficiently long period to enable desulfurization, the catalyst will be overheated at the front end before the rear end reaches the required temperature level for desulfurization. Therefore, rich engine operation has to be interrupted by lean engine operation to enable heat transfer to the rear end. With the application of a lean/rich cycle of 1.5/2s at low load and engine speed (Fig. 7) and 5/2s at high load and engine speed (Fig. 8) a reducing atmosphere at a temperature level of 600-650°C could be found for the LNT when located upstream of the cDPF, Fig. 8. After heating up of the LNT no HC breakthrough could be detected.

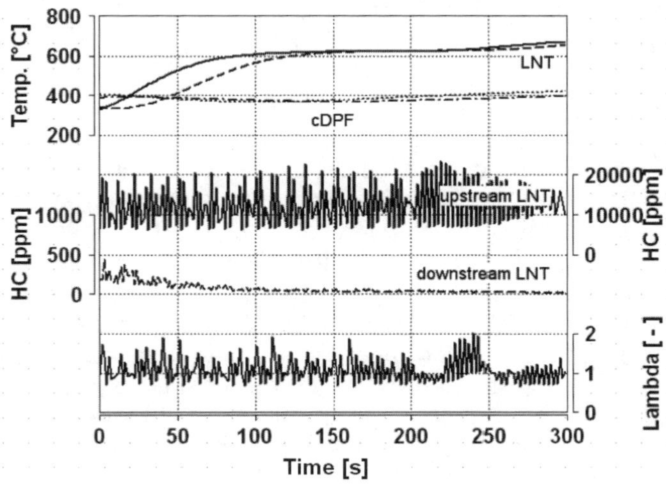

Figure 7: Temperature inside LNT and cDPF, HC emissions up- and downstream of LNT, Lambda at desulfurization sequence at 1500 1/min, 2.5 bar BMEP for System 1

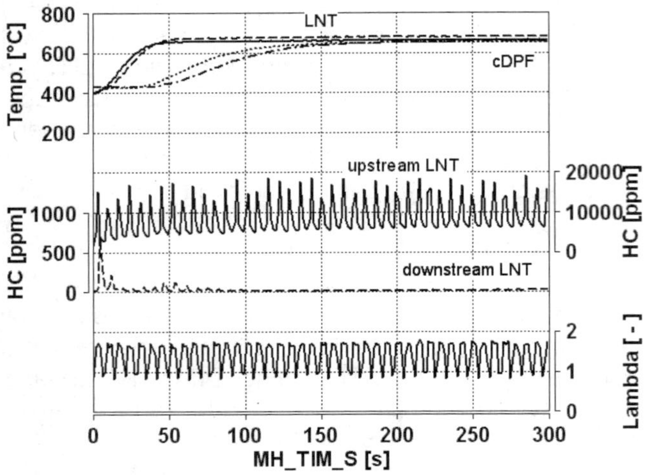

Figure 8: Temperature inside LNT and cDPF, HC emissions up- and downstream of LNT, Lambda during desulfurization sequence at 2400 1/min, 6.5 bar BMEP for System 1

Depending on fuel quality a desulfurization sequence will most probably be required in the same vehicle driving distance as for a cDPF regeneration, so that a desulfurization event can be combined with a cDPF regneration. With activation of a desulfurization sequence directly after a cDPF regeneration a specific heating phase for the desulfurization event and uncontrolled soot oxidation can be avoided and hydrocarbon breakthrough can be significantly reduced.

System 2

The exhaust gas aftertreatment system with oxidation catalyst and cDPF upstream of the LNT, has a significant thermal mass located upstream of the LNT. Therefore, desulfurization becomes very challenging. Some initial tests were performed to investigate the capability of desulfurizing the LNT, however no procedure without any risk of thermal damage to either oxidation catalyst, catalyzed soot filter or LNT could be found. Therefore, the desulfurization of the LNT downstream of the oxidation catalyst and cDPF has to be assessed as extremely critical.

System 3

Because of the small size of the oxidation catalyst upstream of the 4-WC it is possible to overrun the oxidation catalyst with hydrocarbons. The hydrocarbons can then be converted in the 4-WC, so that a high temperature level to enable desulfurization can be achieved directly inside the 4-WWC. A sequence again consisting of a rich and lean phase can easily be achieved for such an exhaust aftertreatment system. In this case the rich phase with a duration of two seconds is interrupted by a four seconds lean phase for both engine operating points. After achieving the desulfurization target temperature, no significant hydrocarbon breakthrough was found. For this system it is also recommended to perform a desulfurization sequence after a cDPF regeneration.

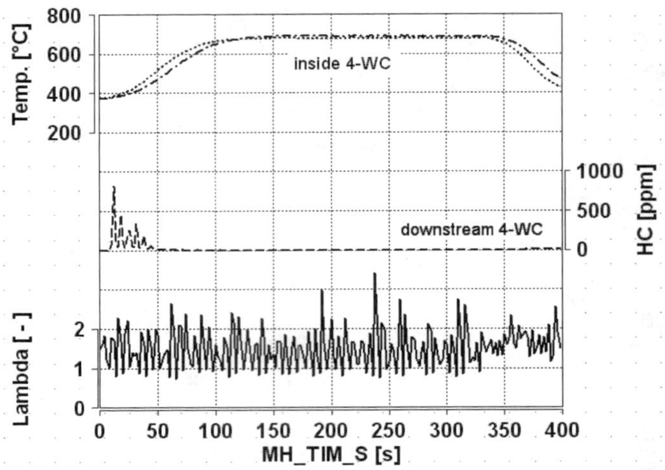

Figure 9: Temperature inside 4-WC, HC emissions downstream 4-WC, Lambda during desulfurization sequence at 1500 1/min, 2.5 bar BMEP for System 3

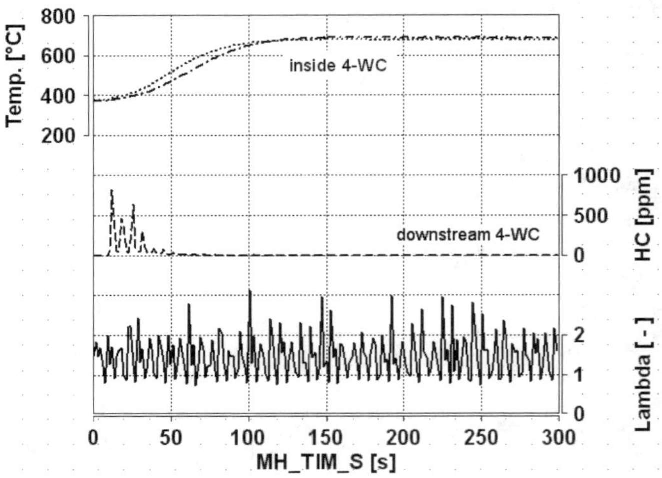

Figure 10: DeSO$_x$ Sequence at 2400 1/min, 6.5 bar BMEP: Temperature inside 4-WC, HC emissions downstream of the 4-WC, Lambda for System 3

SOOT REGENERATION CAPABILITY

During normal engine operation (speed and load) soot oxidation via NO_2 and thermal regeneration occurs partially, however, active soot regeneration has to be available as a "back-up regeneration" mode when a certain soot loading level is exceeded. This active thermal regeneration has to be guaranteed under all driving conditions, especially at low load and temperatures. Therefore, the cDPF regeneration capability was investigated at 1500 1/min and 2.5 bar BMEP. The target temperature inside the cDPF to enable active regeneration is approximately 600°C.

Within these investigations a regeneration strategy utilizing exothermic reaction upstream of the cDPF or 4-WC was applied. This strategy corresponds to a worst case exhaust aftertreatment architecture with an underfloor arrangement including significant heat loss from the exhaust pipe. With a close coupled arrangement, exothermic reaction can be reduced by applying maximum exhaust gas temperatures upstream of the turbine close to the durability limit of the turbine and exhaust manifold. However, a strategy utilizing significant exothermic reaction is still required at low load, where exhaust gas temperatures higher than 650 °C upstream of the turbine cannot be achieved.

System 1

System 1 is characterized by the rather high thermal inertia of the 1.7 l LNT upstream of the cDPF. Therefore, heating phase during active regeneration requires more energy compared to the optimum cDPF arrangement with the oxidation catalyst located upstream of the CDPF (System 2). For the LNT catalyst a long monolith with a length-diameter ratio of 1.39 compared to the oxidation catalyst of System 2 with a length-diameter ratio of 1.89 was used. Additionally, the volume of the LNT is 42% larger than the oxidation catalyst matching well with the engine and CDPF size. Therefore, comparably low space velocity within the LNT is achieved resulting in good hydrocarbon conversion. However, this results in an exothermic reaction which occurs mainly at the front end, of the substrate so that the heat during enforced cDPF regeneration needs to be transferred to the rear end of the LNT. During enforced cDPF regeneration the coating of the LNT has to be prevented from thermal damage.

The first step of the regeneration sequence is to heat up the catalyst above hydrocarbon light-off. This first increase of temperature level is achieved by stepwise shutting off EGR and throttling of intake air, retardation of main injection and application of a close and burning post injection. After hydrocarbon light off of the LNT is achieved, post injection is retarded so that hydrocarbons are converted within the LNT. After achieving sufficient hydrocarbon light off, in the second phase of cDPF regeneration the quantity of late post injection has to be increased to heat up the cDPF up to the regeneration temperature. However, the heat which is produced by an exothermic reaction at the front end of the LNT needs to be transferred to the rear end and to the cDPF. To avoid overheating of the LNT alternating of two post injection quantities is required. The phase with low post injection fuel quantity enables heat transfer to the rear end of the LNT and avoids cooling down of the front end. With an injection system capable of two post injections such a complex calibration can be simplified.

Fig. 11 shows such a regeneration sequence. The filter was loaded with 11.3 g/l soot in a 10 hrs loading cycle. This specific loading is rather high for a cDPF compared to final vehicle application. However, it can be seen, that such a regeneration can be successfully performed at low load.

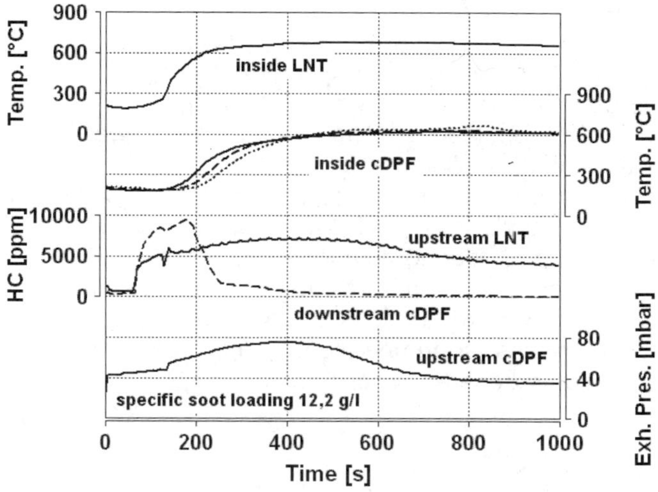

Figure 11: cDPF regeneration at 1500 1/min, 2.5 bar BMEP, specific soot loading 12,2 g/l: Temperature inside LNT and cDPF, HC emissions upstream of LNT and downstream of cDPF, exhaust gas pressure upstream of cDPF for System 1

System 2

System 2 represents an "optimum" oxidation catalyst and cDPF arrangement regarding soot regeneration. The cDPF can be regenerated by a simpler regeneration sequence similar to the procedure mentioned above but without alternating between different post injection fuel quantities.

Fig. 12 shows the regeneration of a cDPF with a typical soot loading of 5.8 g/l. The cDPF was loaded 10 hrs under lean conditions. It can be seen that the cDPF regeneration can be achieved without hydrocarbon breakthrough.

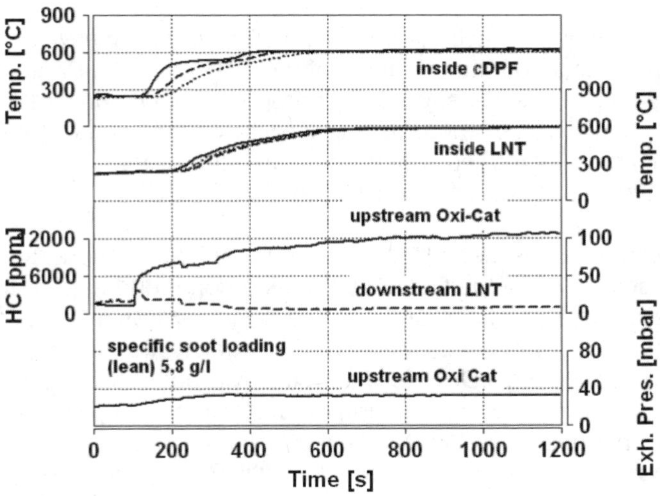

Figure 12: cDPF regeneration at 1500 1/min, 2.5 bar BMEP, specific soot loading (lean) 5.8 g/l: Temperature inside cDPF and LNT, HC emissions upstream oxidation catalyst and downstream LNT, exhaust gas pressure upstream cDPF for System 2

System 3

System 3 consisting of an oxidation catalyst and 4-WC represents again an "optimum" arrangement regarding cDPF regeneration. The 4-WC can be easily regenerated similar to System 2. However, during regeneration temperature peaks inside the 4-WC can still be found. In Fig. 13 a small and in this case uncritical peak during a regeneration of a 4-WC with a 7.4 g/l soot loading can be seen. This loading is relatively high when considered against the need to avoid excessive temperature conditions during un-controlled regeneration conditions

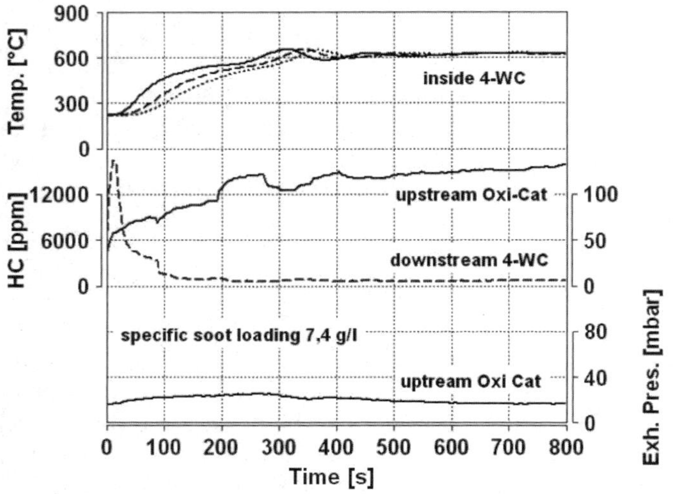

Figure 13: Soot regeneration at 1500 1/min, 2.5 bar BMEP, specific soot loading 7.4 g/l: Temperature inside 4-WC, HC emissions upstream oxidation catalyst and downstream 4-WC, exhaust gas pressure upstream oxidation catalyst for System 3

IMPACT OF LEAN/RICH COMBUSTION

The main interactions between cDPF and LNT are seen in particulate matter loading of the cDPF. Generally, soot loading does not effect NO_x storage and NO_x conversion of the LNT. Combustion sensitivity of rich combustion against increased back pressure are seen, but not investigated within this program.

Since NO_x is desorbed within the LNT at temperatures above ~ 450°C, a $DeNO_x$ purge is recommended prior to each cDPF regeneration. Since temperatures in excess of 600°C are used during cDPF regeneration, lean/rich cycles during cDPF regeneration will not result in any significant NO_x conversion and were not applied. As mentioned previously, the combination of a desulfurization sequence with a cDPF regeneration is recommended, since both sequences require high temperature. To avoid uncontrolled conditions, cDPF regeneration should occur prior the desulfurization. This is relevant for all three investigated exhaust aftertreatment systems.

To investigate the impact of lean/rich cycles on particulate matter loading of the cDPF, a ten hours loading cycle was applied as described previously. This cycle was run under lean and lean/rich (60s/4s) engine operating conditions. The results can be found in following Fig. 14 and will be described in the next paragraphs.

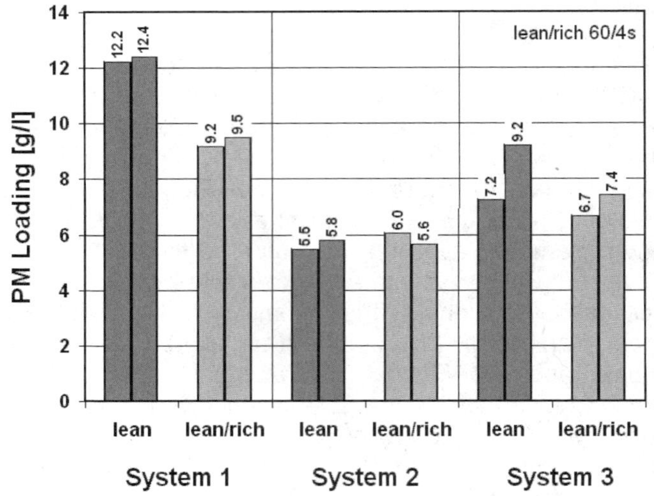

Figure 14: Filter loading after 10 hrs lean and lean/rich (60/4s) cycling

System 1

Lean loading of System 1 results in higher soot loading of approximately 12.3 g/l compared to the lean/rich loading with 9.3 g/l. In both lean and lean/rich loading cycles only limited NO_2 is available for soot oxidation. During lean loading, NO_2 is only available after saturation of the LNT or if it is desorbed at high temperature. At high load (2900 min^{-1}, 8.5 bar BMEP) part of the soot is oxidized with NO_2 or with oxygen. This oxidation is indicated by the decreasing pressure drop across the cDPF, Fig. 16.

Even though during rich combustion particulate emissions are significantly higher compared to lean engine operation, lean/rich cycles result in lower PM loading rates of the cDPF. This effect is based on the generally higher temperatures within the cDPF during lean/rich cycling resulting in improved continuous PM oxidation, Fig. 15. Additionally, lean/rich cycling results in a particulate matter composition containing more volatile organic components. These components are oxidized at lower temperatures resulting in local temperature peaks and hence local soot combustion. Reduced PM loading of the cDPF during lean/rich cycling may also be influenced by improved NO_2 oxidation due to the increased temperature. However, NO_2 is significantly reduced due to NO_x conversion of the LNT.

Pressure drop or pressure upstream of the CDPF is similar for both loading methods (lean and lean/rich) despite peaks during rich combustion.

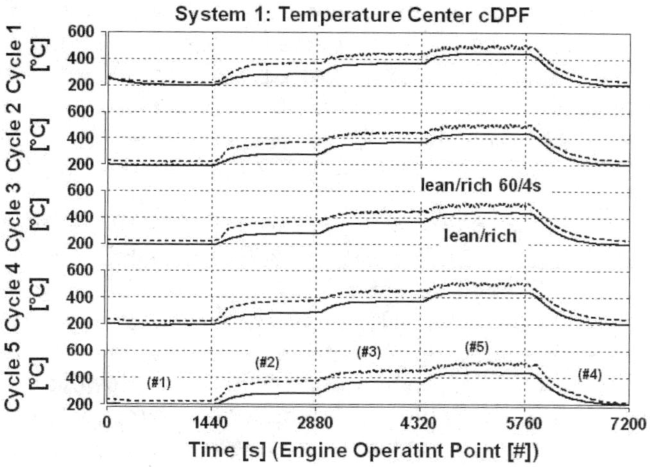

Figure 15: Temperature within cDPF during 10 hrs lean and lean/rich (60/4s) loading for System 1

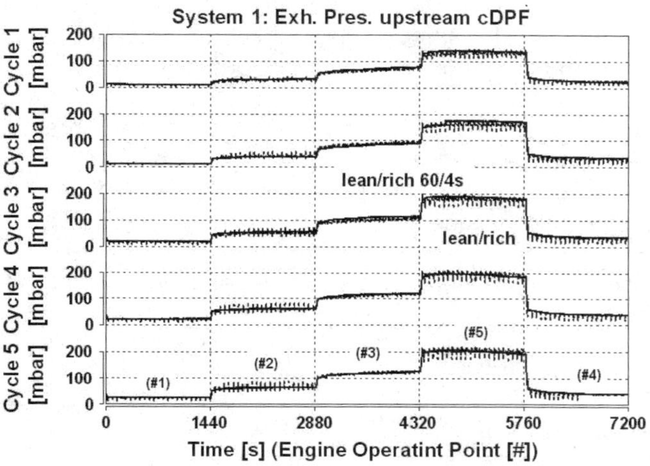

Figure 16: Exhaust gas pressure upstream cDPF during 10 hrs lean and lean/rich (60/4s) loading for System 1

System 2

In this configuration both loading methods result in the same soot loading of ~ 6 g/l. Compared to System 1 it is a reduction of approximately 50-75%. This reduction is based mainly on soot oxidation with NO_2, which is readily available and reduced later by the LNT located downstream of the cDPF. Furthermore, the oxidation catalyst converts most of the soluble organic fraction, so that within the cDPF more or less only dry particulates are stored. The latter is also the reason why lean/rich soot loading levels are the same compared to the original lean loading even though the lean/rich cycling produces higher particulate output levels. The good conversion performance of the oxidation catalyst explains why in this system the lowest accumulated soot mass in the cDPF was found. The temperature within the cDPF is in a similar range to System 1, Fig. 17.

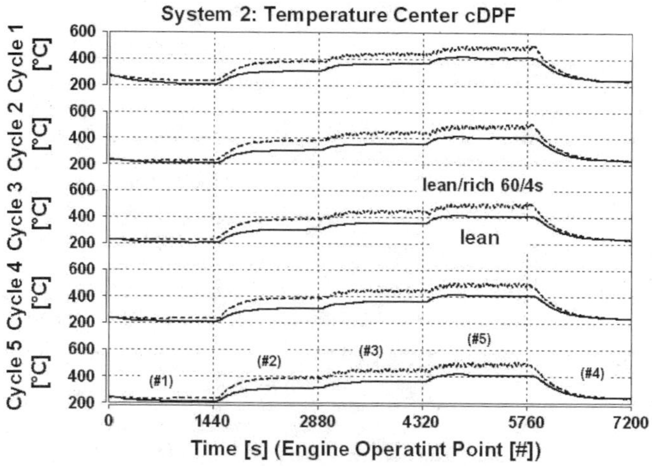

Figure 17: Temperature within cDPF during 10hrs lean and lean/rich (60/4s) loading for System 2

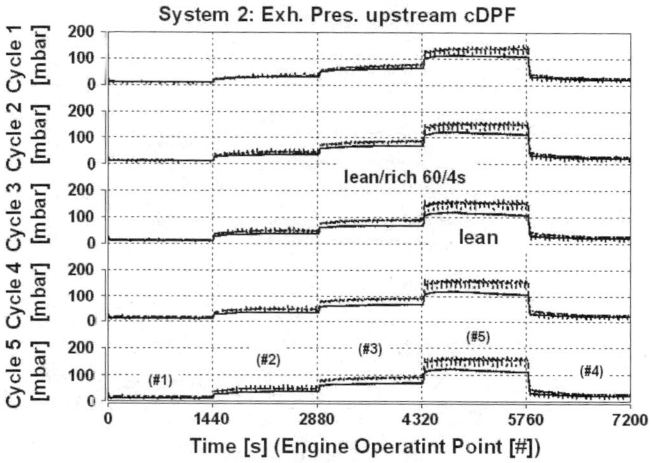

Figure 18: Exhaust gas pressure upstream cDPF during 10hrs lean and lean/rich (60/4s) loading for System 2

System 3

The ten hours loading cycle results in a comparable loading of the 4-WC for both lean and lean/rich loading. Again low loading is mainly achieved by the conversion of the volatile organic components within the oxidation catalyst. Compared to System 2 the loading level is slightly higher, since only limited NO_2 is available for soot oxidation. NO_2 during lean loading is stored directly in the 4-WC and only partly released at the high load operating point or reduced during lean/rich cycling inside the 4-WC. Since both NO_x storage and reduction, and soot oxidation via NO_2 happen directly inside the 4-Way Catalyst, soot oxidation via NO_2 is better compared to System 1. However, compared to System 2 the amount of available NO_2 is reduced either due to NO_x storage or reduction.

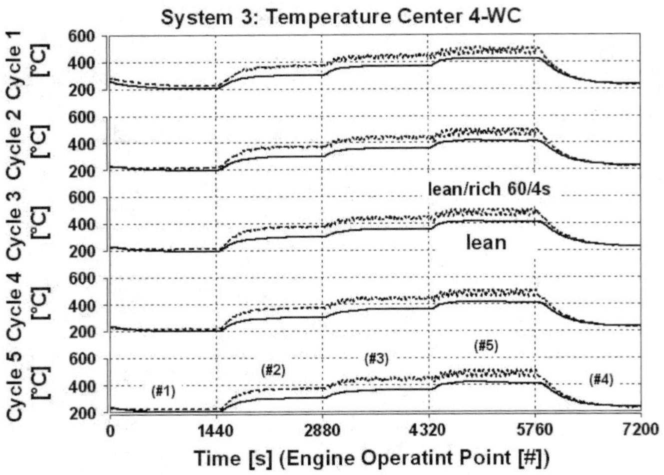

Figure 19: Temperature within 4-WC during 10hrs lean and lean/rich (60/4s) loading for System 3

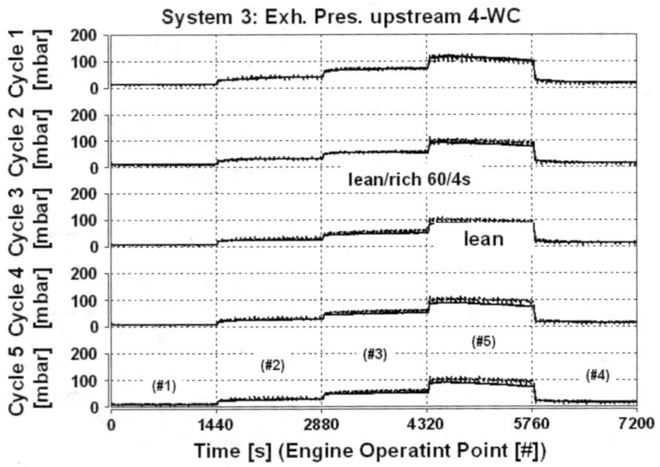

Figure 20: Exhaust gas pressure upstream 4-WC during 10hrs lean and lean/rich (60/4s) loading for System 3

SYSTEM COMPARISON

All three systems show similar NO_x conversion rates under steady state conditions. However, in System 2 with the oxidation catalyst and cDPF upstream, there is a large thermal mass positioned upstream of the LNT. During transient engine operation, long and gradual heating phases have to be considered so that high NO_x conversion rates can be maintained under high load conditions and also the prevention of hydrocarbon breakthrough. Therefore, System 2 is more applicable for small engines in heavier vehicles such as small delivery trucks, heavy passenger cars and SUVs. System 1 and 3 enable a closed coupled arrangement of LNT and 4-WC, respectively, so that they can be applied in low temperature applications with large engines in lighter vehicles, such as luxury passenger cars. With adjustment of the pipe length between the engine and catalyst and pipe insulation, such a system could be also utilized for high temperature applications. However, it has to be taken into consideration, that cDPF regeneration, especially at low load and hence low temperature, has to remain feasible. System 1 applied as underfloor arrangement requires an exothermic hydrocarbon conversion in the LNT positioned directly in front of the cDPF. This strategy requires a very high calibration and control effort for cDPF regeneration. A LNT close coupled to the engine further increases the need to control the underfloor cDPF regeneration.

A remaining open issue for all three systems is the avoidance of hydrocarbon breakthrough. Whereas System 1 has a slight benefit with the cDPF located downstream of the LNT being able to contribute to HC conversion, System 2 and 3 in comparison show improved cDPF regeneration performance. However, for Systems 2 and 3, the maximum permissible soot loading has to be either limited to a lower value compared to System 1, or alternatively the soot oxidation process needs to be controlled to avoid thermal damage of the LNT coating. A reduced maximum soot loading results in shorter cDPF regeneration intervals. However, improved PM oxidation was found for both Systems 2 and 3 when tested over the same cycle, partly compensating for the lower maximum permissible soot loading level.

Regarding desulfurization, System 3 showed the best performance followed by System 1. For System 2 it appears at this stage extremely difficult to find a desulfurization sequence.

The system evaluation is summarized in following table:

	System 1	System 2	System 3
NO_x-Conversion Rate	comparable range for all systems		
HC breakthrough	moderate (still critical)	high	high
Suitability for low temperature application	high	low	high
Desulfurization capability	moderate	low	high
Soot regeneration capability	good	excellent	excellent
Risk of thermal damage of NO_x adsorber during PM Regeneration	low	high	high
Heat calibration effort (PM regeneration)	high	moderate	moderate
Continuous PM oxidation within DPF	low	high	moderate
PM regeneration interval	low	high	moderate

Table 4: System Comparison

From the results shown in the table above, System 3 is the preferred choice. It shows significant benefits regarding cDPF regeneration and desulfurization. The main areas requiring further development for this technology are the avoidance of hydrocarbon breakthrough and prevention of thermal damage of the LNT coating improving system durability.

EXTERNAL FUEL ENRICHMENT

Current HSDI Diesel combustion systems are very well known for their extremely stable combustion under lean engine operation. This combustion is characterized by a leaner than stoichiometric atmosphere in the combustion chamber. Rich combustion can be achieved by two different methods. The first is the reduction of air mass flow and the second, the increase of fuel quantity. Since both of these methods result in deteriorated torque and increased black smoke output, they have to be combined together to achieve a minimum impact on torque and emissions.

To apply internal rich combustion depending on engine load two approaches can be applied. At low load air mass flow reduction via intake throttling is applied. However, this alone would result in reduced cylinder filling and too low compression thus preventing correct combustion to occur. It is therefore necessary to apply EGR to enable complete filling of the cylinder and the ignition point of the fuel. However, such engine operation with EGR rates in the range of 60% and simultaneous combustion retardation result in very unstable engine operation. Since such a calibration is very sensitive to engine aging and contamination and also changes in ambient conditions and production tolerances, new lambda and torque control structures to help compensate are required.

At higher loads, this approach is limited by increasing black smoke and exhaust gas temperatures. Above this level a second approach to achieving rich combustion is employed by the use of intake throttling, but without the application of EGR, down to a limit with acceptable black smoke (Lambda ~ 1.3). Then a late post injection is applied. Depending on the post injection strategy utilized, new lambda and torque control strategies may also be required. Compared to the first strategy, the high load rich combustion strategy is more stable.

Very challenging in general, is the transient behavior when changing from standard lean to rich combustion but also the transient engine operation under rich conditions and the transition between the two rich combustion strategies. One significant problem is oil dilution during rich combustion, especially with application of late post injection. Therefore, internal rich combustion has to be categorized as extremely critical for any LNT application.

EXHAUST PORT FUEL INJECTION

Exhaust port fuel injection was investigated as a method for providing external fuel enrichment. It was decided to inject fuel directly into the exhaust port to make use of the hot exhaust gas to help evaporation of the Diesel fuel, but also to utilize the turbine to homogenize the mixture. A prototype system was installed consisting of a state of the art GDI (Gasoline Direct Injection) injector with a water cooled jacket in the exhaust manifold, opposite the EGR valve, Fig. 21. A GDI injector has a spray pattern which allows fuel injection into the exhaust port either in the direction of the exhaust gas flow (towards EGR valve and turbine) or in the opposite direction (towards exhaust valve). In this study both injection strategies were prepared, however, only the first one was investigated. A remote fuel pump provided the fuel supply. The fuel pressure could be adjusted up to a maximum of 50 bar via a pressure regulation valve. However, it was found that rich combustion generated by exhaust port fuel injection is insensitive to injection pressure in the injector, so a minimum fuel pressure of 10 bar enabling a good spray pattern was applied. Low fuel pressure has the added benefit of reducing wall wetting of the exhaust port. The exhaust port fuel injector was controlled via an control algorithm applied in ASCET™ together with a separate power supply. Both control algorithm and power supply were developed by AVL. The control algorithm is time based but allowed almost cycle to cycle activation of exhaust port fuel injection.

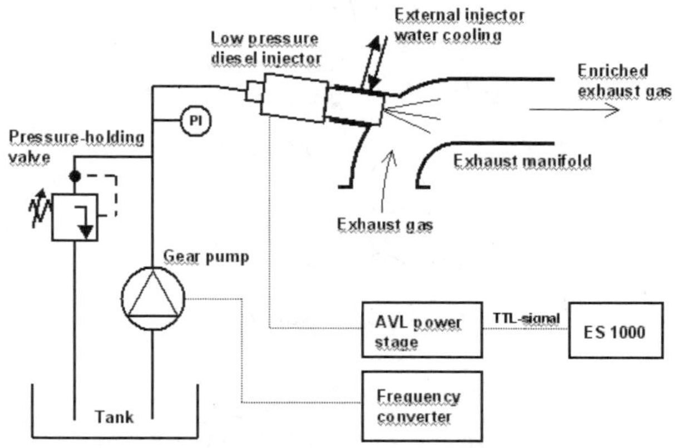

Figure 21: Set-up of the exhaust port fuel injection system

FUEL ENRICHMENT WITH EXHAUST PORT FUEL INJECTION

One pre-requisite for fuel enrichment to be used in combination with external fuel injection is the reduction of the air mass flow, thus reducing the overall fuel consumption required to achieve rich combustion. Intake throttling is therefore applied at high loads and high EGR rates at low load conditions. As mentioned previously, with state of the art combustion systems at low load and high EGR rates, engine operation can become unstable. Therefore, internal rich combustion requires increased pilot injection quantities to stabilize the combustion event. However, with exhaust port fuel injection a significant mass of unburned hydrocarbon is recirculated into the combustion chamber. With application of a large quantity of pilot injection this fuel will be ignited resulting in increased engine load and combustion noise. Therefore, for the application of the exhaust port fuel injection at low load a combustion adjustment using a strategy similar to alternative combustion systems such as HPLI/HCLI (Highly Premixed Late Injection/Homogeneous Charge Late Injection) /4/ was performed. These strategies are characterized by application of one injection only (w/o pilot injection) in combination with high EGR rates.

Such an adjusted combustion, Fig. 22 and Table 5, applied at 1500 1/min, 2,5 bar BMEP results in a significant increase in the carbon monoxide emissions (in this case the CO concentration was beyond the analyzers' range of 5 %), while hydrocarbon was still on a low lever. However, combustion was still leaner than stoichiometric. With activation of the exhaust port fuel injection, engine load increased significantly but this could easily be compensated for by reduction of the main fuel injection quantity. The exhaust port fuel injection also led to a reduction in the high carbon monoxide levels but also a significant increase in hydrocarbon concentration, Table 5. It can also be seen that the adjusted com-bustion with and without exhaust port fuel injection shows increased filter smoke numbers compared with lean combustion. With complete application of a HCLI/HPLI strategy a significant reduction of filter smoke number is expected.

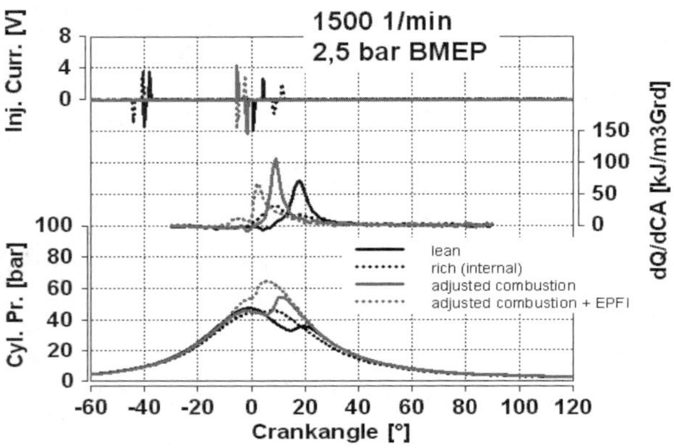

Figure 22: Internal rich combustion and adjusted combustion without and with exhaust port fuel injection at 1500 1/min, 2.5 bar BMEP

1500 1/min 2,5 bar BMEP		lean	rich	adj. com-bustion	adj. com-bustion + EPFI
Air Mass Flow	kg/h	29	18	23	23
EGR	%	36	58	51	54
Throttle	%	5	75	69	70
Lambda	-	1.78	0.71	1.39	0.92
HC	ppm	1072	30669	1494	33935
CO	ppm	2704	24487	50000	39239
O2	%	8.8	3.5	5.7	6.3
FSN	-	0.78	3.7	1.71	2.55

Table 5: Rich combustion results at 1500 1/min, 2.5 bar BMEP

At higher load, where the application of alternative combustion systems is not possible, air mass flow is initially reduced by intake air throttling without EGR up to an acceptable filter smoke number. Afterwards exhaust port injection is activated. Compared with internal rich combustion, a significantly higher amount of hydrocarbons are introduced into the LNT and CO is reduced, however, the order of magnitude is the same compared with internal rich combustion, Fig. 23 and Table 6.

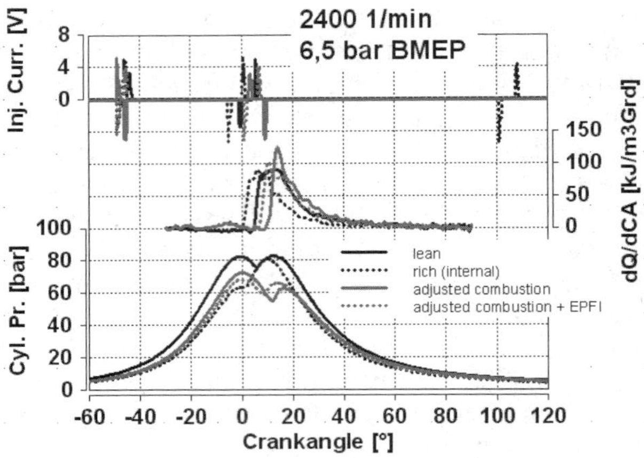

Figure 23: Internal rich combustion and adjusted combustion without and with exhaust port fuel injection at 2400 1/min, 6.5 bar BMEP

2400 1/min 6,5 bar BMEP		lean	rich	adj. com-bustion	adj. com-bustion + EPFI
Air Mass Flow	kg/h	106	90	105	98
EGR	%	16	-	-	-
Throttle	%	5	75	75	75
Lambda	-	1.83	0.92	1.32	0.95
HC	ppm	1901	23062	1093	50000
CO	ppm	451	2539	880	1923
O2	%	9.3	7.9	8.4	8.0
FSN	-	3.4	2.23	2.45	2.74

Table 6: Rich combustion results at 2400 1/min, 6.5 bar BMEP

An improvement in oil dilution is expected when exhaust port fuel injection is compared with internal rich combustion. Since the combustion strategy is similar to alternative combustion and fuel injected directly into the exhaust port does not impinge on the cylinder liner, oil dilution is expected to be in the same range as that achieved with alternative combustion systems. For these alternative combustion systems, no increase in oil dilution levels were found when compared with state of the art combustion systems.

FUEL EVAPORATION

Due to the low temperature of the exhaust gas directly upstream of the exhaust aftertreatment device evaporation of fuel injected at this point is very limited. As an alternative to fuel injection directly upstream of the exhaust aftertreatment device, BOSCH´s fuel evaporation unit was investigated. This evaporator, see Figure 24, consists of an evaporation chamber into which fuel is injected using a state-of-the-art intake port injector. Two glow plugs are heating up the fuel to approximately 600°C, while a third glow plug controls the fuel temperature to approximately 800°C to enable complete evaporation. The evaporated fuel flows out of a small hole at the end of a small pipe which is mounted in the exhaust pipe. To enable homogenous fuel distribution upstream of the exhaust aftertreatment device, the fuel evaporator should be installed 300-500 mm upstream of the exhaust aftertreatment device. A vertical mounting position (+/- 45°) is required to avoid liquid fuel flowing out of the evaporation chamber. The evaporation unit was controlled by an ASCET™ model supplied by BOSCH and implemented in AVL´s control algorithm implemented in ASCET™. The fuel injector of the evaporation unit demands a fuel pressure of 3 bar which was provided by a remote fuel pump.

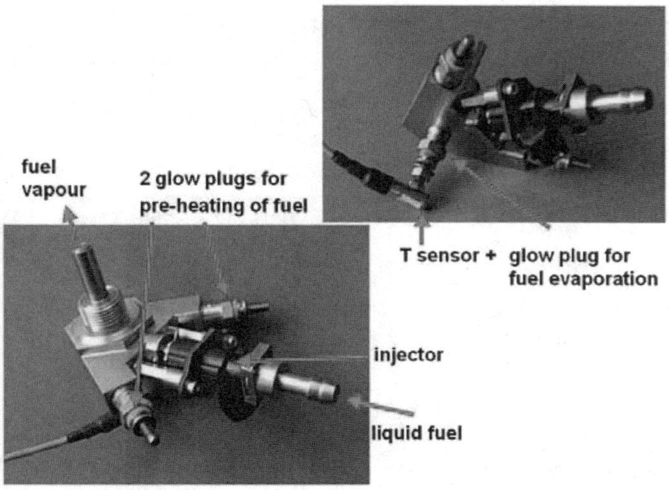

Figure 24: BOSCH fuel evaporation unit

Fuel enrichment with fuel evaporation requires a significant reduction of air mass flow to minimize fuel consumption similar to all other fuel enrichment strategies. This air mass flow reduction was achieved by similar combustion adjustment compared to the exhaust port fuel injection. Activation of fuel evaporation upstream of the exhaust aftertreatment device and hence downstream of the turbine does not effect combustion. Consequently, torque adjustment is not required.

A typical duration of the rich phase for a full stream LNT application is in the range of 3-5s. Prior to injection of the fuel into the evaporation unit, the fuel and chamber have to be heated up to 600 and 800°C, respectively. The quantity of evaporated fuel generated is dependent upon the amount of fuel injected and the heat which is supplied to the fuel. Therefore, transient response of the evaporation unit is slow compared with any internal or external fuel injection system . However, since the rate of change of exhaust mass flow under transient engine operation is also slow, the response time of the fuel evaporation unit is expected to be sufficient.

One significant disadvantage of such a device is the high hydrocarbon and oxygen content of the exhaust gas during fuel enrichment which requires an oxidation catalyst upstream of the LNT. Due to the high hydrocarbon content lambda during regeneration needs to be controlled by the use of a sensor positioned downstream of the oxidation catalyst. The control algorithm needs to compensate for the slow response of the oxidation catalyst whilst also preventing thermal overheating (catalyst aging) by adjusting the required lambda values. Generally, the exhaust gas temperature downstream of the fuel evaporator is low, often below the catalyst light-off temperature resulting in limited hydrocarbon conversion during short rich peaks. After prolonged use of the fuel evaporator unit, significant amounts of condensed Diesel fuel was found on the inner walls of the exhaust pipe upstream of the oxidation catalyst and LNT.

NO_X, HC, CO AND O_2 CONVERSION

With both exhaust port fuel injection and the fuel evaporation unit, NO_x conversion rates were investigated at three engine operating points representing low and mid range engine loads using a lean/rich cycle of 60/4s. A combination of oxidation catalyst positioned upstream of the LNT was selected as the exhaust aftertreatment system for this evaluation.

Fig. 25 shows that with both systems at low load, increased temperature inside the LNT was found compared with internal rich combustion. This temperature increase is the result of the high hydrocarbon content combined with a low space velocity

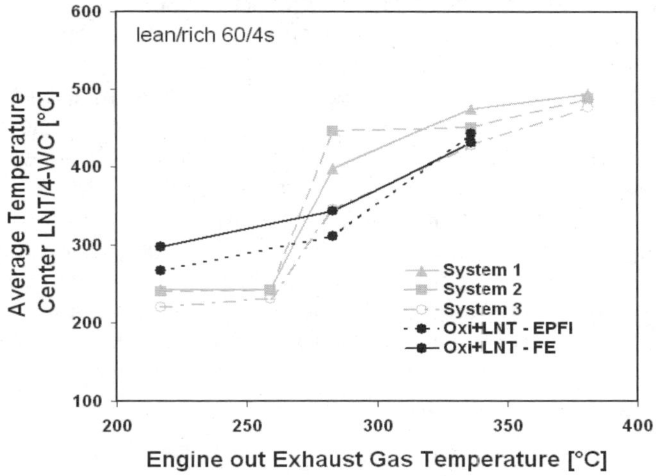

Figure 25: Temperature Center LNT with Exhaust Port Fuel Injection (EPFI) and Fuel Evaporation (FE)

within the substrate, resulting in exothermic hydrocarbon conversion but also high NO_x conversion rates, Fig. 26. The temperature level inside the LNT at mid load indicates that with increased load, the effect of decreasing HC and NO_x conversion rates due to increased space velocity dominates the effect on conversion rate from the higher reaction kinetics due to the increased temperature. At this mid load engine operating point, decreased NO_x conversion has been found to be comparable internal rich combustion. At the high load engine operating point, similar temperatures in the center of the LNT compared with internal rich combustion were found. Obviously, with the higher temperature inside the LNT, the resulting improved conversion performance prevails over the increased space velocity, allowing similar NO_x conversion rates to exhaust port fuel injection and fuel evaporation to be achieved. Generally, it can be seen that with the fuel evaporation unit higher NO_x conversion rates compared with exhaust port fuel injection can be achieved.

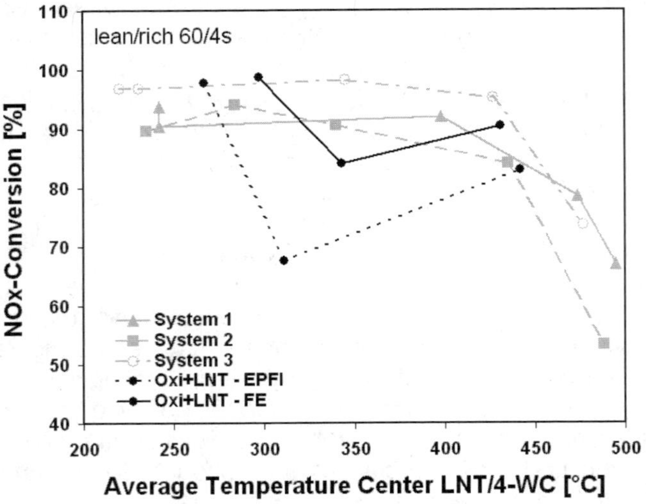

Figure 26: NO_x conversion rates with Exhaust Port Fuel Injection (EPFI) and Fuel Evaporation (FE) with Oxidation Catalyst and LNT

Both exhaust port fuel injection and the fuel evaporation unit are characterized by high hydrocarbon emissions. With exhaust port fuel injection, most of these hydrocarbon are oxidized in the oxidation catalyst and LNT, so that a tailpipe hydrocarbon level in the range of 1000 ppm was found, Fig. 27. The fuel evaporator generally delivers higher hydrocarbon tailpipe emissions in the range of 5000 – 8000 ppm compared with exhaust port fuel injection, Fig. 29. As one exception in favour of the fuel evaporation unit, was the low temperature engine operating point 1500 1/min, 2.5 bar. During this test series a comparably low tailpipe hydrocarbon content in the range of 1000 ppm was detected, see Fig. 28. This improvement in hydrocarbon conversion is also indicated by increased catalyst temperature at this engine operating point.

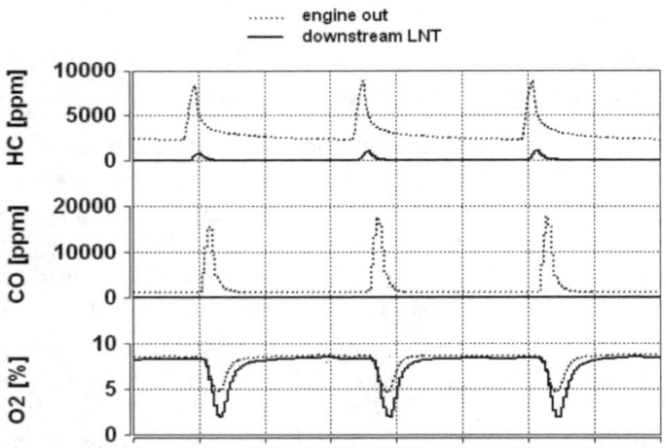

Figure 27: Exhaust Port Fuel Injection: HC-, CO- and O_2-Conversion at 1500 1/min, 2,5 bar BMEP

Tailpipe carbon monoxide emissions of less than 30 ppm were measured for both exhaust port fuel injection and fuel evaporation, which is similar to that of internal rich combustion. Residual oxygen concentrations in the range of 2-3% were also found for both alternative fuel enrichment strategies.

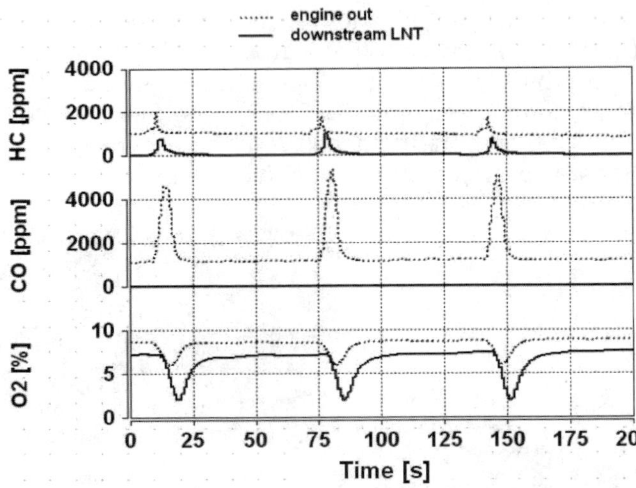

Figure 28: Fuel Evaporation: HC-, CO- and O_2-Conversion at 1500 1/min, 2,5 bar BMEP

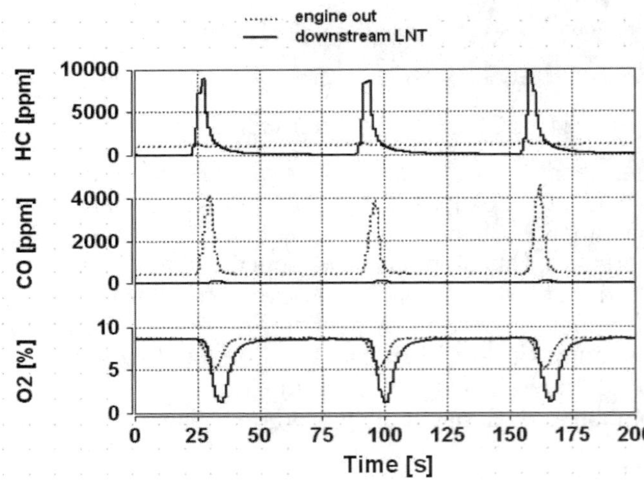

Figure 29: Fuel Evaporation: HC-, CO- and O_2-Conversion at 1900 1/min, 4,3 bar BMEP

SYSTEM COMPARISON

A system comparison of the three fuel enrichment strategies investigated is summarized in following Table 7. The main advantages of rich combustion are higher NO_x conversion rates, suitability for low temperature application, moderate but still critical hydrocarbon breakthroughs and no additional hardware requirements, hence suitability for easy implementation into series production. Disadvantages are a high calibration effort, high oil dilution values, limited transient engine operation under rich conditions and very high production and torque sensitivity.

The external fuel enrichment strategies investigated here are characterized by a low calibration effort and low or even no oil dilution problems. Both systems are relatively insensitive to production tolerances. Torque impact of exhaust port fuel injection is low and can easily be compensated for, and fuel evaporation directly upstream of the exhaust aftertreatment device has zero torque impact. An advantage of exhaust port fuel injection is the very high transient operation capability. In contrast to all the alternative fuel enrichment strategies described here, fuel evaporation directly upstream of the exhaust aftertreatment device has the capability of being used as an add-on or retrofit system. However, the dis-advantages of the fuel evaporator are the very high tailpipe hydrocarbon emissions during fuel enrichment and condensation of Diesel fuel inside the exhaust pipe. Both external fuel enrichment systems require additional hardware and separate fuel supplies, with a fuel pressure in the range of 10 bar for the exhaust port fuel injection potentially being obtained from the common rail pre-supply pump and the 3 bar required for the fuel evaporator obtained from the vehicle fuel supply pump.

	Internal Rich Combustion	Exhaust Port Fuel Injection	Fuel Evaporation
Calibration Effort	high	low	low
Oil Dilution	high	low	low
Suitability for low temperature application	high	high	low
NO_x Conversion	high	moderate	low
Transient Capability	moderate	high	moderate
HC breakthrough	moderate	moderate	high
Sensitivity Against Production Tolerances	high	low	low
Torque Sensitivity	high	moderate	low
Requirement of additional hardware	no	• mid pressure fuel pump (10-30 bar) • Exhaust Port Injector	•(low pressure fuel pump, ~ 3 bar) •Fuel Evaporator
Additional FIE control algorithms	• rich calibration • Lambda control • switch function lean/rich	• rich calibration • Lambda control • transient function lean/rich	•rich calibration •Lambda control
„Add on" capability	no	no	yes

Table 7: Fuel Enrichment System Comparison

For all three fuel enrichment strategies, additional software and control algorithms are required such as additional storage of the maps for rich calibration including ambient correction, control strategies for transients from lean to rich operation and also lambda control.

CONCLUSION

A system comparison of three combined NO_x and PM aftertreatment systems utilizing three different fuel enrichment strategies was performed.

As exhaust aftertreatment system, System 3, the 4-Way-Catalyst is recommended due to its excellent soot regeneration and desulfurization performance. Further-more, this system shows benefits regarding PM loading of the cDPF during lean/rich operation, so that long intervals between regenerations and hence low fuel consumption can be achieved. However this system requires close control of the soot regeneration to avoid excessive temperatures which could lead to catalyst deactivation.

As an alternative to the 4-Way-Catalyst, System 2 (LNT upstream of the cDPF), has to be considered. Risk of thermal damage of this exhaust aftertreatment system is lower compared with the 4-Way-Catalyst and a close coupled arrangement of the LNT is possible. However, disadvantages regarding packaging and high calibration effort would have to be taken into consideration.

As a fuel enrichment strategy for state of the art combustion systems, internal rich combustion has to be considered even with the restricted transient engine operation and the disadvantages regarding oil dilution, and high calibration and control effort. With initial baseline tests of external fuel enrichment systems it was found that the combination of alternative combustion systems such as HCLI/HPLI together with exhaust port fuel injection offer a large potential for an LNT application. With further development, the lack of reduced NO_x conversion rates may be solved.

A common problem of all exhaust aftertreatment systems and fuel enrichment strategies is the unacceptable hydrocarbon breakthrough. Avoidance of these high tailpipe emissions will be one major topic for future LNT and fuel enrichment development. One strategy to overcome this problem may be the application of a dual path system. Such a system would allow lower space velocities, a key factor to significantly increasing HC and NO_2 conversion.

ACKNOWLEDGEMENTS

The authors would like to thank Johnson Matthey for their provision of advanced catalyst and cDPF samples, especially Dr. Paul Phillips and Dr. Phil Blakeman, and BOSCH for the utilization of the fuel evaporator system.

REFERENCES

/1/ Schmidt's Diesel Car Prospects to 2007

/2/ Moser F., et. al., "Development Directions for Passenger Cars and Commercial Vehicle Engines", SIAR 2003, Pune, India

/3/ Ellinger R., et. al., "CO2 Reduction Potential of an Advanced Diesel Combustion Engine in an Advanced Passenger Car Powertrain", EAEC 2003, Paris

/4/ Weissbäck M., et. al., "Alternative Combustion - An Approach for Future HSDI Diesel Engines" MTZ 9/2003

/5/ Pfahl U., et al., "Passenger Car Investigations of NO_x Adsorber and DPF Combination to Fulfill Future Diesel Emission Limits", SAE 2003-01-0043

/6/ Tanaka, T., "Simultaneous Reduction of PM and NO_x – A New Aftertreatment System –" 22nd International Vienna Motor Symposium 2001

/7/ Fujimura T., et. al. "Development towards serial production of a Diesel passenger car with simultaneous reduction system of NO_x and PM for the European Market", 23rd International Vienna Motor Symposium 2002

/8/ McDonald J., Bunker B., "Testing of the Toyota Avensis DPNR at U.S. EPA-NVFEL", SAE 2002-01-2877

/9/ Cartus T., et. al., "The clean and efficient HSDI-diesel using NO_x-adsorber technology", 21st International Vienna Motor Symposium 2000

/10/ Harndorf H., et. al., "Partikelfilter- und NO_x-Speichertechnologie als EURO IV-Option für Dieselmotoren", Forum Partikel Emissionen 2002, Nürnberg

/11/ Sluder, C. S., West B. H., "Effects of Regeneration Conditions NO_x Adsorber Performance", SAE 2002-01-2876

CONTACT

Holger Herrmuth
Exhaust Aftertreatment Systems
Powertrain Engineering

holger.herrmuth@avl.com

AVL LIST GMBH
Hans-List-Platz 1
8020 Graz
Austria

2006-01-0637

DPF SCR Combinations Integrated Systems to Meet Future LDV Emission Limits

Klaus Rusch, Rolf Kaiser and Stefan Hackenberg
ArvinMeritor Light Vehicle Systems

ABSTRACT

This paper describes solutions to handle the complexity of systems including Selective Catalytic Reduction (SCR) and Diesel Particulate Filters (DPF) and possible combinations with additional components. Results of integrated systems with different configurations are also discussed.

A lot of development work was undertaken to develop validated and reliable mathematical models of single components as well as for the complete system. Commencing with DPF soot loading and regeneration models, additional tools are presented to define the best size and loading of oxidation catalysts for enhancing the system performance by generating NO2, as well as models to optimize the size and the dosing logic for the urea SCR.

INTRODUCTION

SCR technology has become the technology of choice for European truck manufacturers to meet future emission limits regarding NO_x. Beginning with EU IV, SCR was introduced on heavy duty trucks into the European market in 2004. In Japan the first serial applications are currently in use and by 2006 it will be the dominant exhaust aftertreatment technology for heavy vehicles.

The worldwide legislations for light vehicles and passenger cars will not only first demand a reduction of Particulate Matter (PM) but also will require a strong reduction of NO_x emissions. Even with next generation of diesel engines the emission control will be a necessity especially at high load and high speed conditions. Fuel economy and CO_2 emission targets will demand that the Diesel engine's efficiency is kept as high as possible. Due to the well known trade-off between NO_x emissions on the one side and fuel consumption on the other side, there is a request for a highly efficient NOx control.

As SCR is being widely introduced on heavy vehicles as a highly efficient NOx control, the technology gets more and more interesting also for light vehicles and passenger cars.

EMISSION TARGETS

Europe, the US and Japan are the leading markets with more and more stringent emission legislation, but China will follow after a short time delay. Limits will enforce exhaust aftertreatment for NO_x as well as for PM for light duty and diesel passenger cars (Fig. 1).

The most challenging emission standards for both light-duty and heavy-duty diesel vehicles are the 2007 and 2010 US standards.

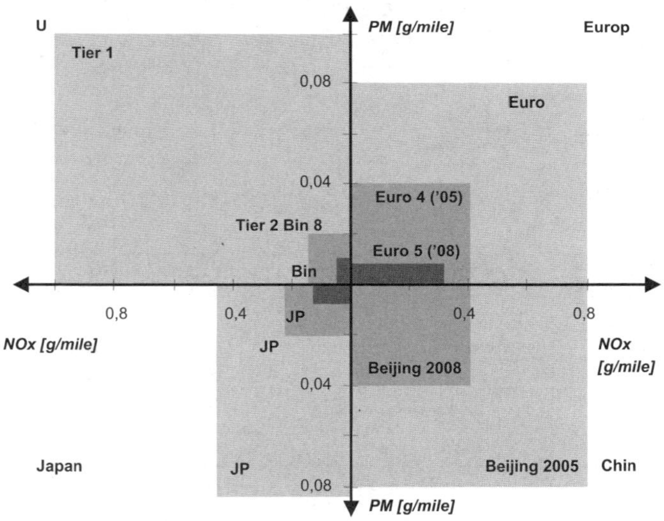

Fig. 1: Worldwide emissions limits for diesel passenger cars

Fig. 2 shows two possible scenarios to reach the stringent US 07 limits for passenger cars (Tier 2 Bin 5) with actual Euro 4 engines. It can be expected that another engine internal reduction of NOx with about 40 % is possible without loosing much of the engine efficiency. To reach the US emission level of 0.05 g/mile (approx. 0.03 g/km) another 75 % reduction will be necessary. With aftertreatment as a stand-alone solution more than 85 % efficiency will be necessary.

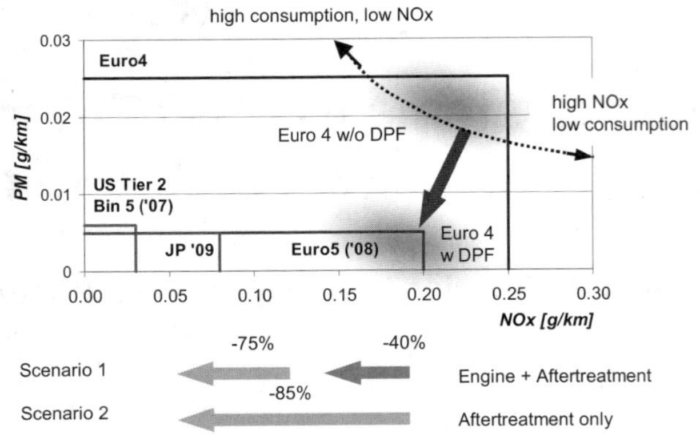

Fig. 2: NO_x reduction scenarios

As described in Fig. 3, with an overall NO_x reduction of 85 % the efficiency of the aftertreatment system has to be much higher during system operation. During the first phase of standard test cycles the emission control system has to stay passive when the exhaust temperatures are low and a proper evaporation of the urea solution is not possible. With a typical Euro 4 vehicle the emission limit for Tier 2 Bin 5 would be exceeded before the SCR system could reach its operating temperature. With a reasonable maximum NO_x reduction during operation of the SCR system of around 90 to 95 % the cumulated engine out NO_x has to be limited to around 1.2 g per NEDC test cycle.

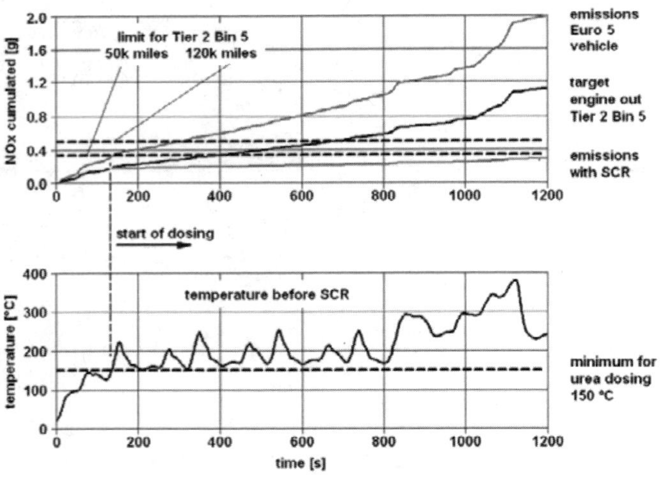

Fig. 3: NO_x emissions (cumulated) with different scenarios during the NEDC

Tab. 1 summarizes the targets for the operating conditions of a system designed to meet the US '07 Tier 2 Bin 5 emission limits.

Engine Out NOx	mg/km	<120
Temperature System Operation	°C	>150
NOx Reduction during Operation	%	≈90

Tab. 1: Target values to reach US Tier 2 Bin 5 limits

COMPARISON OF DIFFERENT NOX CONTROL TECHNOLOGIES

Of the different catalytic technologies currently being developed for Diesel emission control, Lean NO_x Traps (LNT) and SCR are the only competitive technologies. LNT has the advantage that it does not require an additional fuel (urea) on board, and thus does not need any urea infrastructure. On the other hand it is well known that LNT is highly sensitive to fuel sulphur and additionally has the lower temperature window compared to SCR. The competing SCR technology provides high NO_x conversion even at low temperatures, and a wide temperature window which allows a NO_x conversion of about 90 % and more at temperatures of 550 °C or more. As a continuous process it also does not require any regeneration strategy. Aside from the technological aspects investigations of Ford and others also indicate cost advantages of SCR against LNT on both, cost of investment and lifetime usage costs [2].

SCR		LNT	
+	Continuous process	Regeneration leads to fuel penalty	-
+	Insensitive towards sulphur	Very sensitive towards sulphur	-
+	Proven technology	Durability not yet proven	-
+	No PGM necessary	High PGM loading	-
+	Real world technology	Not working at high exhaust temp.	-
-	Needs urea and dosing system	No additional reductant needed	+

Tab. 2: Pros and Cons of SCR versus LNT

Hydrocarbon SCR (HCSCR, HC DeNOx) is at the moment no more than a future option and no real alternative to LNT and urea SCR as neither the temperature window nor the long time stability are sufficient.

SCR TECHNOLOGY

The minimum configuration of an SCR system requires a urea injection and an SCR catalyst. On heavy duty vehicles a NO_x reduction of more than 80 % can be reached with this basic configuration in both, European heavy duty test cycles and real world driving [3]. This compact design allows the packaging of a catalytic converter within the silencer (Fig. 4) minimizing the space requirements of the system.

Fig. 4: Heavy duty silencer with SCR catalyst

As the test cycles for passenger cars in Europe and the US (NEDC and FTP 75) have a much lower mean temperature compared to the European heavy duty tests a more complex layout will be necessary for SCR applications on passenger cars.

With an oxidation catalyst upstream of the SCR the engine out NO is partially converted to NO_2. A well matched oxidation catalyst producing an equimolar ratio of NO_2 and NO can double or triple the system performance at low temperatures. However, as soon as the oxidation catalyst produces an excess of NO_2 the performance at least of vanadia based SCR system is dramatically reduced [1].

As the SCR catalyst itself is very efficient for urea hydrolysis, an additional hydrolysis catalyst may not necessarily lead to better overall performance. Our work shows that it is much more important to achieve a uniform dispersion of urea onto the SCR catalyst. To reach this under the restricted packaging demands on passenger cars it may necessitate the use of mixing elements. To avoid problems with urea deposits on these mixers a hydrolytically active coating seems helpful. An effective urea injection and distribution system is important.

With regards to the integration of all these possible extensions ArvinMeritor benefits from its long experience of heavy duty diesel aftertreatment, and experience in developing SCR systems for the European market. This knowledge base is now being extended for the development of the more complex systems being installed on passenger cars.

SCR DPF COMBINATIONS

The complexity of passenger car systems is in part due to the need to combine NO_x and PM aftertreatment. Even if the SCR technology partially removes particulate matter [4], to fulfil the standards it is increasingly likely that a filter will be explicitly required. The DPF technology has now been widely introduced on passenger cars and it is likely that the actual mass based emission limits will be supplemented by a number based value.

A well known technology for heavy duty vehicles for simultaneous NO_x and PM reduction is the SCRT ™, a combination of SCR and DPF where the DPF is continuously regenerated by NO_2.

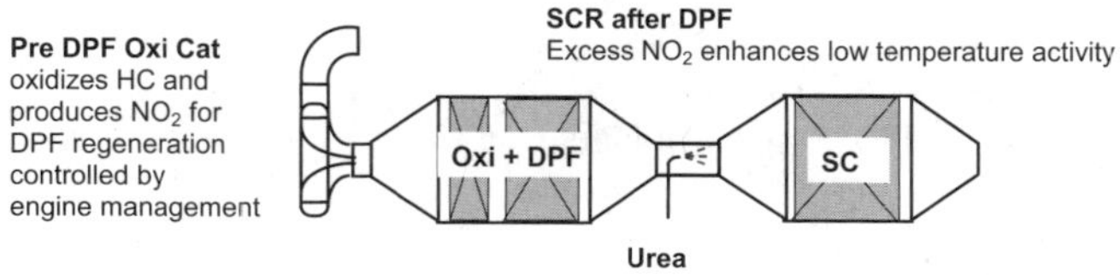

Fig. 5: SCRT system layout

But for an effective regeneration of the DPF with NO_2 only (passive regeneration) the balance of engine out NO_2 to PM emissions has to be higher than it is on today's passenger car engines during most of the driving conditions. Whilst for heavy duty applications SCRT systems are a possible solution, thermal regeneration will remain necessary on passenger cars.

Placing SCR systems downstream of a thermally regenerated DPF it is a necessity to control the DPF regeneration temperature. Recent work of ArvinMeritor has demonstrated that a homogenous flow and soot distribution is one of the key points for effective DPF regeneration. A flow optimized exhaust system with appropriate insulation enables an optimum control of loading and regeneration of a DPF. With an exhaust temperature during DPF regeneration limited to around 700 °C, Zeolite based SCR catalysts are the best choice.

One of ArvinMeritor's Light Vehicle Emissions Technology new products is a fuel vaporizer allowing the introduction of fuel vapour into the exhaust system shortly before an oxidation catalyst or a catalyzed DPF [5], [6].

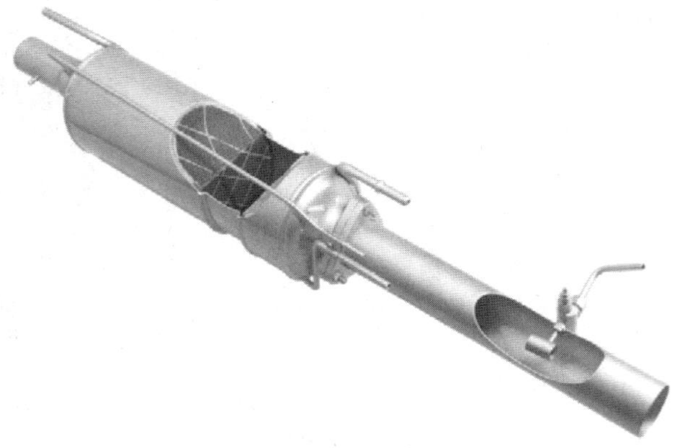

Fig. 6: DPF with fuel vaporizer for regeneration

This device enables other configurations where the SCR is placed upstream of the DPF thus allowing more flexibility for choosing the SCR catalyst and better SCR performance during low load, low speed operation. Urea injection shortly downstream of the turbocharger enables an optimum dispersion of urea onto the SCR catalyst. An oxidation catalyst before the SCR for enhancement of the system efficiency can either be placed before the turbo where the higher temperature can be used to generate NO_2 even at low load conditions or after the turbo where the oxidation catalyst can have an additional soot removal capacity.

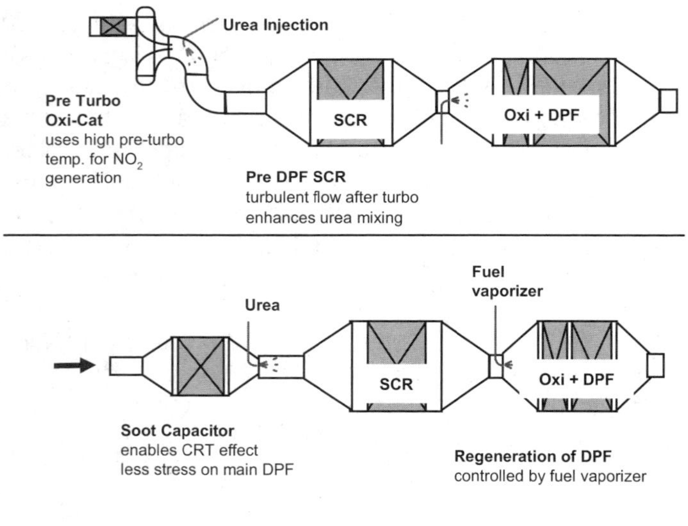

Fig. 7: Exhaust system configurations with SCR before DPF

The latter soot capacitor allows reduction of the main DPF volume so that the overall system displacement can be kept within the confined packaging limits as seen on passenger cars.

UREA INJECTION AND MIXING

Optimum urea mixing and distribution is essential for high system efficiency. Maximum pipe length between the urea dosing valve and the SCR catalyst is achieved when placing the dosing nozzle directly at the outlet cone of a pre oxidation catalyst.

One target for the development of the exhaust system is a minimal pressure drop as this value is directly related to the engine efficiency. Reaching an optimum urea distribution without additional mixing elements is therefore the best possible case. To characterize the mixing and distribution of urea, the NOx reduction efficiency was measured with different mixing elements and without mixers at a matrix of engine operating points (Tab. 3). Mixer #3 is specially designed for urea mixing and is assumed to force an optimum mixture. Thus the values with mixer #3 are set as benchmark.

operating point	engine speed	load	air massflow	exhaust temp.
	rpm	%	kg/h	°C
1	1200	30	100	250
2	2600	35	350	250
3	3800	40	600	250
4	1000	44	115	350
5	2300	45	345	350
6	3200	45	600	350
7	2100	90	510	470
8	2500	90	615	470
9	3800	65	770	470

Tab. 3 Engine operating points for analysis of urea distribution

With the original pipework layout an increase of the system performance could be reached by integrating a mixer indicating that the urea distribution has not been optimal. After optimization of the urea injection point the good performance could be reached even without additional mixing elements.

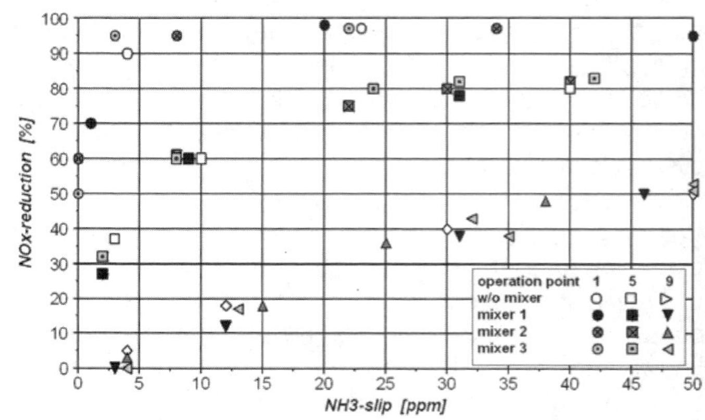

Fig. 8: Comparison of overall system performance with different mixer configurations

Fig. 8 shows the NOx-efficiency versus NH_3-slip for three of the engine operating points with different mixer configurations and without mixer but an optimized placement of the injection nozzle. Significant deviations between the results with and without mixer could no longer be observed, meaning that the mixing efficiency and urea distribution with the optimized nozzle position is good even without mixer.

MODEL BASED SYSTEM INTEGRATION

To handle the complexity of all possible catalyst and DPF combinations a great deal of work was undertaken by ArvinMeritor for model based system integration. Starting with DPF soot loading and regeneration models and measurement tools, a few years ago, enabling the optimization of today's filter systems, additional models for oxidation and SCR catalysts are being developed. As previously explained an oversized oxidation catalyst can lead to an excess of NO_2 and thus extremely reduce the performance of the SCR. ArvinMeritor has a well validated tool to predict the optimum size and PGM loading for the particular application.

Fig. 9 describes the effect of optimizing the size of an oxidation catalyst upstream of the SCR. With the original catalyst a NOx reduction efficiency of 90 % could be reached at 200 °C. But due to a very high NO_2 share at temperatures between 225 and 325 °C the NO_x reduction performance of system dropped down. An improved oxidation catalyst could save most of the low temperature activity but without loosing much performance at elevated temperatures.

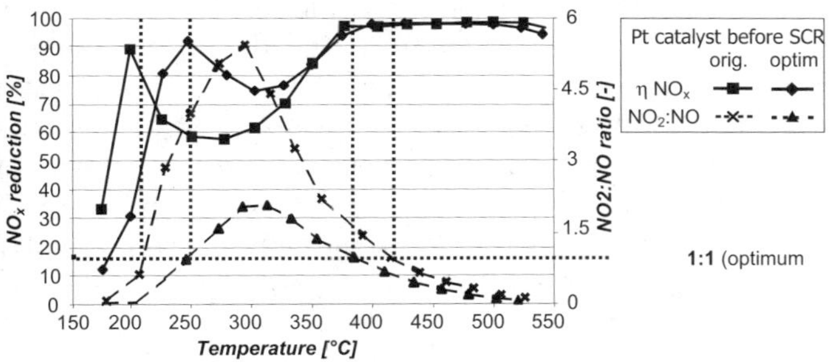

Fig. 9: NOx-reduction performance of a system with different oxidation catalysts

For an optimum combination of DPF and SCR it is necessary to know the effects of one component on the other. Fig. 10 shows the effect of placing a 3 liter SCR catalyst upstream of a DPF on the exhaust temperature before DPF. The dynamic temperature calculation allows to predict the effect of catalysts with different size or wall thickness and thus to optimize the SCR catalyst with respect to the DPF regeneration strategy.

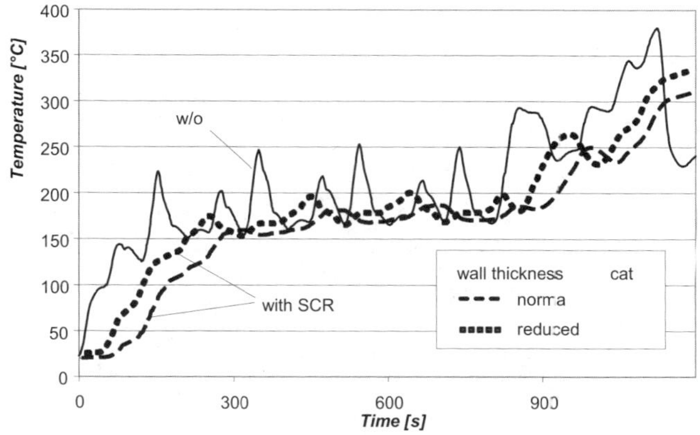

Fig. 10: temperature profile at DPF inlet with different SCR catalysts and w/o SCR

A tool for modelling the SCR reaction has been developed to allow the optimum sizing of the catalyst with respect to the assumed or measured exhaust temperature and emission profile. As the NH_3 slip is very sensitive depending on the catalyst size, efforts have been made to validate the model not only for the correct prediction of NO_x reduction but also of NH_3 conversion.

Fig. 11 shows the calculation of gas phase temperature and NH_3 concentration within the catalyst. Modelled is a transient driving cycle with a sharp temperature increase from 150 to 250 °C where the NH_3 storage capacity of the SCR catalyst drops two orders of magnitude. An overdosing of NH_3 during the start of the cycle leads to a high concentration of stored NH_3 on the catalyst surface. The simulation visualizes the NH_3 wave within the catalyst driven by the temperature wave going trough the catalyst.

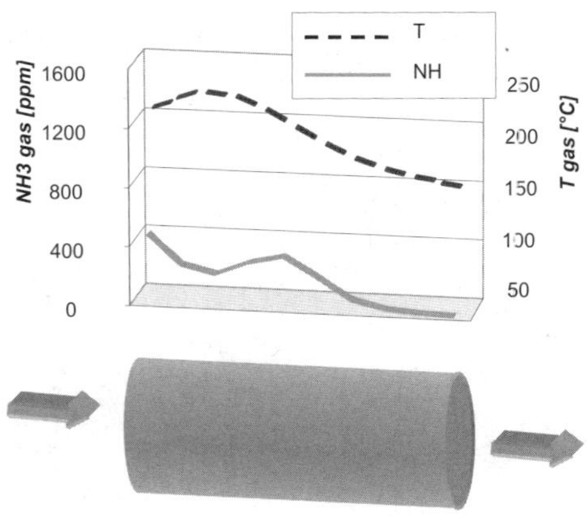

Fig. 11: calculated NH3 concentration and temperature of gas phase within the catalyst

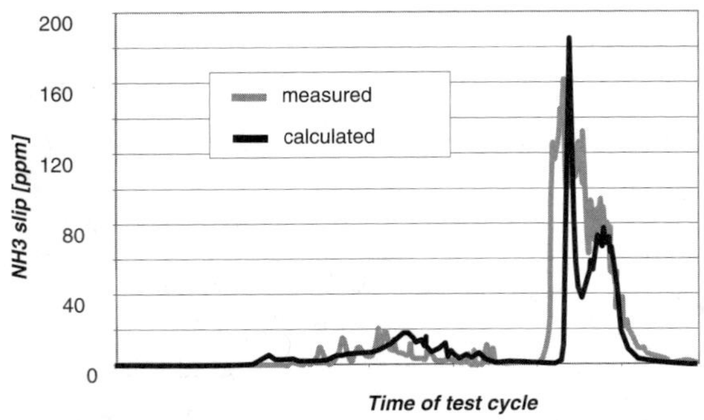

Fig. 12: calculated versus measured NH3 slip during a transient test cycle (due to insufficient dosing strategy)

Regarding the fact that the tool is designed for extremely fast calculation - 1800 sec NEDC cycle takes 10 sec computing time - the predicted and the measured NH_3 slip match very well in height and time (Fig. 12).

Another critical issue of the oxidation catalyst is the light-off behavior. Again, simulation allows finding the optimal balance between catalyst volume and PGM loading. Fig. 13 and Fig. 14 show results calculated for the two oxidation catalysts described in Tab. 4 during the first 250 seconds of a NEDC cycle.

	diamete r	length	volume	Cell density	PGM loading
DOC 1	4"	7.1"	1.5 l	4/400	75 g/ft^3
DOC 2	4"	4.7"	1.0 l	4/400	120 g/ft^3

Tab. 4 Geometric properties and PGM loading of the cordierite oxidation catalysts

As it can be seen from Fig. 14 the temperature increase of the larger oxidation catalyst (DOC 1) is much slower due to its higher thermal mass. As a consequence the light-off for this catalyst is retarded. The CO concentrations in Fig. 13 (between 60 and 80 s) clearly demonstrate this. In this region a small, highly loaded catalyst is advantageous.

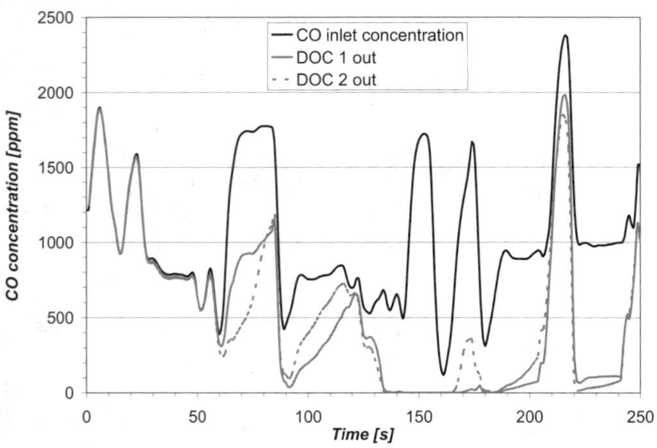

Fig. 13: Calculated CO concentrations after DOC during the first 250s of a NEDC cycle.

On the other hand, the larger catalyst acts as storage of thermal energy and thus is able to balance sudden temperature changes at least partly (see range of 80 to 130 s in Fig. 13 and Fig. 14). This is of interest when constant conditions are requested, for example for urea dosing.

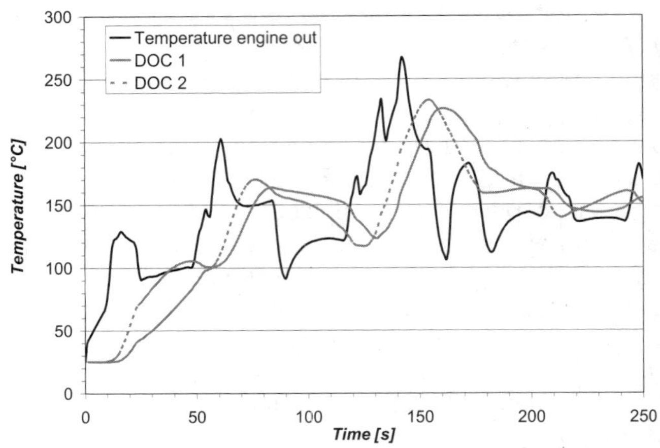

Fig. 14: Calculated temperatures after DOC during the first 250 s of a NEDC cycle.

As urea and flow distribution is a key to high system performance, CFD simulation of urea spray is used to predict the system behaviour. Especially for DPF SCR combinations where limited space and packaging may necessitate active mixing, a model based system optimization is used. A very effective and cost-saving approach to realize the mixing of urea into the main exhaust gas flow is a delta-wing design which extends from the wall into the bulk flow with an inclination of approximately 30°. The functionality of such a vortex mixer is based upon the generation of two counter rotating wake vortices (Fig. 15).

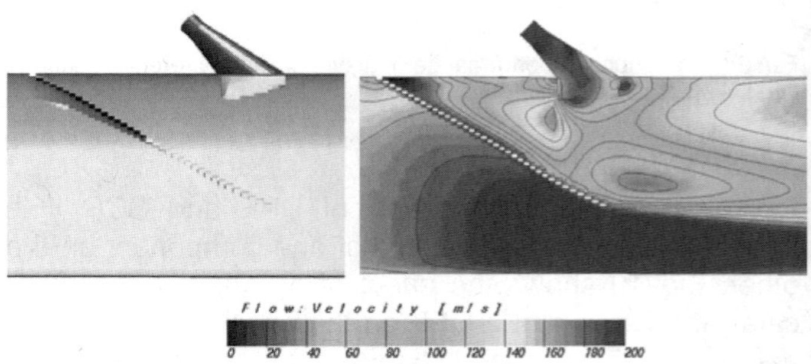

Fig. 15: Vortex mixer, principle and flow field

This secondary flow forms a flow path from the pipe jacket into the bulk flow on which the atomized urea solution is transported in the direction of the energy containing bulk flow of the exhaust gas. By means of this, urea deposits onto the pipe wall are avoided. Besides the secondary flow the vortex mixer generates augmented turbulent kinetic energy (Fig. 16).

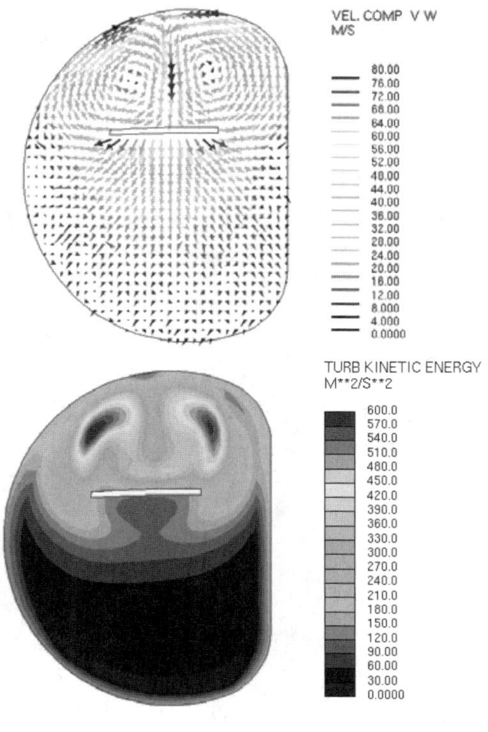

Fig. 16: Vortices and turbulent kinetic energy around a vortex mixer

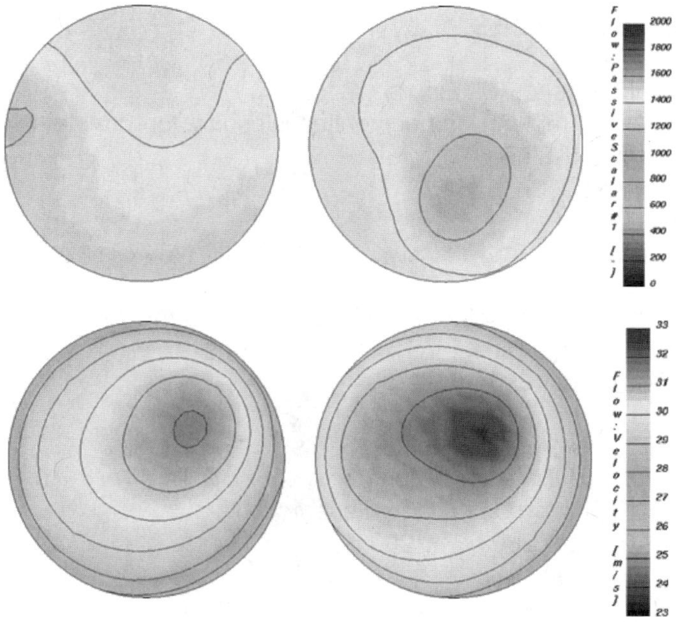

Fig. 17: Urea and flow distribution at catalyst inlet

This increases the turbulent mass transfer and therefore ensures a good mixing of urea and exhaust gas. The positioning of the delta-wing upstream of the urea injection nozzle prevents any urea deposits on the wing. The increase of backpressure of the exhaust system caused by the vortex mixer is low in comparison with other mixer approaches which cover the complete cross section of the pipe. Essential for the optimized functionality of the vortex mixer are, besides the geometry of the delta-wing, the positioning relative to the urea injection nozzle and the inclination angle. By means of 3-dimensional CFD simulation it is possible to design a vortex mixer for all relevant engine operating points in the early concept phase. Fig. 17 shows the CFD results for an optimized SCR system where after a single nozzle the exhaust stream is divided onto a dual leg catalytic converter. It demonstrates that both, uniformity of exhaust gas flow and NH3 distribution can be reached.

CONCLUSIONS

SCR is an effective solution for diesel passenger cars and LDV's to meet future emission limits in the leading markets. For markets with the tightest NOx limits such as the USA in 2010, SCR will be the most efficient system.

The integration of well matched oxidation catalysts and other additional components can enhance the performance at low temperatures. ArvinMeritor's long time experience on heavy duty diesel aftertreatment is the basis for developing smart systems also for passenger cars.

The combination of SCR and DPF needs a precise control of the DPF outlet temperature during regeneration. ArvinMeritor's portfolio includes products such as the fuel vaporizer which allow optimum control of the DPF regeneration.

System integration on the basis of experience and validated models is the key to optimum system performance with minimum complexity and within limited space requirements.

REFERENCES

[1] M. KOEBEL, M. ELSNER, C. SCHÄR
NOx-Verminderung bei mobilen Dieselmotoren mittels Harnstoff-SCR
Zürich 2002

[2] C. LAMBERT, R. HAMMERLE, R. MCGILL, M. KHAIR AND C. SHARP
Technical Advantages of Urea SCR for Light-Duty and Heavy-Duty Diesel Vehicle Applications
SAE Paper 2004-01-1292

[3] L. HOFMANN, K. RUSCH, S. FISCHER, B. LEMIRE
Onboard Emissions Monitoring on a HD Truck with an SCR System Using NOx Sensors
SAE paper 2004-01-1290

[4] K. RUSCH, L. HOFMANN, et al.
PM-Reduction by SCR-Catalyst
SAE paper 1754; Diesel Exhaust Emission Control, March 2003

[5] M. RANALLI, S. SCHMIDT, J. KLEMENT AND A. MAYR :
Diesel Fuel Vaporizer: a Way to Reliable DPF Regeneration. SAE paper 2003-01-50

[6] M. RANALLI, P. KRONER, L. CHIEW,
Diesel Vaporiser: An Innovative Technology for Reducing Complexity and Costs Associated with DPF Regeneration
SAE paper 2005-01-0671

CONTACT

KLAUS RUSCH
ArvinMeritor Emissions Technologies GmbH
Biberbachstr. 9, 86154 Augsburg, Germany
Tel: +49 (0) 821 4103 9364
E-Mail: klaus.rusch@arvinmeritor.com

ROLF KAISER
ArvinMeritor Emissions Technologies GmbH
Biberbachstr. 9, 86154 Augsburg, Germany
Tel: +49 (0) 821 4103 156
E-Mail: rolf.kaiser@arvinmeritor.com

STEFAN HACKENBERG
ArvinMeritor Emissions Technologies GmbH
Biberbachstr. 9, 86154 Augsburg, Germany
Tel: +49 (0) 821 4103 902
E-Mail: stefan.hackenberg@arvinmeritor.com

V. FUEL & LUBRICANTS EFFECTS

2003-01-3179

The Influence of Engine Lubricating Oil on Diesel Nanoparticle Emissions and Kinetics of Oxidation

Heejung Jung, David B. Kittelson and Michael R. Zachariah
University of Minnesota

ABSTRACT

Earlier work [1] shows that kinetics of Diesel soot oxidation is different from that of ethylene diffusion flame soot oxidation [2], possibly due to metals from lube oil. This study investigates the influence of metals on soot oxidation and the exhaust particle emissions using lube oil dosed fuel (2 % by volume). This method does not simulate normal lube oil consumption, but is used as a means of adding metals to particles for oxidation studies. This study also provides insight into the effect of systems that mix lube oil with fuel to minimize oil change service.

The HTO-TDMA (High Temperature Oxidation-Tandem Differential Mobility Analyzer) technique [1] was used to measure the surface specific oxidation rate of Diesel particles over the temperature range 500-750 °C. Diesel particles sampled from the exhaust stream of a Diesel engine were size segregated by differential mobility and oxidized in situ in air in a heated flow tube of known residence time and temperature profile. The change in particle diameter after oxidation took place was measured and converted into the surface specific oxidation rate. The pre-exponential factors increased by about a factor of two, whereas the activation energy determined for the lubrication oil dosed Diesel particles was very close to that of undosed Diesel particles (108 kJ mol^{-1}, [1]). This suggests that the increase of active sites by the presence of Ca (and possibly by other metals included in the additive package of the lube oil) results in the faster oxidation rate.

Particle size measurements show that particle volume emissions, which are roughly proportional to particle mass, decreased by about a factor of two, but that particle number emissions increased by an order of magnitude with lube oil dosed fuel.

INTRODUCTION

The regulations for Diesel particulate emissions are becoming more stringent with a strong emphasis being placed on possible adverse health effects. A lot of work has been done to reduce Diesel emissions by improving aftertreatment devices, such as DPFs (Diesel Particulate Filters) [3, 4], and by enhancing the oxidation of Diesel particulate matters using fuel additives [5, 6]. Although the oxidation rates of Diesel particles have been measured numerous times in the past [7, 8], the influence of engine lube oil metals on the kinetics of Diesel particle oxidation has previously been overlooked. Understanding the role of lube oil in Diesel particle oxidation may be very important in developing new lube oil formulations, as well as new fuel borne catalysts.

It is often assumed that the oxidation kinetics of carbon particles in Diesel exhaust are similar to the kinetics of carbon oxidation in diffusion flame studies [9, 10]. However, a recent study by Higgins et al. [1] showed that the kinetics of oxidation of carbon particles in Diesel exhaust were quite different from that of carbon particles from an ethylene diffusion flame, and from the Nagle and Strickland-Constable kinetic model [11]. Higgins et al [1] suggested this might be due to the presence of metal compounds in Diesel exhaust particles. Engine lube oil is continuously consumed in the combustion chamber. Although oil consumption typically is only about 0.2 % of fuel consumption, it contributes disproportionally to the exhaust particulate matter. The combustion efficiency of lube oil is considerably lower than that of Diesel fuel and lube oil contains metals that form solid particles. Engine lube oil is composed of a base oil and an additive package. In general, the base oil is composed of petroleum-derived mineral oils, whereas the additive package is composed of various chemicals, including metal compounds (such as Ca, Mg, and Zn) [12]. Most of these contribute to lube oil ash. The typical ash content of lube oil is about 1 % [13, 14]. A material balance based on 1 % lube oil ash content, 0.2 % relative lube oil consumption rate, 250 g/kWh fuel consumption rate and 0.13 g/kWh PM (Particulate Matter) (US PM standard for 2004) gives 0.005 g/kWh ash emissions. Under these conditions, the ash would constitute 4 % of the current PM, assuming that all of the ash appeared in the PM standard. Published results show up to 10 % ash by mass [15]. For 2007, the PM standard drops to 0.013 g/kwh. If lube oil ash consumption is assumed at the same level as in the material balance above, it could constitute 40 % of the PM emissions. This suggests growing importance of the lube oil to PM emissions.

It is still not entirely clear which mechanism is the main contributor in oil consumption due to the complexity of the phenomenon. However, recent research shows a clearer picture of oil consumption. Lube oil may leave the cylinder wall by either vaporization or atomization. De Petris et al. [16] showed that oil mist (or oil atomized by a reverse blowby) was a main contributor to oil consumption under their test conditions. Audette III and Wong [17] found that oil vaporization from a cylinder liner is a small contributor to consumption unless more volatile lube oil is used. Lube oil can also get into the combustion chamber by other pathways. It can come through a valve stem seal, a throw-off of accumulated oil above the top ring, etc. Flynn et al. [18] recently described the Diesel combustion process in terms of burning fuel jets that rapidly entrains surrounding combustion chamber air. Some of the lube oil entering the combustion chamber also will be entrained by the burning fuel jets. In this manner, species from the lube oil may have an influence on the kinetics of DPM (Diesel Particulate Matter) formation and oxidation.

Lube oil mixed with Diesel fuel is combusted directly along with the fuel, and is processed somewhat differently by the flame than lube oil consumed by the usual path. Thus, lube oil mixed with fuel is likely to be more involved in the early stages of combustion where soot is being rapidly formed and oxidized. The primary focus of this study is to investigate the role of lube oil on the kinetics of DPM in the exhaust, not in the combustion chamber. We have used fuel dosing as a relatively easy way to add additional lube oil metals to the exhaust PM. It is assumed that metal compounds added in this manner will have a similar influence on oxidation kinetics of the PM in the exhaust and in aftertreatment systems to metals originating from normal oil consumption.

In some cases, lube oil is intentionally mixed with fuel to extend lube oil change intervals. An on-board continuous oil management system called CENTINEL™ was recently developed by Cummins [19, 20]. The system takes a small amount of used oil out of the lubrication system and sends it to the fuel tank, where it is blended with Diesel fuel. At the same time, fresh oil is sent from a makeup tank to the engine, replacing the used oil. This system has been shown not to increase PM mass emissions. However, its influences on unregulated emissions, e.g. metals and aftertreatment systems should be examined. Although our approach of dosing fuel with lube oil is very similar to the CENTINEL™ system, the primary purpose of this work is not to emulate CENTINEL™ system, but rather to estimate the influence of metal species in the lube oil on the kinetics of DPM in the exhaust. However, as a part of this study we have also looked at the influence of lube oil dosing on the exhaust particle size distribution.

In this study the Diesel fuel was dosed with engine lube oil (2 % by volume). This has done to intentionally increase the oil consumption rate and the loading of metal components from lube oil on the DPM with combustion, in order to better observe the effect of lube oil (specifically metal components from lube oil) on the oxidation of Diesel particles downstream of the engine and on the exhaust size distribution. The oxidation rate of 2 % lube oil dosed Diesel particles was compared to that of undosed Diesel particles measured in the previous study [1]. The size distribution was also compared with that for undosed Diesel particles.

EXPERIMENTAL

TEST FUEL AND LUBRICANT

Standard EPA No. 2 on-road Diesel fuel (300-500 ppm sulfur) was used as a base fuel. Two barrels of fuel from the same batch were used to provide consistency in the amount of sulfur content and in chemical composition. One barrel of fuel was used to measure the base size distribution when lube oil was not blended in the fuel. The other barrel was dosed with engine lube oil. Table 1 shows recent fuel analysis for fuel from the same supplier, but not the same batch as used in this study.

The lube oil used in this study was SAE15W-40 (John Deere TY6391). The elemental composition of the same type of lube oil used in this study is shown in Table 2. Calcium is typically the most abundant metal species present in lube oil. Calcium is normally in the form of a sulphonate (e.g. calcium sulphonate) and is used as a detergent to keep engines clean [12]. The lube oil used in this study contains about 2500 ppm Ca. The dosing level of 2 % by volume in the fuel results in about 50 ppm Ca content in the mixed fuel. Although the zinc content of the oil adds another 14 ppm metal to the mixed fuel, Zn has not been studied as a fuel borne oxidation catalyst. Metal contents of 20-100 ppm have typically been used to examine the effect of fuel additives [3, 5, 21, 22]. Also, a dosing level of 2 % lube oil was used to measure the effect of lube oil on Diesel particulate emissions by Kytö et al. [23].

Table 1: Fuel analysis of No.2 on-road Diesel fuel

Property	ASTM	Results
Distillation	D86	498.0 °F 50% Rec. 617.7 °F 90% Rec.
Flash point	D93	71 °C
Heat of combustion	D240	18111 BTU/lb
Kinematic viscosity	D445	2.740 cSt @40 °C
Ash	D482	0.000 mass %
Cetane number	D613	44.9
Spedific gravity	D1298	0.8479 g/ml @ 15 °C
Aromatics by FIA	D1319	27.6 vol. %
Sulfur	D2622	0.0305 mass %
Water and sediment	D2709	0. vol. %
Carbon, hydrogen and nitrogen	D5291	86.2 mass % C 13.00 mass % H
Cloud point	D5773	-29 °C
Pour point	D5949	< -60 °C

Table 2: Lube oil analysis of SAE 15W-40 (John Deere TY6391)

Element	Weight (%)
S	0.260
Ca	0.250
N	0.160
Zn	0.068
P	0.061
B	0.018
Mo	0.011
Mg	0.001

ENGINE

The engine used in this study was a medium-duty, direct injection, 4 cylinder, 4 cycle, mid-90's turbocharged Diesel engine (John Deere T04045TF250) of 4.5 *l* displacement, with a peak power output of 93 kW at 2400 RPM and a peak torque output of 400 N·m at 1400 RPM. The engine was coupled to a dynamometer and operated at 1400 RPM under 75 % load (300 N·m). The exhaust temperature measured at the exhaust manifold downstream of the turbocharger was 530 °C, and the air-fuel ratio was measured as 27 for regular fuel at 1400 RPM under 75 % load (300 N·m).

SAMPLING AND DILUTION

Fig. 1 shows a schematic diagram for the engine bench, sampling system, SMPS (Scanning Mobility Particle Sizer) [24] system and HTO-TDMA system. The exhaust was sampled 25 cm downstream of the turbocharger exit and diluted in a two-stage air ejector variable residence time dilution tunnel, similar to one described by Abdul-Khalek and Kittelson [25].

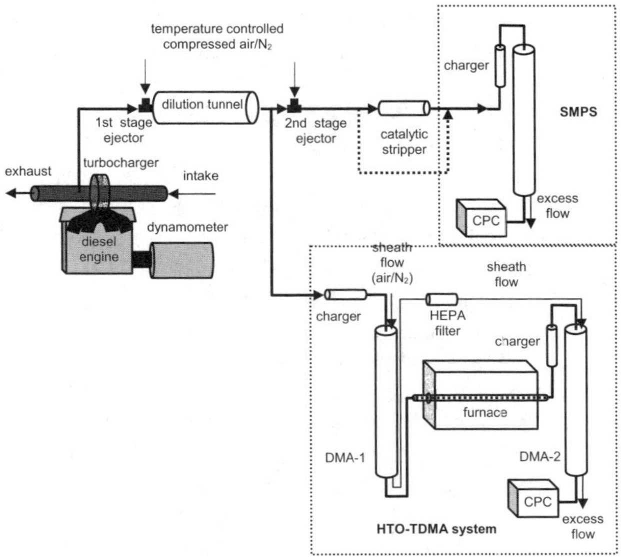

Fig. 1: Schematic diagram of the experimental setup. The bold line indicates the path taken by the Diesel soot particles.

For the oxidation study, particles were extracted after the first stage of dilution, which was operated with a fixed residence time of 0.8 seconds at a dilution ratio of 26-28:1 for 75 % load at 1400 RPM.

The dilution ratio was determined by comparing the NOx concentration in the exhaust gas and diluted exhaust gas after the first stage of dilution after subtracting the background NOx concentration. The first stage dilution air temperature was maintained at a temperature of 32°C at the exit by employing feedback control of a band heater wound around the copper tube for the compressed air to the ejector.

For the size distribution studies, samples were taken after the second stage of dilution to prevent saturation of the SMPS system used for particle sizing. Flow measurements were used to determine the second stage dilution ratio to be 31:1. Thus, the overall dilution ratio after the second stage was thus 810-870:1, calculated by multiplying the dilution ratios from each stage.

SCANNING MOBILITY PARTICLE SIZER (SMPS)

The SMPS measures mobility equivalent particle diameter in the diameter range of 7 to 300 nm during a 4-minute (2 minutes up, 2 minutes down) scan time. This particle measurement system consists of a neutralizer, a mobility section (long column DMA similar to TSI model 3934) with a sampling flow rate of 1 lpm and a sheath flow rate of 10 lpm, a CPC (TSI model 3010) and a computerized control and data acquisition system. This system has been described in more detail by others [25, 26].

HIGH TEMPERATURE OXIDATION-TANDEM DIFFERENTIAL MOBILITY ANALYZER (HTO-TDMA)

The HTO-TDMA method was described in detail in a previous study done by Higgins et al. [1]. Briefly, in the HTO-TDMA method, the sample is sent through a bipolar diffusion charger to establish a known charge distribution. Next, a specific particle size is selected with DMA-1. The DMA (Differential Mobility Analyzer) selects mono-area particles based on electrical mobility, which is related to the drag and charge on a particle, rather than on absolute particle size. The mono-area particles then undergo oxidation in the air while traveling through a quartz flow tube enclosed in a tube furnace of known residence time and temperature profile. The particles are recharged in a bipolar diffusion charger, and the size change resulting from the high temperature processing (which includes oxidation, thermal restructuring and evaporation processes) is measured using DMA-2. Three initial particle sizes of 41, 94 and 130 nm mobility diameters were selected at DMA-1 to match the previous study done by Higgins et al. [1] on Diesel particle oxidation. Furnace temperature settings ranged from room temperature (25-34 °C, depending on the day) to 800 °C.

In order to check for particle size decreases due to thermal effects (evaporation and restructuring), additional experiments were conducted with the air for the first dilution stage and DMA sheath flows replaced with nitrogen (Airgas, industrial grade) in the previous work by Higgins et al [1].

RESULTS AND DISCUSSION

The sampled aerosol stream from the first stage of the dilution tunnel was either introduced to the HTO-TDMA system for the oxidation study, and/or the sampled aerosol stream from the second stage of the dilution tunnel sent to a SMPS system for size distribution measurements as illustrated in Fig. 1.

THE INFLUENCE OF LUBE OIL ON SIZE DISTRIBUTION

Fig. 2 and Table 3 show the effect of lube oil on the number and volume size distributions with a dosing rate of 2 % under a fixed engine condition, 75 % engine load (300 N·m) at 1400 RPM. The data have been fit to a bimodal lognormal distribution. Most of the number is found in the nuclei mode (D_p < ~30nm), most of the volume in the accumulation mode (D_p >~30nm). The volume concentration of accumulation mode particles was reduced by nearly a factor of two while the total number increased by about a factor of 10, mainly as a result of the formation of a large nuclei mode. The same trend was observed by Kytö et al. [23] when they blended 2 % of lube oil with Diesel fuel. They discovered that the dosage of lube oil that gave the lowest particulate mass (meaning the reduction of accumulation mode particles), produced the highest number of particles (which means the appearance of large nuclei mode).

Table 3: Number concentrations and volume concentrations of the nuclei mode and the accumulation mode particles for regular fuel and 2 % lube oil dosed fuel. Base and 2 % indicate regular base fuel and 2 % lube oil dosed fuel, respectively. The units of number and volume concentration are #/cm^3 and μm^3/ cm^3, respectively.

		Total	Nuclei	Accumulation
Base	Number	$3.08 \cdot 10^7$	$6.13 \cdot 10^6$	$2.47 \cdot 10^7$
	Volume	$2.70 \cdot 10^4$	$4.16 \cdot 10^2$	$2.66 \cdot 10^4$
2 %	Number	$3.57 \cdot 10^8$	$3.48 \cdot 10^8$	$8.93 \cdot 10^6$
	Volume	$1.48 \cdot 10^4$	$4.56 \cdot 10^2$	$1.43 \cdot 10^4$

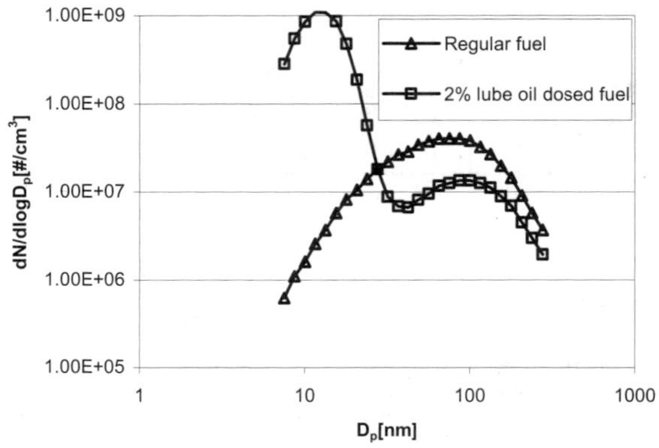

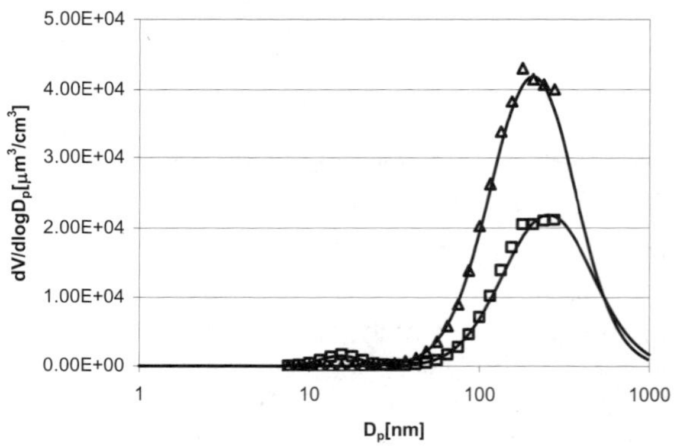

Fig. 2: Size distributions at 1400 RPM, 75 % (300 N·m). (a) (Top) Number weighted size distributions of 2 % lube oil dosed Diesel particles in comparison with base fuel. (b) (Bottom) Volume weighted size distributions of 2 % lube oil dosed Diesel particles in comparison with base fuel.

Calcium is the most abundant species present in the lube oil used in these tests, and is known to suppress the formation of carbonaceous Diesel particles [27] and to enhance their oxidation [27, 28] when it is blended with fuel. Miyamoto et al. [27] concluded that Ca and Ba are the most effective additives for enhancing oxidation and suppressing the formation of soot among the

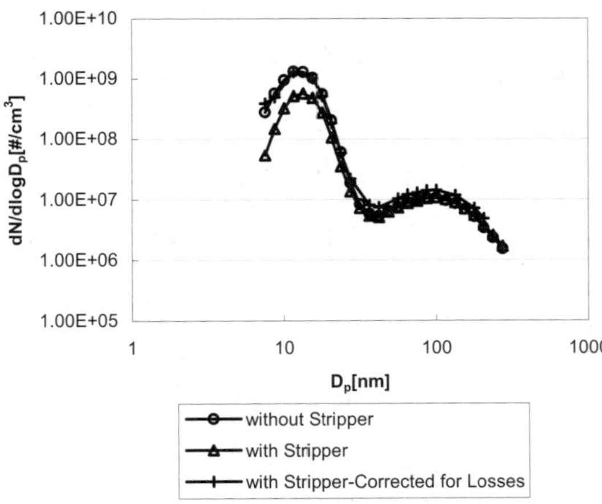

Fig. 3: Size distributions of 2 % lube oil dosed Diesel particles with and without the catalytic stripper at 1400 RPM, 75% load [29].

metallic additives (Ca, Ba, Mg, Fe, Ni, Mn, Cu) they investigated. Other studies [5, 21, 30-32] have shown that metal additives may simultaneously reduce the accumulation mode and increase the nucleation mode. This study shows that metals in lube oil blended with fuel may play a similar role.

When a large nuclei mode is present for Diesel engines operating with typical fuels without additives, the nuclei mode consists primarily of volatile particles [33-38]. However, when metals are present in the fuel, a large solid nuclei mode may form [5, 21, 30-32]. Stenitzer [29] examined Diesel particles with 2 % lube oil dosed fuel from the same engine used in this study using a catalytic stripper operating at 300 °C. The catalytic stripper was designed to remove volatile materials from Diesel particles [34]. If the nuclei mode particles are mainly composed of volatile materials, they will almost disappear completely when passing through the stripper. Fig. 3 compares size distributions for 2 % lube oil dosed fuel with and without the stripper operating at 300 °C [29]. After correcting for transport losses (thermophoretic loss and diffusion loss) in the catalytic stripper, there is no significant reduction in nuclei mode particles. This demonstrates that the nuclei mode particles with lube oil dosed fuel under our test condition are solid.

The formation of a nuclei mode from metal compounds derived from lube oil is not unexpected. Abdul-Khalek et al. [33] performed a chemical equilibrium calculation for both calcium and zinc during Diesel engine combustion to determine the possible fate of metals in lube oil. They found that NASA Program "Flame" [39] predicts that the lube oil metals in the fuel are initially transformed into gas phase compounds at high temperature (~2500K). As the combustion products cool during the expansion stroke to~1500 K, metal compounds transform into solid

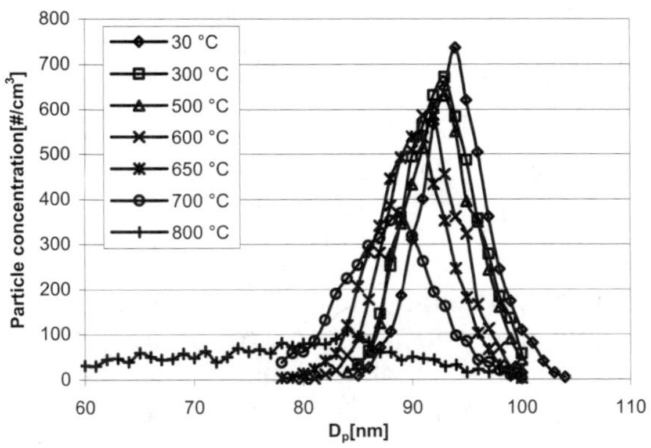

Fig. 4: TDMA Diesel particle oxidation results in air for furnace set temperature of 30-800 °C.: 94 nm initial size with 2 % lube oil dosed fuel.

compounds. When high concentrations of metals are present, some of the compounds nucleate to form a solid nuclei mode. This is consistent with Stenitzer's [29] observation using the catalytic stripper.

RESULTS FOR HTO-TDMA EXPERIMENT

Fig. 4 shows representative HTO-TDMA data for particles with a selected mobility diameter of 94 nm, generated using 2 % lube oil dosed fuel under 75 % engine load (300N·m) at 1400 RPM. Results for other particle sizes selected at the same engine operating condition are similar to those in Fig. 4. Fig. 4 shows that particles shrink as the furnace set temperature increases, with a mobility diameter decrease of 1 nm at 500 °C and 5 nm at 700 °C. The initial shrinkage in size below 500 °C is likely due to the evaporation of semi-volatile materials condensed onto the Diesel particles. In our previous study [1] the total particle size decrease due to non-oxidative effects at 500 °C, was measured using nitrogen as the carrier gas for both the DMA sheath flow and the reactor flow. It amounted to ~1 nm for 75 % load particles using regular Diesel fuel. This is consistent with the amount of shrinkage we observed at 500 °C. Above 500 °C, most of the change in diameter results from particle oxidation. In addition to a decrease in particle size with increasing oven temperature, we also observe a spreading of the particle size distribution believed to be due to the wide distribution in particle residence time in the laminar flow reactors, as well as loss of particles due to thermophoresis.

DETERMINATION OF THE OXIDATION RATE

We used the measured decrease in particle size to obtain kinetic parameters following the procedure detailed in our previous studies [1, 2]. The size decrease rate is modeled using a modified Arrhenius expression,

$$\dot{D}_p = -A_{nm} \cdot T^{1/2} \cdot \exp\left[-\frac{E_a}{RT}\right] \qquad (1)$$

where A_{nm} is an initial-size-dependent pre-exponential factor and E_a is a size-independent activation energy. Values of the A_{nm}'s and E_a are determined by integrating

$$\Delta D_p = \int_0^X \frac{\dot{D}_p(x)}{u(x)} dx \quad (2)$$

where x is the horizontal position in the tube, X is the length of the tube and u is the flow velocity over the heated length of the flow tube. Minimizing the differences between calculated and measured ΔD_p's is done using a non-linear least squares method. The dependence of the size decrease rate and flow velocity on horizontal position is a result of their dependence on temperature, as shown in Eq. (1) and

$$u(x) = \frac{4}{3} u_m \frac{T(x)}{T_o} \quad (3)$$

where u_m is the mean flow velocity calculated from the volume flow rate and the cross-sectional area of the flow tube, and 4/3 u_m is the peak volumetric flow velocity assuming laminar flow. Decreases in size due to oxidation are determined relative to the size of the particles at 500 °C.

COMPARISON WITH PRIOR STUDIES

Fig. 5 shows the experimental size reduction data of the current study with 2 % lube oil dosed fuel (Fig. 5a) compared to our previous work [30] with 25 ppm cerium dosed fuel (Fig. 5b). The size reductions are very similar.

Light-off temperature of an oxidizing particle is defined in our previous study [30]. It is analogous to ignition temperatures in the DPF studies. The light-off temperature is defined as the peak flow tube temperature where the particle shrinks ~1 nm in size from that of a particle where the entire evaporation process has occurred. The light-off temperature of 640~690 °C with 2 % lube oil dosed fuel (Fig 5a) was lower than 840 °C with undosed Diesel particles [1], but higher than 590 °C with cerium dosed fuel, as shown in Fig. 5b. This is qualitatively consistent with observations (Miyamoto et al., [27]) on ignition temperatures of their bulk samples. They found the ignition temperature was lowered as they increased the amount of calcium in the fuel.

The data plotted in Fig. 5a was fitted to the Arrhenius expression of equation (1) to determine a different set of parameters for each of sizes studied. Table 4 shows the fitted parameters. Fig. 6 shows the Arrhenius plot of the surface specific oxidation rates for our measurements compared to other studies. The activation energy of surface specific oxidation rates for our measurements compared to other studies. The activation energy of

Table 4: Arrehinius fit parameters for oxidation rate of particles with 2 % lube oil dosed.

	2% lube oil dosed
E_a [kJ mol^{-1}]	101
A_{41} [10^5 nm K$^{-1/2}$s^{-1}]	1.0
A_{94} [10^5 nm K$^{-1/2}$s^{-1}]	1.2
A_{130} [10^5 nm K$^{-1/2}$s^{-1}]	1.1

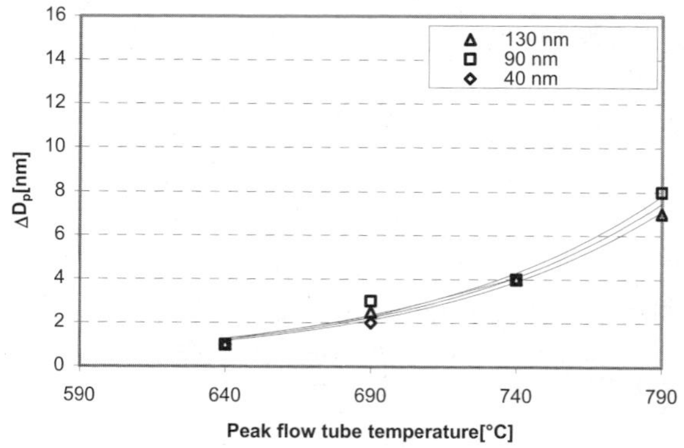

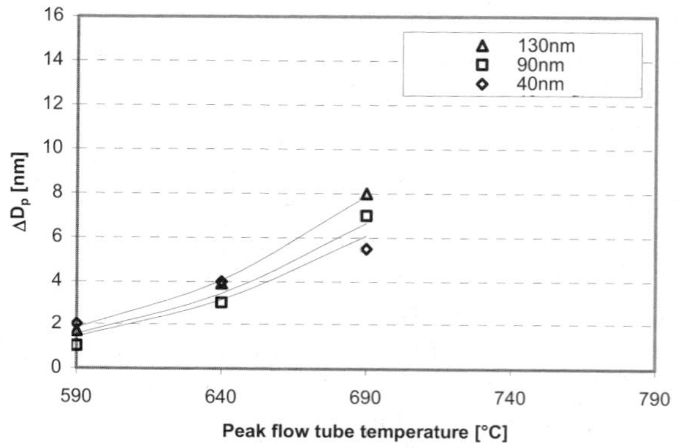

Fig. 5: Particle size change as a function of peak flow tube temperature for the three initial particle sizes. Note: The peak flow tube temperature is approximately 40 °C higher than the furnace set temperature. (a) (Top) 2 % lube oil dosed fuel. (b) (Bottom) 25 ppm cerium dosed fuel [30].

Diesel particle oxidation for 2 % lube oil dosed fuel ('a' in Fig. 6) is 101 kJ mol^{-1}, which is very close to 108 kJ mol^{-1}, the activation energy for particles using (undosed) regular fuel ('b' in Fig. 6) at the same engine operating condition. The frequency factors A_{nm}'s of the Arrehenius equation for the particles with 2 % lube oil-dosed fuel are about two times larger than that of the particles with undosed fuel. It is noteworthy that our previous studies show that the oxidation rate of diffusion flame soot using ethylene ('d' in Fig. 6, [2]) agrees well with the oxidation rate of pyrographite rod ('e' in Fig. 6) by NSC [11], whereas the oxidation rate of the undosed Diesel particle ('b' in Fig. 6, [1]) is not like the NSC (1962) model. Many modelers have used and believed that the NSC [11] model simulates the oxidation of Diesel particles. Our previous study [1] showed the oxidation of Diesel soot is different from the NSC model.

Miyamoto et al. [28] reported oxidation rates of Diesel particles for undosed ('h' in Fig. 6) and calcium dosed fuel ('f' and 'g' in Fig. 6). They found that the beginning of the particle oxidation is rapid ('g' in Fig. 6), whereas the remaining particle oxidation is relatively slow ('f' in Fig. 6) when Ca is blended in the fuel. They found the same trend when other kinds of metal additives (Ba, Ni, Fe) were used. This rapid early stage is similar to that referred to as auto-acceleration by Lahaye et al. [6] and catalytic oxidation of VOF (Volatile Organic Fraction) by Stratakis and Stamatelos [40]. While our current experiments are unable to resolve transients such as described above, we do note that, in general, addition of metals does not seem to change the activation energy, but does increase the oxidation rate significantly. In our experimental setup, suspended particles are oxidized directly, with little heat or mass transfer limitation. Essentially all volatile material is expected to have desorbed before any significant oxidation takes place. Thus, volatiles do not play a role. The auto acceleration observed in the TGA experiments may be due to limitations of heat and mass transfer in bulk soot samples. Catalytic oxidation of VOF is likely associated with mass transfer processes within the TGA sample. The relative change between this study and our previous study [1], due to dosing with Ca (see curve 'a' and 'b' in Fig. 6), is similar to the relative change observed by Miyamoto et al. [28] (see curve 'f' and 'h' in Fig. 6). In both cases, the frequency factor increases, whereas the activation energy is nearly unchanged compared to the undosed case.

DISCUSSION

Our results clearly show that lube oil constituents increase the rate of oxidation of exhaust soot particles. This should help to improve our understanding of soot oxidation in the engine itself, as well as in aftertreatment systems. Clearly, there is a trade-off between the gains associated with lube oil metal catalyzed oxidation of exhaust particles and emissions of metal compounds. Without aftertreatment, these materials are emitted into the atmosphere and might constitute an environmental hazard. Most aftertreatment systems (filters) are very effective at removing solid particles and would trap metal

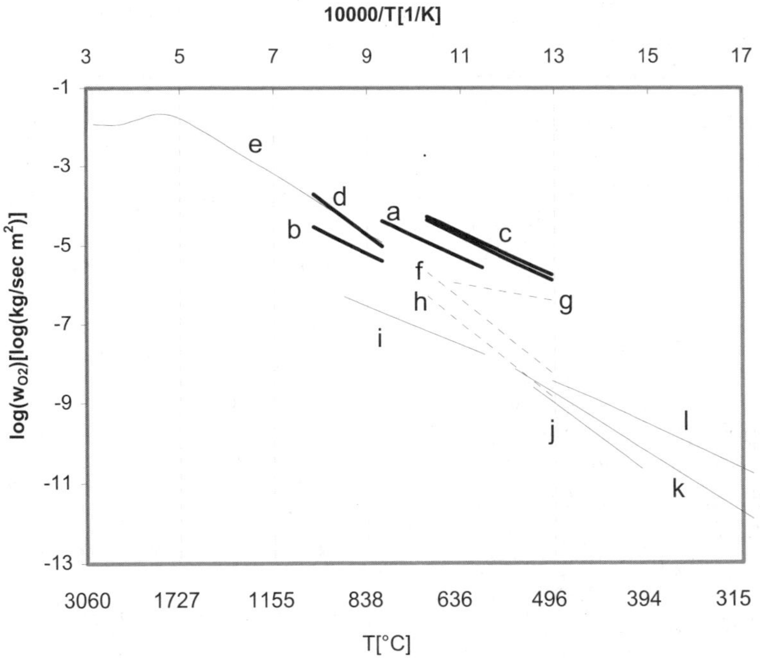

Fig. 6: Arrhenius plot of the surface specific rates of Diesel particle oxidation for the current study and for other relevant prior studies. Previous studies include oxidation rates of various soots (Diffusion flame soot and Carbon black) and pyrographite rod. Rates are presented for 21 % O_2. (a) Current study using 2 % lube oil dosed fuel. (b) Higgins et al. [1] using regular Diesel fuel. (c) Jung et al. [30] using 25 ppm and 100 ppm cerium dosed fuel. (d) Higgins et al. [2] using diffusion flame generated soot. (e) Nagle and Strickland-Constable [11] using a pyrographite rod. (f) Miyamoto et al. [28] TGA of catalyzed (Ca added to fuel) Diesel soot in slow oxidation. (g) Miyamoto et al. [28] TGA of catalyzed (Ca added to fuel) Diesel soot in rapid oxidation. (h) Miyamoto et al. [28] TGA of uncatalyzed Diesel soot. (i) Gilot et al. [41] thermogravimetric analysis (TGA) of carbon black. (j) Neeft et al. [9] flow reactor study of Printex-U flame soot from Degussa AG. (k) Otto et al. [7] TGA of Diesel particles. (l) Ahlström and Odenbrand [8] flow reactor study of Diesel particles. Note: The surface specific rates are obtained for 130 nm initial sizes assuming the density of 1.8 g/cc for (a), (b), (c) and (d).

particles [32]. While this might enhance the oxidation of carbonaceous materials in the filter, the metal compounds remain in the filter leading to gradual plugging. Future oil formulations might be optimized to enhance oxidation while minimizing ash emissions.

A somewhat surprising outcome of this study is the significant reduction of PM volume emissions (which are proportional to mass) with lube oil dosing. These experiments were done with clean lube oil. It would be interesting to examine the effects of aged lube oil, like that used in the CENTINEL™ system, and to further investigate the kinetics of oxidation at a lower level of lube oil dosing.

A disturbing result of lube oil dosing is the formation of a large solid nuclei mode. A number of studies [42, 43] suggest that very small particles, especially solid particles, may be more toxic than larger ones.

CONCLUSION

The main objective of this study was to determine the influence of metals in lube oil on the kinetics of oxidation of soot particles in the exhaust and in aftertreatment systems. This was done by dosing the fuel with 2 % lube oil and comparing the results with previous studies done using undosed fuel. A secondary objective was to examine the influence of dosing the fuel with lube oil on exhaust particle emissions. All testing was done at peak torque speed and 75 % load.

The effect of lube oil metals on the kinetics of Diesel particle oxidation was examined using the HTO-TDMA method. The light-off temperature (which is defined as the peak flow tube temperature to observe ~1 nm size reduction by oxidation) was reduced about 150-200 °C with 2 % lube oil dosed fuel compared to when using regular fuel. The oxidation rate was increased significantly by adding lube oil to the fuel. The activation energy for 2 % lube oil dosed fuel is 101 kJ mol^{-1}, which is very close to the 108 kJ mol^{-1} activation energy for particles using undosed fuel. However, the frequency factors (A_{nm}'s) of the Arrehenius equation with 2 % lube oil-dosed fuel are about twice as large as those for undosed fuel.

The kinetics of oxidation of Diesel particles, especially activation energies are quite different from those of ethylene diffusion flame soot (164 kJ mol^{-1}) and the well-known NSC model (143 kJ mol^{-1}). Our results suggest that Diesel particles are metal catalyzed to some extent, even when using regular undosed Diesel fuels, by the small amount of metals available from normal lube oil consumption. The addition of lube oil in the fuel increases the number of catalytic sites (an increase in frequency factors), but the activation energy is barely lowered, implying that Diesel particles are already catalyzed to some extent by natural consumption of lube oil.

Particle size distribution measurements show that particle volume emissions, which are roughly proportional to particle mass, were decreased by about a factor of two with dosed fuel but that particle number emissions increased by an order of magnitude. Most of the particle number was found in tiny, solid nuclei mode particles (D_p < ~30 nm). These experiments indicate that lube oil metals influence Diesel particle formation, oxidation and the performance of aftertreatment systems. Future lube oil formulations may have to take these effects into consideration.

ACKNOWLEDGMENTS

The authors gratefully acknowledge Deere & Company for providing information about their lube oil, and Cummins Incorporation for providing information about their CENTINEL™ system.

CONTACT

Heejung Jung: heejung@me.umn.edu
David B. Kittelson: kitte001@umn.edu
Michael R. Zachariah: mrz@me.umn.edu

DEFINITIONS, ACRONYMS, ABBREVIATIONS

CPC: Condensation Particle Counter

DMA: Differential Mobility Analyzer

DPF: Diesel Particulate Filter

DPM: Diesel Particulate Matter

HTO-TDMA: High Temperature Oxidation-Tandem Differential Mobility Analyzer

NSC: Nagle Strickland and Constable [11]

PM: Particulate Matter

SMPS: Scanning Mobility Particle Sizer

TDMA: Tandem Differential Mobility Analyzer

TGA: ThermoGravimetric Analyzer

VOF: Volatile Organic Fraction

REFERENCES

1. Higgins, K.J., H. Jung, D.B. Kittelson, J.T. Roberts, and M.R. Zachariah, *Kinetics of Diesel Nanoparticle Oxidation.* Environ. Sci. Technol., 2003. **37**: p. 1949-1954.
2. Higgins, K.J., H. Jung, D.B. Kittelson, J.T. Roberts, and M.R. Zachariah, *Size-Selected Nanoparticle Chemistry: Kinetics of Soot Oxidation.* J. Phys. Chem. A., 2002. **106**: p. 96-103.
3. Pattas, K., N. Kyriakis, Z. Samaras, T. Manikas, A. Mihailidis, W. Mustel, and P. Rouveirolles, SAE TECHNICAL PAPER SERIES, 1998: p. 980543.
4. Quigley, M. and T. Seguelong, SAE TECHNICAL PAPER SERIES, 2002: p. 2002-01-0436.
5. Skillas, G., Z. Qian, U. Baltensperger, U. Matter, and H. Burtscher, *The Influence of Additives on the Size Disribution and Composition of Particles Produced by Diesel Engines.* Combust. Sci. and Tech., 2000. **154**: p. 259-273.
6. Lahaye, J., S. Boehm, P.H. Chambrion, and P. Ehrburger, *Influence of Cerium Oxide on the Formation and Oxidation of Soot.* Combustion and Flame, 1996. **104**: p. 199-207.
7. Otto, K., M.H. Sieg, M. Zinbo, and L. Bartosiewicz, SAE TECHNICAL PAPER SERIES, 1980: p. 800336.
8. Ahlström, A.F. and C.U.I. Odenbrand, Carbon, 1989. **3**: p. 475.
9. Neeft, J.P.A., T.X. Nijhuis, E. Smakman, M. Makkee, and J.A. Moulijin, Fuel, 1997. **76**: p. 1129.
10. Song, H., N. Ladommatos, and H. Zhao, *Diesel soot oxidation under controlled conditions.* SAE TECHNICAL PAPER SERIES, 2001: p. 2001-01-3673.
11. Nagle, J. and R.F. Strickland-Constable. in *Fifth Carbon Conference*. 1962: Pergamon: Oxford.
12. Caines, A. and R. Haycock, *Automotive Lubricants Reference Book*. 1996: SAE.
13. Winsor, R.E., John Deere, Editor. 2003.
14. Thorne, C.J., Castrol, Editor. 2003, Castrol.
15. Farrar-Khan, J.R., G.E. Andrews, P.T. Williams, and K.D. Bartle, SAE TECHNICAL PAPER SERIES, 1992: p. 921649.
16. De Petris, C., V. Giglio, and G. Police, *Some insights on mechanisms of oil consumption.* SAE TECHNICAL PAPER SERIES, 1996: p. 961216.
17. Audette III, W.E. and V.W. Wong, *A model for estimating oil vaporization from the cylinder liner as a contributing mechanism to engine oil consumption.* SAE TECHNICAL PAPER SERIES, 1999: p. 1999-01-1520.
18. Flynn, P.F., R.P. Durrett, G.L. Hunter, A.O. Loye, O.C. Akinyemi, J.E. Dec, and C.K. Westbrook, *Diesel combustion: an integrated view combining laser diagnostics, chemical kinetics and empirical validation.* SAE TECHNICAL PAPER SERIES, 1999: p. 1999-01-0509.
19. Graham, J.P., J.C. Wang, L.A. Roettgen, D.P. Carver, C.I. Black, D.M. Stehouwer, and R.C. Hatton, *Electronically controlled continuous lubricating oil replacement system*. 1998: United States Patent. p. 5,749,339.
20. Cummins, *Centinel*^{TM}. 2003: www.cummins.com/au/pages/en/products/genuinepartsandservices/extendedlifeproducts/centinel.cfm.
21. Kittelson, D.B., D.F. Dolan, R.B. Diver, and E. Aufderheide, *Diesel Exhaust Particle Size Distributions-Fuel and Additive Effects.* SAE TECHNICAL PAPER SERIES, 1978: p. 780787.
22. Stanmore, B., B. J.-F., and P. Gilot, *The ignition and combustion of cerium doped diesel soot.* SAE TECHNICAL PAPER SERIES, 1999: p. 1999-01-0115.
23. Kytö, M., P. Aakko, N.-O. Nylund, and A. Niemi, *Effect of Lubricant on Particulate Emissions of Heavy Duty Diesel Engines.* SAE TECHNICAL PAPER SERIES, 2002: p. 2002-01-2770.
24. Wang, S.C. and R.C. Flagan, Aerosol Sci. and Technol., 1990. **13**: p. 230-240.
25. Abdul-Khalek, I., D.B. Kittelson, and F. Brear, *The Infulence of Dilution Conditions on Diesel Exhaust Particle Size Distribution Measurements.* SAE TECHNICAL PAPER SERIES, 1999: p. 1999-01-1142.
26. Khalek, I.A., D.B. Kittelson, and F. Brear, *Nanoparticle Growth during Dilution and Cooling of Diesel Exhaust: Experimental Investigation and Theoretical Assessment.* SAE TECHNICAL PAPER SERIES, 2000: p. 2000-01-0515.
27. Miyamoto, N., H. Zhixin, A. Harada, H. Ogawa, and T. Murayama, SAE TECHNICAL PAPER SERIES, 1987: p. 871612.
28. Miyamoto, N., H. Zhixin, and O. Hideyuki, SAE TECHNICAL PAPER SERIES, 1988: p. 881224.
29. Stenitzer, M. 2003, Vienna - University of Technology: Vienna.
30. Jung, H., D.B. Kittelson, and M.R. Zachariah, *The Influence of Cerium Additive on Diesel Nanoparticle Emissions and Kinetics of Oxidation.* in preparation for publication, 2003.
31. Kasper, M., K. Sattler, K. Siegmann, U. Matter, and H.C. Siegmann, *The influence of fuel additives on the formation of carbon during combustion.* J. Aerosol. Sci., 1999. **30**: p. 217.
32. Mayer, A., A. Ulrich, J. Czerwinski, U. Matter, and M. Wyser, *Retention of fuel borne catalyst particles by diesel particle filter systems.* SAE Special Publication, 2003(SP-1755): p. 21-27.
33. Abdul-Khalek, I.S., D.B. Kittelson, B.R. Graskow, Q. Wei, and F. Brear, *Diesel Exhaust Particle Size: Measurement Issues and Trends.* SAE TECHNICAL PAPER SERIES, 1998: p. 980525.
34. Abdul-Khalek, I. and D.B. Kittelson, *Real Time Measurement of Volatile and Solid Exhaust Particles Using a Catalytic Stripper.* SAE TECHNICAL PAPER SERIES, 1995: p. 950236.

35. Mayer, A., J. Czerwinski, U. Matter, M. Wyser, P. Scheidegger, D. Kieser, and Weidhofer, *VERT: Diesel Nano-Particulate Emissions: Properties and Reduction Strategies.* SAE TECHNICAL PAPER SERIES, 1998: p. 980539.
36. Mayer, A., J. Czerwinski, and P. Scheidegger, *Trapping efficiency depending on particulate size.* SAE TECHNICAL PAPER SERIES, 1996: p. 960472.
37. Sakurai, H., H.J. Tobias, K. Park, D. Zarling, K.S. Docherty, D.B. Kittelson, P.H. McMurry, and P.J. Ziemann, *On-line measurements of diesel nanoparticle composition and volatility.* Atmospheric Environment, 2003. **37**: p. 1199–1210.
38. Tobias, H.J., D.E. Beving, P.J. Ziemann, H. Sakurai, M. Zuk, P.H. McMurry, D. Zarling, R. Waytulonis, and D.B. Kittelson, *Chemical Analysis of Diesel Engine Nanoparticles Using a Nano-DMA/Thermal Desorption Particle Beam Mass Spectrometer.* Environ. Sci. Technol., 2001. **35**: p. 2233-2243.
39. Gordon, S. and B.J. McBride. 1976, NASA.
40. Stratakis, G.A. and A.M. Stamatelos, *Thermogravimetric analysis of soot emitted by a modern diesel engine run on catalyst-doped fuel.* Combust. Flame, 2003. **132**: p. 157-169.
41. Gilot, P., F. Bonnefoy, F. Marcuccilli, and G. Prado, Combust. Flame, 1993. **95**: p. 87.
42. Pope, C.A., III, M.J. Thun, M.M. Namboodiri, D.W. Dockery, J.S. Evans, F.E. Speizer, and C.W. Heath , Jr., *Particulate Air Pollution as a Predictor of Mortality in a Prospective Study of U.S. Adults.* Am. J. Respir. Crit. Care Med., 1995. **151**: p. 669-674.
43. Dockery, D.W., C.A. Pope III, X. Xu, J.D. Spengler, W.J. H., F. M.E., M.E. Fay, B.G. Ferris, and F.E. Speizer, *An Association between Air Pollution and Mortality in Six U.S. Cities.* New Eng. J. Med., 1993. **329**(24): p. 1753-1759.

2004-01-1989

Fuel and Lubricant Effects on Nucleation Mode Particle Emissions From a Euro III Light Duty Diesel Vehicle

Jon Andersson
Ricardo Consulting Engineers

Hugh Preston, Chris Warrens and Peter Brett
BP

ABSTRACT

The impact of lubricant sulphur and phosphorus levels on the formation of nucleation mode particles was explored in a light duty diesel vehicle operating over the New European Drive Cycle (NEDC). All measurements were undertaken using a Scanning Mobility Particle Sizer (SMPS), sampling from a conventional Constant Volume Sampler (CVS) system. Rigorous sampling system and vehicle conditioning procedures were applied to eliminate oil carry-over and nanoparticle artifact formation.

An initial vehicle selection process was undertaken on vehicles representing three fuel injection strategies, namely; distributor pump, common rail and unit injector. The vehicles met Euro III specifications and were all equipped with oxidation catalysts. Idle and low load stability were key requirements, since these conditions are the most significant in terms of their propensity to generate nucleation mode particles.

The influence of fuel sulphur on nucleation particle generation is well documented and three fuels were tested at idle to confirm the chosen vehicle's response to fuel sulphur. Traditional European fuels exhibited strong nucleation modes, with the 50ppm S fuel showing ~12% greater nucleation mode levels than the 10ppm S fuel. Swedish Class 1 Diesel (SWCL1) fuel gave a very stable nucleation particle mode, but at a lower level. This fuel was selected for the lubricant effect study since it was believed that its reduced final boiling point would not overlap with or interfere with the oil discrimination phase of the work.

A total of 9 oils were tested with sulphur and phosphorus levels ranging from 0 to 0.9 and 0 to 0.1%wt respectively. All the oils were fully formulated having SAE "X"W-30 viscosities and a targeted ACEA A5/B5 performance level. Test order was carried out under strict statistical control.

Oil derived sulphur and phosphorus were both found to increase the production of nucleation mode particles. Little effect of lubricant formulation on accumulation mode particles was observed. Swedish Class 1 (SWCL1) Diesel fuel provided substantial reductions in nucleation mode particles.

INTRODUCTION

Particle measurement processes and emissions from light duty vehicles were investigated in the DETR/ SMMT/ CONCAWE Particulate Research Programme (DSC_PRP)[1,2]. In that programme, the test protocol was well designed and rigorous, to ensure that all vehicles were tested in a directly comparable manner. A conventional CVS-dilution tunnel formed the source for the equivalent sampling of particle and particulates, and results showed that with such a test protocol, both nucleation and accumulation mode particle size distributions could prove highly repeatable.

Data acquired during the DSC_PRP showed that conventional light duty diesel vehicles tend not to produce size distributions with distinct nucleation modes except at idle steady state. Three of the major factors that determine this are the relatively high engine-out carbon concentrations generated by modern NOx limiting calibrations, the absence of sulphur in the fuel and the relatively low levels of lubricant and fuel which survive combustion.

At idle, carbon concentrations in the exhaust are substantially reduced and cooler combustion temperatures enhance the probability of fuel and lubricant survival. These factors lead to a high probability of the formation and growth of a nucleation mode at the dilution ratio, residence time and temperature regime within a conventional CVS-dilution system. Nucleation mode production in real time was also investigated in the

DSC_PRP using SMPS systems set to monitor single particle sizes. This revealed that substantial numbers of 10nm, 15nm and 30nm particles were present in a conventional CVS-dilution system from light diesel exhaust during transient cycles.

In this programme, the tendency of a conventional CVS-dilution system to promote nucleation mode particle formation[3] was exploited in order to investigate differences between the emissions nucleation mode particles from various lubricants at both idle and transient operating conditions.

BACKGROUND

PARTICLE NUMBERS – It has been suggested that a recent reduction in fuel sulphur in Denmark has led to a 50% reduction in ambient particle number concentration[4]. In addition, health effects studies are now beginning to provide the first evidence that there may be links between negative health episodes and the concentration of particle numbers in the ambient air[5].

Over the last decade, European legislation has forced significant cuts in sulphur levels in fuels[6], with typical UK forecourt diesel now showing the sulphur levels of less than 50ppm required for 2005, with typically levels around 30ppm. At these levels of fuel sulphur, the contribution of the lubricant sulphur becomes measureable.

LUBRICANT ISSUES – It has become widely accepted that fuel sulphur can be linked to increased exhaust emissions. This occurs as a direct contribution to particulates and associated chemical compounds and to the direct chemical degradation of catalytic after treatment systems. The link to legislative particulate emissions for lubricant sulphur remains unproven despite a number of studies. In general, it is accepted that fuel sulphur will have to fall below 10ppm before lubricant sulphur contributes noticeably to legislated emissions. The method by which lubricant enters the combustion chamber is not fully understood with uncertain quantities of the formulation surviving the combustion process according to the timing and method of entering the combustion process. It can be appreciated therefore that a simple ratio of fuel consumption at given sulphur content versus oil consumption and sulphur content does not apply. Nucleation mode particulates are not currently part of the emission regulations. It is possible that these may be linked to the presence in the particulate of heavy hydrocarbons. It was therefore decided to investigate this further with a parametric study of special lubricant formulations, which allowed the influence of not only sulphur to be investigated but also the key additive components of phosphorus and zinc.

PROGRAMME OBJECTIVES

The primary objectives of the test programme were two-fold:

- To provide comparative particle size and regulated emissions data between different lubricant chemistries in terms of particulate mass and particle size parameters from idle and New European Drive Cycle (NEDC) transient conditions.
- To process transient particle number data from transient cycles in order to elucidate cycle effects: cruises, idles, accelerations and decelerations on particle production.

EXPERIMENTAL METHODOLOGY

ANALYTICAL MEASUREMENT SYSTEMS

CHASSIS DYNAMOMETER, ENVIRONMENT AND DILUTION SYSTEM – All vehicles were tested on a Schenk GS56 twin roller chassis dynamometer equipped with a 12" diameter stainless steel dilution tunnel. The dilution tunnel has HEPA and charcoal filters at the air intake to ensure low dilution air particle numbers. CVS flow rate was set to 12m^3/min: which was adequate to avoid condensation of water in the dilution tunnel under the highest load conditions. Dilution ratios across an NEDC cycle ranged from ~4:1 to 17:1. A corrugated stainless steel transfer pipe was employed to connect vehicle exhausts to the CVS tunnel. This was approximately 4m in length and was rigorously conditioned prior to the programme and prior to each test day.

The temperature and humidity of the chassis dynamometer area were closely controlled to 25°C +/- 1°C and 40% +/- 3% respectively.

INSTRUMENTATION – The following analytical equipment was employed in both the pre-studies and main test programmes:

Regulated Emissions– Gaseous emissions and particulate masses were measured according to regulated testing procedures[7] and using equipment that is certified by the UK Vehicle Certification Agency (VCA) for type approval purposes:

Hydrocarbon Emissions – were measured by heated flame ionization detector (FID).

Oxides of Nitrogen – were measured using a chemiluminescence analyzer. This method determines NO and NO_2.

Carbon Monoxide and Carbon Dioxide were determined using Non-dispersive Infra-red analyses (NDIR).

Particulate Measurements– Particulate masses were determined according to regulated testing procedures[7]: dilute exhaust gas was drawn through a sample probe equipped with a conical shroud that was situated ten tunnel diameters downstream of the exhaust and dilution air mixing zone. This shroud excludes particles of larger than ~3μm, which are likely to comprise agglomerates re-entrained from the tunnel wall. 47mm Whatman GF/A filters were used to collect particulate mass since these can be subjected to thermogravimetric analysis and elemental carbon levels determined. Sample flow rate through the filters was set to 35 l/min.

Particle Measurements– Twinned model 3934 Scanning Mobility Particle Sizers (SMPS) were employed to determine number weighted particle size distributions. Samples were drawn directly from the CVS tunnel through probes analogous to those used for regulated PM sampling. Since the particle size range measured peaked at less than 500nm, an isokinetic sampling approach was not deemed necessary. Experience gained within the DSC_PRP was employed to give best practice for SMPS measurements[8].

SMPS systems comprise an impactor which removes large particles above the size range of interest and which readily assume multiple charges, a diffusion charger which reduces the charges on the particles to a predictable distribution, an electrostatic classifier which separates out each specific particle size of interest and a condensation particle counter (TSI 3022A) which performs the counting operation. Each particle size distribution takes approximately 2 minutes to acquire. For this reason, in scanning mode the SMPS is unsuitable for determining particle size distributions from transient cycles. However, narrow size ranges can be determined on a second by second basis and these used to continuously trace particle emissions during transient cycles: this is known as the 'size-filter' approach. A comprehensive description of SMPS functionality and the size filter approach has been published previously[9].

Both SMPS systems were configured to measure the size range 7nm to 320nm (based upon sheath and sample flows of 10l/min and 1l/min respectively). The instruments were matched to within 5% absolute number from individual channels and 3% integrated total during simultaneous sampling. Software version 2.4 was employed with both SMPS systems.

VEHICLE TEST METHODOLOGIES

As the final activity on the night prior to the test day, the vehicle, transfer pipe and CVS were preconditioned by 3 consecutive Extra Urban Drive Cycle (EUDC) cycles and then subjected to an overnight soak. As the first tests in the facility the following day, 3 NEDC cycles were driven and the vehicle left to idle for 25 minutes. During the final 15 minutes of this idle period, particle size distributions were acquired using a Scanning Mobility Particle Sizer (SMPS). During this period the catalyst cools to below its normal working temperature. Further 25 minutes steady states at 10kph and 15kph were undertaken subsequently, again with particle size distributions acquired during the final 15 minutes. Regulated gaseous emissions were monitored and particulate filters collected during each steady state period. Details are given in Table 1.

This protocol was employed to ensure that the complete vehicle plus transfer and dilution systems had been 'conditioned' by the fuel and lubricant combination prior to any testing. Data from each vehicle were inspected to determine steady state stability of gaseous emissions, presence and magnitude of a nucleation mode and particulate mass emissions.

Each vehicle was subjected to the entire test protocol twice to indicate day-to-day repeatability.

Fuel Screening Protocol – The fuel screening was undertaken during the full test protocol as described in the following experimental methodology section.

Fuel Screening Test Matrix – the test matrix employed during the fuel screening tests is shown in Table 2.

The main test programme employed a statistical approach, described later, to enable discrimination of lubricant sulphur and phosphorus effects on nucleation mode particles from the Peugeot 406 vehicle fuelled with SWCL1.

Fuel and Oil Change Procedures – the fuel and oil were changed according to strictly defined procedures:

Fuel Change - the fuel tank was drained and filled with the new fuel. The vehicle was driven over a prescribed route for 100 miles on the road and then the oil changed and the vehicle conditioned.

Oil Change - a formal oil change procedure was employed to ensure reproducibility between oils prior to any vehicle conditioning. This included a warm-up procedure to mobilize the oil, a drain and volumetrically controlled flush, and oil filter change and a volumetric controlled fill. A more detailed description of this approach can be found in Appendix 2.

Preconditioning and Sulphur Purge – following the fuel and lubricant fills, the vehicle was subjected to three EUDC cycles and placed in the temperature controlled soak area to stabilize over night. This procedure was always the last testing in the facility on each day during the test programme. This conditioning procedure was selected since it reached higher temperatures than would be observed during the test programme and would enable removal of any sulphate compounds stored in the catalyst and exhaust system. High temperature operation was also required to eliminate low volatility materials stored in the vehicle's exhaust system and transfer pipe and to ensure that prior to testing, the CVS tunnel was exposed to, and conditioned with, exhaust gases and particulates from the test vehicle.

Daily Test Protocol – testing was conducted according to the protocol outlined in Table 1. The timings detailed within this protocol were strictly adhered to which ensured identical between-test cooling profiles from day to day.

Order	Cycle	Description	Duration	Measurements
1	Precon	3*EUDC S Purge	previous night	None
2	Warm-up	Cold start NEDC	1180s	tailpipe raws and bagged dilutes
3	stabilisation	idle	900s +/- 2secs	tailpipe raws and bagged dilutes
4	Transient Cycle#1	Hot start NEDC	1180s	Tailpipe raw regems, bagged dilutes, PM plus transient particle number from sizes 1 and 2
5	stabilisation	idle	900s +/- 2secs	tailpipe raws and bagged dilutes
6	Transient Cycle#2	Hot start NEDC	1180s	Tailpipe raw regems, bagged dilutes, PM plus transient particle number from sizes 3 and 4
7	stabilisation	idle	900s +/- 2secs	tailpipe raws and bagged dilutes
8	Transient Cycle#3	Hot start NEDC	1180s	Tailpipe raw regems, bagged dilutes, PM plus transient particle number from sizes 5 and 6
9	stabilisation	idle	900s +/- 2secs	tailpipe raws and bagged dilutes
10	stabilisation	idle	600s +/- 2secs	tailpipe raws and bagged dilutes
11	Idle Evaluation	idle	1200s	Tailpipe raw regems, bagged dilutes, PM plus full scan particle size distributions

Table 1: Daily Test Protocol

The daily test protocol commenced with a standard cold start NEDC (ECE with 11s idle + EUDC) cycle. This acted as the daily preconditioning for all subsequent tests and provided a link to regulated tests.

The analytical process then commenced with 3 hot start NEDC transient cycles. The detailed transient cycle test procedure and methodology used in this programme is specified in EU Directive 98/69/EC. During each of these NEDC cycles, 2 distinct particle sizes were monitored on a second-by-second basis and the data archived. From the 3 tests, 6 data sets at six different sizes were collected. These data were then processed enabling size distributions to be compiled. These six sizes are shown in Figure 1, and were selected to represent dominant sizes within both nucleation and accumulation modes.

The combination of data from twinned SMPS systems measuring different sizes was considered appropriate due to the close similarity of the two systems (Figure 2), where simultaneous measurements showed minimal differences.

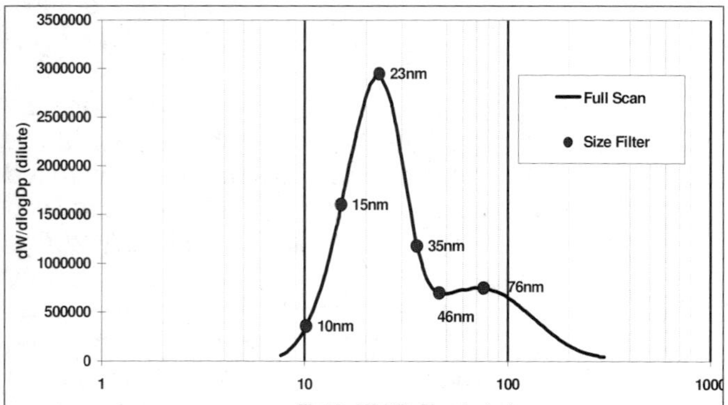

Figure 1: Size-Filtered Particle Measurements

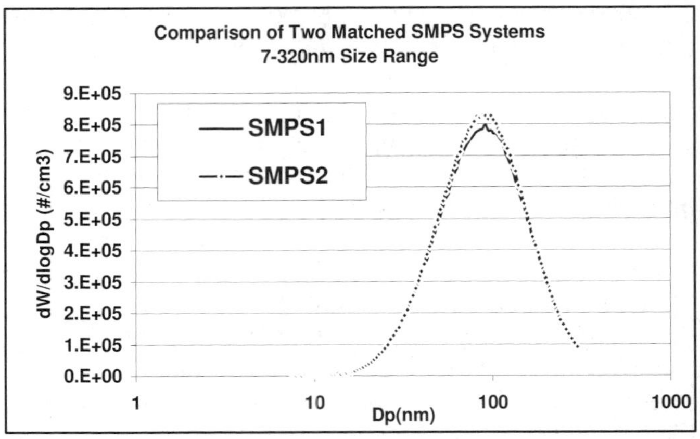

Figure 2: Matched SMPS Systems

Measurement stability in terms of particle size distributions and magnitudes, CO_2 concentrations, dilution ratios and exhaust temperatures at idle were highly consistent. Particle size data at idle are presented in Figure 3.

Measurement Stability – during the fuel screening programme, the repeatability of size distributions measured by SMPS at idle proved to be consistent with that determined for the Peugeot 406 during the vehicle screening pre-study. These data comprised part of the dataset used by the statistician in determination of the measurement variability.

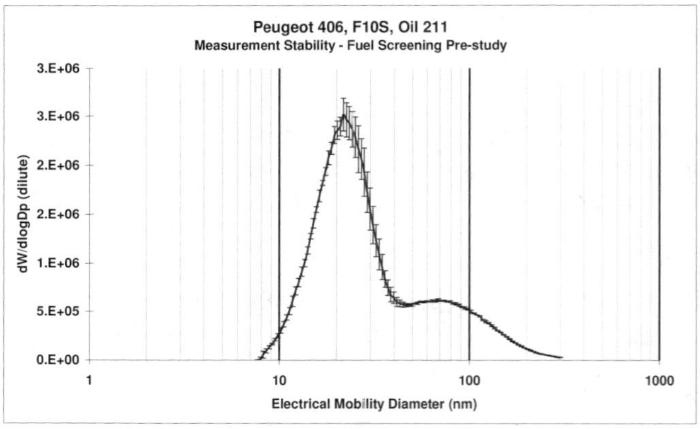

Figure 3: Measurement Stability

Vehicle Operation – the vehicle showed highly stable operation at idle. In general, the dilution ratio was consistent to a coefficient of variance (COV) of 0.5%, while catalyst entry temperature had stabilized at ~77°C prior to sampling. These data are illustrated in Figure 4 and were consistent for all oils.

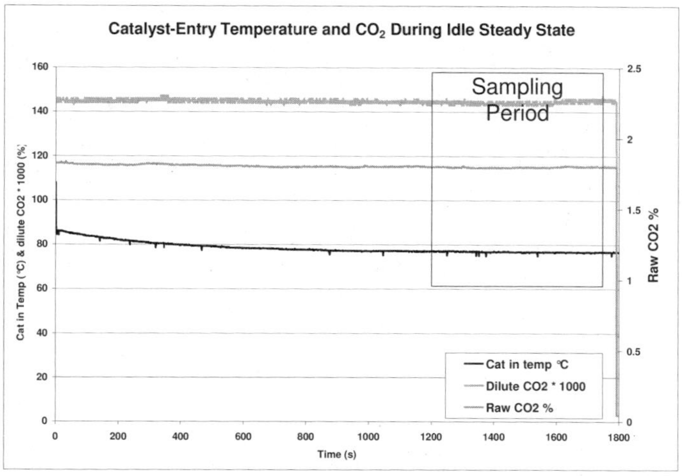

Figure 4: Temperature, dilution and emissions stability at idle

Regulated emissions were monitored from both cold and hot NEDC cycles to ensure consistent vehicle operation. Particulate filters were collected during these tests for comparative mass and, if required, chemical analyses.

Following the transient cycles a warm idle condition was undertaken. During this cycle, high-resolution size distribution measurements were undertaken, with the size range ~7nm to ~320nm examined. Regulated emissions and a PM filter were collected in order to monitor vehicle stability and provide the option for particulate chemical analyses if required.

Between transient cycles and prior to the measured idle, periods of idling were included (see Table 1, steps 3, 5, 7, 9 and 10). These enabled stabilization and standardization of engine, exhaust system and dilution temperatures on a test-to-test basis.

Throughout the test programme, the timing was rigorously controlled with a stopwatch. This enabled extremely high reproducibility of the test protocol on a day-to-day basis.

Test Matrix – the test matrix employed in the main lubricant evaluation programme in shown in Table 2. In this test matrix, 8 oils were each tested once in a sequence prescribed by the programme statistician. The oils were then tested again in an alternative order. Finally, an additional oil; (234A) was tested twice back-to-back.

Order	Test Fuel	Code	Oil No.
1	F10S	190A	1
2	F10S	207E	2
3	F10S	210A	3
4	F10S	201B	4
5	F10S	211C	5
6	F10S	212A	8
7	F10S	178D	6
8	F10S	216D	7
9	F10S	210A	3
10	F10S	207E	2
11	F50S	207E	2
12	SWCL1	207E	2
13	SWCL1	201B	4
14	SWCL1	211C	5
15	SWCL1	190A	1
16	SWCL1	207E	2
17	SWCL1	210A	3
18	SWCL1	201B	4
19	SWCL1	211C	5
20	SWCL1	212A	8
21	SWCL1	178D	6
22	SWCL1	216D	7
23	SWCL1	210A	3
24	SWCL1	212A	8
25	SWCL1	216D	7
26	SWCL1	178D	6
27	SWCL1	190A	1
28	SWCL1	234A	9
29	SWCL1	234A	9

Table 2: Test Matrix, Main programme

Note: F10S = 10ppm sulphur fuel
F50S = 50ppm sulphur fuel

Data Analyses – Particle size distribution data from the idle steady state conditions can be compared on a purely visual basis: due to close control of the laboratory air temperature and humidity and the constant dilution provided by the CVS system, size distributions drawn from the CVS are directly comparable. Therefore, these data are compared graphically as distributions. Peak heights and integrated particle numbers may also be compared without further processing undertaken.

Comparison of transient data is undertaken in two ways:

- Particle production over the entire NEDC cycle is illustrated on a 1Hz transient basis, allowing comparison of oils and lubricants. These data show transient particle number concentration data and do not account for changes in dilution ratio across the cycle.

- Particle data collected over drive cycles can be subdivided into event specific data. The events chosen were; accelerations, cruises, idles and decelerations. In addition, since the final 120kph to idle deceleration is believed to be highly variable, a further category including all decelerations except this was included. For each category, and for each of the six particle sizes measured, an average particle number concentration measurement was determined.

A phased approach to the testing was adopted as follows:

- Phase 1: Vehicle selection
- Phase 2: Fuel sulphur effect validation
- Phase 3: Lubricant testing

RESULTS

PHASE 1 VEHICLE SCREENING – Several Euro III specification vehicles were evaluated in order to determine their tendency to form nucleation mode particles at idle steady state. This evaluation also included establishing the vehicle's stability in terms of regulated emissions, particulate mass and particle size distribution. Three vehicles, including several FIE and 2 and 4-valve combustion systems were selected for screening. These were:

Vehicle	Capacity	Valves/cyl	FIE	Other
VW Passat	1.9	2	Electronic Unit Injector	VGT Intercooler
BMW 320d	2.0	4	High Pressure Distributor Pump	VNT Intercooler
Peugeot 406	2.0	2	Common Rail	TC

Emission testing at idle indicated that the exhaust from the Peugeot 406 had the lowest particulate carbon content, Figure 5, suggesting that this vehicle would be most appropriate for nucleation mode studies.

SMPS testing confirmed this, with the Peugeot 406 having the lowest accumulation mode, and highest nucleation mode. The Peugeot 406 was found to be sensitive to oil anti-foam quality (Figure 6), but good repeatability was achieved when the anti-foam performance of the test oil was improved (Figure 7). Fewer nucleation mode particles were found from any of the vehicles at higher speeds. The Peugeot 406 was therefore selected.

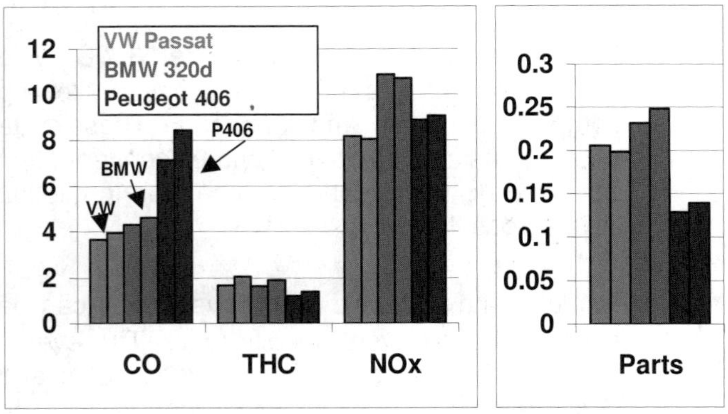

Figure 5: 3 Vehicles' g/h emissions of CO, HC, NOX and PM at idle

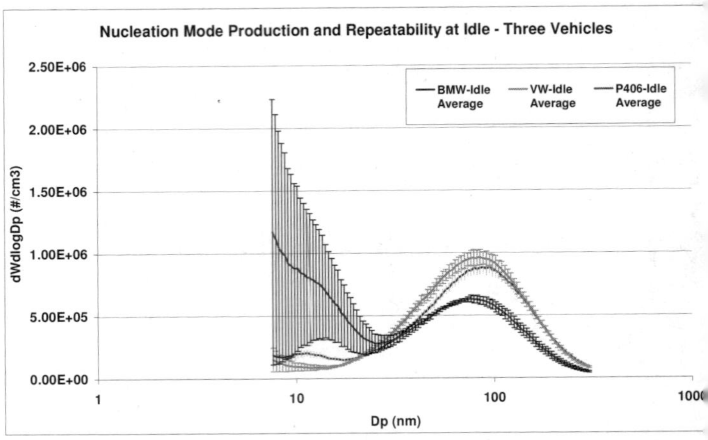

Figure 6: Particle Emissions from 3 Vehicles at Idle

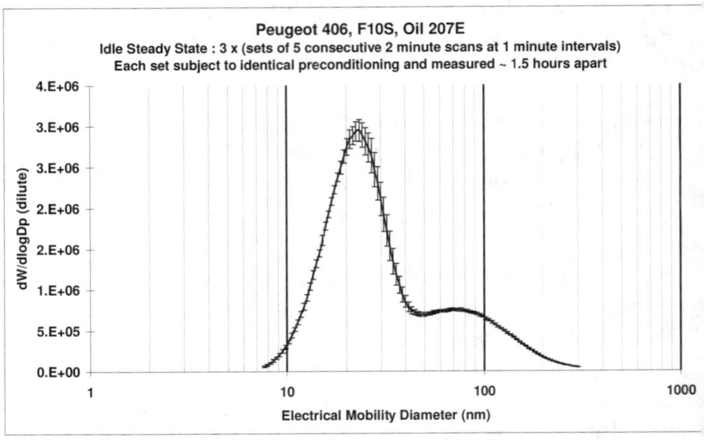

Figure 7: Repeat Analyses of Peugeot 406 with Improved Repeatability

PHASE 2 FUEL SULPHUR VALIDATION STUDIES – Initial tests were conducted on fuels of three sulphur levels (with a single oil) to confirm the engine's expected response. This is shown in (Figure 8).

The three fuels, F10S, F50S and SWCL1 Diesels, were of 10ppm, 50ppm and ~3ppm sulphur content respectively. Otherwise, F10S and F50S were identical (F50S was the same as F10S but doped to 50ppm S). SWCL1 showed many different properties to the other two fuels, including: lower aromatic content, lower density and, in particular, lower T95 (~280°C Vs ~340°C).

Fuel specifications detailing selected physical and chemical properties of both EN590 fuels and the SWCL1 are shown in Appendix 1.

This experiment showed significant nucleation mode production from F10S and F50S, with fewer particles emitted during operation with SWCL1 fuel. This confirmed the vehicle's ability to discriminate between fuels.

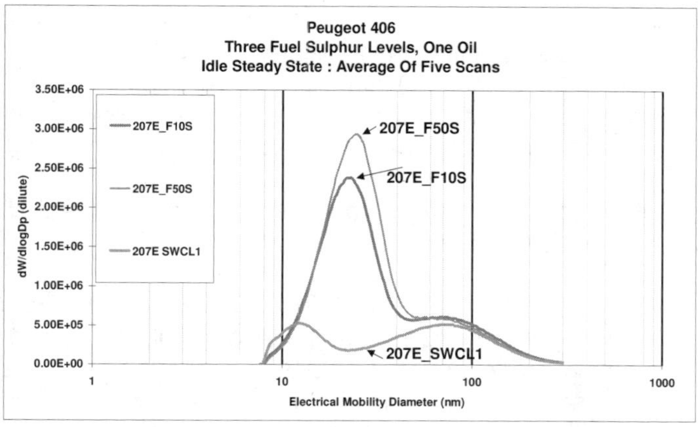

Figure 8: Nucleation Mode Particle Production - 3 Fuel Sulphur Levels

PHASE 3: LUBRICANT STUDIES – A total of 8 SAE 5W-30 lubricants, formulated using Group III (hydrocracked, zero sulphur) base oils were evaluated. A ninth oil was formulated with high sulphur Group I (high sulphur solvent refined mineral) base-oil to give a high sulphur content lubricant. This approach was used, rather than doping with a sulphur compound since this could lead to atypical sulphur consumption by the engine. Phosphorus and sulphur contents were varied by choice of the anti-wear system to give a range of values, as detailed in Table 3.

	Oil code								
Test	190A	207E	210A	201B	211C	178D	216D	212A	234A
Kv40	56.46	58.04	60.57	55.1	65.53	59.77	55.13	53.86	54.37
Kv100	9.77	10	10.24	9.58	9.5	10.22	9.62	9.57	9.46
VI	159	160	157	159	125	160	160	163	158
HTHS	3.04	3.05	3.14	3.02	3.1	2.95	2.98	2.88	3.05
B	0.007	0.003	0.007	0.042	0.038	0.04	0.0005	0.089	0.042
Ca	0.19	0.28	0.18	0.19	0.34	0.18	0.323	0.22	0.181
Mg	0.05	0	0.07	0.03	0.02	0.07	0.013	0.04	0.037
P	0.04	0.1	0.1	0	0.1	0.1	0.043	0	0
S	0.15	0.35	0.25	0.02	0.9	0.04	0.33	0.035	0
Zn	0.04	0.11	0.1	0	0.1	0.04	0.046	0	0

Table 3: Oil Properties

Note: oil 211C was formulated with a high sulphur group I base oil.

A number of novel sulphur- and/or phosphorus-free anti-wear additives were used to supplement the anti-wear performance of those oils containing either reduced or zero zinc dialkyl dithiophosphate (ZDDP).

Tests were conducted upon 8 lubricants with the 10ppm S fuel (F10S). These were tested according to the daily protocol shown in Table 1 and in the order shown in Table 2.

Comparison of Oils tested with Fuel F10S - Size distribution data collected at idle are presented in this section. All data are displayed as measured concentrations (particles/cm^3) – sampled directly from the CVS – in Figure 9.

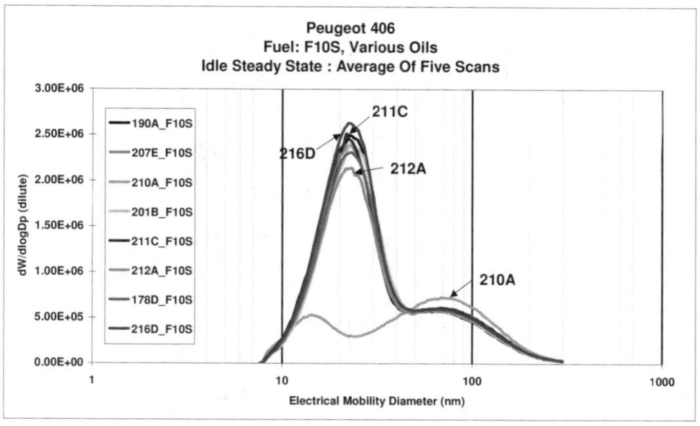

Figure 9: Comparison of Oils - F10S

There was a small reduction in nucleation mode particle numbers associated with oil 212A – the lowest sulphur and zero phosphorus oil tested. The significance of this observation was confirmed by statistical methods. Highest nucleation mode levels were observed from oils 211C and 216D, which were amongst the lubricants with highest sulphur content. However the potential for discrimination between lubricants with F10S and F50S was considered to be low, and an alternative was required.

Consequently, the fuel with very low T95, SWCL1 was chosen. This has a T95 of ~280°C. SWCL1 Diesel also has sulphur levels of <10ppm, lower density and viscosity and a higher H:C ratio than the F10S and F50S fuels, and these differences may also be important. However, the difference (of 60°C) in T95 between the two fuels was clearly the dominant factor.

It was also considered to be advantageous to use SWCL1 fuel for another reason: a fuel with high back-end volatility has less overlap with the front-end distillation of the lubricant. This should better enable discrimination of lubricant effects related to volatility.

Comparison of Oils on Fuel SWCL1 – The fuel was changed to SWCL1 and 100 miles on-road conditioning undertaken. A test was undertaken on oil 207E and this compared with 207E tests on F10S and F50S. These data are shown in Figure 8 and included in Figure 10.

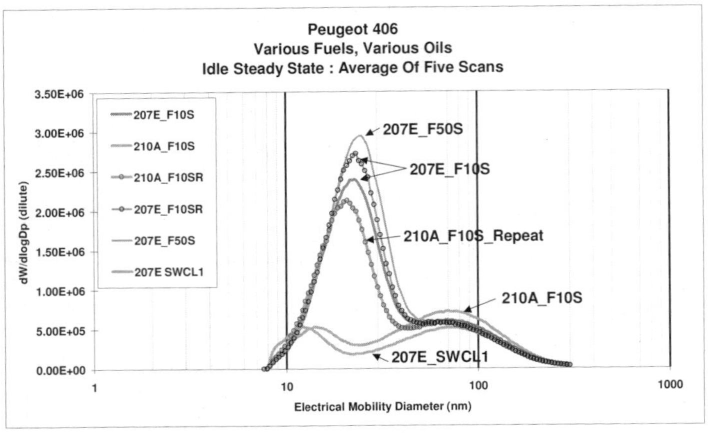

Figure 10: Fuel and Lubricant Effects

This change in fuel promoted a considerable reduction in nucleation mode particle numbers and also a small reduction in accumulation mode particle numbers. Directionally, the reductions in both modes were similar. This is strong evidence of a real reduction in both accumulation mode (carbonaceous) and nucleation mode (volatile) materials with this fuel. The reduction in accumulation mode at idle is probably a consequence of the reduced C:H ratio of the fuel: implying less carbon mass per unit volume injected and the probability of less carbon survival per cycle. The reduction in the nucleation mode volatiles results from the relatively high volatility of the fuel's back-end distillation.

The reduction in nucleation mode levels (7nm to 50nm) with this change in fuel was approximately 75%. With this large reduction in the levels of nucleation mode particles, it was clear that the next stage was to screen extreme oils from the test matrix to evaluate if any oil-to-oil discrimination was feasible. The oils chosen were; 211C (0.9% sulphur and 0.1% phosphorus) and 201B (0.05% sulphur and 0% phosphorus). Shown in Figure 11, this comparison demonstrated (statistically) significant separation between the three oils at the peak of nucleation mode (13nm), and indistinguishable differences between accumulation modes.

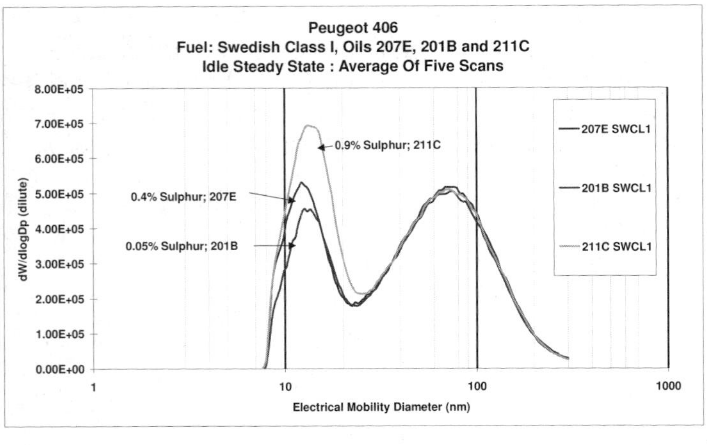

Figure 11: Discrimination between Lubricants on SWCL1 Fuel

The main programme was then commenced, in which nine oils were tested in duplicate, all on SWCL1 fuel. The results of these evaluations were subjected to a rigorous statistical analysis to determine effects of lubricant sulphur, phosphorus and other elements.

In the main test programme, testing was undertaken according to the order shown in Table 1.

PARTICLE SIZE AND NUMBER DATA AT IDLE – From the idle operating condition, particle numbers in both nucleation mode and accumulation mode size regions showed sensitivities to oil. Figure 12 shows oil-averaged data (both repeats of each oil ignoring any test order effects) showing there is a clear separation between oils in the nucleation mode: highest sulphur lubricants (211C, 0.9%; 216D and 207E, 0.4%) show highest particle numbers and lowest sulphur lubricants (234A, zero and 212A, 0.05%) the lowest particle numbers. Accumulation mode differences were smaller and not directly linked with sulphur (or phosphorus) levels.

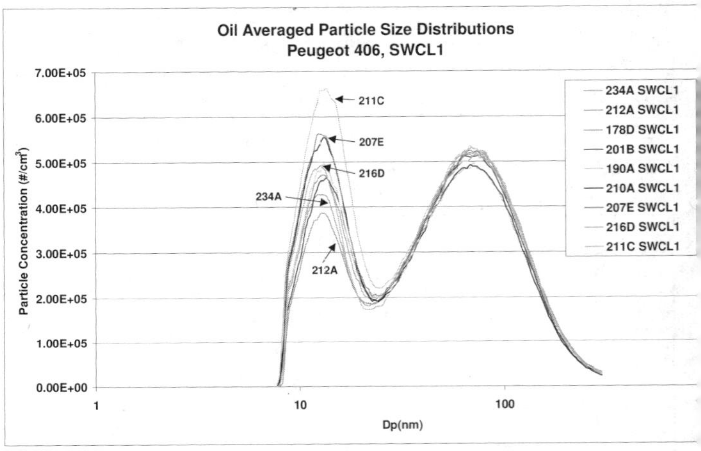

Figure 12: Oil Averaged Particle Size Distributions

TRANSIENT DRIVE CYCLE RESULTS

Both cold and hot start transient cycles, regulated gaseous emissions and particulate mass data were acquired. During hot start transient cycles only, size-filtered particle number measurements were undertaken.

Regulated Emissions Results – During cold start testing, the Peugeot 406 easily achieved Euro III emissions standards for all emissions and with SWCL1 fuel achieved Euro IV emissions levels for CO and PM. Emissions data are summarized in Table 4. The repeatability of the data was generally good, at 16% or below and at 6.4% for both PM and NOx. However, the underlying data quality appeared better than this, and there were apparent oil-to-oil differences within the overall data set.

	Average	COV
CO	0.2015	15.0
CO_2	156.68	1.4
HC	0.0306	16.1
NOx	0.3763	6.4
PM	0.0208	6.4

Table 4: Summarised Cold NEDC Emissions

Cold Start NEDC Cycles Regulated Emissions; Lubricant Effects - Although the test programme was not statistically designed to investigate the effects of lubricants on regulated emissions and particulates, directional differences between lubricants could be discerned. For cold start tests, as Figure 13 and Figure 14 indicate for NOx and PM emissions respectively, some oil-to-oil variations can be seen. These figures show both test order and data grouped in test pairs.

These figures suggest that in general the NOx/PM trade-off seen with conventional diesels is present. However, the lowest NOx is observed with the zero sulphur, zero phosphorus oil (234A), but not the highest PM emissions.

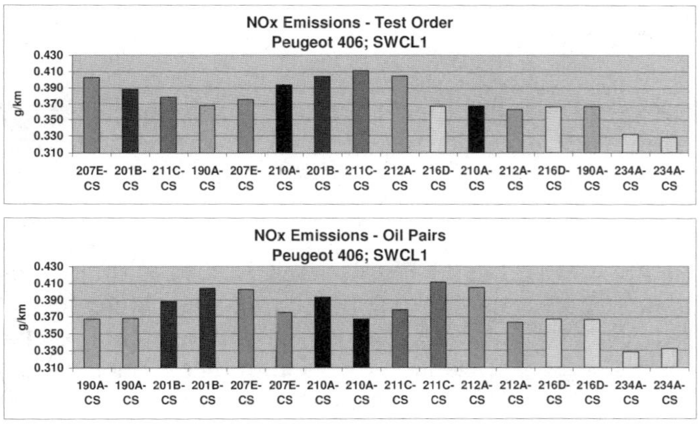

Figure 13: NOx Emissions From Various Oils, Cold Start NEDC Tests

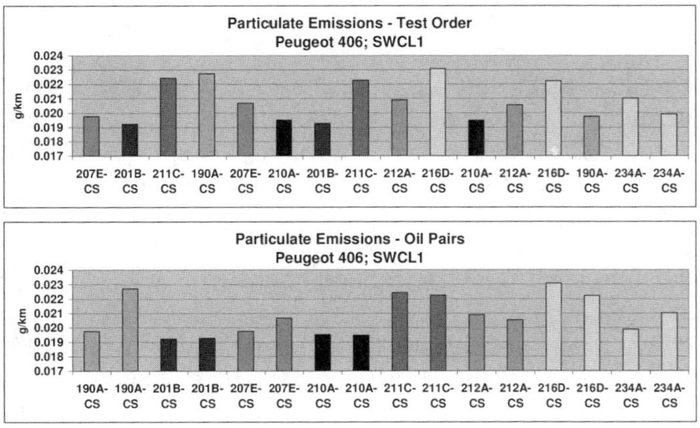

Figure 14: PM Emissions From Various Oils, Cold Start NEDC Tests

No obvious trends were observed with carbon monoxide or hydrocarbon emissions.

Hot Start NEDC Cycles Regulated Emissions; Lubricant Effects – In general, the directional trends observed in the cold start data were repeated in the hot start data. For example, with particulate mass emissions (Figure 15), highest PM emissions from both cold and hot start tests were observed from oils 211C and 216D with lowest emissions from oils 210B and 210A.

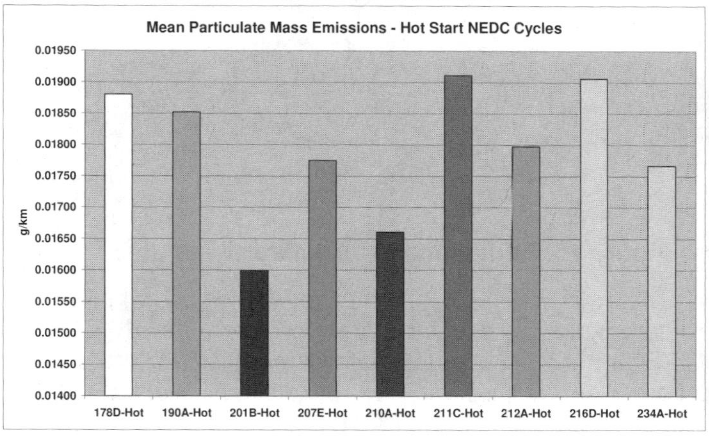

Figure 15: PM Emissions Trends From Hot Start NEDC Tests Reflect Those Of Cold Start Tests

NOx emissions also showed dependency on lubricant formulations, while CO and HC effects were unclear.

Particle Size and Number Measurements – Data from particle measurements taken during hot start transient cycles are described in the following section.

Particle Production Events - With all oils tested, electrical mobilities equivalent to singly charged particles of 10nm and 15nm, 23nm and 35nm plus 46nm and 76nm were monitored during hot start NEDCs 1, 2 and 3 respectively. These particle sizes represent three distinct classes:

- 10nm and 15nm; nucleation mode particles
- 23nm and 35nm; mixed nucleation and accumulation mode
- 46nm and above; dominant accumulation mode

Nucleation mode particles, represented by 10nm data, were primarily produced in response to deceleration events. Particle production proved highly similar between oils, except at the uncontrolled 'foot-off' deceleration from 120kph to idle at the end of the EUDC cycle. Low levels of these particles were also emitted at the start of both ECE and EUDC cycles, though these emissions were small compared with the deceleration related emissions. The close similarity between oils and the variation in the EUDC deceleration can be clearly seen in Figure 16, which illustrates 10nm particle production from 3 oils with the SWCL1 fuel. With the F10S and the F50S fuels, trends were similar though absolute levels varied.

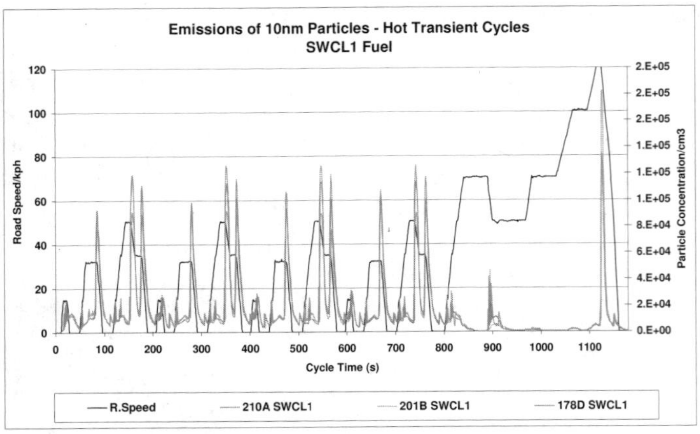

Figure 16: Emissions of 10nm Particles

The emissions of 76nm particles, which are wholly in the carbonaceous accumulation mode from the SWCL1 fuel (Figure 17), showed a quite different trend to the 10nm particles: emissions were clearly highest during accelerations. However, particle numbers remained elevated during cruise conditions, appeared to respond to gear changes and reduced with decelerations. This effect was true with all three fuels tested. Overall, particle production followed the drive trace quite closely and therefore showed a strong synergy with fuelling, and CO_2 emissions.

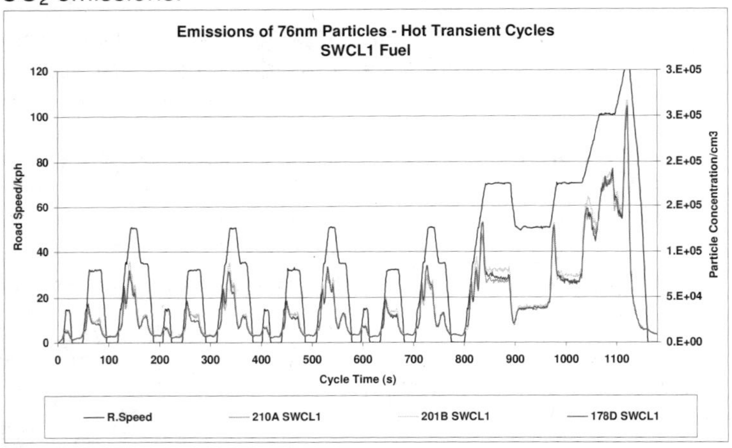

Figure 17: Emissions of 76nm particles

Emissions of 23nm particles (Figure 18) appeared to share characteristics of both 10nm and 76nm emissions. This particle size, and to a lesser extent 36nm, reside within the overlap region between the nucleation and accumulation modes.

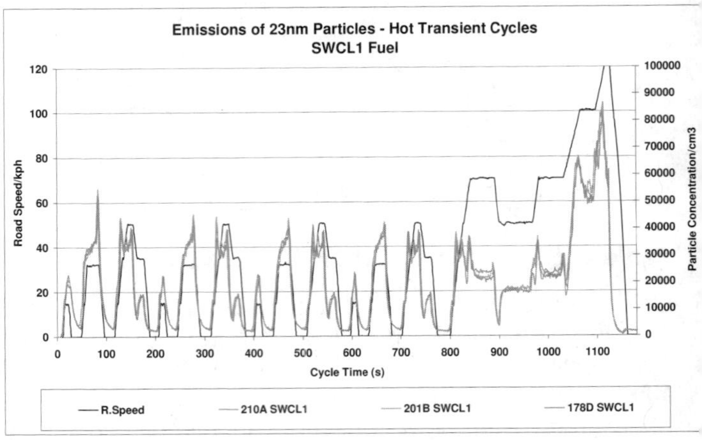

Figure 18: Emissions of 23nm Particles

The transient production of 23nm particles exhibits, as with 76nm, a close replication of the drive trace, but additionally, shows large spikes (as 10nm) coinciding with decelerations. This suggests that although all particles are of 23nm diameter, they are probably of two distinct compositions and origins – both condensed materials and carbonaceous solids.

DISCUSSION

PHASE 1 VEHICLE SELECTION – The principle that every vehicle would have its own characteristic particle emissions was considered, and careful vehicle selection was key to the success of the work.

PHASE 2 FUEL S EFFECT VALIDATION – It has been demonstrated that fuel sulphur, through its conversion by oxidation to sulphate, can be a key element in the formation of a nucleation mode from diesel engines and vehicles[10]. However, other fuel parameters including low back-end volatility may also be substantial influencing factors.

The vehicle was initially filled with UK EN590:2000 specification diesel of 10ppm sulphur. A second UK EN590 fuel was also studied. This fuel (F50S) was identical to the first, except that the sulphur level was increased to 50ppm by doping; this ensured that comparisons between fuels were restricted to sulphur effects. A further fuel, of SWCL1 Diesel specification, was subsequently tested. This showed several differences to the EN590 fuels: <10ppm (~3ppm) sulphur, lower T95 and reduced density.

It was then decided to test the methodology with the 50ppm sulphur fuel (F50S) to evaluate discrimination; but while that showed an increase in the nucleation mode this was relatively small (Figure 9). This test suggested strongly that there was an increase in nucleation mode particle numbers with the 50ppm S fuel. In percentage terms, the increase in nucleation mode particles (in the size range 7 to 50nm) observed from the average of the two results on the 10ppm S fuel to those observed with the 50ppm S fuel was around 12%.

In contrast, the SWCL1 fuel showed substantial reductions in nucleation mode particles that could not be attributed solely to sulphur. Fuel final boiling point and possibly aromatics, or heavy aromatics were considered more likely to be responsible.

PHASE 3 LUBRICANT SCREENING ON F10S - Overall, Figure 9 shows that from all 8 tests, the only clearly different result was observed from oil 210A. This exhibited a noticeable increase in the magnitude of the accumulation mode (peaking at >30nm), and a decrease in the nucleation mode (peak located at <30nm). However, inspection of regulated gaseous emissions revealed there to have been a shift in engine operation on this oil: particulate mass, CO_2, HC and NOx were highest of all oils while CO was lowest. This test was subsequently repeated and the result, which was more consistent with other oils' size distributions, is shown in Figure 10.

At this stage it was considered unlikely that the methodology would discriminate between lubricants adequately, since a relatively small separation was observed between extreme lubricants with sulphur levels varying between 0.035% (212A) and 0.9% (211C).

The influence of other fuel parameters than sulphur to the formation of nucleation mode particles was therefore considered. Other than condensation of hydrocarbons on to sulphate nuclei – the process of heterogeneous nucleation - homogeneous nucleation (where particle droplets spontaneously condense) can occur. This process would be enhanced by incomplete combustion of a large low volatility component fuel. The presence of low volatility material is best represented by a fuel's T95 point – the temperature at which 95% of the fuel has evaporated away during an atmospheric pressure distillation. For the two EN590 fuels, and at 340℃, this was towards the top end of the EN590 specification (360℃). In comparison, the SWCL1 fuel showed a T95 point of ~280℃.

Although in tests SWCL1 produced fewer nucleation mode particles than F10S or F50S, this was potentially favourable for discrimination purposes since relatively small lubricant effects would be proportionally large compared with the fuel background. Consequently, the programme proceeded using SWCL1 as the base fuel.

DATA OUTLIER – As Figure 19 shows, pairs of tests showed consistent results with accumulation modes more repeatable than nucleation mode levels. The one exception was oil 210A, which was highly unrepeatable in the nucleation mode region but showed extremely similar accumulation mode results. In fact, this oil proved to show largely unrepeatable nucleation mode data throughout the programme and may have subtly altered the stability of engine operation. This oil was later discarded as an outlier in the final statistical analyses (Appendix 3).

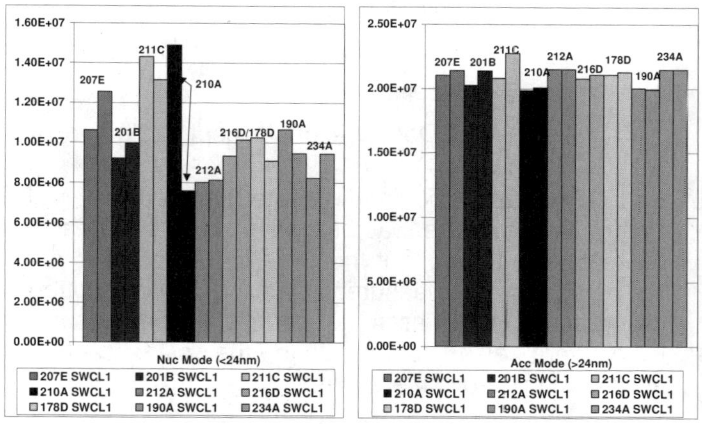

Figure 19: Integrated Nucleation and Accumulation Mode Particle Numbers

Plotting nucleation mode particle number emissions (in the range 7nm to 24nm) Vs. lubricant sulphur content revealed a positive linear relationship (Figure 20). This was confirmed as statistically significant. A positive relationship between particle numbers and phosphorus content was also identified in the statistical analyses. This was made possible from the original statistical design that incorporated phosphorus and sulphur values in such a way that enabled their effects to be decoupled. The effect of phosphorus was larger than the sulphur effect on a simple mass basis.

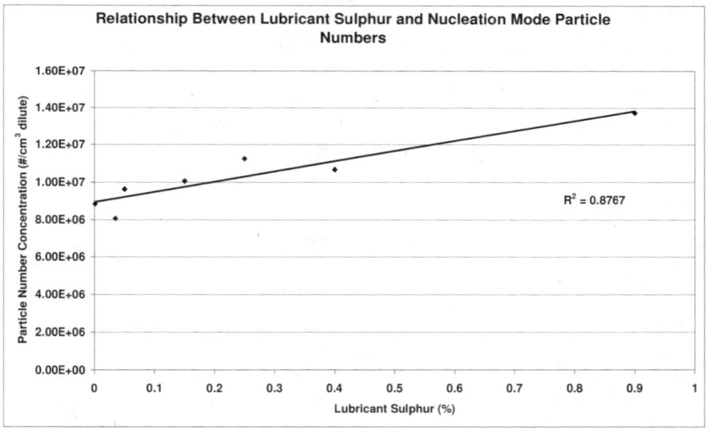

Figure 20: Relationship Between Lubricant Sulphur and Nucleation Mode Particle Numbers

When very abundant, carbonaceous accumulation mode particles tend to suppress nucleation by adsorbing the available volatile materials. This suggests that the increase in nucleation mode particles during decelerations may be a result of the reduction in available carbon, and that the absence of nucleation mode particles during accelerations is due to the high concentrations of carbon under these conditions.

Specific Cycle Events – The NEDC cycle was deconstructed into various categories of driving type. These were:

- Accelerations
- Cruises
- Idles
- Decelerations
- All Decelerations except 120kph to idle

Particle number concentration data from each hot NEDC cycle were averaged across all instances of each of these events throughout the drive cycle, for each of the six particle sizes measured. Data were then plotted in histogram form to estimate particle size distributions. Data were also submitted to the statistician who evaluated oil-to-oil differences.

Accelerations (Figure 21) and cruises showed size distributions with dominant accumulation modes.

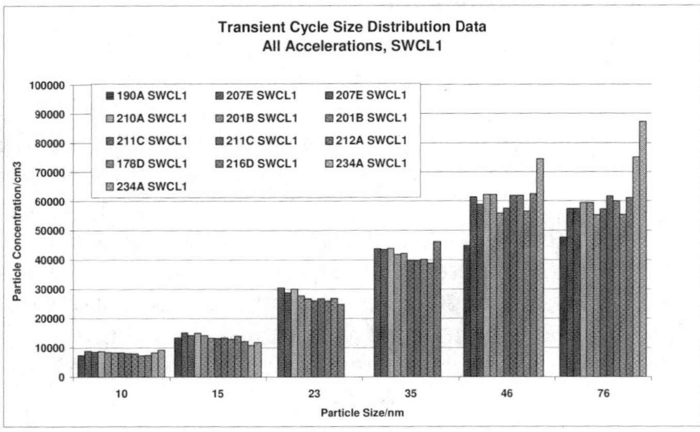

Figure 21: Particle Size Distribution - Accelerations

Hot idling (during the cycle) showed a bimodal distribution (Figure 22) although the nucleation mode was present at or below 10nm.

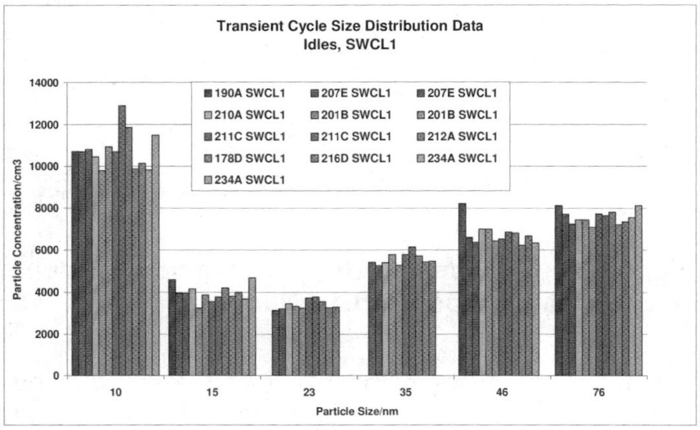

Figure 22: Particle Size Distribution – Idles (Hot)

This distribution agreed well with the stabilized idle data collected with the scanning SMPS (Figure 23) except that due to the elevated temperatures (and consequent reduced condensation) the nucleation mode contained less material and thus smaller particles.

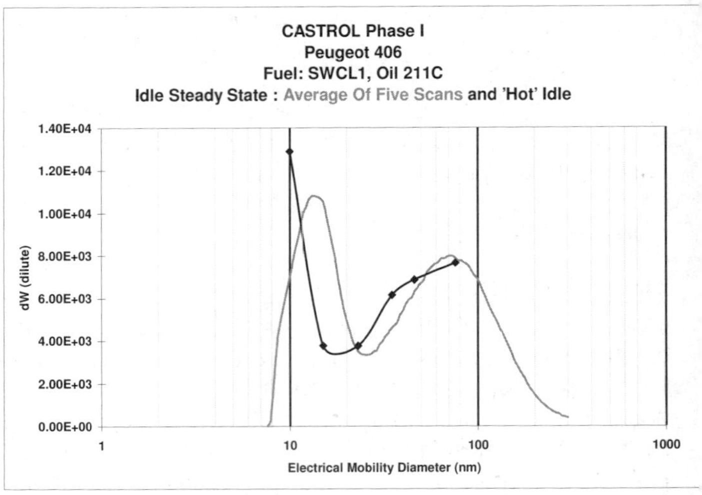

Figure 23: Comparison of Hot and Stabilised Idles

Decelerations also showed dominant nucleation modes (Figure 24), and during these events accumulation mode levels dropped significantly.

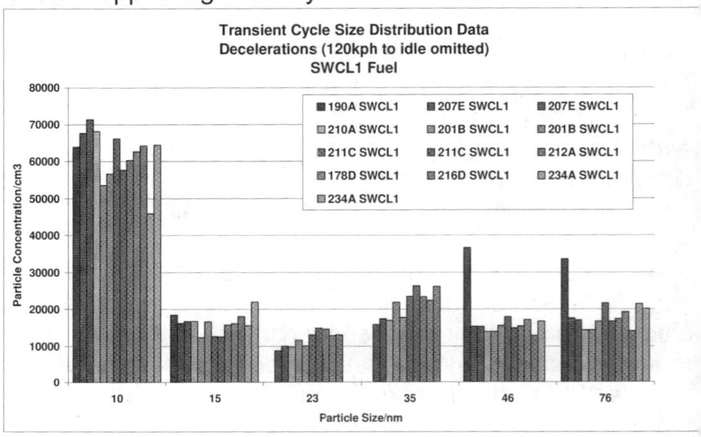

Figure 24: Particle Size Distribution - Decelerations

A number of possible explanations for the nucleation mode particle production during decelerations can be suggested:

- Sudden and major increase in dilution ratio and exhaust cooling accompanying fuel shut-off on the overrun. Exhaust Gas Recirculation (EGR) cut-off and elimination of carbon may promote nucleation.
- Particle precursors may be desorbed from the tunnel walls on decelerations
- Oil films on the cylinder walls being thicker with no combustion pressure and being detached into the exhaust
- Gas temperatures being lower on decelerations leading to lower levels of oxidation of nucleatable materials
- Blow-by gases being drawn into the system and nucleating within the dilution tunnel
- Late injection strategies leading to fuel survival
- One or more of these processes may be implicated in the production of nucleation mode particles, but further work is required in this area.

- Fuel dribble may be contributed from the injector sac volume

CONCLUSIONS

Oil sulphur effects on nucleation mode particle formation were seen at the 99% confidence level, analogous to the fuel sulphur effects.

Unexpectedly, phosphorus levels also had an effect at the 99% confidence level, on nucleation mode particle formation. The phosphorus effect was larger, on a mass basis, than the lubricant sulphur effect.

Fuel T95 had an unexpectedly large effect on nucleation mode particle production. In consequence, lubricant discrimination is significantly improved using a lower final boiling point fuel.

Fuel sulphur effects were clearly seen between 50, 10 and 2 ppm.

Hot idle produces the largest number of nucleation mode particles and is the preferred condition for studying lubricant effects.

Nucleation mode particles, represented by 10nm data, were primarily produced in response to deceleration events under transient cycle operation.

Emissions of accumulation mode particles (represented by 76nm) were clearly highest during accelerations. However, they remained elevated during cruise conditions, appeared to respond to gear changes and reduced with accelerations. Particle production followed the drive trace quite closely and therefore showed a strong synergy with fuelling and CO_2 emissions.

ACKNOWLEDGMENTS

The authors wish to thank Dick Boddy of Statistics for Industry, for his contribution of experimental design and statistical analyses, Angela Morton of Castrol for formulating the oils, Diane Hall of BP for her contribution in discussions prior to commencement of the test programme and Diane Lance, Roger Savill and Ian Lankester of Ricardo Consulting Engineers for their contribution in data acquisition and assistance in writing this paper. The authors would also like to thank the management of Castrol and Ricardo Consulting Engineers for their permission to publish this paper.

REFERENCES

[1] **DETR/SMMT/CONCAWE Particle Research Programme: Light Duty Results**. Barbara G A Wedekind; Jon D Andersson; Diane Hall; Richard Stradling; Chris Barnes; Geoff Wilson. (SAE 2001-01-3577, 2001)

[2] **DETR/SMMT/CONCAWE FINAL REPORT.** Ricardo DP00/2684, J D Andersson (Unrestricted)

3 **LITERATURE STUDY ON TAILPIPE PARTICULATE EMISSION MEASUREMENT FOR DIESEL ENGINES** / Heinz Burtscher. http://www.akpf.org/pub/burtscher_bericht.pdf

[4] **PRONOUNCED DECREASE OF THE AMBIENT PARTICLE NUMBER EMISSIONS FROM DIESEL TRAFFIC IN DENMARK AFTER REDUCTION OF THE SULPHUR CONTENT IN DIESEL FUEL** / Peter Wahlin; Finn Palmgren; Rita Van Dingenen; Frank Raes. (Atmospheric Environment, 2001, Vol.35, Issue 21, 2001, pp3549-3552.)

[5] **RESPIRATORY EFFECTS ARE ASSOCIATED WITH THE NUMBER OF ULTRAFINE PARTICLES** Annette Peters, H Erich Wichmann, Thomas Tuch, Joachim Heinrich & Joachim Heyder (American Journal of Respiratory and Critical Care Medicine, 1997, Vol. 155, pp1376-1383.)

[6] European Directive 98/70/EC

[7] European Directive 98/69/EC

[8] **DETR/SMMT/CONCAWE PARTICLE RESEARCH PROGRAMME: Sampling And Measurement Experiences** Jon D Andersson; Barbara G A Wedekind; Diane Hall; Richard Stradling; Chris Barnes; Geoff Wilson. (SAE 2000-01-2850, 2000)

[9] **DETR/SMMT/CONCAWE Particle Research Programme: Heavy Duty Results**. Barbara G A Wedekind; Jon D Andersson; Diane Hall; Richard Stradling; Chris Barnes; Geoff Wilson. (SAE 2000-01-2851, 2000)

[10] **FORMATION OF DIESEL NANOPARTICLES (SECOND REPORT) - Effect Of Sulfate On Nanoparticle Formation** / Shuichi Kubo; Michio Nakano; Teruaki Kondo; Satoshi Yamazaki; Toshiaki Tanaka. (JSAE 2003 Annual Congress, Yokohama, Japan, May 2003, Session No. 43-03, Measurement & Characterization of Fine Particles II, Paper 20035132.)

[11] **TESTING MARGINAL HOMOGENEITY WITH COCHRAN'SQ;**http://www.id.unizh.ch/software/unix/statmath/sas/sasdoc/stat/chap28/sect42.htm

CONTACT

Jon.Andersson@Ricardo.com

APPENDICES

Appendix 1– Fuel Properties

PROPERTY	UNIT	RESULT SWCL1	RESULT EN590 (10ppm)	RESULT EN590 (50ppm)
Cetane Nº		52.1	54.7	54.7
Cetane Index		51	52.5	52.5
Density @ 15 °C	kg/m³	812	824.6	824.6
Distillation				
T 10%	°C	197.8	202	202
T 50%	°C	226.3	250	250
T 90%	°C	269.3	324	324
T 95%	°C	279.2	340	340
Recovered @ 250 °C	%	75	50	50
Recovered @ 350 °C	%	99.7	98	98
Sulphur content	mg/kg	<10 (~3)	10	50
Poly-Aromatics (Di+Tri)	% vol.	0.4	1.7	1.7
Lubricity	μm	370	400	400
Viscosity @ 40 °C	mm²/s	3.002	2.38	2.38
Cloud		-37	-28	-28
CFPP	°C	-41	-31	-31
Flash	°C	65.5	68	68
Carbon content	%	85.7	87	87
Hydrogen content	%	14.3	13	13
H/C ratio		2.00:1	1.79:1	1.79:1
Calorific value	MJ/kg	43.2	43.06	43.06

Appendix 2– Lubricant change procedure

1. The engine-oil was warmed by low load road driving
2. The vehicle was installed an elevated ramp
3. The sump plug was released and the oil drained. A 1litre sample was collected
4. The oil filter was drained and replaced
5. The ramp was lowered and 4.5 litres of fresh oil was added to the engine.
6. The engine was started and left to idle for a fixed period, during which operating temperature was reached.
7. The oil was drained again.
8. The vehicle was placed upon the raised ramp, the oil filter removed and discarded, and a new filter added. The ramp was lowered.
9. The engine with filled with 4.5 litres of new oil and idled for one minute.
10. The ignition was switched off and the oil left to settle for 5 minutes. The dipstick level was checked to ensure it reached the maximum point.

Appendix 3 - STATISTICAL ANALYSES

Statistical analyses were based upon the relationship between the concentrations in several SMPS size bins at or around the peak of the nucleation mode with the elemental components (specifically sulphur and phosphorus) present in the oils. These sizes were 12nm, 13nm, 14nm, 15nm and 20nm.

Analyses investigating the accumulation mode did not reveal any significant effects. Those data are not discussed here.

METHODOLOGY – The analysis was undertaken using multiple regression, a method of determining the significance and effect of multiple parameters on a response. In this case the response was particle size and the parameters were the concentration of the metals, either singly or combined.

The effectiveness of multiple regression in selecting the relevant parameters is dependent upon the independence of the parameters as represented by the inter-correlations. These are shown in Table 5. It is recognised that there may be some difficulty in interpretation if a correlation is greater than 0.70. Should a correlation be above 0.9 interpretation is required to determine which of two parameters are important should they be statistically significant.

	S	P	Zn	Ca	Mg	B	Other elements	Other metals
Sulphur								
Phosphorus	0.57							
Zinc	0.78	0.86						
Calcium	0.86	0.39	0.67					
Magnesium	-0.53	-0.10	-0.54	-0.76				
Boron	-0.26	-0.43	-0.56	-0.32	0.41			
Other metals	-0.09	-0.26	-0.14	0.05	0.03	-0.50		
Other elements	0.92	0.58	0.74	0.91	-0.49	-0.15	-0.05	

Table 5: Inter-correlation Matrix

The experimental design focused on having as low a correlation as possible between sulphur and phosphorus. All other parameters were a consequence of the necessary additives and as such some high correlations were obtained.

ANALYSIS – The precision of repeat tests on the oils for 13nm particles is given in Table 6.

Oil	Precision (% SD)
212A SWCL1	0.7
201B SWCL1	3.3
216D SWCL1	6.5
211C SWCL1	7.3
178D SWCL1	8.6
190A SWCL1	10.4
207E SWCL1	10.7
210A SWCL1	47.0

Table 6: Measurement Precision - 13nm Particles

The combined standard deviation (SD) of repeat tests excluding 210A SWCL1 was 7.6% compared with 47.0% for 210A. Cochran's test[11] showed that 210A was a significant outlier at a 99% level of confidence.

Analyses were undertaken to determine the effect of certain variables on the number of particles emitted at various sizes of the nucleation mode. Certain of the variables were highly correlated with each other, these were % phosphorus with % zinc (0.86), % sulphur with % calcium (0.86) and zinc (0.78) and % calcium with % magnesium (-0.76): these correlations are determined by the additive packages used in the lubricant formulations.

Table 7 to Table 10 show correlation coefficients for particles sizes between 12nm and 20nm within the nucleation mode. In these tables the following symbols indicate various levels of confidence intervals:

~ significant at a 90% level of confidence

* significant at a 95% level of confidence

** significant at a 99% level of confidence

Size 12 nm	
	Coefficient
% Sulphur % Phosphorus % Other Elements	325.7 ** 759.5 * -696.4 *
RSD (%)	6.1
Multiple r^2	0.90

Table 7: Correlation between 12nm particles and elemental constituents of lubricants

Size 13 nm	
	Coefficient
% Sulphur % Phosphorus % Other Elements	377.7 ** 693.3. ** -843.4 *
RSD (%)	6.9
Multiple r^2	0.88

Table 8: Correlation between 13nm particles and elemental constituents of lubricants

Size 14 nm	
	Coefficient
% Sulphur % Phosphorus % Other Elements	440.8 ** 676.9~ -1110.6 *
RSD (%)	8.6
Multiple r^2	0.85

Table 9: Correlation between 14nm particles and elemental constituents of lubricants

Size 20 nm	
	Coefficient
% Sulphur % Other Elements	260.0** -720.0 *
RSD (%)	13.3
Multiple r^2	0.65

Table 10: Correlation between 20nm particles and elemental constituents of lubricants

CONCLUSIONS – Repeat Tests on 210A SWCL1 were significantly more variable, at a 90% level of confidence, than repeat tests on other oils. Therefore, 210A SWCL1 data were not included in the analysis.

Multiple regression showed that there were strong positive relationships between the number (concentration) of particles and the presence of sulphur, phosphorus and other elements. The relationships held for sizes at the peak of the nucleation mode (between 12 nm and 14 nm), and were present, but not as strong, as low as 10 nm and as high as 20 nm. For this reason, the data for 13nm can be employed to represent the behaviour of the entire nucleation mode. At 13 nm (Table 11) the relationships were:

	% Sulphur	% Phosphorus	% Other Elements
Level of Confidence	0.99	0.99	0.95
Coefficient for treatment at 1% increase in level	378000	693000	-843000
Lower treat rate	0.02%	0.00%	0.262%
Higher treat rate	0.90%	0.10%	0.498%
Effect over the range	332000	69000	-198000

Table 11: Summary of Effects of S and P on Particle Number Emissions (13nm Particles)

This demonstrates that a reduction in nucleation mode particle numbers of greater than 40% is possible with a reduction in lubricant sulphur from 0.9% to 0.02% (Figure 25, oil 211C to 212A). More specifically, Figure 19 addresses solely the concentration of 13nm particles within the size distribution and shows reductions in particle numbers relative to the highest sulphur oil (211C).

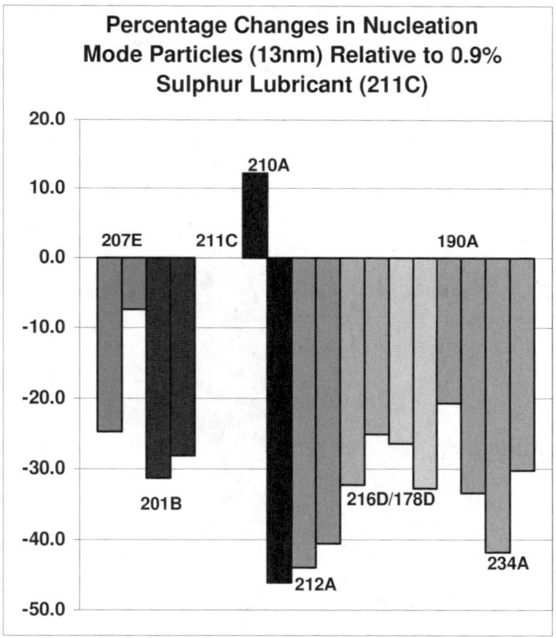

Figure 25: Percentage Changes in 13nm Particles

These data also show that an independent reduction in particle numbers of up to 15% can be achieved by a reduction in lubricant phosphorus from 0.1% to zero.

'Other Elements' is an extra variable for statistical analysis only and comprised zinc, calcium, magnesium, boron and other metals.

The particle number/sulphur relationship gave an exceptionally good fit to the data. At 13 nm the relationship gave a multiple correlation coefficient (r^2) of 0.88 and a standard deviation of 6.9%. This is exactly in line with the standard deviation between the repeat tests used to determine the measurement variability and shows that most of the variability is due to measurement only.

Apart from boron at 10nm, analysis of other elements (Zn, Ca, Mg and other metals) showed no significance. This should not be interpreted as the elements having no effect, especially in the light of other elements being significant. Their lack of significance is probably due to small levels of concentration of many of the elements and their correlation with either sulphur or phosphorus. Since only sulphur and phosphorus were considered in the design matrix, correlations with other elements were due to other circumstances.

The negative sign for other elements could be interpreted in a number of ways, e.g., certain elements are reducing nucleation or it is a feature of the high correlation between calcium and sulphur. Further experimentation would be required to find the true factors causing the significance of other elements.

2006-01-3280

Effect of Biodiesel Blends on Diesel Particulate Filter Performance

Aaron Williams, Robert L. McCormick, R. Robert Hayes and John Ireland
National Renewable Energy Laboratory

Howard L. Fang
Cummins, Inc.

ABSTRACT

Tests of ultra-low sulfur diesel blended with soy-biodiesel at 5% and 20% were conducted using a 2002 model year Cummins ISB engine (with exhaust gas recirculation) that had been retrofitted with a passively regenerated catalyzed diesel particulate filter (DPF). Results show that on average, the DPF balance point temperature (BPT) is 45°C and 112°C lower for B20 blends and neat biodiesel, respectively, than for 2007 certification diesel fuel. Biodiesel causes a measurable increase in regeneration rate at a fixed steady-state condition, even at the 5% blending level. The data show no significant differences in NO_x emissions for these fuels at the steady-state regeneration conditions, suggesting that differences in soot reactivity are responsible for the observed differences in BPT and regeneration rate. Soot from the various fuels was characterized by determining the fuel and lubricant fractions of the soluble organic fraction, elemental and organic carbon content, amorphous carbon/graphitic carbon ratio by Raman spectroscopy, carbon/oxygen ratio by energy dispersive x-ray analysis, and reactivity in oxygen by TGA. Results indicate a much more disordered soot structure, containing higher levels of oxygen as biodiesel is blended into the diesel fuel. The soot produced from biodiesel and blends is much more reactive in oxygen than diesel soot. It is concluded that the lower balance point temperature and higher DPF regeneration rates for biodiesel containing fuels are observed because the soot generated from these blends is more reactive.

INTRODUCTION

Biodiesel is a renewable fuel derived from vegetable oil, animal fat, or waste cooking oil and consists of the methyl esters of fatty acids. It is typically used as a diesel blending component at levels of 20 volume percent or lower. A resource assessment indicates that biodiesel has the potential to displace 5% or more of petroleum diesel over the next decade [1]. A life cycle analysis indicates that biodiesel is a highly renewable fuel, and that use of B20 results in a 19% reduction in life cycle petroleum consumption [2]. The United States Environmental Protection Agency (EPA) has reviewed biodiesel emissions data for heavy-duty engines published up through about 2000 [3]. For transient engine dynamometer testing, the average emission changes for B20 versus petroleum diesel were +2.0% for NO_x, -10.1% for PM, -11.0% for carbon monoxide (CO), and -21.1% for total hydrocarbons (THC). However, closer scrutiny of the available data suggests that increasing NO_x is not necessarily a general phenomenon for biodiesel at the B20 level, and that NO_x emissions depend upon engine duty cycle or mode as well as engine technology [4]. More recently data were published reporting the testing of two newer engines equipped with EGR and meeting the 2004 emission standards, with B20 vs. ultra-low sulfur diesel (ULSD) fuel. This study found, on average, a 25% reduction in PM emissions for B20 [5], but NOx increased by nearly 4 %.

In 2007, new emission standards for on-road heavy-duty diesel engines will begin to phase in. Most relevant to this work, beginning with the 2007 model year the particulate matter emission standard will be lowered from 0.1 g/bhp-h to 0.01 g/bhp-h. This lowering of PM emissions by a factor of 10 is enabled by the introduction of diesel fuel containing a maximum of 15 ppm sulfur (ULSD) beginning in June of 2006 and the use of diesel particulate filters. In a DPF, soot particles are trapped on a filter. In catalyzed DPF systems such as that tested here, the soot is then burned by reaction with NO_2. The role of NO_2 as an oxidizer is critical to catalyzed DPF performance and DPFs typically contain a precious metal catalyst upstream of the ceramic filter, where NO is converted to NO_2. This NO_2 is a more aggressive oxidizer of soot at low temperatures than is oxygen, and thus can control the soot oxidation rate. Thus, a small increase in NO_x emissions (mainly NO) observed for B20 during transient testing could have significant consequences for the performance of B20 with DPFs.

Some published information is available on the performance of DPFs with biodiesel and biodiesel blends. Testing of B20 and several other fuels with various aftertreatment and engine configurations [6] has been reported. The use of a catalyzed DPF produced a factor of 10 reduction in PM emissions relative to the base case. No PM emission advantage was observed for B20 with a DPF installed. No engine-out emissions were reported for B20 in this study. Additionally, 2007 compliant procedures for PM emission measurement do not appear to have been used. However, Boehman and coworkers [7] have recently shown that blending of 20% biodiesel into diesel fuel can significantly lower balance point temperature (BPT). They present results showing that this is not caused by increased availability of NO_2, but by inherent differences in soot reactivity for different fuels. This was confirmed by thermogravimetric analysis wherein soot produced in an engine from different fuels was burned under identical conditions. Soot characterization by electron microscopy suggested that the cause of this increased reactivity is a more highly disordered soot nanostructure for B20 blends such that the soot is more reactive or reactive at lower temperatures. More recent results presented by Boehman [8] suggest that changes in nanostructure are not the cause of increased reactivity, but rather the introduction of highly reactive surface oxygen sites when the soot is produced from B20.

Here we examine the impact of biodiesel and biodiesel blends on balance point temperature, filter regeneration rate, and transient emissions using a modern (2004 compliant) engine and state of the art (2007/2010 compliant) emissions measurement system.

EXPERIMENTAL SETUP

The test setup consisted of a 2002 model year 5.9L 300 hp Cummins ISB, equipped with a diesel particle filter (DPF). Properties of the test engine are shown in Table 1. The engine employs cooled high-pressure EGR, a variable geometry turbocharger, electronic control, and high-pressure common rail direct fuel injection, designed and calibrated to meet the 2004 U.S. heavy-duty emissions standards.

The DPF is a 12 L catalyzed diesel particle filter employing Catalyzed Continuously Regenerating Technology (CCRT™) provided by Johnson Matthey, a passively regenerated system. The CCRT™ filter is a diesel oxidation catalyst followed by a wall-flow catalyzed soot filter. It is used in applications with average exhaust temperatures as low as 200°C - 250°C. The DPF is mounted 152 cm from the engine turbo flange outlet. The DPF was instrumented for inlet and outlet temperatures and pressures as well as differential pressure. Temperatures were measured with K-type thermocouples mounted 8 cm from the face of the pre-catalyst on the inlet side and 8 cm from the face of the DPF on the outlet side. Inlet and outlet pressures as well as differential pressure were measured from the same location. The overall setup and relative location of all instrumentation is illustrated in Figure 1. In addition, engine and emissions sampling systems were instrumented for measurement of all other critical temperatures and pressures.

Table 1. Test engine specifications

	Cummins ISB
Serial Number	56993170
Displacement, L	5.9
Cylinders	6
Rated Power, kW	224 at 2500 rpm
Rated Torque	895 N-m at 1600 rpm
Bore x Stroke	10.2x12 cm
Compression Ratio	16.5:1
Fuel System	Common Rail
Intake Restriction, kPa	4.47
Exhaust Backpressure, kPa	7.95

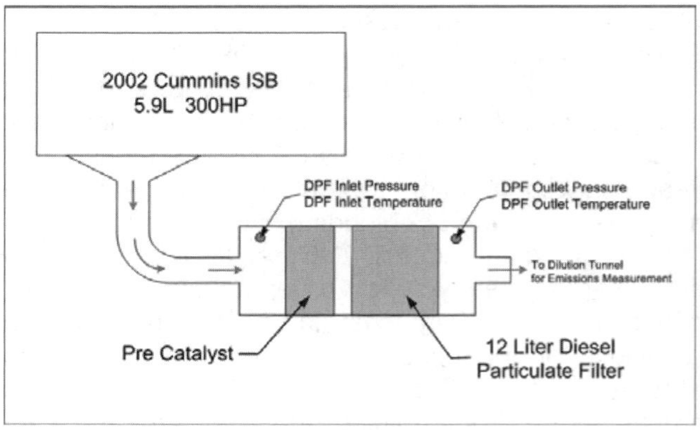

Figure 1. Experimental Setup

The engine was mounted to a DC electric engine dynamometer at NREL's Renewable Fuel and Lubricant (ReFUEL) laboratory. ReFUEL's engine dynamometer is part of a fully functional test cell capable of steady state or transient testing for emissions and fuel consumption. The ReFUEL emission measurement system is based on the full scale dilution method with Constant Volume Sampling (CVS) for mass flow measurement. Gaseous emissions, including CO_2, NO_x, THC and CO are measured continuously. Particulate matter (PM) emissions are measured based on a gravimetric system, in which samples are collected onto 47 mm Teflon membrane filters and then weighed with a microbalance in a clean room environment.

Test fuels included 2007 certification diesel and various biodiesel blends made from a soy biodiesel feedstock. The soy biodiesel was S15 grade and met the requirements of ASTM D6751. Fuel properties for the 2007 certification ULSD and the neat biodiesel are listed in Table A-1 in the Appendix.

PROCEDURES AND RESULTS

ESTIMATION OF PARTICLE LOADING

In addition to fuel effects, performance of the DPF could be affected by the level of particulate loading. Thus, experiments were conducted to understand how to repeatedly preload the DPF with the same amount of particulate, and how to quantify that amount. In all cases the 12 Liter DPF was loaded to approximately 1.5 g/L to approximate typical in-use loading levels. However, the amount of time necessary to produce the total 18 grams of PM at the 2000 rpm 20 ft-lb preload condition is different for each fuel. A series of experiments were conducted to determine the rate of PM production by the engine for each fuel at the preload condition. The results of these experiments were then used to determine the proper load period for all subsequent BPT and regeneration rate testing. In addition, a method of weighing the DPF before and after loading was developed to further confirm the total mass collection.

Particulate Matter Rate Study

Testing was conducted to determine the total output of PM from the engine when operating at the 2000 rpm 20 ft-lb preload condition for a 340 minute period. Without the DPF installed, engine-out PM samples were collected every 20 minutes for the entire 340 minute duration. Samples were collected onto 47 mm Teflon membrane filters at a sample flow rate of 43 L/min for 20 minutes each. The filters were then weighed to determine the total PM production from the engine. This test was conducted for 2007 certification diesel (2007 Cert), B20 and B100, the results of which are shown in Figure 2. Assuming that the DPF captures and stores all of the PM at the preload condition, total collection of 18 grams (1.5 g/L) would take 332, 314 and 239 minutes for B100, B20 and 2007 Cert respectively.

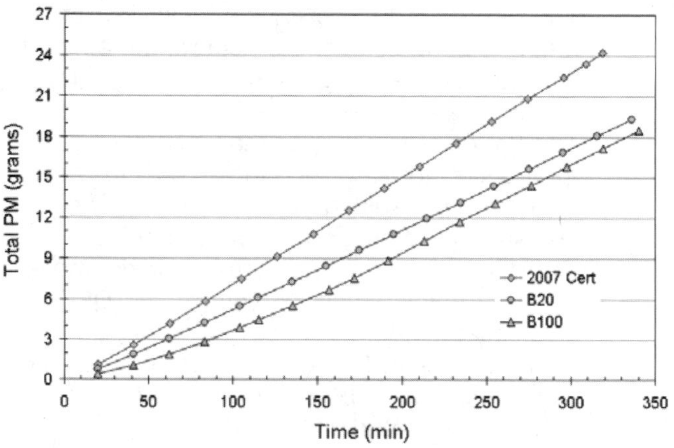

Figure 2. PM rate study results for B100, B20 and 2007 Certification Diesel

DPF Loading by Weight

The preload condition of the DPF was further confirmed by taking fully regenerated (clean) and loaded weights of the DPF. The DPF weighed 17.2 kg compared to the 18 grams of PM being collected, illustrating the degree of accuracy necessary. An electronic balance with a resolution of 0.1 grams was used. Initially the DPF was allowed to equilibrate in an environmental chamber for a minimum of 12 hours before weighing. Temperature and humidity were controlled to 20°C with a relative humidity of 47%. When weighing the DPF during this 12 hour stabilization period, it was found that the DPF would immediately begin gaining weight in the relatively humid environment and eventually stabilize before the twelfth hour. The DPF appeared to absorb moisture even with substrate temperatures as high as 800°F. Overall, the loaded DPF would gain as much as 13 grams of moisture before stabilizing. However, it was discovered that a clean DPF was more hygroscopic than a loaded DPF, gaining as much as 21 grams of moisture. Thus, accurate measurements of loaded mass could not be determined by taking the difference in clean and loaded DPF weights, each of which was taken after equilibration in a humid environment.

The second iteration of a weighing technique consisted of weighing the DPF in a dry room environment. In this case, with relative humidity controlled to below 20%, the DPF still gained weight due to moisture, however, not as quickly. It was decided that the best technique for weighing was to remove the DPF as quickly as possible following engine operation and take immediate measurements in the dry room conditions. This technique yielded repeatable DPF loading measurements between 1.3 and 1.7 g/L compared to the target 1.5 g/L predicted by the PM rate studies.

BALANCE POINT TEMPERATURE

The balance point temperature is defined as the DPF inlet temperature at which the rate of particle oxidation approximately equals the rate of particle collection. At the balance point temperature during steady-state operation, the DPF should not experience a net gain or loss of particulate matter and consequently the differential pressure across the DPF should not change. A series of balance point temperature tests were conducted to compare the operation of a DPF on 2007 certification diesel, B100, and a B20 blend. The balance point temperature was determined with the following test sequence and is illustrated in Figure 3.

1. The DPF was completely regenerated by operating at near rated power (2500 rpm and 575 ft-lbs) for 120 minutes.

2. The DPF was then preloaded with PM at 2000 rpm and 20 ft-lbs over the appropriate amount of time to achieve an approximate 1.5 g/L loading of particulate matter on the DPF.
3. The preloaded DPF was then operated at 1700 rpm while torque was increased to achieve specified exhaust temperatures. The DPF pressure drop was monitored continuously to determine slope of the increase or decrease of the differential pressure across the DPF at a given inlet temperature.
4. This sequence was repeated for 2007 certification diesel, B100 and B20 fuels for determination of BPT at 1700 rpm.

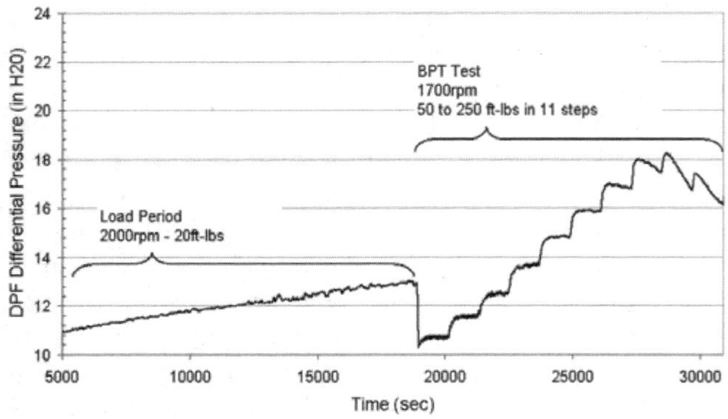

Figure 3. Balance point temperature test DPF differential pressure

The balance point temperature is determined by analyzing the slope of the differential pressure for each of the temperature steps. If the slope is positive (backpressure is increasing) then it is assumed that the DPF is collecting particulate matter. Once the slope becomes negative (backpressure is decreasing) that temperature is above the balance point temperature, as previously defined. The balance point temperature determination is made by plotting the slope of the differential pressure versus the DPF temperature for each of the steps. A linear curve fit is made between the two steps where differential pressure slope transitions from a positive to a negative value. The point where the curve fit crosses the Y-intercept is determined as the best estimate of the balance point temperature. In order to understand variability of the testing method, two repeats of the balance point temperature test were completed for 2007 certification diesel and B20, and three repeats for B100. In the case of 2007 certification fuel, additional torque steps were added in order to move sufficiently beyond the balance point temperature. Figure 4 shows the resulting balance point temperature windows for each of the fuels. On average, the balance point temperature is 45°C lower than 2007 certification diesel for B20 blends and 112°C lower for neat biodiesel.

REGENERATION RATE

The regeneration rate testing is similar to balance point temperature testing in that the DPF is fully regenerated and preloaded to approximately 1.5 g/L. However, rather than stepping through different torques at 1700 rpm, the engine moves directly to a single, relatively high torque (high temperature) operating point. This test method is illustrated in Figure 5. This method provides a potential advantage of comparing regeneration behavior between fuels at the same approximate DPF preload conditions (grams/Liter) and same high load steady-state operating point. This contrasts with the stepped BPT measurement method, which allows for different loading to occur during the "steps" following preload and prior to reaching the BPT.

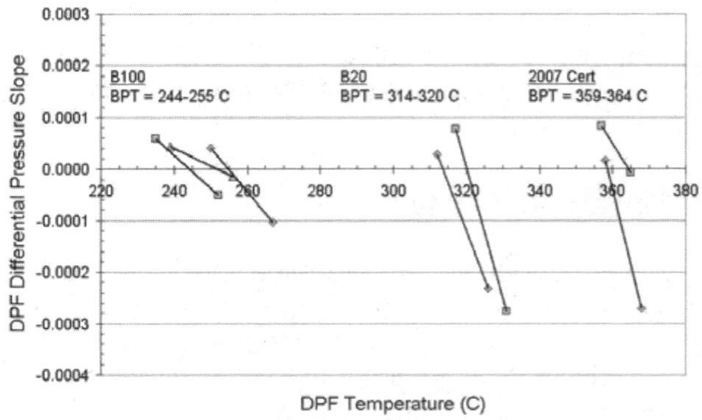

Figure 4. Balance point temperature test results for B100, B20 and 2007 Certification Diesel

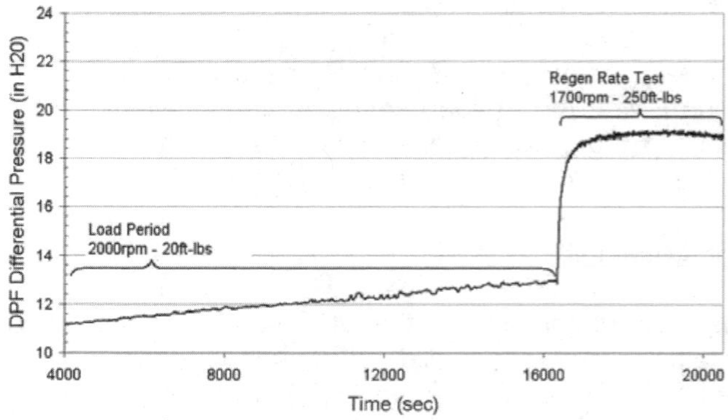

Figure 5. Regeneration rate test DPF differential pressure

Testing was conducted with 2007 certification diesel, B5 and B20 with two repeats for each fuel. Once the DPF had been preloaded the engine was then operated at the higher temperature point (1700 rpm and 250 ft-lbs) for 60 minutes. This allowed enough time for DPF temperature to stabilize. In each case, the DPF temperature at the active regeneration point stabilized between 348°C and 357°C. In general, DPF inlet temperatures were lower with higher blend ratios of biodiesel. DPF differential pressures also varied from test to test, either as a consequence of the slight temperature differences or relative differences in the amount of particle loading. At the stabilized regeneration point, DPF differential pressures ranged between 48.3 cm of H_2O and 52.1 cm of H_2O.

Results for regeneration rates are shown in Figure 6. This plot shows the normalized DPF differential pressure for the final 15 minutes of steady state operation at the active regeneration point. DPF differential pressures are normalized to show relative differences in performance for each of the fuels. As the plot shows, the differential pressure slope for 2007 certification fuel is slightly positive for repeated runs at 1700 rpm, 250 ft-lb operating condition with temperatures around 354°C. This is consistent with previous testing that showed a balance point temperature somewhere between 359°C and 364°C for this fuel. The plot also illustrates that both biodiesel blends at the B5 and B20 level show measurable decreases in DPF differential pressure at the same operating condition with similar amounts of DPF preloading. The slope of the differential pressure decrease increases with biodiesel content. These test results show that when fueled with biodiesel blends, PM appears to measurably oxidize more quickly than when fueled with certification diesel fuel (even down to the B5 level). Increased levels of biodiesel in the fuel appear to increase the rate of DPF regeneration at a given engine operating condition.

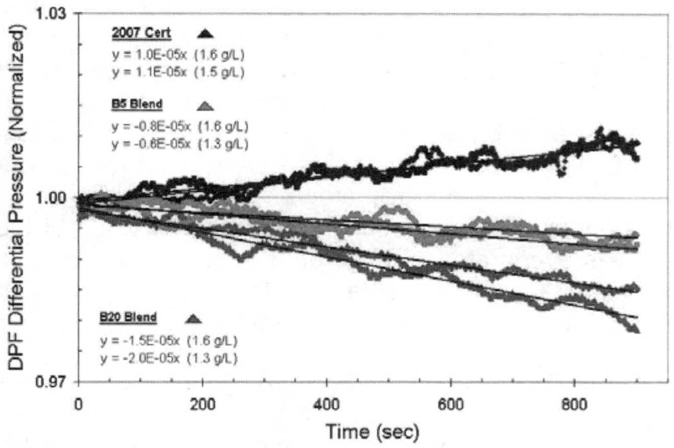

Figure 6. Regeneration rate test results for B20, B5 and 2007 Certification Diesel

AVAILABILITY OF NO_2 FOR DPF REGENERATION

Catalyzed DPFs use NO_2 as the oxidizer of trapped soot. Thus, increased NO_x from biodiesel blends could potentially be the cause of the observed higher regeneration rates. Table 2 shows brake specific NO_x levels at the regeneration condition for each fuel. For this experiment, differences in NO_x levels between 2007 certification diesel and the biodiesel blends are not statistically significant (at alpha = 0.05). This suggests that differences in regeneration rates are <u>not</u> primarily due to differences in the availability of NO_2 as a soot oxidizer, although higher NO_2 availability may play a secondary role for B20.

Table 2. NO_x Levels at Regeneration Condition of 1700 rpm and 250 ft-lb Torque (p-value for comparison of certification fuel NO_x to biodiesel blend NO_x)

	2007 Cert	B5	B20
Run 1 NO_x (g/bhp-hr)	1.96	2.00	2.13
Run 2 NO_x (g/bhp-hr)	2.06	1.93	2.16
Average NO_x	**2.01**	**1.97**	**2.15**
p-value	**--**	**0.445**	**0.078**

The availability of NO_x at each step in the balance point temperature test is illustrated in Figure 7 for each of the three test fuels. B100 NO_x levels (g/bhp-hr) trend higher at temperatures above 280°C. However, at the balance point temperature for B100 (250°C), the availability of NOx is approximately the same as petroleum based diesel or B20. The same is true for NO_x levels at the BPT for B20 (317°C), once again indicating that reactivity of the soot, rather than availability of NO_2, is the primary factor in the determination of the balance point temperature for each test fuel. It should be noted that speciation of NOx for NO and NO_2 components was not performed. Biodiesel blends could also impact the NO/NO_2 split, which may have an impact on efficiency of the DPF. Also of note is the relative difference in the NOx/PM ratio for each test fuel. Because of lower engine out PM for biodiesel and the higher soluble organic fraction, the biodiesel NOx/Soot ratio will be much higher than that of 2007 certification fuel. The impact of this ratio on balance point temperature and regeneration rate was not explored in this study.

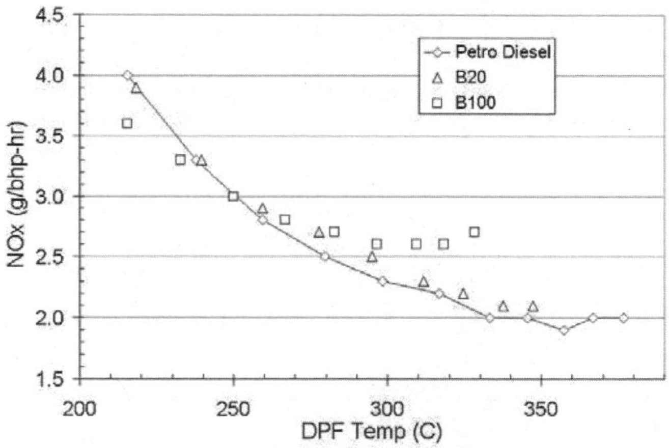

Figure 7. NOx levels at BPT steps for 2007 Certification Diesel, B20 and B100

TRANSIENT EMISSIONS TESTING

Baseline emissions over multiple hot start heavy-duty transient tests were measured for the 2007 certification fuel and a B20 blend. These tests were conducted with and without the DPF for both fuels to confirm operation of the filter and to determine reduction efficiencies of all regulated pollutants. Individual test results for each repeated run are shown in Table A-2. Table 3 shows the percent changes in emissions and fuel economy for each test configuration.

Table 3. Percent change in emissions and fuel economy (nd = non-detectable differences with 95% confidence)

	PM	THC	CO	NOx	BSFC
Cert w/ DPF vs Cert w/o DPF	-99%	-99%	-97%	nd	+1%
B20 w/ DPF vs B20 w/o DPF	-99%	-99%	-98%	nd	+1%
Cert vs B20 w/o DPF	-24%	-50%	-22%	+4%	+3%
Cert vs B20 w/ DPF	-27%	-74%	nd	+4%	+3%

Before installation of the DPF, the B20 blend achieved a 24% PM reduction relative to 2007 Cert, as shown in Figure 8. Following DPF installation, PM emissions were reduced by 90% for 2007 Cert and B20. B20 combined with the DPF still showed a statistically significant ($p<0.05$) reduction in PM of 27% relative to 2007 Cert with DPF.

Without the DPF, biodiesel reduced emissions of both THC and CO by 50% and 22%. In the case of both fuels the catalyzed DPF reduced THC and CO emissions by 97% to 99%, to almost undetectable levels. NO_x emissions increased for the biodiesel blend by 4% on average without the DPF compared to ULSD runs. DPF installation caused NO_x emissions to increase by 1%.

The biodiesel blend produced a 3% ($p<0.001$) increase in brake specific fuel consumption, consistent with the lower energy content of biodiesel, as shown in Table A-1. Installation of the DPF produced a 1% fuel economy penalty for both fuels ($p<0.001$).

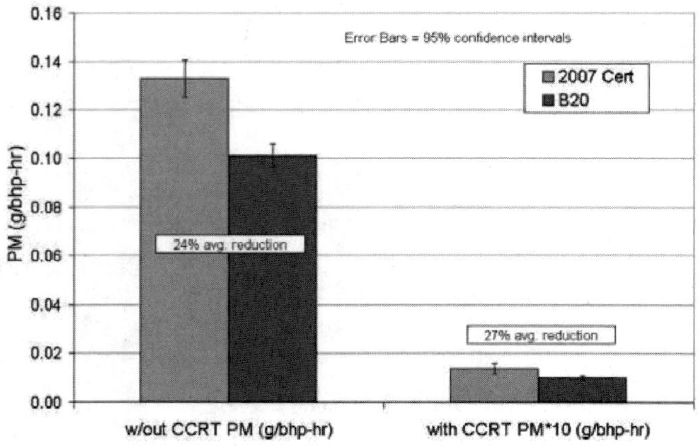

Figure 8. Transient test particulate matter emissions

SOOT CHARACTERIZATION RESULTS

Soot Collection Methods

Particle samples used for soot characterization were collected at the engine preload condition (2000 rpm 20 ft-lbs) for each test fuel. Without the DPF installed, dilute exhaust PM samples were collected onto 47 mm filters using methods consistent with CFR Title 40 Part 86 Subpart N. A butterfly valve was used to simulate exhaust back pressures similar to those created by the DPF. Approximately 46 mg were collected onto Pallflex Quartz filters and 10 mg were collected onto Teflon membrane filters. Bulk soot was collected directly from a preloaded DPF by back-flushing the filter with high pressure air into a collection canister. The DPF was first loaded by operating the engine at the preload condition for the time appropriate to achieve the 1.5 g/L loading. Using this method, approximately 1 gram was collected for 2007 certification diesel and B100.

Fuel and Lubricant Fraction of SOF

The soluble organic fraction (SOF) of PM samples was extracted using supercritical CO_2. The extracted hydrocarbon was delivered into an ODS trap (Agilent) and rinsed with CS_2. The CS_2 solution was injected into a gas chromatograph (HP 5880A) using a packed column (Chromosorb PAW) and a flame ionization detector (FID). The carrier gas was N_2 and the heating rate is set at 10°C/min for the temperature range from 35°C to 350°C. The SOF elutes as two major bands corresponding to fuel at roughly 12 minutes and lubricant as a broad band centered around 17 minutes. The fuel light end components, normally eluting before 10 minutes in the chromatogram, are missing for all samples studied, suggesting that the light portion of the fuel has evaporated. Based on the ratio of the fuel and oil peak areas, we can determine the lubricant fraction in the SOF and the data are shown in Table 4.

PM samples collected on both quartz and Teflon membrane filters were examined for 2007 certification diesel, B20 and B100. For 2007 Cert the major SOF component is lubricant regardless of the filter type. When biodiesel is used, roughly 25% of the SOF can be attributed to the fuel. The contribution of the fuel is even larger when Teflon membrane filters are used.

Table 4. Lubricant Fraction of the PM Soluble Organic Fraction (SOF) Measured by GC

Filter Type/Fuel	Quartz	Teflon Membrane
2007 Cert	95%	96%
B20	71%	78%
B100	75%	50%

Organic Carbon/Elemental Carbon Ratio

The ratio of organic and elemental carbon was measured by a thermal-optical carbon aerosol analyzer. The PM sample is placed in a quartz oven and is heated under helium with a stepped temperature ramp up to 900°C. Under this condition, all organic compounds are desorbed, pyrolyzed, and converted into CO_2. The CO_2 is swept through a heated nickel catalyst and converted into methane which can be quantified with a flame ionization detector. After the initial temperature ramp for organic carbon measurement, the oven is cooled to 600°C and the flow is switched to a helium/oxygen. A second temperature ramp is then initiated in the oxidizing gas stream to oxidize the elemental carbon. The elemental carbon is quantified in the same manner as the organic carbon. The elemental carbon can be also quantified by optical absorption detection.

As shown in Table 5, the OC/EC ratio follows a general trend of B100 > B20 > 2007 Cert. This indicates that the PM generated by biodiesel contains more organic carbon and that organic carbon content is higher with increasing biodiesel content.

Table 5. OC/EC Ratio of Various PM Samples

Fuel	OC $\mu g/cm^2$	EC $\mu g/cm^2$	OC/EC
2007 Cert	206.37	38.11	5.42
2007 Cert	219.07	31.79	6.89
B20	419.94	17.63	23.82
B100	414.54	15.25	27.19

Raman Spectroscopy

Raman spectroscopy is unique in characterizing the long range interactions and the order-disorder effects in carbonaceous materials. Pristine graphite crystals only show two bands in first-order Raman spectrum, at 42 and 1578 cm^{-1}. The 1578 cm^{-1} band is referred to as the graphitic band or G-band. As the graphite crystal is ground to destroy the ordering, a second band, referred as the amorphous band or disordered band, or D-band, begins to appear around 1340 cm^{-1}. For carbonaceous materials such as soot or coal, the D-band is relatively broad and sensitive to the ordering of the carbon structure. Both maturation and oxidation level of PM affect the ordering of the carbon structure. The maturation, in morphological sense, is the collapse of small aromatic domains with 1~4 rings into a configuration containing larger ring number domains. Besides the growth of a two-dimensional domain size, the layer-to-layer orientation also plays a determining role for long range ordering [9, 10].

Two micro-Raman units were used to examine the soot samples: a Dilor XY800 Raman microprobe equipped with Ar^+ laser (514.5nm) and a Rennishaw with a He-Ne laser (632.8nm). Both laser wavelengths were focused on the sample through a microscope with a 50x or 100x objective lens and the scattered light collected through a holographic filter and detected via a CCD detector.

Figure 9 shows a comparison of Raman spectra of PM samples generated from 2007 certification diesel, B20 and B100. The graphite-like G-band is located around 1586 cm^{-1} with a bandwidth of about 100 cm^{-1}. The disordered D-band is located around 1349 cm^{-1} with a much wider bandwidth >250 cm^{-1}. Multiple Raman spectra were taken for each sample at various locations to avoid inhomogeneity in the samples. The average ratio G/D and the corresponding standard deviation are listed in Table 6.

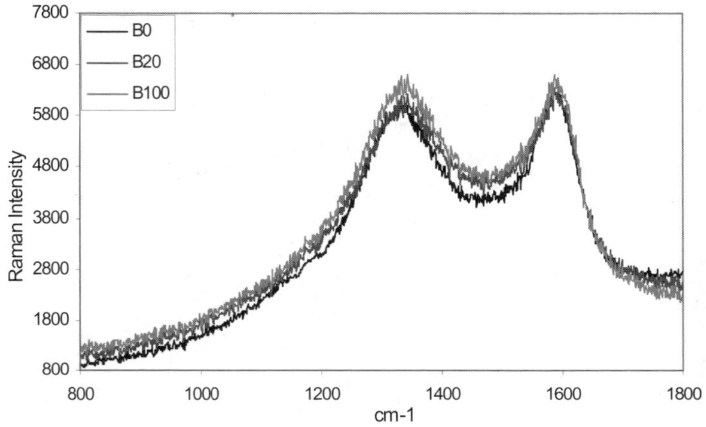

Figure 9. Raman spectra of the PM samples

As shown in Figure 10, the general trend for the G/D ratio is to decrease with biodiesel blend level even though the standard deviation is suddenly increased for PM generated by B100, indicating much greater variation in domain structure. This is consistent with previously published data [7] showing much greater disorder for biodiesel soot. This trend is generally true regardless which Raman unit is being used. The slight ratio variation can be attributed to background difference caused by the different laser excitation wavelengths.

Table 6. G/D Ratio Measured by Rennishaw Micro-Raman with 632.8 nm Excitation

Fuel	G/D	Standard Deviation
2007 Cert (Quartz)	0.836	0.057
B20 (Quartz)	0.756	0.023
B100 (Quartz)	0.586	0.175
2007 Cert (Teflon)	0.823	0.007
B20 (Teflon)	0.836	0.026
B100 (Teflon)	0.66	0.022

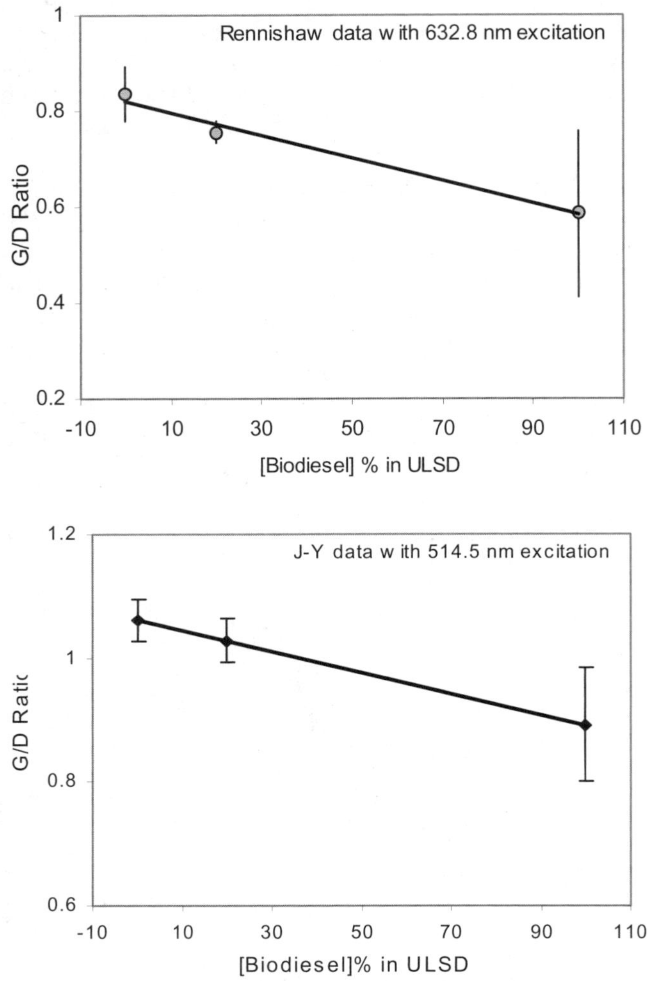

Figure 10. The dependence of the Raman G/D ratio on biodiesel blend level

X-ray Spectroscopy

Energy-dispersive x-ray spectroscopy (EDS) is based on measurement of the characteristic x-rays generated from a sample irradiated with high-energy electrons. EDS measurements were made with a Hitachi 4100 Field Emission scanning electron microscope with a Gresham Sirius EDS detector. The excitation voltage was 5 KV. Carbon to oxygen ratios determined for soot samples produced from 2007 certification diesel and from B100 were 25.34 and 20.34, respectively. This indicates that the B100 soot contains roughly 20% to 25% more oxygen than soot from 2007 certification diesel.

Thermogravimetric Analysis

Thermogravimetric measurement was performed with a Thermal Analysis TGA Q500 unit. All samples were originally collected on quartz filters. The sample was cut into small pieces stacked in the weighing pan. Thus, the majority of the sample weight is due to the quartz substrate. The heating rate was 100°C/min for temperatures below 400°C. The rate was decreased to 30°C/min when temperature reached 400°C (at the location marked ↑ in Figure 11). At 400°C, the purge gas was switched from N_2 to O_2. So at temperatures below 400°C, we observed weight loss from evaporation and devolatilization, while above 400°C weight loss was caused by combustion.

Figure 11 compares TGA results for three PM samples from the different fuels. PM derived from B20 and B100 contains a much greater fraction that is volatile, consistent with the higher organic carbon content. Above 400°C, there is a significant difference in the combustion temperature for these samples. For 2007 certification diesel, combustion occurs between about 650°C and 680°C, while the temperature for B20 is shifted to 550°C-580°C. The B100 sample shows a slightly higher conversion temperature, but the experiment was not precise enough to determine if there is a real difference between B20 and B100.

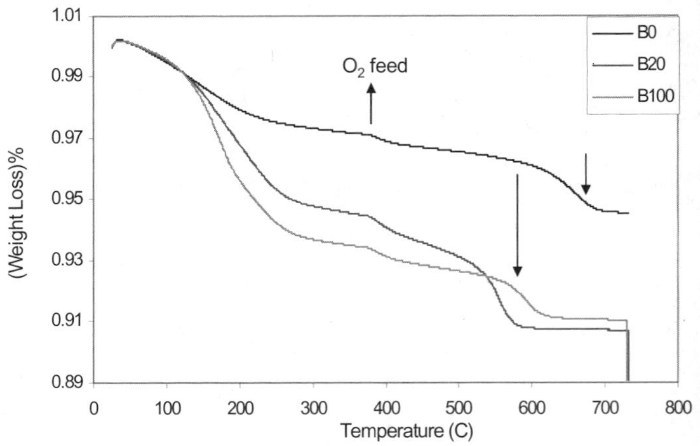

Figure 11. Thermogravimetric analysis results

CONCLUSIONS AND RECOMENDATIONS

Results show that on average, the balance point temperature is 45°C lower than 2007 certification diesel for B20 blends and more than 112°C lower for B100. Filter regeneration rate measurements indicate that biodiesel causes a measurable increase in regeneration rate, even at the 5% blending level. Transient emissions tests show a 24% particulate matter (PM) reduction for B20 without the DPF installed. Installation of the DPF caused PM emissions to drop by more than a factor of 10 for both fuels. Use of B20 with the DPF shows a statistically significant reduction of 27% below the petrodiesel+DPF level. The use of B20 caused a 3% increase in fuel consumption, consistent with the lower energy content of this fuel. Installation of the DPF caused a 1% fuel economy penalty for both 2007 certification diesel and B20.

An examination of NO_x availability at the BPT and DPF regeneration conditions suggests that increased NO_x emissions from biodiesel and biodiesel blends are not responsible for the higher regeneration rate or lower BPT. Detailed chemical and spectroscopic characterization results point to the higher degree of disorder in the soot structure and higher oxygen content as the main factors responsible for the observed effects.

This is confirmed by TGA combustion tests. The results of this study therefore confirm those of Boehman [7, 8] and indicate practical implications in terms of DPF regeneration and the potential for additional PM reduction.

The significant lowering of balance point temperature and increase in regeneration rate might allow passive DPFs to be used in lower temperature engine duty cycles, avoiding or reducing the need for actively regenerated filters and their associated fuel economy penalty. This hypothetical fuel economy benefit has yet to be demonstrated and will be the subject of future research.

ACKNOWLEDGEMENTS

This work was supported by the United States Department of Energy, FreedomCAR and Vehicle Technologies Program, the National Biodiesel Board, and by Cummins, Inc.

REFERENCES

1. Tyson, K.S.; Bozell, J.; Wallace, R.; Petersen, E.; Moens, L. *Biomass Oil Analysis: Research Needs and Recommendations*. NREL/TP-510-34796. Golden, CO: National Renewable Energy Laboratory, June 2004.
2. Sheehan, J.; Camobreco, V.; Duffield, J.; Graboski, M.; Shapouri, H. *An Overview of Biodiesel and Petroleum Diesel Life Cycles*. NREL/TP-580-24772. Golden, CO: National Renewable Energy Laboratory, May 1998.
3. United States Environmental Protection Agency. *A Comprehensive Analysis of Biodiesel Impacts on Exhaust Emissions*. EPA420-P-02-001. Draft Technical Report. 2002.
4. McCormick, R.L., J.D. Ross, M.S. Graboski "Effect of Several Oxygenates on Regulated Emissions from Heavy-Duty Diesel Engines" *Environ. Sci. Technol.*, Vol 31, 1997, p. 1144.
5. McCormick, R.L.; Tennant, C.; Hayes, R.R.; Black, S.; Williams, A.; Ireland, J.; McDaniel, T.; Frailey, M.; Sharp, C.A. *Regulated Emissions from Biodiesel Tested in Heavy-Duty Engines Meeting 2004 Emission Standards*. 2005-01-2200. Society of Automotive Engineers. 2005.
6. Frank, B.P.; Tang, S.; Lanni, T.; Rideout, G.; Beregszaszy, C.; Meyer, N.; Chatterjee, S.; Conway, R.; Lowell, D.; Bush, C.; Evans, J. *A Study of the Effects of Fuel Type and Emission Control Systems on Regulated Gaseous Emissions from Heavy-Duty Diesel Engines*. 2004-01-1085. Society of Automotive Engineers. 2004.
7. Boehman, A.L.; Song, J.; Alam, M. "Impact of Biodiesel Blending on Diesel Soot and the Regeneration of Particulate Filters" *Energy & Fuels*. Vol 19, 2005, p. 1857.
8. Boehman, A.L. "Effect of Alternative Fuels on Soot Properties and Regeneration of Diesel Particulate Filters." *2005 Biodiesel Technical Workshop, November 2005, Denver Colorado*.
9. Kelemen, S.; Fang, H.L. "Maturity Trends in Raman Spectra from Kerogen and Coal." *Energy & Fuels*. Vol. 15, 2001, p. 653.
10. Fang, H.L.; Lance, M. *Influence of Soot Surface Changes on DPF Regeneration*. 2004-01-3043. Society of Automotive Engineers. 2004.

ACRONYMS AND ABBREVIATIONS

ASTM	American Society for Testing and Materials
BSFC	Brake specific fuel consumption
B100	100% biodiesel
B20	20% biodiesel, 80% petrodiesel
B5	5% biodiesel, 95% petrodiesel
CO	Carbon monoxide
CO_2	Carbon dioxide
CVS	Constant volume sampling
DPF	Diesel particle filter
EC	Elemental carbon
EDS	Energy-dispersive x-ray spectroscopy
EPA	U.S. Environmental Protection Agency
FAME	fatty acid methyl esters
g/bhp-hr	grams per brake horsepower hour
NBB	National Biodiesel Board
NO_x	Oxides of nitrogen
NREL	National Renewable Energy Laboratory
OC	Organic carbon
ODS	Octadecylsilane
PM	Particulate matter
SOF	Soluble organic fraction
TGA	Thermogravimetric analysis
THC	Total hydrocarbon
ULSD	Ultra-low sulfur diesel (petrodiesel having 15 ppm or less sulfur)

APPENDIX

Table A-1. Fuel Properties

Property	Method	Units	2007 Certification Diesel	Neat Soy Biodiesel (B100)
Cetane Number	D613		41	--
Kinematic Viscosity 40C	D445	mm^2/s	2.211	4.157
Cloud Point	D2500	°C	--	1
Flash Point	D93	°C	78	152
Total Sulfur	D5453	ppm	9	6
Ash Content	D482	wt%	<0.001	--
Specific Gravity	D4052		0.858	--
Carbon Residue	D524	wt%	0.05	0.04 (neat sample)
Corrosion, Copper strip	D130		1B	1B
Water and Sediment	D2709	vol%	0.01	0.01
Carbon	D5291	wt%	87.27	77.38
Hydrogen	D5291	wt%	13.04	11.86
Aromatics	D1319	%vol	28.2	--
Distillation T90	D86	°C	303	--
Distillation T90	D1160	°C	--	359.6
Net Heat of Combustion	D240	MJ/kg	42.6	37.3
Phosphorus	D4951	ppm	--	<5
Free Glycerin	D6584	wt%	--	0.001
Total Glycerin	D6584	wt%	--	0.236
Acid Value	D664	mg KOH/g	--	0.1
Sulfated Ash	D874	wt%	--	<0.001

Table A-2. Heavy-Duty Transient Test Results (commercial ULSD)

Fuel	Run #	NO_x g/bhp-hr	THC g/bhp-hr	CO g/bhp-hr	PM g/bhp-hr	BSFC g/bhp-hr
Without DPF:						
2007 Cert	1347	2.09	0.185	1.53	0.1557	189.6
2007 Cert	1348	2.14	0.271	1.51	0.1434	189.9
2007 Cert	1349	2.18	0.226	1.56	0.1340	190.7
2007 Cert	1350	2.18	0.240	1.51	0.1141	190.3
2007 Cert	1351	2.19	0.247	1.53	0.1293	189.9
2007 Cert	1354	2.16	0.266	1.52	0.1228	189.7
2007 Cert	1355	2.17	0.281	1.54	0.1294	189.5
2007 Cert	1356	2.17	0.257	1.55	0.1345	189.1
2007 Cert	1357	2.17	0.281	1.51	-	188.8
Mean		*2.16*	*0.250*	*1.53*	*0.1329*	*189.7*
Standard Deviation		*0.03*	*0.03*	*0.02*	*0.01*	*0.56*
COV%		*1.4*	*12.3*	*1.2*	*9.5*	*0.3*
B20	1359	2.19	0.111	1.23	0.1190	193.5
B20	1360	2.26	0.137	1.20	0.0918	195.0
B20	1361	2.28	0.126	1.19	0.0966	195.1
B20	1362	2.29	0.131	1.18	0.0960	195.4
Mean		*2.26*	*0.126*	*1.20*	*0.1009*	*194.8*
Standard Deviation		*0.04*	*0.01*	*0.02*	*0.01*	*0.88*
COV%		*2.0*	*9.1*	*2.0*	*12.2*	*0.5*
With DPF:						
2007 Cert	1308	2.14	0.002	0.02	0.0017	191.4
2007 Cert	1309	2.14	0.004	0.03	0.0015	191.7
2007 Cert	1310	2.13	0.003	0.03	0.0013	190.4
2007 Cert	1311	2.15	0.001	0.04	0.0012	191.4
2007 Cert	1336	2.16	0.001	0.02	0.0022	191.8
2007 Cert	1317	2.16	0.001	0.06	0.0012	191.5
2007 Cert	1338	2.20	0.004	0.04	0.0012	191.9
2007 Cert	1339	2.18	0.003	0.06	0.0011	192.2
2007 Cert	1340	2.19	0.003	0.06	0.0010	191.9
2007 Cert	1341	2.19	0.003	0.06	0.0011	192.3
Mean		*2.16*	*0.002*	*0.04*	*0.0014*	*191.6*
Standard Deviation		*0.03*	*0.00*	*0.02*	*0.00*	*0.52*
COV%		*1.2*	*49.2*	*40.2*	*27.4*	*0.3*
B20	1314	2.26	-0.001	0.02	0.0009	194.1
B20	1315	2.27	0.000	0.03	0.0009	194.7
B20	1316	2.26	-0.001	0.02	0.0009	192.9
B20	1328	2.24	0.004	0.02	-	205.8
B20	1329	2.28	0.000	0.04	0.0010	197.5
B20	1331	2.25	0.000	0.03	0.0012	196.7
B20	1332	2.25	0.000	0.03	0.0011	194.1
B20	1334	2.22	0.000	0.05	0.0009	196.2
Mean		*2.25*	*0.001*	*0.03*	*0.0010*	*196.5*
Standard Deviation		*0.02*	*0.00*	*0.01*	*0.00*	*4.1*
COV%		*0.8*	*360.0*	*32.3*	*10.7*	*2.1*

[a] Negative values indicate emissions below background levels.
[b] 1 g/bhp-h = 1.341 g/kW-h.

2006-01-3443

Effects of Fuel Properties on the Performance of Advanced Diesel NOx Aftertreatment Devices

Shinji Hara
Isuzu Motors Limited

Takashi Kaneko
Nippon Oil Corporation

Toshiaki Kakegawa
Hino Motors Ltd.

Shigeo Senbokuya
Idemitsu Kosan Co., Ltd.

Osamu Nakamura
TonenGeneral Sekiyu K.K.

Masahiko Shibuya
Showa Shell Sekiyu K.K.

Hiroyuki Itoyama
Nissan Motor Co., Ltd.

Masanori Okada
Toyota Motor Corporation

Gen Sugiyama
Japan Automobile Research Institute

Tsutomu Hasegawa
Japan Petroleum Energy Center

ABSTRACT

In the Japan Clean Air Program II (JCAP II) Diesel WG, effects of fuel properties on the performance of two types of diesel NOx emission aftertreatment devices, a Urea-SCR system and a NOx storage reduction (NSR) catalyst system, were examined. For a Urea-SCR system, the NOx emission reduction performance with and without an oxidation catalyst installed in front of the SCR catalyst at low exhaust gas temperature operation was compared. For an NSR catalyst system, the effect of fuel sulfur on both emissions and fuel economy during 50,000 km driving was examined. Furthermore, effects of other fuel properties such as distillation on exhaust emissions were investigated. The results show that sulfur is the influential factor for both devices. Namely, high NOx emission reduction performance of the Urea-SCR system with the oxidation catalyst at low exhaust gas temperature operation is influenced by sulfur. On the other hand, using lower sulfur fuel for the NSR catalyst system, high NOx emission reduction performance is maintained and fuel economy is improved during 50,000 km driving. This study on fuel properties related to the fuel distillation range indicates that fuel composition has an effect on the emission reduction performance of the NSR catalyst system.

INTRODUCTION

"Research program for development of automobile and fuel technologies for better air quality" called Japan Clean Air Program (JCAP) was started in 1997 as a joint research program between the automobile and oil industries, and subsidized by the Ministry of Economy, Trade and Industry. Following the first five-year program called JCAP I, another five-year program called JCAP II was started in 2002. The JCAP II is being conducted to assess the direction of future technologies through the evaluation of automobile and fuel technologies and aims to realize near zero emissions and to reduce carbon dioxide (CO_2) emissions. Furthermore, the JCAP II is aiming to provide a database for environmental and energy policy making through the prediction of air quality improvement by way of Air Quality Model simulation.

Under the JCAP II activity, the Diesel WG has evaluated the effect of fuel properties, in particular the effect of sulfur on exhaust emissions and fuel economy for diesel engines with either a Urea-SCR system or an NSR catalyst. This paper reports on the evaluation results.

EXPERIMENTAL

AFTERTREATMENT DEVICES AND ENGINES

13.1L heavy duty diesel engine was used for the Urea-SCR system, and 4.0L light duty diesel engine was used for the NSR catalyst, respectively.

Table 1 shows the specifications of the test engines, and Table 2 and Table 3 show the specifications of the Urea-SCR system and the NSR system, respectively. For a Urea-SCR catalyst, a vanadium-based SCR catalyst, commonly used as the DeNOx catalyst in fossil-fired combustion plants and stationary diesel engines, was selected. For an NSR catalyst, a combined system with wall-flow type Diesel Particulate Filter (DPF) to reduce NOx and Particulate Matter (PM) simultaneously was selected. Though we have just evaluated another Urea-SCR system with DPF, we conducted the basic test using the Urea-SCR system without DPF in this report.

Table 1. Specifications of test engines

	Urea-SCR system	NSR system
Engine type	In-line 6 (TCI)	In-line 4 (TCI)
Displacement	13.1L	4.0L
Power	283kW/1800rpm	110kW/3000rpm
F.I.E.	Unit injector	Common-rail
EGR	None	Cooled EGR

Table 2. Specifications of Urea-SCR system

	Oxidation catalyst	SCR catalyst
Volume	8.5L	46.4L
Dimensions	Dia.266.7mm L152.4mm	Dia.241.3mm L254mm, 4pieces
Substrate	Cordierite	Extruded
Cell density	400cpsi	300cpsi

Table 3. Specifications of NSR system

	NSR catalyst	DPF (NSR coated)	Oxidation catalyst
Volume	2.2L	5.8L	2.2L
Dimensions	Dia.190.5mm L76.2mm	Dia.190.5mm L203.2mm	Dia.190.5mm L76.2mm
Substrate	Cordierite	Cordierite	Cordierite
Cell density	300cpsi	300cpsi	400cpsi

TEST FUELS

Table 4 shows the properties of the test fuels. To evaluate sulfur effect, we used three fuels with a targeted sulfur content of 0, 10 and 50ppm named S0, S10 and S50. In order to evaluate the effect of distillation characteristics, three fuels with a targeted 90% distillation temperature (T90) of 280, 300 and 330°C named T280, T300 and Base were used. Furthermore, three fuels with a targeted total aromatics content of 5, 10 and 20% named A5, A10 and Base were used.

TEST METHOD

Urea-SCR system

Figure 1 shows a schematic of the Urea-SCR system. In this system, aqueous urea solution is injected into the exhaust pipe, and ammonia is generated, which serves as a reducing agent and reduces NOx. These reactions are expressed in equations (1) to (3). When NO and NO_2 are present in equimolar fraction as shown in equation (4), it is known that the reaction takes place most actively at low temperature ranges [1]. Therefore, the NOx reduction performance at low temperature ranges can be improved by the oxidation catalyst installed in front of the SCR catalyst which acts as NO_2 generator.

$$(NH_2)_2CO + H_2O \rightarrow 2NH_3 + CO_2 \qquad (1)$$

$$4NH_3 + 4NO + O_2 \rightarrow 4N_2 + 6H_2O \qquad (2)$$

$$8NH_3 + 6NO_2 \rightarrow 7N_2 + 12H_2O \qquad (3)$$

$$2NH_3 + NO + NO_2 \rightarrow 2N_2 + 3H_2O \qquad (4)$$

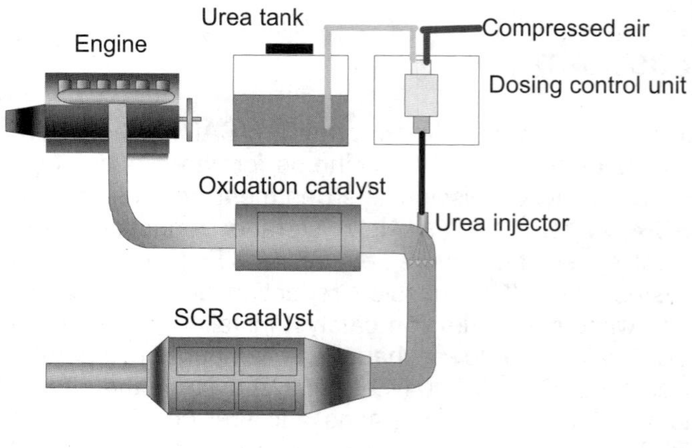

Figure 1. Urea-SCR system

Table 4. Properties of test fuels

Fuel name		Fuels for sulfur effect test			Fuels for properties effect test				
		S0	S10	S50	Base	T300	T280	A10	A5
Density (g/cm³ @15℃)		0.8221	0.8275	0.8271	0.8278	0.8039	0.8013	0.8159	0.8074
Distillation Property (℃)	IBP	196.5	176.0	171.0	174.5	170.0	169.0	185.0	194.0
	10 vol %	216.0	219.5	218.5	219.0	181.0	179.0	223.0	224.0
	50 vol %	261.5	287.5	287.5	273.0	208.0	203.0	281.0	277.0
	90 vol %	301.0	337.0	337.0	330.5	301.0	283.0	335.0	334.0
	EP	313.5	361.0	362.0	344.0	334.0	338.0	361.5	367.0
Kinetic viscosity (mm²/s @30℃)		3.129	3.977	3.949	3.566	1.812	1.679	3.904	3.880
Cetane number		52.2	59.0	58.9	56.6	55.6	56.1	63.0	61.7
Cetane index		57.1	60.5	60.6	57.4	48.4	47.8	65.4	68.8
Total aromatics content (vol %)		12.7	15.6	15.9	18.7	17.2	17.6	9.3	5.4
HPLC Aromatics ring distribution(vol %)	1 ring	10.5	14.7	15.1	16.8	16.9	17.4	9.1	5.3
	2 rings	1.4	0.9	0.8	1.8	0.3	0.2	0.2	0.1
	3 rings+	0.8	0.0	0.0	0.1	0.0	0.0	0.0	0.0
Elements analysis	Sulfur (mass ppm)	<1	7	46	3	6	6	4	9
	Carbon content (mass%)	86.0	85.0	85.0	86.3	85.0	84.9	84.6	84.3
	Hydrogen content (mass%)	14.0	15.0	15.0	13.7	15.0	15.1	15.4	15.7
Net heating value (kJ/kg)		43090	43350	43350	43020	43380	43360	43540	43670
HFRR (μm @60℃)		417	381	377	300	344	274	388	328

In our previous report, the Urea-SCR system is relatively less subject to the effect of sulfur during 10,000 km mileage accumulation [2]. In the accumulation, though the NO_2 generation performance of the oxidation catalyst was somewhat deteriorated, the mileage is not sufficient for evaluating real-world operation.

In this report we conducted a basic test on the Urea-SCR system by means of removing the oxidation catalyst, assuming the activity of the oxidation catalyst was totally poisoned by sulfur in fuel. The result was compared with the case of installing the fresh oxidation catalyst using S10 fuel as shown in Table 4.

NOx storage reduction catalyst

In the NSR system, it is known that the periodical desulfation is necessary to keep the NOx reduction performance, and it is also known that high catalyst temperature during this process causes the deterioration of the NSR catalyst and worsens the fuel economy.

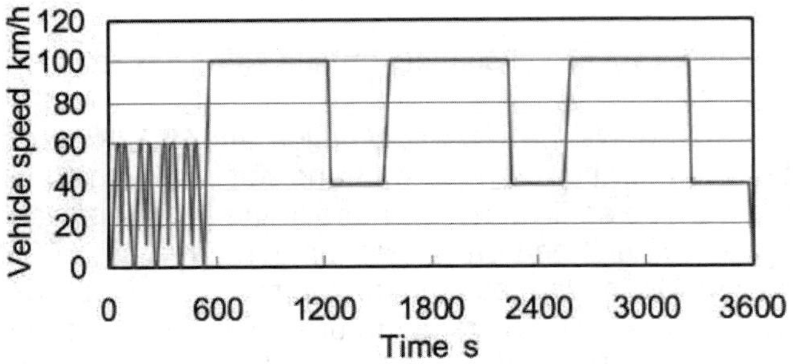

Figure 2. Durability cycle for NSR system

In this report, we conducted a driving test to evaluate the effect of sulfur in fuel on exhaust emissions and fuel economy of diesel engine with the NSR catalyst. The desulfation process is optimally designed for fuel sulfur level. The mileage accumulation of 50,000 km was completed on an engine dynamometer with an average vehicle speed of 73 km/h as shown in Figure 2. For test fuels, three types with varied sulfur levels were prepared

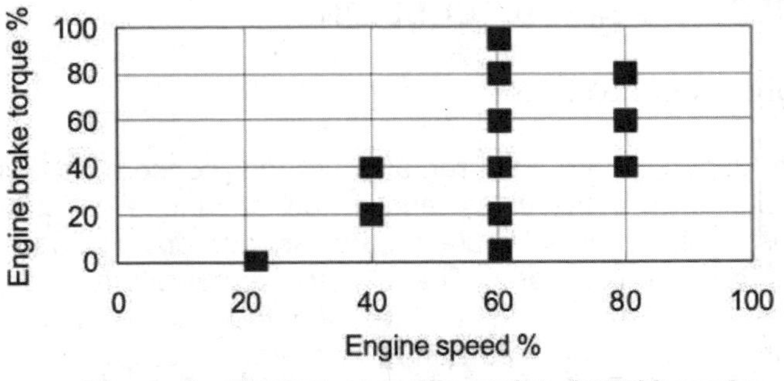

Figure 3. Engine operating points in D13 mode

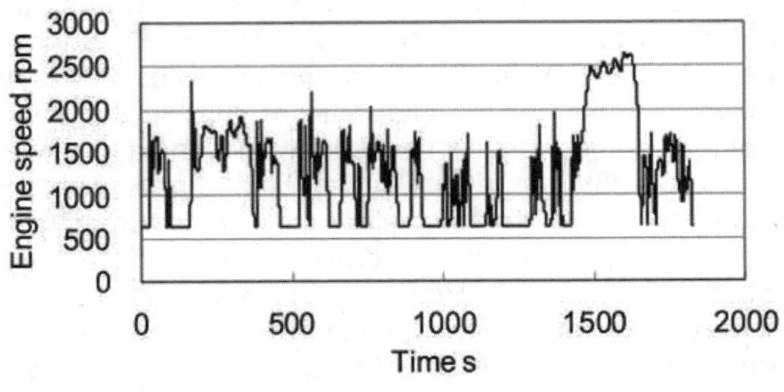

Figure 4. JE05 test cycle

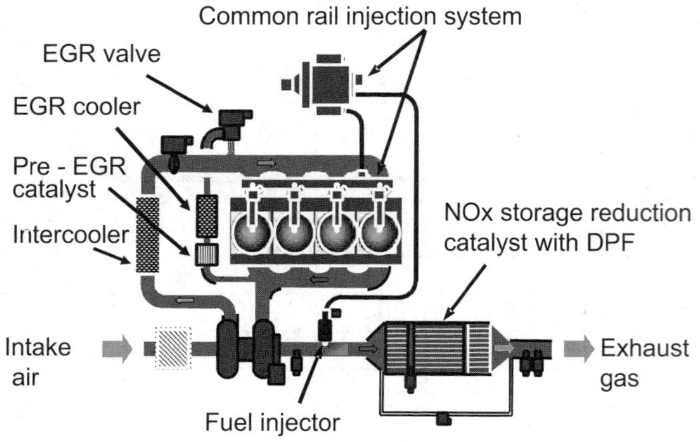

Figure 5. NSR system

as shown in Table 4. An emissions test was carried out 0, 15,000, 30,000 and 50,000 km mileage accumulation. Both the steady state D13 mode and the transient JE05 mode that are shown in Figure 3 and Figure 4 were applied. Fuel consumption during the mileage accumulation was also measured.

To evaluate the effect of other fuel properties (T90 levels and total aromatic contents) on exhaust emissions, we conducted emissions tests using the NSR catalyst that has undergone the mileage accumulation of 50,000 km with S10 fuel.

Figure 5 shows a schematic of the NSR system. A fuel injection valve is provided in the exhaust pipe upstream of the NSR catalyst inlet to spray fuel into the exhaust pipe for the purpose of creating so-called "rich-spike". This can create a fuel-rich circumstance of the catalyst, which is necessary to reducing NO_2 to nitrogen, and also desulfation.

RESULTS AND DISCUSSION

UREA-SCR SYSTEM

In Japan, there still remains strong concern for air pollution in major cities, and a new Japanese emission test cycle JE05 is simulating the average driving pattern in these major cities. The exhaust temperature during the JE05 mode is quite low, and the performance of aftertreatment devices at that low temperature operation is the key issue in evaluation.

Figure 6 shows the NO_2/NOx ratio with and without the oxidation catalyst upstream of the SCR catalyst. Normally, the NOx emitted from engine is mostly NO. As shown in Figure 6, the NO_2 ratio of NOx in the entire engine operation range without the oxidation catalyst was quite low, 25% or less. On the other hand, with the oxidation catalyst, the NO_2 ratio significantly increased, in particular over 50% at an exhaust temperature range of about 400°C or less. However, the NO_2 ratio scarcely increased during high load engine operating condition. It is considered because of chemical equilibrium, by which NO_2 is degraded to NO at high temperature ranges.

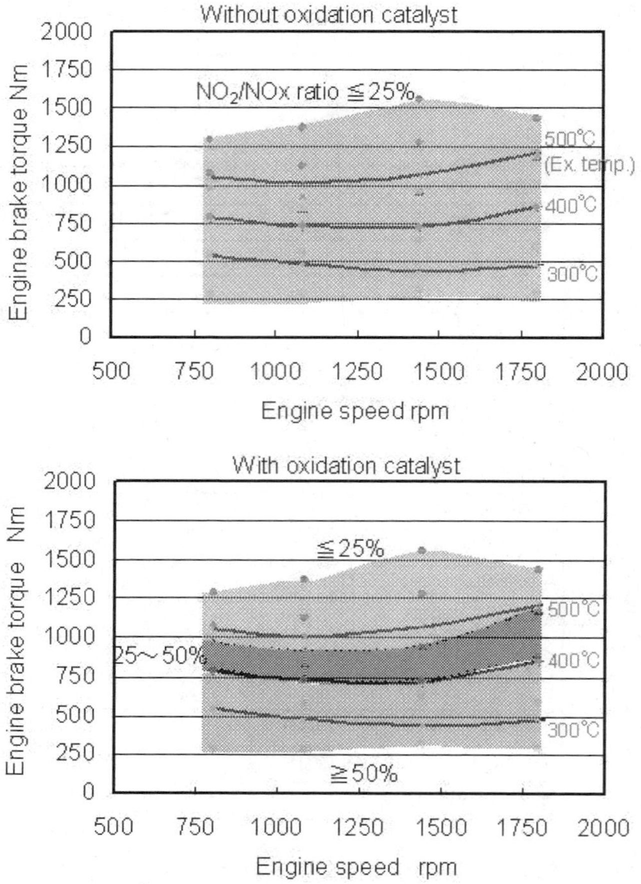

Figure 6. NO_2/NOx ratio

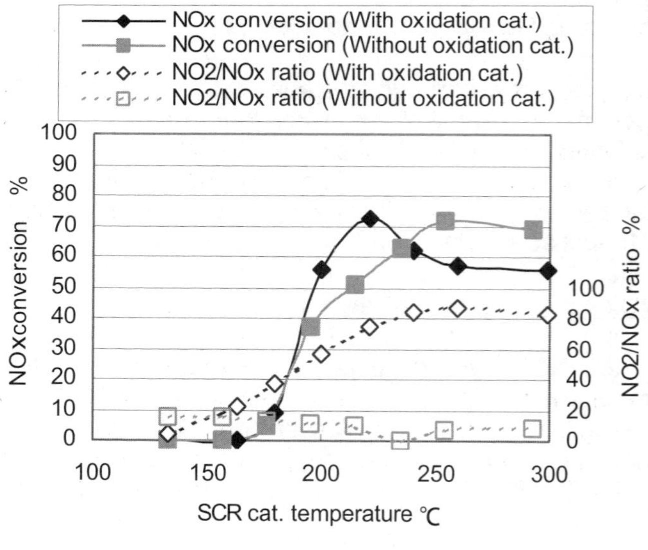

Figure 7. NOx conversion and NO_2/NOx ratio

Figure 7 shows the NOx conversion and NO_2/NOx ratio at low temperature ranges with and without the oxidation catalyst. The catalyst temperature in the figure shows the average of inlet and outlet gas temperatures of the SCR catalyst. During the test, the engine speed and load were regulated to vary the temperature of the SCR catalyst, but space velocity (SV) of the SCR catalyst was kept constant at 8,000 1/hr. With the oxidation catalyst, the NO_2 ratio increased considerably, which indicated the

NOx conversion increased at low temperature range of the catalyst temperature from about 180 to 240°C. The NO_2 ratio in this condition was over 50%, which brought a more active reaction as shown in equation (4), resulting in an increase in NOx conversion. Even the NO_2 ratio exceeded 50%, the NOx conversion tended to increase. The reason for this tendency is yet unidentified, although there may be some differences in the NO/NO_2 ratio in exhaust emissions reactions on the catalyst in actual operation conditions.

Meanwhile, with the oxidation catalyst, the NOx conversion tended to decrease at the temperature range over 240°C, and the NO_2 ratio in this condition was over 80%. According to tests conducted by Saito et al. [3], in a case that reaction of ammonia took place with NO_2 exclusively, NOx conversion decreased in comparison with a case that NO alone exists or a case that both NO and NO_2 exist. Based on this, we suppose that the NOx reduction rate decreased due to the excess NO_2 formed at the temperature range over 240°C in this test.

Based on the test results, when the activity of the oxidation catalyst upstream of the SCR catalyst is decayed because of the effect of sulfur, it is suggested that the NOx reduction performance at low temperature ranges may significantly decrease.

NOX STORAGE REDUCTION CATALYST

Sulfur effect

The 50,000 km mileage accumulation tests have been conducted using S0, S10 and S50 fuels. For the NSR catalyst, lean/rich control is carried out under given conditions regardless of sulfur level, while the frequency of desulfation is regulated taking into account of the sulfur level. The interval between desulfation events is adjusted for those fuel sulfur level. Namely, the interval of desulfation is extended with low sulfur fuel. Table 5 shows the fuel economy obtained in the driving test.

Table 5. Effect of fuel sulfur on fuel economy

Ave. fuel economy for S0	Ave. fuel economy for S10	Ave. fuel economy for S50
6.33 km/L (+1.1 %)	6.26 km/L (Base)	5.97 km/L (-4.6 %)

As a result, average fuel economy improves as the sulfur level is reduced. Provided that S10 fuel is set as the reference fuel, the fuel economy improves by 4.6% with the reduction in sulfur level from S50 to S10, and a further 1.1% improvement with the reduction from S10 to S0.

Figure 8 shows the NOx conversion during the mileage accumulation. Both the D13 and JE05 modes are evaluated. During the mileage accumulation, the NOx storage capacity varies in accordance with SOx accumulation amount on the catalyst. The amount differs by fuel sulfur level and the control of desulfation is set to remove sulfur. In order to eliminate the effect of sulfur accumulated on the catalyst surface, the desulfation was carried out just before the start of the emissions test. The emissions test in the D13 mode was carried out at first, followed by the JE05 mode.

The NOx conversion in the D13 mode showed over 90% at the initial stage of the test even with S50 fuel. In accordance with driving mileage, the NOx conversion gradually decreased, and the tendency is strong in the case of higher sulfur fuel.

The NOx conversion in the JE05 mode also showed a decrease in accordance with the increase in sulfur level, however, the decreasing tendency was relatively small in comparison with that in the D13 mode, which was supposed to occur because the exhaust gas amount in the JE05 mode was lower than that in the D13 mode, so that, though the NOx storage capacity is reduced in higher sulfur fuel case, it was not affected much as in the case of the D13 mode.

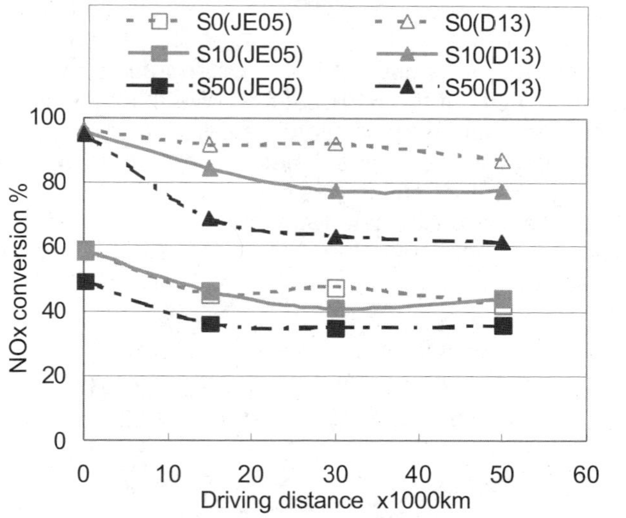

Figure 8. Effect of fuel sulfur on NOx conversion

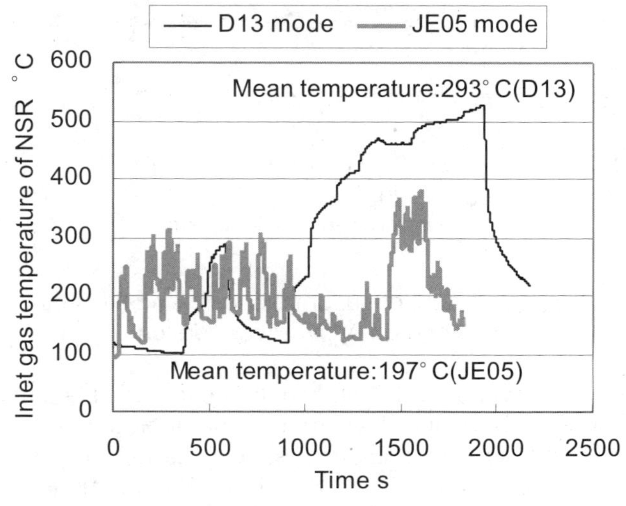

Figure 9. Inlet gas temperature of NSR catalyst in D13 and JE05 mode

The NOx emission reduction performance tended to become nearly stable with every test fuel in the range from 15,000 to 30,000 km of mileage, therefore, the NOx storage capacity was supposed to be lowered suddenly at the initial stage, and then became stable or slow down.

There was a large difference in NOx conversion between the D13 mode and the JE05 mode. This difference was supposed to occur because of the difference in the exhaust gas temperature at the NSR catalyst inlet as shown in Figure 9. The mean inlet gas temperature of the NSR catalyst in the D13 mode was 293°C, while that in the JE05 mode was 197°C, about 100°C lower. It is noted that the activity of the NSR catalyst strongly depends on the performance of NO_2 generation by platinum coated on the catalyst surface, and this activity is extremely decayed under 200°C. In order to improve the NOx conversion of the NSR catalyst at low temperature operation like JE05 mode, it is necessary to improve the NO_2 generation activity of the catalyst at low temperature ranges.

In both test modes, the NOx conversion tended to decrease as the sulfur level increased, which was supposed to be caused by the deterioration in NOx storage capacity of the NSR catalyst. We think there are two mechanisms for the irreversible deterioration. One is that the particle size of sulfated absorbent increased when high level of sulfur was adsorbed [4]. Another is that the dispersion of catalyst metals over the catalyst surface worsens by the heat generated during desulfation.

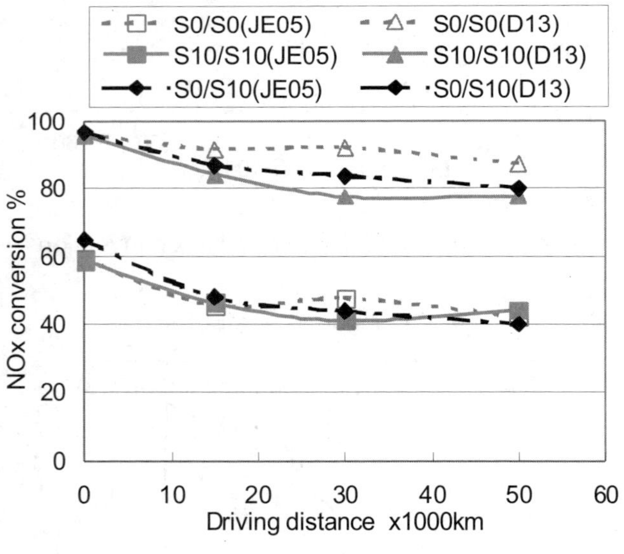

Figure 10. Effect of sulfur poisoning and thermal deterioration on NOx conversion

In this report, in order to clarify both the irreversible sulfur poisoning effect and the thermal deterioration effect, another mileage accumulation test was carried out using S0 fuel and controlling for S10 fuel. The frequency of desulfation was adjusted for S10 fuel, so that, the same thermal deterioration but different sulfur poisoning level could be expected in comparison with the case of S10 fuel/control for S10 fuel combination. The test results are shown in Figure 10. The mark "S0/S10(D13)" in the figure means the test result using S0 fuel, adjusting the frequency of desulfation for S10 fuel and testing in the D13 mode. In comparison with "S0/S10" and "S0/S0", there is a difference in NOx conversion and this difference is considered to arise due to the difference in thermal deterioration. In comparison with "S0/S10" and "S10/S10", there is also a difference in NOx conversion and this difference is considered to arise due to the difference in deterioration by sulfur poisoning, however, this difference is smaller than that in comparison with "S0/S10" and "S0/S0". Therefore, it is predicted that the thermal deterioration because of desulfation is a factor having the most influential impact on the deterioration of the NSR catalyst.

Effect of fuel properties (T90, Aromatics)

The emissions tests in the D13 and JE05 modes were carried out using test fuels with different T90 levels and total aromatics contents. The NOx emission and conversion are shown in Figure 11. The upper graph in the figure indicates the engine-out emissions without the NSR catalyst, while the lower indicates the tail-pipe emissions with the NSR catalyst. The triangle and square in the lower graph indicate the NOx conversion.

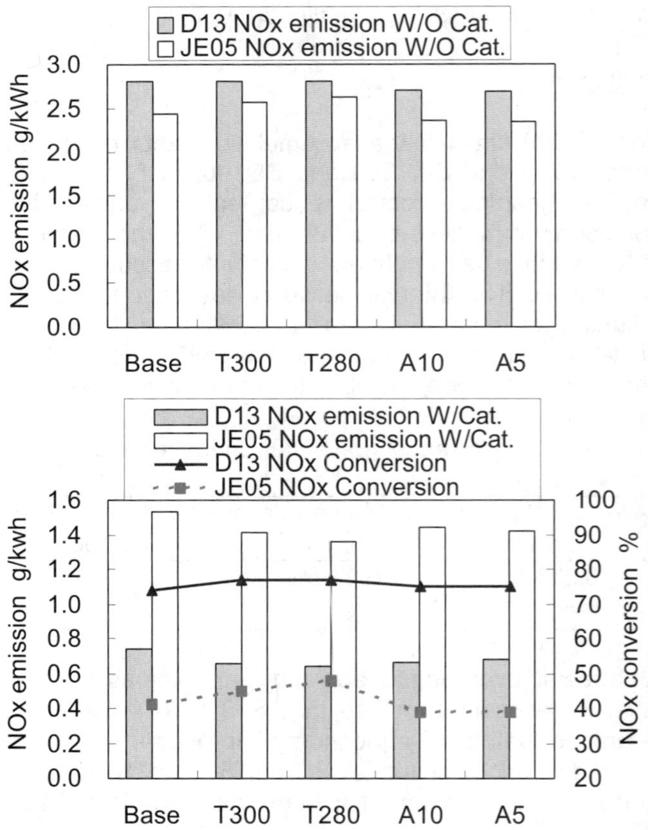

Figure 11. Effect of T90 and total aromatics content on NOx emission and conversion

In terms of the effect of distillation characteristics, it was found that in the case of low T90 fuel named T300 and T280, the NOx conversion increases particularly in the

JE05 mode. It is considered because of the improved evaporability of fuel and the enhanced formation of reduction atmosphere caused by rich-spike operation, and also, it is effective to improve the NOx conversion at low temperature operation like JE05 mode. In terms of the effect of aromatics, the lowering aromatics content reduces the engine-out NOx emission, but effect on NOx conversion is small.

The results on THC emission are shown in Figure 12. In the case of low T90 fuel, the engine-out THC emissions are increased and THC conversion are considerably decreased especially in the JE05 mode. The decrease is supposed to occur because the THC having adsorbed on the NSR catalyst was emitted during acceleration in the JE05 mode.

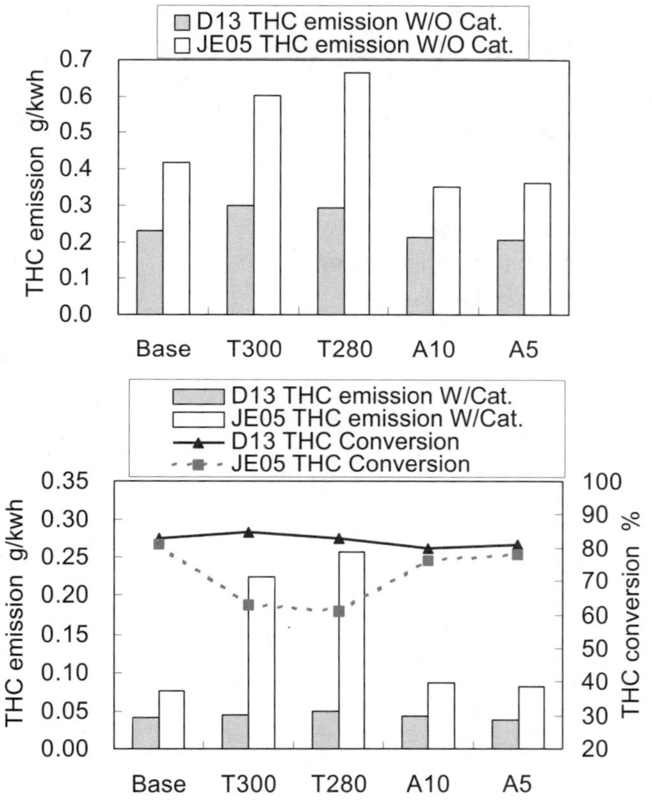

Figure 12. Effect of T90 and total aromatics content on THC emission and conversion

The reason for the increase in the engine-out NOx and THC emissions is considered that the fuel injected during pilot injection failed to ignite and remained unburned. In our previous report, in-cylinder heat release analysis results indicated that the low T90 fuel decreased heat release of pilot fuel [5]. A further analysis is necessary whether the fuel injection quantity during pilot injection reduced or lowered ignitability. The reduction of heat release during pilot injection introduces the extended ignition delay of main injection, the expanded premixed combustion and the extended combustion period.

In terms of the effect of total aromatics content, the lowering aromatics content reduces the engine-out THC emission, but effect on THC conversion is small. It is noticed that in this paper cetane number is kept almost constant as shown in Table 4.

The results on PM emission are shown in Figure 13. It was observed that both T90 and total aromatics content of the fuels affected the engine-out PM emission. In every case, the PM emission reduction performance of the NSR catalyst with the function of DPF added was high, so that the effect of fuel property variation on the tail-pipe PM emission was relatively small.

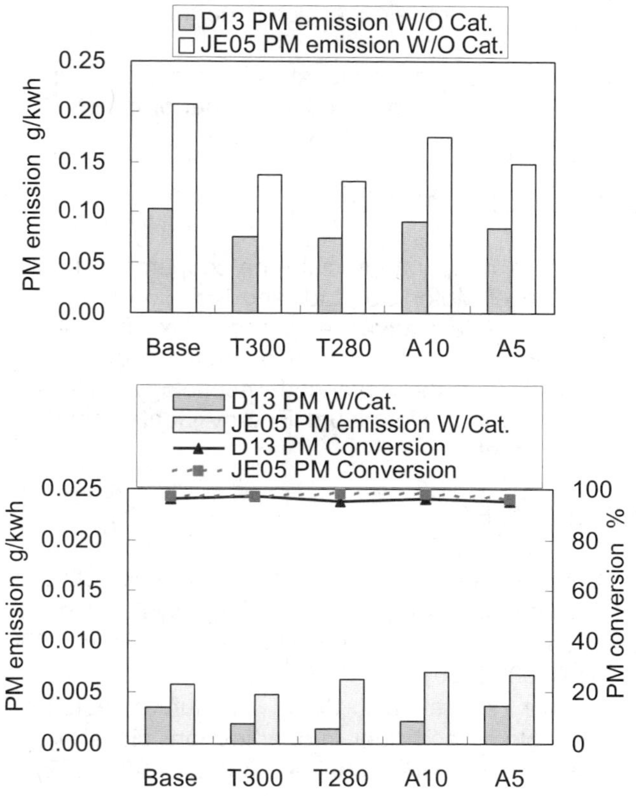

Figure 13. Effect of T90 and total aromatics content on PM emission and conversion

CONCLUSION

The effect of sulfur on NOx conversion and fuel economy was evaluated on diesel engines with both a Urea-SCR system and an NSR catalyst. Furthermore, the effect of distillation and aromatics content on an NSR catalyst was investigated and following conclusions were obtained.

1. For the increase in NOx conversion of the Urea-SCR system at low temperature ranges, the installation of the oxidation catalyst in front of the SCR catalyst is effective.

2. When the activity of the oxidation catalyst is decayed by sulfur poisoning caused by sulfur in fuel, it is suggested that the NOx reduction performance may be considerably lowered at low temperature ranges.

3. With the NSR catalyst, the difference in fuel sulfur levels significantly affects the fuel economy. The

reduction in sulfur from S50 to S10 improves the fuel economy by 4.6%.

4. The sulfur in fuel also greatly affects the NOx reduction performance, and the low sulfur fuel improves the NOx conversion of the NSR catalyst.

5. As to the deterioration of the NSR catalyst in NOx emission reduction performance, it is predicted that the thermal deterioration during the desulfation process is a factor having the most influential impact on the NOx emission reduction performance.

6. The distillation characteristics have an effect on NOx and THC emissions reduction performance of the NSR catalyst.

ACKNOWLEDGMENTS

The authors deeply appreciate the cooperation from the Nissan Diesel Motor Co., Ltd. and the Hino Motors Ltd. provided the latest advanced diesel NOx aftertreatment devices and engines. They also greatly appreciate the cooperation from the Advanced Technology & Research Institute of Japan Petroleum Energy Center conducted the experiments.

REFERENCES

1. Chandler G. R., et al., An Integrated SCR and Continuously Regenerating Trap System to Meet Future NOx and PM Legislation, SAE Paper, 2000-01-0188 (2000)
2. Oyama K., et al., Study on future automobile technology and fuel properties on diesel vehicle emissions, JCAP Technical Report, 2-3-3 (2002)
3. Saito, S. Et al., Development of Urea-SCR System for Commercial Vehicle, Proceedings of JSAE, No.20035165 (2003)
4. Takei Y., et al., Fuel Property requirement for Advanced Technology Engines, SAE Paper, 2000-01-2019 (2000)
5. Oyama K., et al., Study on Current Diesel Vehicle and Fuel Influence on Emissions, JCAP Technical Report, 1-3-1 (2000)

CONTACT

Shinji Hara
Isuzu Motors Limited
shinji_hara@notes.isuzu.co.jp

Takashi Kaneko
Nippon Oil Corporation
takashi.kaneko@eneos.co.jp

Toshiaki Kakegawa
Hino Motors Ltd.
toshiaki.kakegawa@hino.co.jp

Shigeo Senbokuya
Idemitsu Kosan Co., Ltd.
shigeo.sembokuya@si.idemitsu.co.jp

Osamu Nakamura
TonenGeneral Sekiyu K.K.
osamu.nakamura@exxonmobil.com

Masahiko Shibuya
Showa Shell Sekiyu K.K.
masahiko.shibuya@showa-shell.co.jp

Hiroyuki Itoyama
Nissan Motor Co., Ltd.
hiro-ito@mail.nissan.co.jp

Masanori Okada
Toyota Motor Corporation
mokada@engine.tec.toyota.co.jp

Gen Sugiyama
Japan Automobile Research Institute
gsugi@jari.or.jp

Tsutomu Hasegawa
Japan Petroleum Energy Center
hasegawa@atri.pecj.or.jp

VI. Modelling

2005-01-0946

Progress in Diesel Particulate Filter Simulation

Athanasios G. Konstandopoulos, Margaritis Kostoglou, Nickolas Vlachos and Evdoxia Kladopoulou
Aerosol & Particle Technology Laboratory, CERTH/CPERI

ABSTRACT

DPF design, system integration, regeneration control strategy optimization and ash ageing assessment, based on a traditional design of experiments approach becomes very time consuming and costly, due to the high number of tests required. This provides a privileged window of opportunity for the application of simulation tools and hence simulation is increasingly being used for the design of exhaust after-treatment systems with a Diesel Particulate Filter (DPF). DPF behavior depends strongly on the coupling of physico-chemical phenomena occurring over widely disparate spatial and temporal scales and a state-of-the-art simulation approach recognizes and exploits these facts introducing certain assumptions and/or simplifications to derive an accurate but computationally tractable DPF simulation tool, for the needs of industrial users. The present paper summarizes the current state of progress in DPF simulation, from the Aerosol & Particle Technology (APT) Laboratory, addressing consistent and rigorous refinements in a number of areas including: the aerodynamics at the filter and channel inlet, flow, filtration and reaction phenomena at the DPF material microstructure scale, and in coupling/interfacing DPF simulation with general purpose 3-D CFD codes. Where appropriate, experimental results are cited to re-enforce/validate the points brought forward by the theory.

INTRODUCTION

The modern diesel passenger car is increasingly perceived by modern consumers, at least in Europe, as an environmentally friendly and cost-effective transport means building on the first successful market introduction of vehicles equipped with Diesel Particulate Filters (DPFs) and fuel-borne catalyst assisted regeneration in 2001 [1]. The deployment of DPFs also appears necessary for complying with heavy-duty engine emission standards within the next few years while retrofit applications of passive DPF systems for heavy duty vehicles are already wide-spread during the last decade.

Modern trends [2-5] in passenger car emission control systems are now emphasizing on so-called "fit-for-life" solutions overcoming the need for servicing (ash removal) of the DPF during the vehicle lifetime. Robustness and durability of the engine and emission control system is also clearly a first priority in heavy-duty surface transportation. These trends create new challenges to the emission control engineer, who in addition to particulate emission reduction faces the difficult problem of simultaneous reduction of Nitrogen Oxide (NO_x) emissions.

NO_x reduction has relied up to now predominantly on advanced in-cylinder reduction (Exhaust Gas Recirculation, new combustion concepts) approaches, but aftertreatment technologies such as Lean NO_x Traps (LNT), and Lean NO_x Converters (LNC) are actively researched for passenger car application, while urea-based Selective Catalytic Reduction (SCR) systems have already been introduced in the heavy-duty market in Europe [6].

There is little doubt that the next-generation diesel emission control systems will be complex assemblies of chemical reactors and separators (Figure 1), sometimes integrating different functionalities on the same monolithic support to achieve demanding requirements in space and cost, especially in passenger cars [7]. Such complex emission control systems will require for their cost-effective, design, development, system-level integration and optimization advanced simulation tools. In addition, requirements for robust on-board monitoring and control generate the need for efficient algorithms, implementable in computationally limited engine control units (ECUs), which will provide accurate knowledge of the state of emission control system during vehicle operation, and which can be used in control loops for the management of the integrated powertrain-emission control system.

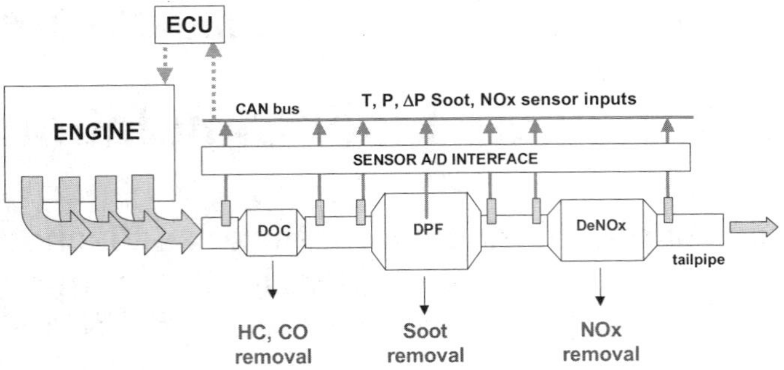

Figure 1. An example of a future integrated soot-NO_x emission control system. DPF stands for Diesel Particulate Filter. Sensory inputs of temperature (T), pressure (P), pressure drop (ΔP), soot and NO_x concentration will be required along the exhaust pipe.

The goal of this paper to present an overview of recent advances in Diesel Particulate Filter simulation technology from the Aerosol and Particle Technology Laboratory (APTL). While the present emphasis is on this highly demanding component of modern diesel emission control systems some of the simulation approaches (especially those pertinent to porous media microstructures) can be fruitfully applied in other catalytic converter systems. The advances in DPF simulation are presented according to a classification scheme based on spatial scales, originally introduced from this laboratory in [8] and exploited in [9-15] to build hierarchically a state-of-the-art DPF simulation tool. Where appropriate experimental results are given to re-enforce the points brought forward by the simulation approach.

THE MANY SCALES OF DPF SYSTEMS

The multi-scale nature of DPFs is by now well appreciated [9-15], and encompasses a span of several orders of magnitude, as shown in Figure 2.

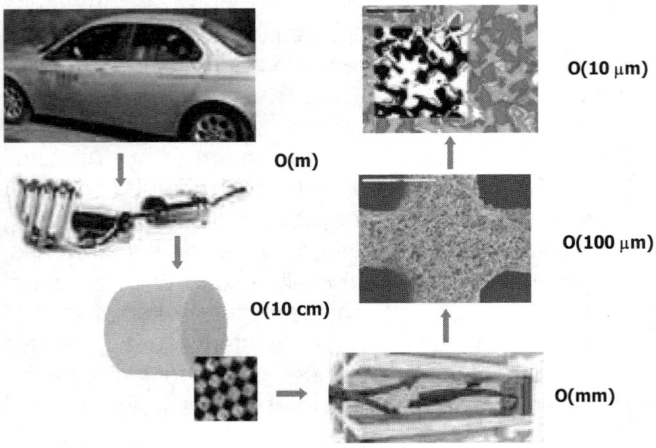

Figure 2. DPF simulation has to face multiple scales, Figure from [9].

This scale disparity has provided a convenient organization scheme for devising a "divide-and-conquer strategy" for the simulation of DPFs, with characteristic examples in [12, 16-19]. In general there are three coexisting length scales that need to be taken into account for the modeling of DPFs. The first scale refers to the phenomena occurring across the soot layer and the porous filter walls, the second scale to the phenomena occurring along the filter channels and the third scale is the macroscopic scale of the entire DPF. In the following sections we discuss selected recent developments from our work at each scale.

FILTER WALL SCALE

In this section we summarize advances in the description of flow, transport and reaction phenomena in the filter wall scale. The filter wall scale is the scale which fuels major material development in the field: new filter structures, multifunctional catalyst coatings, ash interactions [20]. In addition at this scale the structural aspects of the deposited soot particles need to be considered.

CLASSIC FLOW IN POROUS MEDIA DESCRIPTORS

Flow resistance descriptors of DPF walls, experimental protocols for their determination and useful correlations have been previously presented in [21]. At the filter wall scale flow resistance is determined by the Darcy permeability k and the Forchheimer coefficient β of the wall.

$$\Delta P_{wall} = \underbrace{\frac{\mu}{k} u_w w_s}_{Darcy} + \underbrace{\beta \rho u_w^2 w_s}_{Forchheimer} \tag{1}$$

In [21] a generalized correlation spanning 5 orders of magnitude for the permeability of porous DPF walls made of extruded, sintered "grains" was given, which can be stated as:

$$k_{wall} = f_w(\varepsilon) \cdot d_c^2 \cdot SCF(d_c) \tag{2}$$

Where SCF is a factor accounting for slip-flow and free molecular effects (based on the prevailing mean free path and the grain diameter d_c of typical filter media), and the function $f_w(\varepsilon)$ was given in several interchangeable forms [21].

In [21, 22] it was suggested that the Forchheimer coefficient β and the Darcy permeability k were interrelated according to:

$$\beta = \frac{const.}{\varepsilon^{1.5}\sqrt{k}} \tag{3}$$

The value of the constant in Eq. (3) will depend on the actual morphology of the elements making up the porous medium and it was estimated for beds of spheres as 0.143, based on the original Ergun's correlation [23] or as 0.134 for smooth and 0.298 for rough particles of packed beds, based on newer compilations [23]. The Forchheimer contribution to the pressure drop will be negligible unless filters are operated at relatively high filtration velocities. This can occur for some newer filter designs based on sheets of sintered metal materials. From experiments we have performed [24] with granular sintered metal filter media Eq. (3) is shown to be applicable with a value of the constant equal to 0.34 (Figure 3) in good agreement with the value of 0.3 we suggested in [21, 22] by analogy to macroscopic rough granular media [23].

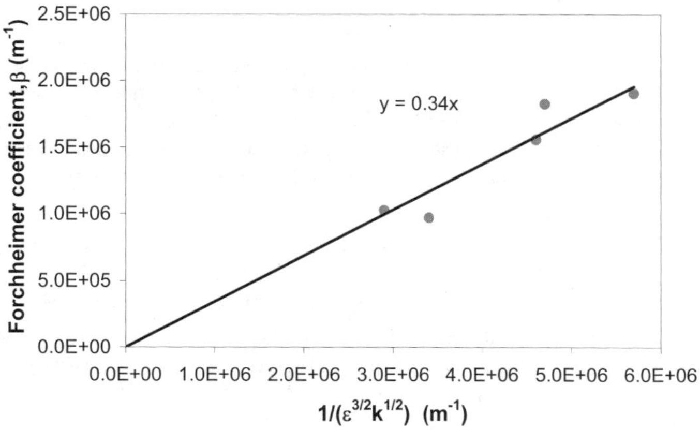

Figure 3. Validity of Eq. (3) for sintered metal granular filters.

We find that Eq. (3) is also applicable for fibrous filter media (Figure 4) with the constant being 16.6. The scatter can be attributed to the fact that the porosity of fibrous filters exhibits a higher local inhomogeneity, therefore sample to sample variations from the nominal specified porosity are expected to be higher.

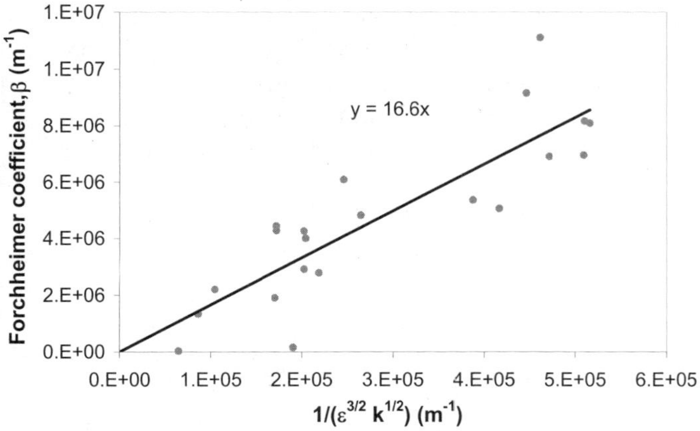

Figure 4. Validity of Eq. (3) for fibrous filters.

MICROFLOW SIMULATION TECHNOLOGY

Experiments and macroscopic continuum flow in porous media approaches have been useful to characterize flow in existing DPF materials. For material design however it is necessary to employ flow simulation techniques in geometrically faithful representations of DPF porous media. The first applications were already introduced in [21, 25] and this line of research is especially important for the development of new filter materials, the optimization of catalyst deposition inside the porous wall and for the design of gradient-functional filter microstructures where multiple functionalities in terms of particle separation and catalyst distribution (for combined gas and particle emission control) can be exploited.

Exploiting advances in computer reconstruction methods (see e.g. [26, 27]) and past experience with discrete particle deposit simulations [28], we have devised algorithmic as well as process-based reconstruction techniques to generate 3-D "digital materials" that are realistic representations of DPF microstructures. We refer to this approach as DPF MicroFlow Simulation (MicroFlowS). MicroFlowS is thus a short name for a computational approach, which combines:

(i) Statistical methods and descriptors (see e.g. [26, 27]) to characterize the microstructure of DPF porous materials (bare as well as coated).

(ii) Computer reconstruction algorithms for the realistic representation of DPF porous media and particle deposit microstructures (see e.g. [26-31]).

(iii) Fluid flow and particle transport, deposition and reaction algorithms in the reconstructed porous materials (employing Lattice Boltzmann (LB) techniques, see e.g. [32]).

Examples of computer reconstructed DPF porous media are given in Figure 5 and encompass all currently available filtration media: extruded ceramic filters (including reaction formed media as cordierite and grain-sintered media as SiC), fibrous filters, foams and sintered metal powder/wiremesh. In addition Figure 5 depicts flow visualization examples for two of the most popular filtration media: extruded ceramic filters and sintered metal media.

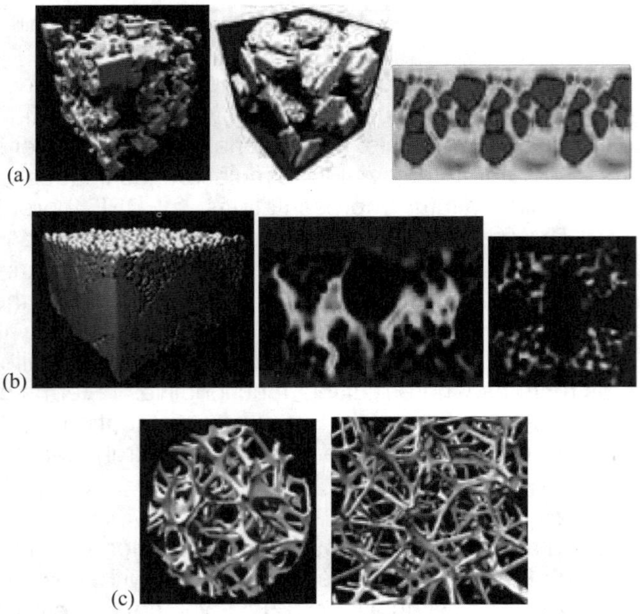

Figure 5. Computer reconstruction of various porous filters. (a): extruded ceramic filters (reaction formed and granular sintered), (b): sintered metal filters (powder and wiremesh), and (c): fibrous filter, foam filter.

A distinct advantage of having "digital materials" at our disposal is the ability to study catalyst deposition processes in these media. Depending on the way the catalyst is applied (e.g. from a solution or from a slurry) the catalyst coating degree of uniformity (Figure 6) is expected to lead to different flow resistance behavior of the DPF.

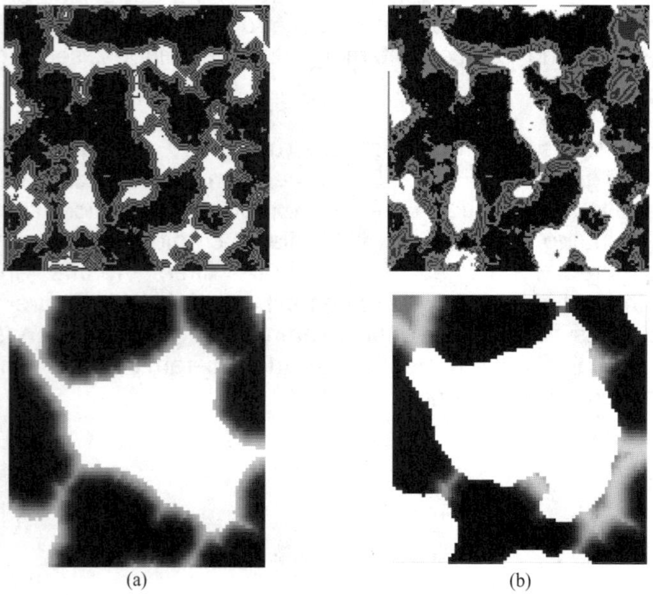

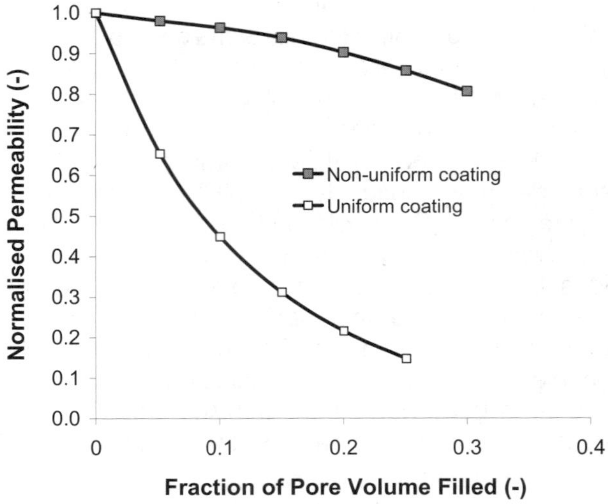

(c)

Figure 6. Simulation of catalytic coating deposition at the same integral quantity of catalyst load. (a): uniform coating, (b): non-uniform coating. The color code (blue to red) assists to visualize the local catalyst coating thickness from the filter surface, (c): effect of coating distribution (uniform vs. non-uniform) on the DPF permeability.

To illustrate the quantitative nature of MicroFlowS we describe the application of the approach to SiC filter structures with pore sizes in the range from 10-90 mm and porosities in the range of 40-50%. At first SEM images of the filter wall are taken at different magnifications, as shown in Figure 7(a), (b), to identify the range of homogeneity of the material. Then a number of polished cross-sections of the material embedded in a special resin are prepared and additional images in the SEM backscattered image mode are obtained, e.g. Figure 7(c). The SEM image is converted into a binary image (with appropriate thresholding algorithms) as shown in Figure 7(d). The binary image is then analyzed with respect to different statistical descriptors (e.g. porosity, autocorrelation function). Subsequently an appropriate reconstruction algorithm is employed (in the particular case the method of [26]) producing the reconstructed "digital material", shown in Figure 7(e). The "real" and "reconstructed" materials have in a statistical sense the same "porosity" and "pore size", as manifested through the very good agreement of their autocorrelation functions [26], shown in Figure 7(f).

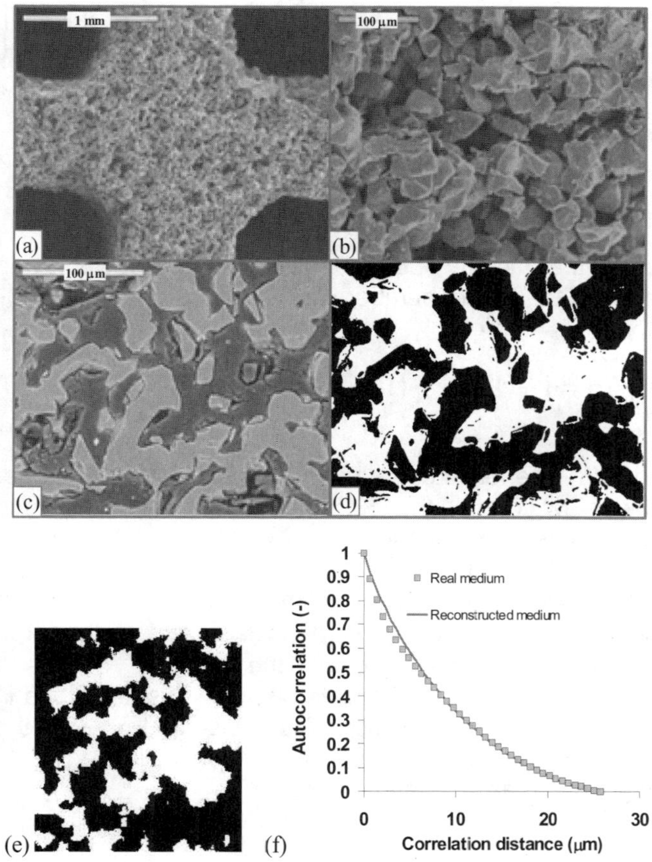

Figure 7. (a), (b): SEM images of filter walls, (c): Backscattered SEM image of a filter wall, (d): Binary thresholded image. The pore space is denoted in black color, (e): Computer reconstruction of filter shown in (d), (f): Comparison of autocorrelation function of reconstructed medium to that of the real filter.

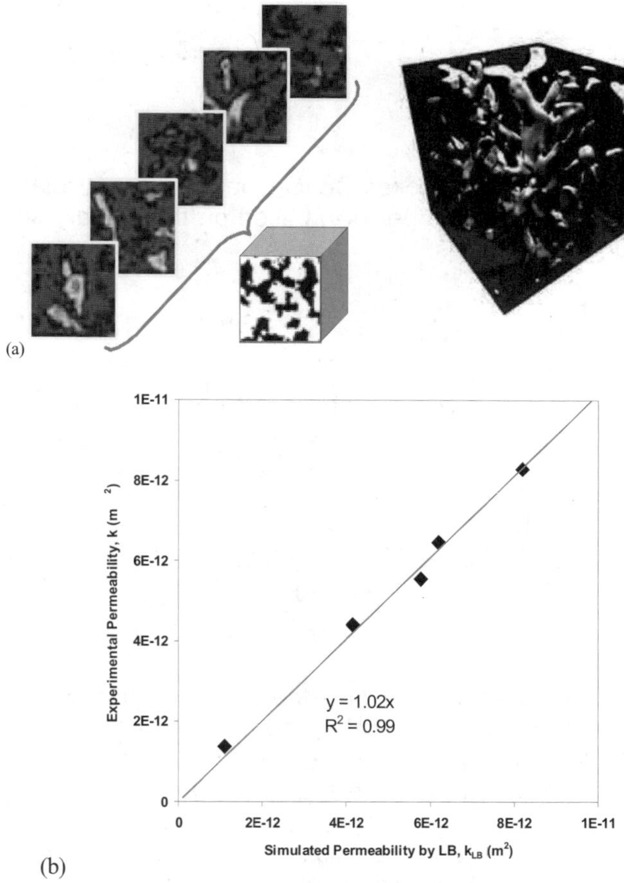

Figure 8. (a): Velocity fields at different sections through the reconstructed filter wall (grey denotes the solid material, violet denotes lowest and red denotes highest velocity), and visualization of flow paths in the reconstructed filter wall and (b): comparison of experimental and simulated filter permeabilities.

The flow resistance behavior of the reconstructed medium is examined by performing three-dimensional flow simulations with the Lattice Boltzmann method [32], and obtaining the permeability of the material [21]. Figure 8(a) depicts a visualization of 3-dimensional flow "tubes" and flow velocity distributions at different cross-sections in a reconstructed filter material. Figure 8(b) shows the comparison of computer simulated and experimental permeabilities obtained with the experimental protocol described in [21].

SOOT DEPOSIT MICROSTRUCTURE

In the literature it has been common practice to assume that the soot layer grown on the filter wall could be described by a uniform density (equivalently porosity), surface area and permeability, which were parameters that had to be tuned according to experiments, leading to widely varying values among different publications, that for the sake of brevity will not be reviewed here. Konstandopoulos et al. [33] demonstrated that during filter loading the microstructure of the soot cake is determined by the convective-diffusive transport of the soot aggregates towards the deposit and it was shown that soot cake packing density (ρ_{soot}) and permeability (k_{soot}) were related to the local value of the dimensionless mass transfer Peclet number, Pe. Furthermore these parameters can be related to the porosity ε and primary particle properties (d_{pr}) of the soot aggregates as follows [33]:

$$\rho_{soot} = \tilde{\rho} \cdot (1-\varepsilon) \quad (4)$$

$$k_{soot} = f(\varepsilon) \cdot d_{pr}^2 \cdot SCF \quad (5)$$

where SCF is the Stokes-Cunningham factor calculated with a Knudsen number (Kn) based on the primary soot particle size (d_{pr}) as:

$$SCF = 1 + Kn \cdot (1.257 + 0.4e^{-1.1/Kn}) \quad (6)$$

and $f(\varepsilon)$ is the Kuwabara function of porosity ε, given by [34]:

$$f(\varepsilon) = \frac{2}{9} \cdot \left[\frac{2 - \frac{9}{5}(1-\varepsilon)^{1/3} - \varepsilon - \frac{1}{5}(1-\varepsilon)^2}{(1-\varepsilon)} \right] \quad (7)$$

These parameters should not therefore be interpreted as constant, intrinsic physical properties of the soot cake, but rather as dynamic properties that depend on the deposit growth mechanism and its history.

Further insight into soot cake properties has been obtained by model experiments with soot aggregates generated by a Combustion Aerosol Standard (CAST) burner (Matter Engineering, Switzerland). The CAST is a quenched diffusion flame gas (propane) burner that allows the stable and controlled generation of soot aggregates over a much larger size range than that found in diesel exhaust. Figure 9 depicts the hydrodynamic resistance factor $(\rho \times k)_{soot}$ of CAST soot deposited cakes on flat disk shaped glass-fiber filters as a function of Peclet number and aggregate mobility diameter d_{ag}.

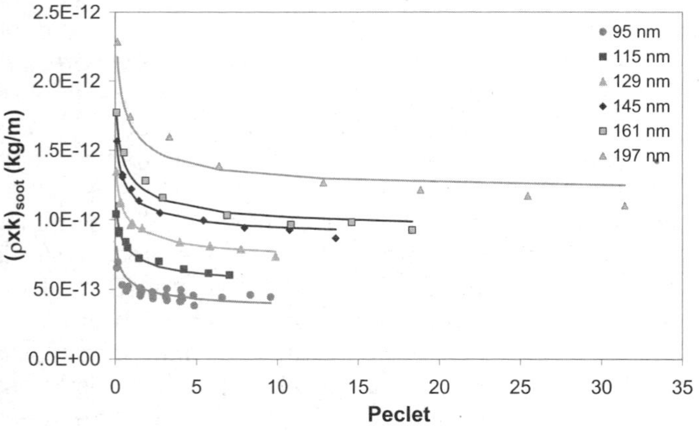

Figure 9. Experimental measurements of CAST soot hydrodynamic resistance factor as a function of Peclet number and aggregate mobility diameter. The continuous lines are plotted using the scaling relation form Eq. (8), (9).

The values of the aggregate mobility diameters shown in Figure 9 are values measured by the Scanning Mobility Particle Sizer (SMPS). The data are consistent with a scaling relation of the form:

$$(\rho \times k)_{soot} = \tilde{\rho} \cdot (1-\varepsilon) \cdot f(\varepsilon) \cdot d_{pr}^2 \cdot SCF \quad (8)$$

where the porosity of the deposits follows a power law in terms of the Peclet number [35]:

$$\varepsilon(Pe) = 1 - (1-\varepsilon_{\infty}) \cdot (1 + \frac{Pe_0}{Pe})^{-n} \quad (9)$$

Eq. (6) generalizes earlier porosity-Peclet number power-law correlations [33] obtained at $Pe > 0.3$ down to the diffusion limited deposition limit. Pe_0 is a characteristic cross-over Pe number defining the scale beyond which the convective mechanism will take over the diffusive mechanism of deposition and ε_{∞} the large Peclet number asymptote of the porosity. ε_{∞} has a dependence on the aggregate size and it is described in a forthcoming publication [36]. Using Eq. (9) the experimental data of Figure 9 can be collapsed on a single curve as shown in Figure 10.

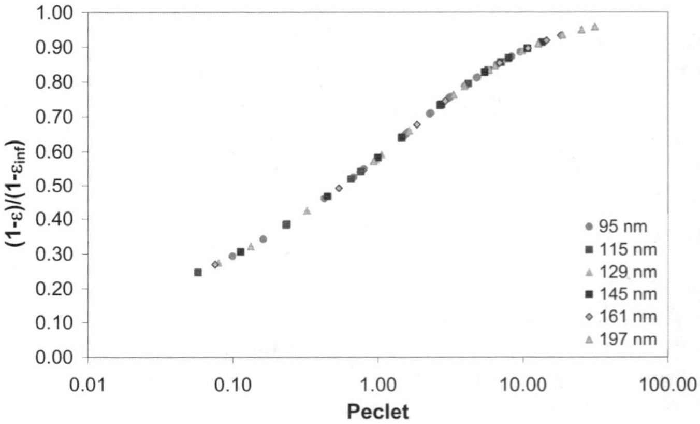

Figure 10.Dependence of the soot cake porosity for different soot aggregate sizes at the prevailing Peclet number.

SOOT DEPOSIT COMPACTION

The model experiments performed with the CAST burner that were reported in the previous section involved the loading of flat disk shaped glass-fiber filters at different Peclet numbers, which was achieved by changing the sampling flow for each case. As experimental pressure drops sometimes reached 300 mbar at the high Peclet number limit, the data had to be analyzed using of our filter pressure drop model accounting for gas compressibility, originally introduced in [37, 38] as shown in Figure 11. Gas compressibility would make the pressure drop signal deviate from a linear behavior however the experimental pressure drop as a function of mass loading remained linear throughout the experimental range of pressure drops. This is an indication that the hydrodynamic resistance factor $(\rho\times k)_{soot}$ of the deposit changed at higher values of the pressure drop, hence deposit compaction was taking place. Accounting for deposit compaction through a power law of the form:

$$\frac{\phi}{\phi_0} = (\frac{\Delta P - \Delta P_{cr}}{\Delta P^*})^{\delta} \qquad (10)$$

with ϕ denoting the solid fraction (1-ε) and ϕ_0 the solid fraction of the deposit in the uncompacted stage. ΔP_{cr} is the critical or yield pressure drop for deposit compaction to set-in and ΔP^* is a scaling constant to make the equation dimensionally correct. This simple deposit compaction model treats the deposit as a Bingham-type of material that remains undeformed below a yield pressure and that deforms with a power law in the post-yield region. The result is shown in Figure 11 as the solid line and illustrates that at high pressure drops the effect of gas compressibility and soot cake compaction are opposing each other.

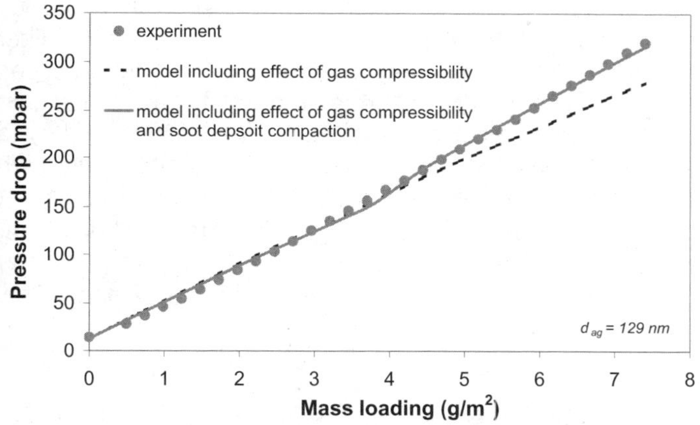

Figure 11. Experimental filter pressure drop as a function of soot mass loading compared with the model taking into account the effect of gas compressibility (dashed line), and the effect of soot deposit compaction (continuous line). The indicative example is given for a soot aggregate size of 129 nm.

Figure 11 gives an indicative example of the effect of gas compressibility and soot deposit compaction on the filter pressure drop for a soot aggregate size of 129 nm, but the same phenomenon was observed for all aggregate sizes at sufficiently high values of the pressure drop. Based on application of the extended pressured drop model that accounts for deposit compaction, we were able to obtain for the first time a quantitative demonstration of the effect of pressure drop on soot deposit compaction. As shown in Figure 12 critical pressure drops for soot cake compaction are setting in the vicinity of 200 mbar causing a gradual increase of the solid fraction (1-ε) of the soot deposit, which can reach 25-30% for some aggregates at high enough values of the pressure drop (> 400 mbar). These pressure drops are non-typical for regular DPF operation but could instantaneously occur in a heavily loaded, undersized filter, say at the start of a forced regeneration event. Our results demonstrate that such phenomena when present can be quantitatively incorporated within the general simulation framework using the simple deposit compaction model mentioned above. ΔP_{cr} and the exponent δ in Eq. (10) have a dependence on aggregate size and morphology as is described in a forthcoming publication [36].

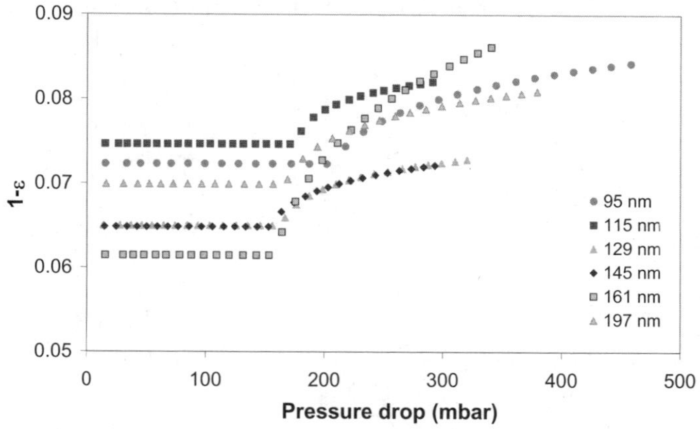

Figure 12. Effect of pressure drop on the solid fraction of soot cake deposits as a function of soot aggregate size.

DEEP-BED FILTRATION

With the advent of more porous wall structures the deep-bed mode of filtration starts to be important and this necessitates a closer look at the soot particle fate inside the wall. As flow in porous media represents a challenging area of fluid mechanics, initial approaches [16-18] have employed so-called unit-cell models, where the porous filter wall is approximated as a collection of "cells", each hosting an object of simple geometry (a sphere for granular filters and a cylinder for fibrous filters).

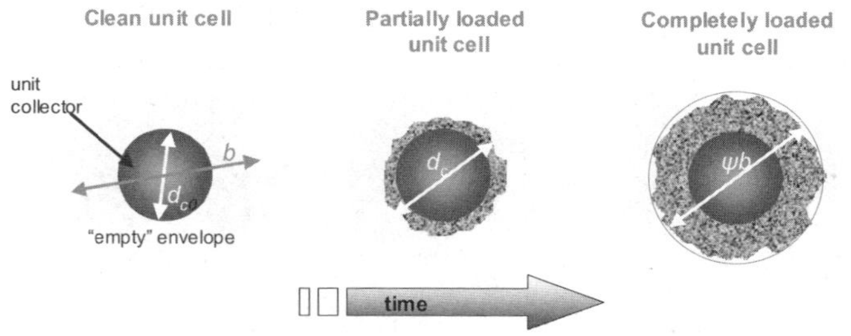

Figure 13. Unit-cell filtration model. The collector size d_c and the empty envelope b are matched to the macroscopic porosity of the filter. The unit-cell blocks when the size of the collector becomes a fraction ψ of b [13, 39].

By employing filtration theory and a local re-computation of the evolving unit-cell geometry due to deposition of particles (Figure 13), a transient filtration model has been derived and tested with very good success against experimental data with ceramic, metallic and fibrous filters [19, 40].

In addition the same unit-cell based filtration model has been successfully employed as a virtual sensor to describe on-line the deposition process of inorganic nanoparticles on different filter media [41]. As seen in Figure 14 the loading of deep bed filters A-C is described equally successfully as the loading of the cake filter D [41].

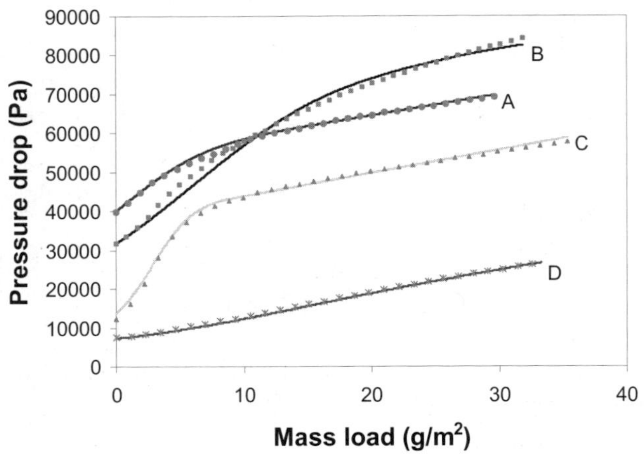

Figure 14. Comparison of simulated pressure drop for different filters (A-D) due to catalyst deposition (solid lines) to the experimental data (points) [41].

The evolution of the non-uniform spatial profile of the inorganic nanoparticles deposited, across the filter wall, can be seen in Figure 15 along with a typical SEM line scan of the distribution of an element of the nanoparticle that serves as a tracer. Both simulated and experimental profiles exhibit a sharp decay along the wall as expected from filtration theory for the particular filter structure employed.

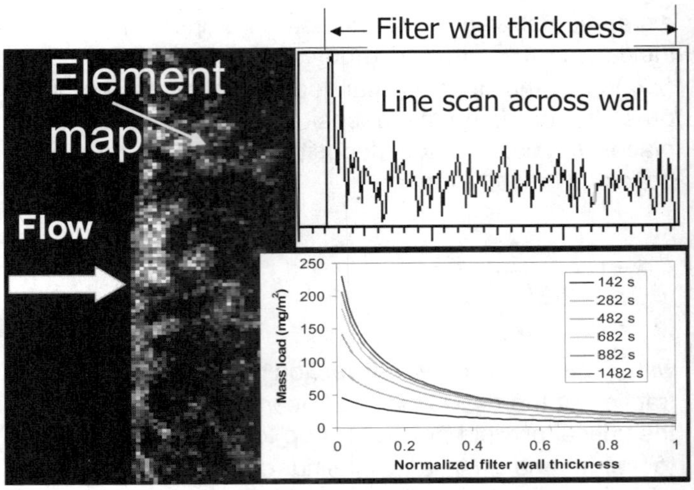

Figure 15. Typical evolution of the spatial deposition of inorganic nanoparticles inside a porous filter wall

Employing our transient filtration unit-cell model for parametric studies we can then obtain the spatial distribution and amount of nanoparticles (soot or inorganic e.g. ash or catalysts) deposited in a porous wall as a function of flow velocity and filter microstructure. This can be achieved for different types of porous materials in terms of porosity, pore size and wall thickness. Table 1 shows the test matrix employed for the parametric study where a wide range of filtration velocities is covered (2-10 cm/s), for a low porous (d_{pore}= 12 μm, ε= 40%) and a high porous material (d_{pore}= 40 μm, ε= 70%), and for a small and a large wall thickness (w_s=0.305 and 0.432 mm corresponding to 12 and 17 mil respectively).

Table 1. Test matrix for parametric study with transient filtration unit-cell model.

filtration velocity (cm/s)	pore size d_{pore} (μm)	porosity ε (%)	wall thickness w_s (mm)
2	12	40	0.305
			0.432
	40	70	0.305
			0.432
4	12	40	0.305
			0.432
	40	70	0.305
			0.432
10	12	40	0.305
			0.432
	40	70	0.305
			0.432

The filtration model can give the mass deposited across the wall thickness at different stages of the deep-bed filtration, which are reported as the utilized capacity of the filter wall, defined as:

$$utilized\ capacity = \frac{V_{solid}}{V_{void}} \qquad (11)$$

where V_{solid} is the total volume of the mass collected inside the filter wall at each stage and V_{void} is the empty space inside the porous wall.

Figure 16 shows the normalized mass distribution inside the filter wall versus the normalized wall thickness as a function of the utilized capacity of the filter wall, for the low porous and the high porous material (small wall thickness) at a filtration velocity of 4 cm/s. The line of the highest utilized capacity gives the state of loading inside the filter wall when the transition from the deep-bed to cake filtration has occurred and there is no more mass entering inside the filter wall.

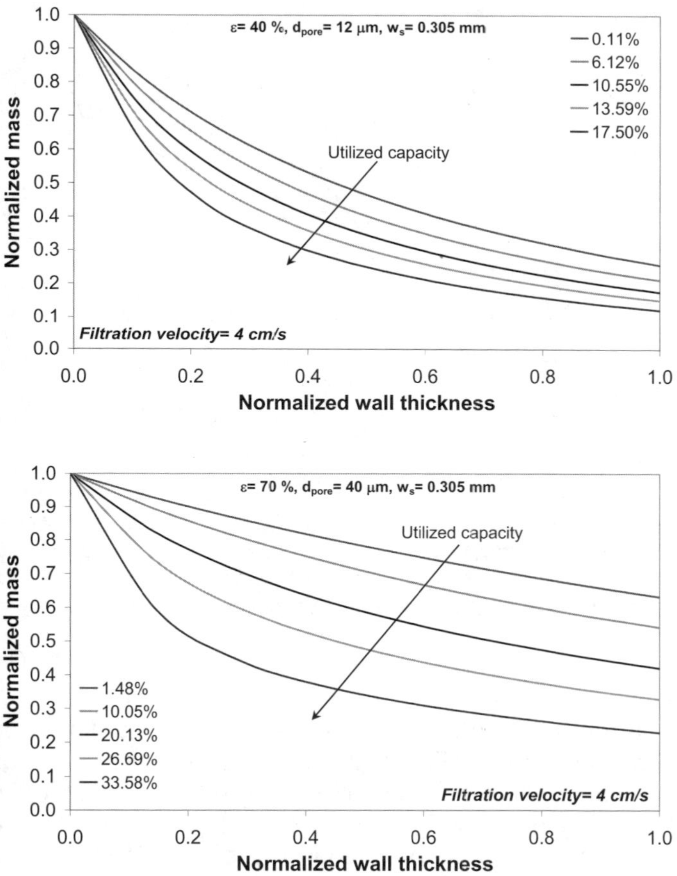

Figure 16. Normalized mass distribution inside the filter wall vs. the normalized wall thickness as a function of the utilized capacity of the filter wall. (top): low porous material, (bottom): high porous material, for a filtration velocity of 4 cm/s, and a wall thickness of 0.305 mm.

This final state of the mass distribution along the filter wall thickness was calculated for all the cases listed in Table 1 and the results are shown in Figure 17.

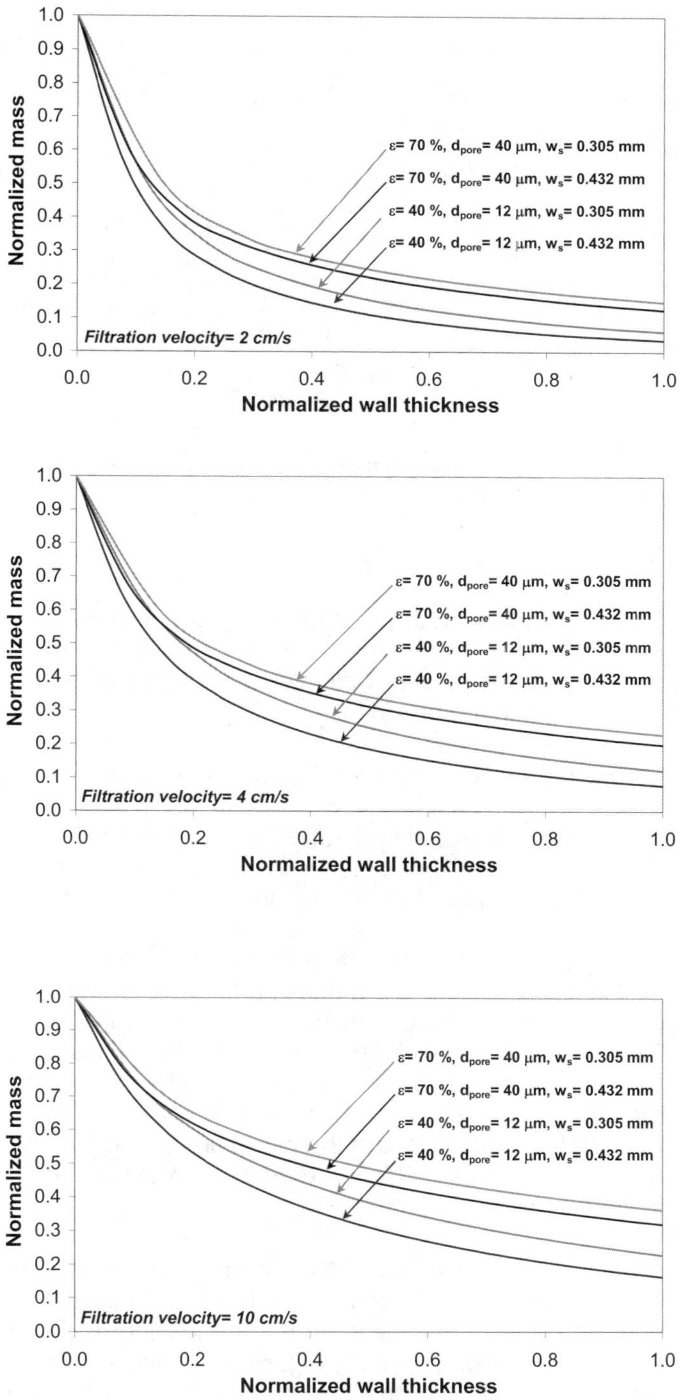

Figure 17. Normalized mass distribution inside the filter wall vs. the normalized wall thickness for the different porous materials. From top to bottom: filtration velocity of 2, 4, 10 cm/s.

Figure 17 shows that a more uniform mass distribution can be achieved at higher filtration velocities, both for the low and the high porous material. Finally, Figure 18 shows the utilized capacity of the filter wall for all the cases in study, computed with a gas temperature of 280 C and a primary and aggregate particle size of 20 and 90 nm, respectively, as a function of the Peclet number.

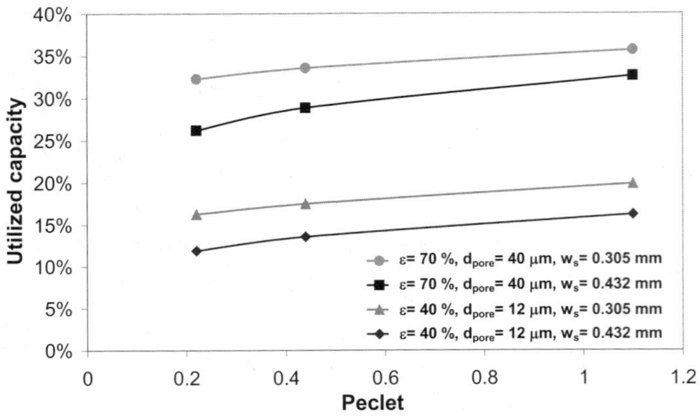

Figure 18. Utilized capacity of the filter wall vs. the Peclet number for all the loading cases in study.

Figure 18 shows that a better usage of the capacity of the filter wall can be achieved for the more porous materials with a smaller wall thickness, with the increase of the Peclet number.

MICROFLOW DEPOSITION IN FILTER WALL

In addition to simulating filtration with unit-cell models, it is possible to simulate soot deposition in the reconstructed "digital materials" previously mentioned.

An example of soot deposition in such a "digital material" (reconstructed with a process-based algorithm that creates and sinters grains from a pre-computed "grain library") is shown in Figure 19(a). As time passes the transition from deep bed to cake filtration as the top regions of the filter are been filled with soot is evident in Figure 19(b). The jumps in the curve are due to the filling of distinct "pores" which the technique allows to be studied in detail.

Over a macroscopic filter area these "incoherent" jumps would average out each other leading to a smooth evolution o the pressure drop as that in Figure 14. The type of simulation shown in Figure 19 is expected to be largely applicable in the near future for industrial use exploiting grid-computing environments [42].

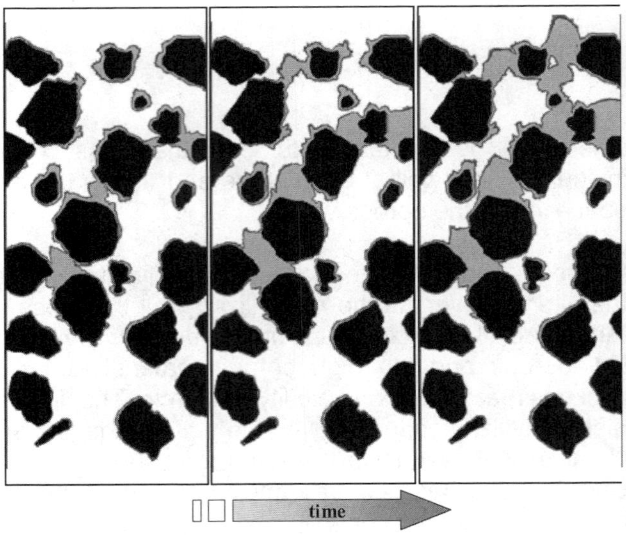

(a)

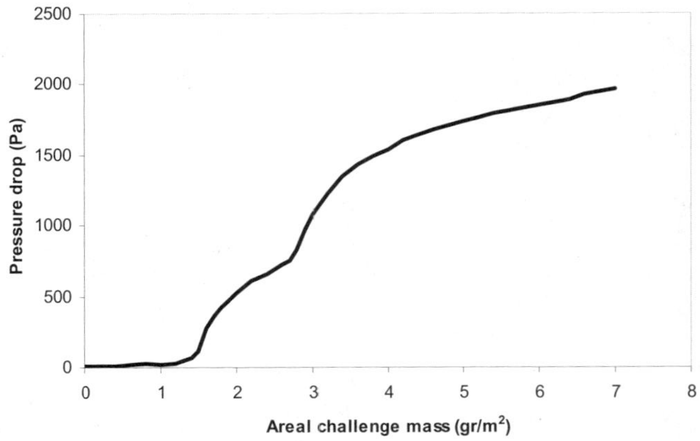

(b)

Figure 19. Simulation of soot deposition on a filter wall. (a): Evolution of soot deposits (gray) in the wall (black is solid, white is pore space) and incipient cake formation, (b): Pressure drop as function of challenge soot mass demonstrating the deep-bed to cake filtration transition.

Current research efforts are concentrating on computationally efficient implementations of the energy equation within the MicroFlowS framework in order to allow realistic simulations of soot particle reaction in the porous structures. A parallel line of development that started in [9] tries to extend continuum models of soot oxidation to account for microstructural effects as shown in the next section.

EFFECT OF MICROSTRUCTURE ON OXIDATION KINETICS

The so-called two-layer model introduced in [12, 14, 43] represents the simplest phenomenological macroscopic description of a catalytic coating (Figure 26). The catalyst coating can interpenetrate/overlap partially with the top part of the filter wall. Hence a spatial "field of catalyst activity" can be assigned to Layer I.

Above the catalyst-influenced layer a normal soot deposit exists that can only react through a thermal oxidation mechanism.

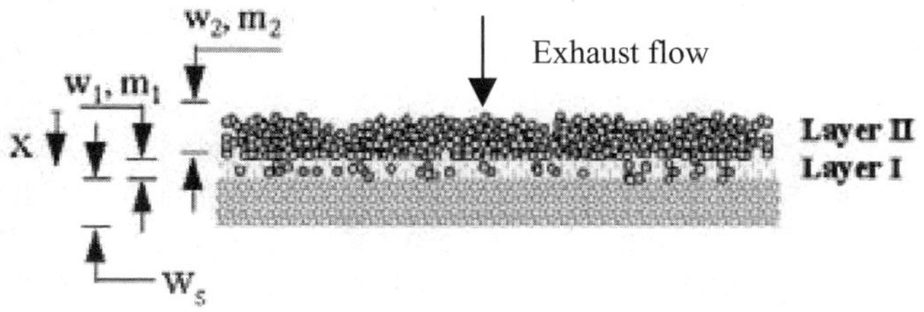

Figure 20. Schematic of the two-layer model of catalytic coating.

A previously developed model of soot oxidation incorporating the soot layer microstructure [9] is combined here with the two-layer model of the catalytic coating. The mathematical model is presented in detail in the Appendix I.

Results are presented for different values of a certain dimensionless microstructural parameter α (See Appendix I) which accounts for the penetration of the thermal layer into the catalytic one, caused by the progressive reaction of the soot which is in contact with the catalyst. Values of α close to 1 mean that the two layers remain separate. As α decreases the degree of penetration increases also leading to enhanced oxidation rates as more soot from the top layer enters the field of activity of the catalyst. This is clearly shown in Figure 21.

For α=1, soot in the catalytic layer is oxidized fast leaving the soot in the thermal layer unreacted. This is a familiar experience with some early catalytic filters. As α decreases the soot from the top layer replaces faster the soot oxidized in catalytic layer increasing the global oxidation rate. The corresponding soot layer thickness evolution is shown in Figure 22. For values of α close to 1 (e.g. 0.9) the catalytic layer is totally depleted from soot at some instances, followed by sudden penetration events from the soot of the thermal layer. These events are clearly shown in the thickness evolution for α=0.9 in Figure 22. For smaller values of α an equilibrium between the soot entering the catalytic layer and the soot oxidized in it, is established leading to a constant global oxidation rate.

The evolution of the dimensionless density profile across the soot layer is shown in Figure 23. The initial gradual replenishment of the soot in the catalytic layer (at t=140s) is followed by sudden penetration events (t=262, t=326) before the establishment of a steady state profile (at t=531, t=778). As regards the non-catalytic (thermal) layer only a gradual reduction of its thickness, accompanied by a very small reduction of its uniform density is observed. Since even this simple microstructural model exhibits so rich dynamic behavior we have also established an experimental program for the study of soot cake microstructures under reactive conditions.

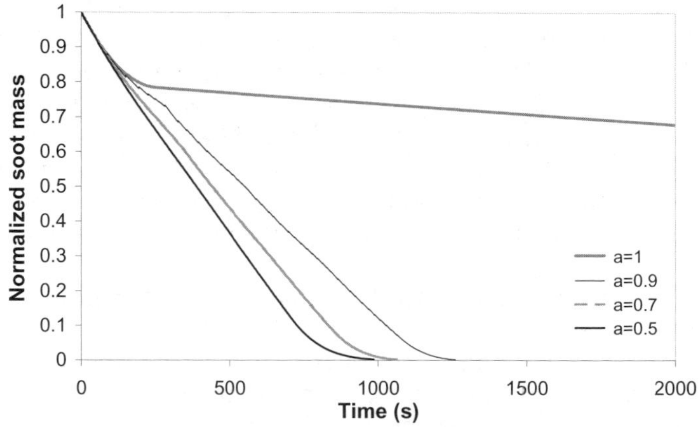

Figure 21. Normalized soot mass evolution for different values of the microstructural parameter α.

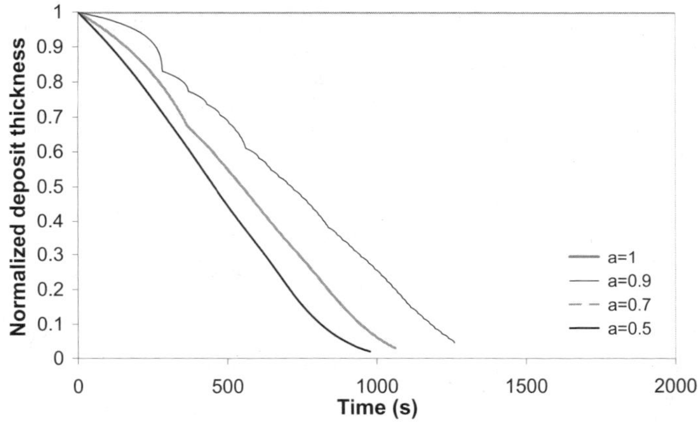

Figure 22. Normalized soot deposit thickness evolution for different values of the microstructural parameter α.

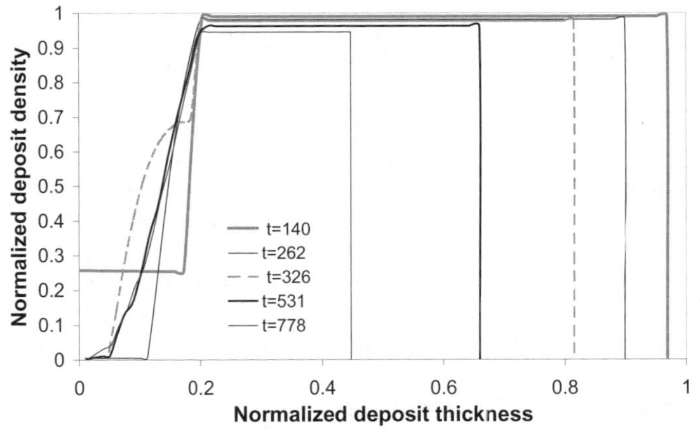

Figure 23. Normalized soot deposit density profile for different times.

This is achieved by performing in-situ permeability experiments on partially reacted soot cakes. An example of such experiments is shown in Figure 24, where the evolution of the soot cake hydrodynamic resistance $(\rho \times k)_{soot}$ as a function of remaining soot mass in the filter is shown. The experiments are performed exposing small filter samples (preloaded with soot from a modern passenger car diesel engine), in a dedicated flow reactor [13] at constant temperature. By cycling between an oxidative (10% O2) exhaust gas and an inert (N2) stream we obtain at each step controlled, partially oxidized soot cakes, which are then analyzed in-situ with the method described in [33] to obtain their hydrodynamic resistance, $(\rho \times k)_{soot}$ by varying the inert gas flow rate and recording the induced pressure drop.

The method has a good resolution since it allows the distinction of $(\rho \times k)_{soot}$ for catalyzed and uncatalyzed filters. Such data in conjunction with the theoretical developments on continuum microstructural models of soot oxidation are expected to improve our description of the relevant phenomena in state-of-the-art DPF simulation tools.

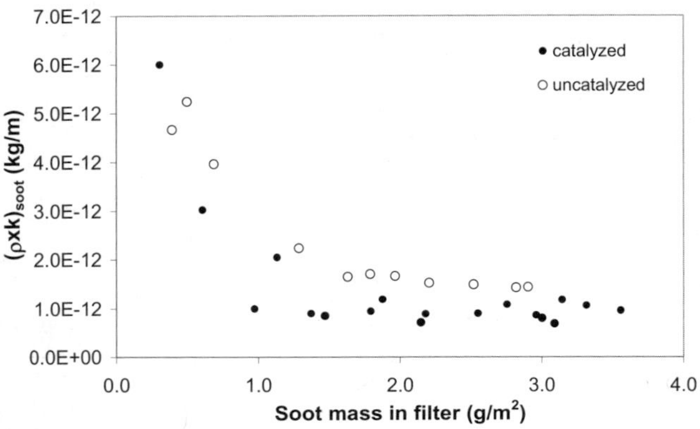

Figure 24. Experimental determination of the change of $(\rho \times k)_{soot}$ during oxidation. At the soot mass load of 1 g/m^2 the soot cake starts to vanish and its hydrodynamic resistance starts to increase rapidly.

THE ROLE OF NO_2 TURNOVER/RECYCLING

The influence of NO_2 on soot oxidation, in conjunction with a highly selective NO to NO_2 oxidation promoting catalytic coating on a DPF was for the first time studied experimentally and theoretically in [13]. At that time NO_2-regenerative technologies were practiced with uncatalyzed filters [44, 45] and it was suggested that combinations of NO_2-regenerative technologies with catalytic filters could lower the dependence of NO_2-regenerative technologies on high engine-out NO_x concentrations, as NO_x emission standards become tighter. This has now become industrial practice [46]

The following reactions occur in presence of an NO oxidation promoting catalyst (such as Pt) in the catalytic layer of the two-layer model [13]:

$$C + NO_2 \rightarrow CO + NO \tag{12}$$

$$NO + \frac{1}{2}O_2 \leftrightarrow NO_2 \tag{13}$$

Let X, Y are the molar concentrations of NO_2 and NO and correspondingly X_o, Y_o are these concentrations in the inlet stream. Z=X+Y is the total NOx concentration. Let k_1 is the reaction rate constant of the soot-NO_2 oxidation reaction and k'_A,k_B the forward and backwards rate constants for the NO oxidation, respectively. For the case of constant oxygen concentration across the soot layer the forward reaction can be considered of pseudo-first order with a modified rate constant $k_A=k_A'[O_2]^{0.5}$.

The reaction of NO_2 with soot has attracted considerable interest in the literature [47, 48] and complex reaction schemes are being developed to account for the influence of H_2O, SO_2, the collaborative effect of O_2-NO_2, presence of various catalysts on the NO to NO_2 oxidation (precious and non-precious metals), inhibition and transient kinetic phenomena [49, 50].

While a detailed kinetics approach may lead in the future to more elaborate rate laws, for our present purposes it is sufficient to consider that the soot-NO_2 and NO oxidation reaction rates as first-order in NO_2, This allows a great simplification and the derivation of an analytic global expression for the NO_2 turnover or recycling factor in the framework of the two-layer model of a catalytic coating can be derived analytically as shown in the Appendix II. The NO_2 turnover or recycling factor R then (shown below from Appendix II) means the number of times an NO_2 molecule is re-used to oxidize soot compared to the absence of a catalytic coating promoting NO into NO_2 oxidation:

$$R = \frac{C_1 C_2 C_3}{1 + C_3 + C_4} + \frac{1}{1 + C_3 + C_4}\left(1 - \frac{C_2 C_3}{1 + C_3 + C_4}\right)\left(1 - e^{-(1+C_3+C_4)C_1}\right) \tag{14}$$

where C_1, C_2, C_3, and C_4 are dimensionless parameters defined as $C_1=k_1w_1/v_w$, $C_2=Z_o/X_i$, $C_3=k_A/k_1$, $C_4=k_B/k_1$ These are determined based on design and operational considerations.

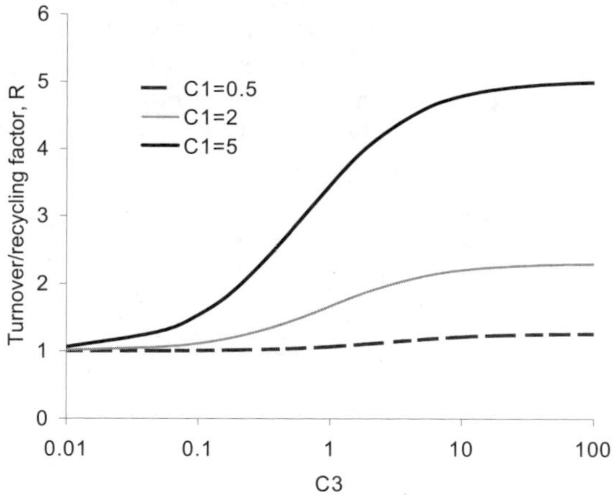

Figure 25. Recycling factor R versus parameter C_3 for several values of parameter C_1.

The recycling factor for the case of an exhaust NO_2 containing stream entering the catalytic layer at equilibrium conditions, is shown in Figure 25 versus the parameter C_3 (affected mainly by the design variable catalyst content which affects k_A) for several values of C_1 (affected by design and operational variables wall velocity, catalyst layer thickness). In case of large C_3 the recycling factor attains its maximum value and therefore it is a useful number to know for design/optimization:

$$R_\infty = C_1 \frac{[NO_2]_{eq}}{[NO_2]_i} \tag{15}$$

Based on the NO_2-soot oxidation data promoted by the catalytic coating tested in [13] an R value of about 3.4 has been computed. This represents a significant enhancement.

FILTER CHANNEL SCALE

The next relevant spatial scale is that of the filter channel. If the inlet conditions to the DPF can be assumed to be spatially (radially) uniform, formulating the DPF simulation model for two representative channels (inlet and outlet) is equivalent to solving it for the entire DPF (assumed radially homogeneous). It is not therefore surprising that this scale claims most of the published works in the literature, starting with Bissett [18] who formulated the first comprehensive model which takes into account the detailed flow distribution in the channel. As the transient behavior of the monolith is slow with respect to the residence time of the exhaust gas as it travels across the filter, the quasi-steady state approximation for the exhaust flow in the channels simplifies considerably the computational burden. Although the flow in the filter channels is actually three-dimensional (3-D), it can be accurately approximated by a (perimeter averaged) one-dimensional (1-D) configuration. This has been confirmed by Konstandopoulos et al. [51, 52] who demonstrated excellent agreement of analytical solutions of a 1-D single channel flow model with exact 3-D CFD results as well as with many experiments for square and triangular shaped channels. New developments in the filter channel scale phenomena simulation include: the study of different geometrical factors and soot loading on inertial inlet/outlet channel losses, the implications of asymmetric channel geometries (for ash tolerance), the modeling of particle entrainment phenomena (for ash and soot) and the refinement of the channel-to-porous wall heat transfer modeling. These are discussed in more detail in the next sections.

INERTIAL LOSSES AT CHANNEL INLET/OUTLET

It has been shown in [22] that CFD calculations or experiments are the only means to evaluate the inertial losses, which result from the contraction of the exhaust gas at the channel entry and its expansion at the channel exit, since estimations on literature correlations (which are not specifically developed for the honeycomb structure of the DPF) are not sufficiently accurate for such estimations. Furthermore the influence of a soot cake on these losses has never before been studied in a rigorous fashion. To improve the inclusion of such phenomena in DPF simulation tools, in the present section we summarize the effect of different cell densities and soot loadings on the DPF contraction/expansion losses coefficient ζ [22].

To this end three-dimensional CFD simulations were performed with a commercial CFD package [53]. Taking advantage of the symmetry of the problem only the physical domain shown in Figure 26 is modeled, which includes part of the inlet, the outlet channel and the wall. The computational domain also includes sufficiently large sections upstream and downstream of the channel in order to obtain well-developed flow. This approach for modeling DPFs has been used and validated in the past [22, 52] where the reader can find more detailed descriptions of the procedure followed for the domain and mesh creation.

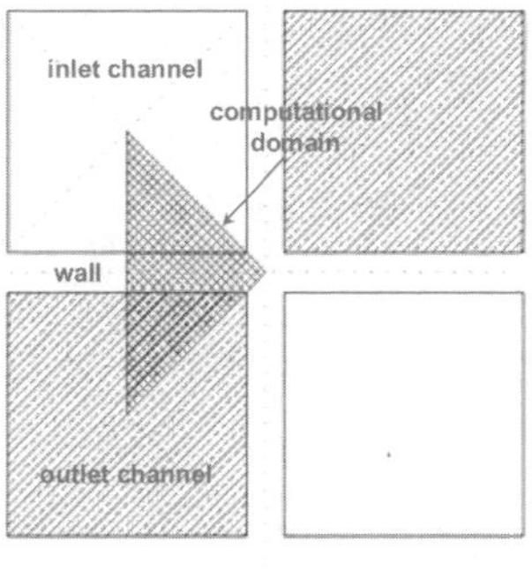

Figure 26. Cross section of filter channels showing the front view of the CFD computational domain.

In order to investigate how the blocking of the channel affects the contraction/expansion losses coefficient ζ various cases were studied in terms of filter cell density, channel size and soot mass loading in the filter. These parameters are shown in Table 2, along with the open frontal area fraction (OAF or ε_A defined as the open channel area over the total filter face area) corresponding to each case. The channel length was fixed to 150 mm (which corresponds to a typical 6 inch DPF), and the channel plug length was initially set to 5 mm.

Table 2. Test matrix for CFD calculations.

cell density (cpsi)	wall thickness (mm)	channel size (mm)	soot layer thickness (μm)	open area fraction (-)
100	0.3356	2.184	0	0.370
			100	0.305
			200	0.247
200		1.440	0	0.322
			100	0.238
			200	0.168
300		1.111	0	0.287
			100	0.193
			200	0.117

The calculations were performed for a Reynolds number based on the channel hydraulic diameter (Re_h) of 1800. This high Reynolds number was chosen since a more accurate determination of ζ can be obtained at these flow conditions (either experimental or computational) keeping in mind that inertial effects are more pronounced at high Re_h. The permeabilities of the filter wall and that of the soot layer were fixed to the representative values of 1×10^{-12} and 3.3×10^{-13} m^2, respectively. The inertial loss coefficient ζ was determined by a single CFD calculation of the pressure distribution along the system, for each case, and by subtraction of the pressure loss due to the porous channel region, and the porous soot layer (when present). The results are shown in Figure 27 as a function of the open area fraction for the different filter cell densities and soot mass loadings.

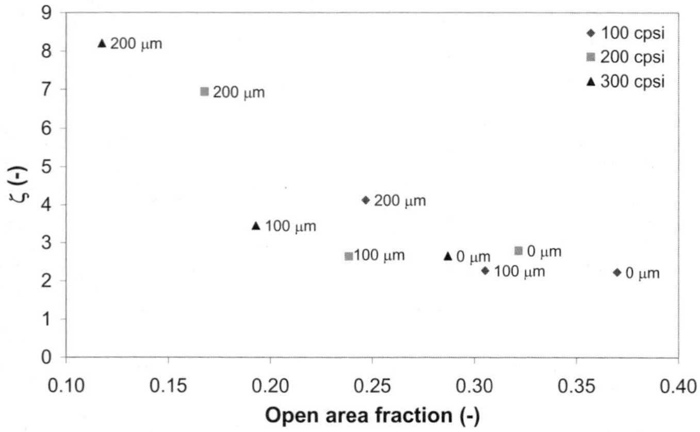

Figure 27. Contraction/expansion losses coefficient ζ vs. open area fraction for different cell densities and soot mass loadings.

The data clearly can be only approximately correlated in terms of the open area fraction as shown in Figure 27. Subsequently we studied the effect of the plug length on the inertial losses coefficient ζ for a clean filter at the three different cell densities listed in Table 2. The plug length value ranged from 5 mm to 15 mm and the results are shown in Figure 28.

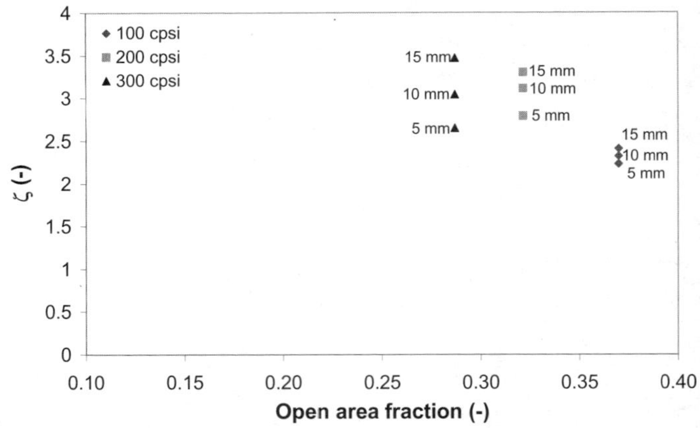

Figure 28. Contraction/expansion losses coefficient ζ vs. open area fraction for different cell densities and plug lengths.

The above illustrate that the task of deriving a theoretical description for ζ requires a substantial amount of computations, but it is not intractable. Pending the derivation of a generalized correlation (that takes into account the combined effect of open area fraction, cell density, soot thickness and pug length) we employ in our simulations tabular interpolations between the data of Figure 27 and Figure 28, which cover a sufficiently large parameter space for practical applications. The advent of other than square channel shapes [52] or asymmetric channel designs for ash tolerance [2-4] poses no problem as it can be handled within the same computational framework. Our current efforts with asymmetric channel geometries (see next section) however, focus on the more important problem of extending our single and multichannel DPF simulation models to account for such effects

ASYMMETRIC CHANNEL GEOMETRIES

The Pe number dependence of the soot cake microstructure has important implications for the morphology of the soot (and ash) deposits in the square channels of DPFs, as illustrated in Figure 29, obtained by a mixed ODE-Monte Carlo simulation [35].

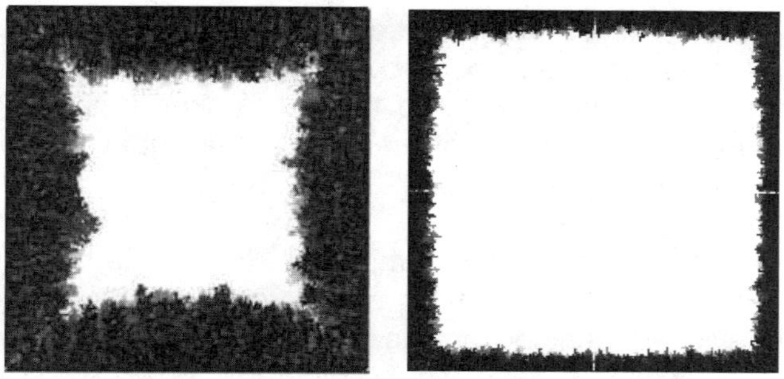

Figure 29. Morphology of simulated soot deposit grown in a square channel for Pe= 0.1 (left) and Pe=100 (right), at the same soot mass load in the filter [35].

The introduction of asymmetric designs [2-4] makes necessary the inclusion of such shape effects on DPF simulation tools employed by the industry. As this inevitably will entail approximations these will have to be assessed with more detailed studies.Soot deposit growth and microstructure evolution is not presently accounted for in general purpose flow solvers. The "hooks" provided in all general purpose flow solvers through User Defined Functions/Subroutines (although often quite awkward) can be used to interface in the solver phenomena occurring at the sub-grid scale and hence study the aforementioned channel shape effects. Our interfacing of local soot deposit growth to large scale CFD simulations is presented in a companion paper [24]. In the present section we provide examples of the current state of our developments with asymmetric filter channel shapes.

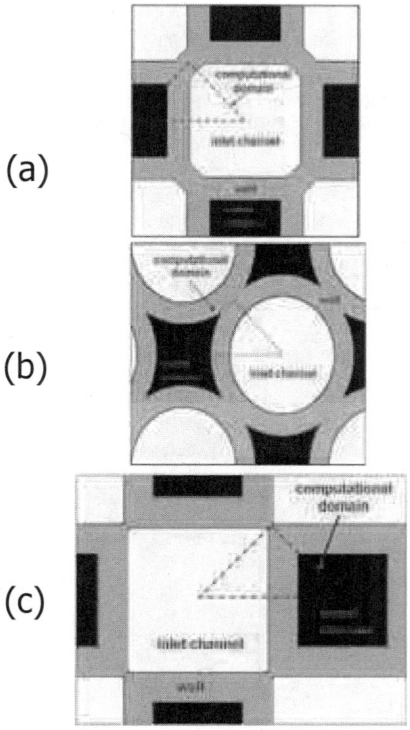

Figure 30. Cross-section of asymmetrical designs showing the front view of the CFD computational domain.

Figure 30 lists schematics of 3 asymmetric channel geometries (A, B, and C) inspired by the designs presented in [2-4] studied (at the same value of open area fraction).

Figure 31 and Figure 32 demonstrate the growth of soot cakes in each structure shown in Figure 30. While a detailed account of soot deposit growth dynamics in asymmetric channel designs will be given elsewhere we can remark that the soot deposits grown bear a very good similarity to those shown on SEM pictures of some of the designs [2]. All filter designs after an initial stage exhibit a nearly uniform soot deposit. This demonstrates the utilization of the entire internal surface area of the inlet channels, as the flow readjusts to transport the soot particles along the path of least resistance (which is not necessarily the shortest geometric path between the inlet and outlet channel, i.e. the wall thickness). This self-regulated behavior is currently advantageously exploited to bring such asymmetric designs within the multichannel simulation framework.

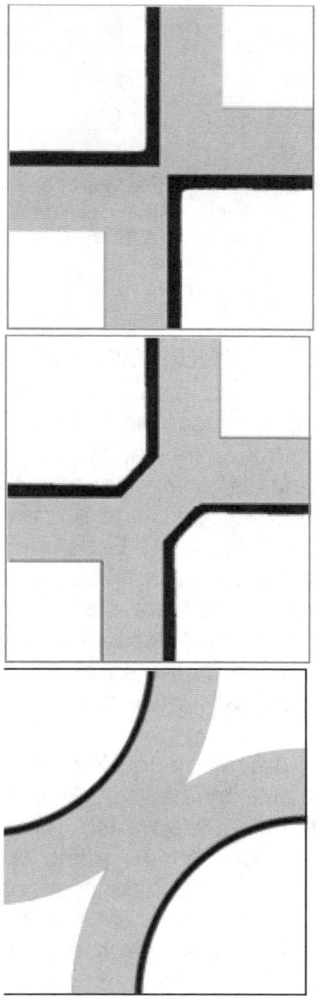

Figure 31. Soot deposits grown on the different channel geometries A, B and C of Figure 30.

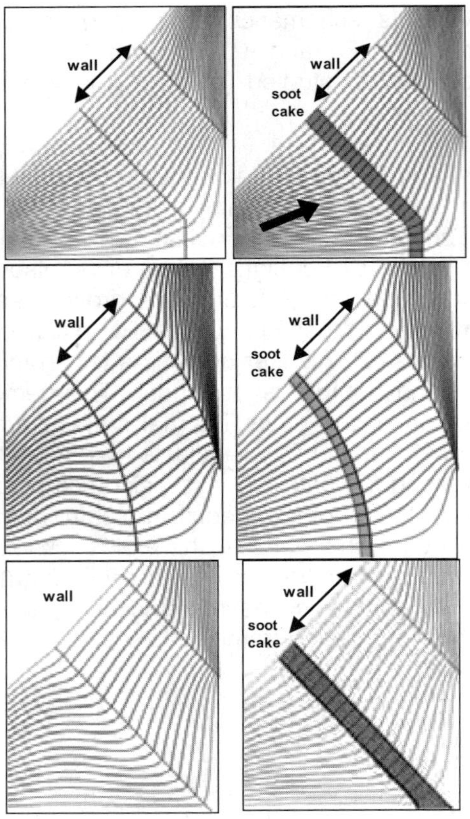

Figure 32. Path lines (spaced at intervals of equal flux) of the wall flow in different asymmetric shape cells in clean (left) and loaded (right) states. From the top, geometries A, B and C of Figure 30 are shown.

ASH AND SOOT ENTRAINMENT PHENOMENA

The mechanistic modeling of ash transport and deposition dynamics was originally introduced in [17]. The same formalism can be applied to account for shear-induced soot particle re-entrainment and further downstream deposition [54]. Experimental studies of ash deposition and transport in filters require extensive and costly engine runs and are scarce in the literature. A dynamic ash transport-deposition model is then an attractive alternative for the assessment of DPF systems with respect to ash accumulation.

The ash simulation model [17, 35] consists of ash transport and ash layer evolution equations describing the interaction between ash deposition and re-entrainment in the channels along with the gas mass balance and momentum balance equations in the inlet and outlet channels of the DPF. Ash re-entrainment is initiated by flow shear stresses along the channel whenever the shear stress in the channel exceeds locally a critical shear stress characteristic of the ash type (as e.g. determined by composition, particle size, etc). The kinetics of ash re-entrainment are determined by the ash stickiness that for a given type of ash depends on the previous filter temperature history (currently accounted by the maximum temperature that the filter wall has experienced), as obtained by a separate filter regeneration simulation [17]. Ash stickiness is described by a critical ash sticking temperature by analogy to the ash fusion temperature employed in studies of ash fouling in heat transfer equipment.

Despite its apparent simplicity, the dynamic ash transport and deposition model can exhibit a rich behavior, depending on the ash quality, and prevailing flow conditions leading to different ash deposition profiles inside DPFs: sticky ash leads to deposition along the walls while non-sticky ash can be transported towards the end of the filter. A comparison to the experimental data of Bardasz et al. [55] in Figure 33 shows good quantitative agreement for the ash layer thickness resulting from different oil consumption protocols. The sulfur content of the oil and whether it is doped into the fuel (to accelerate the ash aging test) determines its "stickiness" and this in turn is described by the model of ash re-entrainment. Optimized filter cell densities accounting for different ash accumulation modes and soot loading of filters have been presented in [56].

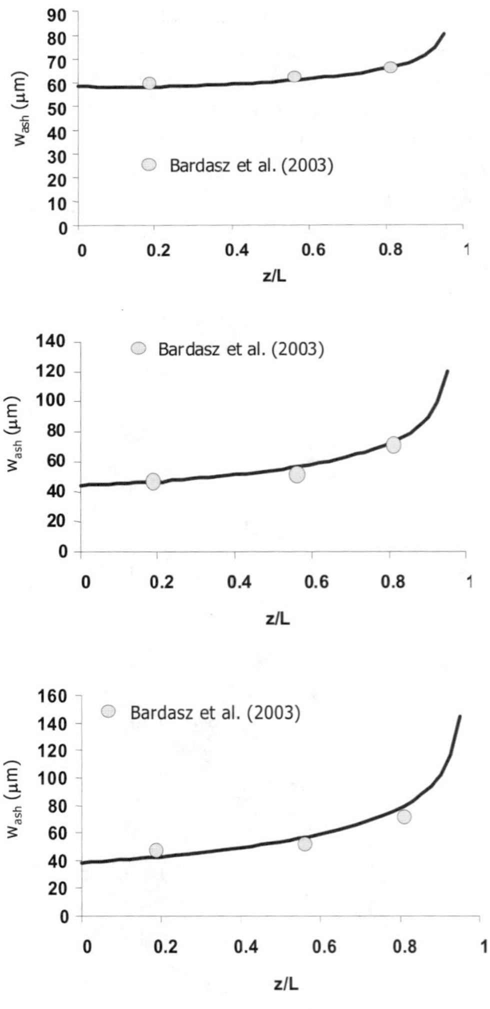

(a)

(b)

(c)

Figure 33. Simulated ash profile along the filter length vs. experimental data of [55]. (a) High sulphur oil doped into fuel, (b) Low sulphur oil doped into fuel, (c) High sulphur oil, regular use undoped into fuel.

The same approach is employed to describe shear-induced transport of soot particles. Based on limited amount of experimental information for such phenomena in the literature we have established a flow cell where soot entrainment from the surface of preloaded filters from the engine exhaust can be studied. Preliminary experiments at ambient conditions reveal that no soot entrainment is observed up to relevant shear rates at the entrance of DPFs. We attribute this to the moisture content in ambient conditions of the soot deposits that due to capillary condensation increases adhesive forces between the particles. In the future experiments at high temperatures are planned to evaluate experimentally the shear-entrained fluxes for soot and ash deposits.

CHANNEL TO POROUS WALL HEAT TRANSFER

The 1-D (perimeter averaged) heat balance equation in a porous channel with suction can be written following the considerations in Appendix III as:

$$a^2\overline{u}_z \frac{dT_c}{dz} = -4\alpha_{th} A(\mathrm{Re}_w)(T_c - T_w) + (1 - B(\mathrm{Re}_w, Pe_T)) 4 a v_w (T_c - T_w) \tag{16}$$

The functions A and B need to be computed for the particular geometry of the DPF channels by solving the complete flow and temperature fields for different values of the wall Reynolds, Re_w and thermal Peclet, Pe_T numbers and fitting the 1-D Eq. (16) to the resulting average perimeter-averaged temperature evolution along the channel. Alternatively, the technique of Raithby and Kinney [57, 58] involving the solution for the self-similar flow field and eigenvalue analysis of the resulting energy equation (of the convection diffusion type) can be followed. For large Pe_T an asymptotic form for B valid in any geometry is given by (Appendix III):

$$B(\mathrm{Re}_w, Pe_T) = 1 - A(\mathrm{Re}_w)/Pe_T \tag{17}$$

Assuming that Nu = A(0) (i.e suction does not influence the temperature profile, [18]) can lead to an inconsistency for for Pe>A(0) since the heat flux is reversed from the hot wall to cold gas in the channel. In [15] the assumption Nu=A(0)+Pe (i.e. B=1) was employed. This is equivalent to assuming that the gas entering the porous wall has a temperature equal to the local mixing-cup temperature of the gas flowing in the channel. This assumption ensures the correct direction of heat flux but overestimates it somewhat since the actual value of B is between 0.5 and 1. A very good assumption consistent with the 1-D flow model is then to use A=A(0) i.e. a Nusselt number applicable to a solid wall but put B=0.5 as an estimation based on Figures 7 and 8 of [57] Accurate calculations of the functions A and B are currently in progress at the APT Laboratory.

ENTIRE FILTER SCALE

THE EFFECTIVE CONDUCTIVITY OF DPFS

The open area fraction ε_A of a square honeycomb structure is:

$$\varepsilon_A = \left(1 - \frac{w_s}{a + w_s}\right)^2 \tag{18}$$

The axial effective conductivity $\lambda_{a,eff}$ of the DPF is given easily as the product of the solid fraction (1- ε_A), the conductivity of the wall material λ_w (this is also an effective conductivity since the wall is porous and it is determined from the wall porosity and intrinsic material thermal conductivity, usually through a power law relation known as Archie's Law).

The computation of the transverse effective conductivity $\lambda_{t,eff}$ is more involved. In this direction the DPF is a periodic medium so its effective conductivity can be found by solving the heat conduction equation in the primary geometric unit (Figure 34)

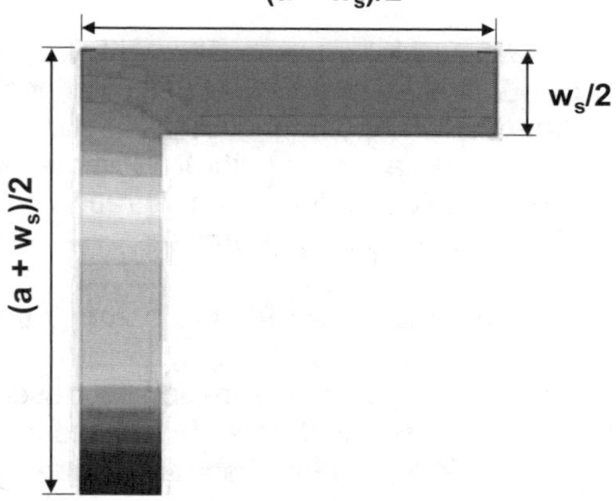

Figure 34. Primary geometric unit for the effective conductivity of DPF.

A unit temperature difference is imposed to the top wall of Figure 14 and the conductance is computed. The boundary conditions are T=1 at the upper side, T=0 at the bottom side and insulation for all the other sides. The gas conductivity is not taken into account here since it is considered through the gas phase heat balance equation. Taking into account the gas conductivity to compute the transverse effective conductivity for the DPF structure is not correct.

The effective conductivity of monoliths with square channels has been approximated by Groppi and Tronconi [59] (using a network of resistances analog) as:

$$\frac{\lambda_{t,eff}}{\lambda_w} = \frac{1-\sqrt{\varepsilon_A}}{1+\varepsilon_A-\sqrt{\varepsilon_A}} \tag{19}$$

A close inspection of the derivation in Groppi and Tronconi [59] reveals that it does not retain the square network symmetry. A larger element size is assigned to the low resistance path than to the high resistance path leading to a 13% underestimation of the total resistance.

Correcting the network analog assigned equal lengths to the two heat transfers paths leads to:

$$\frac{\lambda_{t,eff}}{\lambda_w} = \frac{a+w_s}{a+w_s/2}\left(1+\left(\frac{a+w_s}{a+w_s/2}\right)^2\right)^{-1} = \\ = \frac{2}{1+\sqrt{\varepsilon_A}}\left(1+\left(\frac{2}{1+\sqrt{\varepsilon_A}}\right)^2\right)^{-1} \tag{20}$$

The value of $\lambda_{t,eff}/\lambda_w$ for a typical filter with $w_s/(a+w_s)=0.2$ is: $\lambda_{t,eff}/\lambda_w = 0.212$ (based on Fluent calculations), $\lambda_{t,eff}/\lambda_w = 0.238$ (network approximation), $\lambda_{t,eff}/\lambda_w = 0.211$ (improved network approximation). Commercial filters are at this (or smaller) wall to channel ratios so Eq. (20) can be safely used for design purposes.

Having found a simple way for the computation of the effective conductivity, the validity of the effective homogeneous medium approximation to describe heat transfer in the multichannel scale will be assessed. For this purpose a DPF segment of square shape containing 24x24 channels is considered.

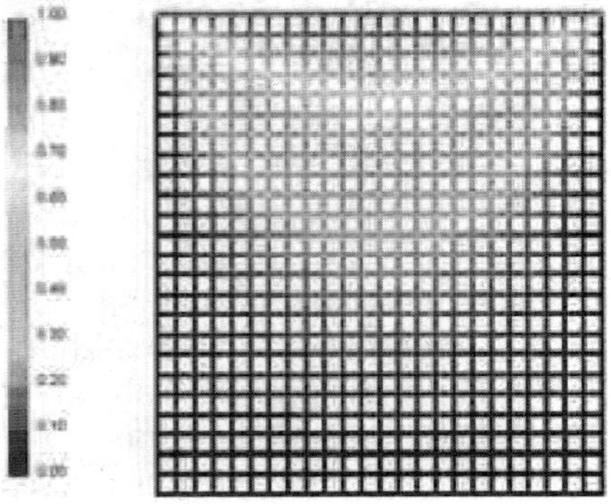

Figure 35. Temperature distribution computed over a DPF segment with a unit temperature difference between the top and the other three sides.

A unit temperature difference between the upper and the other three sides is assumed. The problem is solved using Fluent, with the results shown in Figure 35. Heat fluxes through the right (equal to the left due to symmetry) and bottom sides are shown in Figure 36. The fluxes from Fluent are averages along the side of the unit cell. The approximation is very successful despite the singularity exhibited by the side flux close to the upper corner. Some inconsistencies are observed close to the singularity. This is expected since in that region the macroscopic rate of change of the flux is important on the scale of channel so the local averaging is not successful. The problem was also solved with the effective homogeneous medium theory with effective conductivity given by Eq. (20). The side length of the simulated segment is L=35 mm and the ratio $w_s/(\alpha+w_s)$ is 0.2. The fluxes from the homogeneous medium approximation shown in Figure 36 are computed by:

Side boundary: $$\frac{q\ell}{\lambda_{t,eff}} = 4\sum_{i=1,3,5..}^{\infty}\frac{\sinh(i\pi y)}{\sinh(i\pi)} \tag{21}$$

Bottom boundary: $$\frac{q\ell}{\lambda_{t,eff}} = 4\sum_{i=1,3,5..}^{\infty}\frac{\sin(i\pi x)}{\sinh(i\pi)} \tag{22}$$

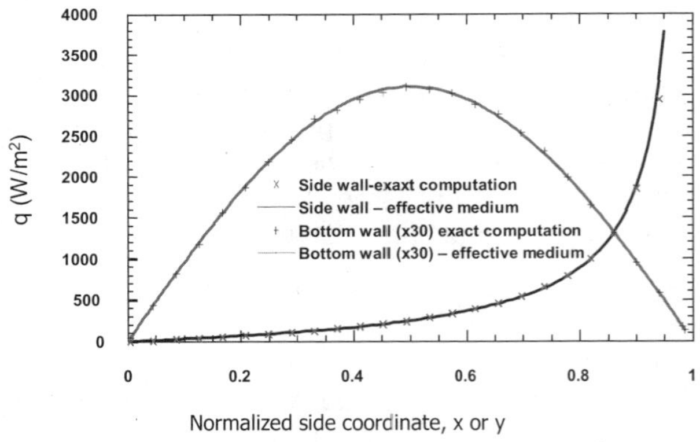

Figure 36. Comparison between approximate and numerically computed heat fluxes on the two sides of a heated DPF segment with 24x24 channels.

With a validated model for the effective thermal conductivity of DPFs it is then possible to study segmented DPF designs. An example showing the influence of the segment gluing material thermal conductivity on the inter-segment heat transfer is shown in Figure 37.

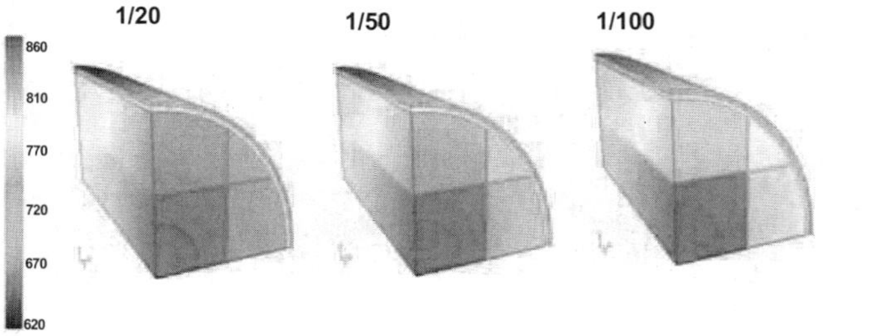

Figure 37. Influence of the segment gluing material thermal conductivity to that of the segment (computed with the effective conductivity approach mentioned in the text) on the inter-segment heat transfer is shown.

MULTICHANNEL PHENOMENA

If the inlet conditions to the trap cannot be assumed to be radially uniform, a single channel description is no longer applicable and the entire DPF needs to be simulated. As a realistic size filter consists of several thousand channels, its direct simulation via the numerical solution of a coupled discrete multichannel problem is an intractable task with the currently available computational resources. An alternative way to deal with the scale-up problem is to employ a continuum model of the filter honeycomb structure.

Rigorous scale homogenization procedures lead to continuum models for the entire DPF [16-18] exploiting (as is common in continuum descriptions) a suitable scale disparity, namely the ratio of the channel hydraulic radius to the entire DPF diameter, to formulate a perturbation expansion of the discrete multichannel equations. This continuum multichannel description of the DPF can accommodate various regeneration methods (thermal, catalytic and NO_2-assisted) and can provide spatiotemporal information of several quantities of interest (e.g. filter temperature, soot mass distribution, flow distribution, etc) as illustrated in Figure 38.

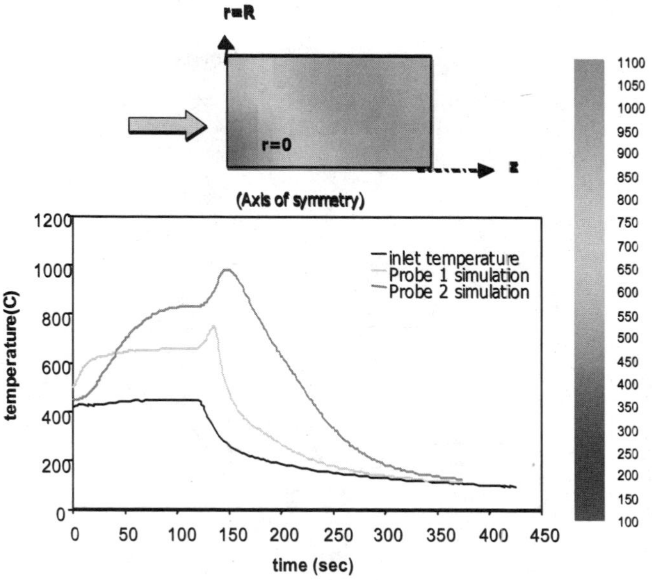

Figure 38. Temporal evolution of temperature probes inserted in a regenerating DPF (bottom) along with the spatial distribution of filter temperature at 145 s, i.e. the moment of exothermic peak (top). Locations (r,z) in mm of temperature probes: Probe 1: (24.25,40), Probe 2: (24.25,142).

In [10, 11] it was demonstrated that during incomplete regeneration radial non-uniformities may develop in a DPF. These will lead to a flow redistribution during subsequent loading. In axisymmetric problems a network analogy [60] suggests itself as an obvious approximation to provide an alternative to CFD but we have not up to now evaluated its accuracy. We have to this end performed CFD calculations with Fluent of the flow distribution at the face of a DPF after incomplete regenerations leading to radial non-uniformities (Figure 39) computed by a multichannel simulation [11] and these are compared to the flow distributions obtained from a network approximation in Figure 40. In the worst case a flux averaged error of 15% is obtained. Current work is focusing on a composite network model, which employs a limited amount of information from CFD simulation of an unloaded DPF in order to improve the above error figure.

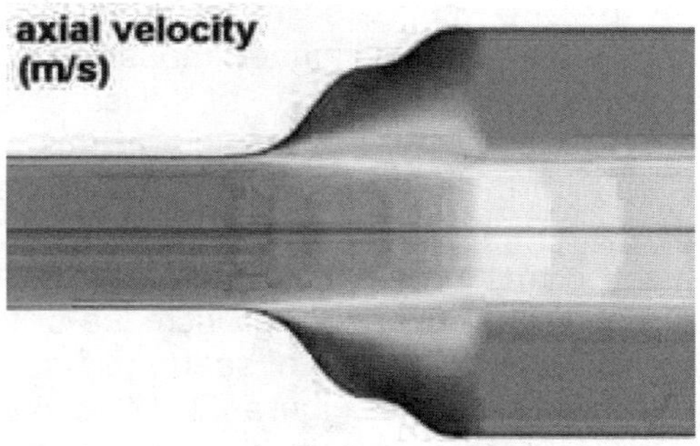

Figure 39. Flow distribution at the entrance of a DPF with radially non-uniform soot distribution, resulting from incomplete regeneration.

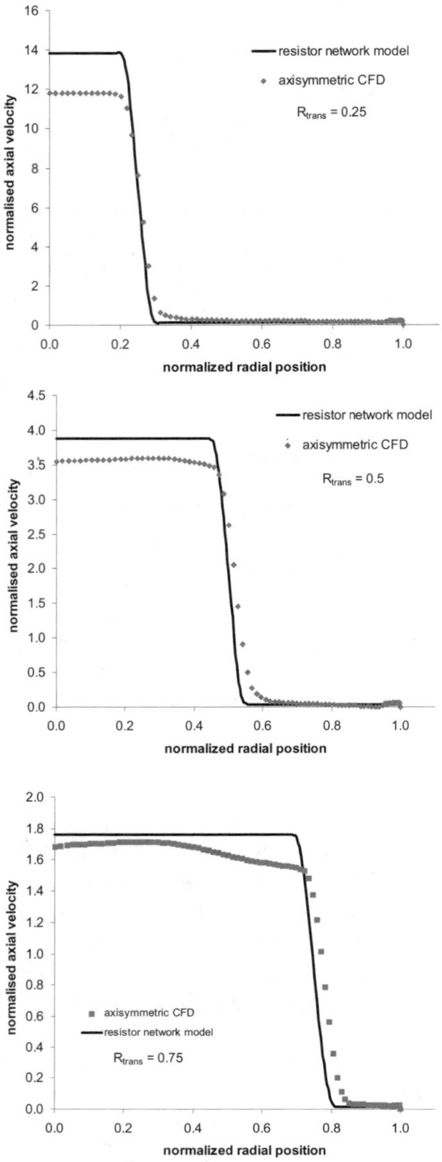

Figure 40. Comparison of flow distribution at the entrance of a DPF with different radially non-uniform soot distributions computed by CFD and a simple network model.

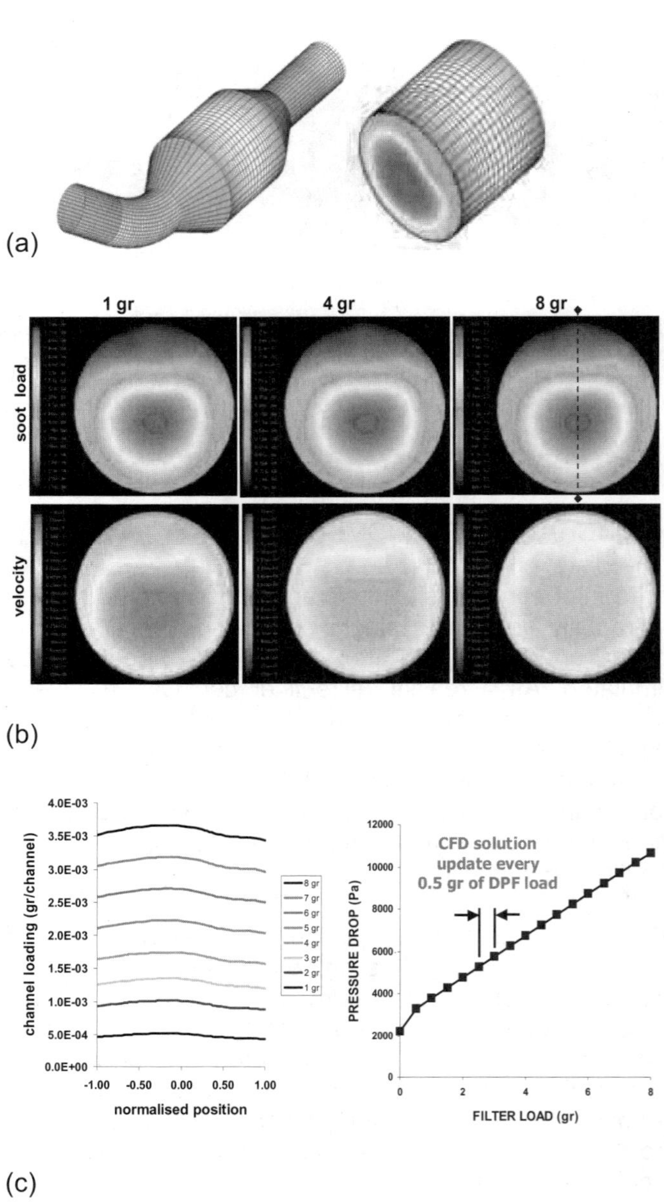

Figure 41. (a) 3-D simulation of a DPF placed downstream of a 45° bend in the exhaust. (b) Soot mass and velocity distribution at the inlet face at different soot loads of the filter. (c) Evolution of radial soot mass profile (along the dotted line) at different soot loads of the filter and resulting pressure drop.

In case of 3-D flow distributions a coupling of the DPF simulation to CFD appears inescapable. We have interfaced our DPF multichannel model with Fluent employing the mechanism of User Defined Functions/Subroutines. As an example of this coupling we show in Figure 41 the coupled CFD-multichannel simulation of a DPF placed downstream of a 45° bend in the exhaust. To make the simulation tractable the CFD solution is updated at every 0.5 gr increase of the filter soot load. From the snapshots of soot mass distribution and inlet velocity shown in Figure 41 it becomes clear that the DPF inlet face velocity field moves in a self-regulated fashion towards a homogeneous distribution, while the soot mass distribution exhibits a stronger "memory" of the initial inlet distribution non-uniformity.

As we proceed to the entire exhaust system scale we face the task of interfacing the DPF behavior to that of other emission control devices in the exhaust (e.g. diesel oxidation catalysts and NO_x reduction devices). An example of a coupled simulation of a Diesel Oxidation Catalyst (DOC) and a DPF in series is shown in Figure 42. We observe how a hydrocarbon pulse injection upstream of the DOC raises the exhaust temperature and causes regeneration of the DPF. Such simulation tools are very useful for the development and optimization of post-injection strategies for DPF regeneration.

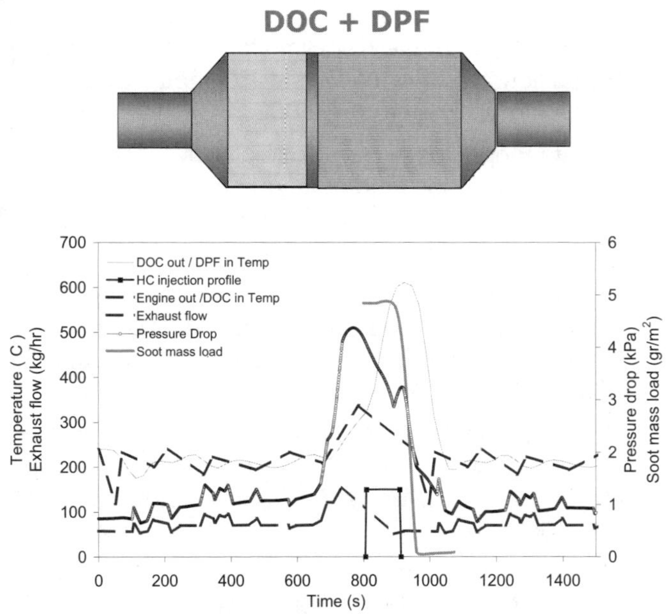

Figure 42. Coupled DOC-DPF simulation. Filter regeneration by hydrocarbon injection upstream of the DOC over transient engine operation.

Closing the section on the entire filter scale we may remark that at the multichannel level the dynamic ash model of [11, 35] can reproduce ash deposition patterns occurring in fuel-additive assisted regeneration [1] as seen in Figure 44 as well as describe the pressure drop evolution due to ash accumulation Figure 44.

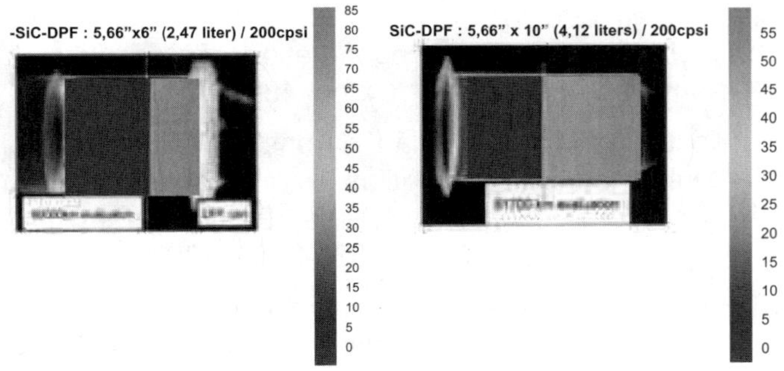

Figure 43. Multichannel dynamic ash model. Simulation of additive ash accumulation at the end of the filter. Comparison with X-ray images of [1].

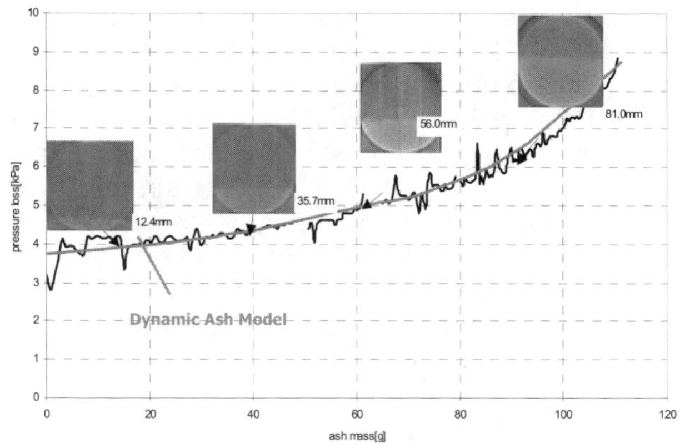

Figure 44. Simulation of the effect of ash accumulation of pressure drop. The experimental data are courtesy of Ibiden Co. Ltd.

CONCLUSION

Experience has proven that a DPF should be treated as an integral part of the powertrain and that it requires a systemic approach for its optimal design and operation, the latter being matched to the engine management strategy. The identification of optimum combinations of emission control components by the process of hardware building and testing entire emission control systems, often requires a prohibitive amount of man power and cost. In addition, deep knowledge of emission control component behavior is required in order to perform an effective system-level integration and optimization. Advanced computational tools are now recognized as tractable alternatives for the rapid assessment of various design options for a particular vehicle platform.

For basic filter material development wall scale models especially the MicroFlowS approach can be applied. During early development of particulate emission control systems single channel models provide a fast and sufficient means for basic DPF design and sizing analysis. In addition, when applicable (i.e. in the absence of radial non-uniformities) they represent a valuable tool for analyzing experimental data collected under well-defined conditions and extraction of relevant physicochemical parameters of the system. Finally, detailed analysis and refinement of the selected DPF system can be achieved with the use of the multichannel models.

In the present work, we have outlined the current state of progress in DPF simulation drawing from our recent work in the area. Employing a convenient spatial scale classification of the various models we have provided a framework for the consistent integration of the information from the various models across all scales. While computing limitations still remain the barrier for the routine deployment of detailed simulations of coupled emission control components at the entire exhaust system scale, we anticipate that in the near future such simulations will be widely employed by the industry, exploiting grid-computing environments. This means that from a research point of view, DPF simulation will focus into providing a deeper understanding and more detailed description of the coupled transport, structural and reaction microphenomena occurring at the wall and pore scales, to materialize the vision for an a priori design of advanced microstructures, hosting multifunctional catalysts for the highly compact and efficient emission control devices of the future.

ACKNOWLEDGMENTS

This work has been supported in part by the European Commission through the (ART-, PSICO-, SYLOC-, STYFF-) DEXA cluster projects as well as projects COMET, IMITEC, FLOWGRID. We thank Dr. E. Kikkinides for sharing his expertise on flow through-porous media reconstruction and Mr. A. Tsakis for his assistance with computations. We are also thankful to all of our colleagues at the APT Lab for their support in carrying out the work presented. Ibiden Co. Ltd. Japan is acknowledged for providing the experimental data in Figure 44.

REFERENCES

1. Blanchard G., Colignon C., Griard C., Rigaudeau C., Salvat O., Seguelong T., “Passenger Car Series Application of a New Diesel Particulate Filter System Using a New Ceria-Based, Fuel- Borne Catalyst: From the Engine Test Bench to European Vehicle Certification”, *SAE Tech. Paper No.* 2002-01-2781, 2004.
2. Ogyu K., Ohno K., Hong S., Komori T., “Ash Storage Capacity Enhancement of Diesel Particulate Filter”, *SAE Tech. Paper No.* 2004-01-0949, 2004.
3. Young D. M., Hickman D. L., Bhatia G., Gunasekaran N., “Ash Storage Concept for Diesel Particulate Filters”, *SAE Tech. Paper No.* 2004-01-0948, 2004.
4. Bardon S., Bouteiller B., Bonnail N., Girot P., Gleize V., Oxarango L., Higelin P., Michelin J., Schuerholz S., Terres F., “Asymmetrical Channels to Increase DPF Lifetime”, *SAE Tech. Paper No.* 2004-01-0950, 2004.
5. Boretto G., Imarisio R., Rellecati P., Barucchi E., Sanguedolce A., “Serial Application of a Catalyzed Particulate Filter on Common Rail DI Diesel Engines For Passenger Cars”, F2004V068, *FISITA World Automotive Congress*, Barcelona, Spain, May 23-27, 2004, 2004.
6. Johnson T. V., “Diesel Emission Control Technology-2003 in Review”, *SAE Tech. Paper No.* 2004-01-0070, 2004.
7. Nakatani K., Hirota S., Takeshima S., Itoh K., Tanaka T., Dohmae K., “Simultaneous PM and NOx Reduction System for Diesel Engines”, *SAE Tech. Paper No.* 2002-01-0957,.2002.
8. Konstandopoulos A. G., Kostoglou M., "Theory and Practice of Soot Filter Operation", *Proceedings of Global Powertrain Congress 99* (invited paper), Powertrain Int., Stuttgart, October 5-7, 1999.
9. Konstandopoulos A. G., Kostoglou M., "Microstructural Aspects of Soot Oxidation in Diesel Particulate Filters", *SAE Tech. Paper No.* 2004-01-0693 (SP-1861), 2004.
10. Konstandopoulos A. G., Kostoglou M., Housiada P., "Spatial Non-Uniformities in Diesel Particulate Trap Regeneration", *SAE Tech. Paper No.* 2001-01-0908 (SP-1582), 2001.
11. Konstandopoulos A. G., Kostoglou M., Housiada P., Vlachos N., Zarvalis D., "Multichannel Simulation of Soot Oxidation in Diesel Particulate Filters", *SAE Tech. Paper No.* 2003-01-0839 (SP-1755), 2003.
12. Konstandopoulos A. G. and Kostoglou M., “Periodically Reversed Flow Regeneration of Diesel Particulate Traps”, *SAE Tech. Paper No.* 1999-01-0469, 1999.
13. Konstandopoulos A. G., Kostoglou M., Skaperdas E., Papaioannou E., Zarvalis D., and Kladopoulou E., “Fundamental Studies of Diesel Particulate Filters: Transient Loading, Regeneration and Aging”, *SAE Tech. Paper No.* 2000-01-1016 (SP-1497), 2000.
14. Konstandopoulos A. G., Kostoglou M., "Reciprocating Flow Regeneration of Soot Filters", *Comb. and Flame*, 121 (3), pp. 488-500, 2000.
15. Kostoglou M., Housiada P., Konstandopoulos A. G., "Multi-channel Simulation of Regeneration in Honeycomb Monolithic Diesel Particulate Filters", *Chem. Eng. Sci.*, 58, pp. 3273-3283, 2003.
16. Konstandopoulos A. G., Kostoglou M. and Housiada P., “Spatial Non-Uniformities in Diesel Particulate Trap Regeneration”, *SAE Tech. Paper No.* 2000-01-0908, 2001.
17. Konstandopoulos A. G., Kostoglou M., Housiada P., Vlachos N. and Zarvalis D., “Multichannel Simulation of Soot Oxidation in Diesel Particulate Filters”, *SAE Tech. Paper No.* 2003-01-0839, 2003.
18. Bissett E. J., “Mathematical Model of the Thermal Regeneration of a Wall-Flow Monolith Diesel Particulate Filter”, *Chem. Eng. Sci.*, 39 (7/8), pp. 1233-1244, 1984.
19. Bissett E. J., Shadman F., “Thermal Regeneration of Diesel Particulate Monolithic Filters”, *AIChE J.*, 31 (5), pp. 753-758, 1985.
20. SAE International, Diesel Exhaust Emission Control, SP-1860, ISBN 0-7680-1396-8, 2004.

21. Konstandopoulos A. G., "Flow Resistance Descriptors for Diesel Particulate Filters: Definitions, Measurements and Testing", *SAE Tech. Paper No.* 2003-01-0846 (SP-1755), 2003.
22. Konstandopoulos A. G., Skaperdas E., Masoudi M., "Inertial Contributions to the Pressure Drop of Diesel Particulate Filters", *SAE Tech. Paper No.* 2001-01-0909 (SP-1582), 2001.
23. Dullien F. A. L., **Porous Media**, Academic Press Inc., San Diego, 1979.
24. Konstandopoulos A. G., Vlachos N., Stavropoulos I., Skopa S., Schumacher U., Woiki D., "Study of a Sintered Metal Diesel Particulate Trap", *SAE Tech. Paper Offer No.* 05FL-106, (submitted), 2005.
25. Muntean M., Rector D., Herling D., Khaleel M., Lessor D., "Lattice-Boltzmann Diesel Particulate Filter Sub-Grid Modelling-A Progress Report", *SAE Tech. Paper No.* 2003-01-0835, 2003.
26. Kikkinides E. S. and Burganos V .N., "Permeation Properties of Three-Dimensional Self-Affine Reconstructions of Porous Materials", *Physical Review E.*, 62(5), pp. 6906, 2000.
27. Torquato S., **Random Heterogeneous Materials: Microstructure and Macroscopic Properties**, Springer-Verlag, New York, 2001.
28. Konstandopoulos A. G., "Deposit Growth Dynamics: Particle Sticking and Scattering Phenomena", *Powder Tech.*, 109 (1-3), pp. 262-277, 2000.
29. Tomadakis M. M., Sotirchos S. V., "Knudsen Diffusivities and Properties of Structures of Unidirectional Fibers", *AIChE J.*, 37, pp. 1175, 1991.
30. Tomadakis M. M., Sotirchos S. V., "Effective Knudsen Diffusivities in Randomly Overlapping Fiber Structures", *AIChE J.*, 37, pp. 74, 1991.
31. Tassopoulos M., "Relationships Between Particle Deposition Mechanism, Deposit Microstructure and Effective Transport Properties", PhD Thesis, Yale University, USA, 1991.
32. Chen S., Doolen G., "Lattice Boltzmann Method for Fluid Flows", *Ann. Rev. Fluid Mech.*, 30, pp. 329, 1998.
33. Konstandopoulos A. G., Skaperdas E., Masoudi M., "Microstructural Properties of Soot Deposits in Diesel Particulate Traps", *SAE Tech. Paper No.* 2002-01-1015 (SP-1673), 2002.
34. Konstandopoulos A. G., Johnson J. H., "Wall-Flow Diesel Particulate Filters-Their Pressure Drop and Collection Efficiency", *SAE Tech. Paper No.* 890405, SAE Trans. 98 sec. 3 (J. Engines), pp. 625-647, 1989.
35. Rodriguez-Perez D., Castillo J. L., Antoranz J. C., Konstandopoulos A. G. and Vlachos N., "Mixed ODE-MC Model for the Soot Cake Formation in a Square Section Filter", *European Aerosol Conference EAC 2004*, Budapest, Hungary, 6-10 Sep, 2004.
36. Konstandopoulos A. G. et al., "Convective-Diffusive Deposition of Fractal-like Aggregates and the Microstructure and Properties of the Resulting Nanostructured Deposits", (submitted), 2004.
37. Skaperdas E., Konstandopoulos A. G., "Prediction of Diesel Particulate Filter Loading Behavior for System Control Applications", CD AUTO 01, *3rd International Conference on Control and Diagnostics in Automotive Applications*, Sestri-Levante, Italy, 4-6 July, 2001.
38. Konstandopoulos A. G., Kladopoulou E., "A Virtual Sensor for On-Board Diagnostics and Control of Diesel Particulate Filters", CDAuto03A2028, CD AUTO 03, *4th International Conference on Control and Diagnostics in Automotive Applications*, Sestri-Levante, Italy, 18-20 June, 2003.
39. Vlachos N., Housiada P., Zarvalis D. and Konstandopoulos A. G., "Transient Filtration and Pressure Drop Behavior of Diesel Particulate Filters", in Particle Loading and Kinetics of Filtration in Fibrous Filters, (Lehmann M. J. and Kasper G, Editors), Karlsruhe (University of Karlsruhe), 13-26, 2002.
40. Zarvalis D., Altiparmakis C., Mei F. and Konstandopoulos A. G., (2003) "Evaluation of a Multi-layer Sintered Metal Fibrous Diesel Particulate Filter", *European Aerosol Conference EAC 2003*, Madrid, Spain, 31 August-5 September, 2003.
41. Karadimitra K., Lorentzou S., Agrafiotis C., Konstandopoulos A. G., "Modeling of Catalytic Particle Synthesis via Spray Pyrolysis & In-Situ Deposition on Porous Materials", PARTEC 2004, *International Conference for Particle Technology*, Nuremberg, Germany, 16 –18 March, 2004.
42. http://www.unizar.es/flowgrid/
43. Konstandopoulos A. G., Kostoglou M., "A Mathematical Model of Soot Oxidation on Catalytically Coated Ceramic Filters", *Advances in Vehicle Control and Safety* (AVCS'98), Amiens, France, July 1-3, 1998.
44. Hawker, P., Myers, N., Huthwohl, G., Vogel, H., Bates, B., Magnusson, L., Bronneberg, P., "Experience with a new particulate trap technology in Europe", *SAE Paper No 970182*, 1997.
45. Lueders, H., Stommel, P., Geckler, S. "Diesel exhaust treatment-new approaches to ultra low emission diesel vehicles", *SAE Paper No. 1999-01-0108*, 1999.
46. Konstandopoulos A. G., Zarvalis D., Papaioannou E., Vlachos N. D., Boretto G., Pidria M. F., Faraldi P., Piacenza O., Prenninger P., Cartus T., Schreier H., Brandstatter W., Wassermayr C., Lepperhof G., Scholz V., Luers B., Schnitzler J., Claussen M.,

Wollmann A., Maly M., Tsotridis G., Vaglieco B. M., Merola S. S., Webster D., Bergeal D., Gorsmann C., Obernosterer H., Fino D., Russo N., Saracco G., Specchia V., Moral N., D'Anna A., D'Alessio A., Zahoransky R., Laile E., Schmidt S., Ranalli M., "The Diesel Exhaust Aftertreatment (DEXA) Cluster: A Systematic Approach to Diesel Particulate Emission Control in Europe", *SAE Tech. Paper No.* 2004-01-0694 (SP-1861), 2004.

47. Jacquot F., Logie V., Brilhac J. F. and Gilot P., "Kinetics of the Oxidation of Carbon Black by NO_2: Influence of the Presence of Water and Oxygen", *Carbon*, 40 (3), pp. 335-343, 2002.
48. Setiabudi A., Makkee M. and Moulijn J. A., "The Role of NO_2 and O_2 in the Accelerated Combustion of Soot in Diesel Exhaust Gases", *Applied Catalysis B: Environmental*, 50 (3), pp. 185-194, 2004.
49. Olsson L., Persson H., Fridell E., Skoglundh M., Andersson B., "A Kinetic Study of NO Oxidation and NO_x Storage on Pt/Al_2O_3 and $Pt/BaO/Al_2O_3$", *J. Phys. Chem. B.*, 105 (29), pp. 6895-6906, 2001.
50. Olsson L., Westerberg B., Persson H., Fridell E., Skoglundh M., and Andersson B., "A Kinetic Study of Oxygen Adsorption/Desorption and NO Oxidation over Pt/Al_2O_3 Catalysts", *J. Phys. Chem. B*, 105, pp. 10433 – 10439, 2001.
51. Konstandopoulos A. G., Skaperdas E., Warren J. and Allanson R., "Optimized Filter Design and Selection Criteria for Continuously Regenerating Diesel Particulate Traps", *SAE Tech. Paper No.* 1999-01-0468, 1999.
52. Konstandopoulos A. G., Vlachos N., Housiada P., Kostoglou M., "Simulation of Triangular-Cell-Shaped, Fibrous Wall-Flow Filters", *SAE Tech. Paper No.* 2003-01-0844 (SP-1755), 2003.
53. Fluent Inc., **Fluent 6 User's Guide**, December 2001.
54. Peters B., "Integrated 1d to 3d Simulation Workflow of Exhaust Aftertreatment Devices", *SAE Tech. Paper No.* 2004-01-1132, 2004.
55. Bardasz E., Mackney D., Britton N., Kleinschek G., Olofsson K., Murray I and Walker A. P., "Investigations of the Interactions between Lubricant-Derived Species and Aftertreatment Systems on a State-of-the-Art Heavy Duty Diesel Engine", *SAE Tech. Paper No.* 2003-01-1963, 2003.
56. Konstandopoulos A. G. and Kladopoulou E., "The Optimum Cell Density for Wall-Flow Monolithic Filters: Effects of Filter Permeability, Soot Cake Structure and Ash Loading", *SAE Tech. Paper No.* 2004-01-1133, 2004.
57. Raithby G., "Laminar Heat Transfer in the Thermal Entrance Region of Circular Tubes and Two-Dimensional Rectangular Ducts with Wall Suction and Injection", *Int. J. Heat Mass Transfer*, 14, pp. 223-243, 1971.
58. Kinney R. B., "Fully Development Frictional and Heat Transfer Characteristics of Laminar Flow in Porous Tubes", *Int. J. Heat Mass Transfer*, 11, pp.1393-1401, 1968.
59. Gropi G. and Tronconi E., "Continuous vs. Discrete Models of Non-adiabatic Monolith Catalysts", *AIChE. J.*, 42 (8), pp. 2382-2387, 1996.
60. Amador C., Gavriilidis A. and. Angeli P, "Flow Distribution in Different Microreactor Scale-Out Geometries and the Effect of Manufacturing Tolerances and Channel Blockage", *Chemical Engineering Journal*, 101 (1-3), pp. 379-390, 2004.

CONTACT

Dr. Athanasios G. Konstandopoulos, Head of Aerosol & Particle Technology Laboratory, CERTH/CPERI, P.O. Box 361, Thermi, Thessaloniki 57001, Greece. e-mail: agk@cperi.certh.gr, web: http://apt.cperi.certh.gr

DEFINITIONS, ACRONYMS, ABBREVIATIONS

***a**:* side length of channel cross section

c_{pg}: specific heat capacity of exhaust gas

c_{ps}: specific heat capacity of soot

c_{pw}: specific heat capacity of filter wall material

E_{cat}: activation energy for catalytic soot oxidation

E'_f: parameter in catalytic CO selectivity equation (11)

E_f: parameter in thermal CO selectivity equation (5)

E_{th}: activation energy for thermal soot oxidation

f_{CO}^{cat} : catalytic CO selectivity

f_{CO}^{th} : thermal CO selectivity

***G**:* exhaust mass flux (with respect to filtration area)

F_{ox}: local soot oxidation rate

k_1: volumetric reaction constant for soot oxidation with NO_2

k_A, k_B: forward and reverse volumetric reaction constants for NO oxidation

K_{cat}: rate coefficient of catalytic soot oxidation

k'_f: parameter in catalytic CO selectivity equation

k_f: parameter in thermal CO selectivity equation

$k_{o,cat}$: preexponential factor for catalytic soot oxidation

$k_{o,th}$: preexponential factor for thermal soot oxidation

K_{th}: rate coefficient of thermal soot oxidation

ℓ: length of segment size

L: slit half width

M_c: molecular weight of soot

M_{CO}: molecular weight of CO

M_{CO2}: molecular weight of CO_2

M_{ox}: molecular weight of oxygen

m_w: filter wall mass per unit filtration area

Pe_T: thermal Peclet number for gas $u_w a/\alpha_{th}$

q: local heat flux

Re_w: wall Reynolds number au_w/ν

R_g: ideal gas density

S: specific surface area (per unit mass) of soot layer

T_c: channel cup mixing temperature

T_o: filter and gas temperature at t=0

$\bar{u}_z$: mean axial channel velocity

u_w: superficial filtration velocity

w: soot layer thickness

w_s: filter wall thickness

X,Y,Z: molar concentration of NO_2, NO and total NOx respectively (Appendix II).

y: oxygen molar fraction in exhaust gases

Y: oxygen mass fraction in exhaust gases

Y_{CO}: CO mass fraction in exhaust gases

Y_{CO2}: CO_2 mass fraction in exhaust gases

Greek letters

α, γ: generalized soot layer structural descriptors ΔH: heat of soot oxidation reaction (per unit mass of soot)

α_{th}: thermal diffusivity of the gas

β: index for the completeness of soot oxidation (Appendix I), Forchheimer coefficient (Equations 1,3)

ΔH_{CO}: specific heat of CO formation

ΔH_{CO2}: specific heat of CO_2 formation

ε_A: open area fraction

θ: parameter in thermal CO selectivity equation

θ': parameter in catalytic CO selectivity equation

$\lambda_{a,eff}$: effective axial thermal conductivity of the filter

$\lambda_{t,eff}$: effective radial thermal conductivity of the filter

λ_w: wall thermal conductivity

ν: kinematic viscosity of the gas

μ: gas viscosity

ρ: soot layer density

ρ_g: exhaust gas density

τ: dimensionless time

-Subscript o denotes the initial value

-Overbar denotes dimensionless quantity

-O() denotes "on the order of"

APPENDIX I

MICROSTRUCTULAR MODEL OF SOOT OXIDATION: THE EFFECT OF CATALYST

In the two-layer model of a catalytic coating, soot particles oxidize when found within the "field of catalyst action". This is modeled with a parameter β denoting the fraction (0-1) of the soot surface in "contact" with the catalyst (meaning within its field of action) and in principle it depends on coating structure and amount of soot loading of the filter. Although in principle β can vary dynamically due to soot deposition conditions and/or flow or reaction induced restructuring of the soot here it is taken as a constant for simplicity. The layer can "store" a certain amount of soot until it is filled-up, which depends on coating structure and filtration velocity. The remaining soot particles form a "queue" on top of the filled-up catalyst-influenced layer, and this second layer can oxidize only thermally.

Soot oxidation in the presence of a catalyst and with a finite selectivity for CO production is described with the following global reactions, one for the thermal and one for the catalytic path (ie oxidation of soot by oxygen transferred from the catalyst by a redox and/or spill-over mechanism). [12, 14]:

$$C+(1-f_{CO}^{th}/2)O_2 \rightarrow f_{CO}^{th}CO+(1-f_{CO}^{th})CO_2 \qquad (23)$$

$$C+(1-f_{CO}^{cat}/2)O_2 \rightarrow f_{CO}^{cat}CO+(1-f_{CO}^{cat})CO_2 \qquad (24)$$

The local soot oxidation rate is:

$$F_{ox}=-[(1-\beta)K_{th}+\beta K_{cat})]S\rho_g\rho Y\frac{M_c}{M_{ox}} \qquad (25)$$

The conservation equations for the different species across the soot layer are:

$$u_w\frac{dY}{dx}=[(1-f_{CO}^{th}/2)(1-\beta)K_{th}+ (1-f_{CO}^{cat}/2)\beta K_{cat})]S\rho Y \qquad (26)$$

$$u_w\frac{dY_{CO}}{dx}=[f_{CO}^{th}(1-\beta)K_{th}+f_{CO}^{cat}\beta K_{cat}]S\rho Y\frac{M_{CO}}{M_{ox}} \qquad (27)$$

$$u_w\frac{dY_{CO2}}{dx}=[(1-f_{CO}^{th})(1-\beta)K_{th}+ (1-f_{CO}^{cat})\beta K_{cat}]S\rho Y\frac{M_{CO2}}{M_{ox}} \qquad (28)$$

A total mass balance for O_2 gives:

$$Y(0)-Y(w)=\frac{1}{2}\frac{M_{ox}}{M_{CO}}Y_{CO}(w)+\frac{M_{ox}}{M_{CO2}}Y_{CO2}(w) \qquad (29)$$

The initial conditions for the above equations are:

$$\text{at } x=w: Y=Y_{in}(t), \quad Y_{co}=Y_{co_2}=0 \qquad (30)$$

The deposit evolution equations according to the microstructural model introduced in [9] with parameters β and γ are:

$$\frac{\partial\rho}{\partial t}=\alpha F_{ox}-\dot{x}\frac{\partial\rho}{\partial x} \qquad (31)$$

$$\dot{x}=\int_0^x(1-\alpha)\frac{F_{ox}}{\rho}dz \qquad (32)$$

$$\frac{S}{S_o}=\left(\frac{\rho_o}{\rho}\right)^{\gamma} \qquad (33)$$

$$(c_{pw}m_w+c_{ps}M)\frac{dT}{dt}=\Delta H_{CO}GY_{CO}(w)\frac{M_c}{M_{CO}}+ +\Delta H_{CO2}GY_{CO2}(w)\frac{M_c}{M_{CO2}}+Gc_{pg}(T_{in}-T) \qquad (34)$$

where:

$$K_{th}=k_{o,th}Te^{-\frac{E_{th}}{R_gT}} \qquad (35)$$

$$K_{cat}=k_{o,cat}Te^{-\frac{E_{cat}}{R_gT}} \qquad (36)$$

$$f_{CO}^{th}=\frac{1}{1+k_fy^{\theta}e^{E_f/R_gT}} \qquad (37)$$

$$f_{CO}^{cat}=\frac{1}{1+k'_fy^{\theta'}e^{E'_f/R_gT}} \qquad (38)$$

The following dimensionless variables are introduced:

$$\bar{T}=T/T_o,\quad \bar{Y}=Y/Y_o,\quad \bar{Y}_{CO}=Y_{CO}/Y_o,$$

$$\bar{Y}_{CO2}=Y_{CO2}/Y_o,\quad \bar{G}=G/G_o,\quad \bar{M}=M/M_o,$$

$$\rho=\rho/\rho_o,\quad \bar{S}=S/S_o,\quad \bar{x}=x/w_o,$$

$$\bar{w}=w/w_o,\quad \bar{\rho}_g=\rho_g/\rho_{go},$$

$$\bar{K}_{th}=\frac{K_{th}}{K_o},\quad \bar{K}_{cat}=\frac{K_{cat}}{K_o},\quad A=\frac{K_ow_o\rho_oS_o}{u_w},$$

$$\tau=\frac{M_c}{M_{ox}}T_oS_o\rho_{go}Y_o,$$

$$C_{1A}=\frac{G_o\Delta H_{CO}}{c_{pw}m_wT_oK_oS_o\rho_{go}}\frac{M_{CO}}{M_c}$$

$$C_{1B} = \frac{G_o \Delta H_{CO2}}{c_{pw} m_w T_o K_o S_o \rho_{go}} \frac{M_{CO2}}{M_c}$$

$$C_2 = \frac{M_{ox} G_o c_{pg}}{M_c c_{pw} m_w K(T_o) S_o \rho_{go} Y_o}$$

$$C_3 = \frac{c_{ps} M_o}{c_{pw} m_w}$$

where T_o is the initial filter temperature, Y_o is the inlet oxygen mass fraction at t=0, G_o is the exhaust mass flux (with respect to filtration area) at t=0, M_o is the initial soot mass per unit filtration area,ρ_{go} is the exhaust density at t=0 (from the ideal gas law), and ρ_o, S_o are reference values for the density and surface area of the deposit respectively. Using the new variables, the Lagrangian form of the problem that must be solved is:

$$\frac{d\overline{Y}}{d\overline{x}} = \frac{1}{\overline{G}(\tau)\overline{T}}[(1 - f_{CO}^{th}/2)(1-\beta)\overline{K}_{th} + \\ +(1 - f_{CO}^{cat}/2)\beta \overline{K}_{cat})]A\overline{\rho}^{1-\gamma}\overline{Y} \quad (39)$$

$$\frac{d\overline{Y}_{CO}}{d\overline{x}} = \frac{1}{\overline{G}(\tau)\overline{T}} \frac{\mathrm{M}_{CO}}{M_{ox}}[f_{CO}^{th}(1-\beta)\overline{K}_{th} + \\ + f_{CO}^{cat}\beta \overline{K}_{cat})]A\overline{\rho}^{1-\gamma}\overline{Y} \quad (40)$$

$$\frac{d\overline{Y}_{CO2}}{d\overline{x}} = \frac{1}{\overline{G}(\tau)\overline{T}} \frac{M_{CO2}}{M_{ox}}[(1 - f_{CO}^{th})(1-\beta)\overline{K}_{th} + \\ +(1 - f_{CO}^{cat})\beta \overline{K}_{cat})]A\overline{\rho}^{1-\gamma}\overline{Y} \quad (41)$$

$$\frac{d\overline{\rho}}{d\tau} = -\alpha \frac{1}{\overline{T}}[(1-\beta)\overline{K}_{th} + \beta \overline{K}_{cat}]\overline{\rho}^{1-\gamma}\overline{Y} \quad (42)$$

$$\frac{d\overline{x}}{d\tau} = -\frac{(1-\alpha)}{\overline{T}}\int_0^{\overline{x}}[(1-\beta)\overline{K}_{th} + \beta \overline{K}_{cat}]\overline{\rho}^{-\gamma}\overline{Y}dz \quad (43)$$

$$(1 + C_3\overline{M})\frac{d\overline{T}}{d\tau} = \overline{G}(\tau)(C_{1A}\overline{Y}_{CO}(w) + \\ + C_{1B}\overline{Y}_{CO2}(w)) + C_2\overline{G}(\tau)(\overline{T}_{in}(\tau) - \overline{T}) \quad (44)$$

with initial and boundary conditions:

$$\overline{Y} = \overline{Y}_{in}(t) \text{ at } \overline{x} = \overline{w} \quad (45)$$

$$\overline{\rho}(x,0) = \overline{\rho}_o(\overline{x}) \quad (46)$$

$$\overline{T}(0) = 1 \quad (47)$$

The above system is discretized across the soot layer leading to the following set of ordinary differential equations:

$$\overline{Y}_i = \exp[-A\frac{1}{\overline{G}\overline{T}}\frac{1}{2}\sum_{j=i+1}^{L}(((1 - f_{CO}^{th}(\overline{Y}_i)/2)(1-\beta_j)\overline{K}_{th} + \\ (1 - f_{CO}^{cat}(\overline{Y}_i)/2)\beta_j K_{cat})\overline{\rho}_j^{1-\gamma} + \\ ((1 - f_{CO}^{th}(\overline{Y}_{i-1})/2)(1-\beta_{j-1})\overline{K}_{th} + \\ (1 - f_{CO}^{cat}(\overline{Y}_{i-1})/2)\beta_{j-1}K_{cat})\overline{\rho}_{j-1}^{1-\gamma})(x_j - x_{j-1})] \quad (48)$$

$$\frac{d\overline{\rho}_i}{d\tau} = -\alpha\frac{1}{\overline{T}}((1-\beta_i)\overline{K}_{th} + \beta_i\overline{K}_{cat})\overline{\rho}_i^{1-\gamma}\overline{Y}_i \quad (49)$$

$$\frac{d\overline{x}_i}{d\tau} = -(1-\alpha)\frac{1}{2\overline{T}}\sum_{j=i+1}^{L}(((1-\beta_j)\overline{K}_{th} + \beta_j\overline{K}_{cat})\overline{\rho}_j^{-\gamma}\overline{Y}_j + \\ ((1-\beta_{j-1})\overline{K}_{th} + \beta_{j-1}\overline{K}_{cat})\overline{\rho}_{j-1}^{-\gamma}\overline{Y}_{j-1})(x_j - x_{j-1}) \quad (50)$$

$$\overline{Y}_{CO,i} = A\frac{1}{\overline{G}\overline{T}}\frac{1}{2}\sum_{j=i+1}^{L}[((f_{CO}^{th}(\overline{Y}_i)(1-\beta_j)\overline{K}_{th} + \\ + f_{CO}^{cat}(\overline{Y}_i)\beta_j K_{cat})\overline{\rho}_j^{1-\gamma}\overline{Y}_j + \\ + (f_{CO}^{th}(\overline{Y}_{i-1})(1-\beta_{j-1})\overline{K}_{th} + \\ + f_{CO}^{cat}(\overline{Y}_{i-1})\beta_{j-1}K_{cat})\overline{\rho}_{j-1}^{1-\gamma}\overline{Y}_{j-1})(x_j - x_{j-1})]\frac{\mathrm{M}_{CO}}{M_{ox}} \quad (51)$$

$$\overline{Y}_{CO2,i} = A\frac{1}{\overline{G}\overline{T}}\frac{1}{2}\sum_{j=i+1}^{L}[(((1 - f_{CO}^{th}(\overline{Y}_i))(1-\beta_j)\overline{K}_{th} + \\ +(1 - f_{CO}^{cat}(\overline{Y}_i))\beta_j K_{cat})\overline{\rho}_j^{1-\gamma}\overline{Y}_j + \\ + (1 - f_{CO}^{th}(\overline{Y}_{i-1}))(1-\beta_{j-1})\overline{K}_{th} + \\ + (1 - f_{CO}^{cat}(\overline{Y}_{i-1}))\beta_{j-1}K_{cat})\overline{\rho}_{j-1}^{1-\gamma}\overline{Y}_{j-1})(x_j - x_{j-1})]\frac{M_{CO2}}{M_{ox}} \quad (52)$$

where $x_i(0)=x_{io}$ are the initial grid points. Each time the density in a grid point goes to zero, this grid point is eliminated and the ODE integrator is reinitialized.

APPENDIX II

NO$_2$ TURNOVER/RECYCLING FACTOR

The two-layer model for soot oxidation with NO$_2$ in a catalytic filter can be written as follows:

Soot layer (thickness w_2)

In this layer only soot oxidation by NO$_2$ happens:

$$u_w \frac{dX}{dz} = -k_1 X \quad (53)$$

with initial condition $X(0)=X_o$

The resulting NO_2 concentration distribution is:

$$X = X_o e^{-\frac{k_1 z}{u_w}} \quad (54)$$

The NO_2 and NO concentrations leaving the soot layer and entering the catalyst layer are X_i and Y_i respectively and they can be computed as:

$$X_i = X_o e^{-\frac{k_1 w_2}{u_w}} \quad \text{and} \quad Y_i = Z_o - X_i \quad (55)$$

Soot + catalyst layer (thickness w_1)

In this layer both soot and NO oxidation take place:

$$u_w \frac{dX}{dz} = -k_1 X + k_A Y - k_B X \quad (56)$$

$$u_w \frac{dY}{dz} = k_1 X - k_A Y + k_B X \quad (57)$$

with initial conditions $X(0)=X_i$, $Y(0)=Y_i$

The resulting NO_2 concentration distribution is:

$$X = \frac{k_A Z_o}{k_1 + k_A + k_B} + \left(X_i - \frac{k_A Z_o}{k_1 + k_A + k_B}\right) e^{-(k_1 + k_A + k_B)\frac{z - w_2}{u_w}} \quad (58)$$

The total reaction rate for each layer is given by integrating the local rate along the respective layer thickness (i.e):

$$\frac{dm_i''}{dt} = -M_c k_1 \int_{w_{i-1}}^{w_i} X dz \quad (59)$$

Where w is obtained from the local value of the soot density ρ_{soot} and the area mass load, m'' and $w_0 = 0$ is taken as the filter surface.

The evolution equation for the two layers in the thin soot deposit limit (used here for analytical convenience) are:

$$\frac{dm_1''}{dt} = -M_c \frac{k_{A1}}{k_1 + k_A + k_B}[k_A Z_o w_1 + u_w (X_i - \frac{k_A Z_o}{k_1 + k_A + k_B})(1 - e^{-\frac{(k_1 + k_A + k_B) w_1}{u_w}})] \quad (60)$$

$$\frac{dm_2''}{dt} = -M_c v_w X_o (1 - e^{-\frac{k_1 w_2}{u_w}}) \quad (61)$$

From the above equations the evolution equation of the catalytic layer can be written in terms of that of a non-catalytic layer multiplied by a recycling factor R given as following:

$$R = \frac{C_1 C_2 C_3}{1 + C_3 + C_4} + \frac{1}{1 + C_3 + C_4}\left(1 - \frac{C_2 C_3}{1 + C_3 + C_4}\right)(1 - e^{-(1 + C_3 + C_4) C_1}) \quad (62)$$

where $C_1 = k_1 w_1 / u_w$, $C_2 = Z_o / X_i$, $C_3 = k_A / k_1$, $C_4 = k_B / k_1$

APPENDIX III

HEAT TRANSFER FROM A CHANNEL WITH UNIFORM SUCTION

In case of a porous wall having small wall thermal Peclet number Pe_T (as is always the case for a DPF as shown by Bissett and Shardman [19]) the problem degenerates to the following modified Graetz problem:

$$\vec{u}\vec{\nabla} T = \alpha_{th} \nabla^2 T \quad (63)$$

with BC's $T=T_i$ at channel inlet and $T=T_w$ at porous wall, where the velocity field is obtained from the solution of the Navier-Stokes equations with the boundary condition of a constant suction velocity v_w. Our goal in developing DPF simulation tools is not the direct solution of Eq. (63) but its replacement with an 1-D equation for the cross-section average (mixing-cup) temperature along the channel. This problem has been solved for the cases of a slit (2-D planar geometry) and a circular pipe. Here we draw some general conclusions for arbitrary channel cross-sections based on the behavior of existing solutions. For the purpose of presentation let us assume a planar geometry with a slit width of 2L. This geometry would be relevant for example for the DPF design studied in [24].

A first necessary condition for the existence of the 1-D analog of Eq. (1) is the existence of a self-similar (asymptotic) velocity profile (itself equivalent to the existence of a 1-D equation for the flow field). This self-similar profile depends only on the wall Reynolds (Re_w) number and has the following form:

$$u_z = (\bar{u}_{zo} - \frac{zu_w}{L}) f(y^2; Re_w) \quad (64)$$

$$u_y = \frac{u_w}{L} \int_0^y f(y^2; Re_w) dy \quad (65)$$

The above self-similar velocity profiles exist only for a Re_w number smaller than a critical value (e.g. 4.6 for a circular pipe). The self-similar velocity profiles must be found from the solution of the Navier-Stokes equations. Then they have to substituted in Eq. (63) which must be solved in order to compute the local Nusselt number $Nu(z)$. The asymptotic Nusselt number Nu_{as} for a pipe flow and constant temperature boundary condition is given by Kinney [58] as a fraction of Re_w and Prandtl (Pr) numbers. The complete $Nu(z)$ curve for the pipe and slit geometries and constant temperature, constant flux boundary conditions were given by Raithy [57]. This author gave Nu_{as} as a function of Re_w and fluid thermal Peclet (Pe_T) number. Both authors solved Eq. (63) via an eigenfunction expansion.

The heat balance for the channel in terms of gas mixing-cup temperature T_c is:

$$2L\rho_g c_p \frac{d\bar{u}_z T_c}{dz} = -2h(T_c - T_w) - 2\rho_g c_p u_w T_w \quad (66)$$

From the Figures 7, 8 of Raithby [57] one can infer that:

$$Nu = A(Re_w) + Pe_T B(Re_w\ Pe) \quad (67)$$

Subsituting in the heat balance leads to:

$$L\bar{u}_z \frac{dT_c}{dz} = -\frac{\alpha_{th}}{2L} A(Re_w)(T_c - T_w) + (1 - B(Re_w, Pe)) u_w (T_c - T_w) \quad (68)$$

The function A stands for the modification to the usual Nusselt number for non-porous wall (the so called Graetz problem) due to the suction-distorted axial velocity profile. Its value in the limit $Re_w \rightarrow 0$ is the well-known asymptotic Nusselt number (3.6568 for pipe). The function B stands for the heat quantity transferred by the fluid stream entering the wall. The temperature of the stream leaving the wall is actually a proper average over the cross section. This averaging process has to do with the flow field (i.e. from where this fluid is coming from) and the temperature field which is imposed by the conduction process and it must be crossed by the transversely flowing fluid. From the Raithby [57] it seems that the function B takes values around 0.5 for small values of Re_w and Pe_T and increases as Pe_T increases for a given Re_w. The large Pe_T asymptote for B is:

$$B(Re_w, Pe_T) = 1 - A(Re_w) / Pe_T \quad (69)$$

The asymptotic form of B given above is valid in every geometry since it results from the requirement for the correct heat flux direction.

The success of the 1-D fluid dynamic model to describe the flow field in the DPF channel [34, 51, 52] is an indication for the existence of a (nearly) self-similar flow field. A necessary condition for the application of the 1-D model for the heat transfer problem as well, is that the wall velocity v_w variation must be small along the characteristic channel length required for establishment of a steady heat transfer pattern (i.e. a length of $a^2\bar{u}_z / \alpha_{th}$). In transferring the above to the case of flow and heat transfer in a DPF channel we may formally write the heat balance as:

$$a^2\bar{u}_z \frac{dT_c}{dz} = -4\alpha_{th} A(Re_w)(T_c - T_w) + (1 - B(Re_w, Pe_T)) 4a u_w (T_c - T_w) \quad (70)$$

It must be noticed that the relative contribution of the conductive to convective heat transfer is given by the ratio:

$$(1 - B) Pe_T / A(0) \quad (71)$$

which can take appreciable values (close to 1) in some practical conditions, so an accurate computation of the functions A and B is desirable and is currently in progress.

2006-01-0467

An Advanced 1D 2-Layer Catalyzed Diesel Particulate Filter Model to Simulate: Filtration by the Wall and Particulate Cake, Oxidation in the Wall and Particulate Cake by NO_2 and O_2, and Regeneration by Heat Addition

Hasan Mohammed, Antonio P. Triana, Song-Lin Yang and John H. Johnson
Michigan Technological University

ABSTRACT

A numerical model to simulate the filtration and regeneration performance of catalyzed diesel particulate filters (CPFs) was developed at Michigan Technological University (MTU). The mathematical formulation of the model and some results are described. The model is a single channel (inlet and outlet) representation of the flow while the thermal and catalytic regeneration framework is based on a 2-layer approach. The 2-layer model can simulate particulate matter (PM) oxidation by thermal and 'catalytic' means of oxidation with O_2.

Several improvements were made to this basic model and are described in this paper. A model to simulate PM oxidation by NO_2/Temperature entering the particulate filter and oxidizing the PM in the two layers of the PM cake was developed. This model can be used to simulate the performance of filters with catalyst washcoats and uncatalyzed filters placed downstream of diesel oxidation catalysts (DOCs), as in the continuously regenerating traps, CRT's®. A sub-model that simulates the production of NO to NO_2 in the washcoat of the CPF, and integrates with the NO_2-PM oxidation model to simulate the higher oxidation rates (due to greater availability of NO_2) in catalyzed continuously regenerating traps, CCRT's®, was developed. An oxidation model to simulate PM oxidation by O_2 and NO_2 in the pores of the filter wall, which can cause the pressure drop of the filter to rapidly decrease with time, is described. The wall oxidation model was able to explain the disproportionate decreases in CPF pressure drop with respect to PM mass in the filter, which was seen at MTU. To ensure proper integration of the wall oxidation model with the overall filtration/oxidation model, a parametric model to simulate the particle filtration by the PM cake layer was developed. The wall and PM cake layer filtration and oxidation models were able to simulate sudden and rapidly decreasing pressure drops with time, with high oxidation rates in and on the filter wall while still predicting high particle filtration efficiencies. The CPF model was also modified to include a heat addition source term in the energy balance of the filter wall. This model can simulate the effect of heat generation inside a filter due to heat addition by external means. Some results from the use of this model are also presented.

A 1-D DOC model previously developed at MTU was coupled to the CPF model using Fortran Mex-Files and Matlab S-functions in Simulink to simulate DOC-CPF system performance . This integrated model can be used to simulate the filtration and oxidation performance of CRT® and CCRT® technologies or other DOC and CPF designs.

INTRODUCTION

PM emissions from diesel engines are one of the major engineering problems to be solved with advanced technology. Future emission regulations in the US, Europe and Japan will most likely require the use of aftertreatment devices to meet the stipulated standards [1]. Upcoming PM emission standards can be met using diesel particulate filter systems, such as a wall-flow monolith. The construction of wall-flow filters is similar to flow through catalyst substrates, but with alternate channels plugged at their ends [2]. The filter physically captures the diesel particles by forcing the exhaust to flow through the porous filter walls into the four outlet channels surrounding each inlet channel. The filtration is done primarily by the filter wall when the filter is relatively clean and by the developing PM cake layer itself when the filter is loaded. The high surface area to volume ratio helps obtain a high filtration efficiency, which means that the deposited PM has to be periodically regenerated so that the backpressure on the engine can be reduced. In spite of particulate filter studies for over twenty years, a system to regenerate the collected PM which is simultaneously efficient, reliable and affordable has yet to be developed. Recently, there has been increased focus on continuously regenerating technologies which uses NO_2 to passively oxidize PM at temperatures as low as 275°C. Such technologies have been implemented in two ways: a DOC upstream of a uncatalyzed DPF, called a CRT®, and a DOC placed upstream of a CPF, called a CCRT®. Passive technologies in conjunction with active regeneration systems can be combined to provide a system that is commercially viable.

The modeling framework developed by Bisset has allowed a single channel flow and thermal regeneration representation of the entire filter [3]. Konstandopoulos et al. built upon that framework and developed the 2-layer model to simulate increased PM oxidation rates with O_2 in filters coated with catalyst washcoats [4, 5]. Konstandopoulos et al. also developed a transient wall filtration model so that the filtration and PM oxidation characteristics of wall flow monoliths could be studied [6]. Models to simulate the oxidation of PM by NO_2 in the cake layer have also been developed [6–9].

A one-dimensional model based on this framework has been developed at MTU. Further, several improvements were made to this model to better simulate the filtration and oxidation characteristics of particulate filters. This paper outlines these improvements in the MTU 1-D CPF model since the first paper published on the MTU Model by Huynh et al. [10, 11] and also validated by Shende [12, 13]. Sample results from MTU studies [9, 14, 15] which used this model are also presented.

BACKGROUND AND MODEL REVIEW

This section contains a review of the MTU CPF modeling effort begun by Huynh et al. [10, 11], which was based on the modeling framework developed by Bisset [3, 16] and Konstandopoulos [4, 6]. Also discussed in this section is a model by Triana [9] to simulate PM oxidation by NO_2 entering an uncatalyzed filter, which was based on the model developed by Konstadopoulos et al. [6]. Figure 1 shows a schematic of the single channel representation of the CPF defining the coordinates used in the mathematical description of the model.

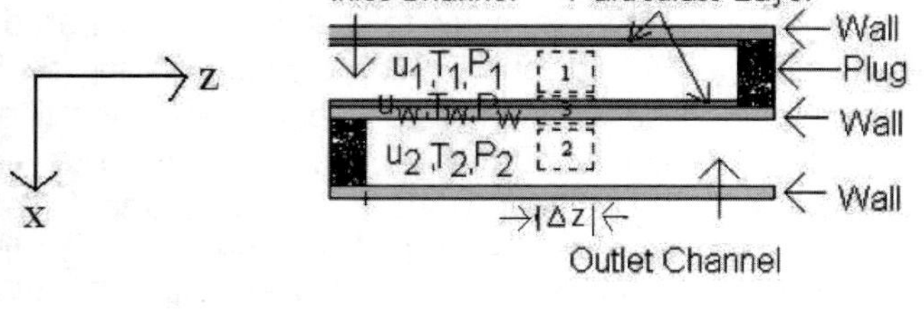

Figure 1: Co-ordinates used in the CPF model, adapted from [10, 11]

The direction along the axial length of filter channel is given by the z coordinate. This length is discretized into control volume units of dimension Δz, for numerical solution purposes. The x coordinate defines the distance through the thickness of the PM cake layer and the filter wall thickness. Further, because filtration by the wall changes its properties along its thickness, it is divided into a number of 'slabs', and the resulting computational domain for the filter wall is shown in Figure 2. The DPF model involves the interaction of four phases: two gas phases for the inlet and outlet channels, respectively, and two solid phases (filter wall and the PM).

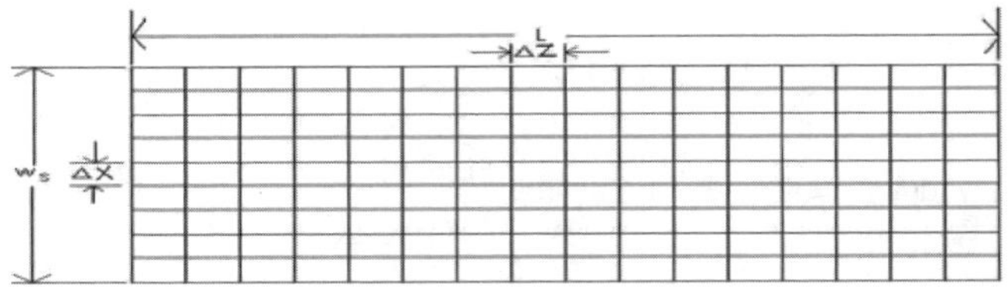

Figure 2: Division of the filter wall into cells for computational purposes [9]

It will be assumed that a single inlet-outlet channel with a porous wall is representative of the thousands of channels in a DPF [3, 4, 6, 10, 16]. This assumption is justified under uniform inlet flow conditions. This assumption reduces the complexity of the 3-D phenomena to a 1-D approximation simplifying the numerical computation. Some of the important assumptions made in the model are:

- A quasi steady-state is assumed, due to the short residence of the exhaust in the inlet and outlet channels. The inlet conditions are transient, but spatially uniform.
- The gas temperatures, species concentrations, gas velocities and properties are an average of the channel cross-section. This means that radial variations within a channel are assumed to be minimal and are neglected.
- The temperature of the porous filter wall and the PM cake layer are assumed to be the same. Additionally, any temperature gradients in the porous filter wall or the PM cake layer in the transverse direction are neglected.
- There is no heat transfer to or from the single channel representation.
- The flow in the inlet and outlet channels is fully developed.
- The oxidation of PM by thermal means, or by the catalyst with O_2 or NO_2, is assumed to be first-order heterogeneous.
- There is no variation in the concentrations of the species of the exhaust gas along the length of the inlet channel.

- Exhaust gas properties vary with temperature and species composition of the gas. The ideal gas equation of state is assumed applicable to the exhaust gas.

Based on the above assumptions, a flow and regeneration model can be formulated. The transient wall filtration model for simulating the depth filtration phase is based on the theory developed by Konstandoupoulos et al., and can be found in References [6, 17]. A review of the wall filtration model is also given in Appendix B. The model can simulate the depth filtration phase by considering the collection of engine-out particles via Brownian diffusion and direct interception by 'unit collectors' in the filter wall.

The flow field model is based on the representative single channel framework of Bisset [3] and has also been used by other researchers [4, 6, 7, 18, 19]. The conservation of mass in the inlet and outlet channel can be described by the following equations [3, 4, 6].

$$\frac{\partial(\rho_1 v_1)}{\partial z} = -F.\frac{4}{D}\rho_w v_w \tag{1}$$

$$\frac{\partial(\rho_2 v_2)}{\partial z} = F.\frac{4}{D}\rho_w v_w \tag{2}$$

The quasi-steady state momentum equation of the exhaust gas in the inlet and outlet channels can be described by [3, 4, 6]:

$$\frac{\partial p_1}{\partial z} + \frac{\partial(\rho_1 v_1^2)}{\partial z} = \frac{-\mu(T_1)v_1}{D^2} \tag{3}$$

$$\frac{\partial p_2}{\partial z} + \frac{\partial(\rho_2 v_2^2)}{\partial z} = \frac{-\mu(T_2)v_2}{D^2} \tag{4}$$

The terms on the right hand side represent the losses due to viscous losses along the length of the channel. F is the friction factor and is equal to 28.454 [17].

An energy balance of the exhaust gas in the inlet and outlet channels results in the following equations [3, 4, 6].

$$C_{p,g}\rho_1 v_1 \frac{\partial T_1}{\partial z} = \frac{4}{D} h_1 (T_w - T_1) \tag{5}$$

$$C_{p,g}\rho_2 v_2 \frac{\partial T_2}{\partial z} = \frac{4}{D} (T_w - T_2) [h_2 + C_{p,g}\rho_2 v_w] \tag{6}$$

The heat transfer coefficients h_1 and h_2 can be obtained from the Nusselt number correlations for square channels [3, 20]. The equations are slightly different than those proposed by Bisset [3], and it is now assumed that the temperature of gas entering the wall is equal to that of the inlet channel gas, T_1. The exhaust gas can be assumed to reach thermal equilibrium with the wall as it passes through the wall, given the high thermal Peclet numbers [3, 16]. Bisset and Shadman also reported that typical length scales over which the gas temperature adjusts to differences with the wall temperature is several times smaller than typical PM cake layer thicknesses [16]. The gas leaves the wall and enters the outlet channel at the wall temperature, T_w.

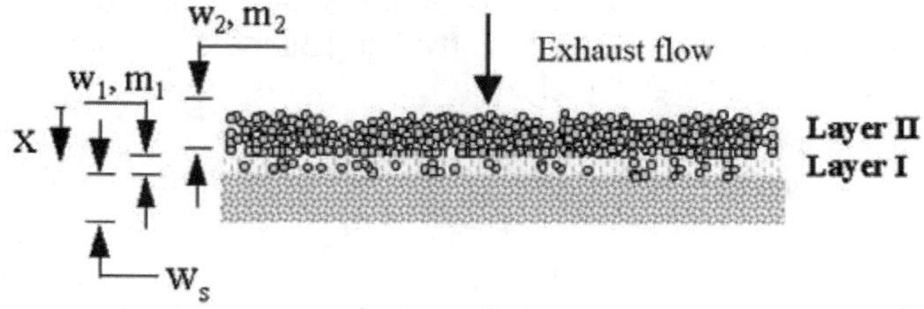

Figure 3: A schematic of the 2-layer formulation for filters coated with catalyst washcoats [4]

PARTICULATE MATTER OXIDATION KINETICS The MTU 1-D CPF oxidation model is based on the 2-layer theory developed by Konstandopoulos et al. [4, 5], which can describe the oxidation of PM by thermal means and by a catalyst washcoat on the filter wall. According to this approach, PM deposited on the surface of catalytically coated filters consist of two layers: the layer closest to the wall benefits from the catalyst coating and is called layer I, while the remaining PM, called layer II, is assumed to be outside the zone of influence of the catalyst. A schematic of the 2-layer approach is shown in Figure 3. This approach was also used by Huynh et al. [10, 11] in the first model formulation at MTU, and results of references [4, 5, 10, 12, 13] have validated this approach. Models of PM oxidation by NO_2 entering the filter based on a single-layer approach have also been developed [6, 8, 21]. A similar approach was implemented and validated in the MTU 1-D model by Triana [9, 22] to simulate the performance of a CRT® at various loads and speeds with and without the upstream DOC.

Thermal oxidation refers to the direct oxidation of the PM by oxygen. While the actual reaction can be complex involving the formation of intermediates and complexes, a global one step heterogeneous equation can be expressed as [3–5, 9, 11]:

$$C + (1 - f_{CO}/2)O_2 \rightarrow f_{CO}CO + (1 - f_{CO})CO_2 \tag{7}$$

where, f_{CO} is the CO selectivity for the PM oxidation by thermal means. α=1-fco/2, which is called the index of completeness of reaction and ranges from 0.5 to 1.0. In the absence of water vapor it has a value of about 0.9 [23]. It has a temperature dependence given by [4, 5, 9, 11]:

$$f_{CO} = \frac{1}{1 + k_{f1}.y_{O_2}^{n1}.exp\left(E_{f1}/\overline{R}T_w\right)} \tag{8}$$

Similarly, a global one step heterogeneous form for the oxidation of PM (with oxygen) in catalytically coated filters is employed for modeling purposes, which is given by [4, 5]:

$$C + (1 - f'_{CO}/2)O_2 \rightarrow f'_{CO}CO + (1 - f'_{CO})CO_2 \tag{9}$$

where f'_{CO} is the CO selectivity for the PM oxidation by catalytic means with a temperature dependence of the form:

$$f'_{CO} = \frac{1}{1 + k_{f2}.y_{O_2}^{n2}.exp\left(E_{f2}/\overline{R}T_w\right)} \tag{10}$$

For oxidation of PM by NO_2, a global one step heterogeneous form for the reaction, involving all the available NO_2 is employed [6–9, 21].

$$C+(2-g_{CO})NO_2 \rightarrow g_{CO}CO+(1-g_{CO})CO_2+(2-g_{CO})NO \quad (11)$$

where g_{CO} is the CO selectivity for the PM oxidation by NO_2 and has a temperature dependence given by [6, 9]:

$$g_{CO} = \frac{1}{1 + k_{f3}.y_{NO_2}^{n3}.exp\left(E_{f3}/\overline{R}T_w\right)} \quad (12)$$

The fact that the reaction

$$2NO_2 + C \rightarrow 2NO + CO_2$$

is the dominant reaction [24–26] means that g_{CO} can be assumed to be close to zero for simpler models.

The oxygen mass fraction, Y_w, entering layer II (if layer I reaches its maximum), or layer I when it is forming, is conserved according to the following equation [3, 4]:

$$\frac{\partial}{\partial x}(\rho_w v_w Y_w) = -S_p \rho_w Y_w k_{th}(T_w)(1 - \frac{f_{co}}{2}) \quad (13)$$

where the specific area of diesel PM is defined as,

$$S_p = A_p.\rho_p \quad (14)$$

The surface area of diesel PM, A_p, is assumed to be around 100 m^2/g [4, 5, 27, 28].

Similarly, the conservation of the NO_2 mass fraction, Y_{NO_2}, entering layer II (if layer I reaches its maximum), or layer I when it is forming, is conserved according to the following equation [6–9]:

$$\frac{\partial}{\partial x}(\rho_w v_w Y_{NO_2}) = -S_p \rho_w Y_{NO_2} k_{NO_2}(T_w)(2 - g_{CO}) \quad (15)$$

Assuming that the exhaust gas density and wall flow velocity remain constant as it flows through the PM cake layer, and integrating Equations 13 and 15 across the thickness of layer II yields the following O_2 [3–5] and NO_2 [6–9] depletion rates in layer II.

$$R_{O_2,2} = \rho_w v_w Y_w.\left(1 - exp\left(\frac{-S_p k_{th}(1-\frac{f_{co}}{2})w_2}{v_w}\right)\right) \quad (16)$$

$$R_{NO_2,2} = \rho_w v_w Y_{NO_2}.\left(1 - exp\left(\frac{-S_p k_{NO_2}(2-g_{CO})w_2}{v_w}\right)\right) \quad (17)$$

The O_2 mass fraction exiting layer II and entering layer I, $Y_{w,1}$, can be obtained from equations 13 and 16.

$$Y_{w,1} = Y_w.exp\left(\frac{-S_p k_{th}(1-\frac{f_{co}}{2})w_2}{v_w}\right) \quad (18)$$

According to the 2-layer theory, the PM in layer I benefits from the presence of the catalyst washcoat and can be oxidized by thermal and catalytic means. It then follows that the O_2 concentration leaving layer II and entering layer I follows two reaction paths: a catalytic path of fraction β, and a thermal path for the remaining O_2 given by (1-β) [4, 5]. Accordingly, the O_2 balance equation for layer I can be written as [4, 5]:

$$\frac{\partial}{\partial x}(\rho_w v_w Y_{w,1}) = -S_p \rho_w Y_{w,1}. \left(k_{th}(T_w)\left(1-\frac{f_{co}}{2}\right)(1-\beta) + k_{cat}(T_w)\left(1-\frac{f_{co'}}{2}\right)\beta\right) \quad (19)$$

The terms k_{th}, k_{cat} and k_{NO_2} refer to the reaction rate constants for PM oxidation by thermal, catalytic and NO_2, respectively. The pre-exponential factors are given by A_{th}, A_{cat} and A_{NO_2}, respectively and the activation energies are given by E_{th}, E_{cat} and E_{NO_2}, respectively. They are assumed to be modified Arrhenius-type functions of the form [3, 4, 6, 9]:

$$k_{th}(T_w) = A_{th}T_w^n exp\left(-E_{th}/(\overline{R}T_w)\right) \quad (20)$$

$$k_{cat}(T_w) = A_{cat}T_w^n exp\left(-E_{cat}/(\overline{R}T_w)\right) \quad (21)$$

$$k_{NO_2}(T_w) = A_{NO_2}.T_w^n.exp\left(-E_{NO_2}/(\overline{R}T_w)\right) \quad (22)$$

Note that the temperature order n can be different for each of the three oxidation mechanisms.

Integrating equation 19 gives the total depletion rate of O_2 in layer I ($R_{O_2,total,1}$) [4, 5].

$$R_{O_2,total,1} = \rho_w v_w Y_{w,1}.\left(1 - exp\left(\frac{-S_p k^* w_1}{v_w}\right)\right) \quad (23)$$

$$k^* = k_{th}(T_w)\left(1-\frac{f_{co}}{2}\right)(1-\beta) + k_{cat}(T_w)\left(1-\frac{f_{co'}}{2}\right)\beta \quad (24)$$

The individual contributions of the thermal ($R_{O_2,th,1}$) and catalytic ($R_{O_2,cat,1}$) oxidation paths in layer I are given by [4, 5]:

$$R_{O_2,th,1} = R_{O_2,total,1}.\left(\frac{k_{th}(T_w)\left(1-\frac{f_{co}}{2}\right)(1-\beta)}{k^*}\right) \quad (25)$$

$$R_{O_2,cat,1} = R_{O_2,total,1}.\left(\frac{k_{cat}(T_w)\left(1-\frac{f_{co'}}{2}\right)\beta}{k^*}\right) \quad (26)$$

The NO_2 mass fraction exiting layer II and entering layer I, $Y_{NO_2,1}$, can be obtained from Equations 15 and 17.

$$Y_{NO_2,1} = Y_{NO_2}.exp\left(\frac{-S_p k_{NO_2}(2-g_{CO})w_2}{v_w}\right) \quad (27)$$

The NO_2 conservation equation in layer I is:

$$\frac{\partial}{\partial x}(\rho_w v_w Y_{NO_2,1}) = -S_p \rho_w Y_{NO_2,1} k_{NO_2}(T_w)(2-g_{CO}) \quad (28)$$

Integrating equation 28 across the thickness of layer I yields the NO_2 depletion rate in layer I.

$$R_{NO_2,1} = \rho_w v_w Y_{NO_2,1} \cdot \left(1 - exp\left(\frac{-S_p k_{NO_2}(2-g_{CO})w_1}{v_w}\right)\right) \quad (29)$$

The evolution of the thicknesses of layer II (w_2) and layer I (w_1), due to oxidation by NO_2 and O_2 can be described by the following equations [4, 6, 9].

$$\frac{\partial}{\delta t}(\rho_p w_2) = -\left(\frac{M_c}{M_{ox}}\left(\frac{R_{O_2,2}}{1-f_{co}/2}\right) + \frac{M_c}{M_{NO2}}\left(\frac{R_{NO_2,2}}{(2-g_{CO})}\right)\right) \quad (30)$$

$$\frac{\partial}{\partial t}(\rho_p w_1) = -\frac{M_c}{M_{ox}}\left(\frac{R_{O_2,th,1}}{1-f_{co}/2} + \frac{R_{O_2,cat,1}}{1-f_{co'}/2}\right) - \frac{M_c}{M_{NO_2}}\left(\frac{R_{NO_2,1}}{(2-g_{CO})}\right) \quad (31)$$

Equations 30 and 31 describe the evolution of PM cake layer, with oxidation by thermal, catalytic and NO_2 assisted means.

The energy due to the reaction is produced in the PM cake layer, conduction in the ceramic wall is so large that wall temperature can be taken to be the same as the PM temperature [3, 16]. The following equation for the energy balance in the wall is solved for the temperature in the wall.

$$(\rho_p C_{pp} w + \rho_s C_{ps} w_s)\frac{\partial T_w}{\partial t} = -h_1(T_1 - T_w) - h_2(T_2 - T_w) + Q^{\cdot\cdot}_{reaction} + Q^{\cdot\cdot}_{conduction} + C_{pg}\rho_w v_w (T_1 - T_w) \quad (32)$$

The energy equation for the filter wall is better described in the section on regeneration by spot heating.

PRESSURE DROP MODEL Inlet channel contraction and outlet channel expansion losses are known to be significant only at very high exhaust flow rates, with relative contributions of order 10^{-2} and thus can be neglected [6, 29]. The pressure loss due to friction in the inlet and outlet channel can be obtained from a solution of the momentum equations (Equations 3 and 4). The pressure drop across the porous ceramic wall and the the PM cake layer is assumed to be given by a Darcy-like formulation [3, 4, 30] and is described by the following equation.

$$\Delta P_{total} = \Delta P_{cake} + \Delta P_{wall} = \frac{\mu}{k_p} v_w w + \frac{\mu}{k_t} v_w w_s \quad (33)$$

IMPROVEMENTS TO 1-D CPF MODEL

This section documents the various improvements made to the MTU 1-D CPF model, in particular a model to simulate PM oxidation in the pores of the filter wall, a model to simulate PM cake layer filtration, a model to take into account the oxidation of NO to NO_2 in the catalyst washcoat of the CPF, and a model to simulate regeneration of the filter by spot heating initiated by external means. They are described in more detail in References [9, 14].

OXIDATION IN THE FILTER WALL Some of the pressure drop data obtained at exhaust temperatures greater than 340°C during the experimental phase of this research [14, 15, 31, 32], showed a rapid decrease in the pressure drop across the CPF even under steady state operating conditions. This is in contrast to the familiar pressure drop curves where a linear PM cake layer filtration regime follows an initial non-linear deep-bed filtration phase [6, 10, 13]. The contrast in such CPF pressure drops can be seen in Figure 4, for the raw exhuast experiments described in References [14, 15, 31]. In Figure 4, the pressure drop across the CPF at 75% load, after about 1.5 hours, was lower than that at 25%, despite the actual flow rates being 1.4 times greater. To simulate these pressure drops, a sub-model that oxidizes PM inside the pores of the filter wall was developed. The approach to PM oxidation inside the filter wall should not only incorporate the oxidation process but also be consistent with the filtration model as explained later.

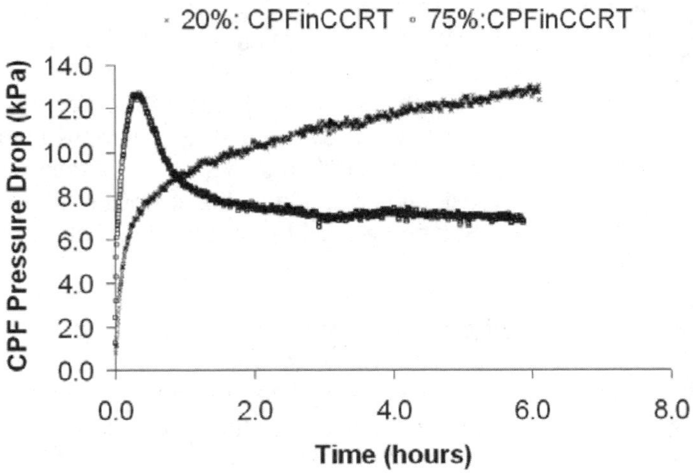

Figure 4: A comparison of steady state CPF pressure drops at exhaust temperatures of 280°C (20% load) and 460°C (75% load) on a Johnson Matthey CCRT®

Starting from clean, the non-linear rise in the CPF pressure drop has been shown to be a result of deep-bed filtration [6, 10, 13]. Further, the simulation results show that there is very little PM mass inside the wall, about 3 grams after which PM cake layer filtration begins [10, 13]. Further, it has also been shown that the deposition of little PM mass inside the filter wall can increase the pressure drop tremendously. The reason is that the porous wall is already fairly solid, about 50% by volume. Hence, even a fairly small deposition of PM in the wall, will further decrease its porosity (and hence permeability) considerably and manifest itself in the pressure drop. The rapid non-linear increase in the CPF pressure drop during the depth filtration phase has been simulated using these mechanisms by References [6, 10, 13, 22]. Just as the deposition of a few grams of PM in the filter wall can increase the CPF pressure drop significantly, it is possible that the oxidation of the same PM can decrease the CPF pressure drop substantially. This could be particularly relevant to CCRT® technologies where the re-oxidation of NO to NO_2 occurs in the catalyst washcoat on the filter wall. It should be noted that there could be other possible explanations for rapidly decreasing pressure drop curves, such as regeneration in parts of the filter, or others as for example reported by Maly et al. [33]. However, the current formulation is spatially homogenous with no need for random or stochastic regeneration models. Data from thermocouples showed that the difference in temperature between the center and periphery of the CPF was only about 10°C, rendering unlikely the scenario of partial filter regeneration.

The O_2 and NO_2 depletion equations apply to PM in the form of a cake layer (Equations 13 and 15). However, inside the filter wall, the PM is dispersed and it is assumed that the mass of the collected particles form a uniform layer around the unit collector causing it to grow in size, but keeping its spherical shape intact [6]. One of the objectives during the development of the wall oxidation model was to be able to use a regeneration framework compatible with that of of the PM cake layer. Because the entire filter wall is divided into 'slabs' and 'cells', the PM mass present in each 'slab' will be lesser, and PM mass present in each cell much lower than the total mass in the filter wall (Figure 2). In addition to being computationally intensive, any oxidation technique that oxidizes the PM cell by cell might also be prone to the propagation of numerical errors.

Given the complexities of the stated problem, it is clear that simplifying assumptions have to be made. It is reasonable to assume that even though there are different masses of PM in each computational cell of the wall, they have the same oxidation rates. However, the total mass oxidized will be related to the initial mass of PM. The approach adopted is to determine oxidation rates in the wall such that if the same mass of PM were present in a cake layer under the same conditions, they would both deplete by the same amounts. To ensure integration with the regeneration framework of Layer I and II, the following technique was formulated: For each axial discretized location, sum all the PM present in each of the 'slabs' of the filter wall (Figure 2) and form a 'virtual' wall cake layer. This ensures that rate depletion of the forms in equations 13 and 15 can be applied. Such an approach also ensures short computational time and is less likely to propagate numerical errors due to 'bulk' oxidation of all PM present in the wall.

The mass of the 'virtual' wall layer, at every discretized location in the axial direction, can be computed by,

$$M_{wall} = \sum_{i=1}^{nlayer} m_i \tag{34}$$

where, m_i is the PM mass present in the ith 'slab'.

To determine the thickness of the virtual wall layer, a PM packing density is required. Here it should be realized that the PM packing density inside the wall used in the filtration model, ρ_{pw} is not a true density, but rather a tool by which all the PM collected by the unit collector is made to uniformly distribute around the unit collector, keeping its spherical shape intact for theoretical purposes. This is the reason for the considerable difference in the relative magnitudes of PM packing densities in the wall and on the wall [6, 9, 11, 12]. Keeping this in mind, and also the aim of keeping oxidation rates in the wall of the same magnitude as on the wall, under the same conditions, the PM cake layer packing density, ρ_p, was used to calculate the thickness of the virtual wall layer.

$$w_{wall} = \frac{M_{wall}}{\rho_p dh \Delta x} \tag{35}$$

dh is the CPF channel width, and Δx is the discretization length in the axial direction.

Equation 35 can be used to calculate the specific area of PM S_p inside the filter wall, which is defined as (same as equation 14):

$$S_p = A_p.\rho_p \tag{36}$$

The O_2 conservation equation for oxidation in the wall can now be written as,

$$\frac{\partial}{\partial x}(\rho_w v_w Y_{w,wall}) = -S_p \rho_w Y_{w,wall} k_{th}(T_w)(1 - \frac{f_{co}}{2}) \tag{37}$$

where $Y_{w,wall}$ is the mass fraction of oxygen entering the wall. It is computed by considering the filter inlet O_2 mass fraction, Y_w, and the depletion in layers I and II (Equations 16 and 23):

$$Y_{w,wall} = Y_w * exp\left(\frac{-S_p k_{th}(1 - \frac{f_{co}}{2})w_2}{v_w}\right) * exp\left(\frac{-S_p k^* w_1}{v_w}\right) \tag{38}$$

Assuming that the exhaust gas velocity and density are constant in the filter wall, the rate of O_2 depletion inside the wall can be obtained from an integration of equation

37 across its thickness,

$$R_{O_2,wall} = \rho_w v_w Y_{w,wall} \cdot \left(1 - exp\left(\frac{-S_p k_{th}\left(1-\frac{f_{co}}{2}\right) w_{wall}}{v_w}\right)\right) \quad (39)$$

In a similar fashion, the NO_2 mass conservation equation inside the filter wall can now be written as,

$$\frac{\partial}{\partial x}(\rho_w v_w Y_{NO_2,wall}) = -S_p \rho_w Y_{NO_2,wall} k_{NO_2}(T_w)(2-g_{CO})) \quad (40)$$

where $Y_{NO_2,wall}$ is the mass fraction of NO_2 entering the filter. It is related to the NO_2 mass fraction entering the filter, Y_{NO_2}, and the depletion in layers I and II.

$$Y_{NO_2,wall} = Y_{NO_2,1} exp\left(\frac{-S_p k_{NO_2}(2-g_{CO})w_1}{v_w}\right) \quad (41)$$

The NO_2 depletion rate inside the wall can be obtained from an integration of Equation 40 across the virtual wall layer thickness.

$$R_{NO_2,wall} = \rho_w v_w Y_{NO_2,wall}\,(1- exp\left(\frac{-S_p k_{NO_2}\,(2-g_{CO})\,w_{wall}}{v_w}\right) \quad (42)$$

The evolution of the the virtual wall layer thickness due to PM oxidation by O_2 and NO_2 can be described by the following equation.

$$\frac{\partial}{\partial t}(\rho_p w_{wall}) = -\frac{M_c}{M_{ox}}\left(\frac{1}{1-f_{co}/2} R_{O_2,wall}\right) - \frac{M_c}{M_{NO_2}}\left(\frac{1}{(2-g_{CO})} R_{NO_2,wall}\right) \quad (43)$$

To determine the rate of PM mass depletion in each of the cells present in the computational domain (redistribute the PM mass in the filter wall), a method of ratio and proportion was adopted. The specific depletion rate of the virtual wall layer in each iteration is given by:

$$\Delta w_{wall} = ((w_{wall})_{n-1} - (w_{wall})_n) / (w_{wall})_{n-1} \quad (44)$$

where, the subscript n refer to the iterations at the nth time step. By ratio and proportion, assuming that all the cells in the wall have the same reaction rates,

$$(m_i)_n = (m_i)_{n-1} * (1 - \Delta w_{wall}) \quad (45)$$

This simple approach of ratio and proportion can easily be verified to conserve mass. The redistributed PM is used in the filtration models to calculate the filtration efficiency and pressure drop across the filter wall (Appendix B).

PARTICULATE CAKE FILTRATION MODEL Use of the wall oxidation model with the wall filtration model of Konstandopoulos et al. [6] resulted in low filtration efficiencies as reported by Triana [9], which has an obvious impact on wall pressure drop computations.

The reason for prediction of low filtration efficiencies was traced to the use of the partition coefficient which is used to filter the engine-out particles before they enter the filter wall (approximating filtration by the PM cake layer) [14]. Oxidation in the wall decreases the loaded unit collector diameters which in turn decreases the partition coefficient resulting in low filtration efficiencies. Thus an improper coupling between the new wall oxidation model and the wall filtration model is seen.

To ensure proper integration of the filtration and oxidation sub models in the overall filter model, a parametric model to simulate filtration by the PM cake layer was developed. To describe filtration by the PM cake layer, the model uses an equation of a form similar to that describing filtration by the wall (Appendix B), thus relating the efficiency of each unit collector and the thickness of the PM cake layer to its filtration efficiency. The filtration efficiency of a single collector in the cake is now a model parameter and is determined during model calibration. The diameter of the cake collector can be assumed to be the aggregate diameter of PM cake deposits [14], usually in the range of 100 - 250 nm [28, 34]. For additional details of this model refer to the thesis of Premchand [35].

The PM cake filtration model assumes that the particles it filters increase its thickness while keeping the diameter of the unit collector and the porosity constant (under steady state operating conditions). In such a case, number of cake collectors in series for a given layer thickness can be calculated by [36],

$$N = \frac{w_{cake}}{dc_{cake}} \quad (46)$$

where dc_{cake} is the diameter of the collector in the PM cake.

The filtration efficiency of a PM cake layer, with unit collector diameter dc_{cake}, and having N such collectors placed in series is given by [35, 37]:

$$\eta_{cake} = A_\eta \cdot \left(1 - e^{-\eta_c \cdot \frac{w_{cake}}{dc_{cake}}}\right) \quad (47)$$

where η_c is the particle collection efficiency of a single collector in the cake layer, and A_η is a parameter which defines the maximum filtration efficiency of the PM cake layer. The parameter A_η was added empirically to prevent Equation 47 from predicting 100% filtration efficiencies at large cake layer thicknesses. In Equation 47, A_η and η_c are input parameters, calibrated to fit pressure drop data, and transient filter outlet particle size distributions PM concentrations. Equation 47 is a relation to calculate the filtration efficiency of the PM cake layer based on its thickness. The PM concentrations entering the filter wall become independent of the oxidation in the filter wall. Thus the PM cake and the filter wall can be viewed as two filters in series.

Equation 47 cannot be used with a 'clean' filter, because in such a case the PM cake layer thickness w_{cake} is equal to zero. Hence the filtration efficiency will be computed to be equal to zero. Hence, an initial starting solution for computational purposes is needed. A good approximation of the clean filtration efficiency is the partition coefficient as defined in Appendix B [6]. With a clean filter, it is logical to assume that the filtration efficiency is related to the size of the collectors in the filter wall as in the partition coefficient. In the first few iterations while simulating clean filters, the partition coefficient is computed before transitioning to Equation 47 to predict the PM cake filtration efficiency.

OXIDATION BY NO_2 PRODUCED IN THE CPF WASHCOAT Research clearly demonstrating the beneficial effect of NO_2 on PM oxidation is well documented [6, 24, 25, 38–40]. More recent research has also firmly established that filters coated with appropriate catalysts can oxidize NO to NO_2 as the exhaust flows through the washcoat. Under such conditions, the rate of PM oxidation can be accelerated due to the re-oxidation of NO→NO_2. This 'recycling' process can happen more than once, and possibly even three times before the exhaust leaves [39–42]. Oxidation of NO→NO_2 in the CPF of the CCRT®, in both CPF-only and CCRT® configurations, was measured during the steady state raw exhaust characterization experiments at MTU [14, 15, 31]. To simulate its effect on PM oxidation, a model to take into account the oxidation of NO to NO_2 in the catalyst washcoat of the CPF was developed.

In the 1-D 2-layer CPF regeneration model framework of Konstandopoulos et al., the catalyst effect with O_2 was limited to layer I [4, 5]. Extending this to filters with catalyst washcoats promoting the oxidation of NO→NO_2, the production of NO_2 can be assumed to occur in layer I which is influenced by the presence of the catalyst. The kinetic effect describing this can be expressed by the following general kinetic expression [6, 38]:

$$\frac{dy_{NO_2}}{dt} = k_{NO} y_{O_2}^{\gamma} y_{NO}^{n} \tag{48}$$

Use of equation 48 to describe the NO→NO_2 oxidation in the CPF washcoat and simulate the higher reaction rates was first proposed in Reference [6]. However, a different numerical procedure for the solution was implemented in this research [14].

Cooper and Thoss reported that the exponents n and γ, representing reaction orders, vary with space velocity [38]. These exponents are reproduced below.

$$\gamma = 0.22 \rightarrow AllSpaceVelocities \tag{49}$$

$$n = 0.5 \rightarrow LowSpaceVelocity \tag{50}$$

$$n = 1.4 \rightarrow HighSpaceVelocity \tag{51}$$

The frequency factor for the NO→NO_2 oxidation reaction, k_{NO}, is assumed to follow a modified Arrhenius function given by [6]:

$$k_{NO}(T_w) = A_{NO}.T_w^n.exp\left(\frac{-E_{NO}}{RT_w}\right) \tag{52}$$

The value of E_{NO} can be taken to be 0.906E+08 (J/kmole) determined experimentally by Konstandopoulos et al. [6] and also used by Triana in modeling the same NO→NO_2 reaction in the DOC [9, 43]. The value of E_{NO} known, the A_{NO} can then be determined from calibration of the model to fit the experimental data.

Oxidation of NO to NO_2 in layer I and oxidation of PM by O_2 and NO_2 results in a gradient of concentrations through the thickness of layer I. Any change in O_2 or NO_2 concentrations affects the PM oxidation rates by thermal, catalytic and NO_2 means. However, the change in O_2 concentrations is expected to be minimal due to the low concentrations of NO and NO_2 relative to O_2 (ppm as compared to %Vol.). Thus, the complexity in layer I kinetics is due to simultaneous depletion and production of NO_2 by the PM and catalyst, respectively. The following numerical solution was devised to solve the problem. To use equation 48 in the regeneration framework of the 1-D 2-layer CPF model, the thickness of layer I was discretized into a number of 'sub-layers' of equal thickness, rather like the division of the filter wall into 'slabs'. This is shown schematically in Figure 5. Depletion of NO_2 (and O_2) takes place within these sub-layers, which results in oxidation of PM present in these sub-layers. This decreases NO_2 and O_2 concentrations, and an increase in NO concentrations, exiting the sub-layer. Between these sub-layers, an increase in NO_2 concentrations (due to NO→NO_2 oxidation) and a corresponding decrease in NO and O_2 concentrations is assumed to take place. Thus the PM in layer I can have higher oxidation rates due to increased availability of NO_2.

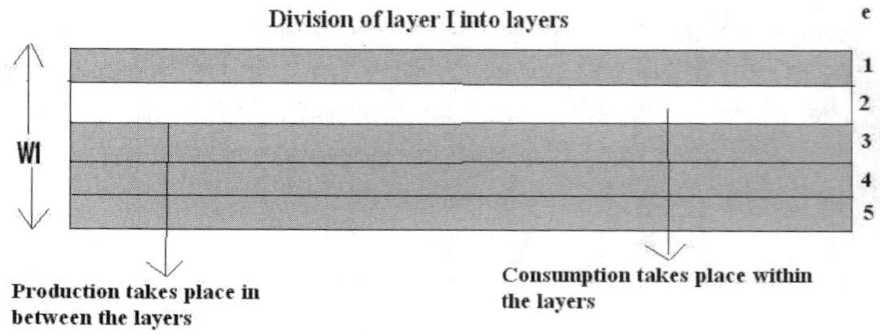

Figure 5: A schematic of the division of layer I into sub-layers

The concentrations of O_2 and NO_2 leaving layer II and entering layer I are calculated using engine-out concentrations and the depletion in layer II given by equations 18 and 27. For the first sub-layer in layer I, the model calculates the rate of PM depletion using concentrations of O_2 and NO_2 leaving layer II and entering layer I. The PM layer thickness shrinks due to oxidation by NO_2 and O_2. The evolution can be described by (similar to equation 31):

$$\frac{\partial}{\partial t}(\rho_p w_e) = -\frac{M_c}{M_{NO2}}\left(\frac{1}{(2-g_{CO})}R_{NO_2,e}\right) - \frac{M_c}{M_{ox}} \cdot \left(\frac{R_{O_2,th,e}}{1-f_{co}/2} + \frac{R_{O_2,cat,e}}{1-f_{co'}/2}\right) \quad (53)$$

where, the subscript *e* refers to the PM layers.

The thermal, catalytic and NO_2 depletion rates in the PM layer are given by (similar to equations 25, 26 and 17 respectively):

$$R_{O_2,th,e} = R_{O_2,total,e} \cdot \left(\frac{k_{th}(T_w)\left(1-\frac{f_{co}}{2}\right)\beta}{k^*}\right) \quad (54)$$

$$R_{O_2,cat,e} = R_{O_2,total,e} \cdot \left(\frac{k_{cat}(T_w)\left(1-\frac{f_{co'}}{2}\right)(1-\beta)}{k^*}\right) \quad (55)$$

$$R_{NO_2,e} = \rho_w v_w Y_{NO_2,1} \cdot \left(1 - exp\left(\frac{-S_p k_{NO_2}(2-g_{CO})w_e}{v_w}\right)\right) \quad (56)$$

where, the total oxygen depletion rate due to thermal and catalytic means, $R_{O_2,total,1}$, is given by (similar to equation 23):

$$R_{O_2,total,e} = \rho_w v_w Y_{w,1}\left(1 - exp\left(\frac{-S_p k^* w_e}{v_w}\right)\right) \quad (57)$$

Equations 53 to 57 can be used to determine the depletion in the thickness of the PM layers due to oxidation of PM by NO_2 and O_2.

The NO_2 and O_2 concentrations (mass fractions), $Y_{NO_2,e}$ and $Y_{w,e}$ leaving the PM layer can be given by (similar to equations 18 and 27, respectively):

$$Y_{NO_2,e} = Y_{NO_2,1}\left(1 - exp\left(\frac{-S_p k_{NO_2}(2-g_{CO})w_e}{v_w}\right)\right) \quad (58)$$

$$Y_{w,e} = Y_{w,1}\left(1 - exp\left(\frac{-S_p k^* w_e}{v_w}\right)\right) \quad (59)$$

where, subscript *e* refers to the mass fractions exiting the PM sub-layers. Note that these decreases in NO_2 and O_2 concentrations are due to consumption of these species by the carbon in the PM sub-layers.

Due to the production of NO_2 from NO (equation 48) by the catalyst in the CPF, the concentrations of NO_2 increase while those of O_2 and NO decrease. These changes are assumed to happen between the sub-layers in layer I. The sub-layer outlet NO mole fraction is calculated from a conservation of nitrogen atoms [6],

$$y_{NO,e} = y_{NO} + y_{NO_2} - y_{NO_2,e} \quad (60)$$

where, y_{NO} and y_{NO_2} are the mole fractions of NO and NO_2 entering the layers. They can also be taken to be equal to the engine-out NO and NO_2 mole fractions as total nitrogen atoms are always conserved. $y_{NO_2,e}$ is the mole fraction of NO_2 leaving the sub-layer and is related to equation 58 by the basic equation relating mass to moles for a given species. The sub-layer outlet O_2 and NO_2 mole fractions can be derived from equations 59 and 58 in a similar fashion, and are then inserted in equation 48 to calculate the production of NO_2 from NO. Equation 48 increases the NO_2 mole fraction entering the next sub-layer, while the O_2 and NO mole fractions decrease. The O_2 concentrations can be calculated by a conservation of O_2 atoms, as follows:[1]

$$y_{O2} = y_{w,e} - 0.5 * (y_{NO2,gen} - y_{NO2,e}) \quad (61)$$

where, $y_{w,e}$ and $y_{NO2,e}$ are the mole fractions of oxygen and NO_2 exiting the PM sub-layer, and are related to equations 59 and 58. $y_{NO2,gen}$ is the increased mole fraction of NO_2 between the PM sub-layers after NO_2 generation by equation 48. Except at conditions of high temperatures (¿500°C) where considerable O_2 can be oxidized, the application of equation 61 in the code is not essential as the decrease in O_2 concentrations is of order $O(10^{-4})$ relative to the decrease in NO_2 concentrations. Nevertheless, equation 61 was implemented in the MTU 1-D 2-layer code.

For the solution of the NO_2 production in equation 48, layer I was divided between 5 to 10 sub-layers, depending on the thickness of layer I. The thickness of layer I is related to the thickness of the catalyst washcoat which is generally between 10μm and 30μm. During coding, an easy and simple method to determine the optimum number of sub-layers is to keep increasing their number until it converges to the single solution. Further increases in the number of sub-layers will only slow down the computation process with no increase in accuracy. Further, for the solution of equation 48, a first-order difference technique was found to be sufficient in accuracy.

The procedure described above is repeated for each of the PM layers. Total depletion of layer I is obtained by summing up the depleted thicknesses of the each of the sub-layers making up layer I.

$$w_{1,dep} = \sum_{i=1}^{e-layers} w_{e,dep} \quad (62)$$

$w_{1,dep}$ is the total thickness of layer I after depletion, and $w_{e,dep}$ are the depleted thicknesses of the each of the sub-layers in layer I.

[1] Refers to the reaction $NO + \frac{1}{2}O_2 \rightarrow NO_2$. Note that the O_2 mole fraction is generally expressed in %Vol. while NO and NO_2 concentrations are expressed in ppm

SIMULATION OF SPOT HEATING The rate of heat accumulation in the filter wall equals the rate of heat transport from the inlet channel minus the rate of heat transport to the outlet channel, minus the rate of heat conduction in the axial direction, plus the rate of heat of reaction. If heat generation is also considered, a heat addition source term is included to account for such effects. Mathematically,

$$\ddot{Q}_{accumulation} = -\ddot{Q}_{conduction} + \ddot{Q}_{inlet\ channel} - \ddot{Q}_{outlet\ channel} + \ddot{Q}_{reaction} + \ddot{Q}_{generation} \qquad (63)$$

The rate of heat accumulation can be expressed as [3, 4]:

$$\ddot{Q}_{accumulation} = \rho_p C_{p,p} w + \rho_s C_{p,s} w_s \frac{\partial T_w}{\partial t} \qquad (64)$$

The heat transported from the inlet channel is [3, 4]:

$$\ddot{Q}_{inlet\ channel} = \rho_w C_{p,w} u_w T_1 - h_1 (T_w - T_1) \qquad (65)$$

The heat transported to the outlet channel is [3, 4]:

$$\ddot{Q}_{outlet\ channel} = \rho_w C_{p,w} u_w T_w - h_2 (T_w - T_2) \qquad (66)$$

The rate of heat conduction in the axial direction, H_{cond}, is given by [3, 4]:

$$\ddot{Q}_{conduction} = -\lambda_p \frac{\partial}{\partial z}\left(w \frac{\delta T_w}{\delta z}\right) - \lambda_s w_s \frac{\partial^2 T_w}{\partial z^2} \qquad (67)$$

The rate of heat addition due to oxidation of PM by thermal, catalytic and NO_2 means is represented by [3, 4, 6, 9]:

$$H_{react} = \frac{\Delta H^{th}}{M_c}\left(\frac{M_c}{M_{ox}} \frac{1}{1 - f_{co}/2} R^{th}_{O_2}\right) + \frac{\Delta H^{cat}}{M_c}\left(\frac{M_c}{M_{ox}} \frac{1}{1 - f_{co'}/2} R_{O_2,cat,1}\right) \qquad (68)$$

The thermal and catalytic heats of reaction depend on their respective CO selectivities, f_{CO} and $f_{CO'}$ [4]:

$$\Delta H^{th} = f_{CO} \Delta H_{CO} + (1 - f_{CO}) \Delta H_{CO_2} \qquad (69)$$

$$\Delta H^{cat} = f_{CO'} \Delta H_{CO} + (1 - f_{CO'}) \Delta H_{CO_2} \qquad (70)$$

Substitution of the above equations in Equation 64 results in an energy equation for the filter wall which is solved to determine the temperature of the filter wall along the length of the filter.

Modifications to incorporate an additional heat source term in the filter wall energy balance were made to simulate the effect of heating a single spot or a combination of strategically located multiple spots along the filter wall by external means to promote active regeneration in local spots inside the porous walls, and by supplying enough energy to diffuse the energy to the remaining spots and oxidize all the solid PM matter. The effect of the heat addition source term is to increase the wall temperature in certain spots on the filter wall channel with the approach of transferring the energy to the rest of the porous wall.The theory and simulation results of this modeling effort are reported in more detail in reference [9].

The rate of heat generated from a local hot spot is [9]:

$$\ddot{Q}_{generation} = \frac{\rho_s w_s}{m_{spot}} (N_{in\ cells} . S) \qquad (71)$$

where S is the absorbed power by a single spot of mass m_{spot}. Assuming that a body of mass m_{spot}, with specific heat of C_p is heated with a distributed heat source S, we have [9],

$$S = m_{spot} C_{p,s} \frac{\partial T}{\partial t} \qquad (72)$$

With a constant heat source S, and a desired temperature increase from $T_{w,i}$ to $T_{w,f}$ in t_o seconds, the spot mass is computed as follows [9]:

$$m_{spot} = \frac{t_o S}{\int_{T_{w,f}}^{T_{w,i}} C_{p,s}(T) dT} \qquad (73)$$

By knowing the magnitude of the heat source, the filter bulk density and geometry (face area, etc), the axial thickness of such spot can be computed [9, 20]. This final term substituted in Equation (64) results in a modified energy equation for the filter wall to simulate the effect of heating a single spot or a combination of strategically located multiple spots along the filter wall by external means to promote active regeneration.

INTEGRATION OF MTU 1-D DOC AND CPF MODELS IN MATLAB/SIMULINK In this section the integration of the MTU 1-D DOC model developed by Triana [9, 43] and the 1-D CPF model in Matlab/Simulink is described. The integration of the two models is better described in the thesis of Triana [9].

The DOC and CPF codes were written in Fortran language, and subsequently in MATLAB/SIMULINK software. Each of the subroutines employed in the DOC and DPF models, shown in figures 6 and 7 respectively, were re-written in the form of MEX-files in MATLAB. MEX-files are MATLAB callable Fortran programs which are dynamically linked subroutines that can be automatically loaded and executed in MATLAB. A MEX-file consists of two main components:

- A computational routine which contains the Fortran code with the numerical computations, and or with the input/output data.
- A gateway routine that interfaces the computational routine with the MATLAB S-functions. The gateway calls the computational routine as a subroutine.

Each subroutine is called in MATLAB inside an S-function which is a computer language description of a SIMULINK

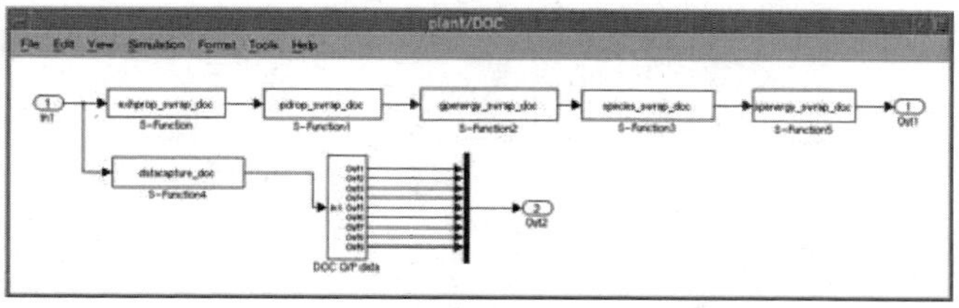

Figure 6: DOC Model in Matlab/Simulink [9]

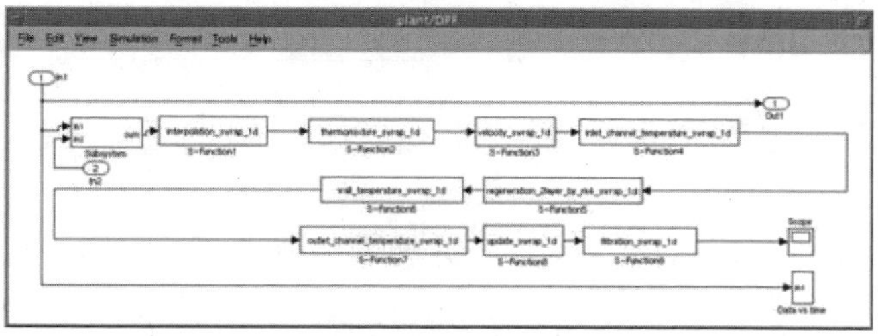

Figure 7: DPF Model in Matlab/Simulink [9]

block diagram (or elementary dynamic system), producing an output. The blocks are interconnected by lines to transfer data from one S-function to the other. The interconnection of functions allows the system to complete all the model features programmed in Fortran language for both the DOC and CPF models. With all the blocks interconnected, the complete DOC or CPF models can be represented by a single block system. The two system blocks (DOC and CPF) were connected by transferring output data from the DOC model as an input to the DPF model. The integrated model can by-pass the DOC model and operate as a DPF model or as a CPF model.

Exhaust gas temperatures, flow velocities, and gaseous emissions species concentrations (mole fractions) at the outlet of the DOC model were used as input data to the DPF model. Since the gas flows from the DOC to the DPF, the mass/momentum and energy of the gas is assumed to be transferred only in the direction of the flow, and the reverse heat and mass transfer in the small gap between the DOC outlet and DPF inlet is neglected. This assumption is particularly valid under most of the normal engine operating conditions as the convective effect is usually much more significant than the heat diffusion in the reverse direction due to the low temperature gradients encountered between the DOC and DPF. However, if a considerable reverse temperature gradient occurs between the DOC and the DPF, heat conduction in the reverse direction cannot be easily neglected and may need to be included in future model development effort. Figure 8 shows a schematic representation in Simulink of the integrated DOC-DPF system blocks.

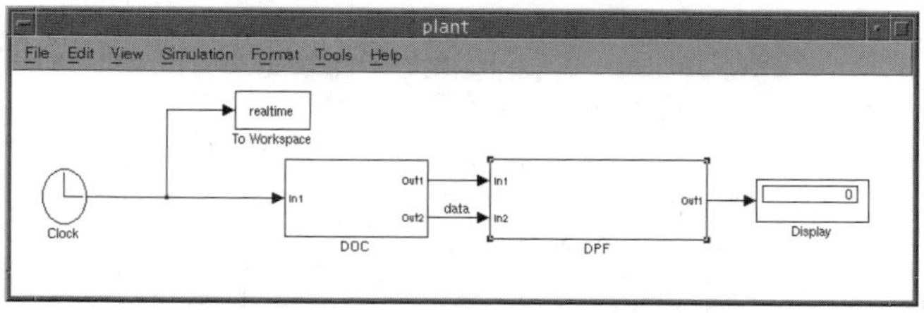

Figure 8: Integrated DOC-DPF Model in Matlab/Simulink [9]

MODEL RESULTS

In this section sample results from the use of the DOC and CPF model are presented. Extensive results from the use of the model at different loads and speeds and device configurations are presented in references [9, 15, 44].

DOC MODEL SIMULATION RESULTS Although not described in this paper, a brief look at the results of the DOC model developed by Triana [9, 43] is shown in this section, as it is part of the overall model used to simulate continuously regenerating technologies.

Table 1 shows a comparison between model predictions and experimentally measured concentrations of HCs, NO, NO_2 and CO upstream and downstream of the DOC. It is seen that all concentrations were predicted within 3 ppm of the experimentally measured values. Since CO concentrations downstream of the DOC were too low to be measured by the Pierburg emissions analyzer, the model was calibrated to predict CO concentrations within 1 ppm downstream of the DOC. The results in Table 1 were taken from the thesis of Reference [14]. In addition, data from a 25% load at rated speed experiment from the dilute emission characterization experiment of reference [31] were also used. The measurements at 20, 40, 60 and 75% load at rated speed (2100 rpm) span a temperature and volumetric flow rate range of 280-465°C and 0.44 and 0.84 actual m^3/sec, respectively, demonstrating good prediction capability across a wide range of temperatures and flowrates.

Table 1: Comparison of DOC model predictions with measurements of HC's, NO_2 and CO [14]

Load↓	Location↓	HC's	NO	NO_2	CO
	Units→	ppmC	ppm	ppm	ppm
20%	US (Expt.)	75	146	36	185
	DS (Expt.)	17	77	105	0
	DS (Model)	17	75	107	0.2
25%	US (Expt.)	81	161	39	177
	DS (Expt.)	18	77	122	0
	DS (Model)	17	77	123	0.3
40%	US (Expt.)	67	200	29	160
	DS (Expt.)	15	92	136	0
	DS (Model)	16	91	138	0.3
60%	US (Expt.)	61	239	21	151
	DS (Expt.)	10	154	103	0
	DS (Model)	10	155	105	0.2
75%	US (Expt.)	56	264	17	145
	DS (Expt.)	11	194	83	0
	DS (Model)	10	199	82	0.3

CCRT® SIMULATION RESULTS In this section, sample results from the use of the model to simulate the performance of a CCRT® and based on the thesis of reference [14], but later modified and presented in reference [15], are shown. Reference [15] also contains a description of the instrumentation, experimental procedures and measurements of the experimental data used to calibrate the CPF model.

Figures 9 shows the experimental and model predicted pressure drops at 75% load steady-state experiments in CPF-only and CCRT® configurations respectively, using ultra low sulfur fuel in a Cummins ISM 2002 heavy duty diesel engine. A good fit between the experimental and model predicted pressure drops can be seen at lending support to the wall oxidation model discussed earlier. The exhaust temperature entering the CPF at 75% load was 460°C, while the mass flow rate was 0.505 kg/sec. The NO_2 concentrations entering the filter were 13 and 63 ppm in CPF-only and CCRT® configurations respectively.

Using the calibrated model, the individual components of the total pressure drop across the CPF were determined and are shown in figure 10 for the 75% load CCRT® configuration. Figure 10 shows that the rapid decrease in pressure drop at the 75% load condition, is due to the decreasing pressure drop across the filter wall, which is a consequence of PM oxidation in the filter wall. This breakdown of the pressure drops in the various parts of the CPF is a new feature of the MTU 1-D 2-layer filter model.

Figure 11 shows the PM mass in the slabs of the filter wall with time for 75% load CCRT® configuration. Slab 1 is the slab closest to the PM cake layer and contains the most PM amongst all the slabs. On the other hand, slab 9 is the last slab in the filter wall after which the exhaust exits the filter and hence contains the least mass as most of the particles are filtered by the PM cake layer and slabs preceding it. The effect of oxidation in the wall is clearly demonstrated in Figure 11 with the PM mass in the filter wall substantially decreasing with time. In Figure 11, note the decrease in PM mass in all the slabs in the filter wall which is a result of the PM cake layer filtration model. If the wall oxidation model was used without the PM cake filtration model, the PM mass in the first slab would be oxidized but be immediately refilled as the results of Triana [9] show, and the model predicted pressure drop would not decrease as rapidly as the experimental data.

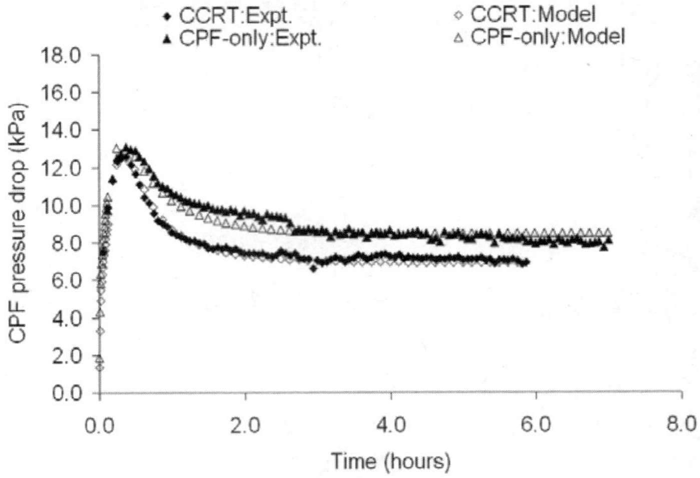

Figure 9: 75% load pressure drop: experimental and model results

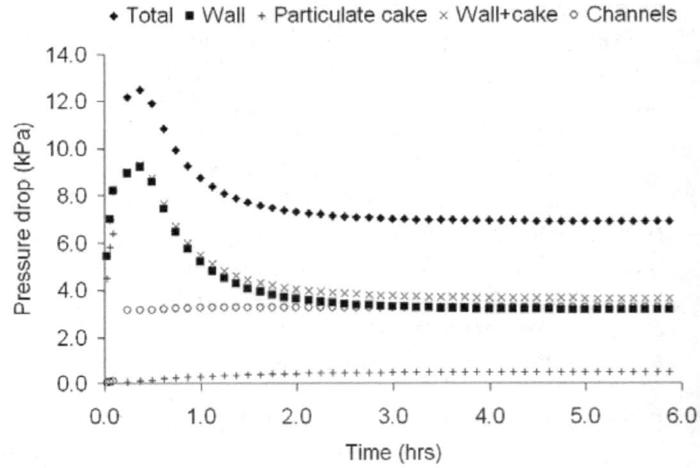

Figure 10: 75% load CCRT®: Components of total CPF pressure drop

The model predicted PM mass evolution with time for the 75% load CPF-only condition is shown in Figure12. It can be seen that most of the PM mass deposited is on the filter wall i.e. the PM cake layer. Good agreement was obtained at all experimental conditions tested, in both CPF-only and CCRT® configurations, and is discussed in more detail in reference [15].

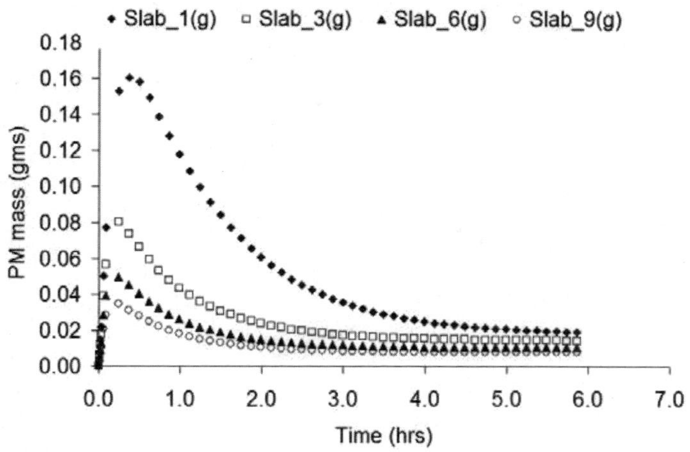

Figure 11: 75% load CCRT®: PM mass in the wall with time

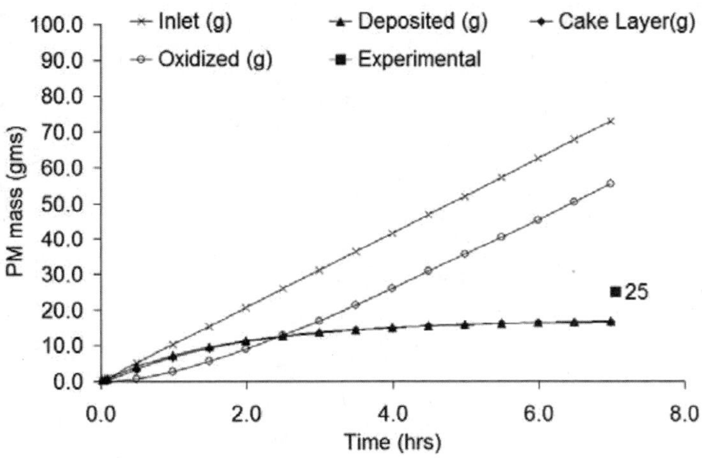

Figure 12: 75% load CPF-only: PM mass evolution with time

A comparison of the relative filtration efficiencies of the filter wall and PM cake layer at 75% load is shown in figure 13. The filtration efficiency of the wall first increases due to deposition of PM in the wall, and then decreases due to oxidation, before stabilizing at about 70%, while that of the PM cake layer keeps increasing with time. The reason is that the efficiency of the filter wall depends on the efficiency of its unit collectors, as the thickness of the filter wall is constant. The efficiency of the unit collectors in the filter wall reaches a maximum when the size of the unit collector loaded with PM equals that of the unit cell. When this happens the unit collectors in the wall cannot grow further in size and their efficiency peaks at this stage. On the other hand, the efficiency of the PM cake layer, in addition to depending on the efficiency of the unit collectors in the PM cake layer, also depends on its own thickness. Thus, if the thickness of the PM cake layer increases, so does its filtration efficiency, as Figure 13 shows.

A comparison of the thicknesses of layer I and layer II for

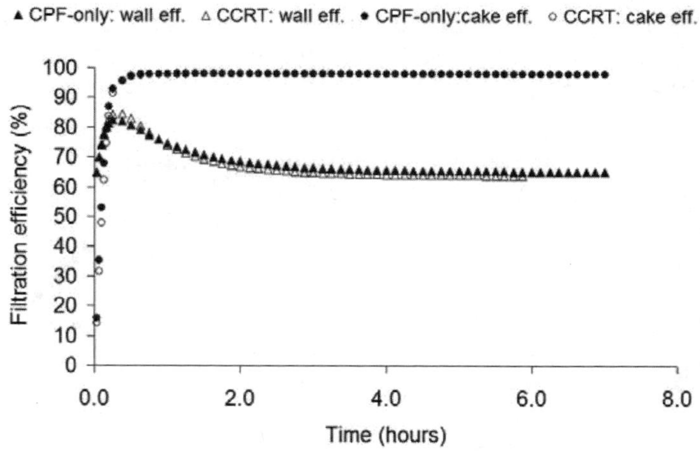

Figure 13: 75% load: Comparison of filtration efficiencies in CPF-only and CCRT® configurations

the 75% load conditions is shown in figure 14. Because of the higher concentrations of NO_2 entering the filter in CCRT® configuration compared to CPF-only configuration, layer I and II form in a shorter time in CCRT® configuration. For the same reason, the thickness of the PM layer in CCRT® configuration, at any given instant, is less than the the thickness of the PM layer in CPF-only configuration. The high oxidation rates at 75% load result in all the PM matter being confined to layer I with no formation of layer II and a non-linear evolution of the layer I thickness.

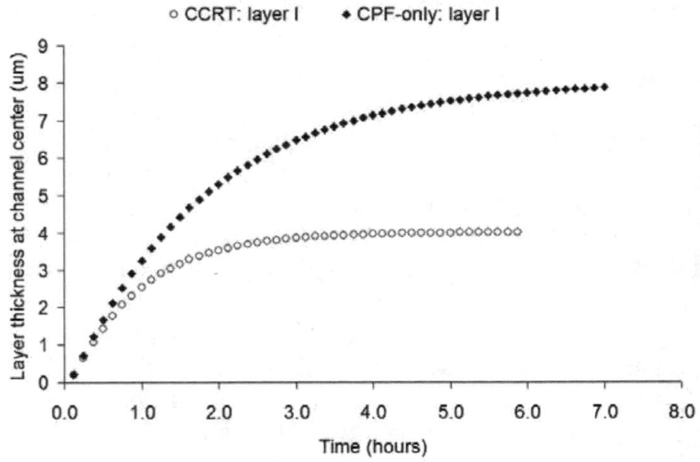

Figure 14: 75% load: comparison of PM layer thickness with time

The PM cake layer and the wall filtration models can be used to predict the particle size distribution downstream of the filter. The result for 75% load CCRT® experiments are shown in Figure 15. It can be seen that the model can predict the 98%+ total filtration efficiencies [31] in spite of the high oxidation rates in the filter wall. Thus, the PM cake layer filtration model, along with the wall filtration model, can model high PM oxidation rates in the wall while still maintaining high filtration efficiencies which were experimentally measured. This was not possible if one uses the wall oxidation model with only the wall filtration model, as found by Triana [9].

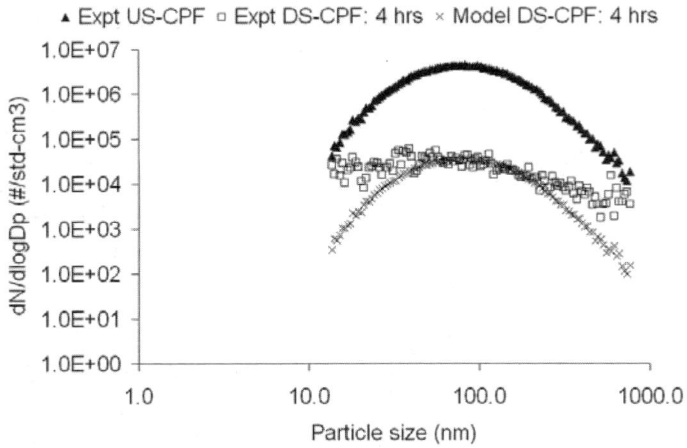

Figure 15: 75% load CCRT®: particle size distribtion comparison

Using the calibrated model, the PM oxidation rates by physical location, i.e., layer I, layer II and filter wall, and by type of oxidation(thermal and NO_2), were determined. These plots are a new feature in the filter model. Figures 16 and 17 shows these results for the 75% load condition respectively. These results should be used in conjunction with Figure 14 for a complete understanding of the physics involved. At 75% load, the thermal oxidation rates, though finite due to exhaust temperatures of 460°C, are one order of magnitude lower than oxidation rates due to NO_2. Because of the exhaust temperatures of 460°C and the NO_2 concentrations, oxidation rates in layer I and in the wall are higher in CCRT® configurations compared to CPF-only configuration. After about 3 hours the total PM oxidation rates asymptotically reach the CPF-inlet PM rates, showing the beneficial effect of NO_2 at high temperatures. Oxidation rates in the wall are one order of magnitude lower than in the PM cake layer because of the smaller mass of PM present in the wall. Further, oxidation rates in the wall reduce after reaching a peak because the developing PM cake layer filters about 98-99% of the particles and only about 1-2% enters the filter wall.

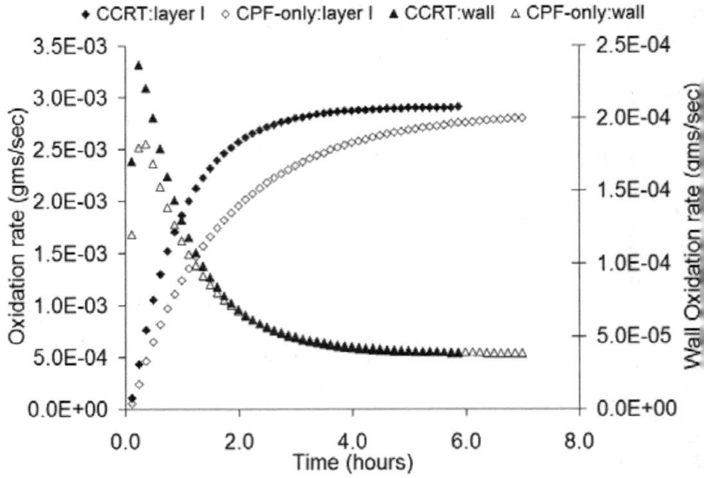

Figure 16: 75% Load: comparison of PM oxidation rate by location

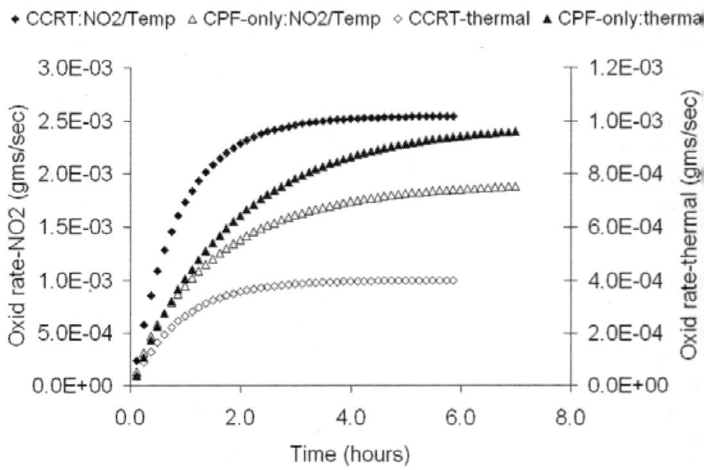

Figure 17: 75% Load: comparison of PM oxidation rate by type

CRT® SIMULATION RESULTS In this section, model results from the simulation of a Johnson Matthey CRT® are described. The engine used for this project was a John Deere 8.1 liter 175 kW 2200 rpm, 6-cylinder inline, turbocharged and after-cooled with a high-pressure common rail fuel injection system and an electronic control unit with variable injection timing. For the DOC experiments no EGR was used, but for the DPF and CR-DPF testing, a cooled low pressure loop EGR strategy was used. For additional details of the test setup, instrumentation, experimental methods and measurements, and DOC and DPF modeling results confer references [9, 22].

Figure 18 shows the pressure drop vs. time characteristics of both the DPF and the DOC-DPF systems at the 2200 rpm 100% load condition, in which model and experimental results are compared. The effect of a higher NO_2/PM ratio (8.6) when using a DOC-DPF system instead a DPF alone (0.3) can be observed. In this case the DOC supplied 155 ppm of NO_2 to the DPF, compared against 10 ppm for the case without the aid of a

DOC. Even when the NO_X to PM ratios are similar, the NO_2/PM ratios clearly make the difference in the pressure drop characteristics of both systems under similar conditions [9, 22].

Figure 19 shows similar results as in Figure 18, but for the 1400 rpm 100% load condition for both DPF and DOC-DPF systems. In this case, the amount of NO_2 in either condition is not significant (¡10ppm), therefore the difference in pressure drop observed between the two loading experiments was mainly due to the difference in temperature (543°C vs. 522°C) [9, 22]. Simulation results from the use of the MTU model at other loads and speeds can be found in References [9, 22].

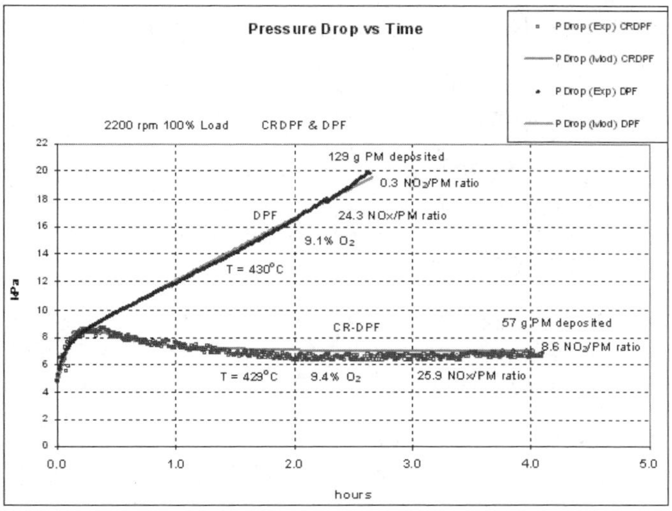

Figure 18: DPF pressure drop comparison between DPF and DOC-DPF operation at 2200 rpm 100% load

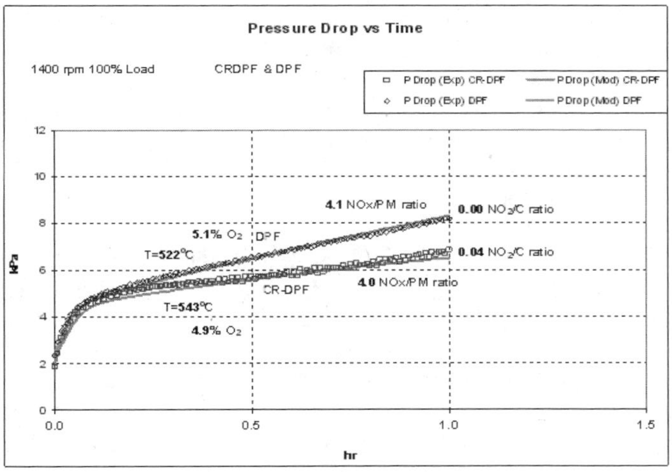

Figure 19: DPF pressure drop comparison between DPF and DOC-DPF operation at 1400 rpm 100% load

SPOT HEATING SIMULATION RESULTS This section discusses sample results from the study of the regeneration model with heat addition (energy generation) inside the filter walls. This energy source was activated only for selected spots along the filter wall length, which represent approximately 1-5% of the total filter wall length. The parameters involved in this study were, spot size, heating time, spot location and power level for the heat generation source. For additional details refer to the thesis of Triana [9].

In the heat addition regeneration study, simulation runs were performed to determine the conditions on which complete regeneration could be achieved in the shortest period of time. As shown in Figure 20, 26 out of the original 29 grams were oxidized during the 4-minute regeneration simulation.

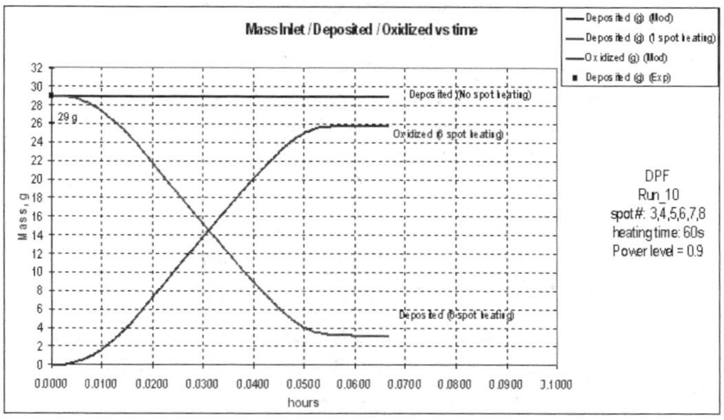

Figure 20: Solid PM mass evolution during spot heating regeneration

Figure 21 shows the PM layer thickness evolution at three different times during the 4-minute regeneration simulation. At 61 seconds, 18% of the PM mass deposited has been oxidized, at 120 seconds, 56% and at 240 seconds, 90%. Figure 21 shows the oxygen consumption during the fast regeneration phase at different times. The plot shows O_2 concentration in the y-axis and spot number in the x-axis. The axial length domain of the wall was discretized into 123 segments. The spots heated were: spots number 3,4,5,6,7, and 8, where spot number 1 is the closest to the front, and spot number 123 is the closest to the end of the filter.

As shown in Figure 22, O_2 is completely depleted at the spot locations where the solid PM is being oxidized due to the high temperature generated during the localized spot heating. As the energy generated is transported to the other segments of the channel wall, the O_2 continues to be depleted as it can be seen by the 10, 20, 30, 60, and 120 s O_2 concentration curves. Another approach was to further increase the number of spots from 6 to 8, and reduce the heating time from 60 to 45 s consuming the same amount of energy input from the heat source. A slight improvement was observed from the previous case. At 61 s, 25% of the PM mass deposited has been oxidized, at 120 s, 60%, and at 240 s, 91%. More detailed results can be found in Reference [9].

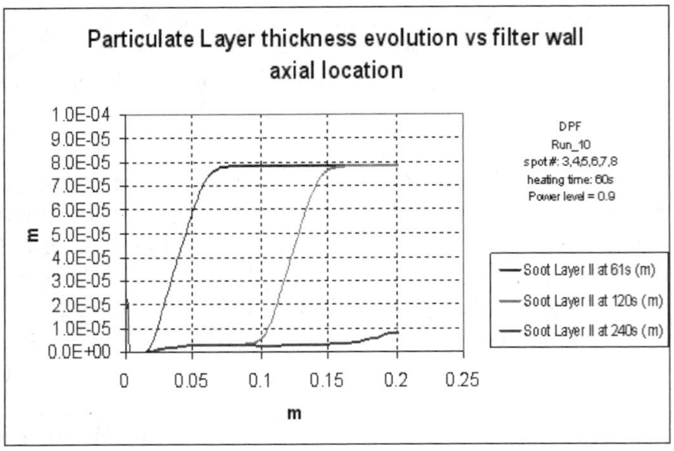

Figure 21: Evolution of solid PM layer thickness (m) along DPF length (m) during spot heating regeneration

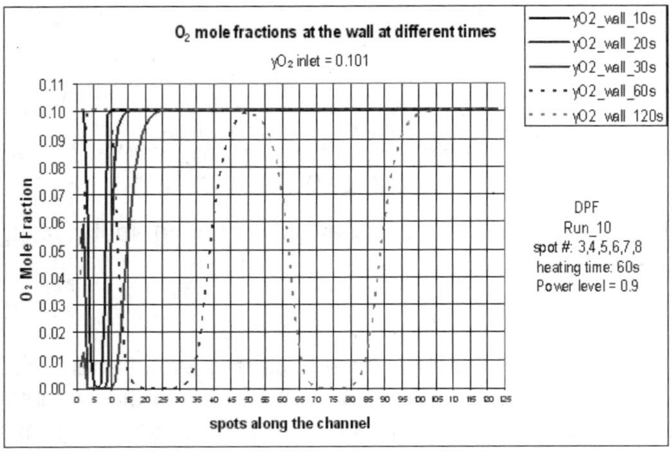

Figure 22: O_2 depletion during spot heating regeneration from front to end of channel

CONCLUSIONS

A single channel one dimensional model to simulate the filtration and PM oxidation performance of catalyzed and uncatalyzed diesel particulate filters was developed at Michigan Technological University. Several improvements were made to the CPF model and were described. The new wall oxidation model coupled to the PM cake layer filtration model shows that oxidation in the pores of the filter wall can explain the disproportionate decrease in the pressure drop across the filter with respect to PM mass. The model to take into account the NO_2 produced by the catalyst in the CPF helped model the higher PM oxidation rates in the CCRT® due to the presence of the catalyst in the CPF. Such models can help evaluate the effectiveness of the catalyst washcoat in the CPF in increasing the PM oxidation rates. The CPF model was also modified to include a heat addition source term in the energy balance of the filter wall. Using this sub-model, simulations to study the regeneration of filters by spot heating from an external source were studied. Th CPF model was integrated with a previously developed DOC model in Matlab/Simulink so that the filtration and oxidation performance of CRT® and CCRT® technologies or other DOC and CPF designs could be simulated.

ACKNOWLEDGMENTS

The authors would like to thanks Cummins Inc, John Deere Co., General Motors and the US Department of Energy for sponsoring this research. The authors would also like to thank Kiran Premchand for sharing his improvements to the PM cake filtration model prior to the publication of his thesis.

REFERENCES

[1] Diesel Net Online Resource. www.dieselnet.com.

[2] John S. Howitt and Max R. Montierth. Cellular Ceramic Diesel Particulate Filter. *SAE:810114*, 1981.

[3] Edward Bisset. Mathematical Model of the Thermal Regeneration of a Wall-Flow Monolith Diesel Particulate Filter. *Chemical Engineering Science*, 39:1232–1244, 1983.

[4] A.G. Konstandopoulos and M. Kostoglou. Periodically Reversed Flow Regeneration of Diesel Particulate Traps. *SAE:1999-01-0469*, 1999.

[5] Athanasios G. Konstandopoulos and M. Kostoglou. Reciprocating Flow Regeneration of Soot Filters. *Combustion and Flame*, 121:488–500, 2000.

[6] Athanasios G. Konstandopoulos, M. Kostoglou, E. Skaperdas, E. Papaionnou, D. Zarvalis, and Evdoxia Kladopoulou. Fundamental Studies of Diesel Particulate Filters: Transient Loading, Regeneration and Aging. *SAE:2000-01-1016*, 2000.

[7] I.P. Kandylas, O.A. Haralampous, and G.C. Koltsakis. Diesel Soot Oxidation with NO_2: Engine Experiments and Simulations. *Ind. Eng. Chem. Res.*, 41:5372–5384, 2002.

[8] I.P. Kandylas and G.C. Koltsakis. NO_2-Assisted Regeneration of Diesel Particulate Filters: A Modeling Study. *Ind. Eng. Chem. Res.*, 41:2115–2123, 2002.

[9] Antonio P. Triana. *Development of Models to Study the Emissions, Flow, Kinetic Characteristics from a Diesel Oxidation Catalyst and Particulate Filter*. PhD thesis, Michigan Technological University, 2005.

[10] T.C. Hunyh, J.H. Johnson, S.L. Yang, S.T. Bagley, and James R. Warner. A One Dimensional Computational Model for Studying the Filtration and Regeneration Characteristics of a Catalyzed Wall Flow Diesel Particulate Filter. *SAE:2003-01-0841*, 2003.

[11] Cuong T. Hunyh. A Study of the Filtration and Regeneration Characteristics of a Catalyzed Wall Flow Diesel Particulate Filter: One Dimensional Model Calibrated and Validated with Experimental Data.

Master's thesis, Michigan Technological University, 2002.

[12] Anand S. Shende. A Study of the Filtration and Particulate Matter Oxidation Characteristics of Two Catalyzed Wall Flow Diesel Particulate Filters: Experimental and 1-D 2-layer Model Results. Master's thesis, Michigan Technological University, 2003.

[13] Anand S. Shende, John H. Johnson, Song L. Yang, Susan T. Bagley, and Abishek M. Thalagavara. The Filtration and Particulate Matter Oxidation Characteristics of a Catalyzed Wall-Flow Diesel Particulate Filter: Experimental and 1-D 2-Layer Model Results. *SAE*, 2005-01-0949.

[14] Mohammed Hasan. The Filtration and Oxidation Characteristics of a Diesel Oxidation Catalyst and a Catalyzed Particulate Filter: Development of a 1-D 2-Layer Model. Master's thesis, Michigan Technological University, 2005.

[15] Hasan Mohammed, Venkata R. Lakkireddy, John H. Johnson, and Susan T. Bagley. An Experimental and Modeling Study of a Diesel Oxidation Catalyst and a Catalyzed Diesel Particulate Filter Using a 1-D 2-Layer Model. *2006-01-0466*, 2006.

[16] Edward Bisset and Farhang Shadman. Thermal Regeneration of Diesel Particulate Monolithic Filters. *A.I.Ch.E*, 31:753–758, 1984.

[17] Athanasios G. Konstandopoulos and John H. Johnson. Wall-Flow Diesel Particulate Filters-Their Pressure Drop and Collection Efficiency. *SAE:890405*, 1989.

[18] Bernhard Peters. Numerical Simulation of Diesel Particulate Filter during Loading and Regeneration. *ASME Internal Combustion Engine Division*, ICES2003-584, 2003.

[19] G.C. Koltsakis and A.M. Stamatelos. Modes of Catalytic Regeneration in Diesel Particulate Filters. *Ind. Eng. Chem. Res.*, 41:2115–2123, 2002.

[20] F.P. Incropera and D.P. DeWitt. *Fundamentals of Heat and Mass Transfer*. John Wiley and Sons, Inc., 2002.

[21] I.P. Kandylas and G.C. Koltsakis. Simulation of Continuously Regenerating Diesel Particulate Filters in Transient Driving Cycles. *Proc. Inst. Mech. Eng. Part D J. Automob. Eng.*, 216(7):591–606, 2002.

[22] Antonio P. Triana, John H. Johnson, Song-Lin Yang, and Kirby J. Baumgard. An Experimental and Computational Study of the Pressure Drop and Regeneration Characteristics of a Diesel Oxidation Catalyst and a Particulate Filter. *SAE: 2006-01-0266*, 2006.

[23] A.F. Ahlstrom and C.I. Odenbrand. Combustion Characteristics of Soot Deposits from Diesel Engines. *Carbon Journal*, 27:475–483, 1989.

[24] F. Jacquot, V. Logie, J.F. Brilhac, and P. Gilot. Kinetics of the Oxidation of Carbon Black by NO_2: Influence of the Presence of Water and Oxygen. *Carbon*, 40:335–343, 2002.

[25] B.A. Lur'e and A.V. Mikhno. Interaction of NO_2 with Soot. *Kinetics and Catalysis*, 38(4):335–343, 1997.

[26] Soren Andersson, Claes Akerlund, and Micael Blomquist. Low Pressure EGR Calibration Strategies for Reliable Diesel Particulate Filter Regeneration on HDD Engines. *SAE:2002-01-2780*, 2002.

[27] M.J. Murphy, L.J. Hillenbrand, D.A. Trayser, and J.H. Wasser. Assessment of Diesel Particulate Control-Direct and Catalytic Oxidation. *SAE:810112*, 1981.

[28] B.R. Stanmore, J.F. Brilhac, and P. Gilot. The Oxidation of Soot: A Review of Experiments, Mechanisms and Models. *Carbon*, 39:2247–2268, 2001.

[29] Athanasios G. Konstandopoulos, E. Skaperdas, and Mansour Masoudi. Inertial Contributions to the Pressure Drop of Diesel Particulate Filters. *SAE:2001-01-0909*, 2001.

[30] Athanasios G. Konstandopoulos. Flow Resistance Descriptors for Diesel Particulate Filters: Definitions, Measurements and Testing. *SAE:2003-01-0846*, 2003.

[31] Venkata R. Lakkireddy. The Effect of an Advanced Oxidation Catalytic Converter and a Catalyzed Particulate Filter on the Emissions from a Heavy Duty Diesel Engine. Master's thesis, Michigan Technological University, 2005.

[32] Venkata R Lakkireddy, Hasan Mohammed, John H Johnson, and Susan T Bagley. The Effect of a Diesel Oxidation Catalyst and a Catalyzed Particulate Filter on the Emissions from a Heavy Duty Diesel Engine. *2006-01-0875*, 2006.

[33] Markus Maly, Michael Claussen, Otto Carlowitz, Peter Kroner, Marco Ranalli, and Stefan Schmidt. Influence of Nitrogen Dioxide Based Regeneration on Soot Distribution. *SAE:2004-01-0823*, 2004.

[34] Athanasios G. Konstandopoulos and E. Skaperdas. Microstructural Properties of Soot Deposits in Diesel Particulate Traps. *SAE:2002-01-1015*, 2002.

[35] Kiran C Premchand. . Master's thesis, Michigan Technological University, 2006.

[36] Kyoo Won Lee and James A. Gieseke. Collection of Aerosol Particles by Packed Beds. *Environmental Science and Technology*, 13(4):466–470, 1979.

[37] Kenneth Wark, Cecil.F.Warner, and Wayne.T.Davis. *Air Pollution - Its Origin and Control*. Addison - Wesley Publishing Company, 2002.

[38] Barry J. Cooper and James E. Thoss. Role of NO in Diesel Particulate Emission Control. *SAE:890404*, 1989.

[39] Andrew P. Walker, Ronny Allansson, Phil G. Blakeman, Barry J. Cooper, Howard Hess, and Peter J. Silcock. Optimizing the Low Temperature Performance and Regeneration Efficiency of the Continuously Regenerating Diesel Particulate Filter (CR-DPF) System. *SAE:2002-01-0428*, 2002.

[40] Ronny Allansson, Andrew P. Walker, Claus Goersmann, Mats Lavenius, Paul Phillips, and Ari J. Uusimaki. The Development and In-Field Performance of Highly Durable Particulate Control Systems. *SAE:2004-01-0072*, 2004.

[41] Agus Setiabudi, Barry A.A.L. van Setten, Michiel Makkee, and Jacob A. Moulijn. The Influence of NO_X on Soot Oxidation Rate: Molten Salt versus Platinum. *Applied Catalysis B: Environmental*, 35:159–166, 2002.

[42] Agus Setiabudi, Michiel Makkee, and Jacob A. Moulijn. An Optimal NO_X Assisted Abatement of Diesel Soot in an Advanced Catalytic Filter Design. *Applied Catalysis B: Environmental*, 42:35–45, 2003.

[43] Antonio P. Triana, John H. Johnson, Song L. Yang, and Kirby J. Baumgard. An Experimental and Numerical Study of the Performance Characteristics of a Diesel Oxidation Catalyst in a Continuously Regenerating Particulate Filter. *SAE:2003-01-3176*, 2003.

[44] Antonio P. Triana, John H. Johnson, and Song-Lin Yang. A Simulation Study of Diesel Particulate Filter Regeneration Initiated by Spot Heating. *SAE 06FL-67*, 2006.

[45] K.W. Lee, L.D. Reed, and J.A. Gieseke. Pressure Drop Across Packed Beds in the Low Knudsen Regime. *Journal of Aerosol Science*, 9:557–565, 1978.

APPENDIX A

PARTICLE SIZE CALCULATION BY MODEL

To understand the deposition and filtration mechanisms occurring in a particulate filter it is necessary to understand how the inlet PM mass is calculated from the particle number distribution, the solid PM matter concentration, the exhaust volumetric flow rate, and the time of loading. In terms of a PM density the inlet mass coming into the filter in a time Δt, is defined as:

$$Mass\_Inlet = \sum_{d_p=1}^{d_p=last} density.V_{act}.\Delta t.partvol(d_p).distup(d_p) \tag{74}$$

where d_p is the particle size diameter, 'density' represents the density of all the particles, assuming that all particle sizes have the same density (particle mass / particle volume), partvol(d_p) is the volume of each particle diameter (particle volume/ particle), distup(d_p) is the particle number distribution (# of particles / actual volume of exhaust), actual V_{act} is the actual exhaust volumetric flow rate at the inlet (actual volume of exhaust / unit time), and Δ t is the discretized time step. The density of each particle is calculated as:

$$density = \frac{C_{in,act}}{totalvol} \tag{75}$$

where $C_{in,act}$ is the inlet solid PM matter concentration (mass of solid PM / actual volume of exhaust) and totalvol is the integration of the entire range of volume distribution (particle volume / actual volume of exhaust), which is then computed as follows:

$$totalvol = \sum_{d_p=1}^{d_p=last} partvol(d_p).distup(d_p) \tag{76}$$

The volume of each particle diameter is calculated assuming the particles are in form of spheres:

$$partvol(d_p) = \frac{4}{3}\pi\left(\frac{d_p}{2}\right)^2 \tag{77}$$

The volume distribution is computed as the product of the volume of each particle times the number of particles at each size diameter, or partvol(dp) * distup(dp). The particle size number distribution, distup(d_p), is measured using a Scanning Mobility Particle Sizer (SMPS) or other instruments. Since the units from the instrument are reported in standard conditions, these are converted to actual conditions in the numerical code. The conversion from standard to actual conditions is performed by assuming ideal gas law to convert the solid PM matter concentration and exhaust volumetric flow rate from standard to actual conditions.

$$\rho_{act} = \frac{P_{act}}{RT_{act}} \tag{78}$$

$$\rho_{std} = \frac{P_{std}}{RT_{std}} \tag{79}$$

$$C_{in,act} = C_{in,std}\frac{\rho_{act}}{\rho_{std}} \tag{80}$$

$$V_{act}^{\cdot} = V_{std}^{\cdot}\frac{\rho_{std}}{\rho_{act}} \tag{81}$$

where P_{std} = 101.3 kPa, T_{std} = 25°C, and R is the exhaust gas constant J/kg-K, calculated from a mass fraction summation of the species present.

$$M_{exh} = \sum_{i=1}^{i=last} y_i M_i \tag{82}$$

Once the incoming PM mass has been defined using equation 74, it is deposited in the DPF wall and PM cake layers. In the filter wall, the particles are filtered by a model which is described in Appendix B.

APPENDIX B

WALL FILTRATION MODEL

In this section a review of model to simulate particle filtration by the wall which was developed by Konstandopoulos et.al [6, 17] is presented.

PM matter filtration in filters made of extruded ceramics like cordierite can be modeled by the theory of filtration of particles by packed beds [17]. According to this view, the granular micro-structure of the ceramic filter wall can be represented by spherical unit cells, each of which contain a unit collector which filters the particles present in the exhaust gas [6, 36]. In formulating the theory of filtration by packed beds, the presence of particles on the fluid flow is neglected, and it is assumed that the individual collectors are so closely packed that the mutual interference effects of neighboring collectors on the flow field have to be considered [36, 45]. The flow field model is the Kuwabara-Happel flow, which is a solution of creeping flow in a system of spheres based on the unit cell model.

By drawing an imaginary spherical boundary at a certain arbitrary distance around the unit collector, the volume fraction α (solidity fraction), related to clean porosity (ϵ_0) and the unit cell diameter (b) by [36]:

$$\alpha = 1 - \epsilon_0 = \left(\frac{d_{c0}}{b}\right)^3 \tag{83}$$

ϵ_0 is the 'clean' filter wall porosity, and d_{c0} is the diameter of the 'clean' spherical unit collector.

The diameter of the clean unit collector, d_{c0}, is related to the filter wall mean pore size, d_{pore}, and filter wall clean porosity, ϵ_0, by [6, 17, 36]:

$$d_{c0} = 1.5\left(\frac{1-\epsilon_0}{\epsilon_0}\right) d_{pore} \tag{84}$$

d_{pore} is the mean pore size of the filter wall. The relation is obtained by assuming that all the void fraction is distributed over cylindrical pores of diameter d_{pore} and that the external area of the collectors matches the area of the surface of these pores [6, 36].

Small particles at low flow rates are primarily collected by Brownian diffusion [36]. As particle size increases, direct interception and gravitational settling become increasingly important. If particle size and flow rates are further increased, particle collection by inertial impaction eventually dominates for particle sizes larger than $1\mu m$ [36]. In the current model, only diffusional and direct interception collection mechanisms have been considered, given that diesel exhaust particles are generally less than $1\mu m$ in size.

The single collector collection efficiency due to diffusional deposition is defined as the ratio of the rate at which particles diffuse to the sphere surface to that at which particles approach a surface with the cross-sectional area of the sphere [6, 36]. In the Kuwabara-Happel flow field this is given by:

$$\eta_D \approx 3.5\left(\frac{\epsilon}{K}\right)^{\frac{1}{3}} Pe^{\frac{-2}{3}} \tag{85}$$

ϵ is the porosity of the filter wall containing PM ('loaded' porosity). K is Kuwabara's hydrodynamic factor given by [6, 36]:

$$K = 2 - \epsilon - \frac{9}{5}(1-\epsilon)^{\frac{1}{3}} - \frac{1}{5}(1-\epsilon)^2 \tag{86}$$

The Peclet number (Pe) is a dimensionless number quantifying the relative magnitudes of convective and diffusive deposition effects as a particle flows around a unit collector. It is given by [6, 36]:

$$Pe = \frac{v_i d_c}{D_p} \tag{87}$$

where, d_c is the diameter of the 'loaded' unit collector, v_i is the undisturbed flow velocity approaching the unit collector, which is related to the wall flow velocity, v_w, by:

$$v_i = \frac{v_w}{\epsilon} \tag{88}$$

D_p is the particle diffusion coefficient given by [6]:

$$D_p = \frac{k_B T C}{3\pi\mu d_p} \tag{89}$$

where k_B is Boltzmann's constant, T is the absolute temperature of the gas, μ is the dynamic viscosity of the exhaust gas and d_p is the size of the diesel particle. C is the Stokes-Cunningham slip correction factor employed to take into account slip conditions present when gas flows through small passages of low porosity media. Slip flow conditions become important when the mean free path of the gas is equal to or higher than the mean pore diameter of the filter wall. It is related to the local Knudsen number by [6]:

$$C = 1 + Kn\left(1.257 + 0.4e^{\frac{-1.1}{Kn}}\right) \tag{90}$$

with the Knudsen number (Kn) defined as,

$$Kn = \frac{2\lambda}{d_{pore}} \tag{91}$$

where λ is the mean free path of the gas molecules, and d_{pore} is the mean pore size of the filter wall,

$$\lambda = \frac{\mu}{P}\sqrt{\frac{\pi \overline{R} T}{2M_{ex}}} \tag{92}$$

In equation 92, P is the absolute pressure of the exhaust gas, T is the absolute temperature of the gas, $\overline{R}$ is the universal gas constant and M_{exh} is the molecular weight of the exhaust gas.

The efficiency of particle collection by a unit collector in a packed bed due to interception, taking into account the interference effect due to surrounding collectors, is given by [6, 36, 45]:

$$\eta_R = 3.5\left(\frac{\epsilon}{K}\right)\frac{R}{(1+R^m)} \tag{93}$$

where R is the particle interception parameter given by (ratio of diameters) [36, 45],

$$R = \frac{d_p}{d_c} \tag{94}$$

and the exponent m is given by [36, 45],

$$m = \frac{3-2\epsilon}{3\epsilon} \tag{95}$$

Assuming that the diffusional and interception mechanisms are independent of each other, the overall collection efficiency of a single collector can then be given as [6, 36]:

$$\eta_{DR} = \eta_D + \eta_R(1-\eta_D) = \eta_D + \eta_R - \eta_D\eta_R \tag{96}$$

Equations (85), (93) and (96) represent the diffusional, interception and total collection efficiency of a single collector. The total particle collection efficiency of the packed bed/filter wall of thickness Δx is related to the total collection efficiency of a single sphere by [6, 36]:

$$E = 1 - exp\left(\frac{-3\,(1-\epsilon)\,\eta_{DR}\Delta x}{2\epsilon d_c}\right) \tag{97}$$

For numerical purposes, equation 97 is not applied to the entire thickness of the filter wall at once. Rather, the thickness of the filter wall is divided into a number of layers called 'slabs'of thickness Δx, and the total filtration efficiency is related to the filtration efficiencies of the 'slabs', as shown in Figure 2.

The exhaust gas is filtered by the wall, according to the local filtration efficiency, given by equation (97) and which causes the local filter wall properties to change. It is assumed that the mass of the particles collected by the filter wall forms a uniform layer around the unit collector causing it to grow in size, but keeping its spherical shape intact. This is done by defining a 'density' of PM in the wall, ρ_{pw} [6]. The total diameter of the PM loaded spherical unit collector can be easily derived by considering the increase in volume of a unit collector by mass m_t packed uniformly around it with density ρ_{pw} [6]:

$$d_c = \left[\left(\frac{d_{c0}}{2}\right)^3 + \frac{3}{4\pi}.\frac{m_t}{\rho_{pw}}\right]^{\frac{1}{3}} \tag{98}$$

where, d_c is the diameter of the 'loaded' unit collector at time t, m_t is the mass around the unit collector at time t and ρ_{pw} is the packing density of PM in the filter wall. Thus it can be seen that ρ_{pw} is not a real PM packing density, but rather a means by which PM filtered by the collector is uniformly distributed around the collector, in order to keep its spherical shape intact for theoretical purposes. This is an important observation which will later be used during the development of the wall oxidation model.

It is easy to show using Equations (83) and (98) that the 'loaded' porosity of the filter wall, ϵ, changes according to the equation:

$$\epsilon_t = 1 - \left(\frac{d_c}{d_{c0}}\right).(1-\epsilon_0) \tag{99}$$

With the 'loaded' unit collector' diameter and 'loaded' porosity known from equations 98 and 99 respectively, the loaded mean pore size of the wall can be computed by rearranging the terms present in equation 83, to obtain:

$$d_{pore} = \frac{2}{3}\left(\frac{\epsilon_t}{1-\epsilon_t}\right)d_c \tag{100}$$

If k_0 is the permeability of the clean filter wall, then the permeability of the 'loaded' filter wall, k_t, is given by [6, 36]:

$$\frac{k_t}{k_0} = \left(\frac{d_{c,t}}{d_{c0}}\right)^2.\left(\frac{K(\epsilon_t)}{K(\epsilon_0)}\right).\left(\frac{1-\epsilon_0}{1-\epsilon_t}\right) \tag{101}$$

K is the Kuwabara hydrodynamic factor defined in equation 86. The term $\frac{K(\epsilon_t)}{K(\epsilon_0)}$ controls the rapidly decreasing permeability of the filter wall during the depth filtration phase due to a decrease in porosity of the filter wall.

During the depth filtration phase, most of the particles are collected inside the filter wall. Eventually, most of the particles are collected by the PM cake layer. On the filter wall, the PM is not dispersed as it is inside the wall and forms a cake like structure. A partition coefficient $\phi(0 \leq \phi \leq 1)$ has been defined to determine the fraction of total upstream filter particles entering the filter wall. It is thus used as a filtration parameter for the PM cake layer itself. The definition can be expressed mathematically as:

$$\phi_t = \frac{d_{c,t} - d_{c0}}{\psi.b^2 - d_{c0}^2} \tag{102}$$

where, $\psi(0 < \psi < 1)$ is a dimensionless 'percolation' factor which determines how close the diameter of a loaded collector can approach that of a unit cell before filtration is completely by the PM cake layer. The percolation factor is used as a control variable to determine the amount of loading (time length of depth-filtration phase) inside the filter wall. A deficiency of this model is that the partition coefficient, which is based on filter wall parameters, is used as a filtration parameter for the PM cake layer even when the PM cake layer itself becomes a filter for the particles.

As explained earlier, filtration in the wall changes the properties of the filter wall along its thickness, and hence the thickness of the filter wall is divided into a number of 'slabs', and the resulting computational domain for the filter wall is shown in Figure 2. In that Figure Δ x represents the thickness of a slab (layer) in the normal direction to the filter wall and Δx the discretized length of the channel in the axial direction. Further, Equation 98 defines the evolution of the unit collector diameter with PM mass deposited on each collector m_t. Since many collectors can be present in each of the discretized cells in Figure 2, the deposited PM mass m_t in each collector is defined as the mass deposited in each discretized cell in the porous wall divided by the number of unit collectors.

$$m_t = \frac{slab\_mass}{num_c ollector} \tag{103}$$

The number of unit collectors is determined as:

$$num\_collector = \frac{\Delta x \Delta y a\,(1-\epsilon_0)}{collector\_vol} \tag{104}$$

where, a is the channel cell size, and collector_vol is the volume of a unit spherical collector, computed as follows:

$$collector\_vol = \frac{4}{3}\pi \left(\frac{d_{c0}}{2}\right)^3 \tag{105}$$

The mass deposited in the each discretized slab of thickness Δy is computed as follows:

$$slab\_mass = \sum_{d_p=1}^{d_p=last} \frac{E.distlay(d_p).density.partvol(d_p).V_{act}\Delta t}{4.N_{cell}.\,(j_{max}-1)} \tag{106}$$

where E is the unit collector efficiency (Equation 97), N_{cell} is the number of open inlet channels in the DPF, j_{max} is the maximum number of divisions into which the axial length has been discretized, and distlay(d_p) is defined in two different forms depending on which slab is considered. For the first slab, the particle number distribution is defined as:

$$distlay(first, d_p) = (1-\phi_t) * distup(d_p) \tag{107}$$

$$distlay_(first, d_p) = (1-E_{cake}) * distup(d_p) \tag{108}$$

Equation 107 is used if only the wall filtration model is used and in this case ϕ_t is the partition coefficient defined by Equation 102. Equation 108 is used if the PM cake filtration model is used along with the wall filtration model, and in this case E_{cake} is the filtration efficiency of the cake defined by Equation 47.

distlay$_($first, $d_p)$ represents the number of particles that have passed through the first filter wall layer (slab). As particles are filtered by the slabs, the number of particles entering the next slab decreases based on an increase in the collection efficiency of the previous slab:

$$distlay(j, d_p) = (1-E) * distlay(j-1, d_p) \tag{109}$$

where, distlay$_(j, d_p)$ represents the number of particles that have passed through the filter wall layer (slab) represented by (j-1).

And at the end of the filter wall (last slab), the downstream particle number distribution is computed by using the solid PM mass of the last slab and the last slab collection efficiency as:

$$distdown(d_p) = (1-E_{last\_slab,d_p}) * distlay(last\_slab, d_p) \tag{110}$$

With the downstream particle number distribution known, the PM matter concentration exiting the DPF over a time Δt can be computed as:

$$Mass\_Outlet = \sum_{k=1}^{k=jmax}\sum_{d_p=1}^{d_p=last} . \frac{density.partvol(d_p).distdown(d_p).V_{act}\Delta t}{4.N_{cell}.\,(j_{max}-1)} \tag{111}$$

From the inlet and outlet solid PM masses, a total PM mass filtration efficiency can be computed as:

$$\eta_{filtration} = \frac{Mass\_Inlet - Mass\_Outlet}{Mass\_Inlet} \tag{112}$$

Knowing the total filtration efficiency, the downstream CPF PM matter concentrations in standard units can be computed as follows:

$$C_{out,std} = C_{in,std}(1-\eta filtration) \tag{113}$$

2007-01-1127

Control Strategies for Peak Temperature Limitation in DPF Regeneration Supported by Validated Modeling

G. C. Koltsakis, O. A. Haralampous and Z. C. Samaras
Aristotle University Thessaloniki

L. Kraemer, F. Heimlich and K. Behnk
IAV GmbH

ABSTRACT

One of the main challenges in developing cost-effective diesel particulate filters is to guarantee a thermally safe regeneration under all possible conditions on the road. Uncontrolled regenerations occur when the soot reaction rate is so high that the cooling effect of the incoming exhaust gas is insufficient to keep the temperature below the required limit for material integrity. These conditions occur when the engine switches to idle while the filter is already hot enough to initiate soot oxidation, typically following engine operation at high torque and speed or active filter regeneration.

The purpose of this work is to investigate engine management techniques to reduce the reaction rate during typical failure mode regenerations. A purely experimental investigation faces many difficulties, especially regarding measurement accuracy, repeatability in filter soot loading, and repeatability in the regeneration protocol. Therefore, the present study is guided by advanced regeneration modeling, which is supported for calibration and validation purposes by careful experimental measurements on the engine bench.

The study covers both uncoated and catalyzed filters. The model used is a previously developed and validated 3-dimensional regeneration model including heterogeneous catalytic reactions for all types of coatings, including NOx storage functionality. The engine management techniques for peak temperature limitation involve engine speed, air path and injection control during the idle mode. Experiments are performed with and without protection measures to verify the model agreement in both conditions. In a next step, the validated model is used to study the acceptable soot mass limit for various calibrations of control measures.

INTRODUCTION

In addition to backpressure constraints, the thermal durability of the Catalyzed Diesel Particulate Filter (CDPF) determines the maximum amount of soot, that can be trapped during driving, and, hence, the regeneration interval. The maximum temperature during filter regeneration depends on a number of parameters comprising the physical exhaust gas conditions (oxygen content, temperature, mass flow etc.) and the filter properties (e.g. length/diameter ratio, position, soot mass loading).

If all DPF regeneration mode parameters are kept constant, the filter peak temperature increases with the soot loading due to exothermal soot oxidation. The integrated catalyst oxidizes the carbon monoxide (CO) produced from the thermal soot oxidation, thus yielding an additional exotherm and reducing the so-called critical soot mass limit of the CDPF. As a consequence, the critical soot mass limit of the next generation DPF is expected to be lower compared to today's technologies, leading to reduced regeneration intervals and fuel consumption penalties.

Furthermore, the future emission standards in Europe (EU6, under discussion) and the US (Tier 2) require a drastic minimization of the NOx emissions. In addition to advanced combustion concepts, promising exhaust aftertreatment technologies are capable of reducing the engine-out NOx emissions between 25 and almost 90% depending on the chosen technology [1]. Whereas the selective catalytic reduction (SCR) of nitrogen oxides is the most efficient measure to reduce the NOx emissions and is expected to be the mid- to long-term solution for future emission standards compliance, the NOx storage catalyst (NSC) technology offers several advantages for the short- to medium-term application: The application of an NSC is independent on any additional reductant infrastructure and can be combined with a DPF as integrated 4-way catalyst as first developed by Toyota [2]. As the required washcoat amount of an NSC is substantially higher compared to an oxidation catalyst, the porosity of the DPF substrate has to be considerably higher compared to other DPF applications to avoid increasing filter back pressures and according fuel consumption penalties. The trade-off between wash coat loading and filter back pressure is optimized by advanced wash coat technologies [3,4] and the development of high porosity Cordierite filter substrates [5,6]

This paper presents a methodology that enables the determination of the critical soot mass limit of catalyzed particulate filters by advanced modeling in combination with experimental validation and enables a reduction of the peak temperatures during uncontrolled regeneration by advanced regeneration control. The demonstrated methodology comprises the following steps to calculate the soot mass limit as a function of all selected parameters:

1. Investigation of application boundaries for idle mode
2. Assessment of polynomial model of idle conditions using Design of Experiments
3. Function development of ECU software to detect critical regeneration conditions and to limit the regeneration progress
4. Calculation of regeneration peak temperatures at severe regeneration conditions by physical modeling
5. Model validation versus selected experiments

EXPERIMENTAL SET-UP

ENGINE TEST BENCH SET-UP

For the investigations hereunder, a 4 cylinder, common-rail Diesel engine of the 2 l class with 110 kW was used. The engine features turbo charging (with variable turbine geometry), charge-air-cooling, cooled exhaust gas recirculation (EGR) and variable swirl control.

The original engine electronic control unit (ECU) was exchanged by a micro auto box (MABx, dSpace), which acts as programmable real-time ECU, in combination with IAV FI^{2RE} (Flexible Injection and Ignition for Rapid Engineering, [7]) to fully control the engine and fuel injection parameters as shown in **Figure 1**. Access to it is realized via graphical user interface (GUI) implemented in the software ControlDesk. The connection to the engine is realized by using the conventional wiring harness. Since – in contrast to regular ECUs – the MABx does not feature the necessary amplifiers to operate the engine's injectors, the IAV FI^{2RE} control unit is added to the system. This control unit enables the user to realize up to five injections with variable timing and quantity (i.e. even rather unconventional injection events are possible). The advantage of that system is the comprehensive access to the very flexible HOCCOS (Homogenous Combustion Control Software) structure and thus, a very high degree of flexibility in implementing new or modifying existing structures in real-time according to the requirements. The utilized software HOCCOS is described in more detail in [8].

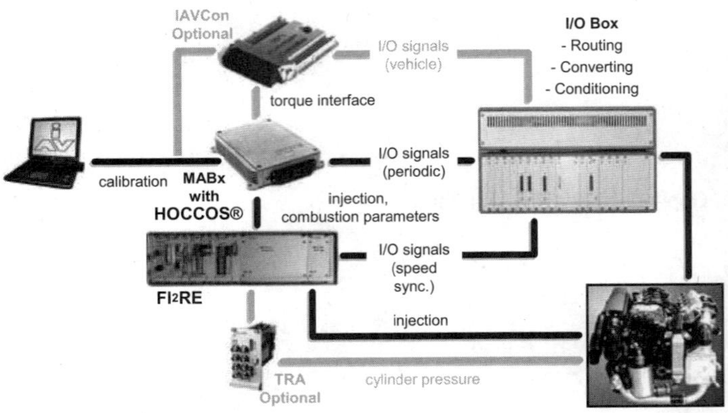

Figure 1: Control Hardware set-up.

The control unit FI^{2RE}, in turn, permits synchronization to different angle-sensing systems. The control unit ensures speed-synchronous control tasks, such as injector control or in-cylinder pressure evaluation. Moreover, it is capable of controlling both conventional solenoid and piezoelectric injection valves, permitting up to five injection events per cycle and cylinder. This control unit package MABx plus FI^{2RE} provides full functionality for process development on the test bench.

The engine test-bench is capable of highly-transient operation. It features an asynchronous dynamometer and state-of-the-art emissions measurement systems (including two HORIBA exhaust gas analyzers for undiluted gas (NO/NOx, CO/CO_2, HC, O_2).

The online monitoring of particulate emissions during various combustion modes depends on the various physico-chemical properties of the emitted soot. A detailed study on the nature of soot originated from rich combustion by applying several established soot measurement techniques has been performed during this study and is described in detail in [9]. It has been shown, that soot produced under rich conditions has different properties than soot produced under lean combustion conditions leading to an overestimation of the soot mass concentration when monitoring the soot emissions by photo acoustic spectroscopy or by black smoke measurement (FSN) compared to chemical filter analysis or laser induced incandescence. Therefore, all stated soot mass concentrations below are determined by gravimetric filter weighing (NOVA partial flow sampler) and laser induced incandescence (ESYTEC), that is described in detail in [10] and the weighing of the (C)DPF itself (see test procedures).

EXHAUST LINE SET-UP

The exhaust line comprises a close-coupled oxidation catalyst, a diesel particulate filter and sample probes/sensors as shown in Figure 2. The close-coupled oxidation catalyst has a volume of 0.7 l. The DPF (or combined DPF/NSC) with a PGM loading of 120 g/ft^3 is located about 1 m after the close-coupled catalyst. A Cordierite wall-flow filter from NGK with 300 cpsi and a volume of 2.4 l was utilized. The Cordierite filters used in this study have a porosity of 59% enabling a high amount of wash coat to be introduced at reasonable back pressure penalties.

In the frame of the present study two cordierite filters were used. The first one is uncoated whereas the second one is coated with a NO_x storage catalyst.

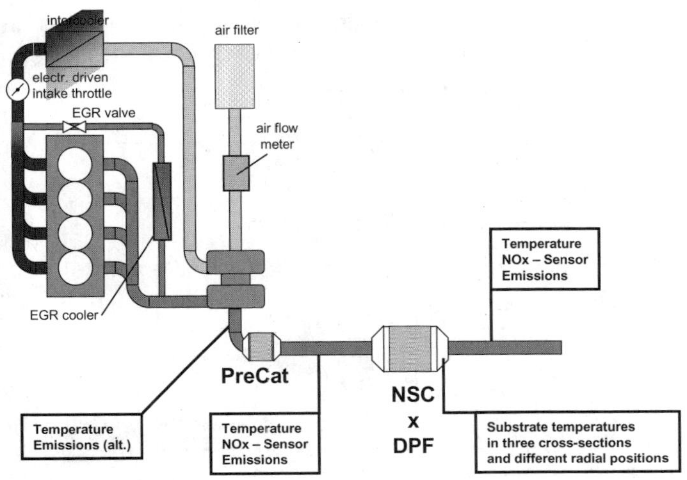

Figure 2: Schematic test bench set-up (standard engine sensors not shown).

To determine the maximum temperature inside the DPF and its location, the filter has been equipped with thermocouples (0.75 mm; NiCrNi). They are located in three radial and axial positions, in center, inlet and outlet positions. Furthermore, the High Speed Surface Temperature Visualization (HSTV) technique has been applied to directly determine the substrate temperature and validate the simulation analysis by thermo graphic measurement. A small window was installed behind the DPF to enable the recording of the surface temperature distribution of the outlet filter area by a thermo camera.

Any necessary bends of the exhaust line are kept as far away as possible from the filter inlet cone. In addition, the filter cone itself was optimized with a relatively small angle such that an even and homogenous flow into the filter can be assumed.

DESIGN OF EXPERIMENTS

The Design of Experiments (DoE) methodology was utilized to characterize the functional correlations of the varied parameters. The DoE method is described in detail in [11] and its implementation into the calibration process is presented in [12].

TEST PROCEDURES

A common particle filter weighing procedure has been established to minimize the measurement error: At first, the filters are regenerated at approx. 615°C (both NOx and soot) before conditioning in an oven at 200° C for one hour. Subsequently, the filters are weighed on a precise balance (uncertainty < 1/100 g) as a function of their temperature as determined by a thermocouple in the middle of the DPF utilizing a heat shield to minimize thermal buoyancy. For any uncoated particle filter, the resulting error margin is in the range of +/- 1 grams of soot. The NO_2 storage capacity of the catalyzed samples has been determined prior to these investigations. The distinction between weight gain due to soot and NOx storage was realized by determining the weight gain of the catalyzed filter by regenerating it and subsequent loading of it with NO_x but without soot (by putting a DPF in front of the 4-way catalyst) and subtracting this mass increase after the equivalent loading procedure with soot.

The DPF loading is done at stationary conditions with approx. 4 g soot/h at low load to keep the exhaust gas temperatures low enough to avoid passive regeneration. During the loading, the differential pressure is monitored to gain some information about the soot loading process and the soot distribution within the DPF. In defined intervals, the loading is stopped and the filter is conditioned and weighed until the desired soot loading is reached.

The regeneration tests are performed in a standardized way: The engine is run at full load for a selected operating point to provide both temperature and sufficient oxygen to start the regeneration. Once the temperature reaches 615°C upstream the DPF corresponding to 600°C in the middle of the filter and a significant differential pressure drop being observed, the engine is quickly (ramp time of 2 seconds) switched into idle. It is then constantly operated in the idle operating point.

CATALYZED DPF MODELING

A modified version of LAT 3-D DPF model [13] has been adopted. For the purpose of the present work, a 2-d axisymmetric approach was sufficient due to the cylindrical shape of the filter and the absence of cement layers. A novel feature of the present model compared to previous publications is the extended reaction scheme which accounts for NO_x storage/release functionality.

The basic equations are described in the Appendix, whereas the extended reaction scheme is presented in Table 1. The inhibition and equilibrium terms are explained in detail in [14].

Table 1: Reaction scheme

Soot reactions	**Rate**
$C+\alpha_1 O_2 \rightarrow 2\left(\alpha_1-\frac{1}{2}\right)CO_2+2(1-\alpha_1)CO$	$k\cdot\left(P_{O_2}\right)^n$
$C+\alpha_2 NO_2 \rightarrow \alpha_2 NO+(2-\alpha_2)CO+(\alpha_2-1)CO_2$	$k\cdot P_{NO_2}$
Catalyzed wall reaction	**Rate**
$CO+\frac{1}{2}O_2 \longrightarrow CO_2$	$\frac{k\cdot x_{CO}\cdot x_{O_2}}{G_2}$
$C_xH_y+\left(x+\frac{y}{4}\right)O_2 \longrightarrow xCO_2+\left(\frac{y}{2}\right)H_2O$	$\frac{k\cdot x_{C_xH_y}\cdot x_{O_2}}{G_1\cdot G_3}$
$NO+\frac{1}{2}O_2 \longrightarrow NO_2$	$\frac{k\cdot x_{NO}\cdot x_{O_2}}{G_4}Eq$
$NO_2 \rightarrow NO+\frac{1}{2}O_2$	$\frac{k\cdot x_{NO_2}}{G_4}Eq$
$BaO+2NO_2+\frac{1}{2}O_2 \rightarrow Ba(NO_3)_2$	$k\cdot c_{NO_2}c_{O_2}\Psi_{cap}\psi_{BaO}$
$Ba(NO_3)_2 \rightarrow BaO+2NO_2+\frac{1}{2}O_2$	$k\cdot\Psi_{cap}\psi_{Ba(NO_3)_2}$
$Ba(NO_3)_2+3CO \rightarrow BaO+2NO+3CO_2$	$k\cdot c_{CO}\Psi_{cap}\psi_{Ba(NO_3)_2}$
$Ba(NO_3)_2+H_2 \rightarrow BaO+2NO_2+H_2O$	$k\cdot c_{H_2}\Psi_{cap}\psi_{Ba(NO_3)_2}$

The extended reaction scheme of the model allows the calculation of both DPF and LNT related functions. As an example, **Figure 3** presents the calculated NO_x emissions downstream the LNT/DPF during a lean/rich cycle at 350° inlet temperature.

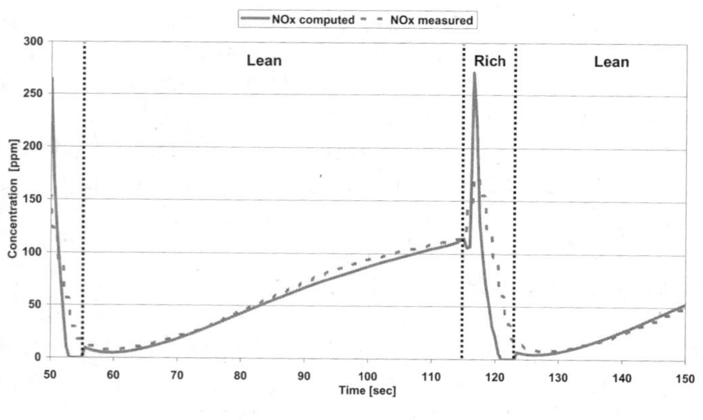

Figure 3: Measured vs simulated NO_x emissions during a lean/rich cycle at 350°C inlet temperature.

Equipped with a coupled transport-reaction scheme, such an improved model is in the position to calculate the complex interactions of the LNT functionality with passive soot regeneration. **Figure 4** shows the calculated NO_2 profiles through the soot layer and catalyzed wall at two different axial positions in the filter for two different instants corresponding to the start and the end of the lean (NO_x storage) phase of the experiment shown in **Figure 3**. As shown in the top graph of **Figure 4**, the NO_2 concentration approaching the soot surface is a function of the axial position due to the convective mass transfer occurring in the channel gas. The reaction of soot with NO_2 is sufficiently fast at this temperature to consume NO_2 in the soot phase. On the other hand, the catalytic coating promotes NO_2 formation and storage on the LNT component. Concurrently, the concentration gradient drives NO_2 back diffusion to the soot layer. It is interesting to note that analogous phenomena occur also at the end of the lean mode, as shown in the bottom graph of the same figure. The differences arise from the fact that the LNT component has been partially saturated, thus affecting the rate of NO_x storage which in turn affects the overall NO_2 profiles in the catalyzed wall and soot layer. A more detailed investigation of the soot/NO_x interactions in such systems is outside the scope of the present paper.

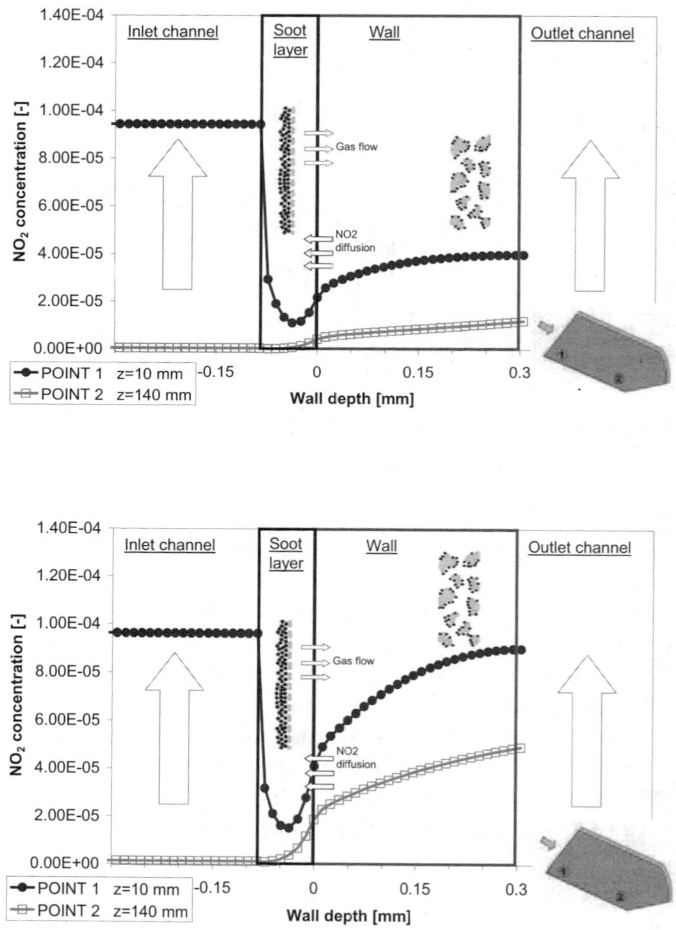

Figure 4: Calculated NO_2 profiles through the soot layer and filter wall at two different positions of the LNT/DPF. Top: At the beginning of the NO_x storage process (lean mode), Bottom: Just before rich mode

RESULTS AND DISCUSION

The most severe regeneration conditions that can occur during the full useful life of a DPF determine the critical soot mass limit. It is common knowledge, that the "unexpected" drop from thermal DPF regeneration (especially when passively initiated at full load running) to idling is the worst condition during DPF operation. The reason is, that during idle, the total exhaust gas flow is very low and the heat enthalpy from the exothermic soot oxidation can not be carried away by the flow media. Furthermore, the oxygen content of the idle flow is quite high compared to normal driving yielding an increase of the soot oxidation rate. To increase the critical soot mass limit, that is defined as the soot mass loading of the DPF that does not induce higher regeneration temperatures at the described worst case regeneration conditions than the DPF durability temperature maximum, the idle conditions have to be optimized with respect to minimized oxygen content and maximized exhaust mass flow. Of course, the DPF layout and composition plays also a major role, but is not within the scope of this work. The conditions which have a direct impact on the regeneration peak temperature are listed in **Figure 5** as well as the indirect impact on these conditions by the engine parameters.

Experimental investigations on the impact of these parameters have been first reported in [15]. It has been shown, that the regeneration peak temperatures could be lowered by several measures. Within the presented work an advanced methodology of the assessment of the critical soot mass limit of catalyzed and non-catalyzed DPF and its maximization by physical modeling and regeneration control is shown.

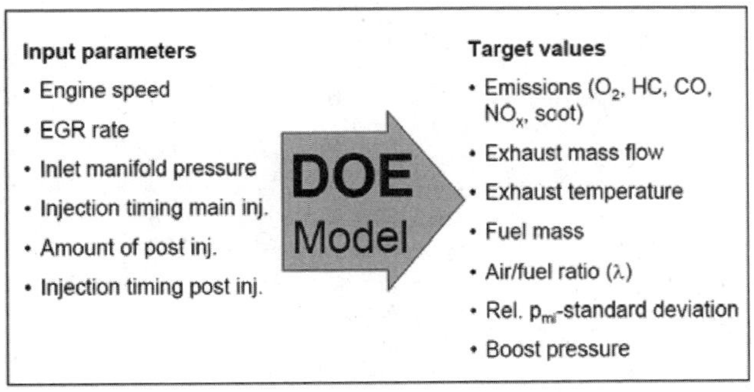

Figure 5: Schematic state diagram of the various operating modes

The DoE methodology was applied to determine a safe idle calibration for the engine during active DPF regeneration. The defined set of experiments was handled by fully controlled automatic engine operation as described in [16]. This procedure yields a DoE polynomial model, that predicts the exhaust gas properties during idle, based on engine operation parameters. The polynomial was then used to determine an idle calibration with minimum oxygen content at maximum flow rate, referred to as "Safe" mode. The exhaust gas properties at "Safe" mode are shown in **Table 2**. Also shown is a reference idle calibration defined as "Standard" mode, which was selected with respect to common serial applications. It should be mentioned that the engine speed of 1400 rpm was selected for demonstration purposes only and is too high for series calibration.

Table 2: Exhaust gas properties during idle

Idle mode		Standard	Safe
O_2	[%]	16.9	8.5
Exhaust flow rate	[kg/h]	45.4	30.3
Engine Speed	[rpm]	800	1400
THC	[ppm]	81	520
$T_{exhaust}$	[°C]	227	259

NON-CATALYZED FILTER

Figure 6 presents the results of the regeneration test at the standard idle calibration with a non-catalyzed filter and an initial soot loading of 5 g/l. The specific test was used to fine tune some of the model parameters regarding soot oxidation kinetics. The top graph presents the pressure drop during the test. The good agreement with the respective model result indicates a reliable prediction of the soot consumption during the full-power mode before switching to idle.

From the calculation, it can be derived that already 2 grams of soot are lost during the full-power mode before switching to idle.

The two graphs in the middle present the temperatures measured at selected positions in the filter in the axial and radial direction. As expected by many previous similar studies [13,17], the maximum temperature occurs near the symmetry axis and close to the exit of the filter. In the specific test the maximum temperature reaches 850°C. The use of 2-d modeling allows for a correct prediction of the temperature differences in the radial direction, which occur as a result of heat losses. The prediction of the regeneration rate differences in the radial direction is important since it affects the transient flow distribution in the complete filter [18].

The agreement between the measured and computed temperature field in the filter guarantees the correct regeneration prediction, a phenomenon which is affected by combined thermal, chemical and mass transfer effects. It may be noted that the remaining soot mass in the filter after the completion of the test is approximately 5 grams which translates to 60% regeneration efficiency.

The bottom graph presents the computed CO emission during the test. Unfortunately, CO measurement was not available for the specific test.

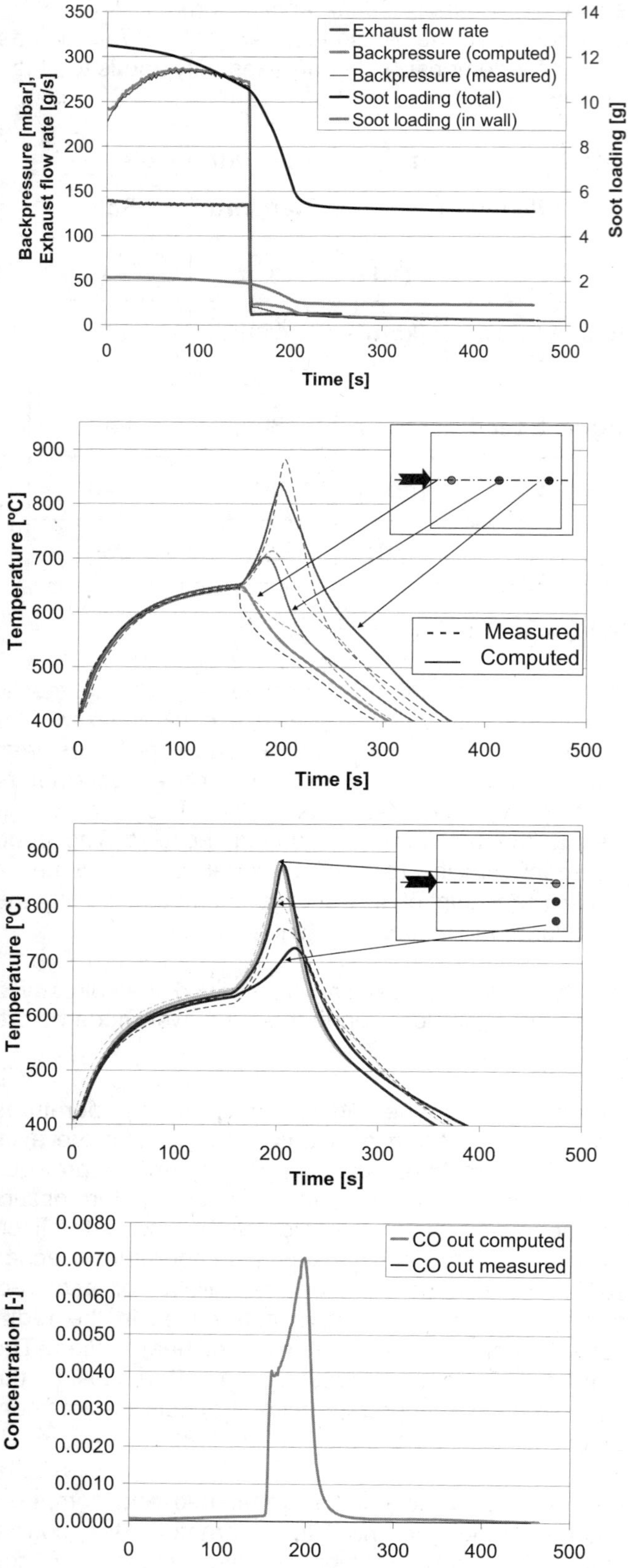

Figure 6: Experimental and simulation results with standard idle mode. Non-catalyzed filter with 5 g/l initial soot loading.

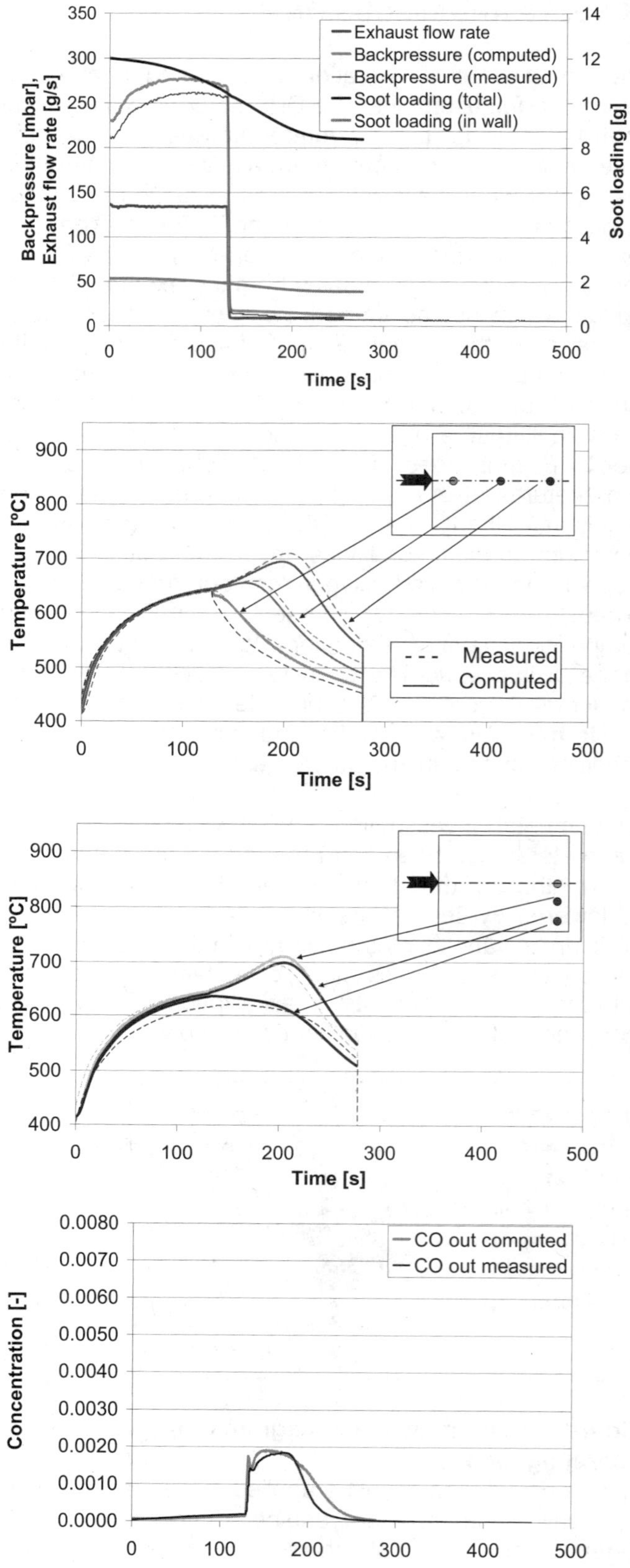

Figure 7: Experimental and simulation results with safe idle mode. Non-catalyzed filter with 5 g/l initial soot loading.

Figure 7 presents the respective measured and computed result for a similar experiment in which the idle mode is set in the "safe mode" with lower oxygen content. From the top graph, it can be seen that the model predicts a significant deceleration of the reaction compared to the standard idle mode, which is mostly due to the reaction rate dependency on O_2. The final predicted soot mass is around 8.5 grams instead of 5 grams in the standard mode.

As a consequence of the slower reaction rates the temperatures in the filter are risen less steeply allowing for a more effective cooling by the incoming exhaust gas. The maximum temperature in the "safe mode" is around 700°C recorded close to the filter exit, 150°C lower compared to the standard idle calibration. Having used the same model calibrations with the standard mode, the agreement regarding the time dependent temperature profiles in the filter is very satisfactory. A further indication of model validity is shown in the bottom graph of **Figure 7**, which shows the measured vs computed CO emission for the specific test.

The above results encourage the use of modeling at different initial conditions. As an example, **Figure 8** presents the model predicted maximum temperatures in the filter as function of initial soot loading for the case of "standard" and "safe mode" calibrations. Obviously, such kind of simulations are helpful to minimize the necessary experimental work and identify critical soot limits with minimum destructive testing. In addition the model can be used for further optimization of filter control strategies, which is currently a subject of on-going work.

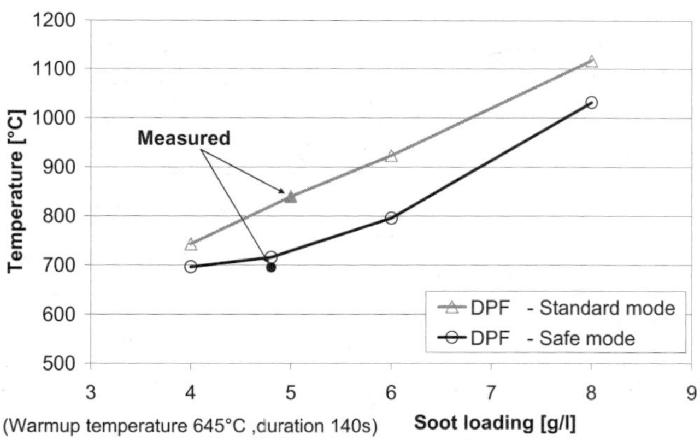

Figure 8: Prediction of maximum temperature in a non-catalyzed filter as function of initial soot loading. The results of the two experiments are also shown.

CATALYZED FILTER

Analogous investigations were performed with a catalyzed filter. The main differences expected in the performance of the two different systems are related to the following parameters:

- The catalyzed filter is expected to oxidize almost completely the CO produced during soot oxidation. This will increase the exothermy and the peak temperatures.
- The catalyzed filter due to a washcoat loading of 120 g/l has approximately 20% higher thermal mass compared to its non-catalyzed counterpart. The additional thermal mass will tend to decrease the peak temperatures.
- Although the catalytic coating does not enable a direct catalysis of soot oxidation, it may enhance the NO_2 based oxidation especially during the transitional operation preceding the idle switch. Consequently, less soot may be available at the time of switching to the critical idle conditions.
- Secondary effects are related to the additional pressure drop of the catalyzed filter, which may influence the flow distribution during regeneration, as well as NO_2 desorption phenomena occurring at elevated temperatures.

From the above, it can be realized that the prediction of the effect of the catalytic coating on the peak temperature during standard and safe mode is not trivial.

In the case of catalyzed filter, the estimation of initial soot loading is more challenging. This is due to the significant amount of NO_x chemisorbed during the soot loading process. For the specific tests this amount is estimated to be 1.5 g/l, which had to be subtracted from the weighted mass gain to obtain the initial soot mass. For the catalyzed filter measurements, the initial soot mass was 6 g/l.

Figure 9 presents the measured vs the computed results for the case of the standard idle mode calibration. In the top graph the calculated soot consumption reveals that approximately 4 grams are consumed during the full-power mode before switching to idle. Half of this soot amount corresponds to the soot trapped within the wall. In fact, the model predicts that the soot trapped in the wall is completely consumed before the idle switch. The soot oxidation within the catalyzed wall is favored due to NO_2 formation and "recycling". These phenomena are probably the most challenging to capture from the modeling point of view [18].

In general, it can be seen that the agreement is very satisfactory taking into consideration the very challenging operating conditions. The peak temperature for both measured and model results reaches 950°C, which may be already too high in terms of catalyst durability.

The two graphs in the bottom present the comparison with respect to CO_2 and O_2 concentrations measured downstream the filter, which are closely linked to the overall reaction rate during regeneration. In the case of the catalyzed filter the CO concentration downstream was negligible.

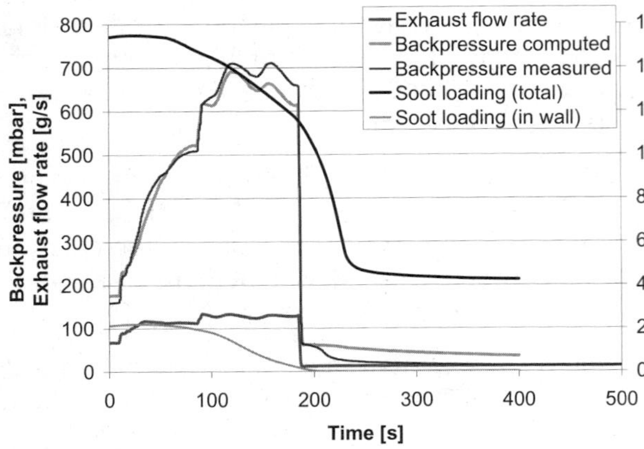

(a)

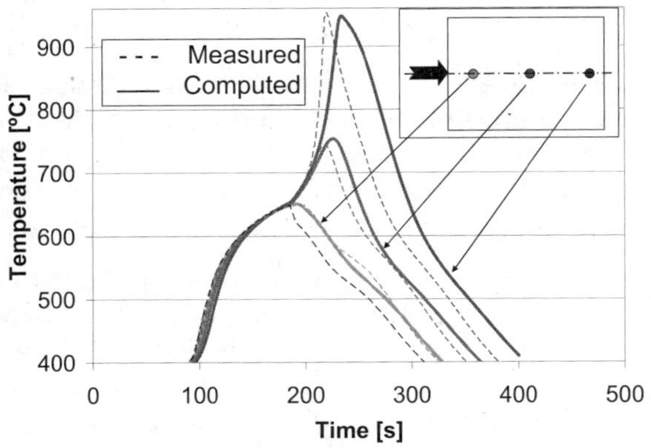

(b)

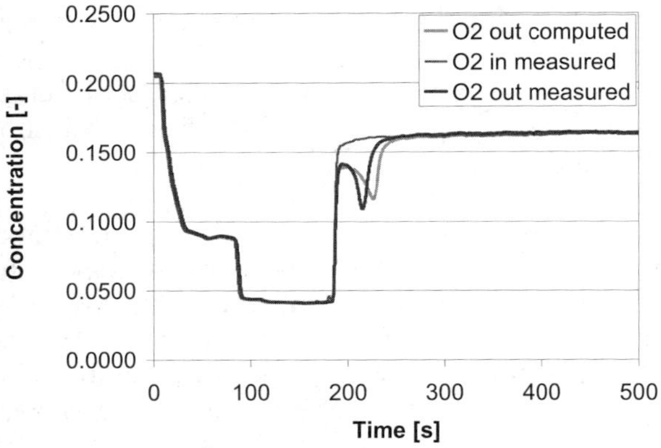

(c)

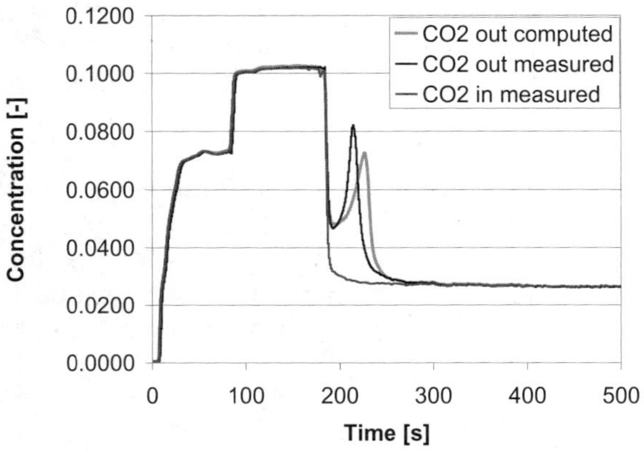

(d)

Figure 9: Experimental and simulation results with standard idle mode. Catalyzed filter with 6 g/l initial soot loading.

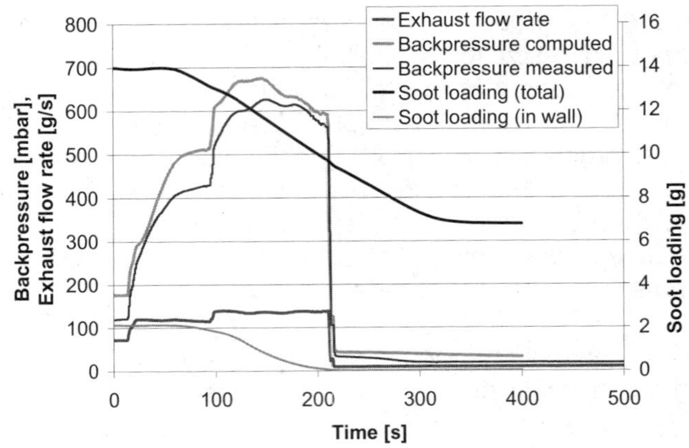

(a)

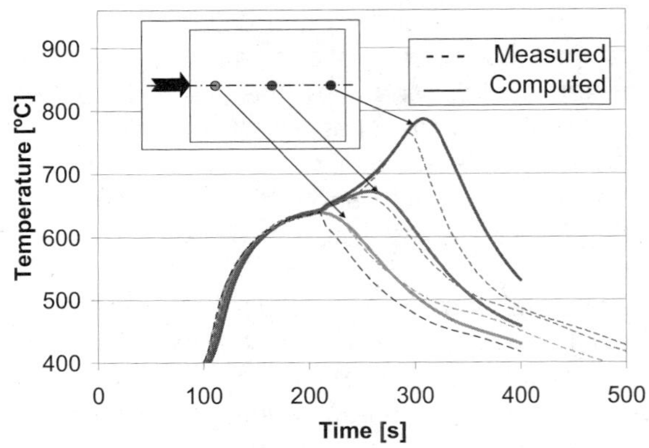

(b)

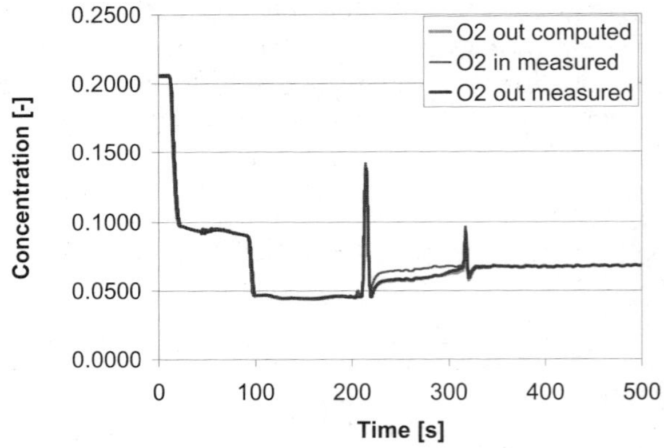

(c)

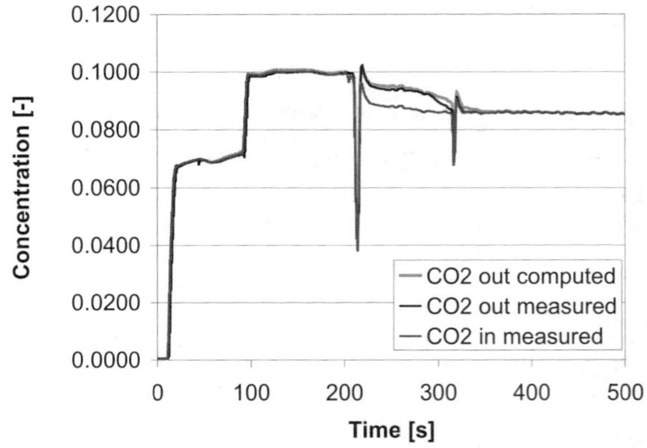

(d)

Figure 10: Experimental and simulation results with safe idle mode. Catalyzed filter with 5.6 g/l initial soot loading.

A similar experiment was performed with the catalyzed filter and a "safe mode" idle calibration. The results are summarized in **Figure 10**. Even though care was taken to obtain an identical initial soot loading, there is a small discrepancy between the two tests of the order of 0.4 g/l.

The model results are again in very good agreement in regards to temperature predictions as well as O_2 and CO_2 downstream the filter. Comparing the O_2 results in the standard and safe mode tests, it is obvious that the overall reaction rate is significantly reduced in the latter case, which is reflected in the respective computed soot mass decrease rate in the top graphs.

By a first examination of the above results, one may conclude that the safe mode is able to decrease the peak temperature by almost 200°C. However, a detailed investigation of the experimental conditions reveals that the test protocol is not identical in both tests. This is not unexpected taking into account the difficulties in obtaining exactly repeatable inlet gas conditions in this kind of engine tests. In the specific case, the maximum inlet gas temperature reached before the idle switch was 650°C in the standard mode compared to 635°C in the safe mode. This small difference may be quite important considering the exponential dependency of the reaction rates on temperature. A lower inlet temperature at the time of idle switch is expected to result in lower peak temperature.

Moreover, the duration of the full-power mode before in the idle switch was 90 s in the case of the standard idle mode compared to 110 s in the safe mode. Consequently, additional soot was consumed at full power leaving less soot at the time of idle switch in the safe mode test. This is again expected to lower the peak temperature.

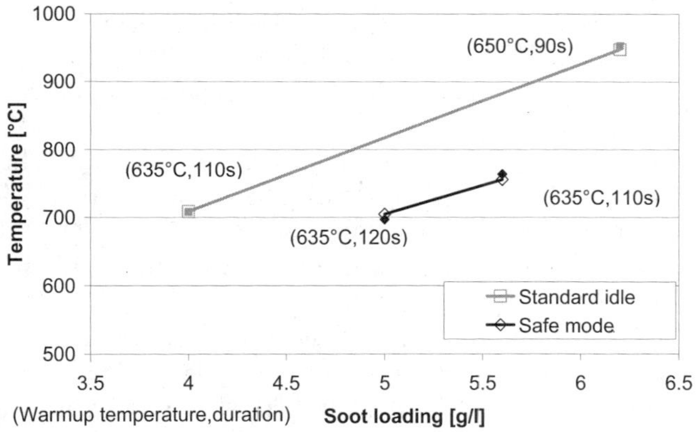

Figure 11: Peak temperatures obtained with the catalyzed filter at different soot loadings and idle modes. Open symbols correspond to model results; closed symbols are measured data.

Two additional experiments were performed with the catalyzed filter at lower initial soot loadings. In all four cases the peak temperatures were correctly predicted by the model, as shown in **Figure 11**. In the parentheses of the same graph, the experimental conditions are summarized regarding exhaust temperature at idle switch and duration of full-power mode.

On the basis the results of **Figure 11**, one cannot safely judge the efficiency of the safe mode calibration, since the peak temperature may be affected by the small differences in the experimental inlet conditions. To overcome this problem, the validated model can be used to create results with "virtual experiments" under identical inlet conditions. More specifically, it was chosen to work with a full-power duration of 110 s and an inlet temperature of 635°C at the time of idle switch. For these specific conditions, two measurements were already available.

A parametric analysis in terms of predicted peak temperature as function of initial soot loading for the case of standard and safe mode is shown in **Figure 12**. Thanks to the inherent ability of the model to work with identical input data, it is possible to make more objective comparisons. In the specific application, it is believed that the result of **Figure 12** gives a more realistic evaluation of the safe mode potential compared to **Figure 11**, which is blurred by the repeatability issues of the test conditions.

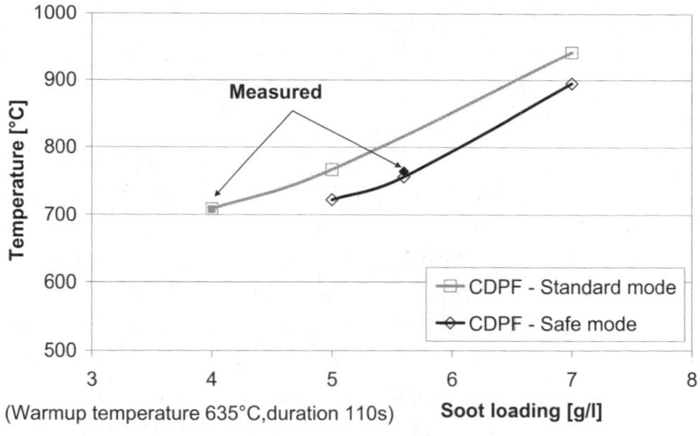

Figure 12: Predicted peak temperature as function of initial soot loading for a catalyzed filter for the standard and safe idle mode. Results of two measurements are also shown.

Finally, a similar parametric investigation with the same gas inlet conditions was performed in the case of a non-catalyzed filter. The results in terms of peak temperature prediction are summarized in **Figure 13**. These results may now be directly compared to the results of **Figure 12** for a fair assessment of the effect of catalyst on maximum soot limit.

From the above results, it can be said that the peak temperatures in the case of the catalyzed filter are expected to be ~ 50 degrees higher at initial soot loadings of the order of 6-7 g/l. The difference becomes smaller at lower soot loadings.

However, such comparisons cannot be directly translated in terms of regeneration frequency. In this case, one must take into account the increased passive regeneration expected in catalyzed filters which will result in lower soot loadings for the same mileage.

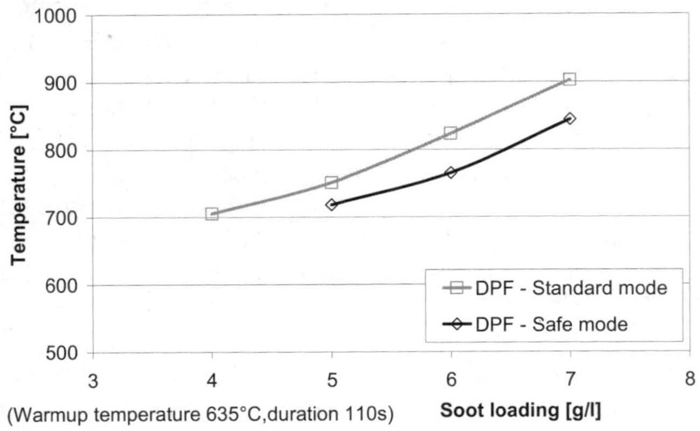

Figure 13: Predicted peak temperature as function of initial soot loading for a non-catalyzed filter for the standard and safe idle mode.

CONCLUSIONS

The presented work features a description of the assessment of the critical soot mass limit of diesel particulate filters by advanced modeling and regeneration control.

It has been shown, that a state-of-the-art physical model is capable of predicting the maximum DPF temperatures during critical DPF regeneration with good agreement to the validation experiments. The input parameters for the DPF modeling procedure can be derived from a polynomial DoE model calibrated by investigations under selected operating points. From the two validated models, the calculation of the critical soot mass limit can be performed for any application within the boundaries of the DoE model on-desk without extensive engine test bench usage. The definition of the methodology to determine and increase the critical soot mass limit of catalyzed and non-catalyzed DPF is schematically shown in **Figure 14**.

In the framework of the present study selected experiments were performed for model validation and for critical soot limit investigations. Apart from the potential to reduce the amount of testing work, modeling proved very useful in evaluating correctly the experimental data. This is important since the experimental conditions are very difficult to duplicate exactly on the engine bench. In the specific case study shown here, the peak temperature during worst-case regeneration condition is not only a function of initial soot loading, but also on the duration and the exhaust gas temperature during heat-up.

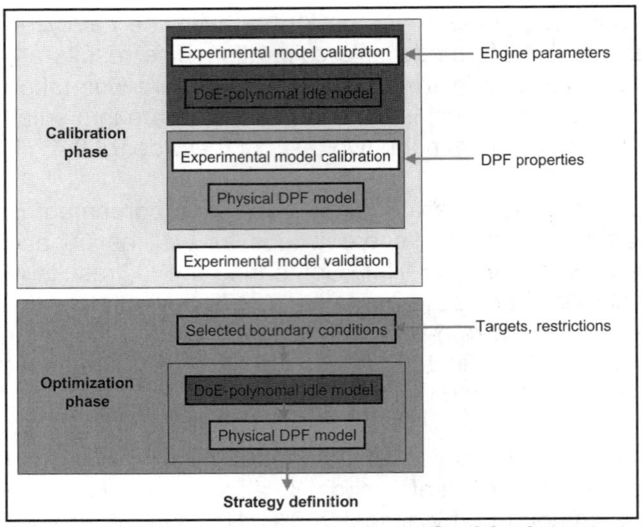

Figure 14: Assessment process of critical soot mass limit.

ACKNOWLEDGMENTS

We would kindly like to thank Dr. S. Stiebels and Mr. A. Punke, Engelhard Technologies, as well as Dr. A. Schäfer-Sindlinger and Mr. C. Vogt, NGK, for providing the state-of-the-art test samples and their helpful contributions.

REFERENCES

1. F. Heimlich, L Kraemer, T. Wombs, J. Manns, K. Behnk, “The Road to EU5 Diesel- Aftertreatment Concepts and Main Development Objectives“, Equip Auto, Paris, 13 Oct. 2005
2. M. Suzuki, N. Tsuzuki, Y. Teramachi: “The new Toyota L-4 direct-injection diesel engine – Toyota D-4D Clean Power Concept -”, 26. Internationales Wiener Motorensymposium, 28. - 29. April 2005.
3. A. Punke, G. Grubert, Yuejin Li, J. Dettling, T. Neubauer: “Catalyzed Soot Filters in Close Coupled Position for Passenger Vehicles", SAE Paper No. 2006-01-1091.
4. T. Mueller, A. Punke, Yuejin Li, J. Dettling: “The Coating Evolution of Catalyzed Soot Filter", ATA International Symposium Diesel Engine: The NOx & PM Emissions Challenge, Monopoli (Bari),13.-15. Oct. 2004.
5. S. Hashimoto, Y. Miyairi, T. Hamanaka, R. Matsubara, T. Harada and S. Miwa: "SiC and Cordierite Diesel Particulate Filters for Low Pressure Drop and Catalyzed, Uncatalyzed Systems", SAE Paper No. 2002-01-0322
6. Y. Kasai, S. Miwa, T. Kuki, K. Senda, Y. Ogura: "New Cordierite Diesel Particulate Filter Material for the Diesel Particulate - NOx Reduction System", SAE Paper No. 2004-01-0953
7. O. Predelli, T. Kracke, W. Schmidt: "Flexible engine controller with integrated cylinder pressure sensing", Sensact, Paris, 7.-8. Dec. 2005.

8. L. Daeubler, R. Gratzke, O. Predelli, A. Rempel: "Flexible Engine Management for Innovative Combustion Systems", Motortechnische Zeitschrift MTZ 09/2005.
9. L. Krämer, K. Behnk, "Assessment of Particle Measurement Techniques and Physico-chemical Soot Properties at Different Diesel Combustion Modes", 10th ETH-Conference on combustion generated nanoparticles, Zürich, 22. August 2006
10. S. Schraml ; C. Heimgaertner, S. Will, A. Leipertz, A. Hemm: "Application of a new soot sensor for exhaust emission control based on time-resolved laser-induced incandescence (TIRE-LII)", SAE Paper Vol. 109, S. 2629-2638, 2000.
11. K. Roepke, R. Gaitzsch, C. Haukap, M. Knaak, C. Knobel, A. Nessler, S. Schaum, U. Schoop, S. Tahl: „DoE- Design of Experiments, Methoden und Anwendungen in der Motorenentwicklung“, Verlag Moderne Industrie, München, 2005.
12. K. Roepke, W. Tschiggfrei, J. Manns, T. Dreher: "The modern Diesel Engine Calibration Proces", SIA Congrès Le Diesel: Aujourd´hui et demain, Lyon, 12. & 13. May 2004.
13. Koltsakis G. C., Haralampous O. A., Margaritis N. K., Samaras Z. C., Vogt C.-D., Ohara E., Watanabe Y. and Mizutani T., “3-Dimensional Modeling of the Regeneration in SiC Particulate Filters”, SAE Transactions: Journal of Fuels & Lubricants, SAE paper 2005-01-0953, Detroit, 2005
14. Koltsakis G. C., Haralampous O. A., Dardiotis C. K., Samaras Z. C., Vogt C.-D., Ohara E., Watanabe Y. and Mizutani T., “Performance of Catalyzed Particulate Filters without Upstream Oxidation Catalyst”, SAE paper 2005-01-0952, Detroit, 2005
15. K. N. Pattas, A. M. Stamatelos, K. N. Kougianos, G. C. Koltsakis: "Trap Protection by Limiting A/F Ratio During Regeneration", SAE paper 950366, SAE transactions, Journal of Engines, pp. 678-688, 1995.
16. K. Roepke, W. Tschiggfrei, J. Manns, T. Dreher: "The modern Diesel Engine Calibration Proces", SIA Congrès Le Diesel: Aujourd´hui et demain, Lyon, 12. & 13. May 2004.
17. Haralampous O. A., Koltsakis, G.C. Samaras Z. C., Vogt C.D., Ohara E., Watanabe Y., Mizutani T.: Modeling and experimental study of uncontrolled regenerations in SiC filters with Fuel Borne catalyst, (SAE 2004-01-0697) SAE International Congress, Detroit, USA, 2004.
18. Haralampous O. A., Koltsakis G. C., Samaras Z.C., “Partial Regenerations in Diesel Particulate Filters”, SAE paper 2003-01-1881, 2003.
19. Haralampous O. A., Kandylas I. P., Koltsakis G.C., Samaras Z. C., “Diesel Particulate Filter Pressure Drop. Part I: Modeling and Experimental Validation”, International Journal of Engine Research, Vol. 5, No 2, pp. 149-162, 2004.
20. Haralampous O. A., Koltsakis G. C.: "Back-diffusion Modeling of NO2 in Catalyzed Diesel Particulate Filters", Industrial & Engineering Chemistry Research, Vol. 43, 4, pp. 875-883, 2004
21. Danckwerts, P. V. Continuous Flow Systems Distribution of Residence Times. Chem. Eng. Sci. 1953, 2, 1

CONTACT

Assistant Professor Grigorios C. Koltsakis,
Laboratory of Applied Thermodynamics,
Aristotle University of Thessaloniki,
Tel: +302310995870,
Fax: +302310996019,
e-mail: grigoris@auth.gr
URL: http://lat.eng.auth.gr

DEFINITIONS

Variables

A	Reaction rate frequency factor
b	Geometric parameter
$c_{j,k}$	Stoichiometric coefficient of species *j* in reaction *k*
c_m	Molecular density, mole/m^3
C_p	Specific heat capacity, J/kgK
d	Hydraulic diameter of clean channel, m
d_i	Hydraulic diameter of respective channel, m
D	Mass diffusivity, m^2/s
E	Activation energy, J/mole
Eq	Chemical equilibrium coefficient, -
f	=*b*/*d*
G	Inhibition term, -
H	Heat transfer per filter volume, W/m^3
h	Convection coefficient, W/m^{2}K
k	Arrhenius term
$k_{i,j}$	Mass transfer coefficient of species *j* in channel *i*, m/s
k_p	Soot layer permeability, m²
k_s	Substrate permeability, m²
p	Pressure, Pa
R	Reaction rate, mole/(m^{3}s)
r	Radial distance, m
S	Sum of heat sources, W/m^3
s_F	Filter specific area, m^{-1}
t	Time, s

T	Temperature, K
v	Velocity, m/s
w	Particulate layer thickness, m
w_s	Channel wall thickness, m
x	Space variable perpendicular to wall surface, m
y_j	Mole fraction of species j, -
z	Axial distance, m

Greek letters

α_1	Constant in channel pressure drop correlation, -
ΔH	Reaction heat, J/mole
Δp	Backpressure, Pa
ε	Emmisivity, -
λ	Thermal conductivity, W/mK
μ	Exhaust gas viscosity, kg/ms
ρ	Density, kg/m^3
σ	Stefan-Boltzmann constant, W/(m^2 K^4)

Subscripts

1s	Inlet channel- soot surface interface ($x = -w$)
2s	Outlet channel- wall surface interface ($x = w_s$)
amb	Ambient
g	Exhaust gas
i	Channel index, (1=inlet channel, 2=outlet channel)
j	Species index
k	Reaction index
r	Radial direction
s	Solid, ceramic substrate
w	Wall-outlet channel interface
z	Axial direction

APPENDIX

The model is based on a radial discretization of the cylindrical filter. Each radial filter segment is represented by a single channel which is solved individually. The link between the radial segments is done by the conductive heat exchange in the radial direction included in the energy balance. Compared to 1-D modeling, this 2-D model allows for the consideration of the effect of heat losses, thermal boundary layer at filter inlet and flow maldistribution as a result of non-uniform regeneration. The model is equipped with intra-layer discretization to account for the coupling between reaction and diffusion phenomena in catalyzed filters.

A schematic of the side and front view of a channel model is given in Figure 15. The governing equations for the conservation of mass, momentum, energy and species are given below.

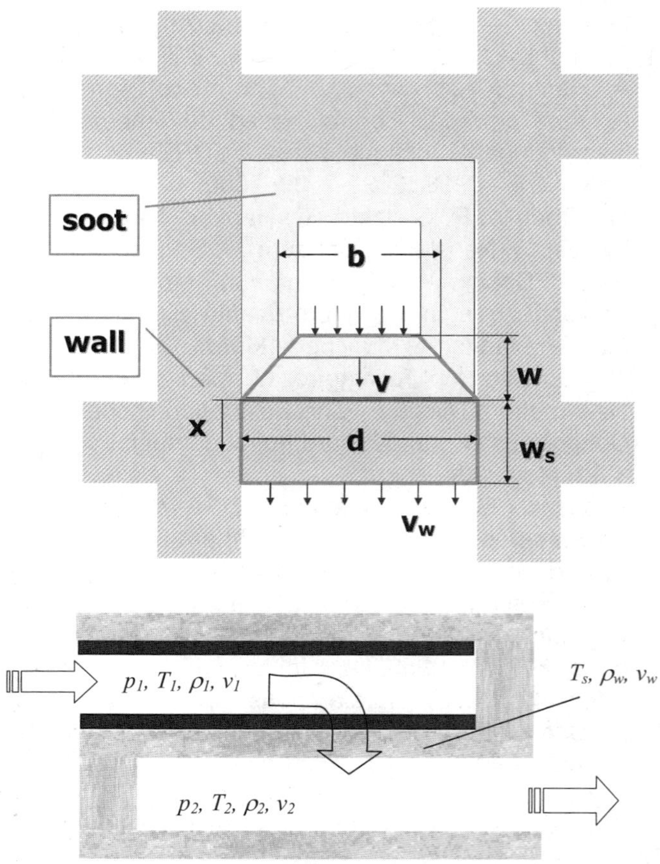

Figure 15: Schematic of channel model. (a) Front view, (b) Side view

MASS-MOMENTUM BALANCE

Conservation of mass of channel gas

$$\frac{\partial}{\partial z}\left(d_i^2 \rho_i v_i\right) = (-1)^i 4 d \rho_w v_w \tag{1}$$

Conservation of axial momentum of channel gas

$$\frac{\partial p_i}{\partial z} + \frac{\partial}{\partial z}\left(\rho_i v_i^2\right) = -\alpha_1 \mu v_i / d_i^2 \tag{2}$$

Pressure drop across the deposit layer and ceramic wall

$$p_1 - p_2 = \Delta p_{soot} + \Delta p_{wall} \tag{3}$$

Based on the analysis presented in [19], the pressure loss through the soot layer and wall is:

$$\Delta p_{soot} = \frac{\Re T}{M_g \bar{p}} \frac{\mu d \rho_w v_w}{2k_p(\bar{p})} \ln\left(\frac{d}{d-2w}\right) \tag{4}$$

$$\Delta p_{wall} = \frac{\mu v_w}{k_s} w_s \tag{5}$$

The solution of the system yields the channel flow field (vw, v1, v2), the pressure field (p1, p2) and the gas temperature field (T1, T2).The inlet flow distribution is calculated every time step based on the downstream flow resistances, as described in [17].

ENERGY BALANCE

Conservation of energy of channel gas

Inlet channel:

$$C_{p,g}\rho_1 v_1\Big|_z \frac{\partial T_1}{\partial z} = h_1 \frac{4}{d_1}(T_s - T_1) \tag{6}$$

Outlet channel:

$$C_{p,g}\rho_2 v_2\Big|_z \frac{\partial T_2}{\partial z} = (h_2 + C_{p,g}\rho_w v_w)\frac{4}{d}(T_s - T_2) \tag{7}$$

Conservation of energy of channel gas

The temperature field in the filter is described by the equation of transient heat conduction with heat sources in axisymmetric co-ordinates:

$$\rho_s \cdot C_{p,s} \frac{\partial T_s}{\partial t} = \lambda_{s,z} \frac{\partial^2 T_s}{\partial x^2} + \lambda_{s,r} \frac{1}{r}\frac{\partial^2 T_s}{\partial r^2} + S \tag{8}$$

The boundary conditions for the heat conduction equation are:

$$\lambda_{s,r} \frac{\partial T_s}{\partial r} = h_{amb}(T_s - T_{amb}) + \varepsilon \cdot \sigma \cdot \left(T_s^{\,4} - T_{amb}^{\,4}\right) \tag{9}$$

at r=R, and

$$\frac{\partial T_s}{\partial r} = 0 \tag{10}$$

at the symmetry line (r=0).

The source term S includes the contribution of the convective heat transfer of the gas flow in the channels and through the wall as well as the exothermic heat release:

$$S = H_{conv} + H_{wall} + H_{react} \tag{11}$$

Reaction exotherm

$$H_{react} = s_F \int_{-w}^{w_s} f_x \sum_k R_k \Delta H_k dx \tag{12}$$

where s_F is the specific filtration area.

Convection of heat due to flow through wall

$$H_{wall} = \rho_w \cdot v_w \cdot s_F \cdot C_{p,g} \cdot (T_1 - T_s) \tag{13}$$

Convection of heat due to flow in channel

$$H_{conv} = h_1 \cdot s_F \cdot (T_1 - T_s) + h_2 \cdot s_F \cdot (T_2 - T_s) \tag{14}$$

SPECIES BALANCE

The governing equation [20,21] for the mass conservation of any species in the soot layer and wall is:

$$v_w \frac{\partial y_j}{\partial x} - D_j \frac{\partial}{\partial x}\left(f_x \frac{\partial y_j}{\partial x}\right) = \frac{f_x}{c_m} \sum_k c_{j,k} R_k \tag{15}$$

Where the geometrical parameter f_x is defined as:

$$f_x = \frac{b(x)}{d} \tag{16}$$

with the following boundary condition for the inlet

$$v_w y_{1s,j} - D_j f_{-w} \frac{\partial y_j}{\partial x}\bigg|_{1s} = v_w y_{1,j} - k_{1,j} \cdot f_{-w} \cdot (y_{1s,j} - y_{1,j}) \tag{17}$$

And for the outlet channel:

$$-D_j \frac{\partial y_j}{\partial x}\bigg|_{2s} = k_{2,j} \cdot (y_{2s,j} - y_{2,j}) \tag{18}$$

The boundary conditions couple the wall with the channel concentrations given in the following mass balances for the inlet and outlet channel:

$$\frac{\partial}{\partial z}(v_1 y_{1,j}) = -v_w y_{1,j} + k_{1,j}(y_{1s,j} - y_{1,j}) \tag{19}$$

(20)

2000-01-1199

Modeling and Control of a NOx Trap Catalyst

Ahmed Ketfi-Cherif, Dirk von Wissel and Stephan Beurthey
RENAULT

Michel Sorine
INRIA

ABSTRACT

An adaptive model based monitoring strategy for a NOx trap catalyst is presented. First, a simplified model of the NOx trap is developed from a physical PDE (Partial Differential Equations) model. A validation on different types of data is given. Then two monitoring strategies are presented: a non-adaptive and an adaptive strategy, which uses a downstream EGO (Exhaust Gas Oxygen) sensor. The basic idea of the adaptive strategy is to put the catalyst and monitoring strategy in a particular condition such that a successive adaptation of three different parameters of the model is possible. The adaptive strategy is compared to the non-adaptive strategy by use of numerical simulations.

INTRODUCTION

Lean burn engines can achieve good fuel economy, however, using classical three-way catalysts with this type of engine decreases HC/CO emissions but not the NOx emission. One way to overcome this drawback is to use another type of catalyst, which stores the NOx in lean conditions and reduces the stored NOx in rich conditions. This type of catalyst is commonly called a NOx trap catalyst, and needs an active monitoring system to achieve good efficiencies on both NOx and HC/CO emissions.

The objective of this paper is the construction of a reduced physical model for a NOx trap catalyst and its application in the design of an on-line regeneration monitoring strategy.

On one hand, there is a large amount of papers on chemical principles and experimental studies of NOx trap catalysts (see references [3, 4, 5, 8, 9, 10] for example). Fewer papers are on modeling and unfortunately, the available models are either incomplete since they describe only adsorption but not regeneration (see [6]), or too complex for control applications (see [7]).

On the other hand, numerous patents have been published on on-line monitoring and on-board diagnostic based on signal processing techniques using temperature and/or gas composition sensors. However, only few applications use phenomenological models (see [2] for example).

In this paper, we introduce a novel closed-loop control strategy based on a reduced physical model. The reduced model is obtained from a full one dimensional PDE description of the basic chemical and physical phenomena. The structure of this description is taken from anterior work of Aimard *et al.* (see [1]). We apply an orthogonal collocation technique to approximate the full PDE model on two collocation points: the input and the output of the catalyst. The result of this approximation is a 3 state ODE (Ordinary Differential Equations)-based model. A separation of time scales and a set of assumptions on the main dependencies yield the reduced physical model.

The reduced model is applied to develop an on-line control strategy of NOx trap catalyst regeneration. This strategy is decomposed in two parts: a basic strategy and an adaptive strategy. The basic strategy uses the reduced model to predict some internal parameters of the NOx trap catalyst and then determines the start and stop of regeneration phases. The adaptive strategy allows adapting the main parameters of the model by the use of the signal of a downstream lambda sensor.

The basic strategy is similar to the open loop strategy proposed in reference [2], however, the performance of this kind of strategy can be deteriorated during the aging of the catalyst or by a large dispersion in catalyst characteristics.

The novel adaptive strategy introduced in this contribution allows the tuning of three main parameters of the reduced model. The main idea of this strategy is to put the catalyst in particular operating conditions for which the lambda sensor signal is influenced by only one parameter of the physical model. A particular order of adaptation allows the on-line tuning of the three parameters. Simultaneous adaptation of the three parameters appears to be impossible due to the cross-dependencies of the measured output on these parameters.

OPERATION OF A NOX TRAP

In our context, we consider a simplified description of a NOx trap operation. Two operation modes exist, storage and regeneration:

STORAGE – In lean conditions, NOx is stored according to a two-step mechanism. In the first step, NO is oxided on Pt-sites, this reaction forms NO_2. In the second step, NO_2 is adsorbed on metal-oxide (MO) sites. This two-step mechanism is described by the following reactions:

$$2NO + O_2 \rightarrow 2NO_2$$

$$2NO_2 + \frac{3}{2}O_2 + MO \rightarrow M(NO_3)_2$$

REGENERATION – The NOx stored is treated in rich conditions. In this conditions, first, the metallic nitrate $M(NO_3)_2$ is decomposed, second, the resulting NO_2 reacts with the reducer species (CO, HC, H_2), according to the following reactions:

$$M(NO_3)_2 \rightarrow 2NO_2 + \frac{3}{2}O_2 + MO$$

$$CH_y + \left(1 + \frac{y}{4}\right)NO_2 \rightarrow CO_2 + \frac{y}{2}H_2O + \left(\frac{1}{2} + \frac{y}{8}\right)N_2$$

$$CO + \frac{1}{2}NO_2 \rightarrow CO_2 + \frac{1}{4}N_2$$

$$H_2 + \frac{1}{2}NO_2 \rightarrow H_2O + \frac{1}{4}N_2$$

Figure 1 gives an illustration of storage and regeneration mechanisms.

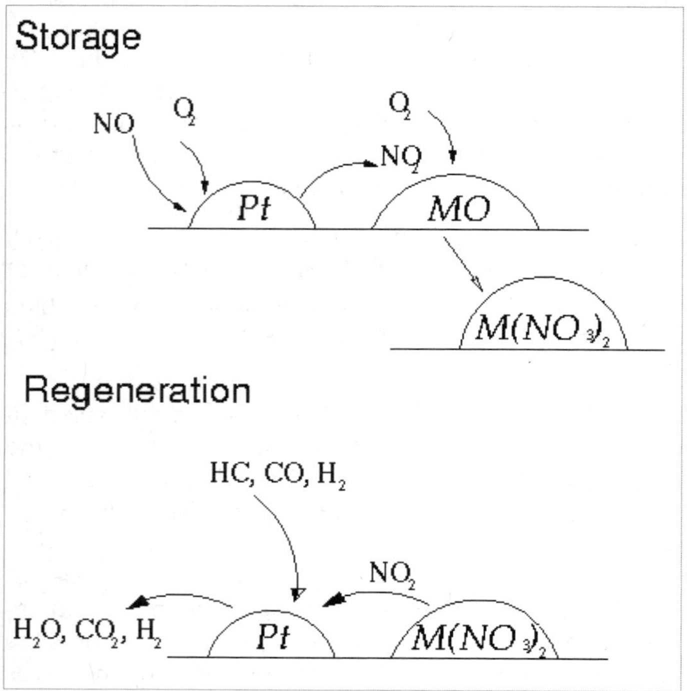

Figure 1. Operation of a NOx trap

MODEL DEVELOPMENT

In this section, we describe the different steps in the development of a NOx trap model sufficiently simple to be used in on-line control. To begin with, we give the physical and chemical laws governing the main phenomena. These laws lead to a one-dimensional PDE based model. Then, we use the orthogonal collocation technique to transform this model to ODE-based one. This last model is next identified and validated. Finally, the structure of the ODE-based model is used to get a more simplified model obtained by keeping only the main dependencies of the different parameters on model inputs.

PHYSICAL AND CHEMICAL PHENOMENA – The two main phenomena taken into account are mass transfer and chemical reactions. These phenomena are mathematically described below.

Mass transfer – We assume that the gaseous species are transferred in the axial direction of the catalyst (denoted by x) with a uniform velocity V_g. The flux density of each species i is given by the following expression:

$$J_i = V_g C_i \tag{1}$$

The adsorbed species are supposed to have zero velocity, which implies for the corresponding flux density that:

$$J_i = 0 \tag{2}$$

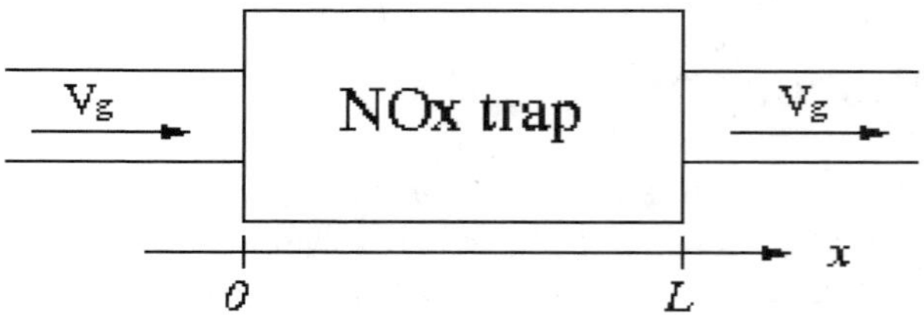

Figure 2. Gas flow in the NOx trap

Chemical reactions – We consider a simplified chemical scheme composed out of three reactions representing NOx storage and regeneration:

$$NOx + \Sigma \rightarrow NOx^* \tag{I}$$

$$NOx^* \rightarrow NOx + \Sigma \tag{II}$$

$$NOx^* + Red \rightarrow \Sigma + a\,H_2O + b\,CO_2 + c\,N_2 \tag{III}$$

In this reactions:

- *NOx* corresponds to (NO+NO_2),
- Σ denotes a free adsorption site
- *NOx** denotes an occupied adsorption site

- *Red* is a fictive species, which replaces ($HC+CO+H_2$), referred to as "equivalent reducer" throughout this paper. Its concentration is calculated using the stoichiometric coefficient of each species in its reaction with NO_2 :

$$C_{Red} = \frac{1}{2}C_{CO} + \frac{1}{2}C_{H2} + \left(1+\frac{y}{4}\right)C_{CHy}$$

The following equation relates the concentrations of free and occupied adsorption sites:

$$C_\Sigma + C_{NOx*} = N_{sc}$$

where N_{sc} is the available concentration of adsorption sites referred to as "NOx storage capacity" in the rest of this paper. Noting by Θ the proportion of occupied sites with respect to total available sites, we can write:

$$C_\Sigma = N_{sc}\,\Theta$$
$$C_{NOx*} = N_{sc}(1-\Theta)$$

In summary, reactions *(I)* and *(II)* represent the storage process which is composed by an adsorption process and desorption process. Reaction *(III)* represents the regeneration process.

PDE BASED DYNAMIC MODEL – The equations of a dynamic transport-reaction model are obtained using the conservation equations:

$$\frac{\partial C_i}{\partial t} = -\frac{\partial J_i}{\partial x} + R_i \tag{3}$$

where R_i is the creation rate of the species i.

Applying (3) for $i \in \{NOx, Red, \Sigma\}$ and using the definition of Θ yield the following system:

$$\begin{cases} \dfrac{\partial C_{NOx}}{\partial t} = -V_g\dfrac{\partial C_{NOx}}{\partial x} - r_1 + r_2 \\ \dfrac{\partial C_{Red}}{\partial t} = -V_g\dfrac{\partial C_{Red}}{\partial x} - r_3 \\ N_{sc}\dfrac{\partial \Theta}{\partial t} = r_1 - r_2 - r_3 \end{cases} \tag{4a}$$

with the following boundary conditions at *x=0*:

$$\begin{cases} C_{NOx}(x=0,t) = C_{NOx}^{in} \\ C_{Red}(x=0,t) = C_{Red}^{in} \end{cases} \tag{4b}$$

where the R_i are rewritten in term of the rates of reactions *(I)*, *(II)* and *(III)* denoted by r_1, r_2, r_3 respectively. To express these last quantities, we adopt the following kinetic laws:

$$\begin{cases} r_1 = k_1(T)C_{NOx}(N_{sc}\Theta)^2 \\ r_2 = k_2(T)(N_{sc}(1-\Theta))^2 \\ r_3 = k_3(T)C_{Red}(N_{sc}\Theta)^2 \end{cases} \tag{4c}$$

where *T* is the temperature of the catalyst which is supposed to be constant with respect to *x* (the catalyst is considered as an isothermal reactor). The choice of second order laws for the reaction rates r_1, r_2, and r_3 is motivated by the fact that each storage site holds two NOx molecules ($M(NO_3)_2$).

The influence of *T* on the kinetic coefficients ki can be expressed by the Arrhenius law:

$$k_i = k_{i0}\exp\left(-\frac{E_i}{RT}\right) \tag{5}$$

MODEL SIMPLIFICATION

<u>1D → 0D reduction</u> – Several methods exist to reduce PDEs to ODEs. Orthogonal collocation is one of the most popular methods used in chemical engineering to achieve this reduction. In this method, the solution of a PDE is approximated using an orthogonal polynomial base:

$$C(x,t) = \sum_{k=1}^{n} P_k(x)\eta_k(t) \tag{6}$$

We want (6) to verify the PDE and its boundary conditions in a set of points $(x_1,\ldots,x_n)$ called "collocation points", which gives a set of ODEs verified by the functions h_k.

In our case, we take only two collocation points, the input and the output of the catalyst *(x=0, x=L)*. The solution of the system (3)-(6) is approximated by:

$$\begin{aligned} C_{NOx}(x,t) &= \left(1-\frac{x}{L}\right)C_{NOx}^{in}(t) + \frac{x}{L}C_{NOx}^{t}(t) \\ C_{Red}(x,t) &= \left(1-\frac{x}{L}\right)C_{Red}^{in}(t) + \frac{x}{L}C_{Red}(t) \\ \Theta(x,t) &= \Theta(t) \end{aligned} \tag{7}$$

$C_{NOx}(t)$ and $C_{Red}(t)$ corresponds to the values of $C_{NOx}(x,t)$ and $C_{Red}(x,t)$ at x=L.

The approximation given by (7) verifies the boundary condition at *x=0*, if we impose that it verifies (3) at the collocation point $x=L$, we obtain the following system of equations:

$$\begin{cases} \frac{dC_{NOx}}{dt} = \frac{V_g}{L}\left(C_{NOx}^{in} - C_{NOx}\right) - r_1 + r_2 \\ \frac{dC_{Red}}{dt} = \frac{V_g}{L}\left(C_{Red}^{in} - C_{Red}\right) - r_3 \\ N_{sc}\frac{d\Theta}{dt} = r_1 - r_2 - r_3 \end{cases} \quad (8)$$

Remark: the previous reduced model can be obtained by considering the catalyst as a CSTR (Continuous Stirred Tank Reactor). However, the orthogonal collocation technique remains more interesting since it can generate higher order approximations.

Time-scale reduction – A further simplification of model (8) is possible if we consider that the dynamics of C_{NOx} and C_{Red} are fast compared to the dynamic of Θ. In chemical engineering, this assumption is called "non-accumulation of gaseous species". The use of this assumption implies the quasi-static conditions:

$$\frac{dC_{NOx}}{dt} = \frac{dC_{Red}}{dt} = 0$$

Furthermore, reactions *(I)* and *(II)* are active only in lean conditions ($\lambda > 1$) and reaction *(III)* is active only in rich conditions ($\lambda < 1$), where λ is the normalized air-to-fuel ratio. This implies that C_{NOx} and C_{red} are functions of Θ in different regimes of λ. By use of this fact and the previous assumption, we get the following reduced model:

$$\lambda > 1 \begin{cases} N_{sc}\frac{d\Theta}{dt} = \frac{V_g}{L}\left(C_{NOx}^{in} - C_{NOx}\right) \\ C_{NOx} = \frac{C_{NOx}^{in} + (k_2 L / V_g)(N_{sc}\Theta)^2}{1 + (k_1 L / V_g)(N_{sc}(1-\Theta))^2} \end{cases}$$

$$\lambda < 1 \begin{cases} N_{sc}\frac{d\Theta}{dt} = \frac{V_g}{L}\left(C_{Red}^{in} - C_{Red}\right) \\ C_{Red} = \frac{C_{Red}^{in}}{1 + (k_3 L / V_g)(N_{sc}\Theta)^2} \end{cases} \quad (9a)$$

MODEL VALIDATION – The identification of model (9a) based on laboratory data yields a good approximation if compared to real engine data.

Data for identification – To identify the kinetic parameters of the simplified model (9a), we used a laboratory set of data including both storage-only and storage/regeneration cycling experiments at different temperatures. The used catalyst has a volume of 0.04l. Figures 3a and 3b give a sample comparison between model and experience.

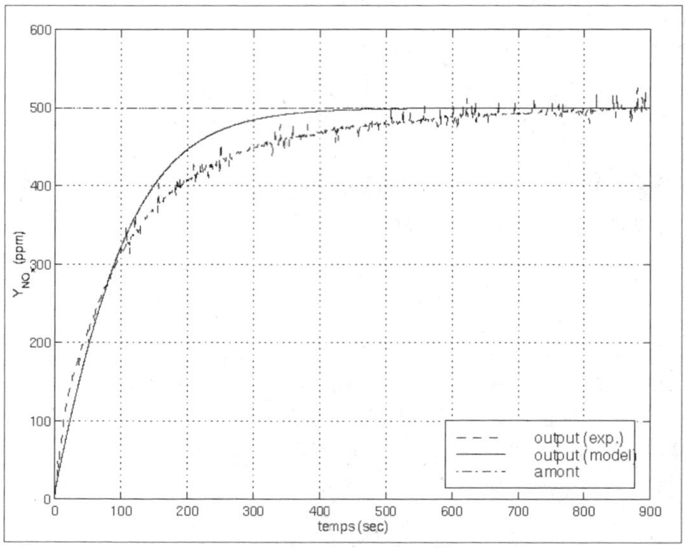

Figure 3a. Comparison between experimental an simulated values of NOx output concentration in storage conditions (laboratory data)

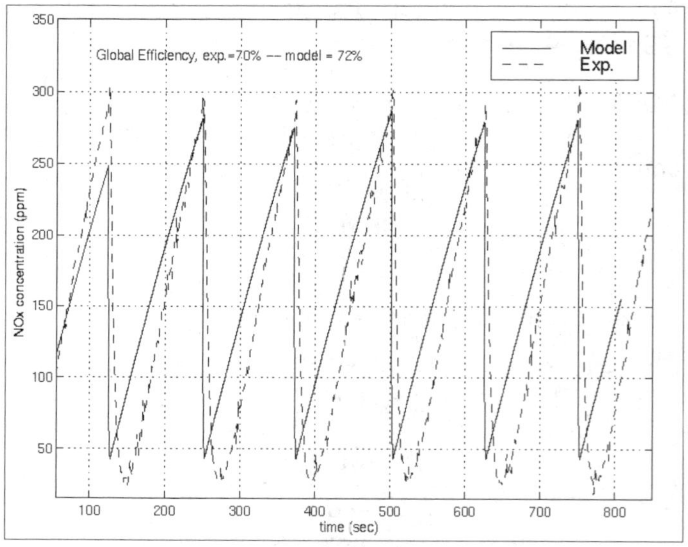

Figure 3b. Comparison between experimental an simulated values of NOx output concentration in storage/regeneration conditions (laboratory data).

Data for validation – Once the parameters are identified, the model is tested on real engine data. The available data corresponds to normalized European cycle (MVEG). We use a 16 valve 4 cylinder 1.6 l Renault engine equipped with a 2.5 l NOx trap catalyst.

Figures 4a and 4b give two comparisons between model and experiments. We superpose first in lean conditions and then in lean/rich conditions measured and simulated cumulative NOx mass at the output of the catalyst.

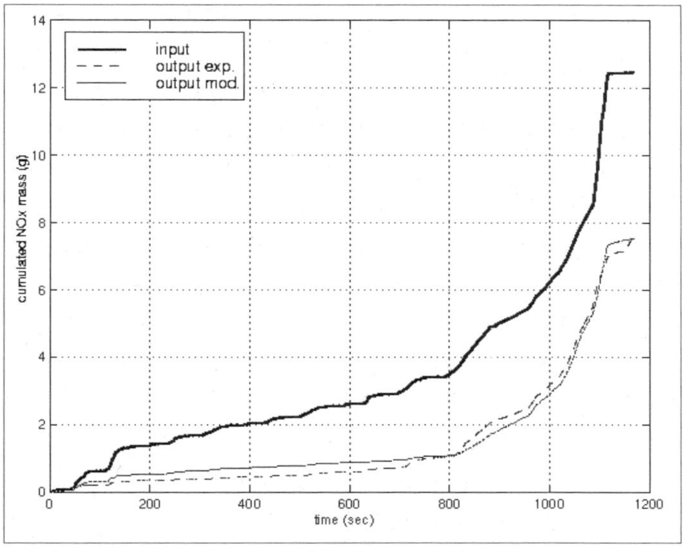

Figure 4a. Comparison between experimental and simulated values of cumulated NOx mass in using engine data (storage condition only).

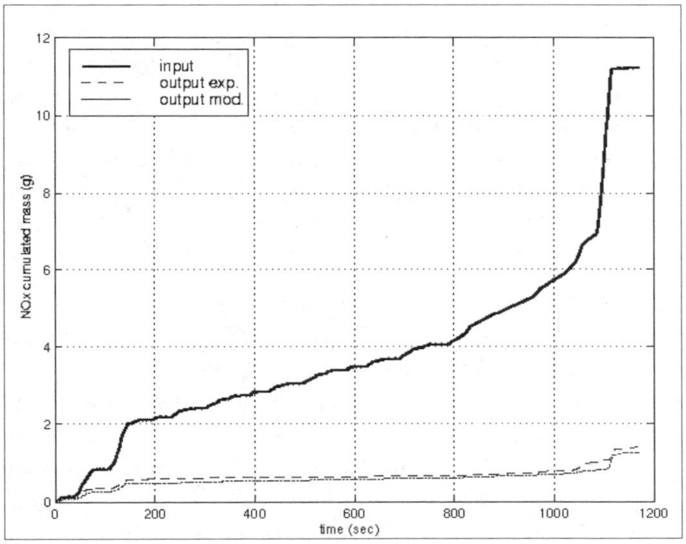

Figure 4b. Comparison between experimental and simulated values of cumulated NOx mass in storage/regeneration conditions using engine data. Five regeneration phases have been performed (at instants t=410, 643, 835, 912 and 993).

A SIMPLIFIED MODEL FOR CONTROL – In this section, we show how to use (9a) to synthesize a even more simplified model which can be used in on-line control strategies. To begin with, we introduce some definitions.

Definitions

- Molar flow rate Q_i: is the number of moles per second of a species *i*. Q_i and C_i are related with the following equation:

$$Q_i = V_g S_{cata} C_i = V_{cata} C_i \frac{V_g}{L}$$

where V_{cata} and S_{cata} are respectively the volume and the cross-sectional area of the catalyst.

- NOx stored quantity *NS*: is the number of moles of NOx stored in the catalyst. It is given by the following equation:

$$NS = V_{cata} N_{sc} \Theta$$

- NOx maximal stored quantity *NSC*: for a given input condition. Due to the competition between adsorption and desorption phenomena, the NOx storage capacity is not necessarily achievable. We can define a maximal stored quantity. C_{NOx}^{in} ding to the model *NSC* depends only on *T* and :

$$NSC(T, C_{NOx}^{in}) = \frac{N_{sc}}{1 + \left(\frac{k_1(T)}{k_2(T) C_{NOx}^{in}} \right)^{1/2}}$$

- Storage efficiency Eff_s: is the ratio between the flow rate of the treated (stored) NOx and that of NOx at the entry of catalyst during lean conditions:

$$\mathrm{Eff_s} = \frac{Q_{NOx}^{in} - Q_{NOx}}{Q_{NOx}^{in}}$$

According to the model, $\mathrm{Eff_s}$ depend on (T,V_g, C_{NOx}^{in}, Θ)

- Regeneration efficiency Eff_p: : it is the ratio between the flow rate of treated reducers and that of reducers at the entry of the catalyst:

$$\mathrm{Eff_p} = \frac{Q_{Red}^{in} - Q_{Red}}{Q_{Red}^{in}}$$

According to the model, Eff_p depend on *(T, V_g,Θ)*

We remind that Q_{NOx} and Q_{Red} are the flow rates of respectively NOx and equivalent reducer at the output of the catalyst.

Using the previous definitions, the model (9a) can be rewritten:

$$\lambda > 1 \begin{cases} \frac{dNS}{dt} = \mathrm{Eff_s} Q_{NOx}^{in} \\ Q_{NOx} = (1 - \mathrm{Eff_s}) Q_{NOx}^{in} \end{cases}$$
$$\lambda < 1 \begin{cases} \frac{dNS}{dt} = -\mathrm{Eff_p} Q_{Red}^{in} \\ Q_{Red} = (1 - \mathrm{Eff_p}) Q_{Red}^{in} \end{cases} \qquad (9b)$$

To further simplify this model (9b), we consider only the main dependencies of Eff_s and Eff_p. Experimentally or by simulation, we can show that, in a first order approximation, these two efficiencies depend only on T and Θ. In addition, NSC can be assumed to depend only on T and the dependencies of Eff_s and Eff_p on Θ can be replaced by a dependency on the ratio NS/NSC.

The model obtained with these assumptions is given in Figure 5.

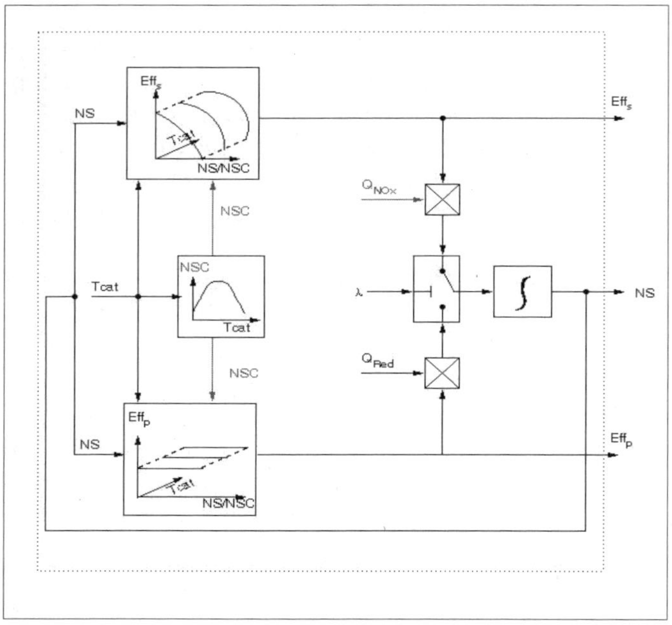

Figure 5. Simplified model structure

ON LINE CONTROL STRATEGY

In this section, we propose a control strategy to achieve desired efficiencies on NOx and HC/CO. This strategy uses the model structure proposed in the previous section. We decompose the strategy into two parts. The first part is called "basic strategy", it supposes that all the parameters of the NOx trap model are well known. The second is called "adaptive strategy", It adapts the main parameters of the catalyst model on a signal delivered by an oxygen sensor placed downstream of the catalyst.

BASIC STRATEGY – The goal of this strategy is to determine the beginning and the end of regeneration phases. The main idea is to estimate the NOx stored quantity and the NOx efficiency, in lean conditions. The regeneration is triggered when the NOx efficiency drops below a threshold S1. It is stopped when the NOx stored quantity NS drops below another threshold S2.

A variation of the thresholds S1 and S2 allows the adaptation of global efficiencies on NOx and HC/CO. On one hand, it is clear that decreasing S1 raises the NOx efficiency. On the other hand, increasing S2 leads to a higher value of the NOx stored quantity at the end of the regeneration phase which implies that HC/CO efficiency increases (see the expression of reaction rate $r_{3\ in}$ (4c)).

The values of S1 and S2 depend on catalyst temperature T and their shape vs. T is chosen to deactivate the strategy at low and high temperatures where the NOx storage capacity is low.

To estimate NOx efficiency and NOx stored quantity, two types of inputs are needed:

- Engine data:
  - NOx flow rate of the engine Q_{NOx}
  - Reducers flow rate Q_{Red}
- NOx trap data
  - NOx Maximal storable quantity NSC
  - Catalyst temperature T
  - Storage efficiency Eff_s
  - Regeneration efficiency Eff_p

Figure 6 illustrates the principle of this strategy.

ADAPTIVE STRATEGY – The inputs of the basic strategy are calibrated off-line using a known engine and catalyst. Unfortunately, due to the manufacturing dispersion and/or aging, often these values are drifting. The role of the adaptive strategy is to correct these inputs on-line using a signal of an oxygen sensor placed downstream of the catalyst. The basic principle of this strategy is illustrated in figure 7.

We assume that Eff_s, Q_{Red} and T are relatively well known and we adapt only Q_{NOx}, NSC and Eff_p. It can be easily shown that simultaneous adaptation of these three parameters is impossible. In fact, the same variation of downstream lambda sensor can be due to an infinite variety of Q_{NOx}, NSC and Eff_p. The idea here is to adapt them successively. For this, we exploit the non-linear characteristics of the catalyst, which allows placing it in conditions such that only one parameter is influenced.

The adaptive strategy modifies the values of Q_{NOx}, Eff_p, NSC by three adaptation factors according to the following equations:

$$Q_{NOx}^{est} = \alpha\, Q_{NOx}$$

$$\mathrm{Eff}_p^{est} = \beta\, \mathrm{Eff}_p$$

$$NSC^{est} = \gamma\, NSC$$

We decompose the adaptive strategy into four steps.

Step 0: NS measurement and initialization – During the regeneration, we can estimate the quantity ΔNS of NOx treated with an efficiency equal to 1. To calculate NS, it is necessary to estimate the residual quantity denoted by NS^{res}, which is performed using the following initialization procedure.

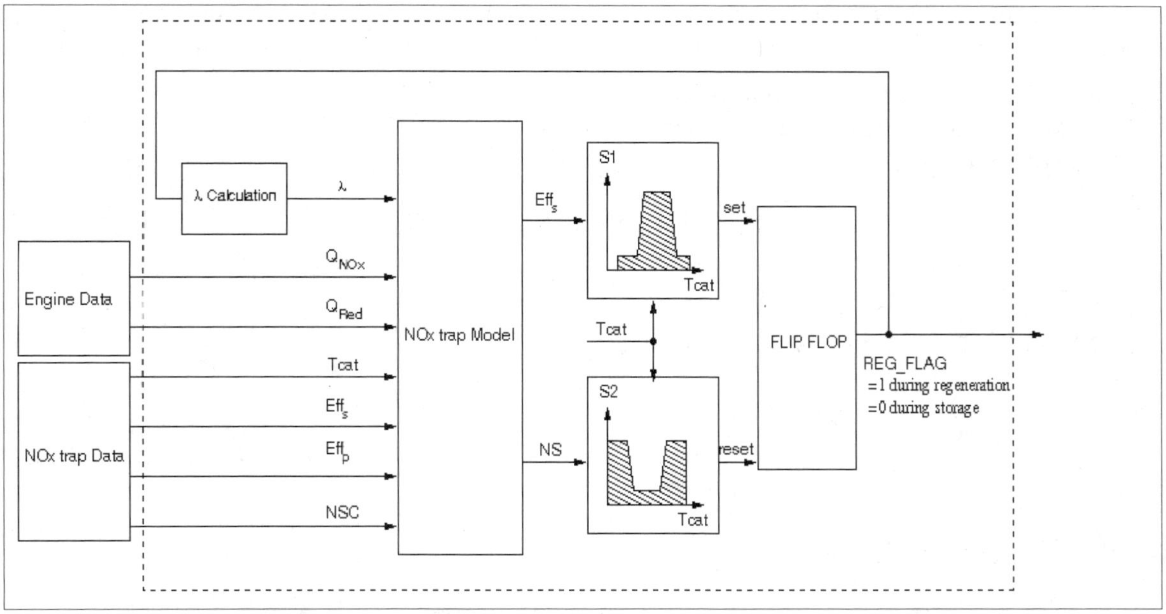

Figure 6. Basic strategy – open loop NOx trap management

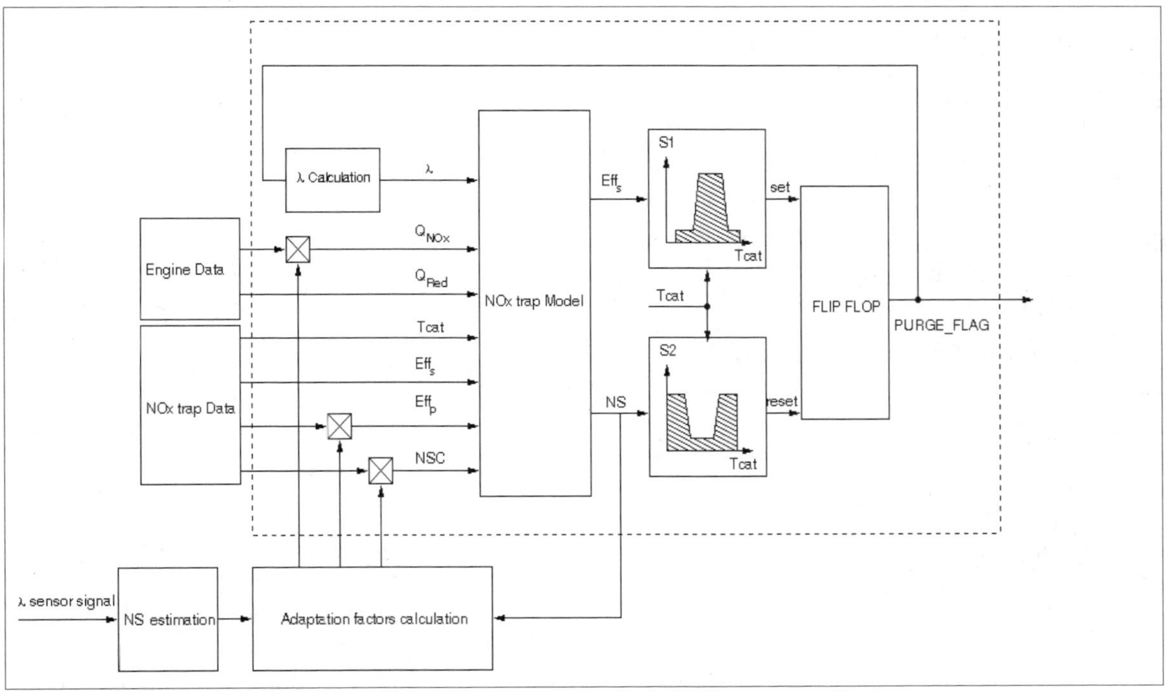

Figure 7. Adaptive strategy

First, we realize a long regeneration to ensure that NS is zero. Next, two successive cycles of storage and regeneration are performed. This procedure is illustrated in Figure 8. It allows calculating ΔNS_1 and ΔNS_2, NS^{res} is then calculated using the formula:

$$NS_{res} = \Delta NS_2 - \Delta NS_1$$

<u>Step1: Adaptation of Q_{NOx}</u> – Adaptation factors are fixed to a values β_0 and γ_0 such that:

$$\beta_0 < 1$$
$$\gamma_0 < 1$$

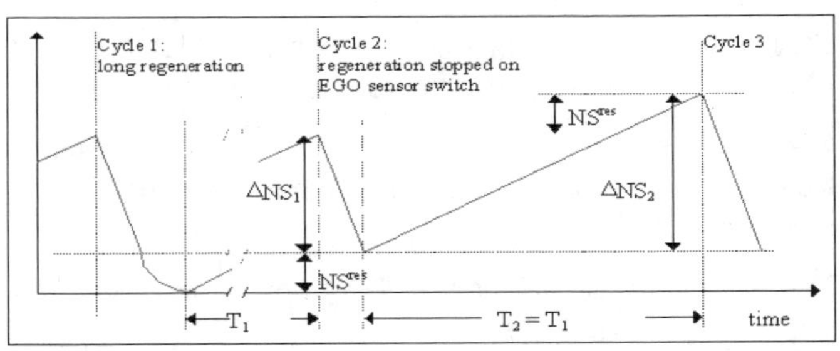

Figure 8. Initialization procedure

The second condition implies that $NSC^{est} < NSC$ which ensures that NS≅0 after each regeneration phase. The first condition ensures that any difference between NS and NS^{est} results from en error on Q_{NOx} value.

The adaptation factor α is modified such that its value decreases if $NS^{est}<NS$ and increases otherwise. The first adaptation phase is stopped when:

$$\left|NS - NS^{est}\right| < S_{\alpha}$$

where S_{α} is a calibrated threshold.

The evolution of NS and NS^{est} during this adaptation is given in figure 9

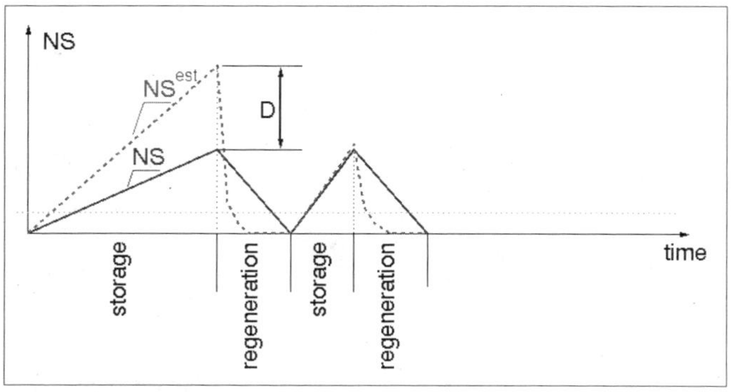

Figure 9. Evolution of NS during Q_{NOx} adaptation

Step2: Adaptation of Eff_p – We fix α to its value resulting from the first adaptation step and γ equal to γ_0.

Adaptation factor β is increased until $NS^{est} > NS$. If this condition is true, the adaptation is terminated by canceling the last increase of β.

The evolution of NS and NS^{est} during this adaptation is given in figure 10.

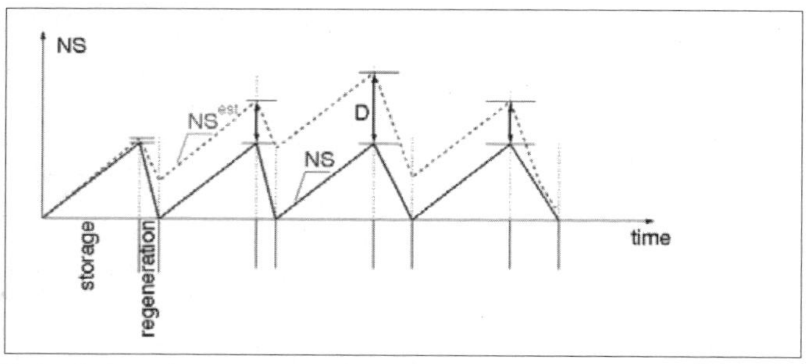

Figure 10. Evolution of NS during Eff_p adaptation

Step 3: Adaptation of NSC – We fix α and β at the values obtained by the two previous adaptation phases.

Next, γ is increased until $NS^{est}>NS$. If this condition is true, the adaptation is terminated by canceling the last increase of γ.

The evolution of NS and NS^{est} during this adaptation phase is given in figure 11.

The three different phases of the adaptation strategy are illustrated by the in Figure 12.

VALIDATION OF THE ADAPTIVE STRATEGY – To validate the adaptive strategy, we have used numerical simulation. We have implemented the basic strategy (including the reduced model) and the adaptive strategy. To represent a real catalyst we have implemented the model (9a).

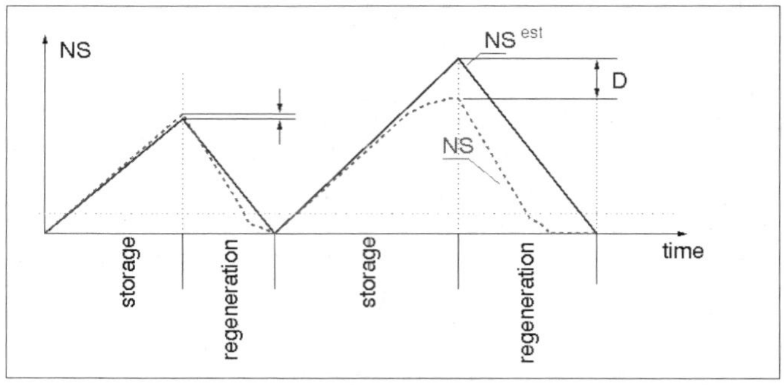

Figure 11. Evolution of NS during NSC adaptation

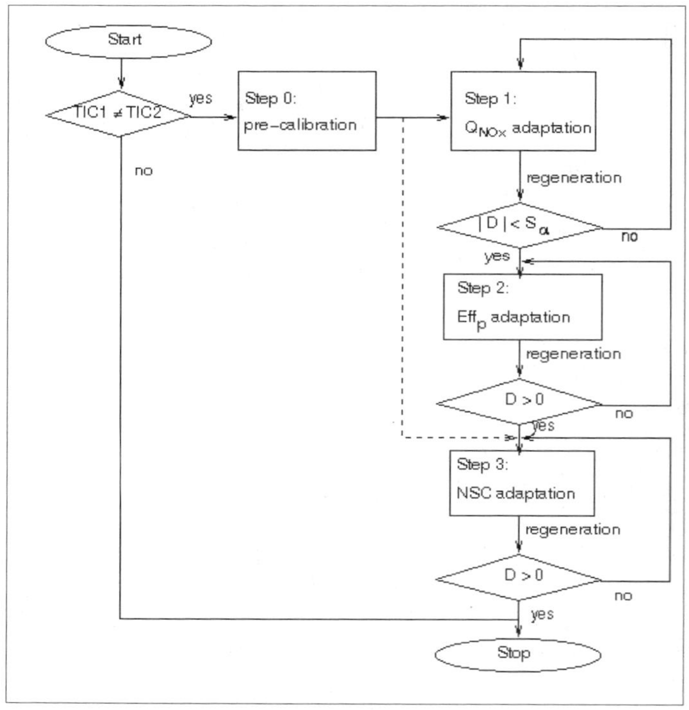

Figure 12. Logical scheme of adaptive strategy

Figures 13 and 14 present numerical results from a test performed on steady state conditions. Figure 13 presents the evolution of adaptation factors during adaptation phases. he adaptation algorithm converges after 11 adsorption/regeneration cycles. Figure 14 illustrates the evolution of output NOx flow rate before and after adaptation. The following table gives the performances of the NOx trap before and after adaptation:

	Before adaptation	**After adaptation**
NOx efficiency	0.90	0.97
HC/CO efficiency	0.13	0.93

The results given below correspond to simulation without any error on measurement. Introducing an error of 10% on NS measurement implies a 4% difference in NOx and HC/CO efficiencies.

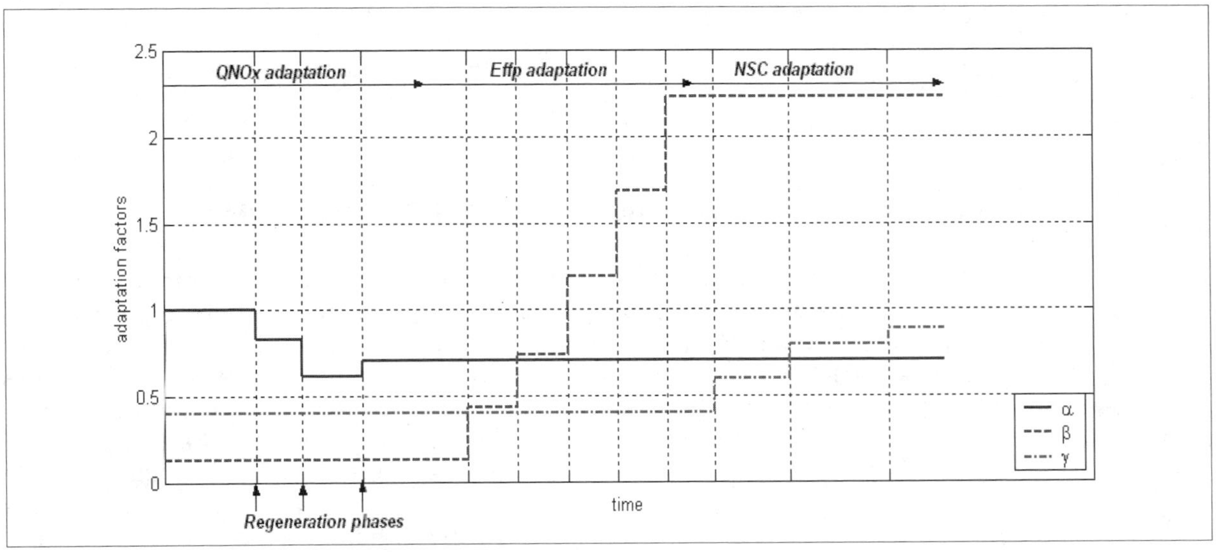

Figure 13. Evolution of adaptation factors

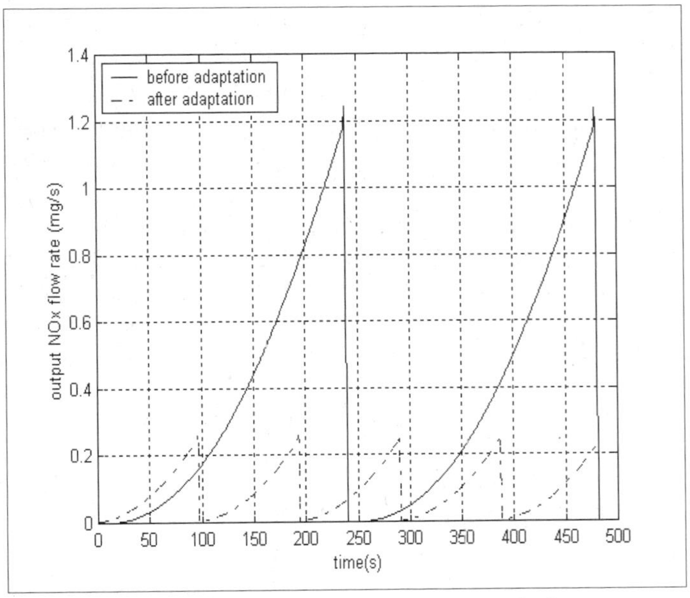

Figure 14. Output NOx flow rate before and after adaptation

CONCLUSION

We have shown that the application of well know reduction techniques on a one dimensional PDE model of a NOx trap catalyst can yield a reduced model with good performances on both NOx and HC/CO efficiencies. The reduced model is sufficiently simple to be integrated in an on-line monitoring strategy. A particular adaptive monitoring strategy using a downstream EGO sensor has been introduced and compared to a non-adaptive one. It has been shown that the adaptive strategy optimizes both NOx and HC/CO efficiencies. The quality of adaptation depends on the accuracy of the measured system output. Special care in the design of the output-manifold is necessary to ensure good observability of the quantity of NOx stored in the catalyst using the output signal of the EGO sensor during regeneration.

REFERENCES

1. F. Aimard, S. Li and M. Sorine, "Mathematical modeling of automotive three way catalytic converters with oxygen storage capacity", Control Eng. Practice, Vol 4, No 8, pp. 1119-1124, 1996
2. S. Beurthey, D. Cozler, D. A. Albert and J.-C. Mollet "Procédé de commande de purge d'un pot catalytique de traitement des gaz d'échappement d'un moteur à combustion interne", French Patent 2772428 A1, 1997.
3. M. S. Brogan and al, "Evaluation of NOx Storage Catalysts as an Effective System for NOx Removal from the Exhaust Gas of Leanburn Gasoline Engines", SAE paper 95490, 1995
4. D. Dou and O. H. Bailey, "Investigation of NOx adsorber Catalyst Deactivation", SAE paper 982594, 1998
5. N. Fekete and al, "Evaluation of NOx storage Catalysts for Lean Burn Gasoline Fueled Passenger Car", SAE paper 970746, 1997
6. J. S. Hepburn, E. Thanasiu, D. A. Dobson and W. L. Watkins, "Experimental and Modeling Investigations of NOx Trap Performance", SAE paper 962051, 1996
7. J. S. Hepburn, T. Kenney, J. MvKenzie, E. Thanasiu and M. Dearth, "Engine Aftertreatment Modeling for Gasoline Direct Injection, SAE paper 982596, 1998
8. T. Kobayashi, T. Yamada and K. Kayano, "Study of NOx Trap Reaction by Thermodynamic Calculation"
9. N. Miyoshi and al, "Development of New Concept Three-Way Catalyst for Automotive Lean-Burn Engines", SAE paper 950809, 1995
10. A. Takami, T. Takemoto, H. Iwakuni, F. Saito and K. Komatsu, "Development of Lean Burn Catalyst", SAE paper 959746, 1995.
11. S. Takeshima and al, "Device for detecting deterioration of NOx absorbent", European Patent Application 0735250 A2,

DEFINITIONS, ACRONYMS, ABBREVIATIONS

EGO sensor: Exhaust Gas Oxygen sensor
PDE: Partial Differential Equation
ODE: Ordinary Differential Equation
NS: NOx Stored quantity
NSC: NOx Storage Capacity

APPENDIX : NS MEASUREMENT USING EGO SENSOR

The estimation of NS is performed during the regeneration using the EGO sensor signal. This sensor switches from low signal to high signal when the downstream gas mixture becomes rich and switches inversely when this mixture becomes lean. In stoichiometric mixture the sensor signal do not change.

The form of EGO sensor signal during regeneration is illustrated in figure 15.

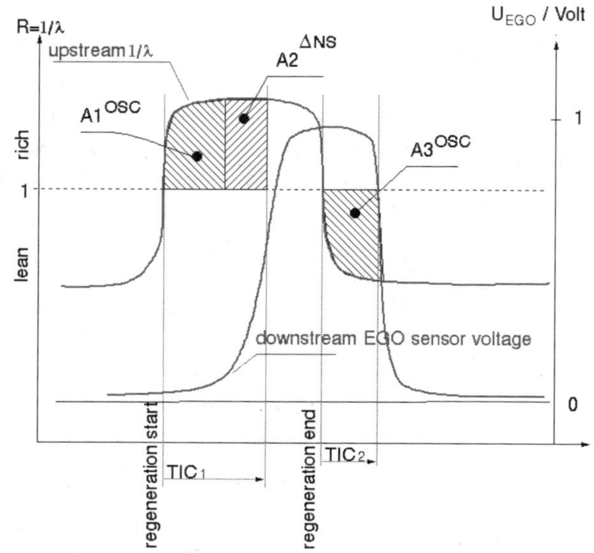

Figure 15. Evolution of lambda sensor signal during regeneration phase

The EGO sensor switches when the downstream gas mixture becomes rich. The time between the start of purge and the EGO sensor switch is called "catalyst integration time", it is denoted by TIC_1. During this time, the reducer species provided at the input of the catalyst are first used to purge the stored oxygen and next used to purge stored NOx. Until the purge efficiency drops below 1, some reducers exhaust from the catalyst and the sensor switches. Therefore, the time TIC_1 allows to calculate the sum of the oxygen storage capacity OSC and a proportion of NOx purged with efficiency equal to 1. This proportion is denoted by ΔNS, and the residual proportion is denoted by NS^{res}.

The EGO sensor switches from high to low signal when the output gas mixture becomes lean. The time between the end of purge and the EGO sensor switch is denoted by TIC_2. During this time, the oxygen provided by the lean input gas mixture is stored in the catalyst. Therefore, TIC_2 allows calculating OSC. Consequently, the values of TIC_1 and TIC_2 allow calculating ΔNS. The value of NS is obtained by adding NS^{res}, and it is given by the following formula:

$$NS^{mes} = TIC_1 Q_{Red} - TIC_2 Q_{O2} + NS^{res}$$

The reducer and the oxygen flow rates, Q_{Red} and Q_{O2} are calculated using engine parameters.

2005-01-0966

Modeling and Control of a Urea-SCR Aftertreatment System

John N. Chi and Herbert F. M. DaCosta
Cummins, Inc.

ABSTRACT

A dynamic system model for simulating the transient performance of a NO_x aftertreatment system using Selective Catalytic Reduction with urea as a reductant (urea-SCR) was developed, calibrated for a heavy-duty engine application, and used to develop a closed loop self-tuning control strategy. The closed loop controller was able to reduce the FTP cycle NO_x emissions from a Cummins heavy-duty engine by 84% while maintaining the mean ammonia slip below 7 ppm and the peak ammonia slip below 55 ppm. The peak ammonia slip occurred during the LA Freeway phase of the FTP cycle.

Components of the urea-SCR aftertreatment system model include a urea dosing system, an exhaust pipe and a fresh vanadia-based SCR catalyst. The urea dosing system model incorporates the evaporation, thermolysis and hydrolysis stages in the conversion of urea to ammonia in the exhaust pipe and on the catalyst. The catalyst model is a 2-dimensional model that incorporates the heat and mass transfer characteristics of a monolith channel, and the chemical kinetics of NO_x conversion by ammonia. The Nusselt number, Sherwood number, and reaction probability are calculated as a function of axial position along the monolith channel. Results from a Cummins heavy-duty engine application were used to calibrate the dynamic system model and parametric studies were carried out to quantify the effect of ammonia storage capacity on NO_x conversion and ammonia slip.

A closed loop self-tuning control strategy with on-line adaptation of the controller gains was designed and implemented on a Cummins heavy-duty urea-SCR aftertreatment system with a rapid prototyping tool. The composite adaptive controller is based on a Model-Reference Adaptive Control (MRAC) system for a first-order plant with composite adaptation law. The controller uses time varying input information for the desired NO_x reduction rate, catalyst inlet exhaust gas temperature, catalyst outlet exhaust gas temperature, catalyst inlet NO_x emissions rate, and catalyst outlet NO_x emissions rate to determine the urea solution dosing rate.

INTRODUCTION

With the increasing demands for low engine emissions, urea-SCR with vanadia-based catalysts is one of the aftertreatment options for reducing NO_x emissions. The schematic of the engine and urea-SCR aftertreatment system presented in this paper is given in Figure 1.

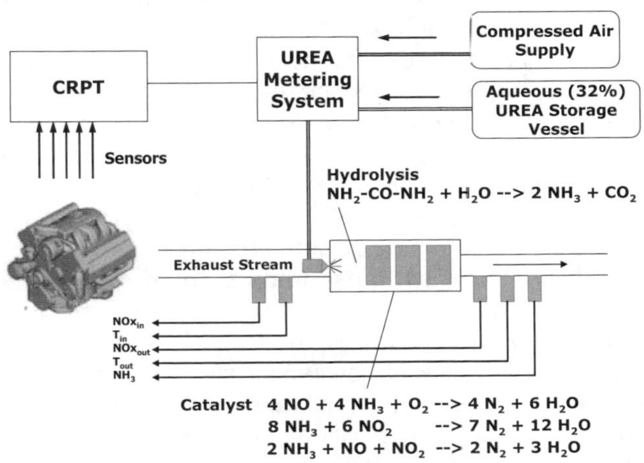

Figure 1: Engine and urea-SCR aftertreatment system

Table 1: Engine Information

Manufacturer	Cummins, Inc.
Model	ISX-450
Features	Non-EGR, VGT
Number of Cylinders	6
Displacement	15.0 liters
Rated Power	450 HP @ 1800 RPM
Peak Torque	1850 Ft-Lbf @ 1200 RPM
Rated Flow Rate	77.8 lbm/min
Base Engine BSNOx	6 grams/hp-hr

The experimental setup consists of a Cummins heavy-duty diesel engine (see Table 1 for details about the engine), an extruded $V_2O_5/WO_3/TiO_2$ catalyst, a urea dosing system, and a Cummins designed Controls Rapid Prototyping Tool developed in-conjunction with Applied Dynamics Incorporated. The after-treatment system was instrumented with exhaust gas temperature sensors at the inlet and outlet of the catalyst, NO_x sensors at the inlet and outlet of the catalyst, and a real-time AltOptronic ammonia analyzer at the tailpipe. Aqueous urea solution was added to the exhaust with compressed air for atomization. The urea solution decomposes to ammonia which reduces engine exhaust oxides of nitrogen (NO_x) to nitrogen (N_2) over the catalyst. When urea is used as a reductant, it is important to understand the details of its roles in the NO_x reduction as it can have serious consequences on the overall process. Model requirements for urea-SCR aftertreatment technology development can be classified into research models, dynamic system models, and embeddable models. Research models will include CFD (computational fluid dynamics) type models which can provide the details to capture the fundamentals of the technology, dynamic system models like the one presented in this paper will include distributed system type models, and embeddable models will include reduced-order models that can be embedded in a controller to run in real time.

The process of modeling engineering systems is a result of a compromise between two opposing trends. In one hand, one needs to use as much mathematical depth as it takes to correctly describe the relevant phenomena. On the other hand, the model needs to be as simple as possible, in order to convey a simple and useful picture of real systems. In the most general case, modeling aftertreatment processes is about mathematically describing the three transport processes occurring on the catalyst surfaces: mass transfer, heat transfer, and momentum transfer. The reaction rates enter as source terms in the equations describing the transport processes. In real systems all the phenomena mentioned above occur simultaneously. However, some assumptions and approximations need to be made during the model development in order to make it less complex and to allow its practical use. Applying that modeling philosophy, the distributed type dynamic system modeling approach was selected for our purposes because it provides the framework for incorporating the mechanisms of chemical reaction kinetics, pore diffusion, and bulk mass transfer characterizing the conversion rate in homogeneous and heterogeneous catalysts. This modeling approach has been demonstrated to be adequate in predicting trends and was used to understand the critical parameters for reducing NO_x while minimizing the amount of ammonia slip in a urea-SCR aftertreatment system during transient operating conditions.

The dynamic system model presented in this paper takes the engine outlet NO_x concentration, exhaust gas mass flow rate, exhaust gas temperature, and urea dosing rate as inputs and predicts the system responses such as NO_x reduction, ammonia slip, and the amount of ammonia stored on the catalyst surface under transient conditions. This paper is organized in two major sections. In the next section, the dynamic models for the different components of the urea-SCR aftertreatment system and the results from parametric studies are presented. That is followed by the control system model used for reducing NO_x while minimizing ammonia slip, and then a summary of the conclusions from the results.

DYNAMIC SYSTEM MODEL

The dynamic system model for a urea-SCR aftertreatment system consists of two main components: a urea dosing system that injects urea solution as droplets into the exhaust pipe and a $V_2O_5/WO_3/TiO_2$ catalyst model.

UREA DOSING SYSTEM MODEL

The urea dosing system model comprises the injection system dynamics as well as the thermodynamic and kinetic components of the urea decomposition. The hydrolysis of urea solution to ammonia and carbon dioxide can be represented by the following chemical reaction:

$$NH_2-CO-NH_2(aq)+H_2O(g)\rightarrow 2NH_3(g)+CO_2(g) \quad (1)$$

The urea dosing system model incorporates the injection system dynamic model, the 4-step mechanism of urea decomposition to ammonia suggested by Koebel et al. [16], and the spatial mixing dynamics in the exhaust pipe upstream of the urea-SCR catalyst. Urea solution from the urea storage tank is introduced as droplets into the exhaust pipe by the actuator. This physical process can be represented by:

$$NH_2-CO-NH_{2(sol)}\rightarrow NH_2-CO-NH_{2(droplets)} \quad (2)$$

The atomization of the urea solution causes another physical transformation, i.e., the evaporation of water from the droplets in hot exhaust yielding solid or molten urea (Koebel et al. [16]). This process can be represented by:

$$NH_2-CO-NH_2(aq)\rightarrow NH_2-CO-NH_2(s) + xH_2O(g) \quad (3)$$

The simple thermolysis mechanism, which assumes that pure urea thermally decomposes to equimolar amounts of ammonia gas and isocyanic acid, described by Koebel et al. [16] was adapted for this model. This simple thermolysis process can be represented by:

$$NH_2-CO-NH_2(s)\rightarrow NH_3(g)+HNCO(g) \quad (4)$$
$$\Delta H^{298}=185.5\ \text{kJ/mol}$$

Alternatively, one can also consider the direct hydrolysis of aqueous urea into ammonia and carbon dioxide i.e., skipping the evaporation process represented by equation (3). In that case, the appropriate kinetic and thermodynamic parameters should be used. The interface representation of the urea dosing system model for the formation of gaseous ammonia, isocyanic acid, and water vapor from urea solution is represented in Figure 2. The inputs to the model include the urea dosing rate, the exhaust gas temperature, and the exhaust gas mass flow rate. The outputs of the model are the ammonia gas concentration, isocyanic acid concentration, and water vapor concentration at the face of the catalyst. Ammonia and isocyanic acid undergo further chemical reactions on the catalyst surface as discussed in the next section.

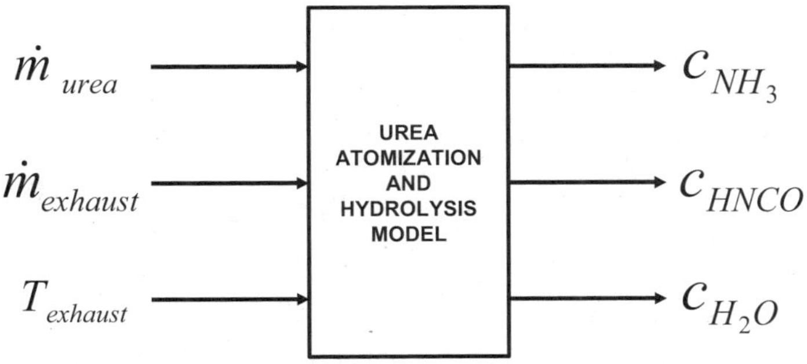

Figure 2: Inputs and Outputs for the Urea Dosing System Model.

The reduced order dynamic system model for the formation of gaseous ammonia and isocyanic acid from urea droplets in the exhaust pipe is given by:

$$\frac{\dot{m}_{NH_3}(s)}{\dot{m}_{urea}(s)} = \frac{1}{\tau s+1}\left(1-e^{-x/L}\right)\eta_{NH_3}(\dot{m},T) \tag{5}$$

$$\frac{\dot{m}_{HNCO}(s)}{\dot{m}_{urea}(s)} = \frac{1}{\tau s+1}\left(1-e^{-x/L}\right)\eta_{HNCO}(\dot{m},T) \tag{6}$$

where τ is the mixing time constant, L is the characteristic mixing length, x is the distance from the urea dosing system to the catalyst face, η is the conversion efficiency.

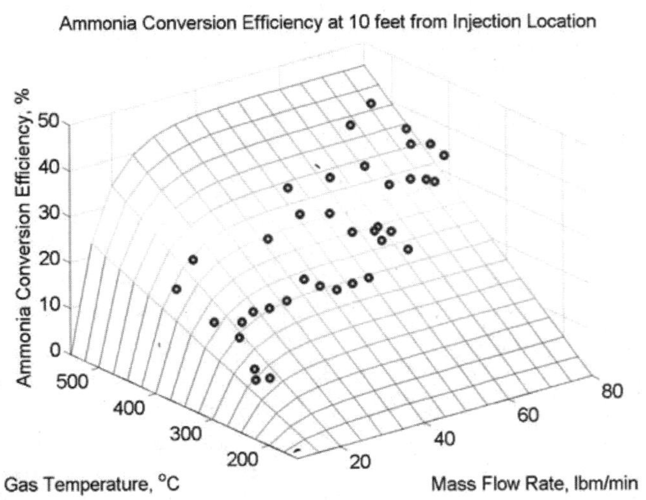

Figure 3: Steady state ammonia conversion efficiency model at an axial location of 10 feet from the injection location.

The steady state conversion efficiency was modeled as a function of the exhaust gas flow rate and the exhaust gas temperature using parametric curves from the stepwise thermolysis of urea study by Fang and DaCosta [7]. Figure 3 shows the steady state ammonia conversion efficiency map used for the urea dosing system model given in equations (5) and (6). The circles represent ammonia conversion efficiency measurements taken at an axial location of 10 feet from the injector. The maximum conversion efficiency observed was 45%. The chemical reaction in equation (4) showing the decomposition of urea into equimolar amounts of ammonia and isocyanic acid implies that the maximum conversion efficiency prior to the catalyst is 50%. Thus the data supports the 4-step mechanism of urea decomposition to ammonia described by Koebel et al. [16] and adopted for the urea dosing system model presented in this paper.

SCR CATALYST MODEL

The SCR catalyst model consists of a thermal component and a chemistry component for a generic channel of a monolith reactor. Figure 4 shows the axial view of a monolith channel used for the SCR catalyst model development.

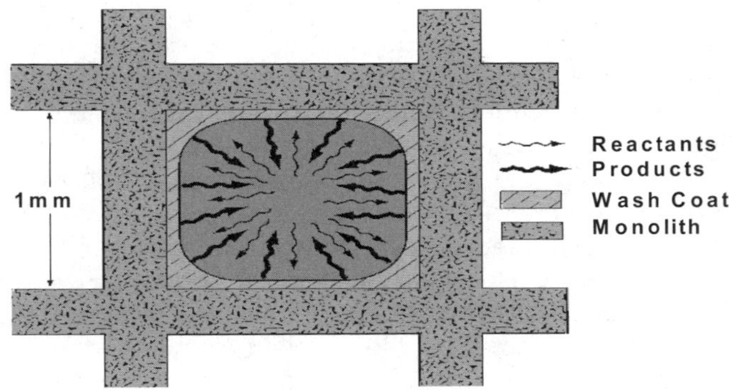

Figure 4: Axial view of a generic monolith channel from Daw et al. [5].

The thermal component models the temperature changes of the exhaust gas and catalyst substrate material resulting from the transfer of heat between the exhaust gas mixture flowing through the catalyst, the catalyst substrate structure, the surroundings, and the energy generated from the chemical reactions taking place at the catalyst surface. The chemistry component models the mass transfer of gas species between the exhaust bulk gas mixture and the catalyst surface/washcoat across the boundary layer, the chemical reactions on the catalyst surface/washcoat, and the diffusion of gas species into the monolith substrate.

Thermal Model

The thermal model is a one-dimensional channel flow model that was obtained by applying an energy balance on a differential control volume of fluid element in the monolith void passage and the corresponding differential element of the solid wall. This thermal model assumes that the solid wall temperature varies only in the axial direction, and heat transfer by radiation was neglected because the temperatures involved correspond to energies that are too low to cause significant changes in molecular states. The resulting one-dimensional partial differential equations from the gas phase energy balance and solid phase energy balance respectively for the catalyst thermal model are:

$$\varepsilon\rho_g C_{p,g}\frac{\partial T_g}{\partial t} = -\varepsilon\rho_g C_{p,g} u\frac{\partial T_g}{\partial x} - hS_v\left(T_g - T_w\right) \quad (7)$$

$$(1-\varepsilon)\rho_w C_{p,w}\frac{\partial T_w}{\partial t} = (1-\varepsilon)\lambda_w\frac{\partial^2 T_w}{\partial x^2} + hS_v\left(T_g - T_w\right) - h_o S_{vo}\left(T_w - T_\infty\right) + \sum_j \Delta H_j R_j S_a \quad (8)$$

where ε is the void fraction of the monolith reactor, ρ_g is the bulk gas density of the exhaust gas, ρ_w is the monolith wall density, $C_{p,g}$ is the specific heat capacity of the exhaust gas, $C_{p,w}$ is the specific heat capacity of the monolith material, λ_w is the thermal conductivity of the monolith material, u is the bulk exhaust gas velocity, S_v is the monolith internal geometric surface area per unit monolith volume, S_{vo} is the monolith external geometric surface area per unit monolith volume, S_a is the catalytic surface area per unit washcoat volume also referred to as washcoat loading or activity factor, h is the local heat transfer coefficient between the exhaust gas mixture and the monolith wall, h_o is an equivalent heat transfer coefficient for the heat loss to the environment, ΔH_j is the enthalpy of reaction for the jth reaction taking place at the catalyst surface, R_j is the reaction rate for the jth reaction taking place on the catalyst surface, T_g is the bulk exhaust gas temperature, T_w is the monolith wall temperature, and T_∞ is the ambient temperature. The inlet and outlet boundary conditions of the monolith were assumed to be adiabatic. The heat transfer coefficient was derived from the following boundary layer transport relation:

$$Nu = \frac{hL}{k_c} = f(\mathrm{Re}, \mathrm{Pr}) \quad (9)$$

where Nu is the Nusselt Number, h is the heat transfer coefficient, L is the characteristic length, k_c is the thermal conductivity, Re is the Reynolds Number, and Pr is the Prandtl Number.

Chemistry Model

The chemistry model has two components: a NO_x reduction model and an ammonia storage model. The chemical reactions incorporated in the NOx reduction model include the hydrolysis of isocyanic acid, the NO/NO_2 reduction reactions, and the ammonia oxidation reactions. The hydrolysis of isocyanic acid to ammonia and carbon dioxide on the SCR catalyst is represented by the following chemical reaction:

$$HNCO(g) + H_2O(g) \rightarrow NH_3(g) + CO_2(g) \quad (10)$$

$$\Delta H^{298} = -95.914 \text{ kJ/mol}$$

Studies by Kleemann et al. [15] have shown that the HNCO hydrolysis reaction on SCR catalysts is mainly limited by external and internal mass transfer. First-order rate expressions were used for this hydrolysis reaction and the rate constants were calibrated for the application presented in this paper.

The three main NO/NO_2 reduction reactions incorporated in the model were as follows:

$$NH_3 + \frac{1}{2}NO + \frac{1}{2}NO_2 \rightarrow N_2 + \frac{3}{2}H_2O \quad (11)$$

$$\Delta H^{298} = -378.534 \text{ kJ/mol}$$

$$NH_3 + NO + \frac{1}{4}O_2 \rightarrow N_2 + \frac{3}{2}H_2O \quad (12)$$

$$\Delta H^{298} = -407.129 \text{ kJ/mol}$$

$$NH_3 + \frac{3}{4}NO_2 \rightarrow \frac{7}{8}N_2 + \frac{3}{2}H_2O \quad (13)$$

$$\Delta H^{298} = -341.664 \text{ kJ/mol}$$

The ammonia reaction in equation (11) requiring equimolar amounts NO and NO_2 is much faster than the ammonia reaction with NO alone in equation (12) which, in turns, is faster than the ammonia reaction with NO_2 alone in equation (13). Most of the reaction rate calibration refinements were made on the ammonia reactions in equations (11) and (12) since diesel exhaust NO_x is predominantly NO (as a rule of thumb, ~90% NO and 10% NO_2).

To capture the catalyst behavior at high temperatures, the following oxidation of ammonia reactions were incorporated in the model:

$$NH_3 + \frac{3}{4}O_2 \rightarrow \frac{1}{2}N_2 + \frac{3}{2}H_2O \quad (14)$$

$$\Delta H^{298} = -316.839 \text{ kJ/mol}$$

$$NH_3 + \frac{5}{4}O_2 \rightarrow NO + \frac{3}{2}H_2O \quad (15)$$

$$\Delta H^{298} = -226.549 \text{ kJ/mol}$$

The selective oxidation of ammonia to N_2 given in equation (14) has been shown to be dominant above 360°C in many SCR catalysts and the non-selective oxidation of ammonia to NO given in equation (15) is favored on vanadia-based catalysts at temperatures above 400°C. The reaction rate expressions for these ammonia oxidation reactions were obtained from open literature and the reaction rate parameters were modified for our system calibration.

Reduction Model

The NO_x reduction model is a one-dimensional (axial direction) model obtained by applying the conservation of mass for each of the chemical species in the exhaust to a differential control volume of fluid element in the monolith reactor void passage and the corresponding differential element of the solid wall. The resulting one-dimensional partial differential equations from gas phase mass balance and solid phase mass balance respectively for the chemical model are:

$$\varepsilon \frac{\partial C_{g,i}}{\partial t} = -\varepsilon u \frac{\partial C_{g,i}}{\partial x} - k_{m,i} S_v \left(C_{g,i} - C_{s,i}\right) \tag{16}$$

$$(1-\varepsilon)\frac{\partial C_{s,i}}{\partial t} = k_{m,i} S_v \left(C_{g,i} - C_{s,i}\right) \pm \sum_j R_{i,j} S_a \tag{17}$$

where ε is the void fraction of the monolith reactor, u is the bulk exhaust gas velocity, S_v is the monolith internal geometric surface area per unit monolith volume, S_a is the catalytic surface area per unit washcoat volume also referred to as washcoat loading/activity factor, $k_{m,i}$ is the boundary layer mass transfer coefficient for the ith species, $R_{i,j}$ is the rate of reaction for the jth reaction taking place on the catalyst surface affecting the ith species, $C_{g,i}$ is the gas concentration of the wall in equilibrium with solid for the ith species, $C_{s,i}$ is the surface concentration for the ith species, and i represents an exhaust gas species. HNCO, NH_3, NO, NO_2, N_2, O_2, CO_2, and H_2O were the eight gas species incorporated in the NO_x reduction model. The mass transfer coefficient for each of the exhaust gas species was estimated from the following boundary layer transport relation:

$$Sh = \frac{k_m L}{D_m} = f(\mathrm{Re}, Sc) \tag{18}$$

where Sh is the Sherwood Number, k_m is the mass transfer coefficient, L is the characteristic length, D_m is the mass diffusion coefficient, Re is the Reynolds Number, and Sc is the Schmidt Number.

Storage Model

The storage model consists of a kinetic model for the NH_3 adsorption-desorption process occurring at the catalyst surface and a one-dimensional (radial direction) model for diffusion/reaction species balance in the catalyst washcoat or substrate. The kinetic model for the adsorption and desorption of ammonia proposed by Tronconi et al. [25] was used and is given as follows:

$$\Omega_{NH_3}(T)\frac{\partial \theta_{NH_3}}{\partial t} = R_{ads,NH_3} - R_{des,NH_3} - S_a R_{NH_3} \tag{19}$$

where Ω_{NH3} is the catalyst NH_3 adsorption capacity, T is the monolith wall temperature, θ_{NH3} is the ammonia surface coverage, $R_{ads,NH3}$ is the catalyst surface NH_3 adsorption rate, $R_{des,NH3}$ is the catalyst surface NH_3 desorption rate, and $R_{rxn,NH3}$ is the cumulative reaction rate of the chemical reactions affecting ammonia.

The species diffusion model was obtained by applying the law of conservation of species to a differential control volume of the monolith reactor solid wall. Assuming constant diffusion coefficients and one-dimensional transfer in the radial direction, the resulting partial differential equations for the species diffusion dynamics are given as follows:

$$\frac{\partial C_{w,i}}{\partial t} = D_{e,i}\frac{\partial^2 C_{w,i}}{\partial z^2} + R_{ads,i} - R_{des,i} \pm S_a R_{rxn,i} \tag{20}$$

where $D_{e,i}$ is the effective diffusion coefficient for the ith species through the catalyst pores, $R_{ads,i}$, is the adsorption rate of the ith species, $R_{des,i}$ is the desorption rate of the ith species, $R_{rxn,i}$ is the cumulative reaction rate of the chemical reactions on the heterogeneous catalyst surface affecting the ith species, and $C_{w,i}$ is the pore concentration of the ith species. NO and NH_3 were the two species incorporated in the NH_3 storage model. A constant species flux boundary condition was assumed at the catalyst surface. For each of the species, a material balance was applied at the catalyst surface using Fick's law to obtain the boundary relation between the species concentration at the gas-solid interface.

Code Structure and Parameter Estimation

The modeling strategy adapted for this paper consisted of breaking down the urea-SCR aftertreatment system into thermal, reduction, and storage components interacting with each other. A chart showing the structure of the urea-SCR aftertreatment model implementation scheme is presented in the appendix. The model component interfaces were defined such that the interactions between the different components are captured through the interface and the series of ordinary differential equations for each generic component can be implemented independently. This approach also provided the added advantage that the fidelity of each of the model components can be independently improved.

The catalyst monolith channel was divided into multiple sequential stages with uniform conditions assumed within each stage. Using axial and radial discretization techniques, the system of partial differential equations (PDEs) for the "average" monolith channel were converted into a system of ordinary differential equations (ODEs) and differential algebraic equations (DAEs). The system of ordinary differential equations and differential algebraic equations were then implemented in a CMEX s-function [22]. The model requires gas and solid physical properties, boundary layer correlations, gas-solid equilibria data, and surface kinetics data.

The correlation used for equation (9) to estimate the local heat transfer coefficient, h, between the exhaust gas and catalyst surface was the empirical relation by Sieder and Tate [10] for laminar heat transfer in tubes given by:

$$Nu_d = 1.86(\mathrm{Re}_d \Pr)^{1/3}\left(\frac{d}{L}\right)^{1/3}\left(\frac{\mu}{\mu_w}\right)^{0.14} \tag{21}$$

where Nu_d is the Nusselt Number (Nu_d = h d / k_c), Re_d is the Reynolds Number (Re_d = u d / v), Pr is the Prandtl Number (Pr = v / α), d is the hydraulic diameter of the monolith channel, L is the length of the monolith channel, k_c is the thermal conductivity of the exhaust gas mixture, u is the exhaust gas velocity, μ is the dynamic viscosity of the exhaust gas mixture evaluated at the gas temperature, ν is the kinematic viscosity of the exhaust gas mixture ($\nu = \mu / \rho$), ρ is the gas density, μ_w is the dynamic viscosity of the exhaust gas mixture evaluated at the catalyst wall temperature, and α is the thermal diffusivity ($\alpha = k_c / C_p \mu$) with C_p being the specific heat capacity of the exhaust gas mixture.

The correlation used for equation (9) to estimate the equivalent heat transfer coefficient for the heat loss to the environment, h_o, was the empirical relation by Churchill and Chu [10] for free convection from horizontal cylinders given by:

$$Nu^{1/2} = 0.60 + 0.387\left\{\frac{Gr\,\mathrm{Pr}}{\left[1+(0.559/\mathrm{Pr})^{9/16}\right]^{16/9}}\right\}^{1/6} \quad (22)$$

where Nu is the Nusselt Number ($Nu = h_o D / k_c$), Gr is the Grashof Number ($Gr = g \beta (T_w - T_\infty) D^3 / \nu^2$), Pr is the Prandtl Number ($Pr = \nu / \alpha$), D is the catalyst diameter, k_c is the thermal conductivity of the catalyst substrate material, g is the gravitational acceleration, β is the inverse of the film temperature i.e., $\beta = 1 / T_{film}$, $T_{film} = (T_w + T_\infty) / 2$, T_w is the catalyst wall temperature, T_∞ is the ambient temperature, ν is the kinematic viscosity of air, and α is the thermal diffusivity of air.

The correlation used for equation (18) to estimate the external mass transfer coefficient between the exhaust gas and catalyst surface for the different species, $k_{m,j}$, was the empirical relation by Uberoi and Pereira [26] for the gas-solid mass transfer in monolith channels given by:

$$Sh = 2.696\left[1 + 0.139 Sc \,\mathrm{Re}\left(\frac{d_h}{L}\right)\right]^{0.81} \quad (23)$$

where Sh is the Sherwood Number ($Sh_j = k_{m,j} d_h / D_j$), Sc is the Schmidt Number ($Sc_j = \nu / D_j$), Re is the Reynolds Number ($Re = u d_h / \nu$), d_h is the monolith hydraulic diameter, D_j is the molecular diffusivity of the jth species, ν is the kinematic viscosity of the exhaust gas mixture, and u is the exhaust gas velocity.

Simplified kinetics for the surface reactions rates from open literature were modified and used for this model. Arrhenius-type function of temperature was used to model the rate expressions for the different chemical reactions as follows:

$$R_j = k_{o,j} \exp\left(\frac{-E_{a,j}}{R_g T_s}\right) \prod_i C_{s,i}^{\alpha_i} f\left(C_{s,i}, T_s, k_{eq,i}\right) \quad (24)$$

where R_j is the rate of the jth reaction, $k_{o,j}$ is the pre-exponential factor for the jth reaction, $E_{a,j}$ is the activation energy for the jth reaction, R_g is the universal gas constant, T_s is the surface temperature (the catalyst surface temperature was assumed to be equal to the monolith wall temperature, T_w), $C_{s,i}$ is the surface concentration of the ith species, α_i is the reaction order, and $k_{eq,i}$ is the Langmuir equilibrium constant for the ith species.

RESULTS AND DISCUSSION

The calibration parameters required for the urea-SCR dynamic system model presented in this paper consists of catalyst geometric parameters, catalyst properties, properties of exhaust gas species, and the reaction rate kinetic parameters. The geometric parameters which include the monolith void fraction, length of monolith, diameter of catalyst, hydraulic diameter of monolith channel, monolith wall thickness, and geometric surface area were obtained from the catalyst manufacturer's spec sheet. The catalyst properties which include the monolith bulk density, specific heat capacity, and thermal conductivity were obtained from engineering handbooks. The properties of exhaust gas species which include the heat and mass transfer coefficients were obtained from correlations reported in open literature. The kinetic parameters for the reaction rate expression, R_j, in equation (24) for the reactions in equations (10)–(15) incorporated in the chemistry model were obtained from open literature and then calibrated for the urea-SCR aftertreatment system using steady state NO_x conversion measurements from the Euro 13 Mode. A comparison of the steady state NO_x conversion efficiency predicted by the calibrated model and the actual test cell measurements from the Euro 13 Mode test for an NSR of 1.0 is summarized in Table 2. A summary of the kinetic parameters is given in Table 3 in the appendix.

Table 2: Euro 13 Mode Steady State Calibration

Mode	Speed (RPM)	Load (%)	Temperature (°C)	Space Velocity (1/hr)	NOx (ppm)	Conversion (%)	
						Actual	Model
1	600	0	120	2081	150	0.0	0.0
2	1058	25	303	4444	749	74.0	85.1
3	1058	50	425	5907	1146	81.8	87.4
4	1058	75	493	7609	1294	80.8	70.7
5	1058	100	515	9532	1250	56.5	51.7
6	1321	25	290	6292	596	67.1	83.8
7	1321	50	368	8462	1045	89.3	92.6
8	1321	75	416	10880	1176	83.1	81.5
9	1321	100	470	13048	1215	78.7	75.1
10	1584	25	270	7871	497	66.2	78.8
11	1584	50	334	10493	778	80.8	93.3
12	1584	75	376	13166	1003	83.3	90.4
13	1584	100	439	14800	1133	71.6	84.9

The steady state results summarized in Table 2 show that even though the model predicted NO_x conversion efficiencies at low loads were higher than the measured values, the model was able to capture the NO_x conversion efficiency trend for a broad range of exhaust gas temperature and catalyst space velocity. More refinement of the kinetic parameters using molecular modeling calculations and the incorporation of multiphase reaction mechanisms can be used to further improve the accuracy of the model predicted NO_x conversion efficiency values. The fact that the calibrated dynamic system model could reasonably predict the relevant NO_x conversion efficiency with sufficient accuracy thus makes it a suitable simulation tool.

Numerical experiments were performed with the calibrated model to identify the critical parameters to be incorporated in a control system model for maximizing NO_x reduction and minimizing ammonia slip under transient operating conditions. In the first set of numerical experiments, the exhaust gas and monolith temperatures were monitored when the urea-SCR system was subjected to a step change in catalyst inlet gas temperature. Figure 5 shows the simulated responses of mid-span gas temperature, catalyst outlet gas temperature, mid-span monolith temperature, and outlet monolith temperature when the urea-SCR system was subjected to a step change in engine exhaust conditions from the Mode 5 steady state condition to Mode 1 steady state condition (see Table 2).

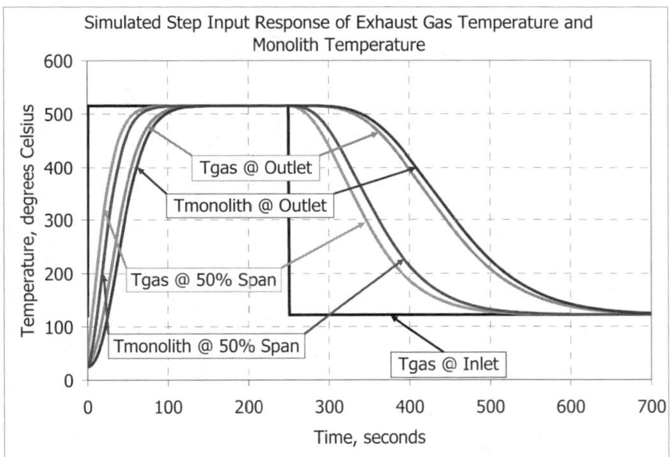

Figure 5: Simulated exhaust gas and monolith wall temperature responses to a step change in engine exhaust gas conditions from Mode 5 to Mode 1.

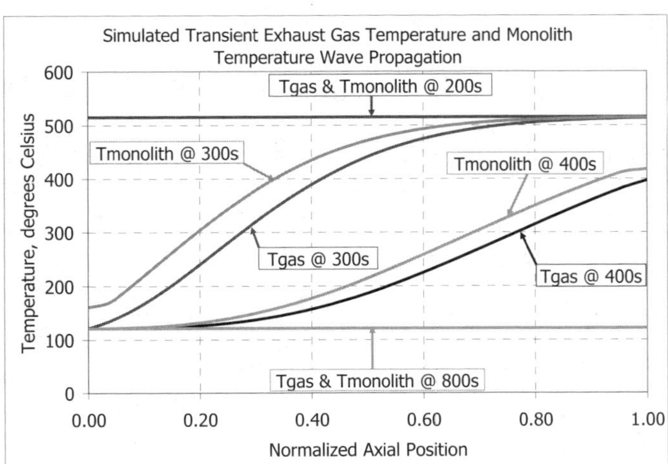

Figure 6: Simulated exhaust gas and monolith wall temperature distributions for a step change in engine exhaust gas conditions from Mode 5 to Mode 1.

Figure 6 shows four snapshots of the corresponding gas and monolith temperature profiles across the catalyst as the urea-SCR system transitions from Mode 5 steady state condition to Mode 1 steady state condition. The simulated transient thermal responses capture the thermal lags in the urea-SCR aftertreatment system, and provide useful information with regards to the relation between the gas and monolith temperatures under transient operating conditions. The control system model described in the next section contains an ammonia slip control algorithm which uses an embedded model of the catalyst bed temperature. This embedded model uses the catalyst inlet and outlet exhaust gas temperature measurements to estimate the catalyst bed temperature, and simulation results from these numerical experiments were used to refine the catalyst bed temperature embedded model.

In the second set of numerical experiments, the NO_x and NH_3 concentrations across the catalyst monolith were monitored when the urea-SCR aftertreatment system was subjected to a step change in urea input. Figure 7 shows the rate of NO/NO_2 reduction reactions, NH_3 oxidation reactions, and corresponding net NO_x reduction rate for Mode 8 steady state exhaust gas concentrations as a function of temperature.

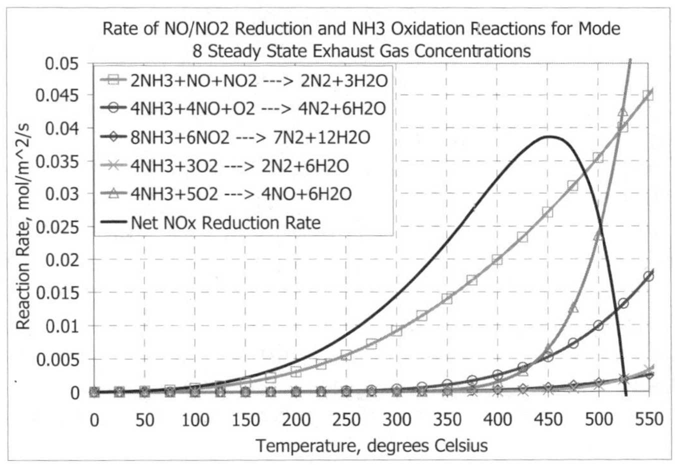

Figure 7: Rate of NO/NO_2 reduction reactions, NH_3 oxidation reactions, and the corresponding net NO_x reduction rate for Mode 8 steady state exhaust gas concentrations (1058 ppm NO, 118 ppm NO_2, 1277 ppm NH_3, 10% O_2, 7.1% CO_2, 5.6% H_2O, and 77.3% N_2) as a function of temperature.

Figure 8 shows the simulated gas-phase NO_x concentration at the catalyst inlet, the gas-phase NO_x concentration at the catalyst outlet, the gas-phase NH_3 concentration at the catalyst inlet and the gas-phase NH_3 concentration at the catalyst outlet (ammonia slip) when the urea-SCR system was subjected to a step urea input equivalent to NSR of 1.1 with the engine operating at Mode 8 steady state condition (Euro 13 Mode Steady State engine exhaust conditions are given in Table 2).

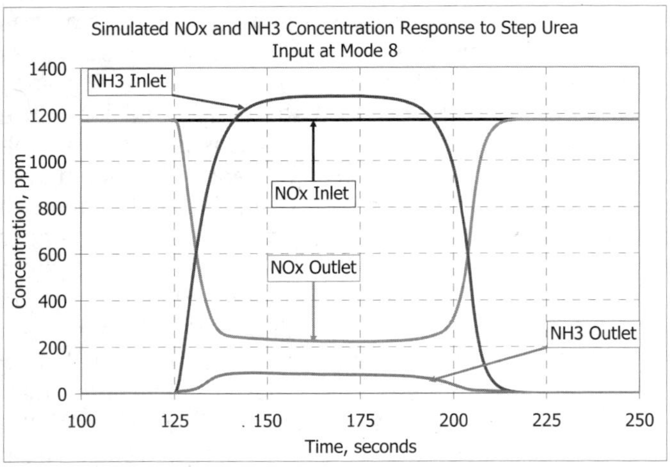

Figure 8: Simulated gas-phase NO_x concentration and gas-phase NH_3 concentration responses to a step change in urea equivalent to NSR of 1.1 at Mode 8 steady state engine exhaust condition.

In the third set of numerical experiments, the peak ammonia slip was monitored when the urea-SCR aftertreatment system was subjected to different inlet exhaust gas temperature ramp rates and a step urea input equivalent to NSR of 1.25. The engine was accelerated from Mode 1 steady state condition to Mode 5 steady state condition at different rates to simulate the different temperature ramp rates that could be encountered during warm-up conditions. A summary of the maximum ammonia slip for the different inlet exhaust gas temperature ramp rates is shown in Figure 9. It is imperative that the ammonia storage and release trade-off which is a localized phenomenon be incorporated in the control strategy. The dynamic system model provides information on the state of the local temperatures and concentrations during highly transient inlet exhaust gas operating conditions. The simulated localized states from these numerical experiments were used to refine the urea dosing strategy in the control system model during highly transient exhaust gas temperature operating conditions.

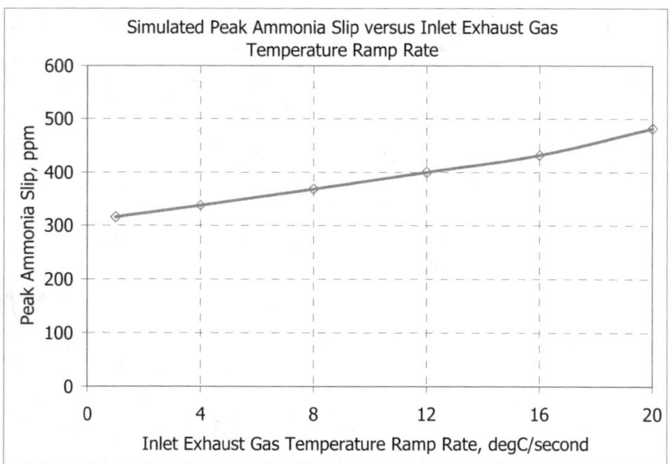

Figure 9: Simulated peak NH_3 slip for different ramp rates of inlet exhaust gas temperature change from Mode 1 to Mode 5.

CONTROL SYSTEM MODEL

The control system algorithm consists of two components: a component for ammonia slip control and a component for NO_x reduction control. These two components are managed such that the amount of NO_x reduction was maximized while minimizing the amount of ammonia slip.

AMMONIA SLIP CONTROL

The amount of ammonia slipping through the catalyst without reducing NO_x was reduced by maintaining the amount of ammonia stored in the catalyst below its saturation level at the operating catalyst bed temperature. This strategy takes advantage of the NO_x reducing ability of the stored ammonia and manages the amount of ammonia stored in the catalyst so that the stored ammonia is used for NO_x reduction instead of thermal desorption. Figure 10 shows a typical catalyst bed temperature dependence of NSR and NO_x conversion efficiencies for three levels of NH_3 slip at a fixed catalyst space velocity.

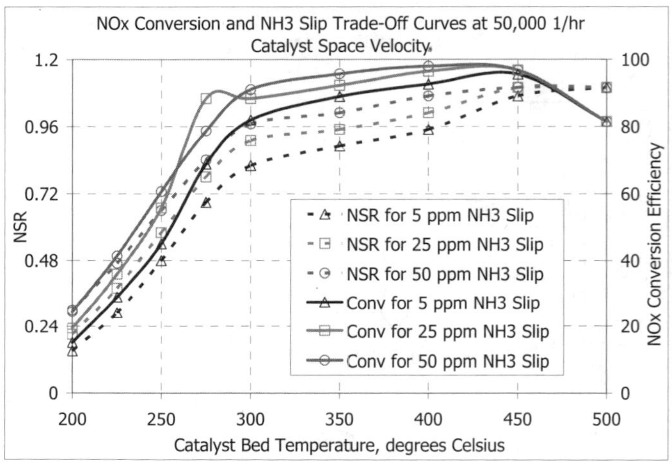

Figure 10: NO_x conversion efficiency and NSR trade-off at 50,000 1/hr space velocity and different catalyst bed temperatures for 5 ppm, 25 ppm, and 50 ppm NH_3 slip.

These maps were embedded in the control algorithm and used to minimize ammonia slip. The ammonia slip control algorithm uses the exhaust gas temperatures at the catalyst inlet and catalyst outlet to estimate the catalyst bed temperature and catalyst bed temperature ramp rate; uses the exhaust gas flow rate and catalyst geometric parameters to estimate the catalyst space velocity; and then determines the desired NO_x reduction to be targeted for minimum ammonia slip based on the capability of the SCR catalyst. The interface for the ammonia slip control algorithm is shown in Figure 11. For a given catalyst bed temperature, space velocity and bed temperature ramp rate, the algorithm limits the maximum desired NO_x reduction target based on the catalyst temperature ramp rate and NO_x conversion versus NH_3 slip trade-off curve by selecting the optimal NSR value for maximum NO_x conversion and minimum NH_3 slip.

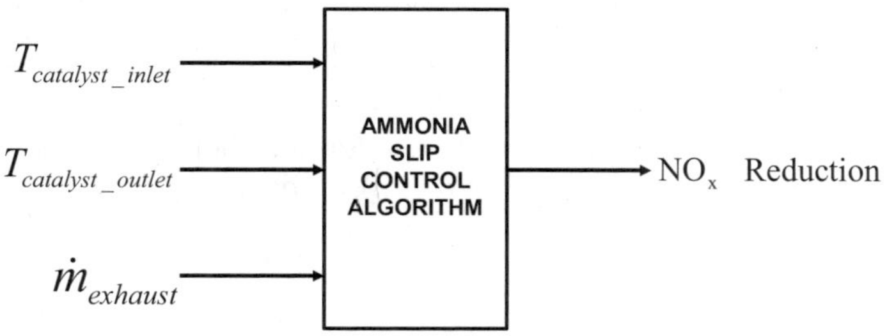

Figure 11: Interface of NH_3 slip control algorithm.

NO_x REDUCTION CONTROL

The desired amount of engine exhaust NO_x reduction was determined by the ammonia slip control algorithm and the NO_x reduction control algorithm was used to determine the urea dosing rate required for achieving the NO_x reduction request. Figure 12 shows a typical system identification response data that was obtained for this urea-SCR aftertreatment system when subjected to step urea of different levels with the engine operating at a steady state mode with exhaust gas temperature of 204°C and catalyst space velocity of 7227 1/hr. The system step response data was then used to model the urea-SCR aftertreatment system as a first order system for control development purposes. The system identification data also shows the inherent non-linearity of the urea-SCR aftertreatment system. By comparing the amount of NO_x reduction obtained with urea step sizes of 30% NSR, 60% NSR, and 90% NSR, it can be seen that the amount of NO_x reduction does not increase linearly with urea injection.

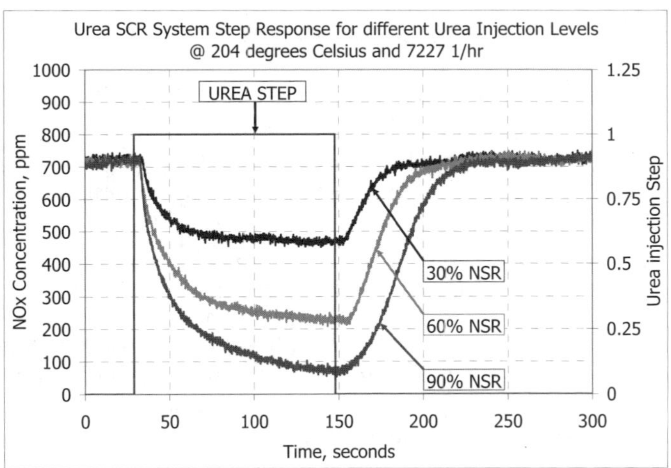

Figure 12: Step Response of urea-SCR aftertreatment system at fixed engine operating mode with exhaust gas temperature of 204°C and catalyst space velocity of 7227 1/hr.

Even though the urea step response exhibits the characteristics of a first order system, the delays, time constants and gains will change at different engine operating modes. The system identification experiments of the urea-SCR aftertreatment system showed that the amount of NO_x reduction can be dynamically modeled as a first-order system whose parameters depend on system operating conditions. This can be characterized by the inlet NO_x concentration, the amount of urea injected, the amount of ammonia stored in the catalyst, the catalyst space velocity, and the exhaust gas temperature. These results imply that the plant dynamics for the urea-SCR aftertreatment system, defined as the transfer function between the urea injected in cc/min and the amount of NO_x reduction in g/hr can be modeled as:

$$\frac{y(s)}{u(s)} = \frac{b_p}{s + a_p} \tag{25}$$

where the parameters $\mathbf{a}_p$ and $\mathbf{b}_p$ depend on the inlet NO_x concentration, the amount of urea injected, the amount of ammonia stored in the catalyst, the catalyst space velocity, and the exhaust gas temperature. For a system with these uncertainties and variability in system dynamics, a gain-scheduling type of control strategy which uses such a system identification model will require four or five-dimensional maps of model parameters at a large number of operating modes.

A system such as the urea-SCR aftertreatment system which can be modeled by equation (25) with constant or slowly-varying uncertain parameters $\mathbf{a}_p$ and $\mathbf{b}_p$ can be controlled by adaptive techniques. The basic idea in adaptive control is to estimate the uncertain plant parameters on-line based on the measured system signals, and use the estimated parameters to compute the control input. One such adaptive controller is the MRAC. A model-reference adaptive control system consists of four parts: a plant model containing unknown parameters, a reference model for compactly specifying the desired output of the control system, a feedback control law containing adjustable parameters, and an adaptation mechanism for updating the adjustable parameters. The structure of the MRAC system used for NO_x reduction control of the urea-SCR aftertreatment system is shown in Figure 13.

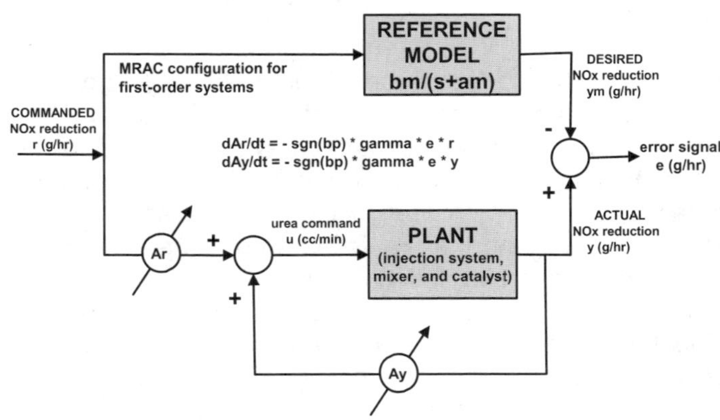

Figure 13: Model-Reference Adaptive Control System for the first-order plant.

The urea-SCR aftertreatment system was modeled to have the structure given in equation (25) with $\mathbf{a}_p$ and $\mathbf{b}_p$ being unknown parameters. The reference signal, **r**, is the amount of commanded NO_x reduction rate in grams/hour. The reference model was used to specify the ideal response of the adaptive control system to the external command. It has the same structure as the assumed plant and provides the ideal plant response that the adaptation mechanism should seek in adjusting the parameters. Thus the reference output, y_m, is the ideal desired NO_x reduction rate in grams/hour. The reference model was given by:

$$\frac{y_m(s)}{r(s)} = \frac{b_m}{s + a_m} \tag{26}$$

The control law used to determine the urea-dosing rate consists of a feedforward component and a feedback component, and is given by:

$$u = \hat{a}_r(t)r + \hat{a}_y(t)y \tag{27}$$

where **u** is the amount of urea in cc/min commanded, $\hat{\mathbf{a}}_r$ is the feedforward gain, $\hat{\mathbf{a}}_y$ is the feedback gain, **r** is the commanded reference NO_x reduction rate in grams/hour, and **y** is the actual NO_x reduction rate in grams/hour.

The adaptation mechanism was used to adjust the parameters in the control law by searching for the parameters such that the plant response under adaptive control becomes the same as the reference model. The adaptation law chosen for the MRAC system is given by:

$$\dot{\hat{a}}_r = -\mathrm{sgn}(b_p)\gamma e r \tag{28}$$

$$\dot{\hat{a}}_y = -\mathrm{sgn}(b_p)\gamma e y \tag{29}$$

where sgn($\mathbf{b}_p$) is the sign of the plant parameter $\mathbf{b}_p$ in equation (25), **e** is the error signal, and γ is the adaptation gain. Even though $\mathbf{b}_p$ is an uncertain plant parameter, sgn($\mathbf{b}_p$) is known to be positive for this system because the plant output, **y**, increases when the input, **u**, is increased within the design catalyst operating conditions, i.e., more urea injection results in more NO_x reduction.

The parameter convergence of the MRAC system was made faster and the tracking errors were made smaller by using a composite adaptation law. The composite adaptation law extracts information about the system parameters from both the tracking error, just as in the MRAC system, and the prediction error. The schematic of the composite adaptation is shown in Figure 14. The main appeal of composite adaptation is that it maintains global stability of the adaptive control system while providing fast parameter convergence and smaller tracking errors.

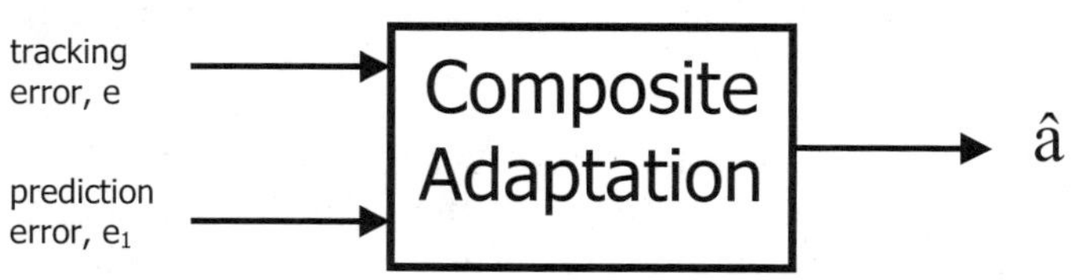

Figure 14: Composite Adaptation: e – tracking error, e_1 – prediction error.

As shown in Figure 14, the composite adaptation law consists of two components: the tracking-error-based adaptation and the prediction-error based estimation. The tracking-error based adaptation component is identical to the adaptation law for the MRAC system given in equations (28) and (29). The prediction-error based estimation laws are given by:

$$\frac{d}{dt}\hat{a} = -Pw^T e_1 \tag{30}$$

where **P** is the estimator gain matrix which can be estimated on-line using various prediction-error-based estimation methods such as the gradient estimation, standard least-squares estimation, least-squares with exponential forgetting etc., $\mathbf{a} = [a_r\ a_y]^T$, $\hat{\mathbf{a}} = [\hat{a}_r\ \hat{a}_y]^T$ is the estimate of **a**, $\mathbf{w} = [y/b_m\ y/(d/dt + a_m)]$, $u_1(t) = u / (d/dt + a_m) = \mathbf{w}(t)\,\mathbf{a}$, and e_1 is the prediction error on u_1 whose estimate is defined as $\hat{e}_1(t) = \mathbf{w}(t)(\hat{\mathbf{a}}(t)-\mathbf{a}) = \mathbf{w}(t)\tilde{\mathbf{a}}(t)$. The least-squares with exponential forgetting technique was used for on-line parameters estimation of the estimator gain matrix, **P**. The parameter update law is given by:

$$\frac{d}{dt}\left[P^{-1}\right] = -\lambda(t)P^{-1} + w^T(t)w(t) \tag{31}$$

where λ(t) is the forgetting factor which was tuned using the following technique: $\lambda(t) = \lambda_o\ (1 - ||\mathbf{P}||/k_o)$, with λ_o being the maximum forgetting rate, ||**P**|| being the magnitude of the estimator gain matrix, and k_o being a pre-specified bound for the magnitude of the estimator gain matrix. The parameter update law in equation (31) for the least-squares with exponential forgetting technique is more efficiently implemented in the following gain update form:

$$\frac{d}{dt}[P] = \lambda(t)P - Pw^T(t)w(t)P \tag{32}$$

The composite adaptation laws for the first order system can be summarized by the following vector equation:

$$\frac{d}{dt}\hat{a} = -P\left[\mathrm{sgn}(b_p)ve + \alpha(t)w^T e_1\right] \tag{33}$$

where $\mathbf{a} = [a_r\ a_y]^T$, $\hat{\mathbf{a}} = [\hat{a}_r\ \hat{a}_y]^T$ is the estimate of **a**, $\mathbf{v} = [r\ y]^T$, e is the error signal $e = y - y_m$, e_1 is the prediction error on $u_1 = \mathbf{w}(t)\,\mathbf{a}$, and α(t) is an adjustable parameter. The composite adaptive control law is the same with the control law for the MRAC system given in equation (27).

The interface for the composite adaptive control algorithm presented in this paper is summarized in Figure 15. There are three inputs to the control algorithm: commanded NO_x reduction rate, engine turbine outlet NO_x emissions rate, and the catalyst outlet NO_x emissions rate. The commanded NO_x reduction rate is the output from the ammonia slip control algorithm. The ammonia slip control algorithm determines the NO_x reduction rate based on an optimization between the need to achieve the desired NO_x reduction rate and the need to prevent or minimize ammonia slip. The engine turbine outlet NO_x emissions rate was determined from the turbine outlet NO_x sensor concentration and the exhaust gas mass flow rate estimate. The catalyst outlet NO_x emissions rate was determined from a NO_x sensor concentration reading and the exhaust gas mass flow rate estimate. The output from the algorithm is the urea dosing rate.

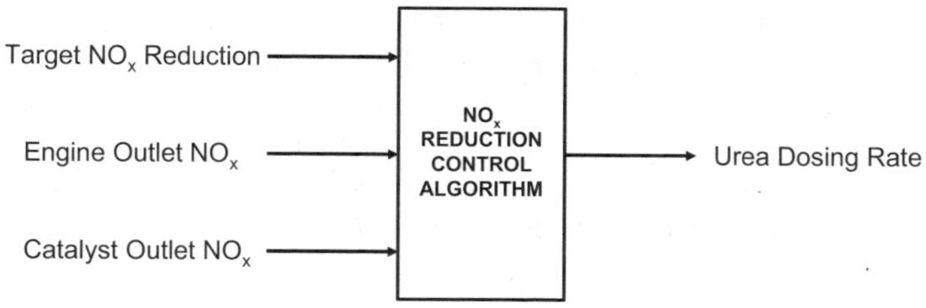

Figure 15: Interface for Composite Adaptive Control Algorithm

RESULTS AND DISCUSSION

The composite adaptive control algorithm presented in this paper was tested for three consecutive heavy-duty hot FTP emission test cycles. Engine outlet and catalyst outlet NO_x sensor traces, and the corresponding urea injection rates from the third test cycle are shown in Figure 16. Most of the urea was injected during the LA Freeway phase of the cycle.

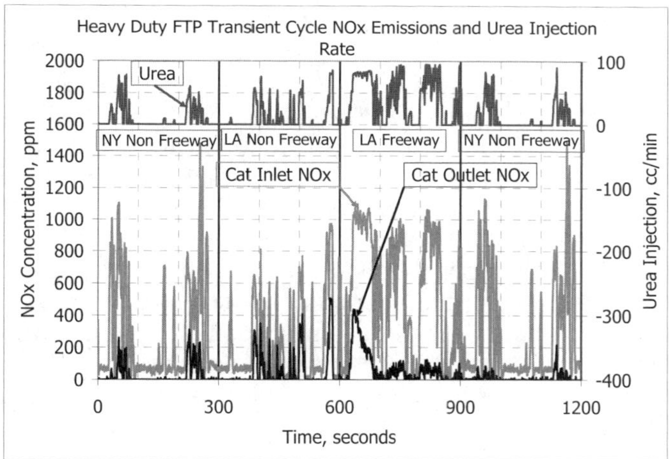

Figure 16: NO_x sensor measurements at the engine outlet (catalyst inlet) and catalyst outlet, and the corresponding urea dosing rates for the third heavy-duty hot FTP cycle using the composite adaptive control algorithm presented in this paper.

The cumulative NO_x emissions for the three consecutive heavy-duty FTP cycles are shown in Figure 17. These results show improvements in the NO_x reduction with successive heavy-duty FTP emission test cycles. More insights can be obtained from a phase breakdown of the NO_x reductions. A breakdown of the NO_x reductions for the four phases is summarized in Figure 18. The percentage of total engine cycle NO_x emissions during the four phases: 13.7% in the first phase (NY Non Freeway), 16.8% in the second phase (LA Non Freeway), 55.6% in the third phase (LA Freeway), and 13.9% in the fourth phase (NY Non Freeway). These results show that the variations in NO_x reduction in successive heavy-duty hot FTP cycle runs came primarily from the first two phases (i.e., the first NY Non Freeway portion and the LA Non Freeway portion). The amount of NO_x reduction in the 3rd and 4th phases are the same or within the measurement system errors for all three cycle runs. The sources of these variations in NO_x reduction for the first and second phases of the heavy-duty hot FTP cycle include the catalyst bed temperature and the amount of ammonia stored in the catalyst.

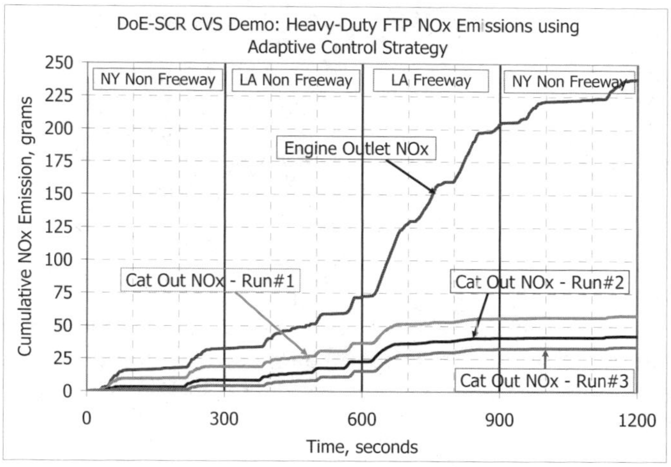

Figure 17: Heavy-duty hot FTP test cycle cumulative NO_x emissions using adaptive control strategy. 75.6%, 82.0%, and 85.6% NO_x Reduction in first, second, and third FTP cycles respectively.

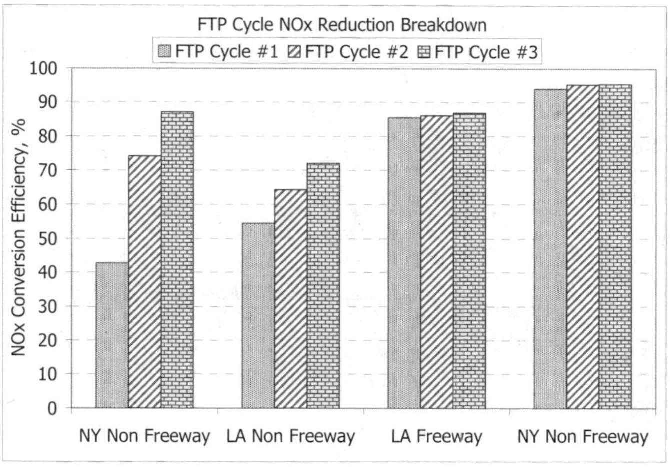

Figure 18: Breakdown of NO_x reduction for the Four Phases of the heavy-duty hot FTP cycle.

Figure 19 shows measurements of the exhaust gas temperature at the catalyst inlet, the exhaust gas temperature at the catalyst outlet, and the ammonia slip for the three heavy-duty hot FTP cycle runs. The catalyst was cleaned (i.e., the ammonia stored in the catalyst was removed) prior to the first heavy-duty hot FTP cycle by running the engine at rated condition for 5 minutes. After cleaning the catalyst, the engine was soaked for 20 minutes before starting the first heavy-duty hot FTP cycle. The engine was soaked for 5 minutes after the first and second cycles before starting the second and third hot FTP cycles respectively. The peak ammonia slip occurred during the LA Freeway phase of the FTP cycle where the catalyst outlet exhaust gas temperature ramp rate was 0.521°C/second.

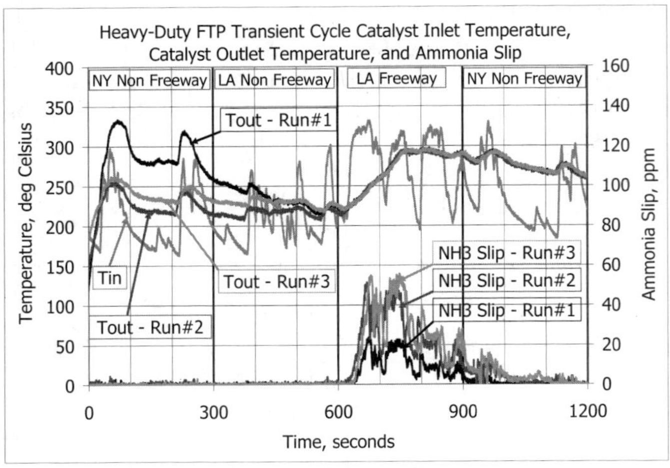

Figure 19: Measurements of catalyst inlet exhaust gas temperature, catalyst outlet exhaust gas temperature, and ammonia slip during heavy-duty hot FTP cycles with adaptive control algorithm.

Another important system behavior that was observed with this urea-SCR aftertreatment system is the phenomenon of ammonia storage. After running three consecutive heavy duty FTP emission test cycles with the composite adaptive algorithm, the engine was then operated at rated condition with the urea injection turned off. At this engine operating mode, the exhaust gas temperature was 371°C and the corresponding catalyst space velocity was 16178 1/hr. Figure 20 shows the measurements that were obtained from NO_x sensors placed at the engine turbine outlet and catalyst outlet.

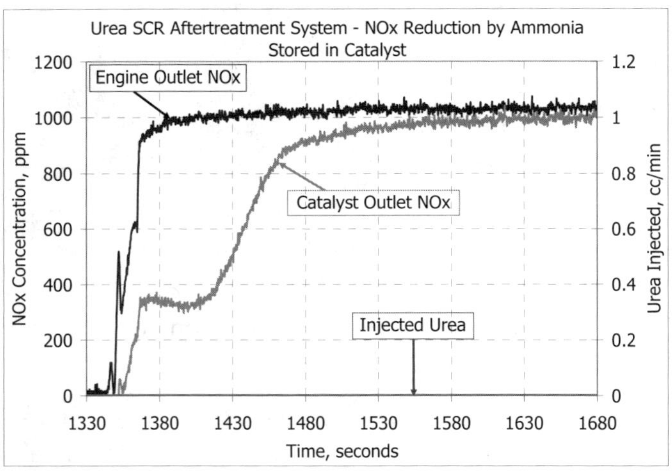

Figure 20: Measured traces of engine turbine outlet and catalyst outlet NO_x concentrations at engine operating mode with exhaust gas temperature of 371°C and catalyst space velocity of 16178 1/hr after three FTP hot cycles with the urea injection turned OFF.

The NO_x concentrations in Figure 20 show NO_x conversion for almost 160 seconds after the urea injection was turned off. This observed NO_x reduction is a result of the ammonia stored in the catalyst from prior urea injections reacting with engine exhaust NO_x. The reaction of NO_x with adsorbed ammonia provides another NO_x reducing ability that if not incorporated in any control design will affect the controller tracking performance since more than commanded NO_x reduction will be achieved at favorable catalyst temperatures. Thus an accurate estimation of the amount of ammonia stored in the catalyst is a very useful quantity in metering the amount of urea injected to the urea-SCR catalyst.

Based on observed system behavior, the injected urea can be divided into four categories. Some urea stays in the exhaust pipe, some urea is converted to ammonia and gets stored in the catalyst, some urea is converted to ammonia and slips through the catalyst, and some urea is converted to ammonia and reacts with engine exhaust NO_x. A good measure of the urea-SCR aftertreatment system efficiency is the fraction of urea delivered by the urea dosing system that actually reduces engine exhaust NO_x. For the three heavy-duty FTP emission test cycle runs using the composite adaptive control algorithm presented in this paper, the total engine exhaust NO_x was 626 grams, and the total tailpipe NO_x emissions was 140 grams. Thus the total engine exhaust NO_x reduced by the urea-SCR aftertreatment system was 486 grams. With the engine exhaust NO_x containing 90% NO and 10% NO_2, the stoichiometric amount of urea required to reduce this 486 grams of engine exhaust NO_x is 328 grams of urea (0.6745 grams of urea per gram of NO_x). This is equivalent to 1009 grams of aqueous urea solution with 32.5% urea concentration by weight. However, measurements of the aqueous urea solution before and after each heavy-duty FTP emissions test cycle showed that a total of 1196 grams of urea was injected during the three heavy-duty FTP emissions test cycles. This implies that 84.4% of the urea injected was used to reduce engine exhaust NO_x. This urea-SCR aftertreatment system efficiency of 84.4% is of the same order of magnitude with the cycle NO_x reduction.

CONCLUSION

The urea-SCR aftertreatment dynamic system model presented in this paper can be calibrated for different systems and is a valuable tool for control system design. This model is capable of predicting the trends in transient performance characteristics such as ammonia storage, ammonia slip, thermal effects, and NO_x reduction in a closed loop system for a fresh vanadia-based catalyst used for selective catalytic reduction of NO_x with urea.

The closed loop controller presented in this paper was able to reduce the FTP cycle NO_x emissions from a Cummins heavy-duty engine by 84% while maintaining the mean ammonia slip below 7 ppm and the peak ammonia slip below 55 ppm. The heavy-duty FTP emission test cycle results show that the catalyst temperature ramp rates and ammonia storage are two parameters that strongly determine the NO_x reduction and ammonia slip by the urea-SCR aftertreatment system. The improvement in NO_x reduction with successive heavy-duty FTP emission test cycle runs occurs primarily in the first NY Non Freeway and LA Non Freeway phases and is due to additional NO_x reduction by the ammonia stored in the catalyst during the previous cycles. The slip in ammonia observed during the heavy-duty FTP emissions test cycle runs is due to ammonia stored in the catalyst during the previous cycles being released as the catalyst temperature ramps up.

ACKNOWLEDGMENTS

The authors would like to thank the Department of Energy (DoE) for supporting this research. The authors would also like to acknowledge Patrick M. Pierz of Fleetguard Emission Solutions, Eric B. Andrews, Jerilyn Tsai, Larry Barnett, and Dr. Howard Fang at Cummins, Inc. for their contributions.

CONTACT

e-mail: john.n.chi@cummins.com

Phone: (812) 377-5345

Fax: (812) 377-7226

REFERENCES

1. http://www.dieselnet.com/
2. Ansell, G. P., Bennett, P. S., Cox, J. P., Frost, J. C., Gray, P. G., Jones, A. –M., Rajaram, R. R., Walker, A. P., Litorell, M., and Smedler, G., "The development of a model capable of predicting diesel lean NO_x catalyst performance under transient conditions", Applied Catalysis B: Environmental 10 (1996) 183-201.
3. Büchner, S., Lardies, S. S., Degen, A., Donnerstag, A., and Held, W., "A Modular Numerical Simulation Tool Predicting Catalytic Converter Light-Off by Improved Modeling of Thermal Management and Conversion Characteristics", SAE Paper No. 2001-01-0940.
4. Chapra, S. C., and Canale, R. P., "Numerical Methods For Engineers", 3rd edition, McGraw-Hill Publishing Company, New York, 1988.
5. Daw, S., Lenox, K., and Chakravarthy, K. Personal Communications, 2001.
6. Dumesic, J. A., Topsøe, N. –Y., Topsøe, H., Chen, Y., and Slabiak, T., "Kinetics of Selective Catalytic Reduction of Nitric Oxide by Ammonia over Vanadia/Titania", Journal of Catalysis 163, 409-417 (1996), Article No. 0342.
7. Fang, H. L., and DaCosta, H. F. M, "Urea thermolysis and NO_x reduction with and without SCR catalysts", Applied Catalysis B: Environmental 46 (2003) 17-34.
8. Gates, B. C., "Catalysis", in K. Othmer (Ed.), *Encyclopedia of Chemical Technology*, 4th, Ed., Volume 5, John Wiley & Sons, New York (NY), (1993), p. 340.
9. Groppi, G., Belloli, A., Tronconi, E., and Forzatti, P., "A Comparison of Lumped and Distributed Models of Monolith Catalytic Combustors", Chemical Engineering Science, Vol. 50, No. 17, pp. 2705-2715, 1995.
10. Holman, J. P., "Heat Transfer", 7th edition, McGraw-Hill Publishing Company, New York, 1990.
11. Incropera, F. P. and DeWitt, D. P., "Fundamentals of Heat and Mass Transfer", 3rd edition, John Wiley & Sons, New York, 1981.
12. Johannessen, T., Larsen, J. H., Chorkendorff, I., Livbjerg, H., and Topsoe, H., "Catalyst dynamics: consequences for classical kinetic descriptions of reactors", *Chem. Eng. J.* **82**, 219-230 (2001).
13. Khodayari, R., and Odenbrand, C. U. I., "Selective catalytic reduction of NO_x: a mathematical model for poison accumulation and conversion performance", Chemical Engineering Science 54 (1999) 1775-1785.
14. Kleemann, M., Elsener, M., Koebel, M., and Wokaun, A., "Investigation of the ammonia adsorption on monolithic SCR catalysts by transient response analysis", Applied Catalysis B: Environmental 27 (2000) 231-242.
15. Kleemann, M., Elsener, M., Koebel, M., and Wokaun, A., "Hydrolysis of Isocyanic Acid on SCR Catalysts", Ind. Eng. Chem. Res. 2000, 39, 4120-4126.
16. Koebel, M., Elsener, M., and Kleemann, M., "Urea-SCR: a promising technique to reduce NO_x emissions from automotive diesel engines", Catalysis Today 59 (2000) 335-345.
17. Kolaczkowski, S. T., "Modelling catalytic combustion in monolith reactors – challenges faced", Catalysis Today 47 (1999) 209-218.
18. Li, Y., and Armor, J. N., "Selective NH_3 oxidation to N_2 in a wet stream", Applied Catalysis B: Environmental 13 (1997) 131-139.
19. Lietti, L., Nova, I., Tronconi, E., and Forzatti, P., "Transient kinetic study of the SCR-DeNO$_x$ reaction", Catalysis Today 45 (1998) 85-92.
20. Nova, I., Lietti, L., Tronconi, E., and Forzatti, P., "Transient response method applied to the kinetic analysis of the DeNO$_x$–SCR reaction", Chemical Engineering Science 56 (2001) 1229-1237.
21. Patankar, S. V., "Numerical Heat Transfer and Fluid Flow (Series in Computational Methods in Mechanics and Thermal Sciences)", Hemisphere Publishing Company, New York, 1980.
22. Simulink Manual, "Writing S-Functions", Version 4, The MathWorks, Inc., Natick, MA. 2003.
23. Slotine, J. –J. E., and Li, W. "Applied Nonlinear Control". Prentice-Hall, 1991.
24. Suárez, S., Jung, S. M., Avila, P., Grange, P., and Blanco, J., "Influence of NH_3 and NO oxidation on the SCR reaction mechanism on copper/nickel and

vanadium oxide catalysts supported on alumina and titania", Catalysis Today 75 (2002) 331-338.
25. Tronconi, E., Lietti, L., Forzatti, P., and Malloggi, S., "Experimental and Theoretical Investigation of the Dynamics of the SCR DeNO$_x$ Reaction", Chemical Engineering Science, Vol. 51, No. 11, pp. 2965-2970, 1996.
26. Uberoi, M., and Pereira, C. J., "External Mass Transfer Coefficient for Monolith Catalysts", Ind. Engng Chem. Res., Vol. 35, pp. 113-116, 1996.

ACRONYMS, ABBREVIATIONS

ADI Applied Dynamics Incorporated

BSNOx Brake Specific NOx Emissions

CRPT Controls Rapid Prototyping Tool

CFD Computational Fluid Dynamics

CO_2 Carbon Dioxide

CVS Constant Volume Sampling

DAE Differential Algebraic Equation

D_m Mass Diffusion Coefficient

DoE Department of Energy

Ea Activation Energy

ECM Engine Control Module

EGR Exhaust Gas Re-circulation

FTP Federal Test Procedure

Gr Grashof Number

h Heat Transfer Coefficient

ΔH^{298} Enthalpy of Formation at 298 K

HNCO Isocyanic Acid

H_2O Water Vapor

k_m Mass Transfer Coefficient

LA Los Angeles

MRAC Model-Reference Adaptive Control

N_2 Nitrogen Gas

NH_3 Ammonia

NO Nitrogen Oxide

NO_2 Nitrogen Dioxide

NO_x Oxides of Nitrogen (NO and NO_2)

NSR Normalized Stoichiometric Ratio

Nu Nusselt Number

NY New York

Ω_{NH3} Ammonia Adsorption Capacity

O_2 Oxygen Gas

ODE Ordinary Differential Equation

PDE Partial Differential Equation

Pr Prandtl Number

Q Net Heat Generation Rate

R Reaction Rate

Re Reynolds Number

Rg Universal Gas Constant

RPM Revolutions Per Minute

Sc Schmidt Number

Sh Sherwood Number

SCR Selective Catalytic Reduction

TiO_2 Titania

VGT Variable Geometry Turbocharger

V_2O_5 Vanadia

WO_3 Tungsta

APPENDIX – REACTION KINETICS

Table 3: Reaction Kinetics	
$HNCO(g)+H_2O(g)\rightarrow NH_3(g)+CO_2(g)$ $\Delta H^{298}=-95.914\text{ kJ/mol}$ $R_{HNCO}=2x10^5 e^{\frac{-3608}{T}}[HNCO][H_2O]$	[HNCO] in mol/m^3 [H_2O] in mol/m^3 R_{HNCO} in mol/m^2/s
$NH_3+\frac{1}{2}NO+\frac{1}{2}NO_2\rightarrow N_2+\frac{3}{2}H_2O$ $\Delta H^{298}=-378.534\text{ kJ/mol}$ $R_{NH_3}=2.53x10^6 e^{\frac{-3007}{T}}[NO][NO_2]\frac{[NH_3]}{1+1.2042x10^{-3}[NH_3]}$	[NH_3] in mol/m^3 [NO] in mol/m^3 [NO_2] in mol/m^3 R_{NH3} in mol/m^2/s
$NH_3+NO+\frac{1}{4}O_2\rightarrow N_2+\frac{3}{2}H_2O$ $\Delta H^{298}=-407.129\text{ kJ/mol}$ $R_{NH_3}=2.36x10^8 e^{\frac{-7151}{T}}[NO]\frac{[NH_3]}{1+1.2042x10^{-3}[NH_3]}$	[NH_3] in mol/m^3 [NO] in mol/m^3 [O_2] in mol/m^3 R_{NH3} in mol/m^2/s
$NH_3+\frac{3}{4}NO_2\rightarrow\frac{7}{8}N_2+\frac{3}{2}H_2O$ $\Delta H^{298}=-341.664\text{ kJ/mol}$ $R_{NH_3}=7.56x10^8 e^{\frac{-8057}{T}}[NO_2]\frac{[NH_3]}{1+1.2042x10^{-3}[NH_3]}$	[NH_3] in mol/m^3 [NO_2] in mol/m^3 R_{NH3} in mol/m^2/s
$NH_3+\frac{3}{4}O_2\rightarrow\frac{1}{2}N_2+\frac{3}{2}H_2O$ $\Delta H^{298}=-316.839\text{ kJ/mol}$ $R_{NH_3}=1.32x10^7 e^{\frac{-15034}{T}}\theta_{NH_3}$	θ_{NH3} is dimensionless R_{NH3} in mol/m^2/s
$NH_3+\frac{5}{4}O_2\rightarrow NO+\frac{3}{2}H_2O$ $\Delta H^{298}=-226.549\text{ kJ/mol}$ $R_{NH_3}=9.11x10^{10} e^{\frac{-14503}{T}}\frac{[NH_3][O_2]}{(1+1.2042x10^{-3}[NH_3])(1+1.5053x10^{-3}[O_2]^{1.25})}$	[NH_3] in mol/m^3 [O_2] in mol/m^3 R_{NH3} in mol/m^2/s
$R_{ads,NH_3}=808[NH_3](1-\theta_{NH_3})$	[NH_3] in mol/m^3 θ_{NH3} is dimensionless $R_{ads,NH3}$ in mol/m^3/s
$R_{des,NH_3}=3.67x10^6 e^{\frac{-12992(1-0.310\theta_{NH_3})}{T}}\theta_{NH_3}$	θ_{NH3} is dimensionless $R_{des,NH3}$ in mol/m^3/s

APPENDIX – MODEL STRUCTURE

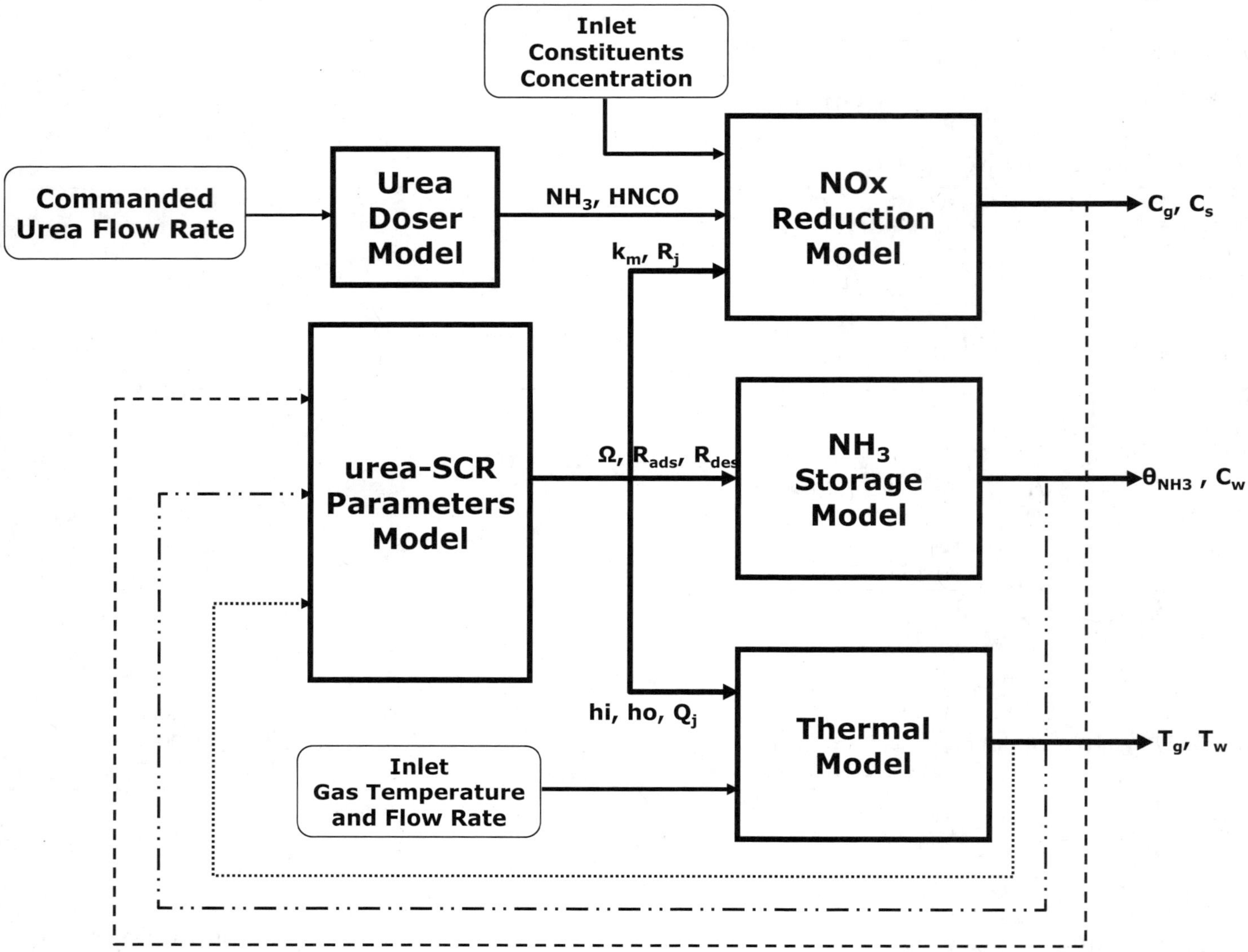

About the Editors

Magdi Khair

Dr. Magdi Khair is an Institute Engineer in the Department of Emissions Research at Southwest Research Institute. He is experienced in the areas of engine testing and exhaust emissions control.

Dr. Khair's prior experience was with AlliedSignal Automotive Catalyst Company, with the development of catalytic aftertreatment for light-duty and heavy-duty diesel engines; Ford New Holland, with primary responsibility for the development of the 6.6- and 7.8-liter midrange diesel engines to meet 1991 emissions standards; Bendix Diesel Engine Controls, where he led the development of advanced electronically controlled diesel fuel injection systems and established several cooperative engineering programs with European and North American engine manufacturers; and Chrysler Corporation, where he converted the Slant Six gasoline engine into an open-chamber, pilot-injected, and electronically controlled diesel engine, supervised a combustion kinetics project, and participated in the design and development of electronic controls for a passenger car turbine engine.

Dr. Khair holds 13 U.S. patents on electronic fuel injection, turbocharging, exhaust gas recirculation, and aftertreatment systems. He also has authored and co-authored numerous papers for SAE International and the American Society of Mechanical Engineers (ASME).

Dr. Khair received a B.S. in Automotive Engineering from Ain Shams University, an M.S. in Thermodynamics from the University of Birmingham, England, an M.B.A. from Michigan State University, and a Ph.D. in Engineering Management from Warren National University.

Federico Millo

Federico Millo is an Associate Professor at the Dipartimento di Energetica, Politecnico di Torino, Italy. He received his master's degree in mechanical engineering at Politecnico di Torino in 1989.

After a short experience in the automotive industry, Mr. Millo joined the faculty at Politecnico di Torino as a Researcher Assistant in 1991. In 1998, he was appointed Associate Professor. His research activity has focused mainly on the analysis of the combustion process in internal combustion engines and of pollutant formation and control in diesel engines. He has published more than 50 articles, most of which are in the proceedings of international congresses and in international journals, and he has been responsible for several research projects with major European automotive companies.

Mr. Millo also has been serving as a technical consultant for the European Commission for the development of a global technical regulation on worldwide harmonized test procedures for emissions regulations. He is a member of SAE International and has been serving as an organizer and chairman for the Engine Modeling session at the annual SAE World Congress, as well as a peer reviewer for several international journals, including the *International Journal of Engine Research* (Institution of Mechanical Engineers [IMechE]), the *Journal of Energy Resources Technology*, and the *Journal of Engineering for Gas Turbines and Power* (ASME).